MW01687710

Get the most out of MyMathLab®

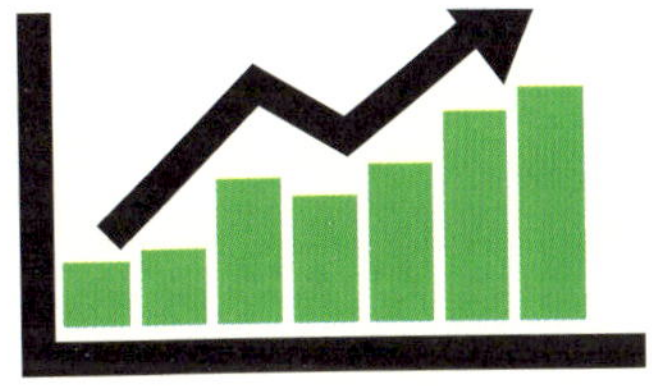

MyMathLab creates personalized experiences to help each student achieve success and provides powerful tools so instructors can create the perfect learning experiences for their courses.

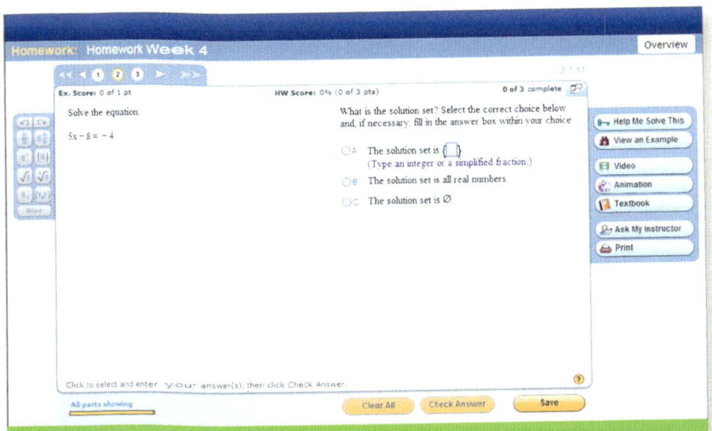

Exercise with feedback and learning aids

Personalized Support *for* Students

- MyMathLab comes with many learning resources—eText, animations, videos, and more—all designed to support you as you complete your assignments.
- Whether you're doing homework or working from the adaptive study plan, you'll receive immediate feedback, so you'll know exactly where you need help.

Data-Driven Reporting *for* Instructors

- MyMathLab's comprehensive online gradebook automatically tracks students' results on tests, quizzes, homework, and in the study plan.
- The Reporting Dashboard makes it easier than ever to identify topics where students are struggling, or specific students who may need extra help.

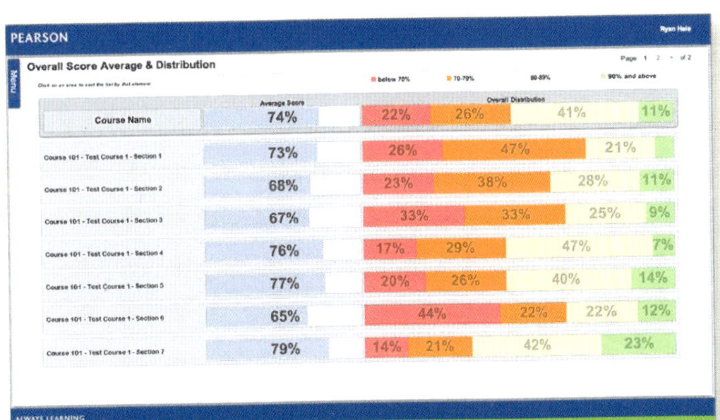

Dashboard

www.mymathlab.com

ANNOTATED INSTRUCTOR'S EDITION

EDITION 12

Intermediate Algebra

MARGARET L. LIAL
American River College

JOHN HORNSBY
University of New Orleans

TERRY MCGINNIS

PEARSON

Boston Columbus Indianapolis New York San Francisco Hoboken Amsterdam
Cape Town Dubai London Madrid Milan Munich Paris Montreal Toronto Delhi
Mexico City São Paulo Sydney Hong Kong Seoul Singapore Taipei Tokyo

Editorial Director: Chris Hoag
Editor in Chief: Michael Hirsch
Editorial Assistant: Chase Hammond
Program Manager: Beth Kaufman
Project Manager: Christine Whitlock
Program Management Team Lead: Karen Wernholm
Project Management Team Lead: Peter Silvia
Media Producer: Shana Siegmund
TestGen Sr. Content Developer: John R. Flanagan
MathXL Executive Content Manager: Rebecca Williams
Marketing Manager: Rachel Ross
Marketing Assistant: Kelly Cross
Senior Author Support/Technology Specialist: Joe Vetere
Rights and Permissions Project Manager: Diahanne Lucas Dowridge
Procurement Specialist: Carol Melville
Associate Director of Design: Andrea Nix
Program Design Lead: Beth Paquin
Text and Cover Design, Production Coordination, Composition, and Illustrations: Cenveo Publisher Services
Cover Image: Boyan Dimitrov/Shutterstock

Library of Congress Cataloging-in-Publication Data
Lial, Margaret L.
Intermediate algebra / Margaret Lial, American River College, John Hornsby, University of New Orleans, Terry McGinnis. – 12th edition.
pages cm
ISBN 978-0-321-96935-4 AIE ISBN: 978-0-321-98337-4 / 0-321-98337-8
1. Algebra–Textbooks. I. Hornsby, John, 1949- II. McGinnis, Terry. III. Title.
QA152.3.L534 2016
512.9–dc23

2013049394

4 2022

PEARSON

www.pearsonhighered.com

ISBN 13: 978-0-321-96935-4
ISBN 10: 0-321-96935-9

To Margaret L. Lial

Margaret L. Lial

On March 16, 2012, the mathematics education community lost one of its most influential members with the passing of our beloved mentor, colleague, and friend Marge Lial. On that day, Marge lost her long battle with ALS. Throughout her illness, Marge showed the remarkable strength and courage that characterized her entire life.

We would like to share a few comments from among the many messages we received from friends, colleagues, and others whose lives were touched by our beloved Marge:

"What a lady"
"A remarkable person"
"Gracious to everyone"
"One of a kind"
"Truly someone special"
"A loss in the mathematical world"
"A great friend"
"Sorely missed but so fondly remembered"
"Even though our crossed path was narrow, she made an impact and I will never forget her."
"There is talent and there is Greatness. Marge was truly Great."
"Her true impact is almost more than we can imagine."

In the world of college mathematics publishing, Marge Lial was a rock star. People flocked to her, and she had a way of making everyone feel like they truly mattered. And to Marge, they did. She and Chuck Miller began writing for Scott Foresman in 1970. As her illness progressed, she told us that she could no longer continue because "just getting from point A to point B" had become too challenging. That's our Marge—she even gave a geometric interpretation to her illness.

It has truly been an honor and a privilege to work with Marge Lial these past twenty years. While we no longer have her wit, charm, and loving presence to guide us, so much of who we are as mathematics educators has been shaped by her influence. We will continue doing our part to make sure that the Lial name represents excellence in mathematics education. And we remember daily so many of the little ways she impacted us, including her special expressions, "Margisms" as we like to call them. She often ended emails with one of them—the single word "Onward."

We conclude with a poem penned by another of Marge's coauthors, Callie Daniels.

Your courage inspires me
Your strength…impressive
Your wit humors me
Your vision…progressive

Your determination motivates me
Your accomplishments pave my way
Your vision sketches images for me
Your influence will forever stay.

Thank you, dearest Marge.
Knowing you and working with you has been a divine gift.

John Hornsby
Terry McGinnis

CONTENTS

3 Systems of Linear Equations 215

4 Exponents, Polynomials, and Polynomial Functions 265

5 Factoring 323

6 Rational Expressions and Functions 365

7 Roots, Radicals, and Root Functions 433

8 Quadratic Equations, Inequalities, and Functions 503

9 Inverse, Exponential, and Logarithmic Functions 581

Additional topics available in MyMathLab®: Solving Systems of Linear Equations by Matrix Methods; Determinants and Cramer's Rule; Piecewise Linear Functions; Conic Sections; Nonlinear Systems of Equations; Second-Degree Inequalities and Systems of Inequalities; Linear Programming; Sequences and Series; The Binomial Theorem

PREFACE

It is with great pleasure that we offer the twelfth edition of *Intermediate Algebra*. We have remained true to the original goal that has guided us over the years—to provide the best possible text and supplements package to help students succeed and instructors teach. This edition faithfully continues that process through enhanced explanations of concepts, new and updated examples and exercises, student-oriented features like Vocabulary Lists, Pointers, Cautions, Problem-Solving Hints, and Now Try Exercises, as well as an extensive package of helpful supplements and study aids.

This text is part of a series that includes the following books, all by Lial, Hornsby, and McGinnis:

- *Beginning Algebra*, Twelfth Edition,
- *Beginning and Intermediate Algebra*, Sixth Edition,
- *Algebra for College Students*, Eighth Edition.

WHAT'S NEW IN THIS EDITION?

We are pleased to offer the following new features:

VOCABULARY LISTS New vocabulary is now given at the beginning of appropriate sections. The list format allows students to preview vocabulary that is introduced in the section and also to review and check-off words they are able to correctly define upon completing a section.

CONCEPT CHECK EXERCISES Each exercise set begins with a set of Concept Check problems that facilitate students' mathematical thinking and conceptual understanding. Problem types include multiple-choice, true/false, matching, completion, and ***What Went Wrong?*** exercises. Many emphasize new vocabulary.

EXTENDING SKILLS EXERCISES These problems, scattered throughout selected exercise sets, expand on section objectives. Some are challenging in nature.

MIXED REVIEW EXERCISES Each chapter review has been expanded to include a dedicated set of Mixed Review Exercises to help students further synthesize concepts.

MARGIN ANSWERS TO REVIEW COMPONENTS To provide immediate reference and enable students to get the most out of review opportunities, answers are included in the margins for Summary Exercises, Chapter Review Exercises, Mixed Review Exercises, Chapter Tests, and Cumulative Review Exercises.

SPECIFIC CONTENT CHANGES include the following:

- **New Chapter R** (former Chapter 1) provides a thorough review of exponents, roots, order of operations, operations and properties of real numbers, and other basic concepts.
- **Application Sections 1.3, 1.4, and 3.3** include new and/or updated problem-solving examples, exercises, and hints. **Section 1.2** provides new health care formulas.
- **The presentation on linear equations in two variables in Sections 2.1–2.3** has been reorganized. Several groups of new exercises make connections between tables, equations, and graphs of linear equations.
- **The introduction to relations and functions in Section 2.5** has a new example and enhanced discussion.
- **Section 7.3** now includes a new objective and discussion, examples, and exercises on equations and graphs of **circles.**
- A new objective and new examples and exercises using calculators to approximate exponentials and logarithms are included in **Sections 9.2 and 9.3. Calculator screens** from the TI-84 have been updated here and throughout the text.
- **Presentations of the following topics have been enhanced and expanded:**

  Solving absolute value inequalities (Section 1.7)
  Graphing linear inequalities in two variables (Section 2.4)
  Applying composition of functions (Section 4.3)
  Factoring by grouping (Section 5.1)
  Determining domains of rational functions (Sections 6.1, 6.4)
  Understanding operations with rational expressions (Sections 6.1, 6.2)
  Solving quadratic inequalities (Section 8.7)
  Finding inverse functions (Section 9.1)
  Solving exponential and logarithmic equations (Section 9.6)

ACKNOWLEDGMENTS

The comments, criticisms, and suggestions of users, nonusers, instructors, and students have positively shaped this textbook over the years, and we are most grateful for the many responses we have received. The feedback gathered for this edition was particularly helpful.

We especially wish to thank the following individuals who provided invaluable suggestions.

Barbara Aaker, *Community College of Denver*
Viola Lee Bean, *Boise State University*
Kim Bennekin, *Georgia Perimeter College*
Dixie Blackinton, *Weber State University*
Tim Caldwell, *Meridian Community College*
Sally Casey, *Shawnee Community College*
Callie Daniels, *St. Charles Community College*
Cheryl Davids, *Central Carolina Technical College*
Chris Diorietes, *Fayetteville Technical Community College*
Sylvia Dreyfus, *Meridian Community College*
Lucy Edwards, *Las Positas College*
LaTonya Ellis, *Bishop State Community College*
Jacqui Fields, *Wake Technical Community College*
Beverly Hall, *Fayetteville Technical Community College*
Sandee House, *Georgia Perimeter College*
Lynette King, *Gadsden State Community College*
Linda Kodama, *Windward Community College*
Ted Koukounas, *Suffolk Community College*
Karen McKarnin, *Allen County Community College*
James Metz, *Kapi'olani Community College*
Jean Millen, *Georgia Perimeter College*
Molly Misko, *Gadsden State Community College*
Jane Roads, *Moberly Area Community College*
Yvonne Sandoval, *Pima Community College, WC*
Melanie Smith, *Bishop State Community College*
Linda Smoke, *Central Michigan University*
Erik Stubsten, *Chattanooga State Technical Community College*
Tong Wagner, *Greenville Technical College*
Sessia Wyche, *University of Texas at Brownsville*

Over the years, we have come to rely on an extensive team of experienced professionals. Our sincere thanks go to these dedicated individuals at Pearson Arts & Sciences, who worked long and hard to make this revision a success: Chris Hoag, Maureen O'Connor, Michael Hirsch, Rachel Ross, Beth Kaufman, Christine Whitlock, Chase Hammond, Kelly Cross, and Shana Siegmund.

Additionally, Shannon d'Hemecourt, Eddie Herring, and Abby Tannebaum did a great job helping us update real data applications. We are also grateful to Kathy Diamond and Marilyn Dwyer of Cenveo, Inc., for their excellent production work; Bonnie Boehme for supplying her copyediting expertise; Aptara for their photo research; and Lucie Haskins for producing another accurate, useful index. Perian Herring, Jack Hornsby, Paul Lorczak, and Sarah Sponholz did a thorough, timely job accuracy checking manuscript and page proofs and Lisa Collette checked the index.

We particularly thank the many students and instructors who have used this textbook over the years. You are the reason we do what we do. It is our hope that we have positively impacted your mathematics journey.

We welcome any comments or suggestions you have via email to math@pearson.com.

John Hornsby
Terry McGinnis

STUDENT SUPPLEMENTS

STUDENT'S SOLUTIONS MANUAL

- Provides detailed solutions to the odd-numbered, section-level exercises and to all Now Try Exercises, Relating Concepts, Summary, Chapter Review, Mixed Review, Chapter Test, and Cumulative Review Exercises

ISBNs: 0-321-96961-8, 978-0-321-96961-3

LIAL VIDEO LIBRARY

The **Lial Video Library,** available in MyMathLab, provides students with a wealth of video resources to help them navigate the road to success. All video resources in the library include optional subtitles in English. The **Lial Video Library** includes the following resources:

- **Section Lecture Videos** offer a new navigation menu that allows students to easily focus on the key examples and exercises that they need to review in each section. Optional Spanish subtitles are available.
- **Solutions Clips** show an instructor working through the complete solutions to selected exercises from the text. Exercises with a solution clip are marked in the text and e-book with a Play Button icon ▶.
- **Quick Review Lectures** provide a short summary lecture of each key concept from the Quick Reviews at the end of every chapter in the text.
- **The Chapter Test Prep Videos** provide step-by-step solutions to all exercises from the Chapter Tests. These videos provide guidance and support when students need it the most: the night before an exam. The Chapter Test Prep Videos are also available on YouTube (searchable using author name and book title).

LIAL VIDEO LIBRARY WORKBOOK

NEW! The **Lial Video Library Workbook** is an unbound, three-hole-punched workbook/note-taking guide that students can use in conjunction with the Lial Video Library. The notebook helps students develop organized notes as they work along with the videos. The notebook includes:

- Guided Examples to be used in conjunction with the Lial Video Library Section Lecture Videos, plus corresponding Now Try Exercises for each text objective.
- Extra practice exercises for every section of the text with ample space for students to show their work.
- Learning objectives and key vocabulary terms for every text section, along with vocabulary practice problems.

ISBNs: 0-321-96968-5, 978-0-321-96968-2

INSTRUCTOR SUPPLEMENTS

ANNOTATED INSTRUCTOR'S EDITION

- Provides "on-page" answers to all text exercises in an easy-to-read margin format, along with helpful Teaching Tips and extensive Classroom Examples

ISBNs: 0-321-98337-8, 978-0-321-98337-4

INSTRUCTOR'S SOLUTIONS MANUAL (Download only)

- Provides complete answers to all text exercises, including all Classroom Examples and Now Try Exercises

ISBNs: 0-321-96969-3, 978-0-321-96969-9

INSTRUCTOR'S RESOURCE MANUAL WITH TESTS

(Download only)

- Contains two diagnostic pretests, four free-response and two multiple-choice test forms per chapter, and two final exams
- Includes a mini-lecture for each section of the text with objectives, key examples, and teaching tips
- Provides a correlation guide from the eleventh to the twelfth edition

ISBNs: 0-321-96960-X, 978-0-321-96960-6

AVAILABLE FOR STUDENTS AND INSTRUCTORS

MYMATHLAB® ONLINE COURSE (access code required)

MyMathLab from Pearson is the world's leading online resource in mathematics, integrating interactive homework, assessment, and media in a flexible, easy-to-use format. It provides **engaging experiences** that personalize, stimulate, and measure learning for each student. And, it comes from an **experienced partner** with educational expertise and an eye on the future.

To learn more about how MyMathLab combines proven learning applications with powerful assessment, visit www.mymathlab.com or contact your Pearson representative.

MYMATHLAB® READY TO GO COURSE (access code required)

These new "Ready to Go" courses provide students with all the same great MyMathLab features, but make it easier for instructors to get started. Each course includes pre-assigned homework and quizzes to make creating a course even simpler. Ask your Pearson representative about the details for this particular course or to see a copy of this course.

MATHXL® ONLINE COURSE (access code required)

MathXL® is the homework and assessment engine that runs MyMathLab. (MyMathLab is MathXL plus a learning management system.)

With MathXL, instructors can:

- Create, edit, and assign online homework and tests using algorithmically generated exercises correlated at the objective level to the textbook.
- Create and assign their own online exercises and import TestGen tests for added flexibility.
- Maintain records of all student work tracked in MathXL's online gradebook.

With MathXL, students can:

- Take chapter tests in MathXL and receive personalized study plans and/or personalized homework assignments based on their test results.
- Use the study plan and/or the homework to link directly to tutorial exercises for the objectives they need to study.
- Access supplemental animations and video clips directly from selected exercises.

MathXL is available to qualified adopters. For more information, visit our website at www.mathxl.com, or contact your Pearson representative.

TESTGEN®

TestGen® (www.pearsoned.com/testgen) enables instructors to build, edit, print, and administer tests using a computerized bank of questions developed to cover all the objectives of the text. TestGen is algorithmically based, allowing instructors to create multiple but equivalent versions of the same question or test with the click of a button. Instructors can also modify test bank questions or add new questions. The software and testbank are available for download from Pearson Education's online catalog.

POWERPOINT® LECTURE SLIDES

- Present key concepts and definitions from the text
- Available for download at www.pearsonhighered.com

ISBNs: 0-321-96970-7, 978-0-321-96970-5

STUDY SKILLS

Using Your Math Textbook

Your textbook is a valuable resource. You will learn more if you fully make use of the features it offers.

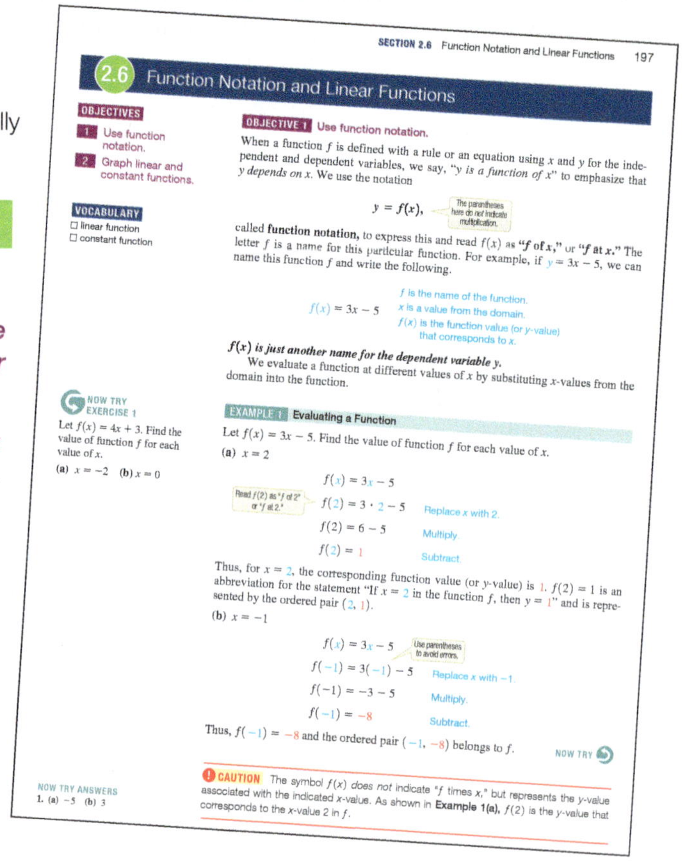
SECTION 2.6 Function Notation and Linear Functions 197

2.6 Function Notation and Linear Functions

OBJECTIVES
1 Use function notation.
2 Graph linear and constant functions.

VOCABULARY
☐ linear function
☐ constant function

OBJECTIVE 1 Use function notation.

When a function f is defined with a rule or an equation using x and y for the independent and dependent variables, we say, "y *is a function of* x" to emphasize that y *depends on* x. We use the notation

$$y = f(x),$$

The parentheses here do not indicate multiplication.

called **function notation,** to express this and read $f(x)$ as "**f of x,**" or "**f at x.**" The letter f is a name for this particular function. For example, if $y = 3x - 5$, we can name this function f and write the following.

$$f(x) = 3x - 5$$

f is the name of the function. x is a value from the domain. $f(x)$ is the function value (or y-value) that corresponds to x.

$f(x)$ is just another name for the dependent variable y.

We evaluate a function at different values of x by substituting x-values from the domain into the function.

NOW TRY EXERCISE 1 Let $f(x) = 4x + 3$. Find the value of function f for each value of x.
(a) $x = -2$ (b) $x = 0$

EXAMPLE 1 Evaluating a Function

Let $f(x) = 3x - 5$. Find the value of function f for each value of x.

(a) $x = 2$

$f(x) = 3x - 5$
$f(2) = 3 \cdot 2 - 5$ Replace x with 2. (Read $f(2)$ as "f of 2" or "f at 2.")
$f(2) = 6 - 5$ Multiply.
$f(2) = 1$ Subtract.

Thus, for $x = 2$, the corresponding function value (or y-value) is 1. $f(2) = 1$ is an abbreviation for the statement "If $x = 2$ in the function f, then $y = 1$" and is represented by the ordered pair $(2, 1)$.

(b) $x = -1$

$f(x) = 3x - 5$ (Use parentheses to avoid errors.)
$f(-1) = 3(-1) - 5$ Replace x with -1.
$f(-1) = -3 - 5$ Multiply.
$f(-1) = -8$ Subtract.

Thus, $f(-1) = -8$ and the ordered pair $(-1, -8)$ belongs to f. NOW TRY

NOW TRY ANSWERS 1. (a) -5 (b) 3

CAUTION The symbol $f(x)$ *does not* indicate "f times x," but represents the y-value associated with the indicated x-value. As shown in **Example 1(a),** $f(2)$ is the y-value that corresponds to the x-value 2 in f.

General Features

Locate each general feature and complete any blanks.

- **Table of Contents** Find this at the front of the text. ***Mark the chapters and sections you will cover, as noted on your course syllabus.***
- **Answer Section** ***Tab this section at the back of the book*** so you can refer to it frequently when doing homework. Answers to odd-numbered section exercises are provided.
- **Glossary** Find this feature after the answer section at the back of the text. It provides an alphabetical list of the key terms found in the text, with definitions and section references. ***Using the glossary, an equation is a statement that*** ________.
- **List of Formulas** Inside the back cover of the text is a helpful list of geometric formulas, along with review information on triangles and angles. Use these for reference throughout the course. ***The formula for finding the volume of a cube is*** ________.

Specific Features

Look through Chapter 1 and give the number of a page that includes an example of each of the following specific features.

- **Objectives** The objectives are listed at the beginning of each section and again within the section as the corresponding material is presented. Once you finish a section, ask yourself if you have accomplished them. ***See page*** ________.
- **Vocabulary List** Important vocabulary is listed at the beginning of each section. You should be able to define these terms when you finish a section. ***See page*** ________.
- **Now Try Exercises** These margin exercises allow you to immediately practice the material covered in the examples and prepare you for the exercises. Check your results using the answers at the bottom of the page. ***See page*** ________.
- **Pointers** These small shaded balloons provide on-the-spot warnings and reminders, point out key steps, and give other helpful tips. ***See page*** ________.
- **Cautions** These provide warnings about common errors that students often make or trouble spots to avoid. ***See page*** ________.
- **Notes** These provide additional explanations or emphasize other important ideas. ***See page*** ________.
- **Problem-Solving Hints** These boxes give helpful tips or strategies to use when you work applications. ***See page*** ________.

R Review of the Real Number System

R.1 Basic Concepts

OBJECTIVES

1. Write sets using set notation.
2. Use number lines.
3. Know the common sets of numbers.
4. Find additive inverses.
5. Use absolute value.
6. Use inequality symbols.

VOCABULARY

- ☐ set
- ☐ elements (members)
- ☐ finite set
- ☐ natural (counting) numbers
- ☐ infinite set
- ☐ whole numbers
- ☐ empty (null) set
- ☐ variable
- ☐ number line
- ☐ integers
- ☐ coordinate
- ☐ graph
- ☐ rational numbers
- ☐ irrational numbers
- ☐ real numbers
- ☐ additive inverse (opposite, negative)
- ☐ signed numbers
- ☐ absolute value
- ☐ equation
- ☐ inequality

OBJECTIVE 1 Write sets using set notation.

A **set** is a collection of objects called the **elements,** or **members,** of the set. In algebra, the elements of a set are usually numbers. Set braces, $\{\ \}$, are used to enclose the elements.

For example, 2 is an element of the set $\{1, 2, 3\}$. Because we can count the number of elements in the set $\{1, 2, 3\}$ and the counting comes to an end, it is a **finite set.**

In our study of algebra, we refer to certain sets of numbers by name. The set

$$N = \{1, 2, 3, 4, 5, 6, \ldots\} \quad \text{Natural (counting) numbers}$$

is the **natural numbers,** or the **counting numbers.** The three dots (*ellipsis* points) show that the list continues in the same pattern indefinitely. We cannot list all of the elements of the set of natural numbers, so it is an **infinite set.**

Including 0 with the set of natural numbers gives the set of **whole numbers.**

$$W = \{0, 1, 2, 3, 4, 5, 6, \ldots\} \quad \text{Whole numbers}$$

The set containing no elements, such as the set of whole numbers less than 0, is the **empty set,** or **null set,** usually written $\varnothing$ or $\{\ \}$.

CAUTION Do not write $\{\varnothing\}$ for the empty set. $\{\varnothing\}$ is a set with one element: $\varnothing$. Use the notation $\varnothing$ or $\{\ \}$ for the empty set.

To indicate that 2 is an element of the set $\{1, 2, 3\}$, we use the symbol $\in$ (read "is an element of").

$$2 \in \{1, 2, 3\}$$

The number 2 is also an element of the set of natural numbers N.

$$2 \in N$$

To show that 0 is *not* an element of set N, we draw a slash through the symbol $\in$.

$$0 \notin N$$

TEACHING TIP
Mention that in language, pronouns are used to stand for persons, places, or objects. In mathematics, variables are used to stand for numbers.

Two sets are equal if they contain exactly the same elements. For example, $\{1, 2\} = \{2, 1\}$. (Order doesn't matter.) However, $\{1, 2\} \neq \{0, 1, 2\}$ ($\neq$ means "is not equal to"), because one set contains the element 0 while the other does not.

A **variable** is a symbol, usually a letter, used to represent a number or to define a set of numbers. For example,

$$\{x \mid x \text{ is a natural number between 3 and 15}\}$$

(read "the set of all elements x such that x is a natural number between 3 and 15") defines the following set.

$$\{4, 5, 6, 7, \ldots, 14\}$$

The notation $\{x \mid x \text{ is a natural number between 3 and 15}\}$ is an example of **set-builder notation.**

TEACHING TIP
We use set-builder notation to specify solutions of dependent equations in **Chapter 3** on systems of equations.

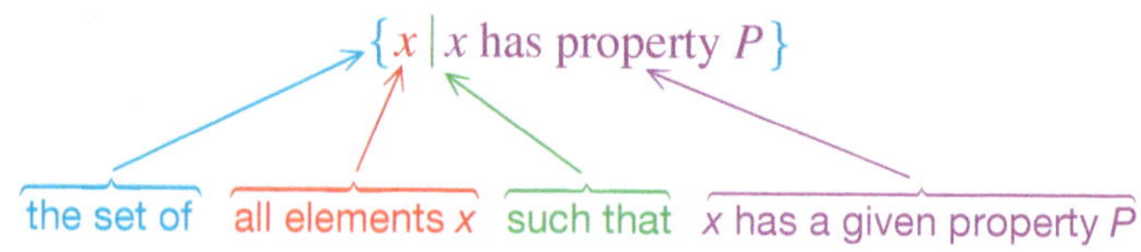

NOW TRY EXERCISE 1
List the elements in the set.

$\{p \mid p \text{ is a natural number less than } 6\}$

EXAMPLE 1 Listing the Elements in Sets

List the elements in each set.

(a) $\{x \mid x \text{ is a natural number less than } 4\}$

The natural numbers less than 4 are 1, 2, and 3. This set is $\{1, 2, 3\}$.

(b) $\{x \mid x \text{ is one of the first five even natural numbers}\}$ is $\{2, 4, 6, 8, 10\}$.

(c) $\{x \mid x \text{ is a natural number greater than or equal to } 7\}$

The set of natural numbers greater than or equal to 7 is an infinite set, written with ellipsis points as

$$\{7, 8, 9, 10, \ldots\}.$$

NOW TRY

CLASSROOM EXAMPLE 1
List the elements in the set.

$\{x \mid x \text{ is a natural number greater than } 12\}$

Answer: $\{13, 14, 15, \ldots\}$

NOW TRY EXERCISE 2
Use set-builder notation to describe the set.

$\{9, 10, 11, 12\}$

EXAMPLE 2 Using Set-Builder Notation to Describe Sets

Use set-builder notation to describe each set.

(a) $\{1, 3, 5, 7, 9\}$

There are often several ways to describe a set in set-builder notation. One way to describe the given set is

$$\{x \mid x \text{ is one of the first five odd natural numbers}\}.$$

(b) $\{5, 10, 15, \ldots\}$

This set can be described as $\{x \mid x \text{ is a positive multiple of } 5\}$. NOW TRY

CLASSROOM EXAMPLE 2
Use set-builder notation to describe the set.

$\{0, 1, 2, 3, 4, 5\}$

Answer: One answer is $\{x \mid x \text{ is a whole number less than } 6\}$.

OBJECTIVE 2 Use number lines.

A good way to picture a set of numbers is to use a **number line.** See **FIGURE 1**.

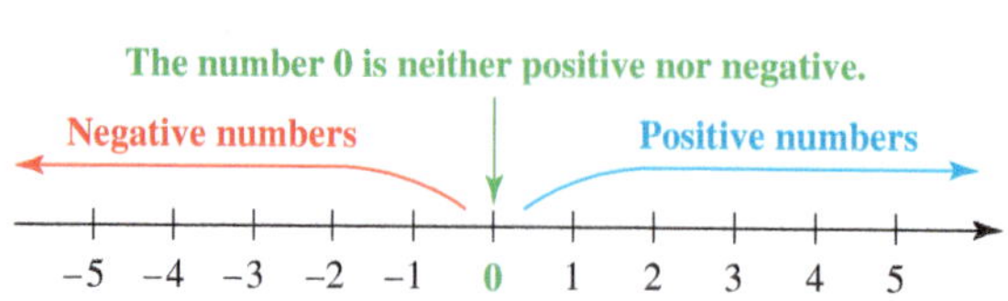

FIGURE 1

To draw a number line, choose any point on the line and label it 0. Then choose any point to the right of 0 and label it 1. Use the distance between 0 and 1 as the scale to locate, and then label, other points.

NOW TRY ANSWERS
1. $\{1, 2, 3, 4, 5\}$
2. One answer is $\{x \mid x \text{ is a natural number between 8 and 13}\}$.

TEACHING TIP
Stress that there may be more than one way to describe a set using set-builder notation. In **Example 1(c)**, for instance, point out that

$\{x \mid x$ is a *natural number greater than* $6\}$

and

$\{x \mid x$ is an *integer greater than or equal to* $7\}$

describe the same set. Mention that the variable used to describe the set can also be different. Work **Exercise 1** in the exercise set as a class to reinforce this concept.

The set of numbers identified on the number line in **FIGURE 1**, including positive and negative numbers and 0, is part of the set of **integers.**

$$I = \{\ldots, -3, -2, -1, 0, 1, 2, 3, \ldots\} \qquad \text{Integers}$$

Each number on a number line is the **coordinate** of the point that it labels, while the point is the **graph** of the number. **FIGURE 2** shows a number line with several points graphed on it.

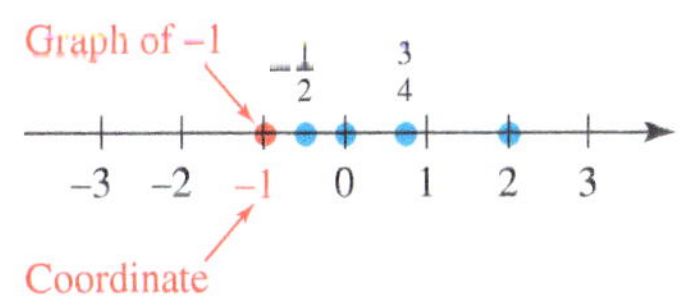

FIGURE 2

The fractions $-\frac{1}{2}$ and $\frac{3}{4}$, graphed on the number line in **FIGURE 2**, are *rational numbers*. A **rational number** can be expressed as the quotient of two integers, with denominator not 0. The set of all rational numbers is written as follows.

$$\left\{\frac{p}{q} \,\middle|\, p \text{ and } q \text{ are integers, where } q \neq 0\right\} \qquad \text{Rational numbers}$$

The set of rational numbers includes the natural numbers, whole numbers, and integers because these numbers can be written as fractions. For example,

$$14 = \frac{14}{1}, \quad -3 = \frac{-3}{1}, \quad \text{and} \quad 0 = \frac{0}{1}.$$

A rational number written as a fraction, such as $\frac{1}{8}$ or $\frac{2}{3}$, can also be expressed as a decimal by dividing the numerator by the denominator.

$$\begin{array}{r} 0.125 \\ 8\overline{)1.000} \\ \underline{8} \\ 20 \\ \underline{16} \\ 40 \\ \underline{40} \\ 0 \end{array} \quad \begin{array}{l} \leftarrow \text{Terminating decimal (rational number)} \\ \\ \\ \\ \\ \\ \\ \leftarrow \text{Remainder is 0.} \end{array}$$

$$\frac{1}{8} = 0.125$$

$$\begin{array}{r} 0.666\ldots \\ 3\overline{)2.000\ldots} \\ \underline{18} \\ 20 \\ \underline{18} \\ 20 \\ \underline{18} \\ 2 \end{array} \quad \begin{array}{l} \leftarrow \text{Repeating decimal (rational number)} \\ \\ \\ \\ \\ \\ \\ \leftarrow \text{Remainder is never 0.} \end{array}$$

$$\frac{2}{3} = 0.\overline{6} \quad \leftarrow \text{A bar is written over the repeating digit(s).}$$

Thus, terminating decimals, such as $0.125 = \frac{1}{8}$, $0.8 = \frac{4}{5}$, and $2.75 = \frac{11}{4}$, and repeating decimals, such as $0.\overline{6} = \frac{2}{3}$ and $0.\overline{27} = \frac{3}{11}$, are rational numbers.

Decimal numbers that neither terminate nor repeat, which include many square roots, are **irrational numbers.**

$$\sqrt{2} = 1.414213562\ldots \quad \text{and} \quad -\sqrt{7} = -2.6457513\ldots \qquad \text{Irrational numbers}$$

NOTE Some square roots, such as $\sqrt{16} = 4$ and $\sqrt{\frac{9}{25}} = \frac{3}{5}$, are rational.

A decimal number such as 0.010010001 . . . has a pattern, but it is irrational because there is no fixed block of digits that repeat. Another irrational number is π. See **FIGURE 3**.

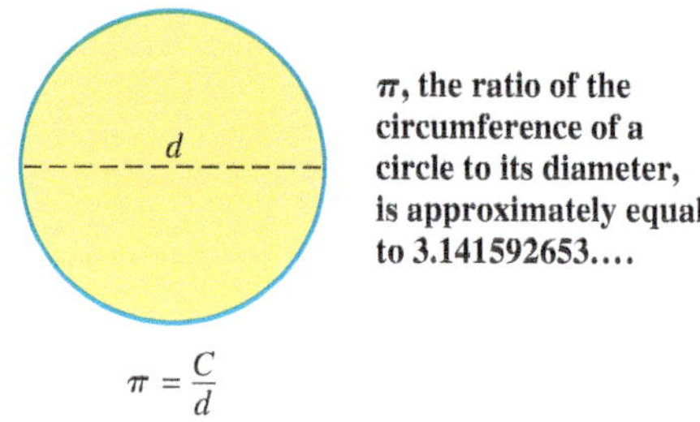

FIGURE 3

Some rational and irrational numbers are graphed on the number line in **FIGURE 4**. The rational numbers together with the irrational numbers make up the set of **real numbers.**

Every point on a number line corresponds to a real number, and every real number corresponds to a point on the number line.

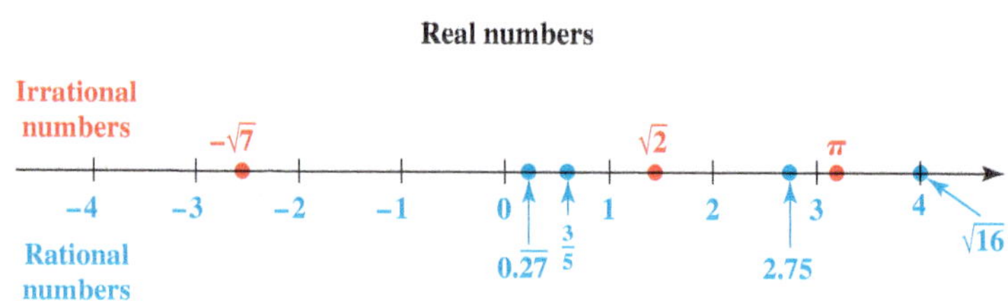

FIGURE 4

OBJECTIVE 3 Know the common sets of numbers.

CLASSROOM EXAMPLE 3
Select all the sets from the following list that apply to each number.
(a) −7 (b) 3.14 (c) $\sqrt{4}$

whole number
integer
rational number
irrational number
real number

Answers: (a) integer, rational number, real number
(b) rational number, real number
(c) $(\sqrt{4} = 2)$; whole number, integer, rational number, real number

Sets of Numbers

Natural numbers, or counting numbers	$\{1, 2, 3, 4, 5, 6, \ldots\}$
Whole numbers	$\{0, 1, 2, 3, 4, 5, 6, \ldots\}$
Integers	$\{\ldots, -3, -2, -1, 0, 1, 2, 3, \ldots\}$
Rational numbers	$\left\{\frac{p}{q} \mid p \text{ and } q \text{ are integers, where } q \neq 0\right\}$ *Examples:* $\frac{4}{1}$, 1.3, $-\frac{9}{2}$ or $-4\frac{1}{2}$, $\frac{16}{8}$ or 2, $\sqrt{9}$ or 3, $0.\overline{6}$
Irrational numbers	$\{x \mid x$ **is a real number that cannot be represented by a terminating or repeating decimal**$\}$ *Examples:* $\sqrt{3}$, $-\sqrt{2}$, π, 0.010010001 . . .
Real numbers	$\{x \mid x$ **is a rational number or an irrational number**$\}$*

*An example of a number that is not real is $\sqrt{-1}$. This number, part of the *complex number system,* is discussed in **Chapter 7.**

TEACHING TIP
Learning the relationships among the sets of real numbers is difficult for students. Use FIGURE 5 and relate each set to a department within a department, as in a large retail store.

FIGURE 5 shows the set of real numbers. ***Every real number is either rational or irrational.*** Notice that the integers are elements of the set of rational numbers and that the whole numbers and natural numbers are elements of the set of integers.

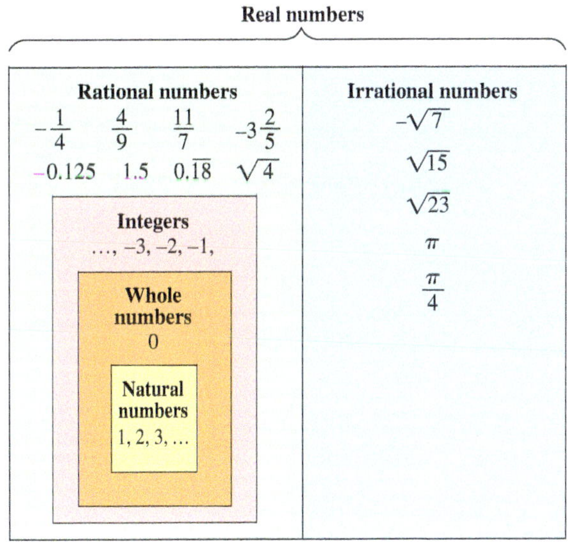

FIGURE 5

CLASSROOM EXAMPLE 4
Decide whether each statement is *true* or *false*. If it is false, tell why.
(a) Some integers are whole numbers.
(b) Every real number is irrational.
Answers: (a) true (b) false; Some real numbers are irrational, but others are rational numbers.

NOW TRY EXERCISE 3
List the numbers in the following set that are elements of each set.

$$\left\{-2.4, -\sqrt{1}, -\frac{1}{2}, 0, 0.\overline{3}, \sqrt{5}, \pi, 5\right\}$$

(a) Whole numbers
(b) Rational numbers

EXAMPLE 3 Identifying Examples of Number Sets

List the numbers in the following set that are elements of each set.

$$\left\{-8, -\sqrt{5}, -\frac{9}{64}, 0, 0.5, \frac{1}{3}, 1.\overline{12}, \sqrt{3}, 2, \pi\right\}$$

(a) Integers
-8, 0, and 2

(b) Rational numbers
$-8, -\frac{9}{64}, 0, 0.5, \frac{1}{3}, 1.\overline{12}$, and 2

(c) Irrational numbers
$-\sqrt{5}$, $\sqrt{3}$, and π

(d) Real numbers
All are real numbers. NOW TRY

NOW TRY EXERCISE 4
Decide whether each statement is *true* or *false*. If it is false, tell why.
(a) All integers are irrational numbers.
(b) Every whole number is an integer.

EXAMPLE 4 Determining Relationships between Sets of Numbers

Decide whether each statement is *true* or *false*. If it is false, tell why.

(a) All irrational numbers are real numbers.

This is true. As shown in **FIGURE 5**, the set of real numbers includes all irrational numbers.

(b) Every rational number is an integer.

This is false. Although some rational numbers are integers, other rational numbers, such as $\frac{2}{3}$ and $-\frac{1}{4}$, are not. NOW TRY

OBJECTIVE 4 Find additive inverses.

In **FIGURE 6**, for each positive number, there is a negative number on the opposite side of 0 that lies the same distance from 0. These pairs of numbers are *additive inverses, opposites,* or *negatives* of each other. For example, 3 and -3 are additive inverses.

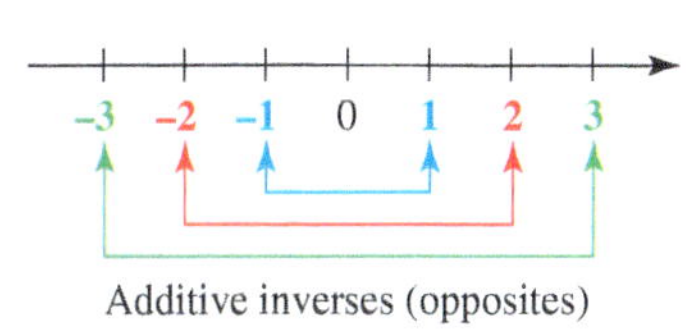

FIGURE 6

NOW TRY ANSWERS
3. (a) $\{0, 5\}$
(b) $\left\{-2.4, -\sqrt{1}, -\frac{1}{2}, 0, 0.\overline{3}, 5\right\}$
4. (a) false; All integers are rational numbers.
(b) true

Additive Inverse

For any real number a, the number $-a$ is the **additive inverse** of a.

We change the sign of a number to find its additive inverse. As we shall see later, the sum of a number and its additive inverse is always 0.

Uses of the Symbol −

The symbol "−" can be used to indicate any of the following.

1. A negative number, as in -9, read "negative 9"
2. The additive inverse of a number, as in "-4 is the additive inverse of 4"
3. Subtraction, as in $12 - 3$, read "12 minus 3"

In the expression $-(-5)$, the symbol "−" is being used in two ways. The first symbol − indicates the additive inverse (or opposite) of -5, and the second indicates a negative number, -5. Because the additive inverse of -5 is 5,

$$-(-5) = 5.$$

▼ Additive Inverses of Signed Numbers

Number	Additive Inverse
6	−6
−4	4
$\frac{2}{3}$	$-\frac{2}{3}$
−8.7	8.7
0	0

The number 0 is its own additive inverse.

−(−a)

For any real number a, $\quad \mathbf{-(-a) = a}$.

Numbers written with positive or negative signs, such as $+4$, $+8$, -9, and -5, are **signed numbers.** A positive number can be called a signed number even though the positive sign is usually left off.

OBJECTIVE 5 Use absolute value.

Geometrically, the **absolute value** of a number a, written $|a|$, is the distance on the number line from 0 to a. For example, the absolute value of 5 is the same as the absolute value of -5 because each number lies five units from 0. See **FIGURE 7**.

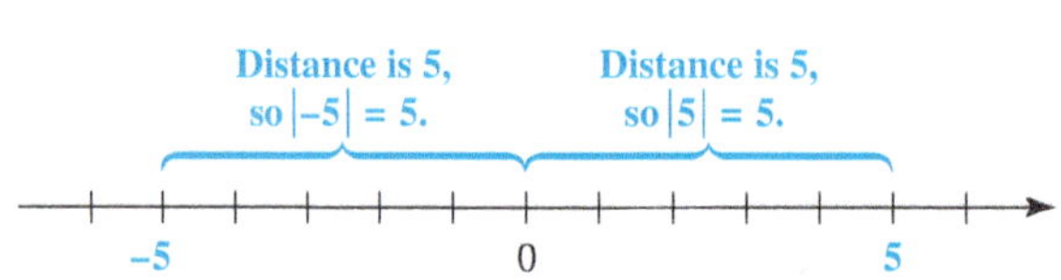

FIGURE 7

TEACHING TIP
Point out that $|7| = 7$, but the additive inverse of 7 is -7. Also, compare these examples.

$$-|-7| \quad \text{and} \quad -(-7)$$

CAUTION ***Because absolute value represents undirected distance, the absolute value of a number is always positive or 0.***

The formal definition of absolute value follows.

TEACHING TIP
Mention that, in general,

$$|a \pm b| \neq |a| \pm |b|.$$

Absolute Value

For any real number a, $\quad |a| = \begin{cases} a & \textbf{if } a \textbf{ is positive or 0} \\ -a & \textbf{if } a \textbf{ is negative.} \end{cases}$

Consider the second part of this definition, $|a| = -a$ if a is negative. If a is a *negative* number, then $-a$, the additive inverse or opposite of a, is a positive number. Thus, $|a|$ is positive. For example, let $a = -3$.

$$|a| = |-3| = -(-3) = 3 \qquad |a| = -a \text{ if } a \text{ is negative.}$$

NOW TRY EXERCISE 5

Evaluate each expression involving absolute value.

(a) $|-7|$ **(b)** $-|-15|$

(c) $|4| - |-4|$

CLASSROOM EXAMPLE 5

Evaluate each expression involving absolute value.

(a) $|-3|$ (b) $-|3|$

(c) $-|-3|$ (d) $|8| + |-1|$

(e) $|8 - 1|$

Answers: (a) 3 (b) -3 (c) -3 (d) 9 (e) 7

EXAMPLE 5 Finding Absolute Value

Evaluate each expression involving absolute value.

(a) $|13| = 13$

(b) $|-2| = -(-2) = 2$

(c) $|0| = 0$

(d) $|-0.75| = 0.75$

(e) $-|8| = -(8) = -8$ Evaluate the absolute value. Then find the additive inverse.

(f) $-|-8| = -(8) = -8$ Work as in part (e); $|-8| = 8$.

(g) $|-2| + |5| = 2 + 5 = 7$ Evaluate each absolute value, and then add.

(h) $-|5 - 2| = -|3| = -3$ Subtract inside the absolute value bars first.

NOW TRY

NOW TRY EXERCISE 6

Refer to the table in **Example 6.** Of the interpreters and translators, postal service clerks, and locomotive firers, which occupation is expected to see the least change (without regard to sign)?

CLASSROOM EXAMPLE 6

Of the physicians assistants and locomotive firers, which will show the greater change (without regard to sign)?

Answer: locomotive firers

EXAMPLE 6 Comparing Rates of Change in Industries

The projected total rates of change in employment (in percent) in some of the fastest-growing and in some of the most rapidly declining occupations from 2012 through 2022 are shown in the table.

Occupation (2012–2022)	*Total Rate of Change (in percent)*
Interpreters and translators	46.1
Personal care aides	48.8
Physicians assistants	38.4
Word processors and typists	−25.1
Postal service clerks	−31.8
Locomotive firers	−42.0

Source: Bureau of Labor Statistics.

What occupation in the table is expected to see the greatest change? The least change?

We want the greatest change, without regard to whether the change is an increase or a decrease. Look for the number in the table with the greatest absolute value. That number is for personal care aides.

$$|48.8| = 48.8$$

Similarly, the least change is for word processors and typists.

$$|-25.1| = 25.1$$

NOW TRY

NOW TRY ANSWERS

5. **(a)** 7 **(b)** -15 **(c)** 0

6. postal service clerks

OBJECTIVE 6 Use inequality symbols.

The statement

$$4 + 2 = 6$$

is an **equation**—a statement that two quantities are equal. The statement

$$4 \neq 6 \quad \text{(read “4 is not equal to 6”)}$$

is an **inequality**—a statement that two quantities are *not* equal. When two numbers are not equal, one must be less than the other. When reading from left to right, the symbol $<$ means “is less than.”

$$8 < 9, \quad -6 < 15, \quad -6 < -1, \quad \text{and} \quad 0 < \frac{4}{3} \qquad \text{All are true.}$$

TEACHING TIP
Emphasize that the “arrowhead” always points to the lesser number.

Reading from left to right, the symbol $>$ means “is greater than.”

$$12 > 5, \quad 9 > -2, \quad -4 > -6, \quad \text{and} \quad \frac{6}{5} > 0 \qquad \text{All are true.}$$

In each case, the symbol “points” toward the lesser number.

The number line in **FIGURE 8** shows the graphs of the numbers 4 and 9. On the graph, 4 is to the *left* of 9, so $4 < 9$. ***The lesser of two numbers is always to the left of the other on a number line.***

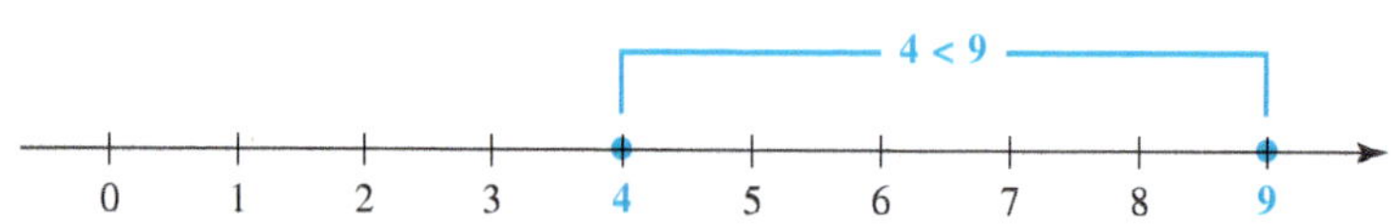

FIGURE 8

Inequalities on a Number Line

On a number line, the following hold.

$a < b$ if a is to the left of b. $\qquad$ $a > b$ if a is to the right of b.

NOW TRY EXERCISE 7
Use a number line to determine whether each statement is *true* or *false*.

(a) $-5 > -1$

(b) $-7 < -6$

EXAMPLE 7 Determining Order on a Number Line

Use a number line to compare -6 and 1 and to compare -5 and -2.

As shown on the number line in **FIGURE 9**, -6 is located to the left of 1. For this reason, $-6 < 1$. Also, $1 > -6$. From **FIGURE 9**, $-5 < -2$, or $-2 > -5$. In each case, the symbol points to the lesser number.

FIGURE 9

NOW TRY

CLASSROOM EXAMPLE 7
Use a number line to determine whether each statement is *true* or *false*.

(a) $-8 > -4$ (b) $-2 < 5$

Answers: (a) false (b) true

The following table summarizes this discussion.

Results about Positive and Negative Numbers

Words	*Symbols*
Every negative number is less than 0.	If a is negative, then $a < 0$.
Every positive number is greater than 0.	If a is positive, then $a > 0$.
0 is neither positive nor negative.	

NOW TRY ANSWERS
7. (a) false **(b)** true

CLASSROOM EXAMPLE 8

Determine whether each statement is *true* or *false*.

(a) $-2 \le -3$

(b) $-1 \ge -9$

(c) $8 \le 8$

(d) $3(4) > 2(6)$

Answers: (a) false (b) true (c) true (d) $12 > 12$; false

In addition to the symbols $\neq$, $<$, and $>$, the symbols $\le$ and $\ge$ are often used.

Inequality Symbols

Symbol	Meaning	Example
$\neq$	is not equal to	$3 \neq 7$
$<$	is less than	$-4 < -1$
$>$	is greater than	$3 > -2$
$\le$	is less than or equal to	$6 \le 6$
$\ge$	is greater than or equal to	$-8 \ge -10$

NOW TRY EXERCISE 8

Determine whether each statement is *true* or *false*.

(a) $-5 \le -6$

(b) $-2 \ge -10$

(c) $0.5 \le 0.5$

(d) $10(6) > 8(7)$

EXAMPLE 8 Using Inequality Symbols

The table shows several uses of inequalities and why each is true.

Inequality	*Why It Is True*
$6 \le 8$	$6 < 8$
$-2 \le -2$	$-2 = -2$
$-9 \ge -12$	$-9 > -12$
$-3 \ge -3$	$-3 = -3$
$6 \cdot 4 \le 5(5)$	$24 < 25$

Notice the reason why $-2 \le -2$ is true. ***With the symbol $\le$, if either the $<$ part or the $=$ part is true, then the inequality is true. This is also the case with the $\ge$ symbol.***

In the last row of the table, recall that the dot in $6 \cdot 4$ indicates the product 6×4, or 24, and $5(5)$ means 5×5, or 25. Thus, the inequality $6 \cdot 4 \le 5(5)$ becomes $24 \le 25$, which is true. NOW TRY

NOW TRY ANSWERS

8. **(a)** false **(b)** true **(c)** true **(d)** $60 > 56$; true

R.1 Exercises

FOR EXTRA HELP

MyMathLab®

Complete solution available in MyMathLab

1. yes

2. **(a)** 7 **(b)** One example is $\frac{1}{2}$. Other answers are possible. **(c)** 0 **(d)** One example is -1. Any negative integer is correct. **(e)** One example is $\sqrt{5}$. Other answers are possible. **(f)** One example is $-\pi$. Other answers are possible.

3. $\{1, 2, 3, 4, 5\}$

4. $\{1, 2, 3, 4, 5, 6, 7, 8\}$

5. $\{5, 6, 7, 8, \ldots\}$

6. $\{9, 10, 11, 12, \ldots\}$

7. $\{\ldots, -1, 0, 1, 2, 3, 4\}$

1. *Concept Check* A student claimed that $\{x \mid x$ is a natural number greater than $3\}$ and $\{y \mid y$ is a natural number greater than $3\}$ actually name the same set, even though different variables are used. Was this student correct?

2. *Concept Check* Give a real number that satisfies each condition.

(a) An integer between 6.75 and 7.75

(b) A rational number between $\frac{1}{4}$ and $\frac{3}{4}$

(c) A whole number that is not a natural number

(d) An integer that is not a whole number

(e) An irrational number between $\sqrt{4}$ and $\sqrt{9}$

(f) An irrational number that is negative

List the elements in each set. ***See Example 1.***

3. $\{x \mid x$ is a natural number less than $6\}$

4. $\{m \mid m$ is a natural number less than $9\}$

5. $\{z \mid z$ is an integer greater than $4\}$

6. $\{y \mid y$ is an integer greater than $8\}$

7. $\{z \mid z$ is an integer less than or equal to $4\}$

8. $\{p \mid p$ is an integer less than $3\}$

9. $\{a \mid a$ is an even integer greater than $8\}$

10. $\{k \mid k$ is an odd integer less than $1\}$

8. $\{\ldots, -2, -1, 0, 1, 2\}$
9. $\{10, 12, 14, 16, \ldots\}$
10. $\{\ldots, -7, -5, -3, -1\}$
11. $\{-4, 4\}$ **12.** $\{-7, 7\}$
13. $\varnothing$ **14.** $\varnothing$

In Exercises 15–18, we give one possible answer.
15. $\{x \mid x$ is an even natural number less than or equal to $8\}$
16. $\{x \mid x$ is an integer between 10 and $15\}$
17. $\{x \mid x$ is a positive multiple of $4\}$
18. $\{x \mid x$ is an integer multiple of $3\}$

19. 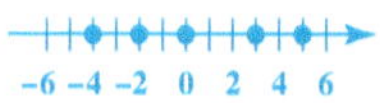

20. 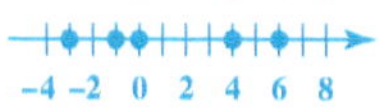

21. 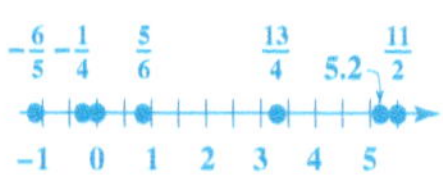

22. 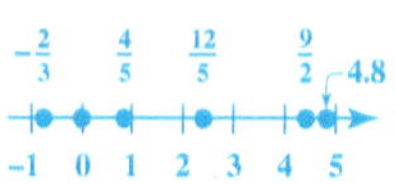

23. **(a)** $8, 13, \frac{75}{5}$ (or 15)
(b) $0, 8, 13, \frac{75}{5}$
(c) $-9, 0, 8, 13, \frac{75}{5}$
(d) $-9, -0.7, 0, \frac{6}{7}, 4.\overline{6}, 8, \frac{21}{2}, 13, \frac{75}{5}$
(e) $-\sqrt{6}, \sqrt{7}$
(f) All are real numbers.

24. **(a)** $5, 17, \frac{40}{2}$ (or 20)
(b) $0, 5, 17, \frac{40}{2}$
(c) $-8, 0, 5, 17, \frac{40}{2}$
(d) $-8, -0.6, 0, \frac{3}{4}, 5, \frac{13}{2}, 17, \frac{40}{2}$
(e) $-\sqrt{5}, \sqrt{3}, \pi$
(f) All are real numbers.

25. false; Some are whole numbers, but negative integers are not.
26. true
27. false; No irrational number is an integer.
28. true
29. true
30. false; No rational number is irrational.
31. true **32.** true
33. true **34.** true

11. $\{p \mid p$ is a number whose absolute value is $4\}$

12. $\{w \mid w$ is a number whose absolute value is $7\}$

13. $\{x \mid x$ is an irrational number that is also rational$\}$

14. $\{r \mid r$ is a number that is both positive and negative$\}$

Use set-builder notation to describe each set. ***See Example 2.*** *(More than one description is possible.)*

15. $\{2, 4, 6, 8\}$ **16.** $\{11, 12, 13, 14\}$

17. $\{4, 8, 12, 16, \ldots\}$ **18.** $\{\ldots, -6, -3, 0, 3, 6, \ldots\}$

Graph the elements of each set on a number line. ***See Objective 2.***

19. $\{-4, -2, 0, 3, 5\}$ **20.** $\{-3, -1, 0, 4, 6\}$

21. $\left\{-\frac{6}{5}, -\frac{1}{4}, 0, \frac{5}{6}, \frac{13}{4}, 5.2, \frac{11}{2}\right\}$ **22.** $\left\{-\frac{2}{3}, 0, \frac{4}{5}, \frac{12}{5}, \frac{9}{2}, 4.8\right\}$

Which elements of each set are ***(a)*** *natural numbers,* ***(b)*** *whole numbers,* ***(c)*** *integers,* ***(d)*** *rational numbers,* ***(e)*** *irrational numbers,* ***(f)*** *real numbers?* ***See Example 3.***

23. $\left\{-9, -\sqrt{6}, -0.7, 0, \frac{6}{7}, \sqrt{7}, 4.\overline{6}, 8, \frac{21}{2}, 13, \frac{75}{5}\right\}$

24. $\left\{-8, -\sqrt{5}, -0.6, 0, \frac{3}{4}, \sqrt{3}, \pi, 5, \frac{13}{2}, 17, \frac{40}{2}\right\}$

Decide whether each statement is true *or* false. *If it is false, tell why.* ***See Example 4.***

25. Every integer is a whole number. **26.** Every natural number is an integer.

27. Every irrational number is an integer. **28.** Every integer is a rational number.

29. Every natural number is a whole number. **30.** Some rational numbers are irrational.

31. Some rational numbers are whole numbers. **32.** Some real numbers are integers.

33. The absolute value of any number is the same as the absolute value of its additive inverse.

34. The absolute value of any nonzero number is positive.

35. ***Concept Check*** Match each expression in parts (a)–(d) with its value in choices A–D. Choices may be used once, more than once, or not at all.

I		II	
(a) $-(-4)$	**(b)** $\lvert -4 \rvert$	**A.** 4	**B.** -4
(c) $-\lvert -4 \rvert$	**(d)** $-\lvert -(-4) \rvert$	**C.** Both A and B	**D.** Neither A nor B

36. ***Concept Check*** For what value(s) of x is $\lvert x \rvert = 4$ true?

Give ***(a)*** *the additive inverse and* ***(b)*** *the absolute value of each number.* ***See Objective 4 and Example 5.***

37. 6 **38.** 9 **39.** -12 **40.** -14 **41.** $\frac{6}{5}$ **42.** 0.16

Evaluate each expression involving absolute value. ***See Example 5.***

43. $\lvert -8 \rvert$ **44.** $\lvert -19 \rvert$ **45.** $\left\lvert \frac{3}{2} \right\rvert$ **46.** $\left\lvert \frac{3}{4} \right\rvert$

47. $-\lvert 5 \rvert$ **48.** $-\lvert 12 \rvert$ **49.** $-\lvert -2 \rvert$ **50.** $-\lvert -6 \rvert$

35. (a) A (b) A (c) B (d) B
36. 4, −4
37. (a) −6 (b) 6
38. (a) −9 (b) 9
39. (a) 12 (b) 12
40. (a) 14 (b) 14
41. (a) $-\frac{6}{5}$ (b) $\frac{6}{5}$
42. (a) −0.16 (b) 0.16
43. 8
44. 19
45. $\frac{3}{2}$
46. $\frac{3}{4}$
47. −5
48. −12
49. −2
50. −6
51. −4.5
52. −12.4
53. 5
54. 30
55. 6
56. 3
57. 0
58. 0
59. (a) Las Vegas; The population increased by 45.4%. (b) Detroit; The population decreased by 3.6%.
60. (a) China; Imports exceeded exports by $28,862 million. (b) Turkey; Exports exceeded imports by $396 million.
61. Pacific Ocean, Indian Ocean, Caribbean Sea, South China Sea, Gulf of California
62. Point Success, Rainier, Matlalcueyetl, Steele, McKinley
63. true
64. false

51. $-|4.5|$ **52.** $-|12.4|$ **53.** $|-2|+|3|$

54. $|-16|+|14|$ **55.** $|-9|-|-3|$ **56.** $|-10|-|-7|$

57. $|-1|+|-2|-|-3|$ **58.** $|-7|+|-3|-|-10|$

Solve each problem. ***See Example 6.***

59. The table shows the percent change in population from 2000 through 2012 for selected metropolitan areas.

Metropolitan Area	*Percent Change*
Las Vegas	45.4
San Francisco	8.0
Chicago	4.7
New Orleans	−8.3
Phoenix	33.1
Detroit	−3.6

Source: U.S. Census Bureau.

(a) Which metropolitan area had the greatest change in population? What was this change? Was it an increase or a decrease?

(b) Which metropolitan area had the least change in population? What was this change? Was it an increase or a decrease?

60. The table gives the net trade balance, in millions of U.S. dollars, for selected U.S. trade partners for October 2013.

Country	*Trade Balance (in millions of dollars)*
India	−1967
China	−28,862
Netherlands	2325
France	−1630
Turkey	396

Source: U.S. Census Bureau.

A negative balance means that imports to the United States exceeded exports from the United States, while a positive balance means that exports exceeded imports.

(a) Which country had the greatest discrepancy between exports and imports? Explain.

(b) Which country had the least discrepancy between exports and imports? Explain.

Sea level refers to the surface of the ocean. The depth of a body of water can be expressed as a negative number, representing average depth in feet below sea level. The altitude of a mountain can be expressed as a positive number, indicating its height in feet above sea level.

Body of Water	*Average Depth in Feet (as a negative number)*	*Mountain*	*Altitude in Feet (as a positive number)*
Pacific Ocean	−14,040	McKinley	20,237
South China Sea	−4802	Point Success	14,164
Gulf of California	−2375	Matlalcueyetl	14,636
Caribbean Sea	−8448	Rainier	14,416
Indian Ocean	−12,800	Steele	16,624

Source: *World Almanac and Book of Facts.*

61. List the bodies of water in order, starting with the deepest and ending with the shallowest.

62. List the mountains in order, starting with the shortest and ending with the tallest.

63. *True* or *false:* The absolute value of the depth of the Pacific Ocean is greater than the absolute value of the depth of the Indian Ocean.

64. *True* or *false:* The absolute value of the depth of the Gulf of California is greater than the absolute value of the depth of the Caribbean Sea.

65. true **66.** true
67. false **68.** false
69. true **70.** true
71. true **72.** true

73. $2 < 6$ **74.** $1 < 5$
75. $4 > -9$ **76.** $1 > -6$
77. $-10 < -5$ **78.** $-12 < -7$
79. $x > 0$ **80.** $x > -3$

81. $7 > y$ **82.** $-4 < 10$
83. $5 \geq 5$ **84.** $-6 \leq -6$
85. $3t - 4 \leq 10$
86. $5x + 4 \geq 21$
87. $5x + 3 \neq 0$
88. $6x + 7 \neq -9$
89. $t \leq -3$ **90.** $r \geq -5$
91. $3x < 4$ **92.** $6x < -2$

93. $-6 < 10$; true
94. $-7 < 5$; true
95. $10 \geq 10$; true
96. $15 \leq 15$; true
97. $-3 \geq -3$; true
98. $-4 \leq -4$; true
99. $-8 > -6$; false
100. $-10 > -4$; false
101. $5 \geq -5$; true

102. true

103. Iowa (IA), Ohio (OH), Pennsylvania (PA)
104. Pennsylvania (PA), Florida (FL)
105. $x < y$
106. $x > y$; $y < x$

Use a number line to determine whether each statement in true *or* false. ***See Example 7.***

65. $-6 < -1$ **66.** $-4 < -2$ **67.** $-4 > -3$ **68.** $-3 > -1$

69. $3 > -2$ **70.** $6 > -3$ **71.** $-3 \geq -3$ **72.** $-5 \leq -5$

Rewrite each statement with $>$ *so that it uses* $<$ *instead. Rewrite each statement with* $<$ *so that it uses* $>$. ***See Example 7.***

73. $6 > 2$ **74.** $5 > 1$ **75.** $-9 < 4$ **76.** $-6 < 1$

77. $-5 > -10$ **78.** $-7 > -12$ **79.** $0 < x$ **80.** $-3 < x$

Use an inequality symbol to write each statement. ***See Example 8.***

81. 7 is greater than y.

82. -4 is less than 10.

83. 5 is greater than or equal to 5.

84. -6 is less than or equal to -6.

85. $3t - 4$ is less than or equal to 10.

86. $5x + 4$ is greater than or equal to 21.

87. $5x + 3$ is not equal to 0.

88. $6x + 7$ is not equal to -9.

89. t is less than or equal to -3.

90. r is greater than or equal to -5.

91. $3x$ is less than 4.

92. $6x$ is less than -2.

Determine whether each statement is true *or* false. ***See Example 8.***

93. $-6 < 7 + 3$ **94.** $-7 < 4 + 1$ **95.** $2 \cdot 5 \geq 4 + 6$

96. $8 + 7 \leq 3 \cdot 5$ **97.** $-|-3| \geq -3$ **98.** $-|-4| \leq -4$

99. $-8 > -|-6|$ **100.** $-10 > -|-4|$ **101.** $|-5| \geq -|-5|$

102. *Concept Check* *True* or *false:* $-|\pi| < -3$ (*Hint:* See **FIGURE 3**.)

The graph shows egg production in millions of eggs in selected states for 2011 and 2012. Use this graph to work Exercises 103–106.

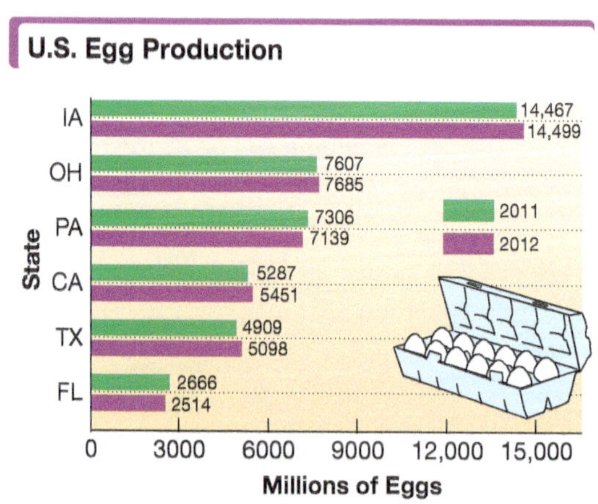

Source: U.S. Department of Agriculture.

103. In 2012, which states had production greater than 6000 million eggs?

104. In which of the states was 2012 egg production less than 2011 egg production?

105. If x represents 2012 egg production for Texas (TX) and y represents 2012 egg production for Ohio (OH), which is true, $x < y$ or $x > y$?

106. If x represents 2011 egg production for Iowa (IA) and y represents 2011 egg production for Pennsylvania (PA), write two inequalities, one with $>$ and one with $<$, that compare the production in these two states.

R.2 Operations on Real Numbers

OBJECTIVES

1. Add real numbers.
2. Subtract real numbers.
3. Find the distance between two points on a number line.
4. Multiply real numbers.
5. Find reciprocals and divide real numbers.

VOCABULARY

- ☐ sum
- ☐ difference
- ☐ product
- ☐ quotient
- ☐ reciprocal (multiplicative inverse)

NOW TRY EXERCISE 1

Find each sum.

(a) $-4 + (-9)$

(b) $-7.25 + (-3.57)$

(c) $-\frac{2}{5} + \left(-\frac{3}{10}\right)$

CLASSROOM EXAMPLE 1

Find each sum.

(a) $-6 + (-15)$

(b) $-1.1 + (-1.2)$

(c) $-\frac{3}{4} + \left(-\frac{5}{8}\right)$

Answers: (a) -21 (b) -2.3 (c) $-\frac{11}{8}$

TEACHING TIP

Perhaps use one of the money problems in **Exercises 125–128** to motivate a discussion of the rules for adding and subtracting signed numbers.

NOW TRY ANSWERS

1. (a) -13 (b) -10.82 (c) $-\frac{7}{10}$

OBJECTIVE 1 Add real numbers.

Recall that the answer to an addition problem is a **sum.**

Adding Real Numbers

Same sign To add two numbers with the *same* sign, add their absolute values. The sum has the same sign as the given numbers.

Examples: $2 + 7 = 9, \quad -2 + (-7) = -9$

Different signs To add two numbers with *different* signs, find the absolute values of the numbers, and subtract the lesser absolute value from the greater. The sum has the same sign as the number with the greater absolute value.

Examples: $-8 + 3 = -5, \quad 15 + (-9) = 6$

EXAMPLE 1 Adding Two Negative Real Numbers

Find each sum.

(a) $-12 + (-8)$

First find the absolute values: $|-12| = 12$ and $|-8| = 8$.

Because -12 and -8 have the *same* sign, add their absolute values.

Both numbers are negative, so the sum will be negative.

$$-12 + (-8)$$
$$= -(12 + 8) \quad \text{Add the absolute values.}$$
$$= -(20)$$
$$= -20$$

(b) $-6 + (-3)$

$$= -(|-6| + |-3|)$$
$$= -(6 + 3)$$
$$= -9$$

(c) $-1.2 + (-0.4)$

$$= -(1.2 + 0.4)$$
$$= -1.6$$

(d) $-\frac{5}{6} + \left(-\frac{1}{3}\right)$

$$= -\left(\frac{5}{6} + \frac{1}{3}\right)$$ Add the absolute values. Both numbers are negative, so the sum will be negative.

$$= -\left(\frac{5}{6} + \frac{2}{6}\right)$$ The least common denominator is 6. $\frac{1 \cdot 2}{3 \cdot 2} = \frac{2}{6}$

$$= -\frac{7}{6}$$ Add numerators. Keep the same denominator.

NOW TRY

NOW TRY EXERCISE 2

Find each sum.

(a) $-15 + 7$

(b) $4.6 + (-2.8)$

(c) $-\frac{5}{9} + \frac{2}{7}$

CLASSROOM EXAMPLE 2

Find each sum.

(a) $3 + (-7)$

(b) $-3 + 7$

(c) $-3.8 + 4.6$

(d) $-\frac{3}{8} + \frac{1}{4}$

Answers: (a) -4 (b) 4 (c) 0.8 (d) $-\frac{1}{8}$

EXAMPLE 2 Adding Real Numbers with Different Signs

Find each sum.

(a) $-17 + 11$

First find the absolute values: $|-17| = 17$ and $|11| = 11$. Because -17 and 11 have *different* signs, subtract their absolute values.

$$17 - 11 = 6$$

The number -17 has a greater absolute value than 11, so the answer is negative.

$$-17 + 11 = -6$$

The sum is negative because $|-17| > |11|$.

(b) $4 + (-1)$

Subtract the absolute values, 4 and 1. Because 4 has the greater absolute value, the sum must be positive.

$$4 + (-1) = 4 - 1 = 3$$

The sum is positive because $|4| > |-1|$.

(c) $-9 + 17$

$$= 17 - 9$$

$$= 8$$

(d) $-2.3 + 5.6$

$$= 5.6 - 2.3$$

$$= 3.3$$

(e) $-16 + 12$

The absolute values are 16 and 12. Subtract the absolute values.

$$-16 + 12 = -(16 - 12) = -4$$

The sum is negative because $|-16| > |12|$.

(f) $-\frac{4}{5} + \frac{2}{3}$

$$= -\frac{12}{15} + \frac{10}{15}$$

The least common denominator is 15. $-\frac{4 \cdot 3}{5 \cdot 3} = -\frac{12}{15}; \frac{2 \cdot 5}{3 \cdot 5} = \frac{10}{15}$

$$= -\left(\frac{12}{15} - \frac{10}{15}\right)$$

Subtract the absolute values. $-\frac{12}{15}$ has the greater absolute value, so the answer will be negative.

$$= -\frac{2}{15}$$

Subtract numerators. Keep the same denominator.

NOW TRY

OBJECTIVE 2 Subtract real numbers.

Recall that the answer to a subtraction problem is a **difference.** Compare the following two statements.

$$6 - 4 = 2$$

$$6 + (-4) = 2$$

Thus,

$$6 - 4 = 6 + (-4).$$

To subtract 4 from 6, we add the additive inverse of 4 to 6. This example suggests the following definition of subtraction.

NOW TRY ANSWERS

2. **(a)** -8 **(b)** 1.8 **(c)** $-\frac{17}{63}$

Subtraction

For all real numbers a and b, the following holds.

$$a - b = a + (-b)$$

To subtract b from a, add the additive inverse (or opposite) of b to a.

Examples: $5 - 12 = 5 + (-12) = -7$, $6 - (-3) = 6 + 3 = 9$

NOW TRY EXERCISE 3

Find each difference.

(a) $-4 - 11$

(b) $-5.67 - (-2.34)$

(c) $\frac{4}{9} - \frac{3}{5}$

CLASSROOM EXAMPLE 3

Find each difference.

(a) $12 - (-5)$

(b) $-11.5 - (-6.3)$

(c) $\frac{3}{4} - \left(-\frac{2}{3}\right)$

Answers: (a) 17 (b) -5.2 (c) $\frac{17}{12}$

EXAMPLE 3 Subtracting Real Numbers

Find each difference.

(a) $6 - 8 = 6 + (-8) = -2$

Change to addition. The additive inverse of 8 is -8.

(b) $-12 - 4 = -12 + (-4) = -16$

Change to addition. The additive inverse of 4 is -4.

(c) $-10 - (-7)$

$= -10 + 7$ The additive inverse of -7 is 7.

$= -3$

(d) $-2.4 - (-8.1)$

$= -2.4 + 8.1$

$= 5.7$

(e) $\frac{5}{6} - \left(-\frac{3}{8}\right)$

$= \frac{5}{6} + \frac{3}{8}$ To subtract $a - b$, add the additive inverse (opposite) of b to a.

$= \frac{20}{24} + \frac{9}{24}$ Write each fraction with the least common denominator, 24.

$= \frac{29}{24}$ Add numerators. Keep the same denominator.

NOW TRY

When working a problem that involves both addition and subtraction, add and subtract in order from left to right. Work inside brackets or parentheses first.

CLASSROOM EXAMPLE 4

Perform the indicated operations.

(a) $-6 - (-2) - 8 - 1$

(b) $-3 - [(-7) + 15] - 6$

Answers: (a) -13 (b) -17

EXAMPLE 4 Adding and Subtracting Real Numbers

Perform the indicated operations.

(a) $15 - (-3) - 5 - 12$

$= (15 + 3) - 5 - 12$ Work in order from left to right.

$= 18 - 5 - 12$ Add inside the parentheses.

$= 13 - 12$ Subtract from left to right.

$= 1$ Subtract.

NOW TRY ANSWERS

3. (a) -15 (b) -3.33 (c) $-\frac{7}{45}$

NOW TRY EXERCISE 4
Perform the indicated operations.

$$-4-(-2-7)-12$$

(b) $-9-[-8-(-4)]+6$

$= -9-[-8+4]+6$ Work inside the brackets.

$= -9-[-4]+6$ Add.

$= -9+4+6$ Add the additive inverse.

$= -5+6$ Add from left to right.

$= 1$ Add. NOW TRY

OBJECTIVE 3 Find the distance between two points on a number line.

The number line in **FIGURE 10** shows several points.

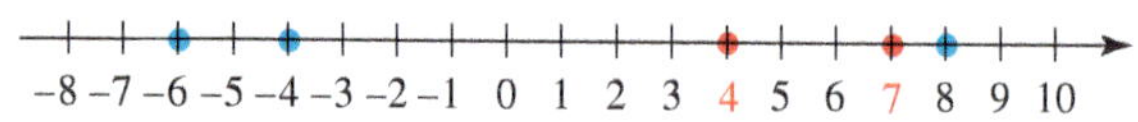

FIGURE 10

To find the distance between the points 4 and 7, we subtract $7-4=3$. Because distance is positive (or 0), we must be careful to subtract in such a way that the answer is positive (or 0). To avoid this problem altogether, we can find the absolute value of the difference. Then the distance between 4 and 7 is found as follows.

$$|7-4| = |3| = 3 \quad \text{or} \quad |4-7| = |-3| = 3 \qquad \text{Distance between 4 and 7}$$

Distance

The **distance** between two points on a number line is the absolute value of the difference of their coordinates.

NOW TRY EXERCISE 5
Find the distance between the points −7 and 12.

CLASSROOM EXAMPLE 5
Find the distance between the points −12 and −1.

Answer: 11

EXAMPLE 5 Finding Distance between Points on the Number Line

Find the distance between each pair of points. Refer to **FIGURE 10.**

(a) 8 and −4

Find the absolute value of the difference of the numbers, taken in either order.

$$|8-(-4)| = 12, \quad \text{or} \quad |-4-8| = 12$$

(b) −4 and −6

$$|-4-(-6)| = 2, \quad \text{or} \quad |-6-(-4)| = 2$$

NOW TRY

OBJECTIVE 4 Multiply real numbers.

Recall that the answer to a multiplication problem is a **product.**

Multiplying Real Numbers

Same sign The product of two numbers with the *same* sign is positive.

Examples: $4(8)=32, \quad -4(-8)=32$

Different signs The product of two numbers with *different* signs is negative.

Examples: $-8(9)=-72, \quad 6(-7)=-42$

NOW TRY ANSWERS
4. −7 **5.** 19

TEACHING TIP
Have students complete these statements to illustrate the multiplication property of 0.

(a) $-197 \cdot 0 =$ ______

(b) ______ $\cdot \frac{8}{9} = 0$

(c) $0 \cdot$ ______ $= 0$

The **multiplication property of 0** states that the product of any real number and 0 is 0.

Multiplication Property of 0

For any real number a, the following hold.

$$a \cdot 0 = 0 \quad \text{and} \quad 0 \cdot a = 0$$

Examples: $12 \cdot 0 = 0, \quad 0 \cdot 12 = 0$

NOW TRY EXERCISE 6

Find each product.

(a) $-3(-10)$

(b) $0.7(-1.2)$

(c) $-\frac{8}{11}(33)$

(d) $14(0)$

EXAMPLE 6 Multiplying Real Numbers

Find each product.

(a) $-3(-9) = 27$

(b) $-0.5(-0.4) = 0.2$

The numbers have the same sign, so the product is positive.

(c) $-\frac{3}{4}\left(-\frac{5}{6}\right)$

$= \frac{15}{24}$ Multiply numerators. Multiply denominators.

$= \frac{5 \cdot 3}{8 \cdot 3}$ Factor to write in lowest terms.

$= \frac{5}{8}$ Divide out the common factor, 3.

(d) $6(-9) = -54$ The numbers have different signs, so the product is negative.

(e) $-0.05(0.3) = -0.015$

(f) $-\frac{5}{8}\left(\frac{12}{13}\right) = -\frac{15}{26}$

(g) $\frac{2}{3}(-6) = -4$ $\quad -6 = -\frac{6}{1}$

NOW TRY

CLASSROOM EXAMPLE 6
Find each product.

(a) $7(-2)$ (b) $-0.9(-15)$

(c) $-\frac{5}{8}(16)$ (d) $-3(0)$

Answers: (a) -14 (b) 13.5
(c) -10 (d) 0

OBJECTIVE 5 Find reciprocals and divide real numbers.

The definition of division uses the concept of a **multiplicative inverse,** or *reciprocal.* Two numbers are *reciprocals* if they have a product of 1.

Reciprocal

The **reciprocal** of a nonzero number a is $\frac{1}{a}$.

Reciprocals

Number	*Reciprocal*	
$-\frac{2}{5}$	$-\frac{5}{2}$	$-\frac{2}{5}\left(-\frac{5}{2}\right) = 1$
-6, or $-\frac{6}{1}$	$-\frac{1}{6}$	$-6\left(-\frac{1}{6}\right) = 1$
$\frac{7}{11}$	$\frac{11}{7}$	$\frac{7}{11}\left(\frac{11}{7}\right) = 1$
0.05	20	$0.05(20) = 1$
0	None	

Reciprocals have a product of 1.

There is no reciprocal for 0 because there is no number that can be multiplied by 0 to give a product of 1.

NOW TRY ANSWERS

6. (a) 30 **(b)** -0.84
(c) -24 **(d)** 0

CAUTION Remember the following.

A number and its additive inverse have opposite signs, such as 3 and −3.

A number and its reciprocal always have the same sign, such as 3 and $\frac{1}{3}$.

The result of dividing one number by another is a **quotient.** For example, we can write the quotient of 45 and 3 as $\frac{45}{3}$, which equals 15. We obtain the same answer if we multiply 45 and $\frac{1}{3}$.

$$45 \div 3 = \frac{45}{3} = 15 \quad \text{and} \quad 45 \cdot \frac{1}{3} = 15$$

This suggests the following definition of division of real numbers.

Division

For all real numbers a and b (where $b \neq 0$), the following holds.

$$\boldsymbol{a \div b = \frac{a}{b} = a \cdot \frac{1}{b}}$$

To divide a by b, multiply a (the **dividend**) by the reciprocal of b (the **divisor**).

TEACHING TIP
To illustrate that division by zero is not allowed, ask students to enter 1 ÷ 0 on their calculators. A calculator returns an error message for an attempted division by zero.

There is no reciprocal for the number 0, so ***division by 0 is undefined.*** For example, $\frac{15}{0}$ is undefined and $-\frac{1}{0}$ is undefined.

CAUTION Although division by 0 is undefined, dividing 0 by a nonzero number gives the quotient 0. For example,

$$\frac{6}{0} \text{ is undefined, but } \frac{0}{6} = 0 \quad (\text{because } 0 \cdot 6 = 0).$$

Be careful when 0 is involved in a division problem.

TEACHING TIP
A student once told one of the authors that she remembered the rule involving division by 0 as follows: "It is *und*efined when 0 is *und*erneath."

Because division is defined as multiplication by the reciprocal, the rules for signs of quotients are the same as those for signs of products.

TEACHING TIP
Restate the rules for addition and subtraction of real numbers presented earlier alongside the rules for multiplication and division.

Dividing Real Numbers

Same sign The quotient of two nonzero real numbers with the *same* sign is positive.

Examples: $\frac{24}{6} = 4, \quad \frac{-24}{-6} = 4$

Different signs The quotient of two nonzero real numbers with *different* signs is negative.

Examples: $\frac{-36}{3} = -12, \quad \frac{36}{-3} = -12$

NOW TRY EXERCISE 7

Find each quotient.

(a) $\frac{-10}{-5}$ (b) $\frac{-\frac{10}{3}}{\frac{3}{8}}$

(c) $-\frac{6}{5} \div \left(-\frac{3}{7}\right)$

EXAMPLE 7 Dividing Real Numbers

Find each quotient.

(a) $\frac{-12}{4} = -3$ The numbers have opposite signs, so the quotient is negative. (b) $\frac{6}{-3} = -2$

(c) $\frac{-\frac{2}{3}}{-\frac{5}{9}}$ This is a *complex fraction* **(Section 6.3).** A complex fraction has a fraction in the numerator, the denominator, or both.

$= -\frac{2}{3}\left(-\frac{9}{5}\right)$ The reciprocal of $-\frac{5}{9}$ is $-\frac{9}{5}$.

$= \frac{18}{15}$ Multiply.

$= \frac{6}{5}$ Write in lowest terms; $\frac{18}{15} = \frac{3 \cdot 6}{3 \cdot 5} = \frac{6}{5}$

(d) $-\frac{9}{14} \div \frac{3}{7}$

$= -\frac{9}{14} \cdot \frac{7}{3}$ Multiply $-\frac{9}{14}$ by the reciprocal of $\frac{3}{7}$.

$= -\frac{63}{42}$ Multiply numerators and multiply denominators.

$= -\frac{3}{2}$ Write in lowest terms. NOW TRY

CLASSROOM EXAMPLE 7

Find each quotient.

(a) $\frac{-15}{-3}$ (b) $\frac{-\frac{3}{8}}{\frac{11}{16}}$

(c) $\frac{3}{4} \div \left(-\frac{7}{16}\right)$

Answers: (a) 5 (b) $-\frac{6}{11}$ (c) $-\frac{12}{7}$

Every fraction has three signs: the sign of the numerator, the sign of the denominator, and the sign of the fraction itself.

TEACHING TIP

Ask students to identify which fractions equal $\frac{-3}{5}$.

A. $\frac{3}{5}$ B. $\frac{3}{-5}$

C. $-\frac{3}{5}$ D. $\frac{-3}{-5}$

Equivalent Forms of a Fraction

The fractions $\frac{-x}{y}$, $\frac{x}{-y}$, and $-\frac{x}{y}$ are equivalent (where $y \neq 0$).

Example: $\frac{-4}{7} = \frac{4}{-7} = -\frac{4}{7}$

The fractions $\frac{x}{y}$ and $\frac{-x}{-y}$ are equivalent (where $y \neq 0$).

Example: $\frac{4}{7} = \frac{-4}{-7}$

NOW TRY ANSWERS

7. (a) 2 (b) $-\frac{80}{9}$ (c) $\frac{14}{5}$

R.2 Exercises

FOR EXTRA HELP

MyMathLab®

Complete solution available in MyMathLab

Concept Check *Complete each statement and give an example.*

1. The sum of two positive numbers is a ___________ number.

2. The sum of two negative numbers is a ___________ number.

3. The sum of a positive number and a negative number is negative if the negative number has the ___________ absolute value.

1. positive; $18 + 6 = 24$
2. negative; $-7 + (-21) = -28$
3. greater; $-14 + 9 = -5$

4. greater; $15 + (-2) = 13$
5. the number with greater absolute value is subtracted from the one with lesser absolute value; $5 - 12 = -7$
6. the number with lesser absolute value is subtracted from the one with greater absolute value; $-15 - (-3) = -12$
7. additive inverses; $4 + (-4) = 0$
8. positive; $-2(-8) = 16$
9. negative; $-5(15) = -75$
10. undefined; zero; $\frac{-17}{0}$ is undefined; $\frac{0}{42} = 0$

11. -19 12. -24
13. 9 14. 1
15. $-\frac{19}{12}$ 16. $-\frac{7}{18}$
17. -1.85 18. 4.342
19. -11 20. -25
21. 21 22. 34
23. -13 24. -15
25. -10.18 26. -6.46
27. $\frac{67}{30}$ 28. $\frac{27}{28}$
29. 14 30. 22
31. -5 32. -1
33. -6 34. 3
35. -11 36. -16
37. 16 38. 22
39. -4 40. 5
41. 8 42. -19
43. 3.018 44. 0.68
45. $-\frac{7}{4}$ 46. $-\frac{1}{8}$
47. $-\frac{7}{8}$ 48. 2
49. 1 50. -3
51. 6 52. 9
53. $\frac{13}{2}$, or $6\frac{1}{2}$ 54. $\frac{13}{2}$, or $6\frac{1}{2}$

55. It is true for multiplication (and division). It is false for addition and false for subtraction when the number to be subtracted has the lesser absolute value. A more precise statement is, "The product or quotient of two negative numbers is positive."

4. The sum of a positive number and a negative number is positive if the positive number has the ____________ absolute value.

5. The difference between two positive numbers is negative if ____________.

6. The difference between two negative numbers is negative if ____________.

7. The sum of a positive number and a negative number is 0 if the numbers are ____________.

8. The product of two numbers with the same sign is ____________.

9. The product of two numbers with different signs is ____________.

10. The quotient formed by any nonzero number divided by 0 is ____________, and the quotient formed by 0 divided by any nonzero number is ____________.

Add or subtract as indicated. ***See Examples 1–3.***

11. $-6 + (-13)$ **12.** $-8 + (-16)$ **13.** $13 + (-4)$

14. $19 + (-18)$ **15.** $-\frac{7}{3} + \frac{3}{4}$ **16.** $-\frac{5}{6} + \frac{4}{9}$

17. $-2.3 + 0.45$ **18.** $-0.238 + 4.58$ **19.** $-6 - 5$

20. $-8 - 17$ **21.** $8 - (-13)$ **22.** $12 - (-22)$

23. $-16 - (-3)$ **24.** $-21 - (-6)$ **25.** $-12.31 - (-2.13)$

26. $-15.88 - (-9.42)$ **27.** $\frac{9}{10} - \left(-\frac{4}{3}\right)$ **28.** $\frac{3}{14} - \left(-\frac{3}{4}\right)$

29. $|-8 - 6|$ **30.** $|-7 - 15|$ **31.** $-|-4 + 9|$

32. $-|-5 + 6|$ **33.** $-2 - |-4|$ **34.** $16 - |-13|$

Perform the indicated operations. ***See Example 4.***

35. $-7 + 5 - 9$ **36.** $-12 + 14 - 18$

37. $6 - (-2) + 8$ **38.** $7 - (-4) + 11$

39. $-9 - 4 - (-3) + 6$ **40.** $-10 - 6 - (-12) + 9$

41. $-8 - (-12) - (2 - 6)$ **42.** $-3 + (-14) + (-6 + 4)$

43. $-0.382 + 4 - 0.6$ **44.** $3 - 2.95 - (-0.63)$

45. $\left(-\frac{5}{4} - \frac{2}{3}\right) + \frac{1}{6}$ **46.** $\left(-\frac{5}{8} + \frac{1}{4}\right) - \left(-\frac{1}{4}\right)$

47. $-\frac{3}{4} - \left(\frac{1}{2} - \frac{3}{8}\right)$ **48.** $\frac{7}{5} - \left(\frac{9}{10} - \frac{3}{2}\right)$

49. $|-11| - |-5| - |7| + |-2|$ **50.** $|-6| + |-3| - |4| - |-8|$

The number line has several points labeled. Find the distance between each pair of points. ***See Example 5.***

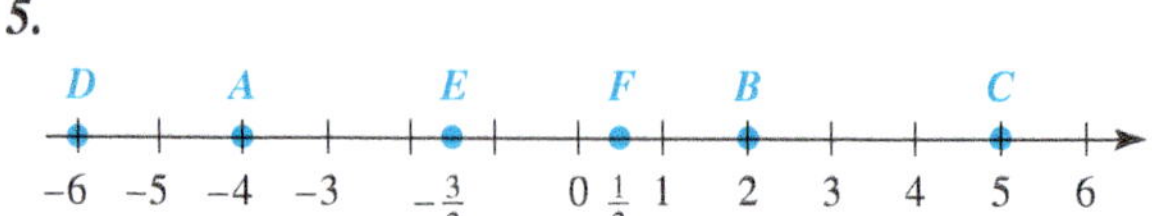

51. *A* and *B* **52.** *A* and *C* **53.** *D* and *F* **54.** *E* and *C*

55. *Concept Check* A statement that is often heard is "Two negatives give a positive." When is this true? When is it not true? Give a more precise statement that conveys this message.

56. Because the reciprocal of any real number is the quotient of 1 (a positive number) and the number, the reciprocal always has the same sign as the number.

57. -35 58. -54
59. 40 60. 80
61. 2 62. 9
63. -12 64. -21
65. $\frac{6}{5}$ 66. $\frac{3}{4}$
67. 1 68. $\frac{7}{2}$
69. 0 70. 0
71. 0 72. 0

73. -7 74. -3
75. 6 76. 5
77. -4 78. -5
79. 0 80. 0
81. undefined 82. undefined
83. $\frac{25}{102}$ 84. $\frac{10}{69}$
85. $-\frac{9}{13}$ 86. -35
87. -2.1 88. -4.5
89. $10{,}000$ 90. 1000

91. $\frac{17}{18}$ 92. $\frac{13}{15}$
93. $\frac{17}{36}$ 94. $\frac{35}{48}$
95. $-\frac{19}{24}$ 96. $-\frac{53}{45}$
97. $-\frac{22}{45}$ 98. $\frac{29}{60}$
99. $-\frac{2}{15}$ 100. $-\frac{21}{100}$
101. $\frac{3}{5}$ 102. $\frac{1}{4}$
103. $-\frac{35}{27}$ 104. $-\frac{10}{3}$
105. $-\frac{1}{3}$ 106. $-\frac{1}{8}$
107. $-\frac{4}{3}$ 108. $-\frac{5}{2}$
109. -12.351 110. -51.382
111. -15.876 112. -1.1948
113. -4.14 114. -4.105
115. 4800 116. 646.8
117. $\frac{1}{3}$ 118. 3

56. *Concept Check* Why must the reciprocal of a nonzero number have the same sign as the number?

Multiply. ***See Example 6.***

57. $5(-7)$ **58.** $6(-9)$ **59.** $-8(-5)$ **60.** $-20(-4)$

61. $-10\left(-\frac{1}{5}\right)$ **62.** $-\frac{1}{2}(-18)$ **63.** $\frac{3}{4}(-16)$ **64.** $\frac{3}{5}(-35)$

65. $-\frac{5}{2}\left(-\frac{12}{25}\right)$ **66.** $-\frac{9}{7}\left(-\frac{21}{36}\right)$ **67.** $-\frac{3}{8}\left(-\frac{24}{9}\right)$ **68.** $-\frac{2}{11}\left(-\frac{77}{4}\right)$

69. $4(0)$ **70.** $\frac{1}{2}(0)$ **71.** $0(-8)$ **72.** $0(-4.5)$

Divide where possible. ***See Example 7.***

73. $\frac{-14}{2}$ **74.** $\frac{-39}{13}$ **75.** $\frac{-24}{-4}$ **76.** $\frac{-45}{-9}$ **77.** $\frac{100}{-25}$

78. $\frac{150}{-30}$ **79.** $\frac{0}{-8}$ **80.** $\frac{0}{-14}$ **81.** $\frac{5}{0}$ **82.** $\frac{13}{0}$

83. $-\frac{10}{17} \div \left(-\frac{12}{5}\right)$ **84.** $-\frac{22}{23} \div \left(-\frac{33}{5}\right)$ **85.** $\frac{\frac{12}{13}}{-\frac{4}{3}}$ **86.** $\frac{\frac{7}{6}}{-\frac{1}{30}}$

87. $\frac{-27.72}{13.2}$ **88.** $\frac{-162.9}{36.2}$ **89.** $\frac{-100}{-0.01}$ **90.** $\frac{-60}{-0.06}$

Exercises 91–120 provide more practice on operations with fractions and decimals. Perform the indicated operations.

91. $\frac{1}{6} - \left(-\frac{7}{9}\right)$ **92.** $\frac{7}{10} - \left(-\frac{1}{6}\right)$ **93.** $-\frac{1}{9} + \frac{7}{12}$ **94.** $-\frac{1}{12} + \frac{13}{16}$

95. $-\frac{3}{8} - \frac{5}{12}$ **96.** $-\frac{11}{15} - \frac{4}{9}$ **97.** $-\frac{7}{30} + \frac{2}{45} - \frac{3}{10}$

98. $-\frac{8}{15} - \frac{3}{20} + \frac{7}{6}$ **99.** $\frac{8}{25}\left(-\frac{5}{12}\right)$ **100.** $\frac{9}{20}\left(-\frac{7}{15}\right)$

101. $\frac{5}{6}\left(-\frac{9}{10}\right)\left(-\frac{4}{5}\right)$ **102.** $\frac{4}{3}\left(-\frac{9}{20}\right)\left(-\frac{5}{12}\right)$ **103.** $\frac{7}{6} \div \left(-\frac{9}{10}\right)$

104. $\frac{12}{5} \div \left(-\frac{18}{25}\right)$ **105.** $\frac{\frac{2}{3}}{-2}$ **106.** $\frac{\frac{3}{4}}{-6}$

107. $\frac{-\frac{8}{9}}{\frac{2}{3}}$ **108.** $\frac{-\frac{15}{16}}{\frac{3}{8}}$ **109.** $-8.6 - 3.751$

110. $-37.8 - 13.582$ **111.** $-4.2(1.4)(2.7)$ **112.** $2.9(-10.3)(0.04)$

113. $-24.84 \div 6$ **114.** $-32.84 \div 8$ **115.** $-2496 \div (-0.52)$

116. $-161.7 \div (-0.25)$ **117.** $\frac{-0.5}{-1.5}$ **118.** $\frac{-1.5}{-0.5}$

119. 70.21 120. −547.006

121. The two quotients are reciprocals by inspection. So once the correct answer to Exercise 117 is known $\left(\frac{1}{3}\right)$, the answer to Exercise 118 is simply its reciprocal (3).

122. (a) 6 (b) 9 (c) 16

123. 112°F 124. 153°F

125. \$30.13 126. \$25.93

127. (a) \$466.02 (b) \$190.68

128. (a) \$222.73 (b) \$179.79

129. (a) 59.53% (b) −36.87% (c) 58.64%

119. $-14.23 + 9.81 + 74.63$ **120.** $-89.416 + 21.32 - 478.91$

121. *Concept Check* How can the answer to **Exercise 118** be determined without showing work if the correct answer to **Exercise 117** is known?

122. *Concept Check* Without showing work, determine the difference between the high and low Fahrenheit temperatures on three consecutive days in Cedar Rapids, Iowa, during the winter of 2014. Express each as a positive number.

(a) *Sunday:* high: 0; low: -6 **(b)** *Monday:* high: -15; low: -24

(c) *Tuesday:* high: -4; low: -20

Solve each problem.

123. The highest temperature ever recorded in Juneau, Alaska, was 90°F. The lowest temperature ever recorded there was −22°F. What is the difference between these two temperatures? (*Source: World Almanac and Book of Facts.*)

124. The highest temperature ever recorded in Illinois was 117°F. The lowest temperature ever recorded there was −36°F. What is the difference between these two temperatures? (*Source: World Almanac and Book of Facts.*)

125. Andrew has \$48.35 in his checking account. He uses his debit card to make purchases of \$35.99 and \$20.00, which overdraws his account. His bank charges his account an overdraft fee of \$28.50. He then deposits his paycheck for \$66.27 from his part-time job at Arby's. What is the balance in his account?

126. Kayla has \$37.60 in her checking account. She uses her debit card to make purchases of \$25.99 and \$19.34, which overdraws her account. Her bank charges her account an overdraft fee of \$25.00. She then deposits her paycheck for \$58.66 from her part-time job at Subway. What is the balance in her account?

127. Ahmad owes \$382.45 on his Visa account. He returns two items costing \$25.10 and \$34.50 for credit. Then he makes purchases of \$45.00 and \$98.17.

(a) How much should his payment be if he wants to pay off the balance on the account?

(b) Instead of paying off the balance, he makes a payment of \$300 and then incurs a finance charge of \$24.66. What is the balance on his account?

128. Charlene owes \$237.59 on her MasterCard account. She returns one item costing \$47.25 for credit and then makes two purchases of \$12.39 and \$20.00.

(a) How much should her payment be if she wants to pay off the balance on the account?

(b) Instead of paying off the balance, she makes a payment of \$75.00 and incurs a finance charge of \$32.06. What is the balance on her account?

129. The graph shows annual returns in percent for class A shares of the Invesco Charter Fund for the years 2007–2013.

(a) Find the sum of the percents for the years shown in the graph.

(b) Find the difference between the returns in 2008 and 2007.

(c) Find the difference between the returns in 2009 and 2008.

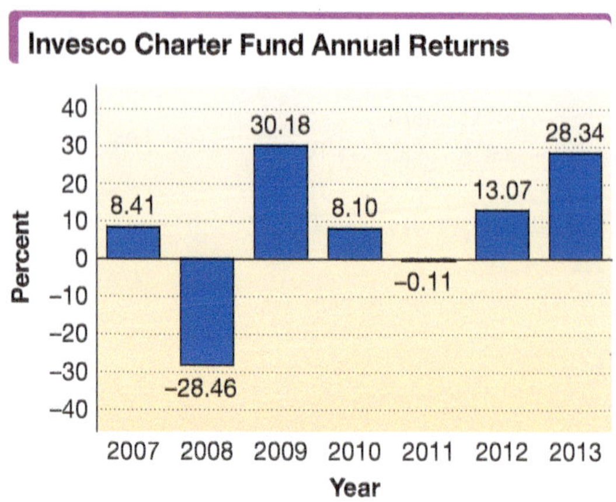

Source: Morningstar, Invesco.

130. (a) −209 thousand dollars
(b) −185 − (−225)
= −185 + 225
= 40;
40 thousand dollars
(c) 124 − (−225)
= 124 + 225
= 349;
349 thousand dollars

131. 1995: −\$164 billion;
2000: \$236 billion;
2005: −\$318 billion;
2012: −\$1087 billion

132. surplus: 2000;
deficit: 1995, 2005, 2012;
A positive difference between receipts and outlays indicates a surplus, while a negative difference indicates a deficit.

130. The graph shows profits and losses in thousands of dollars for a private company for the years 2012 through 2015.

(a) What was the total profit or loss for the years 2012 through 2015?

(b) The company showed an increase from 2012 to 2013, despite still having an overall loss. Illustrate using a subtraction problem how this increase can be determined. What was the increase?

(c) By how much did the company increase between 2012 and 2015? Illustrate using a subtraction problem.

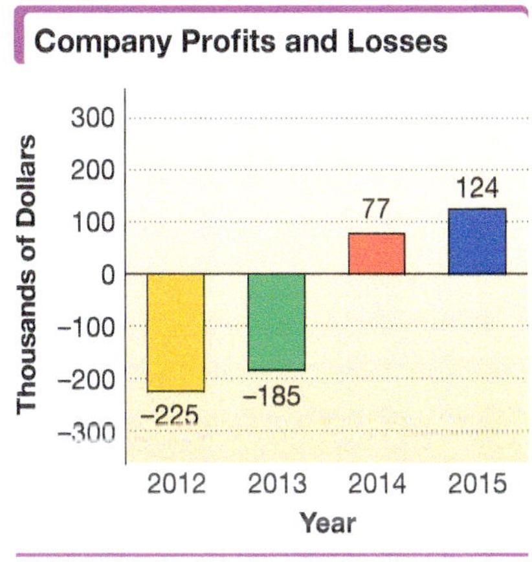

The table shows receipts (income) and outlays (spending) in billions of dollars for the U.S. government in selected years.

Fiscal Year	*Receipts*	*Outlays*
1995	1352	1516
2000	2025	1789
2005	2154	2472
2012	2450	3537

Source: Office of Management and Budget.

131. Find the difference between U.S. government receipts and outlays for each year shown in the table.

132. During which years did the budget show a surplus? A deficit? Use the results from **Exercise 131** to explain the answers.

R.3 Exponents, Roots, and Order of Operations

OBJECTIVES

1. Use exponents.
2. Find square roots.
3. Use the rules for order of operations.
4. Evaluate algebraic expressions for given values of variables.

VOCABULARY

- ☐ factors
- ☐ exponent (power)
- ☐ base
- ☐ exponential expression
- ☐ square root
- ☐ positive (principal) square root
- ☐ negative square root
- ☐ constant
- ☐ algebraic expression

Two (or more) numbers whose product is a third number are **factors** of that third number. For example, 2 and 6 are factors of 12 because $2 \cdot 6 = 12$.

OBJECTIVE 1 Use exponents.

In algebra, we use *exponents* as a way of writing products of repeated factors.

$$\underbrace{2 \cdot 2 \cdot 2 \cdot 2 \cdot 2}_{\text{5 factors of 2}} = 2^5$$

The number 5 shows that 2 is used as a factor 5 times. The number 5 is the *exponent,* and 2 is the *base.*

2^5 ← Exponent
↑ Base

Read 2^5 as "2 to the fifth power," or "2 to the fifth." Multiplying five 2s gives 32.

$$2^5 \quad \text{means} \quad 2 \cdot 2 \cdot 2 \cdot 2 \cdot 2, \quad \text{which equals } 32.$$

Exponential Expression

If a is a real number and n is a natural number, then

$$a^n = \underbrace{a \cdot a \cdot a \cdot \ldots \cdot a}_{n \text{ factors of } a},$$

where n is the **exponent,** a is the **base,** and a^n is an **exponential expression.** Exponents are also called **powers.**

NOW TRY EXERCISE 1

Write using exponents.

(a) $(-3)(-3)(-3)$

(b) $(0.5)(0.5)$

(c) $t \cdot t \cdot t \cdot t \cdot t$

CLASSROOM EXAMPLE 1

Write using exponents.

(a) $\frac{2}{7} \cdot \frac{2}{7} \cdot \frac{2}{7} \cdot \frac{2}{7}$

(b) $(-10)(-10)(-10)$

(c) $y \cdot y \cdot y \cdot y \cdot y \cdot y \cdot y \cdot y$

Answers: (a) $\left(\frac{2}{7}\right)^4$

(b) $(-10)^3$ (c) y^8

EXAMPLE 1 Using Exponential Notation

Write using exponents.

(a) $\underbrace{4 \cdot 4 \cdot 4}_{3 \text{ factors of } 4} = 4^3$

Read 4^3 as "4 **cubed.**"

(b) $\frac{3}{5} \cdot \frac{3}{5} = \left(\frac{3}{5}\right)^2$ 2 factors of $\frac{3}{5}$

Read $\left(\frac{3}{5}\right)^2$ as "$\frac{3}{5}$ **squared.**"

(c) $(-6)(-6)(-6)(-6) = (-6)^4$ 4 factors of -6

Read $(-6)^4$ as "-6 to the fourth power," or "-6 to the fourth."

(d) $(0.3)(0.3)(0.3)(0.3)(0.3) = (0.3)^5$ **(e)** $x \cdot x \cdot x \cdot x \cdot x \cdot x = x^6$

NOW TRY

NOTE In **Example 1,** we used the terms *squared* and *cubed* to refer to powers of 2 and 3, respectively. The term *squared* comes from the figure of a square, which has the same measure for both length and width, as shown in **FIGURE 11(a)**. Similarly, the term *cubed* comes from the figure of a cube, where the length, width, and height have the same measure, as shown in **FIGURE 11(b)**.

(a) $3 \cdot 3$ means 3 squared, or 3^2. **(b)** $6 \cdot 6 \cdot 6$ means 6 cubed, or 6^3.

FIGURE 11

CLASSROOM EXAMPLE 2

Evaluate.

(a) 3^4 (b) $(-3)^2$ (c) -3^2

Answers: (a) 81 (b) 9 (c) -9

EXAMPLE 2 Evaluating Exponential Expressions

Evaluate.

(a) 5^2 means $5 \cdot 5$, which equals 25. 5 is used as a factor 2 times.

5^2 means $5 \cdot 5$, *not* $5 \cdot 2$.

(b) $\left(\frac{2}{3}\right)^3$ means $\frac{2}{3} \cdot \frac{2}{3} \cdot \frac{2}{3}$, which equals $\frac{8}{27}$. $\frac{2}{3}$ is used as a factor 3 times.

(c) 2^6 means $2 \cdot 2 \cdot 2 \cdot 2 \cdot 2 \cdot 2$, which equals 64.

(d) $(-3)^5$ means $(-3)(-3)(-3)(-3)(-3)$, which equals -243. The base is -3.

NOW TRY ANSWERS

1. (a) $(-3)^3$ (b) $(0.5)^2$ (c) t^5

NOW TRY EXERCISE 2

Evaluate.

(a) 7^2 **(b)** $(-7)^2$ **(c)** -7^2

(e) $(-2)^6$ means $(-2)(-2)(-2)(-2)(-2)(-2)$, which equals 64. The base is -2.

(f) -2^6

There are no parentheses. The exponent 6 applies *only* to the number 2, not to -2.

-2^6 means $-(2 \cdot 2 \cdot 2 \cdot 2 \cdot 2 \cdot 2)$, which equals -64. The base is 2.

NOW TRY

Examples 2(d) and (e) suggest the following generalizations.

Sign of an Exponential Expression

The product of an *even* number of negative factors is positive.

Example: $(-2)(-2)(-2)(-2) = 16$

The product of an *odd* number of negative factors is negative.

Example: $(-2)(-2)(-2) = -8$

TEACHING TIP

Show how "pairing" negative factors can determine the sign of the product using the following examples.

$(-2)^5$
$= [(-2)(-2)] \cdot [(-2)(-2)](-2)$
$= (+4)(+4)(-2)$
$= -32$

$(-2)^6$
$= [(-2)(-2)] \cdot [(-2)(-2)][(-2)(-2)]$
$= (+4)(+4)(+4)$
$= +64$

CAUTION As shown in **Examples 2(e)** and **(f),** it is important to distinguish between $-a^n$ and $(-a)^n$.

$-a^n$ means $-1\underbrace{(a \cdot a \cdot a \cdot \ldots \cdot a)}_{n \text{ factors of } a.}$ The base is a.

$(-a)^n$ means $\underbrace{(-a)(-a) \cdot \ldots \cdot (-a)}_{n \text{ factors of } -a}$ The base is $-a$.

Be careful when evaluating an exponential expression with a negative sign.

OBJECTIVE 2 Find square roots.

As we saw in **Example 2(a),** 5 squared (or 5^2) equals 25. The inverse procedure of squaring a number is taking its **square root.** For example, a square root of 25 is 5. Another square root of 25 is -5 because $(-5)^2 = 25$. Thus, 25 has two square roots, 5 and -5.

We write the **positive** or **principal square root** of a number using a **radical symbol** $\sqrt{\ }$. The positive or principal square root of 25 is written

$$\sqrt{25} = 5.$$

The **negative square root** of 25 is written

$$-\sqrt{25} = -5.$$

Because the square of any nonzero real number is positive, the square root of a negative number, such as $\sqrt{-25}$, is not a real number.

CLASSROOM EXAMPLE 3

Find each square root that is a real number.

(a) $-\sqrt{\frac{121}{81}}$ (b) $\sqrt{49}$

(c) $-\sqrt{49}$ (d) $\sqrt{-49}$

Answers: (a) $-\frac{11}{9}$ (b) 7

(c) -7 (d) not a real number

EXAMPLE 3 Finding Square Roots

Find each square root that is a real number.

(a) $\sqrt{36} = 6$ because 6 is positive and $6^2 = 36$.

(b) $\sqrt{0} = 0$ because $0^2 = 0$.

(c) $\sqrt{\frac{9}{16}} = \frac{3}{4}$ because $\left(\frac{3}{4}\right)^2 = \frac{9}{16}$.

NOW TRY ANSWERS

2. (a) 49 (b) 49 (c) -49

NOW TRY EXERCISE 3

Find each square root that is a real number.

(a) $-\sqrt{144}$ (b) $\sqrt{\frac{100}{9}}$

(c) $\sqrt{-144}$

TEACHING TIP

Show how

$5 + 2 \cdot 3$

suggests two possible answers, 21 or 11. However, only 11 is correct according to the rules for order of operations. See **Example 4(a).**

TEACHING TIP

Some students like to use the mnemonic

"Please Excuse My Dear Aunt Sally"

to help remember the rules for order of operations, with the first letters representing **Parentheses, Exponents, Multiply, Divide, Add, Subtract.** Remind them as well that multiplications and divisions should be done *in order from left to right,* followed by additions and subtractions.

CLASSROOM EXAMPLE 4

Simplify.

(a) $5 \cdot 9 + 2 \cdot 4$

(b) $4 - 12 \div 4 \cdot 2$

Answers: (a) 53 (b) −2

NOW TRY EXERCISE 4

Simplify.

$$15 - 3 \cdot 4 + 2$$

NOW TRY ANSWERS

3. **(a)** −12 **(b)** $\frac{10}{3}$ **(c)** not a real number

4. 5

(d) $\sqrt{0.16} = 0.4$ because $(0.4)^2 = 0.16$. **(e)** $\sqrt{100} = 10$ because $10^2 = 100$.

(f) $-\sqrt{100} = -10$ because the negative sign is outside the radical symbol.

(g) $\sqrt{-100}$ is not a real number because the negative sign is inside the radical symbol. No *real number* squared equals −100.

Notice that part (e) is the positive or principal square root of 100, part (f) is the negative square root of 100, and part (g) is the square root of −100, which is not a real number.

NOW TRY

CAUTION The symbol $\sqrt{a}$ is used only for the *positive* square root of *a*. When $a = 0$, $\sqrt{0} = 0$. The symbol $-\sqrt{a}$ is used for the negative square root of *a*.

OBJECTIVE 3 Use the rules for order of operations.

To simplify the following expression, what should we do first—add 5 and 2 or multiply 2 and 3?

$$5 + 2 \cdot 3$$

When an expression involves more than one operation symbol, we use the following rules for **order of operations.**

Order of Operations

If grouping symbols are present, work within them, innermost first (and above and below fraction bars separately), in the following order.

Step 1 Apply all **exponents.**

Step 2 Do any **multiplications** or **divisions** in order from left to right.

Step 3 Do any **additions** or **subtractions** in order from left to right.

If no grouping symbols are present, start with Step 1.

EXAMPLE 4 Using the Rules for Order of Operations

Simplify.

(a) $5 + 2 \cdot 3$

$= 5 + 6$ Multiply.

$= 11$ Add.

(b) $24 \div 3 \cdot 2 + 6$ Multiplications and divisions are done in order from left to right. So we divide first here.

$= 8 \cdot 2 + 6$

$= 16 + 6$ Multiply.

$= 22$ Add.

NOW TRY

NOW TRY EXERCISE 5

Simplify.

(a) $-5^2 + 10 \div 5 - |3 - 7|$

(b) $6 + \frac{2}{3}(-9) - \frac{5}{8} \cdot 16$

CLASSROOM EXAMPLE 5

Simplify.

(a) $(4 + 2) - 3^2 - (8 - 3)$

(b) $6 + \frac{5}{4}(-12) - \frac{2}{3} \cdot 18$

Answers: (a) -8 (b) -21

EXAMPLE 5 Using the Rules for Order of Operations

Simplify.

(a) $10 \div 5 + 2|3 - 4|$ Work inside the absolute value bars first.

$= 10 \div 5 + 2|-1|$ Subtract inside the absolute value bars.

$= 10 \div 5 + 2 \cdot 1$ Take the absolute value.

$= 2 + 2$ Divide first, and then multiply.

$= 4$ Add.

(b) $4 \cdot 3^2 + 7 - (2 + 8)$ Work inside the parentheses first.

$= 4 \cdot 3^2 + 7 - 10$ Add inside the parentheses.

$= 4 \cdot 9 + 7 - 10$ Evaluate the power.

3^2 means $3 \cdot 3$, *not* $3 \cdot 2$.

$= 36 + 7 - 10$ Multiply.

$= 43 - 10$ Add.

$= 33$ Subtract.

(c) $\frac{1}{2} \cdot 4 + (6 \div 3 - 7)$ Work inside the parentheses first.

$= \frac{1}{2} \cdot 4 + (2 - 7)$ Divide inside the parentheses.

$= \frac{1}{2} \cdot 4 + (-5)$ Subtract inside the parentheses.

$= 2 + (-5)$ Multiply.

$= -3$ Add. NOW TRY

CLASSROOM EXAMPLE 6

Simplify.

$$\frac{\frac{1}{2} \cdot 10 - 6 + \sqrt{9}}{\frac{5}{6} \cdot 12 - 3^2 - 1}$$

Answer: undefined

NOW TRY EXERCISE 6

Simplify.

$$\frac{\sqrt{36} - 4 \cdot 3^2}{-2^2 - 8 \cdot 3 + 28}$$

EXAMPLE 6 Using the Rules for Order of Operations

Simplify.

$$\frac{5 - (-2^3)(2)}{6 \cdot \sqrt{9} - 9 \cdot 2}$$ Work separately above and below the fraction bar.

$$= \frac{5 - (-8)(2)}{6 \cdot 3 - 9 \cdot 2}$$ Evaluate the power and the root.

$$= \frac{5 + 16}{18 - 18}$$ Multiply and use the definition of subtraction.

$$= \frac{21}{0}$$ Add and subtract.

Because division by 0 is undefined, the given expression is undefined. NOW TRY

TEACHING TIP

Most calculators follow the order of operations given in this section. Have students work some of the examples on their calculators. Caution them to use the parentheses keys to insert parentheses where they are needed.

NOW TRY ANSWERS

5. **(a)** -27 **(b)** -10

6. undefined

OBJECTIVE 4 Evaluate algebraic expressions for given values of variables.

A **constant** is a fixed, unchanging number.

$$1, \quad 6, \quad -10, \quad \frac{2}{5}, \quad -3.75 \qquad \text{Constants}$$

A collection of constants, variables, operation symbols, and/or grouping symbols is an **algebraic expression.**

$$6ab, \quad 5m - 9n, \quad -2(x^2 + 4y) \qquad \text{Algebraic expressions}$$

Algebraic expressions have different numerical values for different values of the variables. We evaluate such expressions by *substituting* given values for the variables.

NOW TRY EXERCISE 7

Evaluate each expression for $x = -4$, $y = 7$, and $z = 36$.

(a) $3y - 2x$ **(b)** $\dfrac{x^2 - \sqrt{z}}{-3xy}$

CLASSROOM EXAMPLE 7

Evaluate each expression for $w = 4$, $x = -12$, $y = 64$, and $z = -3$.

(a) $\dfrac{5x + z\sqrt{y}}{x - 1}$ (b) $w^2 + 2z^3$

Answers: (a) $\frac{84}{13}$ (b) -38

EXAMPLE 7 Evaluating Algebraic Expressions

Evaluate each expression for $m = -4$, $n = 5$, $p = -6$, and $q = 25$.

(a) $5m - 9n$

$= 5(-4) - 9(5)$ Substitute $m = -4$ and $n = 5$.

Use parentheses around substituted values to avoid errors.

$= -20 - 45$ Multiply.

$= -65$ Subtract.

(b) $\dfrac{m + 2n}{4p}$ Work separately above and below the fraction bar.

$= \dfrac{-4 + 2(5)}{4(-6)}$ Substitute $m = -4$, $n = 5$, and $p = -6$.

$= \dfrac{-4 + 10}{-24}$ Multiply in the numerator. Multiply in the denominator.

$= \dfrac{6}{-24}$ Add in the numerator.

$= -\dfrac{1}{4}$ Write in lowest terms, and rewrite using $\frac{a}{-b} = -\frac{a}{b}$.

(c) $-3m^3 - n^2(\sqrt{q})$

$= -3(-4)^3 - (5)^2(\sqrt{25})$ Substitute $m = -4$, $n = 5$, and $q = 25$.

$= -3(-64) - 25(5)$ Evaluate the powers and the root.

$= 192 - 125$ Multiply.

$= 67$ Subtract.

NOW TRY

NOW TRY ANSWERS

7. (a) 29 (b) $\frac{5}{42}$

R.3 Exercises

FOR EXTRA HELP MyMathLab®

Complete solution available in MyMathLab

Concept Check Decide whether each statement is true or false. If it is false, correct the statement so that it is true.

1. $-7^6 = (-7)^6$

2. $-5^7 = (-5)^7$

3. $\sqrt{25}$ is a positive number.

4. $3 + 5 \cdot 8 = 3 + (5 \cdot 8)$

5. $(-6)^7$ is a negative number.

6. $(-6)^8$ is a positive number.

1. false; $-7^6 = -(7^6)$
2. true
3. true
4. true
5. true
6. true

7. true
8. false; The product is negative.
9. false; The base is 8.
10. true

11. **(a)** 64 **(b)** -64 **(c)** 64 **(d)** -64
12. **(a)** 64 **(a)** -64 **(c)** -64 **(d)** 64

13. 10^4 **14.** 8^3
15. $\left(\frac{3}{4}\right)^5$ **16.** $\left(\frac{3}{2}\right)^2$
17. $(-9)^3$ **18.** $(-9)^4$
19. z^7 **20.** a^5
21. 0^6

22. true

23. 16 **24.** 16
25. 0.021952 **26.** 0.753571
27. $\frac{1}{125}$ **28.** $\frac{1}{1296}$
29. $\frac{256}{625}$ **30.** $\frac{343}{1000}$
31. -125 **32.** -32
33. 256 **34.** 729
35. -729 **36.** -4096
37. -4096 **38.** -1000

39. 9 **40.** 8
41. 13 **42.** 15
43. -20 **44.** -30
45. $\frac{10}{11}$ **46.** $\frac{15}{13}$
47. -0.7 **48.** -0.8
49. not a real number
50. not a real number

51. **(a)** not a real number **(b)** negative
52. **(a)** The grandson's answer was correct. **(b)** The reasoning was *incorrect*. The operation of division must be done first, and then the addition follows. The grandson's "Order of Process rule" is **not correct.** It just happens coincidentally in this problem that he obtained the correct answer the wrong way.

53. 24 **54.** 25
55. 15 **56.** 32
57. 55 **58.** 28

7. The product of 10 positive factors and 10 negative factors is positive.

8. The product of 5 positive factors and 5 negative factors is positive.

9. In the exponential expression -8^5, the base is -8.

10. $\sqrt{a}$ is positive for all positive numbers a.

Concept Check *Evaluate each exponential expression.*

11. **(a)** 8^2 **(b)** -8^2 **(c)** $(-8)^2$ **(d)** $-(-8)^2$

12. **(a)** 4^3 **(b)** -4^3 **(c)** $(-4)^3$ **(d)** $-(-4)^3$

Write each expression using exponents. ***See Example 1.***

13. $10 \cdot 10 \cdot 10 \cdot 10$ **14.** $8 \cdot 8 \cdot 8$ **15.** $\frac{3}{4} \cdot \frac{3}{4} \cdot \frac{3}{4} \cdot \frac{3}{4} \cdot \frac{3}{4}$

16. $\frac{3}{2} \cdot \frac{3}{2}$ **17.** $(-9)(-9)(-9)$ **18.** $(-9)(-9)(-9)(-9)$

19. $z \cdot z \cdot z \cdot z \cdot z \cdot z \cdot z$ **20.** $a \cdot a \cdot a \cdot a \cdot a$ **21.** $0 \cdot 0 \cdot 0 \cdot 0 \cdot 0 \cdot 0$

22. ***Concept Check*** *True* or *false:* If n is odd and a is a positive integer, then $(-a)^n = -a^n$ in all cases.

Evaluate each expression. ***See Example 2.***

23. 4^2 **24.** 2^4 **25.** 0.28^3 **26.** 0.91^3

27. $\left(\frac{1}{5}\right)^3$ **28.** $\left(\frac{1}{6}\right)^4$ **29.** $\left(\frac{4}{5}\right)^4$ **30.** $\left(\frac{7}{10}\right)^3$

31. $(-5)^3$ **32.** $(-2)^5$ **33.** $(-2)^8$ **34.** $(-3)^6$

35. -3^6 **36.** -4^6 **37.** -8^4 **38.** -10^3

Find each square root. If it is not a real number, say so. ***See Example 3.***

39. $\sqrt{81}$ **40.** $\sqrt{64}$ **41.** $\sqrt{169}$ **42.** $\sqrt{225}$

43. $-\sqrt{400}$ **44.** $-\sqrt{900}$ **45.** $\sqrt{\frac{100}{121}}$ **46.** $\sqrt{\frac{225}{169}}$

47. $-\sqrt{0.49}$ **48.** $-\sqrt{0.64}$ **49.** $\sqrt{-36}$ **50.** $\sqrt{-121}$

51. ***Concept Check*** *In these questions, x represents a positive number.*

(a) Is $-\sqrt{-x}$ positive, negative, or not a real number?

(b) Is $-\sqrt{x}$ positive, negative, or not a real number?

52. ***Concept Check*** Frank's grandson was asked to evaluate the following expression.

$$9 + 15 \div 3$$

Frank gave the answer as 8, but his grandson gave the answer as 14. The grandson explained that the answer is 14 because of the "Order of Process rule," which says that when evaluating an expression like this, we proceed from right to left rather than left to right. (*Note:* This is a true story.)

(a) Whose answer was correct for this expression, Frank's or his grandson's?

(b) Was the *reasoning* for the correct answer valid? Explain.

Simplify each expression. ***See Examples 4–6.***

53. $12 + 3 \cdot 4$ **54.** $15 + 5 \cdot 2$ **55.** $6 \cdot 3 - 12 \div 4$

56. $9 \cdot 4 - 8 \div 2$ **57.** $10 + 30 \div 2 \cdot 3$ **58.** $12 + 24 \div 3 \cdot 2$

59. -91 **60.** -42
61. -8 **62.** 17
63. -48 **64.** -85
65. -2 **66.** 8
67. -79 **68.** -5
69. -10 **70.** 23
71. 2 **72.** -1
73. -2 **74.** -9
75. undefined **76.** undefined
77. -7 **78.** $-\frac{32}{5}$
79. -1 **80.** -2
81. 17 **82.** -1
83. -96 **84.** -261
85. $-\frac{15}{238}$ **86.** $\frac{27}{92}$
87. 8 **88.** -1
89. $-\frac{5}{16}$ **90.** $-\frac{15}{16}$
91. $-\frac{11}{4}$, or -2.75
92. $\frac{29}{4}$, or 7.25
93. $-\frac{3}{16}$ **94.** $\frac{29}{32}$
95. \$1572 **96.** \$2434
97. \$3296

59. $-3(5)^2 - (-2)(-8)$

60. $-9(2)^2 - (-3)(-2)$

61. $5 - 7 \cdot 3 - (-2)^3$

62. $-4 - 3 \cdot 5 + 6^2$

63. $-7(\sqrt{36}) - (-2)(-3)$

64. $-8(\sqrt{64}) - (-3)(-7)$

65. $6|4 - 5| - 24 \div 3$

66. $-4|2 - 4| + 8 \cdot 2$

67. $|-6 - 5|(-8) + 3^2$

68. $(-6 - 3)|-2 - 3| \div 9$

69. $6 + \frac{2}{3}(-9) - \frac{5}{8} \cdot 16$

70. $7 - \frac{3}{4}(-8) + 12 \cdot \frac{5}{6}$

71. $-14\left(-\frac{2}{7}\right) \div (2 \cdot 6 - 10)$

72. $-12\left(-\frac{3}{4}\right) - (6 \cdot 5 \div 3)$

73. $\dfrac{(-5 + \sqrt{4})(-2^2)}{-5 - 1}$

74. $\dfrac{(-9 + \sqrt{16})(-3^2)}{-4 - 1}$

75. $\dfrac{2(-5) + (-3)(-2)}{-8 + 3^2 - 1}$

76. $\dfrac{3(-4) + (-5)(-8)}{2^3 - 2 - 6}$

77. $\dfrac{5 - 3\left(\frac{-5 - 9}{-7}\right) - 6}{-9 - 11 + 3 \cdot 7}$

78. $\dfrac{-4\left(\frac{12 - (-8)}{3 \cdot 2 + 4}\right) - 5(-1 - 7)}{-9 - (-7) - [-5 - (-8)]}$

Evaluate each expression for $a = -3$, $b = 64$, and $c = 6$. ***See Example 7.***

79. $3a + \sqrt{b}$

80. $-2a - \sqrt{b}$

81. $\sqrt{b} + c - a$

82. $\sqrt{b} - c + a$

83. $4a^3 + 2c$

84. $-3a^4 - 3c$

85. $\dfrac{2c + a^3}{4b + 6a}$

86. $\dfrac{3c + a^2}{2b - 6c}$

Evaluate each expression for $w = 4$, $x = -\frac{3}{4}$, $y = \frac{1}{2}$, and $z = 1.25$. ***See Example 7.***

87. $wy - 8x$

88. $wz - 12y$

89. $xy + y^4$

90. $xy - x^2$

91. $-w + 2x + 3y + z$

92. $w - 6x + 5y - 3z$

93. $\dfrac{7x + 9y}{w}$

94. $\dfrac{7y - 5x}{2w}$

Residents of Linn County, Iowa, in the Cedar Rapids Community School District can use the expression

$$(v \times 0.5485 - 4850) \div 1000 \times 31.44$$

to determine their property taxes, where v is assessed home value. (Source: The Gazette.) Use the expression to calculate the amount of property taxes to the nearest dollar that the owner of a home with each of the following values would pay. Follow the rules for order of operations.

95. \$100,000

96. \$150,000

97. \$200,000

98. (a) $48 \times 3.2 \times 0.075 \div 190 - 2 \times 0.015$
(b) 0.031
99. 0.035
100. (a) 0.024; 0.023; Increased weight results in lower BACs.
(b) Decreased weight will result in higher BACs; 0.040; 0.053
101. (a) \$28.3 billion
(b) \$42.2 billion
(c) \$56.0 billion
(d) The amount spent on pets almost exactly doubled from 2001 to 2013.
102. (a) 4.69; 7.05; 8.24
(b) The average price of a movie ticket in the United States increased by about 75% from 1997 to 2012.

The Blood Alcohol Concentration (BAC) of a person who has been drinking is given by the expression

number of oz × % alcohol × 0.075 ÷ body weight in lb − hr of drinking × 0.015.

(*Source:* Lawlor, J., *Auto Math Handbook: Calculations, Formulas, Equations and Theory for Automotive Enthusiasts,* Penguin, © 1991.)

Use this formula in Exercises 98–100.

98. Suppose a policeman stops a 190-lb man who, in 2 hr, has ingested four 12-oz beers (48 oz), each having a 3.2% alcohol content.

(a) Substitute the values into the formula, and write the expression for the man's BAC.

(b) Calculate the man's BAC to the nearest thousandth. Follow the rules for order of operations.

99. Find the BAC to the nearest thousandth for a 135-lb woman who, in 3 hr, has drunk three 12-oz beers (36 oz), each having a 4.0% alcohol content.

100. (a) Calculate the BACs in **Exercises 98 and 99** if each person weighs 25 lb more and the rest of the variables stay the same. How does increased weight affect a person's BAC?

(b) Predict how decreased weight would affect the BAC of each person in **Exercises 98 and 99.** Calculate the BACs if each person weighs 25 lb less and the rest of the variables stay the same.

Solve each problem.

101. The amount in billions of dollars that Americans have spent on their pets from 2001 to 2013 can be approximated by substituting a given year for x in the expression

$$2.311x - 4596.$$

(*Source:* American Pet Products Manufacturers Association.) Find the amount spent in each year. Round answers to the nearest tenth.

(a) 2001 **(b)** 2007 **(c)** 2013

(d) How has the amount Americans have spent on their pets changed from 2001 to 2013?

Dotty

102. The average price in dollars of a movie ticket in the United States from 1997 to 2012 can be approximated using the expression

$$0.2363x - 467.2,$$

where x represents the year. (*Source:* National Association of Theater Owners.)

Year	*Average Price (in dollars)*
1997	
2002	5.87
2007	
2012	

(a) Use the expression to complete the table. Round answers to the nearest cent.

(b) How has the average price of a theater ticket in the United States changed from 1997 to 2012?

R.4 Properties of Real Numbers

OBJECTIVES

1. Use the distributive property.
2. Use the identity properties.
3. Use the inverse properties.
4. Use the commutative and associative properties.

VOCABULARY

- ☐ identity element for addition (additive identity)
- ☐ identity element for multiplication (multiplicative identity)
- ☐ term
- ☐ coefficient (numerical coefficient)
- ☐ like terms
- ☐ unlike terms

In **Section R.2** we used the *multiplication property of 0*. There are other basic properties of real numbers that reflect results that occur consistently in mathematics.

OBJECTIVE 1 Use the distributive property.

Notice the following.

$$2(3 + 5) = 2 \cdot 8 = 16$$

and $$2 \cdot 3 + 2 \cdot 5 = 6 + 10 = 16,$$

so $$2(3 + 5) = 2 \cdot 3 + 2 \cdot 5.$$

This idea is illustrated by the divided rectangle in **FIGURE 12**.

Similarly, observe these facts.

$$-4[5 + (-3)] = -4(2) = -8$$

and $$-4(5) + (-4)(-3) = -20 + 12 = -8,$$

so $$-4[5 + (-3)] = -4(5) + (-4)(-3).$$

These two examples are generalized to the set of *all* real numbers as the **distributive property of multiplication with respect to addition,** or simply the **distributive property.**

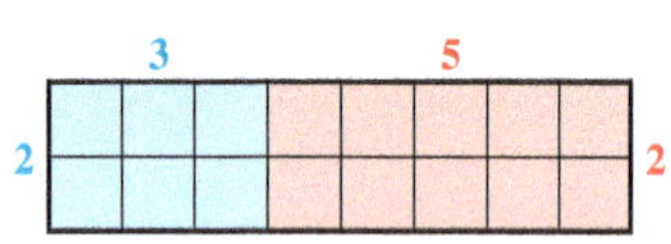

Area of left part is $2 \cdot 3 = 6$.
Area of right part is $2 \cdot 5 = 10$.
Area of total rectangle is $2(3 + 5) = 16$.

FIGURE 12

Distributive Property

For any real numbers a, b, and c, the following hold.

$$a(b + c) = ab + ac \quad \text{and} \quad (b + c)a = ba + ca$$

Examples: $12(4 + 2) = 12 \cdot 4 + 12 \cdot 2$ and $(4 + 2)12 = 4 \cdot 12 + 2 \cdot 12$

The distributive property can also be applied "in reverse."

$$ab + ac = a(b + c) \quad \text{and} \quad ba + ca = (b + c)a$$

Examples: $6 \cdot 8 + 6 \cdot 9 = 6(8 + 9)$ and $8 \cdot 6 + 9 \cdot 6 = (8 + 9)6$

This property can be extended to more than two numbers as well.

$$a(b + c + d) = ab + ac + ad$$

The distributive property provides a way to rewrite a product $a(b + c)$ as a sum $ab + ac$, or a sum as a product.

NOTE When we rewrite $a(b + c)$ as $ab + ac$, we sometimes refer to the process as "removing" or "clearing" parentheses.

NOW TRY EXERCISE 1

Use the distributive property to rewrite each expression.

(a) $-2(3x - y)$

(b) $4k - 12k$

CLASSROOM EXAMPLE 1

Use the distributive property to rewrite each expression.

(a) $-4(p - 5)$

(b) $-6m + 2m$

(c) $2r + 3s$

(d) $5(4p - 2q + r)$

Answers: (a) $-4p + 20$

(b) $-4m$

(c) cannot be rewritten

(d) $20p - 10q + 5r$

TEACHING TIP

Have students work through **Example 1** in groups, and then have the whole class discuss all the results.

EXAMPLE 1 Using the Distributive Property

Use the distributive property to rewrite each expression.

(a) $3(x + y)$

$= 3x + 3y$ — Use the first form of the property to rewrite the given product as a sum.

(b) $-2(5 + k)$

$= -2(5) + (-2)(k)$ — Distributive property

$= -10 - 2k$ — Multiply.

(c) $4x + 8x$

$= (4 + 8)x$ — Use the distributive property in reverse to rewrite the given sum as a product.

$= 12x$ — Add inside the parentheses.

(d) $3r - 7r$

$= 3r + (-7r)$ — Definition of subtraction

$= [3 + (-7)]r$ — Distributive property in reverse

$= -4r$ — Add.

(e) $5p + 7q$ — This expression *cannot* be rewritten as $12pq$.

Because there is no common number or variable here, we cannot use the distributive property to rewrite the expression.

(f) $6(x + 2y - 3z)$

$= 6x + 6(2y) + 6(-3z)$ — Distributive property

$= 6x + 12y - 18z$ — Multiply.

NOW TRY

The distributive property can also be used for subtraction as in **Example 1(d).**

$$a(b - c) = ab - ac$$

OBJECTIVE 2 Use the identity properties.

TEACHING TIP

Point out that when we use an identity property, our result is "identical" to the quantity at the start.

Have students complete these statements to illustrate the identity properties.

(a) $8 + \_\_\_\_ = 8$

(b) $\_\_\_\_ + 0 = 2x$

(c) $-10 \cdot \_\_\_\_ = -10$

(d) $\_\_\_\_ \cdot 1 = \frac{2}{3}$

The number 0 is the only number that can be added to any number to get that number, leaving the identity of the number unchanged. Thus, 0 is the **identity element for addition,** or the **additive identity.**

In a similar way, multiplying any number by 1 leaves the identity of the number unchanged. Thus, 1 is the **identity element for multiplication,** or the **multiplicative identity.** The **identity properties** summarize this discussion.

Identity Properties

For any real number a, the following hold.

$$a + 0 = a \quad \text{and} \quad 0 + a = a$$

Examples: $9 + 0 = 9$ and $0 + 9 = 9$

$$a \cdot 1 = a \quad \text{and} \quad 1 \cdot a = a$$

Examples: $9 \cdot 1 = 9$ and $1 \cdot 9 = 9$

NOW TRY ANSWERS

1. (a) $-6x + 2y$ (b) $-8k$

The identity properties leave the identity of a real number unchanged. Think of a child wearing a costume on Halloween. The child's appearance is changed, but his or her identity is unchanged.

NOW TRY EXERCISE 2
Simplify each expression.
(a) $7x + x$ **(b)** $-(5p - 3q)$

CLASSROOM EXAMPLE 2
Simplify each expression.
(a) $x - 3x$ (b) $-(3 + 4p)$
Answers: (a) $-2x$ (b) $-3 - 4p$

EXAMPLE 2 Using the Identity Property $1 \cdot a = a$

Simplify each expression.

(a) $12m + m$

$= 12m + 1m$ Identity property

$= (12 + 1)m$ Distributive property

$= 13m$ Add inside the parentheses.

(b) $y + y$

$= 1y + 1y$ Identity property

$= (1 + 1)y$ Distributive property

$= 2y$ Add inside the parentheses.

(c) $-(m - 5n)$

$= -1(m - 5n)$ Identity property

$= -1(m) + (-1)(-5n)$ Distributive property

$= -m + 5n$ Multiply.

Multiply *each* term by -1. Be careful with signs.

NOW TRY

OBJECTIVE 3 Use the inverse properties.

The *additive inverse* (or *opposite*) of a number a is $-a$. Additive inverses have a sum of 0 (the additive identity).

$$5 \text{ and } -5, \quad -\frac{1}{2} \text{ and } \frac{1}{2}, \quad -34 \text{ and } 34$$ Additive inverses (sum of 0)

The *multiplicative inverse* (or *reciprocal*) of a number a is $\frac{1}{a}$ (where $a \neq 0$). Multiplicative inverses have a product of 1.

$$5 \text{ and } \frac{1}{5}, \quad -\frac{1}{2} \text{ and } -2, \quad \frac{3}{4} \text{ and } \frac{4}{3}$$ Multiplicative inverses (product of 1)

This discussion leads to the **inverse properties.**

TEACHING TIP
Have students complete these statements to illustrate the inverse properties.
(a) $4 + \_\_\_\_ = 0$
(b) $-7.1 + \_\_\_\_ = 0$
(c) $\_\_\_\_ \cdot \frac{1}{5} = 1$
(d) $-\frac{3}{4} \cdot \_\_\_\_ = 1$

Inverse Properties

For any real number a, the following hold.

$$\boldsymbol{a + (-a) = 0 \quad \text{and} \quad -a + a = 0}$$

Examples: $7 + (-7) = 0$ and $-7 + 7 = 0$

$$\boldsymbol{a \cdot \frac{1}{a} = 1 \quad \text{and} \quad \frac{1}{a} \cdot a = 1 \quad (\text{where } a \neq 0)}$$

Examples: $7 \cdot \frac{1}{7} = 1$ and $\frac{1}{7} \cdot 7 = 1$

NOW TRY ANSWERS
2. (a) $8x$ **(b)** $-5p + 3q$

Terms and Their Coefficients

Term	*Numerical Coefficient*
$-7y$	-7
$34r^3$	34
$-26x^5yz^4$	-26
$-k = -1k$	-1
$r = 1r$	1
$\frac{3x}{8} = \frac{3}{8}x$	$\frac{3}{8}$
$\frac{x}{3} = \frac{1x}{3} = \frac{1}{3}x$	$\frac{1}{3}$

TEACHING TIP
Explain that a term can be made up of factors that are *groupings* of two or more terms, such as

$(2x+1)^2$ and $3(2x+y-z)$.

The inverse properties "undo" addition or multiplication. Think of putting on your shoes when you get up in the morning and then taking them off before you go to bed at night. These are inverse operations that undo each other.

Expressions such as $12m$ and $5n$ from **Example 2** are examples of *terms.* A **term** is a number or the product of a number and one or more variables raised to powers. The numerical factor in a term is the **numerical coefficient,** or just the **coefficient.**

Terms with exactly the same variables raised to exactly the same powers are **like terms.** Otherwise, they are **unlike terms.**

$5p$ and $-21p$	$-6x^2$ and $9x^2$	Like terms
$3m$ and $16x$	$7y^3$ and $-3y^2$	Unlike terms
Different variables	Different exponents on the same variable	

OBJECTIVE 4 Use the commutative and associative properties.

Simplifying expressions as in **Examples 2(a) and (b)** is called **combining like terms.** ***Only like terms may be combined.*** To combine like terms in an expression such as

$$-2m + 5m + 3 - 6m + 8,$$

we need two more properties. From arithmetic, we know that

$$3 + 9 = 12 \quad \text{and} \quad 9 + 3 = 12$$
$$3 \cdot 9 = 27 \quad \text{and} \quad 9 \cdot 3 = 27.$$

The order of the numbers being added or multiplied does not matter. The same answers result. The following computations illustrate this fact.

$$(5+7)+2 = 12+2 = 14$$
$$5+(7+2) = 5+9 = 14$$

$$(5 \cdot 7) \cdot 2 = 35 \cdot 2 = 70$$
$$5 \cdot (7 \cdot 2) = 5 \cdot 14 = 70$$

The grouping of the numbers being added or multiplied does not matter. The same answers result.

These arithmetic examples can be extended to algebra.

Commutative and Associative Properties

For any real numbers a, b, and c, the following hold.

$$\left.\begin{aligned} a + b &= b + a \\ ab &= ba \end{aligned}\right\} \text{Commutative properties}$$

(The *order* of the two terms or factors changes.)

Examples: $9 + (-3) = -3 + 9$ and $9(-3) = (-3)9$

$$\left.\begin{aligned} a + (b + c) &= (a + b) + c \\ a(bc) &= (ab)c \end{aligned}\right\} \text{Associative properties}$$

(The *grouping* among the three terms or factors changes, but the order stays the same.)

Examples: $7 + (8 + 9) = (7 + 8) + 9$ and $7 \cdot (8 \cdot 9) = (7 \cdot 8) \cdot 9$

The commutative properties are used to change the order of the terms or factors in an expression. Think of *commuting* from home to work and then from work to home. The associative properties are used to regroup the terms or factors of an expression. The grouped terms or factors are *associated.*

NOW TRY EXERCISE 3
Simplify.

$-7x + 10 - 3x - 4 + x$

CLASSROOM EXAMPLE 3
Simplify.

$12b - 9 + 4b - 7b + 1$

Answer: $9b - 8$

EXAMPLE 3 Using the Commutative and Associative Properties

Simplify.

$-2m + 5m + 3 - 6m + 8$

$= (-2m + 5m) + 3 - 6m + 8$ Associative property

$= (-2 + 5)m + 3 - 6m + 8$ Distributive property

$= 3m + 3 - 6m + 8$ Add inside parentheses.

The next step would be to add $3m$ and 3, but they are unlike terms. To get $3m$ and $-6m$ together, we use the associative and commutative properties, inserting parentheses and brackets according to the order of operations.

$= [(3m + 3) - 6m] + 8$

$= [3m + (3 - 6m)] + 8$ Associative property

$= [3m + (-6m + 3)] + 8$ Commutative property

$= [(3m + [-6m]) + 3] + 8$ Associative property

$= (-3m + 3) + 8$ Combine like terms.

$= -3m + (3 + 8)$ Associative property

$= -3m + 11$ Add.

In practice, many of these steps are not written down, but it is important to realize that the commutative and associative properties are used whenever the terms in an expression are rearranged to combine like terms. NOW TRY

CLASSROOM EXAMPLE 4
Simplify each expression.

(a) $12b - 9b + 5b - 7b$
(b) $6 - (2x + 7) - 3$
(c) $4m(2n)$

Answers: (a) b (b) $-2x - 4$
(c) $8mn$

EXAMPLE 4 Using the Properties of Real Numbers

Simplify each expression.

(a) $5y - 8y - 6y + 11y$

$= (5 - 8 - 6 + 11)y$ Distributive property

$= 2y$ Subtract and add.

(b) $3x + 4 - 5(x + 1) - 8$

$= 3x + 4 - 5x - 5 - 8$ Distributive property (Be careful with signs.)

$= 3x - 5x + 4 - 5 - 8$ Commutative property

$= -2x - 9$ Combine like terms. Subtract.

(c) $8 - (3m + 2)$

$= 8 - 1(3m + 2)$ Identity property

$= 8 - 3m - 2$ Distributive property

$= 6 - 3m$ Subtract.

NOW TRY ANSWER
3. $-9x + 6$

NOW TRY EXERCISE 4

Simplify each expression.

(a) $-3(t-4)-t+15$

(b) $7x-(4x-2)$

(c) $5x(6y)$

TEACHING TIP

Encourage oral participation by choosing exercises from this section that require students to identify the property being used.

(d) $3x(5)(y)$

$= [3x(5)]y$	Associative property
$= [3(x \cdot 5)]y$	Associative property
$= [3(5x)]y$	Commutative property
$= [(3 \cdot 5)x]y$	Associative property
$= (15x)y$	Multiply.
$= 15(xy)$	Associative property
$= 15xy$	

As previously mentioned, many of these steps are not usually written out. NOW TRY

CAUTION Be careful. The distributive property does not apply in **Example 4(d)** because there is no addition or subtraction involved.

$$(3x)(5)(y) \neq (3x)(5) \cdot (3x)(y)$$

NOW TRY ANSWERS

4. **(a)** $-4t+27$ **(b)** $3x+2$ **(c)** $30xy$

R.4 Exercises

FOR EXTRA HELP

MyMathLab®

Complete solution available in MyMathLab

1. B 2. C
3. A 4. D

5. order 6. grouping
7. same; same 8. like
9. -7 10. product; 0

11. $2m+2p$
12. $3a+3b$
13. $-12x+12y$
14. $-10p+10q$
15. $8k$ 16. $11a$
17. $-2r$ 18. $-2n$
19. cannot be simplified
20. cannot be simplified
21. $8a$ 22. $10s$
23. $-2d+f$ 24. $-3m+n$
25. $x+y$ 26. $3x+4y$

Concept Check *Choose the correct response.*

1. The identity element for addition is

A. $-a$ **B.** 0 **C.** 1 **D.** $\frac{1}{a}$.

2. The identity element for multiplication is

A. $-a$ **B.** 0 **C.** 1 **D.** $\frac{1}{a}$.

3. The additive inverse of a is

A. $-a$ **B.** 0 **C.** 1 **D.** $\frac{1}{a}$.

4. The multiplicative inverse of a, where $a \neq 0$, is

A. $-a$ **B.** 0 **C.** 1 **D.** $\frac{1}{a}$.

Concept Check *Complete each statement.*

5. The commutative property is used to change the ______ of two terms or factors.

6. The associative property is used to change the ______ of three terms or factors.

7. Like terms are terms with the ______ variables raised to the ______ powers.

8. When simplifying an expression, only ______ terms can be combined.

9. The numerical coefficient in the term $-7yz^2$ is ______.

10. The multiplication property of 0 (**Section R.2**) says that the ______ of 0 and any real number is ______.

*Simplify each expression. **See Examples 1 and 2.***

11. $2(m+p)$ **12.** $3(a+b)$ **13.** $-12(x-y)$ **14.** $-10(p-q)$

15. $5k+3k$ **16.** $6a+5a$ **17.** $7r-9r$ **18.** $4n-6n$

19. $-8z+4w$ **20.** $-12k+3r$ **21.** $a+7a$ **22.** $s+9s$

23. $-(2d-f)$ **24.** $-(3m-n)$ **25.** $-(-x-y)$ **26.** $-(-3x-4y)$

27. $-6y+3$ 28. $-6r-5$
29. $p+11$ 30. $-10x-3$
31. $-2k+15$ 32. $9r-11$
33. $-3m+2$ 34. $5a-36$
35. -1 36. $-10x$
37. $2p+7$ 38. $-23m-2$
39. $-6z-39$ 40. $4k-57$

41. $(5+8)x=13x$
42. $(9-6)y=3y$
43. $(5\cdot 9)r=45r$
44. $(-4+12)+8=16$
45. $9y+5x$
46. $7\cdot(-5)=-35$
47. 7 48. $-12x$
49. 0 50. 1
51. $8(-4)+8x=-32+8x$
52. $3x-3y+3z$
53. 0 54. 0

55. Answers will vary. One example of commutativity is washing your face and brushing your teeth. An example of non-commutativity is putting on your socks and putting on your shoes.
56. Answers will vary. One example is waking up and going to sleep.

57. 1900 58. 2700
59. 75 60. 87.5
61. 431 62. 48

Simplify each expression. **See Examples 1–4.**

27. $-12y+4y+3+2y$
28. $-5r-9r+8r-5$
29. $-6p+5-4p+6+11p$
30. $-8x-12+3x-5x+9$
31. $3(k+2)-5k+6+3$
32. $5(r-3)+6r-2r+4$
33. $-2(m+1)-(m-4)$
34. $6(a-5)-(a+6)$
35. $0.25(8+4p)-0.5(6+2p)$
36. $0.4(10-5x)-0.8(5+10x)$
37. $-(2p+5)+3(2p+4)-2p$
38. $-(7m-12)-2(4m+7)-8m$
39. $2+3(2z-5)-3(4z+6)-8$
40. $-4+4(4k-3)-6(2k+8)+7$

Concept Check *Complete each statement so that the indicated property is illustrated. Simplify each answer if possible.*

41. $5x+8x=$ ________ (distributive property)
42. $9y-6y=$ ________ (distributive property)
43. $5(9r)=$ ________ (associative property)
44. $-4+(12+8)=$ ________ (associative property)
45. $5x+9y=$ ________ (commutative property)
46. $-5\cdot 7=$ ________ (commutative property)
47. $1\cdot 7=$ ________ (identity property)
48. $-12x+0=$ ________ (identity property)
49. $-\frac{1}{4}ty+\frac{1}{4}ty=$ ________ (inverse property)
50. $-\frac{9}{8}\left(-\frac{8}{9}\right)=$ ________ (inverse property)
51. $8(-4+x)=$ ________ (distributive property)
52. $3(x-y+z)=$ ________ (distributive property)
53. $0(0.875x+9y-88z)=$ ________ (multiplication property of 0)
54. $0(35t^2-8t+12)=$ ________ (multiplication property of 0)

55. *Concept Check* Give an "everyday" example of a commutative operation and of an operation that is not commutative.

56. *Concept Check* Give an "everyday" example of inverse operations.

The distributive property can be used to mentally perform calculations. For example, calculate $38\cdot 17+38\cdot 3$ *as follows.*

$$38\cdot 17+38\cdot 3$$
$$=38(17+3) \quad \text{Distributive property}$$
$$=38(20) \quad \text{Add inside the parentheses.}$$
$$=760 \quad \text{Multiply.}$$

Use the distributive property to calculate each value mentally.

57. $96\cdot 19+4\cdot 19$
58. $27\cdot 60+27\cdot 40$
59. $58\cdot\frac{3}{2}-8\cdot\frac{3}{2}$
60. $8.75(15)-8.75(5)$
61. $4.31(69)+4.31(31)$
62. $\frac{8}{5}(17)+\frac{8}{5}(13)$

63. No. One example is $7 + (5 \cdot 3) = (7 + 5)(7 + 3)$, which is false.

64. By the distributive property, the sum of two like terms equals the product of the sum of their coefficients and the common variable. For example, $2x + 5x = (2 + 5)x = 7x$.

Write a short answer to each problem.

63. By the distributive property, $a(b + c) = ab + ac$. This property is more completely named the distributive property of multiplication with respect to addition. Is there a distributive property of addition with respect to multiplication? In other words, is

$$a + (b \cdot c) = (a + b)(a + c)$$

true for all real numbers, a, b, and c? To find out, try some sample values of a, b, and c.

64. Explain how the distributive property is used to combine like terms. Give an example.

Relating Concepts **exercises relate topics currently being studied to ones studied earlier, to real-life situations, or to universally accepted truths. In general, these exercises should be worked in numerical order. Answers to *all* of them are given in the answer section of the student text.**

65. associative property
66. associative property
67. commutative property
68. associative property
69. distributive property
70. arithmetic facts

RELATING CONCEPTS For Individual or Group Work **(Exercises 65–70)**

When simplifying the expression $3x + 4 + 2x + 7$ *to* $5x + 11$, *some steps are usually done mentally.* ***Work Exercises 65–70 in order,*** *providing the property that justifies each statement in the given simplification. (These steps could be done in other orders.)*

$3x + 4 + 2x + 7$

65. $= (3x + 4) + (2x + 7)$ ________________

66. $= 3x + (4 + 2x) + 7$ ________________

67. $= 3x + (2x + 4) + 7$ ________________

68. $= (3x + 2x) + (4 + 7)$ ________________

69. $= (3 + 2)x + (4 + 7)$ ________________

70. $= 5x + 11$ ________________

Chapter R Summary

Key Terms

R.1

set
elements (members)
finite set
natural (counting) numbers
infinite set
whole numbers
empty (null) set
variable
number line
integers
coordinate
graph
rational numbers
irrational numbers
real numbers
additive inverse (opposite, negative)
signed numbers
absolute value
equation
inequality

R.2

sum
difference
product
quotient
reciprocal (multiplicative inverse)

R.3

factors
exponent (power)
base
exponential expression
square root
positive (principal) square root
negative square root
constant
algebraic expression

R.4

identity element for addition (additive identity)
identity element for multiplication (multiplicative identity)
term
coefficient (numerical coefficient)
like terms
unlike terms

New Symbols

$\{a, b\}$	set containing the elements a and b
$\varnothing$ or $\{\ \}$	empty set
$\in$	is an element of (a set)
$\notin$	is not an element of
$\neq$	is not equal to
$\{x \mid x$ **has property** $P\}$	set-builder notation
$\lvert x \rvert$	absolute value of x
$<$	is less than
$\leq$	is less than or equal to
$>$	is greater than
$\geq$	is greater than or equal to
a^m	m factors of a
$\sqrt{\ }$	radical symbol
$\sqrt{a}$	positive (principal) square root of a

Test Your Word Power

See how well you have learned the vocabulary in this chapter.

1. The **empty set** is a set
   A. with 0 as its only element
   B. with an infinite number of elements
   C. with no elements
   D. of ideas.

2. A **variable** is
   A. a symbol used to represent an unknown number
   B. a value that makes an equation true
   C. a solution of an equation
   D. the answer in a division problem.

3. The **absolute value** of a number is
   A. the graph of the number
   B. the reciprocal of the number
   C. the opposite of the number
   D. the distance between 0 and the number on a number line.

4. The **reciprocal** of a nonzero number a is
   A. a B. $\frac{1}{a}$ C. $-a$ D. 1.

5. A **factor** is
   A. the answer in an addition problem
   B. the answer in a multiplication problem
   C. one of two or more numbers that are added to get another number
   D. any number that divides evenly into a given number.

6. An **exponential expression** is
   A. a number that is a repeated factor in a product
   B. a number or a variable written with an exponent
   C. a number that shows how many times a factor is repeated in a product
   D. an expression that involves addition.

7. A **term** is
   A. a numerical factor
   B. a number or a product of a number and one or more variables raised to powers
   C. one of several variables with the same exponents
   D. a sum of numbers and variables raised to powers.

8. A **numerical coefficient** is
   A. the numerical factor in a term
   B. the number of terms in an expression
   C. a variable raised to a power
   D. the variable factor in a term.

ANSWERS

1. C; *Example:* The set of whole numbers less than 0 is the empty set, written ∅. **2.** A; *Examples:* a, b, c **3.** D; *Examples:* $|2| = 2$ and $|-2| = 2$ **4.** B; *Examples:* 3 is the reciprocal of $\frac{1}{3}$; $-\frac{5}{2}$ is the reciprocal of $-\frac{2}{5}$. **5.** D; *Example:* 2 and 5 are factors of 10 because both divide evenly (without remainder) into 10. **6.** B; *Examples:* 3^4 and x^{10} **7.** B; *Examples:* $6, \frac{x}{2}, -4ab^2$ **8.** A; *Examples:* The term $8z$ has numerical coefficient 8, and the term $-10x^3y$ has numerical coefficient -10.

Chapter R Test

FOR EXTRA HELP *Step-by-step test solutions are found on the Chapter Test Prep Videos available in* MyMathLab® *or on* YouTube.

View the complete solutions to all Chapter Test exercises in MyMathLab.

1. Graph $\left\{-3, 0.75, \frac{5}{3}, 5, 6.3\right\}$ on a number line.

Let $A = \left\{-\sqrt{6}, -1, -0.5, 0, 3, \sqrt{25}, 7.5, \frac{24}{2}, \sqrt{-4}\right\}$. *Simplify the elements of A as necessary, and then list those elements of A which belong to the specified set.*

2. Whole numbers
3. Integers
4. Rational numbers
5. Real numbers

Decide whether each statement is true *or* false.

6. Every real number is an integer.
7. Every whole number is an integer.

Perform the indicated operations.

8. $-6 + 14 + (-11) - (-3)$
9. $10 - 4 \cdot 3 + 6(-4)$
10. $7 - 4^2 + 2(6) + (-4)^2$
11. $\dfrac{10 - 24 + (-6)}{\sqrt{16}(-5)}$
12. $\dfrac{-2[3 - (-1 - 2) + 2]}{\sqrt{9}(-3) - (-2)}$
13. $\dfrac{8 \cdot 4 - 3^2 \cdot 5 - 2(-1)}{-3 \cdot 2^3 + 24}$

[R.1]

1. 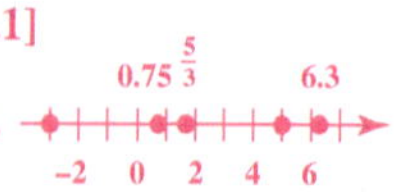

2. $0, 3, \sqrt{25}$ (or 5), $\frac{24}{2}$ (or 12)
3. $-1, 0, 3, \sqrt{25}$ (or 5), $\frac{24}{2}$ (or 12)
4. $-1, -0.5, 0, 3, \sqrt{25}$ (or 5) $7.5, \frac{24}{2}$ (or 12)
5. All are real numbers except $\sqrt{-4}$.
6. false
7. true

[R.2]

8. 0

[R.3]
9. -26 10. 19
11. 1 12. $\frac{16}{7}$
13. undefined

[R.2]
14. 50,395 ft
15. 37,487 ft
16. 1345 ft

[R.3]
17. 14 18. -15
19. not a real number
20. (a) a must be positive.
(b) a must be negative.
(c) a must be 0.
21. $-\frac{6}{23}$

[R.4]
22. $10k - 10$
23. It changes the sign of each term. The simplified form is $7r + 2$.

[R.2, R.4]
24. B
25. D 26. A
27. F 28. C
29. C 30. E

The table shows the heights in feet of some selected mountains and the depths in feet (as negative numbers) of some selected ocean trenches.

Mountain	Height	Trench	Depth
Foraker	17,400	Philippine	−32,995
Wilson	14,252	Cayman	−24,721
Pikes Peak	14,111	Java	−23,376

Source: *World Almanac and Book of Facts.*

14. What is the difference between the height of Mt. Foraker and the depth of the Philippine Trench?

15. What is the difference between the height of Pikes Peak and the depth of the Java Trench?

16. How much deeper is the Cayman Trench than the Java Trench?

Find each square root. If the number is not real, say so.

17. $\sqrt{196}$ **18.** $-\sqrt{225}$ **19.** $\sqrt{-16}$

20. For the expression $\sqrt{a}$, under what conditions will its value be each of the following?

(a) positive **(b)** not real **(c)** 0

21. Evaluate $\frac{8k + 2m^2}{r - 2}$ for $k = -3$, $m = -3$, and $r = 25$.

22. Simplify $-3(2k - 4) + 4(3k - 5) - 2 + 4k$.

23. How does the subtraction sign affect the terms $-4r$ and 6 when

$$(3r + 8) - (-4r + 6)$$

is simplified? What is the simplified form?

Match each statement in Column I with the appropriate property in Column II. Answers may be used more than once.

I	II
24. $6 + (-6) = 0$	**A.** Distributive property
25. $-2 + (3 + 6) = (-2 + 3) + 6$	**B.** Inverse property
26. $5x + 15x = (5 + 15)x$	**C.** Identity property
27. $13 \cdot 0 = 0$	**D.** Associative property
28. $-9 + 0 = -9$	**E.** Commutative property
29. $4 \cdot 1 = 4$	**F.** Multiplication property of 0
30. $(a + b) + c = (b + a) + c$	

STUDY SKILLS

Reading Your Math Textbook

Take time to read each section and its examples before doing your homework. You will learn more and be better prepared to work the exercises your instructor assigns.

Approaches to Reading Your Math Textbook

Student A learns best by listening to her teacher explain things. She "gets it" when she sees the instructor work problems. She previews the section before the lecture, so she knows generally what to expect. **Student A carefully reads the section in her text *AFTER* she hears the classroom lecture on the topic.**

Student B learns best by reading on his own. He reads the section and works through the examples before coming to class. That way, he knows what the teacher is going to talk about and what questions he wants to ask. **Student B carefully reads the section in his text *BEFORE* he hears the classroom lecture on the topic.**

Which of these reading approaches works best for you—that of Student A or Student B?

Tips for Reading Your Math Textbook

- **Turn off your cell phone.** You will be able to concentrate more fully on what you are reading.
- **Survey the material.** Glance over the assigned material to get an idea of the "big picture." Look at the list of objectives to see what you will be learning.
- **Read slowly.** Read only one section—or even part of a section—at a sitting, with paper and pencil in hand.
- **Pay special attention to important information given in colored boxes or set in boldface type.**
- **Study the examples carefully.** Pay particular attention to the blue side comments and any pointer balloons.
- **Do the Now Try exercises in the margin on separate paper as you go.** These mirror the examples and prepare you for the exercise set. The answers are given at the bottom of the page.
- **Make study cards as you read.** Make cards for new vocabulary, rules, procedures, formulas, and sample problems.
- **Mark anything you don't understand. *ASK QUESTIONS*** in class—everyone will benefit. Follow up with your instructor, as needed.

Think through and answer each question.

1. Which two or three reading tips will you try this week?

2. Did the tips you selected improve your ability to read and understand the material? Explain.

1 Linear Equations, Inequalities, and Applications

The concept of balance can be applied to solving *linear equations in one variable*, a topic of this chapter.

1.1 Linear Equations in One Variable

OBJECTIVES

1. Distinguish between expressions and equations.
2. Identify linear equations.
3. Solve linear equations using the addition and multiplication properties of equality.
4. Solve linear equations using the distributive property.
5. Solve linear equations with fractions or decimals.
6. Identify conditional equations, contradictions, and identities.

VOCABULARY

- ☐ linear (first-degree) equation in one variable
- ☐ solution
- ☐ solution set
- ☐ equivalent equations
- ☐ conditional equation
- ☐ identity
- ☐ contradiction

OBJECTIVE 1 Distinguish between expressions and equations.

In our work in **Chapter R,** we reviewed *algebraic expressions.*

$$8x + 9, \quad y - 4, \quad \text{and} \quad \frac{x^3y^8}{z} \qquad \text{Algebraic expressions}$$

Equations and inequalities compare algebraic expressions, just as a balance scale compares the weights of two quantities. Recall that an ***equation*** is a statement that two algebraic expressions are equal. ***An equation always contains an equality symbol, while an expression does not.***

EXAMPLE 1 Distinguishing between Expressions and Equations

Decide whether each of the following is an *expression* or an *equation*.

(a) $3x - 7 = 2$ **(b)** $3x - 7$

In part (a) we have an equation because there is an equality symbol. In part (b), there is no equality symbol, so it is an expression.

Equation: Left side $\overbrace{3x - 7}$ $=$ Right side $\overbrace{2}$

An equation can be *solved.*

Expression: $3x - 7$

An expression **cannot** be solved. It can often be *evaluated* or *simplified.*

NOW TRY

OBJECTIVE 2 Identify linear equations.

A *linear equation in one variable* involves only real numbers and one variable raised to the first power.

Linear Equation in One Variable

A **linear equation in one variable** (here x) can be written in the form

$$Ax + B = C,$$

where A, B, and C are real numbers and $A \neq 0$.

Examples: $x + 1 = -2$, $x - 3 = 5$, and $2x + 5 = 10$ Linear equations

A linear equation is a **first-degree equation** because the greatest power on the variable is 1. Some equations that are not linear (that is, *nonlinear*) follow.

$$x^2 + 3x = 5, \quad \frac{8}{x} = -22, \quad \text{and} \quad \sqrt{x} = 6 \qquad \text{Nonlinear equations}$$

If the variable in an equation can be replaced by a real number that makes the statement true, then that number is a **solution** of the equation. For example, 8 is a solution of the equation $x - 3 = 5$ because replacing x with 8 gives a true statement. An equation is *solved* by finding its **solution set,** the set of all solutions. The solution set of the equation $x - 3 = 5$ is $\{8\}$.

NOW TRY EXERCISE 1

Decide whether each of the following is an *expression* or an *equation*.

(a) $2x + 17 - 3x$

(b) $2x + 17 = 3x$

CLASSROOM EXAMPLE 1

Decide whether each of the following is an *expression* or an *equation*.

(a) $9x + 10 = 0$ (b) $9x + 10$

Answers: (a) equation (b) expression

NOW TRY ANSWERS

1. **(a)** expression **(b)** equation

TEACHING TIP
Students may confuse expressions and equations. Keep reminding them, as we will, of the difference. Stress that an equation has an expression on the left side, an equality symbol, and an expression on the right side.

Equivalent equations are related equations that have the same solution set. To solve an equation, we usually start with the given equation and replace it with a series of simpler equivalent equations. The following are all equivalent because each has solution set $\{3\}$.

$$5x + 2 = 17, \quad 5x = 15, \quad \text{and} \quad x = 3 \qquad \text{Equivalent equations}$$

OBJECTIVE 3 Solve linear equations using the addition and multiplication properties of equality.

We use two important properties of equality to produce equivalent equations.

Addition and Multiplication Properties of Equality

Addition Property of Equality

If A, B, and C represent real numbers, then the equations

$$A = B \quad \text{and} \quad A + C = B + C \quad \text{are equivalent.}$$

That is, the same number may be added to each side of an equation without changing the solution set.

Multiplication Property of Equality

If A, B, and C represent real numbers, where $C \neq 0$, then the equations

$$A = B \quad \text{and} \quad AC = BC \quad \text{are equivalent.}$$

That is, each side of an equation may be multiplied by the same nonzero number without changing the solution set.

Because subtraction and division are defined in terms of addition and multiplication, respectively, the preceding properties can be extended.

The same number may be subtracted from each side of an equation, and each side of an equation may be divided by the same nonzero number, without changing the solution set.

CLASSROOM EXAMPLE 2
Solve.

$$4x + 8x = -9 + 17x - 1$$

Answer: $\{2\}$

EXAMPLE 2 Solving a Linear Equation

Solve $4x - 2x - 5 = 4 + 6x + 3$.

The goal is to isolate x on one side of the equation.

$$4x - 2x - 5 = 4 + 6x + 3$$

$$2x - 5 = 7 + 6x \qquad \text{Combine like terms.}$$

$$2x - 5 + 5 = 7 + 6x + 5 \qquad \text{Add 5 to each side.}$$

$$2x = 12 + 6x \qquad \text{Combine like terms.}$$

$$2x - 6x = 12 + 6x - 6x \qquad \text{Subtract } 6x \text{ from each side.}$$

$$-4x = 12 \qquad \text{Combine like terms.}$$

$$\frac{-4x}{-4} = \frac{12}{-4} \qquad \text{Divide each side by } -4.$$

$$x = -3$$

NOW TRY EXERCISE 2
Solve.

$$5x + 11 = 2x - 13 - 3x$$

CHECK Substitute -3 for x in the original equation.

$$4x - 2x - 5 = 4 + 6x + 3 \quad \text{Original equation}$$
$$4(-3) - 2(-3) - 5 \stackrel{?}{=} 4 + 6(-3) + 3 \quad \text{Let } x = -3.$$
$$-12 + 6 - 5 \stackrel{?}{=} 4 - 18 + 3 \quad \text{Multiply.}$$
$$-11 = -11 \checkmark \quad \text{True}$$

Use parentheses around substituted values to avoid errors.

This is *not* the solution.

The true statement indicates that $\{-3\}$ is the solution set.

NOW TRY

TEACHING TIP
Remind students that only one equality symbol should be used per line in the solution of an equation. Students may have developed some bad habits in earlier courses.

CAUTION In **Example 2,** the equality symbols are aligned in a column. ***Use only one equality symbol in a horizontal line of work when solving an equation.***

Solving a Linear Equation in One Variable

Step 1 **Simplify each side separately.** Use the distributive property as needed.
- Clear any parentheses.
- Clear any fractions or decimals.
- Combine like terms.

Step 2 **Isolate the variable terms on one side.** Use the addition property of equality so that all terms with variables are on one side of the equation and all constants (numbers) are on the other side.

Step 3 **Isolate the variable.** Use the multiplication property of equality to obtain an equation that has just the variable with coefficient 1 on one side.

Step 4 **Check.** Substitute the proposed solution into the *original* equation. If a true statement results, write the solution set. If not, rework the problem.

OBJECTIVE 4 Solve linear equations using the distributive property.

CLASSROOM EXAMPLE 3
Solve.

$$3(4 + x) - x = 5x - 3$$

Answer: $\{5\}$

EXAMPLE 3 Solving a Linear Equation

Solve $2(x - 5) + 3x = x + 6$.

Step 1 Clear parentheses using the distributive property. Then combine like terms.

Be sure to distribute over *all* terms within the parentheses.

$$2(x - 5) + 3x = x + 6$$
$$2x + 2(-5) + 3x = x + 6 \quad \text{Distributive property}$$
$$2x - 10 + 3x = x + 6 \quad \text{Multiply.}$$
$$5x - 10 = x + 6 \quad \text{Combine like terms.}$$

Step 2 Isolate the variable terms on one side and constants on the other.

$$5x - 10 - x = x + 6 - x \quad \text{Subtract } x.$$
$$4x - 10 = 6 \quad \text{Combine like terms.}$$
$$4x - 10 + 10 = 6 + 10 \quad \text{Add 10.}$$
$$4x = 16 \quad \text{Combine like terms.}$$

NOW TRY ANSWER
2. $\{-4\}$

NOW TRY EXERCISE 3
Solve.

$$5(x-4)-12=3-2x$$

Step 3 Use the multiplication property of equality to isolate x on the left.

$$\frac{4x}{4}=\frac{16}{4} \quad \text{Divide by 4.}$$

$$x=4$$

Step 4 Check by substituting 4 for x in the original equation.

CHECK

$$2(x-5)+3x=x+6 \quad \text{Original equation}$$

$$2(4-5)+3(4)\stackrel{?}{=}4+6 \quad \text{Let } x=4.$$

$$2(-1)+12\stackrel{?}{=}10 \quad \text{Simplify.}$$

$$10=10 \ \checkmark \quad \text{True}$$

Always check your work.

The proposed solution 4 checks, so the solution set is $\{4\}$. NOW TRY

NOW TRY EXERCISE 4
Solve.

$$2-3(2+6x)=4(x+1)+36$$

CLASSROOM EXAMPLE 4
Solve.

$$6-(4+x)=8x-2(3x+5)$$

Answer: $\{4\}$

EXAMPLE 4 Solving a Linear Equation

Solve $4(3x-2)=38-2(2x-1)$.

$$4(3x-2)=38-2(2x-1)$$

Step 1

$$4(3x)+4(-2)=38-2(2x)-2(-1) \quad \text{Clear parentheses using the distributive property.}$$

Be careful with signs when distributing.

$$12x-8=38-4x+2 \quad \text{Multiply.}$$

$$12x-8=40-4x \quad \text{Combine like terms.}$$

Step 2

$$12x-8+4x=40-4x+4x \quad \text{Add } 4x.$$

$$16x-8=40 \quad \text{Combine like terms.}$$

$$16x-8+8=40+8 \quad \text{Add 8.}$$

$$16x=48 \quad \text{Combine like terms.}$$

Step 3

$$\frac{16x}{16}=\frac{48}{16} \quad \text{Divide by 16.}$$

$$x=3$$

Step 4 **CHECK**

$$4(3x-2)=38-2(2x-1) \quad \text{Original equation}$$

$$4[3(3)-2]\stackrel{?}{=}38-2[2(3)-1] \quad \text{Let } x=3.$$

$$4[7]\stackrel{?}{=}38-2[5] \quad \text{Work inside the brackets.}$$

$$28=28 \ \checkmark \quad \text{True}$$

The proposed solution 3 checks, so the solution set is $\{3\}$. NOW TRY

OBJECTIVE 5 Solve linear equations with fractions or decimals.

When fractions appear as coefficients in an equation, we multiply each side by the least common denominator (LCD) of all the fractions. This is an application of the multiplication property of equality.

NOW TRY ANSWERS
3. $\{5\}$ **4.** $\{-2\}$

NOW TRY EXERCISE 5

Solve.

$$\frac{x-4}{4}+\frac{2x+4}{8}=5$$

TEACHING TIP

Emphasize that *each term* of the equation must be multiplied by the LCD. It is best to identify how many terms are in the equation before multiplying.

CLASSROOM EXAMPLE 5

Solve.

$$\frac{x+1}{2}+\frac{x+3}{4}=\frac{1}{2}$$

Answer: $\{-1\}$

EXAMPLE 5 Solving a Linear Equation with Fractions

Solve $\frac{x+7}{6}+\frac{2x-8}{2}=-4$.

Step 1	$6\left(\frac{x+7}{6}+\frac{2x-8}{2}\right)=6(-4)$	Eliminate the fractions. Multiply each side by the LCD, 6.
	$6\left(\frac{x+7}{6}\right)+6\left(\frac{2x-8}{2}\right)=6(-4)$	Distributive property
	$x+7+3(2x-8)=-24$	Multiply.
	$x+7+3(2x)+3(-8)=-24$	Distributive property
	$x+7+6x-24=-24$	Multiply.
	$7x-17=-24$	Combine like terms.
Step 2	$7x-17+17=-24+17$	Add 17.
	$7x=-7$	Combine like terms.
Step 3	$\frac{7x}{7}=\frac{-7}{7}$	Divide by 7.
	$x=-1$	
Step 4 CHECK	$\frac{x+7}{6}+\frac{2x-8}{2}=-4$	Original equation
	$\frac{-1+7}{6}+\frac{2(-1)-8}{2}\stackrel{?}{=}-4$	Let $x=-1$.
	$\frac{6}{6}+\frac{-10}{2}\stackrel{?}{=}-4$	Add and subtract in the numerators.
	$1-5\stackrel{?}{=}-4$	Simplify each fraction.
	$-4=-4$ ✓	True

The proposed solution -1 checks, so the solution set is $\{-1\}$. NOW TRY

When an equation involves decimal coefficients, we can clear the decimals by multiplying by a power of 10, such as

$$10^1=10,\quad 10^2=100,\quad \text{and so on,}$$

to obtain an equivalent equation with integer coefficients.

CLASSROOM EXAMPLE 6

Solve.

$$0.02(60)+0.04x=0.03(50+x)$$

Answer: $\{30\}$

EXAMPLE 6 Solving a Linear Equation with Decimals

Solve $0.06x+0.09(15-x)=0.07(15)$.

	$0.06x+0.09(15-x)=0.07(15)$	← Clear parentheses first.
Step 1	$0.06x+0.09(15)+0.09(-x)=0.07(15)$	Distributive property
	$0.06x+1.35-0.09x=1.05$	Multiply.

Now clear the decimals. Because each decimal number is given in hundredths, multiply each term of the equation by 100. A number can be multiplied by 100 by moving the decimal point two places to the right.

NOW TRY ANSWER

5. $\{11\}$

NOW TRY EXERCISE 6

Solve.

$$0.08x - 0.12(x-4) = 0.03(x-5)$$

$$0.06x + 1.35 - 0.09x = 1.05 \quad \text{Multiply each term by 100.}$$

Move decimal points 2 places to the right.

$$6x + 135 - 9x = 105$$

This is an equivalent equation without decimals.

$$-3x + 135 = 105 \quad \text{Combine like terms.}$$

Step 2 $$-3x + 135 - 135 = 105 - 135 \quad \text{Subtract 135.}$$

$$-3x = -30 \quad \text{Combine like terms.}$$

Step 3 $$\frac{-3x}{-3} = \frac{-30}{-3} \quad \text{Divide by } -3.$$

$$x = 10$$

Step 4 CHECK $$0.06x + 0.09(15 - x) = 0.07(15)$$

$$0.06(10) + 0.09(15 - 10) \stackrel{?}{=} 0.07(15) \quad \text{Let } x = 10.$$

$$0.6 + 0.09(5) \stackrel{?}{=} 1.05 \quad \text{Multiply and subtract.}$$

$$0.6 + 0.45 \stackrel{?}{=} 1.05 \quad \text{Multiply.}$$

$$1.05 = 1.05 \ ✓ \quad \text{True}$$

The solution set is $\{10\}$.

NOW TRY

OBJECTIVE 6 Identify conditional equations, contradictions, and identities.

Some equations have one solution, some have no solutions, and others have an infinite number of solutions. The table gives the names of these types of equations.

Solution Sets of Equations

Type of Linear Equation	*Number of Solutions*	*Indication When Solving*
Conditional	One solution; solution set: {a number}	Final line is x = a number. (See **Example 7(a).**)
Identity	Infinitely many solutions; solution set: {all real numbers}	Final line is true, such as $0 = 0$. (See **Example 7(b).**)
Contradiction	No solution; solution set: ∅	Final line is false, such as $-15 = -20$. (See **Example 7(c).**)

CLASSROOM EXAMPLE 7

Solve each equation. Decide whether it is a *conditional equation*, an *identity*, or a contradiction.

(a) $5(x+2) - 2(x+1) = 3x + 1$

(b) $\frac{x+1}{3} + \frac{2x}{3} = x + \frac{1}{3}$

(c) $5(3x+1) = x + 5$

Answers:

(a) ∅; contradiction

(b) {all real numbers}; identity

(c) {0}; conditional equation

EXAMPLE 7 Recognizing Conditional Equations, Identities, and Contradictions

Solve each equation. Decide whether it is a *conditional equation*, an *identity*, or a *contradiction*.

(a) $$5(2x+6) - 2 = 7(x+4)$$

$$10x + 30 - 2 = 7x + 28 \quad \text{Distributive property}$$

$$10x + 28 = 7x + 28 \quad \text{Combine like terms.}$$

$$10x + 28 - 7x - 28 = 7x + 28 - 7x - 28 \quad \text{Subtract } 7x. \text{ Subtract 28.}$$

$$3x = 0 \quad \text{Combine like terms.}$$

$$\frac{3x}{3} = \frac{0}{3} \quad \text{Divide by 3.}$$

$$x = 0 \quad \frac{0}{3} = 0 \text{ (See Section R.2.)}$$

The last line has a variable. The number following "=" is a solution.

NOW TRY ANSWER

6. {9}

NOW TRY EXERCISE 7

Solve each equation. Decide whether it is a *conditional equation*, an *identity*, or a *contradiction*.

(a) $9x - 3(x + 4) = 6(x - 2)$

(b) $-3(2x - 1) - 2x = 3 + x$

(c) $10x - 21 = 2(x - 5) + 8x$

TEACHING TIP

Note the difference between a conditional equation with solution 0 and a contradiction. For example,

$$2x - 10 = -10$$
$$2x = 0$$
$$x = 0$$

is a conditional equation with solution set $\{0\}$. **Example 7(c)** illustrates a contradiction with solution set $\varnothing$.

CHECK

$5(2x + 6) - 2 = 7(x + 4)$	Original equation
$5[2(0) + 6] - 2 \stackrel{?}{=} 7(0 + 4)$	Let $x = 0$.
$5(6) - 2 \stackrel{?}{=} 7(4)$	Multiply and then add.
$28 = 28$ ✓	True

The value 0 checks, so the solution set is $\{0\}$. Because the solution set has only one element, $5(2x + 6) - 2 = 7(x + 4)$ is a conditional equation.

(b)

$5x - 15 = 5(x - 3)$	
$5x - 15 = 5x - 15$	Distributive property
$5x - 15 - 5x + 15 = 5x - 15 - 5x + 15$	Subtract 5x. Add 15.
$0 = 0$ (The variable has disappeared.)	True

The *true* statement $0 = 0$ indicates that the solution set is {all real numbers}. The equation $5x - 15 = 5(x - 3)$ is an identity. Notice that the first step yielded

$$5x - 15 = 5x - 15, \quad \text{which is true for } all \text{ values of } x.$$

We could have identified the equation as an identity at that point.

(c)

$5x - 15 = 5(x - 4)$	
$5x - 15 = 5x - 20$	Distributive property
$5x - 15 - 5x = 5x - 20 - 5x$	Subtract 5x.
$-15 = -20$ (The variable has disappeared.)	False

Because the result, $-15 = -20$, is *false*, the equation has no solution. The solution set is $\varnothing$, so the equation $5x - 15 = 5(x - 4)$ is a contradiction. NOW TRY

NOW TRY ANSWERS

7. (a) {all real numbers}; identity

(b) $\{0\}$; conditional equation

(c) $\varnothing$; contradiction

CAUTION A common error in solving an equation like that in **Example 7(a)** is to think that the equation has no solution and write the solution set as $\varnothing$. This equation has one solution, the number 0, so it is a conditional equation with solution set $\{0\}$.

1.1 Exercises

FOR EXTRA HELP

MyMathLab®

Complete solution available in MyMathLab

1. algebraic expression; does; is not
2. linear equation; $=$; first-degree equation; one
3. true; solution; solution set

Concept Check *Complete each statement. The following key terms may be used once, more than once, or not at all.*

linear equation	solution	algebraic expression	contradiction	all real numbers
solution set	identity	conditional equation	first-degree equation	empty set $\varnothing$

1. A collection of numbers, variables, operation symbols, and grouping symbols, such as $2(8x - 15)$, is a(n) ________. While an equation (*does/does not*) include an equality symbol, there (*is/is not*) an equality symbol in an algebraic expression.

2. A(n) ________ in one variable (here x) can be written in the form $Ax + B\ (=/>/<)\ C$, with $A \neq 0$. Another name for a linear equation is a(n) ________, because the greatest power on the variable is (*one/two/three*).

3. If we let $x = 2$ in the linear equation $2x + 5 = 9$, a (*true/false*) statement results. The number 2 is a(n) ________ of the equation, and $\{2\}$ is the ________.

4. solution set; conditional equation
5. identity; all real numbers
6. contradiction; empty set ∅
7. A, C
8. B is nonlinear because the variable is squared. D is nonlinear because there is a variable in the second denominator.
9. Both sides are evaluated as 30, so 6 is a solution.
10. It is not, because the left side is evaluated as −2 and the right side as −9.
11. equation
12. equation
13. expression
14. expression
15. equation
16. equation
17. A sign error was made when the distributive property was applied. The left side of the second line should be $8x - 4x + 6$. The correct solution is 1.
18. −1; $-5x - (2x - 4) + 5$
$= -5x - 2x + 4 + 5$
$= -7x + 9$
19. $\{-1\}$ 20. $\{5\}$
21. $\{-4\}$ 22. $\{-5\}$
23. $\{-7\}$ 24. $\{3\}$
25. $\{0\}$ 26. $\{0\}$
27. $\{4\}$ 28. $\{2\}$
29. $\left\{-\frac{7}{8}\right\}$ 30. $\left\{-\frac{6}{11}\right\}$
31. ∅; contradiction
32. ∅; contradiction
33. $\left\{-\frac{5}{3}\right\}$ 34. $\{-15\}$
35. $\left\{-\frac{1}{2}\right\}$ 36. $\left\{\frac{4}{3}\right\}$
37. $\{2\}$ 38. $\{-2\}$
39. $\{-2\}$ 40. $\{4\}$

4. A linear equation with one solution in its ________, such as the equation in **Exercise 3,** is a(n) ________.

5. A linear equation with an infinite number of solutions is a(n)________. Its solution set is {________}.

6. A linear equation with no solution is a(n) ________. Its solution set is the ________.

7. *Concept Check* Which equations are linear equations in x?

A. $3x + x - 2 = 0$ **B.** $12 = x^2$ **C.** $9x - 4 = 9$ **D.** $\frac{1}{8}x - \frac{1}{x} = 0$

8. *Concept Check* Which of the equations in **Exercise 7** are nonlinear equations in x? Explain why.

9. *Concept Check* Decide whether 6 is a solution of $3(x + 4) = 5x$ by substituting 6 for x. If it is not a solution, explain why.

10. *Concept Check* Use substitution to decide whether −2 is a solution of the equation $5(x + 4) - 3(x + 6) = 9(x + 1)$. If it is not a solution, explain why.

Decide whether each of the following is an expression *or an* equation. ***See Example 1.***

11. $-3x + 2 - 4 = x$ **12.** $-3x + 2 - 4 - x = 4$

13. $4(x + 3) - 2(x + 1) - 10$ **14.** $4(x + 3) - 2(x + 1) + 10$

15. $-10x + 12 - 4x = -3$ **16.** $-10x + 12 - 4x + 3 = 0$

17. *Concept Check* This work contains a common error.

$8x - 2(2x - 3) = 3x + 7$
$8x - 4x - 6 = 3x + 7$ Distributive property
$4x - 6 = 3x + 7$ Combine like terms
$x = 13$ Subtract $3x$. Add 6.

WHAT WENT WRONG? Give the correct solution.

18. *Concept Check* When clearing parentheses in the expression

$-5x - (2x - 4) + 5$ See Exercise 47, the right side of the equation.

the − sign before the parenthesis acts like a factor representing what number? Clear parentheses and simplify this expression.

Solve each equation, and check the solution. If applicable, tell whether the equation is an identity *or a* contradiction. ***See Examples 2–4 and 7.***

19. $7x + 8 = 1$ **20.** $5x - 4 = 21$ **21.** $5x + 2 = 3x - 6$

22. $9x + 1 = 7x - 9$ **23.** $7x - 5x + 15 = x + 8$ **24.** $2x + 4 - x = 4x - 5$

25. $12w + 15w - 9 + 5 = -3w + 5 - 9$ **26.** $-4x + 5x - 8 + 4 = 6x - 4$

27. $3(2t - 4) = 20 - 2t$ **28.** $2(3 - 2x) = x - 4$

29. $-5(x + 1) + 3x + 2 = 6x + 4$ **30.** $5(x + 3) + 4x - 5 = 4 - 2x$

31. $-2x + 5x - 9 = 3(x - 4) - 5$ **32.** $-6x + 2x - 11 = -2(2x - 3) + 4$

33. $2(x + 3) = -4(x + 1)$ **34.** $4(x - 9) = 8(x + 3)$

35. $3(2x + 1) - 2(x - 2) = 5$ **36.** $4(x - 2) + 2(x + 3) = 6$

37. $2x + 3(x - 4) = 2(x - 3)$ **38.** $6x - 3(5x + 2) = 4(1 - x)$

39. $6x - 4(3 - 2x) = 5(x - 4) - 10$ **40.** $-2x - 3(4 - 2x) = 2(x - 3) + 2$

41. {all real numbers}; identity
42. {all real numbers}; identity
43. $\{-1\}$ **44.** $\left\{-\frac{10}{19}\right\}$
45. $\{7\}$ **46.** $\left\{-\frac{1}{2}\right\}$
47. $\{2\}$ **48.** $\{-6\}$
49. {all real numbers}; identity
50. {all real numbers}; identity
51. $\left\{\frac{3}{2}\right\}$ **52.** $\left\{-\frac{2}{5}\right\}$
53. 12
54. Yes, we will get the correct solution. The coefficients will be larger, but in the end the solution will be the same.
55. **(a)** 10^2, or 100 **(b)** 10^3, or 1000
56. B
57. $\left\{-\frac{18}{5}\right\}$ **58.** $\left\{-\frac{55}{3}\right\}$
59. $\left\{-\frac{5}{6}\right\}$ **60.** $\left\{-\frac{48}{7}\right\}$
61. $\{6\}$ **62.** $\{-20\}$
63. $\{4\}$ **64.** $\{-6\}$
65. $\{3\}$ **66.** $\{9\}$
67. $\{3\}$ **68.** $\{11\}$
69. $\{0\}$ **70.** $\{0\}$
71. $\{35\}$ **72.** $\{15\}$
73. $\{-10\}$ **74.** $\{-18\}$

41. $-2(x+3)-x-4=-3(x+4)+2$
42. $4(2x+7)=2x+25+3(2x+1)$
43. $2[x-(2x+4)+3]=2(x+1)$
44. $4[2x-(3-x)+5]=-(2+7x)$
45. $-[2x-(5x+2)]=2+(2x+7)$
46. $-[6x-(4x+8)]=9+(6x+3)$
47. $-3x+6-5(x-1)=-5x-(2x-4)+5$
48. $4(x+2)-8x-5=-3x+9-2(x+6)$
49. $7[2-(3+4x)]-2x=-9+2(1-15x)$
50. $4[6-(1+2x)]+10x=2(10-3x)+8x$
51. $-[3x-(2x+5)]=-4-[3(2x-4)-3x]$
52. $2[-(x-1)+4]=5+[-(6x-7)+9x]$

Concept Check *Answer each question.*

53. To solve the linear equation

$$\frac{3}{4}x-\frac{1}{3}x=\frac{5}{6}x-5,$$

we multiply each side by the least common denominator of all the fractions in the equation. What is this LCD?

54. Suppose that in solving the equation

$$\frac{1}{3}x+\frac{1}{2}x=\frac{1}{6}x,$$

we begin by multiplying each side by 12, rather than the *least* common denominator, 6. Would we get the correct solution? Explain.

55. To solve a linear equation with decimals, we usually begin by multiplying by a power of 10 so that all coefficients are integers. What is the least power of 10 that will accomplish this goal in each equation?

(a) $0.04x+0.06+0.03x=0.03x+1.46$ **(Exercise 71)**

(b) $0.006x-0.02x+0.03=0.008x+0.25$ **(Exercise 73)**

56. The expression $0.06(10-x)(100)$ is equivalent to which one of the following?

A. $0.06-0.06x$ **B.** $60-6x$ **C.** $6-6x$ **D.** $6-0.06x$

Solve each equation, and check the solution. ***See Examples 5 and 6.***

57. $-\frac{5}{9}x=2$
58. $\frac{3}{11}x=-5$
59. $\frac{6}{5}x=-1$
60. $-\frac{7}{8}x=6$
61. $\frac{x}{2}+\frac{x}{3}=5$
62. $\frac{x}{5}-\frac{x}{4}=1$
63. $\frac{3x}{4}+\frac{5x}{2}=13$
64. $\frac{8x}{3}-\frac{x}{2}=-13$
65. $\frac{x-10}{5}+\frac{2}{5}=-\frac{x}{3}$
66. $\frac{5-x}{6}+\frac{5}{6}=\frac{x}{54}$
67. $\frac{3x-1}{4}+\frac{x+3}{6}=3$
68. $\frac{3x+2}{7}-\frac{x+4}{5}=2$
69. $\frac{4x+1}{3}=\frac{x+5}{6}+\frac{x-3}{6}$
70. $\frac{2x+5}{5}=\frac{3x+1}{2}+\frac{-x+8}{16}$
71. $0.04x+0.06+0.03x=0.03x+1.46$
72. $0.05x+0.08+0.06x=0.07x+0.68$
73. $0.006x-0.02x+0.03=0.008x+0.25$
74. $0.05x-0.1x+0.6=0.04x+2.22$

75. {2000} 76. {100}
77. {25} 78. {400}
79. {40} 80. {60}
81. {3} 82. {14}

75. $0.05x + 0.12(x + 5000) = 940$

76. $0.09x + 0.13(x + 300) = 61$

77. $0.02(50) + 0.08x = 0.04(50 + x)$

78. $0.04(90) + 0.12x = 0.06(460 + x)$

79. $0.05x + 0.10(200 - x) = 0.45x$

80. $0.08x + 0.12(260 - x) = 0.48x$

81. $0.006(x + 2) = 0.007x + 0.009$

82. $0.006(50 - x) = 0.272 - 0.004x$

STUDY SKILLS

Completing Your Homework

You are ready to do your homework ***AFTER*** you have read the corresponding textbook section and worked through the examples and Now Try exercises.

Homework Tips

- **Survey the exercise set.** Glance over the problems that your instructor has assigned to get a general idea of the types of exercises you will be working. Skim directions, and note any references to section examples.
- **Work problems neatly.** Use pencil and write legibly, so others can read your work. Skip lines between steps. Clearly separate problems from each other.
- **Show all your work.** It is tempting to take shortcuts. Include ALL steps.
- **Check your work frequently to make sure you are on the right track.** It is hard to unlearn a mistake. For all odd-numbered problems, answers are given in the back of the book.
- **If you have trouble with a problem, refer to the corresponding worked example in the section.** The exercise directions will often reference specific examples to review. Pay attention to every line of the worked example to see how to get from step to step.
- **If you are having trouble with an even-numbered problem, work the corresponding odd-numbered problem.** Check your answer in the back of the book, and apply the same steps to work the even-numbered problem.
- **Mark any problems you don't understand.** Ask your instructor about them.
- **Do some homework problems every day.** This is a good habit, even if your math class does not meet each day.

Think through and answer each question.

1. What is your instructor's policy regarding homework?
2. Think about your current approach to doing homework. What are you doing well? What improvements could you make?
3. Which one or two homework tips will you try this week? Why?

1.2 Formulas and Percent

OBJECTIVES

1. Solve a formula for a specified variable.
2. Solve applied problems using formulas.
3. Solve percent problems.
4. Solve problems involving percent increase or decrease.
5. Solve problems from the health care industry.

VOCABULARY

- □ mathematical model
- □ formula
- □ percent
- □ percent increase
- □ percent decrease

A **mathematical model** is an equation or inequality that describes a real situation. Models for many applied problems, called *formulas,* already exist. A **formula** is an equation in which variables are used to describe a relationship. For example, the formula for finding the area $\mathcal{A}$* of a triangle is

$$\mathcal{A} = \frac{1}{2}bh.$$

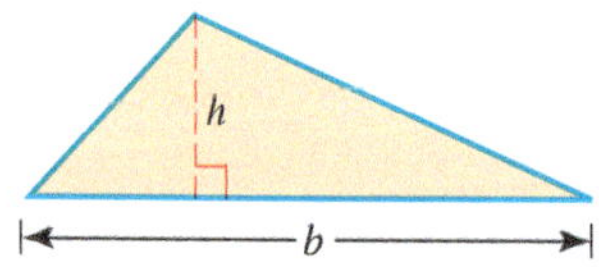

FIGURE 1

Here, b is the length of the base and h is the height. See FIGURE 1. A list of formulas used in algebra is given inside the back cover of this book.

OBJECTIVE 1 Solve a formula for a specified variable.

The formula $I = prt$ says that interest on a loan or investment equals principal (amount borrowed or invested) times rate (percent) times time at interest (in years). To determine how long it will take for an investment at a stated interest rate to earn a predetermined amount of interest, it would help to first solve the formula for t. This process is called **solving for a specified variable** or **solving a literal equation.**

When solving for a specified variable, the key is to treat that variable as if it were the only one. Treat all other variables like numbers (constants).

EXAMPLE 1 Solving for a Specified Variable

Solve the formula $I = prt$ for t.

We solve this formula for t by treating I, p, and r as constants (having fixed values) and treating t as the only variable.

$$prt = I \quad \text{Our goal is to isolate } t.$$

$$(pr)t = I \quad \text{Associative property}$$

$$\frac{(pr)t}{pr} = \frac{I}{pr} \quad \text{Divide by } pr.$$

$$t = \frac{I}{pr}$$

The result is a formula for t, time in years. NOW TRY

NOW TRY EXERCISE 1
Solve the formula $I = prt$ for p.

CLASSROOM EXAMPLE 1
Solve the formula $d = rt$ for r.
Answer: $r = \frac{d}{t}$

Solving for a Specified Variable

Step 1 **Clear any parentheses and fractions.** If the equation contains parentheses, clear them using the distributive property. Clear any fractions by multiplying both sides by the LCD.

Step 2 **Isolate all terms with the specified variable.** Transform so that all terms containing the specified variable are on one side of the equation and all terms without that variable are on the other side.

Step 3 **Isolate the specified variable.** Divide each side by the factor that is the coefficient of the specified variable.

NOW TRY ANSWER
1. $p = \frac{I}{rt}$

*In this book, we use $\mathcal{A}$ to denote area.

NOW TRY EXERCISE 2

Solve the formula for b.

$$P = a + 2b + c$$

CLASSROOM EXAMPLE 2

Solve the formula for L.

$$P = 2L + 2W$$

Answer: $L = \frac{P - 2W}{2}$, or $L = \frac{P}{2} - W$

TEACHING TIP

When solving a formula for a specified variable, encourage students to circle or underline the specified variable. It is helpful to solve a similar equation with numerals side by side with the solution of the formula to show the steps.

EXAMPLE 2 Solving for a Specified Variable

Solve the formula $P = 2L + 2W$ for W.

This formula gives the relationship between perimeter of a rectangle, P, length of the rectangle, L, and width of the rectangle, W. See **FIGURE 2**.

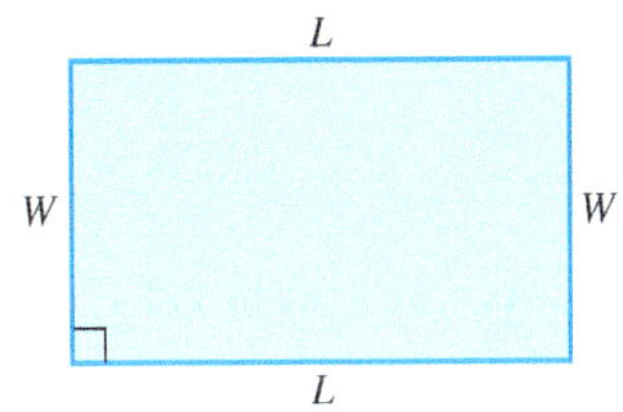

Perimeter, P, the sum of the lengths of the sides of this rectangle, is given by

$$P = 2L + 2W.$$

FIGURE 2

$$P = 2L + 2W$$ Our goal is to isolate W.

Step 1 is not needed here because there are no parentheses or fractions to clear in this formula.

Step 2

$$P - 2L = 2L + 2W - 2L$$ Subtract $2L$.

$$P - 2L = 2W$$ Combine like terms.

Step 3

$$\frac{P - 2L}{2} = \frac{2W}{2}$$ Divide by 2.

$$\frac{P - 2L}{2} = W$$ $\frac{2W}{2} = \frac{2}{2} \cdot W = 1 \cdot W = W$

Although W is isolated, we simplify the right side further.

$$W = \frac{P - 2L}{2}$$ Interchange sides.

$$W = \frac{P}{2} - \frac{2L}{2}$$ $\frac{a - b}{c} = \frac{a}{c} - \frac{b}{c}$

Be careful here. $\frac{P - 2L}{2} \neq P - L$

$$W = \frac{P}{2} - L$$ $\frac{2L}{2} = \frac{2}{2} \cdot L = 1 \cdot L = L$

NOW TRY

CAUTION In Step 3 of **Example 2,** we cannot simplify the fraction by dividing 2 into the term $2L$. The fraction bar serves as a grouping symbol. Thus, the subtraction in the numerator must be done before the division.

$$\frac{P - 2L}{2} \neq P - L$$

CLASSROOM EXAMPLE 3

Solve the equation for x.

$$y = \frac{1}{2}(x + 3)$$

Answer: $x = 2y - 3$

EXAMPLE 3 Solving a Formula Involving Parentheses

The formula

$$s = \frac{1}{2}(a + b + c)$$

is used with another formula, called *Heron's formula,* for finding the area of a triangle. Here, the variable s represents the *semiperimeter* (which is one-half the perimeter), and a, b, and c represent the lengths of the sides of the triangle.

Solve the formula for a.

NOW TRY ANSWER

2. $b = \frac{P - a - c}{2}$

NOW TRY EXERCISE 3
Solve $P = 2(L + W)$ for L.

$$s = \frac{1}{2}(a + b + c)$$

Step 1 $$s = \frac{1}{2}a + \frac{1}{2}b + \frac{1}{2}c$$ Distributive property

$$2s = 2\left(\frac{1}{2}a + \frac{1}{2}b + \frac{1}{2}c\right)$$ Multiply by 2 to clear the fractions.

$$2s = a + b + c$$ Distributive property

Step 2 $$2s - b - c = a + b + c - b - c$$ Subtract b and c.

Step 3 $$2s - b - c = a, \quad \text{or} \quad a = 2s - b - c$$ Combine like terms. Interchange sides.

NOW TRY

In **Examples 1–3,** we solved formulas for specified variables. In **Example 4,** we solve an equation with two variables for one of these variables. This process will be useful when we work with *linear equations in two variables* in **Chapter 2.**

NOW TRY EXERCISE 4
Solve $5x - 6y = 12$ for y.

CLASSROOM EXAMPLE 4
Solve $2x + 7y = 7$ for y.

Answer: $y = -\frac{2}{7}x + 1$

EXAMPLE 4 Solving an Equation for One of the Variables

Solve $3x - 4y = 12$ for y.

$$3x - 4y = 12$$ Our goal is to isolate y.

$$3x - 4y - 3x = 12 - 3x$$ Subtract $3x$.

$$-4y = -3x + 12$$ Combine like terms; commutative property

$$\frac{-4y}{-4} = \frac{-3x + 12}{-4}$$ Divide by -4.

$$y = \frac{-3x}{-4} + \frac{12}{-4}$$ $\frac{a+b}{c} = \frac{a}{c} + \frac{b}{c}$

$$y = \frac{3}{4}x - 3$$ Simplify the expression on the right.

$\frac{-3x}{-4} = \frac{-3}{-4} \cdot \frac{x}{1} = \frac{3}{4}x$

NOW TRY

OBJECTIVE 2 Solve applied problems using formulas.

CLASSROOM EXAMPLE 5
It takes James $\frac{1}{3}$ hr to travel 15 mi. What is his average rate?

Answer: 45 mph

EXAMPLE 5 Finding Average Rate

Phyllis found that on average it took her $\frac{3}{4}$ hr each day to drive a distance of 15 mi to work. What was her average rate (or speed)?

The distance formula $\boldsymbol{d = rt}$ relates d, the distance traveled, r, the rate or speed, and t, the travel time. Solve the formula for r.

$$d = rt$$

$$\frac{d}{t} = \frac{rt}{t}$$ Divide by t.

$$\frac{d}{t} = r, \quad \text{or} \quad r = \frac{d}{t}$$

NOW TRY ANSWERS

3. $L = \frac{P - 2W}{2}$ or $L = \frac{P}{2} - W$,

4. $y = \frac{5}{6}x - 2$

NOW TRY EXERCISE 5

It takes $\frac{1}{2}$ hr for Dorothy to drive 21 mi to work each day. What is her average rate?

We find the desired rate by substituting the given values of d and t into this formula.

$$r = \frac{15}{\frac{3}{4}} \quad \text{Let } d = 15,\ t = \tfrac{3}{4}.$$

$$r = 15 \cdot \frac{4}{3} \quad \text{To divide by } \tfrac{3}{4}, \text{ multiply by its reciprocal, } \tfrac{4}{3}.$$

$$r = 20$$

Her average rate was 20 mph. (That is, at times she may have traveled a little faster or slower than 20 mph, but overall her rate was 20 mph.) NOW TRY

OBJECTIVE 3 Solve percent problems.

An important everyday use of mathematics involves the concept of percent, written with the symbol **%.** The word **percent** means **"per one hundred."** One percent means "one per one hundred" or "one one-hundredth."

$$\mathbf{1\% = 0.01 \quad or \quad 1\% = \frac{1}{100}}$$

Solving a Percent Problem

Let a represent a partial amount of b, the whole amount (or base). Then the following equation can be used to solve a percent problem.

$$\frac{\textbf{partial amount } a}{\textbf{whole amount } b} = \textbf{decimal value (which is converted to a percent)}$$

For example, if a class consists of 50 students and 32 are males, then the percent of males in the class is found as follows.

$$\frac{\text{partial amount } a}{\text{whole amount } b} = \frac{32}{50} \quad \text{Let } a = 32,\ b = 50.$$

$$= \frac{64}{100} \quad \frac{32}{50} \cdot \frac{2}{2} = \frac{64}{100}$$

$$= 0.64 \quad \text{Write as a decimal.}$$

$$= 64\% \quad \text{Convert to a percent.}$$

NOTE When interpreting the "partial amount" in the formula for solving a percent problem, understand that it may represent a quantity that is actually *larger* than the whole amount. In these cases, the percent will be greater than 100%. For example,

25 is 250% of 10.

Here 25 is the partial amount and 10 is the whole amount.

NOW TRY ANSWER

5. 42 mph

NOW TRY EXERCISE 6

Solve each problem.

(a) A 5-L mixture of water and antifreeze contains 2 L of antifreeze. What is the percent of antifreeze in the mixture?

(b) If a savings account earns 2.5% interest on a balance of \$7500 for one year, how much interest is earned?

CLASSROOM EXAMPLE 6

Solve each problem.

(a) A mixture of gasoline and oil contains 20 oz, of which 1 oz is oil. What percent of the mixture is oil?

(b) An automobile salesman earns an 8% commission on every car he sells. How much does he earn on a car that sells for \$12,000?

Answers: (a) 5% (b) \$960

EXAMPLE 6 Solving Percent Problems

(a) A 50-L mixture of acid and water contains 10 L of acid. What is the percent of acid in the mixture?

The given amount of the mixture is 50 L, and the part that is acid is 10 L. Let x represent the percent of acid in the mixture.

$$x = \frac{10}{50} \quad \begin{array}{l}\leftarrow \text{partial amount} \\ \leftarrow \text{whole amount}\end{array}$$

$$x = 0.20 \quad \text{Divide to write as a decimal.}$$

$$x = 20\% \quad 0.20 = \frac{20}{100} = 20\%$$

The mixture is 20% acid.

(b) If \$4780 earns 5% interest in one year, how much interest is earned?

Let x represent the amount of interest earned (that is, the part of the whole amount invested). Because $5\% = 0.05$, the equation is written as follows.

$$\frac{x}{4780} = 0.05 \quad \frac{\text{partial amount } a}{\text{whole amount } b} = \text{decimal value}$$

$$4780 \cdot \frac{x}{4780} = 4780(0.05) \quad \text{Multiply by 4780.}$$

$$x = 239$$

The interest earned is \$239.

NOW TRY

NOW TRY EXERCISE 7

Refer to **FIGURE 3**. How much was spent on vet care? Round the answer to the nearest tenth of a billion dollars.

CLASSROOM EXAMPLE 7

Refer to **FIGURE 3**. How much was spent on pet supplies/medicine? Round the answer to the nearest tenth of a billion dollars.

Answer: \$12.7 billion

EXAMPLE 7 Interpreting Percents from a Graph

In 2012, Americans spent about \$53.3 billion on their pets. Use the graph in **FIGURE 3** to determine how much of this amount was spent on pet food.

Spending on Kitty and Rover

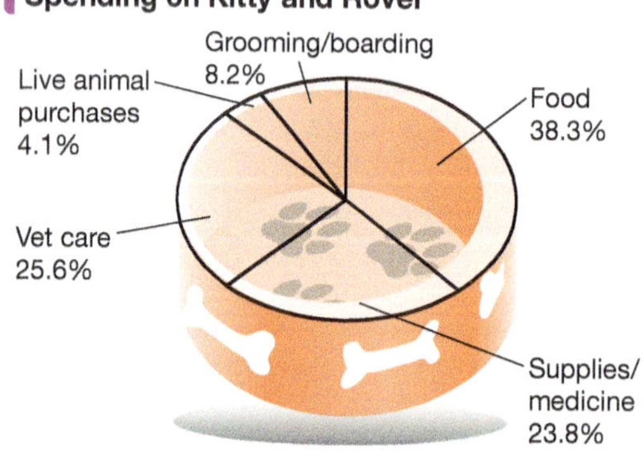

Source: American Pet Products Manufacturers Association Inc.

FIGURE 3

Pythagoras

The graph shows that 38.3% was spent on food. Let $x =$ this amount in billions of dollars.

$$\begin{array}{l}\text{partial amount} \rightarrow \\ \text{whole amount} \rightarrow\end{array} \frac{x}{53.3} = 0.383 \quad 38.3\% = 0.383$$

$$53.3 \cdot \frac{x}{53.3} = 53.3(0.383) \quad \text{Multiply by 53.3.}$$

$$x \approx 20.4 \quad \text{Nearest tenth}$$

Therefore, about \$20.4 billion was spent on pet food.

NOW TRY

NOW TRY ANSWERS

6. (a) 40% **(b)** \$187.50

7. \$13.6 billion

OBJECTIVE 4 Solve problems involving percent increase or decrease.

Percent is often used to express a change in some quantity. Buying an item that has been marked up and getting a raise at a job are applications of **percent increase.** Buying an item on sale and finding population decline are applications of **percent decrease.**

To solve problems of this type, we use the following equation.

$$\textbf{percent change (as a decimal)} = \frac{\textbf{amount of change}}{\textbf{original amount}}$$

Subtract to find this.

NOW TRY EXERCISE 8

Use the percent change equation in each application.

(a) Jane bought a jacket on sale for \$56. The regular price of the jacket was \$80. What was the percent markdown?

(b) When it was time for Horatio to renew the lease on his apartment, the landlord raised his monthly rent from \$650 to \$689. What was the percent increase?

CLASSROOM EXAMPLE 8

Use the percent change equation in each application.

(a) A cost-of-living salary increase caused Keith's monthly salary to go from \$1300 to \$1352. What percent increase was this?

(b) The price of a concert ticket changed from \$54.00 to \$51.30. What percent decrease was this?

Answers: (a) 4% (b) 5%

EXAMPLE 8 Solving Problems about Percent Increase or Decrease

Use the percent change equation in each application.

(a) An electronics store marked up a laptop computer from their cost of \$1200 to a selling price of \$1464. What was the percent markup?

"Markup" is a name for an increase. Let x = the percent increase (as a decimal).

$$\text{percent increase} = \frac{\text{amount of increase}}{\text{original amount}}$$

Subtract to find the *amount* of increase.

$$x = \frac{1464 - 1200}{1200} \quad \text{Substitute the given values.}$$

Use the original cost.

$$x = \frac{264}{1200}$$

$$x = 0.22 \quad \text{Use a calculator.}$$

$$x = 22\% \quad \text{Convert to a percent.}$$

The computer was marked up 22%.

(b) The enrollment at a community college declined from 12,750 during one school year to 11,350 the following year. Find the percent decrease.

Let x = the percent decrease (as a decimal).

$$\text{percent decrease} = \frac{\text{amount of decrease}}{\text{original amount}}$$

Subtract to find the *amount* of decrease.

$$x = \frac{12{,}750 - 11{,}350}{12{,}750} \quad \text{Substitute the given values.}$$

Use the original number.

$$x = \frac{1400}{12{,}750}$$

$$x \approx 0.11 \quad \text{Use a calculator.}$$

$$x = 11\% \quad \text{Convert to a percent.}$$

The college enrollment decreased by about 11%. NOW TRY

CAUTION When calculating a percent increase or decrease, be sure to use the original number (*before* the increase or decrease) as the denominator of the fraction. A common error is to use the final number (*after* the increase or decrease).

NOW TRY ANSWERS

8. **(a)** 30% **(b)** 6%

OBJECTIVE 5 Solve problems from the health care industry.

NOW TRY EXERCISE 9
Use the formula in **Example 9** to find the body surface area, to the nearest hundredth, of a child who weighs 32 lb. Use $k = 14.515$.

CLASSROOM EXAMPLE 9
Use the formula in Example 9 to find the body surface area, to the nearest hundredth, of a child who weighs 20 lb. Use $k = 9.072$.

Answer: 0.44 m^2

EXAMPLE 9 Determining a Child's Body Surface Area

If a child weighs k kilograms, then the child's body surface area S in square meters (m^2) is determined by the formula

$$S = \frac{4k + 7}{k + 90}.$$

What is the body surface area of a child who weighs 40 lb? (*Source:* Hegstad, L., and W. Hayek, *Essential Drug Dosage Calculations,* Fourth Edition, Prentice Hall.)

Tables and web sites for conversion between the metric and English measurement systems are readily available. Using www.metric-conversions.org, we find that 40 lb is approximately 18.144 kg. Use this value for k in the above formula.

$$S = \frac{4(18.144) + 7}{18.144 + 90} \quad \text{Let } k = 18.144 \text{ in the formula.}$$

$$S = 0.74 \quad \text{Use a calculator. Round to the nearest hundredth.}$$

The child's body surface area is 0.74 m^2. NOW TRY

NOW TRY EXERCISE 10
Refer to **Now Try Exercise 9.** Determine the appropriate dose, to the nearest unit, for this child if the usual adult dose is 100 mg.

CLASSROOM EXAMPLE 10
Refer to Classroom Example 10. Determine the appropriate dose, to the nearest unit, for this child if the usual adult dose is 500 mg.

Answer: 129 mg

EXAMPLE 10 Determining a Child's Dose of a Medication

If D represents the usual adult dose of a medication in milligrams, the corresponding child's dose C in milligrams is calculated using the formula

$$C = \frac{\text{body surface area in square meters}}{1.7} \times D.$$

Determine the appropriate dose for a child weighing 40 lb if the usual adult dose is 50 mg. (*Source:* Hegstad, L., and W. Hayek, *Essential Drug Dosage Calculations,* Fourth Edition, Prentice Hall.)

From **Example 9,** the body surface area of a child weighing 40 lb is 0.74 m^2. Use this value in the above formula to find the child's dose.

$$C = \frac{0.74}{1.7} \times 50 \quad \text{Body surface area} = 0.74, \ D = 50$$

$$C = 22 \quad \text{Use a calculator. Round to the nearest unit.}$$

The child's dose is 22 mg. NOW TRY

NOW TRY ANSWERS
9. 0.62 m^2
10. 36 mg

1.2 Exercises

FOR EXTRA HELP MyMathLab®

Complete solution available in MyMathLab

Concept Check *Fill in each blank with the correct response.*

1. A(n) ________ is an equation in which variables are used to describe a relationship.

2. To solve a formula for a specified variable, treat that ________ as if it were the only one and treat all other variables like ________ (numbers).

Concept Check *Work Exercises 3 and 4 to review converting between decimals and percents.*

3. Write each decimal as a percent.

(a) 0.35 **(b)** 0.18 **(c)** 0.02 **(d)** 0.075 **(e)** 1.5

1. formula
2. variable; constants
3. (a) 35% **(b)** 18% **(c)** 2% **(d)** 7.5% **(e)** 150%

4. **(a)** 0.6 **(b)** 0.37 **(c)** 0.08 **(d)** 0.035 **(e)** 2.1

There may be other acceptable forms of the answers in Exercises 5–36.

5. $r = \frac{I}{pt}$ **6.** $t = \frac{d}{r}$

7. **(a)** $b = \frac{\mathcal{A}}{h}$ **(b)** $h = \frac{\mathcal{A}}{b}$

8. **(a)** $W = \frac{\mathcal{A}}{L}$ **(b)** $L = \frac{\mathcal{A}}{W}$

9. $L = \frac{P - 2W}{2}$, or $L = \frac{P}{2} - W$

10. **(a)** $b = P - a - c$ **(b)** $c = P - a - b$

11. **(a)** $W = \frac{V}{LH}$ **(b)** $H = \frac{V}{LW}$

12. **(a)** $h = \frac{2\mathcal{A}}{b}$ **(b)** $b = \frac{2\mathcal{A}}{h}$

13. $r = \frac{C}{2\pi}$ **14.** $h = \frac{V}{\pi r^2}$

15. **(a)** $b = \frac{2\mathcal{A} - hB}{h}$, or $b = \frac{2\mathcal{A}}{h} - B$ **(b)** $B = \frac{2\mathcal{A} - hb}{h}$, or $B = \frac{2\mathcal{A}}{h} - b$

16. $h = \frac{3V}{\pi r^2}$

17. $C = \frac{5}{9}(F - 32)$

18. $F = \frac{9}{5}C + 32$

19. **(a)** $x = \frac{C - B}{A}$ **(b)** $A = \frac{C - B}{x}$

20. **(a)** $x = \frac{y - b}{m}$ **(b)** $m = \frac{y - b}{x}$

4. Write each percent as a decimal.

(a) 60% **(b)** 37% **(c)** 8% **(d)** 3.5% **(e)** 210%

Solve each formula for the specified variable. ***See Examples 1–3.***

5. $I = prt$ for r (simple interest)

6. $d = rt$ for t (distance)

7. $\mathcal{A} = bh$ (area of a parallelogram)

(a) for b **(b)** for h

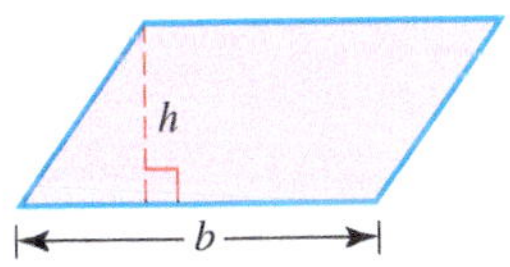

8. $\mathcal{A} = LW$ (area of a rectangle)

(a) for W **(b)** for L

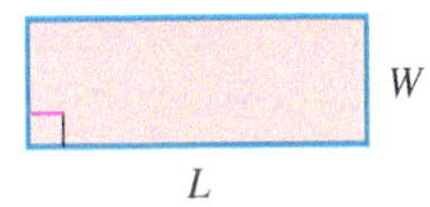

9. $P = 2L + 2W$ for L (perimeter of a rectangle)

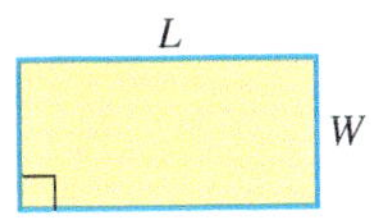

10. $P = a + b + c$ (perimeter of a triangle)

(a) for b **(b)** for c

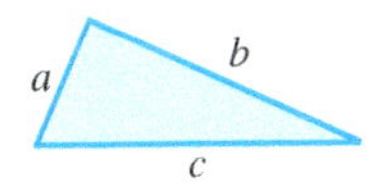

11. $V = LWH$ (volume of a rectangular solid)

(a) for W **(b)** for H

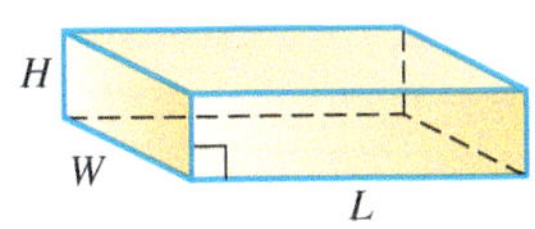

12. $\mathcal{A} = \frac{1}{2}bh$ (area of a triangle)

(a) for h **(b)** for b

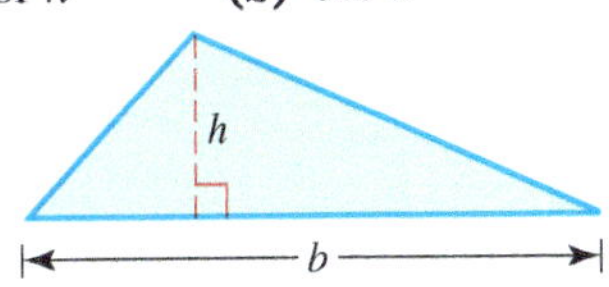

13. $C = 2\pi r$ for r (circumference of a circle)

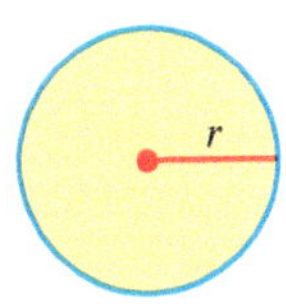

14. $V = \pi r^2 h$ for h (volume of a right circular cylinder)

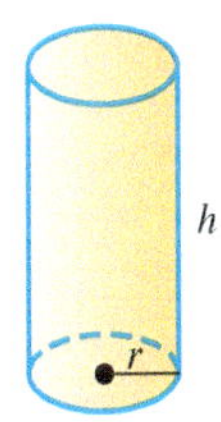

15. $\mathcal{A} = \frac{1}{2}h(b + B)$ (area of a trapezoid)

(a) for b **(b)** for B

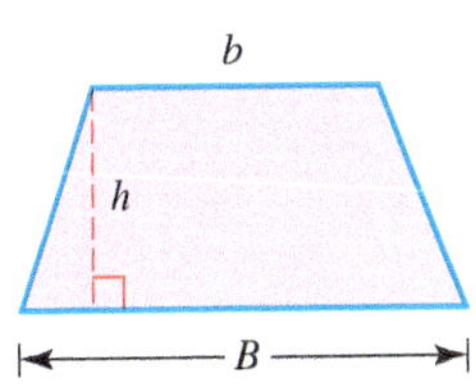

16. $V = \frac{1}{3}\pi r^2 h$ for h (volume of a cone)

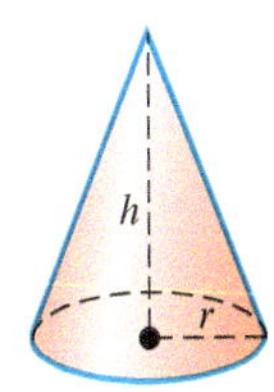

17. $F = \frac{9}{5}C + 32$ for C (Celsius to Fahrenheit)

18. $C = \frac{5}{9}(F - 32)$ for F (Fahrenheit to Celsius)

19. $Ax + B = C$ (linear equation in x)

(a) for x **(b)** for A

20. $y = mx + b$ (slope-intercept form of a linear equation)

(a) for x **(b)** for m

21. $t = \frac{A - P}{Pr}$ **22.** $r = \frac{M - C}{C}$
23. D **24.** D
25. $y = -4x + 1$
26. $y = -3x + 9$
27. $y = \frac{1}{2}x + 3$
28. $y = \frac{1}{5}x + 4$
29. $y = -\frac{4}{9}x + \frac{11}{9}$
30. $y = -\frac{7}{8}x + \frac{11}{8}$
31. $y = \frac{3}{2}x + \frac{5}{2}$
32. $y = \frac{5}{3}x + 4$
33. $y = \frac{6}{5}x - \frac{7}{5}$
34. $y = \frac{8}{3}x - \frac{4}{3}$
35. $y = \frac{3}{2}x - 3$
36. $y = \frac{5}{3}x - 5$
37. 3.140 hr **38.** 2.734 hr
39. 52 mph **40.** 21 hr
41. 113°F **42.** −50°C
43. 230 m

21. $A = P(1 + rt)$ for t (future value for simple interest)

22. $M = C(1 + r)$ for r (markup)

23. *Concept Check* When a formula is solved for a particular variable, several different equivalent forms may be possible. If we solve $\mathcal{A} = \frac{1}{2}bh$ for h, one possible correct answer is

$$h = \frac{2\mathcal{A}}{b}.$$

Which one of the following is *not* equivalent to this?

A. $h = 2\left(\frac{\mathcal{A}}{b}\right)$ **B.** $h = 2\mathcal{A}\left(\frac{1}{b}\right)$ **C.** $h = \frac{\mathcal{A}}{\frac{1}{2}b}$ **D.** $h = \frac{\frac{1}{2}\mathcal{A}}{b}$

24. *Concept Check* The answer to **Exercise 17** is given as $C = \frac{5}{9}(F - 32)$. Which one of the following is *not* equivalent to this?

A. $C = \frac{5}{9}F - \frac{160}{9}$ **B.** $C = \frac{5F}{9} - \frac{160}{9}$ **C.** $C = \frac{5F - 160}{9}$ **D.** $C = \frac{5}{9}F - 32$

Solve each equation for y. **See Example 4.**

25. $4x + y = 1$ **26.** $3x + y = 9$ **27.** $x - 2y = -6$

28. $x - 5y = -20$ **29.** $4x + 9y = 11$ **30.** $7x + 8y = 11$

31. $-3x + 2y = 5$ **32.** $-5x + 3y = 12$ **33.** $6x - 5y = 7$

34. $8x - 3y = 4$ **35.** $\frac{1}{2}x - \frac{1}{3}y = 1$ **36.** $\frac{2}{3}x - \frac{2}{5}y = 2$

Solve each problem. **See Example 5.**

37. Jimmie Johnson won the Daytona 500 (mile) race with a rate of 159.250 mph in 2013. Find his time to the nearest thousandth. (*Source: World Almanac and Book of Facts.*)

38. In 2007, rain shortened the Indianapolis 500 race to 415 mi. It was won by Dario Franchitti, who averaged 151.774 mph. What was his time to the nearest thousandth? (*Source:* www.indy500.com)

39. Nora traveled from Kansas City to Louisville, a distance of 520 mi, in 10 hr. Find her rate in miles per hour.

40. The distance from Melbourne to London is 10,500 mi. If a jet averages 500 mph between the two cities, what is its travel time in hours?

41. As of 2013, the highest temperature ever recorded in Tennessee was 45°C. Find the corresponding Fahrenheit temperature. (*Source:* National Climatic Data Center.)

42. As of 2013, the lowest temperature ever recorded in South Dakota was −58°F. Find the corresponding Celsius temperature. (*Source:* National Climatic Data Center.)

43. The base of the Great Pyramid of Cheops is a square whose perimeter is 920 m. What is the length of each side of this square? (*Source: Atlas of Ancient Archaeology.*)

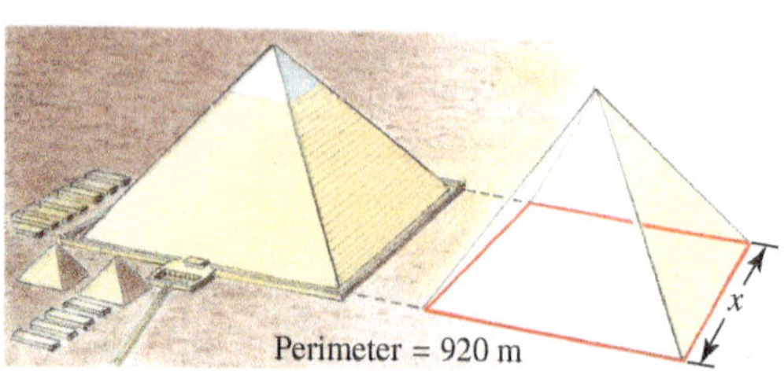

44. 565,722 ft^3
45. (a) 240 in. **(b)** 480 in.
46. (a) 5 in. **(b)** 5π in.
47. 2 in. **48.** 14 in.
49. 75% water; 25% alcohol
50. pure acid: 14 L; pure water: 26 L
51. 3% **52.** 0.75%
53. (a) .558 **(b)** .481 **(c)** .438 **(d)** .315

44. Marina City in Chicago is a complex of two residential towers that resemble corncobs. Each tower has a concrete cylindrical core with a 35-ft diameter and is 588 ft tall. Find the volume of the core of one of the towers to the nearest whole number. (*Hint:* Use the π key on your calculator.) (*Source:* www.architechgallery.com; www.aviewoncities.com)

45. The circumference of a circle is 480π in.

(a) What is its radius?

(b) What is its diameter?

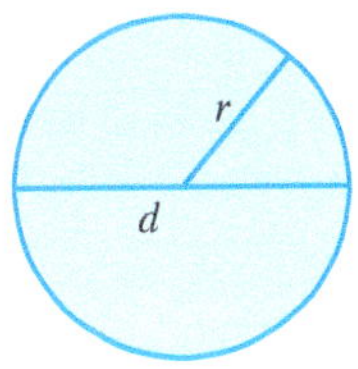

46. The radius of a circle is 2.5 in.

(a) What is its diameter?

(b) What is its circumference?

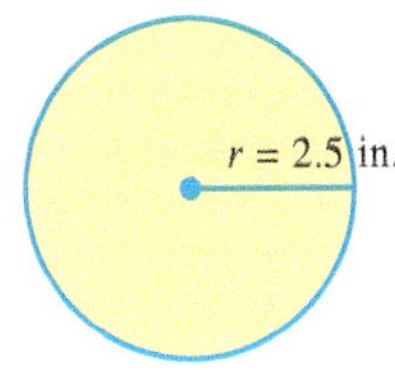

47. A sheet of standard-size copy paper measures 8.5 in. by 11 in. If a ream (500 sheets) of this paper has a volume of 187 in.^{3}, how thick is the ream?

48. Copy paper **(Exercise 47)** also comes in legal size, which has the same width, but is longer than standard size. If a ream of legal-size paper has the same thickness as standard-size paper and a volume of 238 in.^{3}, what is the length of a sheet of legal paper?

Solve each problem. ***See Example 6.***

49. A mixture of alcohol and water contains a total of 36 oz of liquid. There are 9 oz of pure alcohol in the mixture. What percent of the mixture is water? What percent is alcohol?

50. A mixture of acid and water is 35% acid. If the mixture contains a total of 40 L, how many liters of pure acid are in the mixture? How many liters of pure water are in the mixture?

51. A real-estate agent earned \$6300 commission on a property sale of \$210,000. What is her rate of commission?

52. A certificate of deposit for 1 yr pays \$25.50 simple interest on a principal of \$3400. What is the interest rate being paid on this deposit?

In baseball, winning percentage (Pct.) is commonly expressed as a decimal rounded to the nearest thousandth. To find the winning percentage of a team, divide the number of wins (W) *by the total number of games played* (W + L).

53. The final 2013 standings of the American League West division in Major League Baseball are shown in the table. Find the winning percentage of each team.

(a) Texas **(b)** L.A. Angels

(c) Seattle **(d)** Houston

	W	L	Pct.
Oakland	96	66	.593
Texas*	91	72	
L.A. Angels	78	84	
Seattle	71	91	
Houston	51	111	

*Texas played one more game than the others. It was an extra playoff game, which counted in the regular season statistics.

Source: World Almanac and Book of Facts.

54. (a) .580 (b) .556 (c) .457 (d) .407

55. 34% 56. 82%
57. 103.4 million
58. 54.4 million

59. $72,324 60. $43,394
61. $14,465 62. 16%; yes

63. 12% 64. 20%
65. 23.8% 66. 11.9%
67. 49.9%

54. The final 2013 standings of the National League Central division in Major League Baseball are shown in the table. Find the winning percentage of each team.

(a) Pittsburgh **(b)** Cincinnati

(c) Milwaukee **(d)** Chicago Cubs

	W	L	Pct.
St. Louis	97	65	.599
Pittsburgh	94	68	
Cincinnati	90	72	
Milwaukee	74	88	
Chicago Cubs	66	96	

Source: *World Almanac and Book of Facts.*

In 2013, 114.2 million U.S. households owned at least one TV set. (Source: Nielsen Media Research.) Use this to work Exercises 55–58. Round answers to the nearest percent in Exercises 55 and 56, and to the nearest tenth million in Exercises 57 and 58. ***See Example 6.***

55. About 38.4 million U.S. households owned 4 or more TV sets in 2013. What percent of those owning at least one TV set was this?

56. About 94.2 million households that owned at least one TV set in 2013 had a DVD player. What percent of those owning at least one TV set had a DVD player?

57. Of the households owning at least one TV set in 2013, 90.5% received basic cable. How many households received basic cable?

58. Of the households owning at least one TV set in 2013, 47.6% received premium cable. How many households received premium cable?

An average middle-income family will spend $241,080 to raise a child born in 2013 from birth through age 17. The graph shows the percents spent for various categories. Use the graph to answer Exercises 59–62. ***See Example 7.***

The Cost of Parenthood

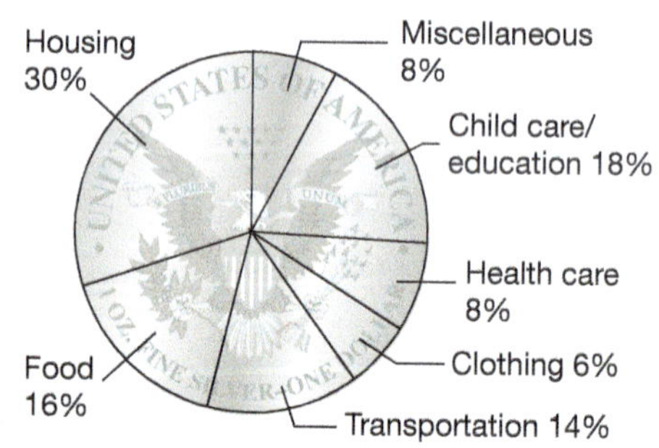

Source: U.S. Department of Agriculture.

59. To the nearest dollar, how much will be spent to provide housing for the child?

60. To the nearest dollar, how much will be spent for child care and education?

61. Use the answer from **Exercise 60** to find how much will be spent on clothing.

62. About $38,500 will be spent for food. To the nearest percent, what percent of the cost of raising a child from birth through age 17 is this? Does the answer agree with the graph?

Solve each problem. ***See Example 8.***

63. After 1 yr on the job, Grady got a raise from $10.50 per hour to $11.76 per hour. What was the percent increase in his hourly wage?

64. Clayton bought a ticket to a rock concert at a discount. The regular price of the ticket was $70.00, but he only paid $56.00. What was the percent discount?

65. Between 2000 and 2012, the population of New Orleans, Louisiana, declined from 484,674 to 369,250. What was the percent decrease to the nearest tenth? (*Source:* U.S. Census.)

66. Between 2000 and 2012, the estimated population of Naperville, Illinois, grew from 128,358 to 143,684. What was the percent increase to the nearest tenth? (*Source:* U.S. Census.)

67. The movie *Lincoln* in Blu-Ray was on sale for $19.52. The list price (full price) of this two-disc combo pack was $39.00. To the nearest tenth, what was the percent discount? (*Source:* www.amazon.com)

68. 39.6%

69. **(a)** 0.79 m^2 **(b)** 0.96 m^2
70. **(a)** 0.59 m^2 **(b)** 0.53 m^2
71. **(a)** 116 mg **(b)** 141 mg
72. **(a)** 174 mg **(b)** 156 mg

68. *The Dark Night Trilogy* in Blu-Ray was on sale for \$31.99. The list price (full price) of this five-disc set was \$52.99. To the nearest tenth, what was the percent discount? (*Source:* www.amazon.com)

In Exercises 69–72, apply the formulas for determining a child's body surface area and a child's dose of a medication. ***See Examples 9 and 10.***

$$S = \frac{4k + 7}{k + 90}$$

Child's body surface area S, in square meters; k = weight in kilograms

$$C = \frac{\text{body surface area in square meters}}{1.7} \times D$$

Child's dose of medication C, in milligrams; D = usual adult dose, in milligrams

69. Find the body surface area, to the nearest hundredth, of a child with each weight.
(a) 20 kg **(b)** 26 kg

70. Find the body surface area, to the nearest hundredth, of a child with each weight.
(a) 30 lb (Use $k = 13.608$.) **(b)** 26 lb (Use $k = 11.793$.)

71. If the usual adult dose of a medication is 250 mg, what is the child's dose, to the nearest unit, for a child with each weight? (*Hint:* Use the results of **Exercise 69.**)
(a) 20 kg **(b)** 26 kg

72. If the usual adult dose of a medication is 500 mg, what is the child's dose, to the nearest unit, for a child with each weight? (*Hint:* Use the results of **Exercise 70.**)
(a) 30 lb **(b)** 26 lb

STUDY SKILLS

Taking Lecture Notes

Study the set of sample math notes given here.

- **Use a new page** for each day's lecture.
- **Include the date and title** of the day's lecture topic.
- **Skip lines and write neatly** to make reading easier.
- **Include cautions and warnings** to emphasize common errors to avoid.
- **Mark important concepts with stars, underlining, circling, boxes, etc.**
- **Use two columns,** which allows an example and its explanation to be close together.
- **Use brackets and arrows** to clearly show steps, related material, etc.

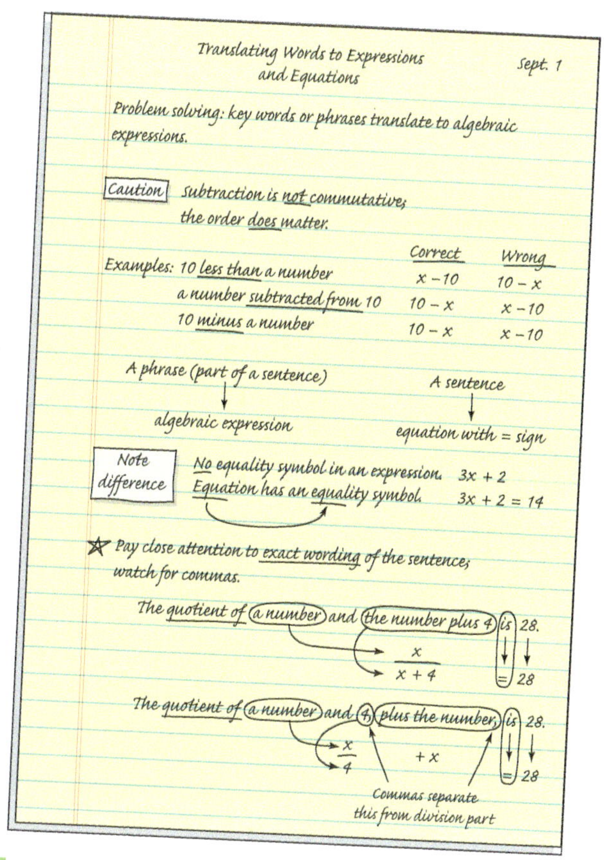

With a partner or in a small group, compare lecture notes. Answer each question.

1. What are you doing to show main points in your notes (such as boxing, using stars or capital letters, etc.)?
2. In what ways do you set off explanations from worked problems and subpoints (such as indenting, using arrows, circling, etc.)?
3. What new ideas did you learn by examining your classmates' notes?
4. What new techniques will you try in your note taking?

1.3 Applications of Linear Equations

OBJECTIVES

1. Translate from words to mathematical expressions.
2. Write equations from given information.
3. Distinguish between simplifying expressions and solving equations.
4. Use the six steps in solving an applied problem.
5. Solve percent problems.
6. Solve investment problems.
7. Solve mixture problems.

OBJECTIVE 1 Translate from words to mathematical expressions.

Producing a mathematical model of a real situation often involves translating verbal statements into mathematical statements.

PROBLEM-SOLVING HINT Usually there are key words and phrases in a verbal problem that translate into mathematical expressions involving addition, subtraction, multiplication, and division. Translations of some commonly used expressions follow.

Use this table for reference when solving applications in this chapter.

TEACHING TIP
As an oral exercise, have the class compile a list of key words and expressions categorized as to arithmetic operations.

Translating from Words to Mathematical Expressions

Verbal Expression	*Mathematical Expression (where x and y are numbers)*
Addition	
The **sum** of a number and 7	$x + 7$
6 **more than** a number	$x + 6$
3 **plus** a number	$3 + x$
24 **added to** a number	$x + 24$
A number **increased by** 5	$x + 5$
The **sum** of two numbers	$x + y$
Subtraction	
2 **less than** a number	$x - 2$
2 **less** a number	$2 - x$
12 **minus** a number	$12 - x$
A number **decreased by** 12	$x - 12$
A number **subtracted from** 10	$10 - x$
10 **subtracted from** a number	$x - 10$
The **difference of** two numbers	$x - y$
Multiplication	
16 **times** a number	$16x$
A number **multiplied by** 6	$6x$
$\frac{2}{3}$ **of** a number (used with fractions and percent)	$\frac{2}{3}x$
$\frac{3}{4}$ **as much as** a number	$\frac{3}{4}x$
Twice (2 times) a number	$2x$
Triple (3 times) a number	$3x$
The **product** of two numbers	xy
Division	
The **quotient** of 8 and a number	$\frac{8}{x}$ $(x \neq 0)$
A number **divided by** 13	$\frac{x}{13}$
The **ratio** of two numbers	$\frac{x}{y}$ $(y \neq 0)$
The **quotient** of two numbers	$\frac{x}{y}$ $(y \neq 0)$

CLASSROOM EXAMPLE 1
Translate each verbal sentence into an equation, using x as the variable.

(a) The sum of a number and 6 is 28.

(b) The product of a number and 7 is twice the number plus 12.

(c) The quotient of a number and 6, added to twice the number, is 7.

Answers: (a) $x + 6 = 28$
(b) $7x = 2x + 12$ (c) $2x + \frac{x}{6} = 7$

CAUTION ***Because subtraction and division are not commutative operations, it is important to correctly translate expressions involving them.*** For example,

"2 less than a number" is translated as $x - 2$, **not** $2 - x$.

"A number subtracted from 10" is expressed as $10 - x$, **not** $x - 10$.

For division, the number *by which* we are dividing is the denominator, and the number *into which* we are dividing is the numerator.

"A number divided by 13" and "13 divided into x" both translate as $\frac{x}{13}$.

"The quotient of x and y" is translated as $\frac{x}{y}$.

CLASSROOM EXAMPLE 2
Decide whether each is an *expression* or an *equation*. Simplify any expressions, and solve any equations.
(a) $5x - 3(x + 2) = 7$
(b) $5x - 3(x + 2)$
Answers: (a) equation; $\left\{\frac{13}{2}\right\}$
(b) expression; $2x - 6$

OBJECTIVE 2 Write equations from given information.

Any words that indicate equality or "sameness," such as *is*, translate as $=$.

NOW TRY EXERCISE 1
Translate each verbal sentence into an equation, using x as the variable.
(a) The quotient of a number and 10 is twice the number.
(b) The product of a number and 5, decreased by 7, is 0.

EXAMPLE 1 Translating Words into Equations

Translate each verbal sentence into an equation, using x as the variable.

Verbal Sentence	*Equation*
Twice a number, decreased by 3, is 42.	$2x - 3 = 42$
The product of a number and 12, decreased by 7, is 105.	$12x - 7 = 105$
The quotient of a number and the number plus 4 is 28.	$\frac{x}{x + 4} = 28$
The quotient of a number and 4, plus the number, is 10.	$\frac{x}{4} + x = 10$

NOW TRY

OBJECTIVE 3 Distinguish between simplifying expressions and solving equations.

An expression translates as a phrase. An equation includes the equality symbol $(=)$, with expressions on both sides, and translates as a sentence.

NOW TRY EXERCISE 2
Decide whether each is an *expression* or an *equation*. Simplify any expressions, and solve any equations.
(a) $3(x - 5) + 2x - 1$
(b) $3(x - 5) + 2x = 1$

EXAMPLE 2 Simplifying Expressions vs. Solving Equations

Decide whether each is an *expression* or an *equation*. Simplify any expressions, and solve any equations.

(a) $2(3 + x) - 4x + 7$

Clear parentheses and combine like terms to **simplify**.

$2(3 + x) - 4x + 7$ — There is no equality symbol. This is an **expression**.

$= 6 + 2x - 4x + 7$ — Distributive property

$= -2x + 13$ — Simplified expression

(b) $2(3 + x) - 4x + 7 = -1$

Find the value of x to **solve**.

$2(3 + x) - 4x + 7 = -1$ — There is an equality symbol. This is an **equation**.

$6 + 2x - 4x + 7 = -1$ — Distributive property

$-2x + 13 = -1$ — Combine like terms.

$-2x = -14$ — Subtract 13.

$x = 7$ — Divide by -2.

The solution set is $\{7\}$. NOW TRY

NOW TRY ANSWERS
1. (a) $\frac{x}{10} = 2x$ **(b)** $5x - 7 = 0$
2. (a) expression; $5x - 16$
(b) equation; $\left\{\frac{16}{5}\right\}$

OBJECTIVE 4 Use the six steps in solving an applied problem.

The following six steps are helpful when solving applied problems.

TEACHING TIP
Mention that in Step 1, the last sentence of the problem usually tells what you are to find.
To illustrate the six problem-solving steps, work through Exercise 27 or 28 in the exercise set with students.

Solving an Applied Problem

Step 1 **Read** the problem carefully. *What information is given? What is to be found?*

Step 2 **Assign a variable** to represent the unknown value. Make a sketch, diagram, or table, as needed. If necessary, express any other unknown values in terms of the variable.

Step 3 **Write an equation** using the variable expression(s).

Step 4 **Solve** the equation.

Step 5 **State the answer.** Label it appropriately. *Does it seem reasonable?*

Step 6 **Check** the answer in the words of the *original* problem.

NOW TRY EXERCISE 3
The length of a rectangle is 2 ft more than twice the width. The perimeter is 34 ft. Find the length and the width of the rectangle.

CLASSROOM EXAMPLE 3
The length of a rectangle is 5 cm more than its width. The perimeter is five times the width. What are the dimensions of the rectangle?

Answer: width: 10 cm; length: 15 cm

EXAMPLE 3 Solving a Perimeter Problem

The length of a rectangle is 1 cm more than twice the width. The perimeter of the rectangle is 110 cm. Find the length and the width of the rectangle.

Step 1 **Read** the problem. *What must be found?* We must find the length and width of the rectangle. *What is given?* The length is 1 cm more than twice the width and the perimeter is 110 cm.

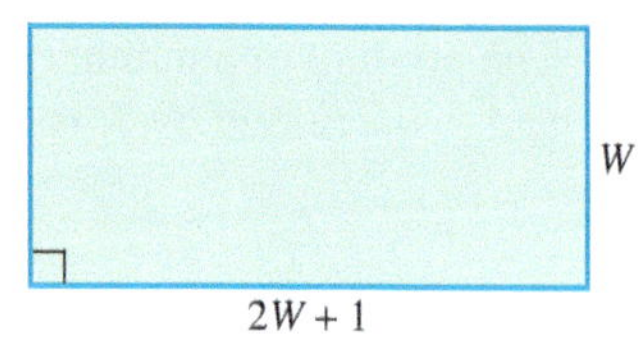

FIGURE 4

Step 2 **Assign a variable.**

Let $W =$ the width.

Then $2W + 1 =$ the length.

Make a sketch, as in **FIGURE 4**.

Step 3 **Write an equation.** Use the formula for the perimeter of a rectangle.

$$P = 2L + 2W \qquad \text{Perimeter of a rectangle}$$

$$110 = 2(2W + 1) + 2W \qquad P = 110 \text{ and } L = 2W + 1$$

Step 4 **Solve.**

$$110 = 4W + 2 + 2W \qquad \text{Distributive property}$$

$$110 = 6W + 2 \qquad \text{Combine like terms.}$$

$$110 - 2 = 6W + 2 - 2 \qquad \text{Subtract 2.}$$

$$108 = 6W \qquad \text{Combine like terms.}$$

$$\frac{108}{6} = \frac{6W}{6} \qquad \text{Divide by 6.}$$

$$18 = W$$

We also need to find the length.

Step 5 **State the answer.** The width of the rectangle is 18 cm and the length is

$$2(18) + 1 = 37 \text{ cm.}$$

Step 6 **Check.** The length, 37 cm, is 1 cm more than 2(18) cm (twice the width). The perimeter is

$$2(37) + 2(18) = 110 \text{ cm}, \quad \text{as required.}$$

NOW TRY

NOW TRY ANSWER
3. width: 5 ft; length: 12 ft

NOW TRY EXERCISE 4

During the 2013 regular NFL season, Peyton Manning of the Denver Broncos threw 16 more touchdown passes than Drew Brees of the New Orleans Saints. Together, these two quarterbacks completed a total of 94 touchdown passes. How many passes did each player complete? (*Source:* www.nfl.com)

CLASSROOM EXAMPLE 4

For the 2013 regular season, the Major League Baseball leaders for home runs were Chris Davis of the Baltimore Orioles and Miguel Cabrera of the Detroit Tigers. These two players had a total of 97 home runs, and Cabrera had 9 fewer home runs than Davis. How many home runs did each player have? (*Source:* www.mlb.com)

Answer: Davis: 53; Cabrera: 44

NOW TRY ANSWER

4. Manning: 55; Brees: 39

EXAMPLE 4 Finding Unknown Numerical Quantities

During the 2013 regular Major League Baseball season, Yu Darvish of the Texas Rangers and Max Scherzer of the Detroit Tigers were the top major league pitchers for strikeouts. The two pitchers had a total of 517 strikeouts. Darvish had 37 more strikeouts than Scherzer. How many strikeouts did each pitcher have? (*Source:* www.mlb.com)

Step 1 **Read** the problem. We are asked to find the number of strikeouts each pitcher had.

Step 2 **Assign a variable** to represent the number of strikeouts for one of the men.

Let $s =$ the number of strikeouts for Max Scherzer.

We must also find the number of strikeouts for Yu Darvish. Because he had 37 more strikeouts than Scherzer,

$s + 37 =$ the number of strikeouts for Darvish.

Step 3 **Write an equation.** The sum of the numbers of strikeouts is 517.

Scherzer's strikeouts	+	Darvish's strikeouts	=	total strikeouts.
↓		↓		↓
s	+	$(s + 37)$	=	517

Step 4 **Solve** the equation.

$$s + (s + 37) = 517$$

$$2s + 37 = 517 \quad \text{Combine like terms.}$$

$$2s + 37 - 37 = 517 - 37 \quad \text{Subtract 37.}$$

$$2s = 480 \quad \text{Combine like terms.}$$

$$\frac{2s}{2} = \frac{480}{2} \quad \text{Divide by 2.}$$

Don't stop here.

$$s = 240$$

Step 5 **State the answer.** The variable s represents the number of strikeouts for Scherzer, so he had 240. Then Darvish had $s + 37$, which is

$240 + 37 = 277$ strikeouts.

Step 6 **Check.** 277 is 37 more than 240, and $240 + 277 = 517$. The conditions of the problem are satisfied, and our answer checks. NOW TRY

CAUTION ***Be sure to answer all the questions asked in the problem.*** In **Example 4,** we were asked for the number of strikeouts for *each* player, so there was extra work in Step 5 in order to find Darvish's number.

PROBLEM-SOLVING HINT In **Example 4,** we chose to let the variable represent the number of strikeouts for Scherzer. Students often ask, "Can I let the variable represent the *other* unknown?" The answer is yes. The equation will be different, but in the end the answers will be the same. Let s represent the number of strikeouts for Darvish, and confirm this.

$s + (s - 37) = 517$ Alternative equation for **Example 4**

OBJECTIVE 5 Solve percent problems.

Recall from **Section 1.2** that percent means "per one hundred," so 5% means 0.05, 14% means 0.14, and so on.

NOW TRY EXERCISE 5

In the fall of 2015, there were 96 Introductory Statistics students at a certain community college, an increase of 700% over the number of Introductory Statistics students in the fall of 1992. How many Introductory Statistics students were there in the fall of 1992?

CLASSROOM EXAMPLE 5

In 2012, Medicare spending in the United States was about $572.5 billion. This was an increase of 4.8% over the total for 2011. What was the approximate Medicare spending in 2011 (to the nearest tenth billion)? (*Source:* U.S. Centers for Medicare & Medicaid Services.)

Answer: $546.3 billion

EXAMPLE 5 Solving a Percent Problem

Total annual health expenditures in the United States were about $2800 billion (or $2.8 trillion) in 2012. This is an increase of 287% over the total for 1990. What were the approximate total health expenditures in billions of dollars in the United States in 1990? (*Source:* U.S. Centers for Medicare & Medicaid Services.)

Step 1 **Read** the problem. We are given that the total health expenditures increased by 287% from 1990 to 2012, and $2800 billion was spent in 2012. We must find the expenditures in 1990.

Step 2 **Assign a variable.** Let x = the total health expenditures for 1990.

$$287\% = 287(0.01) = 2.87,$$

so $2.87x$ represents the expenditures compared to those of 1990.

Step 3 **Write an equation** from the given information.

The expenditures in 1990	+	the increase	=	2800.
↓		↓		↓
x	+	$2.87x$	=	2800

Note the x in $2.87x$.

Step 4 **Solve** the equation. (We do so without clearing the decimal.)

$$1x + 2.87x = 2800 \quad \text{Identity property}$$

$$3.87x = 2800 \quad \text{Combine like terms.}$$

$$x \approx 724 \quad \text{Divide by 3.87.}$$

Step 5 **State the answer.** Total health expenditures in the United States for 1990 were about $724 billion.

Step 6 **Check** that the increase, which is $2800 − $724 = $2076 billion, is about 287% of $724 billion.

NOW TRY

CAUTION Avoid two common errors that occur in solving problems like the one in **Example 5.**

1. Do not try to find 287% of 2800 and subtract that amount from 2800. The 287% should be applied to *the amount in 1990, not the amount in 2012.*
2. Do not write the equation as

$$x + 2.87 = 2800. \quad \text{Incorrect}$$

The percent must be multiplied by some number. In this case, the number is the amount spent in 1990, that is, $2.87x$.

NOW TRY ANSWER

5. 12

OBJECTIVE 6 Solve investment problems.

The investment problems in this chapter deal with *simple interest.* In most real-world applications, *compound interest* (covered in a later chapter) is used.

NOW TRY EXERCISE 6

Gary received a \$20,000 inheritance from his grandfather. He invested some of the money in an account earning 3% annual interest and the remaining amount in an account earning 2.5% annual interest. If the total annual interest earned is \$575, how much is invested at each rate?

CLASSROOM EXAMPLE 6

A man has \$34,000 to invest. He invests some of the money at 2.5% and the balance at 2%. His total annual interest income is \$772.50. Find the amount invested at each rate.

Answer: \$18,500 at 2.5%; \$15,500 at 2%

EXAMPLE 6 Solving an Investment Problem

Thomas invested \$40,000. He put part of the money in an account paying 2% interest and the remainder into stocks paying 3% interest. The total annual income from these investments was \$1020. How much did he invest at each rate?

Step 1 **Read** the problem again. We must find the two amounts.

Step 2 **Assign a variable.**

Let x = the amount invested at 2%;

$40,000 - x$ = the amount invested at 3%.

The formula for interest is $I = prt$. Here the time t is 1 yr. Use a table to organize the given information.

Principal (in dollars)	Rate (as a decimal)	Interest (in dollars)
x	0.02	$0.02x$
$40,000 - x$	0.03	$0.03(40,000 - x)$
40,000	XXXXXX	1020 ← Total

Multiply principal, rate, and time (here, 1 yr) to find interest.

Step 3 **Write an equation.** The last column of the table gives the equation.

Interest at 2% + interest at 3% = total interest.

$$0.02x + 0.03(40,000 - x) = 1020$$

Step 4 **Solve** the equation. Clear the parentheses and then the decimals.

$$0.02x + 1200 - 0.03x = 1020 \quad \text{Distributive property}$$

$$100(0.02x + 1200 - 0.03x) = 100(1020) \quad \text{Multiply by 100.}$$

$$2x + 120,000 - 3x = 102,000 \quad \text{Distributive property; Multiply.}$$

Move decimal points 2 places to the right.

$$-x + 120,000 = 102,000 \quad \text{Combine like terms.}$$

$$-x = -18,000 \quad \text{Subtract 120,000.}$$

$$x = 18,000 \quad \text{Multiply by } -1.$$

Step 5 **State the answer.** Thomas invested \$18,000 of the money at 2%. At 3%, he invested \$40,000 − \$18,000 = \$22,000.

Step 6 **Check.** Find the annual interest at each rate. The sum of these two amounts should total \$1020.

$$0.02(\$18,000) = \$360 \quad \text{and} \quad 0.03(\$22,000) = \$660$$

$$\$360 + \$660 = \$1020, \quad \text{as required.}$$

NOW TRY

NOW TRY ANSWER

6. \$15,000 at 3%; \$5000 at 2.5%

OBJECTIVE 7 Solve mixture problems.

Mixture problems involving rates of concentration can be solved with linear equations.

NOW TRY EXERCISE 7

How many liters of a 20% acid solution must be mixed with 5 L of a 30% acid solution to make a 24% acid solution?

CLASSROOM EXAMPLE 7

How many pounds of candy worth \$8 per lb should be mixed with 100 lb of candy worth \$4 per lb to make a mixture that can be sold for \$7 per lb?

Answer: 300 lb

EXAMPLE 7 Solving a Mixture Problem

A chemist must mix 8 L of a 40% acid solution with some 70% solution to make a 50% solution. How much of the 70% solution should be used?

Step 1 **Read** the problem. The problem asks for the amount of 70% solution to be used.

Step 2 **Assign a variable.**

Let $x =$ the number of liters of 70% solution.

The information in the problem is illustrated in **FIGURE 5**. We use it to complete a table.

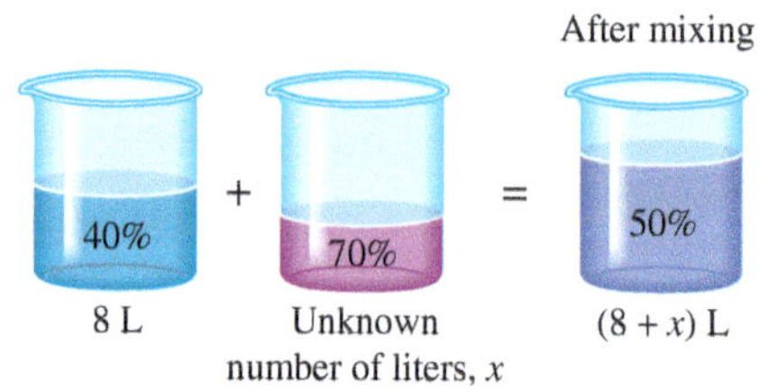

FIGURE 5

Liters of Solution	Percent (as a decimal)	Liters of Pure Acid
8	0.40	$0.40(8) = 3.2$
x	0.70	$0.70x$
$8 + x$	0.50	$0.50(8 + x)$

Sum must equal

The values in the last column were found by multiplying the strengths by the numbers of liters.

Step 3 **Write an equation** using the values in the last column of the table.

Liters of pure acid in 40% solution + liters of pure acid in 70% solution = liters of pure acid in 50% solution.

$$3.2 + 0.70x = 0.50(8 + x)$$

Step 4 **Solve.**

$3.2 + 0.70x = 4 + 0.50x$ Distributive property

$32 + 7x = 40 + 5x$ Multiply by 10 to clear the decimals. (Move decimal points 1 place to the right.)

$2x = 8$ Subtract 32 and $5x$.

$x = 4$ Divide by 2.

Step 5 **State the answer.** The chemist should use 4 L of the 70% solution.

Step 6 **Check.** 8 L of 40% solution plus 4 L of 70% solution is

$$8(0.40) + 4(0.70) = 6 \text{ L of acid.}$$

Similarly, 8 + 4 or 12 L of 50% solution has

$$12(0.50) = 6 \text{ L of acid.}$$

The total amount of pure acid is 6 L both before and after mixing, so the answer checks.

NOW TRY

NOW TRY ANSWER

7. $7\frac{1}{2}$ L

PROBLEM-SOLVING HINT When pure water is added to a solution, water is 0% of the chemical (acid, alcohol, etc.). Similarly, pure chemical is 100% chemical.

NOW TRY EXERCISE 8
How much pure antifreeze must be mixed with 3 gal of 30% antifreeze solution to obtain 40% antifreeze solution?

CLASSROOM EXAMPLE 8
How much water must be added to 20 L of 50% antifreeze solution to reduce it to 40% antifreeze?

Answer: 5 L

EXAMPLE 8 Solving a Mixture Problem When One Ingredient Is Pure

The octane rating of gasoline is a measure of its antiknock qualities. For a standard fuel, the octane rating is the percent of isooctane. How many liters of pure isooctane should be mixed with 200 L of 94% isooctane, referred to as 94 octane, to obtain a mixture that is 98% isooctane?

Step 1 **Read** the problem. We must find the amount of pure isooctane.

Step 2 **Assign a variable.** Let $x =$ the number of liters of pure (100%) isooctane. Complete a table. Recall that $100\% = 100(0.01) = 1$.

Liters of Solution	*Percent (as a decimal)*	*Liters of Pure Isooctane*
x	1	x
200	0.94	0.94(200)
$x + 200$	0.98	$0.98(x + 200)$

Step 3 **Write an equation** using the values in the last column of the table.

$$x + 0.94(200) = 0.98(x + 200)$$ Refer to the table.

Step 4 **Solve.**

$$x + 188 = 0.98x + 196$$ Multiply; distributive property

$$100(x + 188) = 100(0.98x + 196)$$ Multiply by 100 to clear the decimal.

$$100x + 18{,}800 = 98x + 19{,}600$$ Distributive property

$$2x = 800$$ Subtract 98x and 18,800.

$$x = 400$$ Divide by 2.

Step 5 **State the answer.** 400 L of isooctane is needed.

Step 6 **Check** by substituting 400 for x in the equation from Step 3.

$$400 + 0.94(200) \stackrel{?}{=} 0.98(400 + 200)$$ Let $x = 400$.

$$400 + 188 \stackrel{?}{=} 0.98(600)$$ Multiply. Add.

$$588 = 588 \checkmark$$ True

A true statement results, so the answer checks. NOW TRY

NOW TRY ANSWER
8. $\frac{1}{2}$ gal

1.3 Exercises

FOR EXTRA HELP

MyMathLab®

Complete solution available in MyMathLab

1. (a) $x + 15$ **(b)** $15 > x$
2. (a) $x + 5$ **(b)** $5 > x$
3. (a) $x - 8$ **(b)** $8 < x$
4. (a) $x - 6$ **(b)** $6 < x$

Concept Check *In each of the following, translate part (a) as an expression and translate part (b) as an equation or inequality. Use x to represent the number.*

1. (a) 15 more than a number
(b) 15 is more than a number.

2. (a) 5 greater than a number
(b) 5 is greater than a number.

3. (a) 8 less than a number
(b) 8 is less than a number.

4. (a) 6 less than a number
(b) 6 is less than a number.

5. *Concept Check* Which one of the following is *not* a valid translation of "40% of a number," where x represents the number?

A. $0.40x$ **B.** $0.4x$ **C.** $\frac{2x}{5}$ **D.** $40x$

6. *Concept Check* Why is $13 - x$ *not* a correct translation of "13 less than a number?"

Translate each verbal phrase into a mathematical expression using x as the variable. ***See Objective 1.***

7. Twice a number, decreased by 13

8. Triple a number, decreased by 14

9. 12 increased by four times a number

10. 15 more than one-half of a number

11. The product of 8 and 16 less than a number

12. The product of 8 more than a number and 5 less than the number

13. The quotient of three times a number and 10

14. The quotient of 9 and five times a nonzero number

Translate each verbal sentence into an equation, using x as the variable. Then solve the equation. ***See Example 1.***

15. The sum of a number and 6 is -31. Find the number.

16. The sum of a number and -4 is 18. Find the number.

17. If the product of a number and -4 is subtracted from the number, the result is 9 more than the number. Find the number.

18. If the quotient of a number and 6 is added to twice the number, the result is 8 less than the number. Find the number.

19. When $\frac{2}{3}$ of a number is subtracted from 14, the result is 10. Find the number.

20. When 75% of a number is added to 6, the result is 3 more than the number. Find the number.

Decide whether each is an expression *or an* equation. *Simplify any expressions, and solve any equations.* ***See Example 2.***

21. $5(x+3)-8(2x-6)$

22. $-7(x+4)+13(x-6)$

23. $5(x+3)-8(2x-6)=12$

24. $-7(x+4)+13(x-6)=18$

25. $\frac{1}{2}x-\frac{1}{6}x+\frac{3}{2}-8$

26. $\frac{1}{3}x+\frac{1}{5}x-\frac{1}{2}+7$

Complete the six suggested problem-solving steps to solve each problem.

27. In 2012, the corporations securing the most U.S. patents were IBM and Samsung. Together, the two corporations secured a total of 11,500 patents, with Samsung receiving 1414 fewer patents than IBM. How many patents did each corporation secure? (*Source:* U.S. Patent and Trademark Office.)

Step 1 **Read** the problem carefully. We are asked to find ________.

Step 2 **Assign a variable.** Let x = the number of patents that IBM secured. Then

$x - 1414$ = the number of ________.

Step 3 **Write an equation.** ______ + ______ = 11,500

Step 4 **Solve** the equation. x = ______

Step 5 **State the answer.** IBM secured ______ patents. Samsung secured ______ patents.

Step 6 **Check.** The number of Samsung patents was ______ fewer than the number of ________. The total number of patents was 6457 + ______ = ______.

---

5. D
6. $13 - x$ is the translation of "x less than 13." The phrase "13 less than a number" is $x - 13$.
7. $2x-13$
8. $3x-14$
9. $12+4x$
10. $\frac{1}{2}x+15$
11. $8(x-16)$
12. $(x+8)(x-5)$
13. $\frac{3x}{10}$
14. $\frac{9}{5x}\ (x \neq 0)$
15. $x+6=-31$; -37
16. $x+(-4)=18$; 22
17. $x-(-4x)=x+9$; $\frac{9}{4}$
18. $2x+\frac{x}{6}=x-8$; $-\frac{48}{7}$
19. $14-\frac{2}{3}x=10$; 6
20. $6+0.75x=x+3$; 12
21. expression; $-11x+63$
22. expression; $6x-106$
23. equation; $\left\{\frac{51}{11}\right\}$
24. equation; $\left\{\frac{62}{3}\right\}$
25. expression; $\frac{1}{3}x-\frac{13}{2}$
26. expression; $\frac{8}{15}x+\frac{13}{2}$
27. *Step 1:* the number of patents each corporation secured *Step 2:* patents that Samsung secured *Step 3:* x; $x-1414$ *Step 4:* 6457 *Step 5:* 6457; 5043 *Step 6:* 1414; IBM patents; 5043; 11,500

28. *Step 1:* the number of travelers to each country
Step 2: travelers to Canada
Step 3: x; $x - 8.5$
Step 4: 20.1
Step 5: 20.1 million; 11.6 million
Step 6: travelers to Mexico; 8.5 million; Canada; 11.6; 31.7 million

29. width: 165 ft; length: 265 ft
30. width: 100 ft; length: 160 ft
31. 24.34 in. by 29.88 in.
32. length: 14 cm; width: 2 cm
33. 850 mi; 925 mi; 1300 mi
34. 246.75 ft
35. Exxon Mobil: $449.9 billion; Wal-Mart: $469.2 billion

28. In 2011, 8.5 million more U.S. residents traveled to Mexico than to Canada. There was a total of 31.7 million U.S. residents traveling to these two countries. How many traveled to each country? (*Source:* U.S. Department of Commerce.)

Step 1 **Read** the problem carefully. We are asked to find ________.

Step 2 **Assign a variable.** Let x = the number of travelers to Mexico (in millions). Then $x - 8.5$ = the number of ________.

Step 3 **Write an equation.** _____ + _____ = 31.7

Step 4 **Solve** the equation. x = _____

Step 5 **State the answer.** There were _____ travelers to Mexico and _____ travelers to Canada.

Step 6 **Check.** The number of ________ was _____ more than the number of travelers to ________. The total number of travelers was 20.1 + _____ = _____.

Solve each problem. ***See Examples 3 and 4.***

29. The John Hancock Center in Chicago has a rectangular base. The length of the base measures 65 ft less than twice the width. The perimeter of the base is 860 ft. What are the dimensions of the base?

30. The John Hancock Center (**Exercise 29**) tapers as it rises. The top floor is rectangular and has perimeter 520 ft. The width of the top floor measures 20 ft more than one-half its length. What are the dimensions of the top floor?

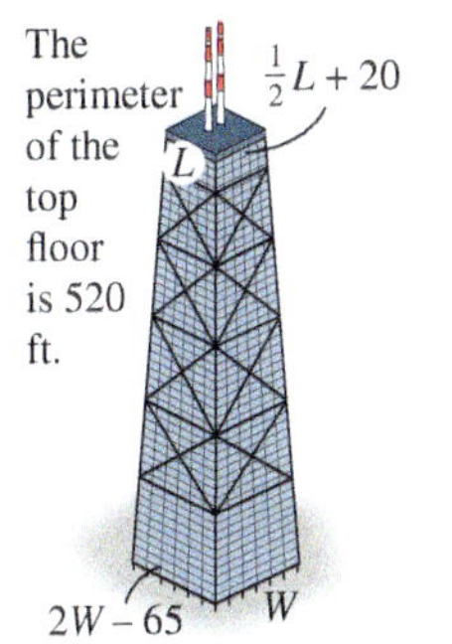

31. Grant Wood painted his most famous work, *American Gothic,* in 1930 on composition board with perimeter 108.44 in. If the rectangular painting is 5.54 in. taller than it is wide, find the dimensions of the painting. (*Source: The Gazette.*)

32. The perimeter of a certain rectangle is 16 times the width. The length is 12 cm more than the width. Find the length and width of the rectangle.

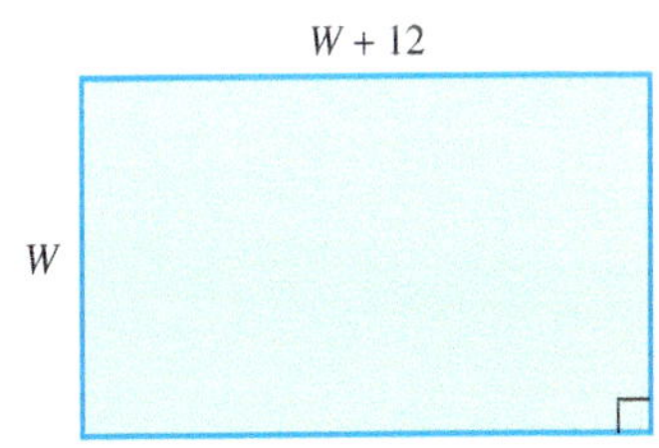

33. The Bermuda Triangle supposedly causes trouble for aircraft pilots. It has a perimeter of 3075 mi. The shortest side measures 75 mi less than the middle side, and the longest side measures 375 mi more than the middle side. Find the lengths of the three sides.

34. The Vietnam Veterans Memorial in Washington, DC, is in the shape of two sides of an isosceles triangle. If the two walls of equal length were joined by a straight line of 438 ft, the perimeter of the resulting triangle would be 931.5 ft. Find the lengths of the two walls. (*Source:* Pamphlet obtained at Vietnam Veterans Memorial.)

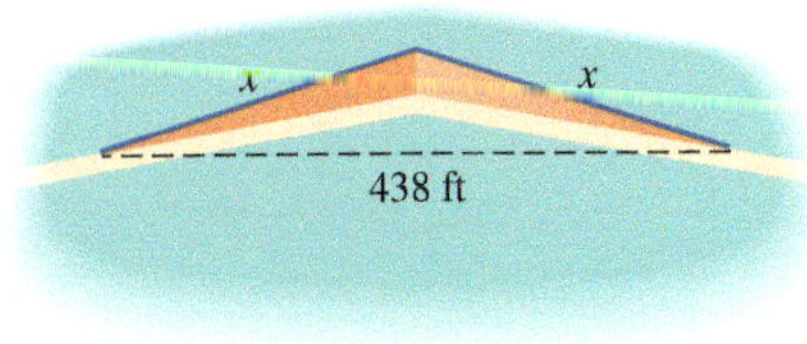

35. The two companies with top revenues in the Fortune 500 list for 2012 were Exxon Mobil and Wal-Mart. Their revenues together totaled $919.1 billion. Exxon Mobil revenues were $19.3 billion less than those of Wal-Mart. What were the revenues of each corporation? (*Source: Fortune* magazine.)

36. *Cats*: 7485; *Les Misérables*: 6680
37. Eiffel Tower: 984 ft; Leaning Tower: 180 ft
38. Yankees: $237.0 million; Dodgers: $236.9 million
39. Obama: 332 votes; Romney: 206 votes
40. Williams: 2654 hits; Hornsby: 2930 hits

41. 49.3% **42.** 1.4%
43. $8265 **44.** $25,687
45. $49.48

36. Two of the longest-running Broadway shows were *Cats* and *Les Misérables*. Together, there were 14,165 performances of these two shows during their Broadway runs. There were 805 fewer performances of *Les Misérables* than of *Cats*. How many performances were there of each show? (*Source:* The Broadway League.)

37. Galileo Galilei conducted experiments involving Italy's famous Leaning Tower of Pisa to investigate the relationship between an object's speed of fall and its weight. The Leaning Tower is 804 ft shorter than the Eiffel Tower in Paris, France. The two towers have a total height of 1164 ft. How tall is each tower? (*Source: World Almanac and Book of Facts.*)

38. In 2013, the New York Yankees and the Los Angeles Dodgers had the highest payrolls in Major League Baseball. The Dodgers' payroll was $0.1 million less than the Yankees' payroll, and the two payrolls totaled $473.9 million. What was the payroll for each team? (*Source: Kansas City Star.*)

39. In the 2012 presidential election, Barack Obama and Mitt Romney together received 538 electoral votes. Obama received 126 more votes than Romney. How many votes did each candidate receive? (*Source: World Almanac and Book of Facts.*)

40. Ted Williams and Rogers Hornsby were two great hitters in Major League Baseball. Together, they had 5584 hits in their careers. Hornsby had 276 more hits than Williams. How many hits did each have? (*Source:* Neft, D. S., and R. M. Cohen, *The Sports Encyclopedia: Baseball,* St. Martins Griffin, New York.)

Rogers Hornsby

Solve each percent problem. ***See Example 5.***

41. In 2012, the number of graduating seniors taking the ACT exam was 1,666,017. In 2002, a total of 1,116,082 graduating seniors took the exam. By what percent did the number increase over this period of time, to the nearest tenth of a percent? (*Source:* ACT.)

42. Composite scores on the ACT exam rose from 20.8 in 2002 to 21.1 in 2012. What percent increase was this, to the nearest tenth of a percent? (*Source:* ACT.)

43. For 1993–1994, the average cost of tuition and fees at public four-year universities in the United States was $2431 for full-time in-state students. Twenty years later, it had increased 240%. To the nearest dollar, what was the approximate cost for 2013–2014? (*Source:* The College Board.)

44. For 1993–1994, the average cost of tuition and fees at private four-year universities in the United States was $9399 for full-time students. Twenty years later, it had increased 173.3%. To the nearest dollar, what was the approximate cost for 2013–2014? (*Source:* The College Board.)

45. In 2013, the average cost of a traditional Thanksgiving dinner for 10, featuring turkey, stuffing, cranberries, pumpkin pie, and trimmings, was $49.04, a decrease of 0.89% over the cost in 2012. What was the cost, to the nearest cent, in 2012? (*Source:* American Farm Bureau.)

46. \$26.75
47. \$225 **48.** \$149,460
49. \$4000 at 3%; \$8000 at 4%
50. \$20,000 at 2%; \$40,000 at 3%
51. \$10,000 at 4.5%; \$19,000 at 3%
52. \$8000 at 3.5%; \$29,000 at 4%
53. \$24,000 **54.** \$8500
55. 5 L **56.** 25 L
57. 4 L **58.** $13\frac{1}{3}$ L

46. Refer to **Exercise 45.** The cost of a traditional Thanksgiving dinner in 2013 was \$49.04, an increase of 83.3% over the cost in 1987 when data was first collected. What was the cost, to the nearest cent, in 1987? (*Source:* American Farm Bureau.)

47. At the end of a day, Lawrence found that the total cash register receipts at the motel where he works were \$2725. This included the 9% sales tax charged. Find the amount of tax.

48. David sold his house for \$159,000. He got this amount knowing that he would have to pay a 6% commission to his agent. What amount did he have after the agent was paid?

Solve each investment problem. ***See Example 6.***

49. Mario earned \$12,000 last year giving tennis lessons. He invested part of the money at 3% simple interest and the rest at 4%. In one year, he earned a total of \$440 in interest. How much did he invest at each rate?

Principal (in dollars)	Rate (as a decimal)	Interest (in dollars)
x	0.03	
	0.04	

50. Sheryl won \$60,000 on a slot machine in Las Vegas. She invested part of the money at 2% simple interest and the rest at 3%. In one year, she earned a total of \$1600 in interest. How much was invested at each rate?

Principal (in dollars)	Rate (as a decimal)	Interest (in dollars)
x	0.02	

51. Jennifer invested some money at 4.5% simple interest and \$1000 less than twice this amount at 3%. Her total annual income from the interest was \$1020. How much was invested at each rate?

52. Piotr invested some money at 3.5% simple interest, and \$5000 more than three times this amount at 4%. He earned \$1440 in annual interest. How much did he invest at each rate?

53. Dan has invested \$12,000 in bonds paying 6%. How much additional money should he invest in a certificate of deposit paying 3% simple interest so that the total return on the two investments will be 4%?

54. Mona received a year-end bonus of \$17,000 from her company and invested the money in an account paying 6.5%. How much additional money should she deposit in an account paying 5% so that the return on the two investments will be 6%?

Solve each problem. ***See Examples 7 and 8.***

55. How many liters of a 10% acid solution must be mixed with 10 L of a 4% solution to obtain a 6% solution?

Liters of Solution	Percent (as a decimal)	Liters of Pure Acid
10	0.04	
x	0.10	
	0.06	

56. How many liters of a 14% alcohol solution must be mixed with 20 L of a 50% solution to obtain a 30% solution?

Liters of Solution	Percent (as a decimal)	Liters of Pure Alcohol
x	0.14	
	0.50	

57. In a chemistry class, 12 L of a 12% alcohol solution must be mixed with a 20% solution to obtain a 14% solution. How many liters of the 20% solution are needed?

58. How many liters of a 10% alcohol solution must be mixed with 40 L of a 50% solution to obtain a 40% solution?

59. 1 gal 60. 2 gal
61. 150 lb 62. 200 oz

63. We cannot expect the final mixture to be worth more than the more expensive of the two ingredients.
64. Let x = the number of liters of 30% acid. Then

$$0.30x + 0.50(15) = 0.60(x + 15).$$

The solution is -5, which is impossible, because the number of liters of 30% acid cannot be negative.

59. How much pure dye must be added to 4 gal of a 25% dye solution to increase the solution to 40%? (*Hint:* Pure dye is 100% dye.)

60. How much water must be added to 6 gal of a 4% insecticide solution to reduce the concentration to 3%? (*Hint:* Water is 0% insecticide.)

61. Randall wants to mix 50 lb of nuts worth \$2 per lb with some nuts worth \$6 per lb to make a mixture worth \$5 per lb. How many pounds of \$6 nuts must he use?

Pounds of Nuts	Cost per Pound (in dollars)	Total Cost (in dollars)

62. Lee Ann wants to mix tea worth 2¢ per oz with 100 oz of tea worth 5¢ per oz to make a mixture worth 3¢ per oz. How much 2¢ tea should be used?

Ounces of Tea	Cost per Ounce (in dollars)	Total Cost (in dollars)

63. *Concept Check* Why is it impossible to mix candy worth \$4 per lb and candy worth \$5 per lb to obtain a final mixture worth \$6 per lb?

64. *Concept Check* Write an equation based on the following problem, solve the equation, and explain why the problem has no solution.

How much 30% acid should be mixed with 15 L of 50% acid to obtain a mixture that is 60% acid?

RELATING CONCEPTS For Individual or Group Work (Exercises 65–68)

Consider each problem.

Problem A Jack has \$800 invested in two accounts. One pays 5% annual interest and the other pays 10% annual interest. The amount of annual interest is the same as he would earn if the entire \$800 was invested at 8.75%. How much does he have invested at each rate?

Problem B Jill has 800 L of acid solution. She obtained it by mixing some 5% acid with some 10% acid. Her final mixture of 800 L is 8.75% acid. How much of each of the 5% and 10% solutions did she use to make her final mixture?

In Problem A, let x represent the amount invested at 5% interest, and in Problem B, let y represent the amount of 5% acid used. ***Work Exercises 65–68 in order.***

65. **(a)** Write an expression in x that represents the amount of money Jack invested at 10% in Problem A.

(b) Write an expression in y that represents the amount of 10% acid solution Jill used in Problem B.

66. **(a)** Write expressions that represent the amount of annual interest Jack earns at 5% and at 10%.

(b) Write expressions that represent the amount of pure acid in Jill's 5% and 10% acid solutions.

67. **(a)** The sum of the two expressions in part (a) of **Exercise 66** must equal the total amount of annual interest earned. Write an equation representing this fact.

(b) The sum of the two expressions in part (b) of **Exercise 66** must equal the amount of pure acid in the final mixture. Write an equation representing this fact.

68. **(a)** Solve Problem A.

(b) Solve Problem B.

(c) Explain the similarities between the processes used in solving Problems A and B.

65. (a) $800 - x$ (b) $800 - y$
66. (a) $0.05x$; $0.10(800 - x)$
(b) $0.05y$; $0.10(800 - y)$
67. (a) $0.05x + 0.10(800 - x) = 800(0.0875)$
(b) $0.05y + 0.10(800 - y) = 800(0.0875)$
68. (a) \$200 at 5%; \$600 at 10%
(b) 200 L of 5% acid; 600 L of 10% acid
(c) The processes are the same. The amounts of money in Problem A correspond to the amounts of solution in Problem B.

1.4 Further Applications of Linear Equations

OBJECTIVES

1. Solve problems about different denominations of money.
2. Solve problems about uniform motion.
3. Solve problems about angles.
4. Solve problems about consecutive integers.

VOCABULARY

☐ consecutive integers
☐ consecutive even integers
☐ consecutive odd integers

NOW TRY EXERCISE 1

Steven has a collection of 52 coins worth \$3.70. His collection contains only dimes and nickels. How many of each type of coin does he have?

CLASSROOM EXAMPLE 1

Mohammed has a box of coins containing only dimes and half-dollars. There are 26 coins, and the total value is \$8.60. How many of each denomination of coin does he have?

Answer: 11 dimes; 15 half-dollars

TEACHING TIP

Mention that consistency is important when working with problems involving money. Either work the problem entirely without decimal points (as numbers of cents) or entirely with decimal points.

NOW TRY ANSWER

1. 22 dimes; 30 nickels

OBJECTIVE 1 Solve problems about different denominations of money.

PROBLEM-SOLVING HINT In problems involving money, use the following basic fact.

number of monetary units of the same kind × denomination = total monetary value

Examples: 30 dimes have a monetary value of 30(\$0.10) = \$3.00.

15 five-dollar bills have a value of 15(\$5) = \$75.

EXAMPLE 1 Solving a Money Denomination Problem

For a bill totaling \$5.65, a cashier received 25 coins consisting of nickels and quarters. How many of each denomination of coin did the cashier receive?

Step 1 **Read** the problem. We must find the number of nickels and the number of quarters the cashier received.

Step 2 **Assign a variable.**

Let x = the number of nickels.

Then $25 - x$ = the number of quarters.

	Number of Coins	Denomination (in dollars)	Value (in dollars)
Nickels	x	0.05	$0.05x$
Quarters	$25 - x$	0.25	$0.25(25 - x)$
			5.65 ← Total

Organize the information in a table.

Step 3 **Write an equation.** Use the values from the last column of the table.

$$0.05x + 0.25(25 - x) = 5.65$$

Step 4 **Solve.** Clear the parentheses and then the decimals.

$0.05x + 6.25 - 0.25x = 5.65$ Distributive property

$100(0.05x + 6.25 - 0.25x) = 100(5.65)$ Multiply by 100.

$5x + 625 - 25x = 565$ Distributive property; Multiply.

Move decimal points 2 places to the right

$-20x = -60$ Subtract 625. Combine like terms.

$x = 3$ Divide by −20.

Step 5 **State the answer.** There are 3 nickels and $25 - 3 = 22$ quarters.

Step 6 **Check.** The cashier has $3 + 22 = 25$ coins. The value is

\$0.05(3) + \$0.25(22) = \$5.65, as required. NOW TRY

CAUTION ***Be sure that your answer is reasonable*** when working problems like **Example 1.** Because we are dealing with a number of coins, the correct answer can neither be negative nor a fraction.

OBJECTIVE 2 Solve problems about uniform motion.

PROBLEM-SOLVING HINT Uniform motion problems use the distance formula $d = rt$. ***When rate (or speed) is given in miles per hour, time must be given in hours. Draw a sketch*** to illustrate what is happening. ***Make a table*** to summarize given information.

NOW TRY EXERCISE 2
Two trains leave a city traveling in opposite directions. One travels at a rate of 80 km per hr and the other at a rate of 75 km per hr. How long will it take before they are 387.5 km apart?

CLASSROOM EXAMPLE 2
Two cars leave the same town at the same time. One travels north at 60 mph and the other south at 45 mph. In how many hours will they be 420 mi apart?

Answer: 4 hr

EXAMPLE 2 Solving a Motion Problem (Opposite Directions)

Two cars leave the same place at the same time, one going east and the other west. The eastbound car averages 40 mph, while the westbound car averages 50 mph. In how many hours will they be 300 mi apart?

Step 1 **Read** the problem. We are looking for the time it takes for the two cars to be 300 mi apart.

Step 2 **Assign a variable.** A sketch shows what is happening in the problem. The cars are going in *opposite* directions. See **FIGURE 6**.

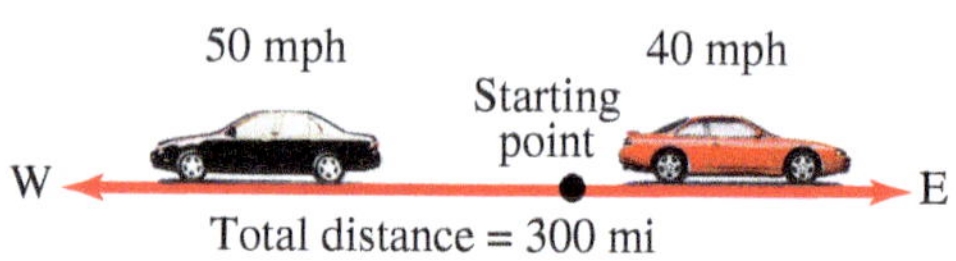

FIGURE 6

Let $x =$ the time traveled by each car.

	Rate	Time	Distance
Eastbound Car	40	x	$40x$
Westbound Car	50	x	$50x$
			300

Fill in each distance by multiplying rate by time, using the formula $d = rt$, or $rt = d$. The sum of the two distances is 300.

Step 3 **Write an equation.** $40x + 50x = 300$

Step 4 **Solve.**

$$90x = 300 \quad \text{Combine like terms.}$$

$$x = \frac{300}{90} \quad \text{Divide by 90.}$$

$$x = \frac{10}{3} \quad \text{Lowest terms}$$

Step 5 **State the answer.** The cars travel $\frac{10}{3} = 3\frac{1}{3}$ hr, or 3 hr, 20 min.

Step 6 **Check.** The eastbound car traveled $40\left(\frac{10}{3}\right) = \frac{400}{3}$ mi. The westbound car traveled $50\left(\frac{10}{3}\right) = \frac{500}{3}$ mi, for a total distance of $\frac{400}{3} + \frac{500}{3} = \frac{900}{3}$, or 300 mi, as required.

NOW TRY

CAUTION It is a common error to write 300 as the distance traveled by each car in **Example 2.** Three hundred miles is the *total* distance traveled.

NOW TRY ANSWER
2. $2\frac{1}{2}$ hr

As in **Example 2,** in general, the equation for a problem involving motion in *opposite* directions is of the following form.

partial distance + partial distance = total distance

NOW TRY EXERCISE 3

Michael can drive to work in $\frac{1}{2}$ hr. When he rides his bicycle, it takes $1\frac{1}{2}$ hours. If his average rate while driving to work is 30 mph faster than his rate while bicycling to work, determine the distance that he lives from work.

CLASSROOM EXAMPLE 3

When Chris drives his car to work, the trip takes $\frac{1}{2}$ hr. When he rides the bus, it takes $\frac{3}{4}$ hr. The average rate of the bus is 12 mph less than his rate when driving. Find the distance he travels to work.

Answer: 18 mi

EXAMPLE 3 Solving a Motion Problem (Same Direction)

Geoff can bike to work in $\frac{3}{4}$ hr. When he takes the bus, the trip takes $\frac{1}{4}$ hr. If the bus travels 20 mph faster than Geoff rides his bike, how far is it to his workplace?

Step 1 **Read** the problem. We must find the distance between Geoff's home and his workplace.

Step 2 **Assign a variable.** Make a sketch to show what is happening. See **FIGURE 7**.

FIGURE 7

The problem asks for a distance, but it is easier here to let x be Geoff's rate when he rides his bike to work. Then the rate of the bus is $(x + 20)$ mph.

For the trip by bike, $\quad d = rt = x \cdot \frac{3}{4} = \frac{3}{4}x.$

For the trip by bus, $\quad d = rt = (x + 20) \cdot \frac{1}{4} = \frac{1}{4}(x + 20).$

	Rate	Time	Distance
Bike	x	$\frac{3}{4}$	$\frac{3}{4}x$
Bus	$x + 20$	$\frac{1}{4}$	$\frac{1}{4}(x + 20)$

Same distance

Step 3 **Write an equation.**

$$\frac{3}{4}x = \frac{1}{4}(x + 20) \quad \text{The distance is the same in each case.}$$

Step 4 **Solve.**

$$\frac{3}{4}x = \frac{1}{4}x + 5 \quad \text{Distributive property}$$

$$4\left(\frac{3}{4}x\right) = 4\left(\frac{1}{4}x + 5\right) \quad \text{Multiply by 4 to clear the fractions.}$$

$$3x = x + 20 \quad \text{Multiply; } 4\left(\tfrac{1}{4}\right) = 1 \text{ and } 1x = x$$

$$2x = 20 \quad \text{Subtract } x.$$

$$x = 10 \quad \text{Divide by 2.}$$

Step 5 **State the answer.** The required distance is

$$d = \frac{3}{4}x = \frac{3}{4}(10) = \frac{30}{4} = 7.5 \text{ mi.}$$

Step 6 **Check** by finding the distance using

$$d = \frac{1}{4}(x + 20) = \frac{1}{4}(10 + 20) = \frac{30}{4} = 7.5 \text{ mi.}$$

The same distance results.

NOW TRY

NOW TRY ANSWER

3. 22.5 mi

As in **Example 3,** the equation for a problem involving motion in the *same* direction is often of the following form.

one distance = other distance

PROBLEM-SOLVING HINT In **Example 3,** it was easier to let the variable represent a quantity other than the one that we were asked to find. It takes practice to learn when this approach works best.

OBJECTIVE 3 Solve problems about angles.

An important result of Euclidean geometry (the geometry of the Greek mathematician Euclid) is as follows.

***The sum of the angle measures of any triangle is* 180°.**

NOW TRY EXERCISE 4

Find the value of x, and determine the measure of each angle.

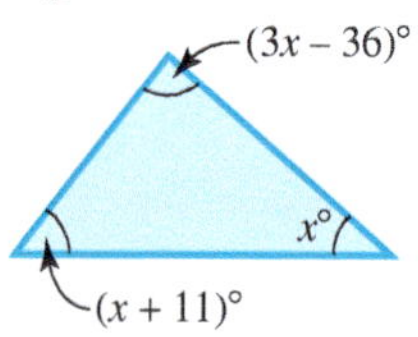

CLASSROOM EXAMPLE 4

Find the value of x, and determine the measure of each angle.

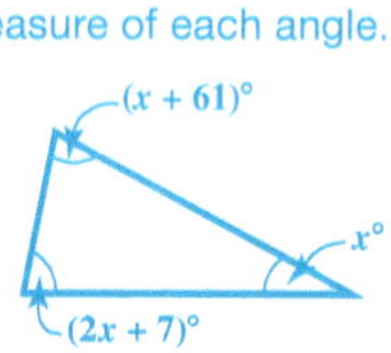

Answer: 28°, 63°, 89°

EXAMPLE 4 Finding Angle Measures

Find the value of x, and determine the measure of each angle in **FIGURE 8.**

Step 1 **Read** the problem. We are asked to find the measure of each angle.

Step 2 **Assign a variable.**

Let $x =$ the measure of one angle.

Step 3 **Write an equation.** The sum of the three measures shown in the figure must be 180°.

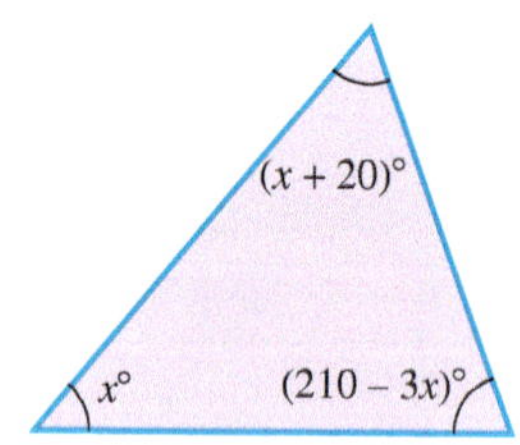

FIGURE 8

$$x + (x + 20) + (210 - 3x) = 180 \quad \text{Measures are in degrees.}$$

Step 4 **Solve.**

$$-x + 230 = 180 \quad \text{Combine like terms.}$$

$$-x = -50 \quad \text{Subtract 230.}$$

$$x = 50 \quad \text{Multiply by } -1.$$

Step 5 **State the answer.** One angle measures 50°. The other two angles measure

$$x + 20 = 50 + 20 = 70^\circ$$

and

$$210 - 3x = 210 - 3(50) = 60^\circ.$$

Step 6 **Check.** Because $50^\circ + 70^\circ + 60^\circ = 180^\circ$ is true, the answers are correct.

NOW TRY

OBJECTIVE 4 Solve problems about consecutive integers.

Recall that the set of integers is $\{\ldots, -3, -2, -1, 0, 1, 2, 3, \ldots\}$. **Consecutive integers** are integers that are next to each other on a number line. Two consecutive integers differ by 1. Some examples of consecutive integers are 4, 5 and −6, −5.

Consecutive even integers always differ by 2. Examples are 4, 6 and −4, −2. Similarly, **consecutive odd integers** also differ by 2, such as 1, 3 and −5, −3.

NOW TRY ANSWER

4. 41°, 52°, 87°

PROBLEM-SOLVING HINT If x = the lesser (least) integer in a consecutive integer problem, then the following apply.

- For two consecutive integers, use $x, \; x + 1$.
- For three consecutive integers, use $x, \; x + 1, \; x + 2$.
- For two consecutive even or odd integers, use $x, \; x + 2$.
- For three consecutive even or odd integers, use $x, \; x + 2, \; x + 4$.

In this book, we list consecutive integers in increasing order.

NOW TRY EXERCISE 5
Find three consecutive integers such that the sum of the first and second is 43 less than three times the third.

CLASSROOM EXAMPLE 5
Find three consecutive integers such that the sum of the first and third is 49 less than three times the second.

Answer: 48, 49, 50

NOW TRY ANSWER
5. 38, 39, 40

EXAMPLE 5 Solving a Consecutive Integer Problem

Find three consecutive integers such that the sum of the first and third, increased by 3, is 50 more than the second.

Step 1 **Read** the problem. We are told to find three consecutive integers satisfying the given conditions.

Step 2 **Assign a variable.**

Let x = the least of the three consecutive integers.

Then $x + 1$ = the second (middle) integer,

and $x + 2$ = the greatest of the three consecutive integers.

Step 3 **Write an equation.**

Sum of the first and third,	increased by 3,	is	50 more than the second.
$x + (x + 2)$	$+3$	$=$	$(x + 1) + 50$

Step 4 **Solve.**

$$2x + 5 = x + 51 \quad \text{Combine like terms.}$$

$$x = 46 \quad \text{Subtract } x \text{ and 5.}$$

Step 5 **State the answer.** The solution is 46, so the first integer is $x = 46$, the second is $46 + 1 = 47$, and the third is $46 + 2 = 48$. The three integers are 46, 47, and 48.

Step 6 **Check.** The sum of the first and third is $46 + 48 = 94$. If this is increased by 3, the result is $94 + 3 = 97$, which is indeed 50 more than the second (47). The answer is correct.

NOW TRY

1.4 Exercises

FOR EXTRA HELP

MyMathLab®

Complete solution available in MyMathLab

1. \$5.70 2. 14 hr
3. 30 mph

Concept Check *Solve each problem.*

1. What amount of money is found in a coin hoard containing 12 dimes and 18 quarters?
2. The distance between Cape Town, South Africa, and Miami is 7700 mi. If a jet averages 550 mph between the two cities, what is its travel time in hours?
3. Tri traveled from Chicago to Des Moines, a distance of 300 mi, in 10 hr. What was his rate in miles per hour?

4. 120 in.
5. The problem asks for the *distance* to the workplace. To find this distance, we must multiply the rate, 10 mph, by the time, $\frac{3}{4}$ hr.
6. Begin by subtracting 36° from 180°. Then find half of this difference to obtain the measure of each of the equal angles.
7. No, the answers must be whole numbers because they represent the numbers of coins.
8. $\frac{1}{6}x$
9. 17 pennies; 17 dimes; 10 quarters
10. 2 nickels; 2 quarters; 4 half-dollars
11. 23 loonies; 14 toonies
12. 57 \$5 bills; 62 \$1 bills
13. 28 \$10 coins; 13 \$20 coins
14. 22 two-cent pieces; 66 three-cent pieces

4. A square has perimeter 160 in. What would be the perimeter of an equilateral triangle whose sides each measure the same length as the side of the square?

Concept Check *Answer each question.*

5. Read over **Example 3** in this section. The solution of the equation is 10. Why is *10 mph* not the answer to the problem?

6. Suppose that we know that two angles of a triangle have equal measures and the third angle measures 36°. How would we find the measures of the equal angles without actually writing an equation?

7. In a problem about the number of coins of different denominations, would an answer that is a fraction be reasonable? Would a negative answer be reasonable?

8. In a motion problem the rate is given as x mph and the time is given as 10 min. What variable expression represents the distance in miles?

Solve each problem. ***See Example 1.***

9. Otis has a box of coins that he uses when he plays poker with his friends. The box contains 44 coins, consisting of pennies, dimes, and quarters. The number of pennies is equal to the number of dimes. The total value is \$4.37. How many of each denomination of coin does he have in the box?

Number of Coins	Denomination (in dollars)	Value (in dollars)	
x	0.01	$0.01x$	
x			
	0.25		
XXXX	XXXX	4.37	← Total

10. Nana found some coins while looking under her sofa pillows. There were equal numbers of nickels and quarters and twice as many half-dollars as quarters. If she found \$2.60 in all, how many of each denomination of coin did she find?

Number of Coins	Denomination (in dollars)	Value (in dollars)	
x	0.05	$0.05x$	
x			
$2x$	0.50		
XXXX	XXXX	2.60	← Total

11. In Canada, \$1 and \$2 bills have been replaced by coins. The \$1 coins are called "loonies" because they have a picture of a loon (a well-known Canadian bird) on the reverse, and the \$2 coins are called "toonies." When Marissa returned home to San Francisco from a trip to Vancouver, she found that she had acquired 37 of these coins, with a total value of 51 Canadian dollars. How many coins of each denomination did she have?

12. Dan works at an ice cream shop. At the end of his shift, he counted the bills in his cash drawer and found 119 bills with a total value of \$347. If all of the bills are \$5 bills and \$1 bills, how many of each denomination were in his cash drawer?

13. Dave collects U.S. gold coins. He has a collection of 41 coins. Some are \$10 coins, and the rest are \$20 coins. If the face value of the coins is \$540, how many of each denomination does he have?

14. In the 19th century, the United States minted two-cent and three-cent pieces. Frances has three times as many three-cent pieces as two-cent pieces, and the face value of these coins is \$2.42. How many of each denomination does she have?

15. 872 adult tickets
16. 315 student tickets; 165 nonstudent tickets

15. In 2014, general admission to the Art Institute of Chicago cost \$23 for adults and \$17 for children and seniors. If \$30,052 was collected from the sale of 1460 general admission tickets, how many adult tickets were sold? (*Source:* www.artic.edu)

16. For a high school production of *A Funny Thing Happened on the Way to the Forum*, student tickets cost \$5 each while nonstudent tickets cost \$8. If 480 tickets were sold and a total of \$2895 was collected, how many tickets of each type were sold?

In Exercises 17–20, find the rate on the basis of the information provided. Use a calculator and round answers to the nearest hundredth. All events were at the 2012 Summer Olympics in London, England. (Source: World Almanac and Book of Facts.)

	Event	Winner	Distance	Time
17.	100-m hurdles, women	Sally Pearson, Australia	100 m	12.35 sec
18.	400-m hurdles, women	Natalya Antyukh, Russia	400 m	52.70 sec
19.	200-m run, men	Usain Bolt, Jamaica	200 m	19.32 sec
20.	400-m run, men	Kirani James, Grenada	400 m	43.94 sec

17. 8.10 m per sec
18. 7.59 m per sec
19. 10.35 m per sec
20. 9.10 m per sec

21. $2\frac{1}{2}$ hr 22. $1\frac{3}{4}$ hr
23. 7:50 P.M. 24. 10:00 A.M.
25. 45 mph 26. 2 mi

Solve each problem. ***See Examples 2 and 3.***

21. Two steamers leave a port on a river at the same time, traveling in opposite directions. Each is traveling 22 mph. How long will it take for them to be 110 mi apart?

	Rate	Time	Distance
First Steamer		t	
Second Steamer	22		
			110

22. A train leaves Kansas City, Kansas, and travels north at 85 km per hr. Another train leaves at the same time and travels south at 95 km per hr. How long will it take before they are 315 km apart?

	Rate	Time	Distance
First Train	85	t	
Second Train			
			315

23. Mulder and Scully are driving to Georgia to investigate "Big Blue," a giant reptile reported in one of the local lakes. Mulder leaves the office at 8:30 A.M. averaging 65 mph. Scully leaves at 9:00 A.M., following the same path and averaging 68 mph. At what time will Scully catch up with Mulder?

	Rate	Time	Distance
Mulder			
Scully			

24. Lois and Clark, two elderly reporters, are covering separate stories and have to travel in opposite directions. Lois leaves the *Daily Planet* building at 8:00 A.M. and travels at 35 mph. Clark leaves at 8:15 A.M. and travels at 40 mph. At what time will they be 140 mi apart?

	Rate	Time	Distance
Lois			
Clark			

25. It took Charmaine 3.6 hr to drive to her mother's house on Saturday morning for a weekend visit. On her return trip on Sunday night, traffic was heavier, so the trip took her 4 hr. Her average rate on Sunday was 5 mph slower than on Saturday. What was her average rate on Sunday?

	Rate	Time	Distance
Saturday			
Sunday			

26. Sharon commutes to her office by train. When she walks to the train station, it takes her 40 min. When she rides her bike, it takes her 12 min. Her average walking rate is 7 mph less than her average biking rate. Find the distance from her house to the train station.

	Rate	Time	Distance
Walking			
Biking			

27. $\frac{1}{2}$ hr 28. 55 mph

29. 60°, 60°, 60°
30. 20°, 30°, 130°
31. 40°, 45°, 95°
32. 28°, 63°, 89°

33. Both measure 122°.
34. Both measure 49°.

35. 64°, 26°
36. 24°, 66°

37. 65°, 115°
38. 40°, 140°

27. Johnny leaves Memphis to visit his cousin Anne, who lives in the town of Hornsby, Tennessee, 80 mi away. He travels at an average rate of 50 mph. One-half hour later, Anne leaves to visit Johnny, traveling at an average rate of 60 mph. How long after Anne leaves will it be before they meet?

28. On an automobile trip, Laura maintained a steady rate for the first two hours. Rush-hour traffic slowed her rate by 25 mph for the last part of the trip. The entire trip, a distance of 125 mi, took $2\frac{1}{2}$ hr. What was her rate during the first part of the trip?

Find the measure of each angle in the triangles shown. ***See Example 4.***

29.

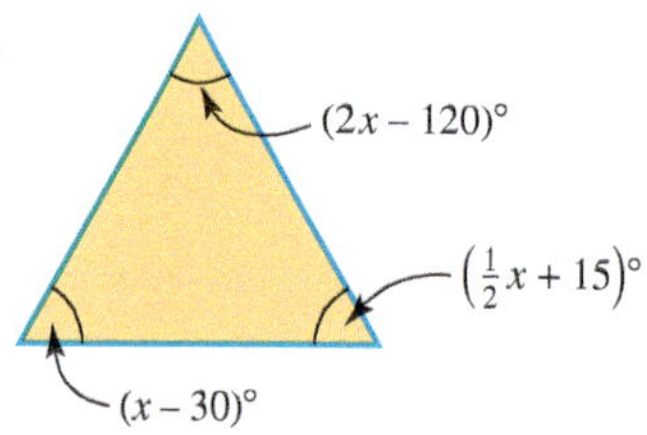

30.

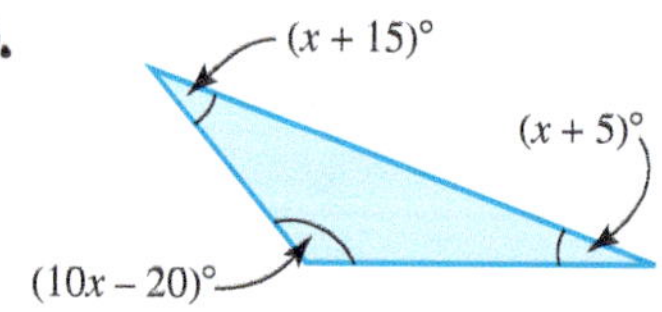

31.

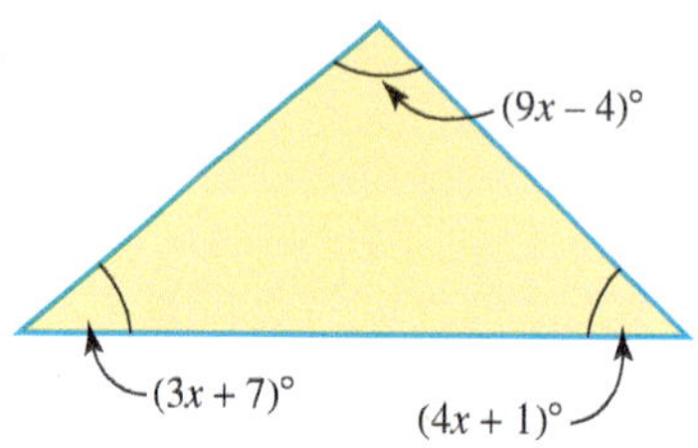

32.

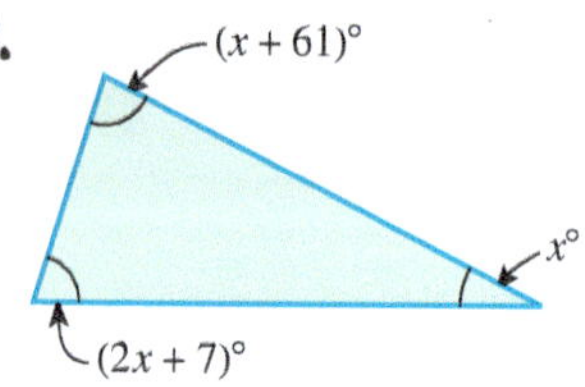

In Exercises 33 and 34, the angles marked with variable expressions are ***vertical angles.*** *It is shown in geometry that vertical angles have equal measures. Find the measure of each angle.*

33.

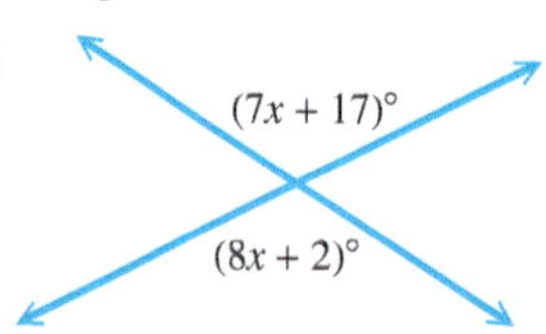

34.

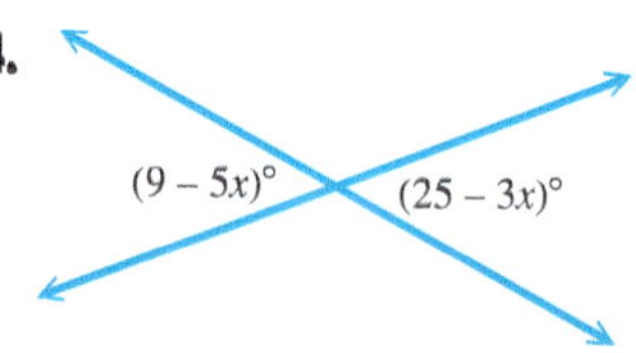

Two angles whose sum is 90° are ***complementary angles.*** *In Exercises 35 and 36, find the measures of the complementary angles shown in the figure.*

35.

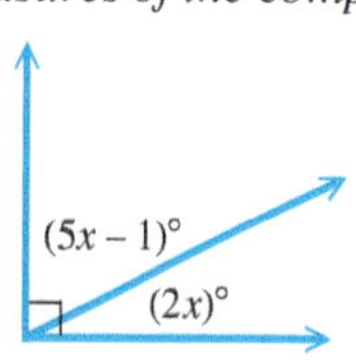

36.

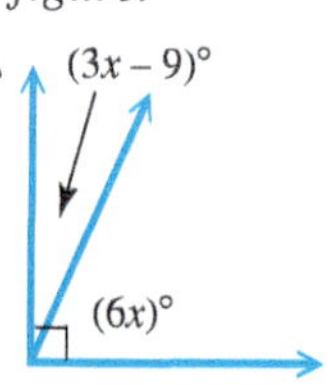

Two angles whose sum is 180° are ***supplementary angles.*** *In Exercises 37 and 38, find the measures of the supplementary angles shown in the figure.*

37.

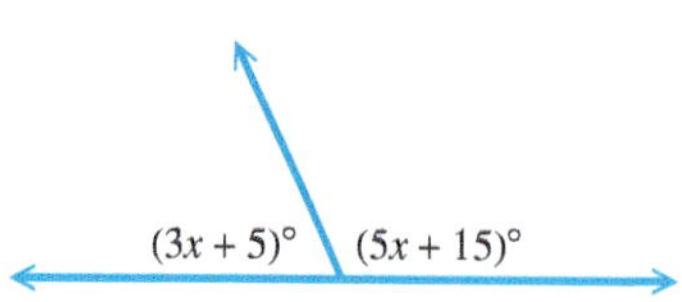

38.

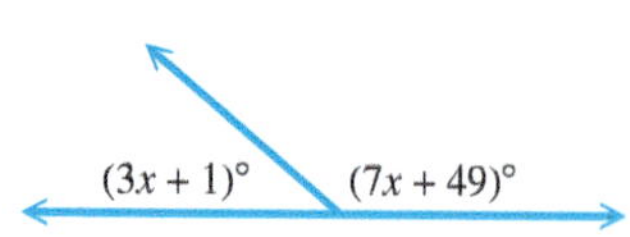

39. 19, 20, 21
40. 37, 38, 39
41. 27, 28, 29, 30
42. 40, 41, 42, 43
43. 61 yr old
44. 69 yr old
45. 28, 30, 32
46. 10, 12, 14
47. 21, 23, 25
48. 11, 13, 15

Solve each problem involving consecutive integers. ***See Example 5.***

39. Find three consecutive integers such that the sum of the first and twice the second is 17 more than twice the third.

40. Find three consecutive integers such that the sum of the first and twice the third is 39 more than twice the second.

41. Find four consecutive integers such that the sum of the first three is 54 more than the fourth.

42. Find four consecutive integers such that the sum of the last three is 86 more than the first.

43. If I add my current age to the age I will be next year on this date, the sum is 103 yr. How old will I be 10 yr from today?

44. If I add my current age to the age I will be next year on this date, the sum is 129 yr. How old will I be 5 yr from today?

45. Find three consecutive *even* integers such that the sum of the least integer and the middle integer is 26 more than the greatest integer.

46. Find three consecutive *even* integers such that the sum of the least integer and the greatest integer is 12 more than the middle integer.

47. Find three consecutive *odd* integers such that the sum of the least integer and the middle integer is 19 more than the greatest integer.

48. Find three consecutive *odd* integers such that the sum of the least integer and the greatest integer is 13 more than the middle integer.

RELATING CONCEPTS For Individual or Group Work **(Exercises 49–52)**

Consider the following two figures. ***Work Exercises 49–52 in order.***

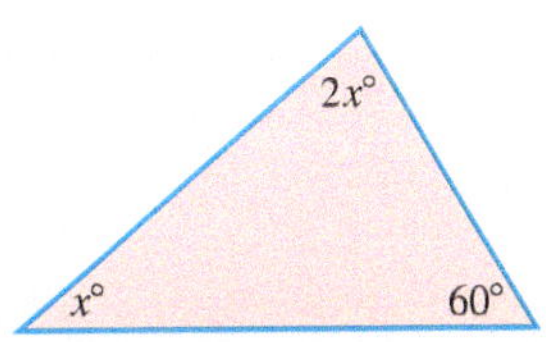

FIGURE A

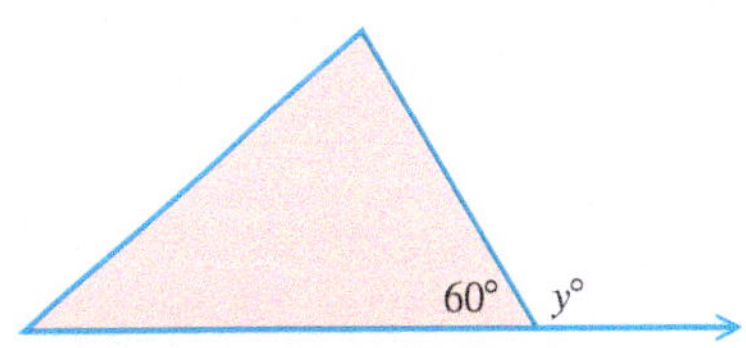

FIGURE B

49. 40°, 80°
50. 120°
51. The sum is equal to the measure of the angle found in Exercise 50.
52. The sum of the measures of angles ① and ② is equal to the measure of angle ③.

49. Solve for the measures of the unknown angles in **FIGURE A.** (*Hint:* What is the sum of the angle measures of any triangle?)

50. Solve for the measure of the unknown angle marked $y°$ in **FIGURE B.** (*Hint:* See **Exercises 37 and 38.**)

51. Add the measures of the two angles found in **Exercise 49.** How does the sum compare to the measure of the angle found in **Exercise 50?**

52. Based on the answers to **Exercises 49–51,** make a conjecture (an educated guess) about the relationship among the angles marked ①, ②, and ③ in the figure below.

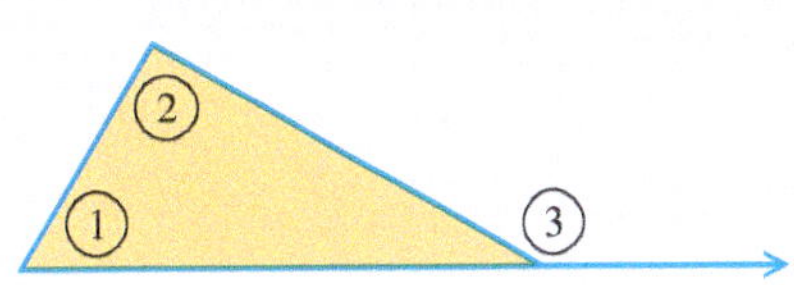

SUMMARY EXERCISES Applying Problem-Solving Techniques

1. length: 8 in.; width: 5 in.
2. length: 60 m; width: 30 m
3. \$399.99
4. \$425
5. \$800 at 4%; \$1600 at 5%
6. \$12,000 at 3%; \$14,000 at 4%
7. Durant: 2280; Bryant: 2133
8. *Titanic:* \$600.8 million; *The Dark Knight:* \$533.3 million
9. 5 hr

Solve each problem.

1. The length of a rectangle is 3 in. more than its width. If the length were decreased by 2 in. and the width were increased by 1 in., the perimeter of the resulting rectangle would be 24 in. Find the dimensions of the original rectangle.

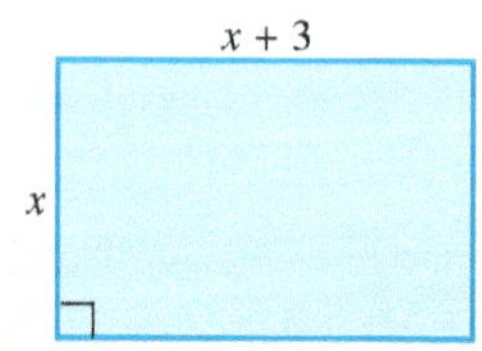

2. A farmer wishes to enclose a rectangular region with 210 m of fencing in such a way that the length is twice the width and the region is divided into two equal parts, as shown in the figure. What length and width should be used?

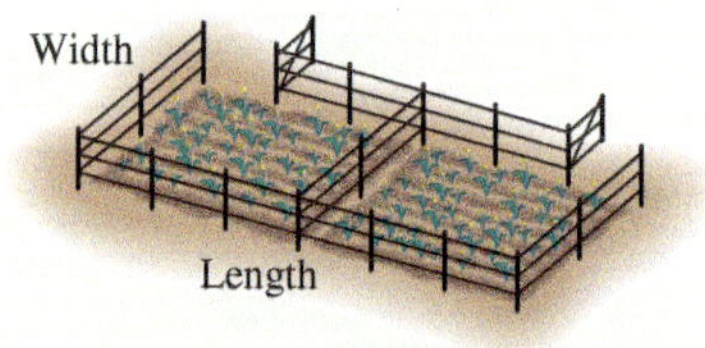

3. After a discount of 20.5%, the sale price for a Samsung Galaxy tablet computer was \$317.99. What was the regular price of the computer to the nearest cent? (*Source*: www.amazon.com)

4. An electronics store offered a new HD television for \$255, the sale price after the regular price was discounted 40%. What was the regular price?

5. An amount of money is invested at 4% annual simple interest, and twice that amount is invested at 5%. The total annual interest is \$112. How much is invested at each rate?

6. An amount of money is invested at 3% annual simple interest, and \$2000 more than that amount is invested at 4%. The total annual interest is \$920. How much is invested at each rate?

7. Kevin Durant of the Oklahoma City Thunder and Kobe Bryant of the Los Angeles Lakers were the leading scorers in the NBA for the 2012–2013 regular season. Together they scored 4413 points, with Bryant scoring 147 fewer points than Durant. How many points did each of them score? (*Source*: www.espn.go.com)

8. Two top-grossing American movies were *Titanic* and *The Dark Knight*. *Titanic* grossed \$67.5 million more than *The Dark Knight*. Together, the two films brought in \$1134.1 million. How much did each movie gross? (*Source:* Rentrak Corporation.)

9. Atlanta and Cincinnati are 440 mi apart. John leaves Cincinnati, driving toward Atlanta at an average rate of 60 mph. Pat leaves Atlanta at the same time, driving toward Cincinnati in her antique auto, averaging 28 mph. How long will it take them to meet?

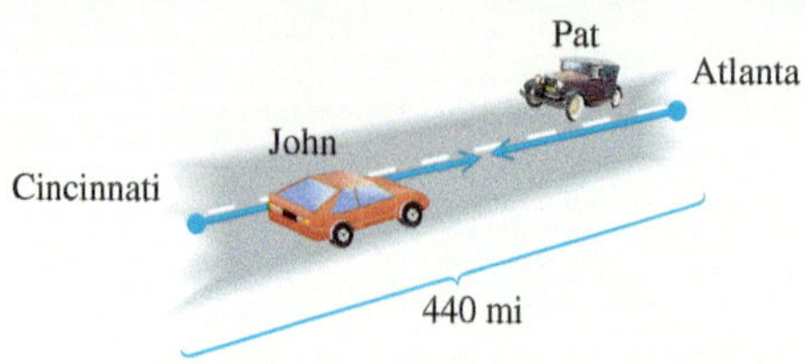

10. $1\frac{1}{2}$ cm
11. $13\frac{1}{3}$ L **12.** $53\frac{1}{3}$ kg
13. 84 fives; 42 tens
14. 1650 tickets at \$9; 810 tickets at \$7
15. 20°, 30°, 130°
16. 107°, 73°
17. 31, 32, 33
18. 9, 11
19. 6 in., 12 in., 16 in.
20. 23 in.

10. Joshua has a sheet of tin 12 cm by 16 cm. He plans to make a box by cutting equal squares out of each of the four corners and folding up the remaining edges. How large a square should he cut so that the finished box will have a length that is 5 cm less than twice the width?

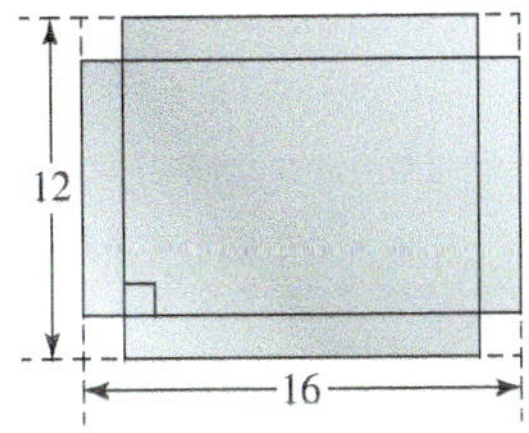

11. A pharmacist has 20 L of a 10% drug solution. How many liters of 5% solution must be added to obtain a mixture that is 8%?

12. A certain metal is 20% tin. How many kilograms of this metal must be mixed with 80 kg of a metal that is 70% tin to obtain a metal that is 50% tin?

Kilograms of Metal	*Percent (as a decimal)*	*Kilograms of Tin*
	0.20	
	0.50	

13. A cashier has a total of 126 bills in fives and tens. The total value of the money is \$840. How many of each denomination of bill does he have?

Number of Bills	*Denomination (in dollars)*	*Value (in dollars)*
XXXXX	XXXXX	840

14. The top-grossing domestic movie in 2013 was *The Hunger Games: Catching Fire*. On opening weekend, one theater showing this movie took in \$20,520 by selling a total of 2460 tickets, some at \$9 and the rest at \$7. How many tickets were sold at each price? (*Source:* www.boxofficemojo.com)

15. Find the measure of each angle.

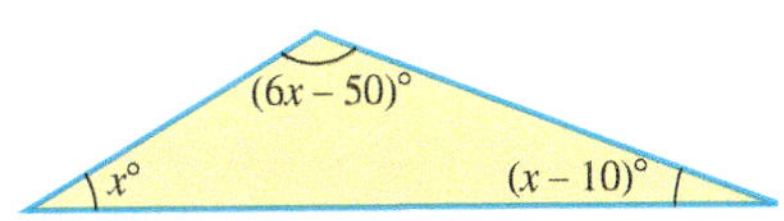

16. Find the measure of each marked angle.

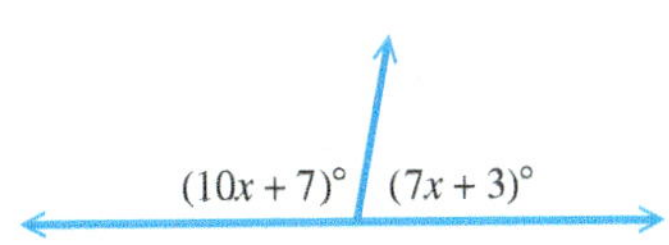

17. The sum of the least and greatest of three consecutive integers is 32 more than the middle integer. What are the three integers?

18. If the lesser of two consecutive odd integers is doubled, the result is 7 more than the greater of the two integers. Find the two integers.

19. The perimeter of a triangle is 34 in. The middle side is twice as long as the shortest side. The longest side is 2 in. less than three times the shortest side. Find the lengths of the three sides.

20. The perimeter of a rectangle is 43 in. more than the length. The width is 10 in. Find the length of the rectangle.

1.5 Linear Inequalities in One Variable

OBJECTIVES

1. Graph intervals on a number line.
2. Solve linear inequalities using the addition property.
3. Solve linear inequalities using the multiplication property.
4. Solve linear inequalities with three parts.
5. Solve applied problems using linear inequalities.

VOCABULARY

- ☐ inequality
- ☐ interval
- ☐ linear inequality in one variable
- ☐ equivalent inequalities
- ☐ three-part inequality

An **inequality** consists of algebraic expressions related by one of the following symbols.

$<$	"is less than"	$\leq$	"is less than or equal to"
$>$	"is greater than"	$\geq$	"is greater than or equal to"

These symbols are read as shown when the inequality is read from left to right.

We ***solve an inequality*** by finding all real number solutions for it. For example, the solution set of $x \leq 2$ includes *all* real numbers that are less than or equal to 2, not just the integers less than or equal to 2.

OBJECTIVE 1 Graph intervals on a number line.

We graph all the real numbers satisfying $x \leq 2$ by placing a square bracket at 2 and drawing an arrow to the left to represent the fact that all numbers less than 2 are also part of the graph. See **FIGURE 9**.

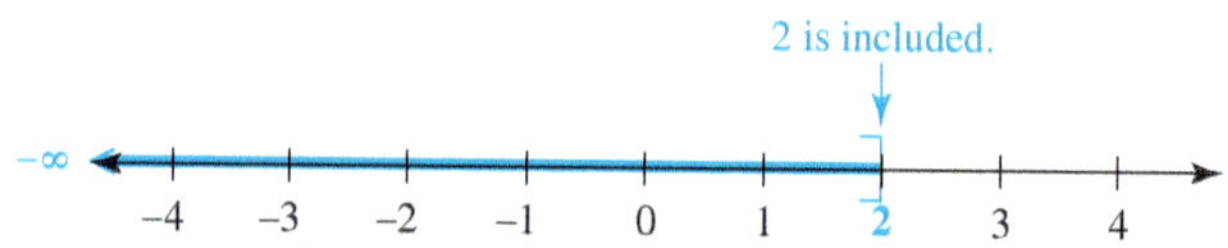

FIGURE 9 The interval $(-\infty, 2]$

The set of numbers less than or equal to 2 is an example of an **interval** on a number line. To write intervals, we use **interval notation,** which includes the **infinity symbols,** ∞ or $-\infty$. We write the interval of *all* numbers less than or equal to 2 as

$$(-\infty, 2]. \quad \text{Interval notation}$$

The negative infinity symbol $-\infty$ does not indicate a number, but shows that the interval includes all real numbers less than 2. On both the number line and in interval notation, the square bracket indicates that 2 is included in the solution set.

TEACHING TIP
Emphasize that parentheses, and not brackets, are written with infinity symbols when using interval notation. Also, mention that other books may use open circles instead of parentheses and solid dots instead of brackets for graphs of inequalities.

Remember the following important concepts regarding interval notation.

- A parenthesis indicates that an endpoint is *not* included.
- A square bracket indicates that an endpoint is included.
- A parenthesis is always used next to an infinity symbol, $-\infty$ or ∞.
- The set of real numbers is written in interval notation as $(-\infty, \infty)$.

CLASSROOM EXAMPLE 1
Write each inequality in interval notation and graph the interval.

(a) $x \geq -3$

(b) $0 < x < 3.5$

Answers:

(a) $[-3, \infty)$

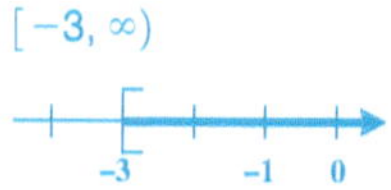

(b) $(0, 3.5)$

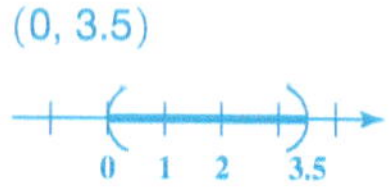

EXAMPLE 1 Using Interval Notation

Write each inequality in interval notation and graph the interval.

(a) $x > -5$

This statement says that x can represent any number greater than -5, but x cannot equal -5. This interval is written $(-5, \infty)$. We graph it by placing a parenthesis at -5 and drawing an arrow to the right, as in **FIGURE 10**. The parenthesis at -5 shows that -5 is *not* part of the graph.

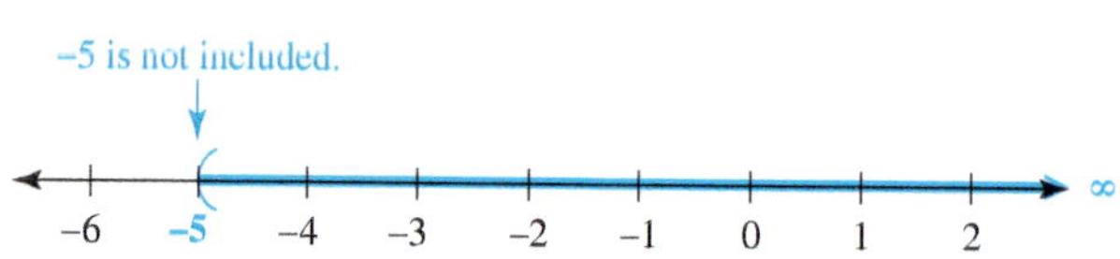

FIGURE 10 The interval $(-5, \infty)$

NOW TRY EXERCISE 1

Write each inequality in interval notation and graph the interval.

(a) $x < -1$

(b) $-4 \leq x < 2$

(b) $-1 \leq x < 3$

This statement is read "-1 is less than or equal to x *and* x is less than 3." We want the set of numbers that are *between* -1 and 3, with -1 included and 3 excluded. In interval notation, we write $[-1, 3)$, using a bracket at -1 because it is part of the graph and a parenthesis at 3 because it is not part of the graph. See **FIGURE 11**.

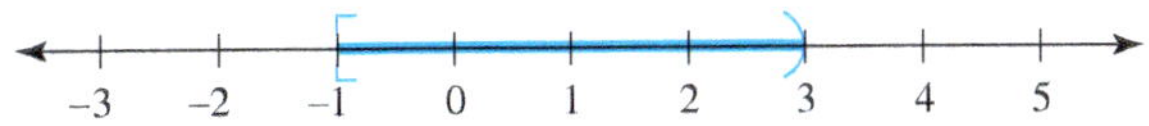

FIGURE 11 The interval $[-1, 3)$

NOW TRY

TEACHING TIP

Refer students to the third column of the table. Point out that the ∞ symbol can never be first in an interval, and $-\infty$ can never be second.

Summary of Types of Intervals

Type of Interval	*Set-Builder Notation*	*Interval Notation*	*Graph*
Open interval	$\{x \mid a < x < b\}$	(a, b)	
Closed interval	$\{x \mid a \leq x \leq b\}$	$[a, b]$	
Half-open (or half-closed) interval	$\{x \mid a \leq x < b\}$	$[a, b)$	
	$\{x \mid a < x \leq b\}$	$(a, b]$	
Disjoint interval*	$\{x \mid x < a \text{ or } x > b\}$	$(-\infty, a) \cup (b, \infty)$	
Infinite interval	$\{x \mid x > a\}$	(a, ∞)	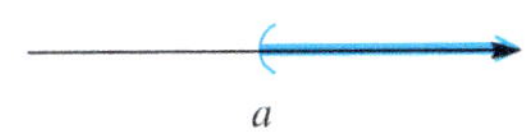
	$\{x \mid x \geq a\}$	$[a, \infty)$	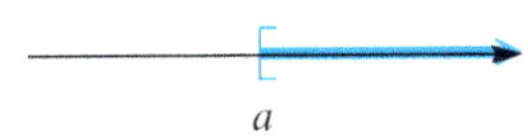
	$\{x \mid x < a\}$	$(-\infty, a)$	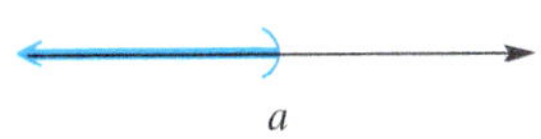
	$\{x \mid x \leq a\}$	$(-\infty, a]$	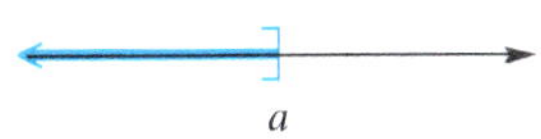
	$\{x \mid x \text{ is a real number}\}$	$(-\infty, \infty)$	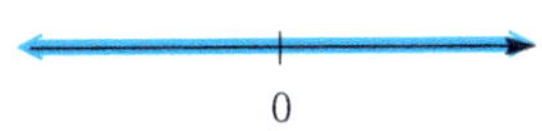

CAUTION A parenthesis is *always* used next to an infinity symbol, $-\infty$ or ∞.

Solving inequalities is similar to solving equations.

Linear Inequality in One Variable

A **linear inequality in one variable** (here x) can be written in the form

$$Ax + B < C, \quad Ax + B \leq C, \quad Ax + B > C, \quad \text{or} \quad Ax + B \geq C,$$

where A, B, and C are real numbers and $A \neq 0$.

Examples: $x + 5 < 2$, $x - 3 \geq 5$, and $2x + 5 \leq 10$ Linear inequalities

NOW TRY ANSWERS

1. (a) $(-\infty, -1)$

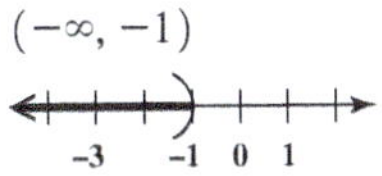

(b) $[-4, 2)$

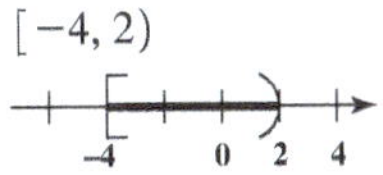

*We will work with disjoint intervals in **Section 1.6** when we study *set operations* and *compound inequalities*.

OBJECTIVE 2 Solve linear inequalities using the addition property.

We solve an inequality by finding all numbers that make the inequality true. Usually, an inequality has an infinite number of solutions. These solutions, like solutions of equations, are found by producing a series of simpler equivalent inequalities. **Equivalent inequalities** are inequalities with the same solution set.

We use two important properties to produce equivalent inequalities.

Addition Property of Inequality

If A, B, and C represent real numbers, then the inequalities

$$\boldsymbol{A < B} \quad \text{and} \quad \boldsymbol{A + C < B + C} \quad \text{are equivalent.*}$$

That is, the same number may be added to each side of an inequality without changing the solution set.

*This also applies to $A \le B$, $A > B$, and $A \ge B$.

NOW TRY EXERCISE 2

Solve $x - 10 > -7$, and graph the solution set.

CLASSROOM EXAMPLE 2

Solve $x - 5 > 1$, and graph the solution set.

Answer: $(6, \infty)$

EXAMPLE 2 Using the Addition Property of Inequality

Solve $x - 7 < -12$, and graph the solution set.

$$x - 7 < -12$$

$$x - 7 + 7 < -12 + 7 \qquad \text{Add 7.}$$

$$x < -5 \qquad \text{Combine like terms.}$$

CHECK Substitute -5 for x in the *equation* $x - 7 = -12$.

$$x - 7 = -12$$

$$-5 - 7 \stackrel{?}{=} -12 \qquad \text{Let } x = -5.$$

$$-12 = -12 \quad ✓ \quad \text{True}$$

The result, a true statement, shows that -5 is the boundary point. Now test a number on each side of -5 to verify that numbers *less than* -5 make the inequality true. We choose -4 and -6.

$$x - 7 < -12$$

$-4 - 7 \stackrel{?}{<} -12$ Let $x = -4$.	$-6 - 7 \stackrel{?}{<} -12$ Let $x = -6$.
$-11 < -12$ False	$-13 < -12$ ✓ True
-4 is *not* in the solution set.	-6 *is* in the solution set.

The check confirms that the interval $(-\infty, -5)$ is the solution set. See **FIGURE 12.**

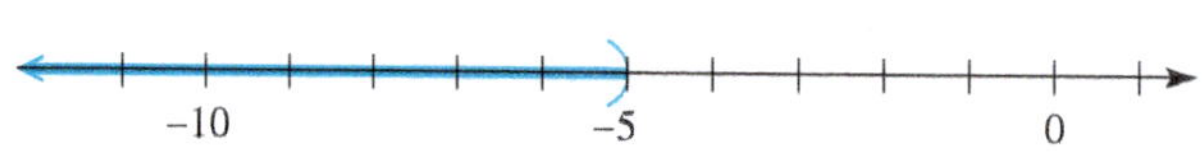

FIGURE 12

NOW TRY

NOW TRY ANSWER

2. $(3, \infty)$

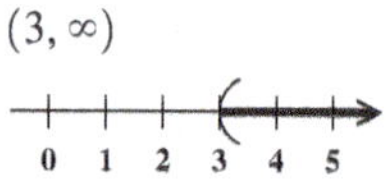

As with equations, the addition property can be used to *subtract* the same number from each side of an inequality.

NOW TRY EXERCISE 3

Solve $4x + 1 \geq 5x$, and graph the solution set.

CLASSROOM EXAMPLE 3

Solve $5x + 3 \geq 4x - 1$, and graph the solution set.

Answer: $[-4, \infty)$

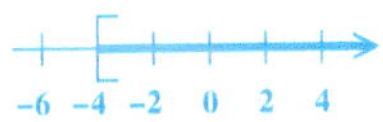

TEACHING TIP

To avoid errors, it is a good idea to have students rewrite an inequality such as $14 \leq x$ as $x \geq 14$ so that the variable is on the left, as in Example 3.

EXAMPLE 3 Using the Addition Property of Inequality

Solve $14 + 2x \leq 3x$, and graph the solution set.

$$14 + 2x \leq 3x$$

$$14 + 2x - 2x \leq 3x - 2x \quad \text{Subtract } 2x.$$

$$14 \leq x \quad \text{Combine like terms.}$$

Be careful. $\quad x \geq 14 \quad$ Rewrite.

The inequality $14 \leq x$ (14 is less than or equal to x) can also be written $x \geq 14$ (x is greater than or equal to 14). ***Notice that in each case the inequality symbol points to the lesser number,* 14.**

CHECK

$$14 + 2x = 3x$$

$$14 + 2(14) \stackrel{?}{=} 3(14) \quad \text{Let } x = 14.$$

$$42 = 42 \; ✓ \quad \text{True}$$

So 14 satisfies the equality part of $\leq$. Choose 10 and 15 as test values.

$$14 + 2x < 3x$$

$14 + 2(10) \stackrel{?}{<} 3(10)$ Let $x = 10$.	$14 + 2(15) \stackrel{?}{<} 3(15)$ Let $x = 15$.
$34 < 30$ False	$44 < 45$ ✓ True
10 is not in the solution set.	15 is in the solution set.

The check confirms that the interval $[14, \infty)$ is the solution set. See **FIGURE 13**.

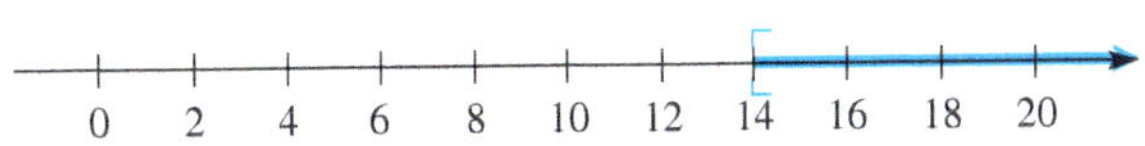

FIGURE 13

NOW TRY

OBJECTIVE 3 Solve linear inequalities using the multiplication property.

Consider the following true statement.

$$-2 < 5$$

Multiply each side by some positive number—for example, 8.

$$-2(8) < 5(8) \quad \text{Multiply by 8.}$$

$$-16 < 40 \quad \text{True}$$

The result is true. Start again with $-2 < 5$, and multiply each side by some negative number—for example, -8.

$$-2(-8) < 5(-8) \quad \text{Multiply by } -8.$$

$$16 < -40 \quad \text{False}$$

The result, $16 < -40$, is false. To make it true, we must change the direction of the inequality symbol.

$$16 > -40 \quad \text{True}$$

As these examples suggest, multiplying each side of an inequality by a *negative* number requires reversing the direction of the inequality symbol. The same is true for dividing by a negative number because division is defined in terms of multiplication.

NOW TRY ANSWER

3. $(-\infty, 1]$

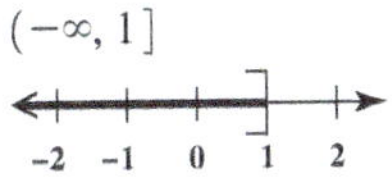

TEACHING TIP
Demonstrate the rule for reversing the inequality symbol using the examples

$-2x < 10$ and $2x < -10$,

emphasizing that in the first one the symbol must be reversed, but in the second it remains unchanged.

Multiplication Property of Inequality

Let A, B, and C represent real numbers, where $C \neq 0$.

(a) If C is ***positive,*** then the inequalities

$$A < B \quad \text{and} \quad AC < BC \quad \text{are equivalent.*}$$

(b) If C is ***negative,*** then the inequalities

$$A < B \quad \text{and} \quad AC > BC \quad \text{are equivalent.*}$$

That is, each side of an inequality may be multiplied (or divided) by the same *positive* number without changing the direction of the inequality symbol. ***If the multiplier is negative, we must reverse the direction of the inequality symbol.***

*This also applies to $A \leq B$, $A > B$, and $A \geq B$.

Because division is defined in terms of multiplication, the multiplication property also applies when dividing each side of an inequality by the same number.

NOW TRY EXERCISE 4
Solve each inequality, and graph the solution set.

(a) $8x \geq -40$

(b) $-20x > -60$

CLASSROOM EXAMPLE 4
Solve each inequality, and graph the solution set.

(a) $4x \leq -100$

(b) $-9x < -81$

Answers:

(a) $(-\infty, -25]$

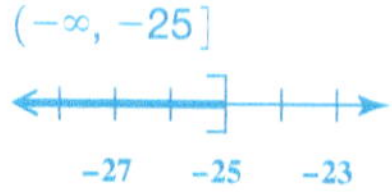

(b) $(9, \infty)$

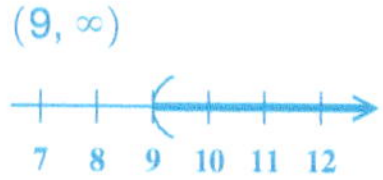

EXAMPLE 4 Using the Multiplication Property of Inequality

Solve each inequality, and graph the solution set.

(a) $5x \leq -30$

Divide each side by 5. ***Because 5 > 0, do not reverse the direction of the inequality symbol.***

$$5x \leq -30$$

$$\frac{5x}{5} \leq \frac{-30}{5} \qquad \text{Divide by 5.}$$

$$x \leq -6$$

Check that the solution set is the interval $(-\infty, -6]$, graphed in **FIGURE 14**.

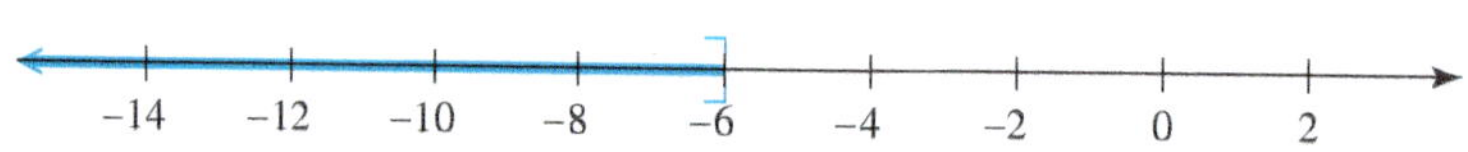

FIGURE 14

(b) $-4x \leq 32$

Divide each side by -4. ***Because −4 < 0, reverse the direction of the inequality symbol.***

$$-4x \leq 32$$

$$\frac{-4x}{-4} \geq \frac{32}{-4} \qquad \text{Divide by } -4. \text{ Reverse the direction of the symbol.}$$

$$x \geq -8$$

Reverse the inequality symbol when dividing by a *negative* number.

Check the solution set. **FIGURE 15** shows the graph of the solution set, $[-8, \infty)$.

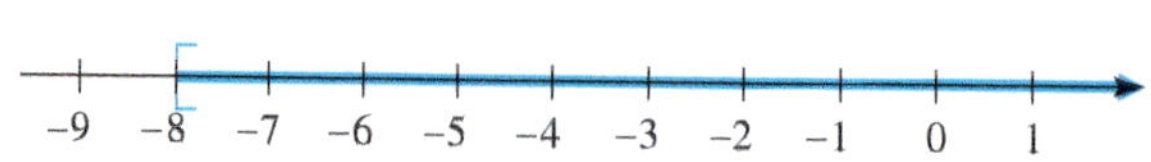

FIGURE 15

NOW TRY

NOW TRY ANSWERS

4. (a) $[-5, \infty)$

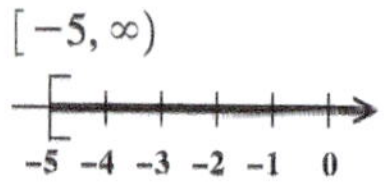

(b) $(-\infty, 3)$

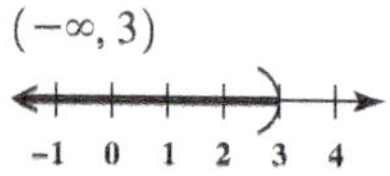

To solve a linear inequality in one variable, use the following steps.

Solving a Linear Inequality in One Variable

Step 1 **Simplify each side separately.** Use the distributive property as needed.

- Clear any parentheses.
- Clear any fractions or decimals.
- Combine like terms.

Step 2 **Isolate the variable terms on one side.** Use the addition property of inequality so that all terms with variables are on one side of the inequality and all constants (numbers) are on the other side.

Step 3 **Isolate the variable.** Use the multiplication property of inequality to obtain an inequality in one of the following forms, where k is a constant (number).

$$\text{variable} < k, \quad \text{variable} \le k, \quad \text{variable} > k, \quad \text{or} \quad \text{variable} \ge k$$

Remember: ***Reverse the direction of the inequality symbol only when multiplying or dividing each side of an inequality by a negative number.***

NOW TRY EXERCISE 5

Solve and graph the solution set.

$$5 - 2(x - 4) \le 11 - 4x$$

EXAMPLE 5 Solving a Linear Inequality

Solve $-3(x + 4) + 2 \ge 7 - x$, and graph the solution set.

Step 1 $-3(x + 4) + 2 \ge 7 - x$

$-3x - 3(4) + 2 \ge 7 - x$ Distributive property

$-3x - 12 + 2 \ge 7 - x$ Multiply.

$-3x - 10 \ge 7 - x$ (*) Combine like terms.

Step 2 $-3x - 10 + x \ge 7 - x + x$ Add x.

$-2x - 10 \ge 7$ Combine like terms.

$-2x - 10 + 10 \ge 7 + 10$ Add 10.

$-2x \ge 17$ Combine like terms.

Step 3 $\dfrac{-2x}{-2} \le \dfrac{17}{-2}$ Divide by -2. Change $\ge$ to $\le$.

Be sure to reverse the direction of the inequality symbol.

$$x \le -\frac{17}{2}$$

FIGURE 16 shows the graph of the solution set, $\left(-\infty, -\frac{17}{2}\right]$.

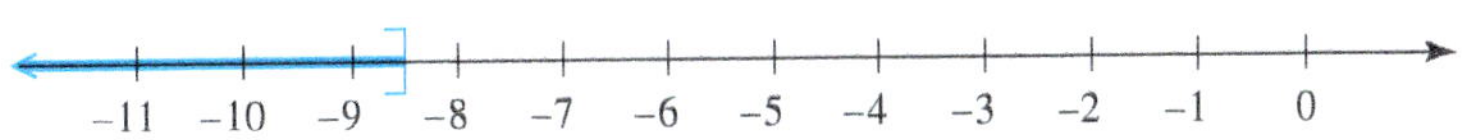

FIGURE 16

NOW TRY

NOW TRY ANSWER

5. $(-\infty, -1]$

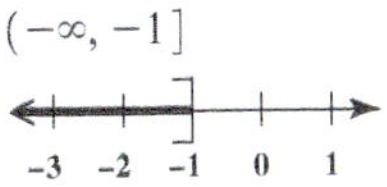

NOTE In Step 2 of **Example 5,** if we add $3x$ (instead of x) to both sides of the inequality, we obtain the following sequence of equivalent inequalities.

$-3x - 10 \geq 7 - x$	See (*) of **Example 5.**
$-3x - 10 + 3x \geq 7 - x + 3x$	Add $3x$.
$-10 \geq 2x + 7$	Combine like terms.
$-10 - 7 \geq 2x + 7 - 7$	Subtract 7.
$-17 \geq 2x$	Combine like terms.
$-\frac{17}{2} \geq x$	Divide by 2.
$x \leq -\frac{17}{2}$	Rewrite. The same solution results.

The symbol points to x in each case.

NOW TRY EXERCISE 6

Solve and graph the solution set.

$$\frac{3}{4}(x-2) + \frac{1}{2} > \frac{1}{5}(x-8)$$

CLASSROOM EXAMPLE 6

Solve and graph the solution set.

$$\frac{1}{4}(x+3) + 2 \leq \frac{3}{4}(x+8)$$

Answer: $\left[-\frac{13}{2}, \infty\right)$

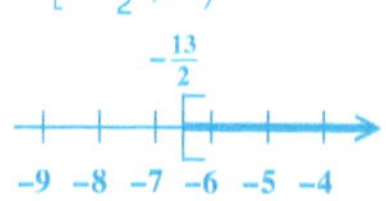

EXAMPLE 6 Solving a Linear Inequality with Fractions

Solve $-\frac{2}{3}(x-3) - \frac{1}{2} < \frac{1}{2}(5-x)$, and graph the solution set.

$$-\frac{2}{3}(x-3) - \frac{1}{2} < \frac{1}{2}(5-x)$$

Step 1	$-\frac{2}{3}x + 2 - \frac{1}{2} < \frac{5}{2} - \frac{1}{2}x$	Clear parentheses.
	$6\left(-\frac{2}{3}x + 2 - \frac{1}{2}\right) < 6\left(\frac{5}{2} - \frac{1}{2}x\right)$	To clear the fractions, multiply by 6, the LCD.
	$6\left(-\frac{2}{3}x\right) + 6(2) + 6\left(-\frac{1}{2}\right) < 6\left(\frac{5}{2}\right) + 6\left(-\frac{1}{2}x\right)$	Distributive property
	$-4x + 12 - 3 < 15 - 3x$	Multiply.
	$-4x + 9 < 15 - 3x$	Combine like terms.
Step 2	$-4x + 9 + 3x < 15 - 3x + 3x$	Add $3x$.
	$-x + 9 < 15$	Combine like terms.
	$-x + 9 - 9 < 15 - 9$	Subtract 9.
	$-x < 6$	Combine like terms.
Step 3	$-1(-x) > -1(6)$	Multiply by -1. Change $<$ to $>$.
	$x > -6$	

Reverse the inequality symbol when multiplying by a *negative* number.

Check that the solution set is $(-6, \infty)$. See the graph in **FIGURE 17**.

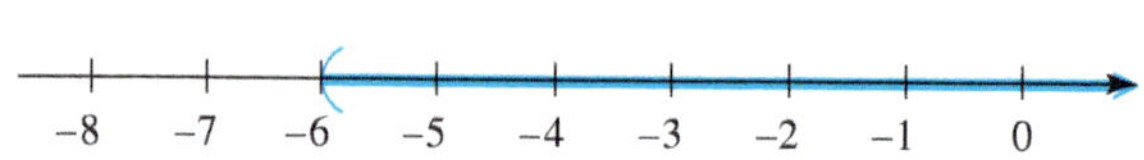

FIGURE 17

NOW TRY

NOW TRY ANSWER

6. $\left(-\frac{12}{11}, \infty\right)$

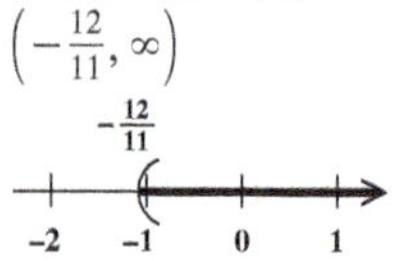

TEACHING TIP
Instructors may want to cover **Objective 4** on three-part inequalities later, in conjunction with **Section 1.7.**

OBJECTIVE 4 Solve linear inequalities with three parts.

Some applications involve a **three-part inequality** such as

$$3 < x + 2 < 8,$$

where $x + 2$ is *between* 3 and 8.

NOW TRY EXERCISE 7
Solve and graph the solution set.

$$-1 < x - 2 < 3$$

EXAMPLE 7 Solving a Three-Part Inequality

Solve $3 < x + 2 < 8$, and graph the solution set.

$$3 < x + 2 < 8$$

$$3 - 2 < x + 2 - 2 < 8 - 2 \quad \text{Subtract 2 from all three parts.}$$

$$1 < x < 6$$

Thus, x must be between 1 and 6 so that $x + 2$ will be between 3 and 8. The solution set, $(1, 6)$, is graphed in **FIGURE 18**.

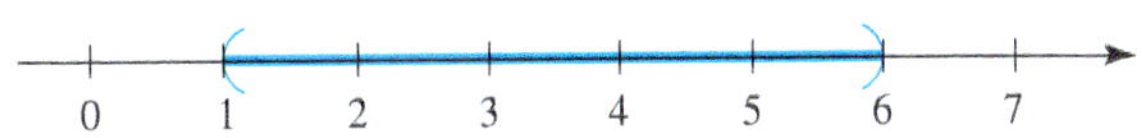

FIGURE 18

NOW TRY

CLASSROOM EXAMPLE 7
Solve and graph the solution set.

$$-4 < x - 2 < 5$$

Answer: $(-2, 7)$

TEACHING TIP
Encourage students to write three-part inequalities so that the numbers are in the same order as on the number line.
Have students rewrite the following three-part inequalities, using order on the number line.
(a) $1 > x > -1$
(b) $16 \geq x \geq 11$
(c) $-2 > x \geq -8$

CAUTION ***In three-part inequalities, the order of the parts is important.*** For example, do not write $8 < x + 2 < 3$ because this would imply that $8 < 3$, a false statement. ***Write three-part inequalities so that the symbols point in the same direction, and both point toward the lesser number.***

NOW TRY EXERCISE 8
Solve and graph the solution set.

$$-2 < -4x - 5 \leq 7$$

EXAMPLE 8 Solving a Three-Part Inequality

Solve $-2 \leq -3x - 1 \leq 5$ and graph the solution set.

$$-2 \leq -3x - 1 \leq 5$$

$$-2 + 1 \leq -3x - 1 + 1 \leq 5 + 1 \quad \text{Add 1 to each part.}$$

$$-1 \leq -3x \leq 6$$

$$\frac{-1}{-3} \geq \frac{-3x}{-3} \geq \frac{6}{-3} \quad \text{Divide each part by } -3. \text{ Reverse the direction of the inequality symbols.}$$

$$\frac{1}{3} \geq x \geq -2$$

$$-2 \leq x \leq \frac{1}{3} \quad \text{Rewrite in the order on the number line.}$$

Check that the solution set is $\left[-2, \frac{1}{3}\right]$, as shown in **FIGURE 19**.

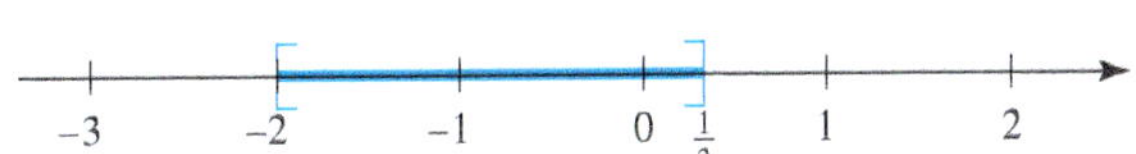

FIGURE 19

NOW TRY

CLASSROOM EXAMPLE 8
Solve and graph the solution set.

$$5 < 3x - 4 < 9$$

Answer: $\left(3, \frac{13}{3}\right)$

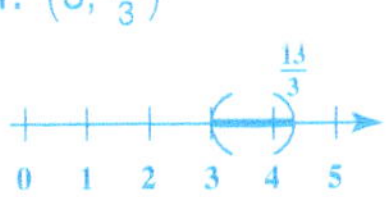

NOW TRY ANSWERS

7. $(1, 5)$

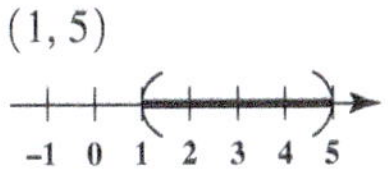

8. $\left[-3, -\frac{3}{4}\right)$

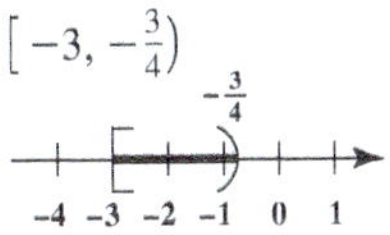

Solution Sets of Equations and Inequalities

Equation or Inequality	*Typical Solution Set*	*Graph of Solution Set*
Linear equation $5x + 4 = 14$	$\{2\}$	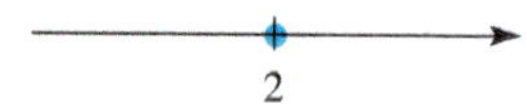
Linear inequality $5x + 4 < 14$	$(-\infty, 2)$	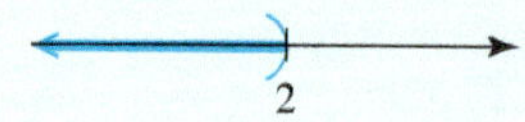
$5x + 4 > 14$	$(2, \infty)$	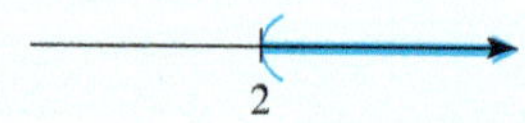
Three-part inequality $-1 \le 5x + 4 \le 14$	$[-1, 2]$	

OBJECTIVE 5 Solve applied problems using linear inequalities.

Words and Phrases That Suggest Inequality

Word Statement	*Interpretation*	*Example*	*Inequality*
a exceeds *b*.	$a > b$	Juan's age *j* exceeds 21 yr.	$j > 21$
a is at least *b*.	$a \ge b$	Juan is at least 21 yr old.	$j \ge 21$
a is no less than *b*.	$a \ge b$	Juan is no less than 21 yr old.	$j \ge 21$
a is at most *b*.	$a \le b$	Mia's age *m* is at most 10 yr.	$m \le 10$
a is no more than *b*.	$a \le b$	Mia is no more than 10 yr old.	$m \le 10$

In **Example 9,** we use the six problem-solving steps from **Section 1.3,** changing Step 3 from

"Write an equation" to "Write an inequality."

NOW TRY EXERCISE 9

A local health club charges a $40 one-time enrollment fee, plus $35 per month for a membership. Sara can spend no more than $355 on this exercise expense. What is the *maximum* number of months that Sara can belong to this health club?

CLASSROOM EXAMPLE 9

A rental company charges $5 to rent a leaf blower, plus $1.75 per hr. Marge can spend no more than $26 to blow leaves from her driveway and pool deck. What is the *maximum* amount of time she can use the rented leaf blower?

Answer: 12 hr

NOW TRY ANSWER

9. 9 months

EXAMPLE 9 Using a Linear Inequality to Solve a Rental Problem

A rental company charges $20 to rent a chain saw, plus $9 per hr. Tom can spend no more than $65 to clear some logs from his yard. What is the maximum amount of time he can use the rented saw?

Step 1 **Read** the problem again.

Step 2 **Assign a variable.** Let $x =$ the number of hours he can rent the saw.

Step 3 **Write an inequality.** He must pay $20, plus $9x, to rent the saw for x hours, and this amount must be *no more than* $65.

Cost of renting	is no more than	65 dollars.
$20 + 9x$	$\le$	65

Step 4 **Solve.**

$$9x \le 45 \quad \text{Subtract 20.}$$

$$x \le 5 \quad \text{Divide by 9.}$$

Step 5 **State the answer.** He can use the saw for a maximum of 5 hr. (Of course, he may use it for less time, as indicated by the inequality $x \le 5$.)

Step 6 **Check.** If Tom uses the saw for 5 hr, he will spend

$$20 + 9(5) = 65 \text{ dollars, the maximum amount.}$$

NOW TRY

NOW TRY EXERCISE 10

Joel has scores of 82, 97, and 93 on his first three exams. What score must he earn on the fourth exam to keep an average of at least 90?

CLASSROOM EXAMPLE 10

Abby has scores of 92, 90, and 84 on her first three tests. What score must she make on her fourth test in order to keep an average of at least 90?

Answer: at least 94

EXAMPLE 10 Finding an Average Test Score

Martha has scores of 88, 86, and 90 on her first three algebra tests. An average score of at least 90 will earn an A in the class. What possible scores on her fourth test will earn her an A average?

Let $x =$ the score on the fourth test. Her average score must be at least 90. To find the average of four numbers, add them and then divide by 4.

$$\underbrace{\frac{88 + 86 + 90 + x}{4}}_{\text{Average}} \underbrace{\geq}_{\text{is at least}} \underbrace{90}_{90.}$$

$$\frac{264 + x}{4} \geq 90 \quad \text{Add the scores.}$$

$$264 + x \geq 360 \quad \text{Multiply by 4.}$$

$$x \geq 96 \quad \text{Subtract 264.}$$

She must score 96 or more on her fourth test.

CHECK $\frac{88 + 86 + 90 + 96}{4} = \frac{360}{4} = 90$, the minimum score. ✓

A score of 96 or more will give an average of at least 90, as required. **NOW TRY**

NOW TRY ANSWER

10. at least 88

1.5 Exercises

FOR EXTRA HELP

MyMathLab®

Complete solution available in MyMathLab

1. D **2.** C **3.** B **4.** A **5.** F **6.** E

7. **(a)** $x < 100$ **(b)** $100 \leq x \leq 129$ **(c)** $130 \leq x \leq 159$ **(d)** $160 \leq x \leq 189$ **(e)** $x \geq 190$

8. **(a)** $x < 100$ **(b)** $100 \leq x \leq 199$ **(c)** $200 \leq x \leq 499$ **(d)** $x \geq 500$

9. $[16, \infty)$
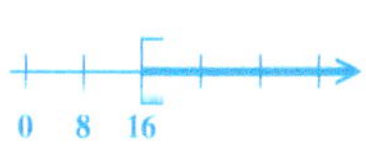

10. $[10, \infty)$
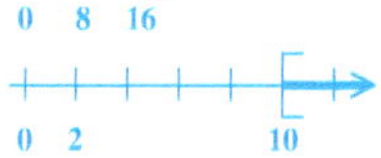

11. $(7, \infty)$
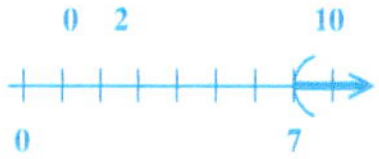

12. $(-\infty, 14)$

13. $(-\infty, -4)$
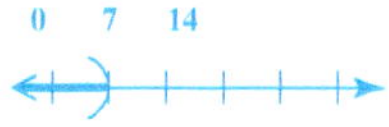

14. $(-5, \infty)$

15. $(-\infty, -40]$

Concept Check *Match each inequality in Column I with the correct graph or interval in Column II.*

I

1. $x \leq 3$

2. $x > 3$

3. $x < 3$

4. $x \geq 3$

5. $-3 \leq x \leq 3$

6. $-3 < x < 3$

II

A.
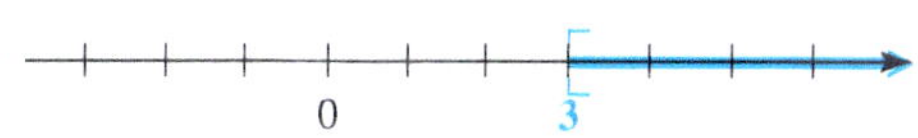

B.
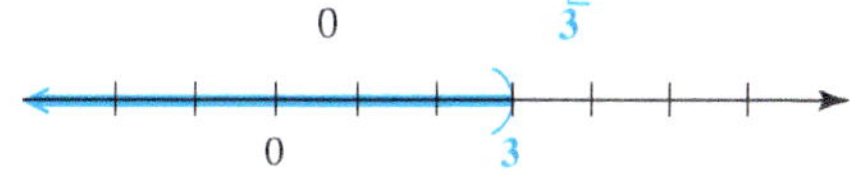

C. $(3, \infty)$

D. $(-\infty, 3]$

E. $(-3, 3)$

F. $[-3, 3]$

Concept Check *Work each problem involving inequalities.*

7. A high level of LDL cholesterol ("bad cholesterol") in the blood increases a person's risk of heart disease. The table shows how LDL levels affect risk.

If x represents the LDL cholesterol number, write a linear inequality or three-part inequality for each category. Use x as the variable.

(a) Optimal

(b) Near optimal/above optimal

(c) Borderline high

(d) High **(e)** Very high

LDL Cholesterol	*Risk Category*
Less than 100	Optimal
100–129	Near optimal/ above optimal
130–159	Borderline high
160–189	High
190 and above	Very high

Source: Cholesterol & Triglycerides Health Center, WebMD, published by WebMD, © 2014.

16. $[-18, \infty)$ 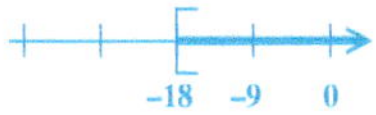

17. $(-\infty, 4]$ 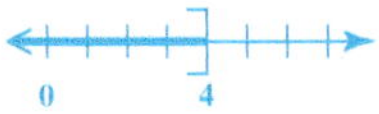

18. $[0.5, \infty)$ 

19. $(-\infty, -10]$ 

20. $(-\infty, -8]$ 

21. $(-\infty, 14)$ 

22. $(7, \infty)$ 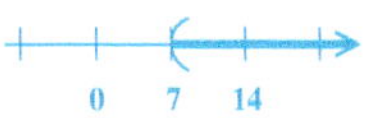

23. $\left(-\infty, -\frac{15}{2}\right)$

24. $\left(-\frac{28}{3}, \infty\right)$ 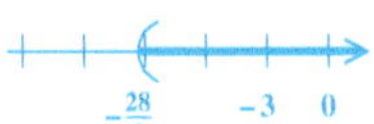

25. $\left[\frac{1}{2}, \infty\right)$ 

26. $[2, \infty)$ 

27. $[2, \infty)$ 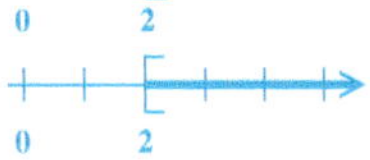

28. $\left[-\frac{1}{2}, \infty\right)$ 

29. $(3, \infty)$ 

30. $[6, \infty)$ 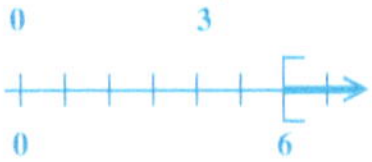

31. $(-\infty, 4)$ 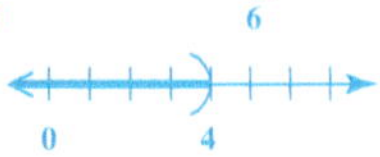

32. $[-3, \infty)$ 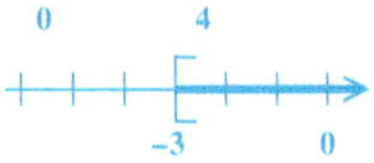

33. $\left(-\infty, \frac{23}{6}\right]$ 

34. $\left(-\infty, \frac{71}{150}\right)$

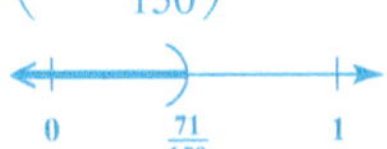

35. $\left(-\infty, \frac{76}{11}\right)$ 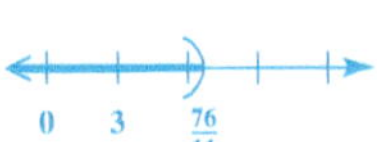

36. $\left(-\infty, \frac{49}{2}\right]$ 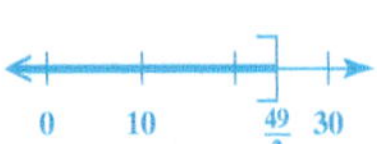

37. $(-\infty, \infty)$ 

38. $(-\infty, \infty)$ 

39. $\emptyset$ **40.** $\emptyset$

41. A; B

8. A high level of triglycerides in the blood also increases a person's risk of heart disease. The table shows how triglyceride levels affect risk.

If x represents the triglycerides number, write a linear inequality or three-part inequality for each category. Use x as the variable.

(a) Normal **(b)** Mildly high

(c) High **(d)** Very high

Triglycerides	*Risk Category*
Less than 100	Normal
100–199	Mildly high
200–499	High
500 or higher	Very high

Source: Cholesterol & Triglycerides Health Center, WebMD, published by WebMD, © 2014.

Solve each inequality. Give the solution set in both interval and graph forms. ***See Examples 1–6.***

9. $x - 4 \geq 12$ **10.** $x - 3 \geq 7$ **11.** $3k + 1 > 22$ **12.** $5x + 6 < 76$

13. $4x < -16$ **14.** $2x > -10$ **15.** $-\frac{3}{4}x \geq 30$ **16.** $-\frac{2}{3}x \leq 12$

17. $-1.3x \geq -5.2$ **18.** $-2.5x \leq -1.25$ **19.** $5x + 2 \leq -48$ **20.** $4x + 1 \leq -31$

21. $\frac{5x - 6}{8} < 8$ **22.** $\frac{3x - 1}{4} > 5$ **23.** $\frac{2x - 5}{-4} > 5$ **24.** $\frac{3x - 2}{-5} < 6$

25. $6x - 4 \geq -2x$ **26.** $2x - 8 \geq -2x$ **27.** $x - 2(x - 4) \leq 3x$

28. $x - 3(x + 1) \leq 4x$ **29.** $-(4 + r) + 2 - 3r < -14$ **30.** $-(9 + x) - 5 + 4x \geq 4$

31. $-3(x - 6) > 2x - 2$ **32.** $-2(x + 4) \leq 6x + 16$

33. $\frac{2}{3}(3x - 1) \geq \frac{3}{2}(2x - 3)$ **34.** $\frac{7}{5}(10x - 1) < \frac{2}{3}(6x + 5)$

35. $-\frac{1}{4}(p + 6) + \frac{3}{2}(2p - 5) < 10$ **36.** $\frac{3}{5}(t - 2) - \frac{1}{4}(2t - 7) \leq 3$

37. $3(2x - 4) - 4x < 2x + 3$ **38.** $7(4 - x) + 5x < 2(16 - x)$

39. $8\left(\frac{1}{2}x + 3\right) < 8\left(\frac{1}{2}x - 1\right)$ **40.** $10\left(\frac{1}{5}x + 2\right) < 10\left(\frac{1}{5}x + 1\right)$

41. *Concept Check* Which one is the graph of $-2 < x$? Of $-x > 2$?

A. 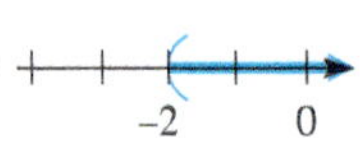 **B.** 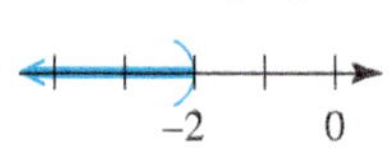 **C.** 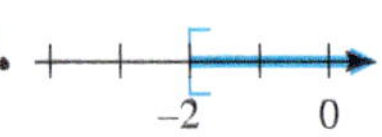 **D.** 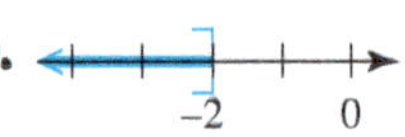

42. *Concept Check* A student solved the following inequality as shown.

$$4x \geq -64$$

$$\frac{4x}{4} \leq \frac{-64}{4}$$

$$x \leq -16 \qquad \text{Solution set: } (-\infty, -16]$$

WHAT WENT WRONG? Give the correct solution set.

Solve each inequality. Give the solution set in both interval and graph forms. ***See Examples 7 and 8.***

43. $-4 < x - 5 < 6$ **44.** $-1 < x + 1 < 8$ **45.** $-9 \leq x + 5 \leq 15$

46. $-4 \leq x + 3 \leq 10$ **47.** $-6 \leq 2x + 4 \leq 16$ **48.** $-15 < 3x + 6 < -12$

49. $-19 \leq 3x - 5 \leq 1$ **50.** $-16 < 3x + 2 < -10$ **51.** $4 \leq -9x + 5 < 8$

52. $4 \leq -2x + 3 < 8$ **53.** $-8 \leq -4x + 2 \leq 6$ **54.** $-12 \leq -6x + 3 \leq 15$

55. $-1 \leq \frac{2x - 5}{6} \leq 5$ **56.** $-3 \leq \frac{3x + 1}{4} \leq 3$

42. Because $4 > 0$, the student should not have reversed the direction of the inequality symbol when dividing by 4. We reverse the symbol only when multiplying or dividing by a *negative* number. The solution set is $[-16, \infty)$.

43. $(1, 11)$

44. $(-2, 7)$

45. $[-14, 10]$

46. $[-7, 7]$

47. $[-5, 6]$

48. $(-7, -6)$

49. $\left[-\frac{14}{3}, 2\right]$
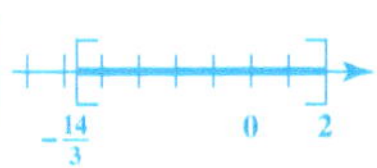

50. $(-6, -4)$

51. $\left(-\frac{1}{3}, \frac{1}{9}\right]$

52. $\left(-\frac{5}{2}, -\frac{1}{2}\right]$

53. $\left[-1, \frac{5}{2}\right]$

54. $\left[-2, \frac{5}{2}\right]$

55. $\left[-\frac{1}{2}, \frac{35}{2}\right]$
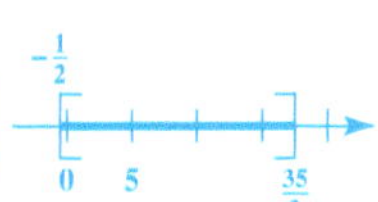

56. $\left[-\frac{13}{3}, \frac{11}{3}\right]$

57. $(0, 1)$ **58.** $(-3, -2)$
59. $(-2, 2)$ **60.** $(-6, 4)$
61. $[3, \infty)$ **62.** $[13, \infty)$
63. $[-9, \infty)$ **64.** $(-\infty, 4]$

65. at least 80 **66.** at least 82
67. 26 months **68.** 167 mi
69. more than 4.4 in.
70. more than 3.3 in.
71. **(a)** 140 to 184 lb
(b) 107 to 141 lb
(c) Answers will vary.

Give, in interval notation, the unknown numbers in each description.

57. A number is between 0 and 1.

58. A number is between -3 and -2.

59. Six times a number is between -12 and 12.

60. Half a number is between -3 and 2.

61. When 1 is added to twice a number, the result is greater than or equal to 7.

62. If 8 is subtracted from a number, then the result is at least 5.

63. One third of a number is added to 6, giving a result of at least 3.

64. Three times a number, minus 5, is no more than 7.

Solve each problem. ***See Examples 9 and 10.***

65. Faith earned scores of 90 and 82 on her first two tests in English literature. What score must she make on her third test to keep an average of 84 or greater?

66. Greg scored 92 and 96 on his first two tests in "Methods in Teaching Mathematics." What score must he make on his third test to keep an average of 90 or greater?

67. Amber is signing up for cell phone service. She must decide between Plan A, which costs \$54.99 per month with a free phone included, and Plan B, which costs \$49.99 per month, but would require her to buy a phone for \$129. Under either plan, Amber does not expect to go over the included number of monthly minutes. After how many months would Plan B be a better deal?

68. Newlyweds Bryce and Lauren need to rent a truck to move their belongings to their new apartment. They can rent a truck of the size they need from U-Haul for \$29.95 per day plus 28 cents per mile or from Budget Truck Rentals for \$34.95 per day plus 25 cents per mile. After how many miles (to the nearest mile) would the Budget rental be a better deal than the U-Haul one?

69. The average monthly precipitation for Dallas, TX, for October, November, and December is 3.47 in. If 2.88 in. falls in October and 3.13 in. falls in November, how many inches must fall in December so that the average monthly precipitation for these months exceeds 3.47 in.? (*Source:* www.weather.com)

70. The average monthly precipitation for Honolulu, HI, for October, November, and December is 3.11 in. If 2.98 in. falls in October and 3.05 in. falls in November, how many inches must fall in December so that the average monthly precipitation for these months exceeds 3.11 in.? (*Source:* www.weather.com)

71. A body mass index (BMI) between 19 and 25 is considered healthy. Use the formula

$$\text{BMI} = \frac{704 \times (\text{weight in pounds})}{(\text{height in inches})^2}$$

to find the weight range w, to the nearest pound, that gives a healthy BMI for each height. (*Source: Washington Post.*)

(a) 72 in. **(b)** 63 in. **(c)** Your height in inches

72. **(a)** 130 to 157 beats per min
(b) 116 to 140 beats per min
(c) Answers will vary.

73. 26 DVDs
74. 921 deliveries

72. To achieve the maximum benefit from exercising, the heart rate, in beats per minute, should be in the target heart rate (THR) zone. For a person aged A, the formula is as follows.

$$0.7(220 - A) \leq \text{THR} \leq 0.85(220 - A)$$

Find the THR to the nearest whole number for each age. (*Source:* Hockey, R. V., *Physical Fitness: The Pathway to Healthful Living,* Times Mirror/Mosby College Publishing.)

(a) 35 **(b)** 55 **(c)** Your age

A product will produce a profit only when the revenue R from selling the product exceeds the cost C of producing it. In Exercises 73 and 74, find the least whole number of units x that must be sold for the business to show a profit for the item described.

73. Peripheral Visions, Inc., finds that the cost of producing x studio-quality DVDs is $C = 20x + 100$, while the revenue produced from them is $R = 24x$ (C and R in dollars).

74. Speedy Delivery finds that the cost of making x deliveries is $C = 3x + 2300$, while the revenue produced from them is $R = 5.50x$ (C and R in dollars).

STUDY SKILLS

Using Study Cards

You may have used "flash cards" in other classes. In math, "study cards" can help you remember terms and definitions, procedures, and concepts. Use study cards to do the following.

- Help you understand and learn the material
- Quickly review when you have a few minutes
- Review before a quiz or test

One of the advantages of study cards is that you learn while you are making them.

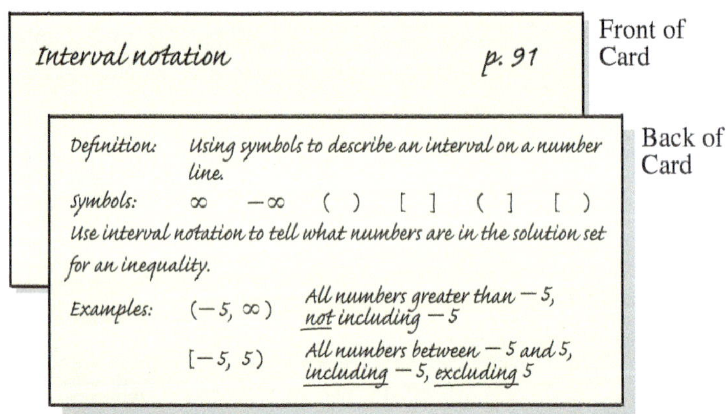

Vocabulary Cards

Put the word and a page reference on the front of the card. On the back, write the definition, an example, any related words, and a sample problem (if appropriate).

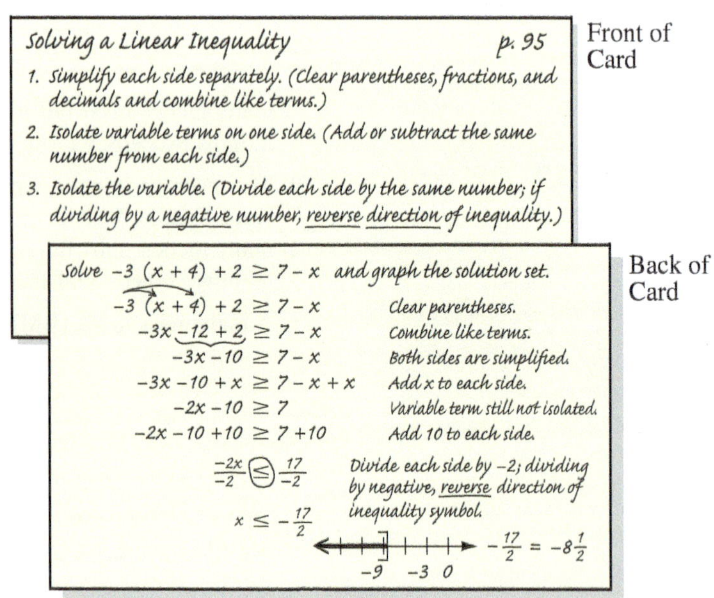

Procedure ("Steps") Cards

Write the name of the procedure on the front of the card. Then write each step in words. On the back of the card, put an example showing each step.

Make a vocabulary card and a procedure card for material you are learning now.

1.6 Set Operations and Compound Inequalities

OBJECTIVES

1. Recognize set intersection and union.
2. Find the intersection of two sets.
3. Solve compound inequalities with the word *and*.
4. Find the union of two sets.
5. Solve compound inequalities with the word *or*.

VOCABULARY

☐ intersection
☐ compound inequality
☐ union

OBJECTIVE 1 Recognize set intersection and union.

Consider the two sets A and B defined as follows.

$$A = \{1, 2, 3\}, \qquad B = \{2, 3, 4\}$$

The set of all elements that belong to both A **and** B, called their *intersection* and symbolized $\boldsymbol{A \cap B}$**,** is given by

$$A \cap B = \{2, 3\}. \quad \text{Intersection}$$

The set of all elements that belong to either A **or** B, or both, called their *union* and symbolized $\boldsymbol{A \cup B}$**,** is given by

$$A \cup B = \{1, 2, 3, 4\}. \quad \text{Union}$$

OBJECTIVE 2 Find the intersection of two sets.

The intersection of two sets is defined with the word *and*.

Intersection of Sets

For any two sets A and B, the **intersection** of A and B, symbolized $A \cap B$, is defined as follows.

$$A \cap B = \{x \mid x \text{ is an element of } A \textbf{ and } x \text{ is an element of } B\}$$

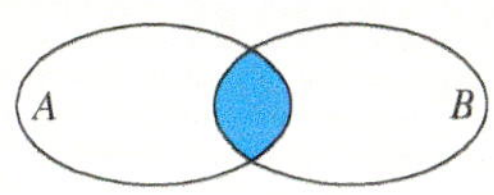

TEACHING TIP
Mention this real-life example of intersection: The intersection of two streets is the region common to *both* streets.

NOW TRY EXERCISE 1
Let $A = \{2, 4, 6, 8\}$ and $B = \{0, 2, 6, 8\}$. Find $A \cap B$.

EXAMPLE 1 Finding the Intersection of Two Sets

Let $A = \{1, 2, 3, 4\}$ and $B = \{2, 4, 6\}$. Find $A \cap B$.

The set $A \cap B$ contains those elements that belong to both A *and* B.

$$\begin{aligned} A \cap B &= \{1, 2, 3, 4\} \cap \{2, 4, 6\} \\ &= \{2, 4\} \end{aligned}$$

NOW TRY

CLASSROOM EXAMPLE 1
Let $A = \{3, 4, 5, 6\}$ and $B = \{5, 6, 7\}$. Find $A \cap B$.
Answer: $\{5, 6\}$

OBJECTIVE 3 Solve compound inequalities with the word *and*.

A **compound inequality** consists of two inequalities linked by a connective word.

$$\begin{aligned} x + 1 \le 9 \quad &\text{and} \quad x - 2 \ge 3 \\ 2x > 4 \quad &\text{or} \quad 3x - 6 < 5 \end{aligned} \qquad \text{Compound inequalities}$$

Solving a Compound Inequality with *and*

Step 1 Solve each inequality individually.

Step 2 Because the inequalities are joined with *and,* the solution set of the compound inequality will include all numbers that satisfy both inequalities in Step 1 (the *intersection* of the solution sets).

NOW TRY ANSWER
1. $\{2, 6, 8\}$

NOW TRY EXERCISE 2

Solve the compound inequality, and graph the solution set.

$$x - 2 \leq 5 \quad \text{and} \quad x + 5 \geq 9$$

CLASSROOM EXAMPLE 2

Solve the compound inequality, and graph the solution set.

$$x + 3 < 1 \quad \text{and} \quad x - 4 > -12$$

Answer: $(-8, -2)$

TEACHING TIP

It is helpful to draw one number line, draw the graphs of the inequalities above the number line, and then graph the intersection on the original number line. The intersection represents the set of points where the two original graphs "overlap." Remind students to pay attention to the graphs of the endpoint(s).

EXAMPLE 2 Solving a Compound Inequality with *and*

Solve the compound inequality, and graph the solution set.

$$x + 1 \leq 9 \quad \text{and} \quad x - 2 \geq 3$$

Step 1 Solve each inequality individually.

$$x + 1 \leq 9 \quad \text{and} \quad x - 2 \geq 3$$

$$x + 1 - 1 \leq 9 - 1 \quad \text{and} \quad x - 2 + 2 \geq 3 + 2$$

$$x \leq 8 \quad \text{and} \quad x \geq 5$$

Step 2 The solution set will include all numbers that satisfy *both* inequalities in Step 1 at the same time. The compound inequality is true whenever $x \leq 8$ and $x \geq 5$ are both true. See the graphs in **FIGURE 20**.

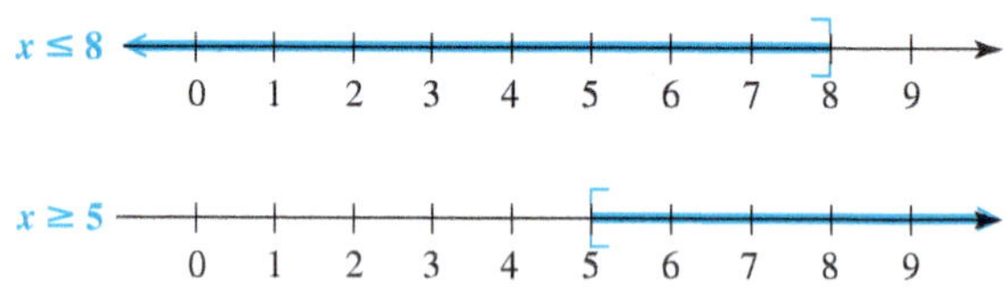

The set of points where the graphs "overlap" represents the intersection.

FIGURE 20

The intersection of the two graphs in **FIGURE 20** is the solution set. **FIGURE 21** shows this solution set, $[5, 8]$.

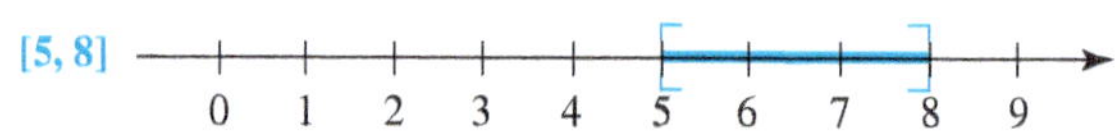

FIGURE 21

NOW TRY

NOW TRY EXERCISE 3

Solve and graph.

$$-4x - 1 < 7 \quad \text{and} \quad 3x + 4 \geq -5$$

CLASSROOM EXAMPLE 3

Solve and graph.

$$2x \leq 4x + 8 \quad \text{and} \quad 3x \geq -9$$

Answer: $[-3, \infty)$

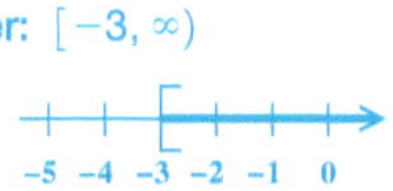

EXAMPLE 3 Solving a Compound Inequality with *and*

Solve the compound inequality, and graph the solution set.

$$-3x - 2 > 5 \quad \text{and} \quad 5x - 1 \leq -21$$

Step 1 Solve each inequality individually.

$$-3x - 2 > 5 \quad \text{and} \quad 5x - 1 \leq -21$$

$$-3x > 7 \quad \text{and} \quad 5x \leq -20$$

$$x < -\frac{7}{3} \quad \text{and} \quad x \leq -4$$

Remember to reverse the direction of the inequality symbol.

The graphs of $x < -\frac{7}{3}$ and $x \leq -4$ are shown in **FIGURE 22**.

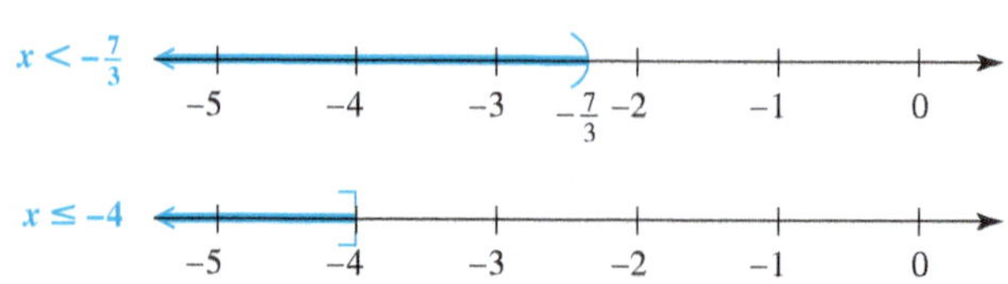

FIGURE 22

Step 2 Now find all values of x that are less than $-\frac{7}{3}$ and also less than or equal to -4. As shown in **FIGURE 23**, the solution set is $(-\infty, -4]$.

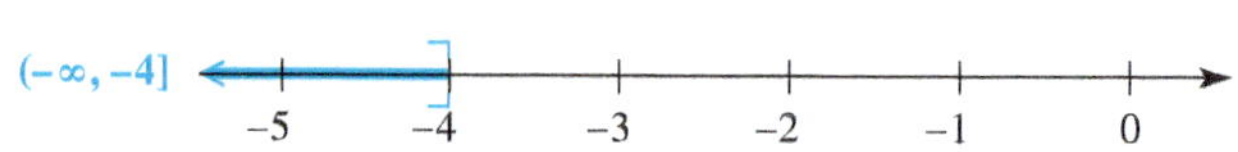

FIGURE 23

NOW TRY

NOW TRY ANSWERS

2. $[4, 7]$

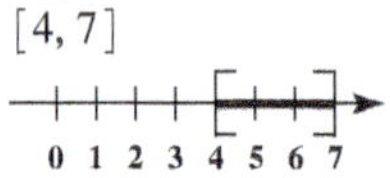

3. $(-2, \infty)$

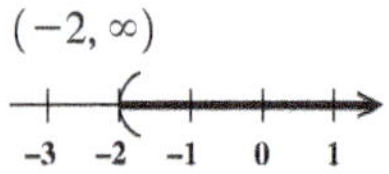

NOW TRY EXERCISE 4

Solve and graph.

$$x - 7 < -12 \quad \text{and} \quad 2x + 1 > 5$$

CLASSROOM EXAMPLE 4

Solve and graph.

$x + 2 > 3$ and $2x + 1 < -3$

Answer: $\varnothing$

EXAMPLE 4 Solving a Compound Inequality with *and*

Solve the compound inequality, and graph the solution set.

$$x + 2 < 5 \quad \text{and} \quad x - 10 > 2$$

Step 1 Solve each inequality individually.

$$x + 2 < 5 \quad \text{and} \quad x - 10 > 2$$
$$x < 3 \quad \text{and} \quad x > 12$$

The graphs of $x < 3$ and $x > 12$ are shown in **FIGURE 24**.

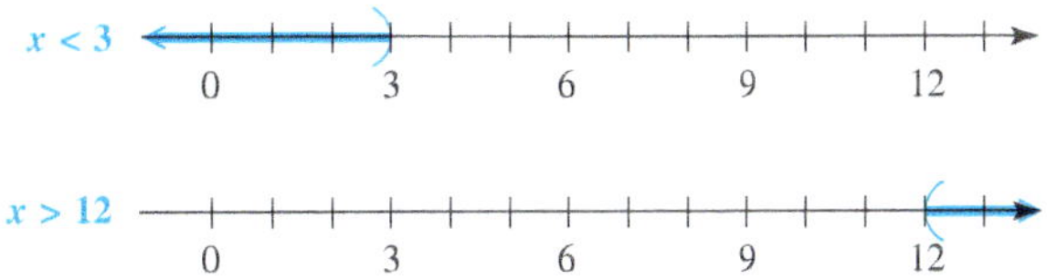

FIGURE 24

Step 2 There is no number that is both less than 3 *and* greater than 12, so the given compound inequality has no solution. The solution set is $\varnothing$. See **FIGURE 25**.

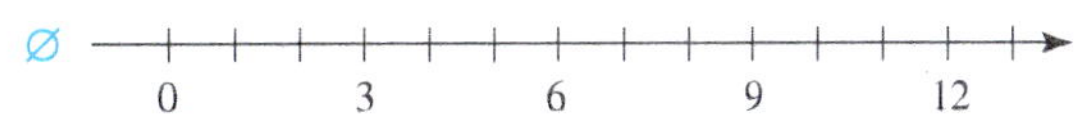

FIGURE 25

NOW TRY

OBJECTIVE 4 Find the union of two sets.

The union of two sets is defined with the word *or*.

Union of Sets

For any two sets A and B, the **union** of A and B, symbolized $A \cup B$, is defined as follows.

$$A \cup B = \{x \mid x \text{ is an element of } A \textbf{ or } x \text{ is an element of } B\}$$

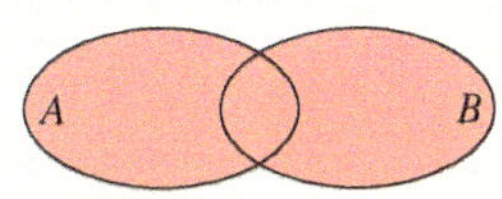

NOW TRY EXERCISE 5

Let $A = \{5, 10, 15, 20\}$ and $B = \{5, 15, 25\}$. Find $A \cup B$.

CLASSROOM EXAMPLE 5

Let $A = \{3, 4, 5, 6\}$ and $B = \{5, 6, 7\}$. Find $A \cup B$.

Answer: $\{3, 4, 5, 6, 7\}$

EXAMPLE 5 Finding the Union of Two Sets

Let $A = \{1, 2, 3, 4\}$ and $B = \{2, 4, 6\}$. Find $A \cup B$.

Begin by listing all the elements of set A: 1, 2, 3, 4. Then list any additional elements from set B. In this case the elements 2 and 4 are already listed, so the only additional element is 6.

$$A \cup B = \{1, 2, 3, 4\} \cup \{2, 4, 6\}$$
$$= \{1, 2, 3, 4, 6\}$$

The union consists of all elements in either A *or* B (or both). NOW TRY

NOTE Although the elements 2 and 4 appeared in both sets A and B in **Example 5,** they are written only once in $A \cup B$.

NOW TRY ANSWERS

4. $\varnothing$

5. $\{5, 10, 15, 20, 25\}$

NOW TRY EXERCISE 6

Solve and graph.

$-12x \le -24 \quad \text{or} \quad x + 9 < 8$

CLASSROOM EXAMPLE 6

Solve and graph.

$x - 1 > 2 \quad \text{or} \quad 3x + 5 < 2x + 6$

Answer: $(-\infty, 1) \cup (3, \infty)$

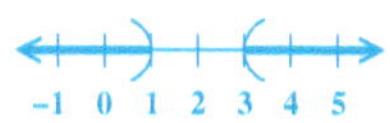

TEACHING TIP

Point out how the two number lines in **FIGURE 26** show the individual solution sets for the two separate parts of the inequality $x < 1$ or $x \ge 3$. Then explain that **FIGURE 27,** the solution set of the union, shows that any point in either of the number lines in **FIGURE 26** is in the final solution set. It is helpful to draw all three of these solution sets one above the other, using the same scale on each.

Loosely speaking, when finding the union of two sets, we "put them together and take whatever is there." Again, remind students to be sure to note whether the endpoints are included.

OBJECTIVE 5 Solve compound inequalities with the word *or*.

Solving a Compound Inequality with *or*

Step 1 Solve each inequality individually.

Step 2 Because the inequalities are joined with *or,* the solution set of the compound inequality includes all numbers that satisfy either one of the two inequalities in Step 1 (the *union* of the solution sets).

EXAMPLE 6 Solving a Compound Inequality with *or*

Solve the compound inequality, and graph the solution set.

$$6x - 4 < 2x \quad \text{or} \quad -3x \le -9$$

Step 1 Solve each inequality individually.

$$6x - 4 < 2x \quad \text{or} \quad -3x \le -9$$

$$4x < 4$$

$$x < 1 \quad \text{or} \quad x \ge 3$$

Remember to reverse the inequality symbol.

The graphs of these two inequalities are shown in **FIGURE 26.**

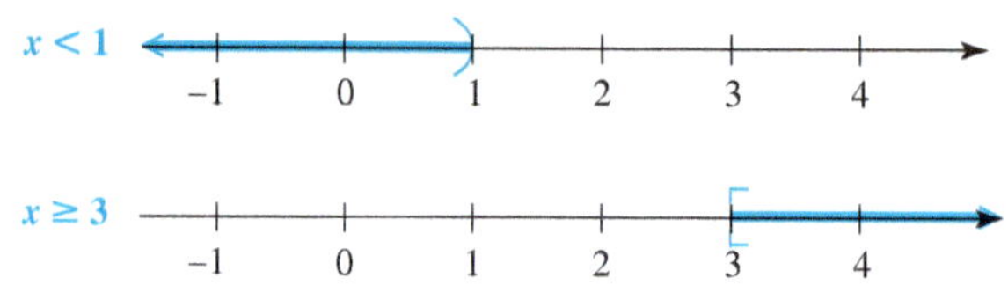

The set of points in either of the graphs represents the union.

FIGURE 26

Step 2 Because the inequalities are joined with *or,* find the union of the two solution sets. The union is the disjoint interval in **FIGURE 27.**

$$(-\infty, 1) \cup [3, \infty)$$

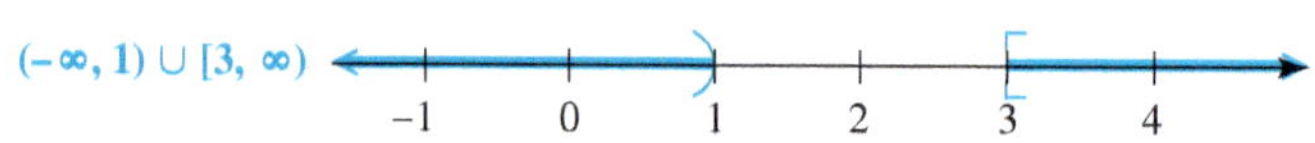

FIGURE 27

Always pay particular attention to the end points of the solution sets and whether parentheses, brackets, or one of each should be used. **NOW TRY**

CAUTION When inequalities are used to write the solution set in **Example 6,** it *must* be written using two separate inequalities.

$$x < 1 \quad \text{or} \quad x \ge 3,$$

Writing $3 \le x < 1$, which translates using *and,* would imply that

$$3 \le 1, \quad \text{which is } \textbf{\textit{FALSE.}}$$

NOW TRY ANSWER

6. $(-\infty, -1) \cup [2, \infty)$

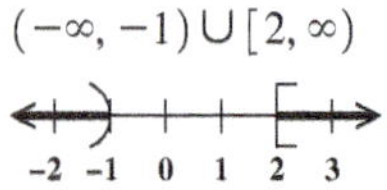

NOW TRY EXERCISE 7

Solve and graph.

$-x + 2 < 6$ or $6x - 8 \geq 10$

CLASSROOM EXAMPLE 7

Solve and graph.

$3x - 2 \leq 13$ or $x + 5 \leq 7$

Answer: $(-\infty, 5]$

EXAMPLE 7 Solving a Compound Inequality with *or*

Solve the compound inequality, and graph the solution set.

$$-4x + 1 \geq 9 \quad \text{or} \quad 5x + 3 \leq -12$$

Step 1 Solve each inequality individually.

$$\begin{aligned} -4x + 1 &\geq 9 \quad &\text{or}\quad 5x + 3 &\leq -12 \\ -4x &\geq 8 \quad &\text{or}\quad 5x &\leq -15 \\ x &\leq -2 \quad &\text{or}\quad x &\leq -3 \end{aligned}$$

The graphs of these two inequalities are shown in **FIGURE 28**.

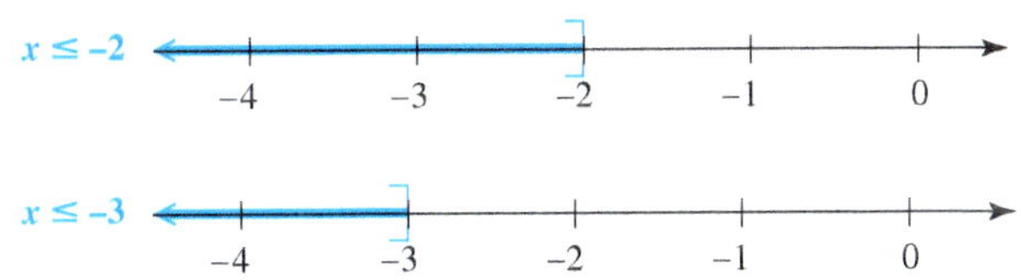

FIGURE 28

Step 2 We take the union to obtain $(-\infty, -2]$. See **FIGURE 29**.

FIGURE 29

NOW TRY

NOW TRY EXERCISE 8

Solve and graph.

$8x - 4 \geq 20$ or $-2x + 1 > -9$

CLASSROOM EXAMPLE 8

Solve and graph.

$3x - 2 \leq 13$ or $x + 5 \geq 7$

Answer: $(-\infty, \infty)$

EXAMPLE 8 Solving a Compound Inequality with *or*

Solve the compound inequality, and graph the solution set.

$$-2x + 5 \geq 11 \quad \text{or} \quad 4x - 7 \geq -27$$

Step 1 Solve each inequality individually.

$$\begin{aligned} -2x + 5 &\geq 11 \quad &\text{or}\quad 4x - 7 &\geq -27 \\ -2x &\geq 6 \quad &\text{or}\quad 4x &\geq -20 \\ x &\leq -3 \quad &\text{or}\quad x &\geq -5 \end{aligned}$$

The graphs of these two inequalities are shown in **FIGURE 30**.

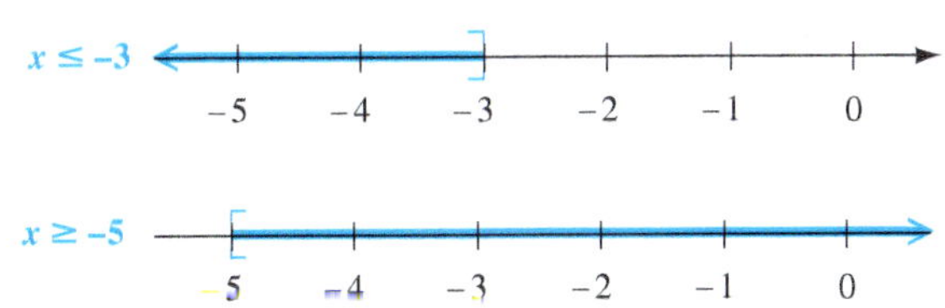

FIGURE 30

Step 2 By taking the union, we obtain every real number as a solution because every real number satisfies at least one of the two inequalities. The set of all real numbers is written in interval notation as $(-\infty, \infty)$ and graphed as in **FIGURE 31**.

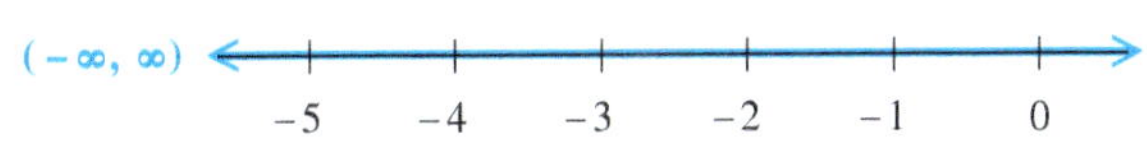

FIGURE 31

NOW TRY

NOW TRY ANSWERS

7. $(-4, \infty)$

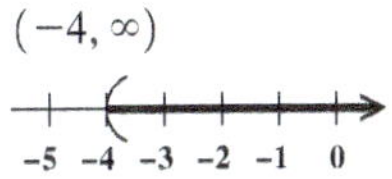

8. $(-\infty, \infty)$

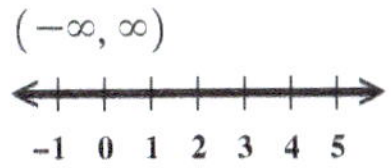

NOW TRY EXERCISE 9

In **Example 9,** list the elements of each set.

(a) The set of countries to which exports were greater than \$100,000 million and from which imports were less than \$400,000 million

(b) The set of countries to which exports were less than \$200,000 million or from which imports were greater than \$250,000 million

CLASSROOM EXAMPLE 9

In Example 9, list the elements of each set.

(a) The set of countries to which exports were greater than \$150,000 million and from which imports were less than \$150,000 million

(b) The set of countries to which exports were between \$50,000 million and \$150,000 million or from which imports were greater than \$300,000 million

Answers: (a) $\varnothing$

(b) {Canada, China, Japan}

NOW TRY ANSWERS

9. (a) {Canada, Mexico}

(b) {Canada, China, Mexico, Japan, Germany}

EXAMPLE 9 Applying Intersection and Union

The five top U.S. trading partners for 2012 are listed in the table. Amounts are in millions of dollars. (*Source:* U.S. Census Bureau.)

Country	*U.S. Exports to Country*	*U.S. Imports from Country*
Canada	292,540	323,937
China	110,484	425,579
Mexico	215,931	277,570
Japan	69,955	146,392
Germany	48,797	108,708

List the elements of the following sets.

(a) The set of countries to which exports were greater than \$200,000 million *and* from which imports were less than \$300,000 million

The only country that satisfies both conditions is Mexico, so the set is

$$\{\text{Mexico}\}.$$

(b) The set of countries to which exports were less than \$100,000 million *or* from which imports were greater than \$200,000 million

Here, any country that satisfies at least one of the conditions is in the set. This set includes all five countries:

$$\{\text{Canada, China, Mexico, Japan, Germany}\}.$$

NOW TRY

1.6 Exercises

FOR EXTRA HELP MyMathLab®

Complete solution available in MyMathLab

Concept Check *Decide whether each statement is* true *or* false. *If it is false, explain why.*

1. The union of the solution sets of $x + 1 = 6$, $x + 1 < 6$, and $x + 1 > 6$ is $(-\infty, \infty)$.

2. The intersection of the sets $\{x \mid x \geq 9\}$ and $\{x \mid x \leq 9\}$ is $\varnothing$.

3. The union of the sets $(-\infty, 7)$ and $(7, \infty)$ is $\{7\}$.

4. The intersection of the sets $(-\infty, 7]$ and $[7, \infty)$ is $\{7\}$.

5. The intersection of the set of rational numbers and the set of irrational numbers is $\{0\}$.

6. The union of the set of rational numbers and the set of irrational numbers is the set of real numbers.

Let $A = \{1, 2, 3, 4, 5, 6\}$, $B = \{1, 3, 5\}$, $C = \{1, 6\}$, *and* $D = \{4\}$. *Find each set. See* ***Examples 1 and 5.***

7. $B \cap A$ **8.** $A \cap B$ **9.** $A \cap D$ **10.** $B \cap C$

11. $B \cap \varnothing$ **12.** $A \cap \varnothing$ **13.** $A \cup B$ **14.** $B \cup D$

1. true
2. false; The intersection is $\{9\}$.
3. false; The union is $(-\infty, 7) \cup (7, \infty)$.
4. true
5. false; The intersection is $\varnothing$.
6. true
7. $\{1, 3, 5\}$, or B
8. $\{1, 3, 5\}$, or B
9. $\{4\}$, or D **10.** $\{1\}$
11. $\varnothing$ **12.** $\varnothing$
13. $\{1, 2, 3, 4, 5, 6\}$, or A
14. $\{1, 3, 4, 5\}$

15. 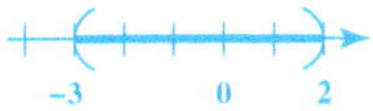

16. 

17. 

18. 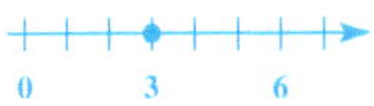

19. $(-3, 2)$ 

20. $(0, 5)$ 

21. $(-\infty, 2]$ 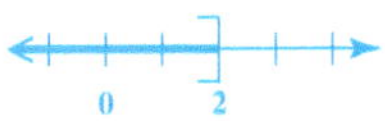

22. $[6, \infty)$ 

23. $\varnothing$ 24. $\varnothing$

25. $[5, 9]$ 

26. $[2, 6]$ 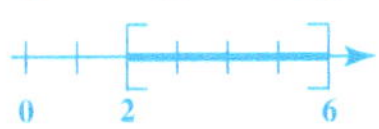

27. $(-3, -1)$ 

28. $(-1, 4)$ 

29. $(-\infty, 4]$ 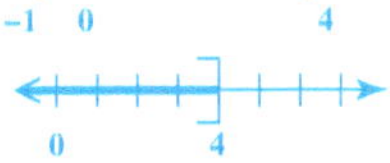

30. $(-\infty, 6]$ 

31. 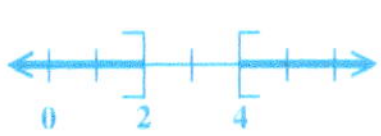

32. 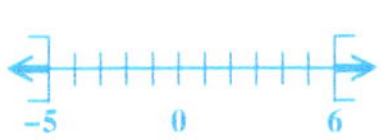

33. 

34. 

35. $(-\infty, 8]$ 

36. $[1, \infty)$ 

37. $[-2, \infty)$ 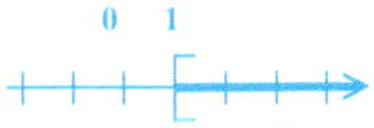

38. $(-\infty, 6]$ 

39. $(-\infty, \infty)$ 

40. $(-\infty, \infty)$ 

41. $(-\infty, -5) \cup (5, \infty)$ 

42. $(-\infty, -2) \cup (2, \infty)$ 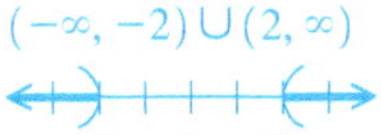

Concept Check *Two sets are specified by graphs. Graph the intersection of the two sets.*

15. 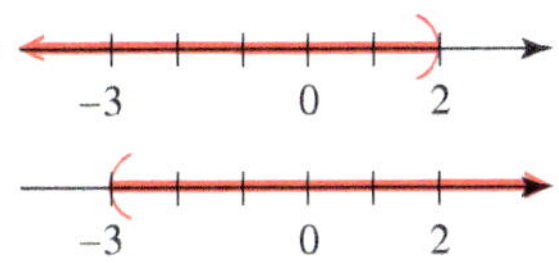

16. 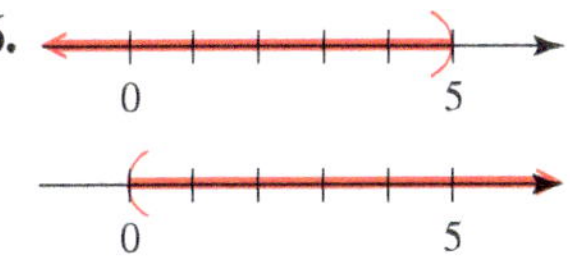

17. 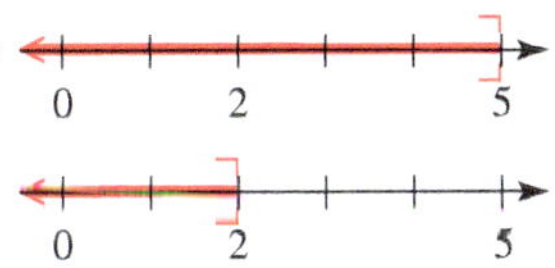

18. 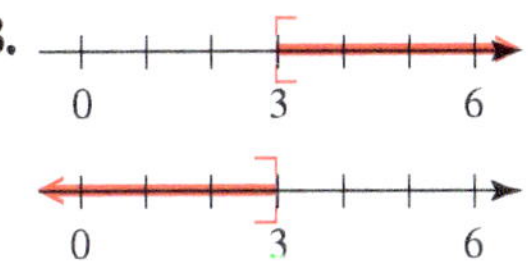

Solve each compound inequality. Give the solution set in both interval and graph forms. ***See Examples 2–4.***

19. $x < 2$ and $x > -3$

20. $x < 5$ and $x > 0$

21. $x \leq 2$ and $x \leq 5$

22. $x \geq 3$ and $x \geq 6$

23. $x \leq 3$ and $x \geq 6$

24. $x \leq -1$ and $x \geq 3$

25. $x - 3 \leq 6$ and $x + 2 \geq 7$

26. $x + 5 \leq 11$ and $x - 3 \geq -1$

27. $-3x > 3$ and $x + 3 > 0$

28. $-3x < 3$ and $x + 2 < 6$

29. $3x - 4 \leq 8$ and $-4x + 1 \geq -15$

30. $7x + 6 \leq 48$ and $-4x \geq -24$

Concept Check *Two sets are specified by graphs. Graph the union of the two sets.*

31. 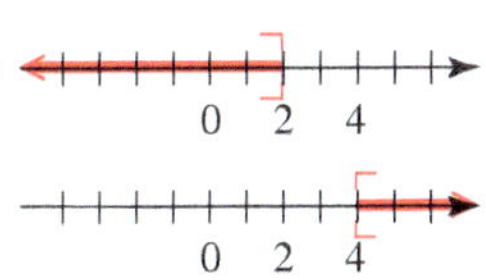

32. 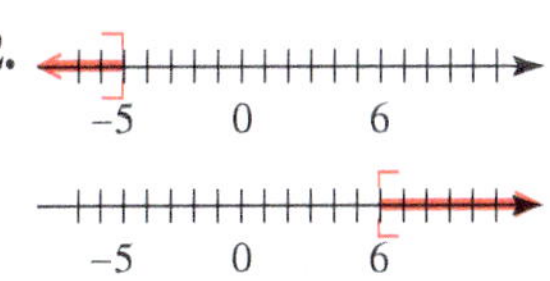

33. 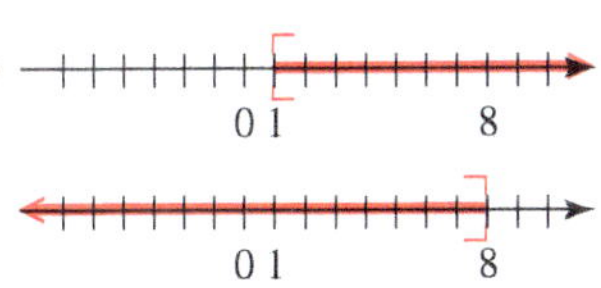

34. 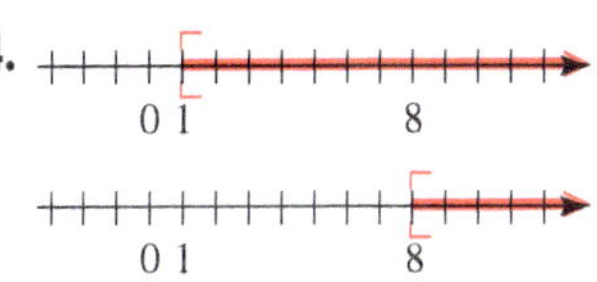

Solve each compound inequality. Give the solution set in both interval and graph forms. ***See Examples 6–8.***

35. $x \leq 1$ or $x \leq 8$

36. $x \geq 1$ or $x \geq 8$

37. $x \geq -2$ or $x \geq 5$

38. $x \leq -2$ or $x \leq 6$

39. $x \geq -2$ or $x \leq 4$

40. $x \geq 5$ or $x \leq 7$

41. $x + 2 > 7$ or $1 - x > 6$

42. $x + 1 > 3$ or $x + 4 < 2$

43. $x + 1 > 3$ or $-4x + 1 > 5$

44. $3x < x + 12$ or $x + 1 > 10$

45. $4x + 1 \geq -7$ or $-2x + 3 \geq 5$

46. $3x + 2 \leq -7$ or $-2x + 1 \leq 9$

Concept Check *Express each set in simplest interval form. (Hint: Graph each set and look for the intersection or union.)*

47. $(-\infty, -1] \cap [-4, \infty)$

48. $[-1, \infty) \cap (-\infty, 9]$

49. $(-\infty, -6] \cap [-9, \infty)$

50. $(5, 11] \cap [6, \infty)$

51. $(-\infty, 3) \cup (-\infty, -2)$

52. $[-9, 1] \cup (-\infty, -3)$

53. $[3, 6] \cup (4, 9)$

54. $[-1, 2] \cup (0, 5)$

43. $(-\infty, -1) \cup (2, \infty)$

44. $(-\infty, 6) \cup (9, \infty)$

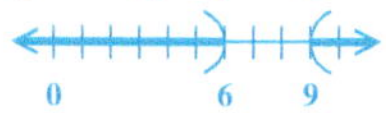

45. $(-\infty, \infty)$

46. $(-\infty, \infty)$

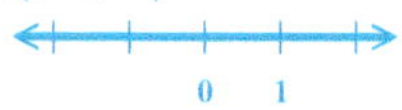

47. $[-4, -1]$ **48.** $[-1, 9]$
49. $[-9, -6]$ **50.** $[6, 11]$
51. $(-\infty, 3)$ **52.** $(-\infty, 1]$
53. $[3, 9)$ **54.** $[-1, 5)$

55. intersection; $(-5, -1)$

56. intersection; $(-1, 7)$

57. union; $(-\infty, 4)$

58. union; $(-\infty, 5)$

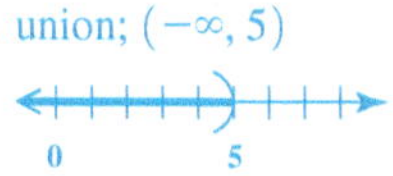

59. union; $(-\infty, 0] \cup [2, \infty)$

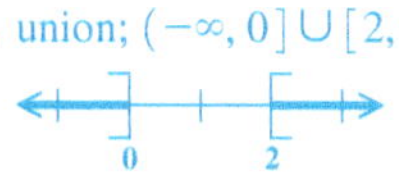

60. intersection; $[-9, -6]$

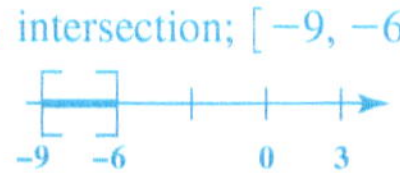

61. intersection; $[4, 12]$

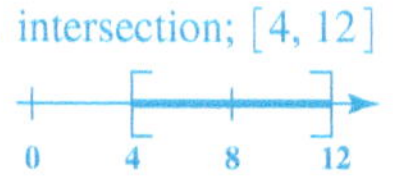

62. union; $(-\infty, -3] \cup [3, \infty)$

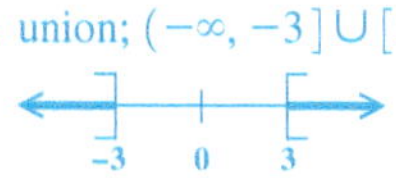

63. {Tuition and fees}
64. {Board rates}
65. {Tuition and fees, Board rates, Dormitory charges}
66. {Tuition and fees}

67. Maria, Joe **68.** none of them
69. none of them **70.** Luigi, Than
71. Maria, Joe **72.** all of them

Solve each compound inequality. Give the solution set in both interval and graph forms. ***See Examples 2–4 and 6–8.***

55. $x < -1$ and $x > -5$

56. $x > -1$ and $x < 7$

57. $x < 4$ or $x < -2$

58. $x < 5$ or $x < -3$

59. $-3x \leq -6$ or $-3x \geq 0$

60. $2x - 6 \leq -18$ and $2x \geq -18$

61. $x + 1 \geq 5$ and $x - 2 \leq 10$

62. $-8x \leq -24$ or $-5x \geq 15$

Average expenses for full-time resident college students at 4-year institutions during the 2011–2012 academic year are shown in the table.

College Expenses (in Dollars)

Type of Expense	*Public Schools (in-state)*	*Private Schools*
Tuition and fees	7701	23,479
Board rates	4052	4591
Dormitory charges	5036	5646

Source: National Center for Education Statistics.

Refer to the table, and list the elements of each set. ***See Example 9.***

63. The set of expenses that are less than \$8000 for public schools *and* are greater than \$15,000 for private schools

64. The set of expenses that are greater than \$4000 for public schools *and* are less than \$5000 for private schools

65. The set of expenses that are less than \$8000 for public schools *or* are greater than \$15,000 for private schools

66. The set of expenses that are greater than \$15,000 *or* are between \$7000 and \$8000

RELATING CONCEPTS For Individual or Group Work **(Exercises 67–72)**

The figures represent the backyards of neighbors Luigi, Maria, Than, and Joe. Find the area and the perimeter of each yard. Suppose that each resident has 150 ft of fencing and enough sod to cover 1400 ft^2 of lawn. Give the name or names of the residents whose yards satisfy each description. ***Work Exercises 67–72 in order.***

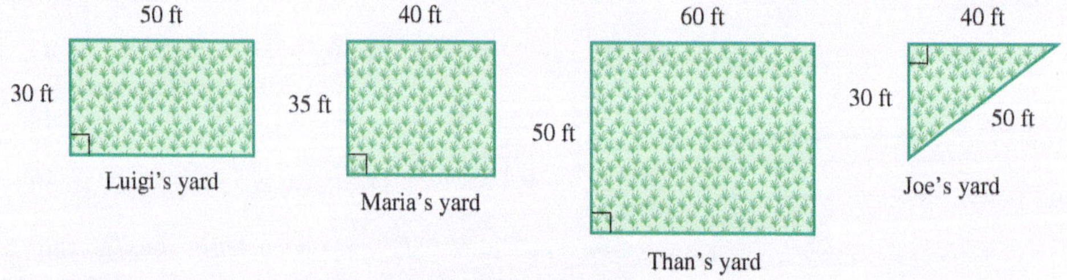

67. The yard can be fenced *and* the yard can be sodded.

68. The yard can be fenced *and* the yard cannot be sodded.

69. The yard cannot be fenced *and* the yard can be sodded.

70. The yard cannot be fenced *and* the yard cannot be sodded.

71. The yard can be fenced *or* the yard can be sodded.

72. The yard cannot be fenced *or* the yard can be sodded.

STUDY SKILLS

Using Study Cards Revisited

We introduced study cards previously. Another type of study card follows.

Practice Quiz Cards

Write a problem with direction words (like *solve, simplify*) on the front of the card, and work the problem on the back. Make one for each type of problem you learn. Use the card when you review for a quiz or test.

Make a practice quiz card for material you are learning now.

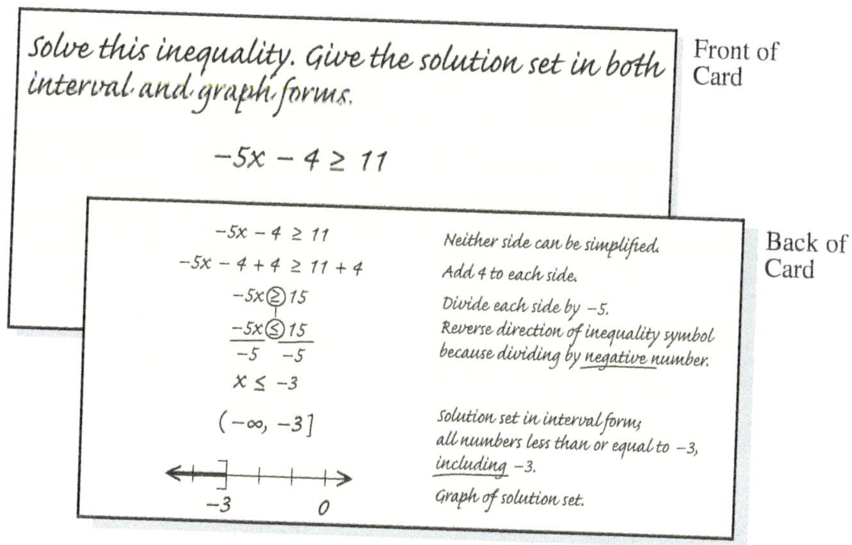

1.7 Absolute Value Equations and Inequalities

OBJECTIVES

1. Use the distance definition of absolute value.
2. Solve equations of the form $|ax + b| = k$, for $k > 0$.
3. Solve inequalities of the form $|ax + b| < k$ and of the form $|ax + b| > k$, for $k > 0$.
4. Solve absolute value equations that involve rewriting.
5. Solve equations of the form $|ax + b| = |cx + d|$.
6. Solve special cases of absolute value equations and inequalities.
7. Solve an application involving relative error.

Suppose the government of a country decides that it will comply with a restriction on greenhouse gas emissions *within* 3 years of 2020. This means that the *difference* between the year it will comply and 2020 is less than 3, *without regard to sign*. We state this mathematically as follows, where x represents the year in which it complies.

$$|x - 2020| < 3 \quad \text{Absolute value inequality}$$

We can intuitively reason that the year must be between 2017 and 2023, and thus $2017 < x < 2023$ makes this inequality true.

OBJECTIVE 1 Use the distance definition of absolute value.

In **Section R.1**, we saw that the absolute value of a number x, written $|x|$, represents the undirected distance from x to 0 on a number line. For example, the solutions of $|x| = 4$ are 4 and -4, as shown in **FIGURE 32**.

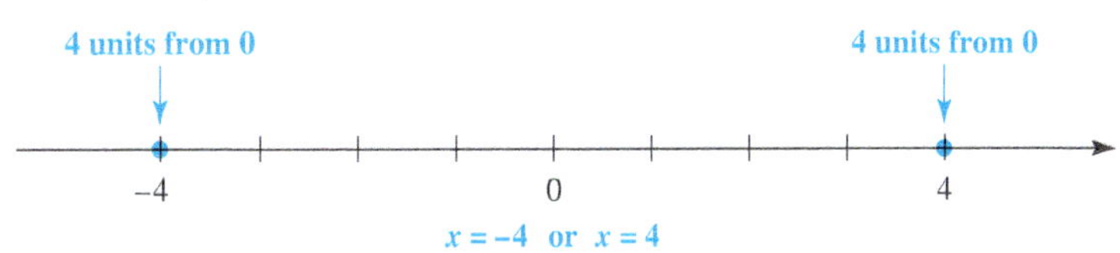

FIGURE 32

Because absolute value represents distance from 0, we interpret the solutions of $|x| > 4$ to be all numbers that are *more* than four units from 0 on a number line. The set $(-\infty, -4) \cup (4, \infty)$ fits this description. **FIGURE 33** on the next page shows the graph of the solution set of $|x| > 4$. The graph consists of two separate intervals, which means $x < -4$ *or* $x > 4$.

VOCABULARY

☐ absolute value equation
☐ absolute value inequality

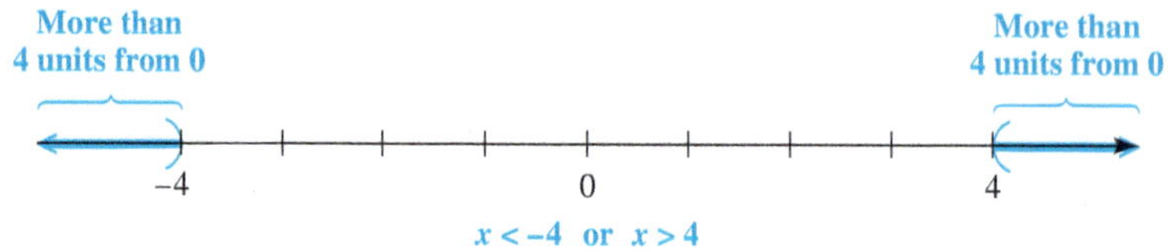

FIGURE 33

The solution set of $|x| < 4$ consists of all numbers that are *less* than 4 units from 0 on a number line. This is represented by all numbers *between* -4 and 4. This set of numbers is given by $(-4, 4)$, as shown in **FIGURE 34**. Here, the graph shows that $-4 < x < 4$, which means $x > -4$ *and* $x < 4$.

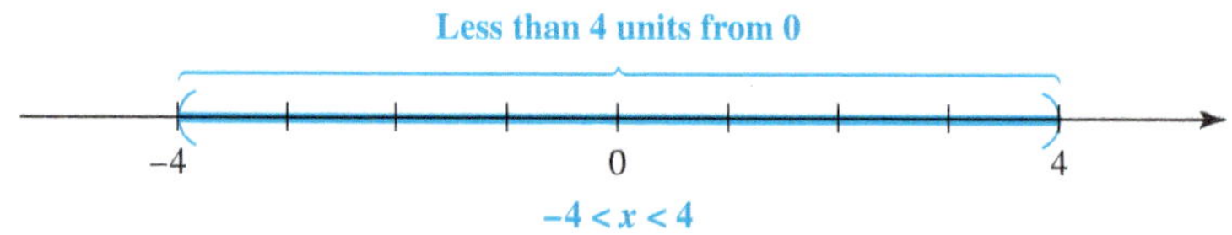

FIGURE 34

Absolute value equations and inequalities generally take the form

$$|ax + b| = k, \qquad |ax + b| > k, \qquad \text{or} \qquad |ax + b| < k,$$

where k is a positive number. From **FIGURES 32–34**, we see that

$|x| = 4$ has the same solution set as $x = -4$ or $x = 4$,

$|x| > 4$ has the same solution set as $x < -4$ or $x > 4$,

$|x| < 4$ has the same solution set as $x > -4$ and $x < 4$.

This is equivalent to $-4 < x < 4$.

Solving Absolute Value Equations and Inequalities

Let k be a positive real number, and p and q be real numbers.

Case 1 To solve $|ax + b| = k$, solve the following compound equation.

$$ax + b = k \quad \textbf{or} \quad ax + b = -k$$

The solution set is usually of the form $\{p, q\}$, which includes two numbers.

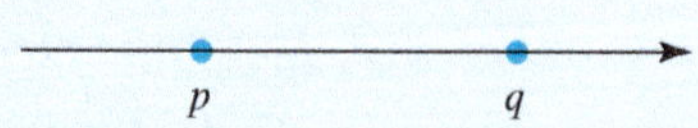

Case 2 To solve $|ax + b| > k$*, solve the following compound inequality.

$$ax + b > k \quad \textbf{or} \quad ax + b < -k$$

The solution set is of the form $(-\infty, p) \cup (q, \infty)$, which is a disjoint interval.

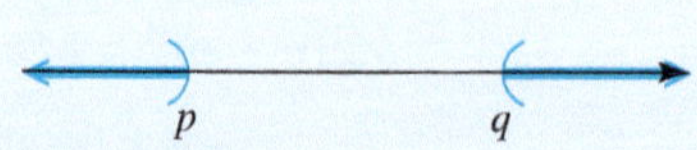

Case 3 To solve $|ax + b| < k$**, solve the following three-part inequality.

$$-k < ax + b < k$$

The solution set is of the form (p, q), which is a single interval.

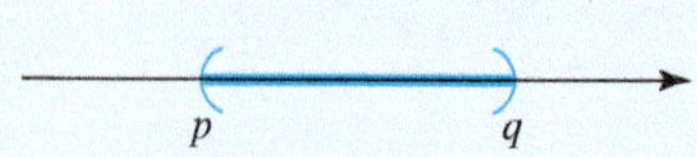

*This also applies to $|ax + b| \geq k$. The solution set *includes* the endpoints, using brackets rather than parentheses.

**This also applies to $|ax + b| \leq k$. The solution set *includes* the endpoints, using brackets rather than parentheses.

NOTE It is acceptable to write the compound statements in Cases 1 and 2 of the preceding box as follows. These forms produce the same results.

$$ax + b = k \quad \text{or} \quad -(ax + b) = k \qquad \text{Alternative for Case 1}$$

$$ax + b > k \quad \text{or} \quad -(ax + b) > k \qquad \text{Alternative for Case 2}$$

OBJECTIVE 2 Solve equations of the form $|ax + b| = k$, for $k > 0$.

Remember that because absolute value refers to distance from the origin, an absolute value equation will have two parts.

NOW TRY EXERCISE 1

Solve $|4x - 1| = 11$.

CLASSROOM EXAMPLE 1

Solve $|3x - 4| = 11$.

Answer: $\left\{-\frac{7}{3}, 5\right\}$

EXAMPLE 1 Solving an Absolute Value Equation (Case 1)

Solve $|2x + 1| = 7$. Graph the solution set.

For $|2x + 1|$ to equal 7, $2x + 1$ must be 7 units from 0 on a number line. This can happen only when $2x + 1 = 7$ or $2x + 1 = -7$. This is Case 1 in the box on the preceding page. Solve this compound equation as follows.

$$2x + 1 = 7 \quad \text{or} \quad 2x + 1 = -7$$

$$2x = 6 \quad \text{or} \quad 2x = -8 \qquad \text{Subtract 1.}$$

$$x = 3 \quad \text{or} \quad x = -4 \qquad \text{Divide by 2.}$$

CHECK $|2x + 1| = 7$

$\lvert 2(3) + 1 \rvert \stackrel{?}{=} 7$ Let $x = 3$.	$\lvert 2(-4) + 1 \rvert \stackrel{?}{=} 7$ Let $x = -4$.
$\lvert 6 + 1 \rvert \stackrel{?}{=} 7$	$\lvert -8 + 1 \rvert \stackrel{?}{=} 7$
$\lvert 7 \rvert \stackrel{?}{=} 7$	$\lvert -7 \rvert \stackrel{?}{=} 7$
$7 = 7$ ✓ True	$7 = 7$ ✓ True

The solution set is $\{-4, 3\}$. The graph is shown in **FIGURE 35.**

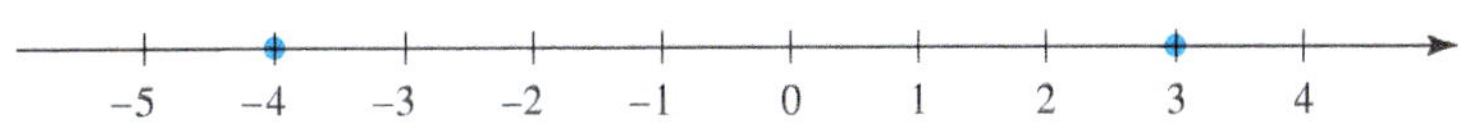

FIGURE 35

NOW TRY

OBJECTIVE 3 Solve inequalities of the form $|ax + b| < k$ and of the form $|ax + b| > k$, for $k > 0$.

CLASSROOM EXAMPLE 2

Solve $|3x - 4| \geq 11$.

Answer: $\left(-\infty, -\frac{7}{3}\right] \cup [5, \infty)$

EXAMPLE 2 Solving an Absolute Value Inequality (Case 2)

Solve $|2x + 1| > 7$. Graph the solution set.

By Case 2 in the box on the preceding page, this absolute value inequality is rewritten as

$$2x + 1 > 7 \quad \text{or} \quad 2x + 1 < -7$$

because $2x + 1$ must represent a number that is *more* than 7 units from 0 on either side of a number line. Now solve the compound inequality.

$$2x + 1 > 7 \quad \text{or} \quad 2x + 1 < -7$$

$$2x > 6 \quad \text{or} \quad 2x < -8 \qquad \text{Subtract 1.}$$

$$x > 3 \quad \text{or} \quad x < -4 \qquad \text{Divide by 2.}$$

NOW TRY ANSWER

1. $\left\{-\frac{5}{2}, 3\right\}$

NOW TRY EXERCISE 2

Solve $|4x - 1| > 11$.

CHECK The excluded endpoints -4 and 3 are correct because from **Example 1** we know that the solutions of the related equation are -4 and 3. We now choose a test point in each of the three intervals $(-\infty, -4)$, $(-4, 3)$, and $(3, \infty)$.

For $(-\infty, -4)$, let $x = -5$.	For $(-4, 3)$, let $x = 0$.	For $(3, \infty)$, let $x = 4$.
$\lvert 2x + 1 \rvert > 7$	$\lvert 2x + 1 \rvert > 7$	$\lvert 2x + 1 \rvert > 7$
$\lvert 2(-5) + 1 \rvert \overset{?}{>} 7$	$\lvert 2(0) + 1 \rvert \overset{?}{>} 7$	$\lvert 2(4) + 1 \rvert \overset{?}{>} 7$
$\lvert -9 \rvert \overset{?}{>} 7$	$\lvert 1 \rvert \overset{?}{>} 7$	$\lvert 9 \rvert \overset{?}{>} 7$
$9 > 7$ ✓ True	$1 > 7$ False	$9 > 7$ ✓ True

The solution set, $(-\infty, -4) \cup (3, \infty)$, is a disjoint interval. See **FIGURE 36.**

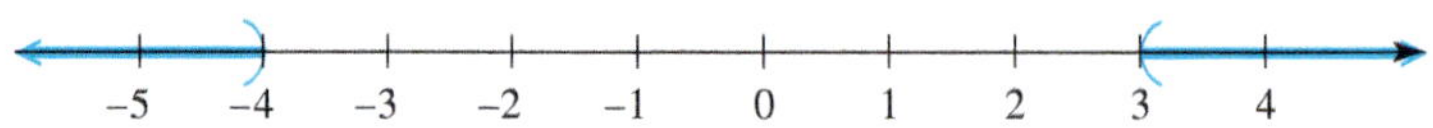

FIGURE 36

NOW TRY

NOW TRY EXERCISE 3

Solve $|4x - 1| < 11$.

CLASSROOM EXAMPLE 3

Solve $|3x - 4| \le 11$.

Answer: $\left[-\frac{7}{3}, 5\right]$

EXAMPLE 3 Solving an Absolute Value Inequality (Case 3)

Solve $|2x + 1| < 7$. Graph the solution set.

The expression $2x + 1$ must represent a number that is less than 7 units from 0 on either side of a number line. That is, $2x + 1$ must be between -7 and 7. As Case 3 in the earlier box shows, this is written as a three-part inequality.

$$-7 < 2x + 1 < 7$$

$$-8 < 2x < 6 \quad \text{Subtract 1 from each part.}$$

$$-4 < x < 3 \quad \text{Divide each part by 2.}$$

Check that the solution set is $(-4, 3)$. The graph is the open interval in **FIGURE 37.**

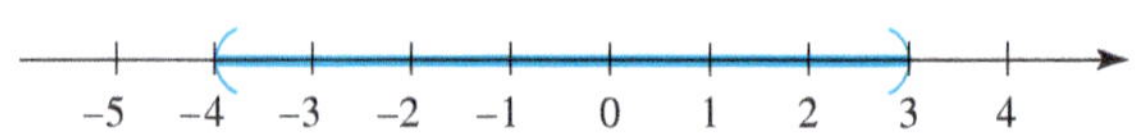

FIGURE 37

NOW TRY

Look back at **FIGURES 35, 36, AND 37,** with the graphs of

$$|2x + 1| = 7, \quad |2x + 1| > 7, \quad \text{and} \quad |2x + 1| < 7.$$

If we find the union of the three sets, we obtain the set of all real numbers. For any value of x, $|2x + 1|$ will satisfy *one and only one* of the following: It is equal to 7, greater than 7, or less than 7.

! CAUTION Remember the following when solving absolute value equations and inequalities.

1. The methods described apply when the constant is alone on one side of the equation or inequality and is *positive.*
2. Absolute value equations $|ax + b| = k$ and inequalities of the form $|ax + b| > k$ translate into "or" compound statements.
3. Absolute value inequalities of the form $|ax + b| < k$ translate into "and" compound statements, which may be written as three-part inequalities.
4. An "or" statement *cannot* be written in three parts. It would be **incorrect** to write $-7 > 2x + 1 > 7$ in **Example 2,** because this would imply that $-7 > 7$, which is *false.*

NOW TRY ANSWERS

2. $\left(-\infty, -\frac{5}{2}\right) \cup (3, \infty)$

3. $\left(-\frac{5}{2}, 3\right)$

NOW TRY EXERCISE 4

Solve $|7 - 4x| \geq 7$.

CLASSROOM EXAMPLE 4

Solve $|20 - 2x| \geq 20$.

Answer: $(-\infty, 0] \cup [20, \infty)$

EXAMPLE 4 Solving an Absolute Value Inequality (Case 2, for $\geq$)

Solve $|5 - 2x| \geq 5$.

Case 2 is applied. Notice that the endpoints are included because equality is part of the symbol $\geq$.

$$5 - 2x \geq 5 \quad \text{or} \quad 5 - 2x \leq -5$$

$$-2x \geq 0 \quad \text{or} \quad -2x \leq -10 \qquad \text{Subtract 5.}$$

$$x \leq 0 \quad \text{or} \quad x \geq 5 \qquad \text{Divide by } -2. \text{ Reverse the direction of the inequality symbols.}$$

Check that the solution set is $(-\infty, 0] \cup [5, \infty)$. See **FIGURE 38**.

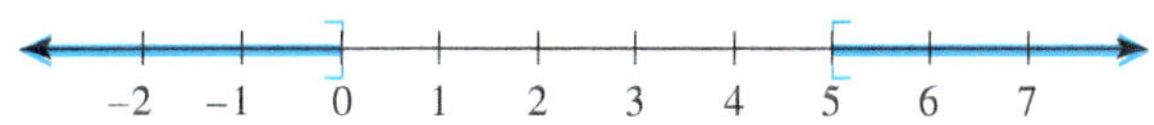

FIGURE 38

NOW TRY

OBJECTIVE 4 Solve absolute value equations that involve rewriting.

NOW TRY EXERCISE 5

Solve $|10x - 2| - 2 = 12$.

CLASSROOM EXAMPLE 5

Solve $|3x + 2| + 4 = 15$.

Answer: $\left\{-\frac{13}{3}, 3\right\}$

EXAMPLE 5 Solving an Absolute Value Equation That Requires Rewriting

Solve $|x + 3| + 5 = 12$.

Isolate the absolute value expression on one side of the equality symbol.

$$|x + 3| + 5 = 12$$

$$|x + 3| + 5 - 5 = 12 - 5 \qquad \text{Subtract 5.}$$

$$|x + 3| = 7 \qquad \text{Combine like terms.}$$

Now use the method shown in **Example 1** for Case 1 to solve $|x + 3| = 7$.

$$x + 3 = 7 \quad \text{or} \quad x + 3 = -7$$

$$x = 4 \quad \text{or} \quad x = -10 \qquad \text{Subtract 3.}$$

CHECK $|x + 3| + 5 = 12$

$\lvert 4 + 3\rvert + 5 \stackrel{?}{=} 12$ Let $x = 4$.	$\lvert -10 + 3\rvert + 5 \stackrel{?}{=} 12$ Let $x = -10$.
$\lvert 7\rvert + 5 \stackrel{?}{=} 12$	$\lvert -7\rvert + 5 \stackrel{?}{=} 12$
$12 = 12$ ✓ True	$12 = 12$ ✓ True

The check confirms that the solution set is $\{-10, 4\}$.

NOW TRY

CLASSROOM EXAMPLE 6

Solve each inequality.

(a) $|x + 2| - 3 > 2$

(b) $|x + 2| - 3 < 2$

Answers: (a) $(-\infty, -7) \cup (3, \infty)$

(b) $(-7, 3)$

NOW TRY EXERCISE 6

Solve each inequality.

(a) $|x - 1| - 4 \leq 2$

(b) $|x - 1| - 4 \geq 2$

EXAMPLE 6 Solving Absolute Value Inequalities That Require Rewriting

Solve each inequality.

(a) $|x + 3| + 5 \geq 12$

$$|x + 3| \geq 7 \qquad \text{(Case 2)}$$

$$x + 3 \geq 7 \quad \text{or} \quad x + 3 \leq -7$$

$$x \geq 4 \quad \text{or} \quad x \leq -10$$

The solution set is $(-\infty, -10] \cup [4, \infty)$.

(b) $|x + 3| + 5 \leq 12$

$$|x + 3| \leq 7 \qquad \text{(Case 3)}$$

$$-7 \leq x + 3 \leq 7$$

$$-10 \leq x \leq 4$$

The solution set is $[-10, 4]$.

NOW TRY

NOW TRY ANSWERS

4. $(-\infty, 0] \cup \left[\frac{7}{2}, \infty\right)$ **5.** $\left\{-\frac{6}{5}, \frac{8}{5}\right\}$

6. (a) $[-5, 7]$

(b) $(-\infty, -5] \cup [7, \infty)$

OBJECTIVE 5 Solve equations of the form $|ax + b| = |cx + d|$.

If two expressions have the same absolute value, they must either be equal or be negatives of each other.

Solving $|ax + b| = |cx + d|$

To solve an absolute value equation of the form

$$|ax + b| = |cx + d|,$$

solve the following compound equation.

$$ax + b = cx + d \quad \textbf{or} \quad ax + b = -(cx + d)$$

NOW TRY EXERCISE 7

Solve

$$|3x - 4| = |5x + 12|.$$

CLASSROOM EXAMPLE 7

Solve $|4x - 1| = |3x + 5|$.

Answer: $\{-\frac{4}{7}, 6\}$

EXAMPLE 7 Solving an Equation Involving Two Absolute Values

Solve $|x + 6| = |2x - 3|$.

This equation is satisfied either if $x + 6$ and $2x - 3$ are equal to each other, or if $x + 6$ and $2x - 3$ are negatives of each other.

$$\begin{aligned} x + 6 &= 2x - 3 && \text{or} & x + 6 &= -(2x - 3) \\ x + 9 &= 2x && \text{or} & x + 6 &= -2x + 3 \\ 9 &= x && \text{or} & 3x &= -3 \\ x &= 9 && \text{or} & x &= -1 \end{aligned}$$

CHECK $|x + 6| = |2x - 3|$

Let $x = 9$.	Let $x = -1$.
$\lvert 9 + 6\rvert \stackrel{?}{=} \lvert 2(9) - 3\rvert$	$\lvert -1 + 6\rvert \stackrel{?}{=} \lvert 2(-1) - 3\rvert$
$\lvert 15\rvert \stackrel{?}{=} \lvert 18 - 3\rvert$	$\lvert 5\rvert \stackrel{?}{=} \lvert -2 - 3\rvert$
$\lvert 15\rvert \stackrel{?}{=} \lvert 15\rvert$	$\lvert 5\rvert \stackrel{?}{=} \lvert -5\rvert$
$15 = 15$ ✓ True	$5 = 5$ ✓ True

The check confirms that the solution set is $\{-1, 9\}$.

NOW TRY

CLASSROOM EXAMPLE 8

Solve each equation.

(a) $|6x + 7| = -5$

(b) $\left|\frac{1}{4}x - 3\right| = 0$

Answers: (a) $\varnothing$ (b) $\{12\}$

OBJECTIVE 6 Solve special cases of absolute value equations and inequalities.

When an absolute value equation or inequality involves a *negative constant or 0* alone on one side, use the properties of absolute value to solve the equation or inequality.

Special Cases of Absolute Value

Case 1 The absolute value of an expression can never be negative—that is, $|a| \geq 0$ for all real numbers a.

Case 2 The absolute value of an expression equals 0 only when the expression is equal to 0.

CLASSROOM EXAMPLE 9

Solve each inequality.

(a) $|x| > -1$

(b) $|x - 10| - 2 \leq -3$

(c) $|x + 2| \leq 0$

Answers: (a) $(-\infty, \infty)$ (b) $\varnothing$ (c) $\{-2\}$

EXAMPLE 8 Solving Special Cases of Absolute Value Equations

Solve each equation.

(a) $|5x - 3| = -4$

See Case 1 in the preceding box. ***The absolute value of an expression can never be negative,*** so there are no solutions for this equation. The solution set is $\varnothing$.

NOW TRY ANSWER

7. $\{-8, -1\}$

NOW TRY EXERCISE 8
Solve each equation.

(a) $|3x - 8| = -2$

(b) $|7x + 12| = 0$

(b) $|7x - 3| = 0$

See Case 2 in the preceding box. The expression $|7x - 3|$ will equal 0 *only* if $7x - 3 = 0$.

$$7x - 3 = 0$$

$$7x = 3 \quad \text{Add 3.}$$

$$x = \frac{3}{7} \quad \text{Divide by 7.}$$

Check by substituting in the original equation.

The solution set $\left\{\frac{3}{7}\right\}$ consists of only one element. NOW TRY

NOW TRY EXERCISE 9
Solve each inequality.

(a) $|x| > -10$

(b) $|4x + 1| + 5 < 4$

(c) $|x - 2| - 3 \le -3$

EXAMPLE 9 Solving Special Cases of Absolute Value Inequalities

Solve each inequality.

(a) $|x| \ge -4$

The absolute value of a number is always greater than or equal to 0. Thus, $|x| \ge -4$ is true for *all* real numbers. The solution set is $(-\infty, \infty)$.

(b)
$$|x + 6| - 3 < -5$$

$$|x + 6| < -2 \quad \text{Add 3 to each side.}$$

There is no number whose absolute value is less than -2, so this inequality has no solution. The solution set is $\emptyset$.

(c)
$$|x - 7| + 4 \le 4$$

$$|x - 7| \le 0 \quad \text{Subtract 4 from each side.}$$

The value of $|x - 7|$ will never be less than 0. However, $|x - 7|$ will equal 0 when $x = 7$. Therefore, the solution set is $\{7\}$. NOW TRY

CLASSROOM EXAMPLE 10
Suppose a machine filling quart orange juice cartons is set for a relative error that is no greater than 0.035. How many ounces may a filled carton contain?

Answer: between 30.88 and 33.12 oz, inclusive

OBJECTIVE 7 Solve an application involving relative error.

Absolute value is used to find the **relative error,** or **tolerance,** in a measurement. If x_t represents the expected measurement and x represents the actual measurement, then the relative error in x equals the absolute value of the difference of x_t and x, divided by x_t.

$$\textbf{relative error in } x = \left|\frac{x_t - x}{x_t}\right|$$

In quality control situations, the relative error must often be less than some predetermined amount.

EXAMPLE 10 Solving an Application Involving Relative Error

Suppose a machine filling *quart* milk cartons is set for a relative error that *is no greater than* 0.05. How many ounces may a filled carton contain?

Here $x_t = 32$ oz (because 1 qt = 32 oz) and the relative error = 0.05. We must find x, given the following condition.

$$\left|\frac{32 - x}{32}\right| \le 0.05 \quad \textit{Is no greater than} \text{ translates as } \le.$$

NOW TRY ANSWERS

8. (a) $\emptyset$ **(b)** $\left\{-\frac{12}{7}\right\}$

9. (a) $(-\infty, \infty)$ **(b)** $\emptyset$ **(c)** $\{2\}$

NOW TRY EXERCISE 10

Suppose a machine filling *quart* milk cartons is set for a relative error that *is no greater than* 0.032. How many ounces may a filled carton contain?

$$\left|\frac{32-x}{32}\right| \le 0.05$$

$$-0.05 \le \frac{32-x}{32} \le 0.05 \qquad \text{(Case 3)}$$

$$-1.6 \le 32 - x \le 1.6 \qquad \text{Multiply by 32.}$$

$$-33.6 \le -x \le -30.4 \qquad \text{Subtract 32.}$$

$$33.6 \ge x \ge 30.4 \qquad \text{Multiply by } -1\text{, and reverse the direction of the inequality symbols.}$$

$$30.4 \le x \le 33.6 \qquad \text{Rewrite.}$$

The filled carton may contain between 30.4 and 33.6 oz, inclusive. NOW TRY

NOW TRY ANSWER

10. between 30.976 and 33.024 oz, inclusive

1.7 Exercises

FOR EXTRA HELP MyMathLab®

Complete solution available in MyMathLab

Concept Check *Match each absolute value equation or inequality in Column I with the graph of its solution set in Column II.*

1.

I	II
$\lvert x\rvert = 5$	**A.** 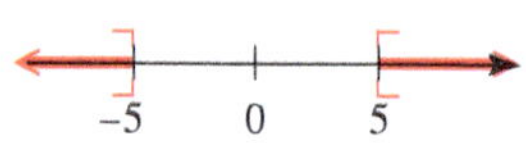
$\lvert x\rvert < 5$	**B.** 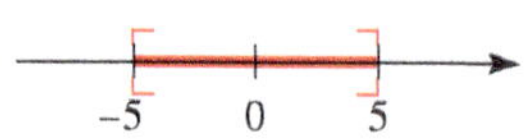
$\lvert x\rvert > 5$	**C.** 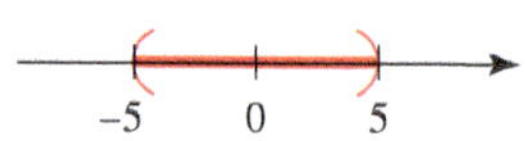
$\lvert x\rvert \le 5$	**D.** 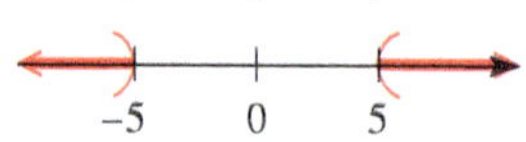
$\lvert x\rvert \ge 5$	**E.** 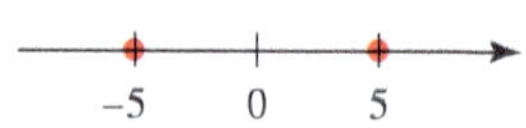

2.

I	II
$\lvert x\rvert = 9$	**A.** 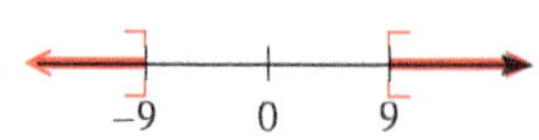
$\lvert x\rvert > 9$	**B.** 
$\lvert x\rvert \ge 9$	**C.** 
$\lvert x\rvert < 9$	**D.** 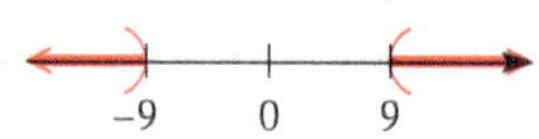
$\lvert x\rvert \le 9$	**E.** 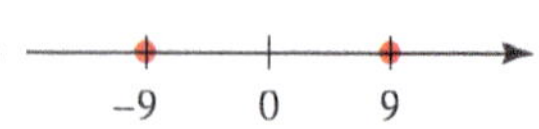

3. *Concept Check* How many solutions will $|ax+b| = k$ have for each situation?

(a) $k = 0$ **(b)** $k > 0$ **(c)** $k < 0$

4. *Concept Check* Explain when to use *and* and when to use *or* for solving an absolute value equation or inequality of the form $|ax+b| = k$, $|ax+b| < k$, or $|ax+b| > k$, where k is a positive number.

Solve each equation. ***See Example 1.***

5. $|x| = 12$ **6.** $|x| = 14$ **7.** $|4x| = 20$ **8.** $|5x| = 30$

9. $|x-3| = 9$ **10.** $|x-5| = 13$ **11.** $|2x-1| = 11$ **12.** $|2x+3| = 19$

13. $|4x-5| = 17$ **14.** $|5x-1| = 21$ **15.** $|2x+5| = 14$ **16.** $|2x-9| = 18$

17. $|-3x+8| = 1$ **18.** $|-6x+5| = 4$ **19.** $\left|12 - \frac{1}{2}x\right| = 6$

20. $\left|14 - \frac{1}{3}x\right| = 8$ **21.** $|0.5x| = 6$ **22.** $|0.3x| = 9$

23. $\left|\frac{1}{2}x + 3\right| = 2$ **24.** $\left|\frac{2}{3}x - 1\right| = 5$ **25.** $\left|1 + \frac{3}{4}x\right| = 7$

26. $\left|2 - \frac{5}{2}x\right| = 14$ **27.** $|0.02x - 1| = 2.50$ **28.** $|0.04x - 3| = 5.96$

Solve each inequality, and graph the solution set. **See Example 2.**

29. $|x| > 3$ **30.** $|x| > 5$ **31.** $|x| \geq 4$

32. $|x| \geq 6$ **33.** $|r + 5| \geq 20$ **34.** $|x + 4| \geq 8$

35. $|5x + 2| > 10$ **36.** $|4x + 1| \geq 21$ **37.** $|3 - x| > 5$

38. $|5 - x| > 3$ **39.** $|-5x + 3| \geq 12$ **40.** $|-2x - 4| \geq 5$

41. *Concept Check* The graph of the solution set of $|2x + 1| = 9$ is given here.

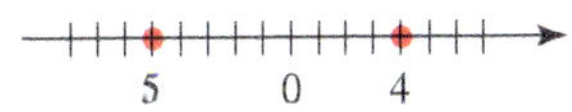

Without actually doing the algebraic work, graph the solution set of each inequality, referring to the graph shown.

(a) $|2x + 1| < 9$ **(b)** $|2x + 1| > 9$

42. *Concept Check* The graph of the solution set of $|3x - 4| < 5$ is given here.

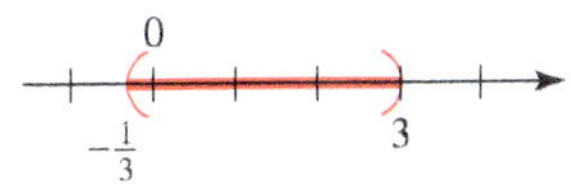

Without actually doing the algebraic work, graph the solution set of the following, referring to the graph shown.

(a) $|3x - 4| = 5$ **(b)** $|3x - 4| > 5$

Solve each inequality, and graph the solution set. **See Example 3.** *(Hint: Compare the answers with those in* **Exercises 29–40.***)*

43. $|x| \leq 3$ **44.** $|x| \leq 5$ **45.** $|x| < 4$

46. $|x| < 6$ **47.** $|r + 5| < 20$ **48.** $|x + 4| < 8$

49. $|5x + 2| \leq 10$ **50.** $|4x + 1| < 21$ **51.** $|3 - x| \leq 5$

52. $|5 - x| \leq 3$ **53.** $|-5x + 3| < 12$ **54.** $|-2x - 4| < 5$

In Exercises 55–78, decide which method of solution applies, and find the solution set. In Exercises 55–66, graph the solution set. **See Examples 1–4.**

55. $|-4 + x| > 9$ **56.** $|-3 + x| > 8$ **57.** $|3x + 2| < 11$

58. $|2x - 1| < 7$ **59.** $|7 + 2x| = 5$ **60.** $|9 - 3x| = 3$

61. $|3x - 1| \leq 11$ **62.** $|2x - 6| \leq 6$ **63.** $|-6x - 6| \leq 1$

64. $|-2x - 6| \leq 5$ **65.** $|-8 + x| \leq 5$ **66.** $|-4 + x| \leq 9$

67. $|10 - 12x| \geq 4$ **68.** $|8 - 10x| \geq 2$ **69.** $|3(x - 1)| = 8$

70. $|7(x - 2)| = 4$ **71.** $|0.1x - 1| > 3$ **72.** $|0.1x + 1| > 2$

73. $|x + 2| = 5 - 2$ **74.** $|x + 3| = 12 - 2$ **75.** $3|x - 6| = 9$

76. $5|x - 4| = 5$ **77.** $|2 - 0.2x| = 2$ **78.** $|5 - 0.5x| = 4$

Solve each equation or inequality. **See Examples 5 and 6.**

79. $|x| - 1 = 4$ **80.** $|x| + 3 = 10$ **81.** $|x + 4| + 1 = 2$

82. $|x + 5| - 2 = 12$ **83.** $|2x + 1| + 3 > 8$ **84.** $|6x - 1| - 2 > 6$

85. $|x + 5| - 6 \leq -1$ **86.** $|x - 2| - 3 \leq 4$

---

37. $(-\infty, -2) \cup (8, \infty)$
38. $(-\infty, 2) \cup (8, \infty)$
39. $\left(-\infty, -\frac{9}{5}\right] \cup [3, \infty)$
40. $\left(-\infty, -\frac{9}{2}\right] \cup \left[\frac{1}{2}, \infty\right)$

41. (a) 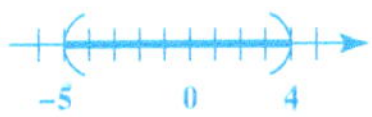

(b) 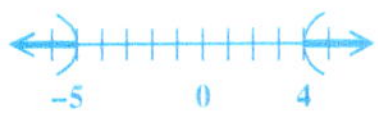

42. (a) 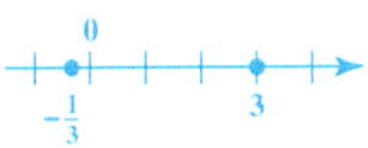

(b) 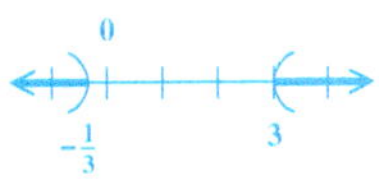

Due to space limitations, we do not give answer graphs for Exercises 43–66.

43. $[-3, 3]$ **44.** $[-5, 5]$
45. $(-4, 4)$ **46.** $(-6, 6)$
47. $(-25, 15)$ **48.** $(-12, 4)$
49. $\left[-\frac{12}{5}, \frac{8}{5}\right]$ **50.** $\left(-\frac{11}{2}, 5\right)$
51. $[-2, 8]$ **52.** $[2, 8]$
53. $\left(-\frac{9}{5}, 3\right)$ **54.** $\left(-\frac{9}{2}, \frac{1}{2}\right)$
55. $(-\infty, -5) \cup (13, \infty)$
56. $(-\infty, -5) \cup (11, \infty)$
57. $\left(-\frac{13}{3}, 3\right)$ **58.** $(-3, 4)$
59. $\{-6, -1\}$ **60.** $\{2, 4\}$
61. $\left[-\frac{10}{3}, 4\right]$ **62.** $[0, 6]$
63. $\left[-\frac{7}{6}, -\frac{5}{6}\right]$ **64.** $\left[-\frac{11}{2}, -\frac{1}{2}\right]$
65. $[3, 13]$ **66.** $[-5, 13]$
67. $\left(-\infty, \frac{1}{2}\right] \cup \left[\frac{7}{6}, \infty\right)$
68. $\left(-\infty, \frac{3}{5}\right] \cup [1, \infty)$
69. $\left\{-\frac{5}{3}, \frac{11}{3}\right\}$ **70.** $\left\{\frac{10}{7}, \frac{18}{7}\right\}$
71. $(-\infty, -20) \cup (40, \infty)$
72. $(-\infty, -30) \cup (10, \infty)$
73. $\{-5, 1\}$ **74.** $\{-13, 7\}$
75. $\{3, 9\}$ **76.** $\{3, 5\}$
77. $\{0, 20\}$ **78.** $\{2, 18\}$
79. $\{-5, 5\}$ **80.** $\{-7, 7\}$
81. $\{-5, -3\}$ **82.** $\{-19, 9\}$
83. $(-\infty, -3) \cup (2, \infty)$
84. $\left(-\infty, -\frac{7}{6}\right) \cup \left(\frac{3}{2}, \infty\right)$

87. $|0.1x - 2.5| + 0.3 \geq 0.8$

88. $|0.5x - 3.5| + 0.2 \geq 0.6$

89. $\left|\frac{1}{2}x + \frac{1}{3}\right| + \frac{1}{4} = \frac{3}{4}$

90. $\left|\frac{2}{3}x + \frac{1}{6}\right| + \frac{1}{2} = \frac{5}{2}$

Solve each equation. ***See Example 7.***

91. $|3x + 1| = |2x + 4|$

92. $|7x + 12| = |x - 8|$

93. $\left|x - \frac{1}{2}\right| = \left|\frac{1}{2}x - 2\right|$

94. $\left|\frac{2}{3}x - 2\right| = \left|\frac{1}{3}x + 3\right|$

95. $|6x| = |9x + 1|$

96. $|13x| = |2x + 1|$

97. $|2x - 6| = |2x + 11|$

98. $|3x - 1| = |3x + 9|$

Solve each equation or inequality. ***See Examples 8 and 9.***

99. $|x| \geq -10$

100. $|x| \geq -15$

101. $|12t - 3| = -8$

102. $|13x + 1| = -3$

103. $|4x + 1| = 0$

104. $|6x - 2| = 0$

105. $|2x - 1| = -6$

106. $|8x + 4| = -4$

107. $|x + 5| > -9$

108. $|x + 9| > -3$

109. $|7x + 3| \leq 0$

110. $|4x - 1| \leq 0$

111. $|5x - 2| = 0$

112. $|7x + 4| = 0$

113. $|x - 2| + 3 \geq 2$

114. $|x - 4| + 5 \geq 4$

115. $|10x + 7| + 3 < 1$

116. $|4x + 1| - 2 < -5$

In Exercises 117–120, determine the number of ounces a filled 32 oz carton may contain for the given relative error. ***See Example 10.***

117. no greater than 0.04

118. no greater than 0.03

119. no greater than 0.025

120. no greater than 0.015

In later courses in mathematics, it is sometimes necessary to find an interval in which x must lie in order to keep y within a given difference of some number. For example, suppose

$$y = 2x + 1$$

and we want y to be within 0.01 *unit of* 4. *This criterion can be written as*

$$|y - 4| < 0.01.$$

Solving this inequality shows that x must lie in the interval (1.495, 1.505) *to satisfy the requirement.*

In Exercises 121–124, find the open interval in which x must lie in order for the given condition to hold.

121. $y = 2x + 1$, and the difference of y and 1 is less than 0.1.

122. $y = 4x - 6$, and the difference of y and 2 is less than 0.02.

123. $y = 4x - 8$, and the difference of y and 3 is less than 0.001.

124. $y = 5x + 12$, and the difference of y and 4 is less than 0.0001.

Work each problem.

125. Dr. Mosely has determined that 99% of the babies he has delivered have weighed x pounds, where

$$|x - 8.3| < 1.5.$$

What range of weights corresponds to this inequality?

126. between and inclusive of -140 and -30 degrees Celsius
127. $|x - 1000| \leq 100$; $900 \leq x \leq 1100$
128. $|x - 7.45| \leq 3.6$; $3.85 \leq x \leq 11.05$

126. The Celsius temperatures x on Mars approximately satisfy the inequality

$$|x + 85| \leq 55.$$

What range of temperatures corresponds to this inequality?

127. The recommended daily intake (RDI) of calcium for females aged 19–50 is 1000 mg. Actual needs vary from person to person. Write this statement as an absolute value inequality, with x representing the RDI, to express the RDI plus or minus 100 mg, and solve the inequality. (*Source:* National Academy of Sciences—Institute of Medicine.)

128. The average clotting time of blood is 7.45 sec, with a variation of plus or minus 3.6 sec. Write this statement as an absolute value inequality, with x representing the time, and solve the inequality.

RELATING CONCEPTS For Individual or Group Work (Exercises 129–132)

The 10 tallest buildings in Houston, Texas, are listed along with their heights.

Building	*Height (in feet)*
JPMorgan Chase Tower	1002
Wells Fargo Plaza	992
Williams Tower	901
Bank of America Center	780
Texaco Heritage Plaza	762
Enterprise Plaza	756
Centerpoint Energy Plaza	741
Continental Center I	732
Fulbright Tower	725
One Shell Plaza	714

Source: *World Almanac and Book of Facts.*

129. 810.5 ft
130. Bank of America Center, Texaco Heritage Plaza
131. Williams Tower, Bank of America Center, Texaco Heritage Plaza, Enterprise Plaza, Centerpoint Energy Plaza, Continental Center I, Fulbright Tower
132. (a) $|x - 810.5| \geq 95$
(b) $x \geq 905.5$ or $x \leq 715.5$
(c) JPMorgan Chase Tower, Wells Fargo Plaza, One Shell Plaza
(d) It makes sense because it includes all buildings *not* listed in the answer to Exercise 131.

*Use this information to **work Exercises 129–132 in order.***

129. To find the average of a group of numbers, we add the numbers and then divide by the number of numbers added. Use a calculator to find the average of the heights.

130. Let k represent the average height of these buildings. If a height x satisfies the inequality

$$|x - k| < t,$$

then the height is said to be within t feet of the average. Using the result from **Exercise 129,** list the buildings that are within 50 ft of the average.

131. Repeat **Exercise 130,** but list the buildings that are within 95 ft of the average.

132. (a) Write an absolute value inequality that describes the height of a building that is *not* within 95 ft of the average.

(b) Solve the inequality from part (a).

(c) Use the result of part (b) to list the buildings that are not within 95 ft of the average.

(d) Confirm that the answer to part (c) makes sense by comparing it with the answer to **Exercise 131.**

SUMMARY EXERCISES Solving Linear and Absolute Value Equations and Inequalities

Solve each equation or inequality. Give the solution set in set notation for equations and in interval notation for inequalities.

1. $4x + 1 = 49$

2. $|x - 1| = 6$

3. $6x - 9 = 12 + 3x$

4. $3x + 7 = 9 + 8x$

5. $|x + 3| = -4$

6. $2x + 1 \leq x$

7. $8x + 2 \geq 5x$

8. $4(x - 11) + 3x = 20x - 31$

9. $2x - 1 = -7$

10. $|3x - 7| - 4 = 0$

11. $6x - 5 \leq 3x + 10$

12. $|5x - 8| + 9 \geq 7$

13. $9x - 3(x + 1) = 8x - 7$

14. $|x| \geq 8$

15. $9x - 5 \geq 9x + 3$

16. $13x - 5 > 13x - 8$

17. $|x| < 5.5$

18. $4x - 1 = 12 + x$

19. $\frac{2}{3}x + 8 = \frac{1}{4}x$

20. $-\frac{5}{8}x \geq -20$

21. $\frac{1}{4}x < -6$

22. $\frac{1}{2} \leq \frac{2}{3}x \leq \frac{5}{4}$

23. $\frac{3}{5}x - \frac{1}{10} = 2$

24. $\frac{x}{6} - \frac{3x}{5} = x - 86$

25. $x + 9 + 7x = 4(3 + 2x) - 3$

26. $6 - 3(2 - x) < 2(1 + x) + 3$

27. $-6 \leq \frac{3}{2} - x \leq 6$

28. $\frac{x}{4} - \frac{2x}{3} = -10$

29. $|5x + 1| \leq 0$

30. $5x - (3 + x) \geq 2(3x + 1)$

31. $-2 \leq 3x - 1 \leq 8$

32. $-1 \leq 6 - x \leq 5$

33. $|7x - 1| = |5x + 3|$

34. $|x + 2| = |x + 4|$

35. $|1 - 3x| \geq 4$

36. $7x - 3 + 2x = 9x - 8x$

37. $-(x + 4) + 2 = 3x + 8$

38. $|x - 1| < 7$

39. $|2x - 3| > 11$

40. $|5 - x| < 4$

41. $|x - 1| \geq -6$

42. $|2x - 5| = |x + 4|$

43. $8x - (1 - x) = 3(1 + 3x) - 4$

44. $8x - (x + 3) = -(2x + 1) - 12$

45. $|x - 5| = |x + 9|$

46. $|x + 2| < -3$

47. $2x + 1 > 5$ or $3x + 4 < 1$

48. $1 - 2x \geq 5$ and $7 + 3x \geq -2$

Answers

1. $\{12\}$ **2.** $\{-5, 7\}$
3. $\{7\}$ **4.** $\left\{-\frac{2}{5}\right\}$
5. $\emptyset$ **6.** $(-\infty, -1]$
7. $\left[-\frac{2}{3}, \infty\right)$ **8.** $\{-1\}$
9. $\{-3\}$ **10.** $\left\{1, \frac{11}{3}\right\}$
11. $(-\infty, 5]$ **12.** $(-\infty, \infty)$
13. $\{2\}$
14. $(-\infty, -8] \cup [8, \infty)$
15. $\emptyset$ **16.** $(-\infty, \infty)$
17. $(-5.5, 5.5)$ **18.** $\left\{\frac{13}{3}\right\}$
19. $\left\{-\frac{96}{5}\right\}$ **20.** $(-\infty, 32]$
21. $(-\infty, -24)$ **22.** $\left[\frac{3}{4}, \frac{15}{8}\right]$
23. $\left\{\frac{7}{2}\right\}$ **24.** $\{60\}$
25. {all real numbers}
26. $(-\infty, 5)$
27. $\left[-\frac{9}{2}, \frac{15}{2}\right]$ **28.** $\{24\}$
29. $\left\{-\frac{1}{5}\right\}$ **30.** $\left(-\infty, -\frac{5}{2}\right]$
31. $\left[-\frac{1}{3}, 3\right]$ **32.** $[1, 7]$
33. $\left\{-\frac{1}{6}, 2\right\}$ **34.** $\{-3\}$
35. $(-\infty, -1] \cup \left[\frac{5}{3}, \infty\right)$
36. $\left\{\frac{3}{8}\right\}$
37. $\left\{-\frac{5}{2}\right\}$ **38.** $(-6, 8)$
39. $(-\infty, -4) \cup (7, \infty)$
40. $(1, 9)$
41. $(-\infty, \infty)$ **42.** $\left\{\frac{1}{3}, 9\right\}$
43. {all real numbers}
44. $\left\{-\frac{10}{9}\right\}$
45. $\{-2\}$ **46.** $\emptyset$
47. $(-\infty, -1) \cup (2, \infty)$
48. $[-3, -2]$

STUDY SKILLS

Reviewing a Chapter

Your textbook provides material to help you prepare for quizzes or tests in this course. Refer to a **Chapter Summary** as you read through the following techniques.

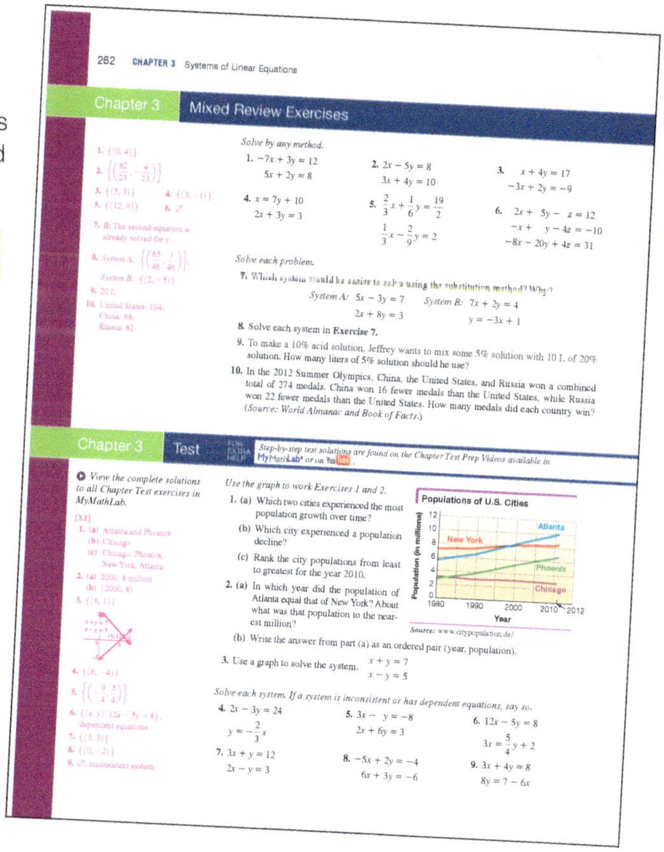

262 CHAPTER 3 Systems of Linear Equations

Chapter 3 Mixed Review Exercises

Solve by any method.

1. $-7x + 3y = 12$, $5x + 2y = 8$
2. $2x - 5y = 8$, $3x + 4y = 10$
3. $x + 4y = 17$, $-3x + 2y = -9$
4. $x = 7y + 10$, $2x + 3y = 3$
5. $\frac{2}{3}x + \frac{1}{6}y = \frac{19}{2}$, $\frac{1}{3}x - \frac{2}{9}y = 2$
6. $2x + 5y - z = 12$, $-x + y - 4z = -10$, $-8x - 20y + 4z = 31$

Solve each problem.

7. Which system would be easier to solve using the substitution method? Why?
   System A: $5x - 3y = 7$, $2x + 8y = 3$ System B: $7x + 2y = 4$, $y = -3x + 1$
8. Solve each system in **Exercise 7.**
9. To make a 10% acid solution, Jeffrey wants to mix some 5% solution with 10 L of 20% solution. How many liters of 5% solution should he use?
10. In the 2012 Summer Olympics, China, the United States, and Russia won a combined total of 274 medals. China won 16 fewer medals than the United States, while Russia won 22 fewer medals than the United States. How many medals did each country win? (*Source: World Almanac and Book of Facts.*)

Answers: 1. $\{(0, 4)\}$ 2. $\left\{\left(\frac{82}{23}, -\frac{4}{23}\right)\right\}$ 3. $\{(5, 3)\}$ 4. $\{(3, -1)\}$ 5. $\{(12, 9)\}$ 6. $\emptyset$ 7. B; The second equation is already solved for y. 8. System A: $\left\{\left(\frac{65}{46}, \frac{1}{46}\right)\right\}$; System B: $\{(2, -5)\}$ 9. 20 L 10. United States: 104; China: 88; Russia: 82

Chapter 3 Test

FOR EXTRA HELP *Step-by-step test solutions are found on the Chapter Test Prep Videos available in* MyMathLab® *or on* YouTube.

View the complete solutions to all Chapter Test exercises in MyMathLab.

Use the graph to work Exercises 1 and 2.

1. (a) Which two cities experienced the most population growth over time?
   (b) Which city experienced a population decline?
   (c) Rank the city populations from least to greatest for the year 2010.
2. (a) In which year did the population of Atlanta equal that of New York? About what was that population to the nearest million?
   (b) Write the answer from part (a) as an ordered pair (year, population).
3. Use a graph to solve the system. $x + y = 7$, $x - y = 5$

Populations of U.S. Cities — Population (in millions); Year; Atlanta, New York, Phoenix, Chicago; 1980, 1990, 2000, 2010, 2012. *Source:* www.citypopulation.de/

Solve each system. If a system is inconsistent or has dependent equations, say so.

4. $2x - 3y = 24$, $y = -\frac{2}{3}x$
5. $3x - y = -8$, $2x + 6y = 3$
6. $12x - 5y = 8$, $3x = \frac{5}{4}y + 2$
7. $3x + y = 12$, $2x - y = 3$
8. $-5x + 2y = -4$, $6x + 3y = -6$
9. $3x + 4y = 8$, $8y = 7 - 6x$

Answers: [3.1] 1. (a) Atlanta and Phoenix (b) Chicago (c) Chicago, Phoenix, New York, Atlanta 2. (a) 2000; 8 million (b) (2000, 8) 3. $\{(6, 1)\}$ 4. $\{(6, -4)\}$ 5. $\left\{\left(-\frac{9}{4}, \frac{5}{4}\right)\right\}$ 6. $\{(x, y) \mid 12x - 5y = 8\}$; dependent equations 7. $\{(3, 3)\}$ 8. $\{(0, -2)\}$ 9. $\emptyset$; inconsistent system

Chapter Reviewing Techniques

- **Review the Key Terms and New Symbols.** Make a study card for each. Include a definition, an example, a sketch (if appropriate), and a section or page reference.
- **Take the Test Your Word Power quiz** to check your understanding of new vocabulary. The answers immediately follow.
- **Read the Quick Review.** Pay special attention to the headings. Study the explanations and examples given for each concept. Try to think about the whole chapter.
- **Reread your lecture notes.** Focus on what your instructor has emphasized in class, and review that material in your text.
- **Look over your homework.** Pay special attention to any trouble spots.
- **Work the Review Exercises.** They are grouped by section.
  - Answers are included in the margin for quick reference.
  - Pay attention to direction words, such as *simplify, solve,* and *evaluate*.
  - Are your answers exact and complete? Did you include the correct labels, such as \$, cm^2, ft, etc.?
  - Make study cards for difficult problems.
- **Work the Mixed Review Exercises.** They are in mixed-up order. Check your answers in the margin.
- **Take the Chapter Test under test conditions.**
  - Time yourself.
  - Use a calculator or notes (if your instructor permits them on tests).
  - Take the test in one sitting.
  - Show all your work.
  - Check your answers in the margin. Section references are provided.

Reviewing a chapter will take some time. Avoid rushing through your review in one night. Use the suggestions over a few days or evenings to better understand the material and remember it longer.

Follow these reviewing techniques to prepare for your next test.

1. How much time did you spend reviewing for your test? Was it enough?
2. How did the reviewing techniques work for you?
3. What will you do differently when reviewing for your next test?

Chapter 1 Summary

Key Terms

1.1
linear (first-degree) equation in one variable
solution
solution set
equivalent equations
conditional equation
identity
contradiction

1.2
mathematical model
formula
percent
percent increase
percent decrease

1.4
consecutive integers
consecutive even integers
consecutive odd integers

1.5
inequality
interval
linear inequality in one variable
equivalent inequalities
three-part inequality

1.6
intersection
compound inequality
union

1.7
absolute value equation
absolute value inequality

New Symbols

%	percent
$1°$	one degree
∞	infinity
$-\infty$	negative infinity
$(-\infty, \infty)$	the set of real numbers
(a, b)	interval notation for $a < x < b$
$[a, b]$	interval notation for $a \leq x \leq b$
$\cap$	set intersection
$\cup$	set union

Test Your Word Power

See how well you have learned the vocabulary in this chapter.

1. An **equation** is
A. an algebraic expression
B. an expression that contains fractions
C. an expression that uses any of the four basic operations or the operation of raising to powers or taking roots on any collection of variables and numbers formed according to the rules of algebra
D. a statement that two algebraic expressions are equal.

2. An **inequality** is
A. a statement that two algebraic expressions are equal
B. a point on a number line
C. an equation with no solutions
D. a statement consisting of algebraic expressions related by $<, \leq, >$, or $\geq$.

3. Interval notation is
A. a point on a number line
B. a special notation for describing a point on a number line
C. a way to use symbols to describe an interval on a number line
D. a notation to describe unequal quantities.

4. The **intersection** of two sets *A* and *B* is the set of elements that belong
A. to both *A* and *B*
B. to either *A* or *B*, or both
C. to either *A* or *B*, but not both
D. to just *A*.

5. The **union** of two sets *A* and *B* is the set of elements that belong
A. to both *A* and *B*
B. to either *A* or *B*, or both
C. to either *A* or *B*, but not both
D. to just *B*.

ANSWERS

1. D; *Examples:* $2a + 3 = 7$, $3y = -8$, $x^2 = 4$ **2.** D; *Examples:* $x < 5$, $7 + 2k \geq 11$, $-5 < 2z - 1 \leq 3$
3. C; *Examples:* $(-\infty, 5]$, $(1, \infty)$, $[-3, 3)$ **4.** A; *Example:* If $A = \{2, 4, 6, 8\}$ and $B = \{1, 2, 3\}$, then $A \cap B = \{2\}$.
5. B; *Example:* Using the sets *A* and *B* from Answer 4, $A \cup B = \{1, 2, 3, 4, 6, 8\}$.

Quick Review

CONCEPTS	EXAMPLES
1.1 Linear Equations in One Variable	
Solving a Linear Equation in One Variable	Solve $4(8-3x)=32-8(x+2)$.
Step 1 Simplify each side separately. • Clear any parentheses. • Clear any fractions or decimals. • Combine like terms.	$32-12x=32-8x-16$ Distributive property $32-12x=16-8x$ Combine like terms.
Step 2 Isolate the variable terms on one side.	$32-12x+12x=16-8x+12x$ Add $12x$. $32=16+4x$ Combine like terms. $32-16=16+4x-16$ Subtract 16. $16=4x$ Combine like terms.
Step 3 Isolate the variable.	$\frac{16}{4}=\frac{4x}{4}$ Divide by 4. $4=x$
Step 4 Check.	The solution set is $\{4\}$. This can be checked by substituting 4 for x in the original equation.
1.2 Formulas and Percent	
Solving a Formula for a Specified Variable (Solving a Literal Equation)	Solve $\mathcal{A}=\frac{1}{2}bh$ for h.
Step 1 Clear any parentheses and fractions.	$\mathcal{A}=\frac{1}{2}bh$
Step 2 Isolate all terms with the specified variable.	$2\mathcal{A}=2\left(\frac{1}{2}bh\right)$ Multiply by 2. $2\mathcal{A}=bh$
Step 3 Isolate the specified variable.	$\frac{2\mathcal{A}}{b}=h$, or $h=\frac{2\mathcal{A}}{b}$ Divide by b.
1.3 Applications of Linear Equations	
Solving an Applied Problem	How many liters of 30% alcohol solution and 80% alcohol solution must be mixed to obtain 100 L of 50% alcohol solution?
Step 1 Read the problem.	
Step 2 Assign a variable.	Let x = number of liters of 30% solution needed. Then $100-x$ = number of liters of 80% solution needed.

Liters of Solution	Percent (as a decimal)	Liters of Pure Alcohol
x	0.30	$0.30x$
$100-x$	0.80	$0.80(100-x)$
100	0.50	$0.50(100)$

CONCEPTS	EXAMPLES
Step 3 Write an equation.	From the last column of the table, the equation is $0.30x+0.80(100-x)=0.50(100)$.

CONCEPTS	EXAMPLES

Step 4 Solve the equation.

$0.30x + 0.80(100 - x) = 0.50(100)$	Equation from Step 3
$0.30x + 80 - 0.80x = 50$	Distributive property; Multiply.
$100(0.30x + 80 - 0.80x) = 100(50)$	Multiply by 100.
$30x + 8000 - 80x = 5000$	Distributive property; Multiply.
$-50x + 8000 = 5000$	Combine like terms.
$-50x = -3000$	Subtract 8000.
$x = 60$	Divide by −50.

Step 5 State the answer.

The solution of the equation is 60.

60 L of 30% solution and 100 − 60 = 40 L of 80% solution are needed.

Step 6 Check.

$0.30(60) + 0.80(100 - 60) = 50$ is true.

1.4 Further Applications of Linear Equations

To solve a uniform motion problem, draw a sketch and make a table. Use the formula

$$d = rt \quad \text{or} \quad rt = d.$$

Two cars start from towns 400 mi apart and travel toward each other. They meet after 4 hr. Find the rate of each car if one travels 20 mph faster than the other.

Let x = rate of the slower car in miles per hour.

Then $x + 20$ = rate of the faster car.

Use the information in the problem and $d = rt$ to complete a table.

	Rate	Time	Distance	
Slower Car	x	4	$4x$	
Faster Car	$x + 20$	4	$4(x + 20)$	
			400	← Total

A sketch shows that the sum of the distances, $4x$ and $4(x + 20)$, must be 400.

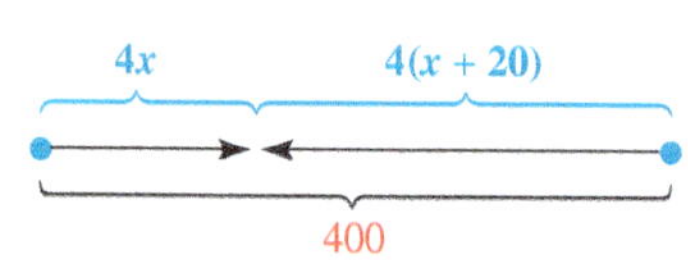

$4x + 4(x + 20) = 400$	
$4x + 4x + 80 = 400$	Distributive property
$8x + 80 = 400$	Combine like terms.
$8x = 320$	Subtract 80.
$x = 40$	Divide by 8.

The slower car travels 40 mph, and the faster car travels 40 + 20 = 60 mph.

$4(40) + 4(40 + 20) = 400$ is true.

CONCEPTS

EXAMPLES

1.5 Linear Inequalities in One Variable

Solving a Linear Inequality in One Variable

Step 1 Simplify each side separately.

Step 2 Isolate the variable terms on one side.

Step 3 Isolate the variable (x) to write the inequality in one of these forms.

$$x < k, \quad x \leq k, \quad x > k, \quad \text{or} \quad x \geq k$$

If an inequality is multiplied or divided by a negative number, the direction of the inequality symbol must be reversed.

Solve each inequality.

$$3(x+2) - 5x \leq 12$$

$$3x + 6 - 5x \leq 12 \quad \text{Distributive property}$$

$$-2x + 6 \leq 12 \quad \text{Combine like terms.}$$

$$-2x + 6 - 6 \leq 12 - 6 \quad \text{Subtract 6.}$$

$$-2x \leq 6 \quad \text{Combine like terms.}$$

$$\frac{-2x}{-2} \geq \frac{6}{-2} \quad \text{Divide by } -2. \text{ Change } \leq \text{ to } \geq.$$

$$x \geq -3$$

The solution set $[-3, \infty)$ is graphed here.

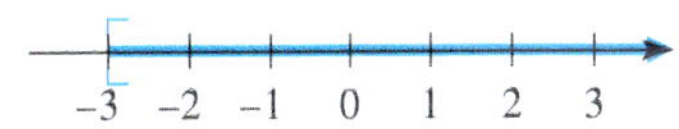

To solve a three-part inequality, work with all three parts at the same time.

$$-4 < 2x + 3 \leq 7$$

$$-4 - 3 < 2x + 3 - 3 \leq 7 - 3 \quad \text{Subtract 3.}$$

$$-7 < 2x \leq 4$$

$$\frac{-7}{2} < \frac{2x}{2} \leq \frac{4}{2} \quad \text{Divide by 2.}$$

$$-\frac{7}{2} < x \leq 2$$

The solution set $\left(-\frac{7}{2}, 2\right]$ is graphed here.

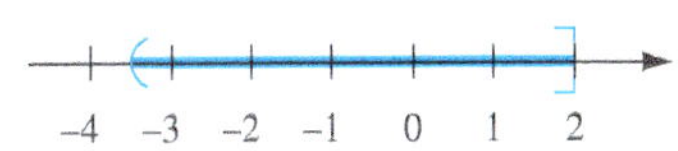

1.6 Set Operations and Compound Inequalities

Solving a Compound Inequality

Step 1 Solve each inequality in the compound inequality individually.

Step 2 If the inequalities are joined with *and,* then the solution set is the intersection of the two individual solution sets.

If the inequalities are joined with *or,* then the solution set is the union of the two individual solution sets.

Solve each compound inequality.

$$x + 1 > 2 \quad \text{and} \quad 2x < 6$$

$$x > 1 \quad \text{and} \quad x < 3$$

The solution set is $(1, 3)$.

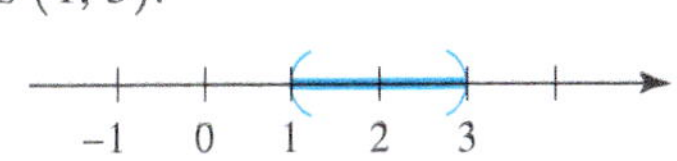

$$2(x+3) - 2 \leq 4 \quad \text{or} \quad -4x \leq -16$$

$$2x + 6 - 2 \leq 4 \quad \text{or} \quad \frac{-4x}{-4} \geq \frac{-16}{-4}$$

$$2x + 4 \leq 4 \quad \text{or} \quad x \geq 4$$

$$x \leq 0$$

The solution set is $(-\infty, 0] \cup [4, \infty)$.

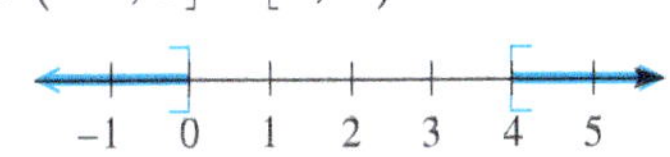

CONCEPTS | **EXAMPLES**

1.7 Absolute Value Equations and Inequalities

Solving Absolute Value Equations and Inequalities

Let k be a positive number.

To solve $|ax + b| = k$, solve the following compound equation.

$$ax + b = k \quad \textbf{or} \quad ax + b = -k$$

Solve each equation or inequality.

$$|x - 7| = 3$$

$$x - 7 = 3 \quad \text{or} \quad x - 7 = -3$$

$$x = 10 \quad \text{or} \quad x = 4 \qquad \text{Add 7.}$$

The solution set is $\{4, 10\}$.

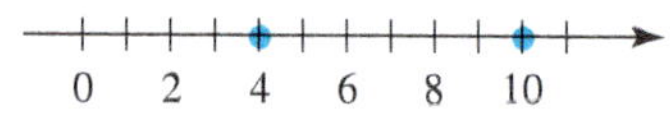

To solve $|ax + b| > k$, solve the following compound inequality.

$$ax + b > k \quad \textbf{or} \quad ax + b < -k$$

$$|x - 7| > 3$$

$$x - 7 > 3 \quad \text{or} \quad x - 7 < -3$$

$$x > 10 \quad \text{or} \quad x < 4 \qquad \text{Add 7.}$$

The solution set is $(-\infty, 4) \cup (10, \infty)$.

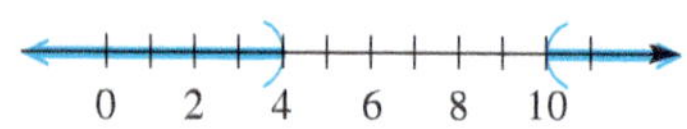

To solve $|ax + b| < k$, solve the following compound inequality.

$$-k < ax + b < k$$

$$|x - 7| < 3$$

$$-3 < x - 7 < 3$$

$$4 < x < 10 \qquad \text{Add 7.}$$

The solution set is $(4, 10)$.

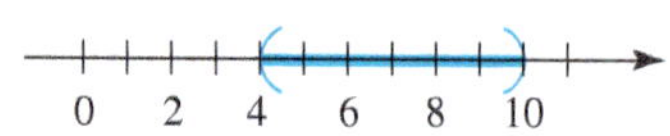

To solve an absolute value equation of the form

$$|ax + b| = |cx + d|,$$

solve the following compound equation.

$$ax + b = cx + d \quad \textbf{or} \quad ax + b = -(cx + d)$$

$$|x + 2| = |2x - 6|$$

$$x + 2 = 2x - 6 \quad \text{or} \quad x + 2 = -(2x - 6)$$

$$x = 8 \quad \text{or} \quad x + 2 = -2x + 6$$

$$3x = 4$$

$$x = \frac{4}{3}$$

The solution set is $\left\{\frac{4}{3}, 8\right\}$.

Chapter 1 Review Exercises

1.1 *Solve each equation.*

1. $-(8 + 3x) + 5 = 2x + 6$

2. $-\frac{3}{4}x = -12$

3. $\frac{2x + 1}{3} - \frac{x - 1}{4} = 0$

4. $5(2x - 3) = 6(x - 1) + 4x$

Answers: **1.** $\left\{-\frac{9}{5}\right\}$ **2.** $\{16\}$ **3.** $\left\{-\frac{7}{5}\right\}$ **4.** $\emptyset$

5. {all real numbers}; identity
6. $\varnothing$; contradiction
7. {0}; conditional equation
8. {all real numbers}; identity
9. $L = \frac{V}{HW}$
10. $b = \frac{2\mathcal{A} - Bh}{h}$, or $b = \frac{2\mathcal{A}}{h} - B$
11. $c = P - a - b - B$
12. $y = -\frac{4}{7}x + \frac{9}{7}$
13. 6 ft **14.** 3.0%
15. 6.5% **16.** 25°
17. \$775.3 billion
18. \$92.0 billion
19. $9 - \frac{1}{3}x$ **20.** $\frac{4x}{x+9}$
21. length: 13 m; width: 8 m
22. 17 in., 17 in., 19 in.

Solve each equation. Then tell whether the equation is a conditional equation, *an* identity, *or a* contradiction.

5. $10x - 3(2x - 4) = 4(x + 3)$

6. $13x - (x + 7) = 12x - 13$

7. $x + 6(x - 1) - (4 - x) = -10$

8. $5 + 3(2x + 6) = 5 + 3(2x + 6)$

1.2 *Solve for the specified variable.*

9. $V = LWH$ for L

10. $\mathcal{A} = \frac{1}{2}h(b + B)$ for b

11. $P = a + b + c + B$ for c

12. $4x + 7y = 9$ for y

1.2, 1.3 *Solve each problem.*

13. A rectangular solid has a volume of 180 ft^3. Its length is 6 ft and its width is 5 ft. Find its height.

14. The number of students attending colleges and universities in the United States dropped from about 20.1 million in the fall of 2011 to 19.5 million in the fall of 2013. To the nearest tenth of a percent, what was the percent decrease? (*Source:* National Student Clearinghouse Research Center.)

15. Find the annual simple-interest rate that Halina is earning if the principal of \$30,000 earns \$7800 interest in 4 yr.

16. If the Fahrenheit temperature is 77°, what is the corresponding Celsius temperature?

In 2012, total U.S. government spending was about \$3540 *billion (or* \$3.54 *trillion). The circle graph shows how the spending was divided.*

2012 U.S. Government Spending

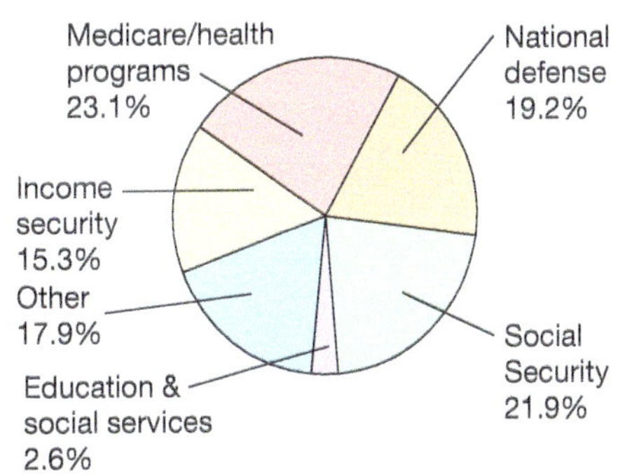

Source: U.S. Office of Management and Budget.

17. About how much was spent on Social Security?

18. About how much did the U.S. government spend on education and social services in 2012?

Translate each verbal phrase into a mathematical expression, using x as the variable.

19. One-third of a number, subtracted from 9

20. The product of 4 and a number, divided by 9 more than the number

Solve each problem.

21. The length of a rectangle is 3 m less than twice the width. The perimeter of the rectangle is 42 m. Find the length and width of the rectangle.

22. In a triangle with two sides of equal length, the third side measures 15 in. less than the sum of the two equal sides. The perimeter of the triangle is 53 in. Find the lengths of the three sides.

23. 12 kg **24.** 30 L
25. 10 L
26. \$10,000 at 6%; \$6000 at 4%
27. 15 dimes; 8 quarters
28. 7 nickels; 12 dimes
29. A **30.** 530 mi
31. 328 mi **32.** 2.2 hr
33. 50 km per hr; 65 km per hr
34. 46 mph
35. 40°, 45°, 95°
36. 150°, 30°

23. A candy clerk has three times as many kilograms of chocolate creams as peanut clusters. The clerk has 48 kg of the two candies altogether. How many kilograms of peanut clusters does the clerk have?

24. How many liters of a 20% solution of a chemical should be mixed with 15 L of a 50% solution to obtain a 30% mixture?

25. How much water should be added to 30 L of a 40% acid solution to reduce it to a 30% solution?

Liters of Solution	Percent (as a decimal)	Liters of Pure Acid
	0.40	
x		
	0.30	

26. Eric invested some money at 6% and \$4000 less than that amount at 4%. Find the amount invested at each rate if his total annual interest income was \$840.

Principal (in dollars)	Rate (as a decimal)	Interest (in dollars)
x	0.06	
	0.04	

1.4 *Solve each problem.*

27. A clerk has \$3.50 in dimes and quarters in her cash drawer. The number of dimes is 1 less than twice the number of quarters. How many of each denomination are there?

28. When Jim emptied his pockets one evening, he found he had 19 nickels and dimes with a total value of \$1.55. How many of each denomination did he have?

29. Which choice is the best *estimate* for the average rate for a trip of 405 mi that lasted 8.2 hr?

A. 50 mph **B.** 30 mph **C.** 60 mph **D.** 40 mph

30. A driver averaged 53 mph and took 10 hr to travel from Memphis to Chicago. What is the distance between Memphis and Chicago?

31. A small plane traveled from Warsaw to Rome, averaging 164 mph. The trip took 2 hr. What is the distance from Warsaw to Rome?

32. A passenger train and a freight train leave a town at the same time and go in opposite directions. They travel at rates of 60 mph and 75 mph, respectively. How long will it take for them to be 297 mi apart?

	Rate	Time	Distance
Passenger Train	60	x	
Freight Train	75	x	

33. Two cars leave small towns 230 km apart at the same time, traveling directly toward one another. One car travels 15 km per hr slower than the other car. They pass one another 2 hr later. What are their rates?

	Rate	Time	Distance
Faster car	x	2	
Slower car	$x - 15$	2	

34. An 85-mi trip to the beach took Susan 2 hr. During the second hour, a rainstorm caused her to average 7 mph less than she traveled during the first hour. Find her average rate for the first hour.

35. Find the measure of each angle in the triangle.

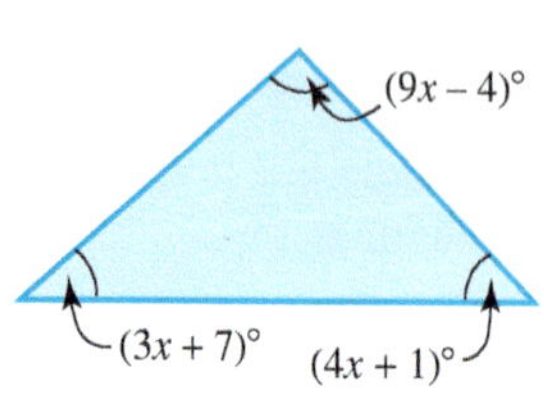

36. Find the measure of each marked angle.

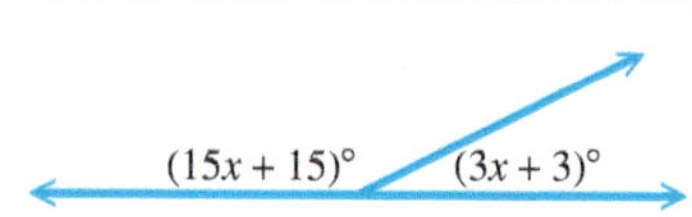

37. $(-9, \infty)$ **38.** $(-\infty, -3]$

39. $\left(\frac{3}{2}, \infty\right)$ **40.** $[-3, \infty)$

41. $[3, 5)$ **42.** $\left(\frac{59}{31}, \infty\right)$

43. **(a)** $x < 6.75$
(b) $6.75 \le x \le 7.25$
(c) $x > 7.25$

44. 38 m or less

45. 31 tickets or less (but at least 15)

46. any score greater than or equal to 61

47. {a, c} **48.** {a}

49. {a, c, e, f, g}

50. {a, b, c, d, e, f, g}

1.5 *Solve each inequality. Give the solution set in interval form.*

37. $-\frac{2}{3}x < 6$

38. $-5x - 4 \ge 11$

39. $\frac{6x + 3}{-4} < -3$

40. $5 - (6 - 4x) \ge 2x - 7$

41. $8 \le 3x - 1 < 14$

42. $\frac{5}{3}(x - 2) + \frac{2}{5}(x + 1) > 1$

Solve each problem.

43. Dr. Paul Donohue writes a syndicated column in which readers question him on a variety of health topics. Reader C. J. wrote, "Many people say they can weigh more because they have a large frame. How is frame size determined?" Here is Dr. Donohue's response:

> *"For a man, a wrist circumference between 6.75 and 7.25 in. [inclusive] indicates a medium frame. Anything above is a large frame and anything below, a small frame."*

(*Source:* Dr. Donohue, © 2004. Used by permission of North America Syndicate.)

Using x to represent wrist circumference in inches, write an inequality or a three-part inequality that represents wrist circumference for a male with the following.

(a) a small frame **(b)** a medium frame **(c)** a large frame

44. The perimeter of a rectangular playground must be no greater than 120 m. One dimension of the playground must be 22 m. Find the possible lengths of the other dimension of the playground.

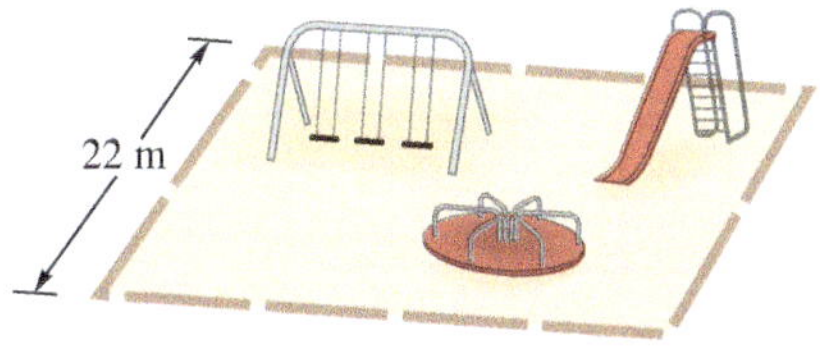

45. A group of college students wants to buy tickets to attend a performance of *Motown the Musical* at the Lunt-Fontanne Theatre in New York City. They can buy student mezzanine seats for a group rate of $51 per ticket for 15 tickets or more. If they have $1600 available to spend, how many tickets can they purchase at this price? (*Source:* www.broadway.com)

46. To pass algebra, a student must have an average of at least 70 on five tests. On the first four tests, a student has scores of 75, 79, 64, and 71. What possible scores on the fifth test would guarantee the student a passing grade in the class?

1.6 *Let* $A = \{a, b, c, d\}$, $B = \{a, c, e, f\}$, *and* $C = \{a, e, f, g\}$. *Find each set.*

47. $A \cap B$

48. $A \cap C$

49. $B \cup C$

50. $A \cup C$

Solve each compound inequality. Give the solution set in both interval and graph forms.

51. $x > 6$ and $x < 9$

52. $x + 4 > 12$ and $x - 2 < 12$

53. $x > 5$ or $x \leq -3$

54. $x \geq -2$ or $x < 2$

55. $x - 4 > 6$ and $x + 3 \leq 10$

56. $-5x + 1 \geq 11$ or $3x + 5 \geq 26$

Express each set in simplest interval form.

57. $(-3, \infty) \cap (-\infty, 4)$

58. $(-\infty, 6) \cap (-\infty, 2)$

59. $(4, \infty) \cup (9, \infty)$

60. $(1, 2) \cup (1, \infty)$

1.7 *Solve each equation.*

61. $|x| = 7$

62. $|x + 2| = 9$

63. $|3x - 7| = 8$

64. $|x - 4| = -12$

65. $|2x - 7| + 4 = 11$

66. $|4x + 2| - 7 = -3$

67. $|3x + 1| = |x + 2|$

68. $|2x - 1| = |2x + 3|$

Solve each inequality.

69. $|x| < 14$

70. $|-x + 6| \leq 7$

71. $|2x + 5| \leq 1$

72. $|x + 1| \geq -3$

73. $|3 - 4x| + 7 < -4$

74. $|-8 - 3x| - 7 > -8$

51. $(6, 9)$

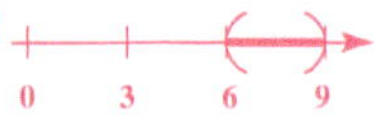

52. $(8, 14)$

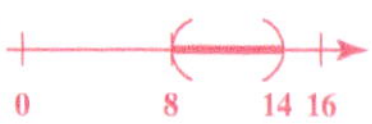

53. $(-\infty, -3] \cup (5, \infty)$

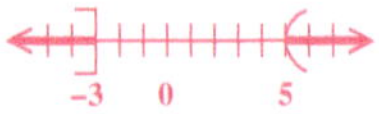

54. $(-\infty, \infty)$

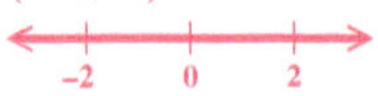

55. $\varnothing$

56. $(-\infty, -2] \cup [7, \infty)$

57. $(-3, 4)$ **58.** $(-\infty, 2)$

59. $(4, \infty)$ **60.** $(1, \infty)$

61. $\{-7, 7\}$ **62.** $\{-11, 7\}$

63. $\left\{-\frac{1}{3}, 5\right\}$ **64.** $\varnothing$

65. $\{0, 7\}$ **66.** $\left\{-\frac{3}{2}, \frac{1}{2}\right\}$

67. $\left\{-\frac{3}{4}, \frac{1}{2}\right\}$ **68.** $\left\{-\frac{1}{2}\right\}$

69. $(-14, 14)$ **70.** $[-1, 13]$

71. $[-3, -2]$ **72.** $(-\infty, \infty)$

73. $\varnothing$ **74.** $(-\infty, \infty)$

Chapter 1 Mixed Review Exercises

Solve.

1. $5 - (6 - 4x) > 2x - 5$

2. $ak + bt = 6r$ for k

3. $x + 4 < 7$ and $x + 5 \geq 3$

4. $\frac{4x + 2}{4} + \frac{3x - 1}{8} = \frac{x + 6}{16}$

5. $|3x + 6| \geq 0$

6. $-5x \geq -10$

7. $|3x + 2| + 4 = 9$

8. $0.05x + 0.03(1200 - x) = 42$

9. $|x + 3| \leq 13$

10. $\frac{3}{4}(x - 2) - \frac{1}{3}(5 - 2x) < -2$

11. $-4 < 3 - 2x < 9$

12. $-0.3x + 2.1(x - 4) \leq -6.6$

13. $|5x - 1| > 14$

14. $x \geq -2$ or $x < 4$

15. $|x - 1| = |2x + 3|$

16. $\frac{3x}{5} - \frac{x}{2} = 3$

17. $|3x - 7| = 4$

18. $5(2x - 7) = 2(5x + 3)$

19. $-5x < -30$ and $-7x > -56$

20. $-5x + 1 \geq 11$ or $3x + 5 \geq 26$

1. $(-2, \infty)$ **2.** $k = \frac{6r - bt}{a}$

3. $[-2, 3)$ **4.** $\{0\}$

5. $(-\infty, \infty)$ **6.** $(-\infty, 2]$

7. $\left\{-\frac{7}{3}, 1\right\}$ **8.** $\{300\}$

9. $[-16, 10]$ **10.** $\left(-\infty, \frac{14}{17}\right)$

11. $\left(-3, \frac{7}{2}\right)$ **12.** $(-\infty, 1]$

13. $\left(-\infty, -\frac{13}{5}\right) \cup (3, \infty)$

14. $(-\infty, \infty)$

15. $\left\{-4, -\frac{2}{3}\right\}$ **16.** $\{30\}$

17. $\left\{1, \frac{11}{3}\right\}$ **18.** $\varnothing$

19. $(6, 8)$

20. $(-\infty, -2] \cup [7, \infty)$

21. 10 ft **22.** 46, 47, 48
23. any amount greater than or equal to \$1100
24. 5 L
25. 1 hr

Solve each problem.

21. A newspaper recycling collection bin is in the shape of a box 1.5 ft wide and 5 ft long. If the volume of the bin is 75 ft^3, find the height.

22. The sum of the first and third of three consecutive integers is 47 more than the second integer. What are the integers?

23. To qualify for a company pension plan, an employee must average at least \$1000 per month in earnings. During the first four months of the year, an employee made \$900, \$1200, \$1040, and \$760. What possible amounts earned during the fifth month will qualify the employee?

24. How many liters of a 20% solution of a chemical should be mixed with 10 L of a 50% solution to obtain a 40% mixture?

25. An automobile averaged 45 mph for the first part of a trip and 50 mph for the second part. If the entire trip took 4 hr and covered 195 mi, for how long was the rate 45 mph?

Chapter 1 Test

FOR EXTRA HELP *Step-by-step test solutions are found on the Chapter Test Prep Videos available in* MyMathLab® *or on* YouTube.

View the complete solutions to all Chapter Test exercises in MyMathLab.

[1.1]
1. $\{-19\}$ **2.** $\{5\}$
3. (a) $\emptyset$; contradiction
(b) {all real numbers}; identity
(c) $\{0\}$; conditional equation
(d) {all real numbers}; identity

[1.2]
4. $h = \frac{3V}{b}$
5. $y = \frac{3}{2}x + 3$

[1.3, 1.4]
6. 2.668 hr
7. 1.25% **8.** 42.9%
9. \$8000 at 1.5%; \$20,000 at 2.5%
10. faster car: 60 mph; slower car: 45 mph
11. 40°, 40°, 100°

Solve each equation.

1. $3(2x - 2) - 4(x + 6) = 3x + 8 + x$ **2.** $0.08x + 0.06(x + 9) = 1.24$

3. Solve each equation. Then tell whether the equation is a *conditional equation,* an *identity,* or a *contradiction.*

(a) $3x - (2 - x) + 4x + 2 = 8x + 3$ **(b)** $\frac{x}{3} + 7 = \frac{5x}{6} - 2 - \frac{x}{2} + 9$

(c) $-4(2x - 6) = 5x + 24 - 7x$ **(d)** $\frac{x + 6}{10} + \frac{x - 4}{15} = \frac{x + 2}{6}$

4. Solve $V = \frac{1}{3}bh$ for h. **5.** Solve $-3x + 2y = 6$ for y.

Solve each problem.

6. The 2013 Indianapolis 500 (mile) race was won by Tony Kanaan, who averaged 187.433 mph. What was his time to the nearest thousandth of an hour? (*Source: World Almanac and Book of Facts.*)

7. A certificate of deposit pays \$456.25 in simple interest for 1 yr on a principal of \$36,500. What is the rate of interest?

8. A total of 160.0 billion items were sent through the U.S. Mail in a recent year. Of these, 68.7 billion items were first-class mail. What percent of the items, to the nearest tenth, were pieces of first-class mail? (*Source:* U.S. Postal Service.)

9. Tyler invested some money at 1.5% simple interest and some at 2.5% simple interest. The total amount of his investments was \$28,000, and the interest he earned during the first year was \$620. How much did he invest at each rate?

10. Two cars leave from the same point at the same time, traveling in opposite directions. One travels 15 mph slower than the other. After 6 hr, they are 630 mi apart. Find the rate of each car.

11. Find the measure of each angle in the triangle.

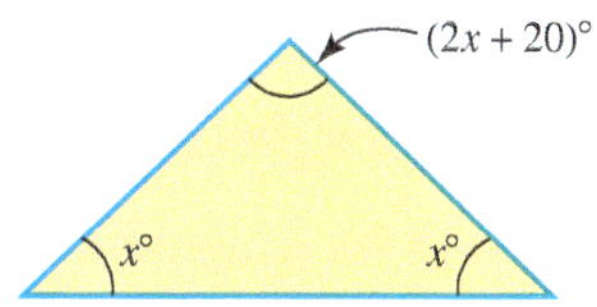

[1.5]

12. $[1, \infty)$ 

13. $(-\infty, 28)$ 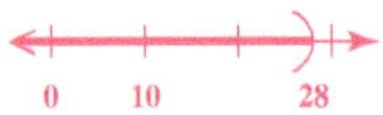

14. $(1, 2)$ 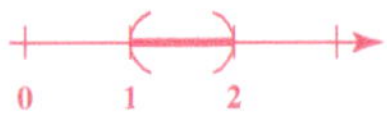

15. $[-3, 3]$ 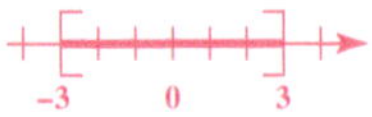

16. C

17. 82 or more

[1.6]

18. **(a)** $\{1, 5\}$ **(b)** $\{1, 2, 5, 7, 9, 12\}$

19. $[2, 9)$

20. $(-\infty, 3) \cup [6, \infty)$

[1.7]

21. $\left[-\frac{5}{2}, 1\right]$

22. $\left(-\infty, -\frac{7}{6}\right) \cup \left(\frac{17}{6}, \infty\right)$

23. $\{1, 5\}$ **24.** $\left(\frac{1}{3}, \frac{7}{3}\right)$

25. $\varnothing$ **26.** $\left\{-\frac{5}{3}, 3\right\}$

27. $\left\{-\frac{5}{7}, \frac{11}{3}\right\}$

28. **(a)** $\varnothing$ **(b)** $(-\infty, \infty)$ **(c)** $\varnothing$

Solve each inequality. Give the solution set in both interval and graph forms.

12. $4 - 6(x + 3) \leq -2 - 3(x + 6) + 3x$

13. $-\frac{4}{7}x > -16$

14. $-1 < 3x - 4 < 2$

15. $-6 \leq \frac{4}{3}x - 2 \leq 2$

16. Which one of the following inequalities is equivalent to $x < -3$?

A. $-3x < 9$ **B.** $-3x > -9$ **C.** $-3x > 9$ **D.** $-3x < -9$

17. Justin must have an average of at least 80 on the four tests in a course to get a B. He had scores of 83, 76, and 79 on the first three tests. What possible scores on the fourth test would guarantee him a B in the course?

18. Let $A = \{1, 2, 5, 7\}$ and $B = \{1, 5, 9, 12\}$. Find each set.

(a) $A \cap B$ **(b)** $A \cup B$

Solve each compound or absolute value equation or inequality.

19. $3x \geq 6$ and $x < 9$

20. $-4x \leq -24$ or $4x < 12$

21. $|4x + 3| \leq 7$

22. $|5 - 6x| > 12$

23. $|3x - 9| = 6$

24. $|-3x + 4| - 4 < -1$

25. $|7 - x| \leq -1$

26. $|3x - 2| + 1 = 8$

27. $|3 - 5x| = |2x + 8|$

28. If $k < 0$, what is the solution set of each of the following?

(a) $|8x - 5| < k$ **(b)** $|8x - 5| > k$ **(c)** $|8x - 5| = k$

2 Linear Equations, Graphs, and Functions

The concept of steepness, or grade, is mathematically interpreted using *slope*, one of the topics of this chapter.

2.1 Linear Equations in Two Variables

OBJECTIVES

1. Interpret a line graph.
2. Plot ordered pairs.
3. Find ordered pairs that satisfy a given equation.
4. Graph lines.
5. Find x- and y-intercepts.
6. Recognize equations of horizontal and vertical lines.
7. Use the midpoint formula.

VOCABULARY

- ☐ ordered pair
- ☐ origin
- ☐ x-axis
- ☐ y-axis
- ☐ rectangular (Cartesian) coordinate system
- ☐ components
- ☐ plot
- ☐ coordinate
- ☐ quadrant
- ☐ table of ordered pairs (table of values)
- ☐ graph of an equation
- ☐ first-degree equation
- ☐ linear equation in two variables
- ☐ x-intercept
- ☐ y-intercept

René Descartes (1596–1650)

OBJECTIVE 1 Interpret a line graph.

The line graph in **FIGURE 1** shows personal spending (in billions of dollars) on medical care in the United States from 2005 through 2012. About how much was spent on medical care in 2010?

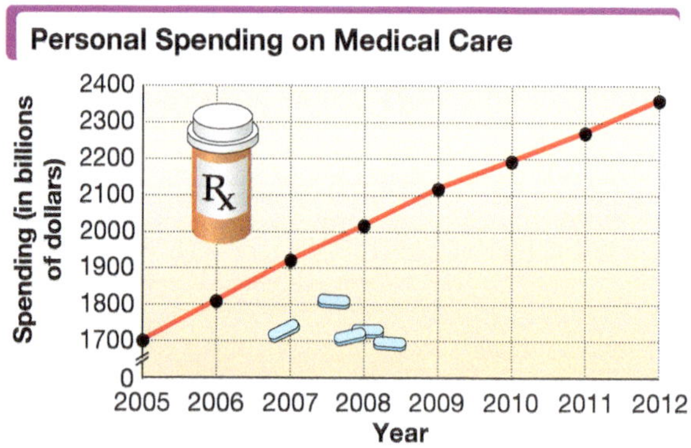

Source: Centers for Medicare & Medicaid Services.

FIGURE 1

The line graph in **FIGURE 1** presents information based on a method for locating a point in a plane developed by René Descartes, a 17th-century French mathematician. According to legend, Descartes was lying in bed ill watching a fly crawl about on the ceiling near a corner of the room. It occurred to him that the location of the fly could be described by determining its distances from the two adjacent walls. (See the figure at the right.)

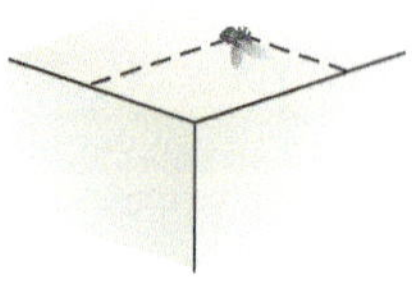

Locating a fly on a ceiling

We use this insight to plot points and graph linear equations in two variables whose graphs are straight lines.

OBJECTIVE 2 Plot ordered pairs.

Each of the pairs of numbers $(3, 2)$, $(-5, 6)$, and $(4, -1)$ is an example of an **ordered pair**—that is, a pair of numbers written within parentheses in which the order of the numbers is important. We graph an ordered pair by using two perpendicular number lines that intersect at their 0 points, as shown in **FIGURE 2**. The common 0 point is the **origin.**

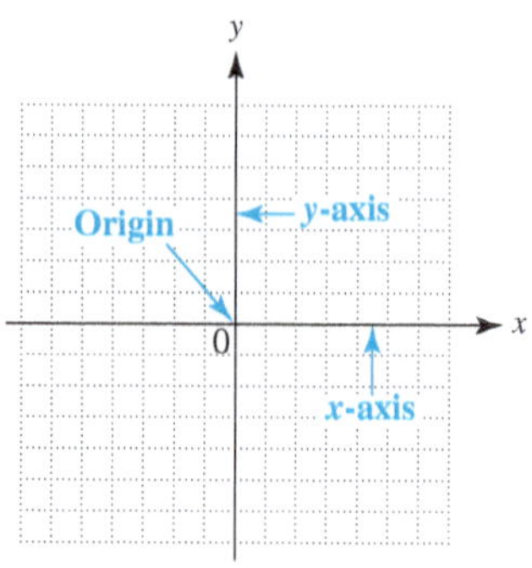

Rectangular coordinate system

FIGURE 2

The position of any point in this coordinate plane is determined by referring to the horizontal number line, or ***x*-axis,** and the vertical number line, or ***y*-axis.** The x-axis and the y-axis make up a **rectangular** (or **Cartesian,** for Descartes) **coordinate system.**

The numbers in an ordered pair (x, y) are its **components.** The first component indicates position relative to the x-axis, and the second component indicates position relative to the y-axis. For example, to locate, or **plot,** the point on the graph that corresponds to the ordered pair $(3, 2)$, we move three units from 0 to the right along the x-axis and then two units up parallel to the y-axis. See **FIGURE 3**. The numbers in an ordered pair are the **coordinates** of the corresponding point.

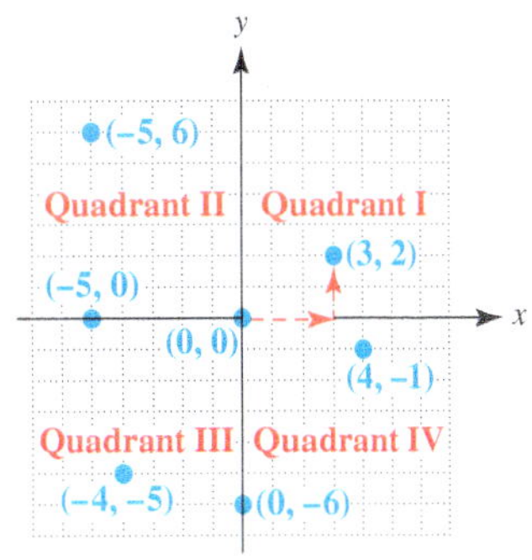

FIGURE 3

We can apply this method of locating ordered pairs to the line graph in **FIGURE 1**. We move along the horizontal axis to a year and then up parallel to the vertical axis to approximate spending for that year. Thus, we can write the ordered pair (2010, 2200) to indicate that in 2010, personal spending on medical care was about $2200 billion.

CAUTION The parentheses used to represent an ordered pair are also used to represent an open interval **(Section 1.5).** The context of the discussion tells whether ordered pairs or open intervals are being represented.

The four regions of the graph, shown in **FIGURE 3**, are **quadrants I, II, III,** and **IV,** reading counterclockwise from the upper right quadrant. ***The points on the x-axis and y-axis do not belong to any quadrant.***

OBJECTIVE 3 Find ordered pairs that satisfy a given equation.

Each solution of an equation with two variables, such as

$$2x + 3y = 6, \quad \text{Equation with two variables } x \text{ and } y$$

includes two numbers, one for each variable. We write the solutions as ordered pairs. ***(If x and y are used as the variables, the x-value is given first.)*** For example, we can show that $(6, -2)$ is a solution of $2x + 3y = 6$ by substitution.

$$2x + 3y = 6$$

$$2(6) + 3(-2) \stackrel{?}{=} 6 \quad \text{Let } x = 6, y = -2.$$

Use parentheses to avoid errors.

$$12 - 6 \stackrel{?}{=} 6 \quad \text{Multiply.}$$

$$6 = 6 \ \checkmark \quad \text{True}$$

Because the ordered pair $(6, -2)$ makes the equation true, it is a solution.

On the other hand, (5, 1) is *not* a solution of the equation $2x + 3y = 6$.

$$2x + 3y = 6$$
$$2(5) + 3(1) \stackrel{?}{=} 6 \quad \text{Let } x = 5, y = 1.$$
$$10 + 3 \stackrel{?}{=} 6 \quad \text{Multiply.}$$
$$13 = 6 \quad \text{False}$$

To find ordered pairs that satisfy an equation, select a number for one of the variables, substitute it into the equation for that variable, and solve for the other variable.

Because any real number could be selected for one variable and would lead to a real number for the other variable, an equation with two variables such as $2x + 3y = 6$ has an infinite number of solutions.

TEACHING TIP
Have students find additional ordered pairs that satisfy the equation

$$2x + 3y = 6$$

in **Example 1.** Stress that no matter what value we select for x (or y), we can always find a corresponding value for y (or x), although some of the numbers are messy.

NOW TRY EXERCISE 1
Complete each ordered pair for $2x - y = 4$.

(0, ___), (___, 0),
(4, ___), (___, 2)

Then write the results as a table of ordered pairs.

CLASSROOM EXAMPLE 1
Complete each ordered pair for $3x - 4y = 12$.

(0, ___), (___, 0),
(___, −2), (−6, ___)

Then write the results as a table of ordered pairs.

Answer:
$(0, -3), (4, 0), \left(\frac{4}{3}, -2\right), \left(-6, -\frac{15}{2}\right)$

x	y
0	−3
4	0
$\frac{4}{3}$	−2
−6	$-\frac{15}{2}$

EXAMPLE 1 Completing Ordered Pairs and Making a Table

In parts (a)–(d), complete each ordered pair for $2x + 3y = 6$. Then, in part (e), write the results as a table of ordered pairs.

(a) (0, ___)

$$2x + 3y = 6$$
$$2(0) + 3y = 6 \quad \text{Let } x = 0.$$
$$3y = 6 \quad \text{Multiply. Add.}$$
$$y = 2 \quad \text{Divide by 3.}$$

The ordered pair is (0, 2).

(b) (___, 0)

$$2x + 3y = 6$$
$$2x + 3(0) = 6 \quad \text{Let } y = 0.$$
$$2x = 6 \quad \text{Multiply. Add.}$$
$$x = 3 \quad \text{Divide by 2.}$$

The ordered pair is (3, 0).

(c) (−3, ___)

$$2x + 3y = 6$$
$$2(-3) + 3y = 6 \quad \text{Let } x = -3.$$
$$-6 + 3y = 6 \quad \text{Multiply.}$$
$$3y = 12 \quad \text{Add 6.}$$
$$y = 4 \quad \text{Divide by 3.}$$

The ordered pair is (−3, 4).

(d) (___, −4)

$$2x + 3y = 6$$
$$2x + 3(-4) = 6 \quad \text{Let } y = -4.$$
$$2x - 12 = 6 \quad \text{Multiply.}$$
$$2x = 18 \quad \text{Add 12.}$$
$$x = 9 \quad \text{Divide by 2.}$$

The ordered pair is (9, −4).

(e) We write a **table of ordered pairs** (or **table of values**) for these results as shown.

x	y	Ordered Pairs
0	2	⟵ Represents (0, 2) from part (a)
3	0	⟵ Represents (3, 0) from part (b)
−3	4	⟵ Represents (−3, 4) from part (c)
9	−4	⟵ Represents (9, −4) from part (d)

NOW TRY

NOW TRY ANSWER
1. (0, −4), (2, 0), (4, 4), (3, 2)

x	y
0	−4
2	0
4	4
3	2

OBJECTIVE 4 Graph lines.

The **graph of an equation** is the set of points corresponding to *all* ordered pairs that satisfy the equation. It gives a "picture" of the equation.

To graph $2x + 3y = 6$, we plot the ordered pairs found in **Objective 3** and **Example 1.** These points, plotted in FIGURE 4(a) on the next page, appear to lie on a straight line. If the infinite number of ordered pairs that satisfy the equation $2x + 3y = 6$ were graphed, they would form the straight line shown in FIGURE 4(b).

TEACHING TIP
Confirm with students that any additional ordered pairs they found that satisfy the equation $2x + 3y = 6$ lie on the line graphed in **FIGURE 4(b)**.

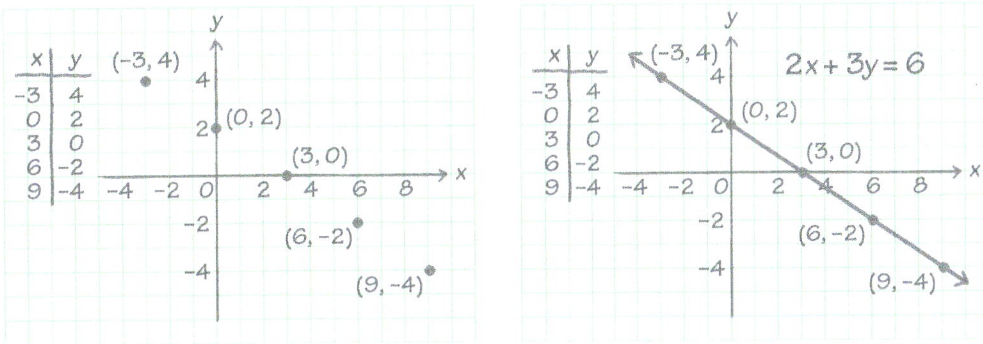

(a) (b)

FIGURE 4

TEACHING TIP
Mention to students that we discussed linear equations in one variable in **Chapter 1**.

The equation $2x + 3y = 6$ is a **first-degree equation** because it has no term with a variable to a power greater than 1.

The graph of any first-degree equation in two variables is a straight line.

Because first-degree equations with two variables have straight-line graphs, they are called ***linear** equations in two variables.*

Linear Equation in Two Variables

A **linear equation in two variables** (here x and y) can be written in the form

$$Ax + By = C,$$

where A, B, and C are real numbers and A and B are not both 0. This form is called **standard form.**

OBJECTIVE 5 Find x- and y-intercepts.

A straight line is determined if any two different points on the line are known. Therefore, finding two different points is sufficient to graph the line.

Two useful points for graphing are the x- and y-intercepts. The **x-intercept** is the point (if any) where the line intersects the x-axis. The **y-intercept** is the point (if any) where the line intersects the y-axis.* See **FIGURE 5**.

- The y-value of the point where the line intersects the x-axis is always 0.
- The x-value of the point where the line intersects the y-axis is always 0.

This suggests a method for finding the x- and y-intercepts.

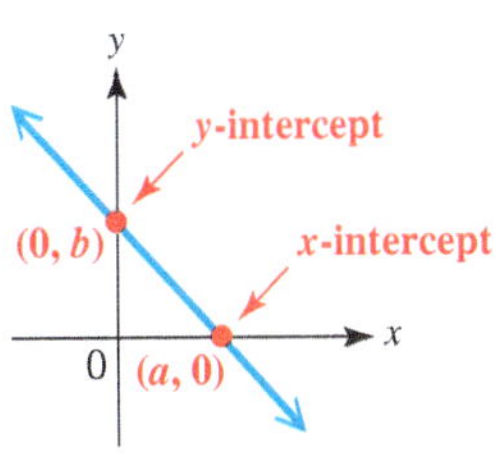

FIGURE 5

Finding Intercepts

When graphing the equation of a line, find the intercepts as follows.

Let $y = 0$ to find the x-intercept.

Let $x = 0$ to find the y-intercept.

*Some texts define an intercept as a number, not a point. For example, "y-intercept $(0, 4)$" would be given as "y-intercept 4."

NOW TRY EXERCISE 2

Find the x- and y-intercepts, and graph the equation.

$$x - 2y = 4$$

CLASSROOM EXAMPLE 2

Find the x- and y-intercepts, and graph the equation.

$$2x - y = 4$$

Answer: x-intercept: $(2, 0)$; y-intercept: $(0, -4)$

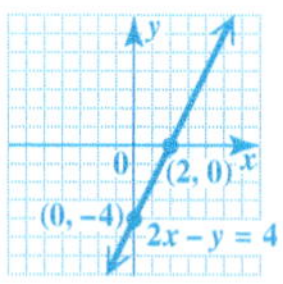

EXAMPLE 2 Finding Intercepts

Find the x- and y-intercepts of $4x - y = -3$, and graph the equation.

To find the x-intercept, let $y = 0$.

$$4x - y = -3$$
$$4x - 0 = -3 \quad \text{Let } y = 0.$$
$$4x = -3 \quad \text{Subtract.}$$
$$x = -\frac{3}{4} \quad \text{Divide by 4.}$$

The x-intercept is $\left(-\frac{3}{4}, 0\right)$.

To find the y-intercept, let $x = 0$.

$$4x - y = -3$$
$$4(0) - y = -3 \quad \text{Let } x = 0.$$
$$-y = -3 \quad \text{Multiply. Subtract.}$$
$$y = 3 \quad \text{Multiply by } -1.$$

The y-intercept is $(0, 3)$.

To guard against errors when graphing the equation, it is a good idea to find a third point. We arbitrarily choose $x = -2$, and substitute this value in the equation to find the corresponding value of y.

$$4x - y = -3$$
$$4(-2) - y = -3 \quad \text{Let } x = -2.$$
$$-8 - y = -3 \quad \text{Multiply.}$$
$$-y = 5 \quad \text{Add 8.}$$
$$y = -5 \quad \text{Multiply by } -1.$$

The ordered pair $(-2, -5)$ lies on the graph. We plot the three ordered pairs and draw a line through them. See **FIGURE 6.**

x	y
$-\frac{3}{4}$	0
0	3
-2	-5

Find a third point as a check.

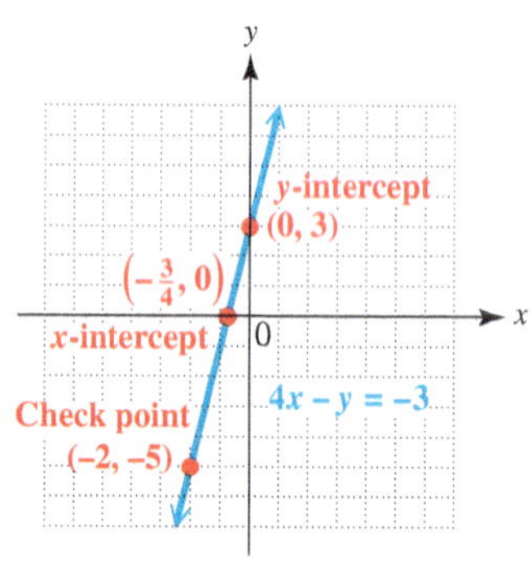

FIGURE 6

A linear equation with both x and y variables will have both x- and y-intercepts. Its graph will be a "slanted" line.

NOW TRY

CLASSROOM EXAMPLE 3

Graph $3x - y = 0$.

Answer:

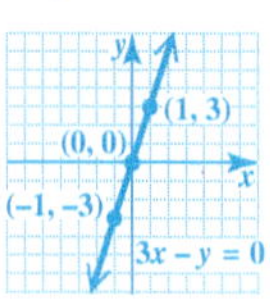

EXAMPLE 3 Graphing a Line That Passes through the Origin

Graph $x + 2y = 0$.

Find the x-intercept.

$$x + 2y = 0$$
$$x + 2(0) = 0 \quad \text{Let } y = 0.$$
$$x + 0 = 0 \quad \text{Multiply.}$$
$$x = 0 \quad x\text{-intercept is } (0, 0).$$

Find the y-intercept.

$$x + 2y = 0$$
$$0 + 2y = 0 \quad \text{Let } x = 0.$$
$$2y = 0 \quad \text{Add.}$$
$$y = 0 \quad y\text{-intercept is } (0, 0).$$

Both intercepts are the *same* point, $(0, 0)$, which means that the graph passes through the origin. To find a second point, we choose any nonzero number for x or y and solve for the other variable. We arbitrarily choose $x = 4$.

NOW TRY ANSWER

2. x-intercept: $(4, 0)$; y-intercept: $(0, -2)$

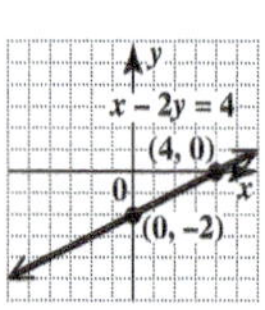

NOW TRY EXERCISE 3

Graph $2x + 3y = 0$.

TEACHING TIP

Students often have trouble finding points for a line passing through the origin. Tell them that a line of the form $Ax + By = 0$ will always pass through $(0, 0)$. A second point with integer coordinates can be found as follows:

1. Find a multiple of the coefficient of y.
2. Substitute this multiple for x.
3. Solve for y.
4. Use the results from Steps 2 and 3 as the second ordered pair.

$$x + 2y = 0$$

Choosing a multiple of 2 for x makes dividing by 2 in the last step easier.

$$4 + 2y = 0 \quad \text{Let } x = 4.$$

$$2y = -4 \quad \text{Subtract 4.}$$

$$y = -2 \quad \text{Divide by 2.}$$

This gives the ordered pair $(4, -2)$. As a final check, substitute 1 for y in the equation and verify that $(-2, 1)$ also lies on the line. The graph is shown in **FIGURE 7**.

This point serves as a check point.

x	y
-2	1
0	0
4	-2

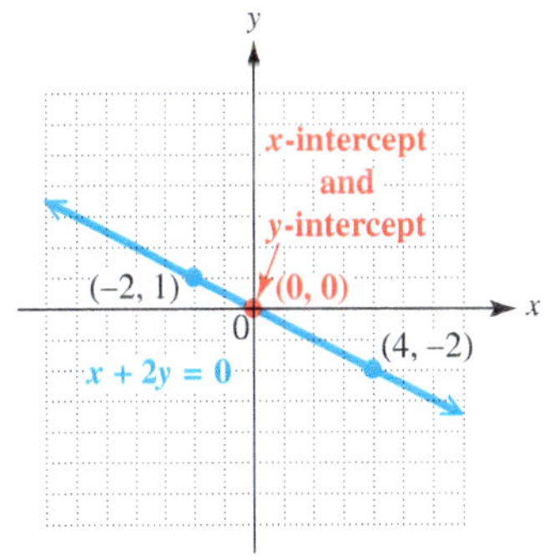

FIGURE 7

NOW TRY

OBJECTIVE 6 Recognize equations of horizontal and vertical lines.

A line parallel to the x-axis will not have an x-intercept. Similarly, a line parallel to the y-axis will not have a y-intercept. This is why we included the phrase "if any" when we introduced intercepts.

NOW TRY EXERCISE 4

Graph each equation.

(a) $y = -2$ (b) $x + 3 = 0$

CLASSROOM EXAMPLE 4

Graph each equation.

(a) $y = 3$ (b) $x + 2 = 0$

Answers:

(a)

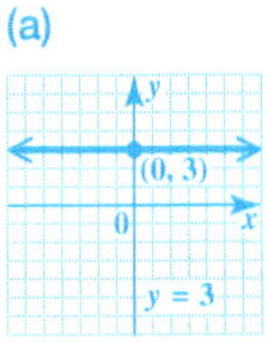

(b)

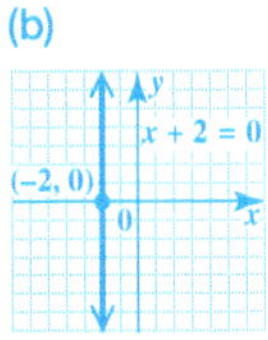

EXAMPLE 4 Graphing Horizontal and Vertical Lines

Graph each equation.

(a) $y = 2$ (This equation can be written as $0x + y = 2$.)

Because y *always* equals 2, there is no value of x corresponding to $y = 0$, and the graph has no x-intercept. One value where $y = 2$ is on the y-axis, so the y-intercept is $(0, 2)$. Plot any two other points with y-coordinate 2, such as $(-1, 2)$ and $(3, 2)$.

The graph is shown in **FIGURE 8**. It is a horizontal line.

x	y
-1	2
0	2
3	2

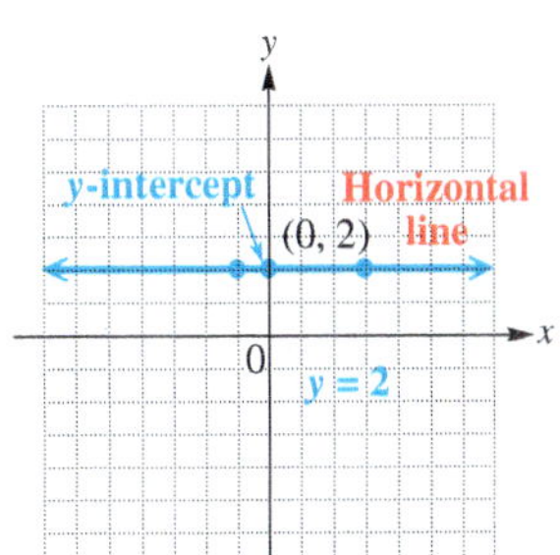

FIGURE 8

x	y
-1	-4
-1	0
-1	5

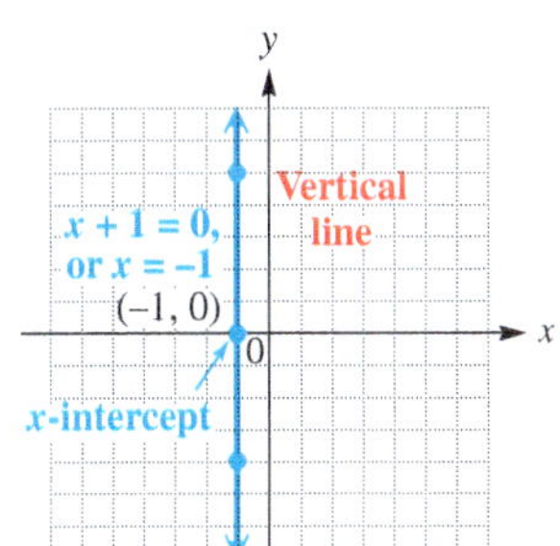

FIGURE 9

(b) $x + 1 = 0$ (This equation can be written as $x = -1$ or $x + 0y = -1$).

Because x *always* equals -1, there is no value of y that makes $x = 0$, and the graph has no y-intercept. One value where $x = -1$ is on the x-axis, so the x-intercept is $(-1, 0)$. Plot any two other points with x-coordinate -1, such as $(-1, -4)$ and $(-1, 5)$.

The graph is shown in **FIGURE 9**. It is a vertical line. NOW TRY

NOW TRY ANSWERS

3.

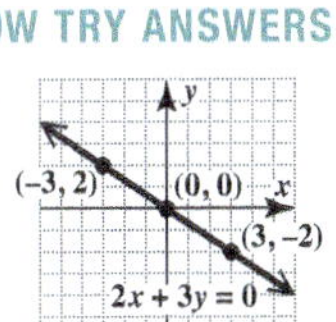

4. (a)

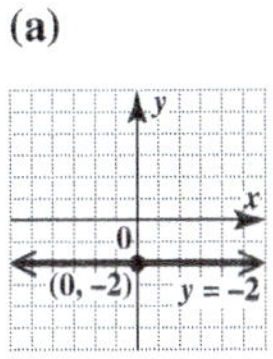

(b)

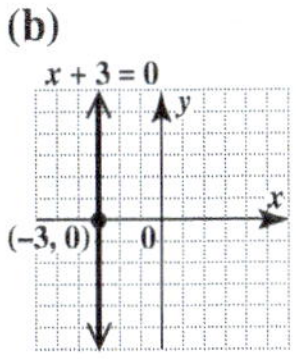

CAUTION A linear equation that has only *one* variable *x* or *y* will have a vertical or horizontal line as its graph.

1. An equation with only the variable *x* will always intersect the *x-axis* and thus will be ***vertical.*** It has the form $x = a$. The vertical line with equation $x = 0$ is the *y*-axis.
2. An equation with only the variable *y* will always intersect the *y-axis* and thus will be ***horizontal.*** It has the form $y = b$. The horizontal line with equation $y = 0$ is the *x*-axis.

OBJECTIVE 7 Use the midpoint formula.

If the coordinates of the endpoints of a line segment are known, then the coordinates of the *midpoint* of the segment can be found.

FIGURE 10 shows a line segment PQ with endpoints $P(-8, 4)$ and $Q(3, -2)$. R is the point with the same *x*-coordinate as P and the same *y*-coordinate as Q. So the coordinates of R are $(-8, -2)$.

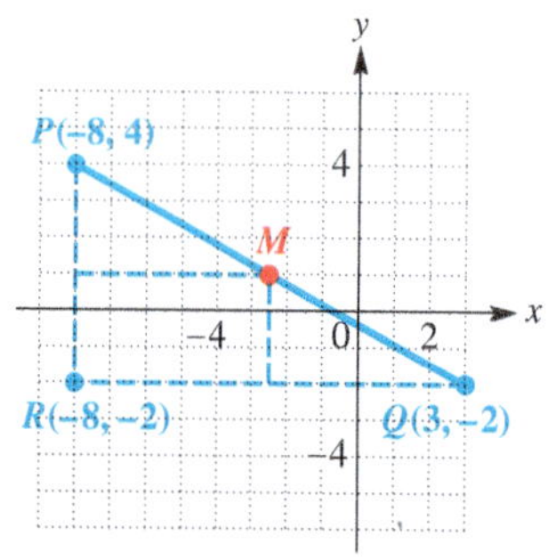

FIGURE 10

The *x*-coordinate of the midpoint M of PQ is the same as the *x*-coordinate of the midpoint of RQ. Because RQ is horizontal, the *x*-coordinate of its midpoint is the *average* (or *mean*) of the *x*-coordinates of its endpoints.

$$\frac{1}{2}(-8 + 3) = -2.5$$

The *y*-coordinate of M is the average (or *mean*) of the *y*-coordinates of the midpoint of PR.

$$\frac{1}{2}(4 + (-2)) = 1$$

The midpoint of PQ is $M(-2.5, 1)$.

This discussion leads to the *midpoint formula.*

TEACHING TIP
Point out the difference between x_2 and x^2. Students sometimes confuse these. Explain how to read them correctly.

Midpoint Formula

The midpoint M of a line segment PQ with endpoints (x_1, y_1) and (x_2, y_2) is found as follows.

$$M = \left(\frac{x_1 + x_2}{2}, \frac{y_1 + y_2}{2}\right)$$

The two nonspecific points (x_1, y_1) and (x_2, y_2) use **subscript notation.** Read (x_1, y_1) as **"*x*-sub-one, *y*-sub-one."**

NOW TRY EXERCISE 5

Find the midpoint of line segment PQ with endpoints $P(2, -5)$ and $Q(-4, 7)$.

CLASSROOM EXAMPLE 5

Find the midpoint of line segment PQ with endpoints $P(-5, 8)$ and $Q(2, 4)$.

Answer: $(-1.5, 6)$

EXAMPLE 5 Finding the Coordinates of a Midpoint

Find the midpoint of line segment PQ with endpoints $P(4, -3)$ and $Q(6, -1)$.

$$\begin{matrix} (x_1, & y_1) & & (x_2, & y_2) \\ \downarrow & \downarrow & & \downarrow & \downarrow \end{matrix}$$

$$P(4, -3) \quad \text{and} \quad Q(6, -1) \qquad \text{Label the points.}$$

$$M = \left(\frac{x_1 + x_2}{2}, \frac{y_1 + y_2}{2}\right) \qquad \text{Midpoint formula}$$

$$= \left(\frac{4+6}{2}, \frac{-3+(-1)}{2}\right) \qquad \text{Substitute.}$$

We are finding the average of the x-coordinates and the average of the y-coordinates.

$$= \left(\frac{10}{2}, \frac{-4}{2}\right) \qquad \text{Add in the numerators.}$$

$$= (5, -2) \leftarrow \text{Midpoint of segment } PQ$$

NOW TRY

NOTE When graphing with a graphing calculator, we "set up" a rectangular coordinate system. In the screen in **FIGURE 11**, which shows the **standard viewing window,** minimum x- and y-values are -10 and maximum x- and y-values are 10. The **scale** on each axis, here 1, determines the distance between the tick marks.

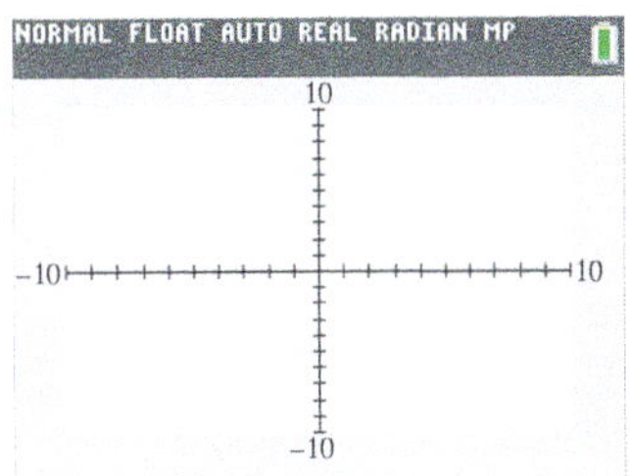

Standard viewing window

FIGURE 11

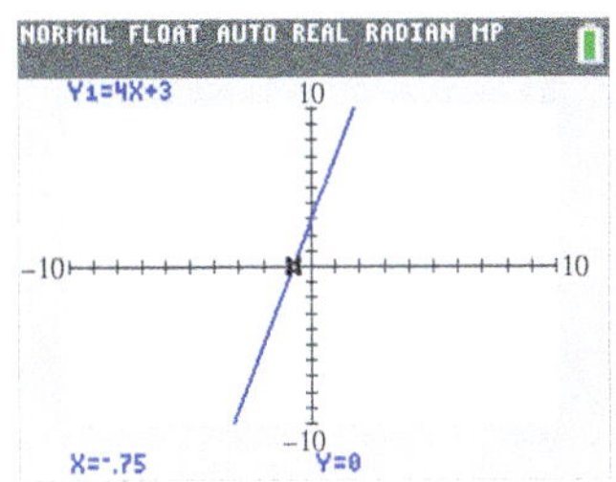

FIGURE 12

To graph an equation such as $4x - y = -3$, we use the intercepts $(-0.75, 0)$ and $(0, 3)$ to determine an appropriate window. Here, we choose the standard viewing window. We solve the equation for y to obtain $y = 4x + 3$ and enter it into the calculator. The graph in **FIGURE 12** also gives the x-intercept at the bottom of the screen.

NOW TRY ANSWER

5. $(-1, 1)$

2.1 Exercises

FOR EXTRA HELP

MyMathLab®

Complete solution available in MyMathLab

1. origin 2. x; y
3. y; x; x; y
4. horizontal; vertical
5. two; 2; 6
6. $y = 0$; $x = 0$

Concept Check *Fill in each blank with the correct response.*

1. The point with coordinates $(0, 0)$ is the _____ of a rectangular coordinate system.

2. For any value of x, the point $(x, 0)$ lies on the _____-axis. For any value of y, the point $(0, y)$ lies on the _____-axis.

3. To find the x-intercept of a line, we let _____ equal 0 and solve for _____. To find the y-intercept, we let _____ equal 0 and solve for _____.

4. The equation $y = 4$ has a _____ line as its graph, while $x = 4$ has a _____ line as its graph.

5. To graph a straight line, we must find a minimum of _____ points. The points (3, _____) and (_____, 4) lie on the graph of $2x - 3y = 0$.

6. The equation of the x-axis is _____. The equation of the y-axis is _____.

7. **(a)** x represents the year; y represents personal spending on medical care in billions of dollars. **(b)** about \$2360 billion **(c)** (2012, 2360) **(d)** 2008

8. **(a)** x represents the year; y represents the percentage of Americans who moved. **(b)** about 11% **(c)** (2013, 11) **(d)** In 1960, the percentage of Americans who moved was about 20%.

9. **(a)** I **(b)** III **(c)** II **(d)** IV **(e)** none **(f)** none

10. **(a)** III **(b)** I **(c)** II **(d)** IV **(e)** none **(f)** none

11. **(a)** I or III **(b)** II or IV **(c)** II or IV **(d)** I or III

12. One of the coordinates must be 0.

13.–22.

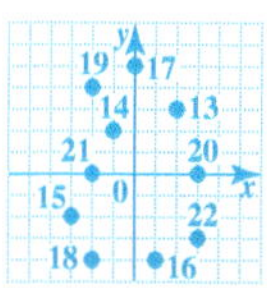

23. **(a)** -4; -3; -2; -1; 0

(b)

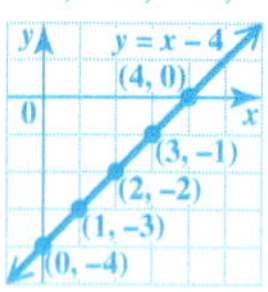

24. **(a)** 3; 4; 5; 6; 7

(b)

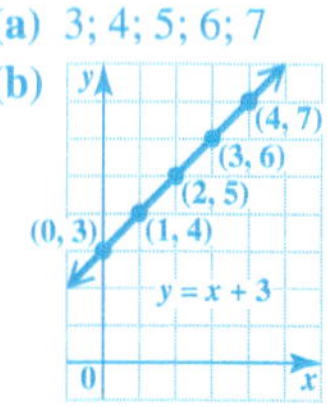

25. **(a)** -3; 3; 2; -1

(b)

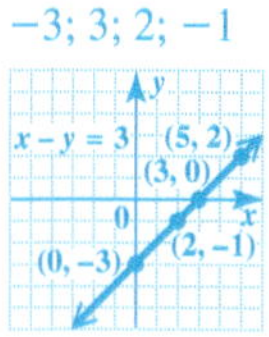

26. **(a)** -5; 5; -4; -2

(b)

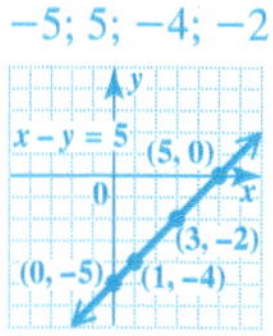

27. **(a)** $\frac{5}{2}$; 5; $\frac{3}{2}$; 1

(b)

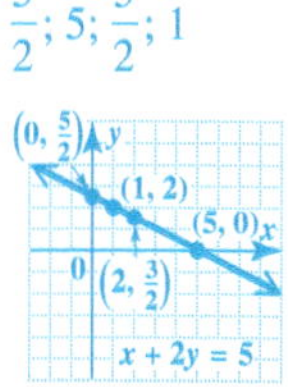

Solve each problem by locating ordered pairs on the graphs. ***See Objective 1.***

7. The graph indicates personal spending in billions of dollars on medical care in the United States.

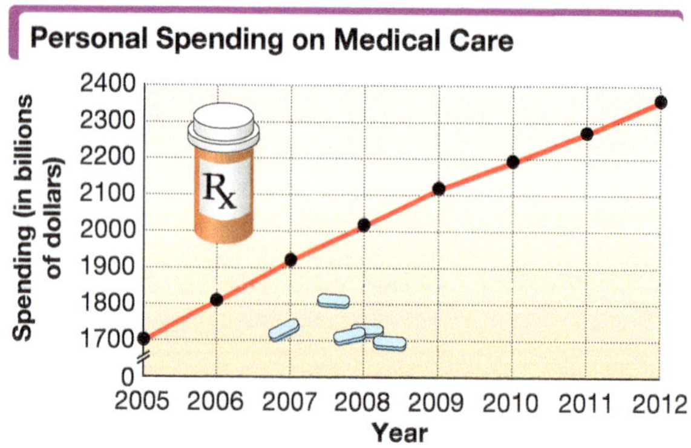

Source: Centers for Medicare & Medicaid Services.

(a) If (x, y) represents a point on the graph, what does x represent? What does y represent?

(b) Estimate spending in 2012.

(c) Write an ordered pair (x, y) that represents approximate spending in 2012.

(d) In what year was spending about \$2000 billion?

8. The graph shows the percentage of Americans who moved in selected years.

(a) If the ordered pair (x, y) represents a point on the graph, what does x represent? What does y represent?

(b) Estimate the percentage of Americans who moved in 2013.

(c) Write an ordered pair (x, y) that gives the approximate percentage of Americans who moved in 2013.

(d) What does the ordered pair (1960, 20) mean in the context of this graph?

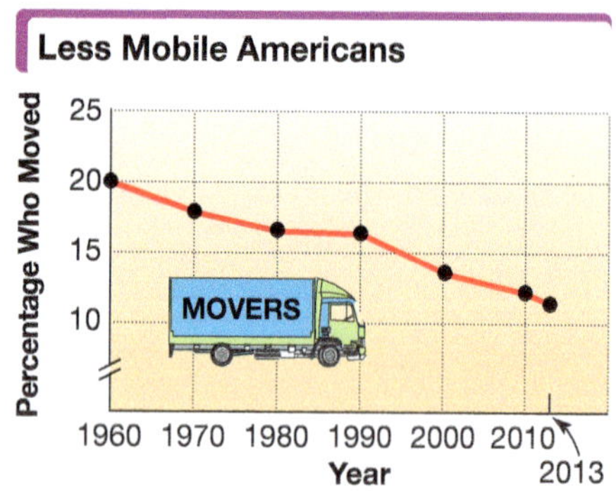

Source: U.S. Census Bureau.

Name the quadrant, if any, in which each point is located. ***See Objective 2.***

9. **(a)** $(1, 6)$ **(b)** $(-4, -2)$ **(c)** $(-3, 6)$ **(d)** $(7, -5)$ **(e)** $(-3, 0)$ **(f)** $(0, -0.5)$

10. **(a)** $(-2, -10)$ **(b)** $(4, 8)$ **(c)** $(-9, 12)$ **(d)** $(3, -9)$ **(e)** $(0, -8)$ **(f)** $(2.3, 0)$

11. *Concept Check* Use the given information to determine the quadrants in which the point (x, y) may lie.

(a) $xy > 0$ **(b)** $xy < 0$ **(c)** $\frac{x}{y} < 0$ **(d)** $\frac{x}{y} > 0$

12. *Concept Check* What must be true about the value of at least one of the coordinates of any point that lies along an axis?

Plot each point in a rectangular coordinate system. ***See Objective 2.***

13. $(2, 3)$ **14.** $(-1, 2)$ **15.** $(-3, -2)$ **16.** $(1, -4)$ **17.** $(0, 5)$

18. $(-2, -4)$ **19.** $(-2, 4)$ **20.** $(3, 0)$ **21.** $(-2, 0)$ **22.** $(3, -3)$

In the following exercises, (a) complete the given table for each equation and then (b) graph the equation. ***See Example 1 and* FIGURE 4.**

23. $y = x - 4$

x	y
0	
1	
2	
3	
4	

24. $y = x + 3$

x	y
0	
1	
2	
3	
4	

25. $x - y = 3$

x	y
0	
	0
5	
2	

26. $x - y = 5$

x	y
0	
	0
1	
3	

27. $x + 2y = 5$

x	y
0	
	0
2	
	2

28. $x + 3y = -5$

x	y
0	
	0
1	
	−1

29. $4x - 5y = 20$

x	y
0	
	0
2	
	−3

30. $6x - 5y = 30$

x	y
0	
	0
3	
	−2

31. $y = -2x + 3$

x	y
0	
1	
2	
3	

32. $y = -3x + 1$

x	y
0	
1	
2	
3	

33. *Concept Check* Consider the patterns formed in the tables for **Exercises 23 and 31.** Fill in each blank with the appropriate number.

(a) In **Exercise 23,** for every increase in x by 1 unit, y increases by ____ unit(s).

(b) In **Exercise 31,** for every increase in x by 1 unit, y decreases by ____ unit(s).

(c) Based on the results in parts (a) and (b), make a conjecture about a similar pattern for $y = 2x + 4$. Then test this conjecture.

34. *Concept Check* Which of the following equations have a graph that is a horizontal line? A vertical line?

A. $x - 6 = 0$ **B.** $x + y = 0$ **C.** $y + 3 = 0$ **D.** $y = -10$ **E.** $x + 1 = 5$

Find the x- and y-intercepts. Then graph each equation. ***See Examples 2–4.***

35. $2x + 3y = 12$ **36.** $5x + 2y = 10$ **37.** $x - 3y = 6$

38. $x - 2y = -4$ **39.** $5x + 6y = -10$ **40.** $3x - 7y = 9$

41. $\frac{2}{3}x - 3y = 7$ **42.** $\frac{5}{7}x + \frac{6}{7}y = -2$ **43.** $y = 5$

44. $y = -3$ **45.** $x = 2$ **46.** $x = -3$

47. $x + 4 = 0$ **48.** $x - 4 = 0$ **49.** $y + 2 = 0$

50. $y - 5 = 0$ **51.** $x + 5y = 0$ **52.** $x - 3y = 0$

53. $2x = 3y$ **54.** $4y = 3x$ **55.** $-\frac{2}{3}y = x$ **56.** $-\frac{3}{4}y = x$

28. (a) $-\frac{5}{3}$; -5; -2; -2

(b)
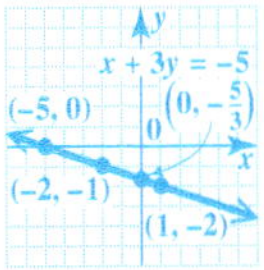

29. (a) -4; 5; $-\frac{12}{5}$; $\frac{5}{4}$

(b)
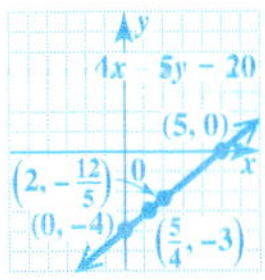

30. (a) -6; 5; $-\frac{12}{5}$; $\frac{10}{3}$

(b)
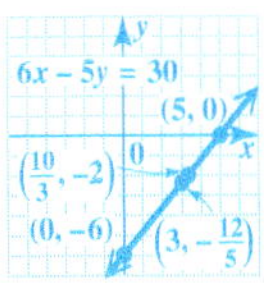

31. (a) 3; 1; −1; −3

(b)
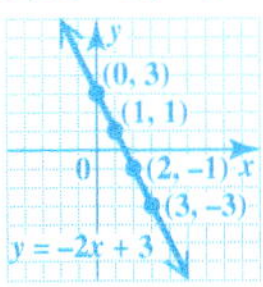

32. (a) 1; −2; −5; −8

(b)
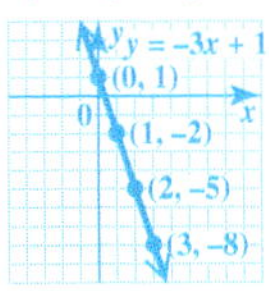

33. (a) 1 (b) 2

(c) For every increase in x by 1 unit, y increases by 2 units.

34. C, D; A, E

Due to space limitations, answers to odd-numbered Exercises 35–59 can be found in the Student Answer Section.

36. (2, 0); (0, 5)

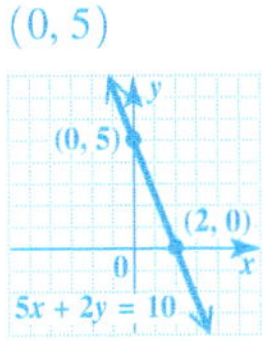

38. (−4, 0); (0, 2)

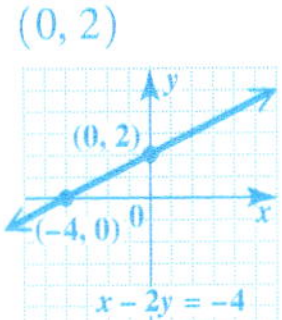

40. (3, 0); $\left(0, -\frac{9}{7}\right)$

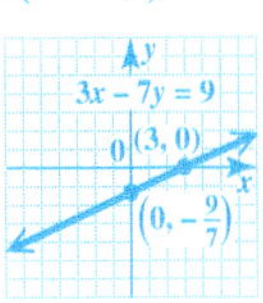

42. $\left(-\frac{14}{5}, 0\right)$; $\left(0, -\frac{7}{3}\right)$

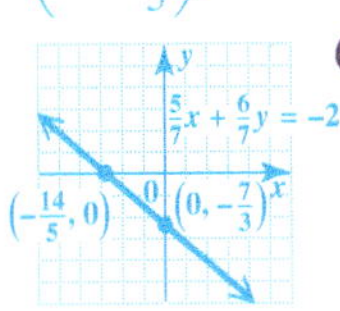

44. none; $(0, -3)$

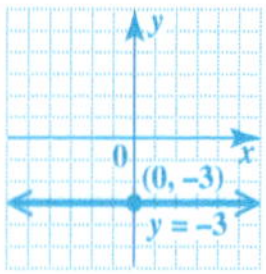

46. $(-3, 0)$; none

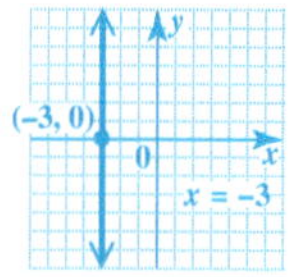

48. $(4, 0)$; none

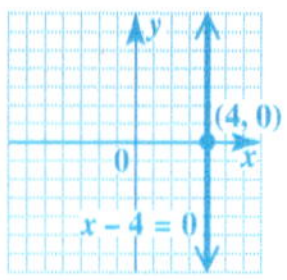

50. none; $(0, 5)$

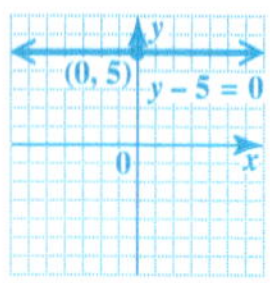

52. $(0, 0)$; $(0, 0)$

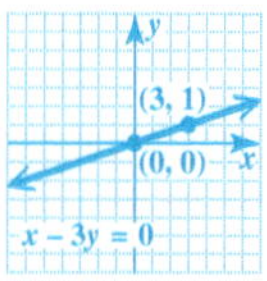

54. $(0, 0)$; $(0, 0)$

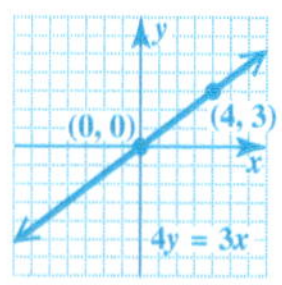

56. $(0, 0)$; $(0, 0)$

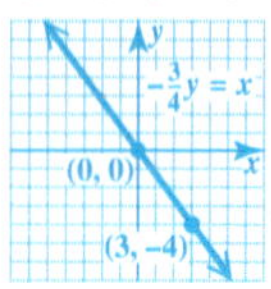

58. **(a)** $(2, 0)$; $(0, 4)$ **(b)** C
(c)

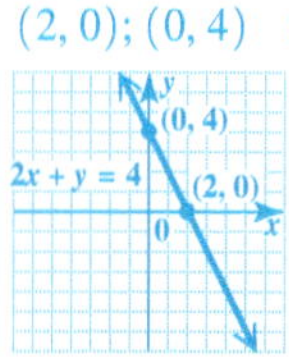

60. **(a)** $(6, 0)$; none **(b)** D
(c)

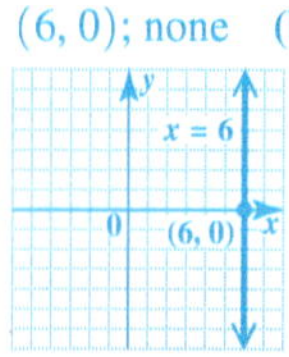

61. C **62.** D
63. B **64.** A

65. $(-5, -1)$ **66.** $(2, 5)$

67. $\left(\frac{9}{2}, -\frac{3}{2}\right)$ **68.** $\left(-\frac{3}{2}, \frac{5}{2}\right)$

69. $\left(0, \frac{11}{2}\right)$ **70.** $\left(\frac{3}{2}, 0\right)$

71. $(2.1, 0.9)$ **72.** $(3.8, 2.6)$

73. $(1, 1)$ **74.** $\left(\frac{7}{2}, \frac{1}{2}\right)$

75. $\left(-\frac{5}{12}, \frac{5}{28}\right)$ **76.** $\left(\frac{11}{20}, -\frac{23}{12}\right)$

77. $Q(11, -4)$ **78.** $Q(3, -4)$
79. $Q(4.5, 0.75)$ **80.** $Q(3.5, 2.25)$

Each table of values gives several points that lie on a line.

(a) What is the x-intercept of the line? The y-intercept?

(b) Which equation in choices A–D corresponds to the given table of values?

(c) Graph the equation.

57.

x	y
−4	−3
−2	0
0	3
2	6

A. $3x + 2y = 6$
B. $3x - 2y = -6$
C. $3x + 2y = -6$
D. $3x - 2y = 6$

58.

x	y
−1	6
0	4
1	2
2	0

A. $2x - y = 4$
B. $2x + y = -4$
C. $2x + y = 4$
D. $2x - y = -4$

59.

x	y
−2	−1
0	−1
2	−1
4	−1

A. $y = -1$
B. $y = 1$
C. $x = 1$
D. $x = -1$

60.

x	y
6	−1
6	0
6	1
6	2

A. $x = -6$
B. $y = 0$
C. $y = 6$
D. $x = 6$

Concept Check *Match each equation with its graph in choices A–D. (Coordinates of the points shown are integers.)*

61. $x + 3y = 3$ **62.** $x - 3y = -3$ **63.** $x - 3y = 3$ **64.** $x + 3y = -3$

A.

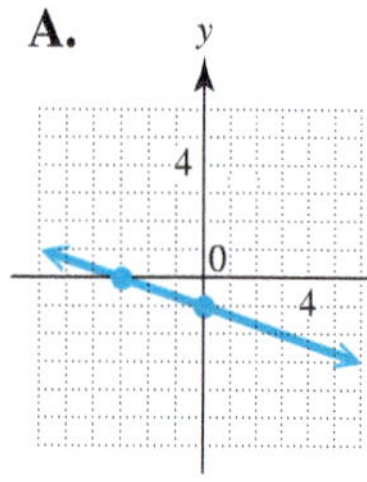

B.

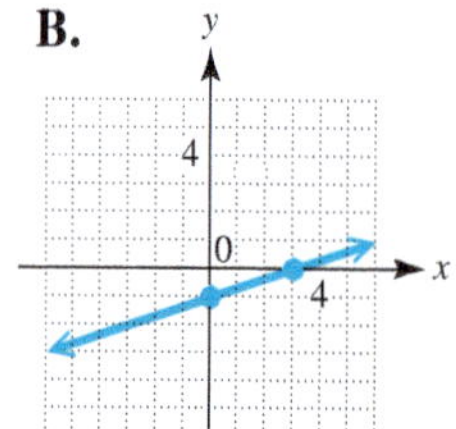

C.

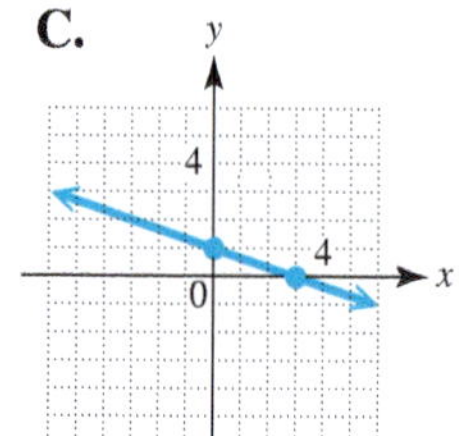

D.

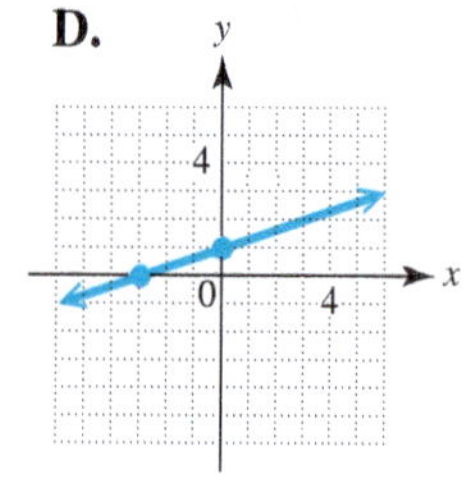

Find the midpoint of each segment with the given endpoints. ***See Example 5.***

65. $(-8, 4)$ and $(-2, -6)$

66. $(5, 2)$ and $(-1, 8)$

67. $(3, -6)$ and $(6, 3)$

68. $(-10, 4)$ and $(7, 1)$

69. $(-9, 3)$ and $(9, 8)$

70. $(4, -3)$ and $(-1, 3)$

71. $(2.5, 3.1)$ and $(1.7, -1.3)$

72. $(6.2, 5.8)$ and $(1.4, -0.6)$

Extending Skills *Find the midpoint of each segment with the given endpoints.*

73. $\left(\frac{1}{2}, \frac{1}{3}\right)$ and $\left(\frac{3}{2}, \frac{5}{3}\right)$

74. $\left(\frac{21}{4}, \frac{2}{5}\right)$ and $\left(\frac{7}{4}, \frac{3}{5}\right)$

75. $\left(-\frac{1}{3}, \frac{2}{7}\right)$ and $\left(-\frac{1}{2}, \frac{1}{14}\right)$

76. $\left(\frac{3}{5}, -\frac{1}{3}\right)$ and $\left(\frac{1}{2}, -\frac{7}{2}\right)$

Extending Skills *Segment PQ has the given coordinates for one endpoint P and for its midpoint M. Find the coordinates of the other endpoint Q. (Hint: Represent Q by (x, y) and write two equations using the midpoint formula, one involving x and the other involving y. Then solve for x and y.)*

77. $P(5, 8)$, $M(8, 2)$

78. $P(7, 10)$, $M(5, 3)$

79. $P(1.5, 1.25)$, $M(3, 1)$

80. $P(2.5, 1.75)$, $M(3, 2)$

STUDY SKILLS

Managing Your Time

Many college students juggle a difficult schedule and multiple responsibilities, including school, work, and family demands.

Time Management Tips

- **Read the syllabus for each class.** Understand class policies, such as attendance, late homework, and make-up tests. Find out how you are graded.
- **Make a semester or quarter calendar.** Put test dates and major due dates for *all* your classes on the *same* calendar. Try using a different color pen for each class.
- **Make a weekly schedule.** After you fill in your classes and other regular responsibilities, block off some study periods. Aim for 2 hours of study for each 1 hour in class.
- **Choose a regular study time and place** (such as the campus library). Routine helps.
- **Keep distractions to a minimum.** Get the most out of the time you have set aside for studying by limiting interruptions. Turn off your cell phone. Take a break from social media. Avoid studying in front of the TV.

- **Make "to-do" lists.** Number tasks in order of importance. Cross off tasks as you complete them.
- **Break big assignments into smaller chunks.** Make deadlines for each smaller chunk so that you stay on schedule.
- **Take breaks when studying.** Do not try to study for hours at a time. Take a 10-minute break each hour or so.
- **Ask for help when you need it.** Talk with your instructor during office hours. Make use of the learning center, tutoring center, counseling office, or other resources available at your school.

Think through and answer each question.

1. Evaluate when and where you are currently studying. Are the places you named quiet and comfortable? Are you studying when you are most alert?
2. How many hours do you have available for studying this week?
3. Which two or three of the above suggestions will you try this week to improve your time management?
4. Once the week is over, evaluate how these suggestions worked. What will you do differently next week?

2.2 The Slope of a Line

OBJECTIVES

1. Find the slope of a line given two points on the line.
2. Find the slope of a line given an equation of the line.
3. Graph a line given its slope and a point on the line.
4. Use slopes to determine whether two lines are parallel, perpendicular, or neither.
5. Solve problems involving average rate of change.

VOCABULARY

- ☐ rise
- ☐ run
- ☐ slope

Slope (steepness) is used in many practical ways. The slope (or *grade*) of a highway is often given as a percent. For example, a 10% $\left(\text{or } \frac{10}{100} = \frac{1}{10}\right)$ slope means that the highway rises 1 unit for every 10 horizontal units. Stairs and roofs have slopes too, as shown in **FIGURE 13**.

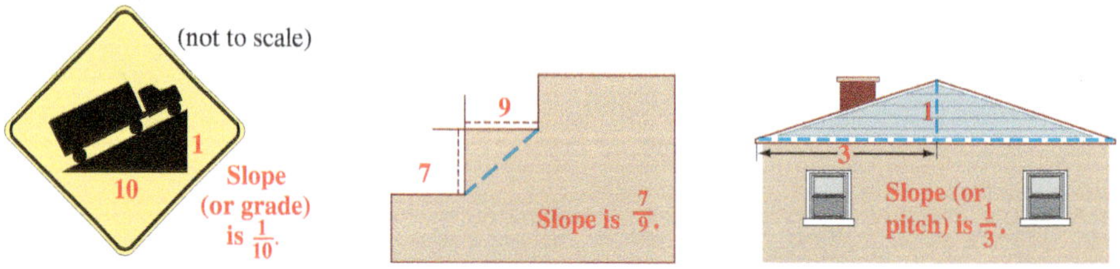

FIGURE 13

Slope is the ratio of vertical change, or **rise,** to horizontal change, or **run.** A simple way to remember this is to think, ***"Slope is rise over run."***

OBJECTIVE 1 Find the slope of a line given two points on the line.

To obtain a formal definition of the slope of a line, we designate two different points on the line as (x_1, y_1) and (x_2, y_2). See **FIGURE 14**.

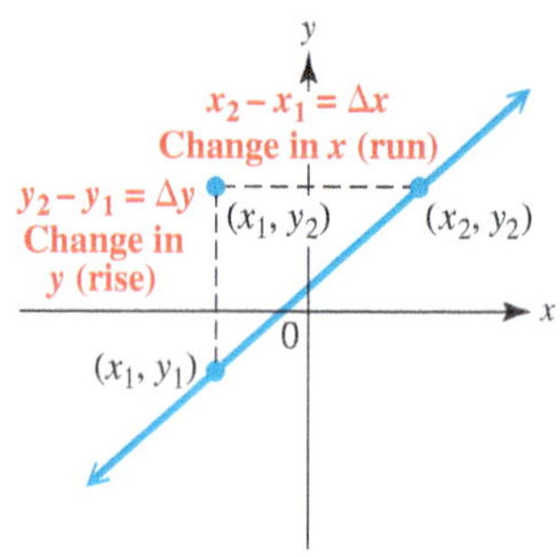

FIGURE 14

As we move along the line in **FIGURE 14** from (x_1, y_1) to (x_2, y_2), the y-value changes (vertically) from y_1 to y_2, an amount equal to $y_2 - y_1$. As y changes from y_1 to y_2, the value of x changes (horizontally) from x_1 to x_2 by the amount $x_2 - x_1$.

NOTE The Greek letter **delta,** Δ, is used in mathematics to denote "change in," so Δy and Δx represent the change in y and the change in x, respectively.

The ratio of the change in y to the change in x $\left(\text{"rise over run," or } \frac{\text{rise}}{\text{run}}\right)$ is the *slope* of the line, with the letter m traditionally used for slope.

Slope Formula

The **slope** m of the line passing through the distinct points (x_1, y_1) and (x_2, y_2) is defined as follows.

$$m = \frac{\textbf{rise}}{\textbf{run}} = \frac{\textbf{change in } y}{\textbf{change in } x} = \frac{\Delta y}{\Delta x} = \frac{y_2 - y_1}{x_2 - x_1} \quad (\textbf{where } x_1 \neq x_2)$$

NOW TRY EXERCISE 1
Find the slope of the line passing through the points $(2, -6)$ and $(-3, 5)$.

CLASSROOM EXAMPLE 1
Find the slope of the line passing through the points $(-6, 9)$ and $(3, -5)$.
Answer: $-\frac{14}{9}$

TEACHING TIP
Stress that in the slope formula the values of x_2 and y_2 are from the *same* ordered pair. The values of x_1 and y_1 are from the other ordered pair.

EXAMPLE 1 Finding the Slope of a Line

Find the slope of the line passing through the points $(2, -1)$ and $(-5, 3)$.

Label the points, and then apply the slope formula.

$$\underset{(x_1,\ y_1)}{(2, -1)} \quad \text{and} \quad \underset{(x_2,\ y_2)}{(-5, 3)}$$

$$\text{slope } m = \frac{y_2 - y_1}{x_2 - x_1} = \frac{3 - (-1)}{-5 - 2} \qquad \text{Substitute.}$$

$$= \frac{4}{-7}, \text{ or } -\frac{4}{7} \qquad \text{Subtract; } \frac{a}{-b} = -\frac{a}{b}$$

The slope is $-\frac{4}{7}$. See **FIGURE 15.**

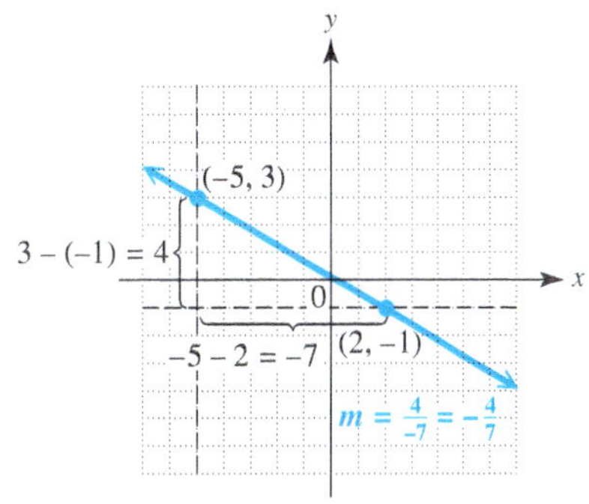

FIGURE 15

The same slope is obtained if we label the points in reverse order. ***It makes no difference which point is identified as (x_1, y_1) or (x_2, y_2).***

$$\underset{(x_2,\ y_2)}{(2, -1)} \quad \text{and} \quad \underset{(x_1,\ y_1)}{(-5, 3)}$$

$$\text{slope } m = \frac{y_2 - y_1}{x_2 - x_1} = \frac{-1 - 3}{2 - (-5)} \qquad \text{Substitute.}$$

y-values are in the numerator, *x*-values in the denominator.

$$= \frac{-4}{7}, \text{ or } -\frac{4}{7} \qquad \text{Subtract; } \frac{-a}{b} = -\frac{a}{b}$$

NOW TRY

Example 1 suggests the following important ideas regarding slope.

1. The slope is the same no matter which point we consider first.

2. Using similar triangles from geometry, we can show that the slope is the same no matter which two different points on the line we choose.

CAUTION *When calculating slope, remember that the change in y (rise) is the numerator and the change in x (run) is the denominator.*

Correct	Incorrect
$\frac{y_2 - y_1}{x_2 - x_1}$	~~$\frac{x_2 - x_1}{y_2 - y_1}$~~ or ~~$\frac{y_2 - y_1}{x_1 - x_2}$~~ or ~~$\frac{y_1 - y_2}{x_2 - x_1}$~~

Be careful to subtract the y-values and the x-values in the same order.

NOW TRY ANSWER
1. $-\frac{11}{5}$

OBJECTIVE 2 Find the slope of a line given an equation of the line.

When an equation of a line is given, one way to find its slope is to use the definition of slope with two different points on the line.

NOW TRY EXERCISE 2

Find the slope of the line $3x - 7y = 21$.

CLASSROOM EXAMPLE 2

Find the slope of the line $3x - 4y = 12$.

Answer: $\frac{3}{4}$

EXAMPLE 2 Finding the Slope of a Line

Find the slope of the line $4x - y = -8$.

The intercepts can be used as the two points needed to find the slope. Let $y = 0$ to find that the x-intercept is $(-2, 0)$. Then let $x = 0$ to find that the y-intercept is $(0, 8)$. Use these two points in the slope formula.

$$\text{slope } m = \frac{y_2 - y_1}{x_2 - x_1} = \frac{8 - 0}{0 - (-2)} \qquad (x_1, y_1) = (-2, 0),\ (x_2, y_2) = (0, 8)$$

$$= \frac{8}{2} \qquad \text{Subtract.}$$

$$= 4 \qquad \text{Divide.}$$

NOW TRY

NOW TRY EXERCISE 3

Find the slope of each line.

(a) $x = 4$ **(b)** $y - 6 = 0$

CLASSROOM EXAMPLE 3

Find the slope of each line.

(a) $y + 3 = 0$ (b) $x = -6$

Answers: (a) 0 (b) undefined

EXAMPLE 3 Applying the Slope Concept to Horizontal and Vertical Lines

Find the slope of each line.

(a) $y = 2$

The graph of $y = 2$ is a horizontal line. See **FIGURE 16**. To find the slope, select two different points on the line, such as $(3, 2)$ and $(-1, 2)$, and apply the slope formula.

$$m = \frac{2 - 2}{-1 - 3} = \frac{0}{-4} = 0 \qquad (x_1, y_1) = (3, 2),\ (x_2, y_2) = (-1, 2)$$

In this case, the *rise* is 0, so the slope is 0.

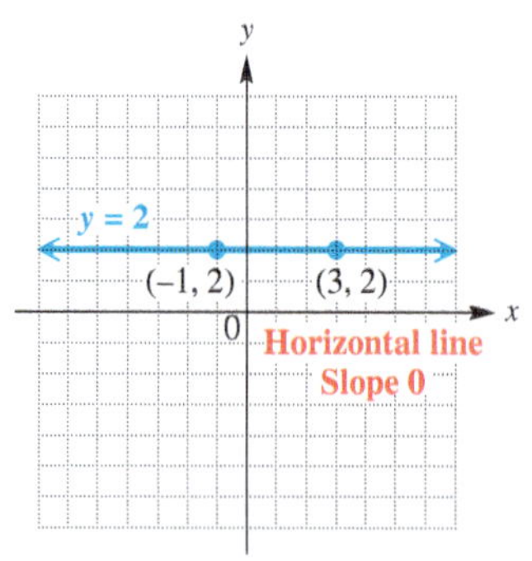

FIGURE 16

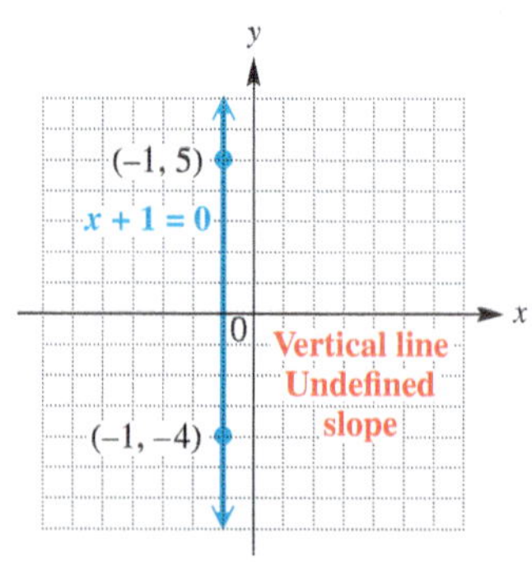

FIGURE 17

(b) $x + 1 = 0$

The graph of $x + 1 = 0$, or $x = -1$, is a vertical line. See **FIGURE 17**. Two points that satisfy the equation $x = -1$ are $(-1, 5)$ and $(-1, -4)$. We use these points and the slope formula.

$$m = \frac{-4 - 5}{-1 - (-1)} = \frac{-9}{0} \qquad (x_1, y_1) = (-1, 5),\ (x_2, y_2) = (-1, -4)$$

Here the *run* is 0. Because division by 0 is undefined, the slope is undefined. This is why the definition of slope includes the restriction that $x_1 \neq x_2$.

NOW TRY

NOW TRY ANSWERS

2. $\frac{3}{7}$

3. **(a)** undefined **(b)** 0

Example 3 illustrates the following important concepts.

TEACHING TIP
Review the cases for the slopes of horizontal and vertical lines. We suggest avoiding the use of *no slope*, as students confuse *no slope* with *0 slope*.

Horizontal and Vertical Lines

- An equation of the form $y = b$ always intersects the y-axis at the point $(0, b)$. A line with this equation is horizontal and has slope 0.
- An equation of the form $x = a$ always intersects the x-axis at the point $(a, 0)$. A line with this equation is vertical and has undefined slope.

The slope of a line can also be found directly from its equation. Consider the equation $4x - y = -8$ from **Example 2.** Solve this equation for y.

$$4x - y = -8 \quad \text{Equation from Example 2}$$
$$-y = -8 - 4x \quad \text{Subtract } 4x.$$
$$-y = -4x - 8 \quad \text{Commutative property}$$
$$y = 4x + 8 \quad \text{Multiply by } -1.$$

Going forward, we combine these steps.

The slope, 4, found with the slope formula in **Example 2** is the same number as the coefficient of x in the equation $y = 4x + 8$. We will see in the next section that this is true in general, *as long as the equation is solved for y.*

NOW TRY EXERCISE 4
Find the slope of the graph of $5x - 4y = 7$.

CLASSROOM EXAMPLE 4
Find the slope of the graph of $3x + 4y = 9$.
Answer: $-\frac{3}{4}$

TEACHING TIP
Find the slope in **Example 4** using two ordered pairs to show students that the same slope results.

EXAMPLE 4 Finding the Slope from an Equation

Find the slope of the graph of $3x - 5y = 8$.

Solve the equation for y.

$$3x - 5y = 8 \quad \text{Solve for } y.$$
$$-5y = -3x + 8 \quad \text{Subtract } 3x.$$
$$\frac{-5y}{-5} = \frac{-3x + 8}{-5} \quad \text{Divide each side by } -5.$$
$$y = \frac{3}{5}x - \frac{8}{5} \quad \frac{a + b}{c} = \frac{a}{c} + \frac{b}{c}$$

$\frac{-3x}{-5} = \frac{-3}{-5} \cdot \frac{x}{1} = \frac{3}{5}x$

The slope is given by the coefficient of x, so the slope is $\frac{3}{5}$. NOW TRY

NOTE We can solve the standard form of a linear equation $Ax + By = C$ (where $B \neq 0$) for y to show that, in general, **the slope of a line in this form is** $-\frac{A}{B}$**.**

$$Ax + By = C \quad \text{Standard form (Section 2.1)}$$
$$By = -Ax + C \quad \text{Subtract } Ax.$$
$$y = -\frac{A}{B}x + \frac{C}{B} \quad \text{Divide each term by } B.$$

The slope is given by the coefficient of x, $-\frac{A}{B}$. In the equation $3x - 5y = 8$ from **Example 4,** $A = 3$ and $B = -5$, so the slope is

$$-\frac{A}{B} = -\frac{3}{-5} = \frac{3}{5}. \quad \text{The same slope results.}$$

NOW TRY ANSWER
4. $\frac{5}{4}$

OBJECTIVE 3 Graph a line given its slope and a point on the line.

NOW TRY EXERCISE 5

Graph the line passing through $(-4, 1)$ that has slope $-\frac{2}{3}$.

CLASSROOM EXAMPLE 5

Graph the line passing through $(-3, -2)$ that has slope $\frac{1}{2}$.

Answer:

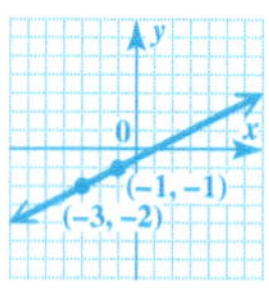

TEACHING TIP

Remind students that graphing a line requires a minimum of two points. Here, the first point is given. From the first point, we use the given slope (a positive or negative change in *y*, followed by a positive change in *x*) to get the second point.

As a check, have students confirm that the graph corresponds to the sign of the slope, as summarized in **FIGURE 20** on the next page.

EXAMPLE 5 Using the Slope and a Point to Graph Lines

Graph each line described.

(a) With slope $\frac{2}{3}$ and y-intercept $(0, -4)$

Begin by plotting the point $P(0, -4)$, as shown in **FIGURE 18.** Then use the geometric interpretation of slope to find a second point.

$$m = \frac{\text{change in } y}{\text{change in } x} = \frac{2}{3} \begin{matrix} \leftarrow \text{rise} \\ \leftarrow \text{run} \end{matrix}$$

We move 2 units *up* from $(0, -4)$ and then 3 units to the *right* to locate another point on the graph, $R(3, -2)$. The line through $P(0, -4)$ and R is the required graph.

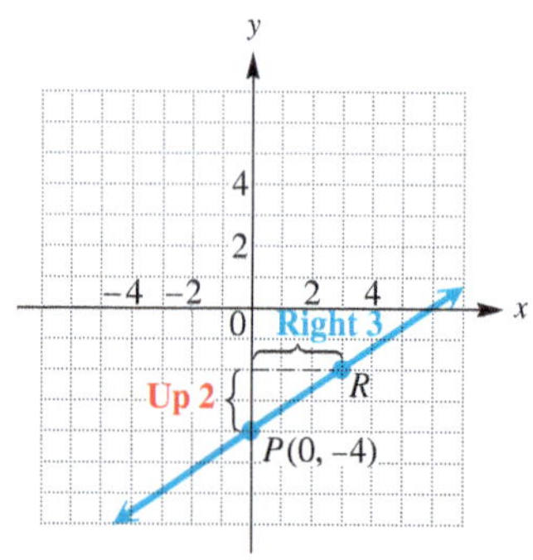

FIGURE 18

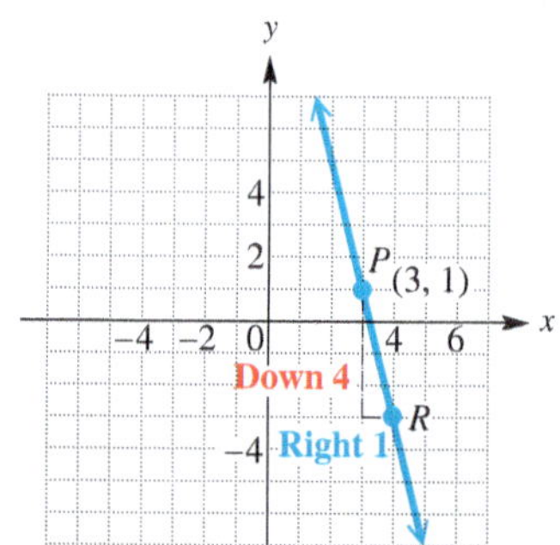

FIGURE 19

(b) Through $(3, 1)$ with slope -4

Start by plotting the point $P(3, 1)$, as shown in **FIGURE 19.** Find a second point R on the line by writing the slope -4 as $\frac{-4}{1}$ and using the geometric interpretation of slope.

$$m = \frac{\text{change in } y}{\text{change in } x} = \frac{-4}{1} \begin{matrix} \leftarrow \text{rise} \\ \leftarrow \text{run} \end{matrix}$$

We move 4 units *down* from $(3, 1)$ and then 1 unit to the *right* to locate this second point $R(4, -3)$. The line through $P(3, 1)$ and R is the required graph.

The slope -4 also could be written as

$$m = \frac{\text{change in } y}{\text{change in } x} = \frac{4}{-1}.$$

In this case, the second point R is located 4 units *up* and 1 unit to the *left.* Verify that this approach also produces the line in **FIGURE 19.**

NOW TRY

In **Example 5(a),** the slope of the line is the *positive* number $\frac{2}{3}$. The graph of the line in **FIGURE 18** slants up (rises) from left to right. The line in **Example 5(b)** has *negative* slope -4. As **FIGURE 19** shows, its graph slants down (falls) from left to right. These facts illustrate the following generalization.

NOW TRY ANSWER

5.

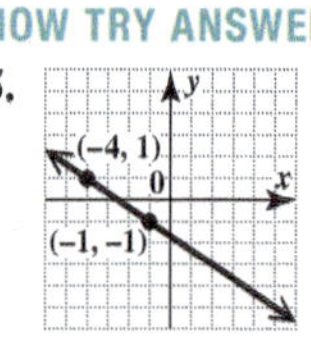

Orientation of a Line in the Plane

A positive slope indicates that the line slants *up* (rises) from left to right.

A negative slope indicates that the line slants *down* (falls) from left to right.

TEACHING TIP
Summarize the four cases for slopes of lines as in **FIGURE 20.**

Positive slope ↗ Negative slope ↘
Undefined slope ↕ Slope 0 ↔

Emphasize that *slope* 0 is not the same as *undefined slope*.

FIGURE 20 summarizes the four cases for slopes of lines.

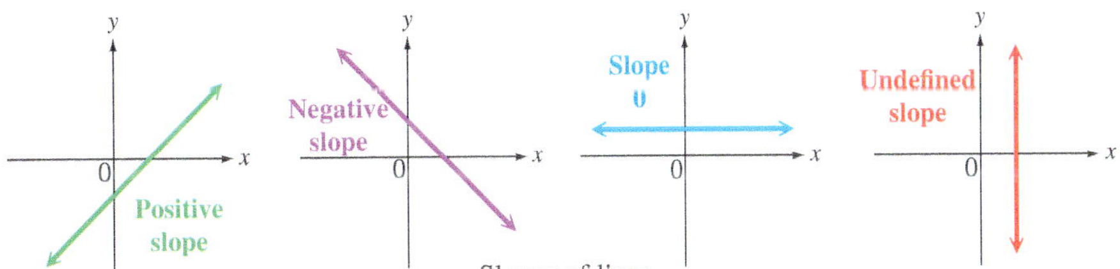

Slopes of lines

FIGURE 20

OBJECTIVE 4 Use slopes to determine whether two lines are parallel, perpendicular, or neither.

Recall that the slope of a line measures its steepness and that parallel lines have equal steepness.

CLASSROOM EXAMPLE 6
Is the line passing through $(-1, 2)$ and $(3, 5)$ parallel to the line passing through $(4, 7)$ and $(8, 10)$?

Answer: yes

Slopes of Parallel Lines

Two nonvertical lines with the same slope are parallel.

Two nonvertical parallel lines have the same slope.

NOW TRY EXERCISE 6
Is the line passing through $(2, 5)$ and $(4, 8)$ parallel to the line passing through $(2, 0)$ and $(-1, -2)$?

EXAMPLE 6 Determining Whether Two Lines Are Parallel

Are the lines L_1, passing through $(-2, 1)$ and $(4, 5)$, and L_2, passing through $(3, 0)$ and $(0, -2)$, parallel?

$$\text{Slope of } L_1: \quad m_1 = \frac{5-1}{4-(-2)} = \frac{4}{6} = \frac{2}{3} \qquad (x_1, y_1) = (-2, 1) \quad (x_2, y_2) = (4, 5)$$

$$\text{Slope of } L_2: \quad m_2 = \frac{-2-0}{0-3} = \frac{-2}{-3} = \frac{2}{3} \qquad (x_1, y_1) = (3, 0) \quad (x_2, y_2) = (0, -2)$$

Because the slopes are equal, the two lines are parallel. NOW TRY

TEACHING TIP
For the *parallel line* case, graph two different lines with y-intercepts $(0, 0)$ and $(0, 2)$, but with the same slope $\frac{2}{3}$.

For the *perpendicular line* case, graph a line with y-intercept $(0, 0)$ and slope $\frac{2}{3}$. Then graph a second line with slope $-\frac{3}{2}$.

To see how the slopes of perpendicular lines are related, consider a nonvertical line with slope $\frac{a}{b}$. If this line is rotated 90°, the vertical change and the horizontal change are interchanged and the slope is $-\frac{b}{a}$, because the horizontal change is now negative. See **FIGURE 21.** Thus, the slopes of perpendicular lines have product -1 and are negative reciprocals of each other.

For example, if the slopes of two lines are $\frac{3}{4}$ and $-\frac{4}{3}$, then the lines are perpendicular because

$$\frac{3}{4}\left(-\frac{4}{3}\right) = -1.$$

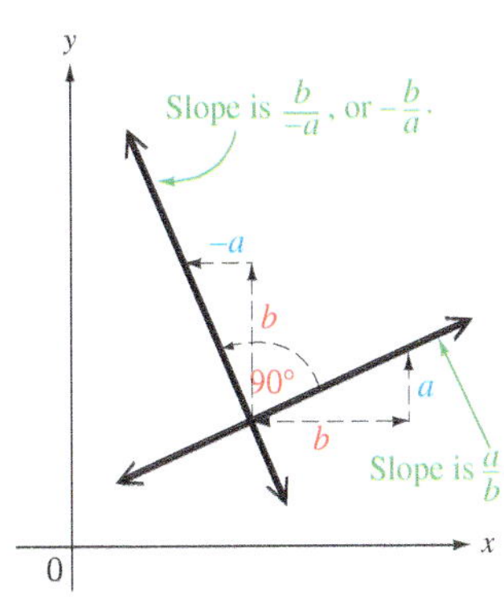

FIGURE 21

NOW TRY ANSWER
6. no

TEACHING TIP
It is helpful to say that perpendicular lines have "different slopes." Then proceed to get more specific.

Slopes of Perpendicular Lines

If neither is vertical, perpendicular lines have slopes that are negative reciprocals—that is, their product is -1. Also, lines with slopes that are negative reciprocals are perpendicular.

A line with slope 0 is perpendicular to a line with undefined slope.

NOW TRY EXERCISE 7
Are the lines with these equations perpendicular?

$$x + 2y = 7$$
$$2x = y - 4$$

CLASSROOM EXAMPLE 7
Are the lines with these equations perpendicular?

$$3x + 5y = 6$$
$$5x - 3y = 2$$

Answer: yes

EXAMPLE 7 Determining Whether Two Lines Are Perpendicular

Are the lines with equations $2y = 3x - 6$ and $2x + 3y = -6$ perpendicular?

Find the slope of each line by solving each equation for y.

$$2y = 3x - 6$$
$$y = \frac{3}{2}x - 3 \quad \text{Divide by 2.}$$
($\frac{3}{2}$ ← Slope)

$$2x + 3y = -6$$
$$3y = -2x - 6 \quad \text{Subtract } 2x.$$
$$y = -\frac{2}{3}x - 2 \quad \text{Divide by 3.}$$
($-\frac{2}{3}$ ← Slope)

The slopes are negative reciprocals because their product is $\frac{3}{2}\left(-\frac{2}{3}\right) = -1$. The lines are perpendicular. NOW TRY

NOTE In **Example 7,** alternatively, we could have found the slope of each line using intercepts and the slope formula. The graph of $2y = 3x - 6$ has x-intercept $(2, 0)$ and y-intercept $(0, -3)$.

$$m = \frac{0 - (-3)}{2 - 0} = \frac{3}{2} \quad \text{The same slope results. See Example 7.}$$

Find the intercepts of the graph of $2x + 3y = -6$ and use them to confirm the slope $-\frac{2}{3}$. Because the slopes are negative reciprocals, the lines are perpendicular.

CLASSROOM EXAMPLE 8
Determine whether the lines with these equations are *parallel, perpendicular,* or *neither.*

$$4x - y = 2$$
$$x - 4y = -8$$

Answer: neither

NOW TRY EXERCISE 8
Determine whether the lines with these equations are *parallel, perpendicular,* or *neither.*

$$2x - y = 4$$
$$2x + y = 6$$

EXAMPLE 8 Determining Whether Two Lines Are Parallel, Perpendicular, or Neither

Determine whether the lines with equations $2x - 5y = 8$ and $2x + 5y = 8$ are *parallel, perpendicular,* or *neither.*

Find the slope of each line by solving each equation for y.

$$2x - 5y = 8$$
$$-5y = -2x + 8 \quad \text{Subtract } 2x.$$
$$y = \frac{2}{5}x - \frac{8}{5} \quad \text{Divide by } -5.$$
($\frac{2}{5}$ ← Slope)

$$2x + 5y = 8$$
$$5y = -2x + 8 \quad \text{Subtract } 2x.$$
$$y = -\frac{2}{5}x + \frac{8}{5} \quad \text{Divide by 5.}$$
($-\frac{2}{5}$ ← Slope)

The slopes, $\frac{2}{5}$ and $-\frac{2}{5}$, are not equal, and they are not negative reciprocals because their product is $-\frac{4}{25}$, *not* -1. Thus, the two lines are *neither* parallel nor perpendicular.

NOW TRY

NOW TRY ANSWERS
7. yes
8. neither

OBJECTIVE 5 Solve problems involving average rate of change.

The slope formula applied to any two points on a line gives the **average rate of change** in y per unit change in x, where the value of y depends on the value of x.

For example, suppose the height of a boy increased from 60 to 68 in. between the ages of 12 and 16, as shown in **FIGURE 22**.

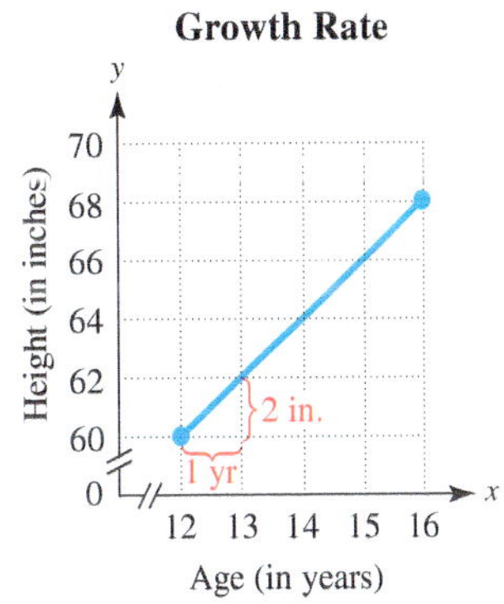

FIGURE 22

$$\frac{\text{Change in height } y \rightarrow 68 - 60}{\text{Change in age } x \rightarrow 16 - 12} = \frac{8}{4} = 2 \text{ in.}$$

Boy's average growth rate (or average change in height) ***per year***

The boy may actually have grown more than 2 in. during some years and less than 2 in. during other years. If we plotted ordered pairs (age, height) for those years and drew a line connecting any two of the points, the average rate of change would likely be slightly different than that found above. However, using the data for ages 12 and 16, the boy's *average* change in height was 2 in. per year over these years.

CLASSROOM EXAMPLE 9

There were approximately 45 million digital cable TV customers in 2010. Using this number for 2010 and the number for 2012 from the graph in **FIGURE 23**, find the average rate of change from 2010 to 2012. How does it compare to the average rate of change found in Example 9?

Answer: 1 million customers per year; It is less than the average rate of change from 2007 to 2012.

NOW TRY EXERCISE 9

There were approximately 40 million digital cable TV customers in 2008. Using this number for 2008 and the number for 2012 from the graph in **FIGURE 23**, find the average rate of change from 2008 to 2012. How does it compare with the average rate of change found in **Example 9?**

EXAMPLE 9 Interpreting Slope as Average Rate of Change

The graph in **FIGURE 23** shows the number of digital cable TV customers in the United States from 2007 to 2012. Find the average rate of change in number of customers per year.

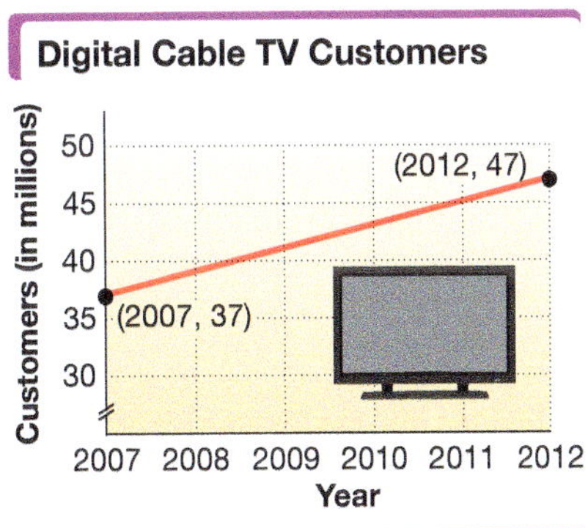

Source: SNL Kagan.

FIGURE 23

To find the average rate of change, we need two pairs of data. From the graph, we have the ordered pairs $(2007, 37)$ and $(2012, 47)$. We use the slope formula.

$$\text{average rate of change} = \frac{47 - 37}{2012 - 2007} = \frac{10}{5} = 2$$

A positive slope indicates an increase.

This means that the number of digital TV customers *increased* by an average of 2 million customers per year from 2007 to 2012.

NOW TRY

NOW TRY ANSWER

9. 1.75 million customers per year; It is less than the average rate of change from 2007 to 2012.

NOW TRY EXERCISE 10

In 2010, sales of digital camcorders in the United States totaled \$1150 million. In 2013, sales totaled \$137 million. Find the average rate of change in sales of digital camcorders per year, to the nearest million dollars. (*Source:* Consumer Electronics Association.)

CLASSROOM EXAMPLE 10

In 2000, 943 million compact discs were sold in the United States. In 2011, 241 million CDs were sold. Find the average rate of change in CDs sold per year, to the nearest million. (*Source:* Recording Industry Association of America.)

Answer: −64 million CDs per year

NOW TRY ANSWER

10. −\$338 million per year

EXAMPLE 10 Interpreting Slope as Average Rate of Change

In 2006, there were 65 million basic cable TV customers in the United States. There were 56 million such customers in 2012. Find the average rate of change in the number of customers per year. (*Source:* SNL Kagan.)

To use the slope formula, we let one ordered pair be (2006, 65) and the other be (2012, 56).

$$\text{average rate of change} = \frac{56-65}{2012-2006} = \frac{-9}{6} = -1.5$$

A negative slope indicates a decrease.

The graph in **FIGURE 24** confirms that the line through the ordered pairs falls from left to right and therefore has negative slope. Thus, the number of basic cable TV customers *decreased* by an average of 1.5 million customers per year from 2006 to 2012.

The negative sign in −1.5 denotes the *decrease*. (We say "The number of customers decreased by 1.5 million per year." It is *incorrect* to say "The number of customers decreased by −1.5 million per year.")

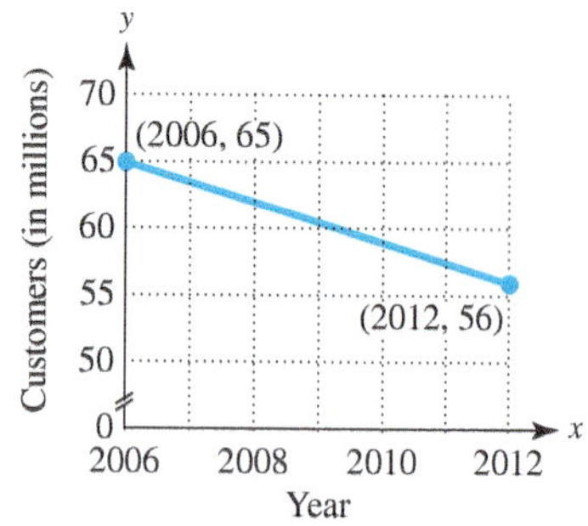

FIGURE 24

NOW TRY

2.2 Exercises

FOR EXTRA HELP

MyMathLab®

Complete solution available in MyMathLab

1. A, B, D, F
2. B, C, E
3. 5 ft

Concept Check *Answer each question.*

1. A hill rises 30 ft for every horizontal 100 ft. Which of the following express its slope (or grade)? (There are several correct choices.)

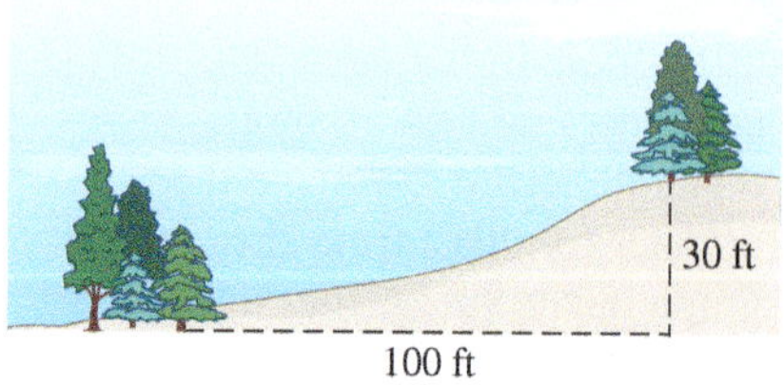

A. 0.3 **B.** $\frac{3}{10}$ **C.** $3\frac{1}{3}$

D. $\frac{30}{100}$ **E.** $\frac{10}{3}$ **F.** 30%

2. If a walkway rises 2 ft for every 24 ft on the horizontal, which of the following express its slope (or grade)? (There are several correct choices.)

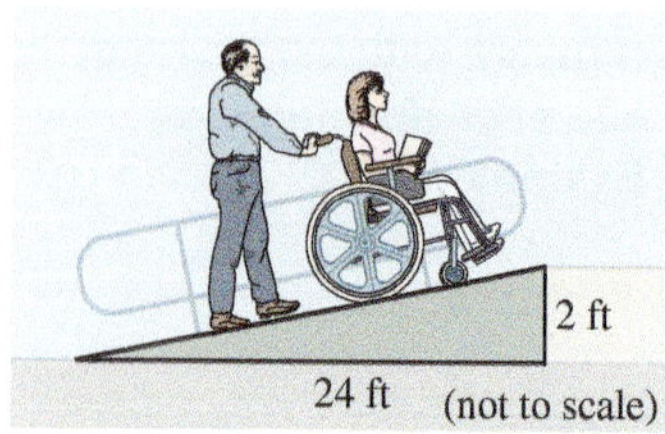

A. 12% **B.** $\frac{2}{24}$ **C.** $\frac{1}{12}$

D. 12 **E.** $8.\overline{3}\%$ **F.** $\frac{24}{2}$

3. A ladder leaning against a wall has slope 3. How many feet in the horizontal direction correspond to a rise of 15 ft?

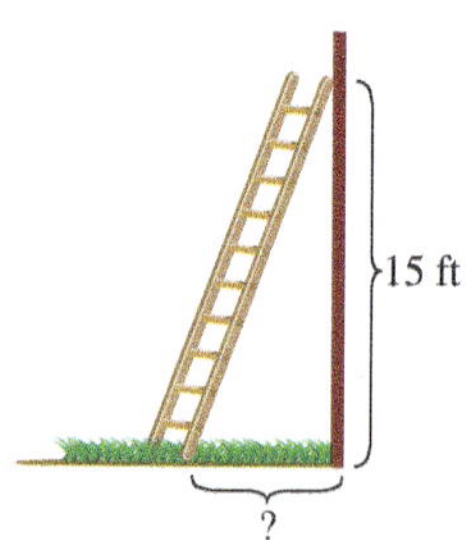

4. 2.5 ft

5. (a) C (b) A (c) D (d) B

6. Answers will vary.

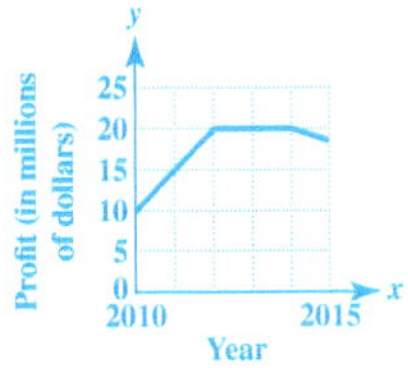

7. 2 8. 0

9. undefined 10. $-\frac{1}{3}$

11. 1 12. -4

13. -1 14. $\frac{7}{4}$

15. (a) *B* (b) *C* (c) *A* (d) *D*

4. A hill has slope 0.05. How many feet in the vertical direction correspond to a run of 50 ft?

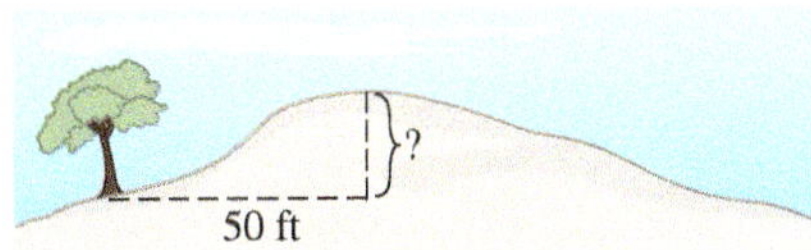

(not to scale)

5. *Concept Check* Match each situation in parts (a)–(d) with the most appropriate graph in choices A–D.

(a) Sales rose sharply during the first quarter, leveled off during the second quarter, and then rose slowly for the rest of the year.

(b) Sales fell sharply during the first quarter and then rose slowly during the second and third quarters before leveling off for the rest of the year.

(c) Sales rose sharply during the first quarter and then fell to the original level during the second quarter before rising steadily for the rest of the year.

(d) Sales fell during the first two quarters of the year, leveled off during the third quarter, and rose during the fourth quarter.

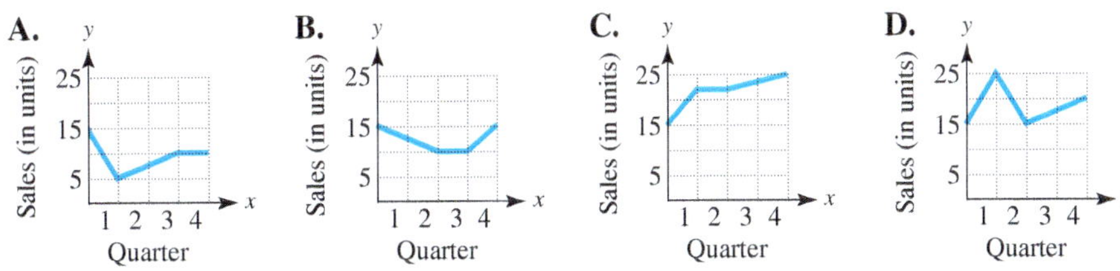

6. *Concept Check* Using the given axes, draw a graph that illustrates the following description.

Profits for a business were \$10 million in 2010. They rose sharply from 2010 through 2012, remained constant from 2012 through 2014, and then fell slowly from 2014 through 2015.

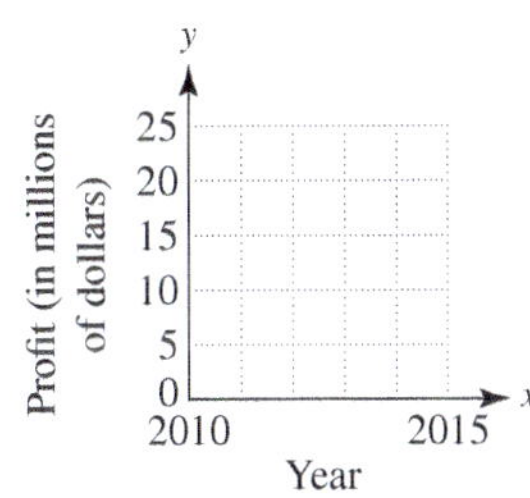

Concept Check *Determine the slope of each line segment in the given figure.*

7. *AB* **8.** *BC* **9.** *CD*

10. *DE* **11.** *EF* **12.** *FG*

13. If *A* and *F* were joined by a line segment in the figure, what would be its slope?

14. If *B* and *D* were joined by a line segment in the figure, what would be its slope?

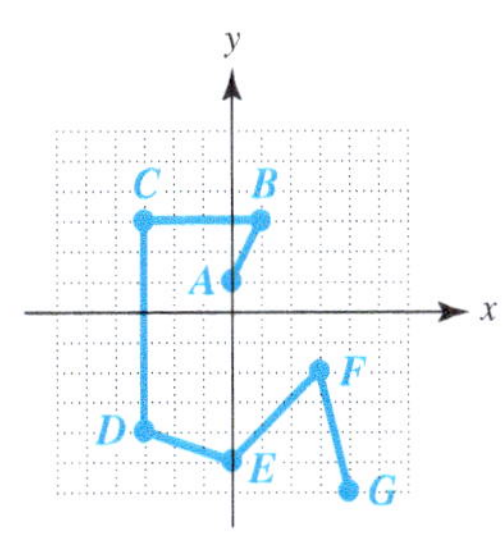

15. *Concept Check* On the basis of the figure shown here, determine which line satisfies the given description.

(a) The line has positive slope.

(b) The line has negative slope.

(c) The line has slope 0.

(d) The line has undefined slope.

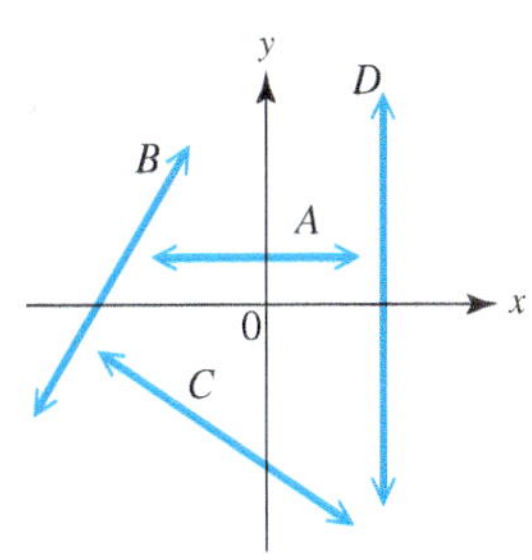

16. B and D are correct. Choice A is wrong because the order of subtraction must be the same in the numerator and denominator. Choice C is wrong because slope is defined as the change in y divided by the change in x.

17. 2 18. $\frac{1}{3}$
19. $\frac{5}{2}$ 20. -2
21. 0 22. 0
23. undefined 24. undefined
25. (a) 8 (b) rises
26. (a) 3 (b) rises
27. (a) $\frac{5}{6}$ (b) rises
28. (a) $\frac{9}{8}$ (b) rises
29. (a) 0 (b) horizontal
30. (a) 0 (b) horizontal
31. (a) $-\frac{1}{2}$ (b) falls
32. (a) $-\frac{1}{3}$ (b) falls
33. (a) undefined (b) vertical
34. (a) undefined (b) vertical
35. (a) -1 (b) falls
36. (a) -3 (b) falls
37. 6 38. 6
39. -3 40. $-\frac{7}{5}$
41. -2 42. -3
43. $\frac{4}{3}$ 44. $\frac{2}{5}$
45. $-\frac{5}{2}$ 46. 2
47. undefined 48. 0

In part (a) of Exercises 49–54, we used the intercepts. Other points can be used.

49. (a) $(4, 0)$ and $(0, -8)$; 2
(b) $y = 2x - 8$; 2
(c) $A = 2, B = -1$; 2
50. (a) $(-2, 0)$ and $(0, 6)$; 3
(b) $y = 3x + 6$; 3
(c) $A = 3, B = -1$; 3
51. (a) $(4, 0)$ and $(0, 3)$; $-\frac{3}{4}$
(b) $y = -\frac{3}{4}x + 3$; $-\frac{3}{4}$
(c) $A = 3, B = 4$; $-\frac{3}{4}$

16. *Concept Check* Which forms of the slope formula are correct? Explain.

A. $m = \dfrac{y_1 - y_2}{x_2 - x_1}$ **B.** $m = \dfrac{y_1 - y_2}{x_1 - x_2}$ **C.** $m = \dfrac{x_2 - x_1}{y_2 - y_1}$ **D.** $m = \dfrac{y_2 - y_1}{x_2 - x_1}$

Evaluate each expression for m, applying the slope formula. ***See Example 1.***

17. $m = \dfrac{6 - 2}{5 - 3}$ **18.** $m = \dfrac{5 - 7}{-4 - 2}$ **19.** $m = \dfrac{4 - (-1)}{-3 - (-5)}$

20. $m = \dfrac{-6 - 0}{0 - (-3)}$ **21.** $m = \dfrac{-5 - (-5)}{3 - 2}$ **22.** $m = \dfrac{-2 - (-2)}{4 - (-3)}$

23. $m = \dfrac{3 - 8}{-2 - (-2)}$ **24.** $m = \dfrac{5 - 6}{-8 - (-8)}$

In the following exercises, **(a)** *find the slope of the line passing through each pair of points, if possible, and* **(b)** *based on the slope, indicate whether the line* rises *from left to right,* falls *from left to right, is* horizontal, *or is* vertical. ***See Examples 1 and 3 and*** **FIGURE 20.**

25. $(-2, -3)$ and $(-1, 5)$ **26.** $(-4, 1)$ and $(-3, 4)$ **27.** $(-4, 1)$ and $(2, 6)$

28. $(-3, -3)$ and $(5, 6)$ **29.** $(2, 4)$ and $(-4, 4)$ **30.** $(-6, 3)$ and $(2, 3)$

31. $(-2, 2)$ and $(4, -1)$ **32.** $(-3, 1)$ and $(6, -2)$ **33.** $(5, -3)$ and $(5, 2)$

34. $(4, -1)$ and $(4, 3)$ **35.** $(1.5, 2.6)$ and $(0.5, 3.6)$ **36.** $(3.4, 4.2)$ and $(1.4, 10.2)$

Extending Skills *Find the slope of the line passing through the given pair of points.* $\left(\textit{Hint: } \dfrac{\frac{a}{b}}{\frac{c}{d}} = \dfrac{a}{b} \div \dfrac{c}{d}\right)$

37. $\left(\frac{1}{6}, \frac{1}{2}\right)$ and $\left(\frac{5}{6}, \frac{9}{2}\right)$ **38.** $\left(\frac{3}{4}, \frac{1}{3}\right)$ and $\left(\frac{5}{4}, \frac{10}{3}\right)$

39. $\left(-\frac{2}{9}, \frac{5}{18}\right)$ and $\left(\frac{1}{18}, -\frac{5}{9}\right)$ **40.** $\left(-\frac{4}{5}, \frac{9}{10}\right)$ and $\left(-\frac{3}{10}, \frac{1}{5}\right)$

Each table of values gives several points that line on a line. Find the slope of the line.

41.

x	y
−1	8
0	6
2	2
3	0

42.

x	y
−3	6
−1	0
0	−3
2	−9

43.

x	y
−6	−4
−3	0
0	4
3	8

44.

x	y
−5	−4
0	−2
5	0
10	2

Use the geometric interpretation of slope ("rise over run") *to find the slope of each line. (Coordinates of the points shown are integers.)*

45.

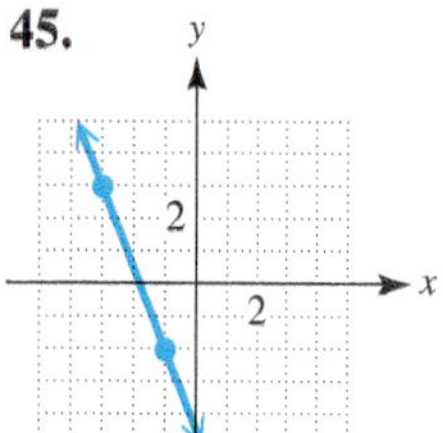

46.

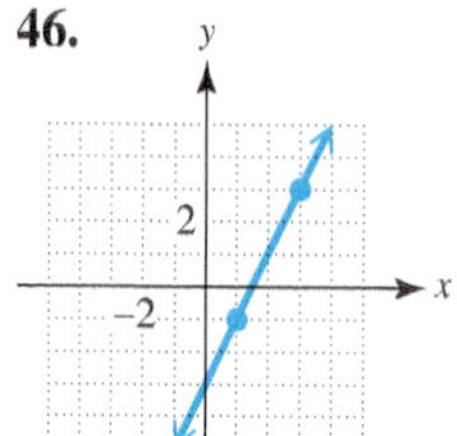

47.

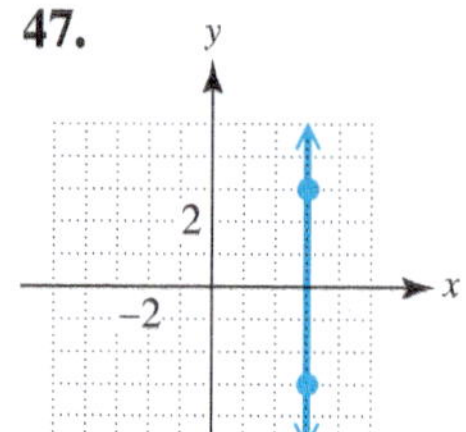

48.

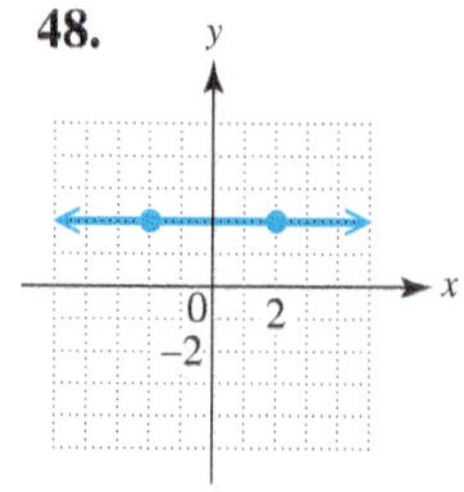

52. (a) $(5, 0)$ and $(0, 6)$; $-\frac{6}{5}$
(b) $y = -\frac{6}{5}x + 6$; $-\frac{6}{5}$
(c) $A = 6, B = 5$; $-\frac{6}{5}$

53. (a) $(-3, 0)$ and $(0, -3)$; -1
(b) $y = -x - 3$; -1
(c) $A = 1, B = 1$; -1

54. (a) $(4, 0)$ and $(0, -4)$; 1
(b) $y = x - 4$; 1
(c) $A = 1, B = -1$; 1

55. $-\frac{1}{2}$

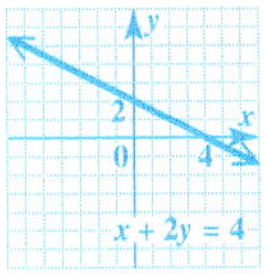

56. $-\frac{1}{3}$

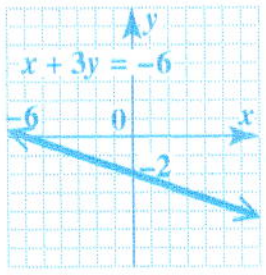

57. $\frac{5}{2}$

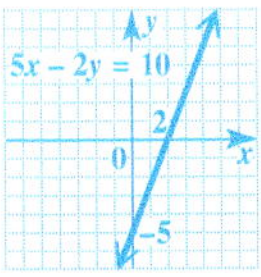

58. 4

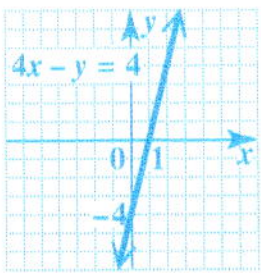

59. 4

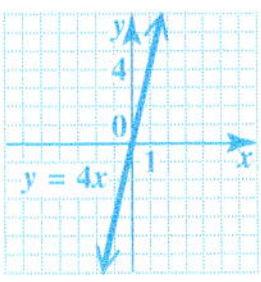

60. -3

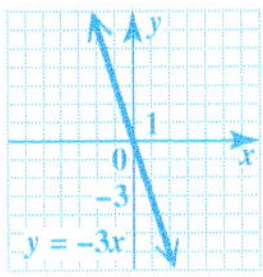

61. undefined

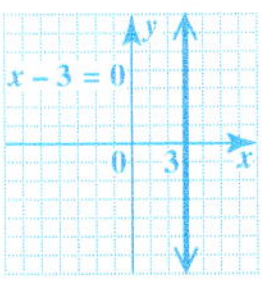

62. undefined

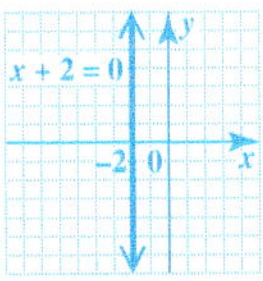

63. 0

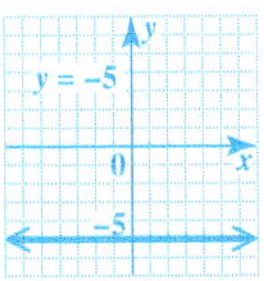

64. 0

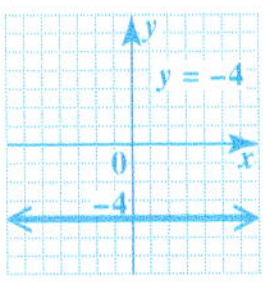

65. 0

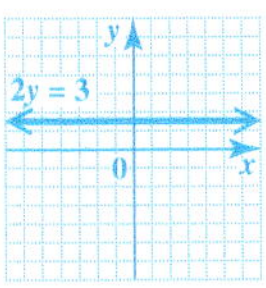

66. undefined

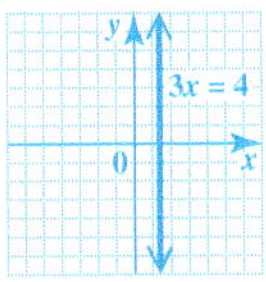

67.

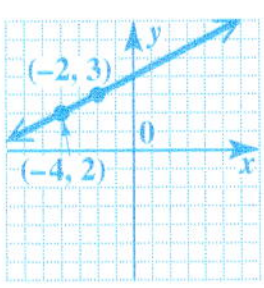

68.

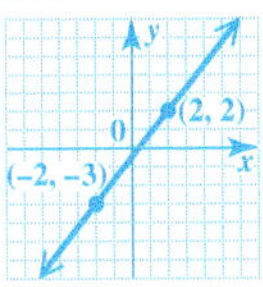

Find the slope of each line in three ways by doing the following.

(a) Give any two points that lie on the line, and use them to determine the slope. ***See Example 2.***

(b) Solve the equation for y, and identify the slope from the equation. ***See Example 4.***

(c) For the form $Ax + By = C$, calculate $-\frac{A}{B}$. ***See the Note following Example 4.***

49. $2x - y = 8$ **50.** $3x - y = -6$ **51.** $3x + 4y = 12$

52. $6x + 5y = 30$ **53.** $x + y = -3$ **54.** $x - y = 4$

Find the slope of each line, and sketch its graph. ***See Examples 1–4.***

55. $x + 2y = 4$ **56.** $x + 3y = -6$ **57.** $5x - 2y = 10$

58. $4x - y = 4$ **59.** $y = 4x$ **60.** $y = -3x$

61. $x - 3 = 0$ **62.** $x + 2 = 0$ **63.** $y = -5$

64. $y = -4$ **65.** $2y = 3$ **66.** $3x = 4$

Graph each line described. ***See Example 5.***

67. Through $(-4, 2)$; $m = \frac{1}{2}$ **68.** Through $(-2, -3)$; $m = \frac{5}{4}$

69. y-intercept $(0, -2)$; $m = -\frac{2}{3}$ **70.** y-intercept $(0, -4)$; $m = -\frac{3}{2}$

71. Through $(-1, -2)$; $m = 3$ **72.** Through $(-2, -4)$; $m = 4$

73. $m = 0$; through $(2, -5)$ **74.** $m = 0$; through $(5, 3)$

75. Undefined slope; through $(-3, 1)$ **76.** Undefined slope; through $(-4, 1)$

77. *Concept Check* If a line has slope $-\frac{4}{9}$, then any line parallel to it has slope ______, and any line perpendicular to it has slope ______.

78. *Concept Check* If a line has slope 0.2, then any line parallel to it has slope ______, and any line perpendicular to it has slope ______.

Determine whether each pair of lines is parallel, perpendicular, *or* neither. ***See Examples 6–8.***

79. The line passing through $(15, 9)$ and $(12, -7)$ and the line passing through $(8, -4)$ and $(5, -20)$

80. The line passing through $(4, 6)$ and $(-8, 7)$ and the line passing through $(-5, 5)$ and $(7, 4)$

81. $x + 4y = 7$ and $4x - y = 3$ **82.** $2x + 5y = -7$ and $5x - 2y = 1$

83. $4x - 3y = 6$ and $3x - 4y = 2$ **84.** $2x + y = 6$ and $x - y = 4$

85. $x = 6$ and $6 - x = 8$ **86.** $3x = y$ and $2y - 6x = 5$

87. $4x + y = 0$ and $5x - 8 = 2y$ **88.** $2x + 5y = -8$ and $6 + 2x = 5y$

89. $2x = y + 3$ and $2y + x = 3$ **90.** $4x - 3y = 8$ and $4y + 3x = 12$

Extending Skills *Solve each problem.*

91. When designing the TD Bank North Garden in Boston, architects designed the ramps leading up to the entrances so that circus elephants would be able to walk up the ramps. The maximum grade (or slope) that an elephant will walk on is 13%. Suppose that such a ramp was constructed with a horizontal run of 150 ft. What would be the maximum vertical rise the architects could use?

69.

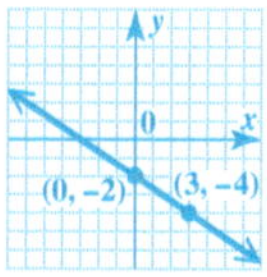

70.

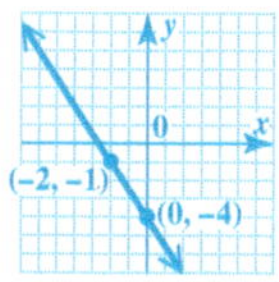

71.

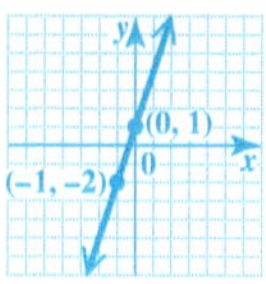

72.

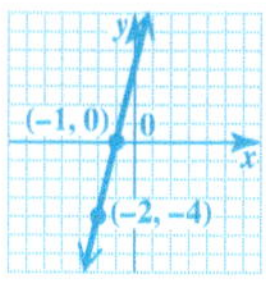

73.

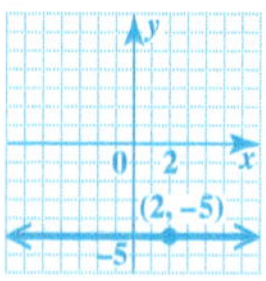

74.

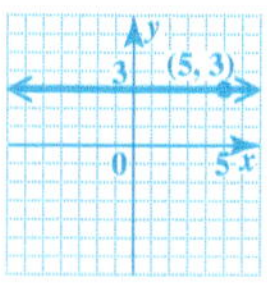

75.

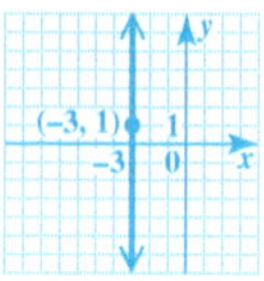

76.

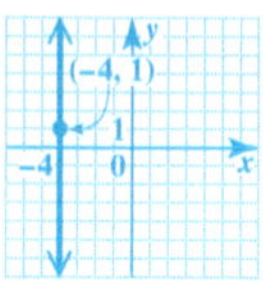

77. $-\frac{4}{9}; \frac{9}{4}$ **78.** 0.2; −5

79. parallel **80.** parallel
81. perpendicular
82. perpendicular
83. neither **84.** neither
85. parallel **86.** parallel
87. neither **88.** neither
89. perpendicular
90. perpendicular

91. 19.5 ft **92.** $\frac{7}{10}$

93. −$4000 per year; The value of the machine is decreasing $4000 per year during these years.

94. $50 per month; The amount saved is increasing $50 per month during these months.

95. 0% per year (or no change); The percent of pay raise is not changing—it is 3% per year during these years.

96. positive; negative

97. (a) In 2012, there were 326 million wireless subscriber connections in the U.S.
(b) 14.2
(c) The number of subscribers increased by an average of 14.2 million per year from 2007 to 2012.

92. The upper deck at U.S. Cellular Field in Chicago has produced, among other complaints, displeasure with its steepness. It is 160 ft from home plate to the front of the upper deck and 250 ft from home plate to the back. The top of the upper deck is 63 ft above the bottom. What is its slope? (Consider the slope as a positive number here.)

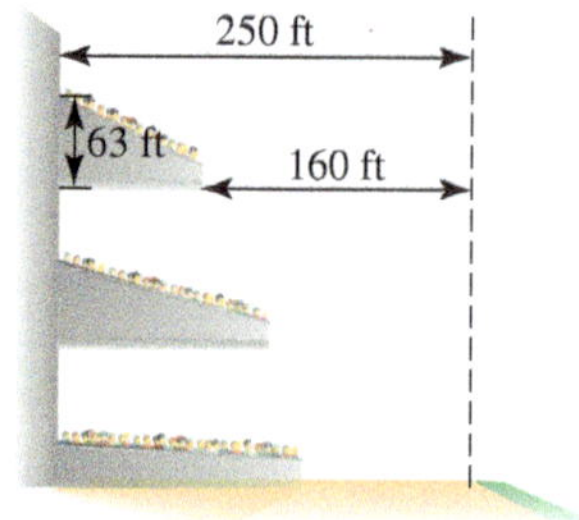

Find and interpret the average rate of change illustrated in each graph. ***See Objective 5 and*** **FIGURE 22.**

93.

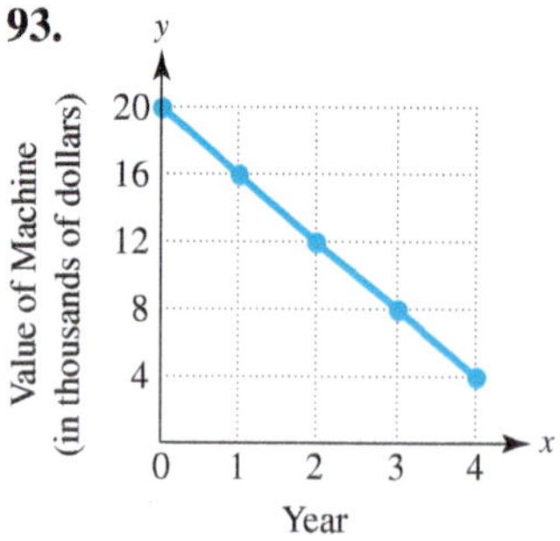

94.

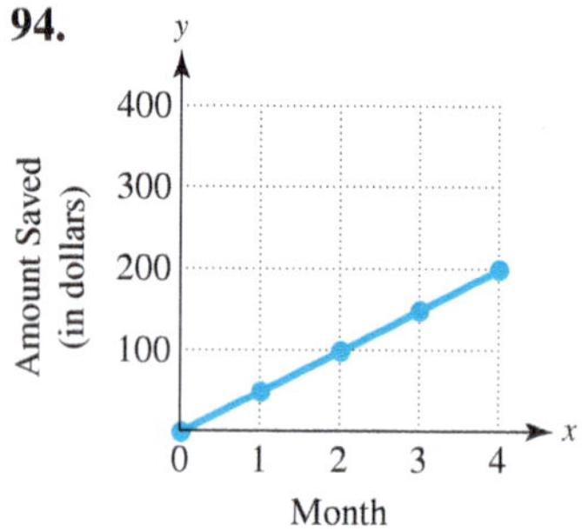

95.

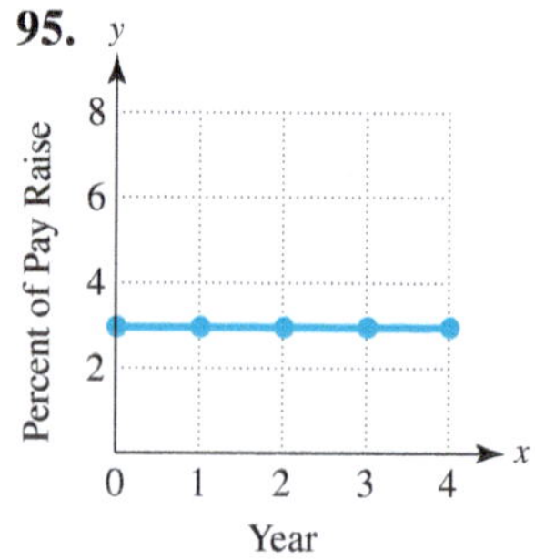

96. *Concept Check* If the graph of a linear equation rises from left to right, then the average rate of change is (*positive / negative*). If the graph of a linear equation falls from left to right, then the average rate of change is (*positive / negative*).

Solve each problem. ***See Examples 9 and 10.***

97. The graph shows the number of wireless subscriber connections (that is, active devices, including smartphones, feature phones, tablets, etc.) in millions in the United States for the years 2007 to 2012.

Wireless Subscriber Connections

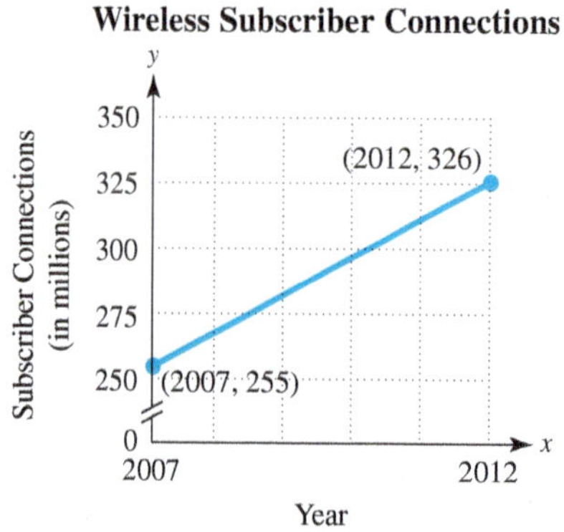

Source: CTIA.

(a) In the context of this graph, what does the ordered pair (2012, 326) mean?

(b) Use the given ordered pairs to find the slope of the line.

(c) Interpret the slope in the context of this problem.

98. The graph shows the percent of households in the United States that were wireless-only households for the years 2007 to 2012.

Wireless-Only Households

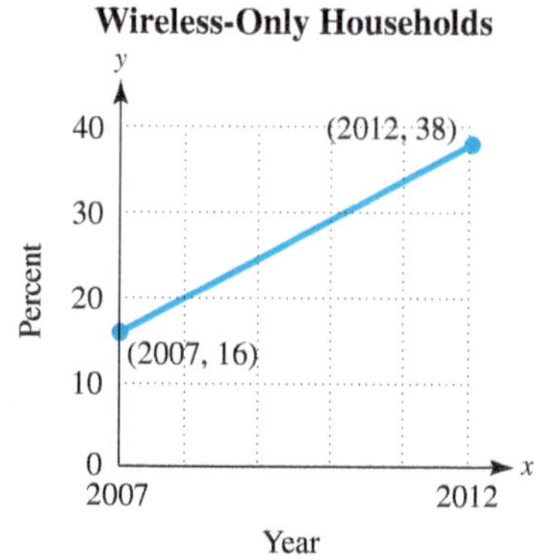

Source: CTIA.

(a) In the context of this graph, what does the ordered pair (2012, 38) mean?

(b) Use the given ordered pairs to find the slope of the line.

(c) Interpret the slope in the context of this problem.

98. (a) In 2012, 38% of U.S. households were wireless-only households.
(b) 4.4
(c) The percent of wireless-only households increased by an average of 4.4% per year from 2007 to 2012.

99. (a) −7 theaters per year
(b) The negative slope means that the number of drive-in theaters decreased by an average of 7 per year from 2005 to 2012.

100. (a) −327 thousand travelers per year
(b) The negative slope means that the number of U.S. travelers to Canada decreased by an average of 327 thousand per year from 2000 to 2011.

101. $0.08 per year; The price of a gallon of gasoline increased by an average of $0.08 per year from 1980 to 2012.

102. $0.17 per year; The price of a movie ticket increased by an average of $0.17 per year from 1990 to 2012.

103. −1859 thousand cameras per year; The number of digital cameras sold decreased by an average of 1859 thousand per year from 2010 to 2013.

104. −$171 million per year; Sales of desktop computers decreased by an average of $171 million per year from 2010 to 2013.

99. The graph shows the number of drive-in movie theaters in the United States from 2005 through 2012.

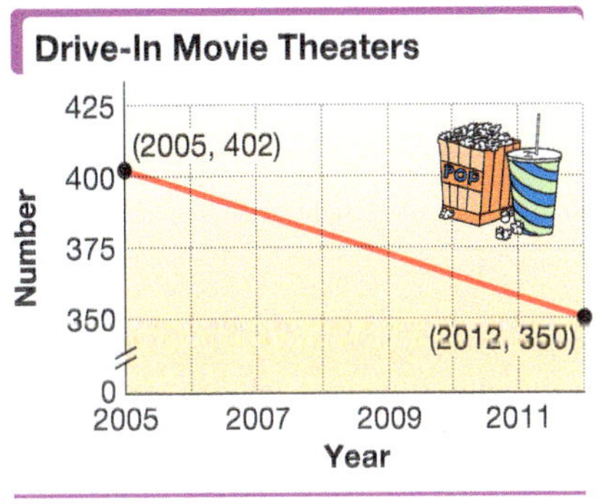

Source: www.drive-ins.com

(a) Use the given ordered pairs to find the average rate of change in the number of drive-in theaters per year during this period. Round the answer to the nearest whole number.

(b) Explain how a negative slope is interpreted in this situation.

100. The graph shows the number of U.S. travelers to Canada (in thousands) from 2000 through 2011.

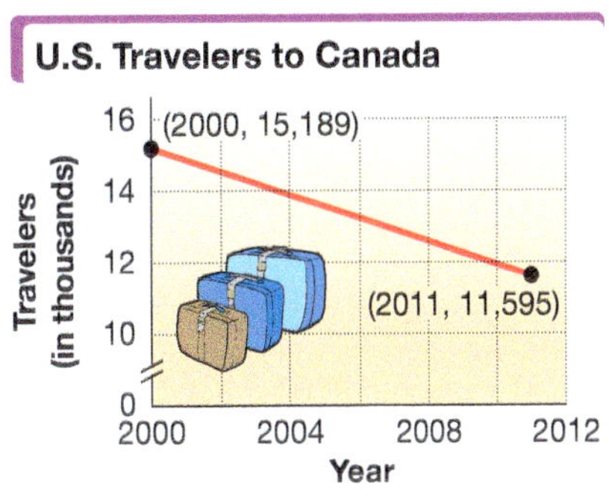

Source: U.S. Department of Commerce.

(a) Use the given ordered pairs to find the average rate of change in the number of U.S. travelers to Canada per year during this period. Round the answer to the nearest thousand.

(b) Explain how a negative slope is interpreted in this situation.

101. The average price of a gallon of gasoline in 1980 was $1.22. In 2012, the average price was $3.70. Find and interpret the average rate of change in the price of a gallon of gasoline per year to the nearest cent. (*Source:* Energy Information Administration.)

102. The average price of a movie ticket in 1990 was $4.23. In 2012, the average price was $7.96. Find and interpret the average rate of change in the price of a movie ticket per year to the nearest cent. (*Source:* Motion Picture Association of America.)

103. In 2010, the number of digital cameras sold in the United States totaled 7246 thousand. There were 1670 thousand sold in 2013. Find and interpret the average rate of change in the number of digital cameras sold per year to the nearest thousand. (*Source:* Consumer Electronics Association.)

104. In 2010, sales of desktop computers in the United States totaled $7390 million. In 2013, sales were $6876 million. Find and interpret the average rate of change in sales of desktop computers per year to the nearest million dollars. (*Source:* Consumer Electronics Association.)

105. Because the slopes of both pairs of opposite sides are equal, the figure is a parallelogram.
106. yes; yes

Extending Skills *Use your knowledge of the slopes of parallel and perpendicular lines.*

105. Show that $(-13, -9)$, $(-11, -1)$, $(2, -2)$, and $(4, 6)$ are the vertices of a parallelogram. (*Hint:* A parallelogram is a four-sided figure with opposite sides parallel.)

106. Is the figure with vertices at $(-11, -5)$, $(-2, -19)$, $(12, -10)$, and $(3, 4)$ a parallelogram? Is it a rectangle? (*Hint:* A rectangle is a parallelogram with a right angle.)

107. $\frac{1}{3}$ 108. $\frac{1}{3}$
109. $\frac{1}{3}$
110. $\frac{1}{3} = \frac{1}{3} = \frac{1}{3}$ is true.
111. collinear
112. not collinear

RELATING CONCEPTS For Individual or Group Work (Exercises 107–112)

Three points that lie on the same straight line are said to be ***collinear.*** *Consider the points* $A(3, 1)$, $B(6, 2)$, *and* $C(9, 3)$. ***Work Exercises 107–112 in order.***

107. Find the slope of segment AB.

108. Find the slope of segment BC.

109. Find the slope of segment AC.

110. If slope of segment AB = slope of segment BC = slope of segment AC, then A, B, and C are collinear. Use the results of **Exercises 107–109** to show that this statement is satisfied.

111. Use the slope formula to determine whether the points $(1, -2)$, $(3, -1)$, and $(5, 0)$ are collinear.

112. Repeat **Exercise 111** for the points $(0, 6)$, $(4, -5)$, and $(-2, 12)$.

2.3 Writing Equations of Lines

OBJECTIVES

1. Write an equation of a line given its slope and y-intercept.
2. Graph a line using its slope and y-intercept.
3. Write an equation of a line given its slope and a point on the line.
4. Write an equation of a line given two points on the line.
5. Write equations of horizontal and vertical lines.
6. Write an equation of a line parallel or perpendicular to a given line.
7. Write an equation of a line that models real data.

VOCABULARY

☐ scatter diagram

OBJECTIVE 1 Write an equation of a line given its slope and y-intercept.

In **Section 2.2,** we found the slope of a line from its equation by solving the equation for y. For example, we found that the slope of the line with equation

$$y = 4x + 8$$

is 4, the coefficient of x. *What does the number 8 represent?*

To answer this question, suppose a line has slope m and y-intercept $(0, b)$. We can find an equation of this line by choosing another point (x, y) on the line, as shown in **FIGURE 25**, and applying the slope formula.

$$m = \frac{y - b}{x - 0} \quad \begin{matrix}\leftarrow \text{Change in } y \\ \leftarrow \text{Change in } x\end{matrix}$$

$$m = \frac{y - b}{x} \quad \text{Subtract.}$$

$$mx = y - b \quad \text{Multiply by } x.$$

$$mx + b = y \quad \text{Add } b.$$

$$\mathbf{y = mx + b} \quad \text{Interchange sides.}$$

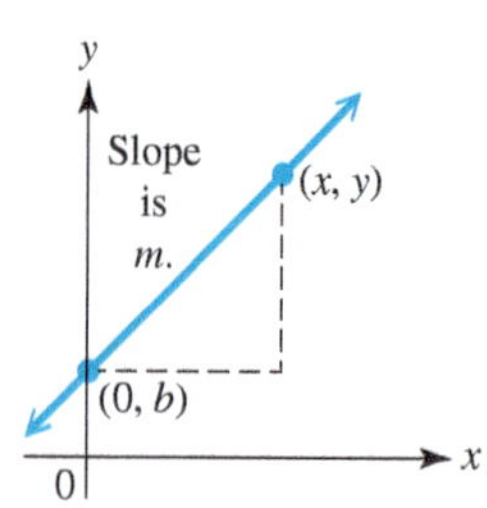

FIGURE 25

This last equation is the *slope-intercept form* of the equation of a line, because we can identify the slope m and y-intercept $(0, b)$ at a glance. In the line with equation $y = 4x + 8$, the number 8 indicates that the y-intercept is $(0, 8)$.

Slope-Intercept Form

The **slope-intercept form** of the equation of a line with slope m and y-intercept $(0, b)$ is

$$y = mx + b.$$

Slope ↑ (pointing to m) ↑ $(0, b)$ is the y-intercept.

NOW TRY EXERCISE 1
Write an equation of the line with slope $\frac{2}{3}$ and y-intercept $(0, 1)$.

CLASSROOM EXAMPLE 1
Write an equation of the line with slope 2 and y-intercept $(0, -3)$.
Answer: $y = 2x - 3$

EXAMPLE 1 Writing an Equation of a Line

Write an equation of the line with slope $-\frac{4}{5}$ and y-intercept $(0, -2)$.

Here, $m = -\frac{4}{5}$ and $b = -2$. Substitute these values into the slope-intercept form.

$$y = mx + b \quad \text{Slope-intercept form}$$

$$y = -\frac{4}{5}x + (-2) \quad \text{Let } m = -\frac{4}{5} \text{ and } b = -2.$$

$$y = -\frac{4}{5}x - 2 \quad \text{Definition of subtraction}$$

NOW TRY

TEACHING TIP
Because lowercase "b" looks like the number "6," write the "b" in $y = mx + b$ in script form to avoid confusion.

NOTE Every linear equation (of a nonvertical line) has a *unique* (one and only one) slope-intercept form. In **Section 2.6,** we introduce *linear functions,* which are defined using slope-intercept form. Also, this form is used when graphing a line with a graphing calculator.

OBJECTIVE 2 Graph a line using its slope and y-intercept.

We first saw this approach in **Example 5(a)** of **Section 2.2.**

CLASSROOM EXAMPLE 2
Graph the line using the slope and y-intercept.

$$x + 2y = -4$$

Answer:

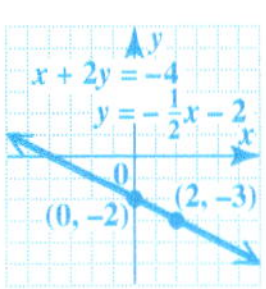

EXAMPLE 2 Graphing Lines Using Slope and y-Intercept

Graph each line using the slope and y-intercept.

(a) $y = 3x - 6$ (In slope-intercept form)

Here, $m = 3$ and $b = -6$. Plot the y-intercept $(0, -6)$. The slope 3 can be interpreted geometrically as follows.

$$m = \frac{\text{rise}}{\text{run}} = \frac{\text{change in } y}{\text{change in } x} = \frac{3}{1}$$

From $(0, -6)$, move 3 units *up* and 1 unit to the *right,* and plot a second point at $(1, -3)$. Join the two points with a straight line. See **FIGURE 26.**

TEACHING TIP
In **Example 2(a),** show that another point on the line can be found by interpreting the slope 3 as $\frac{-3}{-1}$.

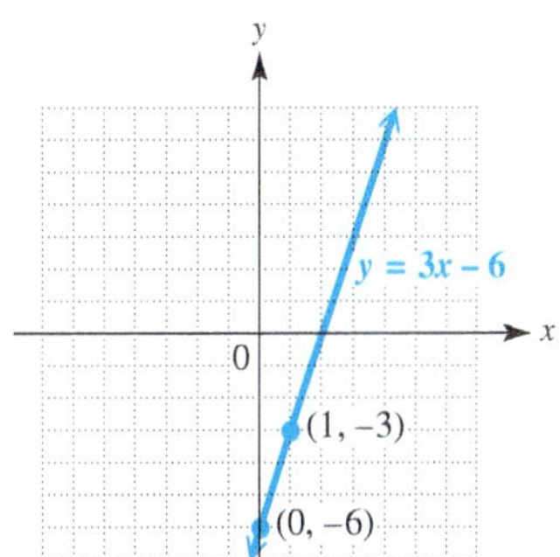

FIGURE 26

NOW TRY ANSWER
1. $y = \frac{2}{3}x + 1$

NOW TRY EXERCISE 2

Graph the line using the slope and y-intercept.

$$4x + 3y = 6$$

(b) $3y + 2x = 9$ (*Not* in slope-intercept form)

Write the equation in slope-intercept form by solving for y.

$$3y + 2x = 9$$

$$3y = -2x + 9 \quad \text{Subtract } 2x.$$

$$y = -\frac{2}{3}x + 3 \quad \text{Divide by 3.}$$

Slope ↑ ↑ y-intercept is $(0, 3)$.

To graph this equation, plot the y-intercept $(0, 3)$. The slope can be interpreted as either $\frac{-2}{3}$ or $\frac{2}{-3}$. Using $\frac{-2}{3}$, begin at $(0, 3)$ and move 2 units *down* and 3 units to the *right* to locate the point $(3, 1)$. The line through these two points is the required graph. See **FIGURE 27**. (Verify that the point obtained using $\frac{2}{-3}$ as the slope is also on this line.)

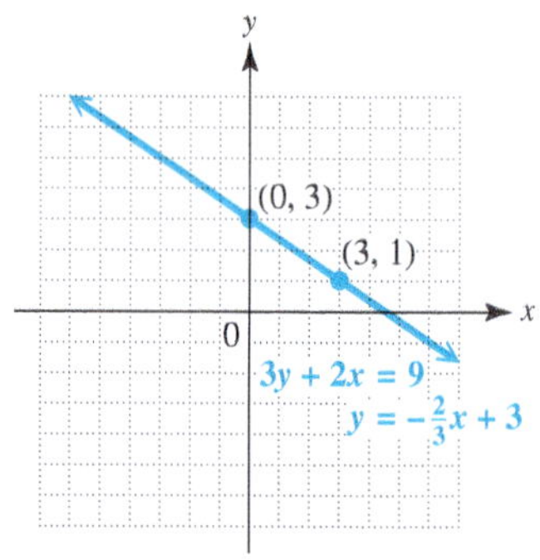

FIGURE 27

NOW TRY

OBJECTIVE 3 Write an equation of a line given its slope and a point on the line.

Let m represent the slope of a line and (x_1, y_1) represent a given point on the line. Let (x, y) represent any other point on the line. See **FIGURE 28**.

$$m = \frac{y - y_1}{x - x_1} \quad \text{Slope formula}$$

$$m(x - x_1) = y - y_1 \quad \text{Multiply each side by } x - x_1.$$

$$\mathbf{y - y_1 = m(x - x_1)} \quad \text{Interchange sides.}$$

This last equation is the *point-slope form* of the equation of a line.

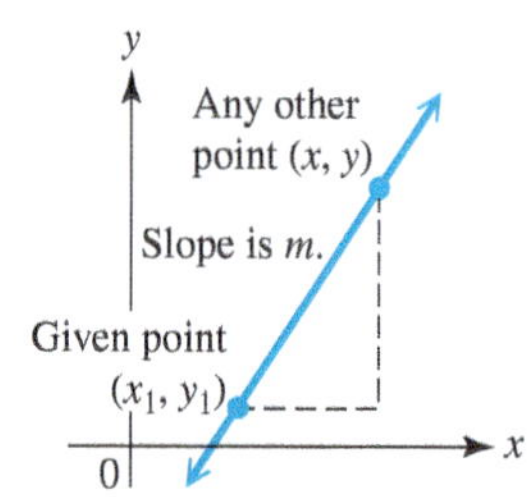

FIGURE 28

Point-Slope Form

The **point-slope form** of the equation of a line with slope m passing through the point (x_1, y_1) is

Slope ↓

$$\mathbf{y - y_1 = m(x - x_1).}$$

↑ Given point ↑

NOW TRY ANSWER

2.

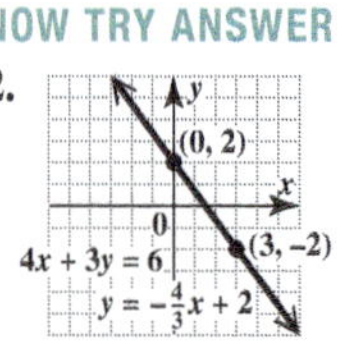

NOW TRY EXERCISE 3
Write an equation of the line with slope $-\frac{1}{5}$ passing through the point $(5, -3)$.

CLASSROOM EXAMPLE 3
Write an equation of the line with slope $\frac{2}{5}$ passing through the point $(3, -4)$.

Answer: $y = \frac{2}{5}x - \frac{26}{5}$

TEACHING TIP
In the point-slope form

$$y - y_1 = m(x - x_1),$$

state that the boldface quantities are the only ones to be replaced. (x and y are left as variables.)

In the slope-intercept form

$$y = mx + b,$$

point out that y, m, and x are replaced (leaving b as the only variable).

EXAMPLE 3 Writing an Equation of a Line Given Its Slope and a Point

Write an equation of the line with slope $\frac{1}{3}$ passing through the point $(-2, 5)$.

Method 1 Use point-slope form with $(x_1, y_1) = (-2, 5)$ and $m = \frac{1}{3}$.

$$y - y_1 = m(x - x_1) \quad \text{Point-slope form}$$

$$y - 5 = \frac{1}{3}[x - (-2)] \quad \text{Let } y_1 = 5,\ m = \tfrac{1}{3}, \text{ and } x_1 = -2.$$

$$y - 5 = \frac{1}{3}(x + 2) \quad \text{Definition of subtraction}$$

$$y - 5 = \frac{1}{3}x + \frac{2}{3} \quad (*)\ \text{Distributive property}$$

$$\text{Slope-intercept form} \rightarrow y = \frac{1}{3}x + \frac{17}{3} \quad 5 = \tfrac{15}{3};\ \text{Add } \tfrac{15}{3}.$$

Method 2 Use slope-intercept form with $(x, y) = (-2, 5)$ and $m = \frac{1}{3}$.

$$y = mx + b \quad \text{Slope-intercept form}$$

$$5 = \frac{1}{3}(-2) + b \quad \text{Let } y = 5,\ m = \tfrac{1}{3}, \text{ and } x = -2.$$

Solve for b.

$$5 = -\frac{2}{3} + b \quad \text{Multiply.}$$

$$\frac{17}{3} = b, \quad \text{or} \quad b = \frac{17}{3} \quad 5 = \tfrac{15}{3};\ \text{Add } \tfrac{2}{3}.$$

Substitute to obtain the equation $y = \frac{1}{3}x + \frac{17}{3}$, as above. NOW TRY

In **Section 2.1,** we defined *standard form* for a linear equation.

$$Ax + By = C \quad \text{Standard form}$$

Here A, B, and C are real numbers and A and B are not both 0. (In most cases in this book, A, B, and C are rational numbers.) For consistency, we give answers so that A, B, and C are integers with greatest common factor 1, and $A \geq 0$. (If $A = 0$, then we give $B > 0$.) For example, the equation in **Example 3** is written in standard form as follows.

$$y - 5 = \frac{1}{3}x + \frac{2}{3} \quad \text{Equation } (*) \text{ from Example 3}$$

$$3y - 15 = x + 2 \quad \text{Multiply each term by 3.}$$

$$-x + 3y = 17 \quad \text{Subtract } x. \text{ Add 15.}$$

$$\text{Standard form} \rightarrow x - 3y = -17 \quad \text{Multiply by } -1.$$

NOTE "Standard form" is not standard among texts. A linear equation can be written in many different, equally correct ways. For example, $2x + 3y = 8$ can be written as

$$2x = 8 - 3y, \quad 3y = 8 - 2x, \quad x + \frac{3}{2}y = 4, \quad \text{and} \quad 4x + 6y = 16.$$

We prefer the standard form $2x + 3y = 8$ over any multiples of each side, such as $4x + 6y = 16$. (To write $4x + 6y = 16$ in this preferred form, divide each side by 2.)

TEACHING TIP
Restate the restrictions on standard form as follows: A, B, and C cannot be fractions and A should be nonnegative. Proceed with several examples. Include slopes of 3 and $-\frac{2}{3}$.

NOW TRY ANSWER
3. $y = -\frac{1}{5}x - 2$

OBJECTIVE 4 Write an equation of a line given two points on the line.

NOW TRY EXERCISE 4

Write an equation of the line passing through the points $(3, -4)$ and $(-2, -1)$. Give the final answer in standard form.

CLASSROOM EXAMPLE 4

Write an equation of the line passing through the points $(-2, 6)$ and $(1, 4)$. Give the final answer in standard form.

Answer: $2x + 3y = 14$

EXAMPLE 4 **Writing an Equation of a Line Given Two Points**

Write an equation of the line passing through the points $(-4, 3)$ and $(5, -7)$. Give the final answer in standard form.

First find the slope using the slope formula.

$$m = \frac{-7 - 3}{5 - (-4)} = -\frac{10}{9}$$

Use either $(-4, 3)$ or $(5, -7)$ as (x_1, y_1) in the point-slope form of the equation of a line. We choose $(-4, 3)$, so $-4 = x_1$ and $3 = y_1$.

$$y - y_1 = m(x - x_1) \quad \text{Point-slope form}$$

$$y - 3 = -\frac{10}{9}[x - (-4)] \quad \text{Let } y_1 = 3,\ m = -\tfrac{10}{9}, \text{ and } x_1 = -4.$$

$$y - 3 = -\frac{10}{9}(x + 4) \quad \text{Definition of subtraction}$$

$$y - 3 = -\frac{10}{9}x - \frac{40}{9} \quad \text{Distributive property}$$

$$9y - 27 = -10x - 40 \quad \text{Multiply each term by 9.}$$

Standard form → $10x + 9y = -13$ Add $10x$. Add 27.

Verify that if $(5, -7)$ were used, the same equation would result. NOW TRY

NOTE Once the slope is found in **Example 4,** the equation of the line could also be determined using Method 2 from **Example 3.**

OBJECTIVE 5 Write equations of horizontal and vertical lines.

A horizontal line has slope 0. Using point-slope form, we can find the equation of a horizontal line through the point (a, b).

$$y - y_1 = m(x - x_1) \quad \text{Point-slope form}$$

$$y - b = 0(x - a) \quad y_1 = b,\ m = 0,\ x_1 = a$$

$$y - b = 0 \quad \text{Multiplication property of 0}$$

Horizontal line → $\boldsymbol{y = b}$ Add b.

Point-slope form does not apply to a vertical line because the slope of a vertical line is undefined. A vertical line through the point (a, b) has equation $\boldsymbol{x = a}$.

Equations of Horizontal and Vertical Lines

A **horizontal line** through the point (a, b) has equation $\boldsymbol{y = b}$.

A **vertical line** through the point (a, b) has equation $\boldsymbol{x = a}$.

NOW TRY ANSWER

4. $3x + 5y = -11$

NOW TRY EXERCISE 5

Write an equation of the line passing through the point $(4, -4)$ that satisfies the given condition.

(a) The line has undefined slope.

(b) The line has slope 0.

CLASSROOM EXAMPLE 5

Write an equation of the line passing through the point $(2, -1)$ that satisfies the given condition.

(a) The line has undefined slope.

(b) The line has slope 0.

Answers: (a) $x = 2$ (b) $y = -1$

CLASSROOM EXAMPLE 6

Write an equation of the line passing through the point $(-8, 3)$ that satisfies the given condition. Give final answers in slope-intercept form.

(a) The line is parallel to the line $2x - 3y = 10$.

(b) The line is perpendicular to the line $2x - 3y = 10$.

Answers:

(a) $y = \frac{2}{3}x + \frac{25}{3}$ (b) $y = -\frac{3}{2}x - 9$

TEACHING TIP

For problems involving parallel and perpendicular lines, begin by writing the point-slope form of a line, leaving y_1, m, and x_1 blank. Fill in x_1 and y_1 from the original point. Solve the given equation for y.

1. If *parallel* lines are required, use the coefficient of x for the slope m.
2. If *perpendicular* lines are required, use the negative reciprocal of the coefficient of x for m.

NOW TRY ANSWERS

5. **(a)** $x = 4$ **(b)** $y = -4$

EXAMPLE 5 **Writing Equations of Horizontal and Vertical Lines**

Write an equation of the line passing through the point $(-3, 3)$ that satisfies the given condition.

(a) The line has slope 0.

Because the slope is 0, this is a horizontal line. A horizontal line through the point (a, b) has equation $y = b$. In $(-3, 3)$, the y-coordinate is 3, so the equation is $y = 3$.

(b) The line has undefined slope.

This is a vertical line because the slope is undefined. A vertical line through the point (a, b) has equation $x = a$. In $(-3, 3)$, the x-coordinate is -3, so the equation is $x = -3$.

Both lines are graphed in **FIGURE 29**.

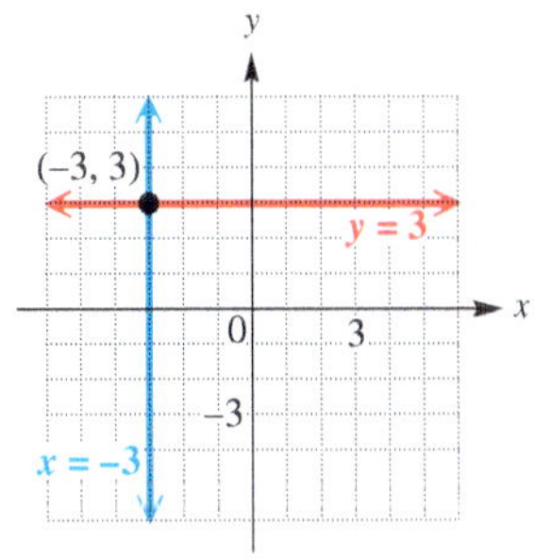

FIGURE 29

NOW TRY

OBJECTIVE 6 Write an equation of a line parallel or perpendicular to a given line.

Recall that parallel lines have the same slope and perpendicular lines have slopes that are negative reciprocals.

EXAMPLE 6 **Writing Equations of Parallel or Perpendicular Lines**

Write an equation of the line passing through the point $(-3, 6)$ that satisfies the given condition. Give final answers in slope-intercept form.

(a) The line is parallel to the line $2x + 3y = 6$.

We can find the slope of the given line by solving for y.

$$2x + 3y = 6$$

$$3y = -2x + 6 \quad \text{Subtract } 2x.$$

$$y = -\frac{2}{3}x + 2 \quad \text{Divide by 3.}$$

(Slope: $-\frac{2}{3}$)

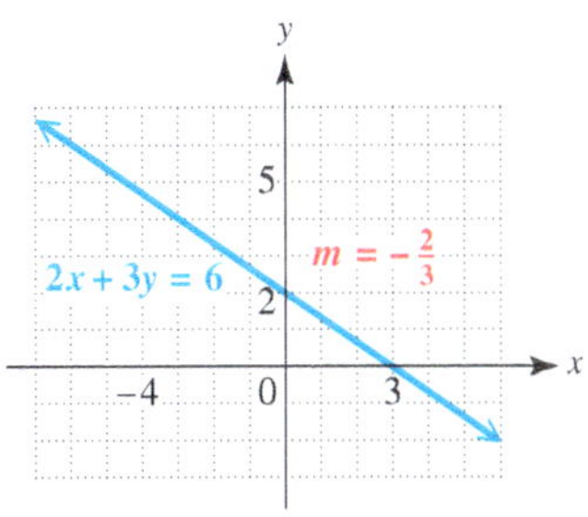

FIGURE 30

The slope of the line is given by the coefficient of x, so $m = -\frac{2}{3}$. See **FIGURE 30**.

The required equation of the line through $(-3, 6)$ and parallel to $2x + 3y = 6$ must also have slope $-\frac{2}{3}$. To find this equation, we use the point-slope form with $(x_1, y_1) = (-3, 6)$ and $m = -\frac{2}{3}$.

$$y - 6 = -\frac{2}{3}[x - (-3)] \quad y_1 = 6,\ m = -\frac{2}{3},\ x_1 = -3$$

$$y - 6 = -\frac{2}{3}(x + 3) \quad \text{Definition of subtraction}$$

$$y - 6 = -\frac{2}{3}x - 2 \quad \text{Distributive property}$$

$$y = -\frac{2}{3}x + 4 \quad \text{Add 6.}$$

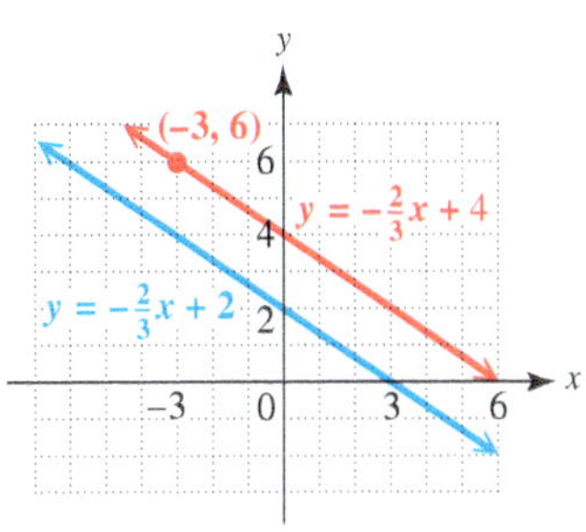

FIGURE 31

We did not clear the fraction because we want the final equation in slope-intercept form—that is, solved for y. Both lines are shown in **FIGURE 31**.

NOW TRY EXERCISE 6

Write an equation of the line passing through the point $(6, -1)$ that satisfies the given condition. Give final answers in slope-intercept form.

(a) The line is parallel to the line $3x - 5y = 7$.

(b) The line is perpendicular to the line $3x - 5y = 7$.

(b) The line is perpendicular to the line $2x + 3y = 6$.

In part (a), we wrote the equation $2x + 3y = 6$ in slope-intercept form.

$$y = -\frac{2}{3}x + 2$$

↑ Slope

To be perpendicular to the line $2x + 3y = 6$, a line must have slope $\frac{3}{2}$, the negative reciprocal of $-\frac{2}{3}$.

We use $(-3, 6)$ and slope $\frac{3}{2}$ in the point-slope form to find the equation of the perpendicular line shown in **FIGURE 32.**

$$y - 6 = \frac{3}{2}[x - (-3)] \qquad y_1 = 6,\ m = \frac{3}{2},\ x_1 = -3$$

$$y - 6 = \frac{3}{2}(x + 3) \qquad \text{Definition of subtraction}$$

$$y - 6 = \frac{3}{2}x + \frac{9}{2} \qquad \text{Distributive property}$$

$$y = \frac{3}{2}x + \frac{21}{2} \qquad \text{Add } 6 = \tfrac{12}{2}.$$

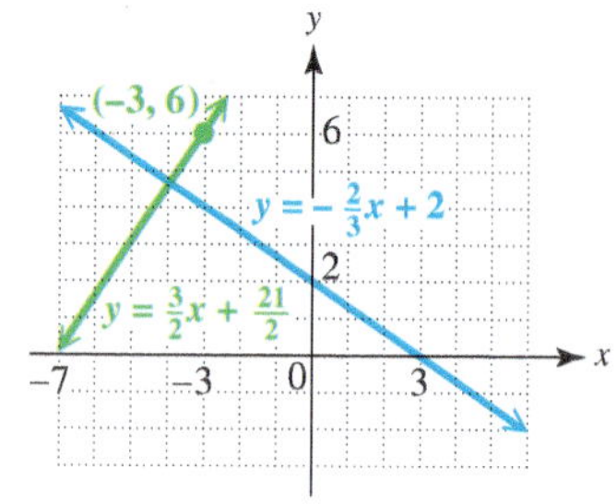

FIGURE 32

Again, we did not clear the fractions because we want the final equation in slope-intercept form.

NOW TRY

TEACHING TIP
Spend some extra time and use this table to ensure that students understand the different forms of linear equations. Work **Exercises 1–6** in the exercise set together as a class.

▼ Summary of Forms of Linear Equations

Equation	*Description*	*When to Use*
$y = mx + b$	**Slope-Intercept Form** Slope is m. y-intercept is $(0, b)$.	The slope and y-intercept can be easily identified and used to quickly graph the equation.
$y - y_1 = m(x - x_1)$	**Point-Slope Form** Slope is m. Line passes through (x_1, y_1).	This form is ideal for finding the equation of a line if the slope and a point on the line or two points on the line are known.
$Ax + By = C$	**Standard Form** (A, B, and C integers, $A \geq 0$) Slope is $-\frac{A}{B}$ $(B \neq 0)$. x-intercept is $\left(\frac{C}{A}, 0\right)$ $(A \neq 0)$. y-intercept is $\left(0, \frac{C}{B}\right)$ $(B \neq 0)$.	The x- and y-intercepts can be found quickly and used to graph the equation. The slope must be calculated.
$y = b$	**Horizontal Line** Slope is 0. y-intercept is $(0, b)$.	If the graph intersects only the y-axis, then y is the only variable in the equation.
$x = a$	**Vertical Line** Slope is undefined. x-intercept is $(a, 0)$.	If the graph intersects only the x-axis, then x is the only variable in the equation.

TEACHING TIP
For some classes, it may be desirable to cover **Objective 7** at a second meeting.

NOW TRY ANSWERS

6. (a) $y = \frac{3}{5}x - \frac{23}{5}$

(b) $y = -\frac{5}{3}x + 9$

OBJECTIVE 7 Write an equation of a line that models real data.

If a given set of data changes at a fairly constant rate, the data may fit a linear pattern, where the rate of change is the slope of the line.

NOW TRY EXERCISE 7

A cell phone plan costs \$100 for the telephone plus \$85 per month for service. Write an equation that gives the cost y in dollars for x months of cell phone service using this plan.

CLASSROOM EXAMPLE 7

There is a flat rate of \$0.20 plus a charge of \$0.10 per minute to make a telephone call. Write an equation that gives the cost y in dollars for a call of x minutes.

Answer: $y = 0.1x + 0.2$

EXAMPLE 7 Writing a Linear Equation to Describe Real Data

A local gasoline station is selling 89-octane gas for \$3.50 per gal.

(a) Write an equation that describes the cost y to buy x gallons of gas.

The total cost is determined by the number of gallons we buy multiplied by the price per gallon (in this case, \$3.50). As the gas is pumped, two sets of numbers spin by: the number of gallons pumped and the cost of that number of gallons.

The table illustrates this situation.

Number of Gallons Pumped	Cost of This Number of Gallons
0	0(\$3.50) = \$ 0.00
1	1(\$3.50) = \$ 3.50
2	2(\$3.50) = \$ 7.00
3	3(\$3.50) = \$10.50
4	4(\$3.50) = \$14.00

If we let x denote the number of gallons pumped, then the total cost y in dollars can be found using the following linear equation.

Total cost → $y = 3.50x$ ← Number of gallons

Theoretically, there are infinitely many ordered pairs (x, y) that satisfy this equation, but here we are limited to nonnegative values for x because we cannot have a negative number of gallons. In this situation, there is also a practical maximum value for x, which varies from one car to another—the size of the gas tank.

(b) A car wash at this gas station costs an additional \$3.00. Write an equation that defines the cost of gas and a car wash.

The cost will be $3.50x + 3.00$ dollars for x gallons of gas and a car wash.

$$y = 3.5x + 3 \qquad \text{Final 0's need not be included.}$$

(c) Interpret the ordered pairs $(5, 20.5)$ and $(10, 38)$ in relation to the equation from part (b).

$(5, 20.5)$ indicates that 5 gal of gas and a car wash cost \$20.50.

$(10, 38)$ indicates that 10 gal of gas and a car wash cost \$38.00.

NOW TRY

NOTE In **Example 7(a),** the ordered pair $(0, 0)$ satisfied the equation $y = 3.50x$, so the linear equation has the form

$$y = mx, \quad \text{where } b = 0.$$

If a realistic situation involves an initial charge b plus a charge per unit, m, as in the equation $y = 3.5x + 3$ in **Example 7(b),** then the equation has the form

$$y = mx + b, \quad \text{where } b \neq 0.$$

NOW TRY ANSWER

7. $y = 85x + 100$

NOW TRY EXERCISE 8

Refer to **Example 8.**

(a) Using the data values for the years 2009 and 2012, write an equation that models the data.

(b) Use the equation from part (a) to approximate the cost of tuition and fees in 2011.

CLASSROOM EXAMPLE 8

The percentage of the U.S. population 25 yr and older with at least a high school diploma is shown in the table for selected years.

Year	Percent
1950	34.3
1960	41.1
1970	55.2
1980	68.6
1990	77.6
2000	84.1
2010	87.1

Source: National Center for Education Statistics.

(a) Let $x = 0$ represent 1950, $x = 10$ represent 1960, and so on. Use the data for 1950 and 2010 to write an equation that models the data.

(b) Use the equation from part (a) to approximate the percentage, to the nearest tenth, of the U.S. population 25 yr and older who were at least high school graduates in 2012.

Answers:

(a) $y = 0.88x + 34.3$

(b) 88.9%

EXAMPLE 8 Writing an Equation of a Line That Models Data

Average annual tuition and fees for in-state students at public two-year colleges are shown in the table for selected years and graphed as ordered pairs of points in the **scatter diagram** in **FIGURE 33**, where $x = 0$ represents 2009, $x = 1$ represents 2010, and so on, and y represents the cost in dollars.

Year	Cost (in dollars)
2009 ($x = 0$)	2787
2010 ($x = 1$)	2938
2011 ($x = 2$)	3074
2012 ($x = 3$)	3216
2013 ($x = 4$)	3264

Source: The College Board.

Tuition and Fees at Two-Year Colleges

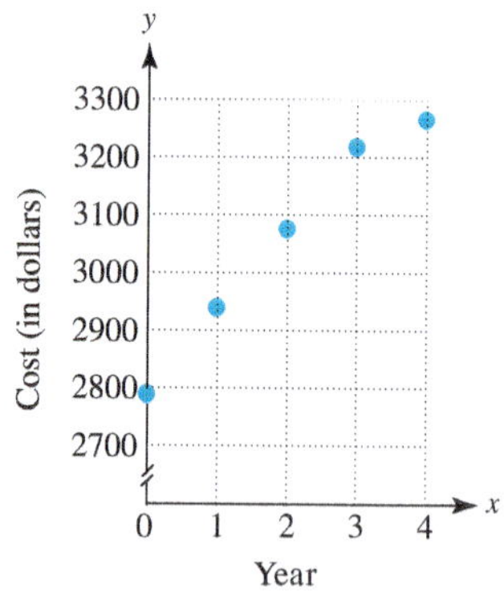

FIGURE 33

(a) Write an equation that models the data.

Because the points in **FIGURE 33** lie approximately on a straight line, we can write a linear equation that models the relationship between year x and cost y. We choose two data points, $(0, 2787)$ and $(4, 3264)$, to find the slope of the line.

$$m = \frac{3264 - 2787}{4 - 0} = \frac{477}{4} = 119.25$$

The slope 119.25 indicates that the cost of tuition and fees increased by about \$119 per year from 2009 to 2013. We use this slope and the y-intercept $(0, 2787)$ to write an equation of the line in slope-intercept form.

$$y = 119.25x + 2787$$

(b) Use the equation from part (a) to approximate the cost of tuition and fees in 2014.

The value $x = 5$ corresponds to the year 2014.

$$y = 119.25x + 2787 \quad \text{Equation from part (a)}$$
$$y = 119.25(5) + 2787 \quad \text{Substitute 5 for } x.$$
$$y = 3383.25 \quad \text{Multiply, and then add.}$$

According to the model, average tuition and fees for in-state students at public two-year colleges in 2014 were about \$3383.

NOW TRY

NOTE Choosing different data points in **Example 8** would result in a slightly different line (particularly in regard to its slope) and, hence, a slightly different equation. However, all such equations should yield similar results. See **Now Try Exercise 8.**

NOW TRY ANSWERS

8. (a) $y = 143x + 2787$
(b) \$3073

NOW TRY EXERCISE 9

Refer to **Example 9.**

(a) Use the ordered pairs (8, 243) and (12, 263) to write an equation that models the data.

(b) Use the equation from part (a) to estimate retail spending on prescription drugs in 2015.

CLASSROOM EXAMPLE 9

Refer to Example 9.

(a) Use the ordered pairs (9, 255) and (11, 262) to write an equation that models the data.

(b) Use the equation from part (a) to estimate retail spending on prescription drugs in 2014.

Answers:

(a) $y = 3.5x + 223.5$

(b) $272.5 billion

EXAMPLE 9 Writing an Equation of a Line That Models Data

Retail spending (in billions of dollars) on prescription drugs in the United States is shown in the graph in **FIGURE 34.**

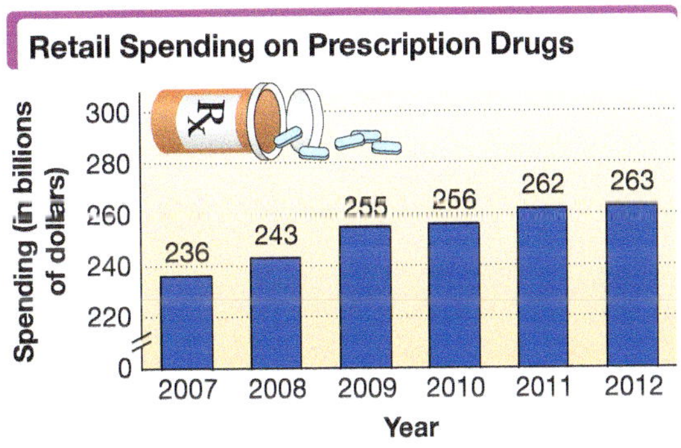

Source: Centers for Medicare & Medicaid Services.

FIGURE 34

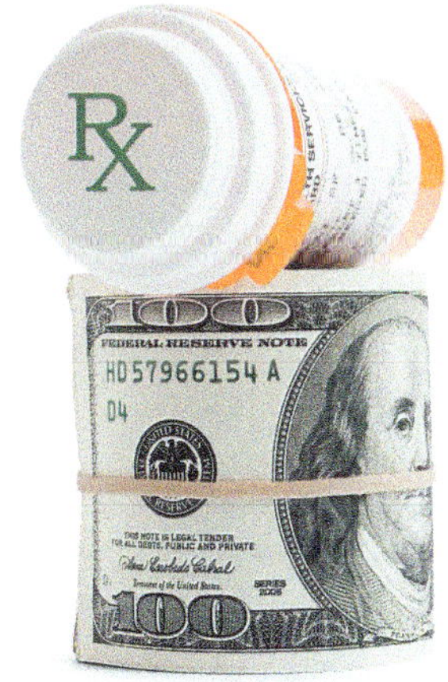

(a) Write an equation that models the data.

The data increase linearly—that is, a straight line through the tops of any two bars in the graph would be close to the top of each bar. To model the relationship between year x and spending on prescription drugs y, we let $x = 7$ represent 2007, $x = 8$ represent 2008, and so on. The given data for 2007 and 2012 can be written as the ordered pairs (7, 236) and (12, 263).

$$m = \frac{263 - 236}{12 - 7} = \frac{27}{5} = 5.4 \quad \text{Find the slope of the line through (7, 236) and (12, 263).}$$

Thus, spending increased by about $5.4 billion per year. To write an equation, we substitute this slope and the point (7, 236) into the point-slope form.

$$y - y_1 = m(x - x_1) \quad \text{Point-slope form}$$

$$y - 236 = 5.4(x - 7) \quad (x_1, y_1) = (7, 236);\ m = 5.4$$

$$y - 236 = 5.4x - 37.8 \quad \text{Distributive property}$$

$$y = 5.4x + 198.2 \quad \text{Add 236.}$$

Either point can be used here. (12, 263) provides the same answer.

Retail spending y (in billions of dollars) on prescription drugs in the United States in year x can be approximated by the equation $y = 5.4x + 198.2$.

(b) Use the equation from part (a) to estimate retail spending on prescription drugs in the United States in 2015. (Assume a constant rate of change.)

Because we let $x = 7$ represent 2007, $x = 15$ represents 2015.

$$y = 5.4x + 198.2 \quad \text{Equation from part (a)}$$

$$y = 5.4(15) + 198.2 \quad \text{Substitute 15 for } x.$$

$$y = 279.2 \quad \text{Multiply, and then add.}$$

About $279 billion was spent on prescription drugs in 2015. NOW TRY

NOW TRY ANSWERS

9. (a) $y = 5x + 203$

(b) $278 billion

2.3 Exercises

FOR EXTRA HELP MyMathLab®

Complete solution available in MyMathLab

1. A 2. C
3. A
4. $y = -3x + 10$
5. $3x + y = 10$
6. $y = \frac{10}{7}x - 10$
7. A 8. D
9. C 10. F
11. H 12. G
13. B 14. E
15. $y = 5x + 15$
16. $y = 2x + 12$
17. $y = -\frac{2}{3}x + \frac{4}{5}$
18. $y = -\frac{5}{8}x - \frac{1}{3}$
19. $y = x - 1$
20. $y = -x - 3$
21. $y = \frac{2}{5}x + 5$
22. $y = -\frac{3}{4}x + 7$
23. $y = \frac{2}{3}x + 1$
24. $y = \frac{5}{2}x - 3$
25. $y = -x - 2$
26. $y = -x + 2$
27. $y = 2x - 4$
28. $y = 3x + 3$
29. $y = -\frac{3}{5}x + 3$
30. $y = -\frac{5}{2}x - 5$
31. **(a)** $y = x + 4$ **(b)** 1 **(c)** (0, 4) **(d)**

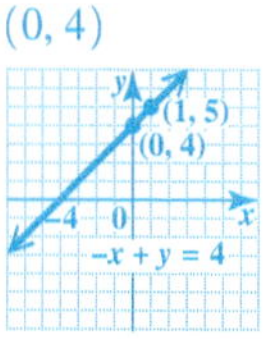

32. **(a)** $y = x + 6$ **(b)** 1 **(c)** (0, 6) **(d)**

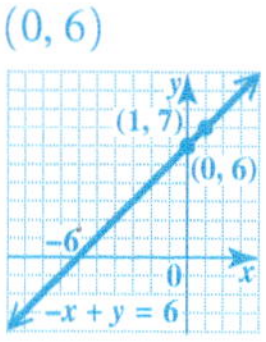

Concept Check *Provide the appropriate response.*

1. The following equations all represent the same line. Which one is in standard form as specified in this section?

A. $3x - 2y = 5$ **B.** $2y = 3x - 5$ **C.** $\frac{3}{5}x - \frac{2}{5}y = 1$ **D.** $3x = 2y + 5$

2. Which equation is in point-slope form?

A. $y = 6x + 2$ **B.** $4x + y = 9$ **C.** $y - 3 = 2(x - 1)$ **D.** $2y = 3x - 7$

3. Which equation in **Exercise 2** is in slope-intercept form?

4. Write the equation $y + 2 = -3(x - 4)$ in slope-intercept form.

5. Write the equation from **Exercise 4** in standard form.

6. Write the equation $10x - 7y = 70$ in slope-intercept form.

Concept Check *Match each equation with the graph that it most closely resembles. (Hint: Determining the signs of m and b will help in each case.)*

7. $y = 2x + 3$

8. $y = -2x + 3$

9. $y = -2x - 3$

10. $y = 2x - 3$

11. $y = 2x$

12. $y = -2x$

13. $y = 3$

14. $y = -3$

A.

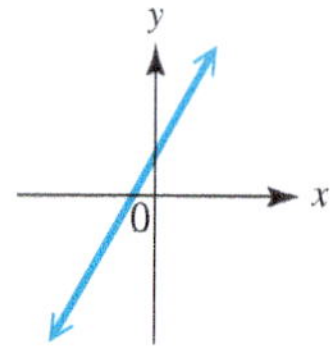

B.

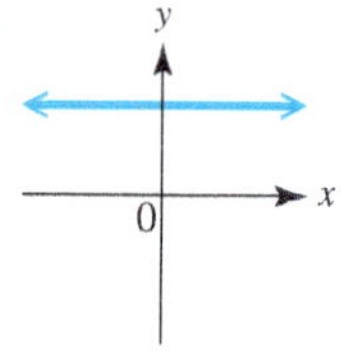

C.

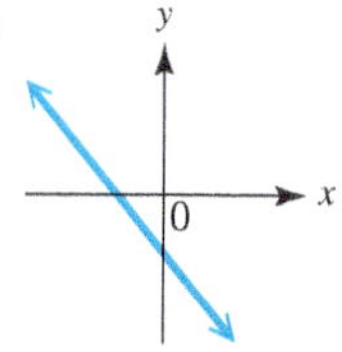

D.

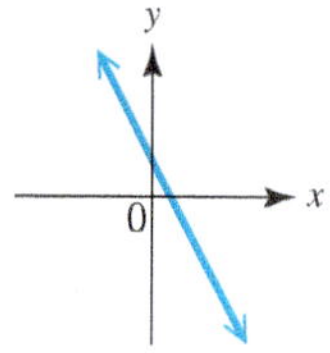

E.

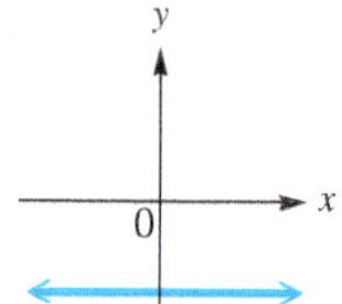

F.

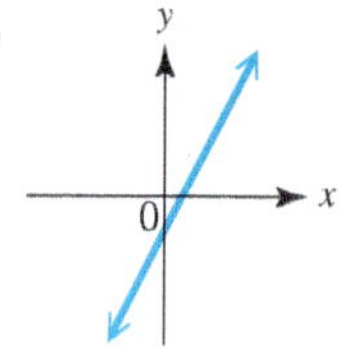

G.

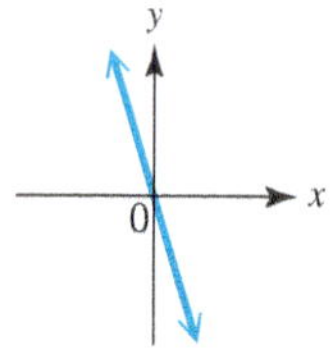

H.

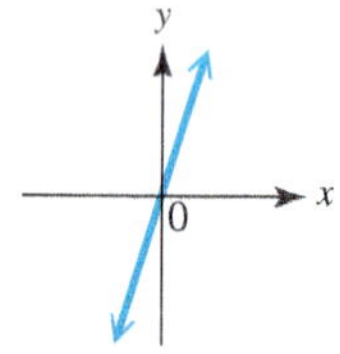

I.

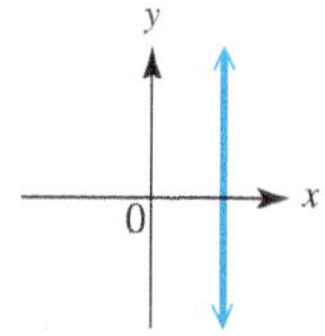

Write an equation in slope-intercept form of the line that satisfies the given conditions. ***See Example 1.***

15. $m = 5;\ b = 15$

16. $m = 2;\ b = 12$

17. $m = -\frac{2}{3};\ b = \frac{4}{5}$

18. $m = -\frac{5}{8};\ b = -\frac{1}{3}$

19. Slope 1; y-intercept $(0, -1)$

20. Slope -1; y-intercept $(0, -3)$

21. Slope $\frac{2}{5}$; y-intercept $(0, 5)$

22. Slope $-\frac{3}{4}$; y-intercept $(0, 7)$

Write an equation in slope-intercept form of the line shown in each graph. (Coordinates of the points shown are integers.)

23.

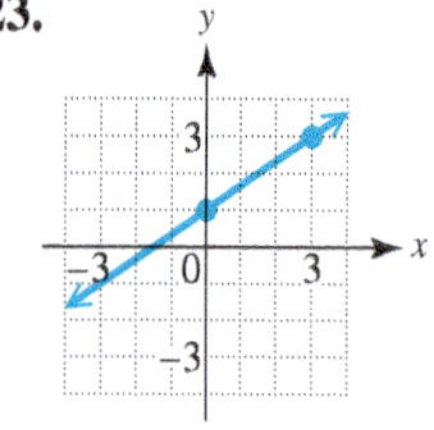

24.

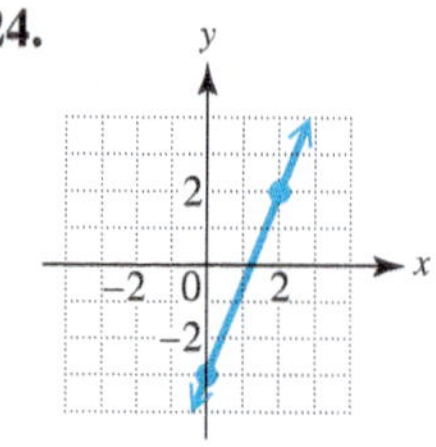

25.

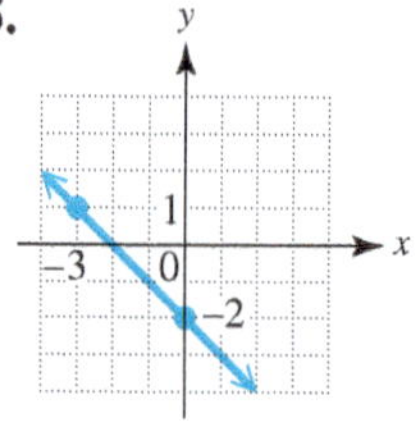

26.

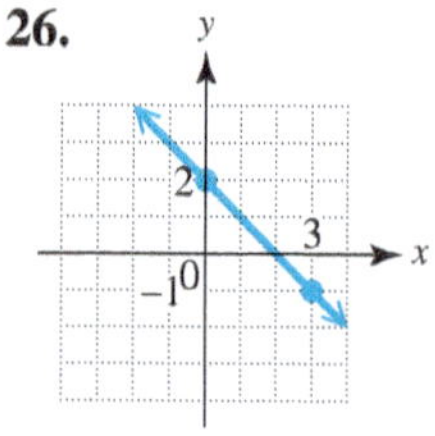

Each table of values gives several points that lie on a line. Write an equation in slope-intercept form of the line.

27.

x	y
−2	−8
0	−4
1	−2
3	2

28.

x	y
−2	−3
0	3
2	9
3	12

29.

x	y
−5	6
0	3
5	0
10	−3

30.

x	y
−4	5
−2	0
0	−5
2	−10

*For each equation, **(a)** write it in slope-intercept form, **(b)** give the slope of the line, **(c)** give the y-intercept, and **(d)** graph the line. **See Example 2.***

31. $-x + y = 4$ **32.** $-x + y = 6$ **33.** $6x + 5y = 30$

34. $3x + 4y = 12$ **35.** $4x - 5y = 20$ **36.** $7x - 3y = 3$

37. $x + 2y = -4$ **38.** $x + 3y = -9$

*Write an equation of the line that satisfies the given conditions. Give the equation **(a)** in slope-intercept form and **(b)** in standard form. **See Example 3 and the discussion on standard form.***

39. Through $(5, 8)$; slope -2 **40.** Through $(12, 10)$; slope 1

41. Through $(-2, 4)$; slope $-\frac{3}{4}$ **42.** Through $(-1, 6)$; slope $-\frac{5}{6}$

43. Through $(-5, 4)$; slope $\frac{1}{2}$ **44.** Through $(7, -2)$; slope $\frac{1}{4}$

45. x-intercept $(3, 0)$; slope 4 **46.** x-intercept $(-2, 0)$; slope -5

47. Through $(2, 6.8)$; slope 1.4 **48.** Through $(6, -1.2)$; slope 0.8

*Write an equation of the line passing through the given points. Give the final answer in standard form. **See Example 4.***

49. $(3, 4)$ and $(5, 8)$ **50.** $(5, -2)$ and $(-3, 14)$ **51.** $(6, 1)$ and $(-2, 5)$

52. $(-2, 5)$ and $(-8, 1)$ **53.** $(2, 5)$ and $(1, 5)$ **54.** $(-2, 2)$ and $(4, 2)$

55. $(7, 6)$ and $(7, -8)$ **56.** $(13, 5)$ and $(13, -1)$

57. $\left(\frac{1}{2}, -3\right)$ and $\left(-\frac{2}{3}, -3\right)$ **58.** $\left(-\frac{4}{9}, -6\right)$ and $\left(\frac{12}{7}, -6\right)$

59. $\left(-\frac{2}{5}, \frac{2}{5}\right)$ and $\left(\frac{4}{3}, \frac{2}{3}\right)$ **60.** $\left(\frac{3}{4}, \frac{8}{3}\right)$ and $\left(\frac{2}{5}, \frac{2}{3}\right)$

*Write an equation of the line that satisfies the given conditions. **See Example 5.***

61. Through $(9, 5)$; slope 0 **62.** Through $(-4, -2)$; slope 0

63. Through $(9, 10)$; undefined slope **64.** Through $(-2, 8)$; undefined slope

65. Through $\left(-\frac{3}{4}, -\frac{3}{2}\right)$; slope 0 **66.** Through $\left(-\frac{5}{8}, -\frac{9}{2}\right)$; slope 0

67. Through $(-7, 8)$; horizontal **68.** Through $(2, -7)$; horizontal

69. Through $(0.5, 0.2)$; vertical **70.** Through $(0.1, 0.4)$; vertical

*Write an equation of the line that satisfies the given conditions. Give the equation **(a)** in slope-intercept form and **(b)** in standard form. **See Example 6.***

71. Through $(7, 2)$; parallel to $3x - y = 8$ **72.** Through $(4, 1)$; parallel to $2x + 5y = 10$

73. Through $(-2, -2)$; parallel to $-x + 2y = 10$

74. Through $(-1, 3)$; parallel to $-x + 3y = 12$

33. (a) $y = -\frac{6}{5}x + 6$ (b) $-\frac{6}{5}$
(c) $(0, 6)$
(d)
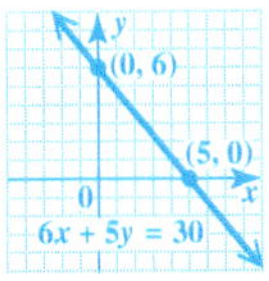

34. (a) $y = -\frac{3}{4}x + 3$ (b) $-\frac{3}{4}$
(c) $(0, 3)$
(d)
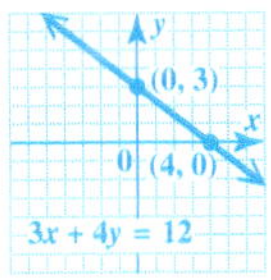

35. (a) $y = \frac{4}{5}x - 4$ (b) $\frac{4}{5}$
(c) $(0, -4)$
(d)
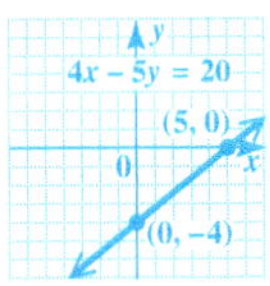

36. (a) $y = \frac{7}{3}x - 1$ (b) $\frac{7}{3}$
(c) $(0, -1)$
(d)
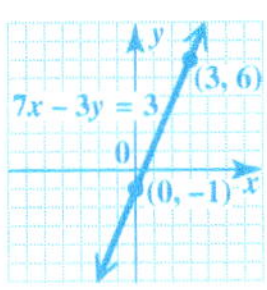

37. (a) $y = -\frac{1}{2}x - 2$ (b) $-\frac{1}{2}$
(c) $(0, -2)$
(d)
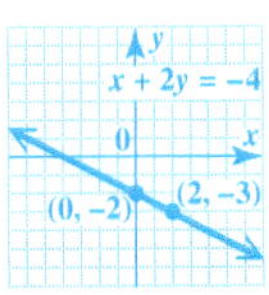

38. (a) $y = -\frac{1}{3}x - 3$ (b) $-\frac{1}{3}$
(c) $(0, -3)$
(d)
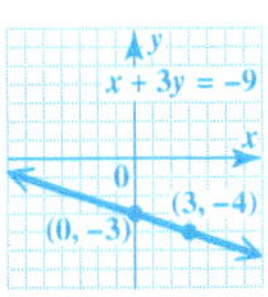

39. (a) $y = -2x + 18$
(b) $2x + y = 18$

40. (a) $y = x - 2$
(b) $x - y = 2$

41. (a) $y = -\frac{3}{4}x + \frac{5}{2}$
(b) $3x + 4y = 10$

42. (a) $y = -\frac{5}{6}x + \frac{31}{6}$
(b) $5x + 6y = 31$

43. (a) $y = \frac{1}{2}x + \frac{13}{2}$
(b) $x - 2y = -13$
44. (a) $y = \frac{1}{4}x - \frac{15}{4}$
(b) $x - 4y = 15$
45. (a) $y = 4x - 12$
(b) $4x - y = 12$
46. (a) $y = -5x - 10$
(b) $5x + y = -10$
47. (a) $y = 1.4x + 4$
(b) $7x - 5y = -20$
48. (a) $y = 0.8x - 6$
(b) $4x - 5y = 30$

49. $2x - y = 2$
50. $2x + y = 8$
51. $x + 2y = 8$
52. $2x - 3y = -19$
53. $y = 5$ 54. $y = 2$
55. $x = 7$ 56. $x = 13$
57. $y = -3$ 58. $y = -6$
59. $2x - 13y = -6$
60. $120x - 21y = 34$

61. $y = 5$ 62. $y = -2$
63. $x = 9$ 64. $x = -2$
65. $y = -\frac{3}{2}$ 66. $y = -\frac{9}{2}$
67. $y = 8$ 68. $y = -7$
69. $x = 0.5$ 70. $x = 0.1$

71. (a) $y = 3x - 19$
(b) $3x - y = 19$
72. (a) $y = -\frac{2}{5}x + \frac{13}{5}$
(b) $2x + 5y = 13$
73. (a) $y = \frac{1}{2}x - 1$
(b) $x - 2y = 2$
74. (a) $y = \frac{1}{3}x + \frac{10}{3}$
(b) $x - 3y = -10$
75. (a) $y = -\frac{1}{2}x + 9$
(b) $x + 2y = 18$
76. (a) $y = \frac{2}{5}x - \frac{39}{5}$
(b) $2x - 5y = 39$
77. (a) $y = 7$ (b) $y = 7$
78. (a) $y = 4$ (b) $y = 4$

79. $y = 45x$;
$(0, 0), (5, 225), (10, 450)$
80. $y = 26x$;
$(0, 0), (5, 130), (10, 260)$
81. $y = 3.75x$;
$(0, 0), (5, 18.75), (10, 37.50)$
82. $y = 4.50x$;
$(0, 0), (5, 22.50), (10, 45.00)$

75. Through $(8, 5)$; perpendicular to $2x - y = 7$

76. Through $(2, -7)$; perpendicular to $5x + 2y = 18$

77. Through $(-2, 7)$; perpendicular to $x = 9$

78. Through $(8, 4)$; perpendicular to $x = -3$

Write an equation in the form $y = mx$ for each situation. Then give the three ordered pairs associated with the equation for x-values 0, 5, and 10. ***See Example 7(a).***

79. x represents the number of hours traveling at 45 mph, and y represents the distance traveled (in miles).

80. x represents the number of t-shirts sold at \$26 each, and y represents the total cost of the t-shirts (in dollars).

81. x represents the number of gallons of gas sold at \$3.75 per gal, and y represents the total cost of the gasoline (in dollars).

82. x represents the number of days a DVD movie is rented at \$4.50 per day, and y represents the total charge for the rental (in dollars).

83. x represents the number of credit hours taken at Kirkwood Community College at \$140 per credit hour, and y represents the total tuition paid for the credit hours (in dollars). (*Source:* www.kirkwood.edu)

84. x represents the number of tickets to a performance of *Jersey Boys* at the Des Moines Civic Center purchased at \$125 per ticket, and y represents the total paid for the tickets (in dollars). (*Source:* Ticketmaster.)

For each situation, do the following.

(a) Write an equation in the form $y = mx + b$.

(b) Find and interpret the ordered pair associated with the equation for $x = 5$.

(c) Answer the question posed in the problem.

See Examples 7(b) and 7(c).

85. A ticket for the Diving Board Tour, featuring Elton John, costs \$149. A parking pass costs \$15. Let x represent the number of tickets and y represent the cost in dollars. How much does it cost for 2 tickets and a parking pass? (*Source:* Ticketmaster.)

86. Resident tuition at Broward College is \$105.90 per credit hour. There is also a \$20 health science application fee. Let x represent the number of credit hours and y represent the cost in dollars. How much does it cost for a student in health science to take 15 credit hours? (*Source:* www.broward.edu)

87. A health club membership costs \$99, plus \$41 per month. Let x represent the number of months and y represent the cost in dollars. How much does the first year's membership cost? (*Source:* Midwest Athletic Club.)

88. An Executive VIP/Gold membership to a health club costs \$159, plus \$57 per month. Let x represent the number of months and y represent the cost in dollars. How much does a one-year membership cost? (*Source:* Midwest Athletic Club.)

89. A wireless plan includes unlimited talk and text plus 2 GB of data for \$95 per month. There is a \$36 activation fee. Let x represent the number of months and y represent the cost in dollars. Over a two-year contract, how much will this plan cost? (*Source*: AT&T.)

90. Another wireless plan includes unlimited talk and text plus 4 GB of data for \$110 per month. There is a \$36 activation fee and a \$99 charge for an Apple iPhone 5c. Let x represent the number of months and y represent the cost in dollars. Over a two-year contract, how much will this plan cost? (*Source*: AT&T.)

91. There is a \$30 fee to rent a chain saw, plus \$6 per day. Let x represent the number of days the saw is rented and y represent the charge to the user in dollars. If the total charge is \$138, for how many days is the saw rented?

92. A rental car costs \$50, plus \$0.45 per mile. Let x represent the number of miles driven and y represent the total charge to the renter in dollars. How many miles was the car driven if the renter paid \$127.85?

Solve each problem. In part (a), give equations in slope-intercept form. ***See Examples 8 and 9.***

93. Total sales of portable media/MP3 players in the United States (in millions of dollars) are shown in the graph, where the year 2010 corresponds to $x = 0$.

(a) Use the ordered pairs from the graph to write an equation that models the data. Interpret the slope in the context of this problem.

(b) Use the equation from part (a) to approximate sales of portable media/MP3 players in the United States in 2011, the year data was unavailable.

Media/MP3 Player Sales

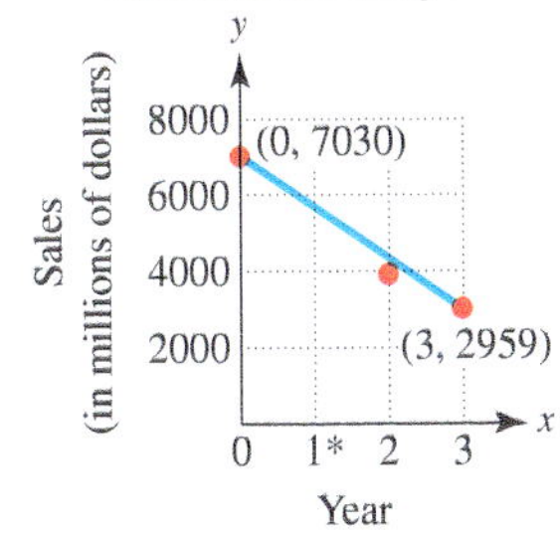

*Data for this year is unavailable.

Source: Consumer Electronics Association.

94. Total sales of smartphones in the United States (in billions of dollars) are shown in the graph, where the year 2010 corresponds to $x = 0$.

(a) Use the ordered pairs from the graph to write an equation that models the data. (Round the slope to the nearest tenth.) Interpret the slope in the context of this problem.

(b) Use the equation from part (a) to approximate smartphone sales in the United States in 2011, the year data was unavailable.

Smartphone Sales

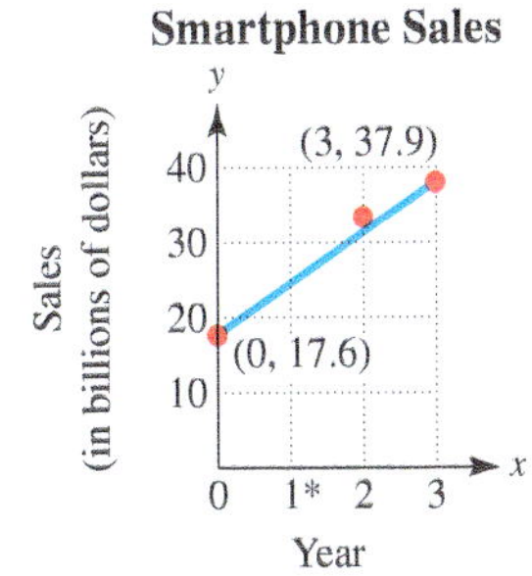

*Data for this year is unavailable.

Source: Consumer Electronics Association.

95. Expenditures for home health care in the United States are shown in the graph.

(a) Use the information given for the years 2008 and 2012, letting $x = 8$ represent 2008 and $x = 12$ represent 2012, and letting y represent spending (in billions of dollars), to write an equation that models the data.

(b) Use the equation from part (a) to approximate the amount spent on home health care in 2011 to the nearest tenth. How does the result compare with the actual value, \$74.0 billion?

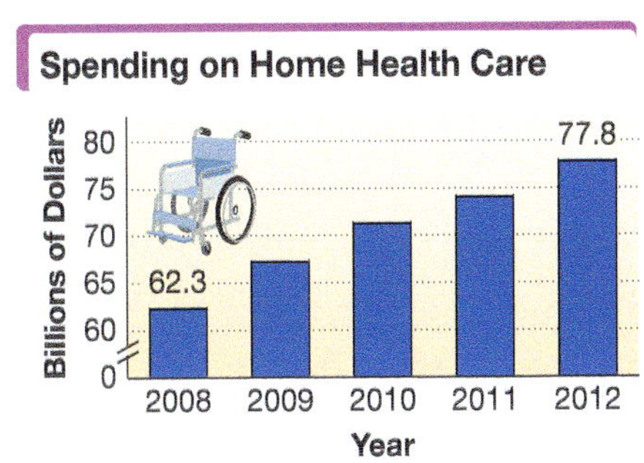

Source: Centers for Medicare & Medicaid Services.

83. $y = 140x$;
$(0, 0), (5, 700), (10, 1400)$

84. $y = 125x$;
$(0, 0), (5, 625), (10, 1250)$

85. (a) $y = 149x + 15$
(b) $(5, 760)$;
The cost for 5 tickets and a parking pass is \$760.
(c) \$313

86. (a) $y = 105.90x + 20$
(b) $(5, 549.5)$;
The cost for 5 credit hours and the application fee is \$549.50.
(c) \$1608.50

87. (a) $y = 41x + 99$
(b) $(5, 304)$;
The cost for a 5-month membership is \$304.
(c) \$591

88. (a) $y = 57x + 159$
(b) $(5, 444)$;
The cost for a 5-month membership is \$444.
(c) \$843

89. (a) $y = 95x + 36$
(b) $(5, 511)$;
The cost of the plan for 5 months is \$511.
(c) \$2316

90. (a) $y = 110x + 135$
(b) $(5, 685)$;
The cost of the plan for 5 months is \$685.
(c) \$2775

91. (a) $y = 6x + 30$
(b) $(5, 60)$;
It costs \$60 to rent the saw for 5 days.
(c) 18 days

92. (a) $y = 0.45x + 50$
(b) $(5, 52.25)$;
The charge for driving 5 mi is \$52.25.
(c) 173 mi

93. (a) $y = -1357x + 7030$;
Sales of portable media/MP3 players in the United States decreased by \$1357 million per year from 2010 to 2013.
(b) \$5673 million

94. (a) $y = 6.8x + 17.6$;
Sales of smartphones in the United States increased by \$6.8 billion per year from 2010 to 2013.
(b) \$24.4 billion

95. (a) $y = 3.875x + 31.3$
(b) $73.9 billion; It is very close to the actual value.

96. (a) $y = -5.44x + 133.98$
(b) 79.6 billion; It is greater than the actual value.

96. The number of pieces of first class mail delivered in the United States is shown in the graph.

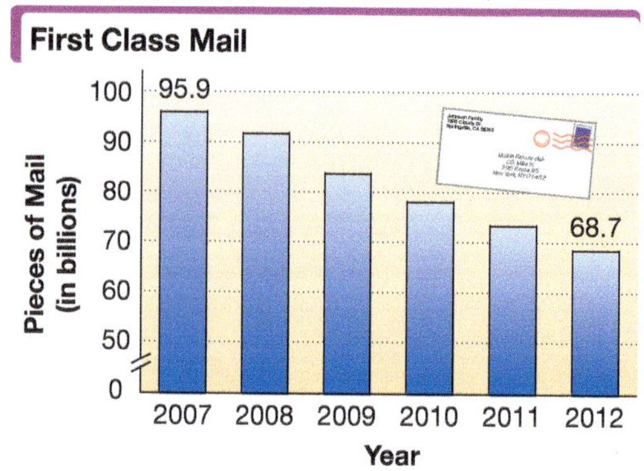

Source: U.S. Postal Service.

(a) Use the information given for the years 2007 and 2012, letting $x = 7$ represent 2007 and $x = 12$ represent 2012, and letting y represent the number of pieces of mail (in billions), to write an equation that models the data.

(b) Use the equation from part (a) to approximate the number of pieces of first class mail delivered in 2010 to the nearest tenth. How does this result compare to the actual value, 78.2 billion?

RELATING CONCEPTS For Individual or Group Work (Exercises 97–104)

In ***Section 1.2,*** *we worked with formulas.* ***Work Exercises 97–104 in order,*** *to see how the formula that relates Celsius and Fahrenheit temperatures is derived.*

97. There is a linear relationship between Celsius and Fahrenheit temperatures.

When $C = 0°$, $F =$ _____°.

When $C = 100°$, $F =$ _____°.

98. Think of ordered pairs of temperatures (C, F), where C and F represent corresponding Celsius and Fahrenheit temperatures. The equation that relates the two scales has a straight-line graph that contains the two points determined in **Exercise 97.** What are these two points?

99. Find the slope of the line described in **Exercise 98.**

100. Use the slope found in **Exercise 99** and one of the two points determined earlier, and write an equation that gives F in terms of C. (*Hint:* Use the point-slope form, where C replaces x and F replaces y.)

101. To obtain another form of the formula, use the equation found in **Exercise 100** and solve for C in terms of F.

102. Use the equation from **Exercise 100** to find the Fahrenheit temperature when $C = 30$.

103. Use the equation from **Exercise 101** to find the Celsius temperature when $F = 50$.

104. For what temperature does $F = C$? (Use the photo to confirm this temperature.)

97. 32; 212
98. $(0, 32)$ and $(100, 212)$
99. $\frac{9}{5}$
100. $F = \frac{9}{5}C + 32$
101. $C = \frac{5}{9}(F - 32)$
102. 86°
103. 10°
104. −40°

STUDY SKILLS

Taking Math Tests

Techniques To Improve Your Test Score	Comments
Come prepared with a pencil, eraser, paper, and calculator, if allowed.	Working in pencil lets you erase, keeping your work neat and readable.
Scan the entire test, note the point values of different problems, and plan your time accordingly.	To do 20 problems in 50 minutes, allow $50 \div 20 = 2.5$ minutes per problem. Spend less time on easier problems.
Do a "knowledge dump" when you get the test. Write important notes, such as formulas, in a corner of the test.	Writing down tips and information that you've learned at the beginning allows you to relax later.
Read directions carefully, and circle any significant words. When you finish a problem, reread the directions. Did you do what was asked?	Pay attention to announcements written on the board or made by your instructor. Ask questions if you don't understand.
Show all your work. Many teachers give partial credit if some steps are correct, even if the final answer is wrong. ***Write neatly.***	If your teacher can't read your writing, you won't get credit for it. If you need more space to work, ask to use extra paper.
Write down anything that might help solve a problem: a formula, a diagram, etc. If you still can't solve it, circle the problem and come back to it later. Do *not* erase anything you wrote down.	If you know even a little bit about the problem, write it down. The answer may come to you as you work on it, or you may get partial credit. Don't spend too long on any one problem.
If you can't solve a problem, make a guess. Do not change it unless you find an obvious mistake.	Have a good reason for changing an answer. Your first guess is usually your best bet.
Check that the answer to an application problem is reasonable and makes sense. Reread the problem to make sure that you have answered the question.	Use common sense. Can the father really be seven years old? Would a month's rent be \$32,140? Remember to label your answer if needed: \$, years, inches, etc.
Check for careless errors. Rework the problem without looking at your previous work. Compare the two answers.	Reworking the problem from the beginning forces you to rethink it. If possible, use a different method to solve the problem.

Think through and answer each question.

1. What two or three tips will you try when you take your next math test?
2. How did the tips you selected work for you when you took your math test?
3. What will you do differently when taking your next math test?

SUMMARY EXERCISES Finding Slopes and Equations of Lines

1. $-\frac{3}{5}$ 2. 0
3. 1 4. $\frac{3}{7}$
5. undefined 6. $-\frac{4}{7}$
7. (a) B (b) F (c) A (d) C (e) E (f) D
8. C is in standard form; A: $4x + y = -7$; B: $3x - 4y = -12$; D: $x + 2y = 0$; E: $3x - y = 5$ F: $5x - 3y = 15$
9. (a) $y = -\frac{5}{6}x + \frac{13}{3}$ (b) $5x + 6y = 26$
10. (a) $y = -8$ (b) $y = -8$
11. (a) $y = -3x + 10$ (b) $3x + y = 10$
12. (a) $y = -\frac{5}{2}x + 2$ (b) $5x + 2y = 4$
13. (a) $y = -\frac{7}{9}$ (b) $9y = -7$
14. (a) $y = \frac{2}{3}x + 8$ (b) $2x - 3y = -24$
15. (a) $y = -\frac{5}{2}x$ (b) $5x + 2y = 0$
16. (a) $y = 3x + 11$ (b) $3x - y = -11$
17. (a) $y = \frac{2}{3}x + \frac{14}{3}$ (b) $2x - 3y = -14$
18. (a) $y = 2x - 10$ (b) $2x - y = 10$

Find the slope of each line, if possible.

1. Through $(3, -3)$ and $(8, -6)$

2. Through $(4, -5)$ and $(-1, -5)$

3. $y = x - 5$

4. $3x - 7y = 21$

5. $x - 4 = 0$

6. $4x + 7y = 3$

7. ***Concept Check*** Match the description in Column I with its equation in Column II.

I	II
(a) Slope -0.5, $b = -2$	**A.** $y = -\frac{1}{2}x$
(b) x-intercept $(4, 0)$, y-intercept $(0, 2)$	**B.** $y = -\frac{1}{2}x - 2$
(c) Passes through $(4, -2)$ and $(0, 0)$	**C.** $x - 2y = 2$
(d) $m = \frac{1}{2}$, passes through $(-2, -2)$	**D.** $y = 2x$
(e) $m = \frac{1}{2}$, passes through the origin	**E.** $x = 2y$
(f) Slope 2, $b = 0$	**F.** $x + 2y = 4$

8. ***Concept Check*** Which equation is written in standard form as specified in the text?

A. $y = -4x - 7$ **B.** $-3x + 4y = 12$ **C.** $x - 6y = 3$

D. $\frac{1}{2}x + y = 0$ **E.** $6x - 2y = 10$ **F.** $3y - 5x = -15$

Write the equations not written in standard form in that form.

For each line described, write an equation of the line ***(a)*** *in slope-intercept form and* ***(b)*** *in standard form.*

9. Through $(-2, 6)$ and $(4, 1)$

10. Through $(5, -8)$, $m = 0$

11. Through $(4, -2)$ with slope -3

12. Through $(4, -8)$ and $(-4, 12)$

13. Through $\left(\frac{3}{4}, -\frac{7}{9}\right)$; perpendicular to $x = \frac{2}{3}$

14. Through $(-3, 6)$ with slope $\frac{2}{3}$

15. Through the origin; perpendicular to $2x - 5y = 6$

16. Through $(-2, 5)$; parallel to $3x - y = 4$

17. Through $(-4, 2)$; parallel to the line through $(3, 9)$ and $(6, 11)$

18. Through $(4, -2)$; perpendicular to the line through $(3, 7)$ and $(5, 6)$

2.4 Linear Inequalities in Two Variables

OBJECTIVES

1. Graph linear inequalities in two variables.
2. Graph the intersection of two linear inequalities.
3. Graph the union of two linear inequalities.

VOCABULARY

☐ linear inequality in two variables
☐ boundary line

OBJECTIVE 1 Graph linear inequalities in two variables.

In **Section 1.5,** we graphed linear inequalities in one variable on a number line. In this section, we graph linear inequalities in two variables on a rectangular coordinate system.

Linear Inequality in Two Variables

A **linear inequality in two variables** (here x and y) can be written in the form

$$Ax + By < C, \quad Ax + By \leq C, \quad Ax + By > C, \quad \text{or} \quad Ax + By \geq C,$$

where A, B, and C are real numbers and A and B are not both 0.

Consider the graph in FIGURE 35. The graph of the line $x + y = 5$ divides the points in the rectangular coordinate system into three sets of points.

1. Those points that lie ***on*** the line itself and satisfy the equation $x + y = 5$, such as $(0, 5)$, $(2, 3)$, and $(5, 0)$;
2. Those points that lie in the region *above* the line and satisfy the inequality $x + y > 5$, such as $(5, 3)$ and $(2, 4)$;
3. Those points that lie in the region *below* the line and satisfy the inequality $x + y < 5$, such as $(0, 0)$ and $(-3, -1)$.

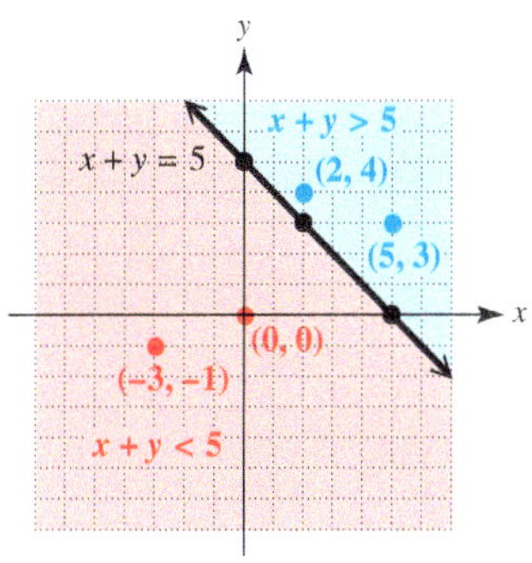

FIGURE 35

The graph of the line $x + y = 5$ is the **boundary line** for the two inequalities

$$x + y > 5 \quad \text{and} \quad x + y < 5.$$

A graph of a linear inequality in two variables is a region in the real number plane that may or may not include the boundary line.

To graph a linear inequality in two variables, follow these steps.

TEACHING TIP
Remind students of this statement:

Equality is the boundary between less than and greater than.

Graphing a Linear Inequality in Two Variables

Step 1 **Draw the graph of the straight line that is the boundary.**

- Make the line solid if the inequality involves $\leq$ or $\geq$.
- Make the line dashed if the inequality involves $<$ or $>$.

Step 2 **Choose a test point.** Choose any point not on the line, and substitute the coordinates of that point in the inequality.

Step 3 **Shade the appropriate region.** Shade the region that includes the test point if it satisfies the original inequality. Otherwise, shade the region on the other side of the boundary line.

NOW TRY EXERCISE 1

Graph $-x + 2y \geq 4$.

CLASSROOM EXAMPLE 1

Graph $x + y \leq 4$.

Answer:

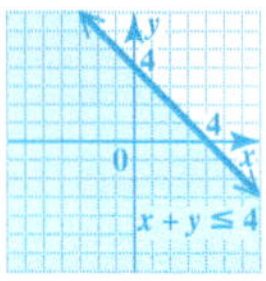

EXAMPLE 1 Graphing a Linear Inequality

Graph $3x + 2y \geq 6$.

Step 1 First graph the boundary line $3x + 2y = 6$, which has intercepts $(2, 0)$ and $(0, 3)$, as shown in **FIGURE 36**.

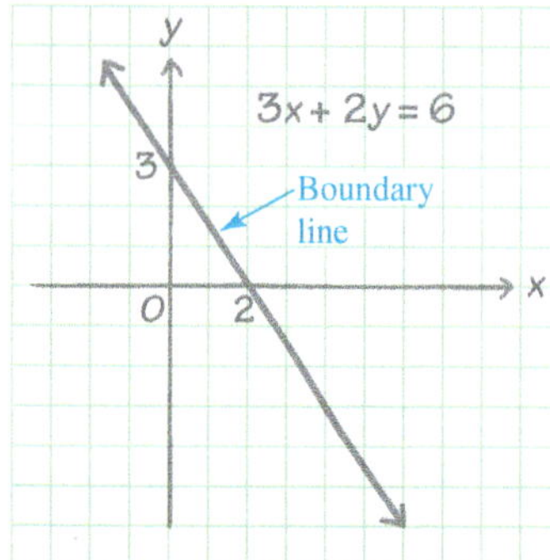

FIGURE 36

Step 2 The graph of the inequality $3x + 2y \geq 6$ includes the points of the boundary line $3x + 2y = 6$ (because the inequality symbol $\geq$ includes equality) and either the points *above* that line or the points *below* it. To decide which, select any point *not* on the boundary line to use as a test point. Substitute the values from the test point for x and y in the inequality.

$$3x + 2y > 6 \quad \text{We are testing the region.}$$

$$3(0) + 2(0) \stackrel{?}{>} 6 \quad \text{Let } x = 0 \text{ and } y = 0.$$

(0, 0) is a convenient test point.

$$0 > 6 \quad \text{False}$$

TEACHING TIP

Remind students that if a boundary line passes through the origin, they must choose a test point other than (0, 0).

Step 3 Because the result is false, $(0, 0)$ does *not* satisfy the inequality. The solution set includes all points in the region on the *other* side of the line. See **FIGURE 37**.

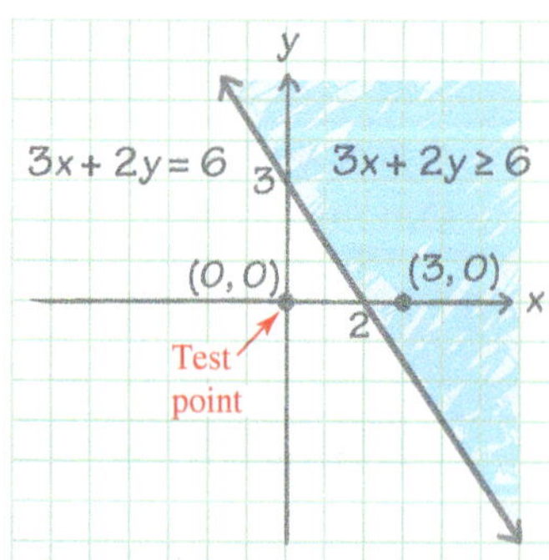

FIGURE 37

As a further check, select a test point in the shaded region, such as $(3, 0)$, and confirm that it does indeed satisfy the inequality. See **FIGURE 37**. NOW TRY

NOW TRY ANSWER

1.

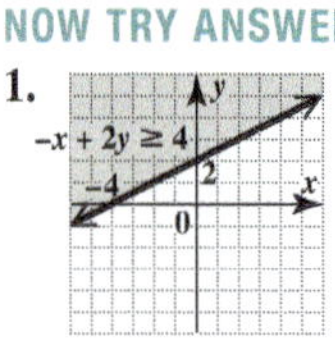

CAUTION When drawing the boundary line in Step 1, be careful to draw a solid line if the inequality includes equality ($\leq$, $\geq$) or a dashed line if equality is not included ($<$, $>$).

TEACHING TIP
It may be helpful to review the addition and multiplication properties of inequality here.

If an inequality is written in the form $y > mx + b$ or $y < mx + b$, then the inequality symbol indicates which region to shade.

If $y > mx + b$, then shade above the boundary line.

If $y < mx + b$, then shade below the boundary line.

This method works only if the inequality is solved for y.

CAUTION A common error in using the method just described is to use the original inequality symbol when deciding which region to shade. ***Be sure to use the inequality symbol found in the inequality** after it is solved for y.*

NOW TRY EXERCISE 2
Graph $3x - 2y < 0$.

CLASSROOM EXAMPLE 2
Graph $x + y > 0$.

Answer:

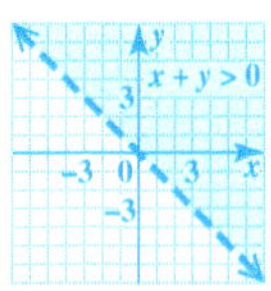

EXAMPLE 2 Graphing a Linear Inequality with Boundary Passing through the Origin

Graph $3x - 4y > 0$.

First graph the boundary line. The x- and y-intercepts are the same point, $(0, 0)$. Thus, this line passes through the origin. Two other points on the line are $(4, 3)$ and $(-4, -3)$. The points of the boundary line do *not* belong to the inequality

$$3x - 4y > 0$$

(because the inequality symbol is $>$, *not* $\geq$). For this reason, the line is dashed. See **FIGURE 38**.

To use the method explained above, we solve the inequality for y.

$$3x - 4y > 0 \quad \text{Original inequality}$$

$$-4y > -3x \quad \text{Subtract } 3x.$$

$$y < \frac{3}{4}x \quad \text{Divide by } -4. \text{ Change } > \text{ to } <.$$

Use this equivalent inequality to decide which region to shade.

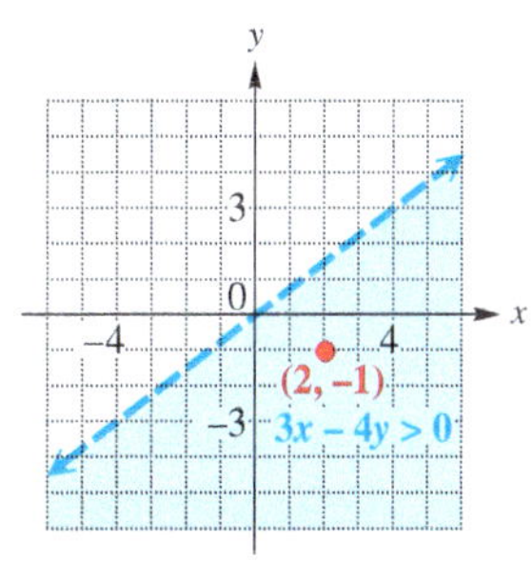

FIGURE 38

Because the *is less than* symbol occurs ***when the original inequality is solved for y,*** shade the region *below* the boundary line.

CHECK As a further check, choose a test point not on the line, which rules out the origin. We choose $(2, -1)$.

$$3x - 4y > 0 \quad \text{Original inequality}$$

$$3(2) - 4(-1) \overset{?}{>} 0 \quad \text{Let } x = 2 \text{ and } y = -1.$$

$$6 + 4 \overset{?}{>} 0 \quad \text{Multiply.}$$

$$10 > 0 \ \checkmark \quad \text{True}$$

This result agrees with the decision to shade below the line. The solution set, graphed in **FIGURE 38**, includes only those points in the shaded region (and *not* those on the line).

NOW TRY

NOW TRY ANSWER

2.

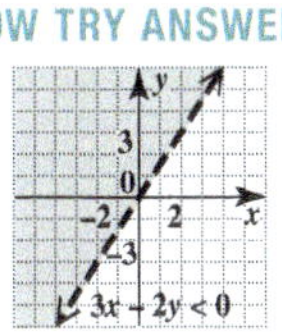

NOW TRY EXERCISE 3

Graph $x + 2 > 0$.

CLASSROOM EXAMPLE 3

Graph $y + 4 \leq 0$.

Answer:

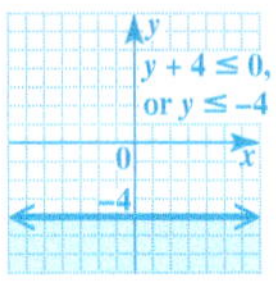

EXAMPLE 3 Graphing a Linear Inequality

Graph $x - 3 < 1$.

We graph $x - 3 = 1$, which is equivalent to $x = 4$, as a dashed vertical line passing through the point $(4, 0)$. To determine which region to shade, we choose $(0, 0)$ as a test point.

$$x - 3 < 1$$
$$0 - 3 \stackrel{?}{<} 1 \quad \text{Let } x = 0.$$
$$-3 < 1 \quad \text{True}$$

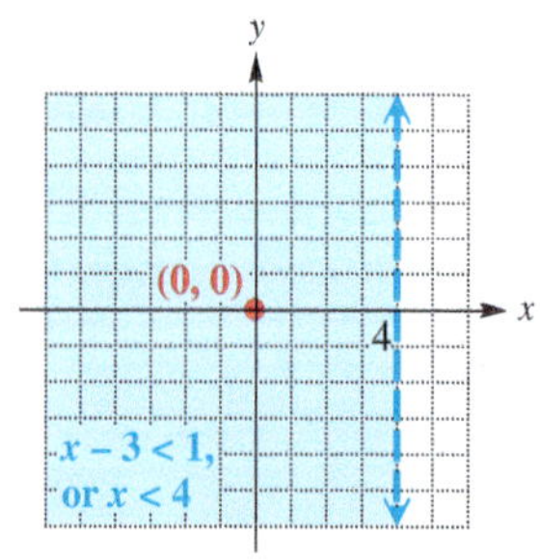

FIGURE 39

Because a true statement results, we shade the region containing $(0, 0)$. See FIGURE 39.

NOW TRY

OBJECTIVE 2 Graph the intersection of two linear inequalities.

A pair of inequalities joined with the word ***and*** is interpreted as the intersection of the solution sets of the inequalities.

The graph of the intersection of two or more inequalities is the region of the plane where all points satisfy all of the inequalities at the same time.

CLASSROOM EXAMPLE 4

Graph $x - y \leq 4$ and $x \geq -2$.

Answer:

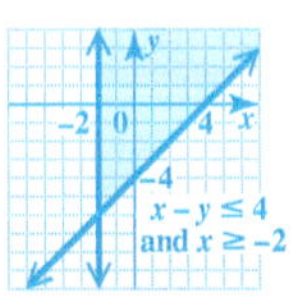

EXAMPLE 4 Graphing the Intersection of Two Inequalities

Graph $2x + 4y \geq 5$ and $x \geq 1$.

To begin, we graph each of the two inequalities $2x + 4y \geq 5$ and $x \geq 1$ separately, as shown in FIGURES 40(a) AND (b). Then we use heavy shading to identify the intersection of the graphs, as shown in FIGURE 40(c).

TEACHING TIP

Suggest that students use two different-colored pencils or highlighters when shading intersections and unions of inequalities so that solution sets can be more easily identified.

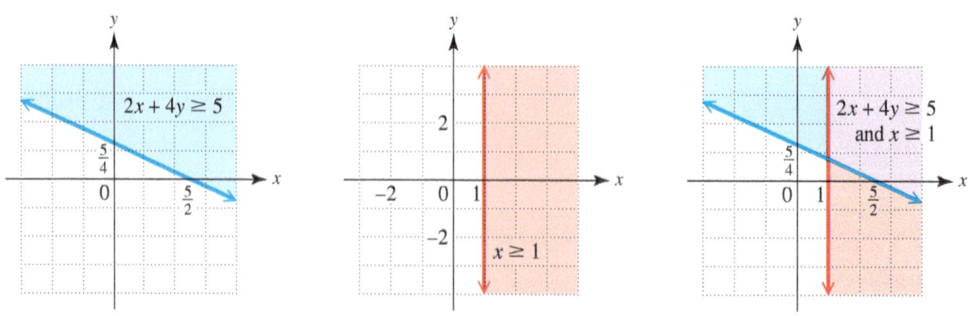

(a) (b) (c)

FIGURE 40

In practice, the graphs in FIGURES 40(a) AND (b) are graphed on the same axes.

NOW TRY ANSWER

3.

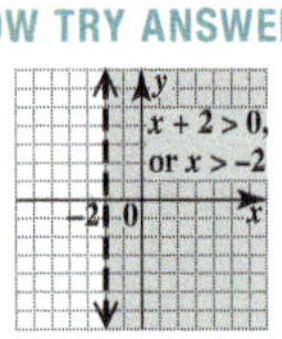

NOW TRY EXERCISE 4

Graph $x + y < 3$ and $y \leq 2$.

CHECK Using **FIGURE 40(c)**, we can choose a test point from each of the four regions formed by the intersection of the boundary lines.

$(2, 1)$,	$(0, 2)$,	$(0, 0)$,	and	$(2, -1)$	Possible test points
↑	↑	↑		↑	
Heavily shaded region	Blue shaded region	Unshaded region		Red shaded region	

Verify that only ordered pairs in the heavily shaded region satisfy *both* inequalities. Ordered pairs in the other regions satisfy only one of the inequalities or neither of them. ✓

NOW TRY

CLASSROOM EXAMPLE 5

Graph $7x - 3y < 21$ or $x > 2$.

Answer:

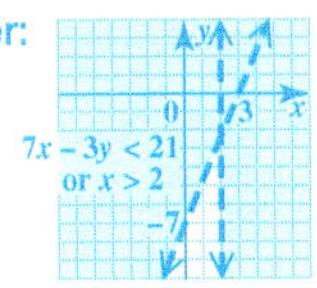

OBJECTIVE 3 Graph the union of two linear inequalities.

When two inequalities are joined by the word ***or***, we must find the union of the graphs of the inequalities.

The graph of the union of two inequalities includes all of the points that satisfy either inequality.

NOW TRY EXERCISE 5

Graph $3x - 5y < 15$ or $x > 4$.

EXAMPLE 5 Graphing the Union of Two Inequalities

Graph $2x + 4y \geq 5$ or $x \geq 1$.

The graphs of the two inequalities are shown in **FIGURES 40(a) AND (b)** in **Example 4** on the preceding page. The graph of the union includes all points in *either* inequality, as shown in **FIGURE 41**.

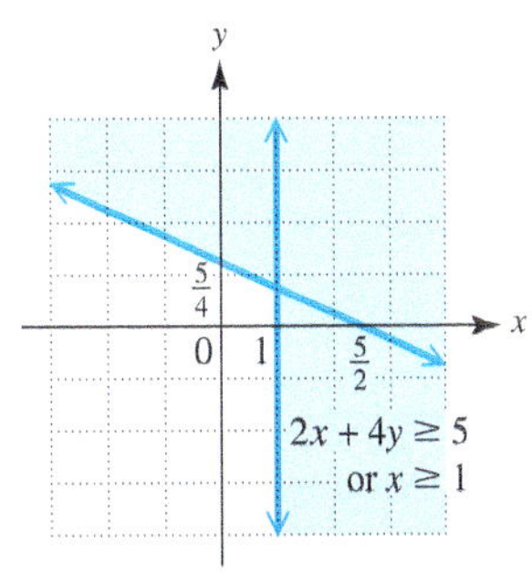

FIGURE 41

NOW TRY

NOW TRY ANSWERS

4.

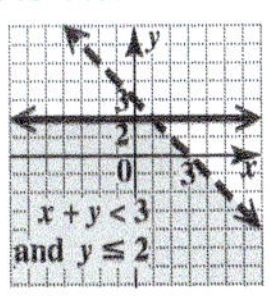

5.

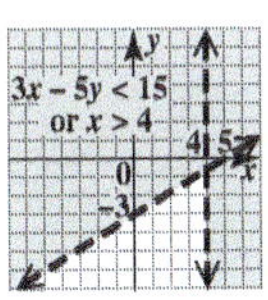

2.4 Exercises

FOR EXTRA HELP

MyMathLab®

Complete solution available in MyMathLab

Concept Check *Decide whether each ordered pair is a solution of the given inequality.*

1. $x - 2y \leq 4$

(a) $(0, 0)$ **(b)** $(2, -1)$

(c) $(7, 1)$ **(d)** $(0, 2)$

2. $x + y > 0$

(a) $(0, 0)$ **(b)** $(-2, 1)$

(c) $(2, -1)$ **(d)** $(-4, 6)$

3. $x - 5 > 0$

(a) $(0, 0)$ **(b)** $(5, 0)$

(c) $(-1, 3)$ **(d)** $(6, 2)$

4. $y \leq 1$

(a) $(0, 0)$ **(b)** $(3, 1)$

(c) $(2, -1)$ **(d)** $(-3, 3)$

1. (a) yes (b) yes (c) no (d) yes
2. (a) no (b) no (c) yes (d) yes
3. (a) no (b) no (c) no (d) yes
4. (a) yes (b) yes (c) yes (d) no

5. solid; below
6. dashed; below
7. dashed; above
8. solid; above

9. $\leq$ 10. $\geq$
11. $>$ 12. $<$

Due to space limitations, odd-numbered graphs can be found in the Student Answer Section.

14.
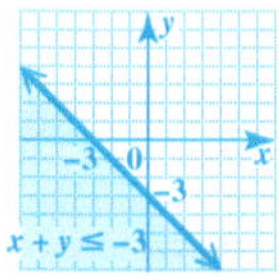

16.
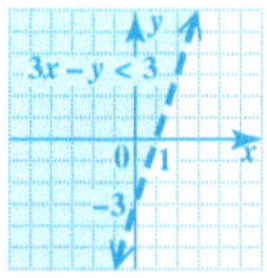

18.
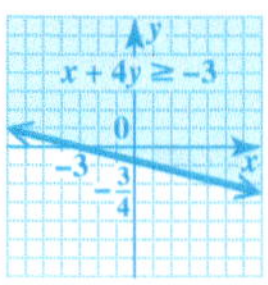

20.
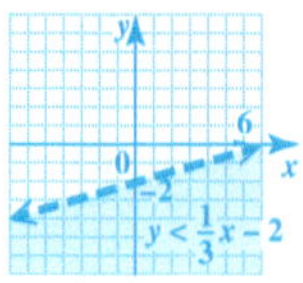

22.
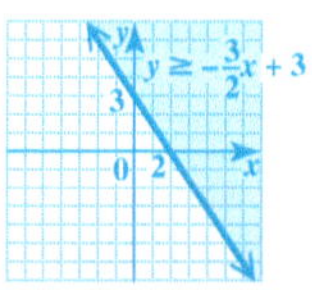

24.
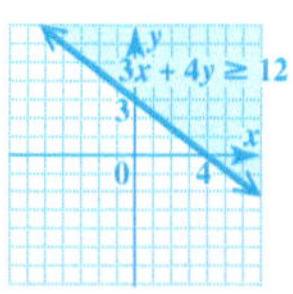

26.
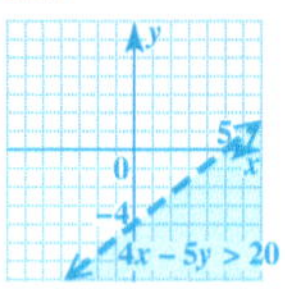

28.
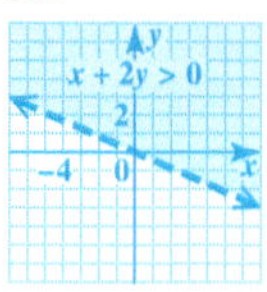

30.
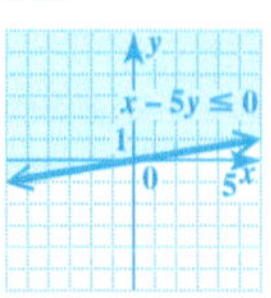

32.
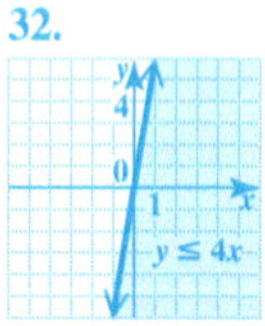

34.
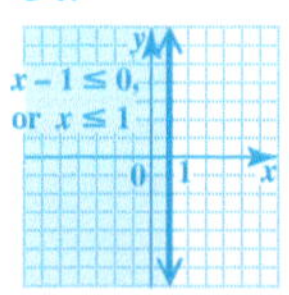

36.
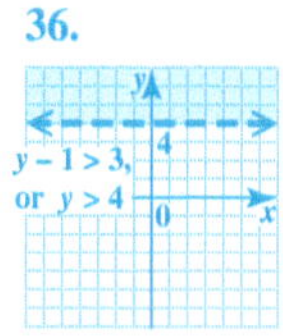

37. 2; $(0, -4)$; $2x - 4$; solid; above; $\geq$; $\geq 2x - 4$

Concept Check *In each statement, fill in the first blank with either* solid *or* dashed. *Fill in the second blank with either* above *or* below.

5. The boundary of the graph of $y \leq -x + 2$ will be a _____ line, and the shading will be _____ the line.

6. The boundary of the graph of $y < -x + 2$ will be a _____ line, and the shading will be _____ the line.

7. The boundary of the graph of $y > -x + 2$ will be a _____ line, and the shading will be _____ the line.

8. The boundary of the graph of $y \geq -x + 2$ will be a _____ line, and the shading will be _____ the line.

Concept Check *Refer to the given graph, and complete each statement with the correct inequality symbol* $<$, $\leq$, $>$, *or* $\geq$.

9. x _____ 4

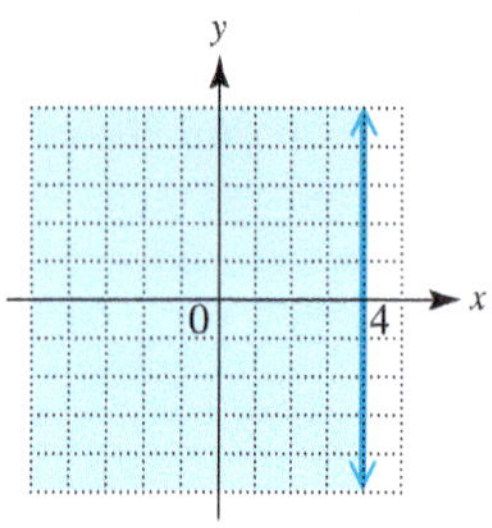

10. y _____ -3

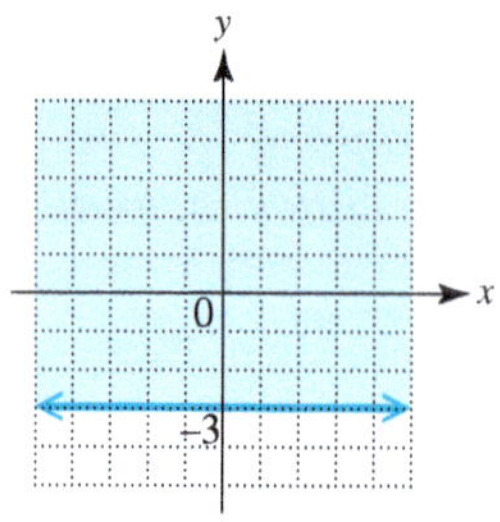

11. y _____ $3x - 2$

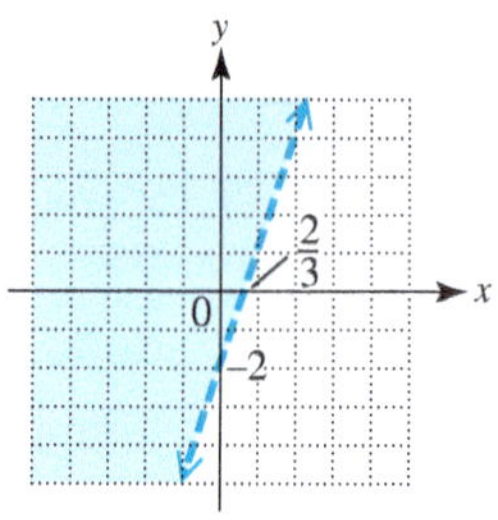

12. y _____ $-x + 3$

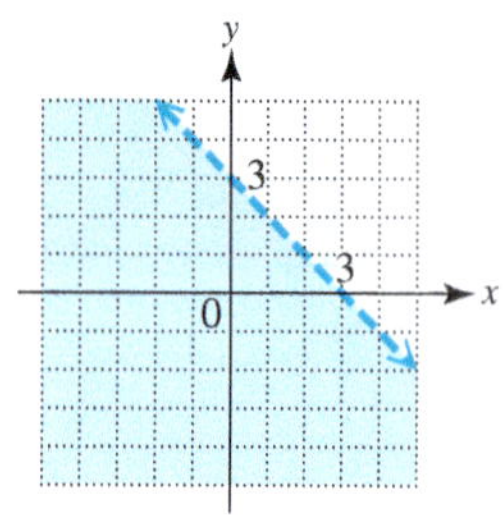

Graph each linear inequality in two variables. ***See Examples 1–3.***

13. $x + y \leq 2$	**14.** $x + y \leq -3$	**15.** $4x - y < 4$
16. $3x - y < 3$	**17.** $x + 3y \geq -2$	**18.** $x + 4y \geq -3$
19. $y < \frac{1}{2}x + 3$	**20.** $y < \frac{1}{3}x - 2$	**21.** $y \geq -\frac{2}{5}x + 2$
22. $y \geq -\frac{3}{2}x + 3$	**23.** $2x + 3y \geq 6$	**24.** $3x + 4y \geq 12$
25. $5x - 3y > 15$	**26.** $4x - 5y > 20$	**27.** $x + y > 0$
28. $x + 2y > 0$	**29.** $x - 3y \leq 0$	**30.** $x - 5y \leq 0$
31. $y < x$	**32.** $y \leq 4x$	**33.** $x + 3 \geq 0$
34. $x - 1 \leq 0$	**35.** $y + 5 < 2$	**36.** $y - 1 > 3$

Extending Skills *Complete each of the following to write an inequality for the graph shown.*

37. Determine the following for the boundary line.

Slope: __________

y-intercept: __________

Equation: $y =$ __________

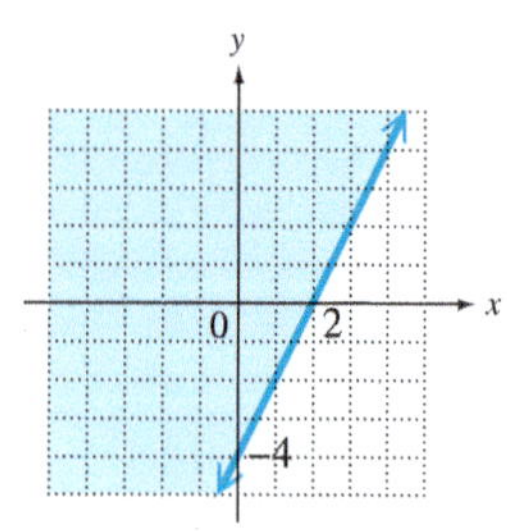

The boundary line here is (*solid / dashed*), and the region (*above / below*) it is shaded.

The inequality symbol to indicate this is ($<$ / $\leq$ / $>$ / $\geq$).

Inequality for the graph: y __________

38. Determine the following for the boundary line.

Slope: __________

y-intercept: __________

Equation: $y =$ __________

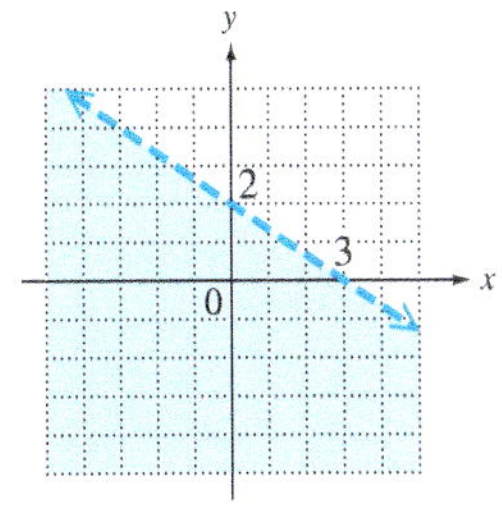

The boundary line here is (*solid* / *dashed*), and the region (*above* / *below*) it is shaded.

The inequality symbol to indicate this is $(< / \le / > / \ge)$.

Inequality for the graph: y __________

Graph each compound inequality. ***See Example 4.***

39. $x + y \le 1$ and $x \ge 1$

40. $x - y \ge 2$ and $x \ge 3$

41. $2x - y \ge 2$ and $y < 4$

42. $3x - y \ge 3$ and $y < 3$

43. $x + y > -5$ and $y < -2$

44. $6x - 4y < 10$ and $y > 2$

Extending Skills *Use the method described in* ***Section 1.7*** *to write each inequality as a compound inequality, and graph its solution set in the rectangular coordinate plane.*

45. $|x| < 3$

46. $|y| < 5$

47. $|x + 1| < 2$

48. $|y - 3| < 2$

Graph each compound inequality. ***See Example 5.***

49. $x - y \ge 1$ or $y \ge 2$

50. $x + y \le 2$ or $y \ge 3$

51. $x - 2 > y$ or $x < 1$

52. $x + 3 < y$ or $x > 3$

53. $3x + 2y < 6$ or $x - 2y > 2$

54. $x - y \ge 1$ or $x + y \le 4$

RELATING CONCEPTS For Individual or Group Work (Exercises 55–60)

Linear programming *is a method for finding the optimal (best possible) solution that meets all the conditions for a problem such as the following.*

A factory can have no more than 200 *workers on a shift, but must have* at least 100 *and must manufacture* at least 3000 *units at minimum cost. How many workers should be on a shift in order to produce the required units at minimal cost?*

Let x represent the number of workers and y represent the number of units manufactured. ***Work Exercises 55–60 in order.***

55. Write three inequalities expressing the problem conditions.

56. Graph the inequalities from **Exercise 55** using the axes at the right, and shade the intersection.

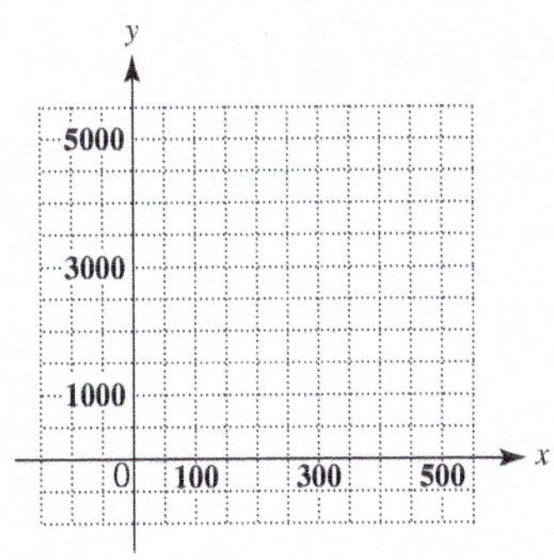

57. The cost per worker is \$50 per day and the cost to manufacture 1 unit is \$100. Write an equation in x, y, and C representing the total daily cost C.

58. Find values of x and y for several points in or on the boundary of the shaded region. Include any "corner points," where C is maximized or minimized.

59. Of the values of x and y found in **Exercise 58,** which ones give the least value when substituted in the cost equation from **Exercise 57?**

60. What does the answer in **Exercise 59** mean in terms of the given problem?

---

38. $-\frac{2}{3}$; $(0, 2)$; $-\frac{2}{3}x + 2$; dashed; below; $<$; $< -\frac{2}{3}x + 2$

40.

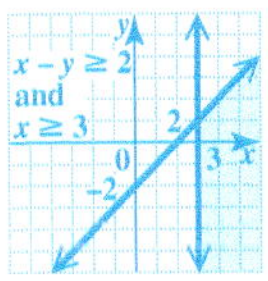

42.

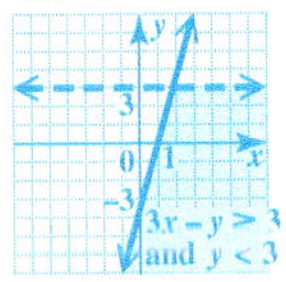

44.

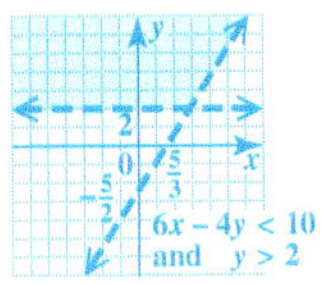

46. $-5 < y < 5$

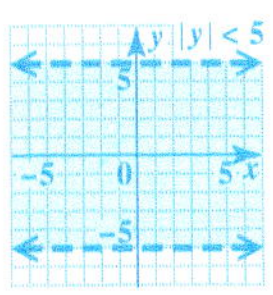

48. $-2 < y - 3 < 2$
$1 < y < 5$

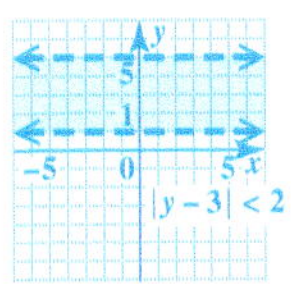

50.

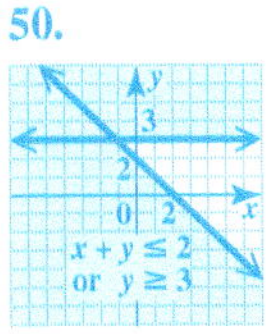

52.

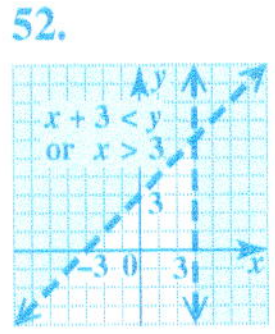

54.

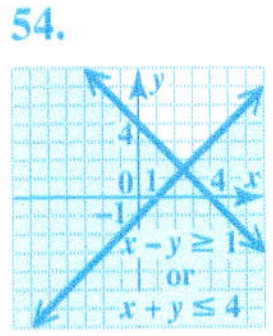

55. $x \le 200$, $x \ge 100$, $y \ge 3000$

56.

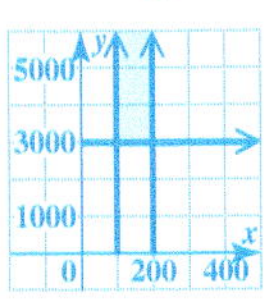

57. $C = 50x + 100y$

58. Some examples are $(100, 5000)$, $(150, 3000)$, and $(150, 5000)$. The corner points are $(100, 3000)$ and $(200, 3000)$.

59. The least value occurs when $x = 100$ and $y = 3000$.

60. The company should use 100 workers and manufacture 3000 units to achieve the least possible cost.

2.5 Introduction to Relations and Functions

OBJECTIVES

1. Define and identify relations and functions.
2. Find the domain and range.
3. Identify functions defined by graphs and equations.

VOCABULARY

- ☐ relation
- ☐ function
- ☐ dependent variable
- ☐ independent variable
- ☐ domain
- ☐ range

OBJECTIVE 1 Define and identify relations and functions.

Consider the relationship illustrated in the following table between number of hours worked and paycheck amount for an hourly worker.

Number of Hours Worked	Paycheck Amount (in dollars)	Ordered Pairs
5	40	⟶ (5, 40)
10	80	⟶ (10, 80)
20	160	⟶ (20, 160)
40	320	⟶ (40, 320)

The data from the table can be represented by a set of ordered pairs.

$$\{(5, 40), (10, 80), (20, 160), (40, 320)\}$$

Number of hours worked ↑ ↑ Paycheck amount in dollars

Each first component of the ordered pairs represents a number of hours worked, and each second component represents the corresponding paycheck amount. Such a set of ordered pairs is a *relation.*

Relation

A **relation** is any set of ordered pairs.

NOW TRY EXERCISE 1

Write the relation as a set of ordered pairs.

Year	Average Gas Price per Gallon (in dollars)
2000	1.56
2005	2.34
2010	2.84
2015	3.39

Source: Energy Information Administration.

EXAMPLE 1 Writing Ordered Pairs for a Relation

Write the relation as a set of ordered pairs.

Number of Gallons of Gas	Cost (in dollars)
0	0
1	3.50
2	7.00
3	10.50
4	14.00

This table is from **Section 2.3, Example 7.**

The data in the table defines a relation between number of gallons of gas and cost and can be written as the following set of ordered pairs.

$$\{(0, 0), (1, 3.50), (2, 7.00), (3, 10.50), (4, 14.00)\}$$

Number of gallons of gas ↑ ↑ Cost in dollars

NOW TRY

NOW TRY ANSWER

1. $\{(2000, 1.56), (2005, 2.34), (2010, 2.84), (2015, 3.39)\}$

A *function* is a special kind of relation.

Function

A **function** is a relation in which, for each distinct value of the first component of the ordered pairs, there is *exactly one value* of the second component.

NOW TRY EXERCISE 2

Determine whether each relation defines a function.

(a) $\{(1, 5), (3, 5), (5, 5)\}$

(b) $\{(-1, -3), (0, 2), (-1, 6)\}$

CLASSROOM EXAMPLE 1

Write the relation as a set of ordered pairs.

x	y
-4	0
0	-2
3	1
5	1

Answer: $\{(-4, 0), (0, -2), (3, 1), (5, 1)\}$

CLASSROOM EXAMPLE 2

Determine whether each relation defines a function.

(a) $\{(0, 3), (-1, 2), (-1, 3)\}$

(b) $\{(5, 4), (6, 4), (7, 4)\}$

Answers: (a) not a function (b) function

TEACHING TIP

Give students additional examples of relations and functions, using different ways to define them. Include some of the applications with graphs or tables from earlier sections.

EXAMPLE 2 Determining Whether Relations Are Functions

Determine whether each relation defines a function.

(a) $F = \{(1, 2), (-2, 4), (3, -1)\}$

Look at the ordered pairs that define this relation.

For $x = 1$, there is only one value of y, 2.

For $x = -2$, there is only one value of y, 4.

For $x = 3$, there is only one value of y, -1.

Relation F is a function—for each distinct x-value, there is *exactly one* y-value.

(b) $G = \{(-2, -1), (-1, 0), (0, 1), (1, 2), (2, 2)\}$

Relation G is also a function. Although the last two ordered pairs have the same y-value (1 is paired with 2, and 2 is paired with 2), this does not violate the definition of a function. The first components (x-values) are distinct, and each is paired with only one second component (y-value).

(c) $H = \{(-4, 1), (-2, 1), (-2, 0)\}$

In relation H, the last two ordered pairs have the ***same*** x-value paired with ***two different*** y-values (-2 is paired with both 1 and 0). H is a relation, but *not* a function. ***In a function, no two ordered pairs have the same first component and different second components.***

Different y-values

Relation $H = \{(-4, 1), (-2, 1), (-2, 0)\}$ Not a function

Same x-value

NOW TRY

Relations may be defined in several different ways.

- **A relation may be defined as a set of ordered pairs. (See Example 2.)**

  Relation $F = \{(1, 2), (-2, 4), (3, -1)\}$ Function

  Relation $H = \{(-4, 1), (-2, 1), (-2, 0)\}$ Not a function

- **A relation may be defined as a correspondence or *mapping*.**

  See FIGURE 42. In the mapping for relation F from **Example 2(a),** 1 is mapped to 2, -2 is mapped to 4, and 3 is mapped to -1. Thus, F is a function—each first component of an ordered pair is paired with exactly one second component.

  In the mapping for relation H from **Example 2(c),** which is *not* a function, the first component -2 is paired with two different second components.

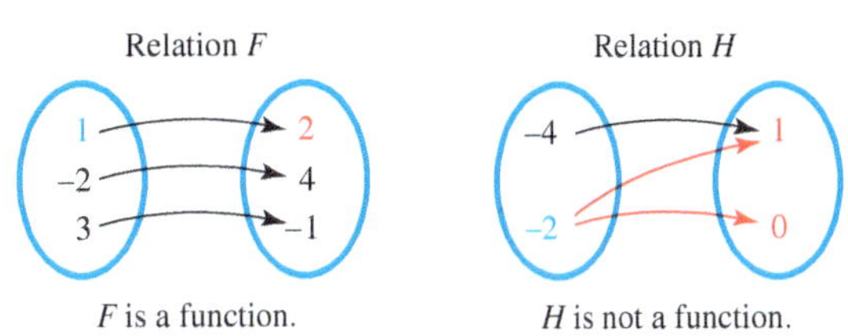

FIGURE 42

NOW TRY ANSWERS

2. (a) function (b) not a function

TEACHING TIP
Work **Exercises 13–16** in the exercise set as a class to reinforce the concepts here.

- **A relation may be defined as a table.**
- **A relation may be defined as a graph.**

FIGURE 43 includes a table and graph for relation F, which is a function, from **Example 2(a).**

x	y
1	2
−2	4
3	−1

Table for relation F

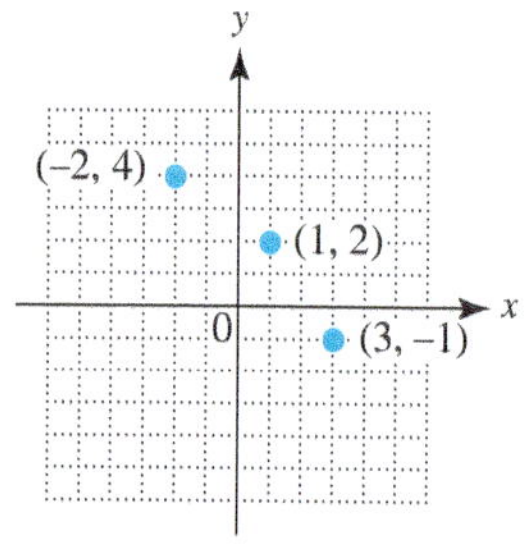

Graph of relation F

FIGURE 43

- **A relation may be defined as an equation (or rule).** The solutions of an equation give an infinite set of ordered pairs. For example, if the value of y is twice the value of x, the equation is

$$y = 2x.$$

The infinite number of ordered-pair solutions (x, y) can be represented by the graph in **FIGURE 44**.

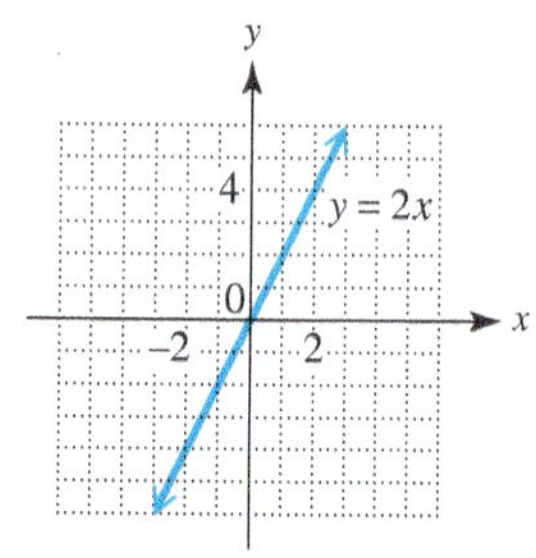

Graph of the relation $y = 2x$

FIGURE 44

In the equation $y = 2x$, the value of y ***depends*** on the value of x. Thus, the variable y is the **dependent variable.** The variable x is the **independent variable.**

Dependent variable → $y = 2x$ ← Independent variable

An equation tells how to determine the value of the dependent variable for a specific value of the independent variable.

TEACHING TIP
Have students give an example of a function from everyday life by completing the blanks.

______ depends on ______,
so ______ is a function of ______.

Example:

The cost of gasoline depends on *the number of gallons*, so *cost* is a function of *number of gallons.*

NOTE An equation that describes the relationship given at the beginning of this section between number of hours worked and paycheck amount is

$$y = 8x. \quad \text{8 represents the hourly rate, \$8.}$$

Paycheck amount y *depends* on number of hours worked x. Thus, paycheck amount is the dependent variable and number of hours worked is the independent variable.

In a function, there is exactly one value of the dependent variable, the second component, for each value of the independent variable, the first component.

NOTE Another way to think of a function relationship is to think of the independent variable as an **input** and the dependent variable as an **output.** This **input-output (function) machine** illustrates the relationship between number of hours worked and paycheck amount.

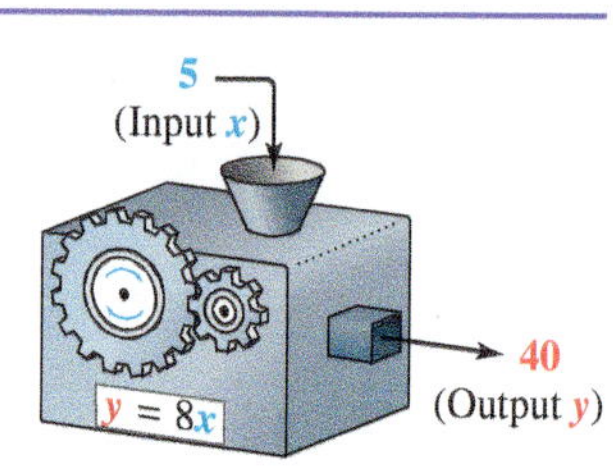

Function machine

OBJECTIVE 2 Find the domain and range.

Domain and Range

For every relation defined by a set of ordered pairs (x, y), there are two important sets of elements.

- The set of all values of the independent variable (x) is the **domain.**
- The set of all values of the dependent variable (y) is the **range.**

EXAMPLE 3 Finding Domains and Ranges of Relations

Give the domain and range of each relation. Decide whether the relation defines a function.

(a) $\{(3, -1), (4, 2), (4, 5), (6, 8)\}$ Only list 4 once.

Domain: $\{3, 4, 6\}$ Set of x-values

Range: $\{-1, 2, 5, 8\}$ Set of y-values

This relation is not a function because the same x-value 4 is paired with two different y-values, 2 and 5.

(b)

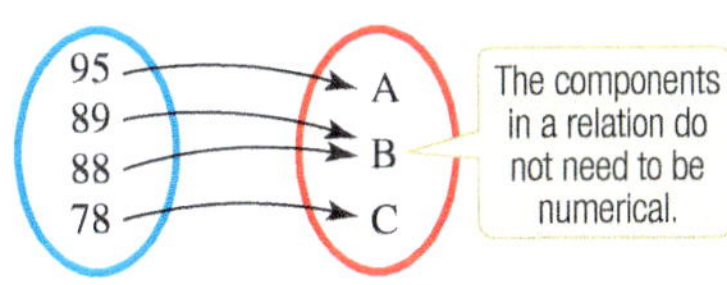

This mapping represents the following set of ordered pairs.

$$\{(95, \text{A}), (89, \text{B}), (88, \text{B}), (78, \text{C})\}$$

Domain: $\{95, 89, 88, 78\}$ Set of first components

Range: $\{\text{A}, \text{B}, \text{C}\}$ Set of second components

The mapping defines a function—each domain value corresponds to exactly one range value.

(c)

x	y
−5	2
0	2
5	2

This table represents the following set of ordered pairs.

$$\{(-5, 2), (0, 2), (5, 2)\}$$

Domain: $\{-5, 0, 5\}$ Set of x-values

Range: $\{2\}$ Set of y-values

The table defines a function—each distinct x-value corresponds to exactly one y-value (even though it is the same y-value). NOW TRY

NOW TRY EXERCISE 3

Give the domain and range of each relation. Decide whether the relation defines a function.

(a) $\{(2, 2), (2, 5), (4, 8)\}$

(b) The table from **Objective 1**

Number of Hours Worked	Paycheck Amount (in dollars)
5	40
10	80
20	160
40	320

CLASSROOM EXAMPLE 3

Give the domain and range of the relation represented by the table in **Example 1.** Does it define a function?

Number of Gallons of Gas	Cost (in dollars)
0	0
1	3.50
2	7.00
3	10.50
4	14.00

Answer: domain: $\{0, 1, 2, 3, 4\}$; range: $\{0, 3.50, 7.00, 10.50, 14.00\}$; Yes, the relation defines a function.

NOW TRY ANSWERS

3. (a) domain: $\{2, 4\}$; range: $\{2, 5, 8\}$; not a function

(b) domain: $\{5, 10, 20, 40\}$; range: $\{40, 80, 160, 320\}$; function

A graph gives a "picture" of a relation and can be used to determine its domain and range.

NOTE Pay particular attention to the use of color to interpret domain and range in **Example 4**—blue for domain and red for range.

NOW TRY EXERCISE 4

Give the domain and range of the relation.

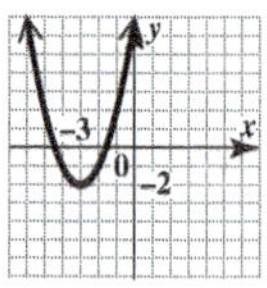

CLASSROOM EXAMPLE 4

Give the domain and range of the relation.

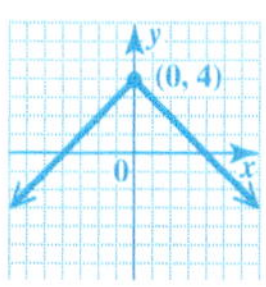

Answer: domain: $(-\infty, \infty)$; range: $(-\infty, 4]$

TEACHING TIP

Use graphs of vertical and horizontal lines, parabolas, hyperbolas, and square root functions to show how to find the domain and range. Give results in interval notation, paying special attention to the use of parentheses and brackets.

EXAMPLE 4 Finding Domains and Ranges from Graphs

Give the domain and range of each relation.

(a)

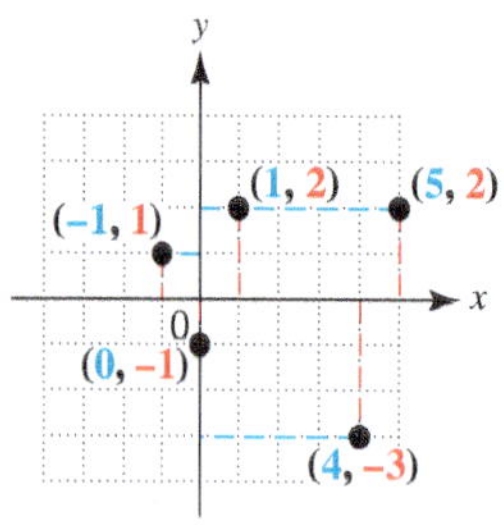

This relation includes the five ordered pairs that are graphed.

$$\{(-1, 1), (0, -1), (1, 2), (4, -3), (5, 2)\}$$

Domain: $\{-1, 0, 1, 4, 5\}$ Set of x-values

Range: $\{1, -1, 2, -3\}$ Set of y-values

Only list 2 once.

(b)

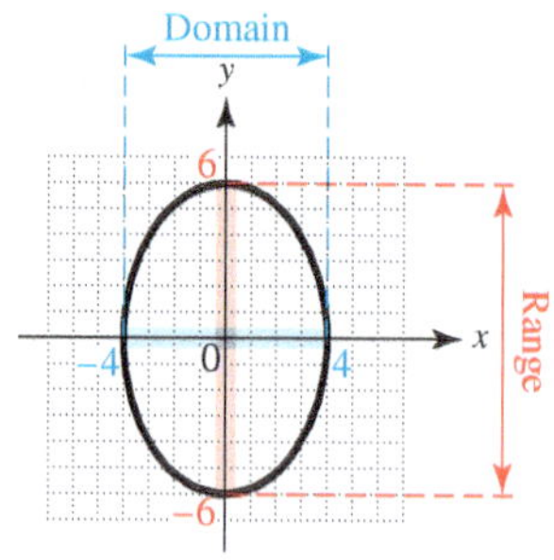

The x-values of the ordered pairs that form the graph include all numbers between -4 and 4, inclusive, as shown in blue. The y-values include all numbers between -6 and 6, inclusive, as shown in red.

Domain: $[-4, 4]$ Use interval notation.

Range: $[-6, 6]$ (Section 1.5)

(c)

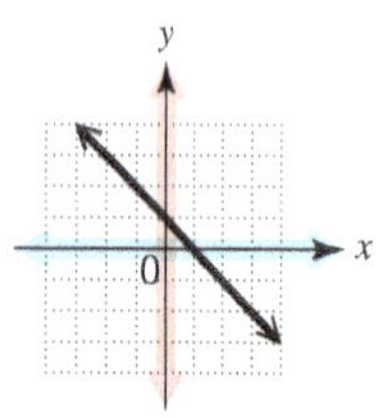

The arrowheads on the graphed line indicate that the line extends indefinitely left and right, as well as up and down. Therefore, both the domain, shown in blue, (the set of x-values) and the range, shown in red, (the set of y-values) include all real numbers.

Domain: $(-\infty, \infty)$ Range: $(-\infty, \infty)$

(d)

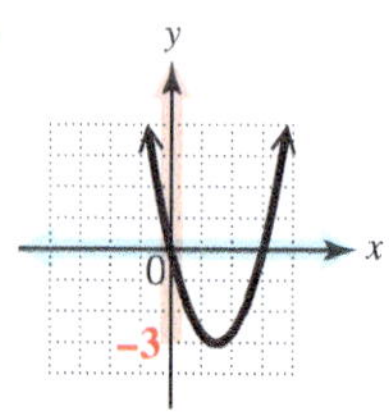

The graphed curve extends indefinitely left and right, as well as upward. The domain, shown in blue, includes all real numbers. Because there is a least y-value, -3, the range, shown in red, includes all numbers greater than or equal to -3.

Domain: $(-\infty, \infty)$ Range: $[-3, \infty)$

NOW TRY

OBJECTIVE 3 Identify functions defined by graphs and equations.

Because each value of x in a function corresponds to only one value of y, any vertical line drawn through the graph of a function must intersect the graph in at most one point. This is the *vertical line test* for a function.

FIGURE 45 on the next page illustrates this test with the graphs of two relations.

NOW TRY ANSWER

4. domain: $(-\infty, \infty)$; range: $[-2, \infty)$

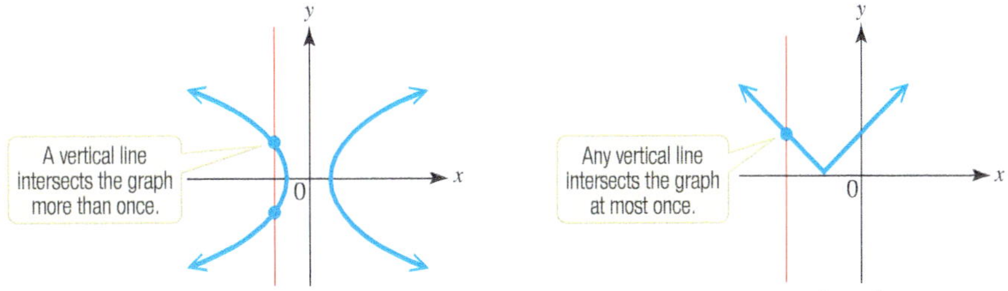

Not a function
The same x-value corresponds to two different y-values.

Function
Each x-value corresponds to only one y-value.

FIGURE 45

Vertical Line Test

If every vertical line intersects the graph of a relation in no more than one point, then the relation represents a function.

NOW TRY EXERCISE 5

Use the vertical line test to determine whether the relation is a function.

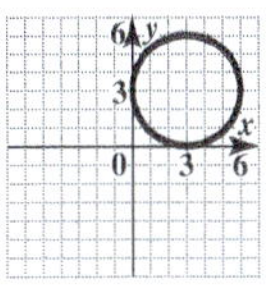

CLASSROOM EXAMPLE 5

Use the vertical line test to determine whether the relation graphed in Classroom Example 4 is a function.

Answer: function

EXAMPLE 5 Using the Vertical Line Test

Use the vertical line test to determine whether each relation graphed in **Example 4** is a function. (We repeat the graphs here.)

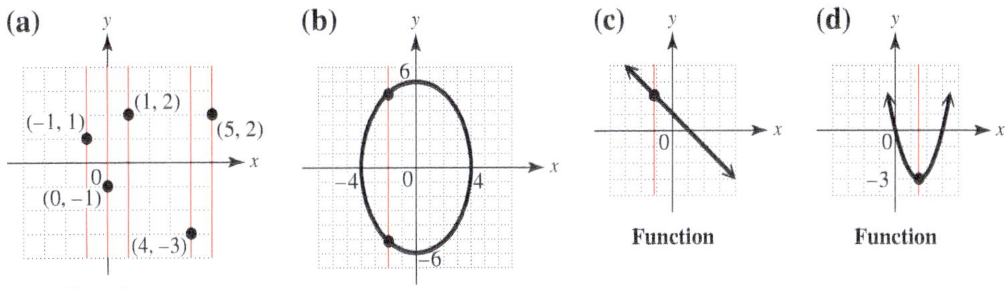

The graphs in (a), (c), and (d) satisfy the vertical line test and represent functions. The graph in (b) fails the vertical line test because a vertical line intersects the graph more than once—that is, the same x-value corresponds to two different y-values. This is not the graph of a function. NOW TRY

NOTE Graphs that do not represent functions are still relations. ***All equations and graphs represent relations, and all relations have a domain and range.***

If a relation is defined by an equation involving a fraction or a radical, apply these guidelines when finding its domain.

1. **Exclude from the domain any values that make the denominator of a fraction equal to 0.**

   Example: The function $y = \frac{1}{x}$ has all real numbers *except* 0 as its domain because division by 0 is undefined.

2. **Exclude from the domain any values that result in an even root of a negative number.**

   Example: The function $y = \sqrt{x}$ has all *nonnegative* real numbers as its domain because the square root of a negative number is not real.

NOW TRY ANSWER

5. not a function

Agreement on Domain

Unless specified otherwise, the domain of a relation is assumed to be all real numbers that produce real numbers when substituted for the independent variable.

NOW TRY EXERCISE 6

Decide whether each relation defines y as a function of x. Give the domain.

(a) $y = 4x - 3$

(b) $y = \sqrt{2x - 4}$

(c) $y = \dfrac{1}{x - 2}$

(d) $y < 3x + 1$

CLASSROOM EXAMPLE 6

Decide whether each relation defines y as a function of x. Give the domain.

(a) $y = -2x + 7$

(b) $y = \sqrt{5x - 6}$

(c) $y^4 = x$

(d) $y \geq 4x + 2$

(e) $y = \dfrac{6}{5 + 3x}$

Answers: (a) yes; $(-\infty, \infty)$

(b) yes; $\left[\frac{6}{5}, \infty\right)$

(c) no; $[0, \infty)$

(d) no; $(-\infty, \infty)$

(e) yes; $\left(-\infty, -\frac{5}{3}\right) \cup \left(-\frac{5}{3}, \infty\right)$

EXAMPLE 6 Identifying Functions from Their Equations

Decide whether each relation defines y as a function of x. Give the domain.

(a) $y = x + 4$

In this equation, y is found by adding 4 to x. Thus, each value of x corresponds to just one value of y, and the relation defines a function. Because x can be any real number, the domain is $(-\infty, \infty)$.

(b) $y = \sqrt{2x - 1}$

For any choice of x in the domain, there is exactly one corresponding value for y. (The radical is a single nonnegative number.) This equation defines a function. The quantity under the radical symbol cannot be negative—that is, $2x - 1$ *must be greater than or equal to* 0.

$$2x - 1 \geq 0$$

$$2x \geq 1 \quad \text{Add 1.}$$

$$x \geq \frac{1}{2} \quad \text{Divide by 2.}$$

The domain of the function is $\left[\frac{1}{2}, \infty\right)$.

(c) $y^2 = x$

The ordered pairs $(16, 4)$ and $(16, -4)$ both satisfy this equation. One value of x, 16, corresponds to two values of y, 4 and -4, so this equation does not define a function. Because x is equal to the square of y, the values of x must always be nonnegative. The domain of the relation is $[0, \infty)$.

(d) $y \leq x - 1$

By definition, y is a function of x if every value of x leads to exactly one value of y. Here, a particular value of x, such as 1, corresponds to many values of y. The ordered pairs

$$(1, 0), \quad (1, -1), \quad (1, -2), \quad (1, -3), \quad \text{and so on}$$

all satisfy the inequality. Thus, this relation does not define a function. Any number can be used for x, so the domain is the set of all real numbers, $(-\infty, \infty)$.

(e) $y = \dfrac{5}{x - 1}$

Given any value of x in the domain, we find y by subtracting 1 and then dividing the result into 5. This process produces exactly one value of y for each value in the domain, so the given equation defines a function.

The domain includes all real numbers except those which make the denominator 0.

$$x - 1 = 0 \quad \text{Set the denominator equal to 0.}$$

$$x = 1 \quad \text{Add 1.}$$

The domain includes all real numbers *except* 1, written $(-\infty, 1) \cup (1, \infty)$.

NOW TRY

NOW TRY ANSWERS

6. (a) yes; $(-\infty, \infty)$

(b) yes; $[2, \infty)$

(c) yes; $(-\infty, 2) \cup (2, \infty)$

(d) no; $(-\infty, \infty)$

In summary, we give three variations of the definition of a function.

Variations of the Definition of a Function

1. A **function** is a relation in which, for each distinct value of the first component of the ordered pairs, there is exactly one value of the second component.
2. A **function** is a set of distinct ordered pairs in which no first component is repeated.
3. A **function** is a correspondence (mapping) or an equation (rule) that assigns exactly one range value to each distinct domain value.

2.5 Exercises

FOR EXTRA HELP MyMathLab®

Complete solution available in MyMathLab

1. relation; ordered pairs
2. function; ordered pairs
3. domain; range
4. does not; domain; range
5. independent variable; dependent variable
6. vertical line test; function; one
7. $\{(2, -2), (2, 0), (2, 1)\}$
8. $\{(-1, -1), (0, -1), (1, -1)\}$
9. $\{(1960, 0.76), (1980, 2.69), (2000, 5.39), (2013, 8.38)\}$
10. $\{(2010, 21.0), (2011, 21.1), (2012, 21.1), (2013, 20.9)\}$
11. $\{(A, 4), (B, 3), (C, 2), (D, 1), (F, 0)\}$
12. $\{(A, 1), (E, 5), (I, 9), (O, 15), (U, 21)\}$

In Exercises 13–15, answers will vary.

13.

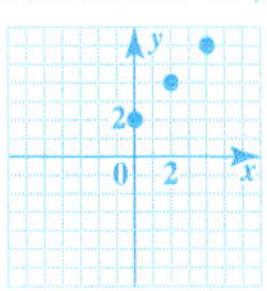

14. $\{(-1, -3), (0, -1), (1, 1), (3, 3)\}$

15.

x	y
-3	-4
-3	1
2	0

16. No, the same x-value, -3, is paired with two different y-values, -4 and 1.

17. function; domain: $\{5, 3, 4, 7\}$; range: $\{1, 2, 9, 6\}$

Concept Check *Complete each statement. Choices may be used more than once.*

function	independent variable	vertical line test	relation
domain	ordered pairs	dependent variable	range

1. A(n) ________ is any set of ________ $\{(x, y)\}$.
2. A(n) ________ is a relation in which, for each distinct value of the first component of the ________, there is exactly one value of the second component.
3. In a relation $\{(x, y)\}$, the ________ is the set of x-values, and the ________ is the set of y-values.
4. The relation $\{(0, -2), (2, -1), (2, -4), (5, 3)\}$ (*does / does not*) define a function. The set $\{0, 2, 5\}$ is its ________, and the set $\{-2, -1, -4, 3\}$ is its ________.
5. Consider the function $d = 50t$, where d represents distance and t represents time. The value of d depends on the value of t, so the variable t is the ________, and the variable d is the ________.
6. The ________ is used to determine whether a graph is that of a function. It says that any vertical line can intersect the graph of a(n) ________ in no more than (*zero / one / two*) point(s).

Write each relation as a set of ordered pairs. ***See Example 1 and Objective 1.***

7.

x	y
2	-2
2	0
2	1

8.

x	y
-1	-1
0	-1
1	-1

9.

Year	Average Movie Ticket Price (in dollars)
1960	0.76
1980	2.69
2000	5.39
2013	8.38

Source: Motion Picture Association of America.

10.

Year	Average ACT Composite Score
2010	21.0
2011	21.1
2012	21.1
2013	20.9

Source: ACT.

11.

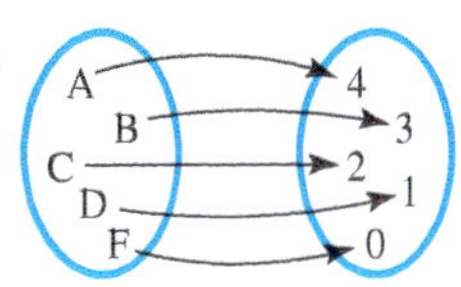

12.

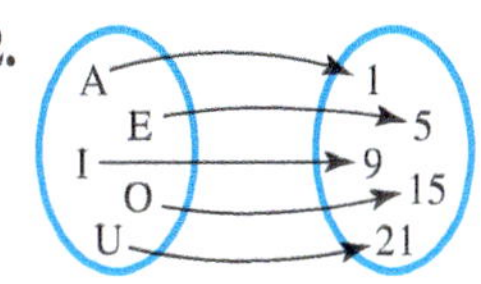

18. function; domain: $\{8, 5, 9, 3\}$; range: $\{0, 4, 3, 8\}$
19. not a function; domain: $\{2, 0\}$; range: $\{4, 2, 5\}$
20. not a function; domain: $\{9, -3\}$; range: $\{-2, 5, 2\}$
21. function; domain: $\{-3, 4, -2\}$; range: $\{1, 7\}$
22. function; domain: $\{-12, -10, 8\}$; range: $\{5, 3\}$
23. not a function; domain: $\{1, 0, 2\}$; range: $\{1, -1, 0, 4, -4\}$
24. function; domain: $\{2, 3, 4, 5\}$; range: $\{5, 7, 9, 11\}$
25. function; domain: $\{2, 5, 11, 17, 3\}$; range: $\{1, 7, 20\}$
26. not a function; domain: $\{1, 2, 3, 5\}$; range: $\{10, 15, 19, 27\}$
27. not a function; domain: $\{1\}$; range: $\{5, 2, -1, -4\}$
28. not a function; domain: $\{-4\}$; range: $\{-4, 0, 4, 8\}$
29. function; domain: $\{4, 2, 0, -2\}$; range: $\{-3\}$
30. function; domain: $\{-3, -1, 1, 3\}$; range: $\{-6\}$
31. function; domain: $\{-2, 0, 3\}$; range: $\{2, 3\}$
32. function; domain: $\{-1, 1, 4\}$; range: $\{-3, 0\}$
33. function; domain: $(-\infty, \infty)$; range: $(-\infty, \infty)$
34. function; domain: $(-\infty, \infty)$; range: $(-\infty, \infty)$
35. not a function; domain: $\{-2\}$; range: $(-\infty, \infty)$
36. function; domain: $(-\infty, \infty)$; range: $\{-2\}$
37. not a function; domain: $(-\infty, 0]$; range: $(-\infty, \infty)$
38. not a function; domain: $[-2, 2]$; range: $[-2, 2]$

Concept Check *Express each relation using a different form. (For example, if the given form is a set of ordered pairs, use a graph.) There is more than one correct way to do this.* ***See Objective 1.***

13. $\{(0, 2), (2, 4), (4, 6)\}$

14.

x	y
−1	−3
0	−1
1	1
3	3

15.
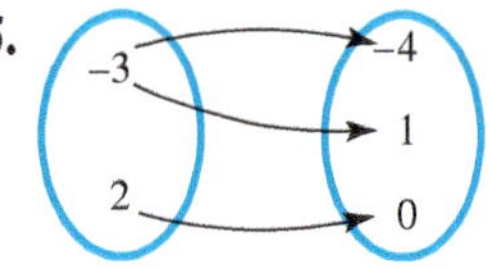

16. *Concept Check* Does the relation given in **Exercise 15** define a function? Why or why not?

Decide whether each relation defines a function, and give the domain and range. ***See Examples 2–5.***

17. $\{(5, 1), (3, 2), (4, 9), (7, 6)\}$

18. $\{(8, 0), (5, 4), (9, 3), (3, 8)\}$

19. $\{(2, 4), (0, 2), (2, 5)\}$

20. $\{(9, -2), (-3, 5), (9, 2)\}$

21. $\{(-3, 1), (4, 1), (-2, 7)\}$

22. $\{(-12, 5), (-10, 3), (8, 3)\}$

23. $\{(1, 1), (1, -1), (0, 0), (2, 4), (2, -4)\}$

24. $\{(2, 5), (3, 7), (4, 9), (5, 11)\}$

25.
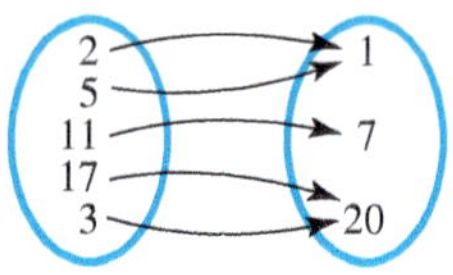

26.
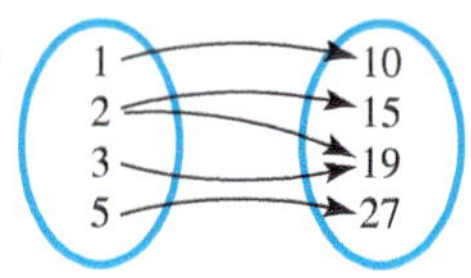

27.

x	y
1	5
1	2
1	−1
1	−4

28.

x	y
−4	−4
−4	0
−4	4
−4	8

29.

x	y
4	−3
2	−3
0	−3
−2	−3

30.

x	y
−3	−6
−1	−6
1	−6
3	−6

31.
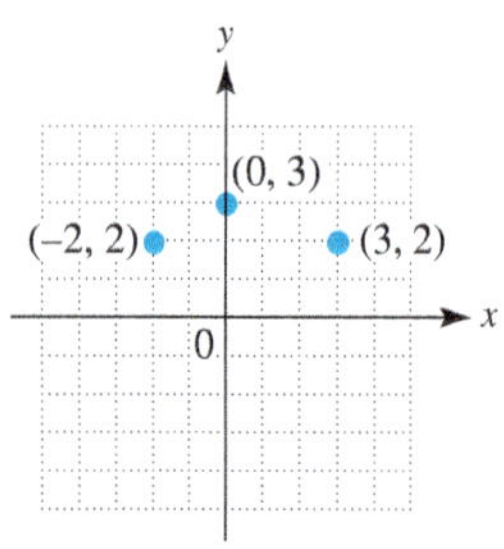

32.
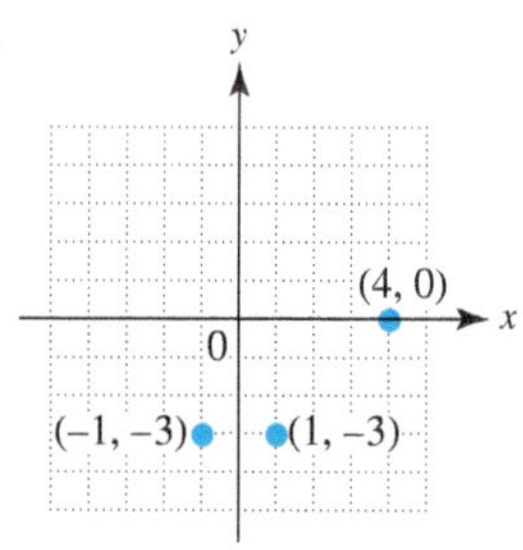

33.
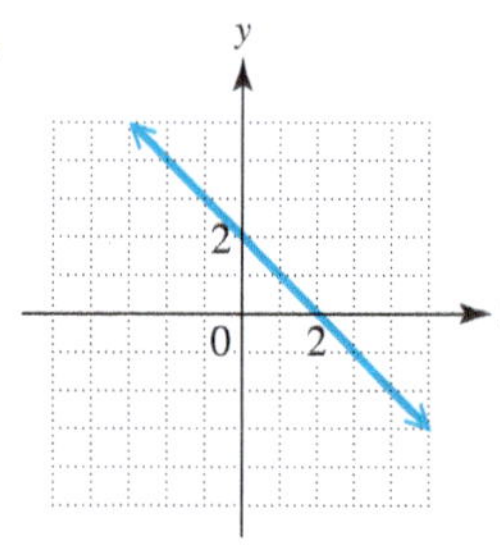

34.
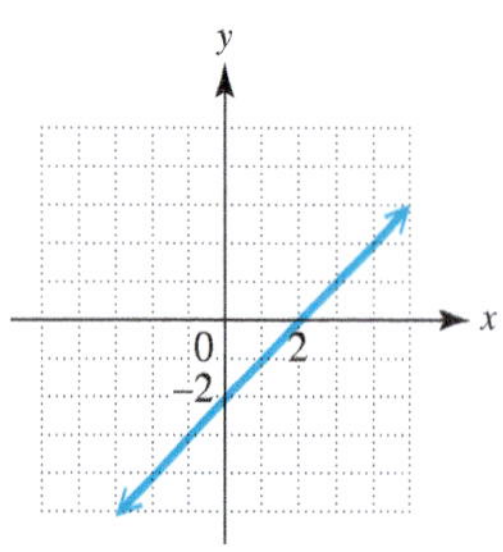

35.
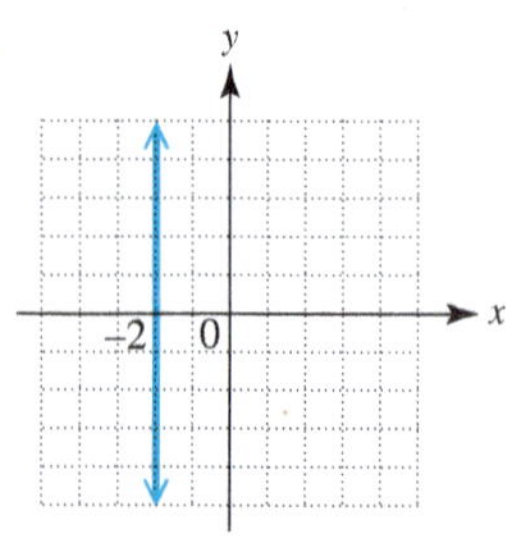

36.
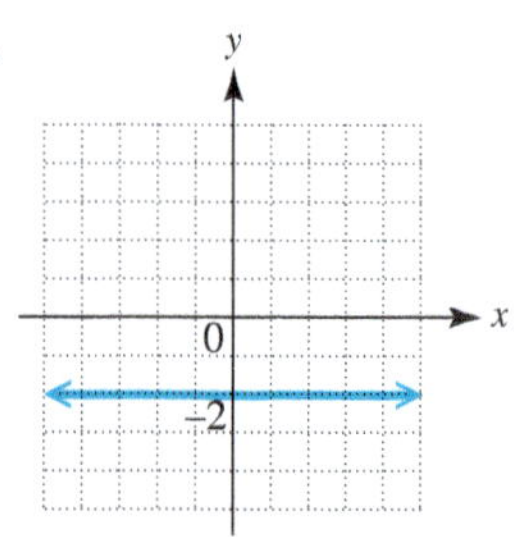

37.
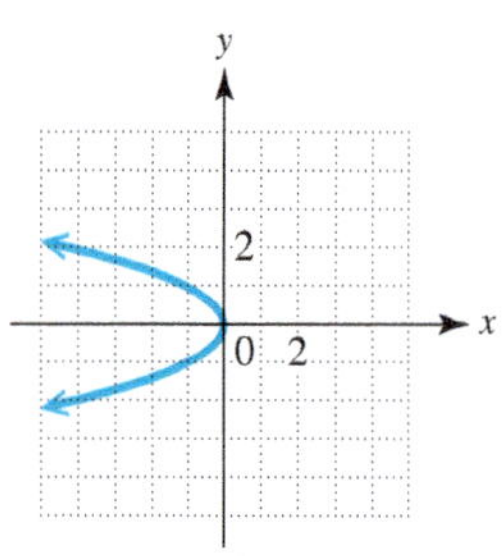

38.
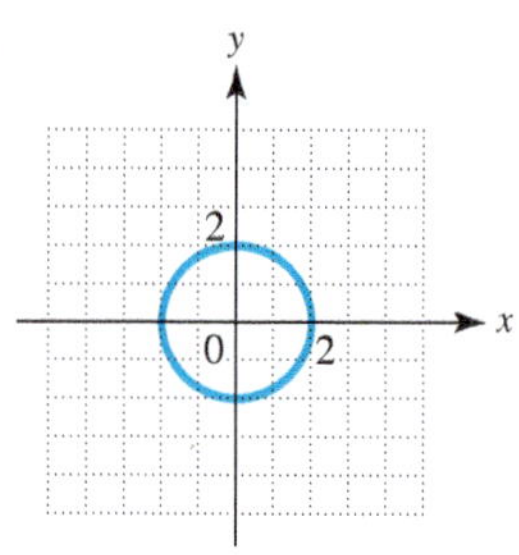

39.
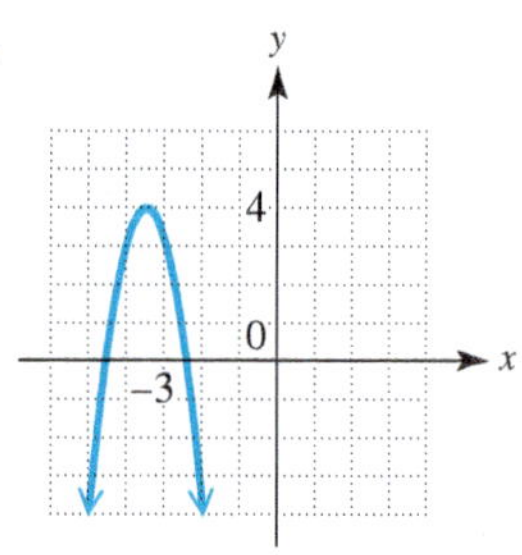

39. function; domain: $(-\infty, \infty)$; range: $(-\infty, 4]$
40. function; domain: $[-2, 2]$; range: $[0, 4]$
41. not a function; domain: $[-4, 4]$; range: $[-3, 3]$
42. not a function; domain: $[3, \infty)$; range: $(-\infty, \infty)$
43. not a function; domain: $(-\infty, \infty)$; range: $[2, \infty)$
44. not a function; domain: $(-\infty, 3]$; range: $(-\infty, \infty)$

45.–50. function; $(-\infty, \infty)$
51. not a function; $[0, \infty)$
52. not a function; $[0, \infty)$
53. not a function; $(-\infty, \infty)$
54. not a function; $(-\infty, \infty)$
55. function; $[0, \infty)$
56. function; $[0, \infty)$
57. function; $[3, \infty)$
58. function; $[7, \infty)$
59. function; $\left[-\frac{1}{2}, \infty\right)$
60. function; $\left[-\frac{9}{2}, \infty\right)$
61. function; $(-\infty, \infty)$
62. function; $(-\infty, \infty)$
63. function; $(-\infty, 0) \cup (0, \infty)$
64. function; $(-\infty, 0) \cup (0, \infty)$
65. function; $(-\infty, 4) \cup (4, \infty)$
66. function; $(-\infty, 2) \cup (2, \infty)$
67. function; $(-\infty, 0) \cup (0, \infty)$
68. function; $(-\infty, 0) \cup (0, \infty)$

69. (a) yes (b) domain: $\{2009, 2010, 2011, 2012, 2013\}$; range: $\{44.0, 43.4, 43.1, 42.9\}$ (c) 42.9; 2010 (d) Answers will vary. Two possible answers are $(2010, 43.4)$ and $(2013, 43.1)$.
70. (a) yes (b) domain: $\{2008, 2009, 2010, 2011, 2012\}$; range: $\{4.4, 4.2, 3.7, 2.8, 2.6\}$ (c) 2.6; 2011 (d) Answers will vary. Two possible answers are $(2008, 4.4)$ and $(2012, 2.6)$.

40.

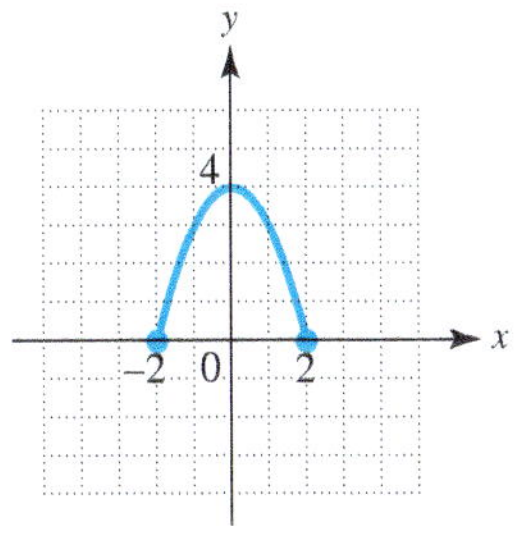

41.

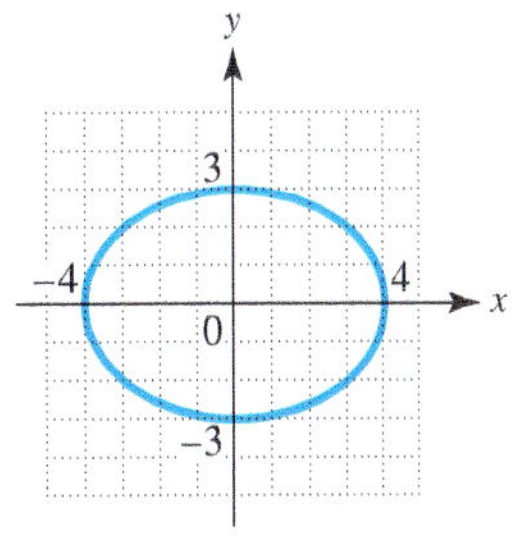

42.

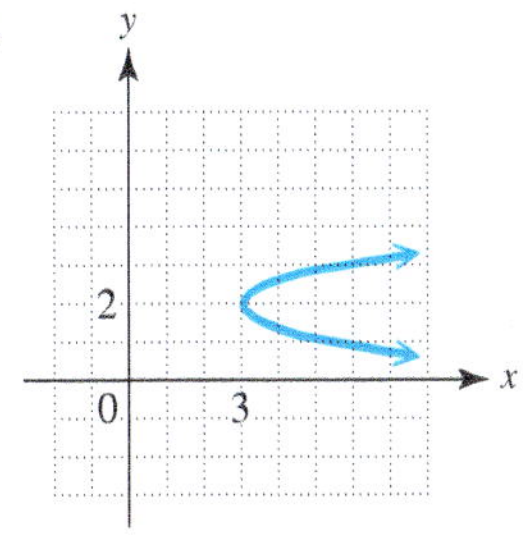

43.

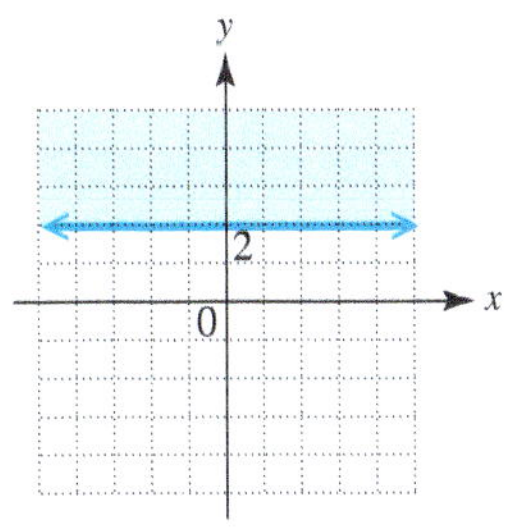

44.

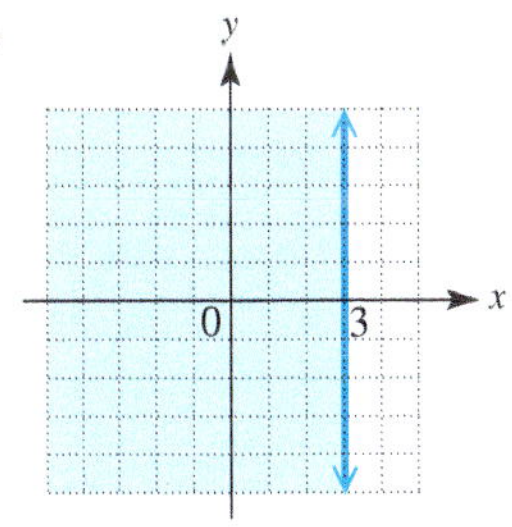

Decide whether each relation defines y as a function of x. (Solve for y first if necessary.) Give the domain. ***See Example 6.***

45. $y = -6x$ **46.** $y = -9x$ **47.** $y = 2x - 6$ **48.** $y = 6x + 8$

49. $y = x^2$ **50.** $y = x^3$ **51.** $x = y^6$ **52.** $x = y^4$

53. $x + y < 4$ **54.** $x - y < 3$ **55.** $y = \sqrt{x}$ **56.** $y = -\sqrt{x}$

57. $y = \sqrt{x - 3}$ **58.** $y = \sqrt{x - 7}$ **59.** $y = \sqrt{4x + 2}$

60. $y = \sqrt{2x + 9}$ **61.** $y = \frac{x + 4}{5}$ **62.** $y = \frac{x - 3}{2}$

63. $y = -\frac{2}{x}$ **64.** $y = -\frac{6}{x}$ **65.** $y = \frac{2}{x - 4}$

66. $y = \frac{7}{x - 2}$ **67.** $xy = 1$ **68.** $xy = 3$

Solve each problem.

69. The table shows the percentage of students at 4-year public colleges who graduated within 5 years.

(a) Does the table define a function?

(b) What are the domain and range?

(c) What is the range element that corresponds to 2012? The domain element that corresponds to 43.4?

(d) Call this function f. Give two ordered pairs that belong to f.

Year	Percentage
2009	44.0
2010	43.4
2011	43.1
2012	42.9
2013	43.1

Source: ACT.

70. The table shows the percentage of full-time college freshmen who said that they frequently smoked cigarettes in the last year.

(a) Does the table define a function?

(b) What are the domain and range?

(c) What is the range element that corresponds to 2012? The domain element that corresponds to 2.8?

(d) Call this function g. Give two ordered pairs that belong to g.

Year	Percentage
2008	4.4
2009	4.2
2010	3.7
2011	2.8
2012	2.6

Source: Higher Education Research Institute.

STUDY SKILLS

Analyzing Your Test Results

An exam is a learning opportunity—learn from your mistakes. After a test is returned, do the following:

- **Note what you got wrong and why you had points deducted.**
- **Figure out how to solve the problems you missed.** Check your textbook or notes, or ask your instructor. Rework the problems correctly.
- **Keep all quizzes and tests that are returned to you.** Use them to study for future tests and the final exam.

Typical Reasons for Errors on Math Tests

These are test taking errors. They are easy to correct if you read carefully, show all your work, proofread, and double-check units and labels.

1. You read the directions wrong.
2. You read the question wrong or skipped over something.
3. You made a computation error.
4. You made a careless error. (For example, you incorrectly copied a correct answer onto a separate answer sheet.)
5. Your answer is not complete.
6. You labeled your answer wrong. (For example, you labeled an answer "ft" instead of "ft^2.")
7. You didn't show your work.

These are test preparation errors. You must practice the kinds of problems that you will see on tests.

8. You didn't understand a concept.
9. You were unable to set up the problem (in an application).
10. You were unable to apply a procedure.

Below are sample charts for tracking your test taking progress. Refer to the tests you have taken so far in your course, and use the charts to find out if you tend to make certain kinds of errors. Check the appropriate box when you've made an error in a particular category.

Test Taking Errors

Test	Read directions wrong	Read question wrong	Computation error	Not exact or accurate	Not complete	Labeled wrong	Didn't show work
1							
2							
3							

Test Preparation Errors

Test	Didn't understand concept	Didn't set up problem correctly	Couldn't apply concept to new situation
1			
2			
3			

What will you do to avoid these kinds of errors on your next test?

2.6 Function Notation and Linear Functions

OBJECTIVES

1. Use function notation.
2. Graph linear and constant functions.

VOCABULARY

☐ linear function
☐ constant function

TEACHING TIP
For the function $y = f(x)$, emphasize that y and $f(x)$ are equivalent.
Review the rules for order of operations before giving examples using function notation.

NOW TRY EXERCISE 1
Let $f(x) = 4x + 3$. Find the value of function f for each value of x.
(a) $x = -2$ **(b)** $x = 0$

CLASSROOM EXAMPLE 1
Let $f(x) = 6x - 2$. Find the value of function f for each value of x.
(a) $x = -3$ (b) $x = 0$
Answers: (a) -20 (b) -2

TEACHING TIP
To reinforce the use of function notation, define $f(x)$ as $2x + 3$. Then ask students to "evaluate" the following.
(a) f(elephant)
(b) f(their own name)
While the answers sound strange, they will help students interpret this unfamiliar notation. (The authors thank Jeff Cole for this illustration.)

NOW TRY ANSWERS
1. (a) -5 (b) 3

OBJECTIVE 1 Use function notation.

When a function f is defined with a rule or an equation using x and y for the independent and dependent variables, we say, "*y is a function of x*" to emphasize that *y depends on x*. We use the notation

$$y = f(x),$$ The parentheses here do *not* indicate multiplication.

called **function notation,** to express this and read $f(x)$ as **"f of x,"** or **"f at x."** The letter f is a name for this particular function. For example, if $y = 3x - 5$, we can name this function f and write the following.

$$f(x) = 3x - 5$$

f is the name of the function.
x is a value from the domain.
$f(x)$ is the function value (or y-value) that corresponds to x.

$f(x)$ is just another name for the dependent variable y.

We evaluate a function at different values of x by substituting x-values from the domain into the function.

EXAMPLE 1 Evaluating a Function

Let $f(x) = 3x - 5$. Find the value of function f for each value of x.

(a) $x = 2$

$f(x) = 3x - 5$

$f(2) = 3 \cdot 2 - 5$ Replace x with 2. (Read $f(2)$ as "f of 2" or "f at 2.")

$f(2) = 6 - 5$ Multiply.

$f(2) = 1$ Subtract.

Thus, for $x = 2$, the corresponding function value (or y-value) is 1. $f(2) = 1$ is an abbreviation for the statement "If $x = 2$ in the function f, then $y = 1$" and is represented by the ordered pair $(2, 1)$.

(b) $x = -1$

$f(x) = 3x - 5$ Use parentheses to avoid errors.

$f(-1) = 3(-1) - 5$ Replace x with -1.

$f(-1) = -3 - 5$ Multiply.

$f(-1) = -8$ Subtract.

Thus, $f(-1) = -8$ and the ordered pair $(-1, -8)$ belongs to f. NOW TRY

CAUTION The symbol $f(x)$ *does not* indicate "f times x," but represents the y-value associated with the indicated x-value. As shown in **Example 1(a),** $f(2)$ is the y-value that corresponds to the x-value 2 in f.

These ideas can be illustrated as follows.

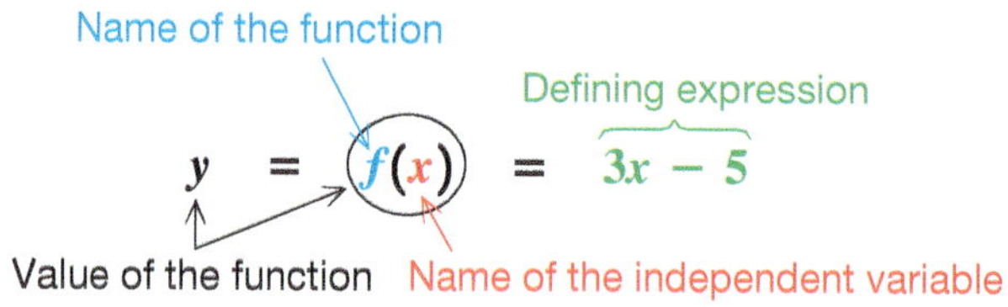

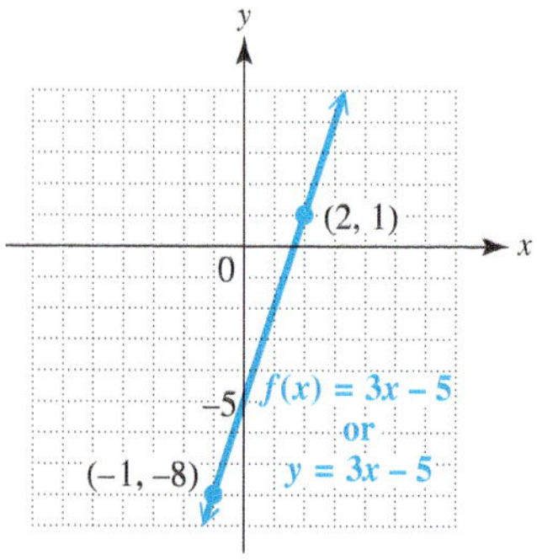

FIGURE 46

NOTE In the function $f(x) = 3x - 5$ in **Example 1,** $f(2) = 1$ and $f(-1) = -8$ correspond to the ordered pairs $(2, 1)$ and $(-1, -8)$. Because the domain of f is $(-\infty, \infty)$—that is, x can be any real number—this function defines an infinite set of ordered pairs whose graph is a line with slope 3 and y-intercept $(0, -5)$. See **FIGURE 46**. This makes sense because $f(x)$ is another name for y, and

$$f(x) = 3x - 5 \quad \text{is equivalent to} \quad y = 3x - 5.$$

As we will see in **Objective 2,** f is a *linear function*.

NOW TRY EXERCISE 2

Let $f(x) = 2x^2 - 4x + 1$. Find the following.

(a) $f(-2)$ **(b)** $f(a)$

CLASSROOM EXAMPLE 2

Let $f(x) = -x^2 + 3x + 3$. Find the following.

(a) $f(-3)$ (b) $f(t)$

Answers: (a) -15
(b) $-t^2 + 3t + 3$

EXAMPLE 2 Evaluating a Function

Let $f(x) = -x^2 + 5x - 3$. Find the following.

(a) $f(4)$

Do *not* read this as "f times 4." Read it as "f of 4," or "f at 4."

$f(x) = -x^2 + 5x - 3$ The base in $-x^2$ is x, *not* $(-x)$.

$f(4) = -4^2 + 5 \cdot 4 - 3$ Replace x with 4.

$f(4) = -16 + 20 - 3$ Apply the exponent. Multiply.

$f(4) = 1$ Add and subtract.

Thus, $f(4) = 1$, and the ordered pair $(4, 1)$ belongs to f.

(b) $f(q)$

$f(x) = -x^2 + 5x - 3$

$f(q) = -q^2 + 5q - 3$ Replace x with q.

The replacement of one variable with another is important in later courses.

NOW TRY

CLASSROOM EXAMPLE 3

Let $g(x) = 5x - 1$. Find and simplify $g(t + 2)$.

Answer: $5t + 9$

Sometimes letters other than f, such as g, h, or capital letters F, G, and H are used to name functions.

NOW TRY EXERCISE 3

Let $g(x) = 8x - 5$. Find and simplify $g(a - 2)$.

EXAMPLE 3 Evaluating a Function

Let $g(x) = 2x + 3$. Find and simplify $g(a + 1)$.

$g(x) = 2x + 3$

$g(a + 1) = 2(a + 1) + 3$ Replace x with $a + 1$.

$g(a + 1) = 2a + 2 + 3$ Distributive property

$g(a + 1) = 2a + 5$ Add.

NOW TRY

NOW TRY ANSWERS

2. (a) 17 **(b)** $2a^2 - 4a + 1$

3. $8a - 21$

NOW TRY EXERCISE 4

Find $f(-1)$ for each function.

(a) $f = \{(-5, -1), (-3, 2), (-1, 4)\}$

(b) $f(x) = x^2 - 12$

CLASSROOM EXAMPLE 4

Find $f(2)$ for each function.

(a) $f = \{(2, 6), (4, 2)\}$

(b) $f(x) = -x^2$

Answers: (a) 6 (b) −4

EXAMPLE 4 Evaluating Functions

For each function, find $f(3)$.

(a) $f(x) = 3x - 7$

$f(3) = 3(3) - 7$ Replace x with 3.

$f(3) = 9 - 7$ Multiply.

$f(3) = 2$ Subtract.

(b)

x	$y = f(x)$
6	−12
3	−6
0	0
−3	6

← Here, $f(3) = -6$.

(c) $f = \{(-3, 5), (0, 3), (3, 1), (6, -1)\}$

We want $f(3)$, the y-value of the ordered pair whose first component is $x = 3$. As indicated by the ordered pair $(3, 1)$, for $x = 3$, $y = 1$. Thus, $f(3) = 1$.

(d)

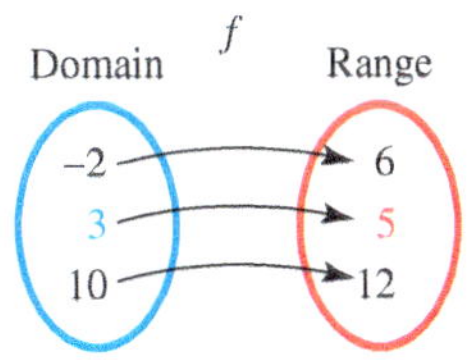

The domain element 3 is paired with 5 in the range, so

$f(3) = 5.$

NOW TRY

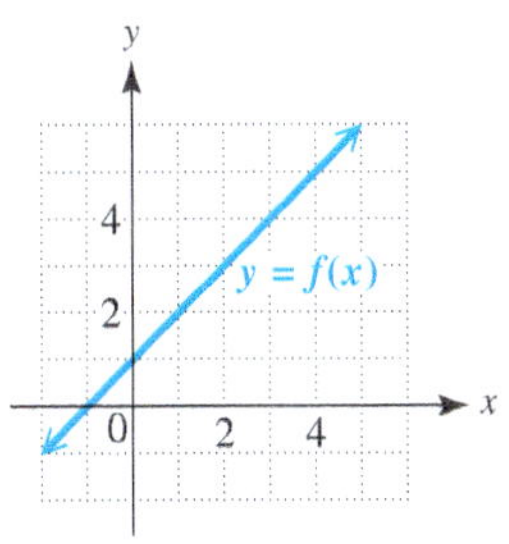

FIGURE 47

EXAMPLE 5 Finding Function Values from a Graph

Refer to the function graphed in **FIGURE 47**.

(a) Find $f(3)$.

Locate 3 on the x-axis. See **FIGURE 48**. Moving up to the graph of f and over to the y-axis gives 4 for the corresponding y-value. Thus, $f(3) = 4$, which corresponds to the ordered pair $(3, 4)$.

(b) Find $f(0)$.

Refer to **FIGURE 48** to see that $f(0) = 1$.

NOW TRY EXERCISE 5

Refer to the function graphed in **FIGURE 47**.

(a) Find $f(-1)$.

(b) For what value of x is $f(x) = 2$?

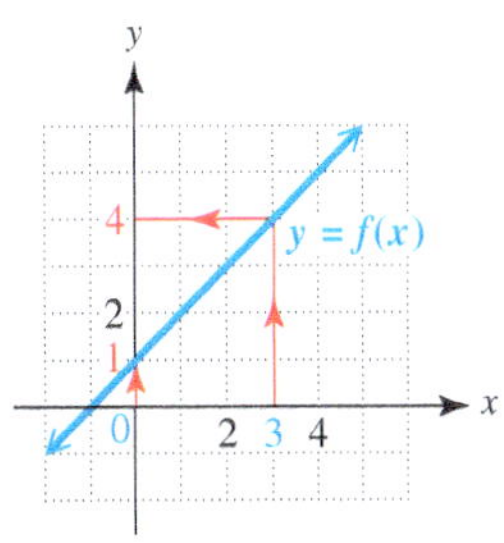

FIGURE 48

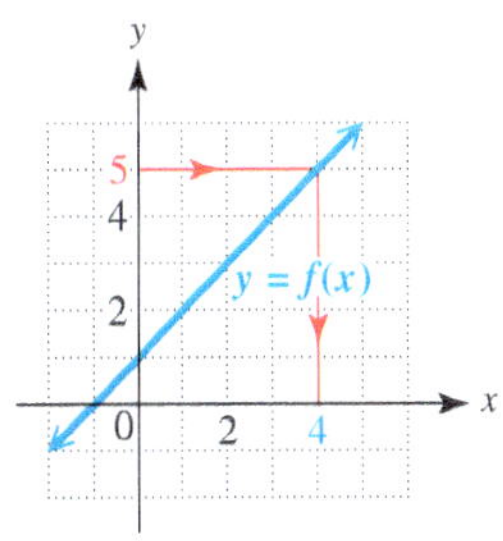

FIGURE 49

CLASSROOM EXAMPLE 5

Refer to the graph of the function.

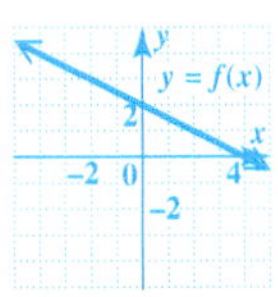

(a) Find $f(2)$.

(b) Find $f(-2)$.

(c) For what value of x is $f(x) = 0$?

Answers: (a) 1 (b) 3 (c) 4

(c) For what value of x is $f(x) = 5$?

Because $f(x) = y$, we want the value of x that corresponds to $y = 5$. See **FIGURE 49**, and locate 5 on the y-axis. Moving across to the graph of f and down to the x-axis gives $x = 4$. Thus, $f(4) = 5$, which corresponds to the ordered pair $(4, 5)$.

NOW TRY

If a function f is defined by an equation in x and y instead of function notation, use the following steps to find $f(x)$.

Writing an Equation Using Function Notation

Step 1 Solve the equation for y.

Step 2 Replace y with $f(x)$.

NOW TRY ANSWERS

4. (a) 4 (b) −11

5. (a) 0 (b) 1

NOW TRY EXERCISE 6

Write the equation using function notation $f(x)$. Then find $f(-3)$ and $f(h)$.

$$-4x^2 + y = 5$$

CLASSROOM EXAMPLE 6

Write the equation using function notation $f(x)$. Then find $f(1)$ and $f(a)$.

$$x^2 - 4y = 3$$

Answer: $f(x) = \frac{x^2 - 3}{4}$; $f(1) = -\frac{1}{2}$; $f(a) = \frac{a^2 - 3}{4}$

EXAMPLE 6 Writing Equations Using Function Notation

Write each equation using function notation $f(x)$. Then find $f(-2)$ and $f(a)$.

(a) $y = x^2 + 1$ ← This equation is already solved for y. (Step 1)

$f(x) = x^2 + 1$ Replace y with $f(x)$. (Step 2)

To find $f(-2)$, let $x = -2$.

$$f(x) = x^2 + 1$$

$f(-2) = (-2)^2 + 1$ Let $x = -2$.

$f(-2) = 4 + 1$ $(-2)^2 = -2(-2)$

$f(-2) = 5$ Add.

To find $f(a)$, let $x = a$.

$$f(a) = a^2 + 1$$

(b) $x - 4y = 5$

Step 1 $-4y = -x + 5$ Subtract x.

$y = \frac{1}{4}x - \frac{5}{4}$ Divide by -4.

Step 2 $f(x) = \frac{1}{4}x - \frac{5}{4}$ Replace y with $f(x)$.

Now find $f(-2)$ and $f(a)$.

$f(-2) = \frac{1}{4}(-2) - \frac{5}{4} = -\frac{7}{4}$ Let $x = -2$.

$f(a) = \frac{1}{4}a - \frac{5}{4}$ Let $x = a$.

NOW TRY

OBJECTIVE 2 Graph linear and constant functions.

Linear equations (except for vertical lines with equations of the form $x = a$) define *linear functions.*

Linear Function

A function f that can be written in the form

$$f(x) = ax + b,$$

where a and b are real numbers, is a **linear function.** The value of a is the slope m of the graph of the function. The domain of a linear function is $(-\infty, \infty)$, unless specified otherwise.

A linear function whose graph is a horizontal line has the form

$f(x) = b$ Constant function

and is a **constant function.** While the range of any nonconstant linear function is $(-\infty, \infty)$, the range of a constant function $f(x) = b$ is $\{b\}$.

NOW TRY ANSWER

6. $f(x) = 4x^2 + 5$; $f(-3) = 41$; $f(h) = 4h^2 + 5$

NOW TRY EXERCISE 7

Graph the function. Give the domain and range.

$$f(x) = \frac{1}{3}x - 2$$

CLASSROOM EXAMPLE 7

Graph the function. Give the domain and range.

$$f(x) = -1.5$$

Answer:

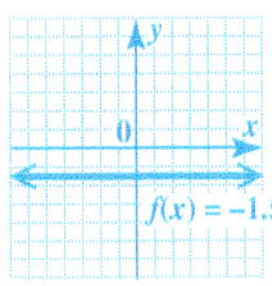

domain: $(-\infty, \infty)$;
range: $\{-1.5\}$

NOW TRY ANSWER

7.

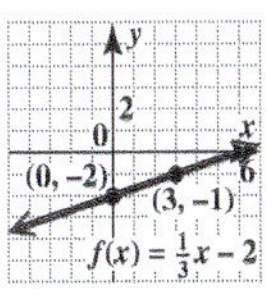

domain: $(-\infty, \infty)$;
range: $(-\infty, \infty)$

EXAMPLE 7 Graphing Linear and Constant Functions

Graph each function. Give the domain and range.

(a) $f(x) = \frac{1}{4}x - \frac{5}{4}$ (from Example 6(b))

Slope ↑ (for $\frac{1}{4}$); ↑ y-intercept is $\left(0, -\frac{5}{4}\right)$.

To graph this function, plot the y-intercept $\left(0, -\frac{5}{4}\right)$. Use the geometric definition of slope as $\frac{\text{rise}}{\text{run}}$ to find a second point on the line. The slope is $\frac{1}{4}$, so we move 1 unit up from $\left(0, -\frac{5}{4}\right)$ and 4 units to the right to the point $\left(4, -\frac{1}{4}\right)$. Draw the straight line through these points. See **FIGURE 50**. The domain and range are both $(-\infty, \infty)$.

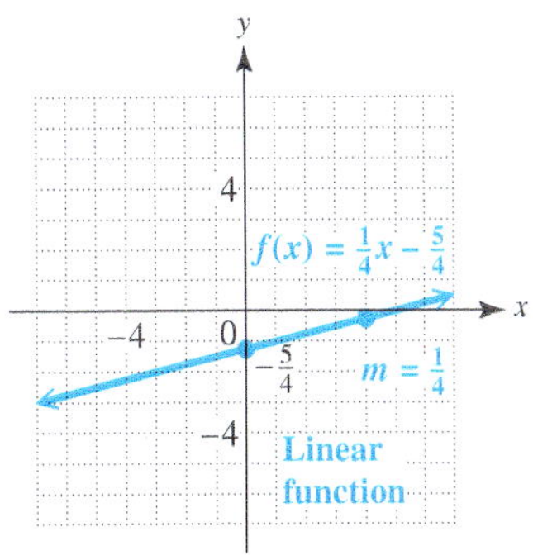

FIGURE 50

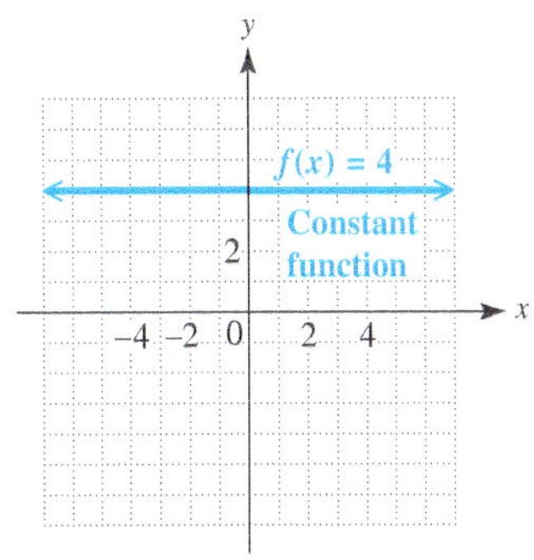

FIGURE 51

(b) $f(x) = 4$

The graph of this constant function is the horizontal line containing all points with y-coordinate 4. See **FIGURE 51**. The domain is $(-\infty, \infty)$ and the range is $\{4\}$.

NOW TRY

2.6 Exercises

FOR EXTRA HELP

MyMathLab®

Complete solution available in MyMathLab

Concept Check *Work each problem.*

1. To emphasize that "y is a function of x" for a given function f, we use function notation and write $y =$ ______. Here, f is the name of the ______, x is a value from the ______, and $f(x)$ is the function value (or y-value) that corresponds to ______. We read $f(x)$ as "______."

2. Choose the correct response.

For a function f, the notation $f(3)$ means ______.

A. the variable f times 3, or $3f$.

B. the value of the dependent variable when the independent variable is 3.

C. the value of the independent variable when the dependent variable is 3.

D. f equals 3.

3. Fill in each blank with the correct response.

The equation $2x + y = 4$ has a straight ______ as its graph. One point that lies on the graph is $(3,$ ____$)$. If we solve the equation for y and use function notation, we have a(n) ______ function $f(x) =$ ______. For this function, $f(3) =$ ____, meaning that the point $($____, ____$)$ lies on the graph of the function.

1. $f(x)$; function; domain; x; f of x (or "f at x")
2. B
3. line; -2; linear; $-2x + 4$; -2; 3; -2

4. Which of the following defines y as a linear function of x?

A. $y = \frac{1}{4}x - \frac{5}{4}$ **B.** $y = \frac{1}{x}$ **C.** $y = x^2$ **D.** $y = \sqrt{x}$

Let $f(x) = -3x + 4$ and $g(x) = -x^2 + 4x + 1$. Find the following. **See Examples 1–3.**

5. $f(0)$ **6.** $g(0)$ **7.** $f(-3)$ **8.** $f(-5)$

9. $g(-2)$ **10.** $g(-1)$ **11.** $g(3)$ **12.** $g(10)$

13. $f(100)$ **14.** $f(-100)$ **15.** $f\left(\frac{1}{3}\right)$ **16.** $f\left(\frac{7}{3}\right)$

17. $g(0.5)$ **18.** $g(1.5)$ **19.** $f(p)$ **20.** $g(k)$

21. $f(-x)$ **22.** $g(-x)$ **23.** $f(x + 2)$ **24.** $f(x - 2)$

25. $f(2t + 1)$ **26.** $f(3t - 2)$ **27.** $g(\pi)$ **28.** $g(t)$

29. $f(x + h)$ **30.** $f(a + b)$ **31.** $g\left(\frac{p}{3}\right)$ **32.** $g\left(\frac{1}{x}\right)$

*For each function, find **(a)** $f(2)$ and **(b)** $f(-1)$.* **See Examples 4, 5(a), and 5(b).**

33. $f = \{(-2, 2), (-1, -1), (2, -1)\}$ **34.** $f = \{(-1, -5), (0, 5), (2, -5)\}$

35. $f = \{(-1, 3), (4, 7), (0, 6), (2, 2)\}$ **36.** $f = \{(2, 5), (3, 9), (-1, 11), (5, 3)\}$

37.

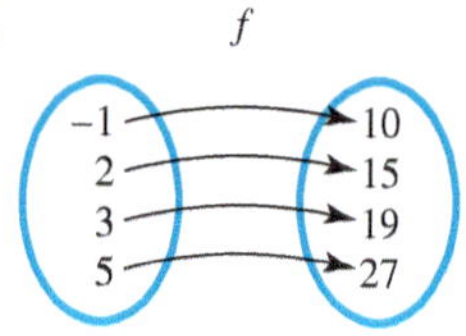

38.

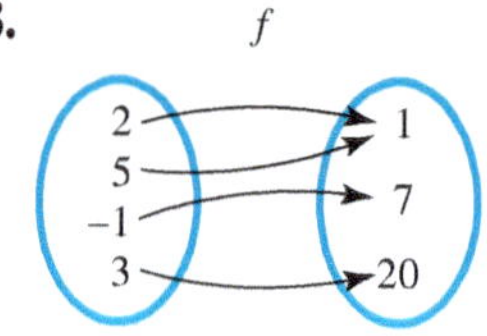

39.

x	$y = f(x)$
2	4
1	1
0	0
−1	1
−2	4

40.

x	$y = f(x)$
8	6
5	3
2	0
−1	−3
−4	−6

41.

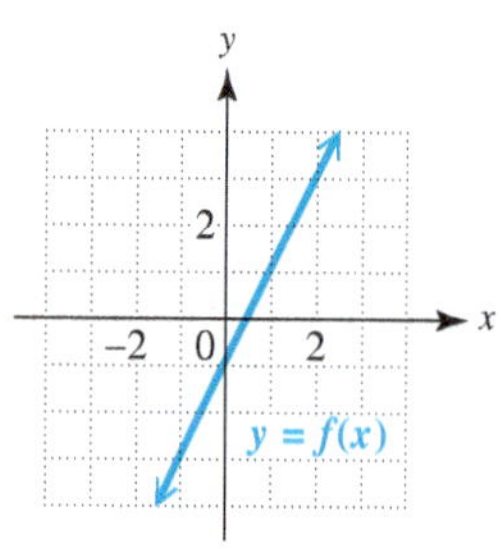

42.

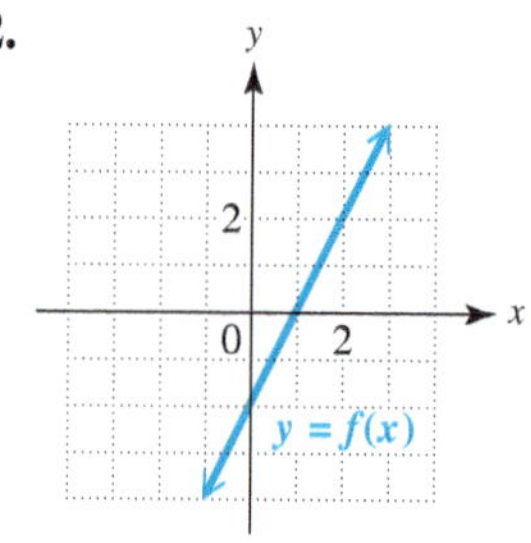

43.

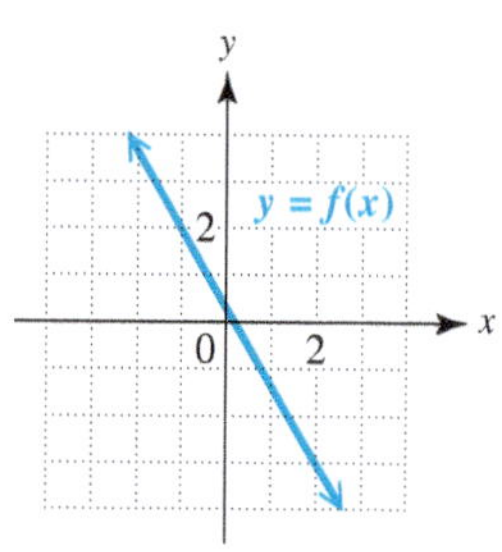

44.

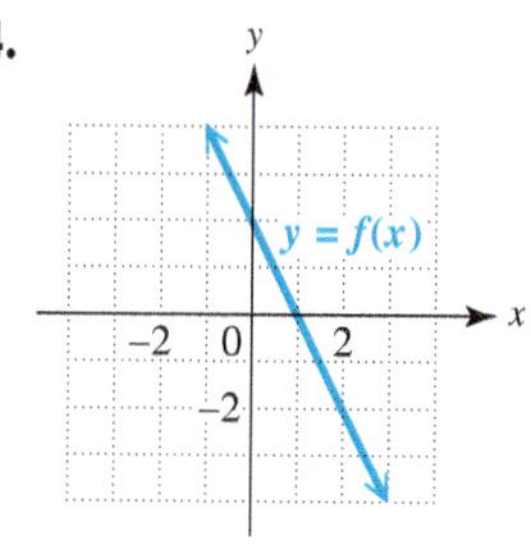

4. A

5. 4 **6.** 1
7. 13 **8.** 19
9. −11 **10.** −4
11. 4 **12.** −59
13. −296 **14.** 304
15. 3 **16.** −3
17. 2.75 **18.** 4.75
19. $-3p + 4$ **20.** $-k^2 + 4k + 1$
21. $3x + 4$ **22.** $-x^2 - 4x + 1$
23. $-3x - 2$ **24.** $-3x + 10$
25. $-6t + 1$ **26.** $-9t + 10$
27. $-\pi^2 + 4\pi + 1$
28. $-t^2 + 4t + 1$
29. $-3x - 3h + 4$
30. $-3a - 3b + 4$
31. $-\frac{p^2}{9} + \frac{4p}{3} + 1$
32. $-\frac{1}{x^2} + \frac{4}{x} + 1$

33. **(a)** −1 **(b)** −1
34. **(a)** −5 **(b)** −5
35. **(a)** 2 **(b)** 3
36. **(a)** 5 **(b)** 11
37. **(a)** 15 **(b)** 10
38. **(a)** 1 **(b)** 7
39. **(a)** 4 **(b)** 1
40. **(a)** 0 **(b)** −3
41. **(a)** 3 **(b)** −3
42. **(a)** 2 **(b)** −4
43. **(a)** −3 **(b)** 2
44. **(a)** −2 **(b)** 4

45. **(a)** 2 **(b)** 0 **(c)** −1
46. **(a)** 3 **(b)** 0 **(c)** 1

47. **(a)** $f(x) = -\frac{1}{3}x + 4$ **(b)** 3
48. **(a)** $f(x) = \frac{1}{4}x - 2$ **(b)** $-\frac{5}{4}$
49. **(a)** $f(x) = 3 - 2x^2$ **(b)** −15
50. **(a)** $f(x) = 2 + 3x^2$ **(b)** 29
51. **(a)** $f(x) = \frac{4}{3}x - \frac{8}{3}$ **(b)** $\frac{4}{3}$
52. **(a)** $f(x) = \frac{2}{5}x + \frac{9}{5}$ **(b)** 3

53. domain: $(-\infty, \infty)$; range: $(-\infty, \infty)$

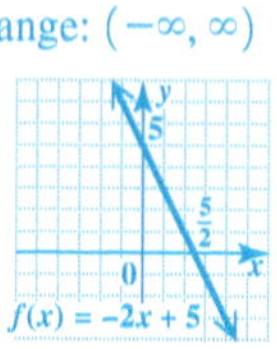

54. domain: $(-\infty, \infty)$; range: $(-\infty, \infty)$

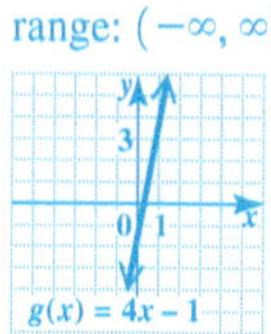

Refer to the given graph. Find the value of x for each value of $f(x)$. ***See Example 5(c).***

45. (a) $f(x) = 3$

(b) $f(x) = -1$

(c) $f(x) = -3$

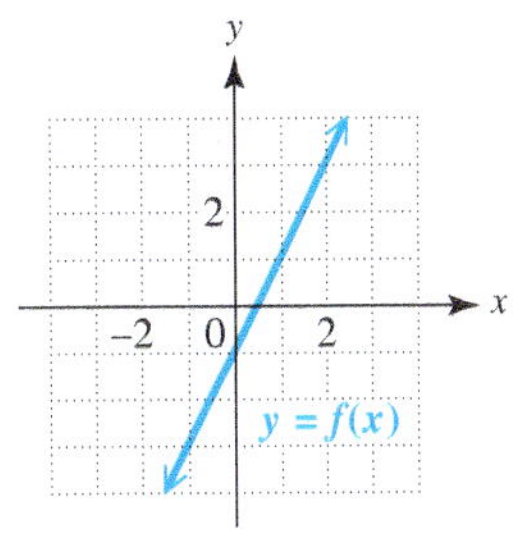

46. (a) $f(x) = 4$

(b) $f(x) = -2$

(c) $f(x) = 0$

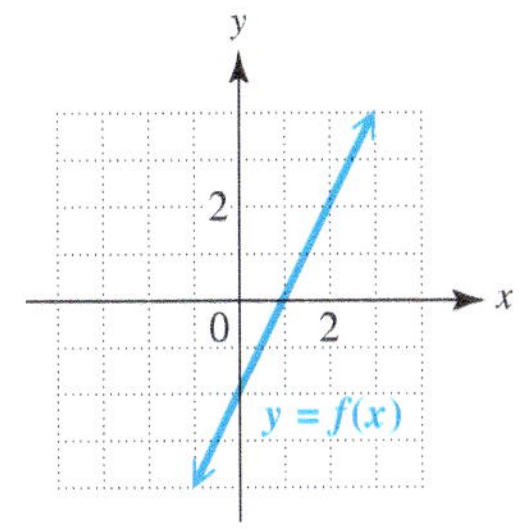

An equation that defines y as a function f of x is given. ***(a)*** *Solve for y in terms of x, and replace y with function notation $f(x)$.* ***(b)*** *Find $f(3)$.* ***See Example 6.***

47. $x + 3y = 12$ **48.** $x - 4y = 8$ **49.** $y + 2x^2 = 3$

50. $y - 3x^2 = 2$ **51.** $4x - 3y = 8$ **52.** $-2x + 5y = 9$

Graph each linear function. Give the domain and range. ***See Example 7.***

53. $f(x) = -2x + 5$ **54.** $g(x) = 4x - 1$ **55.** $h(x) = \frac{1}{2}x + 2$

56. $F(x) = -\frac{1}{4}x + 1$ **57.** $G(x) = 2x$ **58.** $H(x) = -3x$

59. $g(x) = -4$ **60.** $f(x) = 5$ **61.** $f(x) = 0$ **62.** $f(x) = -2.5$

63. Concept Check What is the name that is usually given to the graph in **Exercise 61?**

64. Concept Check Can the graph of a linear function have an undefined slope? Explain.

Solve each problem.

65. A package weighing x pounds costs $f(x)$ dollars to mail to a given location, where

$$f(x) = 3.75x.$$

(a) Evaluate $f(3)$.

(b) Describe what 3 and the value $f(3)$ mean in part (a), using the terms *independent variable* and *dependent variable.*

(c) How much would it cost to mail a 5-lb package? Then write this question and its answer using function notation.

66. A taxicab driver charges $2.50 per mile.

(a) Fill in the table with the correct response for the price $f(x)$ he charges for a trip of x miles.

x	$f(x)$
0	
1	
2	
3	

(b) The linear function that gives a rule for the amount charged is $f(x) =$ ______.

(c) Graph this function for the domain $\{0, 1, 2, 3\}$.

55. domain: $(-\infty, \infty)$; range: $(-\infty, \infty)$

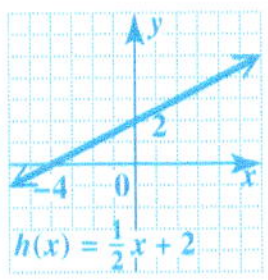

56. domain: $(-\infty, \infty)$; range: $(-\infty, \infty)$

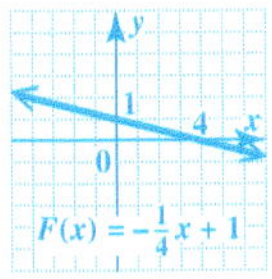

57. domain: $(-\infty, \infty)$; range: $(-\infty, \infty)$

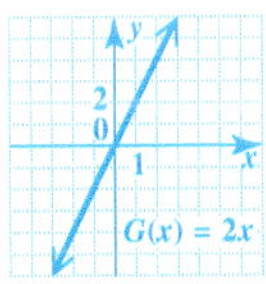

58. domain: $(-\infty, \infty)$; range: $(-\infty, \infty)$

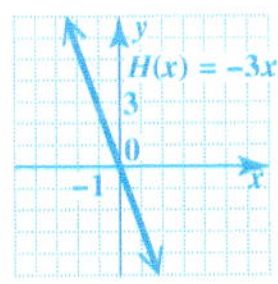

59. domain: $(-\infty, \infty)$; range: $\{-4\}$

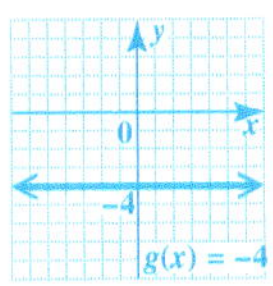

60. domain: $(-\infty, \infty)$; range: $\{5\}$

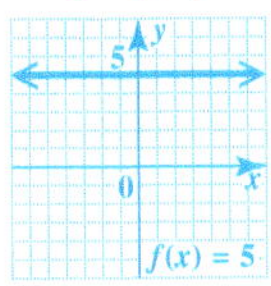

61. domain: $(-\infty, \infty)$; range: $\{0\}$

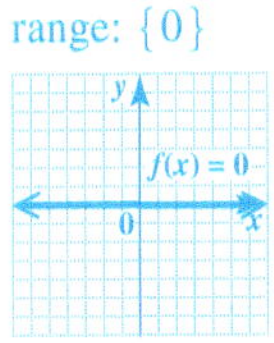

62. domain: $(-\infty, \infty)$; range: $\{-2.5\}$

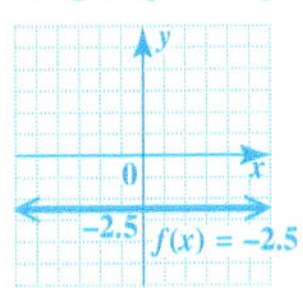

63. x-axis

64. No, because the equation of a line with an undefined slope is $x = a$. The ordered pairs have the form (a, y), where a is a constant and y is a variable. Thus, the number a corresponds to an infinite number of values of y.

65. (a) 11.25 (dollars) **(b)** 3 is the value of the independent variable, which represents a package weight of 3 lb. $f(3)$ is the value of the dependent variable, which represents the cost to mail a 3-lb package. **(c)** \$18.75; $f(5) = 18.75$

66. (a) \$0; \$2.50; \$5.00; \$7.50 **(b)** $2.50x$ **(c)**

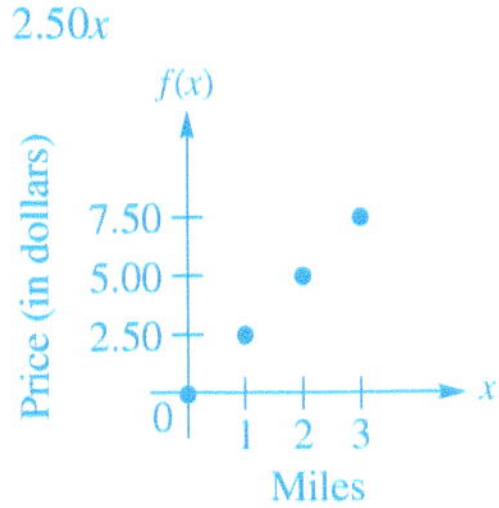

67. **(a)** $f(x) = 12x + 100$
(b) 1600; The cost to print 125 t-shirts is \$1600.
(c) 75; $f(75) = 1000$; The cost to print 75 t-shirts is \$1000.

68. **(a)** $f(x) = 0.50x + 150$
(b) \$275; $f(250) = 275$
(c) 500; It costs \$400 to drive the rental car 500 mi.

69. **(a)** 1.1 **(b)** 5 **(c)** -1.2 **(d)** $(0, 3.5)$ **(e)** $f(x) = -1.2x + 3.5$

70. **(a)** 0.6 **(b)** 3 **(c)** 1.5 **(d)** $(0, -2.4)$ **(e)** $f(x) = 1.5x - 2.4$

71. **(a)** $[0, 100]$; $[0, 3000]$
(b) 25 hr; 25 hr
(c) 2000 gal
(d) $f(0) = 0$; The pool is empty at time 0.
(e) $f(25) = 3000$; After 25 hr, there are 3000 gal of water in the pool.

72. **(a)** For every hour, there is one and only one megawatt reading.
(b) $[0, 24]$
(c) 1200 megawatts
(d) at 18 hr or 6 P.M.; at 4 A.M.
(e) $f(12) = 1800$; At 12 noon, electricity use is 1800 megawatts.

67. To print t-shirts, there is a \$100 set-up fee, plus a \$12 charge per t-shirt. Let x represent the number of t-shirts printed and $f(x)$ represent the total charge.

(a) Write a linear function that models this situation.

(b) Find $f(125)$. Interpret your answer in the context of this problem.

(c) Find the value of x if $f(x) = 1000$. Express this situation using function notation, and interpret it in the context of this problem.

68. Rental on a car is \$150, plus \$0.50 per mile. Let x represent the number of miles driven and $f(x)$ represent the total cost to rent the car.

(a) Write a linear function that models this situation.

(b) How much would it cost to drive 250 mi? Interpret this question and answer, using function notation.

(c) Find the value of x if $f(x) = 400$. Interpret your answer in the context of this problem.

69. The table represents a linear function.

(a) What is $f(2)$?

(b) If $f(x) = -2.5$, what is the value of x?

(c) What is the slope of the line?

(d) What is the y-intercept of the line?

(e) Using your answers from parts (c) and (d), write an equation for $f(x)$.

x	$y = f(x)$
0	3.5
1	2.3
2	1.1
3	−0.1
4	−1.3
5	−2.5

70. The table represents a linear function.

(a) What is $f(2)$?

(b) If $f(x) = 2.1$, what is the value of x?

(c) What is the slope of the line?

(d) What is the y-intercept of the line?

(e) Using your answers from parts (c) and (d), write an equation for $f(x)$.

x	$y = f(x)$
−1	−3.9
0	−2.4
1	−0.9
2	0.6
3	2.1

71. The graph shows water in a swimming pool over time.

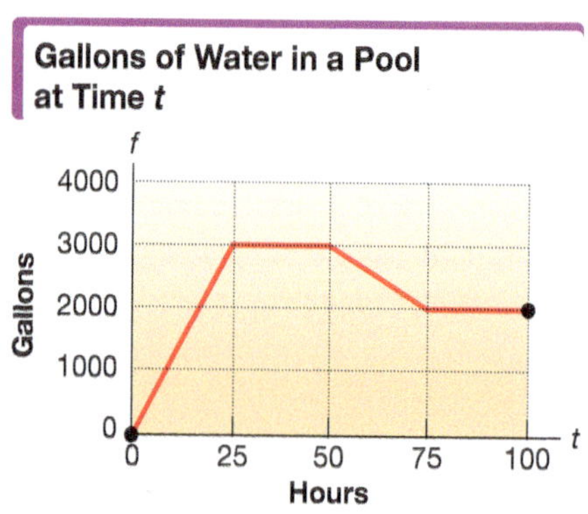

(a) What numbers are possible values of the independent variable? The dependent variable?

(b) For how long is the water level increasing? Decreasing?

(c) How many gallons of water are in the pool after 90 hr?

(d) Call this function f. What is $f(0)$? What does it mean?

(e) What is $f(25)$? What does it mean?

72. The graph shows electricity use on a summer day.

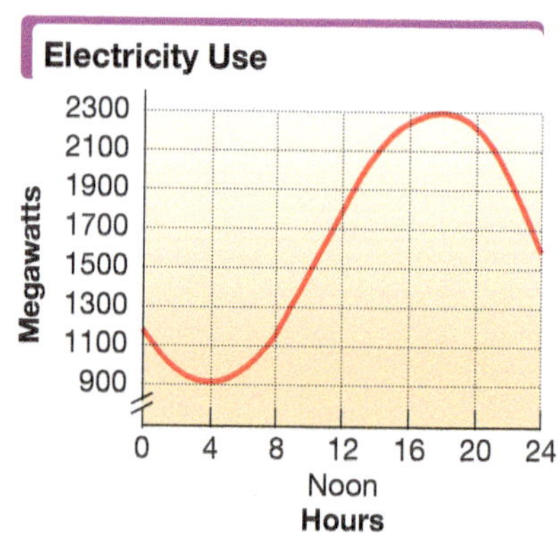

(a) Why is this the graph of a function?

(b) What is the domain?

(c) Estimate the number of megawatts used at 8 A.M.

(d) At what time was the most electricity used? The least electricity?

(e) Call this function f. What is $f(12)$? What does it mean?

73. (a) 194.53 cm (b) 177.29 cm (c) 177.41 cm (d) 163.65 cm

74. (a) 1.83 m^3 (b) 2.86 m^3 (c) 4.11 m^3 (d) 6.43 m^3

73. Forensic scientists use the lengths of certain bones to calculate the height of a person. Two such bones are the tibia (t), the bone from the ankle to the knee, and the femur (r), the bone from the knee to the hip socket. A person's height (h) in centimeters is determined from the lengths of these bones by using the following functions.

For men: $h(r) = 69.09 + 2.24r$ or $h(t) = 81.69 + 2.39t$

For women: $h(r) = 61.41 + 2.32r$ or $h(t) = 72.57 + 2.53t$

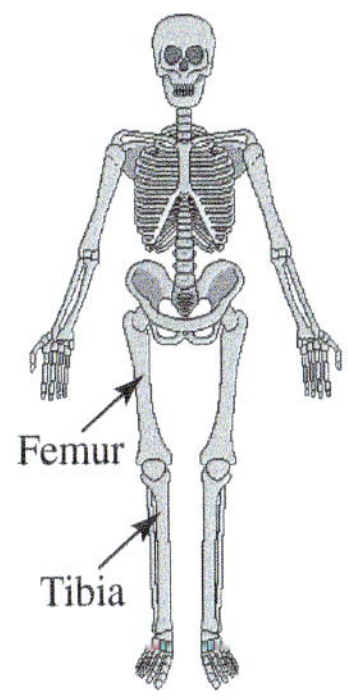

(a) Find the height of a man with a femur measuring 56 cm.

(b) Find the height of a man with a tibia measuring 40 cm.

(c) Find the height of a woman with a femur measuring 50 cm.

(d) Find the height of a woman with a tibia measuring 36 cm.

74. Based on federal regulations, a pool to house sea otters must have a volume that is "the square of the sea otter's average adult length (in meters) multiplied by 3.14 and by 0.91 meter." If x represents the sea otter's average adult length and $f(x)$ represents the volume (in cubic meters) of the corresponding pool size, this formula can be written as the function

$$f(x) = 0.91(3.14)x^2.$$

Find the volume of the pool for each adult sea otter length (in meters). Round answers to the nearest hundredth.

(a) 0.8 **(b)** 1.0 **(c)** 1.2 **(d)** 1.5

RELATING CONCEPTS For Individual or Group Work (Exercises 75–82)

Refer to the straight-line graph and ***work Exercises 75–82 in order.***

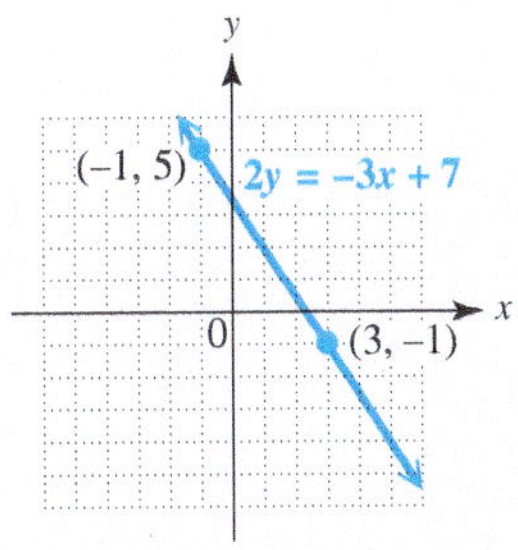

75. By just looking at the graph, how can we tell whether the slope is positive, negative, 0, or undefined?

76. Apply the slope formula to find the slope of the line.

77. What is the slope of any line parallel to the line shown? Perpendicular to the line shown?

78. Find the x-intercept of the graph.

79. Find the y-intercept of the graph.

80. Use function notation to write the equation of the line. Use f to designate the function.

81. Find $f(8)$.

82. If $f(x) = -8$, what is the value of x?

75. Because it falls from left to right, the slope is negative.

76. $-\frac{3}{2}$

77. $-\frac{3}{2}; \frac{2}{3}$

78. $\left(\frac{7}{3}, 0\right)$

79. $\left(0, \frac{7}{2}\right)$

80. $f(x) = -\frac{3}{2}x + \frac{7}{2}$

81. $-\frac{17}{2}$

82. $\frac{23}{3}$

Chapter 2 Summary

Key Terms

2.1
ordered pair
origin
x-axis
y-axis
rectangular (Cartesian) coordinate system
components
plot
coordinate
quadrant
table of ordered pairs (table of values)
graph of an equation
first-degree equation
linear equation in two variables
x-intercept
y-intercept

2.2
rise
run
slope

2.3
scatter diagram

2.4
linear inequality in two variables
boundary line

2.5
relation
function
dependent variable
independent variable
domain
range

2.6
linear function
constant function

New Symbols

(x, y) ordered pair

(x_1, y_1) subscript notation (read "x-sub-one, y-sub-one")

Δ Greek letter delta

m slope

$f(x)$ function notation; function of x (read "f of x" or "f at x")

Test Your Word Power

See how well you have learned the vocabulary in this chapter.

1. An **ordered pair** is a pair of numbers written
A. in numerical order between brackets
B. between parentheses or brackets
C. between parentheses in which order is important
D. between parentheses in which order does not matter.

2. A **linear equation in two variables** is an equation that can be written in the form
A. $Ax + By < C$
B. $ax = b$
C. $y = x^2$
D. $Ax + By = C$.

3. An **intercept** is
A. the point where the x-axis and y-axis intersect
B. a pair of numbers written between parentheses in which order matters
C. one of the regions determined by a coordinate system
D. the point where a graph intersects the x-axis or the y-axis.

4. The **slope** of a line is
A. the measure of the run over the rise of the line
B. the distance between two points on the line
C. the ratio of the change in y to the change in x along the line
D. the horizontal change compared to the vertical change of two points on the line.

5. A **relation** is
A. a set of ordered pairs
B. the ratio of the change in y to the change in x along a line
C. the set of all possible values of the independent variable
D. all the second components of a set of ordered pairs.

6. A **function** is
A. a pair of numbers in an ordered pair
B. a set of ordered pairs in which each x-value corresponds to exactly one y-value
C. a pair of numbers written between parentheses
D. the set of all ordered pairs that satisfy an equation.

7. The **domain** of a relation is
A. the set of all possible values of the dependent variable y
B. a set of ordered pairs
C. the difference between the x-values
D. the set of all possible values of the independent variable x.

8. The **range** of a relation is
A. the set of all possible values of the dependent variable y
B. a set of ordered pairs
C. the difference between the y-values
D. the set of all possible values of the independent variable x.

ANSWERS

1. C; *Examples:* $(0, 3), (3, 8), (4, 0)$ **2.** D; *Examples:* $3x + 2y = 6$, $x = y - 7$ **3.** D; *Example:* In **FIGURE 4(b)** of **Section 2.1,** the x-intercept is $(3, 0)$ and the y-intercept is $(0, 2)$. **4.** C; *Example:* The line through $(3, 6)$ and $(5, 4)$ has slope $\frac{4-6}{5-3} = \frac{-2}{2} = -1$. **5.** A; *Example:* The set $\{(2, 0), (4, 3), (6, 6)\}$ defines a relation. **6.** B; *Example:* The relation given in Answer 5 is a function. **7.** D; *Example:* In the relation in Answer 5, the domain is the set of x-values, $\{2, 4, 6\}$. **8.** A; *Example:* In the relation in Answer 5, the range is the set of y-values, $\{0, 3, 6\}$.

Quick Review

CONCEPTS | **EXAMPLES**

2.1 Linear Equations in Two Variables

Finding Intercepts

To find the x-intercept, let $y = 0$ and solve for x.

To find the y-intercept, let $x = 0$ and solve for y.

Find the intercepts of the graph of $2x + 3y = 12$.

$$2x + 3(0) = 12 \quad \text{Let } y = 0.$$
$$2x = 12$$
$$x = 6$$

The x-intercept is $(6, 0)$.

$$2(0) + 3y = 12 \quad \text{Let } x = 0.$$
$$3y = 12$$
$$y = 4$$

The y-intercept is $(0, 4)$.

Midpoint Formula

The midpoint M of a line segment PQ with endpoints (x_1, y_1) and (x_2, y_2) is found as follows.

$$M = \left(\frac{x_1 + x_2}{2}, \frac{y_1 + y_2}{2}\right)$$

Find the midpoint of the segment with endpoints $(4, -7)$ and $(-10, -13)$.

$$M = \left(\frac{4 + (-10)}{2}, \frac{-7 + (-13)}{2}\right)$$
$$= (-3, -10)$$

2.2 The Slope of a Line

The slope m of the line passing through the points (x_1, y_1) and (x_2, y_2) is defined as follows.

$$\textbf{slope } m = \frac{\textbf{rise}}{\textbf{run}} = \frac{\textbf{change in } y}{\textbf{change in } x} = \frac{\Delta y}{\Delta x} = \frac{y_2 - y_1}{x_2 - x_1}$$

(where $x_1 \neq x_2$)

Find the slope of the graph of $2x + 3y = 12$.

Use the intercepts $(6, 0)$ and $(0, 4)$ and the slope formula.

$$m = \frac{4 - 0}{0 - 6} = \frac{4}{-6} = -\frac{2}{3} \qquad (x_1, y_1) = (6, 0) \quad (x_2, y_2) = (0, 4)$$

A horizontal line has slope 0.

A vertical line has undefined slope.

Parallel lines have equal slopes.

The graph of the horizontal line $y = -5$ has slope $m = 0$.

The graph of the vertical line $x = 3$ has undefined slope.

The lines $y = 2x + 3$ and $4x - 2y = 6$ are parallel. Both have $m = 2$.

$$y = 2x + 3$$

$$4x - 2y = 6 \quad \text{Solve for } y.$$
$$-2y = -4x + 6$$
$$y = 2x - 3$$

Perpendicular lines, neither of which is vertical, **have slopes that are negative reciprocals** (that is, have a product of -1).

The lines $y = 3x - 1$ and $x + 3y = 4$ are perpendicular. Their slopes are negative reciprocals.

$$y = 3x - 1$$

$$x + 3y = 4 \quad \text{Solve for } y.$$
$$3y = -x + 4$$
$$y = -\frac{1}{3}x + \frac{4}{3}$$

Slope gives the **average rate of change** in y per unit change in x, where the value of y depends on the value of x.

The weight of a young child increased from 30 lb to 60 lb between the ages of 3 and 8. What was the child's average change in weight per year over these years?

$$\frac{60 - 30}{8 - 3} = \frac{30}{5} = 6 \text{ lb per yr}$$

CONCEPTS	EXAMPLES

2.3 Writing Equations of Lines

Slope-Intercept Form
$y = mx + b$

$y = 2x + 3$ $m = 2$, y-intercept is $(0, 3)$.

Point-Slope Form
$y - y_1 = m(x - x_1)$

$y - 3 = 4(x - 5)$ $(5, 3)$ is on the line, $m = 4$.

Standard Form
$Ax + By = C$, where A, B, and C are real numbers and A and B are not both 0. (We give A, B, and C integers, with greatest common factor 1 and $A \geq 0$.)

$2x - 5y = 8$ Standard form

Horizontal Line
$y = b$

$y = 4$ Horizontal line

Vertical Line
$x = a$

$x = -1$ Vertical line

2.4 Linear Inequalities in Two Variables

Graphing a Linear Inequality

Step 1 Draw the graph of the straight line that is the boundary. Make the line solid if the inequality involves $\leq$ or $\geq$. Make the line dashed if the inequality involves $<$ or $>$.

Step 2 Choose any point not on the line as a test point. Substitute the coordinates of that point in the inequality.

Step 3 Shade the region that includes the test point if it satisfies the original inequality. Otherwise, shade the region on the other side of the boundary line.

Graph $2x - 3y \leq 6$.

Draw the graph of $2x - 3y = 6$. Use a solid line because of the inclusion of equality in the symbol $\leq$.

Choose $(0, 0)$ as a test point.

$$2(0) - 3(0) \overset{?}{<} 6$$

$$0 < 6 \quad \text{True}$$

Shade the region that includes $(0, 0)$.

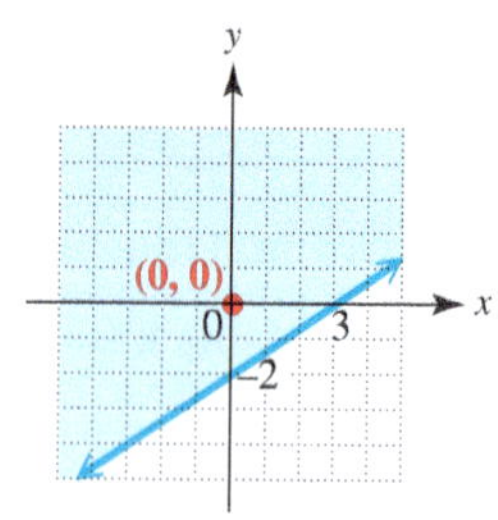

2.5 Introduction to Relations and Functions

A **relation** is any set of ordered pairs. A **function** is a set of ordered pairs such that, for each distinct first component, there is one and only one second component.

The set of first components is the **domain.**

The set of second components is the **range.**

The set of ordered pairs $\{(-1, 4), (0, 6), (1, 4)\}$ defines a function.

Domain: $\{-1, 0, 1\}$ Set of x-values

Range: $\{4, 6\}$ Set of y-values

The equation $y = x^2$ defines a function.

Domain: $(-\infty, \infty)$ Range: $[0, \infty)$

2.6 Function Notation and Linear Functions

To evaluate a function f, where $f(x)$ defines the range value for a given value of x in the domain, substitute the value wherever x appears.

$$f(x) = x^2 - 7x + 12$$

$$f(1) = 1^2 - 7(1) + 12 \quad \text{Let } x = 1.$$

$$f(1) = 6$$

To write an equation that defines a function f in function notation, follow these steps.

Step 1 Solve the equation for y.

Step 2 Replace y with $f(x)$.

Write $2x + 3y = 12$ using function notation for a function f.

$$3y = -2x + 12 \quad \text{Subtract } 2x.$$

$$y = -\frac{2}{3}x + 4 \quad \text{Divide by 3.}$$

$$f(x) = -\frac{2}{3}x + 4 \quad y = f(x)$$

Chapter 2 Review Exercises

1.

x	y
0	5
$\frac{10}{3}$	0
2	2
$\frac{14}{3}$	−2

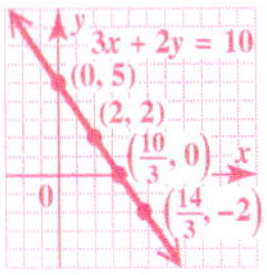

2.

x	y
2	−6
5	−3
3	−5
6	−2

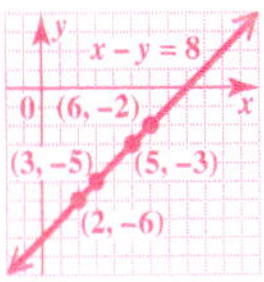

3. (3, 0); (0, −4) 4. $\left(\frac{28}{5}, 0\right)$; (0, 4)

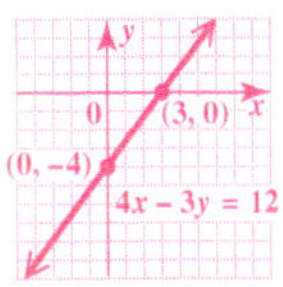

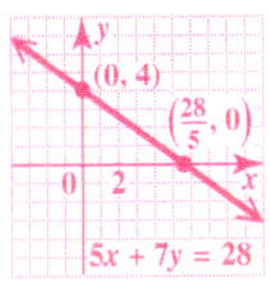

5. (10, 0); (0, 4) 6. (8, 0); (0, −2)

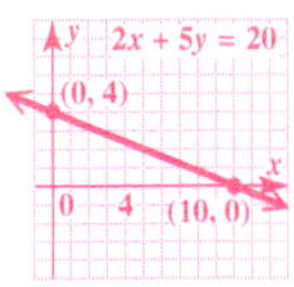

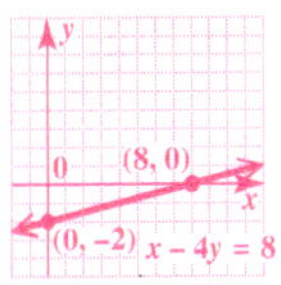

7. (0, 2) 8. $\left(-\frac{9}{2}, \frac{3}{2}\right)$

9. $-\frac{7}{5}$ 10. $-\frac{1}{2}$

11. 2 12. $\frac{3}{4}$

13. undefined 14. $\frac{2}{3}$

15. $-\frac{1}{3}$ 16. undefined

17. $-\frac{1}{3}$ 18. −1

19. positive 20. negative

21. undefined 22. 0

23. 12 ft 24. \$938 per year

2.1 *Complete the table of ordered pairs for each equation. Then graph the equation.*

1. $3x + 2y = 10$

x	y
0	
	0
2	
	−2

2. $x - y = 8$

x	y
2	
	−3
3	
	−2

Find the x- and y-intercepts. Then graph each equation.

3. $4x - 3y = 12$ **4.** $5x + 7y = 28$ **5.** $2x + 5y = 20$ **6.** $x - 4y = 8$

Find the midpoint of each segment with the given endpoints.

7. $(-8, -12)$ and $(8, 16)$ **8.** $(0, -5)$ and $(-9, 8)$

2.2 *Find the slope of each line. (In Exercises 17 and 18, coordinates of the points shown are integers.)*

9. Through $(-1, 2)$ and $(4, -5)$ **10.** Through $(0, 3)$ and $(-2, 4)$

11. $y = 2x + 3$ **12.** $3x - 4y = 5$

13. $x = 5$ **14.** Parallel to $3y = 2x + 5$

15. Perpendicular to $3x - y = 4$ **16.** Through $(-1, 5)$ and $(-1, -4)$

17.

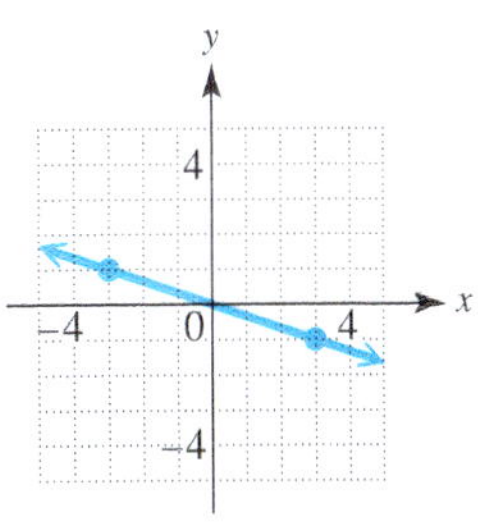

18.

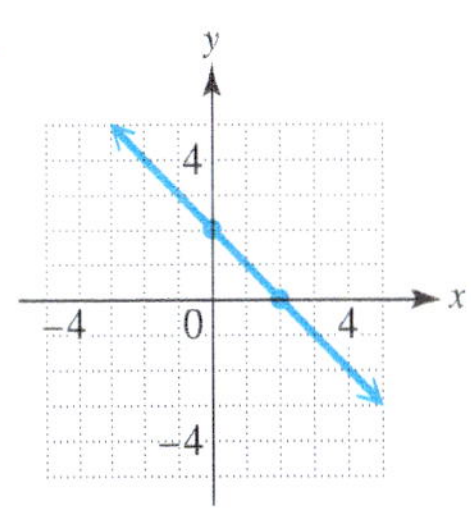

Tell whether the slope of the line is positive, negative, 0, *or* undefined.

19.

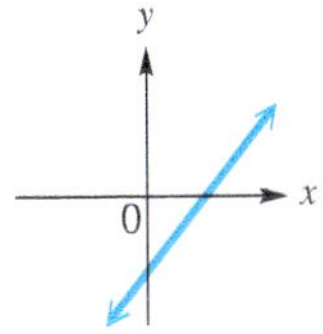

20.

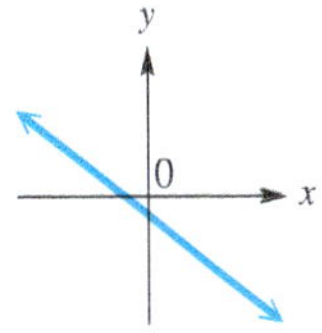

21.

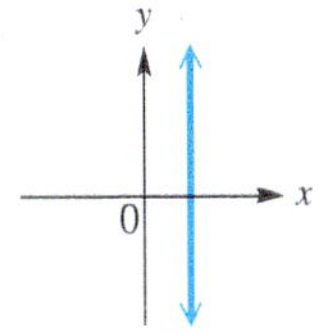

22.

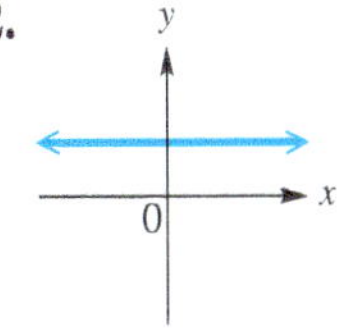

Solve each problem.

23. If the pitch of a roof is $\frac{1}{4}$, how many feet in the horizontal direction correspond to a rise of 3 ft?

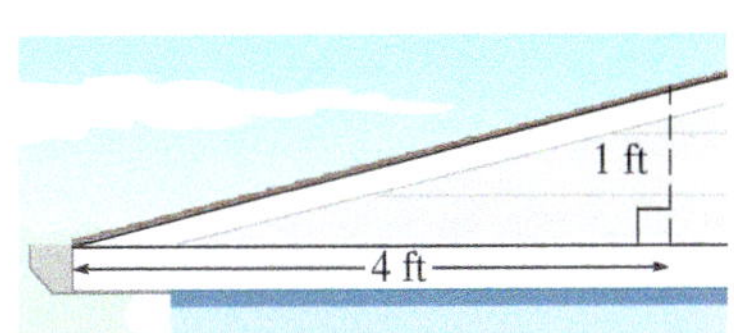

24. In 1980, the median family income in the United States was about \$21,000 per year. In 2012, it was about \$51,017 per year. Find the average rate of change in median family income per year to the nearest dollar during this period. (*Source:* U.S. Census Bureau.)

25. (a) $y = -\frac{1}{3}x - 1$
(b) $x + 3y = -3$
26. (a) $y = -2$
(b) $y = -2$
27. (a) $y = -\frac{4}{3}x + \frac{29}{3}$
(b) $4x + 3y = 29$
28. (a) $y = 3x + 7$
(b) $3x - y = -7$
29. (a) not possible
(b) $x = 2$
30. (a) $y = -9x + 13$
(b) $9x + y = 13$
31. (a) $y = \frac{7}{5}x + \frac{16}{5}$
(b) $7x - 5y = -16$
32. (a) $y = -x + 2$
(b) $x + y = 2$
33. (a) $y = 4x - 29$
(b) $4x - y = 29$
34. (a) $y = -\frac{5}{2}x + 13$
(b) $5x + 2y = 26$
35. $y = 47x + 159$; \$723
36. (a) $y = 38x + 2172$; The revenue from skiing facilities increased by an average of \$38 million per year from 2008 to 2012.
(b) \$2552 million

37.
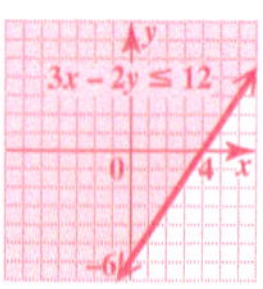

38.
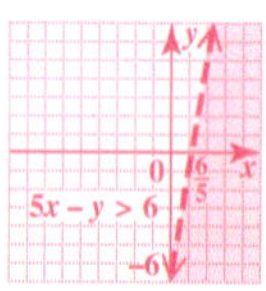

39.
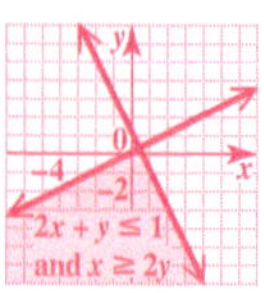

40.
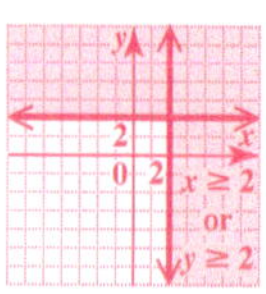

41. not a function;
domain: $\{-4, 1\}$;
range: $\{2, -2, 5, -5\}$
42. function;
domain: $\{9, 11, 4, 17, 25\}$;
range: $\{32, 47, 69, 14\}$
43. function;
domain: $[-4, 4]$;
range: $[0, 2]$
44. not a function;
domain: $(-\infty, 0]$;
range: $(-\infty, \infty)$
45. function;
domain: $(-\infty, \infty)$;
linear function

2.3 *Write an equation of the line that satisfies the given conditions. Give the equation **(a)** in slope-intercept form if possible and **(b)** in standard form.*

25. Slope $-\frac{1}{3}$; y-intercept $(0, -1)$

26. Slope 0; y-intercept $(0, -2)$

27. Slope $-\frac{4}{3}$; through $(2, 7)$

28. Slope 3; through $(-1, 4)$

29. Vertical; through $(2, 5)$

30. Through $(2, -5)$ and $(1, 4)$

31. Through $(-3, -1)$ and $(2, 6)$

32. The line graphed in **Exercise 18**

33. Through $(7, -1)$; parallel to $4x - y = 3$

34. Through $(4, 3)$; perpendicular to $2x - 5y = 7$

Solve each problem.

35. An Executive Regular/Silver membership to a health club costs \$159, plus \$47 per month. Let x represent the number of months and y represent the cost. How much will a one-year membership cost? (*Source:* Midwest Athletic Club.)

36. Revenue for skiing facilities in the United States is shown in the graph.

(a) Use the information given for the years 2008 and 2012, letting $x = 8$ represent 2008 and $x = 12$ represent 2012, and letting y represent revenue (in millions of dollars), to write an equation that models the data. Write the equation in slope-intercept form. Interpret the slope.

(b) Use the equation from part (a) to approximate revenue for skiing facilities in 2010.

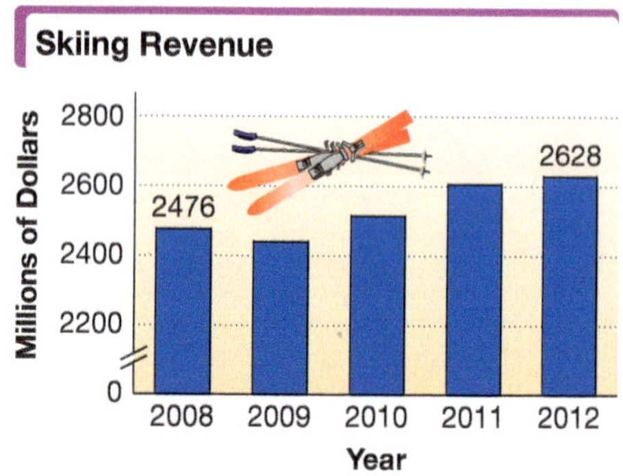

Source: U.S. Department of Commerce.

2.4 *Graph each inequality or compound inequality.*

37. $3x - 2y \le 12$

38. $5x - y > 6$

39. $2x + y \le 1$ and $x \ge 2y$

40. $x \ge 2$ or $y \ge 2$

2.5 *Decide whether each relation defines a function, and give the domain and range.*

41. $\{(-4, 2), (-4, -2), (1, 5), (1, -5)\}$

42.
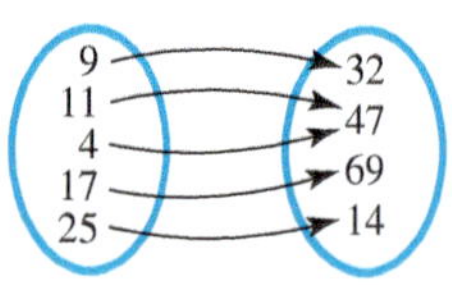

43.
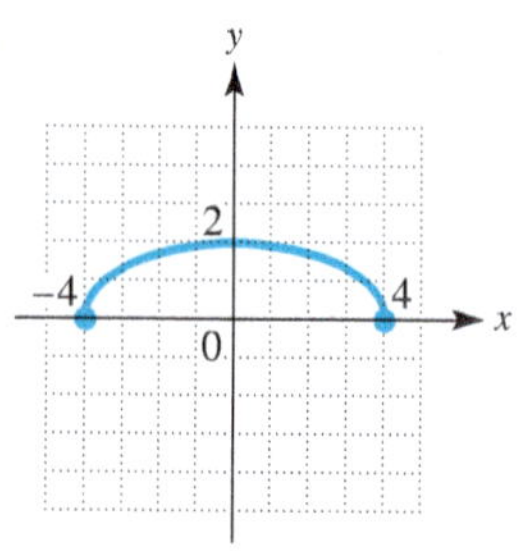

44.
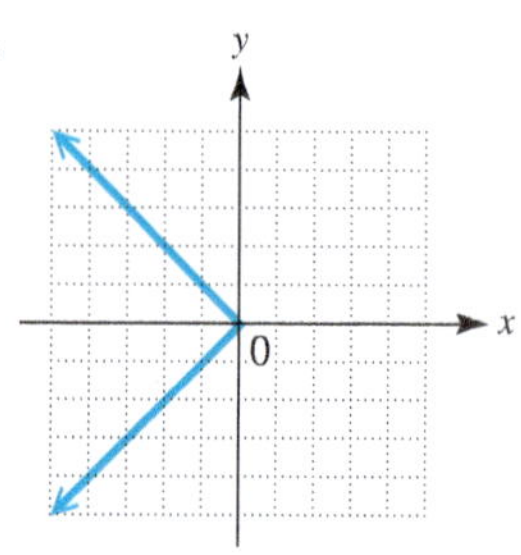

2.5, 2.6 *Decide whether each relation defines y as a function of x. Give the domain. Identify any linear functions.*

45. $y = 3x - 3$

46. $y < x + 2$

47. $y = |x|$

48. $y = \sqrt{4x + 7}$

49. $x = y^2$

50. $y = \frac{7}{x - 6}$

46. not a function;
domain: $(-\infty, \infty)$
47. function;
domain: $(-\infty, \infty)$
48. function;
domain: $\left[-\frac{7}{4}, \infty\right)$
49. not a function;
domain: $[0, \infty)$
50. function;
domain: $(-\infty, 6) \cup (6, \infty)$

51. -6 52. -8.52
53. -8 54. $-2k^2 + 3k - 6$
55. $f(x) = 2x^2$; 18
56. C
57. It is a horizontal line.
58. (a) yes
(b) domain: {1960, 1970, 1980, 1990, 2000, 2010};
range: {69.7, 70.8, 73.7, 75.4, 76.8, 78.7}
(c) Answers will vary. Two possible answers are (1960, 69.7) and (2010, 78.7).
(d) 73.7;
In 1980, life expectancy at birth was 73.7 yr.
(e) 2000

2.6 *Let* $f(x) = -2x^2 + 3x - 6$. *Find the following.*

51. $f(0)$ **52.** $f(2.1)$ **53.** $f\left(-\frac{1}{2}\right)$ **54.** $f(k)$

Solve each problem.

55. The equation $2x^2 - y = 0$ defines y as a function f of x. Write it using function notation, and find $f(3)$.

56. Suppose that $2x - 5y = 7$ defines y as a function f of x. If $y = f(x)$, which one of the following defines the same function?

A. $f(x) = -\frac{2}{5}x + \frac{7}{5}$ **B.** $f(x) = -\frac{2}{5}x - \frac{7}{5}$

C. $f(x) = \frac{2}{5}x - \frac{7}{5}$ **D.** $f(x) = \frac{2}{5}x + \frac{7}{5}$

57. Describe the graph of a constant function.

58. The table shows life expectancy at birth in the United States for selected years.

(a) Does the table define a function?

(b) What are the domain and range?

(c) Call this function f. Give two ordered pairs that belong to f.

(d) Find $f(1980)$. What does this mean?

(e) If $f(x) = 76.8$, what does x equal?

Year	*Life Expectancy at Birth (years)*
1960	69.7
1970	70.8
1980	73.7
1990	75.4
2000	76.8
2010	78.7

Source: National Center for Health Statistics.

Chapter 2 Mixed Review Exercises

1. perpendicular
2. parallel

3. -1.8 lb per year;
Per capita consumption of potatoes decreased by an average of 1.8 lb per year from 2003 to 2011.
4. $y = -1.8x + 46.8$

Determine whether each pair of lines is parallel, perpendicular, *or* neither.

1. $3x + y = 4$ and $3y = x - 6$ **2.** $4x + 3y = 8$ and $6y = 7 - 8x$

The graph shows per capita consumption of potatoes (in pounds) in the United States from 2003 to 2011.

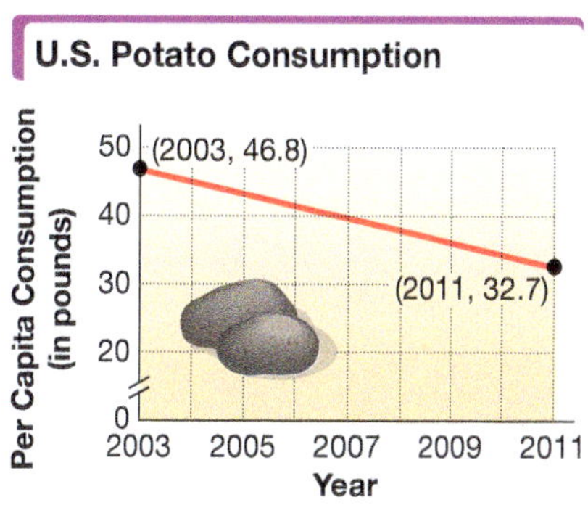

Source: U.S. Department of Agriculture.

3. Use the given ordered pairs to find and interpret the average rate of change in per capita potato consumption per year to the nearest tenth during this period.

4. Write an equation in slope-intercept form that models per capita consumption of potatoes y (in pounds) in year x, where $x = 0$ represents 2003.

Write an equation of a line (in the form specified, if given) that satisfies the given conditions.

5. $m = 3$; y-intercept $(0, 0)$ (slope-intercept form)

6. Through $(0, 3)$ and $(-2, 4)$ (standard form)

7. Through $(2, -3)$; perpendicular to $x = 2$

Solve each problem.

8. Which equations have a graph with just one intercept?

A. $x - 6 = 0$ **B.** $x + y = 0$ **C.** $x - y = 4$ **D.** $y = -4$

9. Which inequality has as its graph a dashed boundary line and shading below the line?

A. $y \geq 4x + 3$ **B.** $y > 4x + 3$ **C.** $y \leq 4x + 3$ **D.** $y < 4x + 3$

10. Refer to the graph of function f.

(a) Find $f(-2)$.

(b) Find $f(0)$.

(c) For what value of x is $f(x) = -3$?

(d) Give the domain and range.

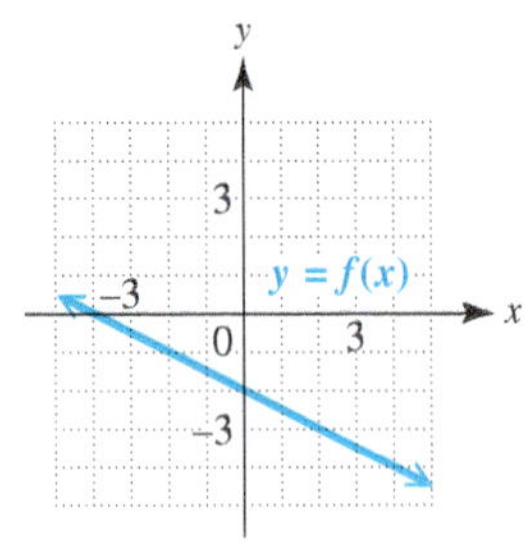

5. $y = 3x$
6. $x + 2y = 6$
7. $y = -3$
8. A, B, D
9. D
10. (a) -1 (b) -2 (c) 2 (d) $(-\infty, \infty)$; $(-\infty, \infty)$

Chapter 2 Test

FOR EXTRA HELP

Step-by-step test solutions are found on the Chapter Test Prep Videos available in MyMathLab® or on YouTube.

View the complete solutions to all Chapter Test exercises in MyMathLab.

1. Complete the table of ordered pairs for the equation $2x - 3y = 12$.

x	y
1	
3	
	-4

Find the x- and y-intercepts. Then graph each equation.

2. $3x - 2y = 20$ **3.** $y = 5$ **4.** $x = 2$

5. Find the slope of the line passing through the points $(6, 4)$ and $(-4, -1)$.

6. Describe how the graph of a line with undefined slope is situated in a rectangular coordinate system.

Determine whether each pair of lines is parallel, perpendicular, *or* neither.

7. $5x - y = 8$ and $5y = -x + 3$ **8.** $2y = 3x + 12$ and $3y = 2x - 5$

9. In 1980, there were 119,000 farms in Iowa. As of 2012, there were 92,200. Find and interpret the average rate of change in the number of farms per year, to the nearest whole number. (*Source:* U.S. Department of Agriculture.)

Write an equation of the line that satisfies the given conditions. Give the equation **(a)** *in slope-intercept form if possible and* **(b)** *in standard form.*

10. Through $(4, -1)$; $m = -5$ **11.** Through $(-3, 14)$; horizontal

12. Through $(-2, 3)$ and $(6, -1)$ **13.** Through $(5, -6)$; vertical

14. Through $(-7, 2)$; parallel to $3x + 5y = 6$

15. Through $(-7, 2)$; perpendicular to $y = 2x$

[2.1]

1. $-\frac{10}{3}$; -2; 0
2. $\left(\frac{20}{3}, 0\right)$; $(0, -10)$

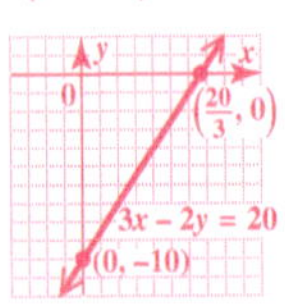

3. none; $(0, 5)$ 4. $(2, 0)$; none

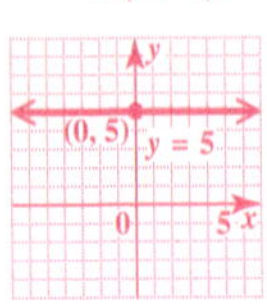

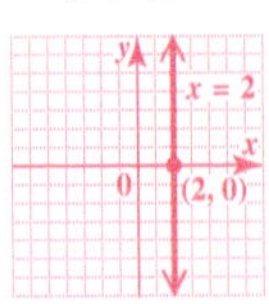

[2.2]

5. $\frac{1}{2}$
6. It is a vertical line.
7. perpendicular
8. neither
9. -838 farms per yr; The number of farms decreased an average of 838 farms per year from 1980 to 2012.

16. Which line has positive slope and negative y-coordinate for its y-intercept?

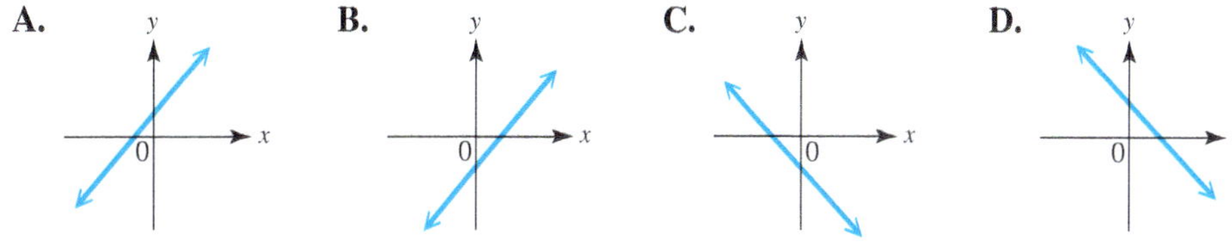

17. A ticket to a rock concert costs $142.75. An advance parking pass costs $45. Let x represent the number of tickets purchased and y represent the total cost in dollars.

(a) Write an equation in slope-intercept form that represents this situation.

(b) How much does it cost for 6 tickets and a parking pass?

Graph each inequality or compound inequality.

18. $3x - 2y > 6$

19. $y < 2x - 1$ and $x - y < 3$

20. Which one of the following is the graph of a function?

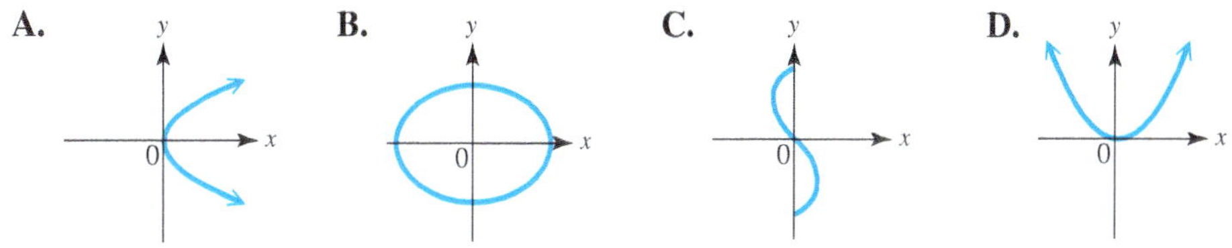

21. Which of the following does not define y as a function of x?

A. $\{(0, 1), (-2, 3), (4, 8)\}$

B. $y = 2x - 6$

C. $y = \sqrt{x + 2}$

D.

x	y
0	1
3	2
0	2
6	3

Give the domain and range of the relation shown in each of the following.

22. Choice A of **Exercise 20**

23. Choice A of **Exercise 21**

24. For $f(x) = -x^2 + 2x - 1$, find $f(1)$ and $f(a)$.

25. Graph the linear function $f(x) = \frac{2}{3}x - 1$. Give the domain and range.

[2.3]

10. **(a)** $y = -5x + 19$ **(b)** $5x + y = 19$

11. **(a)** $y = 14$ **(b)** $y = 14$

12. **(a)** $y = -\frac{1}{2}x + 2$ **(b)** $x + 2y = 4$

13. **(a)** not possible **(b)** $x = 5$

14. **(a)** $y = -\frac{3}{5}x - \frac{11}{5}$ **(b)** $3x + 5y = -11$

15. **(a)** $y = -\frac{1}{2}x - \frac{3}{2}$ **(b)** $x + 2y = -3$

16. B

17. **(a)** $y = 142.75x + 45$ **(b)** $901.50

[2.4]

18.

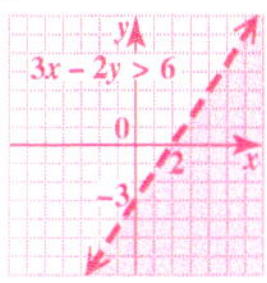

19.

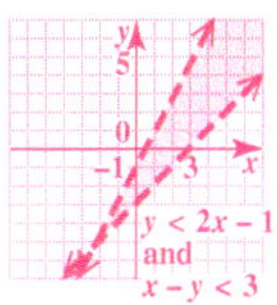

[2.5]

20. D **21.** D

22. domain: $[0, \infty)$; range: $(-\infty, \infty)$

23. domain: $\{0, -2, 4\}$; range: $\{1, 3, 8\}$

[2.6]

24. 0; $-a^2 + 2a - 1$

25.

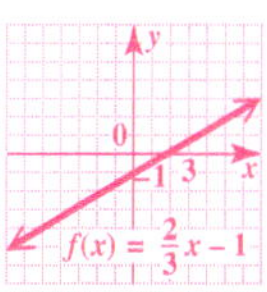

domain: $(-\infty, \infty)$; range: $(-\infty, \infty)$

Chapters R–2 Cumulative Review Exercises

Decide whether each statement is always true, sometimes true, *or* never true. *If the statement is* sometimes true, *give examples for which it is true and for which it is false.*

1. The absolute value of a negative number equals the additive inverse of the number.

2. The sum of two negative numbers is positive.

3. The sum of a positive number and a negative number is 0.

Perform each operation.

4. $-|-2| - 4 + |-3| + 7$

5. $(-0.8)^2$

6. $\sqrt{-64}$

[R.1]

1. always true **2.** never true

3. sometimes true; For example, $3 + (-3) = 0$, but $3 + (-1) = 2$ and $2 \neq 0$.

[R.2]

4. 4

[R.3]

5. 0.64

6. not a real number

[R.4]
7. $4x - 3$
8. $2x^2 + 5x + 4$

[R.3]
9. $-\frac{19}{2}$ **10.** 0
11. -39 **12.** undefined

[1.1]
13. $\left\{\frac{7}{6}\right\}$ **14.** $\{-1\}$

[1.4]
15. 6 in. **16.** 2 hr

[1.5]
17. $\left(-3, \frac{7}{2}\right)$ **18.** $(-\infty, 1]$

[1.6]
19. $(6, 8)$
20. $(-\infty, -2] \cup (7, \infty)$

[1.7]
21. $\{0, 7\}$ **22.** $(-\infty, \infty)$

[2.1]
23. $(4, 0)$; $\left(0, \frac{12}{5}\right)$

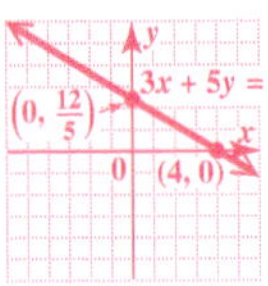

[2.2]
24. **(a)** $-\frac{6}{5}$ **(b)** $\frac{5}{6}$

[2.4]
25.

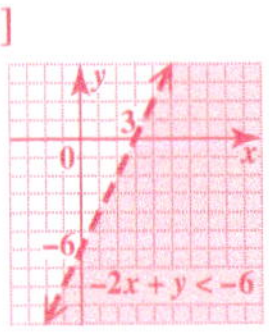

[2.3]
26. **(a)** $y = -\frac{3}{4}x - 1$
(b) $3x + 4y = -4$
27. **(a)** $y = -\frac{4}{3}x + \frac{7}{3}$
(b) $4x + 3y = 7$

[2.5]
28. domain: $\{14, 91, 75, 23\}$;
range: $\{9, 70, 56, 5\}$;
not a function

Simplify each expression.

7. $-(-4x + 3)$

8. $3x^2 - 4x + 4 + 9x - x^2$

9. $\dfrac{(4^2 - 4) - (-1)7}{4 + (-6)}$

10. $\sqrt{25} - 5(-1)^2$

Evaluate each expression for $p = -4$, $q = \frac{1}{2}$, *and* $r = 16$.

11. $-3(2q - 3p)$

12. $\dfrac{\sqrt{r}}{8p + 2r}$

Solve each equation.

13. $2z - 5 + 3z = 4 - (z + 2)$

14. $\dfrac{3x - 1}{5} + \dfrac{x + 2}{2} = -\dfrac{3}{10}$

Solve each problem.

15. If each side of a square were increased by 4 in., the perimeter would be 8 in. less than twice the perimeter of the original square. Find the length of a side of the original square.

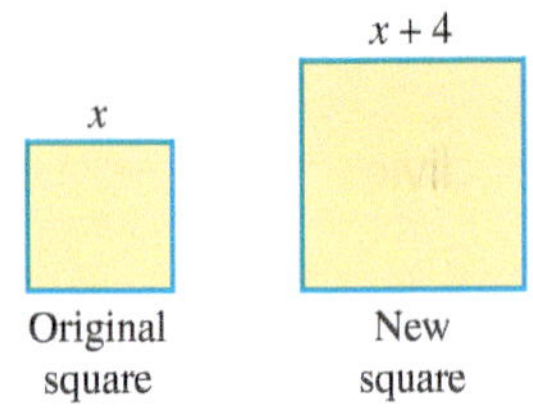

16. Two planes leave the Dallas–Fort Worth airport at the same time. One travels east at 550 mph, and the other travels west at 500 mph. Assuming no wind, how long will it take for the planes to be 2100 mi apart?

Solve.

17. $-4 < 3 - 2k < 9$

18. $-0.3x + 2.1(x - 4) \leq -6.6$

19. $\frac{1}{2}x > 3$ and $\frac{1}{3}x < \frac{8}{3}$

20. $-5x + 1 \geq 11$ or $3x + 5 > 26$

21. $|2k - 7| + 4 = 11$

22. $|3x + 6| \geq 0$

Solve each problem.

23. Find the x- and y-intercepts of the line with equation $3x + 5y = 12$, and graph the line.

24. Consider the points $A(-2, 1)$ and $B(3, -5)$.

(a) Find the slope of the line AB.

(b) Find the slope of a line perpendicular to line AB.

25. Graph the inequality $-2x + y < -6$.

Write an equation of the line that satisfies the given conditions. Give the equation **(a)** *in slope-intercept form if possible and* **(b)** *in standard form.*

26. Slope $-\frac{3}{4}$; y-intercept $(0, -1)$

27. Through $(4, -3)$ and $(1, 1)$

28. Give the domain and range of the relation. Does it define a function?

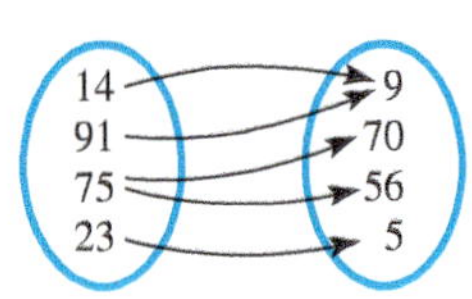

3 Systems of Linear Equations

Just as the *intersection* of two streets consists of the region common to both, a solution of a *system* of two linear equations (represented graphically by lines) is an ordered pair found in the solution sets of *both* of the individual equations.

3.1 Systems of Linear Equations in Two Variables

OBJECTIVES

1. Decide whether an ordered pair is a solution of a linear system.
2. Solve linear systems by graphing.
3. Solve linear systems (with two equations and two variables) by substitution.
4. Solve linear systems (with two equations and two variables) by elimination.
5. Solve special systems.

VOCABULARY

- ☐ system of equations
- ☐ system of linear equations (linear system)
- ☐ solution set of a linear system
- ☐ consistent system
- ☐ independent equations
- ☐ inconsistent system
- ☐ dependent equations

In recent years, the number of Americans aged 7 and over participating in hiking has increased, while the number participating in fishing has decreased. These trends can be seen in the graph in **FIGURE 1**. The two straight-line graphs intersect at the time when the two sports had the *same* number of participants.

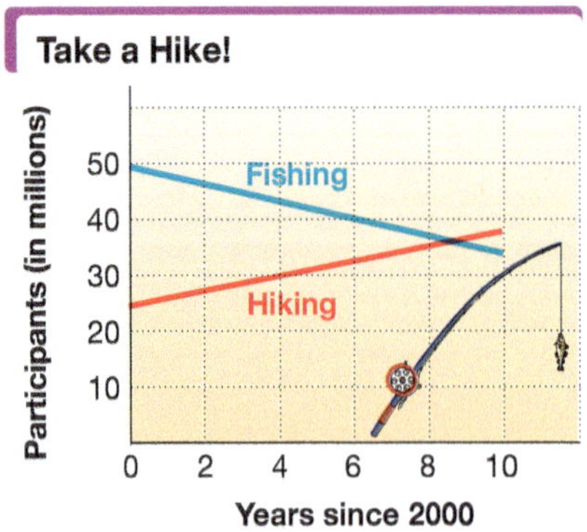

Source: National Sporting Goods Association.

FIGURE 1

$$\begin{aligned}-1.34x + y &= 24.3\\ 1.55x + y &= 49.3\end{aligned}$$ Linear system of equations

(Here, $x = 0$ represents 2000, $x = 1$ represents 2001, and so on. y represents millions of participants.)

As shown beside **FIGURE 1**, we can use a linear equation to model the graph of the number of people who hiked (red graph) and another linear equation to model the graph of the number of people who fished (blue graph). Such a set of equations is a **system of equations**—in this case, a **system of linear equations,** or a **linear system.**

The point where the graphs in **FIGURE 1** intersect is a solution of each of the individual equations. It is also the solution of the linear system of equations.

OBJECTIVE 1 Decide whether an ordered pair is a solution of a linear system.

The **solution set of a linear system** of equations contains all ordered pairs that satisfy all the equations of the system *at the same time.*

CLASSROOM EXAMPLE 1

Decide whether the given ordered pair is a solution of the given system.

$$\begin{aligned}2x + y &= -6\\ x + 3y &= 2\end{aligned}$$

(a) $(-4, 2)$ (b) $(3, -12)$

Answers: (a) solution (b) not a solution

EXAMPLE 1 Deciding Whether an Ordered Pair Is a Solution

Decide whether the given ordered pair is a solution of the given system.

(a) $\begin{aligned}x + y &= 6\\ 4x - y &= 14\end{aligned}$; $(4, 2)$

Replace x with 4 and y with 2 in each equation of the system.

$$x + y = 6$$
$$4 + 2 \stackrel{?}{=} 6$$
$$6 = 6 \quad \checkmark \text{ True}$$

$$4x - y = 14$$
$$4(4) - 2 \stackrel{?}{=} 14$$
$$16 - 2 \stackrel{?}{=} 14$$
$$14 = 14 \quad \checkmark \text{ True}$$

Because $(4, 2)$ makes *both* equations true, $(4, 2)$ is a solution of the system.

NOW TRY EXERCISE 1

Decide whether the ordered pair $(2, 5)$ is a solution of the given system.

(a) $-x + 2y = 8$
$3x - 2y = 0$

(b) $2x - y = -1$
$3x + 2y = 16$

(b) $3x + 2y = 11$; $x + 5y = 36$; $(-1, 7)$

$$3x + 2y = 11$$
$$3(-1) + 2(7) \stackrel{?}{=} 11$$
$$-3 + 14 \stackrel{?}{=} 11$$
$$11 = 11 \quad \checkmark \text{ True}$$

$$x + 5y = 36$$
$$-1 + 5(7) \stackrel{?}{=} 36$$
$$-1 + 35 \stackrel{?}{=} 36$$
$$34 = 36 \quad \text{False}$$

The ordered pair $(-1, 7)$ is not a solution of the system, because it does not make *both* equations true.

NOW TRY

OBJECTIVE 2 Solve linear systems by graphing.

One way to find the solution set of a linear system of equations is to graph each equation and find the point where the graphs intersect.

NOW TRY EXERCISE 2

Solve the system of equations by graphing.

$$3x - 2y = 6$$
$$x - y = 1$$

CLASSROOM EXAMPLE 2

Solve the system of equations by graphing.

$$2x + y = -5$$
$$-x + 3y = 6$$

Answer: $\{(-3, 1)\}$

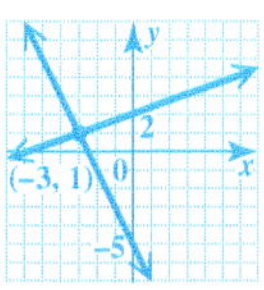

EXAMPLE 2 Solving a System by Graphing

Solve the system of equations by graphing.

$$x + y = 5 \quad (1)$$
$$2x - y = 4 \quad (2)$$

To graph these linear equations, we plot several points for each line.

$x + y = 5$

x	y
0	5
5	0
2	3

The intercepts are a convenient choice.

$2x - y = 4$

x	y
0	−4
2	0
4	4

Find a third ordered pair as a check.

As shown in **FIGURE 2**, the graph suggests that the point of intersection is the ordered pair $(3, 2)$.

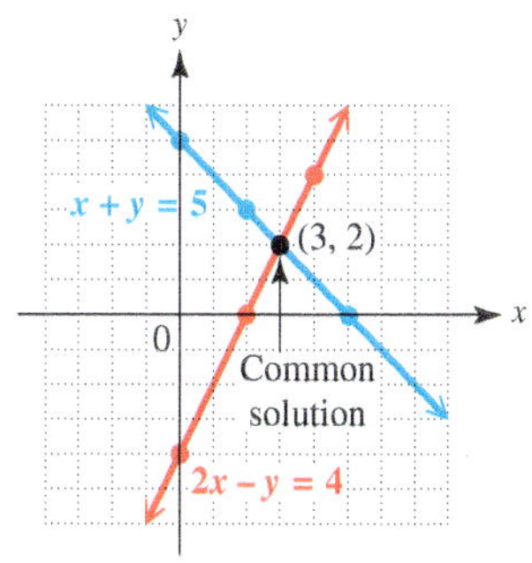

FIGURE 2

TEACHING TIP

Review how to graph lines by determining x- and y-intercepts (where possible). Students *must* use a straightedge to solve systems of equations graphically.

To confirm that $(3, 2)$ is a solution of *both* equations, we check by substituting 3 for x and 2 for y in each equation.

CHECK

$$x + y = 5 \quad (1)$$
$$3 + 2 \stackrel{?}{=} 5$$
$$5 = 5 \quad \checkmark \text{ True}$$

$$2x - y = 4 \quad (2)$$
$$2(3) - 2 \stackrel{?}{=} 4$$
$$6 - 2 \stackrel{?}{=} 4$$
$$4 = 4 \quad \checkmark \text{ True}$$

Because $(3, 2)$ makes both equations true, $\{(3, 2)\}$ is the solution set of the system.

NOW TRY

NOW TRY ANSWERS

1. **(a)** not a solution **(b)** solution
2. $\{(4, 3)\}$

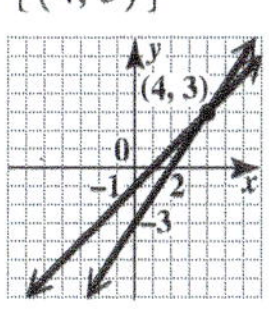

There are three possibilities for the number of elements in the solution set of a linear system in two variables.

Graphs of Linear Systems in Two Variables

Case 1 **The two graphs intersect in a single point.** The coordinates of this point give the only solution of the system. Because the system has a solution, it is **consistent.** The equations are *not* equivalent, so they are **independent.** See FIGURE 3(a).

Case 2 **The graphs are parallel lines.** There is no solution common to both equations, so the solution set is $\emptyset$ and the system is **inconsistent.** Because the equations are *not* equivalent, they are **independent.** See FIGURE 3(b).

Case 3 **The graphs are the same line—that is, they coincide.** Because any solution of one equation of the system is a solution of the other, the solution set is an infinite set of ordered pairs representing the points on the line. This type of system is **consistent** because there is a solution. The equations are equivalent, so they are **dependent.** See FIGURE 3(c).

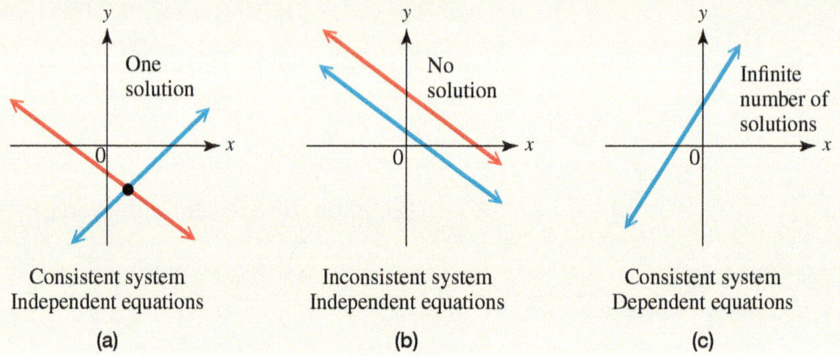

FIGURE 3

OBJECTIVE 3 Solve linear systems (with two equations and two variables) by substitution.

It can be difficult to read exact coordinates, especially if they are not integers, from a graph, so we usually use algebraic methods to solve systems. One such method, the **substitution method,** is well suited for solving linear systems in which one equation is solved or can be easily solved for one variable in terms of the other.

CLASSROOM EXAMPLE 3
Solve the system.

$$5x - 3y = -6$$
$$x = 2 - y$$

Answer: $\{(0, 2)\}$

EXAMPLE 3 **Solving a System by Substitution**

Solve the system.

$$2x - y = 6 \quad (1)$$
$$x = y + 2 \quad (2)$$

Because equation (2) is solved for x, we substitute $y + 2$ for x in equation (1).

$$2x - y = 6 \quad (1)$$
$$2(y + 2) - y = 6 \quad \text{Let } x = y + 2.$$
$$2y + 4 - y = 6 \quad \text{Distributive property}$$
$$y + 4 = 6 \quad \text{Combine like terms.}$$
$$y = 2 \quad \text{Subtract 4.}$$

Be sure to use parentheses here.

NOW TRY EXERCISE 3

Solve the system.

$$x = 3 + 2y$$
$$4x - 3y = 32$$

TEACHING TIP

Emphasize that the solution set found by the substitution method will be the same as the solution set found by graphing. Have students graph the system in Example 3 to see this.

We found y. Now we solve for x by substituting 2 for y in equation (2).

$$x = y + 2 \quad (2)$$
$$x = 2 + 2 \quad \text{Let } y = 2.$$
$$x = 4 \quad \text{Add.}$$

Write the x-value first in the ordered pair.

Thus, $x = 4$ and $y = 2$, giving the ordered pair $(4, 2)$. Check this ordered pair in both equations of the original system.

CHECK

$2x - y = 6$ (1)	$x = y + 2$ (2)
$2(4) - 2 \stackrel{?}{=} 6$	$4 \stackrel{?}{=} 2 + 2$
$8 - 2 \stackrel{?}{=} 6$	$4 = 4$ ✓ True
$6 = 6$ ✓ True	

Because $(4, 2)$ makes both equations true, the solution set is $\{(4, 2)\}$. NOW TRY

Solving a Linear System by Substitution

Step 1 **Solve one equation for either variable.** If one of the equations has a variable term with coefficient 1 or -1, choose it because the substitution method is usually easier.

Step 2 **Substitute** for that variable in the other equation. The result should be an equation with just one variable.

Step 3 **Solve** the equation from Step 2.

Step 4 **Find the other value.** Substitute the result from Step 3 into the equation from Step 1 and solve for the other variable.

Step 5 **Check** the values in *both* of the *original* equations. Then write the solution set as a set containing an ordered pair.

CLASSROOM EXAMPLE 4

Solve the system.

$$4x + y = 5$$
$$2x - 3y = 13$$

Answer: $\{(2, -3)\}$

EXAMPLE 4 Solving a System by Substitution

Solve the system.

$$3x + 2y = 13 \quad (1)$$
$$4x - y = -1 \quad (2)$$

Step 1 First solve one of the equations for either x or y. Because the coefficient of y in equation (2) is -1, it is easiest to solve for y in equation (2).

$$4x - y = -1 \quad (2)$$
$$-y = -1 - 4x \quad \text{Subtract } 4x.$$
$$y = 1 + 4x \quad \text{Multiply by } -1.$$

Step 2 Substitute $1 + 4x$ for y in equation (1).

$$3x + 2y = 13 \quad (1)$$

Substitute in the *other* equation.

$$3x + 2(1 + 4x) = 13 \quad \text{Let } y = 1 + 4x.$$

Step 3 Solve for x.

$$3x + 2 + 8x = 13 \quad \text{Distributive property}$$
$$11x = 11 \quad \text{Combine like terms. Subtract 2.}$$
$$x = 1 \quad \text{Divide by 11.}$$

NOW TRY ANSWER

3. $\{(11, 4)\}$

NOW TRY EXERCISE 4

Solve the system.

$$5x + y = 7$$
$$3x - 2y = 25$$

Step 4 Now find y. From Step 1, $y = 1 + 4x$, so substitute 1 for x.

$$y = 1 + 4(1) \quad \text{Let } x = 1.$$
$$y = 5 \quad \text{Multiply. Add.}$$

Step 5 Check the ordered pair $(1, 5)$ in both equations (1) and (2).

CHECK

$3x + 2y = 13$ (1)	$4x - y = -1$ (2)
$3(1) + 2(5) \stackrel{?}{=} 13$	$4(1) - 5 \stackrel{?}{=} -1$
$3 + 10 \stackrel{?}{=} 13$	$4 - 5 \stackrel{?}{=} -1$
$13 = 13$ ✓ True	$-1 = -1$ ✓ True

The solution set is $\{(1, 5)\}$.

NOW TRY

CLASSROOM EXAMPLE 5

Solve the system.

$$-2x + 5y = 22$$
$$\frac{1}{2}x + \frac{1}{4}y = \frac{1}{2}$$

Answer: $\{(-1, 4)\}$

EXAMPLE 5 Solving a System with Fractional Coefficients

Solve the system.

$$\frac{2}{3}x - \frac{1}{2}y = \frac{7}{6} \quad (1)$$
$$3x - y = 6 \quad (2)$$

This system will be easier to solve if we clear the fractions in equation (1).

$$6\left(\frac{2}{3}x - \frac{1}{2}y\right) = 6\left(\frac{7}{6}\right) \quad \text{Multiply (1) by the LCD, 6.}$$

Remember to multiply *each* term by 6.

$$6 \cdot \frac{2}{3}x - 6 \cdot \frac{1}{2}y = 6 \cdot \frac{7}{6} \quad \text{Distributive property}$$
$$4x - 3y = 7 \quad (3)$$

Now the system consists of equations (2) and (3).

$$3x - y = 6 \quad (2)$$
$$4x - 3y = 7 \quad (3)$$

This equation is equivalent to equation (1).

To use the substitution method, we solve equation (2) for y.

$$3x - y = 6 \quad (2)$$
$$-y = 6 - 3x \quad \text{Subtract } 3x.$$
$$y = 3x - 6 \quad \text{Multiply by } -1. \text{ Rewrite.}$$

Substitute $3x - 6$ for y in equation (3).

$$4x - 3y = 7 \quad (3)$$
$$4x - 3(3x - 6) = 7 \quad \text{Let } y = 3x - 6.$$
$$4x - 9x + 18 = 7 \quad \text{Distributive property}$$

Be careful with signs.

$$-5x + 18 = 7 \quad \text{Combine like terms.}$$
$$-5x = -11 \quad \text{Subtract 18.}$$
$$x = \frac{11}{5} \quad \text{Divide by } -5.$$

NOW TRY ANSWER

4. $\{(3, -8)\}$

NOW TRY EXERCISE 5
Solve the system.

$$\frac{1}{10}x - \frac{3}{5}y = \frac{2}{5}$$
$$-2x + 3y = 1$$

Now find y using $y = 3x - 6$ (equation (2) solved for y).

$$y = 3\left(\frac{11}{5}\right) - 6 \qquad \text{Let } x = \tfrac{11}{5}.$$

$$y = \frac{33}{5} - \frac{30}{5} \qquad 6 = \tfrac{30}{5}$$

$$y = \frac{3}{5} \qquad \text{Subtract fractions.}$$

A check verifies that the solution set is $\left\{\left(\frac{11}{5}, \frac{3}{5}\right)\right\}$. NOW TRY

NOTE If an equation in a system contains decimal coefficients, it is best to first clear the decimals by multiplying by 10, 100, or 1000, depending on the number of decimal places. Then solve the system. For example, we multiply *each side* of the equation

$$0.5x + 0.75y = 3.25$$

by 10^2, or 100, to obtain an equivalent equation

$$50x + 75y = 325.$$

OBJECTIVE 4 Solve linear systems (with two equations and two variables) by elimination.

Another algebraic method, the **elimination method,** involves combining the two equations in a system so that one variable is eliminated. This is done using the following logic.

If $a = b$ and $c = d$, then $a + c = b + d$.

NOW TRY EXERCISE 6
Solve the system.

$$8x - 2y = 5$$
$$5x + 2y = -18$$

CLASSROOM EXAMPLE 6
Solve the system.

$$-2x + 3y = -10$$
$$2x + 2y = 5$$

Answer: $\left\{\left(\frac{7}{2}, -1\right)\right\}$

TEACHING TIP
Point out that if we were to try to solve the system of **Example 6** by the substitution method, fractions would be involved and make the work more difficult.

NOW TRY ANSWERS
5. $\{(-2, -1)\}$
6. $\left\{\left(-1, -\frac{13}{2}\right)\right\}$

EXAMPLE 6 Solving a System by Elimination

Solve the system.

$$2x + 3y = -6 \qquad (1)$$
$$4x - 3y = 6 \qquad (2)$$

Notice that adding the equations together will eliminate the variable y.

$$2x + 3y = -6 \qquad (1)$$
$$\underline{4x - 3y = 6} \qquad (2)$$
$$6x \phantom{{}-3y} = 0 \qquad \text{Add.}$$
$$x = 0 \qquad \text{Solve for } x.$$

To find y, substitute 0 for x in either equation (1) or equation (2).

$$2x + 3y = -6 \qquad (1)$$
$$2(0) + 3y = -6 \qquad \text{Let } x = 0.$$
$$0 + 3y = -6 \qquad \text{Multiply.}$$
$$y = -2 \qquad \text{Add. Divide by 3.}$$

Check by substituting 0 for x and -2 for y in both equations of the original system to verify that $(0, -2)$ satisfies both equations. It does, so the solution set is $\{(0, -2)\}$.

NOW TRY

From **Example 6:**

$$\begin{aligned} 2x + 3y &= -6 \quad (1) \\ 4x - 3y &= 6 \quad (2) \\ \hline 6x \quad &= 0 \end{aligned}$$

By adding the equations in **Example 6** (repeated in the margin), we eliminated the variable y because the coefficients of the y-terms were opposites. In many cases the coefficients will *not* be opposites, and we must transform one or both equations so that the coefficients of one pair of variable terms are opposites.

Solving a Linear System by Elimination

Step 1 **Write both equations in standard form $Ax + By = C$.**

Step 2 **Transform the equations as needed so that the coefficients of one pair of variable terms are opposites.** Multiply one or both equations by appropriate numbers so that the sum of the coefficients of either the x- or y-terms is 0.

Step 3 **Add** the new equations to eliminate a variable. The sum should be an equation with just one variable.

Step 4 **Solve** the equation from Step 3 for the remaining variable.

Step 5 **Find the other value.** Substitute the result from Step 4 into either of the original equations and solve for the other variable.

Step 6 **Check** the values in *both* of the *original* equations. Then write the solution set as a set containing an ordered pair.

CLASSROOM EXAMPLE 7
Solve the system.

$$2x + 3y = 19$$
$$3x - 7y = -6$$

Answer: $\{(5, 3)\}$

TEACHING TIP
Discuss how to decide whether to use the substitution method or the elimination method to solve a given system of equations. Work **Exercise 44** in the exercise set as a class.

EXAMPLE 7 Solving a System by Elimination

Solve the system.

$$5x - 2y = 4 \quad (1)$$
$$2x + 3y = 13 \quad (2)$$

Step 1 Both equations are in standard form.

Step 2 Suppose that we wish to eliminate the variable x. One way to do this is to multiply equation (1) by 2 and equation (2) by -5.

The goal is to have *opposite* coefficients.

$$10x - 4y = 8 \quad \text{2 times each side of equation (1)}$$
$$-10x - 15y = -65 \quad -5 \text{ times each side of equation (2)}$$

Step 3 Now add.

$$\begin{aligned} 10x - 4y &= 8 \\ -10x - 15y &= -65 \\ \hline -19y &= -57 \quad \text{Add.} \end{aligned}$$

Step 4 Solve for y. $\quad y = 3 \quad$ Divide by -19.

Step 5 To find x, substitute 3 for y in either equation (1) or equation (2).

$$\begin{aligned} 2x + 3y &= 13 && (2) \\ 2x + 3(3) &= 13 && \text{Let } y = 3. \\ 2x + 9 &= 13 && \text{Multiply.} \\ 2x &= 4 && \text{Subtract 9.} \\ x &= 2 && \text{Divide by 2.} \end{aligned}$$

NOW TRY EXERCISE 7
Solve the system.

$$2x - 5y = 4$$
$$5x + 2y = 10$$

Step 6 To check, substitute 2 for x and 3 for y in both equations (1) and (2).

CHECK

$$5x - 2y = 4 \quad (1)$$
$$5(2) - 2(3) \stackrel{?}{=} 4$$
$$10 - 6 \stackrel{?}{=} 4$$
$$4 = 4 \ ✓ \ \text{True}$$

$$2x + 3y = 13 \quad (2)$$
$$2(2) + 3(3) \stackrel{?}{=} 13$$
$$4 + 9 \stackrel{?}{=} 13$$
$$13 = 13 \ ✓ \ \text{True}$$

The solution set is $\{(2, 3)\}$. NOW TRY

OBJECTIVE 5 Solve special systems.

As we saw in **FIGURES 3(b) AND (c)**, some systems of linear equations have no solution or an infinite number of solutions.

NOW TRY EXERCISE 8
Solve the system.

$$x - 3y = 7$$
$$-3x + 9y = -21$$

EXAMPLE 8 Solving a System of Dependent Equations

Solve the system.

$$2x - y = 3 \quad (1)$$
$$6x - 3y = 9 \quad (2)$$

We multiply equation (1) by -3 and then add the result to equation (2).

$$-6x + 3y = -9 \quad -3 \text{ times each side of equation (1)}$$
$$6x - 3y = 9 \quad (2)$$
$$0 = 0 \quad \text{True}$$

CLASSROOM EXAMPLE 8
Solve the system.

$$2x + y = 6$$
$$-8x - 4y = -24$$

Answer: $\{(x, y) \mid 2x + y = 6\}$

Adding these equations gives the true statement $0 = 0$. In the original system, we could obtain equation (2) from equation (1) by multiplying equation (1) by 3. Because of this, equations (1) and (2) are equivalent and have the same graph, as shown in **FIGURE 4**. The equations are dependent.

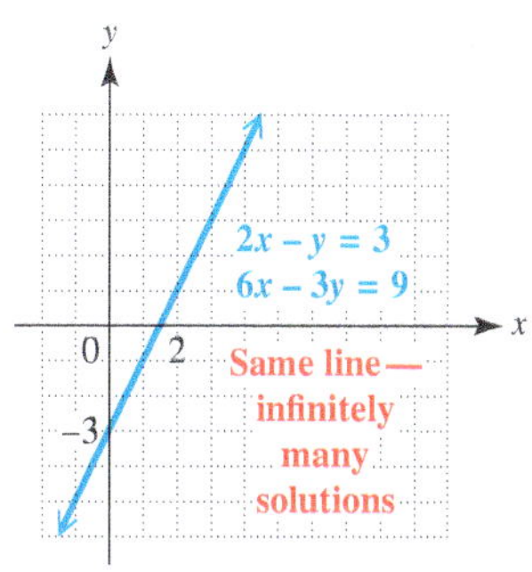

FIGURE 4

TEACHING TIP
Emphasize that the solution set of a system of dependent equations is written in *set-builder notation*, reviewed in Chapter R.

The solution set is the set of all points on the line with equation $2x - y = 3$, written in set-builder notation (**Section R.1**) as

$$\{(x, y) \mid 2x - y = 3\}$$

and read "the set of all ordered pairs (x, y), such that $2x - y = 3$." NOW TRY

NOTE When a system has dependent equations and an infinite number of solutions, as in **Example 8**, either equation of the system could be used to write the solution set. ***In this book, we use the equation in standard form with coefficients that are integers having greatest common factor 1 and positive coefficient of x.***

NOW TRY ANSWERS
7. $\{(2, 0)\}$
8. $\{(x, y) \mid x - 3y = 7\}$

NOW TRY EXERCISE 9
Solve the system.

$$-2x + 5y = 6$$
$$6x - 15y = 4$$

CLASSROOM EXAMPLE 9
Solve the system.

$$4x - 3y = 8$$
$$8x - 6y = 14$$

Answer: $\varnothing$

EXAMPLE 9 Solving an Inconsistent System

Solve the system.

$$x + 3y = 4 \quad (1)$$
$$-2x - 6y = 3 \quad (2)$$

Multiply equation (1) by 2, and then add the result to equation (2).

$$2x + 6y = 8 \quad \text{Equation (1) multiplied by 2}$$
$$-2x - 6y = 3 \quad (2)$$
$$0 = 11 \quad \text{False}$$

The result of the addition step is a false statement, which indicates that the system is inconsistent. As shown in **FIGURE 5**, the graphs of the equations of the system are parallel lines.

There are no ordered pairs that satisfy both equations, so there is no solution for the system. The solution set is $\varnothing$.

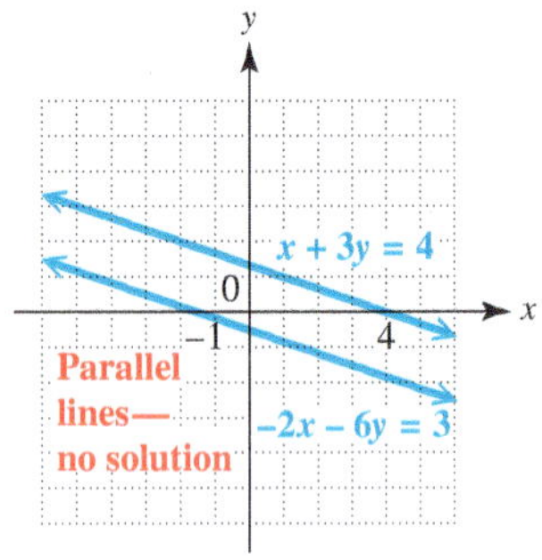

FIGURE 5

NOW TRY

Special Cases of Linear Systems

If both variables are eliminated when a system of linear equations is solved, then the solution sets are determined as follows.

Case 1 There are infinitely many solutions if the resulting statement is *true*. (See **Example 8.**)

Case 2 There is no solution if the resulting statement is *false*. (See **Example 9.**)

Slopes and y-intercepts can be used to decide whether the graphs of a system of equations are parallel lines or whether they coincide.

CLASSROOM EXAMPLE 10
Write each equation in slope-intercept form and then tell how many solutions the system has.

(a) $3x - 6y = 9$
$x - 2y = 3$

(b) $-2x = 5y + 1$
$-4x = 10y + 3$

Answers:
(a) Both are $y = \frac{1}{2}x - \frac{3}{2}$; infinitely many solutions
(b) $y = -\frac{2}{5}x - \frac{1}{5}$; $y = -\frac{2}{5}x - \frac{3}{10}$; no solution

EXAMPLE 10 Using Slope-Intercept Form to Determine the Number of Solutions

Refer to **Examples 8 and 9.** Write each equation in slope-intercept form and then tell how many solutions the system has.

Solve each equation from **Example 8** for y.

$$2x - y = 3 \quad (1)$$
$$-y = -2x + 3 \quad \text{Subtract } 2x.$$
$$y = 2x - 3 \quad \text{Multiply by } -1.$$

Slope (2); y-intercept $(0, -3)$

$$6x - 3y = 9 \quad (2)$$
$$2x - y = 3 \quad \text{Divide by 3.}$$

Solving for y leads to the same result as on the left.

The lines have the same slope and same y-intercept, meaning that they coincide. There are infinitely many solutions.

NOW TRY ANSWER
9. $\varnothing$

NOW TRY EXERCISE 10

Write each equation in slope-intercept form and then tell how many solutions the system has.

(a) $2x - 3y = 3$
$4x - 6y = -6$

(b) $5y = -x - 4$
$-10y = 2x + 8$

Solve each equation from **Example 9** for y.

$x + 3y = 4$	(1)	$-2x - 6y = 3$	(2)
$3y = -x + 4$	Subtract x.	$-6y = 2x + 3$	Add $2x$.
$y = -\frac{1}{3}x + \frac{4}{3}$	Divide by 3.	$y = -\frac{1}{3}x - \frac{1}{2}$	Divide by -6; $\frac{2}{-6} = -\frac{1}{3}$; $\frac{3}{-6} = -\frac{1}{2}$
Slope: $-\frac{1}{3}$; y-intercept $\left(0, \frac{4}{3}\right)$		Slope: $-\frac{1}{3}$; y-intercept $\left(0, -\frac{1}{2}\right)$	

The lines have the same slope, but different y-intercepts, indicating that they are parallel. Thus, the system has no solution.

NOW TRY

NOTE In **Example 2,** we solved the system

$$x + y = 5 \quad (1)$$
$$2x - y = 4 \quad (2)$$

by graphing the two lines and finding their point of intersection. We can also do this with a graphing calculator, as shown in **FIGURE 6**. The two lines were graphed by solving each equation for y.

$y = 5 - x$ Equation (1) solved for y

$y = 2x - 4$ Equation (2) solved for y

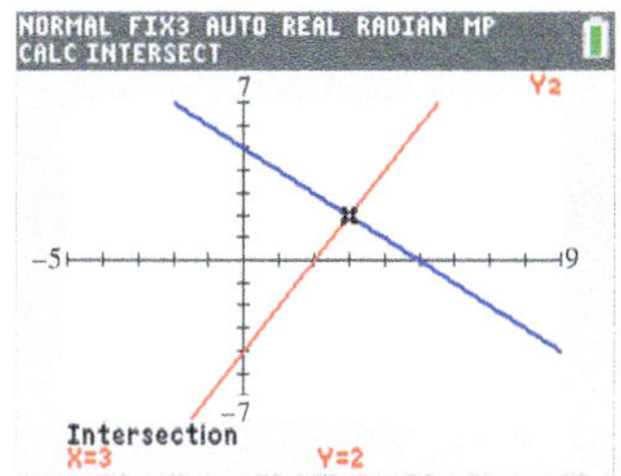

FIGURE 6

The coordinates of their point of intersection are displayed at the bottom of the screen, indicating that the solution set is $\{(3, 2)\}$. (Compare this graph with the one in **FIGURE 2**.)

NOW TRY ANSWERS

10. (a) $y = \frac{2}{3}x - 1$; $y = \frac{2}{3}x + 1$; no solution

(b) Both are $y = -\frac{1}{5}x - \frac{4}{5}$; infinitely many solutions

3.1 Exercises

FOR EXTRA HELP

MyMathLab®

Complete solution available in MyMathLab

1. 4; −3 **2.** pair
3. ∅ **4.** dependent
5. no **6.** one

Concept Check *Complete each statement.*

1. If $(4, -3)$ is a solution of a linear system in two variables, then substituting ________ for x and ________ for y leads to true statements in *both* equations.

2. A solution of a system of independent linear equations in two variables is an ordered ________.

3. If solving a system leads to a false statement such as $0 = 3$, the solution set is ________.

4. If solving a system leads to a true statement such as $0 = 0$, the system has ________ equations.

5. If the two lines forming a system have the same slope and different y-intercepts, the system has (*no* / *one* / *infinitely many*) solution(s).

6. If the two lines forming a system have different slopes, the system has (*no* / *one* / *infinitely many*) solution(s).

7. D; The ordered-pair solution must be in quadrant IV because that is where the graphs of the equations intersect.
8. B; The ordered-pair solution must be on the x-axis, with $x < 0$, because that is where the graphs of the equations intersect.
9. **(a)** B **(b)** C **(c)** A **(d)** D
10. $(0, 0)$

11. solution **12.** solution
13. not a solution
14. not a solution

15. $\{(-2, -3)\}$ **16.** $\{(2, 2)\}$

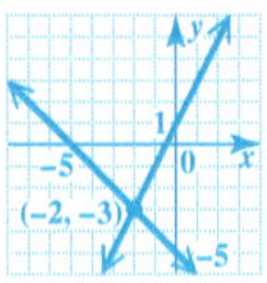

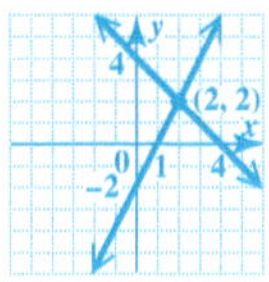

17. $\{(0, 1)\}$ **18.** $\{(0, -2)\}$

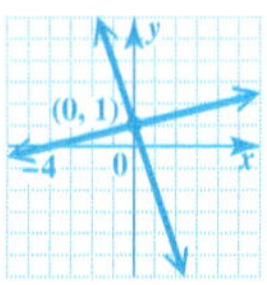

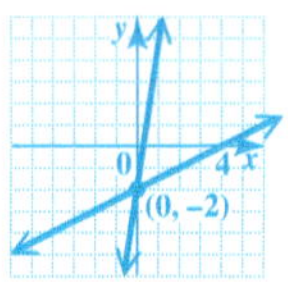

19. $\{(2, 0)\}$ **20.** $\{(-3, 0)\}$

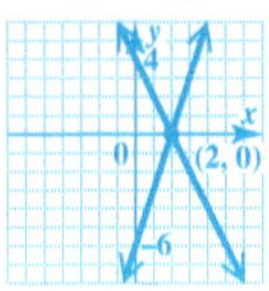

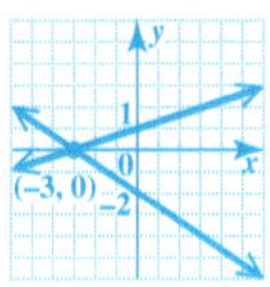

21. $\{(1, 2)\}$ **22.** $\{(-2, -10)\}$
23. $\{(2, 3)\}$ **24.** $\{(-1, 4)\}$
25. $\left\{\left(\frac{22}{9}, \frac{22}{3}\right)\right\}$
26. $\left\{\left(\frac{10}{7}, -\frac{5}{7}\right)\right\}$
27. $\{(5, 4)\}$ **28.** $\{(-1, 3)\}$
29. $\{(1, 3)\}$

7. ***Concept Check*** Which ordered pair could be a solution of the graphed system of equations? Why?

A. $(4, 4)$
B. $(-4, 4)$
C. $(-4, -4)$
D. $(4, -4)$

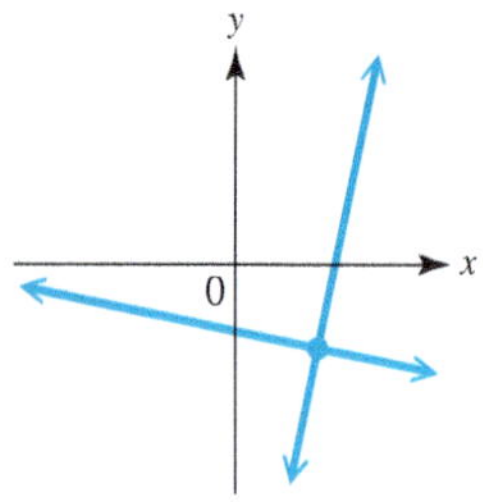

8. ***Concept Check*** Which ordered pair could be a solution of the graphed system of equations? Why?

A. $(4, 0)$
B. $(-4, 0)$
C. $(0, 4)$
D. $(0, -4)$

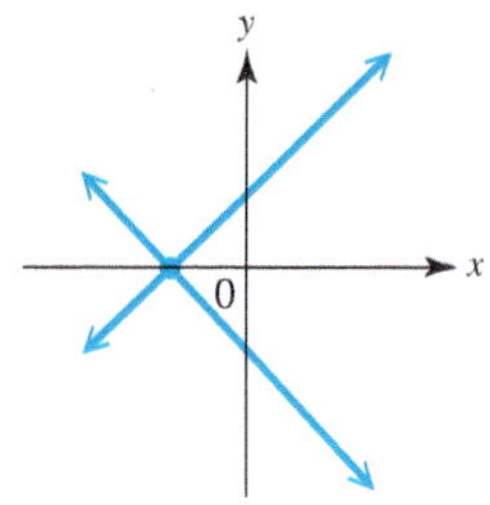

9. ***Concept Check*** Match each system with the correct graph.

(a) $x + y = 6$
$x - y = 0$

(b) $x + y = -6$
$x - y = 0$

(c) $x + y = 0$
$x - y = -6$

(d) $x + y = 0$
$x - y = 6$

A.

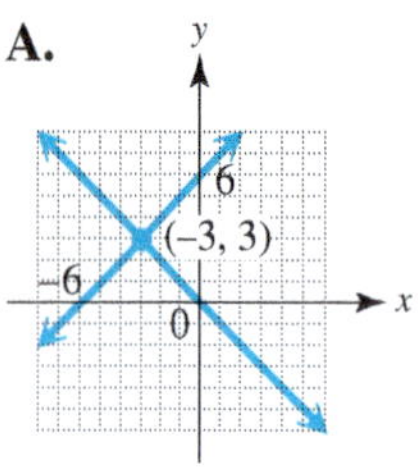

B.

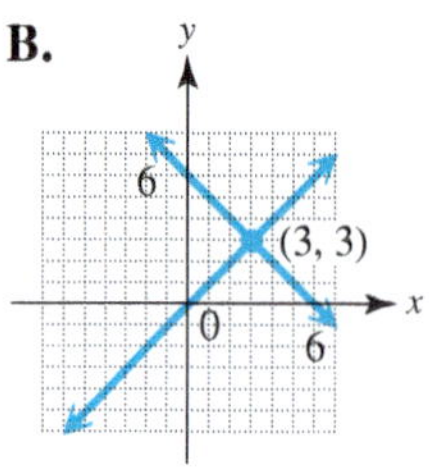

C.

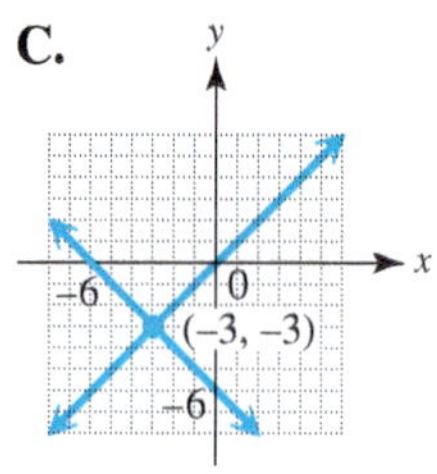

D.

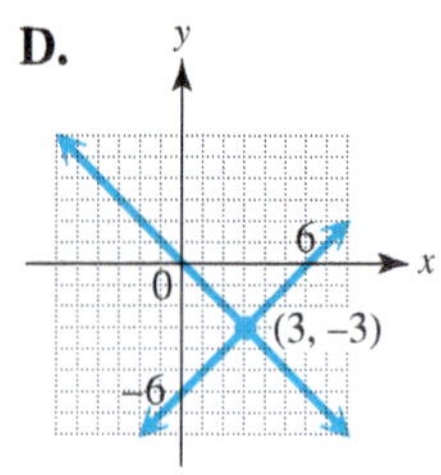

10. ***Concept Check*** If a system of the following form has a single solution, what must that solution be?

$$Ax + By = 0$$
$$Cx + Dy = 0$$

Decide whether the given ordered pair is a solution of the given system. ***See Example 1.***

11. $x - y = 17$; $(8, -9)$
$x + y = -1$

12. $x + y = 6$; $(5, 1)$
$x - y = 4$

13. $3x - 5y = -12$; $(-1, 2)$
$x - y = 1$

14. $2x - y = 8$; $(5, 2)$
$3x + 2y = 20$

Solve each system by graphing. ***See Example 2.***

15. $x + y = -5$
$-2x + y = 1$

16. $x + y = 4$
$2x - y = 2$

17. $x - 4y = -4$
$3x + y = 1$

18. $6x - y = 2$
$x - 2y = 4$

19. $2x + y = 4$
$3x - y = 6$

20. $2x + 3y = -6$
$x - 3y = -3$

Solve each system by substitution. If the system is inconsistent or has dependent equations, say so. ***See Examples 3–5, 8, and 9.***

21. $4x + y = 6$
$y = 2x$

22. $2x - y = 6$
$y = 5x$

23. $-x - 4y = -14$
$y = 2x - 1$

24. $-3x - 5y = -17$
$y = 4x + 8$

25. $3x - 4y = -22$
$-3x + y = 0$

26. $-3x + y = -5$
$3x + 6y = 0$

27. $5x - 4y = 9$
$3 - 2y = -x$

28. $6x - y = -9$
$4 + 7x = -y$

29. $4x - 5y = -11$
$x + 2y = 7$

30. $\{(3, -1)\}$
31. $\left\{\left(-5, -\frac{10}{3}\right)\right\}$
32. $\left\{\left(\frac{2}{7}, \frac{8}{21}\right)\right\}$
33. $\{(2, 6)\}$
34. $\{(-12, -60)\}$
35. $\{(x, y) \mid 2x - y = 0\}$; dependent equations
36. $\{(x, y) \mid x - 3y = 0\}$; dependent equations
37. $\emptyset$; inconsistent system
38. $\emptyset$; inconsistent system
39. $\{(4, 2)\}$ 40. $\{(10, 14)\}$
41. $\{(-8, 4)\}$ 42. $\{(6, -10)\}$
43. $\{(0, 0)\}$
44. (a) Use elimination because the coefficients of the y-terms are opposites. (b) Use substitution because the second equation is solved for y. (c) Use elimination because the equations are in standard form with no coefficients of 1 or -1. Solving by substitution would involve fractions.
45. $\{(2, -4)\}$ 46. $\{(-2, 1)\}$
47. $\{(3, -1)\}$ 48. $\{(1, 1)\}$
49. $\{(2, -3)\}$ 50. $\{(4, -5)\}$
51. $\{(x, y) \mid 7x + 2y = 6\}$; dependent equations
52. $\{(x, y) \mid x - 4y = 2\}$; dependent equations
53. $\left\{\left(\frac{3}{2}, -\frac{3}{2}\right)\right\}$
54. $\left\{\left(\frac{1}{4}, -\frac{1}{2}\right)\right\}$
55. $\emptyset$; inconsistent system
56. $\emptyset$; inconsistent system
57. $\{(0, 0)\}$ 58. $\{(0, 0)\}$
59. $\{(0, -4)\}$ 60. $\{(0, 3)\}$
61. $\left\{\left(6, -\frac{5}{6}\right)\right\}$
62. $\left\{\left(13, -\frac{7}{5}\right)\right\}$

30. $3x - y = 10$, $2x + 5y = 1$

31. $x = 3y + 5$, $x = \frac{3}{2}y$

32. $x = 6y - 2$, $x = \frac{3}{4}y$

33. $\frac{1}{2}x + \frac{1}{3}y = 3$, $-3x + y = 0$

34. $\frac{1}{4}x - \frac{1}{5}y = 9$, $5x - y = 0$

35. $y = 2x$, $4x - 2y = 0$

36. $x = 3y$, $3x - 9y = 0$

37. $x = 5y$, $5x - 25y = 5$

38. $y = -4x$, $8x + 2y = 4$

39. $y = 0.5x$, $1.5x - 0.5y = 5.0$

40. $y = 1.4x$, $0.5x + 1.5y = 26.0$

41. $\frac{1}{2}x - \frac{1}{4}y = -5$, $\frac{1}{8}x + \frac{1}{4}y = 0$

42. $-\frac{1}{3}x + \frac{2}{5}y = -6$, $-\frac{1}{2}x - \frac{3}{2}y = 12$

43. $0.4x + 0.2y = 0$, $\frac{1}{2}x - \frac{1}{3}y = 0$

44. To minimize the amount of work required, tell whether to use the substitution or elimination method to solve each system. Explain. *Do not actually solve.*

(a) $3x + y = -7$, $x - y = -5$ **(b)** $6x - y = 5$, $y = 11x$ **(c)** $3x - 2y = 0$, $9x + 8y = 7$

Solve each system by elimination. If the system is inconsistent or has dependent equations, say so. ***See Examples 6–9.***

45. $-2x + 3y = -16$, $2x - 5y = 24$

46. $6x + 5y = -7$, $-6x - 11y = 1$

47. $2x - 5y = 11$, $3x + y = 8$

48. $-2x + 3y = 1$, $-4x + y = -3$

49. $3x + 4y = -6$, $5x + 3y = 1$

50. $4x + 3y = 1$, $3x + 2y = 2$

51. $7x + 2y = 6$, $-14x - 4y = -12$

52. $x - 4y = 2$, $4x - 16y = 8$

53. $3x + 3y = 0$, $4x + 2y = 3$

54. $8x + 4y = 0$, $4x - 2y = 2$

55. $5x - 5y = 3$, $x - y = 12$

56. $2x - 3y = 7$, $-4x + 6y = 14$

57. $x + y = 0$, $2x - 2y = 0$

58. $3x + 3y = 0$, $-2x - y = 0$

59. $x - \frac{1}{2}y = 2$, $-x + \frac{2}{5}y = -\frac{8}{5}$

60. $\frac{3}{2}x + y = 3$, $\frac{2}{3}x + \frac{1}{3}y = 1$

61. $\frac{1}{2}x + \frac{1}{3}y = \frac{49}{18}$, $\frac{1}{2}x + 2y = \frac{4}{3}$

62. $\frac{1}{5}x + \frac{1}{7}y = \frac{12}{5}$, $\frac{1}{10}x + \frac{1}{3}y = \frac{5}{6}$

Write each equation in slope-intercept form and then tell how many solutions the system has. Do not actually solve. ***See Example 10.***

63. $3x + 7y = 4$
$6x + 14y = 3$

64. $-x + 2y = 8$
$4x - 8y = 1$

65. $2x = -3y + 1$
$6x = -9y + 3$

66. $5x = -2y + 1$
$10x = -4y + 2$

Solve each system by the method of your choice. ***See Examples 3–9.*** *(For Exercises 67–69, see the answers to* ***Exercise 44.****)*

67. $3x + y = -7$
$x - y = -5$

68. $6x - y = 5$
$y = 11x$

69. $3x - 2y = 0$
$9x + 8y = 7$

70. $3x - 5y = 7$
$2x + 3y = 30$

71. $2x + 3y = 10$
$-3x + y = 18$

72. $x + y = 10$
$2x - y = 5$

73. $\frac{1}{2}x - \frac{1}{8}y = -\frac{1}{4}$
$4x - y = -2$

74. $\frac{1}{6}x + \frac{1}{3}y = 8$
$\frac{1}{4}x + \frac{1}{2}y = 12$

75. $0.3x + 0.2y = 0.4$
$0.5x + 0.4y = 0.7$

76. $0.2x + 0.5y = 6$
$0.4x + y = 9$

77. ***Concept Check*** Explain why *System A* would be more difficult to solve by substitution than *System B*.

System A: $3x + 7y = 19$, $8x - 5y = 4$

System B: $x + 7y = 8$, $3x - 2y = 4$

78. ***Concept Check*** Make up a system of the following form, where a, b, c, d, e, and f are consecutive integers. Solve the system.

$$ax + by = c$$
$$dx + ey = f$$

Use each graph provided to work Exercises 79–82.

79. The figure shows graphs that represent supply and demand for a certain brand of low-fat frozen yogurt at various prices per half-gallon (in dollars).

(a) At what price does supply equal demand?

(b) For how many half-gallons does supply equal demand?

(c) What are the supply and demand at a price of \$2 per half-gallon?

80. Sharon compared the monthly payments she would incur for two types of mortgages: fixed rate and variable rate. Her observations led to the following graphs.

(a) For which years would the monthly payment be more for the fixed-rate mortgage than for the variable-rate mortgage?

(b) In what year would the payments be the same? What would those payments be?

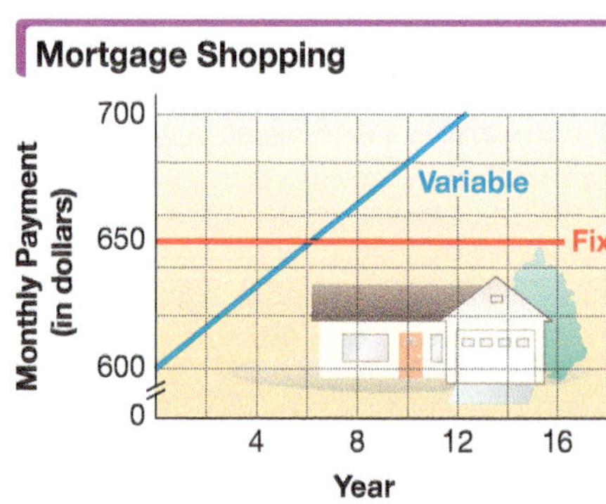

63. $y = -\frac{3}{7}x + \frac{4}{7}$; $y = -\frac{3}{7}x + \frac{3}{14}$; no solution

64. $y = \frac{1}{2}x + 4$; $y = \frac{1}{2}x - \frac{1}{8}$; no solution

65. Both are $y = -\frac{2}{3}x + \frac{1}{3}$; infinitely many solutions

66. Both are $y = -\frac{5}{2}x + \frac{1}{2}$; infinitely many solutions

67. $\{(-3, 2)\}$ **68.** $\{(-1, -11)\}$

69. $\left\{\left(\frac{1}{3}, \frac{1}{2}\right)\right\}$ **70.** $\{(9, 4)\}$

71. $\{(-4, 6)\}$ **72.** $\{(5, 5)\}$

73. $\{(x, y) \mid 4x - y = -2\}$

74. $\{(x, y) \mid x + 2y = 48\}$

75. $\left\{\left(1, \frac{1}{2}\right)\right\}$ **76.** $\emptyset$

77. None of the coefficients in the first system are -1 or 1, and solving by substitution would involve fractions. In the second system, we can solve for x in the first equation to avoid using fractions as coefficients.

78. $\{(-1, 2)\}$

79. **(a)** \$4 **(b)** 300 half-gallons **(c)** supply: 200 half-gallons demand: 400 half-gallons

80. **(a)** years 0 to 6 **(b)** year 6; about \$650

81. (a) 2009–2011
(b) 2011 and 2012
(c) 2010; 13,000
(d) (2010, 13,000)
(e) *The Big Bang Theory* viewership was increasing, *Dancing with the Stars* viewership was decreasing, and *60 Minutes* viewership remained fairly constant.

82. (a) 1977
(b) 25,000
(c) (1977, 25,000)
(d) USSR; 1990; 37,000
(e) The number of nuclear weapons possessed by the USSR decreased during these years; negative

81. The graph shows viewers (in millions) in the United States for three television programs from 2009 through 2012.

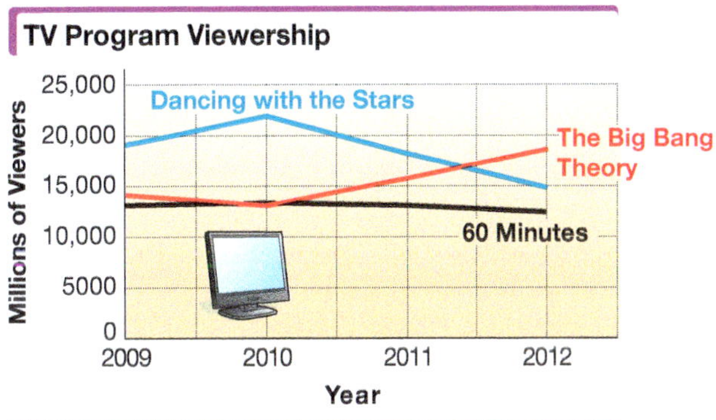

Source: tvbythenumbers.zap2it.com

(a) During what years did the number of viewers for *Dancing with the Stars* exceed that for *The Big Bang Theory?*

(b) Between what two consecutive years did *The Big Bang Theory* first become the most popular program?

(c) In what year were the numbers of viewers of *The Big Bang Theory* and *60 Minutes* approximately equal? What was the approximate number of viewers during that year (to the nearest million)?

(d) Write the answer for part (c) as an ordered pair.

(e) Describe the trends in viewership of the three programs from 2010 through 2012.

82. The graph shows numbers of nuclear weapons possessed by the United States and USSR/Russia from 1950 through 2013.

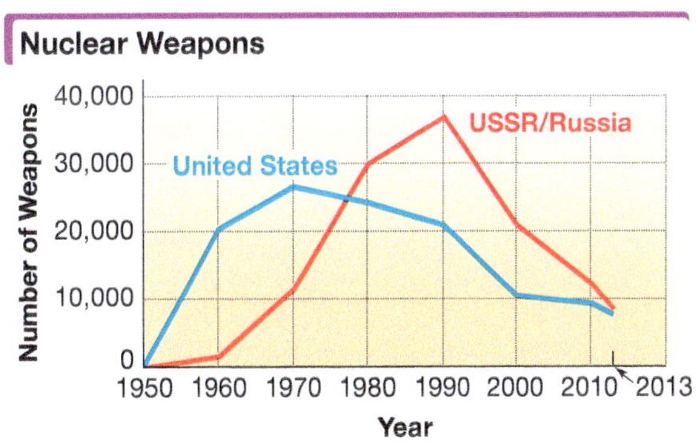

Source: *World Almanac and Book of Facts.*

(a) In what year were the numbers of nuclear weapons approximately equal?

(b) Approximately how many nuclear weapons did each country possess in that year?

(c) Express the point of intersection of the graphs for U.S. and USSR nuclear weapons as an ordered pair of the form (year, number of nuclear weapons).

(d) Over the entire period 1950–2013, which country possessed the greatest number of nuclear weapons at any one particular time? In what year did that maximum occur, and what was the approximate number of nuclear weapons possessed?

(e) Describe the trend in number of nuclear weapons possessed by the USSR from 1990 through 2013. If a straight line were used to approximate its graph, would the line have a slope that is positive, negative, or 0?

83. hiking
84. 2008; 35 million
85. (8.7, 35.9)
86. During year 8 (2008), participation in both sports was about 35.9 million. This is close to the estimate in Exercise 84.

87. $\{(2, 4)\}$ 88. $\{(4, -5)\}$

89. $\left\{\left(\frac{1}{2}, 2\right)\right\}$ 90. $\{(2, 3)\}$

91. $\left\{\left(\frac{c}{a}, 0\right)\right\}$

92. $\left\{\left(\frac{1}{a}, \frac{1}{b}\right)\right\}$

93. $\left\{\left(-\frac{1}{a}, -5\right)\right\}$

94. $\left\{\left(-\frac{3}{5a}, \frac{7}{5}\right)\right\}$

Use the graph shown in **FIGURE 1** *at the beginning of this section (and repeated here) to work Exercises 83–86.*

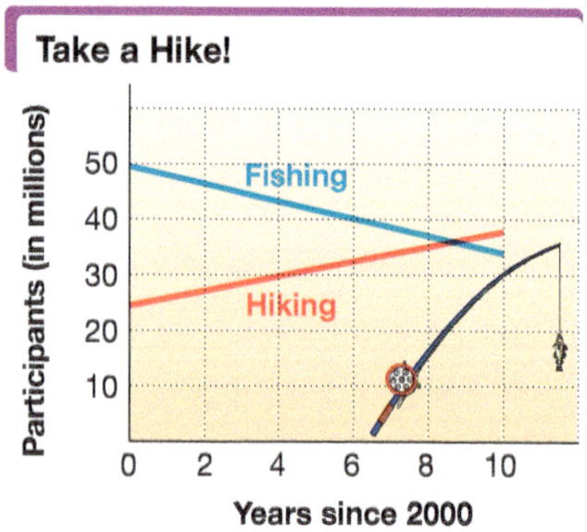

Source: National Sporting Goods Association.

83. Which sport was more popular in 2010?

84. Estimate the year in which participation in the two sports was the same. About how many Americans participated in each sport during that year?

85. If $x = 0$ represents 2000 and $x = 10$ represents 2010, the number of participants y in millions in the two sports can be modeled by the linear equations in the following system.

$$-1.34x + y = 24.3 \quad \text{Hiking}$$
$$1.55x + y = 49.3 \quad \text{Fishing}$$

Solve this system. Express values as decimals rounded to the nearest tenth. Write the solution as an ordered pair of the form (year, number of participants).

86. Interpret the answer for **Exercise 85,** rounding down for the year. How does it compare to the estimate from **Exercise 84?**

Extending Skills *The following systems can be solved by elimination. One way to do this is to let $p = \frac{1}{x}$ and $q = \frac{1}{y}$. Substitute, solve for p and q, and then find x and y.* $\left(\textit{For example, in Exercise 87, } \frac{3}{x} = 3 \cdot \frac{1}{x} = 3p.\right)$

Use this method to solve each system.

87. $\frac{3}{x} + \frac{4}{y} = \frac{5}{2}$
$\frac{5}{x} - \frac{3}{y} = \frac{7}{4}$

88. $\frac{2}{x} - \frac{5}{y} = \frac{3}{2}$
$\frac{4}{x} + \frac{1}{y} = \frac{4}{5}$

89. $\frac{2}{x} + \frac{3}{y} = \frac{11}{2}$
$-\frac{1}{x} + \frac{2}{y} = -1$

90. $\frac{4}{x} - \frac{9}{y} = -1$
$-\frac{7}{x} + \frac{6}{y} = -\frac{3}{2}$

Extending Skills *Solve by any method. Assume that a and b represent nonzero constants.*

91. $ax + by = c$
$ax - 2by = c$

92. $ax + by = 2$
$-ax + 2by = 1$

93. $2ax - y = 3$
$y = 5ax$

94. $3ax + 2y = 1$
$-ax + y = 2$

STUDY SKILLS

Preparing for Your Math Final Exam

Your math final exam is likely to be a comprehensive exam, which means it will cover material from the entire term. **One way to prepare for it now is by working a set of Cumulative Review Exercises** each time your class finishes a chapter. This continual review will help you remember concepts and procedures as you progress through the course.

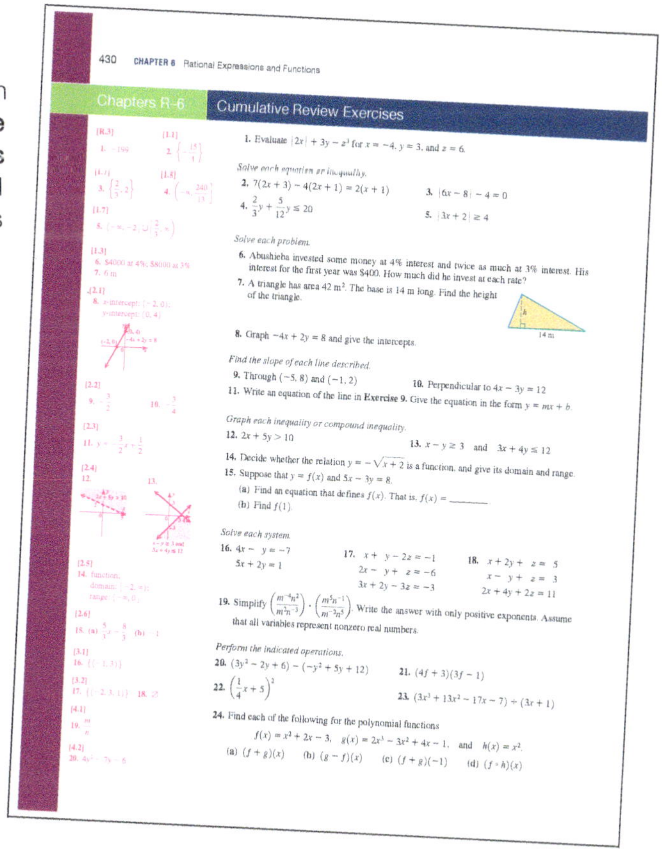
430 CHAPTER 6 Rational Expressions and Functions

Chapters R–6 Cumulative Review Exercises

1. Evaluate $|2x| + 3y - z^3$ for $x = -4$, $y = 3$, and $z = 6$.

Solve each equation or inequality.

2. $7(2x + 3) - 4(2x + 1) = 2(x + 1)$
3. $|6x - 8| - 4 = 0$
4. $\frac{2}{3}y + \frac{5}{12}y \leq 20$
5. $|3x + 2| \geq 4$

Solve each problem.

6. Abushieba invested some money at 4% interest and twice as much at 3% interest. His interest for the first year was $400. How much did he invest at each rate?
7. A triangle has area 42 m^2. The base is 14 m long. Find the height of the triangle.

8. Graph $-4x + 2y = 8$ and give the intercepts.

Find the slope of each line described.

9. Through $(-5, 8)$ and $(-1, 2)$
10. Perpendicular to $4x - 3y = 12$
11. Write an equation of the line in **Exercise 9.** Give the equation in the form $y = mx + b$.

Graph each inequality or compound inequality.

12. $2x + 5y > 10$
13. $x - y \geq 3$ and $3x + 4y \leq 12$
14. Decide whether the relation $y = -\sqrt{x + 2}$ is a function, and give its domain and range.
15. Suppose that $y = f(x)$ and $5x - 3y = 8$.
    (a) Find an equation that defines $f(x)$. That is, $f(x) =$ ______
    (b) Find $f(1)$.

Solve each system.

16. $4x - y = -7$; $5x + 2y = 1$
17. $x + y - 2z = -1$; $2x - y + z = -6$; $3x + 2y - 3z = -3$
18. $x + 2y + z = 5$; $x - y + z = 3$; $2x + 4y + 2z = 11$
19. Simplify $\left(\frac{m^{-4}n^2}{m^2n^{-3}}\right) \cdot \left(\frac{m^5n^{-1}}{m^{-2}n^5}\right)$. Write the answer with only positive exponents. Assume that all variables represent nonzero real numbers.

Perform the indicated operations.

20. $(3y^2 - 2y + 6) - (-y^2 + 5y + 12)$
21. $(4f + 3)(3f - 1)$
22. $\left(\frac{1}{4}x + 5\right)^2$
23. $(3x^3 + 13x^2 - 17x - 7) \div (3x + 1)$
24. Find each of the following for the polynomial functions
    $f(x) = x^2 + 2x - 3$, $g(x) = 2x^3 - 3x^2 + 4x - 1$, and $h(x) = x^2$.
    (a) $(f + g)(x)$ (b) $(g - f)(x)$ (c) $(f + g)(-1)$ (d) $(f \cdot h)(x)$

Final Exam Preparation Suggestions

1. **Figure out the grade you need to earn on the final exam to get the course grade you want.** Check your course syllabus for grading policies, or ask your instructor if you are not sure.

   How many points do you need to earn on your math final exam to get the grade you want?

2. **Create a final exam week plan.** Set priorities that allow you to spend extra time studying. This may mean making adjustments, in advance, in your work schedule or enlisting extra help with family responsibilities.

   What adjustments do you need to make for final exam week? List two or three here.

3. **Use the following suggestions to guide your studying.**
   - **Begin reviewing several days before the final exam.** DON'T wait until the last minute.
   - **Know exactly which chapters and sections will be covered on the exam.**
   - **Divide up the chapters.** Decide how much you will review each day.
   - **Keep returned quizzes and tests. Use them to review.**
   - **Practice all types of problems. Use the Cumulative Review Exercises** at the end of each chapter in your textbook beginning in Chapter 2. All answers, with section references, are given in the margin.
   - **Review or rewrite your notes** to create summaries of important information.
   - **Make study cards for all types of problems.** Carry the cards with you, and review them whenever you have a few minutes.
   - **Take plenty of short breaks (as you study) to reduce physical and mental stress.** Exercising, listening to music, and enjoying a favorite activity are effective stress busters.

   Finally, *DON'T* stay up all night the night before an exam—*get a good night's sleep.*

Which of these suggestions will you use as you study for your math final exam? List two or three here.

3.2 Systems of Linear Equations in Three Variables

OBJECTIVES

1. Understand the geometry of systems of three equations in three variables.
2. Solve linear systems (with three equations and three variables) by elimination.
3. Solve linear systems (with three equations and three variables) in which some of the equations have missing terms.
4. Solve special systems.

VOCABULARY

- ☐ ordered triple
- ☐ focus variable
- ☐ working equation

A solution of an equation in three variables, such as

$$2x + 3y - z = 4, \quad \text{Linear equation in three variables}$$

is an **ordered triple** and is written $\mathbf{(x, y, z)}$. For example, the ordered triple $(0, 1, -1)$ is a solution of the preceding equation, because

$$2(0) + 3(1) - (-1) = 4 \quad \text{is a true statement.}$$

Verify that another solution of this equation is $(10, -3, 7)$.

We now extend the term *linear equation* to equations of the form

$$Ax + By + Cz + \cdots + Dw = K,$$

where not all the coefficients $A, B, C, \ldots, D$ equal 0. For example,

$$2x + 3y - 5z = 7 \quad \text{and} \quad x - 2y - z + 3w = 8$$

are linear equations, the first with three variables and the second with four.

OBJECTIVE 1 Understand the geometry of systems of three equations in three variables.

Consider the solution of a system such as the following.

$$\begin{aligned} 4x + 8y + z &= 2 \\ x + 7y - 3z &= -14 \\ 2x - 3y + 2z &= 3 \end{aligned} \quad \text{System of linear equations in three variables}$$

Theoretically, a system of this type can be solved by graphing. However, the graph of a linear equation with three variables is a *plane,* not a line. Because visualizing a plane requires three-dimensional graphing, the method of graphing is not practical with these systems. However, it does illustrate the number of solutions possible for such systems, as shown in **FIGURE 7**.

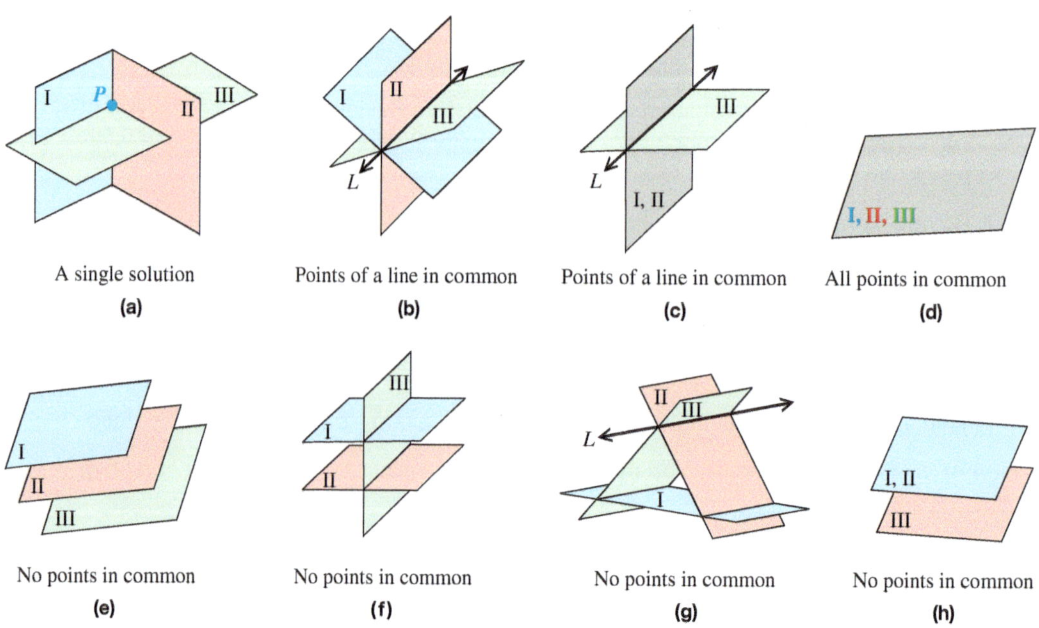

FIGURE 7

TEACHING TIP
Use the walls, floor, and ceiling of a building as examples of intersecting and parallel planes. For example, two perpendicular walls and a ceiling would intersect in a single point (**FIGURE 7(a)**). Three floors of three different levels of a building would not intersect (**FIGURE 7(e)**).

Three pages of this book, which intersect in the spine, illustrate the intersection of three planes in infinitely many points (**FIGURE 7(b)**).

FIGURE 7 illustrates the following cases.

Graphs of Linear Systems in Three Variables

Case 1 **The three planes may meet at a single, common point.** This point is the solution of the system. See **FIGURE 7(a)**.

Case 2 **The three planes may have the points of a line in common.** The infinite set of points that satisfy the equation of the line is the solution of the system. See **FIGURES 7(b) AND (c)**.

Case 3 **The three planes may coincide.** The solution of the system is the set of all points on a plane. See **FIGURE 7(d)**.

Case 4 **The planes may have no points common to all three.** There is no solution of the system. See **FIGURES 7(e)–(h)**.

TEACHING TIP
Outline this procedure as going from *three equations* (in three unknowns), to *two equations* (in two unknowns), to *one equation* (in one unknown). Once this last equation is solved, proceed to substitute the answer into a two-variable equation to find the value of a second unknown. Finally, substitute the two known values into a three-variable equation.

OBJECTIVE 2 Solve linear systems (with three equations and three variables) by elimination.

Because graphing to find the solution set of a system of three equations in three variables is impractical, these systems are solved with an extension of the elimination method from **Section 3.1.**

In the steps that follow, we use the term **focus variable** to identify the first variable to be eliminated in the process. The focus variable will always be present in the **working equation,** which will be used twice to eliminate this variable.

Solving a Linear System in Three Variables*

Step 1 **Select a variable and an equation.** A good choice for the variable, which we call the *focus variable,* is one that has coefficient 1 or −1. Then select an equation, one that contains the focus variable, as the *working equation*.

Step 2 **Eliminate the focus variable.** Use the working equation and one of the other two equations of the original system. The result is an equation in two variables.

Step 3 **Eliminate the focus variable again.** Use the working equation and the remaining equation of the original system. The result is another equation in two variables.

Step 4 **Write the equations in two variables that result from Steps 2 and 3 as a system, and solve it.** Doing this gives the values of two of the variables.

Step 5 **Find the value of the remaining variable.** Substitute the values of the two variables found in Step 4 into the working equation to obtain the value of the focus variable.

Step 6 **Check** the three values in *each* of the *original* equations of the system. Then write the solution set as a set containing an ordered triple.

---

*The authors wish to thank Christine Heinecke Lehmann of Purdue University North Central for her suggestions here.

CLASSROOM EXAMPLE 1
Solve the system.

$$\begin{aligned} x + y + z &= 2 \\ x - y + 2z &= 2 \\ -x + 2y - z &= 1 \end{aligned}$$

Answer: $\{(-1, 1, 2)\}$

EXAMPLE 1 Solving a System in Three Variables

Solve the system.

$$4x + 8y + z = 2 \quad (1)$$
$$x + 7y - 3z = -14 \quad (2)$$
$$2x - 3y + 2z = 3 \quad (3)$$

Step 1 Because z in equation (1) has coefficient 1, we choose z as the focus variable and (1) as the working equation. (Another option would be to choose x as the focus variable—it also has coefficient 1—and use (2) as the working equation.)

Focus variable

$$4x + 8y + z = 2 \quad (1) \leftarrow \text{Working equation}$$

Step 2 Multiply working equation (1) by 3 and add the result to equation (2).

$$12x + 24y + 3z = 6 \quad \text{Multiply each side of (1) by 3.}$$
$$x + 7y - 3z = -14 \quad (2)$$
$$13x + 31y = -8 \quad \text{Add.} \quad (4)$$

Focus variable z was eliminated.

Step 3 Multiply working equation (1) by -2 and add the result to remaining equation (3) to again eliminate focus variable z.

$$-8x - 16y - 2z = -4 \quad \text{Multiply each side of (1) by } -2.$$
$$2x - 3y + 2z = 3 \quad (3)$$
$$-6x - 19y = -1 \quad \text{Add.} \quad (5)$$

Focus variable z was eliminated.

Step 4 Write the equations in two variables that result in Steps 2 and 3 as a system.

Make sure these equations have the same two variables.

$$13x + 31y = -8 \quad (4) \quad \text{The result from Step 2}$$
$$-6x - 19y = -1 \quad (5) \quad \text{The result from Step 3}$$

Now solve this system. We choose to eliminate x.

$$78x + 186y = -48 \quad \text{Multiply each side of (4) by 6.}$$
$$-78x - 247y = -13 \quad \text{Multiply each side of (5) by 13.}$$
$$-61y = -61 \quad \text{Add.}$$
$$y = 1 \quad \text{Divide by } -61.$$

Substitute 1 for y in either equation (4) or (5) to find x.

$$-6x - 19y = -1 \quad (5)$$
$$-6x - 19(1) = -1 \quad \text{Let } y = 1.$$
$$-6x - 19 = -1 \quad \text{Multiply.}$$
$$-6x = 18 \quad \text{Add 19.}$$
$$x = -3 \quad \text{Divide by } -6.$$

Step 5 Now substitute the two values we found in Step 4 in working equation (1) to find the value of the remaining variable, focus variable z.

$$4x + 8y + z = 2 \quad (1)$$
$$4(-3) + 8(1) + z = 2 \quad \text{Let } x = -3 \text{ and } y = 1.$$
$$-4 + z = 2 \quad \text{Multiply, and then add.}$$
$$z = 6 \quad \text{Add 4.}$$

NOW TRY EXERCISE 1
Solve the system.

$$x - y + 2z = 1$$
$$3x + 2y + 7z = 8$$
$$-3x - 4y + 9z = -10$$

TEACHING TIP
Have students work Exercise 6, carrying out the checks for equations (2) and (3) of the system. Stress that the solution must satisfy *all three* of the original equations.

Write the values of *x*, *y*, and *z* in the correct order.

Step 6 It appears that $(-3, 1, 6)$ is the only solution of the system. We must check that this ordered triple satisfies all three original equations of the system. We begin with equation (1).

CHECK

$$4x + 8y + z = 2 \quad (1)$$
$$4(-3) + 8(1) + 6 \stackrel{?}{=} 2 \quad \text{Substitute.}$$
$$-12 + 8 + 6 \stackrel{?}{=} 2 \quad \text{Multiply.}$$
$$2 = 2 \checkmark \quad \text{True}$$

Because $(-3, 1, 6)$ also satisfies equations (2) and (3), $\{(-3, 1, 6)\}$ is the solution set. This is Case 1 as shown in **FIGURE 7(a)** at the beginning of this section.

NOW TRY

OBJECTIVE 3 Solve linear systems (with three equations and three variables) in which some of the equations have missing terms.

If a linear system includes an equation that is missing a term or terms, one elimination step can be omitted.

CLASSROOM EXAMPLE 2
Solve the system.

$$x - y = 6$$
$$2y + 5z = 1$$
$$3x - 4z = 8$$

Answer: $\{(4, -2, 1)\}$

EXAMPLE 2 **Solving a System of Equations with Missing Terms**

Solve the system.

$$6x - 12y = -5 \quad (1) \quad \text{Missing } z$$
$$8y + z = 0 \quad (2) \quad \text{Missing } x$$
$$9x - z = 12 \quad (3) \quad \text{Missing } y$$

Equation (3) is missing the variable y, so one way to begin is to eliminate y again, using equations (1) and (2).

$$12x - 24y = -10 \quad \text{Multiply each side of (1) by 2.}$$
$$24y + 3z = 0 \quad \text{Multiply each side of (2) by 3.}$$
$$12x + 3z = -10 \quad \text{Add. (4)}$$

Leave space for the missing terms.

Use resulting equation (4) in x and z, together with equation (3), $9x - z = 12$, to eliminate z.

$$27x - 3z = 36 \quad \text{Multiply each side of (3) by 3.}$$
$$12x + 3z = -10 \quad (4)$$
$$39x = 26 \quad \text{Add.}$$
$$x = \frac{26}{39} \quad \text{Divide by 39.}$$
$$x = \frac{2}{3} \quad \text{Lowest terms}$$

We can find z by substituting this value for x in equation (3).

$$9x - z = 12 \quad (3)$$
$$9\left(\frac{2}{3}\right) - z = 12 \quad \text{Let } x = \tfrac{2}{3}.$$
$$6 - z = 12 \quad \text{Multiply.}$$
$$z = -6 \quad \text{Subtract 6. Multiply by } -1.$$

NOW TRY ANSWER
1. $\{(2, 1, 0)\}$

NOW TRY EXERCISE 2
Solve the system.

$$3x - z = -10$$
$$4y + 5z = 24$$
$$x - 6y = -8$$

TEACHING TIP
Remind students to be careful to write the values of x, y, and z in the appropriate order in the solution.

We can find y by substituting -6 for z in equation (2).

$$8y + z = 0 \quad (2)$$
$$8y - 6 = 0 \quad \text{Let } z = -6.$$
$$8y = 6 \quad \text{Add 6.}$$
$$y = \frac{6}{8} \quad \text{Divide by 8.}$$
$$y = \frac{3}{4} \quad \text{Lowest terms}$$

Check to verify that the solution set is $\left\{\left(\frac{2}{3}, \frac{3}{4}, -6\right)\right\}$. This is also an example of Case 1.

NOW TRY

NOTE Another way to solve the system in **Example 2** is to begin by eliminating the variable z from equations (2) and (3). The resulting equation together with equation (1) forms a system of two equations in the variables x and y. Try working **Example 2** this way to see that the same solution results.

There are often multiple ways to solve a system of equations. Some ways may involve more work than others.

CLASSROOM EXAMPLE 3
Solve the system.

$$x - y + z = 4$$
$$-3x + 3y - 3z = -12$$
$$2x - 2y + 2z = 8$$

Answer: $\{(x, y, z) \mid x - y + z = 4\}$

OBJECTIVE 4 Solve special systems.

EXAMPLE 3 Solving a System of Dependent Equations with Three Variables

Solve the system.

$$2x - 3y + 4z = 8 \quad (1)$$
$$-x + \frac{3}{2}y - 2z = -4 \quad (2)$$
$$6x - 9y + 12z = 24 \quad (3)$$

Multiplying each side of equation (1) by 3 gives equation (3). Multiplying each side of equation (2) by -6 also gives equation (3). Because of this, the equations are dependent.

When solving a system such as this, attempting to eliminate one variable results in elimination of *all* variables, leading to the true statement $0 = 0$. This indicates that all three equations have the same graph, as illustrated in **FIGURE 7(d)**. This is Case 3. The solution set is written as follows.

$$\{(x, y, z) \mid 2x - 3y + 4z = 8\} \quad \text{Set-builder notation}$$

Although any one of the three equations could be used to write the solution set, we use the equation in standard form with coefficients that are integers with greatest common factor 1, as we did in **Section 3.1.**

NOW TRY

NOW TRY EXERCISE 3
Solve the system.

$$x - 3y + 2z = 10$$
$$-2x + 6y - 4z = -20$$
$$\frac{1}{2}x - \frac{3}{2}y + z = 5$$

CLASSROOM EXAMPLE 4
Solve the system.

$$3x - 5y + 2z = 1$$
$$5x + 8y - z = 4$$
$$-6x + 10y - 4z = 5$$

Answer: $\emptyset$

EXAMPLE 4 Solving an Inconsistent System with Three Variables

Solve the system.

$$2x - 4y + 6z = 5 \quad (1)$$
$$-x + 3y - 2z = -1 \quad (2)$$
$$x - 2y + 3z = 1 \quad (3)$$

Use as the working equation, with focus variable x.

Eliminate the focus variable, x, using equations (1) and (3).

NOW TRY ANSWERS
2. $\{(-2, 1, 4)\}$
3. $\{(x, y, z) \mid x - 3y + 2z = 10\}$

NOW TRY EXERCISE 4
Solve the system.

$$\begin{aligned} x - 5y + 2z &= 4 \\ 3x + y - z &= 6 \\ -2x + 10y - 4z &= 7 \end{aligned}$$

CLASSROOM EXAMPLE 5
Solve the system.

$$\begin{aligned} 2x + 3y - z &= 8 \\ \frac{1}{2}x + \frac{3}{4}y - \frac{1}{4}z &= 2 \\ x + \frac{3}{2}y - \frac{1}{2}z &= -6 \end{aligned}$$

Answer: $\varnothing$

NOW TRY EXERCISE 5
Solve the system.

$$\begin{aligned} x - 3y + 2z &= 4 \\ \frac{1}{3}x - y + \frac{2}{3}z &= 7 \\ \frac{1}{2}x - \frac{3}{2}y + z &= 2 \end{aligned}$$

NOW TRY ANSWERS
4. $\varnothing$
5. $\varnothing$

$$\begin{aligned} -2x + 4y - 6z &= -2 \quad \text{Multiply each side of (3) by } -2. \\ 2x - 4y + 6z &= 5 \quad (1) \\ \hline 0 &= 3 \quad \text{False} \end{aligned}$$

The resulting false statement indicates that equations (1) and (3) have no common solution. Thus, the system is inconsistent and the solution set is $\varnothing$. The graph of this system would show two planes parallel to one another and a third plane which intersects both, as in **FIGURE 7(f)**. This is Case 4. NOW TRY

NOTE If a false statement results when adding as in **Example 4,** it is not necessary to go any further with the solution. Because two of the three planes are parallel, it is not possible for the three planes to have any points in common.

EXAMPLE 5 Solving Another Special System

Solve the system.

$$\begin{aligned} 2x - y + 3z &= 6 \quad (1) \\ x - \frac{1}{2}y + \frac{3}{2}z &= 3 \quad (2) \\ 4x - 2y + 6z &= 1 \quad (3) \end{aligned}$$

Multiplying each side of equation (2) by 2 gives equation (1), so these two equations are dependent. Equations (1) and (3) are not equivalent, however. Multiplying equation (3) by $\frac{1}{2}$ does *not* give equation (1). Instead, we obtain two equations with the same coefficients, but with different constant terms.

The graphs of equations (1) and (3) have no points in common (that is, the planes are parallel). Thus, the system is inconsistent and the solution set is $\varnothing$, as illustrated in **FIGURE 7(h)**. This is another example of Case 4. NOW TRY

3.2 Exercises

FOR EXTRA HELP

MyMathLab®

Complete solution available in MyMathLab

Concept Check *Answer each of the following.*

1. Using your immediate surroundings, give an example of three planes that satisfy the condition.
   (a) They intersect in a single point.
   (b) They do not intersect.
   (c) They intersect in infinitely many points.

2. Suppose that a system has infinitely many ordered triple solutions of the form (x, y, z) such that

$$x + y + 2z = 1.$$

Give three specific ordered triples that are solutions of the system.

3. Explain what the following statement means: "The solution set of the following system is $\{(-1, 2, 3)\}$."

$$\begin{aligned} 2x + y + z &= 3 \\ 3x - y + z &= -2 \\ 4x - y + 2z &= 0 \end{aligned}$$

1. Answers will vary. Some possible answers are (a) two perpendicular walls and the ceiling in a normal room, (b) the floors of three different levels of an office building, (c) three pages of a book (because they intersect in the spine).
2. Answers will vary. Three possibilities are $\left(1, 1, -\frac{1}{2}\right)$, $\left(0, 0, \frac{1}{2}\right)$, and $(2, 5, -3)$.

3. The statement means that when -1 is substituted for x, 2 is substituted for y, and 3 is substituted for z in the three equations, the resulting three statements are true.
4. B
5. 4
6. For equation (2), the result is $-14 = -14$, and for equation (3), the result is $3 = 3$. Both are true.
7. $\{(3, 2, 1)\}$
8. $\{(4, 3, 2)\}$
9. $\{(1, 4, -3)\}$
10. $\{(2, -1, 3)\}$
11. $\{(0, 2, -5)\}$
12. $\{(0, 3, -1)\}$
13. $\{(1, 0, 3)\}$
14. $\{(3, 0, -2)\}$
15. $\left\{\left(1, \frac{3}{10}, \frac{2}{5}\right)\right\}$
16. $\left\{\left(\frac{1}{2}, 1, -\frac{1}{2}\right)\right\}$
17. $\left\{\left(-\frac{7}{3}, \frac{22}{3}, 7\right)\right\}$
18. $\left\{\left(\frac{20}{59}, -\frac{33}{59}, \frac{35}{59}\right)\right\}$
19. $\{(-12, 18, 0)\}$
20. $\{(-15, 0, 16)\}$
21. $\{(0.8, -1.5, 2.3)\}$
22. $\{(-0.4, 2.6, 1.8)\}$
23. $\{(4, 5, 3)\}$
24. $\{(5, 1, -1)\}$
25. $\{(2, 2, 2)\}$

4. The following two equations have a common solution of $(1, 2, 3)$.

$$\begin{aligned} x + y + z &= 6 \\ 2x - y + z &= 3 \end{aligned}$$

Which equation would complete a system of three linear equations in three variables having solution set $\{(1, 2, 3)\}$?

A. $3x + 2y - z = 1$ **B.** $3x + 2y - z = 4$

C. $3x + 2y - z = 5$ **D.** $3x + 2y - z = 6$

5. What constant should replace the question mark in this system so that the solution set is $\{(1, 1, 1)\}$?

$$\begin{aligned} 2x - 3y + z &= 0 \\ -5x + 2y - z &= -4 \\ x + y + 2z &= ? \end{aligned}$$

6. Complete the work of **Example 1** and show that the ordered triple $(-3, 1, 6)$ is also a solution of equations (2) and (3).

$$\begin{aligned} x + 7y - 3z &= -14 \quad \text{Equation (2)} \\ 2x - 3y + 2z &= 3 \quad \text{Equation (3)} \end{aligned}$$

Solve each system. ***See Example 1.***

7. $\begin{aligned} 2x - 5y + 3z &= -1 \\ x + 4y - 2z &= 9 \\ x - 2y - 4z &= -5 \end{aligned}$

8. $\begin{aligned} x + 3y - 6z &= 1 \\ 2x - y + z &= 7 \\ x + 2y + 2z &= 14 \end{aligned}$

9. $\begin{aligned} 3x + 2y + z &= 8 \\ 2x - 3y + 2z &= -16 \\ x + 4y - z &= 20 \end{aligned}$

10. $\begin{aligned} -3x + y - z &= -10 \\ -4x + 2y + 3z &= -1 \\ 2x + 3y - 2z &= -5 \end{aligned}$

11. $\begin{aligned} 2x + 5y + 2z &= 0 \\ 4x - 7y - 3z &= 1 \\ 3x - 8y - 2z &= -6 \end{aligned}$

12. $\begin{aligned} 5x - 2y + 3z &= -9 \\ 4x + 3y + 5z &= 4 \\ 2x + 4y - 2z &= 14 \end{aligned}$

13. $\begin{aligned} x + 2y + z &= 4 \\ 2x + y - z &= -1 \\ x - y - z &= -2 \end{aligned}$

14. $\begin{aligned} x - 2y + 5z &= -7 \\ -2x - 3y + 4z &= -14 \\ -3x + 5y - z &= -7 \end{aligned}$

15. $\begin{aligned} -x + 2y + 6z &= 2 \\ 3x + 2y + 6z &= 6 \\ x + 4y - 3z &= 1 \end{aligned}$

16. $\begin{aligned} 2x + y + 2z &= 1 \\ x + 2y + z &= 2 \\ x - y - z &= 0 \end{aligned}$

17. $\begin{aligned} x + y - z &= -2 \\ 2x - y + z &= -5 \\ -x + 2y - 3z &= -4 \end{aligned}$

18. $\begin{aligned} x + 2y + 3z &= 1 \\ -x - y + 3z &= 2 \\ -6x + y + z &= -2 \end{aligned}$

19. $\begin{aligned} \frac{1}{3}x + \frac{1}{6}y - \frac{2}{3}z &= -1 \\ -\frac{3}{4}x - \frac{1}{3}y - \frac{1}{4}z &= 3 \\ \frac{1}{2}x + \frac{3}{2}y + \frac{3}{4}z &= 21 \end{aligned}$

20. $\begin{aligned} \frac{2}{3}x - \frac{1}{4}y + \frac{5}{8}z &= 0 \\ \frac{1}{5}x + \frac{2}{3}y - \frac{1}{4}z &= -7 \\ -\frac{3}{5}x + \frac{4}{3}y - \frac{7}{8}z &= -5 \end{aligned}$

21. $\begin{aligned} 5.5x - 2.5y + 1.6z &= 11.83 \\ 2.2x + 5.0y - 0.1z &= -5.97 \\ 3.3x - 7.5y + 3.2z &= 21.25 \end{aligned}$

22. $\begin{aligned} 6.2x - 1.4y + 2.4z &= -1.80 \\ 3.1x + 2.8y - 0.2z &= 5.68 \\ 9.3x - 8.4y - 4.8z &= -34.20 \end{aligned}$

Solve each system. ***See Example 2.***

23. $\begin{aligned} 2x - 3y + 2z &= -1 \\ x + 2y + z &= 17 \\ 2y - z &= 7 \end{aligned}$

24. $\begin{aligned} 2x - y + 3z &= 6 \\ x + 2y - z &= 8 \\ 2y + z &= 1 \end{aligned}$

25. $\begin{aligned} 4x + 2y - 3z &= 6 \\ x - 4y + z &= -4 \\ -x + 2z &= 2 \end{aligned}$

26. $\{(4, 4, 4)\}$
27. $\left\{\left(\frac{8}{3}, \frac{2}{3}, 3\right)\right\}$
28. $\left\{\left(\frac{3}{4}, \frac{5}{4}, 2\right)\right\}$
29. $\{(-1, 0, 0)\}$
30. $\{(0, 0, -4)\}$
31. $\{(-4, 6, 2)\}$
32. $\{(2, -3, 1)\}$
33. $\{(-3, 5, -6)\}$
34. $\{(10, -30, 12)\}$
35. $\emptyset$; inconsistent system
36. $\emptyset$; inconsistent system
37. $\{(x, y, z) \mid x - y + 4z = 8\}$; dependent equations
38. $\{(x, y, z) \mid x + 4y - z = 3\}$; dependent equations
39. $\{(3, 0, 2)\}$
40. $\{(-4, 0, 6)\}$
41. $\{(x, y, z) \mid 2x + y - z = 6\}$; dependent equations
42. $\{(x, y, z) \mid x - 4y + z = -5\}$; dependent equations
43. $\{(0, 0, 0)\}$
44. $\{(0, 0, 0)\}$
45. $\emptyset$; inconsistent system
46. $\emptyset$; inconsistent system
47. $\{(2, 1, 5, 3)\}$
48. $\{(4, 1, 2, -1)\}$
49. $\{(-2, 0, 1, 4)\}$
50. $\{(5, -1, 0, 3)\}$

26. $\begin{aligned} 2x + 3y - 4z &= 4 \\ x - 6y + z &= -16 \\ -x + 3z &= 8 \end{aligned}$

27. $\begin{aligned} 2x + y &= 6 \\ 3y - 2z &= -4 \\ 3x - 5z &= -7 \end{aligned}$

28. $\begin{aligned} 4x - 8y &= -7 \\ 4y + z &= 7 \\ -8x + z &= -4 \end{aligned}$

29. $\begin{aligned} -5x + 2y + z &= 5 \\ -3x - 2y - z &= 3 \\ -x + 6y &= 1 \end{aligned}$

30. $\begin{aligned} -4x + 3y - z &= 4 \\ -5x - 3y + z &= -4 \\ -2x - 3z &= 12 \end{aligned}$

31. $\begin{aligned} 7x - 3z &= -34 \\ 2y + 4z &= 20 \\ \frac{3}{4}x + \frac{1}{6}y &= -2 \end{aligned}$

32. $\begin{aligned} 5x - 2z &= 8 \\ 4y + 3z &= -9 \\ \frac{1}{2}x + \frac{2}{3}y &= -1 \end{aligned}$

33. $\begin{aligned} 4x - z &= -6 \\ \frac{3}{5}y + \frac{1}{2}z &= 0 \\ \frac{1}{3}x + \frac{2}{3}z &= -5 \end{aligned}$

34. $\begin{aligned} 5x - z &= 38 \\ \frac{2}{3}y + \frac{1}{4}z &= -17 \\ \frac{1}{5}y + \frac{5}{6}z &= 4 \end{aligned}$

Solve each system. If the system is inconsistent or has dependent equations, say so. ***See Examples 1, 3, 4, and 5.***

35. $\begin{aligned} 2x + 2y - 6z &= 5 \\ -3x + y - z &= -2 \\ -x - y + 3z &= 4 \end{aligned}$

36. $\begin{aligned} -2x + 5y + z &= -3 \\ 5x + 14y - z &= -11 \\ 7x + 9y - 2z &= -5 \end{aligned}$

37. $\begin{aligned} -5x + 5y - 20z &= -40 \\ x - y + 4z &= 8 \\ 3x - 3y + 12z &= 24 \end{aligned}$

38. $\begin{aligned} x + 4y - z &= 3 \\ -2x - 8y + 2z &= -6 \\ 3x + 12y - 3z &= 9 \end{aligned}$

39. $\begin{aligned} x + 5y - 2z &= -1 \\ -2x + 8y + z &= -4 \\ 3x - y + 5z &= 19 \end{aligned}$

40. $\begin{aligned} x + 3y + z &= 2 \\ 4x + y + 2z &= -4 \\ 5x + 2y + 3z &= -2 \end{aligned}$

41. $\begin{aligned} 2x + y - z &= 6 \\ 4x + 2y - 2z &= 12 \\ -x - \frac{1}{2}y + \frac{1}{2}z &= -3 \end{aligned}$

42. $\begin{aligned} 2x - 8y + 2z &= -10 \\ -x + 4y - z &= 5 \\ \frac{1}{8}x - \frac{1}{2}y + \frac{1}{8}z &= -\frac{5}{8} \end{aligned}$

43. $\begin{aligned} x + y - 2z &= 0 \\ 3x - y + z &= 0 \\ 4x + 2y - z &= 0 \end{aligned}$

44. $\begin{aligned} 2x + 3y - z &= 0 \\ x - 4y + 2z &= 0 \\ 3x - 5y - z &= 0 \end{aligned}$

45. $\begin{aligned} x - 2y + \frac{1}{3}z &= 4 \\ 3x - 6y + z &= 12 \\ -6x + 12y - 2z &= -3 \end{aligned}$

46. $\begin{aligned} 4x + y - 2z &= 3 \\ x + \frac{1}{4}y - \frac{1}{2}z &= \frac{3}{4} \\ 2x + \frac{1}{2}y - z &= 1 \end{aligned}$

Extending Skills *Extend the method of this section to solve each system. Express the solution in the form* (x, y, z, w).

47. $\begin{aligned} x + y + z - w &= 5 \\ 2x + y - z + w &= 3 \\ x - 2y + 3z + w &= 18 \\ -x - y + z + 2w &= 8 \end{aligned}$

48. $\begin{aligned} 3x + y - z + 2w &= 9 \\ x + y + 2z - w &= 10 \\ x - y - z + 3w &= -2 \\ -x + y - z + w &= -6 \end{aligned}$

49. $\begin{aligned} 3x + y - z + w &= -3 \\ 2x + 4y + z - w &= -7 \\ -2x + 3y - 5z + w &= 3 \\ 5x + 4y - 5z + 2w &= -7 \end{aligned}$

50. $\begin{aligned} x - 3y + 7z + w &= 11 \\ 2x + 4y + 6z - 3w &= -3 \\ 3x + 2y + z + 2w &= 19 \\ 4x + y - 3z + w &= 22 \end{aligned}$

3.3 Applications of Systems of Linear Equations

OBJECTIVES

1. Solve geometry problems using two variables.
2. Solve money problems using two variables.
3. Solve mixture problems using two variables.
4. Solve distance-rate-time problems using two variables.
5. Solve problems with three variables using a system of three equations.

Although some problems with two unknowns can be solved using just one variable, it is often easier to use two variables and a system of equations. The following problem, which can be solved with a system, appeared in a Hindu work that dates back to about A.D. 850. (See **Exercise 39.**)

> The mixed price of 9 citrons (a lemonlike fruit) and 7 fragrant wood apples is 107; again, the mixed price of 7 citrons and 9 fragrant wood apples is 101. O you arithmetician, tell me quickly the price of a citron and the price of a wood apple here, having distinctly separated those prices well.

PROBLEM-SOLVING HINT When solving an applied problem using two variables, it is a good idea to pick letters that correspond to the descriptions of the unknown quantities. In the example above, we could choose c to represent the number of citrons, and w to represent the number of wood apples.

The following steps are based on the problem-solving method of **Section 1.3.**

Solving an Applied Problem Using a System of Equations

Step 1 **Read** the problem carefully. *What information is given? What is to be found?*

Step 2 **Assign variables** to represent the unknown values. Write down what each variable represents. Make a sketch, diagram, or table, as needed.

Step 3 **Write a system of equations** using both variables.

Step 4 **Solve** the system of equations.

Step 5 **State the answer.** Label it appropriately. *Does it seem reasonable?*

Step 6 **Check** the answer in the words of the *original* problem.

OBJECTIVE 1 Solve geometry problems using two variables.

EXAMPLE 1 Finding the Dimensions of a Soccer Field

A rectangular soccer field may have a width between 50 and 100 yd and a length between 100 and 130 yd. One particular soccer field has a perimeter of 320 yd. Its length measures 40 yd more than its width. What are the dimensions of this field? (*Source:* www.soccer-training-guide.com)

Step 1 **Read** the problem again. We must find the dimensions of the field.

Step 2 **Assign variables.** A sketch may be helpful. See **FIGURE 8** on the next page.

Let L = the length and W = the width.

NOW TRY EXERCISE 1

A rectangular parking lot has a length that is 10 ft more than twice its width. The perimeter of the parking lot is 620 ft. What are the dimensions of the parking lot?

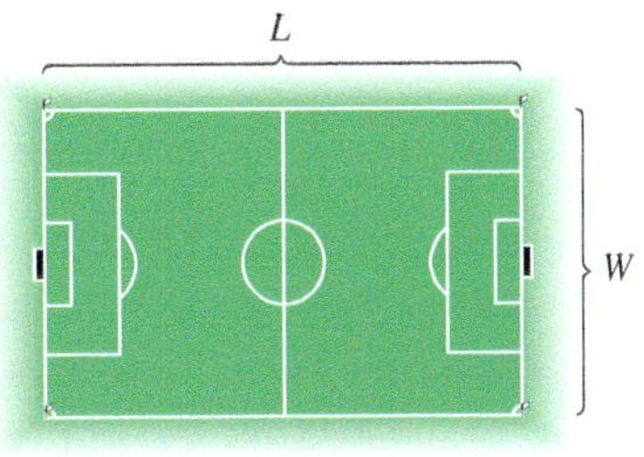

FIGURE 8

CLASSROOM EXAMPLE 1

A rectangular soccer field has perimeter 360 yd. Its length is 20 yd more than its width. What are its dimensions?

Answer: length: 100 yd; width: 80 yd

Step 3 **Write a system of equations.** Because the perimeter is 320 yd, we find one equation by using the perimeter formula.

$$2L + 2W = 320 \qquad 2L + 2W = P$$

For a second equation, use the information given about the width.

$$L = W + 40 \qquad \text{The length is 40 yd more than the width.}$$

These two equations form a system of equations.

$$2L + 2W = 320 \qquad (1)$$
$$L = W + 40 \qquad (2)$$

Step 4 **Solve** the system of equations. Because equation (2), $L = W + 40$, is solved for L, we can substitute $W + 40$ for L in equation (1) and solve for W.

$$2L + 2W = 320 \qquad (1)$$
$$2(W + 40) + 2W = 320 \qquad \text{Let } L = W + 40.$$
$$2W + 80 + 2W = 320 \qquad \text{Distributive property}$$
$$4W + 80 = 320 \qquad \text{Combine like terms.}$$
$$4W = 240 \qquad \text{Subtract 80.}$$
$$W = 60 \qquad \text{Divide by 4.}$$

Be sure to use parentheses around $W + 40$.

Don't stop here.

Let $W = 60$ in the equation $L = W + 40$ to find L.

$$L = 60 + 40 = 100$$

Step 5 **State the answer.** The length is 100 yd, and the width is 60 yd. Both dimensions are within the ranges given in the problem.

Step 6 **Check** using the words of the problem. The answer is correct.

$$2(100) + 2(60) = 320 \qquad \text{The perimeter is 320 yd.}$$
$$100 - 60 = 40 \qquad \text{Length is 40 yd more than width.}$$

NOW TRY

PROBLEM-SOLVING HINT There is often more than one way to write the equations in a system used to solve an application. In **Example 1,** we might write the second equation as

$$W = L - 40. \qquad (2)$$

In this case, we would substitute $L - 40$ for W in equation (1) to obtain

$$2L + 2(L - 40) = 320 \qquad \text{Let } W = L - 40.$$

and solve for L first (instead of W, as in Step 4 above). The *same* answers result.

NOW TRY ANSWER

1. length: 210 ft; width: 100 ft

NOW TRY EXERCISE 2

For the 2013 season at Six Flags St. Louis, two general admission tickets and three tickets for children cost $239.95. One general admission ticket and four tickets for children cost $224.95. Determine the ticket prices for general admission and for children. (*Source:* www.sixflags.com)

CLASSROOM EXAMPLE 2

For the 2013 Major League Baseball and National Football League seasons, based on average ticket prices, three baseball tickets and two football tickets cost $245.52. Two baseball tickets and one football ticket cost $136.50. What were the average ticket prices for the two sports? (*Source:* Team Marketing Report.)

Answer: baseball: $27.48; football: $81.54

OBJECTIVE 2 Solve money problems using two variables.

EXAMPLE 2 Solving a Problem about Ticket Prices

For the 2012–2013 National Hockey League and National Basketball Association seasons, two hockey tickets and one basketball ticket purchased at their average prices cost $173.01. One hockey ticket and two basketball tickets cost $162.99. What were the average ticket prices for the two sports? (*Source:* Team Marketing Report.)

Step 1 **Read** the problem again. There are two unknowns.

Step 2 **Assign variables.**

Let h = the average price for a hockey ticket

and b = the average price for a basketball ticket.

Step 3 **Write a system of equations.** Because two hockey tickets and one basketball ticket cost a total of $173.01, one equation for the system is

$$2h + b = 173.01.$$

By similar reasoning, the second equation is

$$h + 2b = 162.99.$$

These two equations form a system of equations.

$$2h + b = 173.01 \quad (1)$$
$$h + 2b = 162.99 \quad (2)$$

Step 4 **Solve** the system. To eliminate h, multiply equation (2) by -2 and add.

$$2h + b = 173.01 \quad (1)$$
$$-2h - 4b = -325.98 \quad \text{Multiply each side of (2) by } -2.$$
$$-3b = -152.97 \quad \text{Add.}$$
$$b = 50.99 \quad \text{Divide by } -3.$$

To find the value of h, let $b = 50.99$ in equation (2).

$$h + 2b = 162.99 \quad (2)$$
$$h + 2(50.99) = 162.99 \quad \text{Let } b = 50.99.$$
$$h + 101.98 = 162.99 \quad \text{Multiply.}$$
$$h = 61.01 \quad \text{Subtract 101.98.}$$

Step 5 **State the answer.** The average price for one basketball ticket was $50.99. For one hockey ticket, the average price was $61.01.

Step 6 **Check** that these values satisfy the problem conditions.

$$2(\$61.01) + \$50.99 = \$173.01, \quad \text{as required.}$$
$$\$61.01 + 2(\$50.99) = \$162.99, \quad \text{as required.}$$

NOW TRY

NOTE In **Example 2,** we could have solved the system using the substitution method.

NOW TRY ANSWER

2. general admission: $56.99; children: $41.99

OBJECTIVE 3 Solve mixture problems using two variables.

We solved mixture problems in **Section 1.3** using one variable. For many mixture problems we can use more than one variable and a system of equations.

CLASSROOM EXAMPLE 3
A grocer has some \$4-per-lb coffee and some \$8-per-lb coffee that she will mix to make 50 lb of \$5.60-per-lb coffee. How many pounds of each should be used?

Answer: 30 lb at \$4; 20 lb at \$8

EXAMPLE 3 Solving a Mixture Problem

How many ounces each of 5% hydrochloric acid and 20% hydrochloric acid must be combined to obtain 10 oz of solution that is 12.5% hydrochloric acid?

Step 1 **Read** the problem. Two solutions of different strengths are being mixed to obtain a specific amount of a solution with an "in-between" strength.

Step 2 **Assign variables.**

Let $x =$ the number of ounces of 5% solution

and $y =$ the number of ounces of 20% solution.

Use a table to summarize the information from the problem.

Ounces of Solution	*Percent (as a decimal)*	*Ounces of Pure Acid*
x	5% = 0.05	$0.05x$
y	20% = 0.20	$0.20y$
10	12.5% = 0.125	$(0.125)10$

Gives equation (1) (first column) — Gives equation (2) (third column)

Multiply the amount of each solution (given in the first column) by its concentration of acid (given in the second column) to find the amount of acid in that solution (given in the third column).

FIGURE 9 illustrates what is happening in the problem.

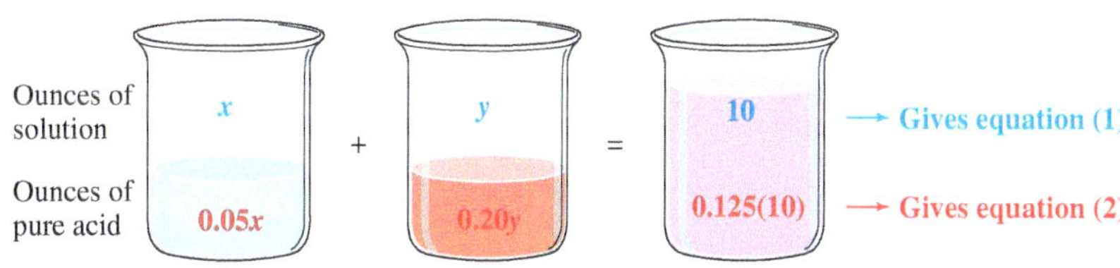

FIGURE 9

Step 3 **Write a system of equations.** When x ounces of 5% solution and y ounces of 20% solution are combined, the total number of ounces is 10.

$$x + y = 10$$

The number of ounces of acid in the 5% solution $(0.05x)$ added to the number of ounces of acid in the 20% solution $(0.20y)$ must equal the total number of ounces of acid in the mixture, which is $(0.125)10$, or 1.25.

$$0.05x + 0.20y = 1.25$$

***Notice that these equations can be quickly determined by reading down the table or using the labels in* FIGURE 9.**

Multiply the equation in red by 100 to clear the decimals and obtain an equivalent system of equations.

$$x + y = 10 \quad (1)$$
$$5x + 20y = 125 \quad (2)$$

This is the system to solve.

NOW TRY EXERCISE 3

How many liters each of a 15% acid solution and a 25% acid solution should be mixed to obtain 30 L of an 18% acid solution?

Step 4 **Solve** the system. To eliminate x, multiply equation (1), $x + y = 10$, by -5.

$$-5x - 5y = -50 \quad \text{Multiply each side of (1) by } -5.$$
$$5x + 20y = 125 \quad (2)$$
$$15y = 75 \quad \text{Add.}$$
$$y = 5 \quad \text{Divide by 15.}$$

Ounces of 20% solution → $y = 5$

Substitute 5 for y in equation (1) to find the value of x.

$$x + y = 10 \quad (1)$$
$$x + 5 = 10 \quad \text{Let } y = 5.$$
$$x = 5 \quad \text{Subtract 5.}$$

Ounces of 5% solution → $x = 5$

Step 5 **State the answer.** The desired mixture will require 5 oz of the 5% solution and 5 oz of the 20% solution.

Step 6 **Check.**

Total amount of solution: $x + y = 5 \text{ oz} + 5 \text{ oz}$

$= 10 \text{ oz}$, as required.

Total amount of acid: 5% of 5 oz + 20% of 5 oz

$= 0.05(5) + 0.20(5)$

$= 1.25 \text{ oz}$

Percent of acid in solution:

Total acid → 1.25; Total solution → 10

$$\frac{1.25}{10} = 0.125, \quad \text{or} \quad 12.5\%, \quad \text{as required.}$$

NOW TRY

OBJECTIVE 4 Solve distance-rate-time problems using two variables.

Motion problems require the distance formula $d = rt$, where d is distance, r is rate (or speed), and t is time.

CLASSROOM EXAMPLE 4

A train travels 600 mi in the same time that a truck travels 520 mi. Find the rate of each vehicle if the train's average rate is 8 mph faster than that of the truck.

Answer: train: 60 mph; truck: 52 mph

EXAMPLE 4 Solving a Motion Problem

A car travels 250 km in the same time that a truck travels 225 km. If the rate of the car is 8 km per hr faster than the rate of the truck, find both rates.

Step 1 **Read** the problem again. Given the distances traveled, we need to find the rate of each vehicle.

Step 2 **Assign variables.**

Let x = the rate of the car,

and y = the rate of the truck.

As in **Example 3,** a table helps organize the information. Fill in the distance for each vehicle, and the variables for the unknown rates.

	d	r	t
Car	250	x	$\frac{250}{x}$
Truck	225	y	$\frac{225}{y}$

To find the expressions for time, we solved the distance formula $d = rt$ for t. Thus, $\frac{d}{r} = t$.

NOW TRY ANSWER

3. 9 L of the 25% acid solution; 21 L of the 15% acid solution

NOW TRY EXERCISE 4
On a bicycle ride, Vann can travel 50 mi in the same amount of time that Ivy can travel 40 mi. Determine both bicyclists' rates, if Vann's rate is 2 mph faster than Ivy's.

Step 3 **Write a system of equations.** The car travels 8 km per hr faster than the truck. Because the two rates are x and y,

$$x = y + 8. \quad (1)$$

Both vehicles travel for the *same* time, so the times must be equal.

$$\text{Time for car} \longrightarrow \frac{250}{x} = \frac{225}{y} \longleftarrow \text{Time for truck}$$

Multiply both sides by xy to obtain an equivalent equation with no variable denominators.

$$xy \cdot \frac{250}{x} = \frac{225}{y} \cdot xy \quad \text{Multiply by the LCD, } xy.$$

$$\frac{250xy}{x} = \frac{225xy}{y}$$

$$250y = 225x \quad \text{Divide out the common factors.} \quad (2)$$

We now have a system of linear equations.

$$x = y + 8 \quad (1)$$

$$250y = 225x \quad (2)$$

Step 4 **Solve** the system by substitution. Replace x with $y + 8$ in equation (2).

$$250y = 225x \quad (2)$$

$$250y = 225(y + 8) \quad \text{Let } x = y + 8.$$

Be sure to use parentheses around $y + 8$.

$$250y = 225y + 1800 \quad \text{Distributive property}$$

$$25y = 1800 \quad \text{Subtract } 225y.$$

$$\text{Truck's rate} \rightarrow y = 72 \quad \text{Divide by 25.}$$

Let $y = 72$ in the equation $x = y + 8$ to find x.

$$\text{Car's rate} \rightarrow x = 72 + 8 = 80$$

Step 5 **State the answer.** The rate of the car is 80 km per hr, and the rate of the truck is 72 km per hr.

Step 6 **Check.**

$$\textit{Car:} \quad t = \frac{d}{r} = \frac{250}{80} = 3.125$$

$$\textit{Truck:} \quad t = \frac{d}{r} = \frac{225}{72} = 3.125$$

Times are equal, as required.

The rate of the car, 80 km per hr, is 8 km per hour greater than that of the truck, 72 km per hr, as required. NOW TRY

PROBLEM-SOLVING HINT When solving a problem as in **Example 4,** where one quantity is compared to another (e.g., the car travels 8 km per hr faster than the truck), be sure to translate correctly in terms of the two variables.

NOW TRY ANSWER
4. Vann: 10 mph; Ivy: 8 mph

NOW TRY EXERCISE 5

In his motorboat, Ed travels 42 mi upstream at top speed in 2.1 hr. Still at top speed, the return trip to the same spot takes only 1.5 hr. Find the rate of Ed's boat in still water and the rate of the current.

EXAMPLE 5 Solving a Motion Problem

While kayaking on the Blackledge River, Rebecca traveled 9 mi upstream (against the current) in 2.25 hr. It only took her 1 hr paddling downstream (with the current) back to the spot where she started. Find Rebecca's kayaking rate in still water and the rate of the current.

Step 1 **Read** the problem. We must find two rates—Rebecca's kayaking rate in still water and the rate of the current.

Step 2 **Assign variables.**

Let x = Rebecca's kayaking rate in still water

and y = the rate of the current.

When the kayak is traveling *against* the current, the current slows it down. The rate of the kayak is the *difference* between its rate in still water and the rate of the current, which is $(x - y)$ mph.

When the kayak is traveling *with* the current, the current speeds it up. The rate of the kayak is the *sum* of its rate in still water and the rate of the current, which is $(x + y)$ mph.

Thus, $x - y =$ the rate of the kayak *against* the current,

and $x + y =$ the rate of the kayak *with* the current.

See **FIGURE 10**. Make a table. Use the formula $d = rt$, or $rt = d$.

FIGURE 10

	r	t	d
Upstream	$x - y$	2.25	$2.25(x - y)$
Downstream	$x + y$	1	$1(x + y)$

The distance is the same in each direction, 9 mi.

Step 3 **Write a system of equations.**

$$2.25(x - y) = 9 \quad \text{Upstream}$$
$$1(x + y) = 9 \quad \text{Downstream}$$

Clear parentheses in each equation, and then divide each term in the first equation by 2.25 to obtain an equivalent system.

$$x - y = 4 \quad (1)$$
$$x + y = 9 \quad (2)$$

Step 4 **Solve.**

$$2x = 13 \quad \text{Add.}$$
$$\text{Rebecca's rate} \rightarrow x = 6.5 \quad \text{Divide by 2.}$$

Substitute 6.5 for x in equation (2) and solve for y.

$$x + y = 9 \quad (2)$$
$$6.5 + y = 9 \quad \text{Let } x = 6.5.$$
$$\text{Rate of current} \rightarrow y = 2.5 \quad \text{Subtract 6.5.}$$

Step 5 **State the answer.** Rebecca's rate in still water was 6.5 mph, and the rate of the current was 2.5 mph.

Step 6 **Check.**

Distance upstream: $2.25(6.5 - 2.5) = 9$

Distance downstream: $1(6.5 + 2.5) = 9$

True statements result.

NOW TRY

CLASSROOM EXAMPLE 5

While canoeing on the Tickfaw River, Shannon traveled 30 mi upstream (against the current) in 5 hr. It took her only 3 hr paddling downstream (with the current) back to the spot where she started. Find Shannon's canoeing rate in still water and the rate of the current.

Answer: canoe: 8 mph; current: 2 mph

NOW TRY ANSWER

5. boat: 24 mph; current: 4 mph

OBJECTIVE 5 Solve problems with three variables using a system of three equations.

PROBLEM-SOLVING HINT If an application has *three* unknown quantities, we can use a system of *three* equations to solve it. We extend the method used for two unknowns.

CLASSROOM EXAMPLE 6
A department store display features three kinds of perfume: Felice, Vivid, and Joy. There are 10 more bottles of Felice than Vivid, and 3 fewer bottles of Joy than Vivid. Each bottle of Felice costs \$8, Vivid costs \$15, and Joy costs \$32. The total value of all the perfume is \$589. How many bottles of each are there?

Answer: 21 bottles of Felice; 11 bottles of Vivid; 8 bottles of Joy

EXAMPLE 6 Solving a Problem Involving Prices

At Panera Bread, a loaf of honey wheat bread costs \$2.95, a loaf of sunflower bread costs \$2.99, and a loaf of French bread costs \$5.79. On a recent day, three times as many loaves of honey wheat bread were sold as sunflower bread. The number of loaves of French bread sold was 5 less than the number of loaves of honey wheat bread sold. Total receipts for these breads were \$87.89. How many loaves of each type of bread were sold? (*Source:* Panera Bread menu.)

Step 1 **Read** the problem again. There are three unknowns in this problem.

Step 2 **Assign variables** to represent the three unknowns.

Let x = the number of loaves of honey wheat bread,

y = the number of loaves of sunflower bread,

and z = the number of loaves of French bread.

Step 3 **Write a system of three equations.** Three times as many loaves of honey wheat bread were sold as sunflower bread.

$$x = 3y, \quad \text{or} \quad x - 3y = 0 \qquad \text{Subtract } 3y. \quad (1)$$

Also, we have the information needed for another equation.

Number of loaves of French	equals	5 less than the number of loaves of honey wheat.
↓	↓	↓
z	$=$	$x - 5$

$$-x + z = -5 \qquad \text{Subtract } x.$$
$$x - z = 5 \qquad \text{Multiply by } -1. \quad (2)$$

Multiplying the cost of a loaf of each kind of bread by the number of loaves of that kind sold and adding gives an equation for the total receipts.

$$2.95x + 2.99y + 5.79z = 87.89$$
$$295x + 299y + 579z = 8789 \qquad \text{Multiply each term by 100 to clear decimals.} \quad (3)$$

Step 4 **Solve** the system of three equations.

$$x - 3y = 0 \quad (1)$$
$$x - z = 5 \quad (2)$$
$$295x + 299y + 579z = 8789 \quad (3)$$

Equation (1) is missing the variable z, so one way to begin is to eliminate z again, using equations (2) and (3).

$$\begin{aligned} 579x - 579z &= 2895 && \text{Multiply (2) by 579.} \\ 295x + 299y + 579z &= 8789 && (3) \\ \hline 874x + 299y &= 11{,}684 && \text{Add.} \quad (4) \end{aligned}$$

NOW TRY EXERCISE 6

At Panera Bread, a loaf of white bread costs \$3.69, a loaf of cheese bread costs \$4.29, and a loaf of whole grain bread costs \$7.69. On a recent day, twice as many loaves of white bread were sold as cheese bread. The number of loaves of whole grain bread sold was 3 less than the number of loaves of white bread sold. Total receipts for these breads were \$139.23. How many loaves of each type of bread were sold? *(Source:* Panera Bread menu.)

Use resulting equation (4) in x and y, together with equation (1), $x - 3y = 0$, to eliminate x.

$$\begin{aligned} -874x + 2622y &= 0 && \text{Multiply (1) by } -874. \\ 874x + 299y &= 11{,}684 && (4) \\ \hline 2921y &= 11{,}684 && \text{Add.} \\ y &= 4 && \text{Divide by 2921.} \end{aligned}$$

We can find x by substituting this value for y in equation (1).

$$\begin{aligned} x - 3y &= 0 && (1) \\ x - 3(4) &= 0 && \text{Let } y = 4. \\ x - 12 &= 0 && \text{Multiply.} \\ x &= 12 && \text{Add 12.} \end{aligned}$$

We can find z by substituting this value for x in equation (2).

$$\begin{aligned} x - z &= 5 && (2) \\ 12 - z &= 5 && \text{Let } x = 12. \\ z &= 7 && \text{Subtract 12. Multiply by } -1. \end{aligned}$$

Thus, $x = 12$, $y = 4$, and $z = 7$.

Step 5 **State the answer.** There were 12 loaves of honey wheat bread, 4 loaves of sunflower bread, and 7 loaves of French bread sold.

Step 6 **Check.** Because $12 = 3 \cdot 4$, the number of loaves of honey wheat bread is three times the number of loaves of sunflower bread. Also, $12 - 7 = 5$, so the number of loaves of French bread is 5 less than the number of loaves of honey wheat bread. Multiply the appropriate cost per loaf by the number of loaves sold and add the results to check that total receipts were \$87.89.

NOW TRY

CLASSROOM EXAMPLE 7

A paper mill makes newsprint, bond, and copy machine paper.

- Each ton of newsprint requires 3 tons of recycled paper and 1 ton of wood pulp.
- Each ton of bond requires 2 tons of recycled paper, 4 tons of wood pulp, and 3 tons of rags.
- Each ton of copy machine paper requires 2 tons of recycled paper, 3 tons of wood pulp, and 2 tons of rags.

The mill has 4200 tons of recycled paper, 5800 tons of wood pulp, and 3900 tons of rags. How much of each kind of paper can be made from these supplies?

Answer: newsprint: 400 tons; bond: 900 tons; copy machine paper: 600 tons

EXAMPLE 7 Solving a Business Production Problem

A company produces three flat screen television sets: models X, Y, and Z.

- Each model X set requires 2 hr of electronics work, 2 hr of assembly time, and 1 hr of finishing time.
- Each model Y requires 1 hr of electronics work, 3 hr of assembly time, and 1 hr of finishing time.
- Each model Z requires 3 hr of electronics work, 2 hr of assembly time, and 2 hr of finishing time.

There are 100 hr available for electronics, 100 hr available for assembly, and 65 hr available for finishing per week. How many of each model should be produced each week if all available time must be used?

Step 1 **Read** the problem again. There are three unknowns.

Step 2 **Assign variables.** Then organize the information in a table.

Let x = the number of model X produced per week,

y = the number of model Y produced per week,

and z = the number of model Z produced per week.

NOW TRY ANSWER

6. 12 loaves of white bread; 6 loaves of cheese bread; 9 loaves of whole grain bread

NOW TRY EXERCISE 7

Katherine has a quilting shop and makes three kinds of quilts: the lone star quilt, the bandana quilt, and the log cabin quilt.

- Each lone star quilt requires 8 hr of piecework, 4 hr of machine quilting, and 2 hr of finishing.
- Each bandana quilt requires 2 hr of piecework, 2 hr of machine quilting, and 2 hr of finishing.
- Each log cabin quilt requires 10 hr of piecework, 5 hr of machine quilting, and 2 hr of finishing.

Katherine allocates 74 hr for piecework, 42 hr for machine quilting, and 24 hr for finishing quilts each month. How many of each type of quilt should be made each month if all available time must be used?

	Each Model X	Each Model Y	Each Model Z	Totals	
Hours of Electronics Work	2	1	3	100	→ Gives equation (1)
Hours of Assembly Time	2	3	2	100	→ Gives equation (2)
Hours of Finishing Time	1	1	2	65	→ Gives equation (3)

Step 3 **Write a system of three equations.** The x model X sets require $2x$ hours of electronics, the y model Y sets require $1y$ (or y) hours of electronics, and the z model Z sets require $3z$ hours of electronics. There are 100 hr available for electronics. This gives one equation.

$$2x + y + 3z = 100$$

By similar reasoning, we write two more equations using the fact that there are 100 hr available for assembly and 65 hr available for finishing.

$$2x + y + 3z = 100 \quad \text{Electronics} \quad (1)$$
$$2x + 3y + 2z = 100 \quad \text{Assembly} \quad (2)$$
$$x + y + 2z = 65 \quad \text{Finishing} \quad (3)$$

Notice that by reading *across* the table, we can easily determine the coefficients and constants in the equations of the system.

Step 4 **Solve** the system of equations (1), (2), and (3). Because x in equation (3) has coefficient 1, we choose x as the focus variable and (3) as the working equation.

$$2x + y + 3z = 100 \quad (1)$$
$$-2x - 2y - 4z = -130 \quad \text{Multiply (3) by } -2.$$
$$-y - z = -30 \quad \text{Add.} \quad (4)$$

Eliminate x again, using equations (2) and (3).

$$2x + 3y + 2z = 100 \quad (2)$$
$$-2x - 2y - 4z = -130 \quad \text{Multiply (3) by } -2.$$
$$y - 2z = -30 \quad \text{Add.} \quad (5)$$

Solve the system of two equations (4) and (5).

$$-y - z = -30 \quad (4)$$
$$y - 2z = -30 \quad (5)$$
$$-3z = -60 \quad \text{Add.}$$
$$z = 20 \quad \text{Divide by } -3.$$

We can find y by substituting this value for z in equation (5).

$$y - 2z = -30 \quad (5)$$
$$y - 2(20) = -30 \quad \text{Let } z = 20.$$
$$y - 40 = -30 \quad \text{Multiply.}$$
$$y = 10 \quad \text{Add 40.}$$

NOW TRY ANSWER

7. lone star quilts: 3; bandana quilts: 5; log cabin quilts: 4

NOW TRY EXERCISE 7
is on the preceding page.

We can find x by substituting the values for y and z in equation (3).

$$x + y + 2z = 65 \quad (3)$$
$$x + 10 + 2(20) = 65 \quad \text{Let } y = 10 \text{ and } z = 20.$$
$$x + 50 = 65 \quad \text{Multiply. Add.}$$
$$x = 15 \quad \text{Subtract 50.}$$

Thus, $x = 15$, $y = 10$, and $z = 20$.

Step 5 **State the answer.** The company should produce 15 model X, 10 model Y, and 20 model Z sets per week.

Step 6 **Check** that these values satisfy the conditions of the problem. NOW TRY

3.3 Exercises

FOR EXTRA HELP MyMathLab®

Complete solution available in MyMathLab

1. (a) 6 oz (b) 15 oz (c) 24 oz (d) 30 oz
2. (a) \$100 (b) \$150 (c) \$200 (d) \$175
3. $\$1.89x$ 4. $\$13.50y$
5. (a) $(10 - x)$ mph (b) $(10 + x)$ mph
6. (a) $25y$ mi (b) $\frac{10}{25}$ hr, or $\frac{2}{5}$ hr (or 24 min)
7. wins: 92; losses: 70
8. wins: 74; losses: 88
9. length: 78 ft; width: 36 ft

Concept Check *Answer each question.*

1. If a container of liquid contains 60 oz of solution, what is the number of ounces of pure acid if the given solution contains the following acid concentrations?

(a) 10% **(b)** 25% **(c)** 40% **(d)** 50%

2. If \$5000 is invested in an account paying simple annual interest, how much interest will be earned during the first year at the following rates?

(a) 2% **(b)** 3% **(c)** 4% **(d)** 3.5%

3. If one pound of turkey costs \$1.89, how much will x pounds cost?

4. If one ticket to the movie *12 Years a Slave* costs \$13.50 and y tickets are sold, how much is collected from the sale?

5. If the rate of a boat in still water is 10 mph, and the rate of the current of a river is x mph, what is the rate of the boat in each case?

(a) The boat is going upstream (that is, against the current, which slows the boat down).

(b) The boat is going downstream (that is, with the current, which speeds the boat up).

6. The swimming rate of a whale is 25 mph.

(a) If the whale swims for y hours, what is its distance?

(b) If the whale travels 10 mi, what is its time?

Solve each problem. ***See Example 1.***

7. During the 2013 Major League Baseball season, the Los Angeles Dodgers played 162 games. They won 22 more games than they lost. What was their win-loss record that year?

8. Refer to **Exercise 7.** During the same 162-game season, the Colorado Rockies lost 14 more games than they won. What was the team's win-loss record?

Team	W	L
L.A. Dodgers	____	____
Arizona	81	70
San Diego	76	86
San Francisco	76	86
Colorado	____	____

Source: Major League Baseball.

9. Venus and Serena measured a tennis court and found that it was 42 ft longer than it was wide and had a perimeter of 228 ft. What were the length and the width of the tennis court?

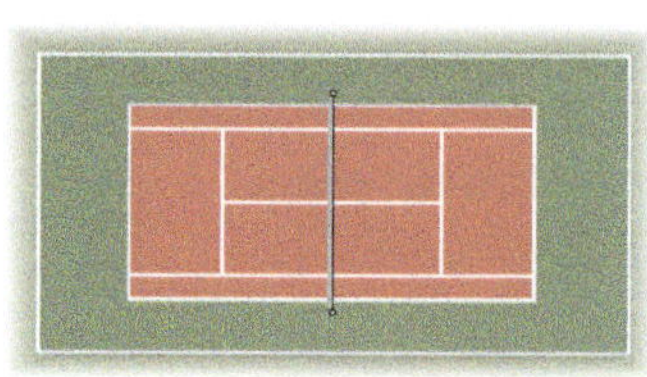

10. length: 94 ft; width: 50 ft

11. AT&T: \$126.4 billion; Verizon: \$115.8 billion

12. Canada: \$292.5 billion; Mexico: \$215.9 billion

13. $x = 40$ and $y = 50$, so the angles measure 40° and 50°.

14. $x = 100$ and $y = 80$, so the angles measure 100° and 80°.

15. hockey: \$354.84; basketball: \$315.66

16. baseball: \$207.80; football: \$459.65

17. ribeye: \$17.29; salmon: \$14.99

10. LeBron and Shaq measured a basketball court and found that the width of the court was 44 ft less than the length. If the perimeter was 288 ft, what were the length and the width of the basketball court?

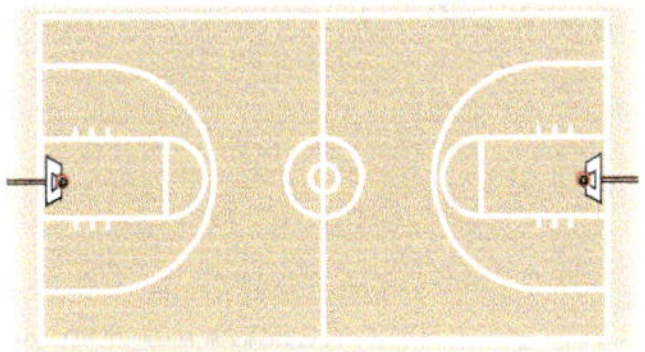

11. In 2012, the two American telecommunication companies with the greatest revenues were AT&T and Verizon. The two companies had combined revenues of \$242.2 billion. AT&T's revenue was \$10.6 billion more than that of Verizon. What was the revenue for each company? (*Source:* Verizon and AT&T Annual Reports.)

12. In 2012, U.S. exports to Canada were \$76.6 billion more than exports to Mexico. Together, exports to these two countries totaled \$508.4 billion. How much were exports to each country? (*Source:* U.S. Census Bureau.)

Find the measures of the angles marked x and y. Remember that **(1)** *the sum of the measures of the angles of a triangle is* 180°, **(2)** *supplementary angles have a sum of* 180°, *and* **(3)** *vertical angles have equal measures.*

13.

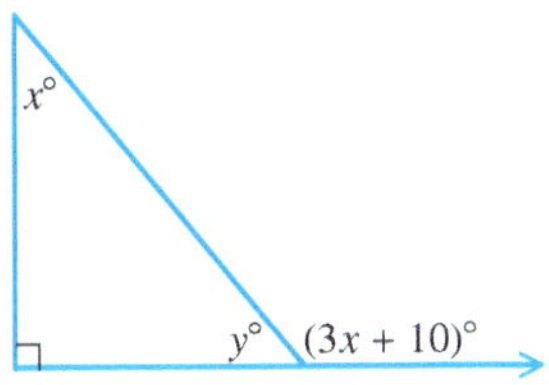

14.

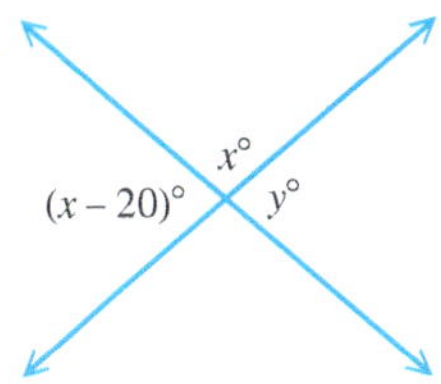

The Fan Cost Index (FCI) represents the cost of four average-price tickets (two adult, two child), four small soft drinks, two small beers, four hot dogs, parking for one car, two game programs, and two souvenir caps to a sporting event. (Source: Team Marketing Report.)

Use the concept of FCI in Exercises 15 and 16. ***See Example 2.***

15. For the 2013 season, the FCI prices for the National Hockey League and the National Basketball Association totaled \$670.50. The hockey FCI was \$39.18 more than that of basketball. What were the FCIs for these sports?

16. In 2013, the FCI prices for Major League Baseball and the National Football League totaled \$667.45. The football FCI was \$251.85 more than that of baseball. What were the FCIs for these sports?

Solve each problem. ***See Example 2.***

17. Leanna is a waitress at Bonefish Grill. During one particular day she sold 15 ribeye steak dinners and 20 grilled salmon dinners, totaling \$559.15. Another day she sold 25 ribeye steak dinners and 10 grilled salmon dinners, totaling \$582.15. How much did each type of dinner cost? (*Source*: Bonefish Grill Menu.)

18. Busch Gardens: \$72; Universal Studios: \$92
19. DVD: \$12.96; Blu-ray: \$19.49
20. New York: \$622; Washington: \$564
21. 6 gal of 25%; 14 gal of 35%
22. 80 L of 15%; 40 L of 33%
23. pure acid: 6 L; 10% acid: 48 L
24. 6 L
25. nuts: 14 kg; cereal: 16 kg
26. 50% juice: 150 L; 30% juice: 50 L

18. Two days at Busch Gardens Williamsburg (Virginia) and 3 days at Universal Studios Florida (Orlando) cost \$420, while 4 days at Busch Gardens and 2 days at Universal Studios cost \$472. (Prices are based on single-day admissions.) What was the cost per day for each park? (*Sources:* Busch Gardens and Universal Studios.)

19. The movie *Saving Mr. Banks* was available in both DVD and Blu-ray formats. The price for 3 DVDs and 2 Blu-ray discs was \$77.86, while the price for 2 DVDs and 3 Blu-ray discs was \$84.39. How much did each format cost?

20. On the basis of average total costs per day for business travel to New York City and Washington, DC (which include a hotel room, car rental, and three meals), 2 days in New York and 3 days in Washington cost \$2936, while 4 days in New York and 2 days in Washington cost \$3616. What was the average cost per day in each city? (*Source: Business Travel News.*)

Solve each problem. ***See Example 3.***

21. How many gallons each of 25% alcohol and 35% alcohol should be mixed to obtain 20 gal of 32% alcohol?

Gallons of Solution	Percent (as a decimal)	Gallons of Pure Alcohol
x	25% = 0.25	
y	35% = 0.35	
20	32% =	

22. How many liters each of 15% acid and 33% acid should be mixed to obtain 120 L of 21% acid?

Liters of Solution	Percent (as a decimal)	Liters of Pure Acid
x	15% = 0.15	
y	33% =	
120	21% =	

23. Pure acid is to be added to a 10% acid solution to obtain 54 L of a 20% acid solution. What amounts of each should be used?

24. A truck radiator holds 36 L of fluid. How much pure antifreeze must be added to a mixture that is 4% antifreeze to fill the radiator with a mixture that is 20% antifreeze?

25. A party mix is made by adding nuts that sell for \$2.50 per kg to a cereal mixture that sells for \$1 per kg. How much of each should be added to obtain 30 kg of a mix that will sell for \$1.70 per kg?

	Number of Kilograms	Price per Kilogram (in dollars)	Value (in dollars)
Nuts	x	2.50	
Cereal	y	1.00	
Mixture		1.70	

26. A fruit drink is made by mixing fruit juices. Such a drink with 50% juice is to be mixed with another drink that is 30% juice to obtain 200 L of a drink that is 45% juice. How much of each should be used?

	Liters of Drink	Percent (as a decimal)	Liters of Pure Juice
50% Juice	x	0.50	
30% Juice	y	0.30	
Mixture		0.45	

27. \$1000 at 2%; \$2000 at 4%
28. \$10,000 at 4%; \$5000 at 3%
29. train: 60 km per hr; plane: 160 km per hr
30. freight train: 50 km per hr; express train: 80 km per hr
31. scooter: 25 mph; bicycle: 10 mph
32. plane: 200 mph; car: 60 mph
33. boat: 21 mph; current: 3 mph
34. plane: 600 mph; wind: 50 mph

27. A total of \$3000 is invested, part at 2% simple interest and part at 4%. If the total annual return from the two investments is \$100, how much is invested at each rate?

Principal (in dollars)	*Rate (as a decimal)*	*Interest (in dollars)*
x	0.02	$0.02x$
y	0.04	$0.04y$
3000	XXXXXX	100

28. An investor will invest a total of \$15,000 in two accounts, one paying 4% annual simple interest and the other 3%. If he wants to earn \$550 annual interest, how much should he invest at each rate?

Principal (in dollars)	*Rate (as a decimal)*	*Interest (in dollars)*
x	0.04	
y	0.03	
15,000	XXXXXX	

Solve each problem. ***See Examples 4 and 5.***

29. A train travels 150 km in the same time that a plane travels 400 km. If the rate of the plane is 20 km per hr less than three times the rate of the train, find both rates.

	r	*t*	*d*
Train	x		150
Plane	y		400

30. A freight train and an express train leave towns 390 km apart, traveling toward one another. The freight train travels 30 km per hr slower than the express train. They pass one another 3 hr later. What are their rates?

	r	*t*	*d*
Freight Train	x	3	
Express Train	y	3	

31. A motor scooter travels 20 mi in the same time that a bicycle travels 8 mi. If the rate of the scooter is 5 mph more than twice the rate of the bicycle, find both rates.

32. A plane travels 1000 mi in the same time that a car travels 300 mi. If the rate of the plane is 20 mph greater than three times the rate of the car, find both rates.

33. In his motorboat, Bill travels upstream at top speed to his favorite fishing spot, a distance of 36 mi, in 2 hr. Returning, he finds that the trip downstream, still at top speed, takes only 1.5 hr. Find the rate of Bill's boat and the rate of the current. Let x = the rate of the boat and y = the rate of the current.

	r	*t*	*d*
Upstream	$x - y$	2	
Downstream	$x + y$		

34. Traveling for 3 hr into a steady head wind, a plane flies 1650 mi. The pilot determines that flying *with* the same wind for 2 hr, he could make a trip of 1300 mi. Find the rate of the plane and the wind speed.

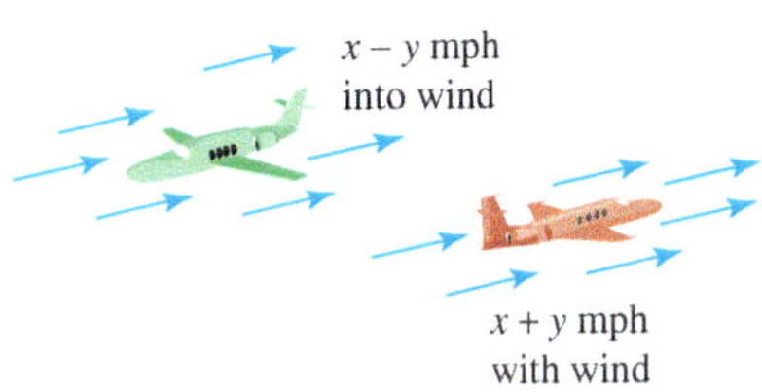

35. \$0.75-per-lb candy: 3 lb; \$1.25-per-lb candy: 7 lb

36. Madonna: \$133.7 million; Bruce Springsteen: \$104.7 million

37. general admission: 76; with student ID: 108

38. cappuccino: \$3.39; house latte: \$3.59

39. 8 for a citron; 5 for a wood apple

40. snow speeder: 2000 mph; wind: 400 mph

41. $x + y + z = 180$; angle measures: 70°, 30°, 80°

42. $x = y - 10$
$x = z - 20$
$x + y + z = 180$;
50°, 60°, 70°

Solve each problem. **See Examples 1–5.**

35. How many pounds of candy that sells for \$0.75 per lb must be mixed with candy that sells for \$1.25 per lb to obtain 10 lb of a mixture that should sell for \$1.10 per lb?

36. The top-grossing tour on the North American concert circuit for 2012 was Madonna, followed by Bruce Springsteen and the E Street Band. Together, they took in \$238.4 million from ticket sales. If Springsteen took in \$29 million less than Madonna, how much did each band generate? (*Source:* Pollstar.)

37. Tickets to a production of *A Midsummer Night's Dream* at Broward College cost \$5 for general admission or \$4 with a student ID. If 184 people paid to see a performance and \$812 was collected, how many of each type of ticket were sold?

38. At a business meeting at Panera Bread, the bill for two cappuccinos and three house lattes was \$17.55. At another table, the bill for one cappuccino and two house lattes was \$10.57. How much did each type of beverage cost? (*Source:* Panera Bread menu.)

39. The mixed price of 9 citrons and 7 fragrant wood apples is 107; again, the mixed price of 7 citrons and 9 fragrant wood apples is 101. O you arithmetician, tell me quickly the price of a citron and the price of a wood apple here, having distinctly separated those prices well. (*Source:* Hindu work, A.D. 850.)

40. Braving blizzard conditions on the planet Hoth, Luke Skywalker sets out in his snow speeder for a rebel base 4800 mi away. He travels into a steady head wind and makes the trip in 3 hr. Returning, he finds that the trip back, now with a tailwind, takes only 2 hr. Find the rate of Luke's snow speeder and the speed of the wind.

	r	*t*	*d*
Into Head Wind			
With Tailwind			

Solve each problem. **See Examples 6 and 7.** *(In Exercises 41–44, remember that the sum of the measures of the angles of a triangle is 180°.)*

41. In the triangle below, $z = x + 10$ and $x + y = 100$. Determine a third equation involving x, y, and z, and then find the measures of the three angles.

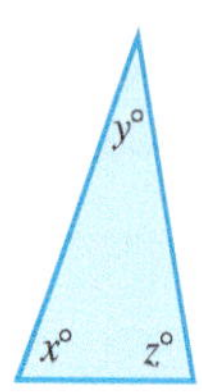

42. In the triangle below, x is 10 less than y and 20 less than z. Write a system of equations and find the measures of the three angles.

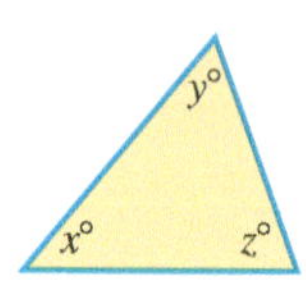

43. first: 20°; second: 70°; third: 90°
44. largest: 84°; middle: 70°; smallest: 26°
45. shortest: 12 cm; middle: 25 cm; longest: 33 cm
46. shortest: 10 in.; middle: 20 in.; longest: 26 in.
47. gold: 13; silver: 11; bronze: 9
48. Independent: 42%; Republican: 25%; Democrat: 31%
49. general admission: 1170; courtside: 985; bench: 130
50. "up close": \$40; "in the middle": \$30; "far out": \$20
51. bookstore A: 140; bookstore B: 280; bookstore C: 380
52. model A: 42; model B: 24; model C: 19

43. In a certain triangle, the measure of the second angle is 10° greater than three times the first. The third angle measure is equal to the sum of the measures of the other two. Find the measures of the three angles.

44. The measure of the largest angle of a triangle is 12° less than the sum of the measures of the other two. The smallest angle measures 58° less than the largest. Find the measures of the angles.

45. The perimeter of a triangle is 70 cm. The longest side is 4 cm less than the sum of the other two sides. Twice the shortest side is 9 cm less than the longest side. Find the length of each side of the triangle.

46. The perimeter of a triangle is 56 in. The longest side measures 4 in. less than the sum of the other two sides. Three times the shortest side is 4 in. more than the longest side. Find the lengths of the three sides.

47. In the 2014 Winter Olympics in Sochi, Russia, host Russia earned 4 more gold medals than bronze. The number of silver medals earned was 7 less than twice the number of bronze medals. Russia earned a total of 33 medals. How many of each kind of medal did Russia earn? (*Source:* www.sochi.com)

48. In a random sample of Americans of voting age conducted recently, 17% more people identified themselves as Independents than as Republicans, while 6% fewer people identified themselves as Republicans than as Democrats. Of those sampled, 2% did not identify with any of the three categories. What percent of the people in the sample identified themselves with each of the three political affiliations? (*Source:* Gallup, Inc.)

49. Tickets for a Harlem Globetrotters show cost \$28 general admission, \$43 courtside, or \$173 bench seats. Nine times as many general admission tickets were sold as bench tickets, and the number of general admission tickets sold was 55 more than the sum of the number of courtside tickets and bench tickets. Sales of all three kinds of tickets totaled \$97,605. How many of each kind of ticket were sold? (*Source:* www.harlemglobetrotters.com)

50. Three kinds of tickets are available for a rock concert: "up close," "in the middle," and "far out." "Up close" tickets cost \$10 more than "in the middle" tickets. "In the middle" tickets cost \$10 more than "far out" tickets. Twice the cost of an "up close" ticket is \$20 more than three times the cost of a "far out" ticket. Find the price of each kind of ticket.

51. A wholesaler supplies college t-shirts to three college bookstores: A, B, and C. The wholesaler recently shipped a total of 800 t-shirts to the three bookstores. Twice as many t-shirts were shipped to bookstore B as to bookstore A, and the number shipped to bookstore C was 40 less than the sum of the numbers shipped to the other two bookstores. How many t-shirts were shipped to each bookstore?

52. An office supply store sells three models of computer desks: A, B, and C. In one month, the store sold a total of 85 computer desks. The number of model B desks was five more than the number of model C desks. The number of model A desks was four more than twice the number of model C desks. How many of each model did the store sell that month?

53. first chemical: 50 kg; second chemical: 400 kg; third chemical: 300 kg

54. 5% solution: 1 oz; 20% solution: 4 oz; water: 5 oz

55. wins: 30; losses: 12; overtime losses: 6

56. wins: 22; losses: 22; overtime losses: 4

57. box of fish: 8 oz; box of bugs: 2 oz; box of worms: 5 oz

58. 25 oz

53. A plant food is to be made from three chemicals. The mix must include 60% of the first and second chemicals. The second and third chemicals must be in the ratio of 4 to 3 by weight. How much of each chemical is needed to make 750 kg of the plant food?

54. How many ounces of 5% hydrochloric acid, 20% hydrochloric acid, and water must be combined to obtain 10 oz of solution that is 8.5% hydrochloric acid if the amount of water used must equal the total amount of the other two solutions?

The National Hockey League uses a point system to determine team standings. A team is awarded 2 points for a win (W), 0 *points for a loss in regulation play* (L), *and* 1 *point for an overtime loss* (OTL). *Use this information to solve each problem.*

55. During the 2012–2013 NHL regular season, the Anaheim Ducks played 48 games. Their wins and overtime losses resulted in a total of 66 points. They had 6 more losses in regulation play than overtime losses. How many wins, losses, and overtime losses did they have that year?

Team	GP	W	L	OTL	Points
Anaheim	48	____	____	____	66
Los Angeles	48	27	16	5	59
San Jose	48	25	16	7	57
Phoenix	48	21	18	9	51
Dallas	48	____	____	____	48

Source: *World Almanac and Book of Facts.*

56. During the same NHL regular season, the Dallas Stars also played 48 games. Their wins and overtime losses resulted in a total of 48 points. They had 4 more total losses (in regulation play and overtime) than wins. How many wins, losses, and overtime losses did they have that year?

The following exercises are based on the "Bait Box Puzzle," featured in a recent "Ask Marilyn" column in Parade Magazine.

57. There are three kinds of bait boxes—those containing fish, those containing bugs, and those containing worms. Although the bait boxes look alike, each kind has a different weight. Use the information in **FIGURE A** to write a system of three equations, and solve it to determine the weight of each kind of bait box.

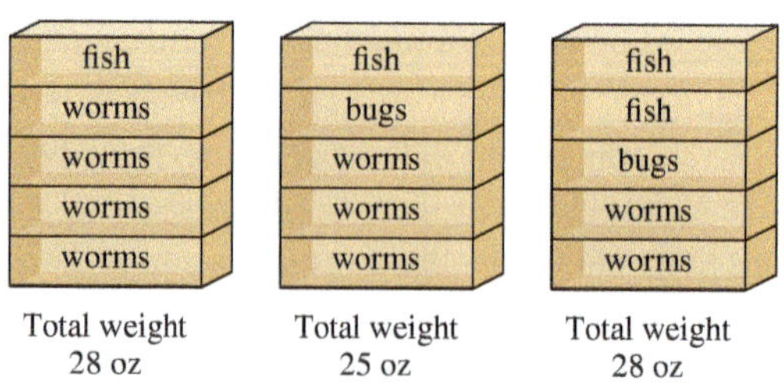

FIGURE A

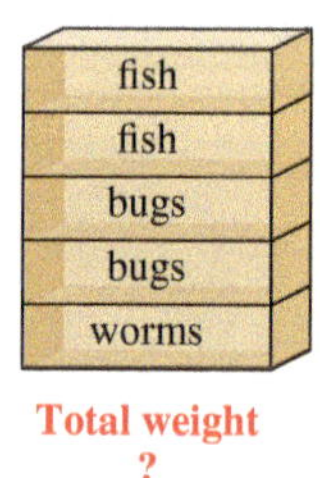

FIGURE B

58. Write an equation that describes the situation depicted in **FIGURE B**. Use the same three variables used in **Exercise 57.** Then determine the total weight of the bait boxes in **FIGURE B**.

Chapter 3 Summary

Key Terms

3.1

system of equations
system of linear equations (linear system)
solution set of a linear system
consistent system
independent equations
inconsistent system
dependent equations

3.2

ordered triple
focus variable
working equation

New Symbol

(x, y, z) ordered triple

Test Your Word Power

See how well you have learned the vocabulary in this chapter.

1. A **system of equations** consists of
   A. at least two equations with different variables
   B. two or more equations that have an infinite number of solutions
   C. two or more equations that are to be solved at the same time
   D. two or more inequalities that are to be solved.

2. The **solution set of a system of equations** is
   A. all ordered pairs that satisfy one equation of the system
   B. all ordered pairs that satisfy all the equations of the system at the same time
   C. any ordered pair that satisfies one or more equations of the system
   D. the set of values that make all the equations of the system false.

3. An **inconsistent system** is a system of equations
   A. with one solution
   B. with no solution
   C. with an infinite number of solutions
   D. that have the same graph.

4. **Dependent equations**
   A. have different graphs
   B. have no solution
   C. have one solution
   D. are different forms of the same equation.

5. A **consistent system** has
   A. no solution
   B. a solution
   C. exactly two solutions
   D. exactly three solutions.

ANSWERS

1. C; *Example:* $\begin{aligned}3x - y &= 3\\ 2x + y &= 7\end{aligned}$ **2.** B; *Example:* The ordered pair $(2, 3)$ satisfies both equations of the system in Answer 1, so $\{(2, 3)\}$ is the solution set of the system. **3.** B; *Example:* The equations of two parallel lines form an inconsistent system. Their graphs never intersect, so the system has no solution. **4.** D; *Example:* The equations $4x - y = 8$ and $8x - 2y = 16$ are dependent because their graphs are the same line. **5.** B; *Example:* The system in Answer 1 is a consistent system.

Quick Review

CONCEPTS	EXAMPLES
3.1 Systems of Linear Equations in Two Variables	
Solving a Linear System by Substitution	Solve by substitution.
Step 1 Solve one of the equations for either variable.	$4x - y = 7$ (1) $3x + 2y = 30$ (2) Solve for y in equation (1) to obtain $y = 4x - 7$.

CONCEPTS	EXAMPLES
Step 2 Substitute for that variable in the other equation. The result should be an equation with just one variable.	Substitute $4x - 7$ for y in equation (2), and solve for x. $3x + 2y = 30$ (2) $3x + 2(4x - 7) = 30$ Let $y = 4x - 7$. $3x + 8x - 14 = 30$ Distributive property
Step 3 Solve the equation from Step 2.	$11x = 44$ Combine like terms. Add 14. $x = 4$ Divide by 11.
Step 4 Find the value of the other variable by substituting the result from Step 3 into the equation from Step 1.	Substitute 4 for x in the equation $y = 4x - 7$ to find y. $y = 4(4) - 7$ $y = 9$ Multiply. Subtract.
Step 5 Check the values in *both* of the *original* equations. Then write the solution set as a set containing an ordered pair.	Check to verify that $\{(4, 9)\}$ is the solution set.
Solving a Linear System by Elimination	Solve by elimination.
Step 1 Write both equations in standard form.	$5x + y = 2$ (1) $2x - 3y = 11$ (2)
Step 2 Transform the equations as needed so that the coefficients of one pair of variable terms are opposites.	To eliminate y, multiply equation (1) by 3 and add the result to equation (2).
Step 3 Add the new equations. The sum should be an equation with just one variable.	$15x + 3y = 6$ 3 times equation (1) $2x - 3y = 11$ (2) $17x = 17$ Add.
Step 4 Solve the equation from Step 3.	$x = 1$ Divide by 17.
Step 5 Find the value of the other variable by substituting the result from Step 4 into either of the original equations.	Let $x = 1$ in equation (1), and solve for y. $5(1) + y = 2$ $y = -3$ Multiply. Subtract 5.
Step 6 Check the values in *both* of the *original* equations. Then write the solution set as a set containing an ordered pair.	Check to verify that $\{(1, -3)\}$ is the solution set.
If the result of the addition step (Step 3) is a false statement, such as $0 = 4$, the graphs are parallel lines and *there is no solution. The solution set is* $\varnothing$.	$x - 2y = 6$ $-x + 2y = -2$ $0 = 4$ Solution set: $\varnothing$
If the result is a true statement, such as $0 = 0$, the graphs are the same line, and an *infinite number of ordered pairs are solutions. The solution set is written in set-builder notation as* $\{(x, y) \mid \_\_\_\_\_\_\}$, *where a form of the equation is written in the blank.*	$x - 2y = 6$ $-x + 2y = -6$ $0 = 0$ Solution set: $\{(x, y) \mid x - 2y = 6\}$

CONCEPTS

EXAMPLES

3.2 Systems of Linear Equations in Three Variables

Solving a Linear System in Three Variables

Step 1 Select a focus variable, preferably one with coefficient 1 or -1, and a working equation.

Solve the system.

$$x + 2y - z = 6 \quad (1)$$
$$x + y + z = 6 \quad (2)$$
$$2x + y - z = 7 \quad (3)$$

We choose z as the focus variable and (2) as the working equation.

Step 2 Eliminate the focus variable, using the working equation and one of the equations of the system.

Add equations (1) and (2).

$$2x + 3y = 12 \quad (4)$$

Step 3 Eliminate the focus variable again, using the working equation and the remaining equation of the system.

Add equations (2) and (3).

$$3x + 2y = 13 \quad (5)$$

Step 4 Solve the system of two equations in two variables formed by the equations from Steps 2 and 3.

Use equations (4) and (5) to eliminate x.

$$-6x - 9y = -36 \quad \text{Multiply (4) by } -3.$$
$$6x + 4y = 26 \quad \text{Multiply (5) by 2.}$$
$$-5y = -10 \quad \text{Add.}$$
$$y = 2 \quad \text{Divide by } -5.$$

To find x, substitute 2 for y in equation (4).

$$2x + 3(2) = 12 \quad \text{Let } y = 2 \text{ in (4).}$$
$$2x + 6 = 12 \quad \text{Multiply.}$$
$$2x = 6 \quad \text{Subtract 6.}$$
$$x = 3 \quad \text{Divide by 2.}$$

Step 5 Find the value of the remaining variable.

Substitute 3 for x and 2 for y in working equation (2).

$$x + y + z = 6 \quad (2)$$
$$3 + 2 + z = 6 \quad \text{Substitute.}$$
$$z = 1 \quad \text{Subtract 5.}$$

Step 6 Check the values in *each* of the *original* equations of the system. Then write the solution set as a set containing an ordered triple.

A check of the ordered triple $(3, 2, 1)$ confirms that the solution set is $\{(3, 2, 1)\}$.

3.3 Applications of Systems of Linear Equations

Use the six-step problem-solving method.

Step 1 Read the problem carefully.

Step 2 Assign variables.

Step 3 Write a system of equations.

Step 4 Solve the system.

Step 5 State the answer.

Step 6 Check.

The perimeter of a rectangle is 18 ft. The length is 3 ft more than twice the width. What are the dimensions of the rectangle?

Let x = the length and y = the width. Write a system of equations.

$$2x + 2y = 18 \quad \text{From the perimeter formula}$$
$$x = 2y + 3 \quad \text{Length is 3 ft more than twice the width.}$$

Solve to find that

$$x = 7 \quad \text{and} \quad y = 2.$$

The length is 7 ft, and the width is 2 ft. The answer checks because

$$2(7) + 2(2) = 18 \quad \text{and} \quad 2(2) + 3 = 7, \quad \text{as required.}$$

Chapter 3 Review Exercises

1. (a) 2004–2005
   (b) 30%
2. $\{(2, 2)\}$

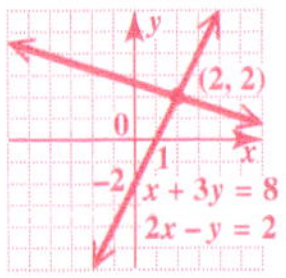

3. D
4. Answers will vary.
   (a)

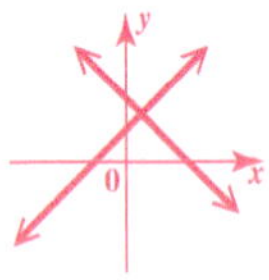

   (b)

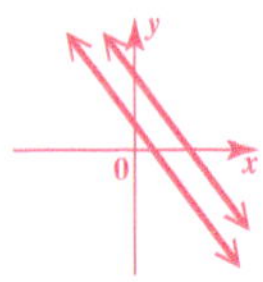

   (c)

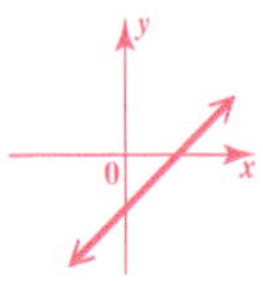

5. $\left\{\left(-\frac{8}{9}, -\frac{4}{3}\right)\right\}$
6. $\{(0, 4)\}$
7. $\{(2, 4)\}$
8. $\{(2, 2)\}$
9. $\{(0, 1)\}$
10. $\{(-1, 2)\}$
11. $\{(-6, 3)\}$
12. $\{(x, y) \mid 3x - y = -6\}$; dependent equations
13. $\emptyset$; inconsistent system
14. The two lines have the same slope, 3, but the y-intercepts, $(0, 2)$ and $(0, -4)$, are different. Therefore, the lines are parallel, do not intersect, and have no common solution.

3.1

1. The graph shows trends during the years 2002 through 2013 relating broadband vs. dial-up Internet users in the United States.
   (a) Between what years did the percentage of broadband users equal that for dial-up users?
   (b) When the percentage of broadband internet users was equal to that for dial-up internet users, approximately what was that number?

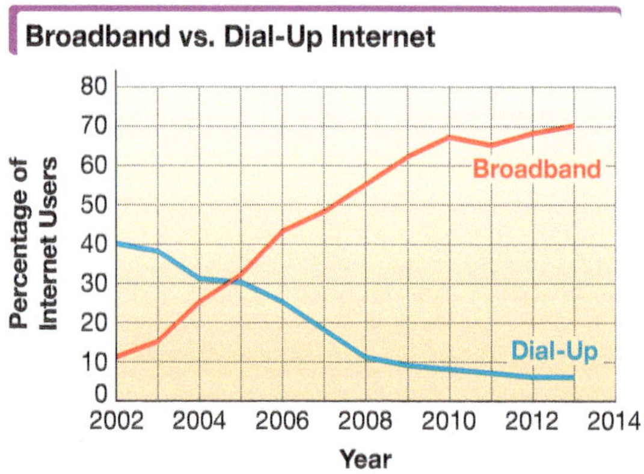

Source: Pew Research Center, © 2013. Used by permission of Pew Research Center.

2. Solve the system by graphing.

$$x + 3y = 8$$
$$2x - y = 2$$

3. Which ordered pair is *not* a solution of the equation $3x + 2y = 6$?
   A. $(2, 0)$ **B.** $(0, 3)$ **C.** $(4, -3)$ **D.** $(3, -2)$

4. Suppose that two linear equations are graphed on the same set of coordinate axes. Sketch what the graph might look like if the system has the given description.
   (a) The system has a single solution.
   (b) The system has no solution.
   (c) The system has infinitely many solutions.

Solve each system by the substitution method. If a system is inconsistent or has dependent equations, say so.

5. $3x + y = -4$
   $x = \frac{2}{3}y$

6. $9x - y = -4$
   $y = x + 4$

7. $-5x + 2y = -2$
   $x + 6y = 26$

Solve each system of equations by the elimination method. If a system is inconsistent or has dependent equations, say so.

8. $5x + y = 12$
   $2x - 2y = 0$

9. $x - 4y = -4$
   $3x + y = 1$

10. $6x + 5y = 4$
    $-4x + 2y = 8$

11. $\frac{1}{6}x + \frac{1}{6}y = -\frac{1}{2}$
    $x - y = -9$

12. $-3x + y = 6$
    $y = 6 + 3x$

13. $5x - 4y = 2$
    $-10x + 8y = 7$

14. Without doing any algebraic work, but answering on the basis of knowledge of the graphs of the two lines, explain why the following system has $\emptyset$ as its solution set.

$$y = 3x + 2$$
$$y = 3x - 4$$

15. $\{(1, 2, 3)\}$
16. $\{(1, -5, 3)\}$
17. $\emptyset$; inconsistent system
18. length: 200 ft; width: 85 ft
19. New York Yankees: \$51.55; Boston Red Sox: \$53.38
20. plane: 300 mph; wind: 20 mph
21. gold: 24; silver: 26; bronze: 32
22. \$2-per-lb nuts: 30 lb; \$1-per-lb candy: 70 lb
23. 85°, 60°, 35°
24. \$40,000 at 10%; \$100,000 at 6%; \$140,000 at 5%
25. 5 L of 8%; 3 L of 20%; none of 10%
26. Mantle: 54; Maris: 61; Berra: 22

3.2 *Solve each system. If a system is inconsistent or has dependent equations, say so.*

15. $\begin{aligned} 4x - y &= 2 \\ 3y + z &= 9 \\ x + 2z &= 7 \end{aligned}$

16. $\begin{aligned} 2x + 3y - z &= -16 \\ x + 2y + 2z &= -3 \\ -3x + y + z &= -5 \end{aligned}$

17. $\begin{aligned} 3x - y - z &= -8 \\ 4x + 2y + 3z &= 15 \\ -6x + 2y + 2z &= 10 \end{aligned}$

3.3 *Solve each problem.*

18. A regulation National Hockey League ice rink has perimeter 570 ft. The length of the rink is 30 ft longer than twice the width. What are the dimensions of an NHL ice rink? (*Source:* www.nhl.com)

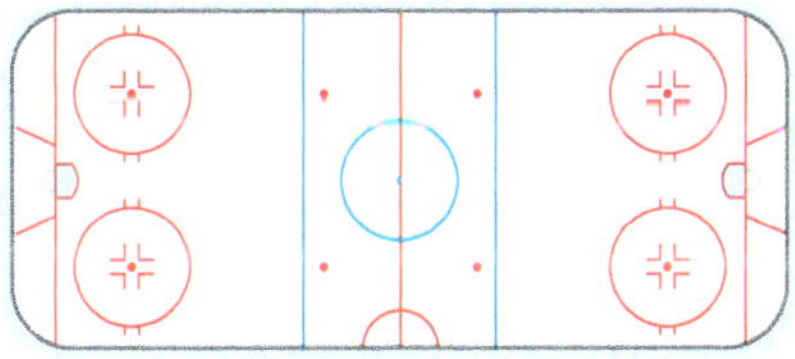

19. During a recent Major League Baseball season, two New York Yankees tickets and three Boston Red Sox tickets purchased at their average prices cost \$263.24. Three Yankees tickets and two Red Sox tickets cost \$261.41. Find the average ticket price for a Yankees ticket and a Red Sox ticket. (*Source:* Team Marketing Report.)

20. A plane flies 560 mi in 1.75 hr traveling with the wind. The return trip later against the same wind takes the plane 2 hr. Find the rate of the plane and the wind speed. Let $x =$ the rate of the plane and $y =$ the wind speed.

	r	*t*	*d*
With Wind	$x + y$	1.75	
Against Wind		2	

21. In the 2012 Summer Olympics in London, England, Russia earned 8 fewer gold medals than bronze. The number of silver medals earned was 38 less than twice the number of bronze medals. Russia earned a total of 82 medals. How many of each kind of medal did Russia earn? (*Source: World Almanac and Book of Facts.*)

22. For Valentine's Day, Ms. Sweet will mix some \$2-per-lb nuts with some \$1-per-lb chocolate candy to obtain 100 lb of mix, which she will sell at \$1.30 per lb. How many pounds of each should she use?

	Number of Pounds	*Price per Pound (in dollars)*	*Value (in dollars)*
Nuts	*x*		
Chocolate	*y*		
Mixture	100		

23. The sum of the measures of the angles of a triangle is 180°. The largest angle measures 10° less than the sum of the other two. The measure of the middle-sized angle is the average of the other two. Find the measures of the three angles.

24. Noemi sells real estate. On three recent sales, she made 10% commission, 6% commission, and 5% commission. Her total commissions on these sales were \$17,000, and she sold property worth \$280,000. If the 5% sale amounted to the sum of the other two, what were the three sales prices?

25. How many liters each of 8%, 10%, and 20% hydrogen peroxide should be mixed together to obtain 8 L of 12.5% solution if the amount of 8% solution used must be 2 L more than the amount of 20% solution used?

26. In the great baseball year of 1961, Yankee teammates Mickey Mantle, Roger Maris, and Yogi Berra combined for 137 home runs. Mantle hit 7 fewer than Maris, and Maris hit 39 more than Berra. What were the home run totals for each player? (*Source:* www.mlb.com)

Chapter 3 Mixed Review Exercises

1. $\{(0, 4)\}$
2. $\left\{\left(\frac{82}{23}, -\frac{4}{23}\right)\right\}$
3. $\{(5, 3)\}$
4. $\{(3, -1)\}$
5. $\{(12, 9)\}$
6. $\varnothing$
7. B; The second equation is already solved for y.
8. *System A:* $\left\{\left(\frac{65}{46}, \frac{1}{46}\right)\right\}$; *System B:* $\{(2, -5)\}$
9. 20 L
10. United States: 104; China: 88; Russia: 82

Solve by any method.

1. $-7x + 3y = 12$
$5x + 2y = 8$

2. $2x - 5y = 8$
$3x + 4y = 10$

3. $x + 4y = 17$
$-3x + 2y = -9$

4. $x = 7y + 10$
$2x + 3y = 3$

5. $\frac{2}{3}x + \frac{1}{6}y = \frac{19}{2}$
$\frac{1}{3}x - \frac{2}{9}y = 2$

6. $2x + 5y - z = 12$
$-x + y - 4z = -10$
$-8x - 20y + 4z = 31$

Solve each problem.

7. Which system would be easier to solve using the substitution method? Why?

System A: $5x - 3y = 7$
$2x + 8y = 3$

System B: $7x + 2y = 4$
$y = -3x + 1$

8. Solve each system in **Exercise 7.**

9. To make a 10% acid solution, Jeffrey wants to mix some 5% solution with 10 L of 20% solution. How many liters of 5% solution should he use?

10. In the 2012 Summer Olympics, China, the United States, and Russia won a combined total of 274 medals. China won 16 fewer medals than the United States, while Russia won 22 fewer medals than the United States. How many medals did each country win? (*Source: World Almanac and Book of Facts.*)

Chapter 3 Test

FOR EXTRA HELP *Step-by-step test solutions are found on the Chapter Test Prep Videos available in* MyMathLab® *or on* YouTube™.

View the complete solutions to all Chapter Test exercises in MyMathLab.

[3.1]
1. (a) Atlanta and Phoenix
   (b) Chicago
   (c) Chicago, Phoenix, New York, Atlanta
2. (a) 2000; 8 million
   (b) (2000, 8)
3. $\{(6, 1)\}$

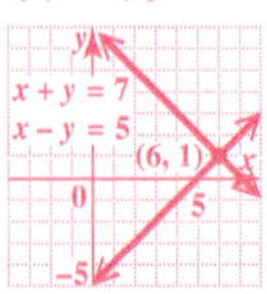

4. $\{(6, -4)\}$
5. $\left\{\left(-\frac{9}{4}, \frac{5}{4}\right)\right\}$
6. $\{(x, y) \mid 12x - 5y = 8\}$; dependent equations
7. $\{(3, 3)\}$
8. $\{(0, -2)\}$
9. $\varnothing$; inconsistent system

Use the graph to work Exercises 1 and 2.

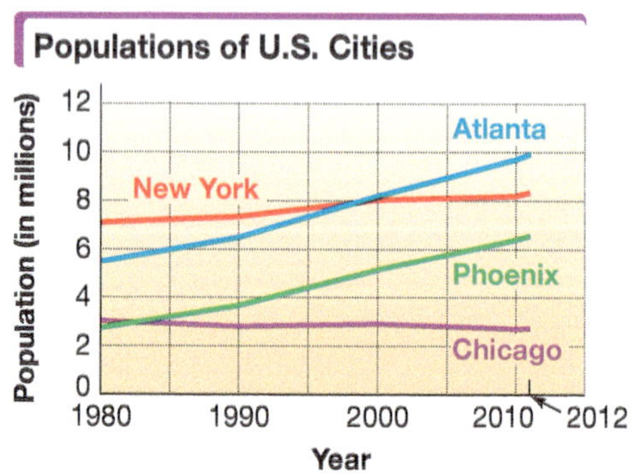

Source: "City Population," by Thomas Brinkhoff, © 2013.

1. (a) Which two cities experienced the most population growth over time?

(b) Which city experienced a population decline?

(c) Rank the city populations from least to greatest for the year 2010.

2. (a) In which year did the population of Atlanta equal that of New York? About what was that population to the nearest million?

(b) Write the answer from part (a) as an ordered pair (year, population).

3. Use a graph to solve the system.
$x + y = 7$
$x - y = 5$

Solve each system. If a system is inconsistent or has dependent equations, say so.

4. $2x - 3y = 24$
$y = -\frac{2}{3}x$

5. $3x - y = -8$
$2x + 6y = 3$

6. $12x - 5y = 8$
$3x = \frac{5}{4}y + 2$

7. $3x + y = 12$
$2x - y = 3$

8. $-5x + 2y = -4$
$6x + 3y = -6$

9. $3x + 4y = 8$
$8y = 7 - 6x$

[3.2]

10. $\left\{\left(-\frac{2}{3}, \frac{4}{5}, 0\right)\right\}$

11. $\{(3, -2, 1)\}$

[3.3]

12. *Iron Man 3:* \$409 million; *Man of Steel:* \$291 million

13. 45 mph, 75 mph

14. 20% solution: 4 L; 50% solution: 8 L

15. AC adaptor: \$8; rechargeable flashlight: \$15

16. Orange Pekoe: 60 oz; Irish Breakfast: 30 oz; Earl Grey: 10 oz

10. $\begin{aligned} 3x + 5y + 3z &= 2 \\ 6x + 5y + z &= 0 \\ 3x + 10y - 2z &= 6 \end{aligned}$

11. $\begin{aligned} 4x + y + z &= 11 \\ x - y - z &= 4 \\ y + 2z &= 0 \end{aligned}$

Solve each problem.

12. Two top-grossing super hero films, *Iron Man 3* and *Man of Steel,* earned \$700 million together. If *Man of Steel* grossed \$118 million less than *Iron Man 3,* how much did each film gross? (*Source*: www.the-numbers.com)

13. Two cars start from points 420 mi apart and travel toward each other. They meet after 3.5 hr. Find the average rate of each car if one travels 30 mph slower than the other.

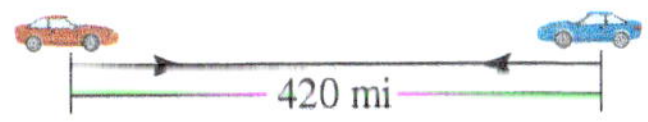

14. A chemist needs 12 L of a 40% alcohol solution. She must mix a 20% solution and a 50% solution. How many liters of each will be required to obtain what she needs?

Liters of Solution	Percent (as a decimal)	Liters of Pure Alcohol

15. A local electronics store will sell seven AC adaptors and two rechargeable flashlights for \$86, or three AC adaptors and four rechargeable flashlights for \$84. What is the price of a single AC adaptor and a single rechargeable flashlight?

16. The owner of a tea shop wants to mix three kinds of tea to make 100 oz of a mixture that will sell for \$0.83 per oz. He uses Orange Pekoe, which sells for \$0.80 per oz, Irish Breakfast, for \$0.85 per oz, and Earl Grey, for \$0.95 per oz. If he wants to use twice as much Orange Pekoe as Irish Breakfast, how much of each kind of tea should he use?

Chapters R–3 Cumulative Review Exercises

[R.3]

1. 81
2. −81
3. −81
4. 0.7
5. −0.7
6. It is not a real number.
7. −199
8. 455

[R.4]

9. commutative property

[1.1] 10. $\left\{-\frac{15}{4}\right\}$

[1.7] 11. $\left\{\frac{2}{3}, 2\right\}$

[1.2] 12. $x = \frac{d - by}{a}$

[1.1] 13. $\{11\}$

[1.5] 14. $\left(-\infty, \frac{240}{13}\right]$

[1.7] 15. $\left[-2, \frac{2}{3}\right]$ 16. $(-\infty, \infty)$

[1.6] 17. $(-\infty, \infty)$

Evaluate each expression if possible.

1. $(-3)^4$ **2.** -3^4 **3.** $-(-3)^4$

4. $\sqrt{0.49}$ **5.** $-\sqrt{0.49}$ **6.** $\sqrt{-0.49}$

Evaluate for $x = -4$, $y = 3$, *and* $z = 6$.

7. $|2x| + 3y - z^3$ **8.** $-5(x^3 - y^3)$

9. Which property of real numbers justifies the statement $5 + (3 \cdot 6) = 5 + (6 \cdot 3)$?

Solve each equation.

10. $7(2x + 3) - 4(2x + 1) = 2(x + 1)$ **11.** $|6x - 8| = 4$

12. $ax + by = d$ for x **13.** $0.04x + 0.06(x - 1) = 1.04$

Solve each inequality.

14. $\frac{2}{3}x + \frac{5}{12}x \le 20$ **15.** $|3x + 2| \le 4$

16. $|12t + 7| \ge 0$ **17.** $2x + 3 > 5 \quad \text{or} \quad x - 1 \le 6$

[1.2]
18. 2010; 1813; 62.8%; 57.2%

[1.3, 1.4]
19. pennies: 35; nickels: 29; dimes: 30
20. 46°, 46°, 88°

[2.1]
21. $y = 6$ **22.** $x = 4$

[2.2]
23. $-\frac{4}{3}$ **24.** $\frac{3}{4}$

[2.3]
25. $4x + 3y = 10$

[2.2]
26.

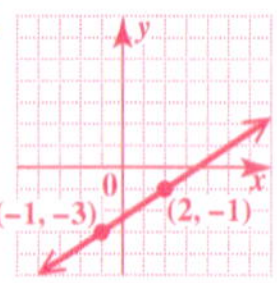

[2.4]
27.

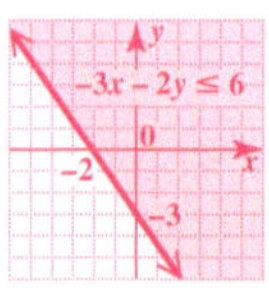

[2.6]
28. (a) -6
(b) $a^2 + 3a - 6$

[3.1]
29. $\{(3, -3)\}$
30. $\{(x, y) \mid x - 3y = 7\}$

[3.2]
31. $\{(5, 3, 2)\}$

[3.1]
32. (a) $x = 8$, or 800 items; \$3000
(b) about \$400

Solve each problem.

18. A survey measured public recognition of some popular contemporary advertising slogans. Complete the results shown in the table if 2500 people were surveyed.

Slogan (product or company)	*Percent Recognition (nearest tenth of a percent)*	*Actual Number Who Recognized Slogan (nearest whole number)*
Please Don't Squeeze the . . . (Charmin)	80.4%	
The Breakfast of Champions (Wheaties)	72.5%	
The King of Beers (Budweiser)		1570
Like a Good Neighbor (State Farm)		1430

Source: Department of Integrated Marketing Communications, Northwestern University.

19. A jar contains only pennies, nickels, and dimes. The number of dimes is one more than the number of nickels, and the number of pennies is six more than the number of nickels. How many of each denomination are in the jar, if the total value is \$4.80?

20. Two angles of a triangle have the same measure. The measure of the third angle is 4° less than twice the measure of each of the equal angles. Find the measures of the three angles.

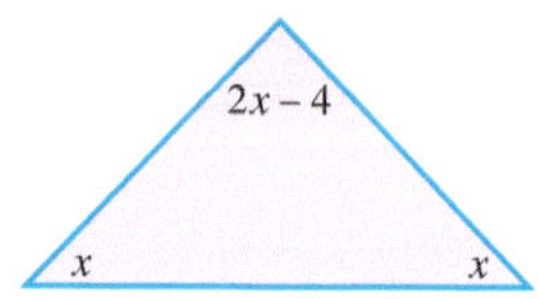

Measures are in degrees.

In Exercises 21–25, point A has coordinates $(-2, 6)$ *and point B has coordinates* $(4, -2)$.

21. What is the equation of the horizontal line through A?

22. What is the equation of the vertical line through B?

23. What is the slope of line AB?

24. What is the slope of a line perpendicular to line AB?

25. What is the standard form of the equation of line AB?

26. Graph the line having slope $\frac{2}{3}$ and passing through the point $(-1, -3)$.

27. Graph the inequality $-3x - 2y \le 6$.

28. Given that $f(x) = x^2 + 3x - 6$, find **(a)** $f(-3)$ and **(b)** $f(a)$.

Solve by any method.

29. $-2x + 3y = -15$
$4x - y = 15$

30. $x - 3y = 7$
$2x - 6y = 14$

31. $x + y + z = 10$
$x - y - z = 0$
$-x + y - z = -4$

32. The graph shows a company's costs to produce computer parts and the revenue from the sale of computer parts.

(a) At what production level does the cost equal the revenue? What is the revenue at that point?

(b) Profit is revenue less cost. Estimate the profit on the sale of 1100 parts.

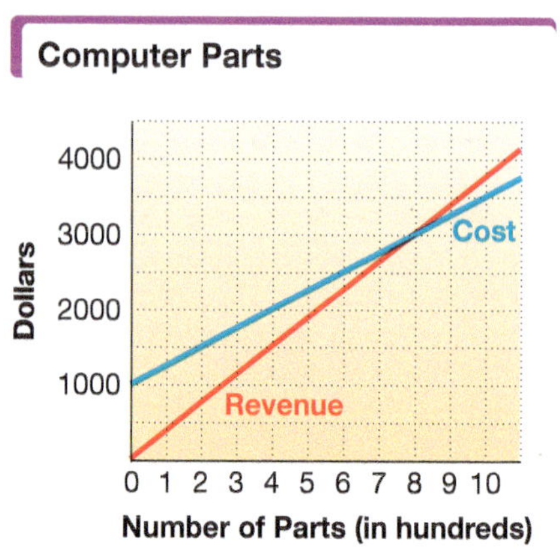

4 Exponents, Polynomials, and Polynomial Functions

Large numbers, such as world population, can be written using *scientific notation,* one of the topics of this chapter.

4.1 Integer Exponents and Scientific Notation

OBJECTIVES

1. Use the product rule for exponents.
2. Define 0 and negative exponents.
3. Use the quotient rule for exponents.
4. Use the power rules for exponents.
5. Simplify exponential expressions.
6. Use the rules for exponents with scientific notation.

VOCABULARY

- ☐ exponent (power)
- ☐ base
- ☐ exponential expression

Recall that we use exponents to write products of repeated factors.

$$2^5 \quad \text{is defined as} \quad 2 \cdot 2 \cdot 2 \cdot 2 \cdot 2, \quad \text{which equals} \quad 32.$$

In 2^5, the number 5 is the **exponent** (or **power**). It indicates that the **base** 2 appears as a factor five times. The quantity 2^5 is an **exponential expression.** We read 2^5 as **"2 to the fifth power"** or **"2 to the fifth."**

OBJECTIVE 1 Use the product rule for exponents.

Consider the product $2^5 \cdot 2^3$, which can be simplified as follows.

$$5 + 3 = 8$$

$$2^5 \cdot 2^3 = (2 \cdot 2 \cdot 2 \cdot 2 \cdot 2)(2 \cdot 2 \cdot 2) = 2^8$$

This result suggests the **product rule for exponents.**

Product Rule for Exponents

If m and n are natural numbers and a is any real number, then

$$a^m \cdot a^n = a^{m+n}.$$

That is, when multiplying powers of like bases, keep the same base and add the exponents.

CLASSROOM EXAMPLE 1

Apply the product rule, if possible, in each case.

(a) $m^8 \cdot m^6$ (b) $m^5 \cdot p^4$

(c) $(-5p^4)(-9p^5)$

(d) $(-3x^2y^3)(7xy^4)$

Answers: (a) m^{14}

(b) The product rule does not apply.

(c) $45p^9$ (d) $-21x^3y^7$

To see that the product rule is true, use the definition of an exponent.

$$a^m = \underbrace{a \cdot a \cdot a \cdot \ldots \cdot a}_{a \text{ appears as a factor } m \text{ times.}} \qquad a^n = \underbrace{a \cdot a \cdot a \cdot \ldots \cdot a}_{a \text{ appears as a factor } n \text{ times.}}$$

From this,

$$a^m \cdot a^n = \underbrace{a \cdot a \cdot a \cdot \ldots \cdot a}_{m \text{ factors}} \cdot \underbrace{a \cdot a \cdot a \cdot \ldots \cdot a}_{n \text{ factors}}$$

$$= \underbrace{a \cdot a \cdot a \cdot \ldots \cdot a}_{(m+n) \text{ factors}}$$

$$a^m \cdot a^n = a^{m+n}.$$

NOW TRY EXERCISE 1

Apply the product rule, if possible, in each case.

(a) $8^5 \cdot 8^4$

(b) $r^7 \cdot r$

(c) $p^2 \cdot q^2$

(d) $(5x^4y^7)(-7xy^3)$

EXAMPLE 1 Using the Product Rule for Exponents

Apply the product rule for exponents, if possible, in each case.

(a) $3^4 \cdot 3^7$

$= 3^{4+7}$ Keep the same base.

$= 3^{11}$

(b) $5^3 \cdot 5$

$= 5^3 \cdot 5^1$

$= 5^{3+1}$

$= 5^4$

(c) $y^3 \cdot y^8 \cdot y^2$

$= y^{3+8+2}$

$= y^{13}$

(d) $(5y^2)(-3y^4)$

$= 5(-3)y^2y^4$ Commutative property

$= -15y^{2+4}$ Multiply; product rule

$= -15y^6$

(e) $(7p^3q)(2p^5q^2)$

$= 7(2)p^3p^5q^1q^2$

$= 14p^{3+5}q^{1+2}$

$= 14p^8q^3$

(f) $x^2 \cdot y^4$ The product rule does not apply because the bases are not the same.

NOW TRY

NOW TRY ANSWERS

1. (a) 8^9 (b) r^8
   (c) The product rule does not apply.
   (d) $-35x^5y^{10}$

TEACHING TIP
Show this example.

$4^2 \cdot 5^3 \neq (4 \cdot 5)^{2+3}$

CAUTION Be careful not to multiply the bases. In **Example 1(a),**

$$3^4 \cdot 3^7 = 3^{11}, \quad \textit{not} \quad 9^{11}.$$

Keep the same base and add the exponents.

OBJECTIVE 2 Define 0 and negative exponents.

Consider the following, where the product rule is extended to whole numbers.

$$4^2 \cdot 4^0 = 4^{2+0} = 4^2$$

For the product rule to hold, 4^0 must equal 1, so we define a^0 this way for any nonzero real number a.

Zero Exponent

If a is any nonzero real number, then

$$\mathbf{a^0 = 1.}$$

The expression 0^0 is undefined.*

NOW TRY EXERCISE 2
Evaluate.
(a) 5^0 **(b)** $(-5x)^0 \quad (x \neq 0)$
(c) -5^0 **(d)** $10^0 - 9^0$

EXAMPLE 2 Using 0 as an Exponent

Evaluate.

(a) $6^0 = 1$

(b) $(-6)^0 = 1$ — Here the base is −6.

(c) $-6^0 = -(6)^0 = -1$ — The base is 6, not −6.

(d) $-(-6)^0 = -1$

(e) $5^0 + 12^0$
$= 1 + 1$
$= 2$

(f) $(8k)^0 = 1 \quad (k \neq 0)$
Any nonzero quantity raised to the zero power equals 1. NOW TRY

TEACHING TIP
In **Example 2,** students may try to evaluate $(-6)^0$ in the same manner as -6^0. Emphasize the difference.

CLASSROOM EXAMPLE 2
Evaluate.
(a) 29^0 (b) $(-29)^0$
(c) -29^0 (d) $8^0 - 15^0$
Answers: (a) 1 (b) 1
(c) −1 (d) 0

To define a negative exponent, we extend the product rule, as follows.

$$8^2 \cdot 8^{-2} = 8^{2+(-2)} = 8^0 = 1$$

Because their product is 1, it appears that 8^{-2} is the *reciprocal* of 8^2. But we know that $\frac{1}{8^2}$ is the reciprocal of 8^2, and a number can have only one reciprocal. Therefore, $\mathbf{8^{-2}}$ **must equal** $\mathbf{\frac{1}{8^2}}$.

We can generalize this result.

Negative Exponent

For any natural number n and any nonzero real number a,

$$\mathbf{a^{-n} = \frac{1}{a^n}.}$$

With this definition, the expression a^n is meaningful for any integer exponent n and any nonzero real number a.

NOW TRY ANSWERS
2. **(a)** 1 **(b)** 1 **(c)** −1 **(d)** 0

*In advanced courses, 0^0 is called an *indeterminate form.*

TEACHING TIP
Caution students that changing the sign of the exponent does not change the sign of the base.

CAUTION ***A negative exponent does not indicate that an expression represents a negative number. Negative exponents lead to reciprocals.***

$$3^{-2} = \frac{1}{3^2} = \frac{1}{9} \quad \text{Not negative} \qquad -3^{-2} = -\frac{1}{3^2} = -\frac{1}{9} \quad \text{Negative}$$

The base is 3 in both cases. 3^{-2} represents the reciprocal of 3^2.

NOW TRY EXERCISE 3
Write with only positive exponents.
(a) 9^{-4}
(b) $(3y)^{-6}$ $(y \neq 0)$
(c) $-4k^{-3}$ $(k \neq 0)$
(d) Evaluate $4^{-1} + 6^{-1}$.

CLASSROOM EXAMPLE 3
Write with only positive exponents.
(a) 6^{-5}
(b) $(2x)^{-4}$ $(x \neq 0)$
(c) $-7p^{-4}$ $(p \neq 0)$
(d) Evaluate $4^{-1} - 2^{-1}$.

Answers: (a) $\frac{1}{6^5}$ (b) $\frac{1}{(2x)^4}$ (c) $-\frac{7}{p^4}$ (d) $-\frac{1}{4}$

EXAMPLE 3 Using Negative Exponents

In parts (a)–(f), write with only positive exponents.

(a) $2^{-3} = \frac{1}{2^3}$ — This is the *reciprocal* of 2^3.

(b) $6^{-1} = \frac{1}{6^1} = \frac{1}{6}$ — This is the *reciprocal* of 6^1, or 6.

(c) $(5z)^{-3} = \frac{1}{(5z)^3}$ $(z \neq 0)$

The base is $5z$.
(Notice the parentheses around it.)

(d) $5z^{-3} = 5\left(\frac{1}{z^3}\right) = \frac{5}{z^3}$ $(z \neq 0)$

The base is z, *not* $5z$.

(e) $-m^{-2} = -\frac{1}{m^2}$ $(m \neq 0)$

(What is the base here?)

(f) $(-m)^{-2} = \frac{1}{(-m)^2}$ $(m \neq 0)$

(What is the base here?)

In parts (g) and (h), evaluate.

(g) $3^{-1} + 4^{-1}$

$= \frac{1}{3} + \frac{1}{4}$ Definition of negative exponent

$= \frac{4}{12} + \frac{3}{12}$ $\frac{1}{3} \cdot \frac{4}{4} = \frac{4}{12}$; $\frac{1}{4} \cdot \frac{3}{3} = \frac{3}{12}$

$= \frac{7}{12}$ $\frac{a}{c} + \frac{b}{c} = \frac{a+b}{c}$

(h) $5^{-1} - 2^{-1}$

$= \frac{1}{5} - \frac{1}{2}$ Definition of negative exponent

$= \frac{2}{10} - \frac{5}{10}$ Find a common denominator.

$= -\frac{3}{10}$ $\frac{a}{c} - \frac{b}{c} = \frac{a-b}{c}$

NOW TRY

CLASSROOM EXAMPLE 4
Evaluate.
(a) $\frac{1}{4^{-3}}$ (b) $\frac{3^{-3}}{9^{-1}}$

Answers: (a) 64 (b) $\frac{1}{3}$

CAUTION Note that $3^{-1} + 4^{-1} \neq (3+4)^{-1}$. The expression $3^{-1} + 4^{-1}$ simplifies to $\frac{7}{12}$, as shown in **Example 3(g).** The expression $(3+4)^{-1}$ simplifies to 7^{-1}, which equals $\frac{1}{7}$. Similar reasoning can be applied to part (h).

NOW TRY EXERCISE 4
Evaluate.
(a) $\frac{1}{5^{-3}}$ (b) $\frac{10^{-2}}{2^{-5}}$

EXAMPLE 4 Using Negative Exponents

Evaluate.

(a) $\frac{1}{2^{-3}} = \frac{1}{\frac{1}{2^3}} = 1 \div \frac{1}{2^3} = 1 \cdot \frac{2^3}{1} = 2^3 = 8$

Multiply by the reciprocal of the divisor.

$\frac{1}{2^{-3}}$ represents the reciprocal of 2^{-3}. Because $2^{-3} = \frac{1}{8}$, the reciprocal is 8, which agrees with the final answer.

(b) $\frac{2^{-3}}{3^{-2}} = \frac{\frac{1}{2^3}}{\frac{1}{3^2}} = \frac{1}{2^3} \div \frac{1}{3^2} = \frac{1}{2^3} \cdot \frac{3^2}{1} = \frac{3^2}{2^3} = \frac{9}{8}$

NOW TRY

NOW TRY ANSWERS
3. (a) $\frac{1}{9^4}$ (b) $\frac{1}{(3y)^6}$ (c) $-\frac{4}{k^3}$ (d) $\frac{5}{12}$
4. (a) 125 (b) $\frac{8}{25}$

TEACHING TIP
Challenge students to predict some of the more common incorrect answers to **Examples 3 and 4,** and then point out why they are incorrect. Show them why, for example,

$$\frac{2^{-1}+3^{-1}}{6^{-1}} \neq \frac{6}{2+3}.$$

Example 4 suggests the following generalizations.

Special Rules for Negative Exponents

If $a \neq 0$ and $b \neq 0$, then

$$\frac{1}{a^{-n}} = a^n \quad \text{and} \quad \frac{a^{-n}}{b^{-m}} = \frac{b^m}{a^n}.$$

OBJECTIVE 3 Use the quotient rule for exponents.

We simplify a quotient, such as $\frac{a^8}{a^3}$, in much the same way as a product. (In all quotients of this type, assume that the denominator is not 0.) Consider this example.

$$\frac{a^8}{a^3} = \frac{a \cdot a \cdot a \cdot a \cdot a \cdot a \cdot a \cdot a}{a \cdot a \cdot a} = a \cdot a \cdot a \cdot a \cdot a = a^5$$

Notice that $8 - 3 = 5$. In the same way, we simplify $\frac{a^3}{a^8}$.

$$\frac{a^3}{a^8} = \frac{a \cdot a \cdot a}{a \cdot a \cdot a \cdot a \cdot a \cdot a \cdot a \cdot a} = \frac{1}{a^5} = a^{-5}$$

Here, $3 - 8 = -5$. These examples suggest the **quotient rule for exponents.**

CLASSROOM EXAMPLE 5
Apply the quotient rule, if possible, and write each result with only positive exponents.

(a) $\frac{m^8}{m^{13}}$ $(m \neq 0)$ (b) $\frac{5^{-6}}{5^{-8}}$

(c) $\frac{x^3}{y^5}$ $(y \neq 0)$

Answers: (a) $\frac{1}{m^5}$ (b) 5^2, or 25
(c) The quotient rule does not apply.

Quotient Rule for Exponents

If a is any nonzero real number and m and n are integers, then

$$\frac{a^m}{a^n} = a^{m-n}.$$

That is, when dividing powers of like bases, keep the same base and subtract the exponent of the denominator from the exponent of the numerator.

NOW TRY EXERCISE 5
Apply the quotient rule, if possible, and write each result with only positive exponents.

(a) $\frac{t^8}{t^2}$ $(t \neq 0)$ **(b)** $\frac{4^5}{4^{-2}}$

(c) $\frac{m^4}{n^3}$ $(n \neq 0)$

EXAMPLE 5 Using the Quotient Rule for Exponents

Apply the quotient rule for exponents, if possible, and write each result with only positive exponents.

Numerator exponent; Denominator exponent

(a) $\frac{3^7}{3^2} = 3^{7-2} = 3^5$

Subtraction symbol

(b) $\frac{p^6}{p^2} = p^{6-2} = p^4 \quad (p \neq 0)$

(c) $\frac{k^7}{k^{12}} = k^{7-12} = k^{-5} = \frac{1}{k^5} \quad (k \neq 0)$

(d) $\frac{2^7}{2^{-3}} = 2^{7-(-3)} = 2^{7+3} = 2^{10}$

Use parentheses to avoid errors.

(e) $\frac{8^{-2}}{8^5} = 8^{-2-5} = 8^{-7} = \frac{1}{8^7}$

(f) $\frac{6}{6^{-1}} = \frac{6^1}{6^{-1}} = 6^{1-(-1)} = 6^2$

(g) $\frac{z^{-5}}{z^{-8}} = z^{-5-(-8)} = z^3 \quad (z \neq 0)$

Be careful with signs.

(h) $\frac{a^3}{b^4} \quad (b \neq 0)$

This expression cannot be simplified further.

The quotient rule does not apply because the bases are different.

NOW TRY

NOW TRY ANSWERS
5. (a) t^6 **(b)** 4^7
(c) The quotient rule does not apply.

TEACHING TIP
Students have difficulty using the quotient rule with negative exponents. You may wish to show that **Example 5(g)** can also be worked by first converting to positive exponents and then using the quotient rule.
Be sure that an exponent of 1 is included when no exponent is shown, as in **Example 5(f)**.

TEACHING TIP
When showing examples of the power rules, include
$(x+y)^2 \neq x^2+y^2$.

OBJECTIVE 4 Use the power rules for exponents.

We can simplify $(3^4)^2$ as follows.

$$(3^4)^2 = 3^4 \cdot 3^4 = 3^{4+4} = 3^8$$

Notice that $4 \cdot 2 = 8$. This example suggests the first **power rule for exponents.** The other two power rules can be demonstrated similarly.

Power Rules for Exponents

If a and b are real numbers and m and n are integers, then

(a) $(a^m)^n = a^{mn}$, **(b)** $(ab)^m = a^m b^m$, and **(c)** $\left(\frac{a}{b}\right)^m = \frac{a^m}{b^m}$ $(b \neq 0)$.

That is,

(a) To raise a power to a power, multiply exponents.

(b) To raise a product to a power, raise each factor to that power.

(c) To raise a quotient to a power, raise the numerator and the denominator to that power.

NOW TRY EXERCISE 6
Simplify using the power rules.
(a) $(9x)^3$ **(b)** $(-2m^3)^4$
(c) $\left(\frac{3x^2}{y^3}\right)^3$ $(y \neq 0)$

CLASSROOM EXAMPLE 6
Simplify using the power rules.
(a) $(r^5)^4$ (b) $(-3y^5)^2$
(c) $\left(\frac{3}{4}\right)^3$ (d) $\left(\frac{-3n^4}{m^2}\right)^3$ $(m \neq 0)$
Answers: (a) r^{20} (b) $9y^{10}$
(c) $\frac{27}{64}$ (d) $-\frac{27n^{12}}{m^6}$

EXAMPLE 6 Using the Power Rules for Exponents

Simplify using the power rules.

(a) $(p^8)^3$
$= p^{8 \cdot 3}$ Power rule (a)
$= p^{24}$

(b) $(3y)^4$
$= 3^4y^4$ Power rule (b)
$= 81y^4$

(c) $\left(\frac{2}{3}\right)^4$
$= \frac{2^4}{3^4}$ Power rule (c)
$= \frac{16}{81}$

(d) $(6p^7)^2$
$= 6^2(p^7)^2$ Power rule (b)
$= 6^2p^{7 \cdot 2}$ Power rule (a)
$= 36p^{14}$ Square 6. Multiply exponents.

(e) $\left(\frac{-2m^5}{z}\right)^3$
$= \frac{(-2m^5)^3}{z^3}$ Power rule (c)
$= \frac{(-2)^3m^{5 \cdot 3}}{z^3}$ Power rule (b)
$= -\frac{8m^{15}}{z^3}$ $(z \neq 0)$ Simplify.

NOW TRY

The reciprocal of a^n is $\frac{1}{a^n}$, which equals $\left(\frac{1}{a}\right)^n$. Also, a^n and a^{-n} are reciprocals.

$$a^n \cdot a^{-n} = a^n \cdot \frac{1}{a^n} = 1 \quad \text{Reciprocals have product 1.}$$

Thus, because both $\left(\frac{1}{a}\right)^n$ and a^{-n} are reciprocals of a^n, the following is true.

$$a^{-n} = \left(\frac{1}{a}\right)^n \quad a^{-n} \text{ represents the reciprocal of } a^n.$$

NOW TRY ANSWERS
6. (a) $729x^3$ (b) $16m^{12}$
(c) $\frac{27x^6}{y^9}$

This discussion can be generalized.

Special Rules for Negative Exponents, Continued

If $a \neq 0$ and $b \neq 0$ and n is an integer, then

$$a^{-n} = \left(\frac{1}{a}\right)^n \quad \text{and} \quad \left(\frac{a}{b}\right)^{-n} = \left(\frac{b}{a}\right)^n.$$

That is, any nonzero number raised to the negative nth power is equal to the reciprocal of that number raised to the nth power.

NOW TRY EXERCISE 7

Write with only positive exponents and then evaluate.

(a) $\left(\frac{1}{3}\right)^{-2}$ (b) $\left(\frac{5}{3}\right)^{-3}$

CLASSROOM EXAMPLE 7

Write with only positive exponents and then evaluate.

(a) $\left(\frac{2}{3}\right)^{-4}$ (b) $\left(\frac{1}{2x}\right)^{-5}$ $(x \neq 0)$

Answer: (a) $\frac{81}{16}$ (b) $32x^5$

EXAMPLE 7 Using Negative Exponents with Fractions

Write with only positive exponents and then evaluate.

(a) $\left(\frac{1}{2}\right)^{-4}$

$= 2^4$ $\quad \left(\frac{1}{a}\right)^n = a^{-n}$

$= 16$

(b) $\left(\frac{3}{7}\right)^{-2}$

$= \left(\frac{7}{3}\right)^2$ $\quad \left(\frac{a}{b}\right)^{-n} = \left(\frac{b}{a}\right)^n$

$= \frac{49}{9}$

(c) $\left(\frac{4x}{5}\right)^{-3}$ $(x \neq 0)$

$= \left(\frac{5}{4x}\right)^3$ $\quad \left(\frac{a}{b}\right)^{-n} = \left(\frac{b}{a}\right)^n$

$= \frac{5^3}{4^3x^3}$ $\quad$ Power rules (c) and (b)

$= \frac{125}{64x^3}$ $\quad$ Apply the exponents.

In each case, applying the negative exponent involves changing the base to its reciprocal and changing the sign of the exponent. NOW TRY

Summary of Definitions and Rules for Exponents

For all integers m and n and all real numbers a and b, the following hold.

Product Rule $\quad a^m \cdot a^n = a^{m+n}$

Quotient Rule $\quad \frac{a^m}{a^n} = a^{m-n} \quad (a \neq 0)$

Zero Exponent $\quad a^0 = 1 \quad (a \neq 0)$

Negative Exponent $\quad a^{-n} = \frac{1}{a^n} \quad (a \neq 0)$

Power Rules $\quad$ (a) $(a^m)^n = a^{mn}$ $\quad$ (b) $(ab)^m = a^m b^m$

(c) $\left(\frac{a}{b}\right)^m = \frac{a^m}{b^m} \quad (b \neq 0)$

Special Rules for Negative Exponents $\quad \frac{1}{a^{-n}} = a^n \quad (a \neq 0)$ $\quad \frac{a^{-n}}{b^{-m}} = \frac{b^m}{a^n} \quad (a, b \neq 0)$

$a^{-n} = \left(\frac{1}{a}\right)^n \quad (a \neq 0)$ $\quad \left(\frac{a}{b}\right)^{-n} = \left(\frac{b}{a}\right)^n \quad (a, b \neq 0)$

NOW TRY ANSWERS

7. (a) 9 (b) $\frac{27}{125}$

OBJECTIVE 5 Simplify exponential expressions.

NOW TRY EXERCISE 8

Simplify. Assume that all variables represent nonzero real numbers.

(a) $x^{-8} \cdot x \cdot x^4$

(b) $(5^{-3})^2$

(c) $\frac{p^2q^{-4}}{p^{-2}q^{-1}}$

(d) $\left(\frac{2x^2}{y^2}\right)^3\left(\frac{-5x^{-2}}{y}\right)^{-2}$

EXAMPLE 8 Using the Definitions and Rules for Exponents

Simplify each expression so that no negative exponents appear in the final result. Assume that all variables represent nonzero real numbers.

(a) $3^2 \cdot 3^{-5}$

$= 3^{2+(-5)}$ Product rule

$= 3^{-3}$ Add exponents.

$= \frac{1}{3^3}$ $a^{-n} = \frac{1}{a^n}$

$= \frac{1}{27}$ $3^3 = 27$

(b) $x^{-3} \cdot x^{-4} \cdot x^2$

$= x^{-3+(-4)+2}$ Product rule

$= x^{-5}$ Add exponents.

$= \frac{1}{x^5}$ $a^{-n} = \frac{1}{a^n}$

(c) $(4^{-2})^{-5}$

$= 4^{(-2)(-5)}$ Power rule (a)

$= 4^{10}$ Multiply exponents.

(d) $(x^{-4})^6$

$= x^{(-4)6}$ Power rule (a)

$= x^{-24}$ Multiply exponents.

$= \frac{1}{x^{24}}$ $a^{-n} = \frac{1}{a^n}$

(e) $\frac{x^{-4}y^2}{x^2y^{-5}}$

$= \frac{x^{-4}}{x^2} \cdot \frac{y^2}{y^{-5}}$

$= x^{-4-2} \cdot y^{2-(-5)}$ Quotient rule

$= x^{-6}y^7$ Subtract exponents.

$= \frac{y^7}{x^6}$ $a^{-n} = \frac{1}{a^n}$

(f) $(2^3x^{-2})^{-2}$

$= (2^3)^{-2} \cdot (x^{-2})^{-2}$ Power rule (b)

$= 2^{-6}x^4$ Power rule (a)

$= \frac{x^4}{2^6}$ $a^{-n} = \frac{1}{a^n}$

$= \frac{x^4}{64}$ $2^6 = 64$

(g) $\left(\frac{3x^2}{y}\right)^2\left(\frac{4x^3}{y^{-2}}\right)^{-1}$

$= \frac{3^2(x^2)^2}{y^2} \cdot \frac{y^{-2}}{4x^3}$ Combination of rules

$= \frac{9x^4}{y^2} \cdot \frac{y^{-2}}{4x^3}$ $3^2 = 9$; Power rule (a)

$= \frac{9}{4}x^{4-3}y^{-2-2}$ Quotient rule

$= \frac{9}{4}x^1y^{-4}$ Subtract exponents.

$= \frac{9x}{4y^4}$ $a^{-n} = \frac{1}{a^n}$

(h) $\left(\frac{-4m^5n^4}{24mn^{-7}}\right)^{-2}$

$= \left(\frac{m^{5-1}n^{4-(-7)}}{-6}\right)^{-2}$ Quotient rule; divide coefficients.

$= \left(\frac{m^4n^{11}}{-6}\right)^{-2}$ Subtract exponents.

$= \frac{(m^4)^{-2}(n^{11})^{-2}}{(-6)^{-2}}$ Power rules (b) and (c)

$= \frac{m^{-8}n^{-22}}{(-6)^{-2}}$ Power rule (a)

$= \frac{(-6)^2}{m^8n^{22}}$ $\frac{a^{-n}}{b^{-m}} = \frac{b^m}{a^n}$ (The sign on −6 does *not* change in this step.)

$= \frac{36}{m^8n^{22}}$ $(-6)^2 = 36$

NOW TRY

CLASSROOM EXAMPLE 8

Simplify. Assume that all variables represent nonzero real numbers.

(a) $(4^2)^{-5}$

(b) $x^{-4} \cdot x^{-6} \cdot x^8$

(c) $\frac{(m^2n)^{-2}}{m^{-3}n}$

(d) $\left(\frac{2y}{x^3}\right)^2\left(\frac{4y}{x}\right)^{-1}$

Answers: (a) $\frac{1}{4^{10}}$ (b) $\frac{1}{x^2}$ (c) $\frac{1}{mn^3}$ (d) $\frac{y}{x^5}$

TEACHING TIP

Explain the distinction between the expressions

$(xy)^3$ and xy^3.

Emphasize that in a term such as $3x^2$, x is the base of the exponent 2, while in $(3x)^2$, the base is $3x$.

NOW TRY ANSWERS

8. (a) $\frac{1}{x^3}$ (b) $\frac{1}{5^6}$ (c) $\frac{p^4}{q^3}$ (d) $\frac{8x^{10}}{25y^4}$

NOTE There is often more than one way to simplify expressions involving negative exponents.

In **Example 8(e),** we began by using the quotient rule. At the right, we simplify the same expression using one of the special rules for exponents. The final result is the same.

$$\frac{x^{-4}y^2}{x^2y^{-5}} \quad \text{Example 8(e)}$$
$$= \frac{y^5y^2}{x^4x^2} \quad \text{Use } \frac{a^{-n}}{b^{-m}} = \frac{b^m}{a^n}.$$
$$= \frac{y^7}{x^6} \quad \text{Product rule}$$

TEACHING TIP
Show how the

EXP or EE

keys work on scientific calculators. Many calculators have an eight-digit display, so they cannot handle extremely large or extremely small numbers unless the numbers are entered in scientific notation.

OBJECTIVE 6 Use the rules for exponents with scientific notation.

The number of one-celled organisms that will sustain a whale for a few hours is 400,000,000,000,000, and the shortest wavelength of visible light is approximately 0.0000004 m. It is often simpler to write such numbers in *scientific notation.*

In scientific notation, a number is written with the decimal point after the first nonzero digit and multiplied by a power of 10.

Scientific Notation

A number is written in **scientific notation** when it is expressed in the form

$$a \times 10^n,$$

where $1 \le |a| < 10$ **and** n **is an integer.**

$1 \le |a| < 10$ — It is customary to use × rather than · for multiplication.

Examples: $8500 = 8.5 \times 1000 = 8.5 \times 10^3$ — In scientific notation

0.230×10^4 ($a < 1$) $\quad 46.5 \times 10^{-3}$ ($a > 10$) — *Not* in scientific notation

To write a positive number in scientific notation, follow these steps.

Converting a Positive Number to Scientific Notation

Step 1 **Position the decimal point.** Place a caret, ^, to the right of the first nonzero digit, where the decimal point will be placed.

Step 2 **Determine the numeral for the exponent.** Count the number of digits from the decimal point to the caret. This number gives the absolute value of the exponent on 10.

Step 3 **Determine the sign for the exponent.** Decide whether multiplying by 10^n should make the result of Step 1 greater or less.

- The exponent should be positive to make the result greater.
- The exponent should be negative to make the result less.

It is helpful to remember the following.

$$\textbf{For } n \ge 1, \quad 10^{-n} < 1 \quad \textbf{and} \quad 10^n \ge 10.$$

To convert a negative number to scientific notation, temporarily ignore the negative sign and go through the steps in the box above. Then attach a negative sign to the result.

NOW TRY EXERCISE 9

Write each number in scientific notation.

(a) 7,560,000,000

(b) −0.000000245

CLASSROOM EXAMPLE 9

Write each number in scientific notation.

(a) 29,800,000

(b) 0.0000000503

Answers: (a) 2.98×10^7

(b) 5.03×10^{-8}

EXAMPLE 9 Writing Numbers in Scientific Notation

Write each number in scientific notation.

(a) 820,000

Step 1 Place a caret to the right of the 8 (the first nonzero digit) to mark the new location of the decimal point.

$$8_{\wedge}20,000$$

Step 2 Count from the decimal point, which is understood to be after the last 0, to the caret.

$$8.20,000. \leftarrow \text{Decimal point}$$

Count 5 places.

Step 3 Because 8.2 is to be made greater, the exponent on 10 is positive.

$$820,000 = 8.2 \times 10^5$$

(b) 0.0000072

$0.000007_{\wedge}2$ Place a caret to the right of the first nonzero digit.

$0.000007.2$ Count from left to right.

6 places

Because 7.2 is to be made less, the exponent on 10 is negative.

$$0.0000072 = 7.2 \times 10^{-6}$$

(c) $-0.0004_{\wedge}62 = -4.62 \times 10^{-4}$ Remember the negative sign.

Count 4 places.

NOW TRY

CLASSROOM EXAMPLE 10

Write each number in standard notation.

(a) 2.51×10^3

(b) -6.8×10^{-4}

Answers: (a) 2510

(b) −0.00068

Converting a Positive Number from Scientific Notation

Multiplying a positive number by a positive power of 10 makes the number greater, so move the decimal point to the right if n is positive in 10^n.

Multiplying a positive number by a negative power of 10 makes the number less, so move the decimal point to the left if n is negative in 10^n.

If n is 0, leave the decimal point where it is in 10^n.

NOW TRY EXERCISE 10

Write each number in standard notation.

(a) 4.45×10^{10}

(b) -5.9×10^{-5}

EXAMPLE 10 Converting from Scientific Notation to Standard Notation

Write each number in standard notation.

(a) 6.93×10^7

6.9300000 Attach 0's as necessary.

7 places

We moved the decimal point 7 places to the right. (We had to attach five 0's.)

$$6.93 \times 10^7 = 69,300,000$$ Insert commas.

Standard notation

(b) 4.7×10^{-3}

$.004.7$ Move the decimal point 3 places to the left. Attach 0's as necessary.

3 places

$$4.7 \times 10^{-3} = 0.0047$$ Attach a leading zero.

(c) $-1.083 \times 10^0 = -1.083 \times 1 = -1.083$

NOW TRY

NOW TRY ANSWERS

9. **(a)** 7.56×10^9
(b) -2.45×10^{-7}

10. **(a)** 44,500,000,000
(b) −0.000059

NOTE *When converting from scientific notation to standard notation, use the exponent to determine the number of places and the direction in which to move the decimal point.*

NOW TRY EXERCISE 11

Evaluate.

$$\frac{0.00063 \times 400{,}000}{1400 \times 0.000003}$$

CLASSROOM EXAMPLE 11

Evaluate.

$$\frac{200{,}000 \times 0.0003}{0.06}$$

Answer: 1000

EXAMPLE 11 Using Scientific Notation in Computation

Evaluate.

$$\frac{1{,}920{,}000 \times 0.0015}{0.000032 \times 45{,}000}$$

$= \dfrac{1.92 \times 10^6 \times 1.5 \times 10^{-3}}{3.2 \times 10^{-5} \times 4.5 \times 10^4}$ Express all numbers in scientific notation.

$= \dfrac{1.92 \times 1.5 \times 10^6 \times 10^{-3}}{3.2 \times 4.5 \times 10^{-5} \times 10^4}$ Commutative property

$= \dfrac{1.92 \times 1.5 \times 10^3}{3.2 \times 4.5 \times 10^{-1}}$ Product rule

$= \dfrac{1.92 \times 1.5}{3.2 \times 4.5} \times 10^4$ Quotient rule

$= 0.2 \times 10^4$ Simplify.

Don't stop here.

$= (2 \times 10^{-1}) \times 10^4$ Write 0.2 in scientific notation.

$= 2 \times 10^3$ Product rule

$= 2000$ Standard notation

NOW TRY

NOW TRY EXERCISE 12

The speed of light is approximately 30,000,000,000 cm per sec. How long will it take light to travel 1.2×10^{15} cm?

CLASSROOM EXAMPLE 12

The distance to the sun is about 9.3×10^7 mi. How long would it take a rocket traveling at 3.2×10^3 mph to reach the sun?

Answer: about 2.9×10^4 hr

EXAMPLE 12 Using Scientific Notation to Solve Problems

In 1990, national health care expenditures in the United States were \$724 billion. By 2012, health care expenditures were four times this amount. (*Source:* Centers for Medicare & Medicaid Services.)

(a) Write the amount of health care expenditures in 1990 using scientific notation.

724 billion

$= 724 \times 10^9$ 1 billion = 1,000,000,000 = 10^9

$= (7.24 \times 10^2) \times 10^9$ Write 724 in scientific notation.

$= 7.24 \times 10^{11}$ Product rule; $10^2 \times 10^9 = 10^{2+9}$

In 1990, health care expenditures were \$$7.24 \times 10^{11}$.

(b) What were health care expenditures in 2012?

$(7.24 \times 10^{11}) \times 4$ Multiply the result in part (a) by 4.

$= (4 \times 7.24) \times 10^{11}$ Commutative and associative properties

$= 28.96 \times 10^{11}$ Multiply.

$= (2.896 \times 10^1) \times 10^{11}$ Write 28.96 in scientific notation.

$= 2.896 \times 10^{12}$ Product rule; $10^1 \times 10^{11} = 10^{12}$

Expenditures in 2012 were about \$2,896,000,000,000 (almost \$3 trillion).

NOW TRY

NOW TRY ANSWERS

11. 60,000

12. 4.0×10^4 sec, or 40,000 sec

4.1 Exercises

FOR EXTRA HELP

MyMathLab®

Complete solution available in MyMathLab

1. incorrect; $(ab)^2 = a^2b^2$
2. correct
3. incorrect; $\left(\frac{4}{a}\right)^3 = \frac{4^3}{a^3}$
4. incorrect; $y^2 \cdot y^6 = y^8$
5. correct
6. incorrect; $xy^0 = x \cdot 1 = x$
7. Do not multiply the bases. $4^5 \cdot 4^2 = 4^7$
8. Do not divide the bases. $\frac{6^5}{3^2} = \frac{7776}{9} = 864$
9. 13^{12}
10. 11^{10}
11. 8^{10}
12. 12^7
13. x^{17}
14. y^{15}
15. $-27w^8$
16. $-15x^6$
17. $18x^3y^8$
18. $24s^7t^6$
19. The product rule does not apply.
20. The product rule does not apply.
21. (a) B (b) C (c) B (d) C
22. (a) D (b) E (c) B (d) B
23. 1
24. 1
25. -1
26. -1
27. 1
28. 1
29. 2
30. 2
31. 0
32. 0
33. -2
34. -2
35. (a) B (b) D (c) B (d) D
36. (a) C (b) D (c) D (d) C

Concept Check *Decide whether each expression has been simplified correctly. If not, correct it.*

1. $(ab)^2 = ab^2$

2. $(5x)^3 = 5^3x^3$

3. $\left(\frac{4}{a}\right)^3 = \frac{4^3}{a} \quad (a \neq 0)$

4. $y^2 \cdot y^6 = y^{12}$

5. $x^3 \cdot x^4 = x^7$

6. $xy^0 = 0 \quad (y \neq 0)$

7. ***Concept Check*** A friend evaluated $4^5 \cdot 4^2$ as 16^7. **WHAT WENT WRONG?** Give the correct answer.

8. ***Concept Check*** Another friend evaluated $\frac{6^5}{3^2}$ as 2^3, which equals 8. **WHAT WENT WRONG?** Give the correct answer.

Apply the product rule for exponents, if possible, in each case. ***See Example 1.***

9. $13^4 \cdot 13^8$ **10.** $11^6 \cdot 11^4$ **11.** $8^9 \cdot 8$ **12.** $12 \cdot 12^6$

13. $x^3 \cdot x^5 \cdot x^9$ **14.** $y^4 \cdot y^5 \cdot y^6$ **15.** $(-3w^5)(9w^3)$ **16.** $(-5x^2)(3x^4)$

17. $(2x^2y^5)(9xy^3)$ **18.** $(8s^4t)(3s^3t^5)$ **19.** $r^2 \cdot s^4$ **20.** $p^3 \cdot q^2$

Match the expression in Column I with its equivalent expression in Column II. Choices may be used once, more than once, or not at all. ***See Example 2.***

21.

I	II
(a) 9^0	A. 0
(b) -9^0	B. 1
(c) $(-9)^0$	C. -1
(d) $-(-9)^0$	D. 9
	E. -9

22.

I	II
(a) $8x^0$	A. 0
(b) $-8x^0$	B. 1
(c) $(8x)^0$	C. -1
(d) $(-8x)^0$	D. 8
	E. -8

Evaluate. Assume that all variables represent nonzero real numbers. ***See Example 2.***

23. 15^0 **24.** 19^0 **25.** -8^0 **26.** -10^0

27. $(-25)^0$ **28.** $(-30)^0$ **29.** $3^0 + (-3)^0$ **30.** $5^0 + (-5)^0$

31. $-3^0 + 3^0$ **32.** $-5^0 + 5^0$ **33.** $-4^0 - m^0$ **34.** $-8^0 - k^0$

Match the expression in Column I with its equivalent expression in Column II. Choices may be used once, more than once, or not at all. ***See Example 3.***

35.

I	II
(a) 5^{-2}	A. 25
(b) -5^{-2}	B. $\frac{1}{25}$
(c) $(-5)^{-2}$	C. -25
(d) $-(-5)^{-2}$	D. $-\frac{1}{25}$

36.

I	II
(a) 4^{-3}	A. 64
(b) -4^{-3}	B. -64
(c) $(-4)^{-3}$	C. $\frac{1}{64}$
(d) $-(-4)^{-3}$	D. $-\frac{1}{64}$

Write each expression with only positive exponents. Assume that all variables represent nonzero real numbers. ***See Examples 3(a)–(f).***

37. 5^{-4} **38.** 7^{-2} **39.** 3^{-5} **40.** 8^{-3}

41. 9^{-1} **42.** 14^{-1} **43.** $(4x)^{-2}$ **44.** $(5t)^{-3}$

45. $4x^{-2}$ **46.** $5t^{-3}$ **47.** $-a^{-3}$ **48.** $-b^{-4}$

49. $(-a)^{-4}$ **50.** $(-b)^{-6}$ **51.** $(-3x)^{-3}$ **52.** $(-2x)^{-5}$

Evaluate. ***See Examples 3(g) and (h).***

53. $5^{-1}+6^{-1}$ **54.** $2^{-1}+8^{-1}$ **55.** $8^{-1}-3^{-1}$

56. $6^{-1}-4^{-1}$ **57.** $-2^{-1}-4^{-1}$ **58.** $-3^{-1}-9^{-1}$

Evaluate. ***See Example 4.***

59. $\frac{1}{4^{-2}}$ **60.** $\frac{1}{3^{-3}}$ **61.** $\frac{2^{-2}}{3^{-3}}$ **62.** $\frac{3^{-3}}{2^{-2}}$

Apply the quotient rule for exponents, if possible, and write each result with only positive exponents. Assume that all variables represent nonzero real numbers. ***See Example 5.***

63. $\frac{4^8}{4^6}$ **64.** $\frac{5^9}{5^7}$ **65.** $\frac{x^{12}}{x^8}$ **66.** $\frac{y^{14}}{y^{10}}$

67. $\frac{r^7}{r^{10}}$ **68.** $\frac{y^8}{y^{12}}$ **69.** $\frac{6^4}{6^{-2}}$ **70.** $\frac{7^5}{7^{-3}}$

71. $\frac{6^{-3}}{6^7}$ **72.** $\frac{5^{-4}}{5^2}$ **73.** $\frac{7}{7^{-1}}$ **74.** $\frac{8}{8^{-1}}$

75. $\frac{r^{-3}}{r^{-6}}$ **76.** $\frac{s^{-4}}{s^{-8}}$ **77.** $\frac{x^3}{y^2}$ **78.** $\frac{y^5}{t^3}$

Simplify using the power rules. Assume that all variables represent nonzero real numbers. ***See Example 6.***

79. $(x^3)^6$ **80.** $(y^5)^4$ **81.** $\left(\frac{3}{5}\right)^3$ **82.** $\left(\frac{4}{3}\right)^2$

83. $(4t)^3$ **84.** $(5t)^4$ **85.** $(-6x^2)^3$ **86.** $(-2x^5)^5$

87. $\left(\frac{-4m^2}{t}\right)^3$ **88.** $\left(\frac{-5n^4}{r^2}\right)^3$ **89.** $\left(\frac{-s^3}{t^5}\right)^4$ **90.** $\left(\frac{-2a^4}{b^5}\right)^6$

Match the expression in Column I with its equivalent expression in Column II. Choices may be used once, more than once, or not at all. ***See Example 7.***

	I	II
91.	(a) $\left(\frac{1}{3}\right)^{-1}$	A. $\frac{1}{3}$
	(b) $\left(-\frac{1}{3}\right)^{-1}$	B. 3
	(c) $-\left(\frac{1}{3}\right)^{-1}$	C. $-\frac{1}{3}$
	(d) $-\left(-\frac{1}{3}\right)^{-1}$	D. -3

	I	II
92.	(a) $\left(\frac{2}{5}\right)^{-2}$	A. $\frac{25}{4}$
	(b) $\left(-\frac{2}{5}\right)^{-2}$	B. $-\frac{25}{4}$
	(c) $-\left(\frac{2}{5}\right)^{-2}$	C. $\frac{4}{25}$
	(d) $-\left(-\frac{2}{5}\right)^{-2}$	D. $-\frac{4}{25}$

Answers

37. $\frac{1}{5^4}$, or $\frac{1}{625}$ 38. $\frac{1}{7^2}$, or $\frac{1}{49}$
39. $\frac{1}{3^5}$, or $\frac{1}{243}$ 40. $\frac{1}{8^3}$, or $\frac{1}{512}$
41. $\frac{1}{9}$ 42. $\frac{1}{14}$
43. $\frac{1}{(4x)^2}$, or $\frac{1}{16x^2}$
44. $\frac{1}{(5t)^3}$, or $\frac{1}{125t^3}$
45. $\frac{4}{x^2}$ 46. $\frac{5}{t^3}$
47. $-\frac{1}{a^3}$ 48. $-\frac{1}{b^4}$
49. $\frac{1}{(-a)^4}$, or $\frac{1}{a^4}$
50. $\frac{1}{(-b)^6}$, or $\frac{1}{b^6}$
51. $\frac{1}{(-3x)^3}$, or $-\frac{1}{27x^3}$
52. $\frac{1}{(-2x)^5}$, or $-\frac{1}{32x^5}$
53. $\frac{11}{30}$ 54. $\frac{5}{8}$
55. $-\frac{5}{24}$ 56. $-\frac{1}{12}$
57. $-\frac{3}{4}$ 58. $-\frac{4}{9}$
59. 16 60. 27
61. $\frac{27}{4}$ 62. $\frac{4}{27}$
63. 4^2, or 16 64. 5^2, or 25
65. x^4 66. y^4
67. $\frac{1}{r^3}$ 68. $\frac{1}{y^4}$
69. 6^6 70. 7^8
71. $\frac{1}{6^{10}}$ 72. $\frac{1}{5^6}$
73. 7^2, or 49 74. 8^2, or 64
75. r^3 76. s^4
77. The quotient rule does not apply.
78. The quotient rule does not apply.
79. x^{18} 80. y^{20}
81. $\frac{27}{125}$ 82. $\frac{16}{9}$
83. $64t^3$ 84. $625t^4$
85. $-216x^6$ 86. $-32x^{25}$
87. $-\frac{64m^6}{t^3}$ 88. $-\frac{125n^{12}}{r^6}$
89. $\frac{s^{12}}{t^{20}}$ 90. $\frac{64a^{24}}{b^{30}}$
91. (a) B (b) D (c) D (d) B
92. (a) A (b) A (c) B (d) B

93. 64 **94.** 25
95. $\frac{27}{8}$ **96.** $\frac{8}{27}$
97. $\frac{25}{16}$ **98.** $\frac{16}{25}$
99. $\frac{81}{16t^4}$ **100.** $\frac{64}{27z^3}$
101. $16x^2$ **102.** $125x^3$
103. $\frac{32}{x^5}$ **104.** $\frac{256}{t^4}$
105. $\frac{1}{3}$ **106.** $\frac{1}{4^2}$, or $\frac{1}{16}$
107. $\frac{1}{a^5}$ **108.** $\frac{1}{k^4}$
109. 5^6 **110.** 4^8
111. $\frac{1}{x^{12}}$ **112.** $\frac{1}{x^{18}}$
113. $\frac{1}{k^2}$ **114.** $\frac{1}{x^7}$
115. $-4r^6$ **116.** $-2m^5$
117. $\frac{625}{a^{10}}$ **118.** $\frac{9}{p^{11}}$
119. $\frac{z^4}{x^3}$ **120.** $\frac{y^6}{z^{12}}$
121. $-14k^3$ **122.** $-\frac{15}{a^4}$
123. $\frac{p^4}{5}$ **124.** $\frac{m^3}{9}$
125. $\frac{m^8}{n^{11}}$ **126.** $\frac{b^{10}}{a^{12}}$
127. $\frac{1}{2pq}$ **128.** $-\frac{2}{x^4}$
129. $\frac{4}{a^2}$ **130.** $2k^5$
131. $\frac{1}{6y^{13}}$ **132.** $\frac{2m^{10}}{9}$
133. $\frac{4k^5}{m^2}$ **134.** $\frac{s^2}{81r^4}$
135. $\frac{2k^5}{3}$ **136.** $\frac{m}{196}$
137. $\frac{8}{3pq^{10}}$ **138.** $\frac{27a^9}{64z^{15}}$
139. $\frac{y^9}{8}$ **140.** $\frac{m^4}{27}$
141. $\frac{n^{10}}{25m^{18}}$ **142.** $\frac{x^{20}}{64y^{14}}$
143. $-\frac{125y^3}{x^{30}}$ **144.** $-\frac{27a^6}{b^{18}}$
145. $\frac{4k^{17}}{125}$ **146.** $\frac{36}{r^{14}}$
147. $-\frac{3}{32m^8p^4}$ **148.** $\frac{15}{8y^5z^3}$
149. $\frac{2}{3y^4}$ **150.** $\frac{1}{54m^{11}x^9}$
151. $\frac{3p^8}{16q^{14}}$ **152.** $\frac{108b^9}{25a^{14}}$

Write with only positive exponents and then evaluate. Assume that all variables represent nonzero real numbers. ***See Example 7.***

93. $\left(\frac{1}{4}\right)^{-3}$ **94.** $\left(\frac{1}{5}\right)^{-2}$ **95.** $\left(\frac{2}{3}\right)^{-3}$ **96.** $\left(\frac{3}{2}\right)^{-3}$

97. $\left(\frac{4}{5}\right)^{-2}$ **98.** $\left(\frac{5}{4}\right)^{-2}$ **99.** $\left(\frac{2t}{3}\right)^{-4}$ **100.** $\left(\frac{3z}{4}\right)^{-3}$

101. $\left(\frac{1}{4x}\right)^{-2}$ **102.** $\left(\frac{1}{5x}\right)^{-3}$ **103.** $\left(\frac{x}{2}\right)^{-5}$ **104.** $\left(\frac{t}{4}\right)^{-4}$

Simplify each expression so that no negative exponents appear in the final result. Assume that all variables represent nonzero real numbers. ***See Examples 1–8.***

105. $3^5 \cdot 3^{-6}$ **106.** $4^4 \cdot 4^{-6}$ **107.** $a^{-3}a^2a^{-4}$ **108.** $k^{-5}k^{-3}k^4$

109. $(5^{-3})^{-2}$ **110.** $(4^{-4})^{-2}$ **111.** $(x^{-4})^3$ **112.** $(x^{-3})^6$

113. $(k^2)^{-3}k^4$ **114.** $(x^3)^{-4}x^5$ **115.** $-4r^{-2}(r^4)^2$ **116.** $-2m^{-1}(m^3)^2$

117. $(5a^{-1})^4(a^2)^{-3}$ **118.** $(3p^{-4})^2(p^3)^{-1}$ **119.** $(z^{-4}x^3)^{-1}$ **120.** $(y^{-2}z^4)^{-3}$

121. $7k^2(-2k)(4k^{-5})^0$ **122.** $3a^2(-5a^{-6})(-2a)^0$ **123.** $\frac{(p^{-2})^0}{5p^{-4}}$

124. $\frac{(m^4)^0}{9m^{-3}}$ **125.** $\frac{m^2n^{-8}}{m^{-6}n^3}$ **126.** $\frac{a^{-9}b^6}{a^3b^{-4}}$

127. $\frac{(3pq)q^2}{6p^2q^4}$ **128.** $\frac{(-8xy)y^3}{4x^5y^4}$ **129.** $\frac{4a^5(a^{-1})^3}{(a^{-2})^{-2}}$

130. $\frac{12k^{-2}(k^{-3})^{-4}}{6k^5}$ **131.** $\frac{(-y^{-4})^2}{6(y^{-5})^{-1}}$ **132.** $\frac{2(-m^{-1})^{-4}}{9(m^{-3})^2}$

133. $\frac{(2k)^2m^{-5}}{(km)^{-3}}$ **134.** $\frac{(3rs)^{-2}}{3^2r^2s^{-4}}$ **135.** $\left(\frac{3k^{-2}}{k^4}\right)^{-1} \cdot \frac{2}{k}$

136. $\left(\frac{7m^{-2}}{m^{-3}}\right)^{-2} \cdot \frac{m^3}{4}$ **137.** $\left(\frac{2p}{q^2}\right)^3\left(\frac{3p^4}{q^{-4}}\right)^{-1}$ **138.** $\left(\frac{5z^3}{2a^2}\right)^{-3}\left(\frac{8a^{-1}}{15z^{-2}}\right)^{-3}$

139. $\frac{2^2y^4(y^{-3})^{-1}}{2^5y^{-2}}$ **140.** $\frac{3^{-1}m^4(m^2)^{-1}}{3^2m^{-2}}$ **141.** $\left(\frac{5m^4n^{-3}}{m^{-5}n^2}\right)^{-2}$

142. $\left(\frac{8x^{-6}y^3}{x^4y^{-4}}\right)^{-2}$ **143.** $\left(\frac{-3x^4y^6}{15x^{-6}y^7}\right)^{-3}$ **144.** $\left(\frac{-4a^3b^2}{12a^5b^{-4}}\right)^{-3}$

Extending Skills *Simplify each expression. Assume that all variables represent nonzero real numbers.*

145. $\frac{(2k)^2k^3}{k^{-1}k^{-5}}(5k^{-2})^{-3}$ **146.** $\frac{(3r^2)^2r^{-5}}{r^{-2}r^3}(2r^{-6})^2$

147. $\frac{(2m^2p^3)^2(4m^2p)^{-2}}{(-3mp^4)^{-1}(2m^3p^4)^3}$ **148.** $\frac{(-5y^3z^4)^2(2yz^5)^{-2}}{10(y^4z)^3(3y^3z^2)^{-1}}$

149. $\frac{(-3y^3x^3)(-4y^4x^2)(x^2)^{-4}}{18x^3y^2(y^3)^3(x^3)^{-2}}$ **150.** $\frac{(2m^3x^2)^{-1}(3m^4x)^{-3}}{(m^2x^3)^3(m^2x)^{-5}}$

151. $\left(\frac{p^2q^{-1}}{2p^{-2}}\right)^2 \cdot \left(\frac{p^3 \cdot 4q^{-2}}{3q^{-5}}\right)^{-1} \cdot \left(\frac{pq^{-5}}{q^{-2}}\right)^3$ **152.** $\left(\frac{a^6b^{-2}}{2a^{-2}}\right)^{-1} \cdot \left(\frac{6a^{-2}}{5b^{-4}}\right)^2 \cdot \left(\frac{2b^{-1}a^2}{3b^{-2}}\right)^{-1}$

153. after; power; a; 10^n
154. **(a)** not in scientific notation; $16.8 > 10$; 1.68×10^6
(b) in scientific notation
(c) not in scientific notation; $0.2 < 1$; 2×10^{-3}

155. 5.3×10^2 **156.** 1.6×10^3
157. 8.3×10^{-1} **158.** 7.2×10^{-3}
159. 6.92×10^{-6}
160. 8.75×10^{-1}
161. -3.85×10^4
162. -9.76×10^8

163. $\$1 \times 10^9$ (or $\$10^9$); $\$1 \times 10^{12}$ (or $\$10^{12}$); $\$3.8 \times 10^{12}$; 2.57891×10^5
164. 5×10; 3.65482×10^8; 6.77×10^5; 2.271343×10^9

165. 72,000 **166.** 891
167. 0.00254 **168.** 0.000542
169. −60,000 **170.** −9000
171. 0.000012 **172.** 0.0000027

173. 0.06 **174.** 0.004
175. 0.0000025 **176.** 0.000002
177. 200,000 **178.** 1,000,000
179. 3000 **180.** 200,000

181. 7.5×10^9 **182.** 9.6×10^7
183. 4×10^{17} **184.** 1.25×10^{13}

153. ***Concept Check*** In scientific notation, a number is written with a decimal point (*before*/*after*) the first nonzero digit and multiplied by a ________ of 10. A number written in scientific notation is expressed in the form ________ ×________, where $1 \leq |a| < 10$ and n is an integer.

154. ***Concept Check*** Tell whether each number is written in scientific notation. If it is not, tell why and write the number in scientific notation.

(a) 16.8×10^5 **(b)** 6×10^{-9} **(c)** 0.2×10^{-2}

Write each number in scientific notation. ***See Example 9.***

155. 530 **156.** 1600 **157.** 0.830 **158.** 0.0072

159. 0.00000692 **160.** 0.875 **161.** −38,500 **162.** −976,000,000

Write the boldfaced numbers in each problem in scientific notation. ***See Example 9.***

163. The annual U.S. budget first passed **$1,000,000,000** in 1917. In 1987, it exceeded **$1,000,000,000,000** for the first time. The budget request for fiscal-year 2014 was **$3,800,000,000,000.** If stacked in dollar bills, this amount would stretch **257,891** mi. (*Source:* Office of Management and Budget.)

164. The largest of the **50** United States is Alaska, with land area of about **365,482,000** acres, while the smallest is Rhode Island, with land area of about **677,000** acres. The total land area of the United States is about **2,271,343,000** acres. (*Source:* General Services Administration.)

Write each number in standard notation. ***See Example 10.***

165. 7.2×10^4 **166.** 8.91×10^2 **167.** 2.54×10^{-3} **168.** 5.42×10^{-4}

169. -6×10^4 **170.** -9×10^3 **171.** 1.2×10^{-5} **172.** 2.7×10^{-6}

Evaluate. Give answers in standard notation. ***See Example 11.***

173. $\dfrac{12 \times 10^4}{2 \times 10^6}$ **174.** $\dfrac{16 \times 10^5}{4 \times 10^8}$ **175.** $\dfrac{3 \times 10^{-2}}{12 \times 10^3}$

176. $\dfrac{5 \times 10^{-3}}{25 \times 10^2}$ **177.** $\dfrac{0.05 \times 1600}{0.0004}$ **178.** $\dfrac{0.003 \times 40{,}000}{0.00012}$

179. $\dfrac{20{,}000 \times 0.018}{300 \times 0.0004}$ **180.** $\dfrac{840{,}000 \times 0.03}{0.00021 \times 600}$

Calculators can express numbers in scientific notation. The displays often use notation such as

2.5E5 *to represent* 2.5×10^5.

Similarly, 2.5E⁻3 *represents* 2.5×10^{-3}. *They can also perform operations with numbers entered in scientific notation.*

Predict the result the calculator will give for each screen. (Use the usual scientific notation to write the answers.)

181.
```
(1.5E12)*(5E-3)
```

182.
```
(3.2E-5)*(3E12)
```

183.
```
(8.4E14)/(2.1E-3
)
```

184.
```
(2.5E10)/(2E-3)
```

185. **(a)** 3.174×10^8
(b) $\$1 \times 10^{12}$ (or $\$10^{12}$)
(c) \$3151
186. **(a)** 3.174×10^8
(b) 1.727×10^{13}
(c) \$54,411
187. 30,500, or 3.05×10^4
188. \$40,045
189. 300 sec
190. approximately $3.2 \times 10^4 = 32{,}000$ hr (about 3.7 yr)
191. approximately 5.87×10^{12} mi
192. 500 sec, which is approximately 8.3 min
193. **(a)** 5.16×10^2
(b) 998 mi^2
194. **(a)** 9.212×10
(b) 51,020 km^2

Solve each problem. ***See Example 12.***

185. In 2014, the population of the United States was approximately 317.4 million. (*Source:* U.S. Census Bureau.)

(a) Write the population using scientific notation.

(b) Write \$1 trillion, that is, \$1,000,000,000,000, using scientific notation.

(c) Using the answers from parts (a) and (b), calculate how much each person in the United States in the year 2014 would have had to contribute in order to make someone a trillionaire. Write this amount in standard notation to the nearest dollar.

186. In 2014, the national debt of the United States was about \$17.27 trillion. The U.S. population at that time was approximately 317.4 million. (*Source:* U.S. Census Bureau.)

(a) Write the population using scientific notation.

(b) Write the amount of the national debt using scientific notation.

(c) Using the answers for parts (a) and (b), calculate how much debt this is per American. Write this amount in standard notation to the nearest dollar.

187. In 2013, the population of India was about 1.22×10^{10}, which was 4.0×10^5 times the population of Monaco. What was the population of Monaco? (*Source: World Almanac and Book of Facts.*)

188. In the early years of the Powerball® Lottery, a player would choose five numbers from 1 through 49 and one number from 1 through 42. It can be shown that there are about 8.009×10^7 different ways to do this. Suppose that a group of 2000 persons decided to purchase tickets for all these numbers and each ticket cost \$1.00. How much should each person have expected to pay? (*Source:* www.powerball.com)

189. The speed of light is approximately 3×10^{10} cm per sec. How long will it take light to travel 9×10^{12} cm?

190. The average distance from Earth to the sun is 9.3×10^7 mi. How long would it take a rocket, traveling at 2.9×10^3 mph, to reach the sun?

191. A *light-year* is the distance that light travels in one year. Find the number of miles in a light-year if light travels 1.86×10^5 mi per sec.

192. Use the information given in the previous two exercises to find the number of minutes necessary for light from the sun to reach Earth.

193. In 2013, the population of Luxembourg was approximately 5.15×10^5. The population density was 516 people per square mile. (*Source: The World Factbook.*)

(a) Write the population density in scientific notation.

(b) To the nearest square mile, what is the area of Luxembourg?

194. In 2013, the population of Costa Rica was approximately 4.70×10^6. The population density was 92.12 people per square kilometer. (*Source: The World Factbook.*)

(a) Write the population density in scientific notation.

(b) To the nearest square kilometer, what is the area of Costa Rica?

4.2 Adding and Subtracting Polynomials

OBJECTIVES

1. Know the basic definitions for polynomials.
2. Add and subtract polynomials.

VOCABULARY

- ☐ term
- ☐ algebraic expression
- ☐ polynomial
- ☐ numerical coefficient (coefficient)
- ☐ degree of a term
- ☐ polynomial in *x*
- ☐ descending powers
- ☐ leading term
- ☐ leading coefficient
- ☐ trinomial
- ☐ binomial
- ☐ monomial
- ☐ degree of a polynomial
- ☐ like terms
- ☐ negative of a polynomial

TEACHING TIP
Take extra time to make sure students master the vocabulary in this section. Give plenty of examples.

OBJECTIVE 1 Know the basic definitions for polynomials.

Recall that a **term** is a number (constant), a variable, or the product of a number and one or more variables raised to powers.

Examples: $4x,\quad \frac{1}{2}m^5 \left(\text{or } \frac{m^5}{2}\right),\quad -7z^9,\quad 6x^2z,\quad \frac{5}{3x^2},\quad 3\sqrt{x},\quad \text{and}\quad 9$ Terms

A term or a sum of two or more terms is an **algebraic expression.** The simplest kind of algebraic expression is a *polynomial.*

Polynomial

A **polynomial** is a term or a finite sum of terms of the form ax^n, where a is a real number, x is a variable, and the exponent n is a whole number.

Examples: $12x^9,\quad 3t-5,\quad \text{and}\quad 4m^3-5m^2+8$ Polynomials in *x*, *t*, and *m*

Even though an expression such as $3t-5$ involves subtraction, it can be written as a sum of terms, here $3t+(-5)$.

Some algebraic expressions that are *not* polynomials follow.

Examples: $x^{-1}+3x^{-2},\quad \sqrt{9-x},\quad \text{and}\quad \frac{1}{x}$ Not polynomials

The first has negative integer exponents, the second involves a variable under a radical, and the third has a variable in the denominator.

For each term ax^n of a polynomial, a is the **numerical coefficient,** or just the **coefficient** of x^n, and the exponent n is the **degree of the term.**

Terms and Their Coefficients and Degrees

Term ax^n	Numerical Coefficient	Degree
$12x^9$	12	9
$3x$, or $3x^1$	3	1
-6, or $-6x^0$	-6	0
$-x^4$, or $-1x^4$	-1	4
$\frac{x^2}{3}=\frac{1x^2}{3}=\frac{1}{3}x^2$	$\frac{1}{3}$	2

Any nonzero constant has degree 0.

NOTE The number 0 has no degree because 0 times a variable to any power is 0.

A polynomial containing only the variable x is a **polynomial in *x*.** A polynomial in one variable is written in **descending powers** of the variable if the exponents on the variable in the terms decrease from left to right.

$$x^5-6x^2+12x-5$$

Descending powers of *x*

Think of $12x$ as $12x^1$ and -5 as $-5x^0$.

When written in descending powers of the variable, the greatest-degree term is written first and is the **leading term** of the polynomial. Its coefficient is the **leading coefficient.**

NOW TRY EXERCISE 1

Write the polynomial in descending powers of the variable. Then give the leading term and the leading coefficient.

$$-2x^3 - 2x^5 + 4x^2 + 7 - x$$

CLASSROOM EXAMPLE 1

Write the polynomial in descending powers of the variable. Then give the leading term and the leading coefficient.

$$-3z^4 + 2z^3 + z^5 - 6z$$

Answer: $z^5 - 3z^4 + 2z^3 - 6z$; z^5; 1

EXAMPLE 1 Writing Polynomials in Descending Powers

Write each polynomial in descending powers of the variable. Then give the leading term and the leading coefficient.

(a) $y - 6y^3 + 8y^5 - 9y^4 + 12$ is written as $8y^5 - 9y^4 - 6y^3 + y + 12$.

(b) $-2 + m + 6m^2 - m^3$ is written as $-m^3 + 6m^2 + m - 2$.

Each leading term is shown in color. In part (a), the leading coefficient is 8, and in part (b) it is -1.

NOW TRY

Some polynomials with a specific number of terms are given special names.

- A polynomial with exactly three terms is a **trinomial.**
- A two-term polynomial is a **binomial.**
- A single-term polynomial is a **monomial.**

Although many polynomials contain only one variable, polynomials may have more than one variable. The degree of a term with more than one variable is the sum of the exponents on the variables. The **degree of a polynomial** is the greatest degree of all of its terms.

Polynomials and Their Degrees

Type of Polynomial	*Example*	*Degree*
Monomial	7	0 $(7 = 7x^0)$
	$5x^3y^7$	10 $(3 + 7 = 10)$
Binomial	$6 + 2x^3$	3
	$11y + 8$	1 $(y = y^1)$
Trinomial	$t^2 + 11t + 4$	2
	$-3 + 2k^5 + 9z^4$	5
	$x^3y^9 + 12xy^4 + 7xy$	12 (The terms have degrees 12, 5, and 2, and 12 is the greatest.)

NOTE If a polynomial in a single variable is written in descending powers of that variable, the degree of the polynomial will be the degree of the leading term.

CLASSROOM EXAMPLE 2

Identify each polynomial as a *monomial, binomial, trinomial,* or *none of these*. Give the degree.

(a) $a^4b^2 - ab^6$

(b) -100

Answers: (a) binomial; 7
(b) monomial; 0

NOW TRY EXERCISE 2

Identify each polynomial as a *monomial, binomial, trinomial,* or *none of these*. Give the degree.

(a) $-4x^3 + 10x^5 + 7$

(b) $-6m^2n^5$

NOW TRY ANSWERS

1. $-2x^5 - 2x^3 + 4x^2 - x + 7$; $-2x^5$; -2

2. **(a)** trinomial; 5
(b) monomial; 7

EXAMPLE 2 Classifying Polynomials

Identify each polynomial as a *monomial, binomial, trinomial,* or *none of these.* Give the degree.

(a) $-x^2 + 5x + 1$ This is a trinomial of degree 2.

(b) $\frac{3}{4}xy^4$ $\left(\text{or } \frac{3}{4}x^1y^4\right)$ This is a monomial of degree 5 (because $1 + 4 = 5$).

(c) $7m^9 + 18m^{14}$ This is a binomial of degree 14.

(d) $p^4 - p^2 - 6p - 5$ Polynomials of four terms or more do not have special names, so *none of these* is the answer that applies here. This polynomial has degree 4.

NOW TRY

OBJECTIVE 2 Add and subtract polynomials.

We use the distributive property to simplify polynomials by combining like terms.

$$x^3 + 4x^2 + 5x^2 - 1$$
$$= x^3 + (4 + 5)x^2 - 1 \quad \text{Distributive property}$$
$$= x^3 + 9x^2 - 1 \quad \text{Add.}$$

The terms in a polynomial such as $4x + 5x^2$ cannot be combined. ***Only terms containing exactly the same variables raised to the same powers may be combined.*** Recall that such terms are **like terms.**

CLASSROOM EXAMPLE 3
Combine like terms.
(a) $2z^4 + 3x^4 + z^4 - 9x^4$
(b) $3t + 4r - 4t - 8r$
(c) $5x^2z - 3x^3z^2 + 8x^2z + 12x^3z^2$
Answers: (a) $3z^4 - 6x^4$
(b) $-t - 4r$ (c) $13x^2z + 9x^3z^2$

CAUTION *Only like terms can be combined.*

NOW TRY EXERCISE 3
Combine like terms.
(a) $2x^2 - 8x^2 + x^2$
(b) $3p^3 - 2q + p^3 - 5q$
(c) $-x^2t + 4x^2t + 3xt^2 - 7xt^2$

EXAMPLE 3 Combining Like Terms

Combine like terms.

(a) $-5y^3 + 8y^3 - y^3$
$$= (-5 + 8 - 1)y^3 \quad \text{Distributive property}$$
$$= 2y^3 \quad \text{Add and subtract.}$$

(b) $6x + 5y - 9x + 2y$
$$= 6x - 9x + 5y + 2y \quad \text{Commutative property}$$
$$= -3x + 7y \quad \text{Combine like terms.}$$

Because $-3x$ and $7y$ are unlike terms, no further simplification is possible.

(c) $5x^2y - 6xy^2 + 9x^2y + 13xy^2$
$$= 5x^2y + 9x^2y - 6xy^2 + 13xy^2 \quad \text{Commutative property}$$
$$= 14x^2y + 7xy^2 \quad \text{Combine like terms.}$$
NOW TRY

TEACHING TIP
When using the vertical method shown in **Example 4,** be aware that some students may erroneously add both the coefficients and the exponents of the like terms.

CLASSROOM EXAMPLE 4
Add.
(a) $(-5p^3 + 6p^2) + (8p^3 - 12p^2)$
(b)
$$\begin{array}{r} -6r^5 + 2r^3 - r^2 \\ \underline{8r^5 - 2r^3 + 5r^2} \end{array}$$
Answers: (a) $3p^3 - 6p^2$
(b) $2r^5 + 4r^2$

Adding Polynomials

To find the sum of two polynomials, combine like terms.

NOW TRY EXERCISE 4
Add.
$$(7x^2 - 9x + 4) + (x^3 - 3x^2 - 5)$$

EXAMPLE 4 Adding Polynomials

Add $(3a^5 - 9a^3 + 4a^2) + (-8a^5 + 8a^3 + 2)$.

$$(3a^5 - 9a^3 + 4a^2) + (-8a^5 + 8a^3 + 2)$$
$$= 3a^5 - 8a^5 - 9a^3 + 8a^3 + 4a^2 + 2 \quad \text{Commutative and associative properties}$$
$$= -5a^5 - a^3 + 4a^2 + 2 \quad \text{Combine like terms.}$$

Alternatively, we can add these two polynomials vertically.

$$\begin{array}{r} 3a^5 - 9a^3 + 4a^2 + 0 \\ \underline{-8a^5 + 8a^3 + 0a^2 + 2} \\ -5a^5 - a^3 + 4a^2 + 2 \end{array}$$

Place like terms in columns. Using placeholders may help.
Add in columns.
The sum is the same.
NOW TRY

NOW TRY ANSWERS
3. (a) $-5x^2$ **(b)** $4p^3 - 7q$
(c) $3x^2t - 4xt^2$
4. $x^3 + 4x^2 - 9x - 1$

CLASSROOM EXAMPLE 5
Subtract.
(a) $(p^4 + p^3 + 5) - (3p^4 + 5p^3 + 2)$
(b) $2k^3 - 3k^2 - 2k + 5$
$\underline{4k^3 + 6k^2 - 5k + 8}$

Answers: (a) $-2p^4 - 4p^3 + 3$
(b) $-2k^3 - 9k^2 + 3k - 3$

NOW TRY EXERCISE 5
Subtract.

$$(2y^2 - 7y - 4) - (8y^2 - 2y + 10)$$

TEACHING TIP
To emphasize how the operation of subtraction can also be performed vertically, write the problem in **Example 5** on the board as shown below.

$$\begin{array}{r} -6m^2 - 8m + 5 \\ \underline{-(-5m^2 + 7m - 8)} \end{array}$$

NOW TRY ANSWER
5. $-6y^2 - 5y - 14$

In **Section R.2,** we defined subtraction of real numbers as follows.

$$a - b = a + (-b)$$

That is, we add the first number and the negative (or opposite) of the second. We define the **negative of a polynomial** as that polynomial with the sign of every coefficient changed.

Subtracting Polynomials

To find the difference of two polynomials, add the first polynomial (minuend) and the negative (or opposite) of the *second* polynomial (subtrahend).

EXAMPLE 5 Subtracting Polynomials

Subtract $(-6m^2 - 8m + 5) - (-5m^2 + 7m - 8)$.

Change every sign in the second polynomial (subtrahend) and add.

$$\begin{aligned} &(-6m^2 - 8m + 5) - (-5m^2 + 7m - 8) \\ &= -6m^2 - 8m + 5 + 5m^2 - 7m + 8 && \text{Definition of subtraction} \\ &= -6m^2 + 5m^2 - 8m - 7m + 5 + 8 && \text{Rearrange terms.} \\ &= -m^2 - 15m + 13 && \text{Combine like terms.} \end{aligned}$$

CHECK $(-m^2 - 15m + 13) + (-5m^2 + 7m - 8) = -6m^2 - 8m + 5$ ✓

(Difference: $-m^2 - 15m + 13$; Subtrahend: $-5m^2 + 7m - 8$; Minuend: $-6m^2 - 8m + 5$)

Alternatively, we can subtract these two polynomials vertically.

$$\begin{array}{r} -6m^2 - 8m + 5 \\ \underline{-5m^2 + 7m - 8} \end{array}$$

Write the subtrahend below the minuend, lining up like terms in columns.

Change all the signs in the subtrahend and add.

$$\begin{array}{rl} -6m^2 - 8m + 5 & \\ \underline{+5m^2 - 7m + 8} & \text{Change all signs.} \\ -m^2 - 15m + 13 & \text{Add in columns.} \end{array}$$

The difference is the same.

NOW TRY

4.2 Exercises

FOR EXTRA HELP

MyMathLab®

● *Complete solution available in MyMathLab*

1. *Concept Check* Which polynomial is a trinomial in descending powers, having degree 6?

A. $5x^6 - 4x^5 + 12$ B. $6x^5 - x^6 + 4$

C. $2x + 4x^2 - x^6$ D. $4x^6 - 6x^4 + 9x^2 - 8$

2. *Concept Check* Give an example of a polynomial of four terms in the variable x, having degree 5, written in descending powers, and lacking a fourth-degree term.

Give the numerical coefficient and the degree of each term. ***See Objective 1.***

3. $7z$
4. $3r$
5. $-15p^2$
6. $-27k^3$
7. x^4
8. y^6
9. $\frac{t}{6}$
10. $\frac{m}{4}$
11. 8
12. 2
13. $-x^3$
14. $-y^9$

Answers:
1. A
2. Answers will vary. One example is $7x^5 + 2x^3 - 6x^2 + 9x$.
3. 7; 1
4. 3; 1
5. −15; 2
6. −27; 3
7. 1; 4
8. 1; 6
9. $\frac{1}{6}$; 1
10. $\frac{1}{4}$; 1
11. 8; 0
12. 2; 0
13. −1; 3
14. −1; 9

15. $2x^3 - 3x^2 + x + 4$; $2x^3$; 2
16. $3q^4 + q^2 - 2q + 1$; $3q^4$; 3
17. $p^7 - 8p^5 + 4p^3$; p^7; 1
18. $y^4 - 2y^3 + 3y^2$; y^4; 1
19. $-3m^4 - m^3 + 10$; $-3m^4$; -3
20. $-8x^2 - x + 4$; $-8x^2$; -8

21. monomial; 0 **22.** monomial; 0
23. binomial; 1 **24.** binomial; 1
25. binomial; 8 **26.** binomial; 5
27. monomial; 6 **28.** monomial; 4
29. trinomial; 3 **30.** trinomial; 2
31. none of these; 5
32. none of these; 4

33. $8z^4$ **34.** $6r^5$
35. $7m^3$ **36.** $6p^4$
37. $5x$ **38.** 0
39. already simplified
40. already simplified
41. $-t + 13s$ **42.** $5p - 8q$
43. $8k^2 + 2k - 7$
44. $-2x^2 + 2x - 6$
45. $-2n^4 - n^3 + n^2$
46. $q^3 + 8q^2 - 4q$
47. $-2ab^2 + 20a^2b$
48. $-m^2n - 5mn^2$
49. $3m + 11$ **50.** $2a + 1$
51. $-p - 4$ **52.** $-8x - 12$

53. $8x^2 + x - 2$
54. $6k^3 - 3k^2 - 2k$
55. $-t^4 + 2t^2 - t + 5$
56. $-4p^3 + 10p^2 + 5p - 5$
57. $5y^3 - 3y^2 + 5y + 1$
58. $2x^5 - x^4 - 2x^3 + 1$
59. $r + 13$ **60.** $-d + 8$
61. $-2a^2 - 2a - 7$
62. $-2q^4 - 7q^2 + 15$
63. $-3z^5 + z^2 + 7z$
64. $t^3 - 5t^2 - t$
65. $12p - 4$ **66.** $19m + 3$
67. $-9p^2 + 11p - 9$
68. $3y^3 + 12y - 1$
69. $5a + 18$ **70.** $-5b + 14$
71. $14m^2 - 13m + 6$
72. $-7z^2 + 7z - 3$
73. $13z^2 + 10z - 3$

Write each polynomial in descending powers of the variable. Then give the leading term and the leading coefficient. ***See Example 1.***

15. $2x^3 + x - 3x^2 + 4$ **16.** $q^2 + 3q^4 - 2q + 1$ **17.** $4p^3 - 8p^5 + p^7$

18. $3y^2 + y^4 - 2y^3$ **19.** $10 - m^3 - 3m^4$ **20.** $4 - x - 8x^2$

Identify each polynomial as a monomial, binomial, trinomial, *or* none of these. *Give the degree.* ***See Example 2.***

21. 25 **22.** 15 **23.** $7m - 22$ **24.** $6x + 15$

25. $-7y^6 + 11y^8$ **26.** $12k^2 - 9k^5$ **27.** $-mn^5$ **28.** $-a^3b$

29. $-5m^3 + 6m - 9m^2$ **30.** $4z^2 - 11z + 2$

31. $-6p^4q - 3p^3q^2 + 2pq^3 - q^4$ **32.** $8s^3t - 4s^2t^2 + 2st^3 + 9$

Combine like terms. ***See Example 3.***

33. $5z^4 + 3z^4$ **34.** $8r^5 - 2r^5$ **35.** $-m^3 + 2m^3 + 6m^3$

36. $3p^4 + 5p^4 - 2p^4$ **37.** $x + x + x + x + x$ **38.** $z - z - z + z$

39. $m^4 - 3m^2 + m$ **40.** $5a^5 + 2a^4 - 9a^3$ **41.** $5t + 4s - 6t + 9s$

42. $8p - 9q - 3p + q$ **43.** $2k + 3k^2 + 5k^2 - 7$ **44.** $4x^2 + 2x - 6x^2 - 6$

45. $n^4 - 2n^3 + n^2 - 3n^4 + n^3$ **46.** $2q^3 + 3q^2 - 4q - q^3 + 5q^2$

47. $3ab^2 + 7a^2b - 5ab^2 + 13a^2b$ **48.** $6m^2n - 8mn^2 + 3mn^2 - 7m^2n$

49. $4 - (2 + 3m) + 6m + 9$ **50.** $8 - (3a + 4) + 5a - 3$

51. $6 + 3p - (2p + 1) - (2p + 9)$ **52.** $-8 + 4x - (x - 1) - (11x + 5)$

Add or subtract as indicated. ***See Examples 4 and 5.***

53. $(5x^2 + 7x - 4) + (3x^2 - 6x + 2)$ **54.** $(4k^3 + k^2 + k) + (2k^3 - 4k^2 - 3k)$

55. $(6t^2 - 4t^4 - t) + (3t^4 - 4t^2 + 5)$ **56.** $(3p^2 + 2p - 5) + (7p^2 - 4p^3 + 3p)$

57. $(y^3 + 3y + 2) + (4y^3 - 3y^2 + 2y - 1)$ **58.** $(2x^5 - 2x^4 + x^3 - 1) + (x^4 - 3x^3 + 2)$

59. $(3r + 8) - (2r - 5)$ **60.** $(2d + 7) - (3d - 1)$

61. $(2a^2 + 3a - 1) - (4a^2 + 5a + 6)$ **62.** $(q^4 - 2q^2 + 10) - (3q^4 + 5q^2 - 5)$

63. $(z^5 + 3z^2 + 2z) - (4z^5 + 2z^2 - 5z)$ **64.** $(5t^3 - 3t^2 + 2t) - (4t^3 + 2t^2 + 3t)$

65. Add.
$$\begin{array}{r} 21p - 8 \\ \underline{-9p + 4} \end{array}$$

66. Add.
$$\begin{array}{r} 15m - 9 \\ \underline{4m + 12} \end{array}$$

67. Add.
$$\begin{array}{r} -12p^2 + 4p - 1 \\ \underline{3p^2 + 7p - 8} \end{array}$$

68. Add.
$$\begin{array}{r} -6y^3 + 8y + 5 \\ \underline{9y^3 + 4y - 6} \end{array}$$

69. Subtract.
$$\begin{array}{r} 12a + 15 \\ \underline{7a - 3} \end{array}$$

70. Subtract.
$$\begin{array}{r} -3b + 6 \\ \underline{2b - 8} \end{array}$$

71. Subtract.
$$\begin{array}{r} 6m^2 - 11m + 5 \\ \underline{-8m^2 + 2m - 1} \end{array}$$

72. Subtract.
$$\begin{array}{r} -4z^2 + 2z - 1 \\ \underline{3z^2 - 5z + 2} \end{array}$$

73. Add.
$$\begin{array}{r} 12z^2 - 11z + 8 \\ 5z^2 + 16z - 2 \\ \underline{-4z^2 + 5z - 9} \end{array}$$

74. $-m^3 + 8m^2 - 8m$
75. $10y^3 - 7y^2 + 5y + 8$
76. $-7r^8 + 5r^6 - r^5 + 5$
77. $-5a^4 - 6a^3 + 9a^2 - 11$
78. $-m^4 - m^3 + 8m^2 - 2m$
79. $3y^2 - 4y + 2$
80. $5z^2 - 7x + 2m$
81. $-4m^2 + 4n^2 - 7n$
82. $-9m^3 - 6m^2 + 6m$
83. $y^4 - 4y^2 - 4$
84. $-19p + 14$
85. $10z^2 - 16z$
86. $20k$
87. $12x^2 + 8x + 5$
88. $4x^2 + 16x + 2$

74. Add.
$$\begin{array}{r} -6m^3 + 2m^2 + 5m \\ 8m^3 + 4m^2 - 6m \\ \underline{-3m^3 + 2m^2 - 7m} \end{array}$$

75. Add.
$$\begin{array}{r} 6y^3 - 9y^2 \qquad + 8 \\ \underline{4y^3 + 2y^2 + 5y \qquad} \end{array}$$

76. Add.
$$\begin{array}{r} -7r^8 + 2r^6 - r^5 \qquad \\ \underline{3r^6 \qquad + 5} \end{array}$$

77. Subtract.
$$\begin{array}{r} -5a^4 \qquad + 8a^2 - 9 \\ \underline{6a^3 - a^2 + 2} \end{array}$$

78. Subtract.
$$\begin{array}{r} -2m^3 + 8m^2 \qquad \\ \underline{m^4 - m^3 \qquad + 2m} \end{array}$$

Extending Skills *Perform the indicated operations.*

79. Subtract $4y^2 - 2y + 3$ from $7y^2 - 6y + 5$.

80. Subtract $-(-4x + 2z^2 + 3m)$ from $[(2z^2 - 3x + m) + (z^2 - 2m)]$.

81. $(-4m^2 + 3n^2 - 5n) - [(3m^2 - 5n^2 + 2n) + (-3m^2) + 4n^2]$

82. $[-(4m^2 - 8m + 4m^3) - (3m^2 + 2m + 5m^3)] + m^2$

83. $[-(y^4 - y^2 + 1) - (y^4 + 2y^2 + 1)] + (3y^4 - 3y^2 - 2)$

84. $[2p - (3p - 6)] - [(5p - (8 - 9p)) + 4p]$

85. $-[3z^2 + 5z - (2z^2 - 6z)] + [(8z^2 - [5z - z^2]) + 2z^2]$

86. $5k - (5k - [2k - (4k - 8k)]) + 11k - (9k - 12k)$

Find the perimeter of each figure. Express it as a polynomial in descending powers of the variable x.

87.

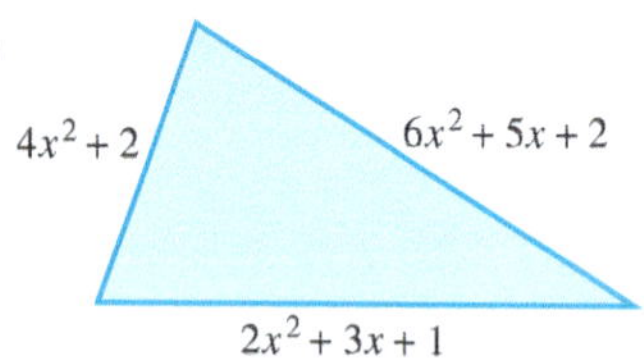

88.

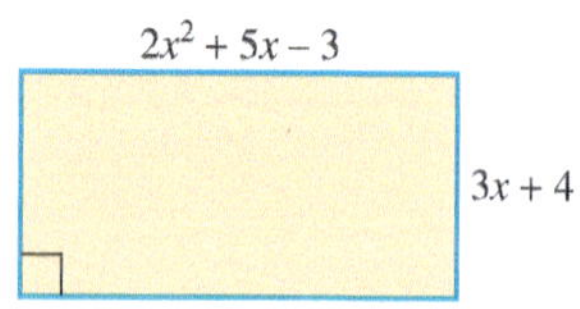

4.3 Polynomial Functions, Graphs, and Composition

OBJECTIVES

1. Recognize and evaluate polynomial functions.
2. Use a polynomial function to model data.
3. Add and subtract polynomial functions.
4. Find the composition of functions.
5. Graph basic polynomial functions.

OBJECTIVE 1 Recognize and evaluate polynomial functions.

In **Chapter 2,** we studied linear (first-degree polynomial) functions $f(x) = ax + b$. Now we consider more general polynomial functions.

Polynomial Function

A **polynomial function of degree *n*** is defined by

$$f(x) = a_n x^n + a_{n-1}x^{n-1} + \cdots + a_1 x + a_0,$$

for real numbers $a_n, a_{n-1}, \ldots, a_1$, and a_0, where $a_n \neq 0$ and n is a whole number.

Another way of describing a polynomial function is to say that it is a function defined by a polynomial in one variable, consisting of one or more terms. It is usually written in descending powers of the variable, and its degree is the degree of the polynomial that defines it.

We can evaluate a polynomial function $f(x)$ at different values of the variable x.

VOCABULARY

- ☐ polynomial function
- ☐ composite function (composition of functions)
- ☐ identity function
- ☐ squaring function
- ☐ cubing function

EXAMPLE 1 Evaluating Polynomial Functions

Let $f(x) = 4x^3 - x^2 + 5$. Find each value.

(a) $f(3)$

Read this as "f of 3," not "f times 3."

$f(x) = 4x^3 - x^2 + 5$ Given function

$f(3) = 4(3)^3 - 3^2 + 5$ Substitute 3 for x.

$f(3) = 4(27) - 9 + 5$ Apply the exponents.

$f(3) = 108 - 9 + 5$ Multiply.

$f(3) = 104$ Subtract, and then add.

Thus, $f(3) = 104$ and the ordered pair $(3, 104)$ belongs to f.

(b) $f(-4)$

$f(x) = 4x^3 - x^2 + 5$ Use parentheses.

$f(-4) = 4(-4)^3 - (-4)^2 + 5$ Let $x = -4$.

$f(-4) = 4(-64) - 16 + 5$ Be careful with signs.

$f(-4) = -256 - 16 + 5$ Multiply.

$f(-4) = -267$ Subtract, and then add.

So, $f(-4) = -267$. The ordered pair $(-4, -267)$ belongs to f. NOW TRY

NOW TRY EXERCISE 1
Let $f(x) = x^3 - 2x^2 + 7$. Find $f(-3)$.

CLASSROOM EXAMPLE 1
Let $f(x) = -x^2 + 5x - 11$. Find $f(-4)$.

Answer: -47

TEACHING TIP
Use **CLASSROOM EXAMPLE 1** to reinforce evaluating $-x^2$ for $x = -4$. Explain carefully why the result for this term is -16, and not 16.

The capital letter P is sometimes used for polynomial functions. The function

$$P(x) = 4x^3 - x^2 + 5$$

yields the same ordered pairs as the function f in **Example 1.**

OBJECTIVE 2 Use a polynomial function to model data.

EXAMPLE 2 Using a Polynomial Model to Approximate Data

The number of students enrolled in public schools (grades pre-K–12) in the United States during the years 2000 through 2011 can be modeled by the polynomial function

$$P(x) = -0.02675x^2 + 0.4873x + 47.26,$$

where $x = 0$ corresponds to the year 2000, $x = 1$ corresponds to 2001, and so on, and $P(x)$ is in millions. Use this function to approximate the number of public school students in 2011. (*Source:* National Center for Education Statistics.)

Here $x = 11$ corresponds to 2011, so we find $P(11)$.

$P(x) = -0.02675x^2 + 0.4873x + 47.26$

$P(11) = -0.02675(11)^2 + 0.4873(11) + 47.26$ Let $x = 11$.

$P(11) \approx 49.38$ Evaluate.

There were about 49.38 million public school students in 2011. NOW TRY

NOW TRY EXERCISE 2
Use the function in **Example 2** to approximate the number of public school students in 2010 (to two decimal places).

CLASSROOM EXAMPLE 2
Use the function in **Example 2** to approximate the number of public school students in 2005 (to two decimal places).

Answer: 49.03 million

NOW TRY ANSWERS

1. -38
2. 49.46 million

OBJECTIVE 3 Add and subtract polynomial functions.

The operations of addition, subtraction, multiplication, and division are also defined for functions. For example, the graph in **FIGURE 1** shows dollars (in billions) spent for general science and for space/other technologies over a 20-year period.

$G(x)$ represents dollars spent for general science.

$S(x)$ represents dollars spent for space/other technologies.

$T(x)$ represents total expenditures for these two categories.

Science and Space Spending

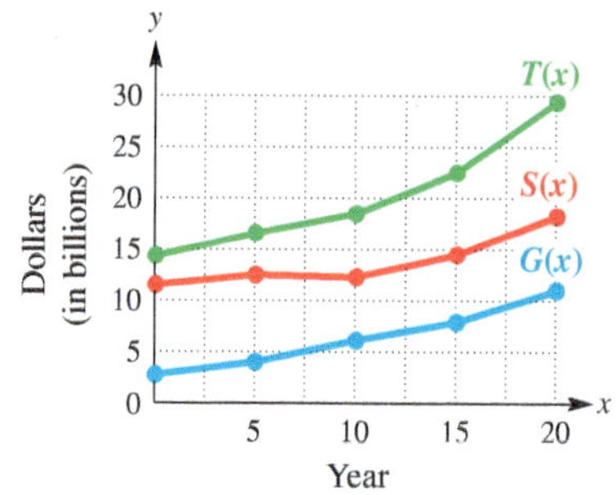

Source: U.S. Office of Management and Budget.

FIGURE 1

The total expenditures function can be found by *adding* the spending functions for the two individual categories.

$$T(x) = G(x) + S(x)$$

As another example, businesses use the equation "profit equals revenue minus cost," which can be written using function notation.

$$P(x) = R(x) - C(x)$$

$P(x)$: Profit function; $R(x)$: Revenue function; $C(x)$: Cost function. x is the number of items produced and sold.

The profit function is found by *subtracting* the cost function from the revenue function.

We define the following **operations on functions.**

Adding and Subtracting Functions

If $f(x)$ and $g(x)$ define functions, then

$$(f + g)(x) = f(x) + g(x) \quad \text{Sum function}$$

and

$$(f - g)(x) = f(x) - g(x). \quad \text{Difference function}$$

In each case, the domain of the new function is the intersection of the domains of $f(x)$ and $g(x)$.

CLASSROOM EXAMPLE 3

For $f(x) = 3x^2 + 8x - 6$ and $g(x) = -4x^2 + 4x - 8$, find each of the following.

(a) $(f + g)(x)$ (b) $(f - g)(x)$

Answers: (a) $-x^2 + 12x - 14$ (b) $7x^2 + 4x + 2$

EXAMPLE 3 Adding and Subtracting Functions

Find each of the following for the polynomial functions f and g as defined.

$$f(x) = x^2 - 3x + 7 \quad \text{and} \quad g(x) = -3x^2 - 7x + 7$$

(a) $(f + g)(x)$ — This notation does *not* indicate the distributive property.

$= f(x) + g(x)$ Use the definition.

$= (x^2 - 3x + 7) + (-3x^2 - 7x + 7)$ Substitute.

$= -2x^2 - 10x + 14$ Add the polynomials.

NOW TRY EXERCISE 3

For $f(x) = x^3 - 3x^2 + 4$

and $g(x) = -2x^3 + x^2 - 12$,

find each of the following.

(a) $(f + g)(x)$

(b) $(f - g)(x)$

NOW TRY EXERCISE 4

For $f(x) = x^2 - 4$

and $g(x) = -6x^2$,

find each of the following.

(a) $(f + g)(x)$

(b) $(f - g)(-4)$

CLASSROOM EXAMPLE 4

For $f(x) = 18x^2 - 24x$

and $g(x) = 3x$,

find each of the following.

(a) $(f + g)(x)$ and $(f + g)(-1)$

(b) $(f - g)(x)$ and $(f - g)(1)$

Answers: (a) $18x^2 - 21x$; 39

(b) $18x^2 - 27x$; -9

NOW TRY ANSWERS

3. (a) $-x^3 - 2x^2 - 8$

(b) $3x^3 - 4x^2 + 16$

4. (a) $-5x^2 - 4$

(b) 108

(b) $(f - g)(x)$

$= f(x) - g(x)$ — Use the definition.

$= (x^2 - 3x + 7) - (-3x^2 - 7x + 7)$ — Substitute.

$= (x^2 - 3x + 7) + (3x^2 + 7x - 7)$ — Change subtraction to addition.

$= 4x^2 + 4x$ — Add. NOW TRY

EXAMPLE 4 Adding and Subtracting Functions

Find each of the following for the polynomial functions f and g as defined.

$$f(x) = 10x^2 - 2x \quad \text{and} \quad g(x) = 2x$$

(a) $(f + g)(2)$

$= f(2) + g(2)$ — Use the definition.

$f(x) = 10x^2 - 2x$ $g(x) = 2x$

This is a key step.

$= [10(2)^2 - 2(2)] + 2(2)$ — Substitute.

$= [40 - 4] + 4$ — Order of operations

$= 40$ — Subtract, and then add.

Alternative method: $(f + g)(x)$ — Find $(f + g)(x)$.

$= f(x) + g(x)$ — Use the definition.

$= (10x^2 - 2x) + 2x$ — Substitute.

$= 10x^2$ — Combine like terms.

$(f + g)(2)$ — Now find $(f + g)(2)$.

$= 10(2)^2$ — $(f + g)(x) = 10x^2$; Substitute.

$= 40$ — The result is the same.

(b) $(f - g)(x)$ and $(f - g)(1)$

$(f - g)(x)$

$= f(x) - g(x)$ — Use the definition.

$= (10x^2 - 2x) - 2x$ — Substitute.

$= 10x^2 - 4x$ — Combine like terms.

$(f - g)(1)$ — Now find $(f - g)(1)$.

$= 10(1)^2 - 4(1)$ — $(f - g)(x) = 10x^2 - 4x$; Substitute.

$= 6$ — Perform the operations. NOW TRY

Confirm that $f(1) - g(1)$ gives the same result.

OBJECTIVE 4 Find the composition of functions.

The diagram in **FIGURE 2** on the next page shows a function g that assigns, to each element x of set X, some element y of set Y. Suppose that a function f takes each element of set Y and assigns a value z of set Z. Then f and g together assign an element x in X to an element z in Z.

The result of this process is a new function h that takes an element x in X and assigns it an element z in Z.

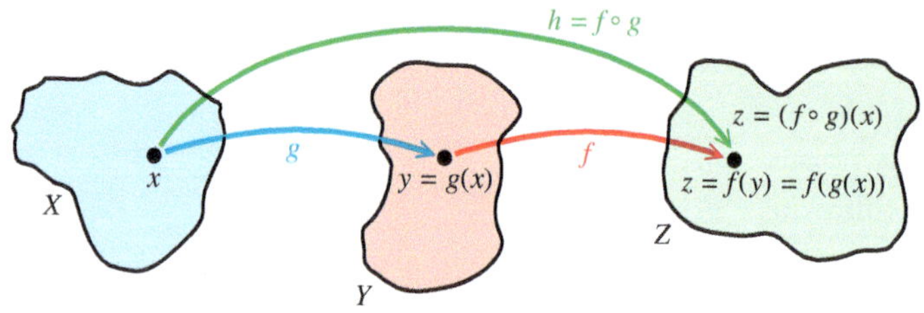

FIGURE 2

Function h is the *composition* of functions f and g, written $\boldsymbol{f \circ g}$.

Composition of Functions

The **composite function,** or **composition,** of functions f and g is defined by

$$(f \circ g)(x) = f(g(x)),$$

for all x in the domain of g such that $g(x)$ is in the domain of f.

Read $f \circ g$ as "f of g" (or "f compose g").

As a real-life example of how composite functions occur, consider the following.

A \$40 pair of blue jeans is on sale for 25% off. If we purchase the jeans before noon, the retailer offers an additional 10% off. What is the final sale price of the blue jeans?

We might be tempted to say that the blue jeans are 25% + 10% = 35% off and calculate \$40(0.35) = \$14, giving a final sale price of

\$40 − \$14 = \$26. This is not correct.

To find the correct final sale price, we must first find the price after taking 25% off, and then take an additional 10% off *that* price.

\$40(0.25) = \$10, giving a sale price of \$40 − \$10 = \$30. Take 25% off original price.

\$30(0.10) = \$3, giving a ***final sale price*** of \$30 − \$3 = \$27. Take additional 10% off.

This is the idea behind composition of functions.

CLASSROOM EXAMPLE 5
Let $f(x) = x - 4$ and $g(x) = x^2$.
Find $(f \circ g)(3)$.
Answer: 5

NOW TRY EXERCISE 5
Let $f(x) = 3x + 7$ and $g(x) = x - 2$.
Find $(f \circ g)(7)$.

EXAMPLE 5 Finding a Composite Function

Let $f(x) = x^2$ and $g(x) = x + 3$. Find $(f \circ g)(4)$.

$(f \circ g)(4)$ — Evaluate the "inside" function value first.

$= f(g(4))$ — Definition of composition

$= f(4 + 3)$ — Use the rule for $g(x)$; $g(4) = 4 + 3$.

$= f(7)$ — Add.

$= 7^2$ — Now evaluate the "outside" function. Use the rule for $f(x)$; $f(7) = 7^2$.

$= 49$ — Square 7.

In this composition, g is the innermost "operation" and acts on x (here 4) first. Then the output value of g (here 7) becomes the input (domain) value of f. NOW TRY

NOW TRY ANSWER
5. 22

If we interchange the order of functions f and g, the composition $g \circ f$, read "g of f" (or "g compose f"), is defined by

$$(g \circ f)(x) = g(f(x)), \quad \text{for all } x \text{ in the domain of } f \text{ such that } f(x) \text{ is in the domain of } g.$$

NOW TRY EXERCISE 6

As in **Now Try Exercise 5,** let $f(x) = 3x + 7$ and $g(x) = x - 2$. Find $(g \circ f)(7)$.

CLASSROOM EXAMPLE 6

As in **CLASSROOM EXAMPLE 5,** let $f(x) = x - 4$ and $g(x) = x^2$. Find $(g \circ f)(3)$.

Answer: 1

EXAMPLE 6 Finding a Composite Function

Find $(g \circ f)(4)$ for the functions $f(x) = x^2$ and $g(x) = x + 3$ from **Example 5.**

$(g \circ f)(4)$	Evaluate the "inside" function value first.
$= g(f(4))$	Definition of composition
$= g(4^2)$	Use the rule for $f(x)$; $f(4) = 4^2$.
$= g(16)$	Square 4.
$= 16 + 3$	Use the rule for $g(x)$; $g(16) = 16 + 3$. (Now evaluate the "outside" function.)
$= 19$	Add.

In this composition, f is the innermost "operation" and acts on x (again 4) first. Then the output value of f (here 16) becomes the input (domain) value of g. NOW TRY

We see in **Examples 5 and 6** that

$$(f \circ g)(4) \neq (g \circ f)(4) \quad \text{because} \quad 49 \neq 19.$$

In general,

$$(f \circ g)(x) \neq (g \circ f)(x).$$

CLASSROOM EXAMPLE 7

Let $f(x) = 3x + 6$ and $g(x) = x^3$.

Find each of the following.

(a) $(f \circ g)(2)$ (b) $(g \circ f)(x)$

Answers: (a) 30 (b) $(3x + 6)^3$

EXAMPLE 7 Finding Composite Functions

Let $f(x) = 4x - 1$ and $g(x) = x^2 + 5$. Find each of the following.

(a) $(f \circ g)(2)$

$= f(g(2))$	Definition of composition
$= f(2^2 + 5)$	$g(x) = x^2 + 5$
$= f(9)$	Work inside the parentheses.
$= 4(9) - 1$	$f(x) = 4x - 1$
$= 35$	Multiply, and then subtract.

(b) $(f \circ g)(x)$

$= f(g(x))$	Use $g(x)$ as the input for function f.
$= 4(g(x)) - 1$	Use the rule for $f(x)$; $f(x) = 4x - 1$.
$= 4(x^2 + 5) - 1$	$g(x) = x^2 + 5$
$= 4x^2 + 20 - 1$	Distributive property
$= 4x^2 + 19$	Combine like terms.

NOW TRY ANSWER

6. 26

NOW TRY EXERCISE 7

Let $f(x) = x - 5$
and $g(x) = -x^2 + 2$.
Find each of the following.
(a) $(g \circ f)(-1)$
(b) $(f \circ g)(x)$

(c) Find $(f \circ g)(2)$ again, this time using the rule obtained in part (b).

$$(f \circ g)(x) = 4x^2 + 19 \qquad \text{From part (b)}$$
$$(f \circ g)(2) = 4(2)^2 + 19 \qquad \text{Let } x = 2.$$
$$= 4(4) + 19 \qquad \text{Square 2.}$$
$$= 16 + 19 \qquad \text{Multiply.}$$
$$\text{Same result as in part (a)} \rightarrow \quad = 35 \qquad \text{Add.}$$

NOW TRY

OBJECTIVE 5 Graph basic polynomial functions.

Recall from **Section 2.5** that each input (or x-value) of a function results in one output (or y-value). The set of input values (for x) defines the domain of the function, and the set of output values (for y) defines the range.

The simplest polynomial function is the **identity function $f(x) = x$,** graphed in **FIGURE 3**. This function pairs each real number with itself.

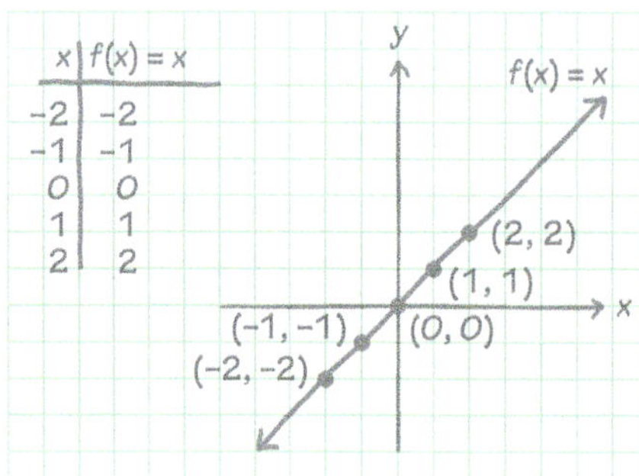

Identity function

$f(x) = x$

Domain: $(-\infty, \infty)$
Range: $(-\infty, \infty)$

FIGURE 3

NOTE A *linear function* **(Section 2.6)** is a specific kind of polynomial function.

Another polynomial function, the **squaring function $f(x) = x^2$,** is graphed in **FIGURE 4**. For this function, every real number is paired with its square. The graph of the squaring function is a *parabola.*

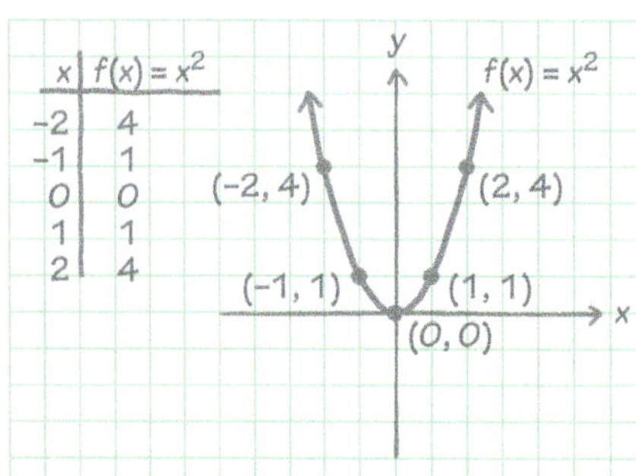

Squaring function

$f(x) = x^2$

Domain: $(-\infty, \infty)$
Range: $[0, \infty)$

FIGURE 4

NOW TRY ANSWERS
7. (a) -34 **(b)** $-x^2 - 3$

The **cubing function** $f(x) = x^3$ is graphed in **FIGURE 5**. This function pairs every real number with its cube.

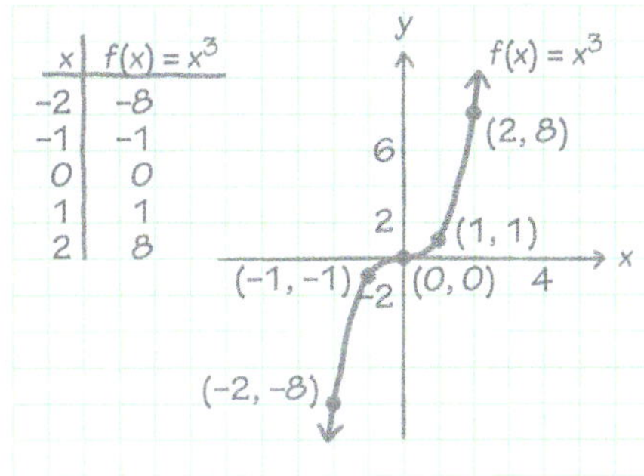

Cubing function

$f(x) = x^3$

Domain: $(-\infty, \infty)$

Range: $(-\infty, \infty)$

FIGURE 5

NOW TRY EXERCISE 8

Graph $f(x) = x^2 - 4$. Give the domain and range.

CLASSROOM EXAMPLE 8

Graph $f(x) = -2x^2$. Give the domain and range.

Answer:

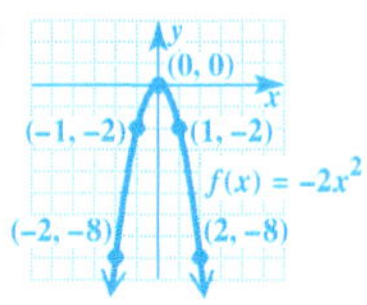

domain: $(-\infty, \infty)$;
range: $(-\infty, 0]$

EXAMPLE 8 Graphing Variations of Polynomial Functions

Graph each function by creating a table of ordered pairs. Give the domain and range of each function by observing its graph.

(a) $f(x) = 2x$

To find each range value, multiply the domain value by 2. Plot the points and join them with a straight line. See **FIGURE 6**. Both the domain and the range are $(-\infty, \infty)$.

x	$f(x) = 2x$
−2	−4
−1	−2
0	0
1	2
2	4

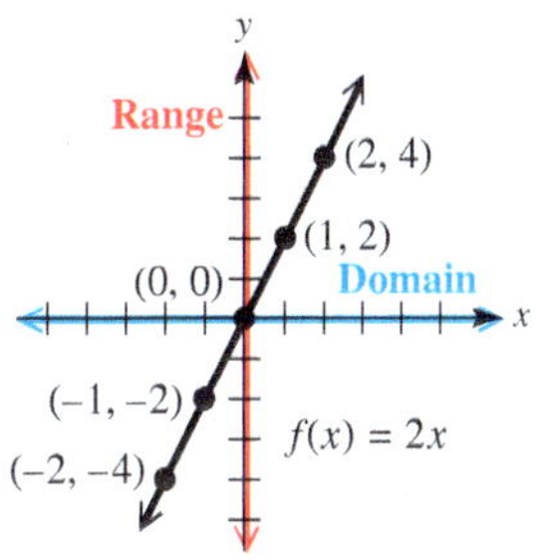

FIGURE 6

x	$f(x) = -x^2$
−2	−4
−1	−1
0	0
1	−1
2	−4

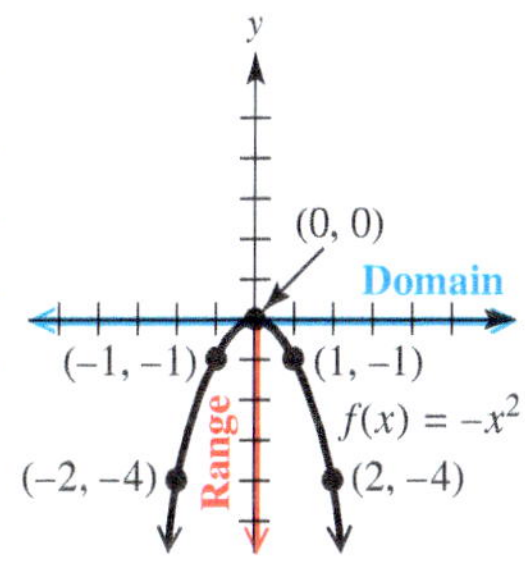

FIGURE 7

(b) $f(x) = -x^2$

For each input x, square it and then take its opposite. Plotting and joining the points gives a parabola that opens down. It is a *reflection* of the graph of the squaring function across the x-axis. See the table and **FIGURE 7**. The domain is $(-\infty, \infty)$ and the range is $(-\infty, 0]$.

(c) $f(x) = x^3 - 2$

For this function, cube the input and then subtract 2 from the result. The graph is that of the cubing function *shifted* 2 units down. See the table and **FIGURE 8**. The domain and range are both $(-\infty, \infty)$.

x	$f(x) = x^3 - 2$
−2	−10
−1	−3
0	−2
1	−1
2	6

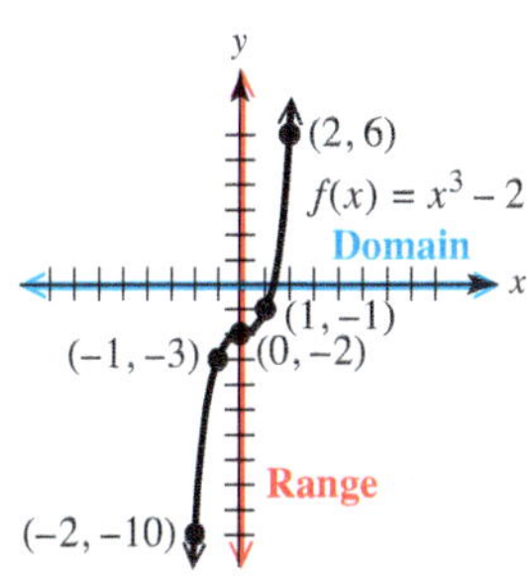

FIGURE 8

NOW TRY

NOW TRY ANSWER

8.

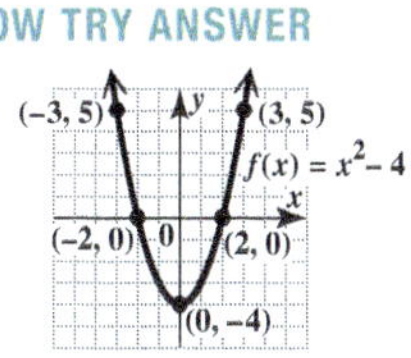

domain: $(-\infty, \infty)$;
range: $[-4, \infty)$

4.3 Exercises

FOR EXTRA HELP MyMathLab®

Complete solution available in MyMathLab

1. polynomial; one; terms; powers
2. A, D
3. (a) -10 (b) 8 (c) -4
4. (a) 7 (b) 1 (c) 5
5. (a) 8 (b) -10 (c) 0
6. (a) -4 (b) 14 (c) 0
7. (a) 8 (b) 2 (c) 4
8. (a) 2 (b) -10 (c) -4
9. (a) 7 (b) 1 (c) 1
10. (a) -3 (b) 9 (c) -5
11. (a) 8 (b) 74 (c) 6
12. (a) 5 (b) 71 (c) -1
13. (a) -11 (b) 4 (c) -8
14. (a) 11 (b) -1 (c) 11
15. (a) 1.0 million lb (b) 56.3 million lb (c) 158.8 million lb
16. (a) 11.5% (b) 5.5% (c) 13.2%

1. *Concept Check* A polynomial function is a function defined by a ________ in (*one / two / three*) variable(s), consisting of one or more (*factors / terms*) and usually written in descending ________ of the variable.

2. *Concept Check* Which of the following are *not* polynomial functions?

A. $P(x) = x^{-2} - 2x$ **B.** $f(x) = \frac{1}{2}x^2 + x - 1$

C. $g(x) = -4x + 1.5$ **D.** $p(x) = x^3 - x^2 - \frac{5}{x}$

For each polynomial function, find **(a)** $f(-1)$, **(b)** $f(2)$, *and* **(c)** $f(0)$. ***See Example 1.***

3. $f(x) = 6x - 4$ **4.** $f(x) = -2x + 5$ **5.** $f(x) = x^2 - 7x$

6. $f(x) = x^2 + 5x$ **7.** $f(x) = x^2 - 3x + 4$ **8.** $f(x) = x^2 - 5x - 4$

9. $f(x) = 2x^2 - 4x + 1$ **10.** $f(x) = 3x^2 + x - 5$ **11.** $f(x) = 5x^4 - 3x^2 + 6$

12. $f(x) = 4x^4 + 2x^2 - 1$ **13.** $f(x) = -x^2 + 2x^3 - 8$ **14.** $f(x) = -x^2 - x^3 + 11$

Solve each problem. ***See Example 2.***

15. Imports of Fair Trade Certified™ coffee into the United States during the years 2000 through 2012 can be modeled by the polynomial function

$$P(x) = 0.6558x^2 + 5.278x + 1.048,$$

where $x = 0$ represents 2000, $x = 1$ represents 2001, and so on, and $P(x)$ is in millions of pounds. Use this function to approximate the amount of Fair Trade coffee imported into the United States (to the nearest tenth) in each given year. (*Source:* Fair Trade USA.)

(a) 2000 **(b)** 2006 **(c)** 2012

16. The U.S. population that was foreign-born during the years 1930 through 2011 can be modeled by the polynomial function

$$P(x) = 0.0042x^2 - 0.3187x + 11.49,$$

where $x = 0$ represents 1930, $x = 10$ represents 1940, and so on, and $P(x)$ is percent. Use this function to approximate the percent of the U.S. population that was foreign-born (to the nearest tenth) in each given year. (*Source:* U.S. Census Bureau.)

(a) 1930 **(b)** 1970 **(c)** 2011

17. (a) \$280.6 billion
(b) \$568.5 billion
(c) \$711.4 billion

18. (a) \$450.9 billion
(b) \$745.5 billion
(c) \$1052.7 billion

19. (a) $8x - 3$ (b) $2x - 17$

20. (a) $2x + 3$ (b) $-10x - 1$

21. (a) $-x^2 + 12x - 12$
(b) $9x^2 + 4x + 6$

22. (a) $-x^2 - 7x + 22$
(b) $7x^2 - 11x - 2$

23. $f(x)$ and $g(x)$ can be any two polynomials that have a sum of $3x^3 - x + 3$, such as $f(x) = 3x^3 + 1$ and $g(x) = -x + 2$.

24. $f(x)$ and $g(x)$ can be any two polynomials whose difference is $-x^2 + x - 5$, such as $f(x) = 2x^2 + 3x - 2$ and $g(x) = 3x^2 + 2x + 3$.

25. $x^2 + 2x - 9$ 26. $x^2 - 2x - 9$
27. 6 28. 6
29. $x^2 - x - 6$ 30. $x^2 + x - 12$
31. 6 32. -10
33. -33 34. 13
35. 0 36. 0
37. $-\frac{9}{4}$ 38. -2
39. $-\frac{9}{2}$ 40. $-\frac{15}{4}$

41. (a) $P(x) = 8.49x - 50$
(b) \$799

42. (a) $P(x) = 20.7x - 75$
(b) \$960

17. The amount spent by Americans on domestic travel during the years 1990 through 2012 can be modeled by the polynomial function

$$P(x) = -0.00942x^3 + 0.4041x^2 + 15.25x + 280.6,$$

where $x = 0$ represents 1990, $x = 1$ represents 1991, and so on, and $P(x)$ is in billions of dollars. Use this function to approximate the amount spent by Americans on domestic travel (to the nearest tenth) in each given year. (*Source:* U.S. Travel Association.)

(a) 1990 **(b)** 2005 **(c)** 2012

18. World tourism receipts during the years 2000 through 2012 can be modeled by the polynomial function

$$P(x) = -0.4499x^3 + 8.274x^2 + 15.65x + 450.9,$$

where $x = 0$ represents 2000, $x = 1$ represents 2001, and so on, and $P(x)$ is in billions of dollars. Use this function to approximate world tourism receipts (to the nearest tenth) in each given year. (*Source:* World Tourism Association.)

(a) 2000 **(b)** 2006 **(c)** 2012

For each pair of functions, find **(a)** $(f + g)(x)$ *and* **(b)** $(f - g)(x)$. ***See Example 3.***

19. $f(x) = 5x - 10, \quad g(x) = 3x + 7$

20. $f(x) = -4x + 1, \quad g(x) = 6x + 2$

21. $f(x) = 4x^2 + 8x - 3, \quad g(x) = -5x^2 + 4x - 9$

22. $f(x) = 3x^2 - 9x + 10, \quad g(x) = -4x^2 + 2x + 12$

Concept Check *Find two polynomial functions defined by $f(x)$ and $g(x)$ such that each statement is true.*

23. $(f + g)(x) = 3x^3 - x + 3$ **24.** $(f - g)(x) = -x^2 + x - 5$

Let $f(x) = x^2 - 9$, $g(x) = 2x$, and $h(x) = x - 3$. Find each of the following. ***See Example 4.***

25. $(f + g)(x)$ **26.** $(f - g)(x)$ **27.** $(f + g)(3)$ **28.** $(f - g)(-3)$

29. $(f - h)(x)$ **30.** $(f + h)(x)$ **31.** $(f - h)(-3)$ **32.** $(f + h)(-2)$

33. $(g + h)(-10)$ **34.** $(g - h)(10)$ **35.** $(g - h)(-3)$ **36.** $(g + h)(1)$

37. $(g + h)\left(\frac{1}{4}\right)$ **38.** $(g + h)\left(\frac{1}{3}\right)$ **39.** $(g + h)\left(-\frac{1}{2}\right)$ **40.** $(g + h)\left(-\frac{1}{4}\right)$

Solve each problem. ***See Objective 3.***

41. The cost in dollars to produce x t-shirts is $C(x) = 2.5x + 50$. The revenue in dollars from sales of x t-shirts is $R(x) = 10.99x$.

(a) Write and simplify a function P that gives profit in terms of x.

(b) Find the profit if 100 t-shirts are produced and sold.

42. The cost in dollars to produce x baseball caps is $C(x) = 4.3x + 75$. The revenue in dollars from sales of x caps is $R(x) = 25x$.

(a) Write and simplify a function P that gives profit in terms of x.

(b) Find the profit if 50 caps are produced and sold.

43. B **44.** D
45. A **46.** C

47. 6 **48.** 125
49. 83 **50.** 35
51. 53 **52.** −6
53. 13 **54.** 29
55. $2x^2 + 11$ **56.** $2x - 7$
57. $2x - 2$ **58.** $x^2 - 1$
59. $\frac{97}{4}$ **60.** $-\frac{3}{4}$
61. 8 **62.** $\frac{23}{2}$

63. 1 **64.** 9
65. 9 **66.** 3
67. 1 **68.** 12

69. $(f \circ g)(x) = 63{,}360x$; It computes the number of inches in x miles.
70. $(f \circ g)(x) = 1440x$; It computes the number of minutes in x days.
71. **(a)** $s = \frac{x}{4}$ **(b)** $y = \frac{x^2}{16}$ **(c)** 2.25

Concept Check *Let $f(x) = x^2$ and $g(x) = 2x - 1$. Match each expression in Column I with the description of how to evaluate it in Column II.*

I	II
43. $(f \circ g)(5)$	**A.** Square 5. Take the result and square it.
44. $(g \circ f)(5)$	**B.** Double 5 and subtract 1. Take the result and square it.
45. $(f \circ f)(5)$	**C.** Double 5 and subtract 1. Take the result, double it, and subtract 1.
46. $(g \circ g)(5)$	**D.** Square 5. Take the result, double it, and subtract 1.

Let $f(x) = x^2 + 4$, $g(x) = 2x + 3$, and $h(x) = x - 5$. Find each value or expression. ***See Examples 5–7.***

47. $(h \circ g)(4)$ **48.** $(f \circ g)(4)$ **49.** $(g \circ f)(6)$ **50.** $(h \circ f)(6)$

51. $(f \circ h)(-2)$ **52.** $(h \circ g)(-2)$ **53.** $(f \circ g)(0)$ **54.** $(f \circ h)(0)$

55. $(g \circ f)(x)$ **56.** $(g \circ h)(x)$ **57.** $(h \circ g)(x)$ **58.** $(h \circ f)(x)$

59. $(f \circ h)\left(\frac{1}{2}\right)$ **60.** $(h \circ f)\left(\frac{1}{2}\right)$ **61.** $(f \circ g)\left(-\frac{1}{2}\right)$ **62.** $(g \circ f)\left(-\frac{1}{2}\right)$

Extending Skills *The tables give some selected ordered pairs for functions f and g.*

x	3	4	6	8
$f(x)$	1	3	9	2

x	2	7	1	9
$g(x)$	3	6	9	12

Tables like these can be used to evaluate composite functions. For example, to evaluate $(g \circ f)(6)$, use the first table to find $f(6) = 9$. Then use the second table to find

$$(g \circ f)(6) = g(f(6)) = g(9) = 12.$$

Find each of the following.

63. $(f \circ g)(2)$ **64.** $(f \circ g)(7)$ **65.** $(g \circ f)(3)$

66. $(g \circ f)(8)$ **67.** $(f \circ f)(4)$ **68.** $(g \circ g)(1)$

Solve each problem. ***See Objective 4.***

69. The function $f(x) = 12x$ computes the number of inches in x feet, and the function $g(x) = 5280x$ computes the number of feet in x miles. Find and simplify $(f \circ g)(x)$. What does it compute?

70. The function $f(x) = 60x$ computes the number of minutes in x hours, and the function $g(x) = 24x$ computes the number of hours in x days. Find and simplify $(f \circ g)(x)$. What does it compute?

71. The perimeter x of a square with sides of length s is given by the formula $x = 4s$.

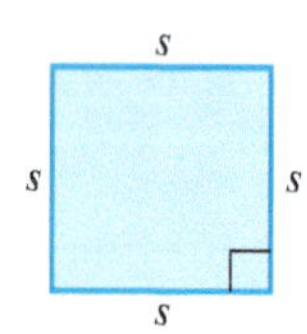

(a) Solve for s in terms of x.

(b) If y represents the area of this square, write y as a function of the perimeter x.

(c) Use the composite function of part (b) to find the area of a square with perimeter 6.

72. (a) $s = \frac{x}{3}$ (b) $y = \frac{x^2\sqrt{3}}{36}$
(c) $4\sqrt{3}$

73. $(\mathcal{A} \circ r)(t) = 4\pi t^2$; This is the area of the circular layer as a function of time.

74. $(\mathcal{A} \circ r)(t) = 16\pi t^2$; This is the area of the circular oil slick as a function of time.

75. 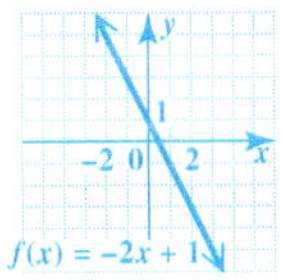

domain: $(-\infty, \infty)$; range: $(-\infty, \infty)$

76. 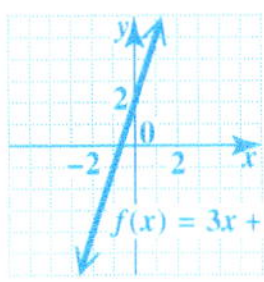

domain: $(-\infty, \infty)$; range: $(-\infty, \infty)$

77. 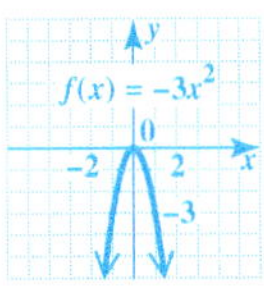

domain: $(-\infty, \infty)$; range: $(-\infty, 0]$

78. 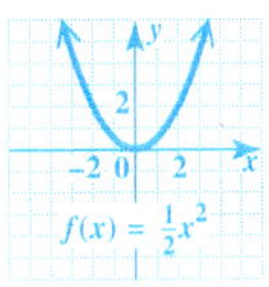

domain: $(-\infty, \infty)$; range: $[0, \infty)$

79. 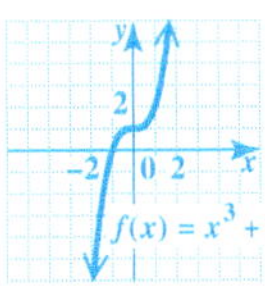

domain: $(-\infty, \infty)$; range: $(-\infty, \infty)$

80. 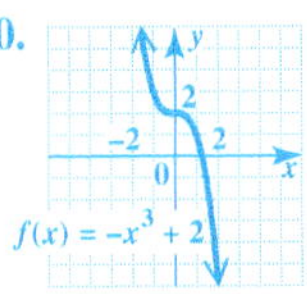

domain: $(-\infty, \infty)$; range: $(-\infty, \infty)$

72. The perimeter x of an equilateral triangle with sides of length s is given by the formula $x = 3s$.

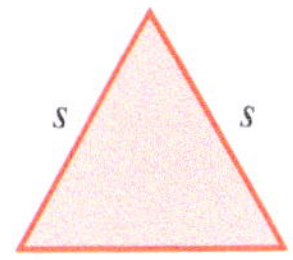

(a) Solve for s in terms of x.

(b) The area y of an equilateral triangle with sides of length s is given by the formula $y = \frac{s^2\sqrt{3}}{4}$. Write y as a function of the perimeter x.

(c) Use the composite function of part (b) to find the area of an equilateral triangle with perimeter 12.

73. When a thermal inversion layer is over a city (as happens often in Los Angeles), pollutants cannot rise vertically, but are trapped below the layer and must disperse horizontally.

Assume that a factory smokestack begins emitting a pollutant at 8 A.M and that the pollutant disperses horizontally over a circular area. Suppose that t represents the time, in hours, since the factory began emitting pollutants ($t = 0$ represents 8 A.M.), and assume that the radius of the circle of pollution is $r(t) = 2t$ miles. Let $\mathcal{A}(r) = \pi r^2$ represent the area of a circle of radius r. Find and interpret $(\mathcal{A} \circ r)(t)$.

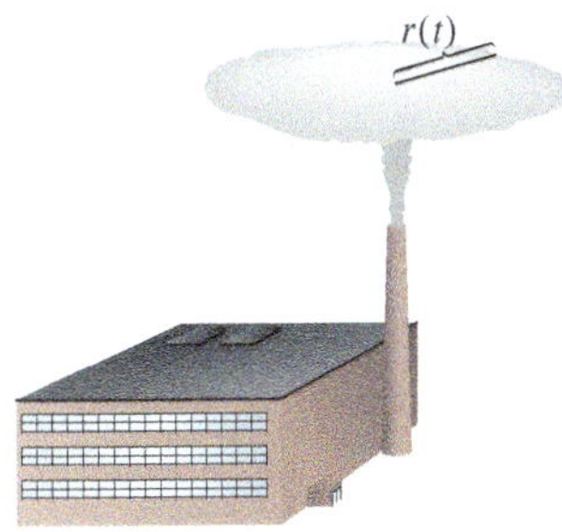

74. An oil well is leaking, with the leak spreading oil over the surface as a circle. At any time t, in minutes, after the beginning of the leak, the radius of the circular oil slick on the surface is $r(t) = 4t$ feet. Let $\mathcal{A}(r) = \pi r^2$ represent the area of a circle of radius r. Find and interpret $(\mathcal{A} \circ r)(t)$.

Graph each polynomial function. Give the domain and range. ***See Example 8.***

75. $f(x) = -2x + 1$

76. $f(x) = 3x + 2$

77. $f(x) = -3x^2$

78. $f(x) = \frac{1}{2}x^2$

79. $f(x) = x^3 + 1$

80. $f(x) = -x^3 + 2$

4.4 Multiplying Polynomials

OBJECTIVES

1. Multiply terms.
2. Multiply any two polynomials.
3. Multiply binomials.
4. Find the product of a sum and difference of two terms.
5. Find the square of a binomial.
6. Multiply polynomial functions.

OBJECTIVE 1 Multiply terms.

EXAMPLE 1 Multiplying Monomials

Find each product.

(a) $3x^4(5x^3)$

$= 3 \cdot 5 \cdot x^4 \cdot x^3$ Commutative and associative properties

$= 15x^{4+3}$ Multiply; product rule for exponents

$= 15x^7$ Add the exponents.

(b) $-4a^3(3a^5)$

$= -4(3)a^3 \cdot a^5$

$= -12a^8$

(c) $2m^2z^4(8m^3z^2)$

$= 2(8)m^2 \cdot m^3 \cdot z^4 \cdot z^2$

$= 16m^5z^6$

NOW TRY

NOW TRY EXERCISE 1
Find the product.

$-3s^2t(15s^3t^4)$

OBJECTIVE 2 Multiply any two polynomials.

EXAMPLE 2 Multiplying Polynomials

Find each product.

(a) $-2(8x^3 - 9x^2)$

$= -2(8x^3) - 2(-9x^2)$ Distributive property (Be careful with signs.)

$= -16x^3 + 18x^2$ Multiply.

(b) $5x^2(-4x^2 + 3x - 2)$

$= 5x^2(-4x^2) + 5x^2(3x) + 5x^2(-2)$ Distributive property

$= -20x^4 + 15x^3 - 10x^2$ Multiply.

(c) $(3x - 4)(2x^2 + x)$ Distributive property; Multiply each term of $2x^2 + x$ by $3x - 4$.

Treat $3x - 4$ as a single expression.

$= (3x - 4)(2x^2) + (3x - 4)(x)$

$= 3x(2x^2) + (-4)(2x^2) + (3x)(x) + (-4)(x)$ Distributive property

$= 6x^3 - 8x^2 + 3x^2 - 4x$ Multiply.

$= 6x^3 - 5x^2 - 4x$ Combine like terms.

(d) $2x^2(x + 1)(x - 3)$

$= 2x^2[(x + 1)(x) + (x + 1)(-3)]$ Distributive property

$= 2x^2[x^2 + x - 3x - 3]$ Distributive property

$= 2x^2(x^2 - 2x - 3)$ Combine like terms.

$= 2x^4 - 4x^3 - 6x^2$ Distributive property

NOW TRY

NOW TRY EXERCISE 2
Find each product.

(a) $3k^3(-2k^5 + 3k^2 - 4)$

(b) $5x(2x - 1)(x + 4)$

CLASSROOM EXAMPLE 1
Find the product.

$8k^3y(9ky^3)$

Answer: $72k^4y^4$

CLASSROOM EXAMPLE 2
Find each product.

(a) $-2r(9r - 5)$

(b) $(2k - 5m)(3k + 2m)$

Answer: (a) $-18r^2 + 10r$

(b) $6k^2 - 11km - 10m^2$

NOW TRY ANSWERS

1. $-45s^5t^5$
2. (a) $-6k^8 + 9k^5 - 12k^3$
 (b) $10x^3 + 35x^2 - 20x$

NOW TRY EXERCISE 3

Find the product.

$$\begin{array}{r} 3t^2 - 5t + 4 \\ \underline{t - 3} \end{array}$$

CLASSROOM EXAMPLE 3

Find each product.

(a) $\begin{array}{r} 4x - 3y \\ \underline{3x - y} \end{array}$

(b) $\begin{array}{r} 5a^3 - 6a^2 + 2a - 3 \\ \underline{2a - 5} \end{array}$

Answers:

(a) $12x^2 - 13xy + 3y^2$

(b) $10a^4 - 37a^3 + 34a^2 - 16a + 15$

EXAMPLE 3 Multiplying Polynomials Vertically

Find each product.

(a) $(5a - 2b)(3a + b)$

$$\begin{array}{rrl} & 5a - 2b & \\ & \underline{3a + b} & \text{Write the factors vertically.} \\ & 5ab - 2b^2 & \leftarrow \text{Multiply } b(5a - 2b). \\ \underline{15a^2 - 6ab} & & \leftarrow \text{Multiply } 3a(5a - 2b). \\ 15a^2 - ab - 2b^2 & & \text{Combine like terms.} \end{array}$$

(b) $(3m^3 - 2m^2 + 4)(3m - 5)$

$$\begin{array}{r} 3m^3 - 2m^2 + 4 \\ \underline{3m - 5} \\ -15m^3 + 10m^2 \qquad - 20 \\ \underline{9m^4 - 6m^3 \qquad + 12m \qquad} \\ 9m^4 - 21m^3 + 10m^2 + 12m - 20 \end{array}$$

Be sure to write like terms in columns.

$\leftarrow$ Multiply $-5(3m^3 - 2m^2 + 4)$.

$\leftarrow$ Multiply $3m(3m^3 - 2m^2 + 4)$.

Combine like terms.

NOW TRY

NOTE We can use a rectangle to model polynomial multiplication.

$$(5a - 2b)(3a + b) \qquad \text{See Example 3(a).}$$

Label a rectangle with each term, as shown below on the left. Then put the product of each pair of monomials in the appropriate box, as shown on the right.

	$3a$	b
$5a$		
$-2b$		

	$3a$	b
$5a$	$15a^2$	$5ab$
$-2b$	$-6ab$	$-2b^2$

$(5a - 2b)(3a + b)$

$= 15a^2 + 5ab - 6ab - 2b^2$ Add the four monomial products.

$= 15a^2 - ab - 2b^2$ The result is the same as in **Example 3(a).**

TEACHING TIP

Warn students that while the mechanics of multiplication presented in this section are simple, errors often occur due to carelessness with signs.

OBJECTIVE 3 Multiply binomials.

There is a shortcut method for finding the product of two binomials.

$(3x - 4)(2x + 3)$

$= 3x(2x + 3) - 4(2x + 3)$ Distributive property

$= 3x(2x) + 3x(3) - 4(2x) - 4(3)$ Distributive property again

$= 6x^2 + 9x - 8x - 12$ Multiply.

Before combining like terms to find the simplest form of the answer, we check the origin of each of the four terms in the sum

$$6x^2 + 9x - 8x - 12.$$

NOW TRY ANSWER

3. $3t^3 - 14t^2 + 19t - 12$

From the discussion on the preceding page, multiplying

$$(3x - 4)(2x + 3)$$

using the distributive property gave

$$6x^2 + 9x - 8x - 12.$$

The term $6x^2$ is the product of the two *first* terms of the binomials.

$$(3x - 4)(2x + 3) \qquad 3x(2x) = 6x^2 \qquad \text{First terms}$$

To obtain $9x$, the *outer* terms are multiplied.

$$(3x - 4)(2x + 3) \qquad 3x(3) = 9x \qquad \text{Outer terms}$$

The term $-8x$ comes from the *inner* terms.

$$(3x - 4)(2x + 3) \qquad -4(2x) = -8x \qquad \text{Inner terms}$$

Finally, -12 comes from the *last* terms.

$$(3x - 4)(2x + 3) \qquad -4(3) = -12 \qquad \text{Last terms}$$

The product is found by adding these four results.

$$(3x - 4)(2x + 3)$$
$$= 6x^2 + 9x + (-8x) + (-12) \qquad \text{FOIL method}$$
$$= 6x^2 + x - 12 \qquad \text{Combine like terms.}$$

TEACHING TIP
Emphasize that the FOIL method is an extension of the distributive property and applies only to multiplying binomials.

To keep track of the order of multiplying terms, we use the initials FOIL, where the letters refer to the positions of the terms (First, Outer, Inner, Last). The above steps can be done as follows.

Compact form:

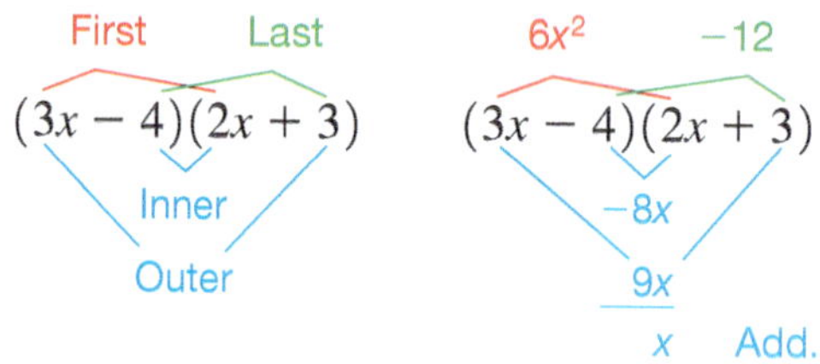

Try to do as many of these steps as possible mentally.

CAUTION ***The FOIL method applies only to multiplying two binomials.***

CLASSROOM EXAMPLE 4
Use the FOIL method to find each product.
(a) $(5r - 3)(2r - 5)$
(b) $(4y - z)(2y + 3z)$
Answers:
(a) $10r^2 - 31r + 15$
(b) $8y^2 + 10yz - 3z^2$

TEACHING TIP
To prepare for factoring, use examples of the FOIL method with the same and different signs in the binomial factors. Discuss how these signs affect the signs of the middle and last terms of the resulting trinomial.

EXAMPLE 4 Using the FOIL Method

Use the FOIL method to find each product.

(a) $(4m - 5)(3m + 1)$

First terms	$(4m - 5)(3m + 1)$	$4m(3m) = 12m^2$
Outer terms	$(4m - 5)(3m + 1)$	$4m(1) = 4m$
Inner terms	$(4m - 5)(3m + 1)$	$-5(3m) = -15m$
Last terms	$(4m - 5)(3m + 1)$	$-5(1) = -5$

$$(4m - 5)(3m + 1)$$
$$\quad \text{F} \qquad \text{O} \qquad \text{I} \qquad \text{L}$$
$$= 12m^2 + 4m - 15m - 5$$
$$= 12m^2 - 11m - 5 \qquad \text{Combine like terms.}$$

NOW TRY EXERCISE 4

Use the FOIL method to find each product.

(a) $(2x + 1)(7x + 2)$

(b) $(3p - k)(5p + 4k)$

Compact form: $(4m - 5)(3m + 1)$ — First: $12m^2$, Last: -5; Outer: $4m$, Inner: $-15m$; Add: $-11m$.

Combine these results to obtain $12m^2 - 11m - 5$.

(b) $(5x - 4)(2x - 5)$

First, Outer, Inner, Last

$= 5x(2x) + 5x(-5) - 4(2x) - 4(-5)$ FOIL method

$= 10x^2 - 25x - 8x + 20$ Multiply.

$= 10x^2 - 33x + 20$ Combine like terms.

(c) $(6a + 5b)(3a - 4b)$

First, Outer, Inner, Last

$= 18a^2 - 24ab + 15ab - 20b^2$ FOIL method

$= 18a^2 - 9ab - 20b^2$ Combine like terms.

(d) $(2k + 3z)(5k + 3z)$

$= 10k^2 + 6kz + 15kz + 9z^2$ FOIL method

$= 10k^2 + 21kz + 9z^2$ Combine like terms.

NOW TRY

TEACHING TIP

Students should recognize the special products and make use of the patterns shown. Gaining skill now will help them factor these special forms later.

OBJECTIVE 4 Find the product of a sum and difference of two terms.

The product of a sum and difference of the same two terms occurs frequently.

$(x + y)(x - y)$

$= x^2 - xy + xy - y^2$ FOIL method

$= x^2 - y^2$ Combine like terms.

CLASSROOM EXAMPLE 5

Find each product.

(a) $(m + 5)(m - 5)$

(b) $(x - 4y)(x + 4y)$

(c) $4y^2(y + 7)(y - 7)$

Answers: (a) $m^2 - 25$

(b) $x^2 - 16y^2$

(c) $4y^4 - 196y^2$

Product of a Sum and Difference of Two Terms

The **product of the sum and difference of the two terms *x* and *y*** is the difference of the squares of the terms.

$$(x + y)(x - y) = x^2 - y^2$$

NOW TRY EXERCISE 5

Find each product.

(a) $(x + 8)(x - 8)$

(b) $(3x - 7y)(3x + 7y)$

(c) $5k(2k - 3)(2k + 3)$

EXAMPLE 5 Multiplying a Sum and Difference of Two Terms

Find each product.

(a) $(p + 7)(p - 7)$

$= p^2 - 7^2$

$= p^2 - 49$

(b) $(2r + 5)(2r - 5)$

$= (2r)^2 - 5^2$

$= 2^2r^2 - 25$ $(ab)^2 = a^2b^2$ (Be careful squaring $2r$.)

$= 4r^2 - 25$

(c) $(6m + 5n)(6m - 5n)$

$= (6m)^2 - (5n)^2$

$= 36m^2 - 25n^2$

(d) $2x^3(x + 3)(x - 3)$

$= 2x^3(x^2 - 9)$

$= 2x^5 - 18x^3$

NOW TRY

NOW TRY ANSWERS

4. (a) $14x^2 + 11x + 2$

(b) $15p^2 + 7kp - 4k^2$

5. (a) $x^2 - 64$

(b) $9x^2 - 49y^2$

(c) $20k^3 - 45k$

OBJECTIVE 5 Find the square of a binomial.

To find the square of a binomial sum $x + y$—that is, $(x + y)^2$—multiply $x + y$ by itself.

$$\begin{aligned} &(x+y)^2 \\ &= (x+y)(x+y) && a^2 = a \cdot a \\ &= x^2 + xy + xy + y^2 && \text{FOIL method} \\ &= x^2 + 2xy + y^2 && \text{Combine like terms.} \end{aligned}$$

A similar result is true for the square of a binomial difference.

TEACHING TIP
Verbalize the rule for squaring a binomial as follows.
1. Square the first term.
2. Add twice the product of the two terms.
3. Add the square of the last term.

Square of a Binomial

The **square of a binomial** is the sum of the square of the first term, twice the product of the two terms, and the square of the last term.

$$(x + y)^2 = x^2 + 2xy + y^2$$
$$(x - y)^2 = x^2 - 2xy + y^2$$

NOW TRY EXERCISE 6
Find each product.
(a) $(y - 10)^2$
(b) $(4x + 5y)^2$

CLASSROOM EXAMPLE 6
Find each product.
(a) $(t + 9)^2$
(b) $(2m + 5)^2$
(c) $(3k - 2n)^2$

Answers: (a) $t^2 + 18t + 81$
(b) $4m^2 + 20m + 25$
(c) $9k^2 - 12kn + 4n^2$

EXAMPLE 6 Squaring Binomials

Find each product.

(a) $(m + 7)^2$

$$\begin{aligned} &= m^2 + 2 \cdot m \cdot 7 + 7^2 && (x+y)^2 = x^2 + 2xy + y^2 \\ &= m^2 + 14m + 49 && \text{Multiply. Apply the exponent.} \end{aligned}$$

(b) $(p - 4)^2$

$$\begin{aligned} &= p^2 - 2 \cdot p \cdot 4 + 4^2 && (x-y)^2 = x^2 - 2xy + y^2 \\ &= p^2 - 8p + 16 && \text{Multiply. Apply the exponent.} \end{aligned}$$

(c) $(2p + 3v)^2$

$$\begin{aligned} &= (2p)^2 + 2(2p)(3v) + (3v)^2 \\ &= 4p^2 + 12pv + 9v^2 \end{aligned}$$

(d) $(3r - 5s)^2$

$$\begin{aligned} &= (3r)^2 - 2(3r)(5s) + (5s)^2 \\ &= 9r^2 - 30rs + 25s^2 \end{aligned}$$

NOW TRY

CAUTION As the products in the formula for the square of a binomial show,

$$(x + y)^2 \neq x^2 + y^2.$$

More generally, $(x + y)^n \neq x^n + y^n$ (where $n \neq 1$).

CLASSROOM EXAMPLE 7
Find each product.
(a) $[(x - y) + z][(x - y) - z]$
(b) $(p + 2q)^3$
(c) $(x + 2)^4$

Answers:
(a) $x^2 - 2xy + y^2 - z^2$
(b) $p^3 + 6p^2q + 12pq^2 + 8q^3$
(c) $x^4 + 8x^3 + 24x^2 + 32x + 16$

EXAMPLE 7 Multiplying More Complicated Binomials

Find each product.

(a) $[(3p - 2) + 5q][(3p - 2) - 5q]$

$$\begin{aligned} &= (3p - 2)^2 - (5q)^2 && \text{Product of a sum and difference of terms} \\ &= 9p^2 - 12p + 4 - 25q^2 && \text{Square both quantities.} \end{aligned}$$

(b) $[(2z + r) + 1]^2$

$$\begin{aligned} &= (2z + r)^2 + 2(2z + r)(1) + 1^2 && \text{Square of a binomial} \\ &= 4z^2 + 4zr + r^2 + 4z + 2r + 1 && \text{Square again; distributive property} \end{aligned}$$

NOW TRY ANSWERS
6. (a) $y^2 - 20y + 100$
(b) $16x^2 + 40xy + 25y^2$

NOW TRY EXERCISE 7

Find each product.

(a) $[(4x - y) + 2][(4x - y) - 2]$

(b) $(y - 3)^4$

(c) $(x + y)^3$

This does *not* equal $x^3 + y^3$.

$$= (x + y)^2(x + y) \qquad a^3 = a^2 \cdot a$$

$$= (x^2 + 2xy + y^2)(x + y) \qquad \text{Square } x + y.$$

$$= x^2(x + y) + 2xy(x + y) + y^2(x + y) \qquad \text{Distributive property}$$

$$= x^3 + x^2y + 2x^2y + 2xy^2 + xy^2 + y^3 \qquad \text{Distributive property}$$

$$= x^3 + 3x^2y + 3xy^2 + y^3 \qquad \text{Combine like terms.}$$

(d) $(2a + b)^4$

$$= (2a + b)^2(2a + b)^2 \qquad a^4 = a^2 \cdot a^2$$

$$= (4a^2 + 4ab + b^2)(4a^2 + 4ab + b^2) \qquad \text{Square } 2a + b \text{ twice.}$$

$$= 4a^2(4a^2 + 4ab + b^2) + 4ab(4a^2 + 4ab + b^2) + b^2(4a^2 + 4ab + b^2) \qquad \text{Distributive property}$$

$$= 16a^4 + 16a^3b + 4a^2b^2 + 16a^3b + 16a^2b^2 + 4ab^3 + 4a^2b^2 + 4ab^3 + b^4 \qquad \text{Distributive property again}$$

$$= 16a^4 + 32a^3b + 24a^2b^2 + 8ab^3 + b^4 \qquad \text{Combine like terms.}$$

NOW TRY

OBJECTIVE 6 Multiply polynomial functions.

In **Section 4.3,** we added and subtracted functions. Functions can be multiplied.

Multiplying Functions

If $f(x)$ and $g(x)$ define functions, then

$$(fg)(x) = f(x) \cdot g(x). \qquad \text{Product function}$$

The domain of the product function is the intersection of the domains of $f(x)$ and $g(x)$.

CAUTION Write the product $f(x) \cdot g(x)$ as $(fg)(x)$, ***not*** $f(g(x))$, which indicates the composition of functions f and g. (See **Section 4.3.**)

CLASSROOM EXAMPLE 8

For $f(x) = 3x + 1$ and $g(x) = 2x - 5$, find $(fg)(x)$ and $(fg)(2)$.

Answer: $6x^2 - 13x - 5$; -7

EXAMPLE 8 Multiplying Polynomial Functions

For $f(x) = 3x + 4$ and $g(x) = 2x^2 + x$, find $(fg)(x)$ and $(fg)(-1)$.

$(fg)(x)$

This notation indicates function multiplication.

$$= f(x) \cdot g(x) \qquad \text{Use the definition.}$$

$$= (3x + 4)(2x^2 + x) \qquad \text{Substitute.}$$

$$= 6x^3 + 3x^2 + 8x^2 + 4x \qquad \text{FOIL method}$$

$$= 6x^3 + 11x^2 + 4x \qquad \text{Combine like terms.}$$

NOW TRY ANSWERS

7. (a) $16x^2 - 8xy + y^2 - 4$

(b) $y^4 - 12y^3 + 54y^2 - 108y + 81$

NOW TRY EXERCISE 8
For $f(x) = 3x^2 - 1$ and $g(x) = 8x + 7$, find $(fg)(x)$ and $(fg)(-2)$.

$(fg)(-1)$	$(fg)(x) = 6x^3 + 11x^2 + 4x$
$= 6(-1)^3 + 11(-1)^2 + 4(-1)$	Let $x = -1$ in $(fg)(x)$.
$= -6 + 11 - 4$ (Be careful with signs.)	Apply the exponents. Multiply.
$= 1$	Add and subtract.

Another way to find $(fg)(-1)$ is to find $f(-1)$ and $g(-1)$ and multiply the results. Verify that $f(-1) \cdot g(-1) = 1$. This follows from the definition. NOW TRY

NOW TRY ANSWER
8. $24x^3 + 21x^2 - 8x - 7$; -99

4.4 Exercises

FOR EXTRA HELP MyMathLab®

Complete solution available in MyMathLab

Concept Check *Match each product in Column I with the correct polynomial in Column II.*

I	II
1. $(2x - 5)(3x + 4)$	**A.** $6x^2 + 23x + 20$
2. $(2x + 5)(3x + 4)$	**B.** $6x^2 + 7x - 20$
3. $(2x - 5)(3x - 4)$	**C.** $6x^2 - 7x - 20$
4. $(2x + 5)(3x - 4)$	**D.** $6x^2 - 23x + 20$

Find each product. ***See Examples 1–3.***

5. $-8m^3(3m^2)$ **6.** $-4p^2(5p^4)$ **7.** $14x^2y^3(-2x^5y)$ **8.** $5m^3n^4(-4m^2n^5)$

9. $3x(-2x + 5)$ **10.** $5y(-6y + 1)$ **11.** $-q^3(2 + 3q)$ **12.** $-3a^4(4 + a)$

13. $6k^2(3k^2 + 2k + 1)$ **14.** $5r^3(2r^2 + 3r + 4)$ **15.** $(2t + 3)(3t^2 - 4t - 1)$

16. $(4z + 2)(z^2 - 3z - 5)$ **17.** $m(m + 5)(m - 8)$ **18.** $p(p + 4)(p - 6)$

19. $4z(2z + 1)(3z - 4)$ **20.** $2y(2y + 1)(8y - 3)$ **21.** $4x^3(x - 3)(x + 2)$

22. $2y^5(y - 8)(y + 2)$ **23.** $(2y + 3)(3y - 4)$ **24.** $(2m + 6)(5m - 3)$

25. $\begin{array}{r} 5m - 3n \\ \underline{5m + 3n} \end{array}$ **26.** $\begin{array}{r} 2k + 6q \\ \underline{2k - 6q} \end{array}$ **27.** $\begin{array}{r} -b^2 + 3b + 3 \\ \underline{2b + 4} \end{array}$

28. $\begin{array}{r} -r^2 - 4r + 8 \\ \underline{3r - 2} \end{array}$ **29.** $\begin{array}{r} 2z^3 - 5z^2 + 8z - 1 \\ \underline{4z + 3} \end{array}$ **30.** $\begin{array}{r} 3z^4 - 2z^3 + z - 5 \\ \underline{2z - 5} \end{array}$

31. $\begin{array}{r} 2p^2 + 3p + 6 \\ \underline{3p^2 - 4p - 1} \end{array}$ **32.** $\begin{array}{r} 5y^2 - 2y + 4 \\ \underline{2y^2 + y + 3} \end{array}$

Use the FOIL method to find each product. ***See Example 4.***

33. $(m + 5)(m - 8)$ **34.** $(p + 4)(p - 6)$ **35.** $(4k + 3)(3k - 2)$

36. $(5w + 2)(2w - 5)$ **37.** $(3x - 2)(5x - 1)$ **38.** $(4x - 1)(6x - 7)$

39. $(5x + 2y)(4x + 3y)$ **40.** $(3x + 4y)(2x + 3y)$ **41.** $(z - w)(3z + 4w)$

42. $(s - t)(2s + 5t)$ **43.** $(6c - d)(2c + 3d)$ **44.** $(2m - n)(3m + 5n)$

Find each product. ***See Example 5.***

45. $(x + 9)(x - 9)$ **46.** $(z + 6)(z - 6)$ **47.** $(2p - 3)(2p + 3)$

48. $(3x - 8)(3x + 8)$ **49.** $(5m - 1)(5m + 1)$ **50.** $(6y - 3)(6y + 3)$

1. C 2. A 3. D 4. B
5. $-24m^5$ 6. $-20p^6$
7. $-28x^7y^4$ 8. $-20m^5n^9$
9. $-6x^2 + 15x$ 10. $-30y^2 + 5y$
11. $-2q^3 - 3q^4$
12. $-12a^4 - 3a^5$
13. $18k^4 + 12k^3 + 6k^2$
14. $10r^5 + 15r^4 + 20r^3$
15. $6t^3 + t^2 - 14t - 3$
16. $4z^3 - 10z^2 - 26z - 10$
17. $m^3 - 3m^2 - 40m$
18. $p^3 - 2p^2 - 24p$
19. $24z^3 - 20z^2 - 16z$
20. $32y^3 + 4y^2 - 6y$
21. $4x^5 - 4x^4 - 24x^3$
22. $2y^7 - 12y^6 - 32y^5$
23. $6y^2 + y - 12$
24. $10m^2 + 24m - 18$
25. $25m^2 - 9n^2$ 26. $4k^2 - 36q^2$
27. $-2b^3 + 2b^2 + 18b + 12$
28. $-3r^3 - 10r^2 + 32r - 16$
29. $8z^4 - 14z^3 + 17z^2 + 20z - 3$
30. $6z^5 - 19z^4 + 10z^3 + 2z^2 - 15z + 25$
31. $6p^4 + p^3 + 4p^2 - 27p - 6$
32. $10y^4 + y^3 + 21y^2 - 2y + 12$
33. $m^2 - 3m - 40$
34. $p^2 - 2p - 24$
35. $12k^2 + k - 6$
36. $10w^2 - 21w - 10$
37. $15x^2 - 13x + 2$
38. $24x^2 - 34x + 7$
39. $20x^2 + 23xy + 6y^2$
40. $6x^2 + 17xy + 12y^2$
41. $3z^2 + zw - 4w^2$
42. $2s^2 + 3st - 5t^2$
43. $12c^2 + 16cd - 3d^2$
44. $6m^2 + 7mn - 5n^2$
45. $x^2 - 81$ 46. $z^2 - 36$
47. $4p^2 - 9$ 48. $9x^2 - 64$
49. $25m^2 - 1$ 50. $36y^2 - 9$

51. $(3a + 2c)(3a - 2c)$ **52.** $(5r + 4s)(5r - 4s)$ **53.** $(4m + 7n^2)(4m - 7n^2)$

54. $(2k^2 + 6h)(2k^2 - 6h)$ **55.** $3y(5y^3 + 2)(5y^3 - 2)$ **56.** $4x(3x^3 + 4)(3x^3 - 4)$

Find each product. ***See Example 6.***

57. $(y - 5)^2$ **58.** $(a - 3)^2$ **59.** $(x + 1)^2$ **60.** $(t + 2)^2$

61. $(2p + 7)^2$ **62.** $(3z + 8)^2$ **63.** $(4n - 3m)^2$ **64.** $(5r - 7s)^2$

Extending Skills *The factors in the following exercises involve fractions or decimals. Apply the methods of this section, and find each product.*

65. $\left(k - \frac{5}{7}p\right)^2$ **66.** $\left(q - \frac{3}{4}r\right)^2$ **67.** $(0.2x - 1.4y)^2$

68. $(0.3x - 1.6y)^2$ **69.** $\left(4x - \frac{2}{3}\right)\left(4x + \frac{2}{3}\right)$ **70.** $\left(3t - \frac{5}{4}\right)\left(3t + \frac{5}{4}\right)$

71. $(0.2x + 1.3)(0.5x - 0.1)$ **72.** $(0.1y + 2.1)(0.5y - 0.4)$

73. $\left(3w + \frac{1}{4}z\right)(w - 2z)$ **74.** $\left(5r + \frac{2}{3}y\right)(r - 5y)$

Find each product. ***See Example 7.***

75. $[(5x + 1) + 6y]^2$ **76.** $[(3m - 2) + p]^2$

77. $[(2a + b) - 3]^2$ **78.** $[(4k + h) - 4]^2$

79. $[(2a + b) - 3][(2a + b) + 3]$ **80.** $[(m + p) - 5][(m + p) + 5]$

81. $[(2h - k) + j][(2h - k) - j]$ **82.** $[(3m - y) + z][(3m - y) - z]$

83. $(y + 2)^3$ **84.** $(z - 3)^3$ **85.** $(5r - s)^3$

86. $(x + 3y)^3$ **87.** $(q - 2)^4$ **88.** $(r + 3)^4$

Extending Skills *Find each product.*

89. $(2a + b)(3a^2 + 2ab + b^2)$ **90.** $(m - 5p)(m^2 - 2mp + 3p^2)$

91. $(4z - x)(z^3 - 4z^2x + 2zx^2 - x^3)$ **92.** $(3r + 2s)(r^3 + 2r^2s - rs^2 + 2s^3)$

93. $(m^2 - 2mp + p^2)(m^2 + 2mp - p^2)$ **94.** $(3 + x + y)(-3 + x - y)$

95. $ab(a + b)(a + 2b)(a - 3b)$ **96.** $mp(m - p)(m - 2p)(2m + p)$

In Exercises 97–100, two expressions are given. Replace x with 3 and y with 4 to show that, in general, the two expressions are not equivalent.

97. $(x + y)^2$; $x^2 + y^2$ **98.** $(x + y)^3$; $x^3 + y^3$

99. $(x + y)^4$; $x^4 + y^4$ **100.** $(x + y)^5$; $x^5 + y^5$

Find the area of each figure. Express it as a polynomial in descending powers of the variable x. Refer to the formulas given inside the back cover of this book if necessary.

101.

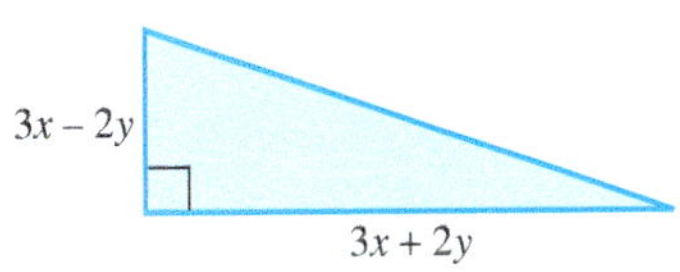

102.

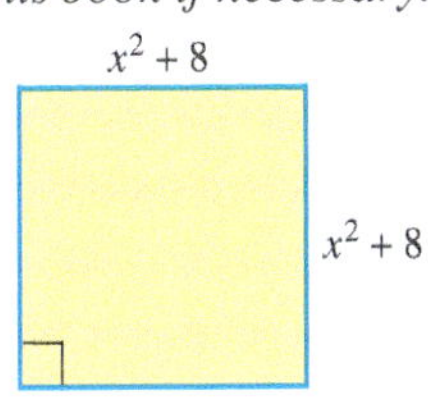

Answers:

51. $9a^2 - 4c^2$ 52. $25r^2 - 16s^2$
53. $16m^2 - 49n^4$ 54. $4k^4 - 36h^2$
55. $75y^7 - 12y$ 56. $36x^7 - 64x$
57. $y^2 - 10y + 25$
58. $a^2 - 6a + 9$
59. $x^2 + 2x + 1$
60. $t^2 + 4t + 4$
61. $4p^2 + 28p + 49$
62. $9z^2 + 48z + 64$
63. $16n^2 - 24nm + 9m^2$
64. $25r^2 - 70rs + 49s^2$
65. $k^2 - \frac{10}{7}kp + \frac{25}{49}p^2$
66. $q^2 - \frac{3}{2}qr + \frac{9}{16}r^2$
67. $0.04x^2 - 0.56xy + 1.96y^2$
68. $0.09x^2 - 0.96xy + 2.56y^2$
69. $16x^2 - \frac{4}{9}$ 70. $9t^2 - \frac{25}{16}$
71. $0.1x^2 + 0.63x - 0.13$
72. $0.05y^2 + 1.01y - 0.84$
73. $3w^2 - \frac{23}{4}wz - \frac{1}{2}z^2$
74. $5r^2 - \frac{73}{3}ry - \frac{10}{3}y^2$
75. $25x^2 + 10x + 1 + 60xy + 12y + 36y^2$
76. $9m^2 - 12m + 4 + 6mp - 4p + p^2$
77. $4a^2 + 4ab + b^2 - 12a - 6b + 9$
78. $16k^2 + 8kh + h^2 - 32k - 8h + 16$
79. $4a^2 + 4ab + b^2 - 9$
80. $m^2 + 2mp + p^2 - 25$
81. $4h^2 - 4hk + k^2 - j^2$
82. $9m^2 - 6my + y^2 - z^2$
83. $y^3 + 6y^2 + 12y + 8$
84. $z^3 - 9z^2 + 27z - 27$
85. $125r^3 - 75r^2s + 15rs^2 - s^3$
86. $x^3 + 9x^2y + 27xy^2 + 27y^3$
87. $q^4 - 8q^3 + 24q^2 - 32q + 16$
88. $r^4 + 12r^3 + 54r^2 + 108r + 81$
89. $6a^3 + 7a^2b + 4ab^2 + b^3$
90. $m^3 - 7m^2p + 13mp^2 - 15p^3$
91. $4z^4 - 17z^3x + 12z^2x^2 - 6zx^3 + x^4$
92. $3r^4 + 8r^3s + r^2s^2 + 4rs^3 + 4s^4$
93. $m^4 - 4m^2p^2 + 4mp^3 - p^4$
94. $-9 - 6y + x^2 - y^2$
95. $a^4b - 7a^2b^3 - 6ab^4$
96. $2m^4p - 5m^3p^2 + m^2p^3 + 2mp^4$

97. 49; 25; 49 ≠ 25
98. 343; 91; 343 ≠ 91
99. 2401; 337; 2401 ≠ 337
100. 16,807; 1267; 16,807 ≠ 1267

101. $\frac{9}{2}x^2 - 2y^2$
102. $x^4 + 16x^2 + 64$
103. $15x^2 - 2x - 24$
104. $x^3 + \frac{7}{2}x^2 + 7x + 6$

105. $10x^2 - 2x$
106. $18x^2 - 24x$
107. $2x^2 - x - 3$
108. $4x^2 - 23x - 35$
109. $8x^3 - 27$
110. $27x^3 + 64$

111. $2x^3 - 18x$
112. $x^3 - 3x^2 - 9x + 27$
113. -20 **114.** 16
115. $2x^2 - 6x$ **116.** 32
117. 36 **118.** 20
119. $\frac{35}{4}$ **120.** $\frac{160}{27}$
121. $\frac{1859}{64}$ **122.** $\frac{3584}{125}$

123. $a - b$
124. $\mathcal{A} = s^2$; $(a - b)^2$
125. $(a - b)b$, or $ab - b^2$; $2ab - 2b^2$
126. b^2
127. a^2; a
128. $a^2 - (2ab - 2b^2) - b^2 = a^2 - 2ab + b^2$
129. **(a)** They must be equal.
(b) $(a - b)^2 = a^2 - 2ab + b^2$
130.

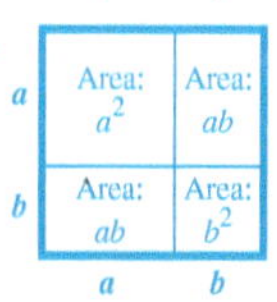

The large square is made up of two smaller squares and two congruent rectangles. The sum of the areas is

$$a^2 + 2ab + b^2.$$

Because $(a + b)^2$ must represent the same quantity, they must be equal. Thus,

$$(a + b)^2 = a^2 + 2ab + b^2.$$

103.

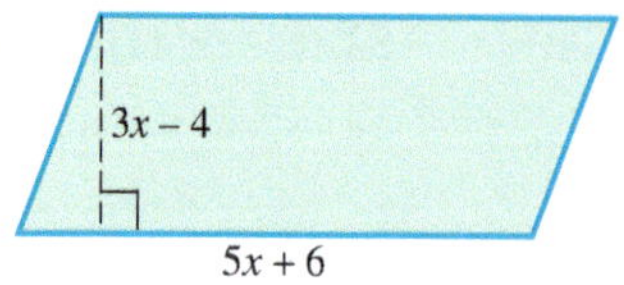

104.

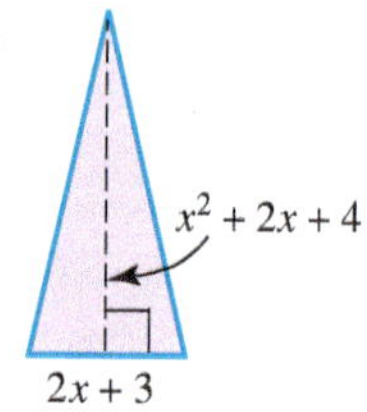

For each pair of functions, find $(fg)(x)$. ***See Example 8.***

105. $f(x) = 2x, \quad g(x) = 5x - 1$
106. $f(x) = 3x, \quad g(x) = 6x - 8$
107. $f(x) = x + 1, \quad g(x) = 2x - 3$
108. $f(x) = x - 7, \quad g(x) = 4x + 5$
109. $f(x) = 2x - 3, \quad g(x) = 4x^2 + 6x + 9$
110. $f(x) = 3x + 4, \quad g(x) = 9x^2 - 12x + 16$

Let $f(x) = x^2 - 9$, $g(x) = 2x$, and $h(x) = x - 3$. Find each of the following. ***See Example 8.***

111. $(fg)(x)$
112. $(fh)(x)$
113. $(fg)(2)$
114. $(fh)(1)$
115. $(gh)(x)$
116. $(fh)(-1)$
117. $(gh)(-3)$
118. $(fg)(-2)$
119. $(fg)\left(-\frac{1}{2}\right)$
120. $(fg)\left(-\frac{1}{3}\right)$
121. $(fh)\left(-\frac{1}{4}\right)$
122. $(fh)\left(-\frac{1}{5}\right)$

RELATING CONCEPTS For Individual or Group Work **(Exercises 123–130)**

Consider the figure. ***Work Exercises 123–130 in order.***

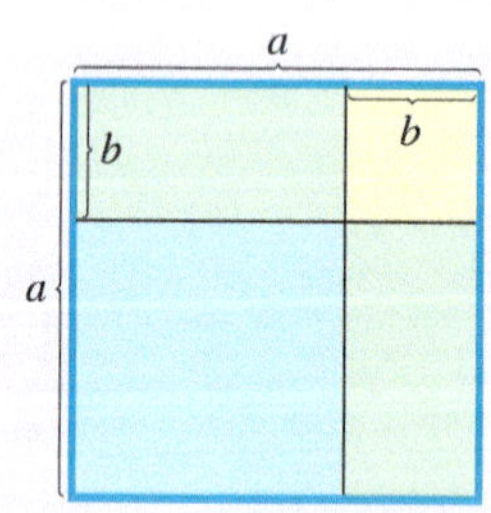

123. What is the length of each side of the blue square in terms of a and b?

124. What is the formula for the area of a square? Use the formula to write an expression, in the form of a product, for the area of the blue square.

125. Each green rectangle has an area of ______. Therefore, the total area in green is represented by the polynomial ______.

126. The yellow square has an area of ______.

127. The area of the entire colored region is represented by ______, because each side of the entire colored region has length ______.

128. The area of the blue square is equal to the area of the entire colored region, minus the total area of the green squares, minus the area of the yellow square. Write this as a simplified polynomial in a and b.

129. **(a)** What must be true about the expressions for the area of the blue square found in **Exercises 124 and 128?**

(b) Write an equation based on the answer in part (a).

130. Draw a figure and give a similar proof for $(a + b)^2 = a^2 + 2ab + b^2$.

4.5 Dividing Polynomials

OBJECTIVES

1 Divide a polynomial by a monomial.
2 Divide a polynomial by a polynomial of two or more terms.
3 Divide polynomial functions.

OBJECTIVE 1 Divide a polynomial by a monomial.

Recall that a monomial is a single term, such as 3, $5m^2$, or x^2y^2.

Dividing a Polynomial by a Monomial

To divide a polynomial by a monomial, divide each term of the polynomial by the monomial.

$$\frac{a+b}{c} = \frac{a}{c} + \frac{b}{c} \quad (\text{where } c \neq 0)$$

Then write each quotient in lowest terms.

NOW TRY EXERCISE 1

Divide.

(a) $\dfrac{16x^3 - 8x^2 - 4}{4}$

(b) $\dfrac{81y^4 - 54y^3 + 18y}{9y^3}$

CLASSROOM EXAMPLE 1

Divide.

(a) $\dfrac{10x^2 - 25x + 35}{5}$

(b) $\dfrac{4x^4 - 7x^3 + 12x^2}{4x^3}$

(c) $\dfrac{6a^2b^4 - 9a^3b^3 + 4a^3b^4}{a^3b^4}$

Answers: (a) $2x^2 - 5x + 7$ (b) $x - \frac{7}{4} + \frac{3}{x}$ (c) $\frac{6}{a} - \frac{9}{b} + 4$

EXAMPLE 1 Dividing Polynomials by Monomials

Divide.

(a) $\dfrac{15x^2 - 12x + 3}{3}$ — Do *not* divide out the 3's.

$= \dfrac{15x^2}{3} - \dfrac{12x}{3} + \dfrac{3}{3}$ — Divide *each* term by 3.

$= 5x^2 - 4x + 1$ — Write in lowest terms.

CHECK $3(5x^2 - 4x + 1) = 15x^2 - 12x + 3$ ✓

(Divisor: 3; Quotient: $5x^2 - 4x + 1$; Original polynomial (Dividend): $15x^2 - 12x + 3$)

(b) $\dfrac{5m^3 - 9m^2 + 10m}{5m^2}$

$= \dfrac{5m^3}{5m^2} - \dfrac{9m^2}{5m^2} + \dfrac{10m}{5m^2}$ — Divide each term by $5m^2$. (Think: $\frac{10m}{5m^2} = \frac{10}{5}m^{1-2} = 2m^{-1} = \frac{2}{m}$)

$= m - \dfrac{9}{5} + \dfrac{2}{m}$ — Simplify each term. Use the quotient rule for exponents.

CHECK $5m^2\left(m - \dfrac{9}{5} + \dfrac{2}{m}\right) = 5m^3 - 9m^2 + 10m$ ✓ — Divisor × Quotient = Original polynomial

The result $m - \frac{9}{5} + \frac{2}{m}$ is not a polynomial because the last term has a variable in its denominator. The quotient of two polynomials need not be a polynomial.

(c) $\dfrac{8xy^2 - 9x^2y + 6x^2y^2}{x^2y^2}$

$= \dfrac{8xy^2}{x^2y^2} - \dfrac{9x^2y}{x^2y^2} + \dfrac{6x^2y^2}{x^2y^2}$ — Divide each term by x^2y^2.

$= \dfrac{8}{x} - \dfrac{9}{y} + 6$ — $\frac{a^m}{a^n} = a^{m-n}$

NOW TRY

NOW TRY ANSWERS

1. (a) $4x^3 - 2x^2 - 1$
(b) $9y - 6 + \dfrac{2}{y^2}$

NOW TRY EXERCISE 2

Divide.

$$\frac{3x^2 - 4x - 15}{x - 3}$$

CLASSROOM EXAMPLE 2

Divide.

$$\frac{2k^2 + 17k + 30}{k + 6}$$

Answer: $2k + 5$

TEACHING TIP

Use an arithmetic example to show the basic steps in long division.

$$25\overline{)5860}$$

Identify the divisor, dividend, and quotient, and review how to check a result.

TEACHING TIP

Point out that the number of terms in the second line (the line being subtracted) is the same as the number of terms in the divisor.

It helps to have students show the sign change on the second line, and then circle the original sign.

OBJECTIVE 2 Divide a polynomial by a polynomial of two or more terms.

This process is similar to that for dividing whole numbers.

EXAMPLE 2 Dividing a Polynomial by a Polynomial

Divide $\frac{2x^2 + x - 10}{x - 2}$.

Make sure each polynomial is written in descending powers of the variables.

$$x - 2\overline{)2x^2 + x - 10}$$ ← Write as if dividing whole numbers.

Divide the first term of the dividend $2x^2 + x - 10$ by the first term of the divisor $x - 2$. Here $\frac{2x^2}{x} = 2x$.

$$\begin{array}{r} 2x \\ x - 2\overline{)2x^2 + x - 10} \end{array}$$ ← Result of $\frac{2x^2}{x}$

Multiply $x - 2$ and $2x$, and write the result below $2x^2 + x - 10$.

$$\begin{array}{r} 2x \\ x - 2\overline{)2x^2 + x - 10} \\ 2x^2 - 4x \end{array}$$ ← $2x(x - 2) = 2x^2 - 4x$

Now subtract by mentally changing the signs on $2x^2 - 4x$ and *adding*.

$$\begin{array}{r} 2x \\ x - 2\overline{)2x^2 + x - 10} \\ \underline{2x^2 - 4x} \\ 5x \end{array}$$ ← Subtract. The difference is $5x$.

To subtract, add the opposite.
$2x^2 + x$
$\oplus \ominus 2x^2 \oplus 4x$

Bring down -10 and continue by dividing $5x$ by x.

$$\begin{array}{r} 2x + 5 \\ x - 2\overline{)2x^2 + x - 10} \\ \underline{2x^2 - 4x} \\ 5x - 10 \\ \underline{5x - 10} \\ 0 \end{array}$$

← $\frac{5x}{x} = 5$
← Bring down -10.
← $5(x - 2) = 5x - 10$
← Subtract. The difference is 0.

Think: $5x - 10$
$\oplus \ominus 5x \oplus 10$

CHECK Multiply $x - 2$ (the divisor) and $2x + 5$ (the quotient).

$$(x - 2)(2x + 5)$$
$$= 2x^2 + 5x - 4x - 10 \quad \text{FOIL method}$$
$$= 2x^2 + x - 10 \quad \text{Combine like terms.}$$

The result is $2x^2 + x - 10$ (the dividend). ✓

NOW TRY

! CAUTION Remember to do the following when dividing a polynomial by a polynomial of two or more terms.

1. Be sure the terms in both polynomials are written in descending powers.
2. Write any missing terms with 0 placeholders.

NOW TRY ANSWER

2. $3x + 5$

NOW TRY EXERCISE 3

Divide

$2x^3 - 12x - 10$ by $x - 4$.

EXAMPLE 3 Dividing a Polynomial with a Missing Term

Divide $3x^3 - 2x + 5$ by $x - 3$.

Add a term with 0 coefficient as a placeholder for the missing x^2-term.

$$x - 3\overline{)3x^3 + 0x^2 - 2x + 5}$$

Missing term ($0x^2$)

← Descending powers of the variable

Start with $\frac{3x^3}{x} = 3x^2$.

$$\begin{array}{r} 3x^2 \phantom{{}+0x^2-2x+5} \\ x - 3\overline{)3x^3 + 0x^2 - 2x + 5} \\ \underline{3x^3 - 9x^2} \phantom{{}-2x+5} \end{array}$$

← $\frac{3x^3}{x} = 3x^2$

← $3x^2(x - 3)$

Subtract by mentally changing the signs on $3x^3 - 9x^2$ and adding.

$$\begin{array}{r} 3x^2 \phantom{{}-2x+5} \\ x - 3\overline{)3x^3 + 0x^2 - 2x + 5} \\ \underline{3x^3 - 9x^2} \phantom{{}-2x+5} \\ 9x^2 \phantom{{}-2x+5} \end{array}$$

Add the opposite.

← Subtract.

Bring down the next term.

$$\begin{array}{r} 3x^2 \phantom{{}-2x+5} \\ x - 3\overline{)3x^3 + 0x^2 - 2x + 5} \\ \underline{3x^3 - 9x^2} \phantom{{}-2x+5} \\ 9x^2 - 2x \phantom{{}+5} \end{array}$$

← Bring down $-2x$.

In the next step, $\frac{9x^2}{x} = 9x$.

$$\begin{array}{r} 3x^2 + 9x \phantom{{}+5} \\ x - 3\overline{)3x^3 + 0x^2 - 2x + 5} \\ \underline{3x^3 - 9x^2} \phantom{{}-2x+5} \\ 9x^2 - 2x \phantom{{}+5} \\ \underline{9x^2 - 27x} \phantom{{}+5} \\ 25x + 5 \end{array}$$

← $\frac{9x^2}{x} = 9x$

Add the opposite.

← $9x(x - 3)$

← Subtract. Bring down 5.

Finally, $\frac{25x}{x} = 25$.

$$\begin{array}{r} 3x^2 + 9x + 25 \\ x - 3\overline{)3x^3 + 0x^2 - 2x + 5} \\ \underline{3x^3 - 9x^2} \phantom{{}-2x+5} \\ 9x^2 - 2x \phantom{{}+5} \\ \underline{9x^2 - 27x} \phantom{{}+5} \\ 25x + 5 \\ \underline{25x - 75} \\ 80 \end{array}$$

← $\frac{25x}{x} = 25$

Add the opposite.

← $25(x - 3)$

← Remainder

Write the remainder, 80, as the numerator of a fraction, $\frac{80}{x-3}$.

$$3x^2 + 9x + 25 + \frac{80}{x - 3}$$

Be sure to add $\frac{\text{remainder}}{\text{divisor}}$. Don't forget the + sign.

CHECK Multiply $x - 3$ (the divisor) and $3x^2 + 9x + 25$ (the quotient), and then add 80 (the remainder). The result is $3x^3 - 2x + 5$ (the dividend). ✓

NOW TRY

NOW TRY ANSWER

3. $2x^2 + 8x + 20 + \frac{70}{x - 4}$

NOW TRY EXERCISE 4
Divide
$2x^4 + 8x^3 + 2x^2 - 5x - 3$
by $2x^2 - 2$.

CLASSROOM EXAMPLE 4
Divide
$4m^4 - 23m^3 + 16m^2 - 4m - 1$
by $m^2 - 5m$.
Answer:
$4m^2 - 3m + 1 + \frac{m-1}{m^2 - 5m}$

EXAMPLE 4 Dividing by a Polynomial with a Missing Term

Divide $6r^4 + 9r^3 + 2r^2 - 8r + 7$ by $3r^2 - 2$.

$$\begin{array}{r} 2r^2 + 3r + 2 \\ 3r^2 + 0r - 2 \overline{)\, 6r^4 + 9r^3 + 2r^2 - 8r + 7} \\ \underline{6r^4 + 0r^3 - 4r^2} \qquad\qquad \\ 9r^3 + 6r^2 - 8r \qquad \\ \underline{9r^3 + 0r^2 - 6r} \qquad \\ 6r^2 - 2r + 7 \\ \underline{6r^2 + 0r - 4} \\ -2r + 11 \end{array}$$

Missing term ($0r$)

Stop when the degree of the remainder is less than the degree of the divisor.

$-2r + 11$ ← Remainder

The degree of the remainder, $-2r + 11$, is less than the degree of the divisor, $3r^2 - 2$, so the division process is finished. The result is written as follows.

$$2r^2 + 3r + 2 + \frac{-2r + 11}{3r^2 - 2} \qquad \text{Quotient} + \frac{\text{remainder}}{\text{divisor}}$$

NOW TRY

NOW TRY EXERCISE 5
Divide
$6m^3 - 8m^2 - 5m - 6$
by $3m - 6$.

CLASSROOM EXAMPLE 5
Divide
$8x^3 + 21x^2 - 2x - 24$
by $4x + 8$.
Answer: $2x^2 + \frac{5}{4}x - 3$

EXAMPLE 5 Finding a Quotient with a Fractional Coefficient

Divide $2p^3 + 5p^2 + p - 2$ by $2p + 2$.

$\frac{3p^2}{2p} = \frac{3}{2}p$

$$\begin{array}{r} p^2 + \frac{3}{2}p - 1 \\ 2p + 2 \overline{)\, 2p^3 + 5p^2 + p - 2} \\ \underline{2p^3 + 2p^2} \qquad\qquad \\ 3p^2 + p \qquad \\ \underline{3p^2 + 3p} \qquad \\ -2p - 2 \\ \underline{-2p - 2} \\ 0 \end{array}$$

The remainder is 0, so the quotient is $p^2 + \frac{3}{2}p - 1$. NOW TRY

OBJECTIVE 3 Divide polynomial functions.

Dividing Functions

If $f(x)$ and $g(x)$ define functions, then

$$\left(\frac{f}{g}\right)(x) = \frac{f(x)}{g(x)}. \qquad \text{Quotient function}$$

The domain of the quotient function is the intersection of the domains of $f(x)$ and $g(x)$, excluding any values of x for which $g(x) = 0$.

CLASSROOM EXAMPLE 6
For $f(x) = 2x^2 + 17x + 30$ and $g(x) = 2x + 5$, find $\left(\frac{f}{g}\right)(x)$ and $\left(\frac{f}{g}\right)(-1)$.
Answer: $x + 6, x \neq -\frac{5}{2}$; 5

EXAMPLE 6 Dividing Polynomial Functions

For $f(x) = 2x^2 + x - 10$ and $g(x) = x - 2$, find $\left(\frac{f}{g}\right)(x)$ and $\left(\frac{f}{g}\right)(-3)$.

$$\left(\frac{f}{g}\right)(x) = \frac{f(x)}{g(x)} = \frac{2x^2 + x - 10}{x - 2}$$

NOW TRY ANSWERS
4. $x^2 + 4x + 2 + \frac{3x + 1}{2x^2 - 2}$
5. $2m^2 + \frac{4}{3}m + 1$

NOW TRY EXERCISE 6

For $f(x) = 8x^2 + 2x - 3$ and $g(x) = 2x - 1$, find $\left(\frac{f}{g}\right)(x)$ and $\left(\frac{f}{g}\right)(8)$.

NOW TRY ANSWER

6. $4x + 3$, $x \neq \frac{1}{2}$; 35

This quotient, found in **Example 2,** is $2x + 5$. Thus,

$$\left(\frac{f}{g}\right)(x) = 2x + 5, \quad x \neq 2. \leftarrow$$ 2 is not in the domain. It causes denominator $g(x) = x - 2$ to equal 0.

$$\left(\frac{f}{g}\right)(-3) = 2(-3) + 5 = -1$$ Let $x = -3$ in $\left(\frac{f}{g}\right)(x) = 2x + 5$.

Verify that the same value is found by evaluating $\frac{f(-3)}{g(-3)}$. NOW TRY

4.5 Exercises

FOR EXTRA HELP

MyMathLab®

Complete solution available in MyMathLab

Concept Check *Complete each statement.*

1. We find the quotient of two monomials using the _____ rule for _____.
2. When dividing polynomials that are not monomials, first write them in _____ powers.
3. If a polynomial in a division problem has a missing term, insert a term with coefficient equal to _____ as a placeholder.
4. To check a division problem, multiply the _____ by the quotient. Then add the _____.

Divide. ***See Example 1.***

5. $\dfrac{15x^3 - 10x^2 + 5}{5}$
6. $\dfrac{27m^4 - 18m^3 - 9}{9}$
7. $\dfrac{9y^2 + 12y - 15}{3y}$
8. $\dfrac{80r^2 - 40r + 10}{10r}$
9. $\dfrac{15m^3 + 25m^2 + 30m}{5m^3}$
10. $\dfrac{64x^3 - 72x^2 + 12x}{8x^3}$
11. $\dfrac{4m^2n^2 - 21mn^3 + 18mn^2}{14m^2n^3}$
12. $\dfrac{24h^2k + 56hk^2 - 28hk}{16h^2k^2}$
13. $\dfrac{8wxy^2 + 3wx^2y + 12w^2xy}{4wx^2y}$
14. $\dfrac{12ab^2c + 10a^2bc + 18abc^2}{6a^2bc}$

Complete the division. ***See Example 2.***

15.
$$\begin{array}{r} r^2 \\ 3r - 1\overline{)3r^3 - 22r^2 + 25r - 6} \\ \underline{3r^3 - r^2} \\ -21r^2 \end{array}$$

16.
$$\begin{array}{r} 3b^2 \\ 2b - 5\overline{)6b^3 - 7b^2 - 4b - 40} \\ \underline{6b^3 - 15b^2} \\ 8b^2 \end{array}$$

Divide. ***See Examples 2–5.***

17. $\dfrac{y^2 + y - 20}{y + 5}$
18. $\dfrac{y^2 + 3y - 18}{y + 6}$
19. $\dfrac{q^2 + 4q - 32}{q - 4}$
20. $\dfrac{q^2 + 2q - 35}{q - 5}$
21. $\dfrac{3t^2 + 17t + 10}{3t + 2}$
22. $\dfrac{2k^2 - 3k - 20}{2k + 5}$
23. $\dfrac{p^2 + 2p + 20}{p + 6}$
24. $\dfrac{x^2 + 11x + 16}{x + 8}$
25. $\dfrac{3m^3 + 5m^2 - 5m + 1}{3m - 1}$
26. $\dfrac{8z^3 - 6z^2 - 5z + 3}{4z + 3}$
27. $\dfrac{m^3 - 2m^2 - 9}{m - 3}$
28. $\dfrac{p^3 + 3p^2 - 4}{p + 2}$

Answers:

1. quotient; exponents
2. descending
3. 0
4. divisor; remainder
5. $3x^3 - 2x^2 + 1$
6. $3m^4 - 2m^3 - 1$
7. $3y + 4 - \frac{5}{y}$
8. $8r - 4 + \frac{1}{r}$
9. $3 + \frac{5}{m} + \frac{6}{m^2}$
10. $8 - \frac{9}{x} + \frac{3}{2x^2}$
11. $\frac{2}{7n} - \frac{3}{2m} + \frac{9}{7mn}$
12. $\frac{3}{2k} + \frac{7}{2h} - \frac{7}{4hk}$
13. $\frac{2y}{x} + \frac{3}{4} + \frac{3w}{x}$
14. $\frac{2b}{a} + \frac{5}{3} + \frac{3c}{a}$
15. $r^2 - 7r + 6$
16. $3b^2 + 4b + 8$
17. $y - 4$
18. $y - 3$
19. $q + 8$
20. $q + 7$
21. $t + 5$
22. $k - 4$
23. $p - 4 + \frac{44}{p + 6}$
24. $x + 3 + \frac{-8}{x + 8}$
25. $m^2 + 2m - 1$
26. $2z^2 - 3z + 1$
27. $m^2 + m + 3$
28. $p^2 + p - 2$

29. $x^2 + 2x - 3 + \frac{6}{4x + 1}$

30. $2z^2 - 4z + 1 + \frac{-10}{5z - 3}$

31. $2x - 5 + \frac{-4x + 5}{3x^2 - 2x + 4}$

32. $4m - 3 + \frac{4m + 5}{2m^2 - 3m + 6}$

33. $x^2 + x + 3$

34. $x^2 + 2x - 6$

35. $2x^2 - x - 5$

36. $2x^2 + 5x + 5$

37. $3x^2 + 6x + 11 + \frac{26}{x - 2}$

38. $3k^2 + 6k + 21 + \frac{28}{k - 2}$

39. $2k^2 + 3k - 1$

40. $3k^2 + 4k + 1$

41. $z^2 + 3$ **42.** $p^2 + 6$

43. $2y^2 + 2$ **44.** $2t^2 + 4$

45. $p^2 + p + 1$

46. $4a^2 - 2a + 1$

47. $x^2 - 4x + 2 + \frac{9x - 4}{x^2 + 3}$

48. $3t^2 + 5t + 7 + \frac{12t + 37}{t^2 - 5}$

49. $p^2 + \frac{5}{2}p + 2 + \frac{-1}{2p + 2}$

50. $x^2 + \frac{1}{3}x + 2 + \frac{-2}{3x + 3}$

51. $\frac{3}{2}a - 10 + \frac{77}{2a + 6}$

52. $\frac{5}{4}t + 1 + \frac{-5}{4t + 12}$

53. $\frac{2}{3}x - 1$ **54.** $\frac{1}{2}m + 1$

55. $\frac{3}{4}a - 2 + \frac{1}{4a + 3}$

56. $\frac{3}{5}q + 1 + \frac{-1}{5q - 2}$

57. $(2p + 7)$ feet

58. $(m^2 + 3m + 4)$ mph

59. -13; -13; They are the same, which suggests that when $P(x)$ is divided by $x - r$, the result is $P(r)$. Here, $r = -1$.

60. $Q(x) = 2x^2 - 3x + 5$; $R(x) = 3$

61. $5x - 1$; 0 **62.** $6x - 8$; 0

63. $2x - 3$; -1 **64.** $4x + 5$; 7

65. $4x^2 + 6x + 9$; $\frac{3}{2}$

---

29. $(4x^3 + 9x^2 - 10x + 3) \div (4x + 1)$

30. $(10z^3 - 26z^2 + 17z - 13) \div (5z - 3)$

31. $\dfrac{6x^3 - 19x^2 + 14x - 15}{3x^2 - 2x + 4}$

32. $\dfrac{8m^3 - 18m^2 + 37m - 13}{2m^2 - 3m + 6}$

33. $(x^3 + 2x - 3) \div (x - 1)$

34. $(x^3 + 5x^2 - 18) \div (x + 3)$

35. $(2x^3 - 11x^2 + 25) \div (x - 5)$

36. $(2x^3 + 3x^2 - 5) \div (x - 1)$

37. $(3x^3 - x + 4) \div (x - 2)$

38. $(3k^3 + 9k - 14) \div (k - 2)$

39. $\dfrac{4k^4 + 6k^3 + 3k - 1}{2k^2 + 1}$

40. $\dfrac{9k^4 + 12k^3 - 4k - 1}{3k^2 - 1}$

41. $(2z^3 - 5z^2 + 6z - 15) \div (2z - 5)$

42. $(3p^3 + p^2 + 18p + 6) \div (3p + 1)$

43. $\dfrac{6y^4 + 4y^3 + 4y - 6}{3y^2 + 2y - 3}$

44. $\dfrac{8t^4 + 6t^3 + 12t - 32}{4t^2 + 3t - 8}$

45. $\dfrac{p^3 - 1}{p - 1}$

46. $\dfrac{8a^3 + 1}{2a + 1}$

47. $(x^4 - 4x^3 + 5x^2 - 3x + 2) \div (x^2 + 3)$

48. $(3t^4 + 5t^3 - 8t^2 - 13t + 2) \div (t^2 - 5)$

49. $(2p^3 + 7p^2 + 9p + 3) \div (2p + 2)$

50. $(3x^3 + 4x^2 + 7x + 4) \div (3x + 3)$

51. $(3a^2 - 11a + 17) \div (2a + 6)$

52. $(5t^2 + 19t + 7) \div (4t + 12)$

Extending Skills *Divide.*

53. $\left(2x^2 - \frac{7}{3}x - 1\right) \div (3x + 1)$

54. $\left(m^2 + \frac{7}{2}m + 3\right) \div (2m + 3)$

55. $\left(3a^2 - \frac{23}{4}a - 5\right) \div (4a + 3)$

56. $\left(3q^2 + \frac{19}{5}q - 3\right) \div (5q - 2)$

Solve each problem.

57. Suppose that the volume of a box is

$$(2p^3 + 15p^2 + 28p) \text{ cubic feet.}$$

The height is p feet and the length is $(p + 4)$ feet. Find an expression in p that represents the width.

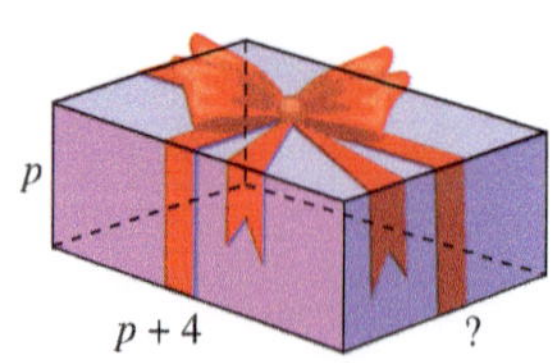

58. Suppose that a minivan travels

$$(2m^3 + 15m^2 + 35m + 36) \text{ miles}$$

in $(2m + 9)$ hours. Find an expression in m that represents the rate of the van in mph.

59. For $P(x) = x^3 - 4x^2 + 3x - 5$, find $P(-1)$. Then divide $P(x)$ by $D(x) = x + 1$. Compare the remainder with $P(-1)$. What do these results suggest?

60. *Concept Check* Let $P(x) = 4x^3 - 8x^2 + 13x - 2$ and $D(x) = 2x - 1$. Use division to find polynomials $Q(x)$ and $R(x)$ such that

$$P(x) = Q(x) \cdot D(x) + R(x).$$

66. $9x^2 - 12x + 16$; $-\frac{4}{3}$
67. $\frac{x^2-9}{2x}$, $x \neq 0$
68. $x + 3$, $x \neq 3$
69. $-\frac{5}{4}$ 70. 4
71. $\frac{x-3}{2x}$, $x \neq 0$
72. $\frac{2x}{x-3}$, $x \neq 3$
73. 0 74. $\frac{1}{2}$
75. $-\frac{35}{4}$ 76. $-\frac{9}{4}$
77. $\frac{7}{2}$ 78. $\frac{3}{2}$

For each pair of functions, find $\left(\frac{f}{g}\right)(x)$ and give any x-values that are not in the domain of the quotient function. ***See Example 6.***

61. $f(x) = 10x^2 - 2x$, $g(x) = 2x$

62. $f(x) = 18x^2 - 24x$, $g(x) = 3x$

63. $f(x) = 2x^2 - x - 3$, $g(x) = x + 1$

64. $f(x) = 4x^2 - 23x - 35$, $g(x) = x - 7$

65. $f(x) = 8x^3 - 27$, $g(x) = 2x - 3$

66. $f(x) = 27x^3 + 64$, $g(x) = 3x + 4$

Let $f(x) = x^2 - 9$, $g(x) = 2x$, and $h(x) = x - 3$. Find each of the following. ***See Example 6.***

67. $\left(\frac{f}{g}\right)(x)$ **68.** $\left(\frac{f}{h}\right)(x)$ **69.** $\left(\frac{f}{g}\right)(2)$ **70.** $\left(\frac{f}{h}\right)(1)$

71. $\left(\frac{h}{g}\right)(x)$ **72.** $\left(\frac{g}{h}\right)(x)$ **73.** $\left(\frac{h}{g}\right)(3)$ **74.** $\left(\frac{g}{h}\right)(-1)$

75. $\left(\frac{f}{g}\right)\left(\frac{1}{2}\right)$ **76.** $\left(\frac{f}{g}\right)\left(\frac{3}{2}\right)$ **77.** $\left(\frac{h}{g}\right)\left(-\frac{1}{2}\right)$ **78.** $\left(\frac{h}{g}\right)\left(-\frac{3}{2}\right)$

Chapter 4 Summary

Key Terms

4.1
exponent (power)
base
exponential expression

4.2
term
algebraic expression
polynomial
numerical coefficient (coefficient)
degree of a term
polynomial in x
descending powers
leading term
leading coefficient
trinomial
binomial
monomial
degree of a polynomial
like terms
negative of a polynomial

4.3
polynomial function
composite function (composition of functions)
identity function
squaring function
cubing function

New Symbols

$(f \circ g)(x) = f(g(x))$ composite function

Test Your Word Power

See how well you have learned the vocabulary in this chapter.

1. A **polynomial** is an algebraic expression made up of
- **A.** a term or a finite product of terms with positive coefficients and exponents
- **B.** the sum of two or more terms with whole number coefficients and exponents
- **C.** the product of two or more terms
- **D.** a term or a finite sum of terms with real number coefficients and whole number exponents.

2. A **monomial** is a polynomial with
- **A.** only one term
- **B.** exactly two terms
- **C.** exactly three terms
- **D.** more than three terms.

3. A **binomial** is a polynomial with
- **A.** only one term
- **B.** exactly two terms
- **C.** exactly three terms
- **D.** more than three terms.

4. A **trinomial** is a polynomial with
- **A.** only one term
- **B.** exactly two terms
- **C.** exactly three terms
- **D.** more than three terms.

5. The **FOIL** method is used to
- **A.** add two binomials
- **B.** add two trinomials
- **C.** multiply two binomials
- **D.** multiply two trinomials.

ANSWERS

1. D; *Example:* $5x^3 + 2x^2 - 7$ **2.** A; *Examples:* -4, $2x^3$, $15a^2b$ **3.** B; *Example:* $3t^3 + 5t$ **4.** C; *Example:* $2a^2 - 3ab + b^2$
5. C; *Example:* $(m + 4)(m - 3) = m(m) - 3m + 4m + 4(-3) = m^2 + m - 12$
F O I L

Quick Review

CONCEPTS	EXAMPLES
4.1 Integer Exponents and Scientific Notation	
Definitions and Rules for Exponents For all integers m and n and all real numbers a and b, the following hold.	Apply the rules for exponents.
Product Rule $\quad a^m \cdot a^n = a^{m+n}$	$3^4 \cdot 3^2 = 3^6$
Quotient Rule $\quad \dfrac{a^m}{a^n} = a^{m-n} \quad (a \neq 0)$	$\dfrac{2^5}{2^3} = 2^2$
Zero Exponent $\quad a^0 = 1 \quad (a \neq 0)$	$27^0 = 1, \quad (-5)^0 = 1$
Negative Exponent $\quad a^{-n} = \dfrac{1}{a^n} \quad (a \neq 0)$	$5^{-2} = \dfrac{1}{5^2}$
Power Rules (a) $(a^m)^n = a^{mn}$ (b) $(ab)^m = a^m b^m$	$(6^3)^4 = 6^{12}, \quad (5p)^4 = 5^4 p^4$
(c) $\left(\dfrac{a}{b}\right)^n = \dfrac{a^n}{b^n} \quad (b \neq 0)$	$\left(\dfrac{2}{3}\right)^5 = \dfrac{2^5}{3^5}$
Special Rules for Negative Exponents $\dfrac{1}{a^{-n}} = a^n \quad (a \neq 0) \qquad \dfrac{a^{-n}}{b^{-m}} = \dfrac{b^m}{a^n} \quad (a, b \neq 0)$	$\dfrac{1}{3^{-2}} = 3^2, \qquad \dfrac{5^{-3}}{4^{-6}} = \dfrac{4^6}{5^3}$
$a^{-n} = \left(\dfrac{1}{a}\right)^n \quad (a \neq 0) \qquad \left(\dfrac{a}{b}\right)^{-n} = \left(\dfrac{b}{a}\right)^n \quad (a, b \neq 0)$	$4^{-3} = \left(\dfrac{1}{4}\right)^3, \qquad \left(\dfrac{4}{7}\right)^{-2} = \left(\dfrac{7}{4}\right)^2$
Scientific Notation A number is written in scientific notation when it is expressed in the form $a \times 10^n,$ **where $1 \leq \lvert a \rvert < 10$ and n is an integer.**	Write 23,500,000,000 in scientific notation. $23{,}500{,}000{,}000 = 2.35 \times 10^{10}$ Write 4.3×10^{-6} in standard notation. $4.3 \times 10^{-6} = 0.0000043$
4.2 Adding and Subtracting Polynomials	
Add or subtract polynomials by combining like terms.	Subtract. $\quad (5x^4 + 3x^2) - (7x^4 + x^2 - x)$ $= 5x^4 + 3x^2 - 7x^4 - x^2 + x$ $= -2x^4 + 2x^2 + x$
4.3 Polynomial Functions, Graphs, and Composition	
Adding and Subtracting Functions If $f(x)$ and $g(x)$ define functions, then $(f + g)(x) = f(x) + g(x)$ and $\quad (f - g)(x) = f(x) - g(x).$	Let $f(x) = x^2$ and $g(x) = 2x + 1$. Find $(f + g)(x)$ and $(f - g)(x)$. $(f + g)(x) = f(x) + g(x) = x^2 + 2x + 1$ $(f - g)(x) = f(x) - g(x) = x^2 - (2x + 1) = x^2 - 2x - 1$
Composition of f and g $(f \circ g)(x) = f(g(x))$ $(g \circ f)(x) = g(f(x))$	Let $f(x) = x^2$ and $g(x) = 2x + 1$. Find $(f \circ g)(x)$ and $(g \circ f)(x)$. $(f \circ g)(x) = f(g(x)) = f(2x + 1) = (2x + 1)^2$ $(g \circ f)(x) = g(f(x)) = g(x^2) = 2x^2 + 1$

CONCEPTS | **EXAMPLES**

Graphs of Basic Polynomial Functions

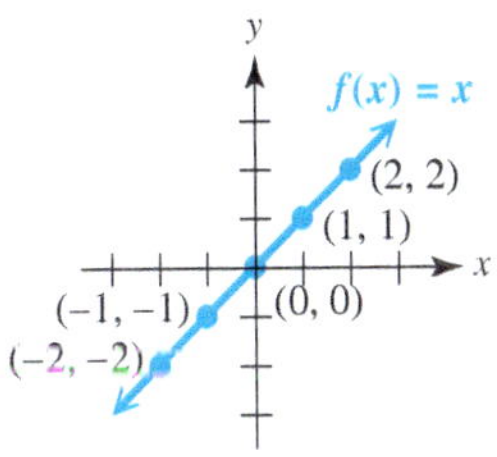

Identity function
$f(x) = x$
Domain: $(-\infty, \infty)$
Range: $(-\infty, \infty)$

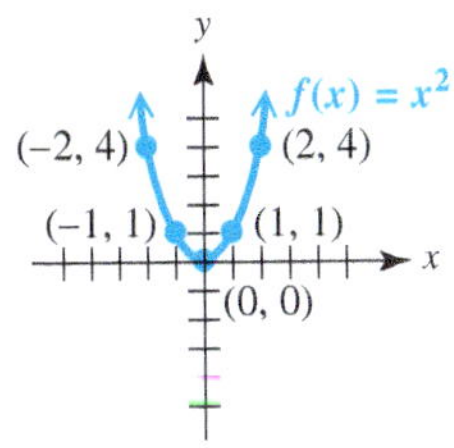

Squaring function
$f(x) = x^2$
Domain: $(-\infty, \infty)$
Range: $[0, \infty)$

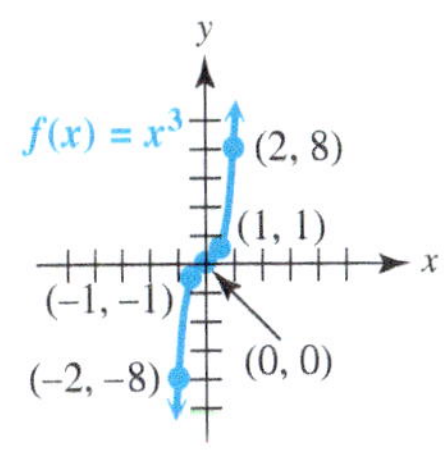

Cubing function
$f(x) = x^3$
Domain: $(-\infty, \infty)$
Range: $(-\infty, \infty)$

4.4 Multiplying Polynomials

To multiply two polynomials, multiply each term of one by each term of the other.

Multiply.

$$(x^3 + 3x)(4x^2 - 5x + 2)$$
$$= x^3(4x^2 - 5x + 2) + 3x(4x^2 - 5x - 2)$$
$$= 4x^5 - 5x^4 + 2x^3 + 12x^3 - 15x^2 - 6x$$
$$= 4x^5 - 5x^4 + 14x^3 - 15x^2 + 6x$$

To multiply two binomials, use the **FOIL method.** Multiply the **First** terms, the **Outer** terms, the **Inner** terms, and the **Last** terms. Then add these products.

$$(2x + 3)(x - 7)$$
$$= 2x(x) + 2x(-7) + 3x + 3(-7) \quad \text{FOIL method}$$
$$= 2x^2 - 14x + 3x - 21 \quad \text{Multiply.}$$
$$= 2x^2 - 11x - 21 \quad \text{Combine like terms.}$$

Special Products

$$(x + y)(x - y) = x^2 - y^2$$
$$(x + y)^2 = x^2 + 2xy + y^2$$
$$(x - y)^2 = x^2 - 2xy + y^2$$

$$(3m + 8)(3m - 8)$$
$$= 9m^2 - 64$$

$$(5a + 3b)^2$$
$$= 25a^2 + 30ab + 9b^2$$

$$(2k - 1)^2$$
$$= 4k^2 - 4k + 1$$

Multiplying Functions
If $f(x)$ and $g(x)$ define functions, then

$$(fg)(x) = f(x) \cdot g(x).$$

Let $f(x) = x^2$ and $g(x) = 2x + 1$. Find $(fg)(x)$.

$$(fg)(x)$$
$$= f(x) \cdot g(x)$$
$$= x^2(2x + 1)$$
$$= 2x^3 + x^2$$

4.5 Dividing Polynomials

Dividing by a Monomial
To divide a polynomial by a monomial, divide each term in the polynomial by the monomial, and then write each fraction in lowest terms.

Divide.

$$\frac{2x^3 - 4x^2 + 6x - 8}{2x}$$
$$= \frac{2x^3}{2x} - \frac{4x^2}{2x} + \frac{6x}{2x} - \frac{8}{2x}$$
$$= x^2 - 2x + 3 - \frac{4}{x}$$

CONCEPTS	EXAMPLES
Dividing by a Polynomial Use the "long division" process. The process ends when the remainder is 0 or when the degree of the remainder is less than the degree of the divisor.	Find $(m^3 - m^2 + 2m + 5) \div (m + 1)$. $$\begin{array}{r} m^2 - 2m + 4 \\ m + 1 \overline{) m^3 - m^2 + 2m + 5} \\ \underline{m^3 + m^2 } \\ -2m^2 + 2m \\ \underline{-2m^2 - 2m } \\ 4m + 5 \\ \underline{4m + 4} \\ 1 \end{array}$$ $1 \leftarrow$ Remainder The answer is $m^2 - 2m + 4 + \frac{1}{m + 1}$.
Dividing Functions If $f(x)$ and $g(x)$ define functions, then $$\left(\frac{f}{g}\right)(x) = \frac{f(x)}{g(x)}, \quad \text{(where } g(x) \neq 0\text{)}.$$	Let $f(x) = x^2$ and $g(x) = 2x + 1$. Find $\left(\frac{f}{g}\right)(x)$. $$\left(\frac{f}{g}\right)(x) = \frac{f(x)}{g(x)} = \frac{x^2}{2x + 1}, \quad x \neq -\frac{1}{2}$$

Chapter 4 Review Exercises

4.1 *Simplify. Write answers with only positive exponents. Assume that all variables represent nonzero real numbers.*

1. 4^3 **2.** $\left(\frac{1}{3}\right)^4$ **3.** $(-5)^3$

4. $\frac{2}{(-3)^{-2}}$ **5.** $\left(\frac{2}{3}\right)^{-4}$ **6.** $\left(\frac{5}{4}\right)^{-2}$

7. $5^{-1} - 6^{-1}$ **8.** $2^{-1} + 4^{-1}$ **9.** $-3^0 + 3^0$

10. $(3^{-4})^2$ **11.** $(x^{-4})^{-2}$ **12.** $(xy^{-3})^{-2}$

13. $(z^{-3})^3 z^{-6}$ **14.** $(5m^{-3})^2(m^4)^{-3}$ **15.** $\frac{(3r)^2 r^4}{r^{-2} r^{-3}}(9r^{-3})^{-2}$

16. $\left(\frac{5z^{-3}}{z^{-1}}\right)\frac{5}{z^2}$ **17.** $\left(\frac{6m^{-4}}{m^{-9}}\right)^{-1}\left(\frac{m^{-2}}{16}\right)$ **18.** $\left(\frac{3r^5}{5r^{-3}}\right)^{-2}\left(\frac{9r^{-1}}{2r^{-5}}\right)^3$

19. $(-3x^4y^3)(4x^{-2}y^5)$ **20.** $\frac{6m^{-4}n^3}{-3mn^2}$ **21.** $\frac{(5p^{-2}q)(4p^5q^{-3})}{2p^{-5}q^5}$

22. Explain the difference between the expressions $(-6)^0$ and -6^0.

Write each number in scientific notation.

23. 13,450 **24.** 0.0000000765 **25.** 0.138

26. The total population of the United States was recently estimated at **308,700,000.** Of this amount, about **53,000** Americans were centenarians, that is, age **100** or older. Write the three boldfaced numbers using scientific notation. (*Source:* U.S. Census Bureau.)

Answers

1. 64
2. $\frac{1}{81}$
3. -125
4. 18
5. $\frac{81}{16}$
6. $\frac{16}{25}$
7. $\frac{1}{30}$
8. $\frac{3}{4}$
9. 0
10. $\frac{1}{3^8}$
11. x^8
12. $\frac{y^6}{x^2}$
13. $\frac{1}{z^{15}}$
14. $\frac{25}{m^{18}}$
15. $\frac{r^{17}}{9}$
16. $\frac{25}{z^4}$
17. $\frac{1}{96m^7}$
18. $\frac{2025}{8r^4}$
19. $-12x^2y^8$
20. $-\frac{2n}{m^5}$
21. $\frac{10p^8}{q^7}$
22. In $(-6)^0$, the base is -6 and the expression simplifies to 1. In -6^0, the base is 6 and the expression simplifies to -1.
23. 1.345×10^4
24. 7.65×10^{-8}
25. 1.38×10^{-1}
26. 3.087×10^8; 5.3×10^4; 1×10^2 (or 10^2)

27. 1,210,000
28. 0.0058
29. 2×10^{-4}; 0.0002
30. 1.5×10^{3}; 1500
31. 4.1×10^{-5}; 0.000041
32. 2.7×10^{-2}; 0.027
33. (a) 20,000 hr
(b) 833 days
34. 14, 5 35. −1; 2
36. $\frac{1}{10}$; 1 37. 504; 8
38. (a) $11k^3 - 3k^2 + 9k$
(b) trinomial (c) 3
39. (a) $9m^7 + 14m^6$
(b) binomial (c) 7
40. (a) $-5y^4 + 3y^3 + 7y^2 - 2y$
(b) none of these (c) 4
41. (a) $-7q^5r^3$ (b) monomial
(c) 8
42. One example is
$x^5 + 2x^4 - x^2 + x + 2$.
43. $-x^2 - 3x + 1$
44. $-5y^3 - 4y^2 + 6y - 12$
45. $6a^3 - 4a^2 - 16a + 15$
46. $8y^2 - 9y + 5$
47. (a) −11 (b) 4 (c) 7
48. (a) $5x^2 - x + 5$
(b) $-5x^2 + 5x + 1$
(c) 11 (d) −9
49. (a) 167 (b) 1495
(c) 20 (d) 42
(e) $75x^2 + 220x + 160$
(f) $15x^2 + 10x + 2$

Write each number in standard notation.

27. 1.21×10^6 **28.** 5.8×10^{-3}

Evaluate. Give answers in both scientific notation and standard notation.

29. $\dfrac{16 \times 10^4}{8 \times 10^8}$ **30.** $\dfrac{6 \times 10^{-2}}{4 \times 10^{-5}}$

31. $\dfrac{0.0000000164}{0.0004}$ **32.** $\dfrac{0.0009 \times 12{,}000{,}000}{400{,}000}$

33. The planet Mercury has an average distance from the sun of 3.6×10^7 mi, while the average distance from Venus to the sun is 6.7×10^7 mi.

(a) How long would it take a spacecraft traveling at 1.55×10^3 mph to travel from Venus to Mercury? Give the answer in hours, in standard notation.

(b) Use the answer from part (a) to find the number of days it would take the spacecraft to travel from Venus to Mercury, to the nearest whole number.

4.2 *Give the numerical coefficient and the degree of each term.*

34. $14p^5$ **35.** $-z^2$ **36.** $\dfrac{x}{10}$ **37.** $504p^3r^5$

*For each polynomial, **(a)** write it in descending powers, **(b)** identify it as a* monomial, binomial, trinomial, *or* none of these, *and **(c)** give its degree.*

38. $9k + 11k^3 - 3k^2$ **39.** $14m^6 + 9m^7$

40. $-5y^4 + 3y^3 + 7y^2 - 2y$ **41.** $-7q^5r^3$

42. Give an example of a polynomial in the variable x such that the polynomial has degree 5, is lacking a third-degree term, and is in descending powers of the variable.

Add or subtract as indicated.

43. Add.
$$\begin{array}{r} 3x^2 - 5x + 6 \\ \underline{-4x^2 + 2x - 5} \end{array}$$

44. Subtract.
$$\begin{array}{r} -5y^3 \qquad + 8y - 3 \\ \underline{4y^2 + 2y + 9} \end{array}$$

45. $(4a^3 - 9a + 15) - (-2a^3 + 4a^2 + 7a)$ **46.** $(3y^2 + 2y - 1) + (5y^2 - 11y + 6)$

4.3 *Work each problem.*

47. For the polynomial function $f(x) = -2x^2 + 5x + 7$, find each value.

(a) $f(-2)$ **(b)** $f(3)$ **(c)** $f(0)$

48. Find each of the following for the polynomial functions
$$f(x) = 2x + 3 \quad \text{and} \quad g(x) = 5x^2 - 3x + 2.$$

(a) $(f + g)(x)$ **(b)** $(f - g)(x)$ **(c)** $(f + g)(-1)$ **(d)** $(f - g)(-1)$

49. Find each of the following for the polynomial functions
$$f(x) = 3x^2 + 2x - 1 \quad \text{and} \quad g(x) = 5x + 7.$$

(a) $(g \circ f)(3)$ **(b)** $(f \circ g)(3)$ **(c)** $(f \circ g)(-2)$

(d) $(g \circ f)(-2)$ **(e)** $(f \circ g)(x)$ **(f)** $(g \circ f)(x)$

50. **(a)** 95,408
(b) 117,556
(c) 130,792

51.

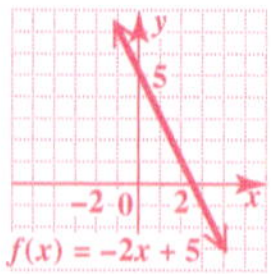

domain: $(-\infty, \infty)$; range: $(-\infty, \infty)$

52.

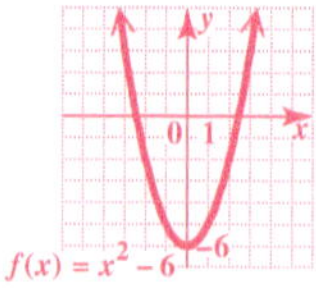

domain: $(-\infty, \infty)$; range: $[-6, \infty)$

53.

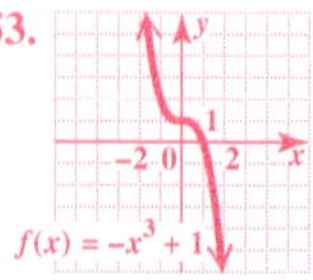

domain: $(-\infty, \infty)$; range: $(-\infty, \infty)$

54. $-12k^3 - 42k$
55. $2x^3 + 22x^2 + 56x$
56. $15m^2 - 7m - 2$
57. $6w^2 - 13wt + 6t^2$
58. $10p^4 + 30p^3 - 8p^2 - 24p$
59. $3q^3 - 13q^2 - 14q + 20$
60. $36r^4 - 1$
61. $16m^2 + 24m + 9$
62. $9t^2 - 12ts + 4s^2$
63. $y^2 - 3y + \frac{5}{4}$
64. $x^2 - 4x + 6$
65. $p^2 + 6p + 9 + \frac{54}{2p - 3}$
66. $p^2 + 3p - 6$
67. **(a)** $36x^3 - 9x^2$ **(b)** -45
68. **(a)** $4x - 1,\ x \neq 0$ **(b)** 7

50. The number of twin births in the United States during the years 1990 through 2011 can be modeled by the polynomial function

$$P(x) = -21.769x^3 + 626.67x^2 - 1875.0x + 95{,}408,$$

where $x = 0$ represents 1990, $x = 1$ represents 1991, and so on. Use this function to approximate number of twin births (to the nearest whole number) in each given year. (*Source:* National Center for Health Statistics.)

(a) 1990 **(b)** 2000 **(c)** 2011

Graph each polynomial function. Give the domain and range.

51. $f(x) = -2x + 5$ **52.** $f(x) = x^2 - 6$ **53.** $f(x) = -x^3 + 1$

4.4 *Find each product.*

54. $-6k(2k^2 + 7)$ **55.** $2x(x + 4)(x + 7)$ **56.** $(3m - 2)(5m + 1)$

57. $(3w - 2t)(2w - 3t)$ **58.** $(2p^2 + 6p)(5p^2 - 4)$ **59.** $(3q^2 + 2q - 4)(q - 5)$

60. $(6r^2 - 1)(6r^2 + 1)$ **61.** $(4m + 3)^2$ **62.** $(3t - 2s)^2$

4.5 *Divide.*

63. $\dfrac{4y^3 - 12y^2 + 5y}{4y}$ **64.** $\dfrac{x^3 - 9x^2 + 26x - 30}{x - 5}$

65. $\dfrac{2p^3 + 9p^2 + 27}{2p - 3}$ **66.** $\dfrac{5p^4 + 15p^3 - 33p^2 - 9p + 18}{5p^2 - 3}$

4.4, 4.5 *Find each of the following for the polynomial functions*

$$f(x) = 12x^2 - 3x \quad \textit{and} \quad g(x) = 3x.$$

67. **(a)** $(fg)(x)$ **(b)** $(fg)(-1)$ **68.** **(a)** $\left(\frac{f}{g}\right)(x)$ **(b)** $\left(\frac{f}{g}\right)(2)$

Chapter 4 Mixed Review Exercises

1. **(a)** A **(b)** G **(c)** C **(d)** C **(e)** A **(f)** E **(g)** B **(h)** H **(i)** F **(j)** I

Solve each problem.

1. Match each expression (a)–(j) in Column I with its equivalent expression A–J in Column II. Choices may be used once, more than once, or not at all.

I		II	
(a) 4^{-2}	**(b)** -4^2	**A.** $\frac{1}{16}$	**B.** 0
(c) 4^0	**(d)** $(-4)^0$	**C.** 1	**D.** $-\frac{1}{16}$
(e) $(-4)^{-2}$	**(f)** -4^0	**E.** -1	**F.** $\frac{5}{16}$
(g) $-4^0 + 4^0$	**(h)** $-4^0 - 4^0$	**G.** -16	**H.** -2
(i) $4^{-2} + 4^{-1}$	**(j)** 4^2	**I.** 16	**J.** none of these

2. 103,363 mi^2

3. $\frac{y^4}{36}$ 4. $\frac{1}{125}$

5. -9 6. $-\frac{1}{5z^9}$

7. $21p^9 + 7p^8 + 14p^7$

8. $4x^2 - 36x + 81$

9. $\frac{1}{16y^{18}}$

10. $8x + 1 + \frac{5}{x-3}$

11. $8x^2 - 10x - 3$

12. $\frac{1250z^7x^6}{9}$

13. $9m^2 - 30mn + 25n^2 - p^2$

14. $2y^2x + \frac{3y^3}{2x} + \frac{5x^2}{2}$

15. $-3k^2 + 4k - 7$

2. In 2013, the estimated population of New Zealand was 4.365×10^6. The population density was 42.23 people per square mile. What is the area of New Zealand to the nearest square mile? (*Source: The World Factbook.*)

Perform the indicated operations and then simplify. Write answers with only positive exponents. Assume that all variables represent nonzero real numbers.

3. $\frac{6^{-1}y^3(y^2)^{-2}}{6y^{-4}(y^{-1})}$ **4.** 5^{-3} **5.** $-(-3)^2$

6. $\frac{(-z^{-2})^3}{5(z^{-3})^{-1}}$ **7.** $7p^5(3p^4 + p^3 + 2p^2)$ **8.** $(2x - 9)^2$

9. $(y^6)^{-5}(2y^{-3})^{-4}$ **10.** $\frac{8x^2 - 23x + 2}{x - 3}$ **11.** $(4x + 1)(2x - 3)$

12. $\frac{(5z^2x^3)^2(2zx^2)^{-1}}{(-10zx^{-3})^{-2}(3z^{-1}x^{-4})^2}$ **13.** $[(3m - 5n) + p][(3m - 5n) - p]$

14. $\frac{20y^3x^3 + 15y^4x + 25yx^4}{10yx^2}$ **15.** $(2k - 1) - (3k^2 - 2k + 6)$

Chapter 4 Test

FOR EXTRA HELP *Step-by-step test solutions are found on the Chapter Test Prep Videos available in* MyMathLab® *or on* YouTube.

View the complete solutions to all Chapter Test exercises in MyMathLab.

[4.1]

1. (a) C (b) A (c) D (d) A (e) E (f) F (g) B (h) G (i) I (j) C

2. $\frac{4x^7}{9y^{10}}$

3. $\frac{6}{r^{14}}$ 4. $\frac{16}{9p^{10}q^{28}}$

5. $\frac{16}{x^6y^{16}}$

6. 0.00000091

7. 3×10^{-4}; 0.0003

[4.3]

8. (a) -18 (b) $-2x^2 + 12x - 9$ (c) $-2x^2 - 2x - 3$ (d) -7

1. Match each expression (a)–(j) in Column I with its equivalent expression A–J in Column II. Choices may be used once, more than once, or not at all.

I		II	
(a) 7^{-2}	**(b)** 7^0	**A.** 1	**B.** $\frac{1}{9}$
(c) -7^0	**(d)** $(-7)^0$	**C.** $\frac{1}{49}$	**D.** -1
(e) -7^2	**(f)** $7^{-1} + 2^{-1}$	**E.** -49	**F.** $\frac{9}{14}$
(g) $(7 + 2)^{-1}$	**(h)** $\frac{7^{-1}}{2^{-1}}$	**G.** $\frac{2}{7}$	**H.** 0
(i) 7^2	**(j)** $(-7)^{-2}$	**I.** 49	**J.** none of these

Simplify. Write answers with only positive exponents. Assume that all variables represent nonzero real numbers.

2. $(3x^{-2}y^3)^{-2}(4x^3y^{-4})$ **3.** $\frac{36r^{-4}(r^2)^{-3}}{6r^4}$

4. $\left(\frac{4p^2}{q^4}\right)^3\left(\frac{6p^8}{q^{-8}}\right)^{-2}$ **5.** $(-2x^4y^{-3})^0(-4x^{-3}y^{-8})^2$

6. Write 9.1×10^{-7} in standard notation.

7. Evaluate. Give the answer in both scientific notation and standard notation.

$$\frac{2{,}500{,}000 \times 0.00003}{0.05 \times 5{,}000{,}000}$$

9. (a) 23 (b) $3x^2 + 11$
(c) $9x^2 + 30x + 27$

10.

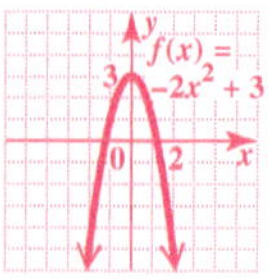

domain: $(-\infty, \infty)$; range: $(-\infty, 3]$

11.

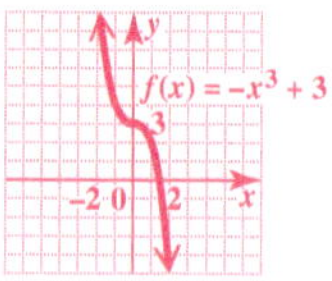

domain: $(-\infty, \infty)$; range: $(-\infty, \infty)$

[4.4]
12. $10x^2 - x - 3$
13. $6m^3 - 7m^2 - 30m + 25$
14. $36x^2 - y^2$
15. $9k^2 + 6kq + q^2$
16. $4y^2 - 9z^2 + 6zx - x^2$

[4.5]
17. $4p - 8 + \frac{6}{p}$
18. $x^2 + 4x + 4$

[4.2]
19. $x^3 - 2x^2 - 10x - 13$

[4.4, 4.5]
20. (a) $x^3 + 4x^2 + 5x + 2$
(b) 0
(c) $x + 2, \quad x \neq -1$
(d) 0

8. Find each of the following for the polynomial functions

$$f(x) = -2x^2 + 5x - 6 \quad \text{and} \quad g(x) = 7x - 3.$$

(a) $f(4)$ **(b)** $(f + g)(x)$ **(c)** $(f - g)(x)$ **(d)** $(f - g)(-2)$

9. Find each of the following for the polynomial functions

$$f(x) = 3x + 5 \quad \text{and} \quad g(x) = x^2 + 2.$$

(a) $(f \circ g)(-2)$ **(b)** $(f \circ g)(x)$ **(c)** $(g \circ f)(x)$

Graph each polynomial function. Give the domain and range.

10. $f(x) = -2x^2 + 3$

11. $f(x) = -x^3 + 3$

Perform the indicated operations.

12. $(5x - 3)(2x + 1)$

13. $(2m - 5)(3m^2 + 4m - 5)$

14. $(6x + y)(6x - y)$

15. $(3k + q)^2$

16. $[2y + (3z - x)][2y - (3z - x)]$

17. $\dfrac{16p^3 - 32p^2 + 24p}{4p^2}$

18. $(x^3 + 3x^2 - 4) \div (x - 1)$

19. $(4x^3 - 3x^2 + 2x - 5) - (3x^3 + 11x + 8) + (x^2 - x)$

20. Find each of the following for the polynomial functions

$$f(x) = x^2 + 3x + 2 \quad \text{and} \quad g(x) = x + 1.$$

(a) $(fg)(x)$ **(b)** $(fg)(-2)$ **(c)** $\left(\frac{f}{g}\right)(x)$ **(d)** $\left(\frac{f}{g}\right)(-2)$

Chapters R–4 Cumulative Review Exercises

[R.1]
1. (a) A, B, C, D, F
(b) B, C, D, F
(c) D, F (d) C, D, F
(e) E, F (f) D, F

[R.3]
2. 32
3. $-\frac{1}{72}$ 4. 0

[1.1]
5. $\{-65\}$
6. {all real numbers}

[1.2]
7. $t = \frac{A - p}{pr}$

[1.5]
8. $(-\infty, 6)$

[1.7]
9. $\left\{-\frac{1}{3}, 1\right\}$
10. $\left(-\infty, -\frac{8}{3}\right] \cup [2, \infty)$

1. Match each number in Column I with the choice or choices of sets of numbers in Column II to which the number belongs.

I		II	
(a) 34	**(b)** 0	**A.** Natural numbers	**B.** Whole numbers
(c) 2.16	**(d)** $-\sqrt{36}$	**C.** Integers	**D.** Rational numbers
(e) $\sqrt{13}$	**(f)** $-\frac{4}{5}$	**E.** Irrational numbers	**F.** Real numbers

Evaluate.

2. $9 \cdot 4 - 16 \div 4$

3. $\left(\frac{1}{3}\right)^2 - \left(\frac{1}{2}\right)^3$

4. $-|8 - 13| - |-4| + |-9|$

Solve.

5. $-5(8 - 2z) + 4(7 - z) = 7(8 + z) - 3$

6. $3(x + 2) - 5(x + 2) = -2x - 4$

7. $A = p + prt$ for t

8. $2(m + 5) - 3m + 1 > 5$

9. $|3x - 1| = 2$

10. $|3z + 1| \geq 7$

[1.2, 1.4]

11. 1650; 32%; 21%; 700

12. 15°, 35°, 130°

[2.2]

13. $-\frac{4}{3}$ **14.** 0

[2.3]

15. **(a)** $y = -4x + 15$

(b) $4x + y = 15$

16. **(a)** $y = 4x$ **(b)** $4x - y = 0$

[2.1]

17.

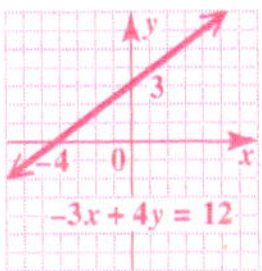

[2.4]

18.

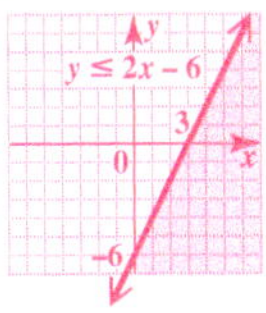

19.

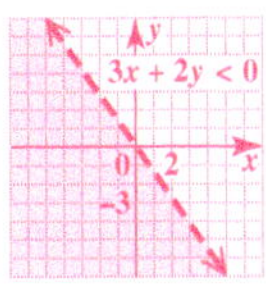

[2.2, 2.3]

20. **(a)** -0.48 gal per yr; Per capita consumption of whole milk decreased by an average of 0.48 gal per year from 1970 to 2011.

(b) $y = -0.48x + 25.3$

(c) 10.9 gal

11. A survey polled Internet users age 12 and older about their most frequent daily Internet activities. Complete the results shown in the table if 5000 such users were surveyed.

Internet Activity	*Percent*	*Number*
Look for news	33%	
Check email		1600
Find or check a fact		1050
Play games	14%	

Source: The 2013 Digital Future Report.

12. Find the measure of each angle of the triangle.

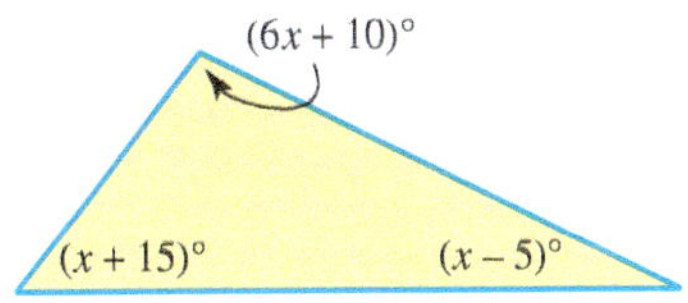

Find the slope of each line described.

13. Through $(-4, 5)$ and $(2, -3)$

14. Through $(4, 5)$; horizontal

Write an equation of each line that satisfies the given conditions. Give the equation ***(a)*** *in slope-intercept form and* ***(b)*** *in standard form.*

15. Through $(4, -1)$; $m = -4$

16. Through $(0, 0)$ and $(1, 4)$

Graph each equation or inequality.

17. $-3x + 4y = 12$

18. $y \leq 2x - 6$

19. $3x + 2y < 0$

20. Per capita consumption of whole milk in the United States (in gallons) is shown in the graph, where $x = 0$ represents 1970.

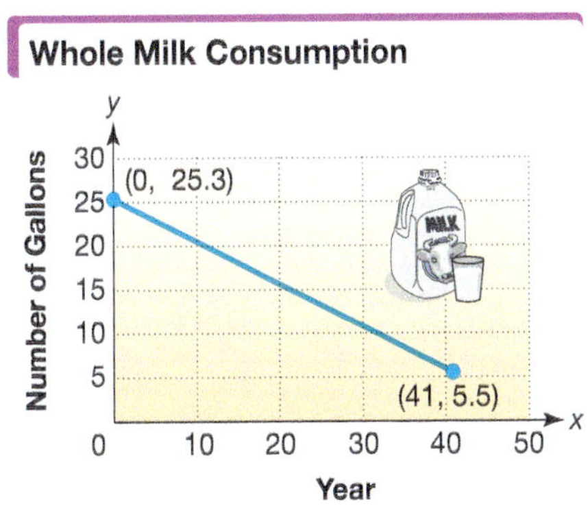

Source: U.S. Department of Agriculture.

(a) Use the given ordered pairs to find the average rate of change in per capita consumption of whole milk (in gallons, to the nearest hundredth) per year during this period. Interpret the answer.

(b) Use the answer from part (a) to write an equation of the line in slope-intercept form that models per capita consumption of whole milk y (in gallons).

(c) Use the equation from part (b) to approximate per capita consumption of whole milk in 2000.

[2.5]
21. domain: $\{-4, -1, 2, 5\}$; range: $\{-2, 0, 2\}$; function

[2.6]
22. -9

[3.1]
23. $\{(3, 2)\}$ **24.** $\varnothing$

[3.2]
25. $\{(1, 0, -1)\}$

[3.3]
26. length: 42 ft; width: 30 ft
27. 15% solution: 6 L; 30% solution: 3 L

[4.1]
28. $\dfrac{8m^9n^3}{p^6}$
29. $\dfrac{y^7}{x^{13}z^2}$
30. $\dfrac{m^6}{8n^9}$

[4.2]
31. $2x^2 - 5x + 10$

[4.4]
32. $x^3 + 8y^3$
33. $15x^2 + 7xy - 2y^2$

[4.5]
34. $4xy^4 - 2y + \dfrac{1}{x^2y}$
35. $m^2 - 2m + 3$

21. Give the domain and range of the relation

$$\{(-4, -2), (-1, 0), (2, 0), (5, 2)\}.$$

Does this relation define a function?

22. If $g(x) = -x^2 - 2x + 6$, find $g(3)$.

Solve each system.

23. $3x - 4y = 1$
$2x + 3y = 12$

24. $3x - 2y = 4$
$-6x + 4y = 7$

25. $x + 3y - 6z = 7$
$2x - y + z = 1$
$x + 2y + 2z = -1$

Solve each problem.

26. The Star-Spangled Banner that flew over Fort McHenry during the War of 1812 had a perimeter of 144 ft. Its length measured 12 ft more than its width. Find the dimensions of this flag, which is displayed in the Smithsonian Institution's Museum of American History in Washington, DC. (*Source:* National Park Service brochure.)

27. Agbe needs 9 L of a 20% solution of alcohol. Agbe has a 15% solution on hand, as well as a 30% solution. How many liters of the 15% solution and the 30% solution should Agbe mix to obtain the 20% solution needed?

Simplify. Write answers with only positive exponents. Assume that all variables represent positive real numbers.

28. $\left(\dfrac{2m^3n}{p^2}\right)^3$

29. $\dfrac{x^{-6}y^3z^{-1}}{x^7y^{-4}z}$

30. $(2m^{-2}n^3)^{-3}$

Perform the indicated operations.

31. $(3x^2 - 8x + 1) - (x^2 - 3x - 9)$

32. $(x + 2y)(x^2 - 2xy + 4y^2)$

33. $(3x + 2y)(5x - y)$

34. $\dfrac{16x^3y^5 - 8x^2y^2 + 4}{4x^2y}$

35. $\dfrac{m^3 - 3m^2 + 5m - 3}{m - 1}$

5 Factoring

Determining the maximum area that can be enclosed by a fence of fixed perimeter is accomplished by solving a *quadratic equation*, one of the topics covered in this chapter on *factoring*.

5.1 Greatest Common Factors and Factoring by Grouping

OBJECTIVES

1. Factor out the greatest common factor.
2. Factor by grouping.

VOCABULARY

- ☐ factoring
- ☐ greatest common factor (GCF)

Writing a polynomial as the product of two or more simpler polynomials is called **factoring** the polynomial. For example, consider the following.

$$3x(5x-2)=15x^2-6x \quad \text{Multiplying}$$

$$15x^2-6x=3x(5x-2) \quad \text{Factoring}$$

Notice that both multiplying and factoring use the distributive property, but in opposite directions. ***Factoring "undoes," or reverses, multiplying.***

OBJECTIVE 1 Factor out the greatest common factor.

The first step in factoring a polynomial is to find the *greatest common factor* for the terms of the polynomial. The product of the greatest common numerical factor and each variable factor of least degree common to every term in a polynomial is the **greatest common factor (GCF)** of the terms of the polynomial.

For example, the greatest common factor for $8x+12$ is 4, because 4 is the greatest factor that *divides into* both $8x$ and 12.

$$8x+12$$

$$=4(2x)+4(3) \quad \text{Factor 4 from each term.}$$

$$=4(2x+3) \quad \text{Distributive property, } ab+ac=a(b+c)$$

CHECK Multiply $4(2x+3)$ to obtain $8x+12$. ✓ Original polynomial

Using the distributive property in this way is called **factoring out the greatest common factor.**

CLASSROOM EXAMPLE 1

Factor out the greatest common factor.

(a) $7k+28$
(b) $32m+24$
(c) $8a-9$
(d) $5z+5$

Answers:
(a) $7(k+4)$
(b) $8(4m+3)$
(c) There is no common factor other than 1.
(d) $5(z+1)$

EXAMPLE 1 Factoring Out the Greatest Common Factor

Factor out the greatest common factor.

(a) $9z-18$

$$=9\cdot z-9\cdot 2 \quad \text{GCF} = 9\text{; Factor 9 from each term.}$$

$$=9(z-2) \quad \text{Distributive property}$$

CHECK Multiply $9(z-2)$ to obtain $9z-18$. ✓ Original polynomial

(b) $56m+35p$

$$=7(8m+5p)$$

(c) $2y+5$

There is no common factor other than 1.

(d) $12+24z$

$$=12\cdot 1+12\cdot 2z \quad \text{Identity property}$$

$$=12(1+2z) \quad \text{12 is the GCF.}$$

Remember to write the 1.

CHECK $12(1+2z)$

$$=12(1)+12(2z) \quad \text{Distributive property}$$

$$=12+24z \ ✓ \quad \text{Original polynomial}$$

NOW TRY

NOW TRY EXERCISE 1

Factor out the greatest common factor.

(a) $54m-45$
(b) $2k-7$

NOW TRY ANSWERS

1. **(a)** $9(6m-5)$
   (b) There is no common factor other than 1.

EXAMPLE 2 Factoring Out the Greatest Common Factor

Factor out the greatest common factor.

(a) $9x^2 + 12x^3$

The numerical part of the GCF is 3, the greatest integer that divides into both 9 and 12. The least exponent that appears on x is 2. The GCF is $3x^2$.

$$9x^2 + 12x^3$$
$$= 3x^2(3) + 3x^2(4x) \quad \text{GCF} = 3x^2$$
$$= 3x^2(3 + 4x) \quad \text{Distributive property}$$

(b) $32p^4 - 24p^3 + 40p^5$

$$= 8p^3(4p) + 8p^3(-3) + 8p^3(5p^2) \quad \text{GCF} = 8p^3$$
$$= 8p^3(4p - 3 + 5p^2) \quad \text{Distributive property}$$

(c) $3k^4 - 15k^7 + 24k^9$

Remember the 1.

$$= 3k^4(1 - 5k^3 + 8k^5) \quad \text{GCF} = 3k^4$$

(d) $24m^3n^2 - 18m^2n + 6m^4n^3$

$$= 6m^2n(4mn) + 6m^2n(-3) + 6m^2n(m^2n^2) \quad \text{GCF} = 6m^2n$$
$$= 6m^2n(4mn - 3 + m^2n^2) \quad \text{Distributive property}$$

(e) $25x^2y^3 + 30y^5 - 15x^4y^7$

$$= 5y^3(5x^2 + 6y^2 - 3x^4y^4)$$

In each case, check the factored form by multiplying. NOW TRY

EXAMPLE 3 Factoring Out a Binomial Factor

Factor out the greatest common factor.

(a) $(x - 5)(x + 6) + (x - 5)(2x + 5)$ — The greatest common factor is $x - 5$.

$$= (x - 5)[(x + 6) + (2x + 5)] \quad \text{Factor out } x - 5.$$
$$= (x - 5)(3x + 11) \quad \text{Combine like terms inside the brackets.}$$

(b) $z^2(m + n) + x^2(m + n)$

$$= (m + n)(z^2 + x^2) \quad \text{Factor out the common factor } (m + n).$$

(c) $p(r + 2s)^2 - q(r + 2s)^3$

$$= (r + 2s)^2[p - q(r + 2s)] \quad \text{Factor out the common factor } (r + 2s)^2.$$
$$= (r + 2s)^2(p - qr - 2qs)$$

Be careful with signs.

(d) $(p - 5)(p + 2) - (p - 5)(3p + 4)$

$$= (p - 5)[(p + 2) - (3p + 4)] \quad \text{Factor out the common factor } p - 5.$$
$$= (p - 5)[p + 2 - 3p - 4] \quad \text{Distributive property}$$
$$= (p - 5)[-2p - 2] \quad \text{Combine like terms.}$$
$$= (p - 5)[-2(p + 1)] \quad \text{Look for a common factor.}$$
$$= -2(p - 5)(p + 1) \quad \text{Commutative property}$$

NOW TRY

NOW TRY EXERCISE 2

Factor out the greatest common factor.

(a) $16y^4 + 8y^3$

(b) $14p^2 - 9p^3 + 6p^4$

(c) $9x^3y^2 - 3x^2y^2 + 6x^2y^3$

CLASSROOM EXAMPLE 2

Factor out the greatest common factor.

(a) $25x^4 + 30x^3$

(b) $100m^5 - 50m^4 + 25m^3$

(c) $5m^4x^3 + 15m^5x^6 - 20m^4x^6$

Answers:

(a) $5x^3(5x + 6)$

(b) $25m^3(4m^2 - 2m + 1)$

(c) $5m^4x^3(1 + 3mx^3 - 4x^3)$

CLASSROOM EXAMPLE 3

Factor out the greatest common factor.

(a) $(a + 2)(a - 3) + (a + 2)(a + 6)$

(b) $(y - 1)(y + 3) - (y - 1)(y + 4)$

(c) $k^2(a + 5b) + m^2(a + 5b)^2$

Answers:

(a) $(a + 2)(2a + 3)$

(b) $(y - 1)(-1)$, or $-y + 1$

(c) $(a + 5b)(k^2 + m^2a + 5m^2b)$

NOW TRY EXERCISE 3

Factor out the greatest common factor.

(a) $(3x - 2)(x + 1) + (3x - 2)(2x - 5)$

(b) $z(a - b)^3 - 2w(a - b)^2$

TEACHING TIP

Include with your examples one that cannot be factored and one with a common numerical factor, but no common variable factor in all terms.

NOW TRY ANSWERS

2. (a) $8y^3(2y + 1)$
(b) $p^2(14 - 9p + 6p^2)$
(c) $3x^2y^2(3x - 1 + 2y)$

3. (a) $(3x - 2)(3x - 4)$
(b) $(a - b)^2(za - zb - 2w)$

NOW TRY EXERCISE 4

Factor $-4y^5 - 3y^3 + 8y$ in two ways.

CLASSROOM EXAMPLE 4

Factor $-6r^2 + 5r$ in two ways.

Answer: $-r(6r - 5)$; $r(-6r + 5)$

EXAMPLE 4 Factoring Out a Negative Common Factor

Factor $-a^3 + 3a^2 - 5a$ in two ways.

First, a could be used as the common factor.

$$-a^3 + 3a^2 - 5a$$
$$= a(-a^2) + a(3a) + a(-5) \quad \text{Factor out } a.$$
$$= a(-a^2 + 3a - 5) \quad \text{Distributive property}$$

CHECK Multiply $a(-a^2 + 3a - 5)$ to obtain $-a^3 + 3a^2 - 5a$. ✓

Because of the leading negative sign, $-a$ could also be used as the common factor.

$$-a^3 + 3a^2 - 5a$$
$$= -a(a^2) + (-a)(-3a) + (-a)(5) \quad \text{Factor out } -a.$$
$$= -a(a^2 - 3a + 5) \quad \text{Distributive property}$$

CHECK Multiply $-a(a^2 - 3a + 5)$ to obtain $-a^3 + 3a^2 - 5a$. ✓

Sometimes there may be a reason to prefer one of these forms, but either is correct.

NOW TRY

NOTE The answer section in this book will usually give the factored form where the common factor has a positive coefficient.

OBJECTIVE 2 Factor by grouping.

Sometimes the *individual terms* of a polynomial have a greatest common factor of 1, but it still may be possible to factor the polynomial by using a process called **factoring by grouping.** ***We usually factor by grouping when a polynomial has more than three terms.***

NOW TRY EXERCISE 5

Factor.

$$3m - 3n + xm - xn$$

CLASSROOM EXAMPLE 5

Factor.

$$6p - 6q + rp - rq$$

Answer: $(p - q)(6 + r)$

EXAMPLE 5 Factoring by Grouping

Factor $ax - ay + bx - by$.

Group the terms in pairs so that each pair has a common factor.

$$ax - ay + bx - by$$

Terms with common factor a — Terms with common factor b

$$= (ax - ay) + (bx - by)$$
$$= a(x - y) + b(x - y) \quad \text{Factor each group.}$$
$$= (x - y)(a + b) \quad \text{The common factor is } x - y.$$

By the commutative property, $(a + b)(x - y)$ is also correct.

CHECK $(x - y)(a + b)$

$$= xa + xb - ya - yb \quad \text{Multiply using the FOIL method.}$$
$$= ax + bx - ay - by \quad \text{Commutative property}$$
$$= ax - ay + bx - by \text{ ✓} \quad \text{Original polynomial}$$

NOW TRY

NOW TRY ANSWERS

4. $y(-4y^4 - 3y^2 + 8)$; $-y(4y^4 + 3y^2 - 8)$

5. $(m - n)(3 + x)$

NOW TRY EXERCISE 6

Factor.

$$ab - 7a - 5b + 35$$

CLASSROOM EXAMPLE 6

Factor.

$$xy - 2y - 4x + 8$$

Answer: $(x - 2)(y - 4)$

TEACHING TIP

Show several examples of factoring by grouping, and then mention that the binomial factors found after the first step must be identical. If they are not, try a different grouping or try factoring out -1 in one of the groups.

EXAMPLE 6 Factoring by Grouping

Factor $3x - 3y - ax + ay$.

$$3x - 3y - ax + ay$$ Pay close attention here.

$$= (3x - 3y) + (-ax + ay)$$ Group the terms.

$$= 3(x - y) + a(-x + y)$$ Factor out 3, and factor out a.

The factors $(x - y)$ and $(-x + y)$ are opposites. If we factor out $-a$ instead of a in the second group, we obtain the common binomial factor $(x \quad y)$. So we start over.

$$(3x - 3y) + (-ax + ay)$$

$$= 3(x - y) - a(x - y)$$ Be careful with signs.

$$= (x - y)(3 - a)$$ Factor out $x - y$.

CHECK $(x - y)(3 - a)$

$$= 3x - ax - 3y + ay$$ Multiply using the FOIL method.

$$= 3x - 3y - ax + ay \checkmark$$ Original polynomial

NOW TRY

NOTE In **Example 6,** a different grouping would lead to the factored form $(a - 3)(y - x)$. Verify by multiplying that this is also correct.

Factoring by Grouping

Step 1 **Group terms.** Collect the terms into groups so that each group has a common factor.

Step 2 **Factor within the groups.** Factor out the common factor in each group.

Step 3 **Factor the entire polynomial.** If each group now has a common factor, factor it out. If not, try a different grouping.

Always check the factored form by multiplying.

NOW TRY EXERCISE 7

Factor.

$$3ax - 6xy - a + 2y$$

CLASSROOM EXAMPLE 7

Factor.

$$kn + mn - k - m$$

Answer: $(k + m)(n - 1)$

EXAMPLE 7 Factoring by Grouping

Factor $6ax + 12bx + a + 2b$.

$$6ax + 12bx + a + 2b$$

$$= (6ax + 12bx) + (a + 2b)$$ Group the terms.

Now factor $6x$ from the first group, and use the identity property of multiplication to introduce the factor 1 in the second group.

Remember to write the 1.

$$= 6x(a + 2b) + 1(a + 2b)$$ Factor each group.

$$= (a + 2b)(6x + 1)$$ Factor out $a + 2b$.

CHECK $(a + 2b)(6x + 1)$

$$= 6ax + a + 12bx + 2b$$ Use the FOIL method.

$$= 6ax + 12bx + a + 2b \checkmark$$ Original polynomial

NOW TRY

NOW TRY ANSWERS

6. $(b - 7)(a - 5)$

7. $(a - 2y)(3x - 1)$

NOW TRY EXERCISE 8

Factor.

$2kp^2 + 6 - 3p^2 - 4k$

CLASSROOM EXAMPLE 8

Factor.

$10x^2y^2 - 18 + 15y^2 - 12x^2$

Answer: $(2x^2 + 3)(5y^2 - 6)$

EXAMPLE 8 Rearranging Terms before Factoring by Grouping

Factor $p^2q^2 - 10 - 2q^2 + 5p^2$.

Neither the first two terms nor the last two terms have a common factor except 1. We can rearrange and group the terms as follows.

$$p^2q^2 - 10 - 2q^2 + 5p^2$$

$$= (p^2q^2 - 2q^2) + (5p^2 - 10) \quad \text{Rearrange and group the terms.}$$

$$= q^2(p^2 - 2) + 5(p^2 - 2) \quad \text{Factor out the common factors.}$$

Don't stop here.

$$= (p^2 - 2)(q^2 + 5) \quad \text{Factor out } p^2 - 2\text{. Use parentheses.}$$

CHECK $(p^2 - 2)(q^2 + 5)$

$$= p^2q^2 + 5p^2 - 2q^2 - 10 \quad \text{Use the FOIL method.}$$

$$= p^2q^2 - 10 - 2q^2 + 5p^2 \checkmark \quad \text{Original polynomial}$$

NOW TRY

CAUTION In **Example 8,** do not stop at the step

$$q^2(p^2 - 2) + 5(p^2 - 2).$$

This expression is *not in factored form* because it is a *sum* of two terms, $q^2(p^2 - 2)$ and $5(p^2 - 2)$, not a *product.*

NOW TRY EXERCISE 9

Factor.

$12wy - 24xy + 4wz - 8xz$

CLASSROOM EXAMPLE 9

Factor.

$14ac - 42bc + 21bd - 7ad$

Answer: $7(a - 3b)(2c - d)$

EXAMPLE 9 Factoring Out a GCF before Factoring by Grouping

Factor $10ax - 5ay + 10bx - 5by$.

Always start by factoring out any greatest common factor from the terms.

$$10ax - 5ay + 10bx - 5by$$

$$= 5(2ax - ay + 2bx - by) \quad \text{Factor out the GCF, 5.}$$

$$= 5[(2ax - ay) + (2bx - by)] \quad \text{Group the terms inside the brackets.}$$

$$= 5[a(2x - y) + b(2x - y)] \quad \text{Factor out the common factors.}$$

$$= 5[(2x - y)(a + b)] \quad \text{Factor out } 2x - y.$$

$$= 5(2x - y)(a + b) \quad \text{Write without the brackets.}$$

CHECK $5(2x - y)(a + b)$

$$= 5(2ax + 2bx - ay - by) \quad \text{Use the FOIL method.}$$

$$= 10ax + 10bx - 5ay - 5by \quad \text{Distributive property}$$

$$= 10ax - 5ay + 10bx - 5by \checkmark \quad \text{Original polynomial}$$

NOW TRY

NOTE Verify that the result in **Example 9** can also be obtained by grouping the terms of the polynomial

$$2ax - ay + 2bx - by \quad \text{as} \quad (2ax + 2bx) + (-ay - by).$$

NOW TRY ANSWERS

8. $(p^2 - 2)(2k - 3)$

9. $4(w - 2x)(3y + z)$

5.1 Exercises

FOR EXTRA HELP

MyMathLab®

Complete solution available in MyMathLab

1. C
2. The polynomial was not factored *completely*. The binomial factor can be factored further as $2(xy^2 - 2)$, so the completely factored form is $4xy^3(xy^2 - 2)$.
3. $3m$
4. $2a$
5. $8xy$
6. $5mn$
7. $3(r + t)^2$
8. $z^2(m + n)^4$
9. $12(m - 5)$
10. $15(r - 3)$
11. $4(1 + 5z)$
12. $9(1 + 3x)$
13. cannot be factored
14. cannot be factored
15. $8k(k^2 + 3)$
16. $9z(z^3 + 9)$
17. $-2p^2q^4(2p + q)$
18. $-3z^3w^2(z^2 + 6w^2)$
19. $7x^3(1 + 5x - 2x^2)$
20. $6k^3(1 - 6k - 8k^2)$
21. $2t^3(5t^2 - 1 - 2t)$
22. $3p^2(2p - 1 - 3p^2)$
23. $5ac(3ac^2 - 5c + 1)$
24. $3yz^3(5y^2 - 9yz + 1)$
25. $16zn^3(zn^3 + 4n^4 - 2z^2)$
26. $5r^2s^2(rs^3 + 2 - 3r^2)$
27. $7ab(2a^2b + a - 3a^4b^2 + 6b^3)$
28. $12km^2(m - 2k^2 + 3km^2 - 5k^3m)$
29. $(m - 4)(2m + 5)$
30. $(z - 5)(2z + 17)$
31. $11(2z - 1)$
32. $-12(3x + 2)$
33. $(2 - x)^2(1 + 2x)$
34. $(5 - x)^2(7 - 2x)$
35. $(3 - x)(6 + 2x - x^2)$
36. $(t - s)(2 + 4t - 4s - t^2 + 2ts - s^2)$
37. $20z(2z + 1)(3z + 4)$
38. $2(a + 2b)^2(3 - 2a - 4b + 6a^2 + 24ab + 24b^2)$
39. $5(m + p)^2(m + p - 2 - 3m^2 - 6mp - 3p^2)$
40. $-3a(p + q)(3a + a^2p + a^2q - 2p^2 - 4pq - 2q^2)$
41. $r(-r^2 + 3r + 5)$; $-r(r^2 - 3r - 5)$
42. $t(-t^3 + 8t^2 - 12)$; $-t(t^3 - 8t^2 + 12)$
43. $12s^4(-s + 4)$; $-12s^4(s - 4)$
44. $16y^3(-y + 4)$; $-16y^3(y - 4)$

1. *Concept Check* Which choice is an example of a polynomial in factored form?

A. $3x^2y^3 + 6x^2(2x + y)$ **B.** $5(x + y)^2 - 10(x + y)^3$

C. $(-2 + 3x)(5y^2 + 4y + 3)$ **D.** $(3x + 4)(5x - y) - (3x + 4)(2x - 1)$

2. *Concept Check* When directed to factor the polynomial $4x^2y^5 - 8xy^3$ completely, a student wrote

$$2xy^3(2xy^2 - 4).$$

The teacher did not give the student full credit. **WHAT WENT WRONG?** Give the correct answer.

Concept Check *To prepare for the exercises that follow, find the greatest common factor for each list of terms.*

3. $9m^3,\ 3m^2,\ 15m$ **4.** $4a^2,\ 6a,\ 2a^3$

5. $16xy^3,\ 24x^2y^2,\ 8x^2y$ **6.** $10m^2n^2,\ 25mn^3,\ 50m^2n$

7. $6m(r + t)^2,\ 3p(r + t)^4$ **8.** $7z^2(m + n)^4,\ 9z^3(m + n)^5$

Factor out the greatest common factor. ***See Examples 1–4.***

9. $12m - 60$ **10.** $15r - 45$ **11.** $4 + 20z$

12. $9 + 27x$ **13.** $8y - 15$ **14.** $7x - 40$

15. $8k^3 + 24k$ **16.** $9z^4 + 81z$ **17.** $-4p^3q^4 - 2p^2q^5$

18. $-3z^5w^2 - 18z^3w^4$ **19.** $7x^3 + 35x^4 - 14x^5$ **20.** $6k^3 - 36k^4 - 48k^5$

21. $10t^5 - 2t^3 - 4t^4$ **22.** $6p^3 - 3p^2 - 9p^4$ **23.** $15a^2c^3 - 25ac^2 + 5ac$

24. $15y^3z^3 - 27y^2z^4 + 3yz^3$ **25.** $16z^2n^6 + 64zn^7 - 32z^3n^3$

26. $5r^3s^5 + 10r^2s^2 - 15r^4s^2$ **27.** $14a^3b^2 + 7a^2b - 21a^5b^3 + 42ab^4$

28. $12km^3 - 24k^3m^2 + 36k^2m^4 - 60k^4m^3$ **29.** $(m - 4)(m + 2) + (m - 4)(m + 3)$

30. $(z - 5)(z + 7) + (z - 5)(z + 10)$ **31.** $(2z - 1)(z + 6) - (2z - 1)(z - 5)$

32. $(3x + 2)(x - 4) - (3x + 2)(x + 8)$ **33.** $5(2 - x)^2 - 2(2 - x)^3$

34. $2(5 - x)^3 - 3(5 - x)^2$ **35.** $4(3 - x)^2 - (3 - x)^3 + 3(3 - x)$

36. $2(t - s) + 4(t - s)^2 - (t - s)^3$

37. $15(2z + 1)^3 + 10(2z + 1)^2 - 25(2z + 1)$

38. $6(a + 2b)^2 - 4(a + 2b)^3 + 12(a + 2b)^4$

39. $5(m + p)^3 - 10(m + p)^2 - 15(m + p)^4$

40. $-9a^2(p + q) - 3a^3(p + q)^2 + 6a(p + q)^3$

Factor each polynomial twice. First use a common factor with a positive coefficient, and then use a common factor with a negative coefficient. ***See Example 4.***

41. $-r^3 + 3r^2 + 5r$ **42.** $-t^4 + 8t^3 - 12t$ **43.** $-12s^5 + 48s^4$

44. $-16y^4 + 64y^3$ **45.** $-2x^5 + 6x^3 + 4x^2$ **46.** $-5a^3 + 10a^4 - 15a^5$

Factor by grouping. ***See Examples 5–9.***

47. $mx + qx + my + qy$ **48.** $2k + 2h + jk + jh$

49. $10m + 2n + 5mk + nk$

50. $3ma + 3mb + 2ab + 2b^2$

51. $4 - 2q - 6p + 3pq$

52. $20 + 5m + 12n + 3mn$

53. $p^2 - 4zq + pq - 4pz$

54. $r^2 - 9tw + 3rw - 3rt$

55. $2xy + 3y + 2x + 3$

56. $7ab + 35bc + a + 5c$

57. $m^3 + 4m^2 - 6m - 24$

58. $2a^3 + a^2 - 14a - 7$

59. $-3a^3 - 3ab^2 + 2a^2b + 2b^3$

60. $-16m^3 + 4m^2p^2 - 4mp + p^3$

61. $4 + xy - 2y - 2x$

62. $10ab - 21 - 6b + 35a$

63. $8 + 9y^4 - 6y^3 - 12y$

64. $x^3y^2 - 3 - 3y^2 + x^3$

65. $1 - a + ab - b$

66. $2ab^2 - 8b^2 + a - 4$

67. $2mx - 6qx + 2my - 6qy$

68. $12 - 6q - 18p + 9pq$

69. $4a^3 - 4a^2b^2 + 8ab - 8b^3$

70. $5x^3 + 15x^2y^2 - 5xy - 15y^3$

71. $2x^3y^2 + x^2y^2 - 14xy^2 - 7y^2$

72. $3m^2n^3 + 15m^2n - 2m^2n^2 - 10m^2$

Extending Skills *Factor out the variable that is raised to the lesser exponent. (For example, in* **Exercise 73,** *factor out* m^{-5}.*)*

73. $3m^{-5} + m^{-3}$ **74.** $k^{-2} + 2k^{-4}$ **75.** $3p^{-3} + 2p^{-2}$ **76.** $-5q^{-3} + 8q^{-2}$

45. $2x^2(-x^3 + 3x + 2)$; $-2x^2(x^3 - 3x - 2)$
46. $5a^3(-1 + 2a - 3a^2)$; $-5a^3(1 - 2a + 3a^2)$
47. $(m + q)(x + y)$
48. $(k + h)(2 + j)$
49. $(5m + n)(2 + k)$
50. $(a + b)(3m + 2b)$
51. $(2 - q)(2 - 3p)$
52. $(4 + m)(5 + 3n)$
53. $(p + q)(p - 4z)$
54. $(r + 3w)(r - 3t)$
55. $(2x + 3)(y + 1)$
56. $(a + 5c)(7b + 1)$
57. $(m + 4)(m^2 - 6)$
58. $(2a + 1)(a^2 - 7)$
59. $(a^2 + b^2)(-3a + 2b)$
60. $(4m^2 + p)(-4m + p^2)$
61. $(y - 2)(x - 2)$
62. $(5a - 3)(2b + 7)$
63. $(3y - 2)(3y^3 - 4)$
64. $(x^3 - 3)(y^2 + 1)$
65. $(1 - a)(1 - b)$
66. $(a - 4)(2b^2 + 1)$
67. $2(m - 3q)(x + y)$
68. $3(2 - q)(2 - 3p)$
69. $4(a^2 + 2b)(a - b^2)$
70. $5(x^2 - y)(x + 3y^2)$
71. $y^2(2x + 1)(x^2 - 7)$
72. $m^2(3n - 2)(n^2 + 5)$
73. $m^{-5}(3 + m^2)$
74. $k^{-4}(k^2 + 2)$
75. $p^{-3}(3 + 2p)$
76. $q^{-3}(-5 + 8q)$

5.2 Factoring Trinomials

OBJECTIVES

1. Factor trinomials when the coefficient of the second-degree term is 1.
2. Factor trinomials when the coefficient of the second-degree term is not 1.
3. Use an alternative method for factoring trinomials.
4. Factor by substitution.

VOCABULARY

☐ prime polynomial

TEACHING TIP
Emphasize the importance of this section as it relates to subsequent sections and the next chapter. A great deal of practice is required to become proficient in factoring.

OBJECTIVE 1 Factor trinomials when the coefficient of the second-degree term is 1.

We begin by finding the product of $x + 3$ and $x - 5$.

$$(x + 3)(x - 5)$$
$$= x^2 - 5x + 3x - 15 \quad \text{FOIL method}$$
$$= x^2 - 2x - 15 \quad \text{Combine like terms.}$$

By this result, the factored form of $x^2 - 2x - 15$ is $(x + 3)(x - 5)$.

Multiplication →

Factored form → $(x + 3)(x - 5) = x^2 - 2x - 15$ ← Product

← Factoring

Because multiplying and factoring are operations that "undo" each other, factoring trinomials involves using the FOIL method in reverse. As shown here, the x^2-term comes from multiplying x and x, and -15 comes from multiplying 3 and -5.

Product of x and x is x^2.

$$(x + 3)(x - 5) = x^2 - 2x - 15$$

Product of 3 and -5 is -15.

We find the $-2x$ in $x^2 - 2x - 15$ by multiplying the outer terms, multiplying the inner terms, and adding the results.

Outer terms: $x(-5) = -5x$

$$(x + 3)(x - 5)$$

Inner terms: $3 \cdot x = 3x$

Add to obtain $-2x$.

Factoring $x^2 + bx + c$

Step 1 **Find pairs whose product is *c*.** Find all pairs of integers whose product is c, the third term of the trinomial.

Step 2 **Find pairs whose sum is *b*.** Choose the pair whose sum is b, the coefficient of the middle term.

If there are no such integers, the polynomial cannot be factored. A polynomial that cannot be factored with integer coefficients is a **prime polynomial.**

Examples: $x^2 + x + 2,\quad x^2 - x - 1,\quad 2x^2 + x + 7$ Prime polynomials

NOW TRY EXERCISE 1

Factor each trinomial.

(a) $t^2 - t - 30$

(b) $w^2 + 12w + 32$

CLASSROOM EXAMPLE 1

Factor each trinomial.

(a) $a^2 + 9a + 20$

(b) $b^2 - 7b + 10$

Answers:

(a) $(a + 4)(a + 5)$

(b) $(b - 5)(b - 2)$

EXAMPLE 1 Factoring Trinomials in $x^2 + bx + c$ Form

Factor each trinomial.

(a) $x^2 + 2x - 35$

Step 1 Find pairs of integers whose product is -35.	***Step 2*** Write sums of those pairs of integers.
$35(-1)$	$35 + (-1) = 34$
$-35(1)$	$-35 + 1 = -34$
$7(-5)$	$7 + (-5) = 2$ ← Coefficient of the middle term
$-7(5)$	$-7 + 5 = -2$

The integers 7 and -5 have the necessary product and sum.

$$x^2 + 2x - 35 \quad \text{factors as} \quad (x + 7)(x - 5).$$

Multiply to check.

(b) $r^2 + 8r + 12$

Look for two integers with a product of 12 and a sum of 8. Of all pairs having a product of 12, only the pair 6 and 2 has a sum of 8.

$$r^2 + 8r + 12 \quad \text{factors as} \quad (r + 6)(r + 2).$$

By the commutative property, it is equally correct to write $(r + 2)(r + 6)$. ***Check by using the FOIL method to multiply the factored form.***

NOW TRY

CLASSROOM EXAMPLE 2

Factor $t^2 + 3t - 5$.

Answer: prime

NOW TRY EXERCISE 2

Factor $m^2 + 12m - 11$.

EXAMPLE 2 Recognizing a Prime Polynomial

Factor $m^2 + 6m + 7$.

Look for two integers whose product is 7 and whose sum is 6. Only two pairs of integers, 7 and 1 and -7 and -1, give a product of 7. Neither of these pairs has a sum of 6, so $m^2 + 6m + 7$ cannot be factored with integer coefficients and is prime.

NOW TRY

NOW TRY ANSWERS

1. **(a)** $(t - 6)(t + 5)$
   (b) $(w + 4)(w + 8)$
2. prime

NOW TRY EXERCISE 3
Factor $a^2 + ab - 20b^2$.

CLASSROOM EXAMPLE 3
Factor $p^2 - 5pq + 6q^2$.
Answer: $(p - 3q)(p - 2q)$

EXAMPLE 3 Factoring a Multivariable Trinomial

Factor $x^2 + 6ax - 16a^2$.

This trinomial is in the form $x^2 + bx + c$, where $b = 6a$ and $c = -16a^2$.

Step 1 Find pairs of expressions whose product is $-16a^2$.	***Step 2*** Write sums of those pairs of expressions.
$16a(-a)$	$16a + (-a) = 15a$
$-16a(a)$	$-16a + a = -15a$
$8a(-2a)$	$8a + (-2a) = 6a$ ← Coefficient of the middle term
$-8a(2a)$	$-8a + 2a = -6a$
$-4a(4a)$	$-4a + 4a = 0$

The expressions $8a$ and $-2a$ have the necessary product and sum.

$$x^2 + 6ax - 16a^2 \quad \text{factors as} \quad (x + 8a)(x - 2a).$$

CHECK $(x + 8a)(x - 2a)$

$= x^2 - 2ax + 8ax - 16a^2$ FOIL method

$= x^2 + 6ax - 16a^2$ ✓ Original polynomial

NOW TRY

NOW TRY EXERCISE 4
Factor $5w^3 - 40w^2 + 60w$.

CLASSROOM EXAMPLE 4
Factor $3a^3 + 12a^2 - 15a$.
Answer: $3a(a - 1)(a + 5)$

EXAMPLE 4 Factoring a Trinomial with a Common Factor

Factor $16y^3 - 32y^2 - 48y$.

$16y^3 - 32y^2 - 48y$

$= 16y(y^2 - 2y - 3)$ Factor out the GCF, $16y$.

To factor $y^2 - 2y - 3$, look for two integers whose product is -3 and whose sum is -2. The necessary integers are -3 and 1.

$= 16y(y - 3)(y + 1)$ Factor the trinomial.

Remember to include the GCF, $16y$.

NOW TRY

TEACHING TIP
Students may be helped by the following visual approach to **Objective 2.** To factor $3x^2 + 7x + 2$, form a lattice, using $7x = 1x + 6x$.

$3x^2$	$1x$
$6x$	2

Now factor out the common factor in each column and row.

	1st factor: $3x$	$+1$
2nd factor: x	$3x^2$	$1x$
$+2$	$6x$	2

The factored form is

$(3x + 1)(x + 2)$.

! CAUTION ***When factoring, always look for a common factor first. Remember to write the common factor as part of the answer.***

OBJECTIVE 2 Factor trinomials when the coefficient of the second-degree term is not 1.

We can use a generalization of the method shown in **Objective 1** to factor a trinomial of the form $ax^2 + bx + c$, where $a \neq 1$.

To factor $3x^2 + 7x + 2$, for example, we first identify the values a, b, and c.

$$\begin{array}{c} ax^2 + bx + c \\ \downarrow \quad \downarrow \quad \downarrow \\ 3x^2 + 7x + 2, \end{array} \quad \text{so} \quad a = 3, \quad b = 7, \quad c = 2$$

The product ac is $3 \cdot 2 = 6$, so we must find integers having a product of 6 and a sum of 7 (because the middle term has coefficient 7). The necessary integers are 1 and 6, so we write $7x$ as $1x + 6x$, or $x + 6x$.

NOW TRY ANSWERS
3. $(a + 5b)(a - 4b)$
4. $5w(w - 6)(w - 2)$

$$\begin{aligned}
&3x^2 + 7x + 2\\
&= 3x^2 + \underbrace{x + 6x}_{x + 6x = 7x} + 2\\
&= (3x^2 + x) + (6x + 2) && \text{Group the terms.}\\
&= x(3x + 1) + 2(3x + 1) && \text{Factor by grouping.}\\
&= (3x + 1)(x + 2) && \text{Factor out the common factor.}
\end{aligned}$$

Check by multiplying.

NOW TRY EXERCISE 5

Factor $8y^2 - 10y - 3$.

CLASSROOM EXAMPLE 5

Factor $6k^2 - 19k + 10$.

Answer: $(2k - 5)(3k - 2)$

EXAMPLE 5 Factoring a Trinomial in $ax^2 + bx + c$ Form

Factor $12r^2 - 5r - 2$.

Here $a = 12$, $b = -5$, and $c = -2$, so the product ac is -24. The two integers whose product is -24 and whose sum b is -5, are 3 and -8.

$$\begin{aligned}
&12r^2 - 5r - 2\\
&= 12r^2 + 3r - 8r - 2 && \text{Write } -5r \text{ as } 3r - 8r.\\
&= 3r(4r + 1) - 2(4r + 1) && \text{Factor by grouping.}\\
&= (4r + 1)(3r - 2) && \text{Factor out the common factor.}
\end{aligned}$$

Check by multiplying.

NOW TRY

OBJECTIVE 3 Use an alternative method for factoring trinomials.

This method involves trying repeated combinations and using the FOIL method.

CLASSROOM EXAMPLE 6

Factor each trinomial.

(a) $10x^2 + 17x + 3$

(b) $6r^2 + 13r - 5$

Answers:

(a) $(5x + 1)(2x + 3)$

(b) $(2r + 5)(3r - 1)$

EXAMPLE 6 Factoring Trinomials in $ax^2 + bx + c$ Form

Factor each trinomial.

(a) $3x^2 + 7x + 2$

The goal is to find the correct numbers to put in the blanks.

$$(\_\_\_\_ x + \_\_\_\_)(\_\_\_\_ x + \_\_\_\_)$$

Addition signs are used because all the signs in the polynomial indicate addition. The first two expressions have a product of $3x^2$, so they must be $3x$ and x.

$$(3x + \_\_\_\_)(x + \_\_\_\_)$$

The product of the two last terms must be 2, so the numbers must be 2 and 1. There is a choice. The 2 could be placed with the $3x$ or with the x. Only one of these choices will give the correct middle term, $7x$.

$3x$	$6x$
$(3x + 2)(x + 1)$	$(3x + 1)(x + 2)$
$2x$	x
$3x + 2x = 5x$	$6x + x = 7x$
Wrong middle term	Correct middle term

Therefore, $3x^2 + 7x + 2$ factors as $(3x + 1)(x + 2)$. (Compare to the solution obtained by factoring by grouping at the top of this page.)

NOW TRY ANSWER

5. $(2y - 3)(4y + 1)$

NOW TRY EXERCISE 6

Factor $10r^2 + 19r + 6$.

(b) $12r^2 - 5r - 2$

To reduce the number of trials, we note that the terms of the trinomial have greatest common factor 1. This means that neither of its factors can have a common factor except 1. We should keep this in mind as we choose factors. We try 4 and 3 for the two first terms.

$$(4r \_\_\_\_)(3r \_\_\_\_)$$

The factors of -2 are -2 and 1 or -1 and 2. We try both possibilities.

$(4r - 2)(3r + 1)$	$(4r - 1)(3r + 2)$ (outer: $8r$; inner: $-3r$)
Wrong: $4r - 2$ has a factor of 2, which cannot be correct, because 2 is not a factor of $12r^2 - 5r - 2$.	$8r - 3r = 5r$ Wrong middle term

The middle term on the right is $5r$, instead of the $-5r$ that is needed. We obtain $-5r$ by interchanging the signs of the second terms in the factors.

$$(4r + 1)(3r - 2)$$

(outer: $-8r$; inner: $3r$)

$-8r + 3r = -5r$

Correct middle term

Thus, $12r^2 - 5r - 2$ factors as $(4r + 1)(3r - 2)$. (Compare to **Example 5.**)

CHECK $(4r + 1)(3r - 2)$

$= 12r^2 - 8r + 3r - 2$ FOIL method

$= 12r^2 - 5r - 2$ ✓ Original polynomial

NOW TRY

TEACHING TIP

Emphasize the concept discussed in the Note with students.

NOTE As shown in **Example 6(b),** if the terms of a polynomial have greatest common factor 1, then none of the terms of its factors can have a common factor (except 1). Remembering this will eliminate some potential factors.

Factoring $ax^2 + bx + c$ (GCF of a, b, c is 1)

Step 1 **Find pairs whose product is a.** Write all pairs of integer factors of a, the coefficient of the second-degree term.

Step 2 **Find pairs whose product is c.** Write all pairs of integer factors of c, the last term.

Step 3 **Choose inner and outer terms.** Use the FOIL method and various combinations of the factors from Steps 1 and 2 until the necessary middle term is obtained.

If no such combinations exist, the trinomial is prime.

NOW TRY ANSWER

6. $(2r + 3)(5r + 2)$

NOW TRY EXERCISE 7
Factor $12a^2 - 19ab - 21b^2$.

CLASSROOM EXAMPLE 7
Factor $6m^2 + 7mn - 5n^2$.

Answer: $(3m + 5n)(2m - n)$

EXAMPLE 7 Factoring a Multivariable Trinomial

Factor $18m^2 - 19mx - 12x^2$.

The terms of the polynomial have greatest common factor 1. Follow the steps in the preceding box to factor the trinomial. There are many possible factors of both 18 and -12. Try 6 and 3 for 18 and -3 and 4 for -12.

$(6m - 3x)(3m + 4x)$	$(6m + 4x)(3m - 3x)$
Wrong: common factor	Wrong: common factors

Because 6 and 3 do not work as factors of 18, try 9 and 2 instead, with 3 and -4 as factors of -12.

$(9m + 3x)(2m - 4x)$	$(9m - 4x)(2m + 3x)$ (outer: $27mx$; inner: $-8mx$)
Wrong: common factors	$27mx + (-8mx) = 19mx$ Wrong middle term

The result on the right differs from the correct middle term only in sign, so interchange the signs of the second terms in the factors.

$$18m^2 - 19mx - 12x^2 \quad \text{factors as} \quad (9m + 4x)(2m - 3x)$$

Check by multiplying.

NOW TRY

NOW TRY EXERCISE 8
Factor $-8x^2 + 22x - 15$.

CLASSROOM EXAMPLE 8
Factor $-2p^2 - 5p + 12$.

Answer: $-(2p - 3)(p + 4)$

EXAMPLE 8 Factoring a Trinomial in $ax^2 + bx + c$ Form ($a < 0$)

Factor $-3x^2 + 16x + 12$.

While we could factor directly, it is helpful to first factor out -1 so that the coefficient of the x^2-term is positive.

$$-3x^2 + 16x + 12$$
$$= -1(3x^2 - 16x - 12) \quad \text{Factor out } -1.$$
$$= -1(3x + 2)(x - 6) \quad \text{Factor the trinomial.}$$
$$= -(3x + 2)(x - 6) \quad -1a = -a$$

NOW TRY

NOTE The factored form in **Example 8** can be written in other ways, such as

$$(-3x - 2)(x - 6) \quad \text{and} \quad (3x + 2)(-x + 6).$$

Verify that these both give the original trinomial when multiplied.

CLASSROOM EXAMPLE 9
Factor $4m^3 + 2m^2 - 6m$.

Answer: $2m(2m + 3)(m - 1)$

NOW TRY EXERCISE 9
Factor $12y^3 + 33y^2 - 9y$.

EXAMPLE 9 Factoring a Trinomial with a Common Factor

Factor $16y^3 + 24y^2 - 16y$.

$$16y^3 + 24y^2 - 16y$$
$$= 8y(2y^2 + 3y - 2) \quad \text{GCF} = 8y$$
$$= 8y(2y - 1)(y + 2) \quad \text{Factor the trinomial.}$$

Remember the common factor.

NOW TRY

NOW TRY ANSWERS

7. $(4a + 3b)(3a - 7b)$
8. $-(4x - 5)(2x - 3)$
9. $3y(4y - 1)(y + 3)$

OBJECTIVE 4 Factor by substitution.

NOW TRY EXERCISE 10
Factor.

$3(a+2)^2 - 11(a+2) - 4$

EXAMPLE 10 Factoring a Polynomial by Substitution

Factor $2(x+3)^2 + 5(x+3) - 12$.

The binomial $x+3$ appears to powers 2 and 1, so we let a substitution variable represent $x+3$. We may choose any letter we wish except x. We choose t.

$$2(x+3)^2 + 5(x+3) - 12$$

$= 2t^2 + 5t - 12$ Let $t = x + 3$.

$= (2t-3)(t+4)$ Factor. (Don't stop here.)

$= [2(x+3) - 3][(x+3) + 4]$ Replace t with $x+3$. (Be sure to make the final substitution.)

$= (2x + 6 - 3)(x + 7)$ Simplify.

$= (2x+3)(x+7)$ NOW TRY

CLASSROOM EXAMPLE 10
Factor.

$8(z+5)^2 - 2(z+5) - 3$

Answer: $(4z+17)(2z+11)$

NOW TRY EXERCISE 11
Factor $6x^4 + 11x^2 + 3$.

EXAMPLE 11 Factoring a Trinomial in $ax^4 + bx^2 + c$ Form

Factor $6y^4 + 7y^2 - 20$.

The variable y appears to powers in which the greater exponent is twice the lesser exponent. We can let a substitution variable represent the variable to the lesser power.

$$6y^4 + 7y^2 - 20$$

$= 6(y^2)^2 + 7y^2 - 20$ $y^4 = (y^2)^2$

$= 6t^2 + 7t - 20$ Let $t = y^2$.

$= (3t-4)(2t+5)$ Factor. (Don't stop here. Replace t with y^2.)

$= (3y^2-4)(2y^2+5)$ Replace t with y^2. NOW TRY

CLASSROOM EXAMPLE 11
Factor $6r^4 - 13r^2 + 5$.

Answer: $(3r^2 - 5)(2r^2 - 1)$

TEACHING TIP
Point out that this is a good time to learn to use the substitution method as it will appear later in equation-solving techniques.

NOTE Some students feel comfortable factoring polynomials like the one in **Example 11** directly, without using the substitution method. We try 3 and 2 as factors of 6 and -4 and 5 as factors of -20.

$$6y^4 + 7y^2 - 20$$

$= (3y^2 - 4)(2y^2 + 5)$ $3y^2 \cdot 2y^2 = 6y^4$; $-4(5) = 20$

Outer: $15y^2$; Inner: $-8y^2$

$15y^2 - 8y^2 = 7y^2$
Correct middle term

NOW TRY ANSWERS
10. $(3a+7)(a-2)$
11. $(3x^2+1)(2x^2+3)$

5.2 Exercises

FOR EXTRA HELP

MyMathLab®

Complete solution available in MyMathLab

Concept Check *Choose the correct response.*

1. Which is *not* a valid way of starting the process of factoring the trinomial $12x^2 + 29x + 10$?

A. $(12x \quad)(x \quad)$ B. $(4x \quad)(3x \quad)$
C. $(6x \quad)(2x \quad)$ D. $(8x \quad)(4x \quad)$

1. D

2. A
3. C 4. B
5. The factor $(4x + 10)$ can be factored further into $2(2x + 5)$, giving the *completely* factored form as $2(2x + 5)(x - 2)$.
6. They are both correct. In each case, the product is $-4x^2 - 29x + 24$.
7. $x + 3$ 8. $y + 9$
9. $m - 7$ 10. $n - 8$
11. $r - 5$ 12. $x + 8$
13. $x - 2a$ 14. $m - 5n$
15. $2x - 3$ 16. $2z - 1$
17. $2u + v$ 18. $4p + q$
19. $(y - 3)(y + 10)$
20. $(z + 6)(z - 4)$
21. $(p + 8)(p + 7)$
22. $(k - 5)(k - 6)$
23. prime 24. prime
25. $(a + 5b)(a - 7b)$
26. $(z + 3w)(z + 5w)$
27. $(a - 6b)(a - 3b)$
28. $(k - 4h)(k - 7h)$
29. prime 30. prime
31. $-(6m - 5)(m + 3)$
32. $-(3y + 2)(5y - 9)$
33. $(5x - 6)(2x + 3)$
34. $(4k + 7)(2k + 5)$
35. $(4k + 3)(5k + 8)$
36. $(9z - 1)(3z + 5)$
37. $(3a - 2b)(5a - 4b)$
38. $(7c + 2d)(2c - 3d)$
39. $(6m - 5)^2$ 40. $(5r - 9)^2$
41. prime 42. prime
43. $(2xz - 1)(3xz + 4)$
44. $(4mn - 3)(2mn - 1)$
45. $3(4x + 5)(2x + 1)$
46. $2(6x - 1)(3x + 2)$
47. $-5(a + 6)(3a - 4)$
48. $-2(3a + 7)(2a - 3)$
49. $-11x(x - 6)(x - 4)$
50. $-9k(k + 7)(k - 3)$
51. $2xy^3(x - 12y)^2$
52. $6mn^2(m - 5n)^2$
53. $6a(a - 3)(a + 5)$
54. $3m^2(m + 6)(m - 4)$
55. $13y(y + 4)(y - 1)$
56. $4p(p + 8)(p - 2)$

2. Which is the completely factored form of the trinomial $2x^6 - 5x^5 - 3x^4$?

A. $x^4(2x + 1)(x - 3)$ **B.** $x^4(2x - 1)(x + 3)$
C. $(2x^5 + x^4)(x - 3)$ **D.** $x^3(2x^2 + x)(x - 3)$

3. Which is the completely factored form of the trinomial $4x^2 - 4x - 24$?

A. $4(x - 2)(x + 3)$ **B.** $4(x + 2)(x + 3)$
C. $4(x + 2)(x - 3)$ **D.** $4(x - 2)(x - 3)$

4. Which is *not* a factored form of the trinomial $-x^2 + 16x - 60$?

A. $(x - 10)(-x + 6)$ **B.** $(-x - 10)(x + 6)$
C. $(-x + 10)(x - 6)$ **D.** $-1(x - 10)(x - 6)$

5. ***Concept Check*** When a student was given the polynomial $4x^2 + 2x - 20$ to factor completely on a test, the student lost some credit when her answer was

$$(4x + 10)(x - 2).$$

WHAT WENT WRONG? Give the correct answer.

6. ***Concept Check*** When factoring the polynomial

$$-4x^2 - 29x + 24,$$

Terry obtained $(-4x + 3)(x + 8)$. Johnny wrote $(4x - 3)(-x - 8)$. Who is correct?

Complete each factoring. ***See Examples 1 and 3–8.***

7. $x^2 + 8x + 15 = (x + 5)(\_\_\_\_)$

8. $y^2 + 11y + 18 = (y + 2)(\_\_\_\_)$

9. $m^2 - 10m + 21 = (m - 3)(\_\_\_\_)$

10. $n^2 - 14n + 48 = (n - 6)(\_\_\_\_)$

11. $r^2 - r - 20 = (r + 4)(\_\_\_\_)$

12. $x^2 + 4x - 32 = (x - 4)(\_\_\_\_)$

13. $x^2 + ax - 6a^2 = (x + 3a)(\_\_\_\_)$

14. $m^2 - 3mn - 10n^2 = (m + 2n)(\_\_\_\_)$

15. $4x^2 - 4x - 3 = (2x + 1)(\_\_\_\_)$

16. $6z^2 - 11z + 4 = (3z - 4)(\_\_\_\_)$

17. $12u^2 + 10uv + 2v^2 = 2(3u + v)(\_\_\_\_)$

18. $16p^2 - 4pq - 2q^2 = 2(2p - q)(\_\_\_\_)$

Factor each trinomial. ***See Examples 1–9.***

19. $y^2 + 7y - 30$ **20.** $z^2 + 2z - 24$ **21.** $p^2 + 15p + 56$

22. $k^2 - 11k + 30$ **23.** $m^2 - 11m + 60$ **24.** $p^2 - 12p - 27$

25. $a^2 - 2ab - 35b^2$ **26.** $z^2 + 8zw + 15w^2$ **27.** $a^2 - 9ab + 18b^2$

28. $k^2 - 11hk + 28h^2$ **29.** $x^2y^2 + 12xy + 18$ **30.** $p^2q^2 - 5pq - 18$

31. $-6m^2 - 13m + 15$ **32.** $-15y^2 + 17y + 18$ **33.** $10x^2 + 3x - 18$

34. $8k^2 + 34k + 35$ **35.** $20k^2 + 47k + 24$ **36.** $27z^2 + 42z - 5$

37. $15a^2 - 22ab + 8b^2$ **38.** $14c^2 - 17cd - 6d^2$ **39.** $36m^2 - 60m + 25$

40. $25r^2 - 90r + 81$ **41.** $40x^2 + xy + 6y^2$ **42.** $15p^2 + 24pq + 8q^2$

43. $6x^2z^2 + 5xz - 4$ **44.** $8m^2n^2 - 10mn + 3$ **45.** $24x^2 + 42x + 15$

46. $36x^2 + 18x - 4$ **47.** $-15a^2 - 70a + 120$ **48.** $-12a^2 - 10a + 42$

49. $-11x^3 + 110x^2 - 264x$ **50.** $-9k^3 - 36k^2 + 189k$

51. $2x^3y^3 - 48x^2y^4 + 288xy^5$ **52.** $6m^3n^2 - 60m^2n^3 + 150mn^4$

53. $6a^3 + 12a^2 - 90a$ **54.** $3m^4 + 6m^3 - 72m^2$

55. $13y^3 + 39y^2 - 52y$ **56.** $4p^3 + 24p^2 - 64p$

57. $3p(2p-1)^2$
58. $5t(3t+2)^2$

59. $(6p^3-r)(2p^3-5r)$
60. $(2y^3+3x)(y^3+2x)$
61. $(5k+4)(2k+1)$
62. $(2m-15)(2m-7)$
63. $(3m+3p+5)(m+p-4)$
64. $(4x-4y+1)(x-y-6)$

65. $(a+b)^2(a-3b)(a+2b)$
66. $(m-p)^2(m+3p)(m-4p)$
67. $(p+q)^2(p+3q)$
68. $(5-y)(2k-5)(k-1)$
69. $(z-x)^2(z+2x)$
70. $(r-s)^2(r+6s)$

71. $(p^2-8)(p^2-2)$
72. $(k^2+1)(k^2+9)$
73. $(2x^2+3)(x^2-6)$
74. $(3z^2-1)(2z^2+1)$
75. $(4x^2+3)(4x^2+1)$
76. $(3r^2+1)(3r^2+2)$

57. $12p^3-12p^2+3p$ **58.** $45t^3+60t^2+20t$

Factor each trinomial. ***See Example 10.***

59. $12p^6-32p^3r+5r^2$ **60.** $2y^6+7xy^3+6x^2$

61. $10(k+1)^2-7(k+1)+1$ **62.** $4(m-5)^2-4(m-5)-15$

63. $3(m+p)^2-7(m+p)-20$ **64.** $4(x-y)^2-23(x-y)-6$

Extending Skills *Factor each trinomial. (Hint: Factor out the GCF first.)*

65. $a^2(a+b)^2-ab(a+b)^2-6b^2(a+b)^2$

66. $m^2(m-p)^2-mp(m-p)^2-12p^2(m-p)^2$

67. $p^2(p+q)+4pq(p+q)+3q^2(p+q)$

68. $2k^2(5-y)-7k(5-y)+5(5-y)$

69. $z^2(z-x)-zx(x-z)-2x^2(z-x)$

70. $r^2(r-s)-5rs(s-r)-6s^2(r-s)$

Factor each trinomial. ***See Example 11.***

71. p^4-10p^2+16 **72.** k^4+10k^2+9 **73.** $2x^4-9x^2-18$

74. $6z^4+z^2-1$ **75.** $16x^4+16x^2+3$ **76.** $9r^4+9r^2+2$

5.3 Special Factoring

OBJECTIVES

1. Factor a difference of squares.
2. Factor a perfect square trinomial.
3. Factor a difference of cubes.
4. Factor a sum of cubes.

VOCABULARY

- ☐ difference of squares
- ☐ perfect square trinomial
- ☐ difference of cubes
- ☐ sum of cubes

TEACHING TIP
Mention that perfect square variable terms will have exponents that are even (divisible by 2).
Work **Exercise 1** in the exercise set as a class to give students practice in identifying differences of squares.

OBJECTIVE 1 Factor a difference of squares.

The special products introduced in **Section 4.4** are used in reverse when factoring. Recall that the product of the sum and difference of two terms leads to a **difference of squares.**

Factoring a Difference of Squares

$$x^2-y^2=(x+y)(x-y)$$

EXAMPLE 1 Factoring Differences of Squares

Factor each polynomial.

(a) t^2-36

$=t^2-6^2$ $36=6^2$

$=(t+6)(t-6)$ Factor the difference of squares.

(b) $4a^2-64$ ← Always check for a common factor first.

$=4(a^2-16)$ Factor out the common factor, 4.

$=4(a+4)(a-4)$ Factor the difference of squares.

$$\begin{array}{ccccc} x^2 & - & y^2 & = & (x+y)(x-y) \\ \downarrow & & \downarrow & & \downarrow\ \ \downarrow\ \ \downarrow\ \ \downarrow \end{array}$$

(c) $16m^2-49p^2=(4m)^2-(7p)^2=(4m+7p)(4m-7p)$

NOW TRY EXERCISE 1

Factor each polynomial.

(a) $4m^2 - 25n^2$

(b) $9x^2 - 729$

(c) $(a + b)^2 - 25$

(d) $v^4 - 1$

CLASSROOM EXAMPLE 1

Factor each polynomial.

(a) $p^2 - 100$

(b) $2x^2 - 18$

(c) $9a^2 - 16b^2$

(d) $(m + 3)^2 - 49z^2$

(e) $y^4 - 16$

Answers:

(a) $(p + 10)(p - 10)$

(b) $2(x + 3)(x - 3)$

(c) $(3a + 4b)(3a - 4b)$

(d) $(m + 3 + 7z)(m + 3 - 7z)$

(e) $(y^2 + 4)(y + 2)(y - 2)$

TEACHING TIP

Multiply out several examples of the form

$$(x + y)^2 \quad \text{and} \quad (x - y)^2$$

to establish the pattern. State that factoring is a reversal of this process. Some trinomials, such as $x^2 + 5x + 4$, may appear to be perfect square trinomials, but require the methods of **Section 5.2** instead.

Work **Exercise 3** in the exercise set as a class to give students practice in identifying perfect square trinomials.

$$x^2 - y^2 = (x + y)(x - y)$$

(d) $81k^2 - (a + 2)^2 = (9k)^2 - (a + 2)^2 = (9k + a + 2)(9k - (a + 2))$

$$= (9k + a + 2)(9k - a - 2)$$

We could have used the method of substitution here.

(e) $x^4 - 81$

$= (x^2 + 9)(x^2 - 9)$ Factor the difference of squares.

$= (x^2 + 9)(x + 3)(x - 3)$ Factor the difference of squares again.

NOW TRY

CAUTION *Assuming that the greatest common factor is 1, it is not possible to factor (with real numbers) a sum of squares* such as $x^2 + 9$ in **Example 1(e).** In particular, $x^2 + y^2 \neq (x + y)^2$, as shown next.

OBJECTIVE 2 Factor a perfect square trinomial.

Two other special products from **Section 4.4** lead to the following rules for factoring.

Factoring a Perfect Square Trinomial

$$x^2 + 2xy + y^2 = (x + y)^2$$
$$x^2 - 2xy + y^2 = (x - y)^2$$

Because the trinomial $x^2 + 2xy + y^2$ is the square of $x + y$, it is called a **perfect square trinomial.** In this pattern, both the first and the last terms of the trinomial must be perfect squares. In the factored form $(x + y)^2$, twice the product of the first and the last terms must give the middle term of the trinomial.

$4m^2 + 20m + 25$	$p^2 - 8p + 64$
Perfect square trinomial; $4m^2 = (2m)^2$, $25 = 5^2$, and $2(2m)(5) = 20m$.	Not a perfect square trinomial; middle term would have to be $16p$ or $-16p$.

EXAMPLE 2 Factoring Perfect Square Trinomials

Factor each polynomial.

(a) $144p^2 - 120p + 25$

Here, $144p^2 = (12p)^2$ and $25 = 5^2$. The sign on the middle term is $-$, so if $144p^2 - 120p + 25$ is a perfect square trinomial, the factored form will have to be

$$(12p - 5)^2.$$

Determine twice the product of the two terms to see if this is correct.

$$2(12p)(-5) = -120p \longleftarrow \text{Desired middle term}$$

This is the middle term of the given trinomial.

$$144p^2 - 120p + 25 \quad \text{factors as} \quad (12p - 5)^2.$$

NOW TRY ANSWERS

1. (a) $(2m + 5n)(2m - 5n)$

(b) $9(x + 9)(x - 9)$

(c) $(a + b + 5)(a + b - 5)$

(d) $(v^2 + 1)(v + 1)(v - 1)$

NOW TRY EXERCISE 2

Factor each polynomial.

(a) $a^2 + 12a + 36$

(b) $16x^2 - 56xy + 49y^2$

(c) $y^2 - 16y + 64 - z^2$

CLASSROOM EXAMPLE 2

Factor each polynomial.

(a) $49z^2 - 14z + 1$

(b) $9a^2 + 48ab + 64b^2$

(c) $x^2 - 2x + 1 - y^2$

Answers:

(a) $(7z - 1)^2$

(b) $(3a + 8b)^2$

(c) $(x - 1 + y)(x - 1 - y)$

(b) $4m^2 + 20mn + 49n^2$

If this is a perfect square trinomial, it will equal $(2m + 7n)^2$. By the pattern in the box, if multiplied out, this squared binomial has a middle term of

$$2(2m)(7n) = 28mn, \quad \textit{which does not equal} \quad 20mn.$$

Verify that this trinomial cannot be factored by the methods of the previous section either. It is prime.

(c) $(r + 5)^2 + 6(r + 5) + 9$

$= [(r + 5) + 3]^2$ — $2(r + 5)(3) = 6(r + 5)$ is the middle term.

$= (r + 8)^2$ — Add.

(d) $m^2 - 8m + 16 - p^2$

There are four terms, so we will use factoring by grouping. The first three terms here form a perfect square trinomial.

$m^2 - 8m + 16 - p^2$

$= (m^2 - 8m + 16) - p^2$ — Group the terms of the perfect square trinomial.

$= (m - 4)^2 - p^2$ — Factor the perfect square trinomial.

$= (m - 4 + p)(m - 4 - p)$ — Factor the difference of squares. NOW TRY

NOTE Perfect square trinomials can be factored by the general methods shown earlier for other trinomials. The patterns given here provide "shortcuts."

OBJECTIVE 3 Factor a difference of cubes.

A **difference of cubes,** $x^3 - y^3$, can be factored as follows.

Factoring a Difference of Cubes

$$x^3 - y^3 = (x - y)(x^2 + xy + y^2)$$

CLASSROOM EXAMPLE 3

Factor each polynomial.

(a) $x^3 - 1000$

(b) $8k^3 - y^3$

(c) $27m^3 - 64n^3$

Answers:

(a) $(x - 10)(x^2 + 10x + 100)$

(b) $(2k - y)(4k^2 + 2ky + y^2)$

(c) $(3m - 4n) \cdot (9m^2 + 12mn + 16n^2)$

Check by showing that the product of $x - y$ and $x^2 + xy + y^2$ is $x^3 - y^3$.

EXAMPLE 3 Factoring Differences of Cubes

Factor each polynomial.

$$x^3 - y^3 = (x - y)(x^2 + x \cdot y + y^2)$$

(a) $m^3 - 8 = m^3 - 2^3 = (m - 2)(m^2 + m \cdot 2 + 2^2)$

$= (m - 2)(m^2 + 2m + 4)$

CHECK $(m - 2)(m^2 + 2m + 4)$

$= m^3 + 2m^2 + 4m - 2m^2 - 4m - 8$ — Distributive property

$= m^3 - 8$ ✓ — Combine like terms.

NOW TRY ANSWERS

2. (a) $(a + 6)^2$

(b) $(4x - 7y)^2$

(c) $(y - 8 + z)(y - 8 - z)$

NOW TRY EXERCISE 3

Factor each polynomial.

(a) $t^3 - 1$

(b) $125a^3 - 8b^3$

TEACHING TIP

Mention that perfect cube variable terms will have exponents that are divisible by 3. Students should memorize the basic form for factoring the sum or difference of cubes. Each factorization consists of a binomial and a trinomial factor. Point out the similar patterns in the two factorizations.

Work **Exercise 2** in the exercise set as a class to give students practice in identifying sums or differences of cubes.

(b) $27x^3 - 8y^3$

$= (3x)^3 - (2y)^3$ Difference of cubes

$= (3x - 2y)[(3x)^2 + (3x)(2y) + (2y)^2]$ Factor.

$= (3x - 2y)(9x^2 + 6xy + 4y^2)$

$(3x)^2 = 3^2x^2$, *not* $3x^2$.
$(2y)^2 = 2^2y^2$, *not* $2y^2$.

(c) $1000k^3 - 27n^3$

$= (10k)^3 - (3n)^3$ Difference of cubes

$= (10k - 3n)[(10k)^2 + (10k)(3n) + (3n)^2]$ Factor.

$= (10k - 3n)(100k^2 + 30kn + 9n^2)$ Multiply. NOW TRY

OBJECTIVE 4 Factor a sum of cubes.

While the binomial $x^2 + y^2$ cannot be factored with real numbers, a **sum of cubes,** such as $x^3 + y^3$, is factored as follows.

Factoring a Sum of Cubes

$$x^3 + y^3 = (x + y)(x^2 - xy + y^2)$$

To verify this result, find the product of $x + y$ and $x^2 - xy + y^2$.

NOTE In a sum or difference of cubes, the following are true.

- In the binomial factor, the sign of the second term is *always the same* as the sign of the second term in the original polynomial.
- In the trinomial factor, the first and last terms are *always positive.*
- In the trinomial factor, the sign of the middle term is *the opposite of* the sign of the second term in the binomial factor.

CLASSROOM EXAMPLE 4

Factor each polynomial.

(a) $8p^3 + 125$

(b) $64m^3 + 125n^3$

(c) $2x^3 + 2000$

(d) $(a - 4)^3 + b^3$

Answers:

(a) $(2p + 5)(4p^2 - 10p + 25)$

(b) $(4m + 5n) \cdot (16m^2 - 20mn + 25n^2)$

(c) $2(x + 10)(x^2 - 10x + 100)$

(d) $(a - 4 + b) \cdot (a^2 - 8a + 16 - ab + 4b + b^2)$

EXAMPLE 4 Factoring Sums of Cubes

Factor each polynomial.

(a) $r^3 + 27$

$= r^3 + 3^3$ Sum of cubes

$= (r + 3)(r^2 - 3r + 3^2)$ Factor.

$= (r + 3)(r^2 - 3r + 9)$ This trinomial cannot be factored further.

(b) $27z^3 + 125$

$= (3z)^3 + 5^3$ Sum of cubes

$= (3z + 5)[(3z)^2 - (3z)(5) + 5^2]$ Factor.

$= (3z + 5)(9z^2 - 15z + 25)$ Multiply.

(c) $125t^3 + 216s^6$

$= (5t)^3 + (6s^2)^3$ Sum of cubes

$= (5t + 6s^2)[(5t)^2 - (5t)(6s^2) + (6s^2)^2]$ Factor.

$= (5t + 6s^2)(25t^2 - 30ts^2 + 36s^4)$ Multiply.

NOW TRY ANSWERS

3. **(a)** $(t - 1)(t^2 + t + 1)$

(b) $(5a - 2b)(25a^2 + 10ab + 4b^2)$

NOW TRY EXERCISE 4

Factor each polynomial.

(a) $1000 + z^3$

(b) $81a^6 + 3b^3$

(c) $(x - 3)^3 + y^3$

(d) $3x^2 + 192$

$= 3(x^3 + 64)$ Factor out the common factor.

$= 3(x^3 + 4^3)$ Write as a sum of cubes.

Remember the common factor.

$= 3(x + 4)(x^2 - 4x + 16)$ Factor.

(e) $(x + 2)^3 + t^3$

$= [(x + 2) + t][(x + 2)^2 - (x + 2)t + t^2]$ Sum of cubes

$= (x + 2 + t)(x^2 + 4x + 4 - xt - 2t + t^2)$ Multiply. NOW TRY

CAUTION A common error when factoring $x^3 + y^3$ or $x^3 - y^3$ is to think that the xy-term has a coefficient of 2. Because there is no coefficient of 2, expressions of the form $x^2 + xy + y^2$ and $x^2 - xy + y^2$ usually cannot be factored further.

NOW TRY ANSWERS

4. (a) $(10 + z) \cdot (100 - 10z + z^2)$
(b) $3(3a^2 + b) \cdot (9a^4 - 3a^2b + b^2)$
(c) $(x - 3 + y) \cdot (x^2 - 6x + 9 - xy + 3y + y^2)$

Summary of Special Types of Factoring	
Difference of Squares	$x^2 - y^2 = (x + y)(x - y)$
Perfect Square Trinomial	$x^2 + 2xy + y^2 = (x + y)^2$
	$x^2 - 2xy + y^2 = (x - y)^2$
Difference of Cubes	$x^3 - y^3 = (x - y)(x^2 + xy + y^2)$
Sum of Cubes	$x^3 + y^3 = (x + y)(x^2 - xy + y^2)$

5.3 Exercises

FOR EXTRA HELP MyMathLab®

Complete solution available in MyMathLab

Concept Check *Work each problem.*

1. Which of the following binomials are differences of squares?

A. $64 - k^2$ **B.** $2x^2 - 25$ **C.** $k^2 + 9$ **D.** $4z^4 - 49$

2. Which of the following binomials are sums or differences of cubes?

A. $64 + r^3$ **B.** $125 - p^6$ **C.** $9x^3 + 125$ **D.** $(x + y)^3 - 1$

3. Which of the following trinomials are perfect squares?

A. $x^2 - 8x - 16$ **B.** $4m^2 + 20m + 25$

C. $9z^4 + 30z^2 + 25$ **D.** $25p^2 - 45p + 81$

4. Of the twelve polynomials listed in **Exercises 1–3,** which ones can be factored by the methods of this section?

5. The binomial $4x^2 + 64$ is an example of a sum of two squares that can be factored. Under what conditions can a sum of two squares be factored?

6. Insert the correct signs in the blanks.

(a) $8 + m^3$
$= (2 \_\_ m)(4 \_\_ 2m \_\_ m^2)$

(b) $n^3 - 1$
$= (n \_\_ 1)(n^2 \_\_ n \_\_ 1)$

1. A, D
2. A, B, D
3. B, C
4. 1A, 1D, 2A, 2B, 2D, 3B, 3C
5. A sum of two squares can be factored if the binomial terms have a common factor greater than 1.
6. (a) +; −; + (b) −; +; +
7. $(p + 4)(p - 4)$
8. $(k + 3)(k - 3)$
9. $(5x + 2)(5x - 2)$
10. $(6m + 5)(6m - 5)$
11. $2(3a + 7b)(3a - 7b)$
12. $2(4c + 7d)(4c - 7d)$
13. $4(4m^2 + y^2)(2m + y)(2m - y)$
14. $3(9x^2 + t^2)(3x + t)(3x - t)$
15. $(y + z + 9)(y + z - 9)$

16. $(h+k+3)(h+k-3)$
17. $(4+x+3y)(4-x-3y)$
18. $(8+r+2t)(8-r-2t)$
19. $(p^2+16)(p+4)(p-4)$
20. $(a^2+25)(a+5)(a-5)$
21. $(k-3)^2$ 22. $(x+5)^2$
23. $(2z+w)^2$ 24. $(3y+z)^2$
25. $(4m-1+n)(4m-1-n)$
26. $(5c-2+d)(5c-2-d)$
27. $(2r-3+s)(2r-3-s)$
28. $(3a-4+b)(3a-4-b)$
29. $(x+y-1)(x-y+1)$
30. $(2+k-h)(2-k+h)$
31. $2(7m+3n)^2$
32. $5(4z-w)^2$
33. $(p+q+1)^2$
34. $(x+y+3)^2$
35. $(a-b+4)^2$
36. $(m-n+2)^2$
37. $(x-3)(x^2+3x+9)$
38. $(y-4)(y^2+4y+16)$
39. $(6-t)(36+6t+t^2)$
40. $(8-m)(64+8m+m^2)$
41. $(x+4)(x^2-4x+16)$
42. $(r+7)(r^2-7r+49)$
43. $(10+y)(100-10y+y^2)$
44. $(9+x)(81-9x+x^2)$
45. $(2x+1)(4x^2-2x+1)$
46. $(3y+1)(9y^2-3y+1)$
47. $(5x-6)(25x^2+30x+36)$
48. $(2w-5)(4w^2+10w+25)$
49. $(x-2y)(x^2+2xy+4y^2)$
50. $(z-5p)(z^2+5zp+25p^2)$
51. $(4g-3h)(16g^2+12gh+9h^2)$
52. $(3a-2b)(9a^2+6ab+4b^2)$
53. $(7p+5q)(49p^2-35pq+25q^2)$
54. $(8t+3s)(64t^2-24ts+9s^2)$
55. $3(2n+3p)(4n^2-6np+9p^2)$
56. $2(5x+2y)(25x^2-10xy+4y^2)$
57. $(y+z+4)(y^2+2yz+z^2-4y-4z+16)$
58. $(p-q+5)(p^2-2pq+q^2-5p+5q+25)$
59. $(m^2-5)(m^4+5m^2+25)$
60. $(3r^2+1)(9r^4-3r^2+1)$
61. $(3-10x^3)(9+30x^3+100x^6)$
62. $(4-9p^3)(16+36p^3+81p^6)$
63. $(5y^2+z)(25y^4-5y^2z+z^2)$
64. no
65. $(5p+2q)(25p^2-10pq+4q^2+5p-2q)$
66. $(3x+y)(9x^2-3xy+y^2+3x-y)$
67. $(3a-4b)(9a^2+12ab+16b^2+5)$
68. $(10k-m)(100k^2+10km+m^2+2)$

Factor each polynomial. ***See Examples 1–4.***

7. p^2-16 8. k^2-9 9. $25x^2-4$
10. $36m^2-25$ 11. $18a^2-98b^2$ 12. $32c^2-98d^2$
13. $64m^4-4y^4$ 14. $243x^4-3t^4$ 15. $(y+z)^2-81$
16. $(h+k)^2-9$ 17. $16-(x+3y)^2$ 18. $64-(r+2t)^2$
19. p^4-256 20. a^4-625 21. k^2-6k+9
22. $x^2+10x+25$ 23. $4z^2+4zw+w^2$ 24. $9y^2+6yz+z^2$
25. $16m^2-8m+1-n^2$ 26. $25c^2-20c+4-d^2$ 27. $4r^2-12r+9-s^2$
28. $9a^2-24a+16-b^2$ 29. x^2-y^2+2y-1 30. $-k^2-h^2+2kh+4$
31. $98m^2+84mn+18n^2$ 32. $80z^2-40zw+5w^2$
33. $(p+q)^2+2(p+q)+1$ 34. $(x+y)^2+6(x+y)+9$
35. $(a-b)^2+8(a-b)+16$ 36. $(m-n)^2+4(m-n)+4$
37. x^3-27 38. y^3-64 39. $216-t^3$
40. $512-m^3$ 41. x^3+64 42. r^3+343
43. $1000+y^3$ 44. $729+x^3$ 45. $8x^3+1$
46. $27y^3+1$ 47. $125x^3-216$ 48. $8w^3-125$
49. x^3-8y^3 50. z^3-125p^3 51. $64g^3-27h^3$
52. $27a^3-8b^3$ 53. $343p^3+125q^3$ 54. $512t^3+27s^3$
55. $24n^3+81p^3$ 56. $250x^3+16y^3$ 57. $(y+z)^3+64$
58. $(p-q)^3+125$ 59. m^6-125 60. $27r^6+1$
61. $27-1000x^9$ 62. $64-729p^9$ 63. $125y^6+z^3$

64. ***Concept Check*** Consider $(x-y)^2-25$. To factor this polynomial, is the first step $x^2-2xy+y^2-25$ correct?

Extending Skills *In some cases, the method of factoring by grouping can be combined with the methods of special factoring discussed in this section. Consider this example.*

$8x^3+4x^2+27y^3-9y^2$

$=(8x^3+27y^3)+(4x^2-9y^2)$ Associative and commutative properties

$=(2x+3y)(4x^2-6xy+9y^2)+(2x+3y)(2x-3y)$ Factor within groups.

$=(2x+3y)[(4x^2-6xy+9y^2)+(2x-3y)]$ Factor out the greatest common factor, $2x+3y$.

$=(2x+3y)(4x^2-6xy+9y^2+2x-3y)$ Combine like terms.

In problems such as this, how we choose to group in the first step is essential to factoring correctly. If we reach a "dead end," then we should group differently and try again.

Use the method just described to factor each polynomial.

65. $125p^3+25p^2+8q^3-4q^2$ 66. $27x^3+9x^2+y^3-y^2$
67. $27a^3+15a-64b^3-20b$ 68. $1000k^3+20k-m^3-2m$

69. $(t-3)(2t+1)(4t^2-2t+1)$
70. $(y+1)^2(y^2-y+1)$
71. $(8m-9n)(8m+9n-64m^2-72mn-81n^2)$
72. $5(x+y)(2x-2y+x^2-xy+y^2)$

69. $8t^4-24t^3+t-3$

70. y^4+y^3+y+1

71. $64m^2-512m^3-81n^2+729n^3$

72. $10x^2+5x^3-10y^2+5y^3$

RELATING CONCEPTS For Individual or Group Work **(Exercises 73–78)**

The binomial x^6-y^6 may be considered either as a difference of squares or a difference of cubes. ***Work Exercises 73–78 in order.***

73. Factor x^6-y^6 by first factoring as a difference of squares. Then factor further by considering one of the factors as a sum of cubes and the other factor as a difference of cubes.

74. Based on the answer in **Exercise 73,** fill in the blank with the correct factors so that x^6-y^6 is factored completely.

$$x^6-y^6=(x-y)(x+y)\_\_\_\_\_\_\_\_\_\_\_\_$$

75. Factor x^6-y^6 by first factoring as a difference of cubes. Then factor further by considering one of the factors as a difference of squares.

76. Based on the answer in **Exercise 75,** fill in the blank with the correct factor so that x^6-y^6 is factored.

$$x^6-y^6=(x-y)(x+y)\_\_\_\_\_\_\_\_\_\_\_\_$$

77. Notice that the factor written in the blank in **Exercise 76** is a fourth-degree polynomial, while the two factors written in the blank in **Exercise 74** are both second-degree polynomials. What must be true about the product of the two factors written in the blank in **Exercise 74?** Verify this.

78. If we have a choice of factoring as a difference of squares or a difference of cubes, how should we start to more easily obtain the completely factored form of the polynomial? Base the answer on the results in **Exercises 73–77.**

73. $(x^3-y^3)(x^3+y^3)$; $(x-y)(x^2+xy+y^2)\cdot(x+y)(x^2-xy+y^2)$
74. $(x^2+xy+y^2)(x^2-xy+y^2)$
75. $(x^2-y^2)(x^4+x^2y^2+y^4)$; $(x-y)(x+y)(x^4+x^2y^2+y^4)$
76. $x^4+x^2y^2+y^4$
77. The product must equal $x^4+x^2y^2+y^4$. Multiply $(x^2+xy+y^2)(x^2-xy+y^2)$ to verify this.
78. Start by factoring as a difference of squares.

5.4 A General Approach to Factoring

OBJECTIVE

1 Factor any polynomial.

A polynomial is *completely factored* when it is written as a product of prime polynomials with integer coefficients.

TEACHING TIP
This summary section is designed to give students additional practice in applying the various methods of factoring.

Factoring a Polynomial

Step 1 **Factor out any common factor.**

Step 2 **If the polynomial is a binomial,** check to see if it is a difference of squares, a difference of cubes, or a sum of cubes.

If the polynomial is a trinomial, check to see if it is a perfect square trinomial. If it is not, factor as in **Section 5.2.**

If the polynomial has more than three terms, try to factor by grouping.

Step 3 **If any of the factors can be factored further, do so.**

Step 4 **Check the factored form by multiplying.**

OBJECTIVE 1 Factor any polynomial.

We use the various methods of factoring introduced in **Sections 5.1–5.3.**

NOW TRY EXERCISE 1

Factor each polynomial.

(a) $13x + 39$

(b) $21x^3y^2 - 27x^2y^4$

(c) $8y(m - n) - 5(m - n)$

CLASSROOM EXAMPLE 1

Factor each polynomial.

(a) $10t + 150$

(b) $2x^3 + 10x^2 - 4x$

(c) $12m(p - q) - 7n(p - q)$

Answers:

(a) $10(t + 15)$

(b) $2x(x^2 + 5x - 2)$

(c) $(p - q)(12m - 7n)$

EXAMPLE 1 Factoring Out a Common Factor

Factor each polynomial.

(a) $9p + 45$

$= 9(p + 5)$ GCF $= 9$

(b) $8m^2p^2 + 4mp$

$= 4mp(2mp + 1)$

(c) $5x(a + b) - y(a + b)$

$= (a + b)(5x - y)$ Factor out $(a + b)$.

NOW TRY

Factoring a Binomial

For a **binomial** (two terms), check for the following patterns.

Difference of squares	$x^2 - y^2 = (x + y)(x - y)$
Difference of cubes	$x^3 - y^3 = (x - y)(x^2 + xy + y^2)$
Sum of cubes	$x^3 + y^3 = (x + y)(x^2 - xy + y^2)$

NOW TRY EXERCISE 2

Factor each binomial if possible.

(a) $4a^2 - 49b^2$

(b) $9x^2 + 100$

(c) $27v^3 - 1000$

CLASSROOM EXAMPLE 2

Factor each binomial if possible.

(a) $36x^2 - y^2$

(b) $4t^2 + 1$

(c) $125x^3 - 27y^3$

Answers:

(a) $(6x + y)(6x - y)$

(b) prime

(c) $(5x - 3y) \cdot$
$(25x^2 + 15xy + 9y^2)$

EXAMPLE 2 Factoring Binomials

Factor each binomial if possible.

(a) $64m^2 - 9n^2$

$= (8m)^2 - (3n)^2$ Difference of squares

$= (8m + 3n)(8m - 3n)$ $x^2 - y^2 = (x + y)(x - y)$

(b) $8p^3 - 27$

$= (2p)^3 - 3^3$ Difference of cubes

$= (2p - 3)[(2p)^2 + (2p)(3) + 3^2]$ $x^3 - y^3 = (x - y)(x^2 + xy + y^2)$

$= (2p - 3)(4p^2 + 6p + 9)$ $(2p)^2 = 2^2p^2 = 4p^2$

(c) $1000m^3 + 1$

$= (10m)^3 + 1^3$ Sum of cubes

$= (10m + 1)[(10m)^2 - (10m)(1) + 1^2]$ $x^3 + y^3 = (x + y)(x^2 - xy + y^2)$

$= (10m + 1)(100m^2 - 10m + 1)$ $(10m)^2 = 10^2m^2 = 100m^2$

(d) $25m^2 + 121$ is prime. It is a *sum* of squares, with no common factor (except 1).

NOW TRY

NOTE Although the binomial $25m^2 + 625$ is a sum of squares, it can be factored because the greatest common factor of the terms is *not* 1.

$$25m^2 + 625 = 25(m^2 + 25)$$ Factor out the common factor 25.

This sum of squares cannot be factored further.

NOW TRY ANSWERS

1. **(a)** $13(x + 3)$
   (b) $3x^2y^2(7x - 9y^2)$
   (c) $(m - n)(8y - 5)$
2. **(a)** $(2a + 7b)(2a - 7b)$
   (b) prime
   (c) $(3v - 10)(9v^2 + 30v + 100)$

Factoring a Trinomial

For a **trinomial** (three terms), decide whether it is a perfect square trinomial of either of these forms.

$$x^2 + 2xy + y^2 = (x + y)^2 \quad \text{or} \quad x^2 - 2xy + y^2 = (x - y)^2$$

If not, use the methods of **Section 5.2.**

NOW TRY EXERCISE 3

Factor each trinomial.

(a) $25x^2 - 90x + 81$

(b) $7x^2 - 7xy - 84y^2$

(c) $12m^2 + 5m - 28$

CLASSROOM EXAMPLE 3

Factor each trinomial.

(a) $16m^2 + 56m + 49$

(b) $8t^2 - 13t + 5$

(c) $6x^2 - 3x - 63$

Answers:

(a) $(4m + 7)^2$

(b) $(8t - 5)(t - 1)$

(c) $3(2x - 7)(x + 3)$

EXAMPLE 3 Factoring Trinomials

Factor each trinomial.

(a) $p^2 + 10p + 25$

$= (p + 5)^2$ Perfect square trinomial

(b) $49z^2 - 42z + 9$

$= (7z - 3)^2$ Perfect square trinomial

(c) $y^2 - 5y - 6$

$= (y - 6)(y + 1)$ The numbers -6 and 1 have a product of -6 and a sum of -5.

(d) $2k^2 - k - 6$

$= (2k + 3)(k - 2)$ Use either method from **Section 5.2.**

(e) $28z^2 + 6z - 10$

$= 2(14z^2 + 3z - 5)$ Factor out the common factor.

$= 2(7z + 5)(2z - 1)$ Factor the trinomial.

Remember the common factor.

NOW TRY

If a polynomial has more than three terms, consider factoring by grouping.

CLASSROOM EXAMPLE 4

Factor each polynomial.

(a) $p^3 - 2pq^2 + p^2q - 2q^3$

(b) $9x^2 + 24x + 16 - y^2$

(c) $64a^3 + 16a^2 + b^3 - b^2$

Answers:

(a) $(p + q)(p^2 - 2q^2)$

(b) $(3x + 4 + y)(3x + 4 - y)$

(c) $(4a + b)(16a^2 - 4ab + b^2 + 4a - b)$

TEACHING TIP

Include this example:

Factor $x^2 + 2xy + y^2 - 1$.

Students often have difficulty grouping properly in factoring such a polynomial. Explain that the first three terms form a perfect square trinomial, and then a difference of squares will result.

NOW TRY ANSWERS

3. **(a)** $(5x - 9)^2$
(b) $7(x + 3y)(x - 4y)$
(c) $(4m + 7)(3m - 4)$

EXAMPLE 4 Factoring Polynomials with More Than Three Terms

Factor each polynomial.

(a) $xy^2 - y^3 + x^3 - x^2y$

$= (xy^2 - y^3) + (x^3 - x^2y)$ Group the terms.

$= y^2(x - y) + x^2(x - y)$ Factor each group.

$= (x - y)(y^2 + x^2)$ $x - y$ is a common factor.

(b) $20k^3 + 4k^2 - 45k - 9$

$= (20k^3 + 4k^2) - (45k + 9)$ Be careful with signs.

$= 4k^2(5k + 1) - 9(5k + 1)$ Factor each group.

$= (5k + 1)(4k^2 - 9)$ $5k + 1$ is a common factor.

$= (5k + 1)(2k + 3)(2k - 3)$ Difference of squares

(c) $4a^2 + 4a + 1 - b^2$

$= (4a^2 + 4a + 1) - b^2$ Associative property

$= (2a + 1)^2 - b^2$ Perfect square trinomial

$= (2a + 1 + b)(2a + 1 - b)$ Difference of squares

NOW TRY EXERCISE 4

Factor each polynomial.

(a) $5a^3 + 5a^2b - ab^2 - b^3$

(b) $9u^2 - 48u + 64 - v^2$

(c) $x^3 - 9y^2 - 27y^3 + x^2$

(d) $8m^3 + 4m^2 - n^3 - n^2$

$= (8m^3 - n^3) + (4m^2 - n^2)$ Rearrange and group the terms.

(Difference of cubes; Difference of squares)

$= (2m - n)(4m^2 + 2mn + n^2) + (2m - n)(2m + n)$ Factor each group.

$= (2m - n)(4m^2 + 2mn + n^2 + 2m + n)$ Factor out the common factor $2m - n$. NOW TRY

NOW TRY ANSWERS

4. (a) $(a + b)(5a^2 - b^2)$
(b) $(3u - 8 + v)(3u - 8 - v)$
(c) $(x - 3y)(x^2 + 3xy + 9y^2 + x + 3y)$

NOTE Observe the final two lines in **Example 4(d).** Students often ask, *"What happened to the other* $(2m - n)$*?"* Think of the variable a in the expression $ax + ay$, which is factored as $a(x + y)$. The binomial $(2m - n)$ is handled just like a in this simpler example.

5.4 Exercises

FOR EXTRA HELP

MyMathLab®

Complete solution available in MyMathLab

1. (a) B (b) D (c) A (d) A, C (e) A, B
2. (a) C (b) E (c) B (d) B (e) A, B
3. $3p^2(p - 6)(p + 5)$
4. $(k^2 + 4)(k + 2)(k - 2)$
5. $3pq(a + 6b)(a - 5b)$
6. $(7z + 4)(7z - 4)$
7. prime
8. $3mn(3m + 2n)(2m - n)$
9. $(6b + 1)(b - 3)$
10. $(k - 8)(k + 2)$
11. $(x - 10)(x^2 + 10x + 100)$
12. $(3t - 7u)(2t + 11u)$
13. $(p + 2)(4 + m)$
14. $8(5p - 4r)$

Concept Check *In Exercises 1 and 2, match each polynomial in Column I with the method or methods for factoring it in Column II. The choices in Column II may be used once, more than once, or not at all.*

I	II
1. (a) $49x^2 - 81y^2$	**A.** Factor out the GCF.
(b) $125z^6 + 1$	**B.** Factor a difference of squares.
(c) $88r^2 - 55s^2$	**C.** Factor a difference of cubes.
(d) $64a^3 - 8b^9$	**D.** Factor a sum of cubes.
(e) $50x^2 - 128y^4$	**E.** The polynomial is prime.

I	II
2. (a) $ab - 5a + 3b - 15$	**A.** Factor out the GCF.
(b) $z^2 - 3z + 6$	**B.** Factor a perfect square trinomial.
(c) $x^2 - 12x + 36 - 4p^2$	**C.** Factor by grouping.
(d) $r^2 - 24r + 144$	**D.** Factor into two distinct binomials.
(e) $2y^2 + 36y + 162$	**E.** The polynomial is prime.

The following exercises are of mixed variety. Factor each polynomial. ***See Examples 1–4.***

3. $3p^4 - 3p^3 - 90p^2$

4. $k^4 - 16$

5. $3a^2pq + 3abpq - 90b^2pq$

6. $49z^2 - 16$

7. $225p^2 + 256$

8. $18m^3n + 3m^2n^2 - 6mn^3$

9. $6b^2 - 17b - 3$

10. $k^2 - 6k - 16$

11. $x^3 - 1000$

12. $6t^2 + 19tu - 77u^2$

13. $4(p + 2) + m(p + 2)$

14. $40p - 32r$

15. $9m(m - 5 + 2m^2)$
16. $(2k + 7r)^2$
17. $2(3m - 10)(9m^2 + 30m + 100)$
18. $(n + 5)(m - 2)$
19. $(3m - 5n)^2$
20. $(2a + 1)(a - 4)$
21. $(k - 9)(q + r)$
22. $7(2k - 5)(4k^2 + 10k + 25)$
23. $16z^2x(zx - 2)$ **24.** prime
25. $(x + 7)(x - 5)$
26. $(3 + a - b)(3 - a + b)$
27. $(25 + x^2)(5 - x)(5 + x)$
28. $(2m + 5n)(m - 3n)$
29. $(p + 1)(p^2 - p + 1)$
30. $4y^2z^3(12 - 7yz)$
31. $(8m + 25)(8m - 25)$
32. $(7z + 2k)(2z - k)$
33. $6z(2z^2 - z + 3)$
34. $9(5k + 2r)(5k - 2r)$
35. $16(4b + 5c)(4b - 5c)$
36. $(z - 5p)(z + 4p)$
37. $8(4 + 5z)(16 - 20z + 25z^2)$
38. $(8m + 5n)(8m - 5n)$
39. $(5r - s)(2r + 5s)$
40. $(4k + q)(3k - 5q)$
41. $4pq(2p + q)(3p + 5q)$
42. $8x^2(4 + 2x - 3x^3)$
43. $3(4k^2 + 9)(2k + 3)(2k - 3)$
44. $(7x + 5q)(2x - 5q)$
45. $(m - n)(m^2 + mn + n^2 + m + n)$
46. $(4x + y)(16x^2 - 4xy + y^2 - 4x + y)$
47. $(x - 2m - n)(x + 2m + n)$
48. $(2r - s - t)(2r + s + t)$
49. $6p^3(3p^2 - 4 + 2p^3)$
50. prime
51. $2(x + 4)(x - 5)$
52. $3(9x^3 - y^3)$
53. $8mn$ **54.** $9(k + 1)^2$
55. $2(5p + 9)(5p - 9)$
56. $(y + 5)(y - 2)$
57. $4rx(3m^2 + mn + 10n^2)$
58. $(9p - 5r)(2p + 7r)$
59. $(7a - 4b)(3a + b)$
60. $(x - y + 2)(x - y - 2)$
61. prime **62.** $(5r + 2s - 3)^2$
63. $(p + 8q - 5)^2$
64. $(z^2 - 5)(z + 2)(z - 2)$
65. $(7m^2 + 1)(3m^2 - 5)$
66. $(x - 27)(x^2 - 3xy + 3y^2 + 27x - 81y + 729)$
67. $(2r - t)(r^2 - rt + 19t^2)$
68. $(4x + 3)(4x - 3)(x + 2)$
69. $(x + 3)(x^2 + 1)(x + 1)(x - 1)$
70. $(1 + x^8)(1 + x^4)(1 + x^2)(1 + x)(1 - x)$ **71.** $(m + n - 5)(m - n + 1)$ **72.** $(x^2 + 4)(1 + y)$

15. $9m^2 - 45m + 18m^3$
16. $4k^2 + 28kr + 49r^2$
17. $54m^3 - 2000$
18. $mn - 2n + 5m - 10$
19. $9m^2 - 30mn + 25n^2$
20. $2a^2 - 7a - 4$
21. $kq - 9q + kr - 9r$
22. $56k^3 - 875$
23. $16z^3x^2 - 32z^2x$
24. $9r^2 + 100$
25. $x^2 + 2x - 35$
26. $9 - a^2 + 2ab - b^2$
27. $625 - x^4$
28. $2m^2 - mn - 15n^2$
29. $p^3 + 1$
30. $48y^2z^3 - 28y^3z^4$
31. $64m^2 - 625$
32. $14z^2 - 3zk - 2k^2$
33. $12z^3 - 6z^2 + 18z$
34. $225k^2 - 36r^2$
35. $256b^2 - 400c^2$
36. $z^2 - zp - 20p^2$
37. $512 + 1000z^3$
38. $64m^2 - 25n^2$
39. $10r^2 + 23rs - 5s^2$
40. $12k^2 - 17kq - 5q^2$
41. $24p^3q + 52p^2q^2 + 20pq^3$
42. $32x^2 + 16x^3 - 24x^5$
43. $48k^4 - 243$
44. $14x^2 - 25xq - 25q^2$
45. $m^3 + m^2 - n^3 - n^2$
46. $64x^3 + y^3 - 16x^2 + y^2$
47. $x^2 - 4m^2 - 4mn - n^2$
48. $4r^2 - s^2 - 2st - t^2$
49. $18p^5 - 24p^3 + 12p^6$
50. $k^2 - 6k + 16$
51. $2x^2 - 2x - 40$
52. $27x^3 - 3y^3$
53. $(2m + n)^2 - (2m - n)^2$
54. $(3k + 5)^2 - 4(3k + 5) + 4$
55. $50p^2 - 162$
56. $y^2 + 3y - 10$
57. $12m^2rx + 4mnrx + 40n^2rx$
58. $18p^2 + 53pr - 35r^2$
59. $21a^2 - 5ab - 4b^2$
60. $x^2 - 2xy + y^2 - 4$
61. $x^2 - y^2 - 4$
62. $(5r + 2s)^2 - 6(5r + 2s) + 9$
63. $(p + 8q)^2 - 10(p + 8q) + 25$
64. $z^4 - 9z^2 + 20$
65. $21m^4 - 32m^2 - 5$
66. $(x - y)^3 - (27 - y)^3$
67. $(r + 2t)^3 + (r - 3t)^3$
68. $16x^3 + 32x^2 - 9x - 18$
69. $x^5 + 3x^4 - x - 3$
70. $1 - x^{16}$
71. $m^2 - 4m + 4 - n^2 + 6n - 9$
72. $x^2 + 4 + x^2y + 4y$

5.5 Solving Equations by the Zero-Factor Property

OBJECTIVES

1. Learn and use the zero-factor property.
2. Solve applied problems that require the zero-factor property.
3. Solve a formula for a specified variable, where factoring is necessary.

VOCABULARY

☐ quadratic equation
☐ double solution

In **Chapter 1,** we developed methods for solving linear, or first-degree, equations. Solving higher degree polynomial equations requires other methods, one of which involves factoring.

OBJECTIVE 1 Learn and use the zero-factor property.

The **zero-factor property** is a special property of the number 0.

Zero-Factor Property

If two numbers have a product of 0, then at least one of the numbers must be 0.

If $\boldsymbol{ab = 0}$, then either $\boldsymbol{a = 0}$ or $\boldsymbol{b = 0}$.

To prove the zero-factor property, we first assume that $a \neq 0$. (If a does equal 0, then the property is proved already.) If $a \neq 0$, then $\frac{1}{a}$ exists, and both sides of $ab = 0$ can be multiplied by $\frac{1}{a}$.

$$\frac{1}{a} \cdot ab = \frac{1}{a} \cdot 0 \quad \text{Multiply each side of } ab = 0 \text{ by } \tfrac{1}{a}.$$

$$b = 0 \quad \text{Multiply.}$$

Thus, if $a \neq 0$, then $b = 0$, and the property is proved.

CAUTION If $ab = 0$, then $a = 0$ or $b = 0$. However, if $ab = 6$, for example, it is not necessarily true that $a = 6$ or $b = 6$. In fact, it is very likely that *neither* $a = 6$ *nor* $b = 6$. ***The zero-factor property works only for a product equal to* 0.**

NOW TRY EXERCISE 1

Solve.

$$(x + 5)(4x - 7) = 0$$

CLASSROOM EXAMPLE 1

Solve.

$$(8x + 3)(2x + 1) = 0$$

Answer: $\left\{-\frac{3}{8}, -\frac{1}{2}\right\}$

EXAMPLE 1 Using the Zero-Factor Property to Solve an Equation

Solve $(x + 6)(2x - 3) = 0$.

Here, the product of $x + 6$ and $2x - 3$ is 0. By the zero-factor property, the following must hold true.

$$x + 6 = 0 \quad \text{or} \quad 2x - 3 = 0 \quad \text{Zero-factor property}$$

$$x = -6 \quad \text{or} \quad 2x = 3 \quad \text{Solve each of these equations.}$$

$$x = \frac{3}{2}$$

CHECK

$$(x + 6)(2x - 3) = 0$$

$$(-6 + 6)[2(-6) - 3] \stackrel{?}{=} 0 \quad \text{Let } x = -6.$$

$$0(-15) \stackrel{?}{=} 0$$

$$0 = 0 \quad ✓ \text{ True}$$

$$(x + 6)(2x - 3) = 0$$

$$\left(\frac{3}{2} + 6\right)\left(2 \cdot \frac{3}{2} - 3\right) \stackrel{?}{=} 0 \quad \text{Let } x = \tfrac{3}{2}.$$

$$\frac{15}{2}(0) \stackrel{?}{=} 0$$

$$0 = 0 \quad ✓ \text{ True}$$

Both values check, so the solution set is $\left\{-6, \frac{3}{2}\right\}$. NOW TRY

NOW TRY ANSWER

1. $\left\{-5, \frac{7}{4}\right\}$

Because the product $(x + 6)(2x - 3)$ equals $2x^2 + 9x - 18$, the equation of **Example 1** has a term with a squared variable and is an example of a *quadratic equation. **A quadratic equation has degree 2.***

TEACHING TIP
Tell students that quadratic equations are discussed in more detail in **Chapter 8**.

Quadratic Equation

A **quadratic equation** (in x here) can be written in the form

$$ax^2 + bx + c = 0,$$

where a, b, and c are real numbers, and $a \neq 0$. The given form is called **standard form.**

TEACHING TIP
Emphasize that 0 must be isolated on one side before the zero-factor property can be used.

Solving a Quadratic Equation Using the Zero-Factor Property

Step 1 **Write the equation in standard form**—that is, with all terms on one side of the equality symbol in descending powers of the variable and 0 on the other side.

Step 2 **Factor** completely.

Step 3 **Apply the zero-factor property.** Set each factor with a variable equal to 0.

Step 4 **Solve** the resulting equations.

Step 5 **Check** each value in the *original* equation. Write the solution set.

CLASSROOM EXAMPLE 2
Solve each equation.
(a) $3x^2 - x = 4$
(b) $16m^2 + 24m + 9 = 0$
Answers: (a) $\left\{-1, \frac{4}{3}\right\}$ (b) $\left\{-\frac{3}{4}\right\}$

EXAMPLE 2 Solving Quadratic Equations

Solve each equation.

(a) $2x^2 + 3x = 2$

Step 1 $2x^2 + 3x - 2 = 0$ Standard form

Step 2 $(2x - 1)(x + 2) = 0$ Factor.

Step 3 $2x - 1 = 0$ or $x + 2 = 0$ Zero-factor property

Step 4 $2x = 1$ or $x = -2$ Solve each equation.

$x = \frac{1}{2}$

Step 5 *Check* each value in the original equation.

CHECK

$2x^2 + 3x = 2$

$2\left(\frac{1}{2}\right)^2 + 3\left(\frac{1}{2}\right) \stackrel{?}{=} 2$ Let $x = \frac{1}{2}$.

$2\left(\frac{1}{4}\right) + \frac{3}{2} \stackrel{?}{=} 2$

$\frac{1}{2} + \frac{3}{2} \stackrel{?}{=} 2$

$2 = 2$ ✓ True

$2x^2 + 3x = 2$

$2(-2)^2 + 3(-2) \stackrel{?}{=} 2$ Let $x = -2$.

$2(4) - 6 \stackrel{?}{=} 2$

$8 - 6 \stackrel{?}{=} 2$

$2 = 2$ ✓ True

Because both values check, the solution set is $\left\{-2, \frac{1}{2}\right\}$. (We give solutions from least to greatest.)

NOW TRY EXERCISE 2

Solve each equation.

(a) $7x = 3 - 6x^2$

(b) $16x^2 + 40x + 25 = 0$

(b)

$$4x^2 - 4x + 1 = 0 \quad \text{Standard form}$$

We could factor as $(2x - 1)(2x - 1)$.

$$(2x - 1)^2 = 0 \quad \text{Factor.}$$

$$2x - 1 = 0 \quad \text{Zero-factor property}$$

$$2x = 1 \quad \text{Add 1.}$$

$$x = \frac{1}{2} \quad \text{Divide by 2.}$$

There is only one *distinct* solution, called a **double solution,** because the trinomial is a perfect square. The solution set is $\{\frac{1}{2}\}$. NOW TRY

NOW TRY EXERCISE 3

Solve $3x^2 + 12x = 0$.

CLASSROOM EXAMPLE 3

Solve $2x^2 + 24x = 0$.

Answer: $\{-12, 0\}$

EXAMPLE 3 Solving a Quadratic Equation (Missing Constant Term)

Solve $4x^2 - 20x = 0$.

This quadratic equation has a missing constant term. Comparing it with the standard form $ax^2 + bx + c = 0$ shows that $c = 0$. The zero-factor property can still be used.

$$4x^2 - 20x = 0$$

$$4x(x - 5) = 0 \quad \text{Factor.}$$

Set each *variable* factor equal to 0.

$$4x = 0 \quad \text{or} \quad x - 5 = 0 \quad \text{Zero-factor property}$$

$$x = 0 \quad \text{or} \quad x = 5 \quad \text{Solve each equation.}$$

CHECK

$4x^2 - 20x = 0$		$4x^2 - 20x = 0$
$4(0)^2 - 20(0) \stackrel{?}{=} 0$ Let $x = 0$.		$4(5)^2 - 20(5) \stackrel{?}{=} 0$ Let $x = 5$.
$0 - 0 = 0$ ✓ True		$100 - 100 = 0$ ✓ True

The solution set is $\{0, 5\}$. NOW TRY

CLASSROOM EXAMPLE 4

Solve $5x^2 - 80 = 0$.

Answer: $\{-4, 4\}$

CAUTION Remember to include 0 as a solution in **Example 3.**

NOW TRY EXERCISE 4

Solve $4x^2 - 100 = 0$.

EXAMPLE 4 Solving a Quadratic Equation (Missing Linear Term)

Solve $3x^2 - 108 = 0$.

$$3x^2 - 108 = 0$$

$$3(x^2 - 36) = 0 \quad \text{Factor out 3.}$$

The factor 3 does *not* lead to a solution.

$$3(x + 6)(x - 6) = 0 \quad \text{Factor } x^2 - 36.$$

$$x + 6 = 0 \quad \text{or} \quad x - 6 = 0 \quad \text{Zero-factor property}$$

$$x = -6 \quad \text{or} \quad x = 6 \quad \text{Solve each equation.}$$

Check that the solution set is $\{-6, 6\}$. NOW TRY

TEACHING TIP

In **Example 4,** point out that we could begin solving the equation by dividing both sides by 3, a constant factor. This eliminates the factor 3 in the second step of the solution. Stress that we cannot divide both sides by a *variable* factor (see **Example 3**), or we will lose a solution.

CAUTION The factor 3 in **Example 4** is not a ***variable*** factor, so it does *not* lead to a solution of the equation. However, the factor x in **Example 3** is a variable factor and leads to the solution 0.

NOW TRY ANSWERS

2. **(a)** $\{-\frac{3}{2}, \frac{1}{3}\}$ **(b)** $\{-\frac{5}{4}\}$

3. $\{0, -4\}$ **4.** $\{-5, 5\}$

NOW TRY EXERCISE 5

Solve.

$$(x+3)(2x-1) = 4(x+4)-4$$

CLASSROOM EXAMPLE 5

Solve.

$(x+6)(x-2) = 2+x-10$

Answer: $\{-4, 1\}$

EXAMPLE 5 Solving an Equation That Requires Rewriting

Solve $(2x+1)(x+1) = 2(1-x)+6$.

$$(2x+1)(x+1) = 2(1-x)+6$$

$$2x^2+3x+1 = 2-2x+6 \quad \text{Multiply on each side.}$$

$$2x^2+3x+1 = 8-2x \quad \text{Add on the right.}$$

Write in standard form.

$$2x^2+5x-7 = 0 \quad \text{Add } 2x\text{. Subtract 8.}$$

$$(2x+7)(x-1) = 0 \quad \text{Factor.}$$

$$2x+7=0 \quad \text{or} \quad x-1=0 \quad \text{Zero-factor property}$$

$$x = -\frac{7}{2} \quad \text{or} \quad x = 1 \quad \text{Solve each equation.}$$

CHECK

$$(2x+1)(x+1) = 2(1-x)+6$$

$$\left[2\left(-\frac{7}{2}\right)+1\right]\left(-\frac{7}{2}+1\right) \stackrel{?}{=} 2\left[1-\left(-\frac{7}{2}\right)\right]+6 \quad \text{Let } x = -\tfrac{7}{2}.$$

$$(-7+1)\left(-\frac{5}{2}\right) \stackrel{?}{=} 2\left(\frac{9}{2}\right)+6 \quad \text{Simplify; } 1 = \tfrac{2}{2}.$$

$$(-6)\left(-\frac{5}{2}\right) \stackrel{?}{=} 9+6$$

$$15 = 15 \ \checkmark \quad \text{True}$$

Check that 1 is also a solution. The solution set is $\left\{-\frac{7}{2}, 1\right\}$.

NOW TRY

The zero-factor property can be extended to solve certain polynomial equations of degree 3 or greater, as shown in the next example.

NOW TRY EXERCISE 6

Solve $12x = 2x^3 + 5x^2$.

CLASSROOM EXAMPLE 6

Solve $3x^3 + x^2 = 4x$.

Answer: $\left\{-\frac{4}{3}, 0, 1\right\}$

EXAMPLE 6 Solving an Equation of Degree 3

Solve $-x^3 + x^2 = -6x$.

$$-x^3+x^2 = -6x$$

$$-x^3+x^2+6x = 0 \quad \text{Add } 6x \text{ to each side.}$$

$$x^3-x^2-6x = 0 \quad \text{Multiply each side by } -1.$$

$$x(x^2-x-6) = 0 \quad \text{Factor out } x.$$

$$x(x-3)(x+2) = 0 \quad \text{Factor the trinomial.}$$

Remember to set x equal to 0.

$$x=0 \quad \text{or} \quad x-3=0 \quad \text{or} \quad x+2=0 \quad \text{Extend the zero-factor property to the three } \textit{variable} \text{ factors.}$$

$$x=3 \quad \text{or} \quad x=-2$$

Check that the solution set is $\{-2, 0, 3\}$.

NOW TRY

NOW TRY ANSWERS

5. $\left\{-3, \frac{5}{2}\right\}$ 6. $\left\{-4, 0, \frac{3}{2}\right\}$

OBJECTIVE 2 Solve applied problems that require the zero-factor property.

An application may lead to a quadratic equation.

NOW TRY EXERCISE 7

The height of a triangle is 1 ft less than twice the length of the base. The area is 14 ft^2. What are the measures of the base and the height?

CLASSROOM EXAMPLE 7

The length of a room is 2 m less than three times the width. The area of the room is 96 m^2. Find the width of the room.

Answer: 6 m

EXAMPLE 7 Using a Quadratic Equation in an Application

A piece of sheet metal is in the shape of a parallelogram. The longer sides of the parallelogram are each 8 m longer than the distance between them. The area of the piece is 48 m^2. Find the length of the longer sides and the distance between them.

Step 1 **Read** the problem again. There will be two answers.

Step 2 **Assign a variable.**

Let x = the distance between the longer sides

and $x + 8$ = the length of each longer side. (See **FIGURE 1**.)

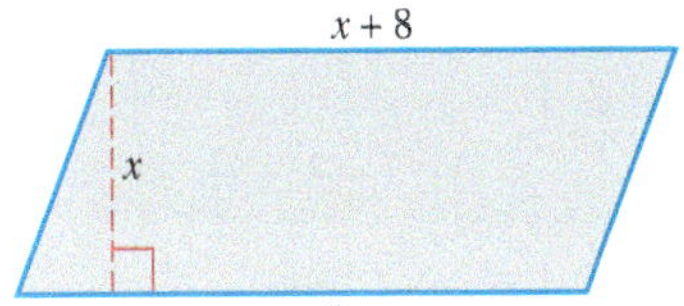

FIGURE 1

Step 3 **Write an equation.** The area of a parallelogram is given by $\mathcal{A} = bh$, where b is the length of the longer side and h is the distance between the longer sides. Here, $b = x + 8$ and $h = x$.

$$\mathcal{A} = bh \quad \text{Formula for area of a parallelogram}$$

$$48 = (x + 8)x \quad \text{Let } \mathcal{A} = 48,\ b = x + 8,\ h = x.$$

Step 4 **Solve.**

$$48 = x^2 + 8x \quad \text{Distributive property}$$

$$x^2 + 8x - 48 = 0 \quad \text{Standard form}$$

$$(x + 12)(x - 4) = 0 \quad \text{Factor.}$$

$$x + 12 = 0 \quad \text{or} \quad x - 4 = 0 \quad \text{Zero-factor property}$$

$$x = -12 \quad \text{or} \quad x = 4 \quad \text{Solve each equation.}$$

Step 5 **State the answer.** ***The distance between the longer sides cannot be negative, so reject*** -12 ***as an answer.*** The only possible answer is 4, so the required distance is 4 m. The length of the longer sides is $4 + 8 = 12$ m.

Step 6 **Check.** The length of the longer sides is 8 m more than the distance between them, and the area is

$$4 \cdot 12 = 48 \text{ m}^2, \quad \text{as required.}$$

The answer checks. NOW TRY

CAUTION When an application leads to a quadratic equation, a solution of the equation may not satisfy the physical requirements of the problem, as in **Example 7.** Reject such solutions as answers.

NOW TRY ANSWER

7. base: 4 ft; height: 7 ft

A function defined by a quadratic polynomial is a ***quadratic function.*** Quadratic functions are used to describe the height a falling object or a projected object reaches in a specific time. The next example uses such a function.

NOW TRY EXERCISE 8
Refer to **Example 8.** After how many seconds will the rocket be 192 ft above the ground?

CLASSROOM EXAMPLE 8
How long will it take the rocket in **Example 8** to reach a height of 256 ft?

Answer: 4 sec

EXAMPLE 8 Using a Quadratic Function in an Application

If a small rocket is launched vertically upward from ground level with an initial velocity of 128 ft per sec, then its height in feet after t seconds (if air resistance is neglected) can be modeled by the function

$$h(t) = -16t^2 + 128t.$$

After how many seconds will the rocket be 220 ft above the ground?

We let $h(t) = 220$ and solve for t.

$$220 = -16t^2 + 128t \qquad \text{Let } h(t) = 220.$$

$$16t^2 - 128t + 220 = 0 \qquad \text{Standard form}$$

$$4t^2 - 32t + 55 = 0 \qquad \text{Divide by 4.}$$

$$(2t - 5)(2t - 11) = 0 \qquad \text{Factor.}$$

$$2t - 5 = 0 \quad \text{or} \quad 2t - 11 = 0 \qquad \text{Zero-factor property}$$

$$t = 2.5 \quad \text{or} \quad t = 5.5 \qquad \text{Solve each equation.}$$

The rocket will reach a height of 220 ft twice: on its way up at 2.5 sec and again on its way down at 5.5 sec.

NOW TRY

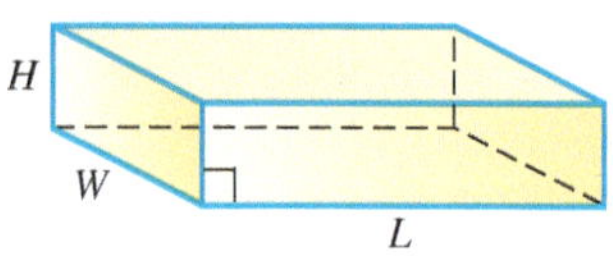

Rectangular solid
$S = 2HW + 2LW + 2LH$

FIGURE 2

OBJECTIVE 3 Solve a formula for a specified variable, where factoring is necessary.

A rectangular solid has the shape of a box, but is solid. See **FIGURE 2.** The surface area of any solid three-dimensional figure is the total area of its surface. For a rectangular solid, the surface area S is

$$S = 2HW + 2LW + 2LH. \qquad H, W, \text{ and } L \text{ represent height, width, and length.}$$

NOW TRY EXERCISE 9
Solve the formula for H.

$$S = 2HW + 2LW + 2LH$$

CLASSROOM EXAMPLE 9
Solve the formula for W.

$$S = 2HW + 2LW + 2LH$$

Answer: $W = \frac{S - 2LH}{2H + 2L}$

EXAMPLE 9 Using Factoring to Solve for a Specified Variable

Solve the formula $S = 2HW + 2LW + 2LH$ for L.

To solve for the length L, treat L as the only variable and treat all other variables as constants.

$$S = 2HW + 2LW + 2LH \qquad \text{We must isolate the } L\text{-terms.}$$

$$S - 2HW = 2LW + 2LH \qquad \text{Subtract } 2HW.$$

$$S - 2HW = L(2W + 2H) \qquad \text{Factor out } L. \quad (\text{This is a key step.})$$

$$\frac{S - 2HW}{2W + 2H} = L, \quad \text{or} \quad L = \frac{S - 2HW}{2W + 2H} \qquad \text{Divide by } 2W + 2H.$$

NOW TRY

CAUTION *In* Example 9, *we must write the expression so that the specified variable is a factor. Then we can divide by its coefficient in the final step.*

NOW TRY ANSWERS
8. 2 sec and 6 sec
9. $H = \frac{S - 2LW}{2W + 2L}$

NOTE In **Section 4.3,** we saw that the graph of $f(x) = x^2$ is a parabola. In general, the graph of

$$f(x) = ax^2 + bx + c \quad (\text{where } a \neq 0)$$

is a parabola, and the x-intercepts of its graph give the real number solutions of the equation $ax^2 + bx + c = 0$.

A graphing calculator can locate these x-intercepts (called **zeros** of the function) for $f(x) = 2x^2 + 3x - 2$. Notice that this quadratic expression was found on the left side of the equation in **Example 2(a)** earlier in this section, where the equation was written in standard form. The x-intercepts (zeros) given with the graphs in **FIGURE 3** are the same as the solutions found in **Example 2(a).**

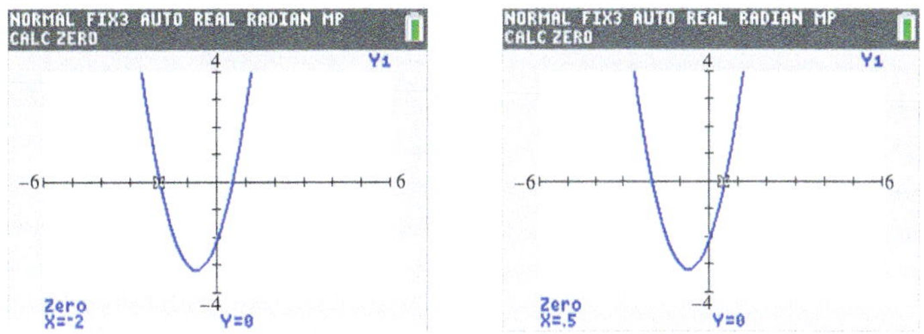

FIGURE 3

5.5 Exercises

FOR EXTRA HELP

MyMathLab®

● *Complete solution available in MyMathLab*

1. D; The polynomial is not factored.
2. D
3. $\{-10, 5\}$
4. $\{-7, -3\}$
5. $\left\{-\frac{8}{3}, \frac{5}{2}\right\}$
6. $\left\{-\frac{5}{2}, \frac{4}{3}\right\}$
7. $\{-2, 5\}$
8. $\{-4, 3\}$
9. $\{-6, -3\}$
10. $\{8, 10\}$
11. $\left\{-\frac{1}{2}, 4\right\}$
12. $\left\{-\frac{3}{2}, 1\right\}$
13. $\left\{-\frac{1}{3}, \frac{4}{5}\right\}$
14. $\left\{-3, -\frac{1}{3}\right\}$
15. $\{-4, 0\}$
16. $\{0, 4\}$
17. $\{0, 6\}$
18. $\{0, 9\}$
19. $\{-2, 2\}$
20. $\{-3, 3\}$
21. $\{-3, 3\}$
22. $\{-2, 2\}$
23. $\{3\}$
24. $\{-4\}$
25. $\left\{-\frac{4}{3}\right\}$
26. $\left\{\frac{5}{2}\right\}$
27. $\{-4, 2\}$
28. $\{-5, -1\}$

Concept Check *Choose the correct response.*

1. Which equation is *not* in proper form for using the zero-factor property? Tell why it is not in proper form.

   A. $(x + 2)(x - 6) = 0$ B. $x(3x - 7) = 0$

   C. $3t(t + 8)(t - 9) = 0$ D. $y(y - 3) + 6(y - 3) = 0$

2. Without actually solving, determine which equation has 0 in its solution set.

   A. $4x^2 - 25 = 0$ B. $x^2 + 2x - 3 = 0$

   C. $6x^2 + 9x + 1 = 0$ D. $x^3 + 4x^2 = 3x$

Solve each equation. ***See Examples 1–5.***

3. $(x + 10)(x - 5) = 0$
4. $(x + 7)(x + 3) = 0$
5. ● $(2k - 5)(3k + 8) = 0$
6. $(3q - 4)(2q + 5) = 0$
7. $x^2 - 3x - 10 = 0$
8. $x^2 + x - 12 = 0$
9. $x^2 + 9x + 18 = 0$
10. $x^2 - 18x + 80 = 0$
11. ● $2x^2 = 7x + 4$
12. $2x^2 = 3 - x$
13. $15x^2 - 7x = 4$
14. $3x^2 + 3 = -10x$
15. ● $4p^2 + 16p = 0$
16. $2t^2 - 8t = 0$
17. $6x^2 - 36x = 0$
18. $3x^2 - 27x = 0$
19. ● $4p^2 - 16 = 0$
20. $9z^2 - 81 = 0$
21. $-3x^2 + 27 = 0$
22. $-2x^2 + 8 = 0$
23. $-x^2 = 9 - 6x$
24. $-x^2 - 8x = 16$
25. $9x^2 + 24x + 16 = 0$
26. $4x^2 - 20x + 25 = 0$
27. $(x - 3)(x + 5) = -7$
28. $(x + 8)(x - 2) = -21$

29. $\left\{-\frac{1}{2}, 6\right\}$ 30. $\left\{-\frac{1}{3}, 5\right\}$

31. $\{-3, 4\}$ 32. $\{-6, -5\}$

33. $\left\{-5, -\frac{1}{5}\right\}$ 34. $\left\{-\frac{4}{3}, 5\right\}$

35. $\{1, 6\}$ 36. $\{-4, 7\}$

37. $\left\{-\frac{1}{2}, 0, 5\right\}$ 38. $\left\{-\frac{1}{3}, 0, \frac{5}{2}\right\}$

39. $\{-1, 0, 3\}$ 40. $\{0, 2, 4\}$

41. $\left\{-\frac{4}{3}, 0, \frac{4}{3}\right\}$ 42. $\left\{-\frac{8}{5}, 0, \frac{8}{5}\right\}$

43. $\left\{-\frac{5}{2}, -1, 1\right\}$

44. $\left\{-7, -\frac{1}{2}, 7\right\}$

45. $\{-3, 3, 6\}$ 46. $\{-2, 2, 3\}$

47. By dividing each side by a variable expression, she lost the solution 0. The solution set is $\left\{-\frac{4}{3}, 0, \frac{4}{3}\right\}$.

48. By dividing each side by a variable expression, he lost the solution 6. The solution set is $\{-3, 3, 6\}$.

49. $\left\{-\frac{1}{2}, 6\right\}$ 50. $\left\{-\frac{15}{8}, -1\right\}$

51. $\left\{-\frac{2}{3}, \frac{4}{15}\right\}$ 52. $\left\{-\frac{5}{2}, -1\right\}$

53. $\left\{-\frac{3}{2}, \frac{1}{2}\right\}$ 54. $\left\{-1, -\frac{1}{4}\right\}$

55. width: 16 ft; length: 20 ft

56. mirror: 7 ft; painting: 9 ft

29. $(2x + 1)(x - 3) = 6x + 3$

30. $(3x + 2)(x - 3) = 7x - 1$

31. $2x^2 - 12 - 4x = x^2 - 3x$

32. $3x^2 + 9x + 30 = 2x^2 - 2x$

33. $(5x + 1)(x + 3) = -2(5x + 1)$

34. $(3x + 1)(x - 3) = 2 + 3(x + 5)$

35. $(x + 3)(x - 6) = (2x + 2)(x - 6)$

36. $(2x + 1)(x + 5) = (x + 11)(x + 3)$

Solve each equation. ***See Example 6.***

37. $2x^3 - 9x^2 - 5x = 0$

38. $6x^3 - 13x^2 - 5x = 0$

39. $x^3 - 2x^2 = 3x$

40. $x^3 - 6x^2 = -8x$

41. $9x^3 = 16x$

42. $25x^3 = 64x$

43. $2x^3 + 5x^2 - 2x - 5 = 0$

44. $2x^3 + x^2 - 98x - 49 = 0$

45. $x^3 - 6x^2 - 9x + 54 = 0$

46. $x^3 - 3x^2 - 4x + 12 = 0$

47. *Concept Check* A student tried to solve the equation in **Exercise 41** by first dividing each side by x, obtaining

$$9x^2 = 16.$$

She then solved the resulting equation by the zero-factor property to obtain the solution set $\left\{-\frac{4}{3}, \frac{4}{3}\right\}$. ***WHAT WENT WRONG?*** Give the correct solution set.

48. *Concept Check* A student tried to solve the equation in **Exercise 45** by first writing it as

$$x^3 - 6x^2 = 9x - 54$$

$$x^2(x - 6) = 9(x - 6)$$

and then dividing both sides by $x - 6$, to obtain $x^2 = 9$. He then gave the solution set as $\{-3, 3\}$. ***WHAT WENT WRONG?*** Give the correct solution set.

Extending Skills *Solve each equation. (Hint: In Exercises 49–52, use the substitution of variable method.)*

49. $2(x - 1)^2 - 7(x - 1) - 15 = 0$

50. $4(2x + 3)^2 - (2x + 3) - 3 = 0$

51. $5(3x - 1)^2 + 3 = -16(3x - 1)$

52. $2(x + 3)^2 = 5(x + 3) - 2$

53. $(2x - 3)^2 = 16x^2$

54. $9x^2 = (5x + 2)^2$

Solve each problem. ***See Examples 7 and 8.***

55. A garden has an area of 320 ft^2. Its length is 4 ft more than its width. What are the dimensions of the garden?

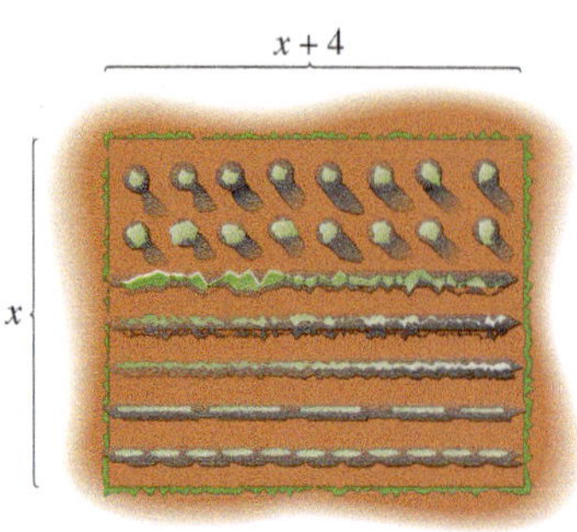

56. A square mirror has sides measuring 2 ft less than the sides of a square painting. If the difference between their areas is 32 ft^2, find the lengths of the sides of the mirror and the painting.

57. base: 12 ft; height: 5 ft
58. base: 8 m; height: 11 m
59. 50 ft by 100 ft
60. width: 100 ft; length: 300 ft
61. -6 and -5 or 5 and 6
62. -9 and -8 or 8 and 9
63. length: 15 in.; width: 9 in.
64. Each side measures 7 in.
65. 5 sec
66. **(a)** 3 sec and 5 sec
(b) 1 sec and 7 sec
67. $6\frac{1}{4}$ sec **68.** 6 sec

57. The base of a parallelogram is 7 ft more than the height. If the area of the parallelogram is 60 ft^2, what are the measures of the base and the height?

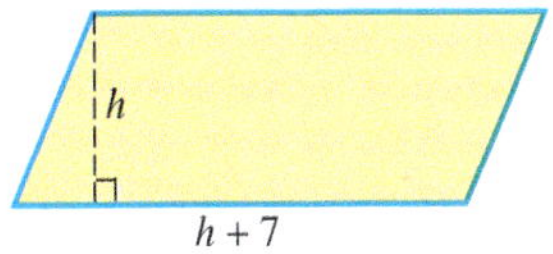

58. A sign has the shape of a triangle. The length of the base is 3 m less than the height. What are the measures of the base and the height if the area is 44 m^2?

59. A farmer has 300 ft of fencing and wants to enclose a rectangular area of 5000 ft^2. What dimensions should she use?

60. A rectangular landfill has an area of 30,000 ft^2. Its length is 200 ft more than its width. What are the dimensions of the landfill?

61. Find two consecutive integers such that the sum of their squares is 61.

62. Find two consecutive integers such that their product is 72.

63. A box with no top is to be constructed from a piece of cardboard whose length measures 6 in. more than its width. The box is to be formed by cutting squares that measure 2 in. on each side from the four corners and then folding up the sides. If the volume of the box will be 110 in.^{3}, what are the dimensions of the piece of cardboard?

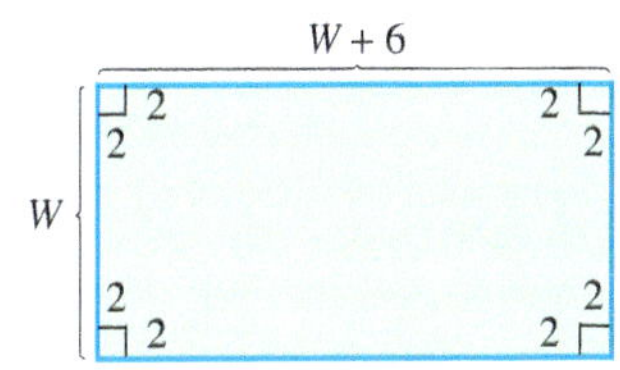

64. The surface area of the box with open top shown in the figure is 161 in.^{2}. Find the dimensions of the base. (*Hint:* The surface area of the box is modeled by the function $S(x) = x^2 + 16x$.)

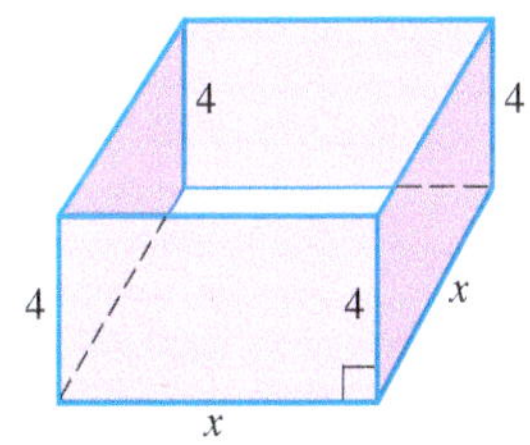

65. If an object is projected upward with an initial velocity of 64 ft per sec from a height of 80 ft, then its height in feet t seconds after it is projected is modeled by the function

$$f(t) = -16t^2 + 64t + 80.$$

How long after it is projected will it hit the ground? (*Hint:* When it hits the ground, its height is 0 ft.)

66. Refer to **Example 8.** After how many seconds will the rocket be

(a) 240 ft above the ground? **(b)** 112 ft above the ground?

67. If a baseball is dropped from a helicopter 625 ft above the ground, then its distance in feet from the ground t seconds later is modeled by the function

$$f(t) = -16t^2 + 625.$$

How long after it is dropped will it hit the ground?

68. If a rock is dropped from a building 576 ft high, then its distance in feet from the ground t seconds later is modeled by the function

$$f(t) = -16t^2 + 576.$$

How long after it is dropped will it hit the ground?

69. L appears on both sides of the equation.
70. A
71. $F = \frac{k}{d - D}$ 72. $m = \frac{Mv}{v - V}$
73. $r = \frac{-2k - 3y}{a - 1}$, or $r = \frac{2k + 3y}{1 - a}$
74. $p = \frac{7 + 4s}{t - 7}$, or $p = \frac{-7 - 4s}{7 - t}$
75. $y = \frac{-x}{w - 3}$, or $y = \frac{x}{3 - w}$
76. $t = \frac{4}{c + 2}$, or $t = \frac{-4}{-c - 2}$

69. *Concept Check* A student solved the formula $S = 2HW + 2LW + 2LH$ for L as follows. **WHAT WENT WRONG?**

$$S = 2HW + 2LW + 2LH$$
$$S - 2LW - 2HW = 2LH$$
$$\frac{S - 2LW - 2HW}{2H} = L$$

70. *Concept Check* Which is the correct result when solving the following equation for t?

$$2t + c = kt$$

A. $t = \frac{-c}{2 - k}$ **B.** $t = \frac{c - kt}{-2}$

C. $t = \frac{2t + c}{k}$ **D.** $t = \frac{kt - c}{2}$

Solve each equation for the specified variable. **See Example 9.**

71. $k = dF - DF$ for F **72.** $Mv = mv - Vm$ for m **73.** $2k + ar = r - 3y$ for r

74. $4s + 7p = tp - 7$ for p **75.** $w = \frac{3y - x}{y}$ for y **76.** $c = \frac{-2t + 4}{t}$ for t

Chapter 5 Summary

Key Terms

5.1
factoring
greatest common factor (GCF)

5.2
prime polynomial

5.3
difference of squares
perfect square trinomial
difference of cubes
sum of cubes

5.5
quadratic equation
double solution

Test Your Word Power

See how well you have learned the vocabulary in this chapter.

1. **Factoring** is
A. a method of multiplying polynomials
B. the process of writing a polynomial as a product of prime factors
C. the answer in a multiplication problem
D. a way to add the terms of a polynomial.

2. The **greatest common factor** of a polynomial is
A. the least integer that divides evenly into all its terms
B. the least expression that is a factor of all its terms
C. the greatest expression that is a factor of all its terms
D. the variable that is common to all its terms.

3. A **difference of squares** is a binomial
A. that can be factored as the difference of two cubes
B. that cannot be factored
C. that is squared
D. that can be factored as the product of the sum and difference of two terms.

4. A **perfect square trinomial** is a trinomial
A. that can be factored as the square of a binomial
B. that cannot be factored
C. that is multiplied by a binomial
D. where all terms are perfect squares.

5. A **quadratic equation** is a polynomial equation of
A. degree one
B. degree two
C. degree three
D. degree four.

ANSWERS

1. B; *Example:* $x^2 - 5x - 14$ factors as $(x - 7)(x + 2)$. **2.** C; *Example:* The greatest common factor of $8x^2$, $22xy$, and $16x^3y^2$ is $2x$. **3.** D; *Example:* $b^2 - 49$ is the difference of the squares b^2 and 7^2. It can be factored as $(b + 7)(b - 7)$. **4.** A; *Example:* $a^2 + 2a + 1$ is a perfect square trinomial. Its factored form is $(a + 1)^2$. **5.** B; *Examples:* $x^2 - 3x + 2 = 0$, $x^2 - 9 = 0$, and $2x^2 = 6x + 8$

Quick Review

CONCEPTS	EXAMPLES
5.1 Greatest Common Factors and Factoring by Grouping	
Factoring Out the Greatest Common Factor Use the distributive property to write the given polynomial as a product of two factors, one of which is the greatest common factor of the terms of the polynomial.	Factor $4x^2y - 50xy^2$. $4x^2y - 50xy^2$ $= 2xy(2x - 25y)$ The greatest common factor is $2xy$.
Factoring by Grouping *Step 1* Group the terms so that each group has a common factor. *Step 2* Factor out the common factor in each group. *Step 3* If the groups now have a common factor, factor it out. If not, try a different grouping. ***Always check the factored form by multiplying.***	Factor by grouping. $5a - 5b - ax + bx$ $= (5a - 5b) + (-ax + bx)$ Group the terms. $= 5(a - b) - x(a - b)$ Factor out 5 and $-x$. $= (a - b)(5 - x)$ Factor out $a - b$.
5.2 Factoring Trinomials	
To factor a trinomial, choose factors of the first term and factors of the last term. Then place them within a pair of parentheses of this form. $(\quad\quad)(\quad\quad)$ Try various combinations of the factors until the correct middle term of the trinomial is found.	Factor $15x^2 + 14x - 8$. The factors of 15 are 5 and 3, and 15 and 1. The factors of -8 are -4 and 2, 4 and -2, -1 and 8, and 1 and -8. Various combinations lead to the correct factorization. $15x^2 + 14x - 8$ $= (5x - 2)(3x + 4)$ Check by multiplying.
5.3 Special Factoring	
Difference of Squares $x^2 - y^2 = (x + y)(x - y)$	Factor. $4m^2 - 25n^2$ $= (2m)^2 - (5n)^2$ $= (2m + 5n)(2m - 5n)$
Perfect Square Trinomials $x^2 + 2xy + y^2 = (x + y)^2$ $x^2 - 2xy + y^2 = (x - y)^2$	$9y^2 + 6y + 1$ $= (3y + 1)^2$ $16p^2 - 56p + 49$ $= (4p - 7)^2$
Difference of Cubes $x^3 - y^3 = (x - y)(x^2 + xy + y^2)$	Factor. $8 - 27a^3$ $= (2 - 3a)(4 + 6a + 9a^2)$
Sum of Cubes $x^3 + y^3 = (x + y)(x^2 - xy + y^2)$	$64z^3 + 1$ $= (4z + 1)(16z^2 - 4z + 1)$

CONCEPTS	EXAMPLES
5.5 Solving Equations by the Zero-Factor Property	
Zero-Factor Property If two numbers have a product of 0, then at least one of the numbers must be 0. If $ab = 0$, then either $a = 0$ or $b = 0$.	If $(2x - 1)(x + 4) = 0$, then by the zero-factor property $2x - 1 = 0$ or $x + 4 = 0$.
Solving a Quadratic Equation Using the Zero-Factor Property. *Step 1* Rewrite the equation if necessary so that one side is 0. *Step 2* Factor the polynomial. *Step 3* Set each factor equal to 0. *Step 4* Solve each equation from Step 3. *Step 5* Check each value, and write the solution set.	Solve. $2x^2 + 5x = 3$ $2x^2 + 5x - 3 = 0$ Standard form $(2x - 1)(x + 3) = 0$ Factor. $2x - 1 = 0$ or $x + 3 = 0$ Zero-factor property $x = \frac{1}{2}$ or $x = -3$ Solve each equation. A check verifies that the solution set is $\left\{-3, \frac{1}{2}\right\}$.

Chapter 5 Review Exercises

5.1 *Factor out the greatest common factor.*

1. $12p^2 - 6p$

2. $21x^2 + 35x$

3. $12q^2b + 8qb^2 - 20q^3b^2$

4. $6r^3t - 30r^2t^2 + 18rt^3$

5. $(x + 3)(4x - 1) - (x + 3)(3x + 2)$

6. $(z + 1)(z - 4) + (z + 1)(2z + 3)$

Factor by grouping.

7. $4m + nq + mn + 4q$

8. $x^2 + 5y + 5x + xy$

9. $2m + 6 - am - 3a$

10. $x^2 + 3x - 3y - xy$

5.2 *Factor completely.*

11. $3p^2 - p - 4$

12. $6k^2 + 11k - 10$

13. $12r^2 - 5r - 3$

14. $10m^2 + 37m + 30$

15. $10k^2 - 11kh + 3h^2$

16. $9x^2 + 4xy - 2y^2$

17. $24x - 2x^2 - 2x^3$

18. $6b^3 - 9b^2 - 15b$

19. $y^4 + 2y^2 - 8$

20. $2k^4 - 5k^2 - 3$

21. $p^2(p + 2)^2 + p(p + 2)^2 - 6(p + 2)^2$

22. $3(r + 5)^2 - 11(r + 5) - 4$

23. When asked to factor $x^2y^2 - 6x^2 + 5y^2 - 30$, a student gave the following incorrect answer.

$$x^2(y^2 - 6) + 5(y^2 - 6)$$

WHAT WENT WRONG? What is the correct answer?

Answers:

1. $6p(2p - 1)$
2. $7x(3x + 5)$
3. $4qb(3q + 2b - 5q^2b)$
4. $6rt(r^2 - 5rt + 3t^2)$
5. $(x + 3)(x - 3)$
6. $(z + 1)(3z - 1)$
7. $(m + q)(4 + n)$
8. $(x + y)(x + 5)$
9. $(m + 3)(2 - a)$
10. $(x + 3)(x - y)$
11. $(3p - 4)(p + 1)$
12. $(3k - 2)(2k + 5)$
13. $(3r + 1)(4r - 3)$
14. $(2m + 5)(5m + 6)$
15. $(2k - h)(5k - 3h)$
16. prime
17. $2x(4 + x)(3 - x)$
18. $3b(2b - 5)(b + 1)$
19. $(y^2 + 4)(y^2 - 2)$
20. $(2k^2 + 1)(k^2 - 3)$
21. $(p + 2)^2(p + 3)(p - 2)$
22. $(3r + 16)(r + 1)$

23. It is not factored because there are two terms: $x^2(y^2 - 6)$ and $5(y^2 - 6)$. The correct answer is $(y^2 - 6)(x^2 + 5)$.

24. $p + 1$

25. $(4x + 5)(4x - 5)$

26. $(3t + 7)(3t - 7)$

27. $(6m - 5n)(6m + 5n)$

28. $(x + 7)^2$

29. $(3k - 2)^2$

30. $(r + 3)(r^2 - 3r + 9)$

31. $(5x - 1)(25x^2 + 5x + 1)$

32. $(m + 1)(m^2 - m + 1) \cdot (m - 1)(m^2 + m + 1)$

33. $(x^4 + 1)(x^2 + 1)(x + 1) \cdot (x - 1)$

34. $(x + 3 + 5y)(x + 3 - 5y)$

35. $2b(3a^2 + b^2)$

36. $(x + 1)(x - 1)(x - 2) \cdot (x^2 + 2x + 4)$

37. $\{4\}$ **38.** $\left\{-1, -\frac{2}{5}\right\}$

39. $\{2, 3\}$ **40.** $\{-4, 2\}$

41. $\left\{-\frac{5}{2}, \frac{10}{3}\right\}$ **42.** $\left\{-\frac{3}{2}, \frac{1}{3}\right\}$

43. $\left\{-\frac{3}{2}, -\frac{1}{4}\right\}$ **44.** $\{-3, 3\}$

45. $\left\{-\frac{3}{2}, 0\right\}$ **46.** $\left\{\frac{1}{2}, 1\right\}$

47. $\{1, 4\}$ **48.** $\left\{-\frac{7}{2}, 0, 4\right\}$

49. $\{-3, -2, 2\}$

50. $\left\{-2, -\frac{6}{5}, 3\right\}$

51. 3 ft

52. length: 60 ft; width: 40 ft

53. 16 sec

54. 1 sec and 15 sec

55. 8 sec

56. The rock reaches a height of 240 ft once on its way up and once on its way down.

57. $k = \frac{-3s - 2t}{b - 1}$, or $k = \frac{3s + 2t}{1 - b}$

58. $w = \frac{7}{z - 3}$, or $w = \frac{-7}{3 - z}$

24. If the area of this rectangle is represented by

$$4p^2 + 3p - 1,$$

what is the width in terms of p?

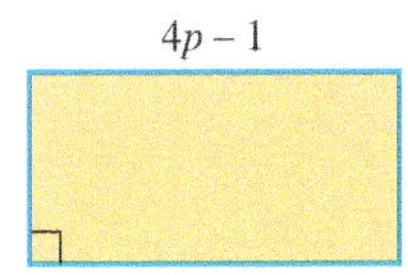

5.3 *Factor completely.*

25. $16x^2 - 25$ **26.** $9t^2 - 49$ **27.** $36m^2 - 25n^2$

28. $x^2 + 14x + 49$ **29.** $9k^2 - 12k + 4$ **30.** $r^3 + 27$

31. $125x^3 - 1$ **32.** $m^6 - 1$ **33.** $x^8 - 1$

34. $x^2 + 6x + 9 - 25y^2$ **35.** $(a + b)^3 - (a - b)^3$ **36.** $x^5 - x^3 - 8x^2 + 8$

5.5 *Solve each equation.*

37. $x^2 - 8x + 16 = 0$ **38.** $(5x + 2)(x + 1) = 0$ **39.** $x^2 - 5x + 6 = 0$

40. $x^2 + 2x = 8$ **41.** $6x^2 = 5x + 50$ **42.** $6x^2 + 7x = 3$

43. $8x^2 + 14x + 3 = 0$ **44.** $-4x^2 + 36 = 0$

45. $6x^2 + 9x = 0$ **46.** $(2x + 1)(x - 2) = -3$

47. $(x - 1)(x + 3) = (2x - 1)(x - 1)$ **48.** $2x^3 - x^2 - 28x = 0$

49. $-x^3 - 3x^2 + 4x + 12 = 0$ **50.** $(x + 2)(5x^2 - 9x - 18) = 0$

Solve each problem.

51. A triangular wall brace has the shape of a right triangle. One of the perpendicular sides is 1 ft longer than twice the other. The area enclosed by the triangle is 10.5 ft^2. Find the shorter of the perpendicular sides.

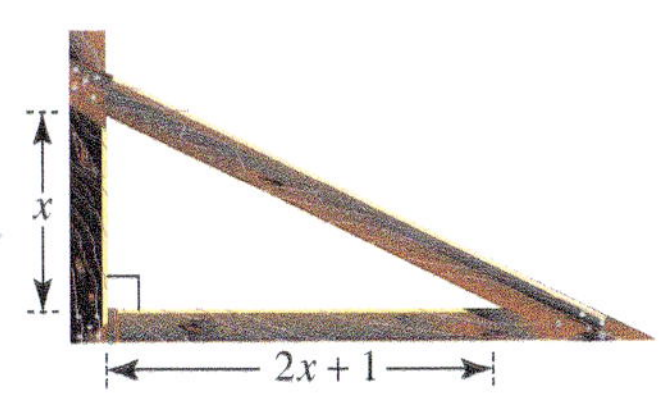

The area is 10.5 ft^2.

52. A rectangular parking lot has a length 20 ft more than its width. Its area is 2400 ft^2. What are the dimensions of the lot?

The area is 2400 ft^2.

A rock is projected directly upward from ground level. After t seconds, its height (if air resistance is neglected) is modeled by the function

$$f(t) = -16t^2 + 256t.$$

53. After how many seconds will the rock return to the ground?

54. After how many seconds will it be 240 ft above the ground?

55. After how many seconds does the rock reach its maximum height, 1024 ft?

56. Why does the question in **Exercise 54** have two answers?

Solve each equation for the specified variable.

57. $3s + bk = k - 2t$ for k

58. $z = \frac{3w + 7}{w}$ for w

Chapter 5 Mixed Review Exercises

Factor completely.

1. $16 - 81k^2$
2. $30a + am - am^2$
3. $9x^2 + 13xy - 3y^2$
4. $8 - a^3$
5. $25z^2 - 30zm + 9m^2$
6. $15y^3 + 20y^2$

Solve.

7. $5x^2 - 17x = 12$
8. $x^3 - x = 0$
9. The length of a rectangular picture frame is 2 in. longer than its width. The area enclosed by the frame is 48 in.^{2}. What is the width?
10. When Europeans arrived in America, many Native Americans of the Northeast lived in *longhouses* that sheltered several related families. The rectangular floor area of a typical Huron longhouse was about 2750 ft^2. The length was 85 ft greater than the width. What were the dimensions of the floor?

Answers:

1. $(4 + 9k)(4 - 9k)$
2. $a(6 - m)(5 + m)$
3. prime
4. $(2 - a)(4 + 2a + a^2)$
5. $(5z - 3m)^2$
6. $5y^2(3y + 4)$
7. $\left\{-\frac{3}{5}, 4\right\}$
8. $\{-1, 0, 1\}$
9. 6 in.
10. width: 25 ft; length: 110 ft

Chapter 5 Test

FOR EXTRA HELP *Step-by-step test solutions are found on the Chapter Test Prep Videos available in* MyMathLab® *or on* YouTube.

View the complete solutions to all Chapter Test exercises in MyMathLab.

Factor.

1. $11z^2 - 44z$
2. $10x^2y^5 - 5x^2y^3 - 25x^5y^3$
3. $3x + by + bx + 3y$
4. $-2x^2 - x + 36$
5. $6x^2 + 11x - 35$
6. $4p^2 + 3pq - q^2$
7. $16a^2 + 40ab + 25b^2$
8. $x^2 + 2x + 1 - 4z^2$
9. $a^3 + 2a^2 - ab^2 - 2b^2$
10. $9k^2 - 121j^2$
11. $y^3 - 216$
12. $6k^4 - k^2 - 35$
13. $27x^6 + 1$
14. Which one of the following is *not* a factored form of $-x^2 - x + 12$?
   A. $(3 - x)(x + 4)$ **B.** $-(x - 3)(x + 4)$
   C. $(-x + 3)(x + 4)$ **D.** $(x - 3)(-x + 4)$

Solve each equation.

15. $3x^2 + 8x = -4$
16. $3x^2 - 5x = 0$
17. $5m(m - 1) = 2(1 - m)$
18. $ar + 2 = 3r - 6t$ for r

Solve each problem.

19. The area of the rectangle shown is 40 in.^{2}. Find the length and the width of the rectangle.

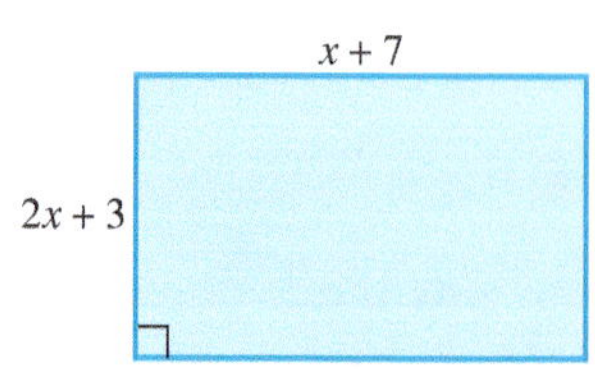

The area is 40 in.^{2}.

20. A ball is projected upward from ground level. After t seconds, its height in feet is modeled by the function

$$f(t) = -16t^2 + 96t.$$

After how many seconds will it reach a height of 128 ft?

Answers:

[5.1–5.4]

1. $11z(z - 4)$
2. $5x^2y^3(2y^2 - 1 - 5x^3)$
3. $(x + y)(3 + b)$
4. $-(2x + 9)(x - 4)$
5. $(3x - 5)(2x + 7)$
6. $(4p - q)(p + q)$
7. $(4a + 5b)^2$
8. $(x + 1 + 2z)(x + 1 - 2z)$
9. $(a + b)(a - b)(a + 2)$
10. $(3k + 11j)(3k - 11j)$
11. $(y - 6)(y^2 + 6y + 36)$
12. $(2k^2 - 5)(3k^2 + 7)$
13. $(3x^2 + 1)(9x^4 - 3x^2 + 1)$

[5.2]

14. D

[5.5]

15. $\left\{-2, -\frac{2}{3}\right\}$
16. $\left\{0, \frac{5}{3}\right\}$
17. $\left\{-\frac{2}{5}, 1\right\}$
18. $r = \frac{-2 - 6t}{a - 3}$, or $r = \frac{2 + 6t}{3 - a}$
19. length: 8 in.; width: 5 in.
20. 2 sec and 4 sec

Chapters R–5 Cumulative Review Exercises

[R.4]
1. $-2m + 6$
2. $2x^2 + 5x + 4$

[R.3]
3. 10
4. undefined

[1.1]
5. $\left\{\frac{7}{6}\right\}$
6. $\{-1\}$

[1.5]
7. $\left(-\frac{1}{2}, \infty\right)$

[1.6]
8. $(2, 3)$
9. $(-\infty, 2) \cup (3, \infty)$

[1.7]
10. $\left\{-\frac{16}{5}, 2\right\}$
11. $(-11, 7)$
12. $(-\infty, -2] \cup [7, \infty)$

[1.4]
13. 2 hr

[2.2]
14. 0
15. -1

[2.1]
16.

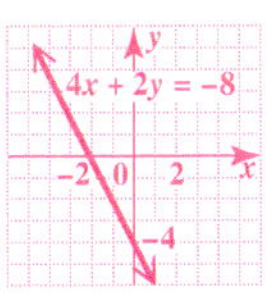

[2.6]
17. -1
18. $\left(-\frac{7}{2}, 0\right)$
19. $(0, 7)$

[3.1]
20. $\{(1, 5)\}$
21. $\varnothing$

[3.2]
22. $\{(1, 1, 0)\}$

Simplify.

1. $-2(m - 3)$

2. $3x^2 - 4x + 4 + 9x - x^2$

Evaluate for $p = -4$, $q = -2$, and $r = 5$.

3. $\dfrac{5p + 6r^2}{p^2 + q - 1}$

4. $\dfrac{\sqrt{r}}{p + 2q}$

Solve.

5. $2x - 5 + 3x = 4 - (x + 2)$

6. $\dfrac{3x - 1}{5} + \dfrac{x + 2}{2} = -\dfrac{3}{10}$

7. $3 - 2(x + 3) < 4x$

8. $2x + 4 < 10$ and $3x - 1 > 5$

9. $2x + 4 > 10$ or $3x - 1 < 5$

10. $|5x + 3| - 10 = 3$

11. $|x + 2| < 9$

12. $|2x - 5| \geq 9$

13. Two planes leave the Dallas–Fort Worth airport at the same time. One travels east at 550 mph, and the other travels west at 500 mph. Assuming no wind, how long will it take for the planes to be 2100 mi apart?

Plane	r	t	d
Eastbound	550	x	
Westbound	500	x	
			← Total

14. What is the slope of the line shown here?

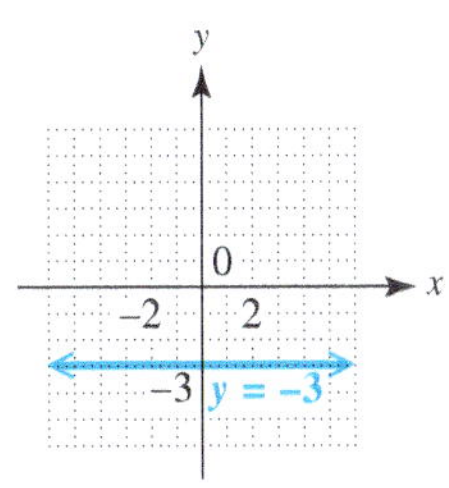

15. Find the slope of the line passing through the points $(-4, 8)$ and $(-2, 6)$.

16. Graph $4x + 2y = -8$.

Use the function $f(x) = 2x + 7$ to find each of the following.

17. $f(-4)$

18. The x-intercept of its graph

19. The y-intercept of its graph

Solve each system.

20. $3x - 2y = -7$
$2x + 3y = 17$

21. $y = 5x + 6$
$-10x = -2y - 12$

22. $2x + 3y - 6z = 5$
$8x - y + 3z = 7$
$3x + 4y - 3z = 7$

Perform the indicated operations. In Exercises 23 and 26, assume that variables represent nonzero real numbers.

23. $(3x^2y^{-1})^{-2}(2x^{-3}y)^{-1}$

24. $(7x + 3y)^2$

25. $(3x^3 + 4x^2 - 7) - (2x^3 - 8x^2 + 3x)$

26. $\dfrac{2x^4 + x^3 + 7x^2 + 2x + 6}{x^2 + 2}$

Factor.

27. $16w^2 + 50wz - 21z^2$

28. $4x^2 - 4x + 1 - y^2$

29. $100x^4 - 81$

30. $8p^3 + 27$

Solve.

31. $(x - 1)(2x + 3)(x + 4) = 0$

32. $9x^2 = 6x - 1$

33. A sign is to have the shape of a triangle with a height 3 ft greater than the length of the base. How long should the base be if the area is to be 14 ft^2?

34. A game board has the shape of a rectangle. The longer sides are each 2 in. longer than the distance between them. The area of the board is 288 in.^{2}. Find the length of the longer sides and the distance between them.

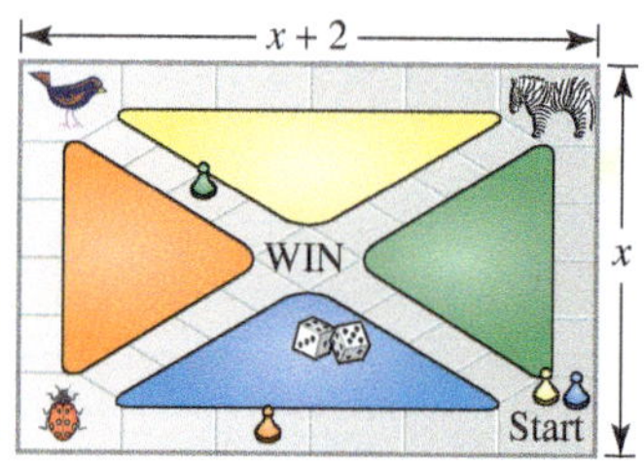

[4.1]
23. $\dfrac{y}{18x}$

[4.4]
24. $49x^2 + 42xy + 9y^2$

[4.2]
25. $x^3 + 12x^2 - 3x - 7$

[4.5]
26. $2x^2 + x + 3$

[5.1–5.4]
27. $(2w + 7z)(8w - 3z)$
28. $(2x - 1 + y)(2x - 1 - y)$
29. $(10x^2 + 9)(10x^2 - 9)$
30. $(2p + 3)(4p^2 - 6p + 9)$

[5.5]
31. $\left\{-4, -\dfrac{3}{2}, 1\right\}$
32. $\left\{\dfrac{1}{3}\right\}$
33. 4 ft
34. longer sides: 18 in.; distance between: 16 in.

6 Rational Expressions and Functions

Ratios and *proportions*, topics covered in this chapter and used in comparisons of quantities, involve *rational expressions*.

6.1 Rational Expressions and Functions; Multiplying and Dividing

OBJECTIVES

1 Define rational expressions.
2 Define rational functions and give their domains.
3 Write rational expressions in lowest terms.
4 Multiply rational expressions.
5 Find reciprocals of rational expressions.
6 Divide rational expressions.

VOCABULARY

☐ rational expression
☐ rational function
☐ reciprocal

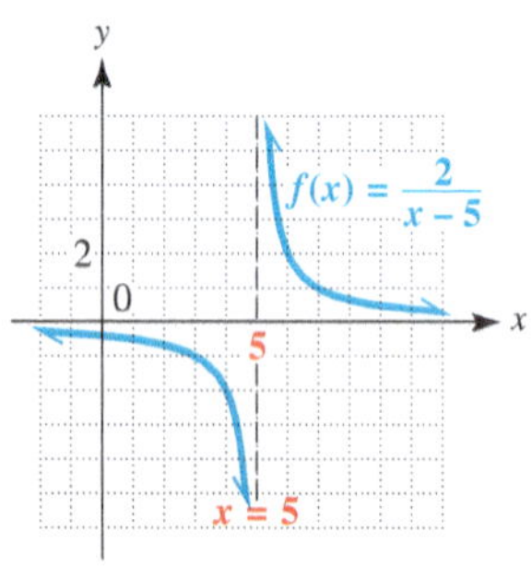

FIGURE 1

CLASSROOM EXAMPLE 1
Give the domain of each rational function using set-builder notation and interval notation.

(a) $f(x) = \frac{x+6}{x^2 - x - 6}$

(b) $f(x) = \frac{3+2x}{5}$

Answers:
(a) $\{x \mid x \text{ is a real number}, x \neq -2, 3\}$; $(-\infty, -2) \cup (-2, 3) \cup (3, \infty)$
(b) $\{x \mid x \text{ is a real number}\}$; $(-\infty, \infty)$

OBJECTIVE 1 Define rational expressions.

In arithmetic, a rational number is the quotient of two integers, with the denominator not 0. In algebra, a **rational expression,** or *algebraic fraction,* is the quotient of two polynomials, again with the denominator not 0.

$$\frac{x}{y}, \quad \frac{-a}{4}, \quad \frac{m+4}{m-2}, \quad \frac{8x^2 - 2x + 5}{4x^2 + 5x}, \quad \text{and} \quad x^5 \left(\text{or } \frac{x^5}{1}\right)$$ Rational expressions

Rational expressions are elements of the set

$$\left\{ \frac{P}{Q} \,\middle|\, P \text{ and } Q \text{ are polynomials, where } Q \neq 0 \right\}.$$

OBJECTIVE 2 Define rational functions and give their domains.

Rational Function

A **rational function** is defined by a quotient of polynomials and has the form

$$f(x) = \frac{P(x)}{Q(x)}, \quad \text{where } Q(x) \neq 0.$$

The domain of a rational function includes all real numbers except those that make $Q(x)$—that is, the denominator—equal to 0.

For example, the domain of the rational function

$$f(x) = \frac{2}{x-5} \leftarrow \text{Cannot equal 0}$$

includes all real numbers except 5 because 5 would make the denominator equal to 0. See FIGURE 1. The graph does not exist when $x = 5$. (It does not intersect the dashed vertical line with equation $x = 5$. This line is an *asymptote.*)

EXAMPLE 1 Finding Domains of Rational Functions

Give the domain of each rational function.

(a) $f(x) = \frac{3}{7x - 14}$

Values of x that make the denominator 0 must be *excluded* from the domain. To find these values, set the denominator equal to 0 and solve the resulting equation.

$$7x - 14 = 0$$
$$7x = 14 \quad \text{Add 14.}$$
$$x = 2 \quad \text{Divide by 7.}$$

The number 2 cannot be used as a replacement for x. The domain of f, which includes all real numbers *except* 2, can be written using set-builder or interval notation.

Set-builder notation: $\{x \mid x \text{ is a real number}, x \neq 2\}$

Interval notation: $(-\infty, 2) \cup (2, \infty)$

NOW TRY EXERCISE 1

Give the domain of each rational function using set-builder notation and interval notation.

(a) $\frac{2x-1}{x^2-4x-5}$

(b) $\frac{15}{2x^2+1}$

(b) $g(x) = \frac{3+x}{x^2-4x+3}$ Values that make the denominator 0 must be *excluded.*

$x^2 - 4x + 3 = 0$ Set the denominator equal to 0.

$(x-1)(x-3) = 0$ Factor.

$x - 1 = 0$ or $x - 3 = 0$ Zero-factor property

$x = 1$ or $x = 3$ Solve each equation.

Domain: $\{x \mid x \text{ is a real number}, x \neq 1, 3\}$ Set-builder notation

$(-\infty, 1) \cup (1, 3) \cup (3, \infty)$ Interval notation

(c) $h(x) = \frac{8x+2}{3}$

The denominator, 3, can never be 0, so the domain of h includes all real numbers.

Domain: $\{x \mid x \text{ is a real number}\}$ Set-builder notation

$(-\infty, \infty)$ Interval notation

(d) $f(x) = \frac{2}{x^2+4}$

Setting $x^2 + 4$ equal to 0 leads to $x^2 = -4$. There is no real number whose square is -4. Therefore, as in part (c), any real number can be used as a replacement for x.

Domain: $\{x \mid x \text{ is a real number}\}$ Set-builder notation

$(-\infty, \infty)$ Interval notation NOW TRY

TEACHING TIP

Good factoring skill is required for this and subsequent sections. Review examples of the factoring techniques introduced in Chapter 5.

OBJECTIVE 3 Write rational expressions in lowest terms.

In arithmetic, we use the **fundamental property of rational numbers** to write a fraction in lowest terms by dividing out any common factors in the numerator and denominator.

Fundamental Property of Rational Numbers

If $\frac{a}{b}$ is a rational number and if c is any nonzero real number, then

$$\frac{a}{b} = \frac{ac}{bc}.$$

That is, the numerator and denominator of a rational number may either be multiplied or divided by the same *nonzero number* without changing the value of the rational number.

Example: $\frac{15}{20} = \frac{3 \cdot 5}{4 \cdot 5} = \frac{3}{4} \cdot 1 = \frac{3}{4}$

Because $\frac{c}{c}$ is equivalent to 1, the fundamental property is based on the identity property of multiplication.

We write a rational expression in lowest terms similarly.

Writing a Rational Expression in Lowest Terms

Step 1 **Factor** both numerator and denominator to find their greatest common factor (GCF).

Step 2 **Apply the fundamental property.** Divide out common factors.

NOW TRY ANSWERS

1. (a) $\{x \mid x \text{ is a real number}, x \neq -1, 5\}$; $(-\infty, -1) \cup (-1, 5) \cup (5, \infty)$

(b) $\{x \mid x \text{ is a real number}\}$; $(-\infty, \infty)$

NOW TRY EXERCISE 2

Write each rational expression in lowest terms.

(a) $\dfrac{3a^2 - 7a + 2}{a^2 + 2a - 8}$

(b) $\dfrac{t^3 + 8}{t + 2}$

(c) $\dfrac{am - bm + an - bn}{am + bm + an + bn}$

CLASSROOM EXAMPLE 2

Write each rational expression in lowest terms.

(a) $\dfrac{y^2 + 2y - 3}{y^2 - 3y + 2}$

(b) $\dfrac{y + 2}{y^2 + 4}$

(c) $\dfrac{1 + p^3}{1 + p}$

Answers: (a) $\dfrac{y+3}{y-2}$

(b) It is already in lowest terms.

(c) $1 - p + p^2$

EXAMPLE 2 Writing Rational Expressions in Lowest Terms

Write each rational expression in lowest terms.

(a) $\dfrac{(x + 5)(x + 2)}{(x + 2)(x - 3)}$ $\dfrac{x+2}{x+2} = 1$

$= \dfrac{(x + 5)(x + 2)}{(x - 3)(x + 2)}$ Commutative property

$= \dfrac{x + 5}{x - 3}$ Fundamental property

(b) $\dfrac{a^2 - a - 6}{a^2 + 5a + 6}$

$= \dfrac{(a - 3)(a + 2)}{(a + 3)(a + 2)}$ Factor the numerator. Factor the denominator.

$= \dfrac{a - 3}{a + 3}$ $\dfrac{a+2}{a+2} = 1$; Fundamental property

(c) $\dfrac{y^2 - 4}{2y + 4}$

$= \dfrac{(y + 2)(y - 2)}{2(y + 2)}$ Factor the difference of squares in the numerator. Factor the denominator.

$= \dfrac{y - 2}{2}$ $\dfrac{y+2}{y+2} = 1$; Fundamental property

(d) $\dfrac{x^3 - 27}{x - 3}$

$= \dfrac{(x - 3)(x^2 + 3x + 9)}{x - 3}$ Factor the difference of cubes in the numerator.

$= x^2 + 3x + 9$ Fundamental property

(e) $\dfrac{pr + qr + ps + qs}{pr + qr - ps - qs}$

$= \dfrac{(pr + qr) + (ps + qs)}{(pr + qr) - (ps + qs)}$ Group the terms. Be careful with signs.

$= \dfrac{r(p + q) + s(p + q)}{r(p + q) - s(p + q)}$ Factor within the groups.

$= \dfrac{(p + q)(r + s)}{(p + q)(r - s)}$ Factor by grouping.

$= \dfrac{r + s}{r - s}$ Fundamental property

(f) $\dfrac{8 + k}{16}$ Be careful. The numerator cannot be factored.

This expression cannot be simplified further and is in lowest terms.

NOW TRY

TEACHING TIP

Be sure students understand that they are dividing out the greatest common factor.

Emphasize the difference between Examples 2(a) and 2(f). Students often try to simplify in Example 2(f).

NOW TRY ANSWERS

2. (a) $\dfrac{3a - 1}{a + 4}$

(b) $t^2 - 2t + 4$

(c) $\dfrac{a - b}{a + b}$

TEACHING TIP
Refer students to the Caution, and replace y with a number and then evaluate the fraction. Show how the result differs from y and $y - 1$. (For example, let $y = 5$.)

CAUTION ***When using the fundamental property of rational numbers, only common factors may be divided out.*** For example,

$$\frac{y-2}{2} \neq y \quad \text{and} \quad \frac{y-2}{2} \neq y - 1.$$

The 2 in $y - 2$ is **not** a *factor* of the numerator.

The expression $\frac{y-2}{2}$ indicates that the *entire* numerator is being divided by 2, not just certain terms. It is already in lowest terms, although it could be written in the equivalent form $\frac{y}{2} - \frac{2}{2}$, or $\frac{y}{2} - 1$.

Factor before writing a fraction in lowest terms.

Look again at the rational expression from **Example 2(b).**

$$\frac{a^2 - a - 6}{a^2 + 5a + 6}, \quad \text{or} \quad \frac{(a-3)(a+2)}{(a+3)(a+2)}$$

TEACHING TIP
In Example 3(a), rewriting

$3 - m$ as $-1(m - 3)$

can be justified as "factoring out a factor of -1." Multiply out $-1(m - 3)$ to check this result.
Use this opportunity to explain the distinction between factoring out -1 and multiplying by $\frac{-1}{-1}$ to change the form of a fraction.

In this expression, a can take any value *except* -3 or -2, because these values make the denominator 0. In the simplified expression $\frac{a-3}{a+3}$, a cannot equal -3. Thus,

$$\frac{a^2 - a - 6}{a^2 + 5a + 6} = \frac{a-3}{a+3}, \quad \text{for all values of } a \text{ except } -3 \text{ or } -2.$$

From now on, such statements of equality will be made with the understanding that they apply only to those real numbers which make neither denominator equal 0. We will no longer state such restrictions.

NOW TRY EXERCISE 3
Write each rational expression in lowest terms.

(a) $\frac{a - 10}{10 - a}$ **(b)** $\frac{81 - y^2}{y - 9}$

CLASSROOM EXAMPLE 3
Write each rational expression in lowest terms.

(a) $\frac{r-1}{1-r}$ (b) $\frac{p-2}{4-p^2}$

Answers: (a) -1

(b) $-\frac{1}{2+p}$

(There are other correct forms.)

EXAMPLE 3 Writing Rational Expressions in Lowest Terms

Write each rational expression in lowest terms.

(a) $\frac{m-3}{3-m}$ Here, the numerator and denominator are opposites.

To write this expression in lowest terms, write the denominator as $-1(m - 3)$.

$$\frac{m-3}{3-m} = \frac{m-3}{-1(m-3)} = \frac{1}{-1} = -1$$

Alternatively, we could write the numerator as $-1(3 - m)$ and obtain the same result.

$$\frac{m-3}{3-m} = \frac{-1(3-m)}{3-m} = \frac{-1}{1} = -1$$

(b)

$$\frac{r^2 - 16}{4 - r}$$

$$= \frac{(r+4)(r-4)}{4-r} \qquad \text{Factor the difference of squares in the numerator.}$$

$$= \frac{(r+4)(r-4)}{-1(r-4)} \qquad \text{Factor out } -1 \text{ in the denominator to write } 4 - r \text{ as } -1(r-4).$$

Distinduish to check.
Distribute to check.
$-1(r - 4)$
$= -1 \cdot r - 1(-4)$
$= -r + 4$, or $4 - r$

$$= \frac{r+4}{-1} \qquad \text{Fundamental property}$$

$$= -(r+4), \quad \text{or} \quad -r - 4 \qquad \text{Lowest terms}$$

NOW TRY

NOW TRY ANSWERS
3. (a) -1
(b) $-(y + 9)$, or $-y - 9$

As shown in **Example 3,** the quotient $\frac{a}{-a}$ (where $a \neq 0$) can be simplified.

$$\frac{a}{-a} = \frac{a}{-1(a)} = \frac{1}{-1} = -1$$

The quotient $\frac{-a}{a}$ (again where $a \neq 0$) can be simplified similarly.

The following statement summarizes these results.

Quotient of Opposites

In general, if the numerator and the denominator of a rational expression are opposites, then the expression equals -1.

Based on this result, the following are true statements.

$$\frac{q-7}{7-q} = -1 \quad \text{and} \quad \frac{-5a+2b}{5a-2b} = -1$$

Numerator and denominator in each expression are opposites.

However, the following expression cannot be simplified further.

$$\frac{r-2}{r+2}$$

Numerator and denominator are *not* opposites.

OBJECTIVE 4 Multiply rational expressions.

Recall that if $\frac{a}{b}$ and $\frac{c}{d}$ are rational numbers (where $b \neq 0$ and $d \neq 0$), then

$$\frac{a}{b} \cdot \frac{c}{d} = \frac{ac}{bd}.$$

Multiply numerators.
Multiply denominators.

Example: $\frac{2}{3} \cdot \frac{4}{5} = \frac{2 \cdot 4}{3 \cdot 5} = \frac{8}{15}$

Sometimes it is possible to factor and then divide out any common factors before multiplying.

$$\frac{7}{10} \cdot \frac{5}{9} = \frac{7}{2 \cdot 5} \cdot \frac{5}{3 \cdot 3} = \frac{7}{18}$$

We multiply rational expressions similarly, using the following steps.

Multiplying Rational Expressions

Step 1 **Factor** all numerators and denominators as completely as possible.

Step 2 **Apply the fundamental property.**

Step 3 **Multiply** remaining factors in the numerators and remaining factors in the denominators. Leave the denominator in factored form.

Step 4 **Check** to be sure the product is in lowest terms.

NOW TRY EXERCISE 4

Multiply.

(a) $\frac{8t^2}{t^2-4} \cdot \frac{3t+6}{9t}$

(b) $\frac{m^2+2m-15}{m^2-5m+6} \cdot \frac{m^2-4}{m^2+5m}$

CLASSROOM EXAMPLE 4

Multiply.

(a) $\frac{c^2+2c}{c^2-4} \cdot \frac{c^2-4c+4}{c^2-c}$

(b) $\frac{m^2-16}{m+2} \cdot \frac{1}{m+4}$

Answers: (a) $\frac{c-2}{c-1}$ (b) $\frac{m-4}{m+2}$

TEACHING TIP

Remind students to factor all numerators and denominators first when working problems like those in **Examples 4 and 5.**

NOW TRY ANSWERS

4. (a) $\frac{8t}{3(t-2)}$ (b) $\frac{m+2}{m}$

EXAMPLE 4 Multiplying Rational Expressions

Multiply.

(a) $\frac{5p-5}{p} \cdot \frac{3p^2}{10p-10}$

$= \frac{5(p-1)}{p} \cdot \frac{3p \cdot p}{2 \cdot 5(p-1)}$ Factor.

$= \frac{5(p-1)}{5(p-1)} \cdot \frac{p}{p} \cdot \frac{3p}{2}$ Commutative property

$= 1 \cdot 1 \cdot 1 \cdot \frac{3p}{2}$ Fundamental property

In practice, this step is usually done mentally.

$= \frac{3p}{2}$ Lowest terms

(b) $\frac{k^2+2k-15}{k^2-4k+3} \cdot \frac{k^2-k}{k^2+k-20}$

$= \frac{(k+5)(k-3)}{(k-3)(k-1)} \cdot \frac{k(k-1)}{(k+5)(k-4)}$ Factor.

$= \frac{k}{k-4}$ Fundamental property; Multiply.

(c) $(p-4) \cdot \frac{3}{5p-20}$

$= \frac{p-4}{1} \cdot \frac{3}{5p-20}$ Write $p-4$ as $\frac{p-4}{1}$.

$= \frac{p-4}{1} \cdot \frac{3}{5(p-4)}$ Factor.

$= \frac{3}{5}$ Fundamental property; Multiply.

(d) $\frac{x^2+2x}{x+1} \cdot \frac{x^2-1}{x^3+x^2}$

$= \frac{x(x+2)}{x+1} \cdot \frac{(x+1)(x-1)}{x^2(x+1)}$ Factor.

$= \frac{(x+2)(x-1)}{x(x+1)}$ Fundamental property; Multiply.

The final expression can be left in factored form. It is in lowest terms.

(e) $\frac{x-6}{x^2-12x+36} \cdot \frac{x^2-3x-18}{x^2+7x+12}$

$= \frac{x-6}{(x-6)^2} \cdot \frac{(x+3)(x-6)}{(x+3)(x+4)}$ Factor.

Remember to include 1 in the numerator when all other factors are eliminated.

$= \frac{1}{x+4}$ Fundamental property; Multiply.

NOW TRY

Reciprocals of Rational Expressions

Rational Expression	Reciprocal
3, or $\frac{3}{1}$	$\frac{1}{3}$
$\frac{5}{k}$	$\frac{k}{5}$
$\frac{m^2 - 9m}{2}$	$\frac{2}{m^2 - 9m}$
$\frac{0}{4}$	undefined

Reciprocals have a product of 1. Recall that 0 has no reciprocal.

OBJECTIVE 5 Find reciprocals of rational expressions.

When dividing rational numbers, we use reciprocals. The rational numbers $\frac{a}{b}$ and $\frac{c}{d}$ are reciprocals of each other if they have a product of 1.

The **reciprocal** of a rational expression is defined in the same way. ***Two rational expressions are reciprocals of each other if they have a product of 1.***

Finding the Reciprocal

To find the reciprocal of a nonzero rational expression, interchange the numerator and denominator of the expression. (See the table.)

OBJECTIVE 6 Divide rational expressions.

Recall that if $\frac{a}{b}$ and $\frac{c}{d}$ are rational numbers (where $b \neq 0$, $c \neq 0$, and $d \neq 0$), then

$$\frac{a}{b} \div \frac{c}{d} = \frac{a}{b} \cdot \frac{d}{c} = \frac{ad}{bc}.$$ To divide by a fraction, multiply by its reciprocal.

Example: $\frac{3}{4} \div \frac{2}{3} = \frac{3}{4} \cdot \frac{3}{2} = \frac{9}{8}$

Dividing rational expressions is similar to dividing rational numbers.

Dividing Rational Expressions

To divide two rational expressions, *multiply* the first expression (the *dividend*) by the reciprocal of the second expression (the *divisor*).

CLASSROOM EXAMPLE 5

Divide.

(a) $\frac{5p + 2}{6} \div \frac{15p + 6}{5}$

(b) $\frac{q^2 + 2q}{5 + q} \div \frac{4 - q^2}{3q - 6}$

Answers: (a) $\frac{5}{18}$ (b) $-\frac{3q}{5 + q}$

EXAMPLE 5 Dividing Rational Expressions

Divide.

(a) $\frac{2z}{9} \div \frac{5z^2}{18}$

$= \frac{2z}{9} \cdot \frac{18}{5z^2}$ Multiply by the reciprocal of the divisor.

$= \frac{2z}{9} \cdot \frac{2 \cdot 9}{5z \cdot z}$ Factor.

$= \frac{4}{5z}$ Fundamental property; Multiply.

(b) $\frac{8k - 16}{3k} \div \frac{3k - 6}{4k^2}$

$= \frac{8k - 16}{3k} \cdot \frac{4k^2}{3k - 6}$ Multiply by the reciprocal.

$= \frac{8(k - 2)}{3k} \cdot \frac{4k \cdot k}{3(k - 2)}$ Factor.

$= \frac{32k}{9}$ Fundamental property; Multiply.

NOW TRY EXERCISE 5

Divide.

(a) $\frac{16k^2}{5} \div \frac{3k}{10}$

(b) $\frac{3k^2+5k-2}{9k^2-1} \div \frac{4k^2+8k}{k^2-7k}$

(c) $\frac{5m^2+17m-12}{3m^2+7m-20} \div \frac{5m^2+2m-3}{15m^2-34m+15}$

$= \frac{5m^2+17m-12}{3m^2+7m-20} \cdot \frac{15m^2-34m+15}{5m^2+2m-3}$ Definition of division

$= \frac{(5m-3)(m+4)}{(3m-5)(m+4)} \cdot \frac{(5m-3)(3m-5)}{(5m-3)(m+1)}$ Factor.

$= \frac{5m-3}{m+1}$ Fundamental property; Multiply. NOW TRY

NOW TRY ANSWERS

5. (a) $\frac{32k}{3}$ (b) $\frac{k-7}{4(3k+1)}$

6.1 Exercises

FOR EXTRA HELP

MyMathLab®

Complete solution available in MyMathLab

Concept Check *Complete each statement.*

1. A rational number such as $\frac{3}{4}$ is the quotient of two ________, with denominator not ____. A ________ expression such as $\frac{2x^2}{2x+1}$ is the quotient of two ________, with denominator not ____.

2. A function of the form $f(x) = \frac{P(x)}{\_\_\_\_}$, where $P(x)$ and $Q(x)$ are polynomials and $Q(x) \neq$ ____, is a ________ function. The domain includes all ________ except any values of x that make the (*numerator/denominator*) equal to 0.

Give the domain of each rational function using ***(a)*** *set-builder notation and* ***(b)*** *interval notation. See Example 1.*

3. $f(x) = \frac{x}{x-7}$

4. $f(x) = \frac{x}{x+3}$

5. $f(x) = \frac{6x-5}{7x+1}$

6. $f(x) = \frac{8x-3}{2x+7}$

7. $f(x) = \frac{12x+3}{x}$

8. $f(x) = \frac{9x+8}{x}$

9. $f(x) = \frac{3x+1}{2x^2+x-6}$

10. $f(x) = \frac{2x+4}{3x^2+11x-42}$

11. $f(x) = \frac{x+2}{14}$

12. $f(x) = \frac{x-9}{26}$

13. $f(x) = \frac{2x^2-3x+4}{3x^2+8}$

14. $f(x) = \frac{9x^2-8x+3}{4x^2+1}$

Concept Check *As review, multiply or divide the rational numbers as indicated. Write answers in lowest terms.*

15. $\frac{4}{21} \cdot \frac{7}{10}$

16. $\frac{5}{9} \cdot \frac{12}{25}$

17. $\frac{3}{8} \div \frac{5}{12}$

18. $\frac{5}{6} \div \frac{14}{15}$

19. $\frac{2}{3} \div \frac{8}{9}$

20. $\frac{3}{8} \div \frac{9}{14}$

1. integers; 0; rational; polynomials; 0
2. $Q(x)$; 0; rational; real numbers; denominator
3. (a) $\{x \mid x \text{ is a real number}, x \neq 7\}$ (b) $(-\infty, 7) \cup (7, \infty)$
4. (a) $\{x \mid x \text{ is a real number}, x \neq -3\}$ (b) $(-\infty, -3) \cup (-3, \infty)$
5. (a) $\{x \mid x \text{ is a real number}, x \neq -\frac{1}{7}\}$ (b) $(-\infty, -\frac{1}{7}) \cup (-\frac{1}{7}, \infty)$
6. (a) $\{x \mid x \text{ is a real number}, x \neq -\frac{7}{2}\}$ (b) $(-\infty, -\frac{7}{2}) \cup (-\frac{7}{2}, \infty)$
7. (a) $\{x \mid x \text{ is a real number}, x \neq 0\}$ (b) $(-\infty, 0) \cup (0, \infty)$
8. (a) $\{x \mid x \text{ is a real number}, x \neq 0\}$ (b) $(-\infty, 0) \cup (0, \infty)$
9. (a) $\{x \mid x \text{ is a real number}, x \neq -2, \frac{3}{2}\}$ (b) $(-\infty, -2) \cup (-2, \frac{3}{2}) \cup (\frac{3}{2}, \infty)$
10. (a) $\{x \mid x \text{ is a real number}, x \neq -6, \frac{7}{3}\}$ (b) $(-\infty, -6) \cup (-6, \frac{7}{3}) \cup (\frac{7}{3}, \infty)$
11. (a) $\{x \mid x \text{ is a real number}\}$ (b) $(-\infty, \infty)$
12. (a) $\{x \mid x \text{ is a real number}\}$ (b) $(-\infty, \infty)$
13. (a) $\{x \mid x \text{ is a real number}\}$ (b) $(-\infty, \infty)$

14. **(a)** $\{x \mid x \text{ is a real number}\}$
(b) $(-\infty, \infty)$

15. $\frac{2}{15}$ **16.** $\frac{4}{15}$

17. $\frac{9}{10}$ **18.** $\frac{25}{28}$

19. $\frac{3}{4}$ **20.** $\frac{7}{12}$

21. **(a)** C **(b)** A **(c)** D **(d)** B **(e)** E **(f)** F

22. numerator: x^2, $4x$; denominator: x, 4; x

23. B, E, F **24.** D

25. B **26.** B, D

27. x **28.** y

29. $\frac{x-3}{x+5}$ **30.** $\frac{x-1}{2x+3}$

31. $\frac{x+3}{2x(x-3)}$ **32.** $\frac{y(y+8)}{3(y-8)}$

33. It is already in lowest terms.

34. It is already in lowest terms.

35. $\frac{6}{7}$ **36.** $\frac{5}{3}$

37. $\frac{z}{6}$ **38.** $\frac{x}{8}$

39. $\frac{t-3}{3}$ **40.** $\frac{m+5}{4}$

41. $\frac{2}{t-3}$ **42.** $\frac{5}{s+5}$

43. $\frac{x-3}{x+1}$ **44.** $\frac{y-7}{y-1}$

45. $\frac{4x+1}{4x+3}$ **46.** $\frac{6x-5}{4x-5}$

47. $a^2 - ab + b^2$

48. $r^2 + rs + s^2$

49. $\frac{c+6d}{c-d}$ **50.** $\frac{s+3t}{s+4t}$

51. $\frac{a+b}{a-b}$ **52.** $\frac{y+w}{y-w}$

21. *Concept Check* Rational expressions can often be written in lowest terms in seemingly different ways. For example, although the expressions

$$\frac{y-3}{-5} \quad \text{and} \quad \frac{-y+3}{5}$$

look different, we can obtain the second expression by multiplying the first by 1 in the form $\frac{-1}{-1}$. To practice recognizing equivalent rational expressions, match the expressions in parts (a)–(f) with their equivalents in choices A–F.

(a) $\frac{x-3}{x+4}$ **(b)** $\frac{x+3}{x-4}$ **(c)** $\frac{x-3}{x-4}$ **(d)** $\frac{x+3}{x+4}$ **(e)** $\frac{3-x}{x+4}$ **(f)** $\frac{x+3}{4-x}$

A. $\frac{-x-3}{4-x}$ **B.** $\frac{-x-3}{-x-4}$ **C.** $\frac{3-x}{-x-4}$ **D.** $\frac{-x+3}{-x+4}$ **E.** $\frac{x-3}{-x-4}$ **F.** $\frac{-x-3}{x-4}$

22. *Concept Check* Identify the two *terms* in the numerator and the two *terms* in the denominator of the rational expression $\frac{x^2+4x}{x+4}$, and write it in lowest terms.

Concept Check *Answer each question.*

23. Which rational expressions are equivalent to $-\frac{x}{y}$?

A. $\frac{-x}{-y}$ **B.** $\frac{x}{-y}$ **C.** $\frac{x}{y}$ **D.** $-\frac{x}{-y}$ **E.** $\frac{-x}{y}$ **F.** $-\frac{-x}{-y}$

24. Which rational expression can be simplified?

A. $\frac{x^2+2}{x^2}$ **B.** $\frac{x^2+2}{2}$ **C.** $\frac{x^2+y^2}{y^2}$ **D.** $\frac{x^2-5x}{x}$

25. Which rational expression is *not* equivalent to $\frac{x-3}{4-x}$?

A. $\frac{3-x}{x-4}$ **B.** $\frac{x+3}{4+x}$ **C.** $-\frac{3-x}{4-x}$ **D.** $-\frac{x-3}{x-4}$

26. Which two rational expressions are equivalent to -1?

A. $\frac{2x+3}{2x-3}$ **B.** $\frac{2x-3}{3-2x}$ **C.** $\frac{2x+3}{3+2x}$ **D.** $\frac{2x+3}{-2x-3}$

Write each rational expression in lowest terms. ***See Example 2.***

27. $\frac{x^2(x+1)}{x(x+1)}$ **28.** $\frac{y^3(y-4)}{y^2(y-4)}$ **29.** $\frac{(x+4)(x-3)}{(x+5)(x+4)}$

30. $\frac{(2x+7)(x-1)}{(2x+3)(2x+7)}$ **31.** $\frac{4x(x+3)}{8x^2(x-3)}$ **32.** $\frac{5y^2(y+8)}{15y(y-8)}$

33. $\frac{3x+7}{3}$ **34.** $\frac{4x-9}{4}$ **35.** $\frac{6m+18}{7m+21}$

36. $\frac{5r-20}{3r-12}$ **37.** $\frac{3z^2+z}{18z+6}$ **38.** $\frac{2x^2-5x}{16x-40}$

39. $\frac{t^2-9}{3t+9}$ **40.** $\frac{m^2-25}{4m-20}$ **41.** $\frac{2t+6}{t^2-9}$

42. $\frac{5s-25}{s^2-25}$ **43.** $\frac{x^2+2x-15}{x^2+6x+5}$ **44.** $\frac{y^2-5y-14}{y^2+y-2}$

45. $\dfrac{8x^2 - 10x - 3}{8x^2 - 6x - 9}$

46. $\dfrac{12x^2 - 4x - 5}{8x^2 - 6x - 5}$

47. $\dfrac{a^3 + b^3}{a + b}$

48. $\dfrac{r^3 - s^3}{r - s}$

49. $\dfrac{2c^2 + 2cd - 60d^2}{2c^2 - 12cd + 10d^2}$

50. $\dfrac{3s^2 - 9st - 54t^2}{3s^2 - 6st - 72t^2}$

51. $\dfrac{ac - ad + bc - bd}{ac - ad - bc + bd}$

52. $\dfrac{2xy + 2xw + y + w}{2xy + y - 2xw - w}$

Write each rational expression in lowest terms. ***See Example 3.***

53. $\dfrac{7 - b}{b - 7}$

54. $\dfrac{r - 13}{13 - r}$

55. $\dfrac{x^2 - y^2}{y - x}$

56. $\dfrac{m^2 - n^2}{n - m}$

57. $\dfrac{x^2 - 4}{2 - x}$

58. $\dfrac{x^2 - 81}{9 - x}$

59. $\dfrac{(a - 3)(x + y)}{(3 - a)(x - y)}$

60. $\dfrac{(8 - p)(x + 2)}{(p - 8)(x - 2)}$

61. $\dfrac{5k - 10}{20 - 10k}$

62. $\dfrac{7x - 21}{63 - 21x}$

63. $\dfrac{a^2 - b^2}{a^2 + b^2}$

64. $\dfrac{p^2 + q^2}{p^2 - q^2}$

Multiply or divide as indicated. ***See Examples 4 and 5.***

65. $\dfrac{(x + 2)(x + 1)}{(x + 3)(x - 2)} \cdot \dfrac{(x + 3)(x + 4)}{(x + 2)(x + 1)}$

66. $\dfrac{(x + 3)(x - 4)}{(x - 4)(x + 2)} \cdot \dfrac{(x + 5)(x - 6)}{(x + 3)(x - 6)}$

67. $\dfrac{(2x + 3)(x - 4)}{(x + 8)(x - 4)} \div \dfrac{(x - 4)(x + 2)}{(x - 4)(x + 8)}$

68. $\dfrac{(6x + 5)(x - 3)}{(x - 1)(x - 3)} \div \dfrac{(2x + 7)(x + 9)}{(x - 1)(x + 9)}$

69. $\dfrac{4x}{8x + 4} \cdot \dfrac{14x + 7}{6}$

70. $\dfrac{12x - 20}{5x} \cdot \dfrac{6}{9x - 15}$

71. $\dfrac{p^2 - 25}{4p} \cdot \dfrac{2}{5 - p}$

72. $\dfrac{a^2 - 1}{4a} \cdot \dfrac{2}{1 - a}$

73. $(7k + 7) \div \dfrac{4k + 4}{5}$

74. $(8y - 16) \div \dfrac{3y - 6}{10}$

75. $(z^2 - 1) \cdot \dfrac{1}{1 - z}$

76. $(y^2 - 4) \cdot \dfrac{8}{2 - y}$

77. $\dfrac{4x - 20}{5x} \div \dfrac{2x - 10}{7x^3}$

78. $\dfrac{3x + 12}{4x} \div \dfrac{4x + 16}{6x^2}$

79. $\dfrac{m^2 - 49}{m + 1} \div \dfrac{7 - m}{m}$

80. $\dfrac{k^2 - 4}{3k^2} \div \dfrac{2 - k}{11k}$

81. $\dfrac{12x - 10y}{3x + 2y} \cdot \dfrac{6x + 4y}{10y - 12x}$

82. $\dfrac{9s - 12t}{2s + 2t} \cdot \dfrac{3s + 3t}{4t - 3s}$

83. $\dfrac{x^2 - 25}{x^2 + x - 20} \cdot \dfrac{x^2 + 7x + 12}{x^2 - 2x - 15}$

84. $\dfrac{t^2 - 49}{t^2 + 4t - 21} \cdot \dfrac{t^2 + 8t + 15}{t^2 - 2t - 35}$

85. $\dfrac{a^3 - b^3}{a^2 - b^2} \div \dfrac{2a - 2b}{2a + 2b}$

86. $\dfrac{x^3 + y^3}{2x + 2y} \div \dfrac{x^2 - y^2}{2x - 2y}$

87. $\dfrac{8x^3 - 27}{2x^2 - 18} \cdot \dfrac{2x + 6}{8x^2 + 12x + 18}$

88. $\dfrac{64x^3 + 1}{4x^2 - 100} \cdot \dfrac{4x + 20}{64x^2 - 16x + 4}$

89. $\dfrac{a^3 - 8b^3}{a^2 - ab - 6b^2} \cdot \dfrac{a^2 + ab - 12b^2}{a^2 + 2ab - 8b^2}$

90. $\dfrac{p^3 - 27q^3}{p^2 + pq - 12q^2} \cdot \dfrac{p^2 - 2pq - 24q^2}{p^2 - 5pq - 6q^2}$

91. $\dfrac{6x^2 + 5x - 6}{12x^2 - 11x + 2} \div \dfrac{4x^2 - 12x + 9}{8x^2 - 14x + 3}$

92. $\dfrac{8a^2 - 6a - 9}{6a^2 - 5a - 6} \div \dfrac{4a^2 + 11a + 6}{9a^2 + 12a + 4}$

53. -1 **54.** -1
55. $-(x + y)$, or $-x - y$
56. $-(m + n)$, or $-m - n$
57. $-(x + 2)$, or $-x - 2$
58. $-(x + 9)$, or $-x - 9$
59. $-\dfrac{x + y}{x - y}$ (There are other answers.)
60. $\dfrac{x + 2}{x - 2}$ (There are other answers.)
61. $-\dfrac{1}{2}$ **62.** $-\dfrac{1}{3}$
63. It is already in lowest terms.
64. It is already in lowest terms.
65. $\dfrac{x + 4}{x - 2}$ **66.** $\dfrac{x + 5}{x + 2}$
67. $\dfrac{2x + 3}{x + 2}$ **68.** $\dfrac{6x + 5}{2x + 7}$
69. $\dfrac{7x}{6}$ **70.** $\dfrac{8}{5x}$
71. $-\dfrac{p + 5}{2p}$ (There are other answers.)
72. $-\dfrac{a + 1}{2a}$ (There are other answers.)
73. $\dfrac{35}{4}$ **74.** $\dfrac{80}{3}$
75. $-(z + 1)$, or $-z - 1$
76. $-8(y + 2)$, or $-8y - 16$
77. $\dfrac{14x^2}{5}$ **78.** $\dfrac{9x}{8}$
79. $\dfrac{-m(m + 7)}{m + 1}$ (There are other answers.)
80. $\dfrac{-11(k + 2)}{3k}$ (There are other answers.)
81. -2 **82.** $-\dfrac{9}{2}$
83. $\dfrac{x + 4}{x - 4}$ **84.** $\dfrac{t + 3}{t - 3}$
85. $\dfrac{a^2 + ab + b^2}{a - b}$
86. $\dfrac{x^2 - xy + y^2}{x + y}$
87. $\dfrac{2x - 3}{2(x - 3)}$ **88.** $\dfrac{4x + 1}{4(x - 5)}$
89. $\dfrac{a^2 + 2ab + 4b^2}{a + 2b}$
90. $\dfrac{p^2 + 3pq + 9q^2}{p + q}$
91. $\dfrac{2x + 3}{2x - 3}$ **92.** $\dfrac{3a + 2}{a + 2}$

93. $\frac{3a^3b^2}{4}$ 94. $\frac{t^2}{16s^3}$
95. $\frac{27}{2mn^7}$ 96. $\frac{b^2}{a^6}$
97. $\frac{k+5p}{2k+5p}$ 98. $\frac{4c+3d}{4c-d}$
99. $(k-1)(k-2)$
100. $\frac{5(x-2)}{9}$

Extending Skills *Multiply or divide as indicated.*

93. $\frac{5a^4b^2}{16a^2b} \div \frac{25a^2b}{60a^3b^2}$

94. $\frac{s^3t^2}{10s^2t^4} \div \frac{8s^4t^2}{5t^6}$

95. $\frac{(-3mn)^2 \cdot 64(m^2n)^3}{16m^2n^4(mn^2)^3} \div \frac{24(m^2n^2)^4}{(3m^2n^3)^2}$

96. $\frac{(-4a^2b^3)^2 \cdot 9(a^2b^4)^2}{(2a^2b^3)^4 \cdot (3a^3b)^2} \div \frac{(ab)^4}{(a^2b^3)^2}$

97. $\frac{3k^2+17kp+10p^2}{6k^2+13kp-5p^2} \div \frac{6k^2+kp-2p^2}{6k^2-5kp+p^2}$

98. $\frac{16c^2+24cd+9d^2}{16c^2-16cd+3d^2} \div \frac{16c^2-9d^2}{16c^2-24cd+9d^2}$

99. $\left(\frac{6k^2-13k-5}{k^2+7k} \div \frac{2k-5}{k^3+6k^2-7k}\right) \cdot \frac{k^2-5k+6}{3k^2-8k-3}$

100. $\left(\frac{2x^3+3x^2-2x}{3x-15} \div \frac{2x^3-x^2}{x^2-3x-10}\right) \cdot \frac{5x^2-10x}{3x^2+12x+12}$

6.2 Adding and Subtracting Rational Expressions

OBJECTIVES

1. Add and subtract rational expressions with the same denominator.
2. Find a least common denominator.
3. Add and subtract rational expressions with different denominators.

VOCABULARY

☐ least common denominator (LCD)

OBJECTIVE 1 Add and subtract rational expressions with the same denominator.

Recall that if $\frac{a}{b}$ and $\frac{c}{b}$ are rational numbers (where $b \neq 0$), then

$$\frac{a}{b} + \frac{c}{b} = \frac{a+c}{b} \quad \text{and} \quad \frac{a}{b} - \frac{c}{b} = \frac{a-c}{b}.$$

Add or subtract the numerators.
Keep the same denominator.

Examples: $\frac{4}{7} + \frac{1}{7} = \frac{5}{7}$ and $\frac{4}{7} - \frac{1}{7} = \frac{3}{7}$

Adding and subtracting rational expressions is similar.

Adding or Subtracting Rational Expressions

Step 1 **If the denominators are the same,** add or subtract the numerators. Place the result over the common denominator.

If the denominators are different, find the least common denominator and write all rational expressions with this LCD. Add or subtract the numerators. Place the result over the common denominator.

Step 2 **Simplify.** Write all answers in lowest terms.

EXAMPLE 1 Adding and Subtracting Rational Expressions (Same Denominators)

Add or subtract as indicated.

(a) $\frac{3y}{5} + \frac{x}{5}$

$= \frac{3y+x}{5}$ ← Add the numerators.
← Keep the common denominator.

NOW TRY EXERCISE 1

Add or subtract as indicated.

(a) $\frac{5}{3x} + \frac{2}{3x}$

(b) $\frac{x^2}{x-3} - \frac{9}{x-3}$

(c) $\frac{2}{x^2+x-2} + \frac{x}{x^2+x-2}$

CLASSROOM EXAMPLE 1

Add or subtract as indicated.

(a) $\frac{7x}{9} + \frac{y}{9}$

(b) $\frac{8}{3x^3} - \frac{14}{3x^3}$

(c) $\frac{r}{r^2-t^2} + \frac{t}{r^2-t^2}$

(d) $\frac{6}{x^2+3x-18} + \frac{x}{x^2+3x-18}$

Answers: (a) $\frac{7x+y}{9}$ (b) $-\frac{2}{x^3}$ (c) $\frac{1}{r-t}$ (d) $\frac{1}{x-3}$

(b) $\frac{7}{2r^2} - \frac{11}{2r^2}$

$= \frac{7-11}{2r^2}$ Subtract the numerators. Keep the common denominator.

$= \frac{-4}{2r^2}$

$= -\frac{2}{r^2}$ Write in lowest terms.

(c) $\frac{m}{m^2-p^2} + \frac{p}{m^2-p^2}$

$= \frac{m+p}{m^2-p^2}$ Add the numerators. Keep the common denominator.

$= \frac{m+p}{(m+p)(m-p)}$ Factor.

$= \frac{1}{m-p}$ Fundamental property

Remember to write 1 in the numerator.

(d) $\frac{4}{x^2+2x-8} + \frac{x}{x^2+2x-8}$

$= \frac{4+x}{x^2+2x-8}$ Add.

$= \frac{4+x}{(x-2)(x+4)}$ Factor.

$= \frac{1}{x-2}$ Fundamental property

NOW TRY

OBJECTIVE 2 Find a least common denominator.

We add or subtract rational numbers with different denominators by first writing them with a common denominator, usually the **least common denominator (LCD).**

TEACHING TIP

Explain to students the importance of developing the ability to determine the LCD before they add and subtract fractions.

Finding the Least Common Denominator

Step 1 **Factor** each denominator.

Step 2 **Find the least common denominator.** The LCD is the product of all of the different factors from each denominator, with each factor raised to the *greatest* power that occurs in any denominator.

Example: To find the LCD for $\frac{7}{15}$ and $\frac{5}{12}$, we factor each denominator.

$$15 = 3 \cdot 5 \quad \text{and} \quad 12 = 2 \cdot 2 \cdot 3 = 2^2 \cdot 3$$

The LCD is the product of the factors 2, 3, and 5, each raised to its *greatest* power.

$$\text{LCD} = 2^2 \cdot 3 \cdot 5 = 60$$

We use a similar procedure to find the LCD of two or more rational expressions.

NOW TRY ANSWERS

1. (a) $\frac{7}{3x}$ (b) $x+3$ (c) $\frac{1}{x-1}$

NOW TRY EXERCISE 2

Find the LCD for each group of denominators.

(a) $15m^3n$, $10m^2n$

(b) t, $t - 8$

(c) $3x^2 + 9x - 30$, $x^2 - 4$, $x^2 + 10x + 25$

CLASSROOM EXAMPLE 2

Find the LCD for each group of denominators.

(a) $10a^3b^5$, $15a^2b^6$

(b) z, $z + 6$

(c) $m^2 - 16$, $m^2 + 8m + 16$

(d) $x^2 - 2x + 1$, $x^2 - 4x + 3$, $4x - 4$

Answers: (a) $30a^3b^6$

(b) $z(z + 6)$

(c) $(m + 4)^2(m - 4)$

(d) $4(x - 3)(x - 1)^2$

NOW TRY ANSWERS

2. **(a)** $30m^3n$ **(b)** $t(t - 8)$

(c) $3(x - 2)(x + 2)(x + 5)^2$

EXAMPLE 2 Finding Least Common Denominators

Suppose that the given expressions are denominators of fractions. Find the LCD for each group of denominators.

(a) $5xy^2$, $2x^3y$

$$5xy^2 = 5 \cdot x \cdot y^2$$
$$2x^3y = 2 \cdot x^3 \cdot y$$

Each denominator is already factored.

Use each factor, raised to the *greatest* power that occurs in any denominator.

$$\text{LCD} = 5 \cdot 2 \cdot x^3 \cdot y^2$$

← Greatest exponent on y is 2.

$$= 10x^3y^2$$

Greatest exponent on x is 3.

(b) $k - 3$, k Each denominator is already factored.

The LCD must be divisible by *both* $k - 3$ and k.

$$\text{LCD} = k(k - 3)$$

Don't forget the factor k.

It is usually best to leave a least common denominator in factored form.

(c) $y^2 - 2y - 8$, $y^2 + 3y + 2$

$$y^2 - 2y - 8 = (y - 4)(y + 2)$$
$$y^2 + 3y + 2 = (y + 2)(y + 1)$$

Factor.

$$\text{LCD} = (y - 4)(y + 2)(y + 1)$$

(d) $8z - 24$, $5z^2 - 15z$

$$8z - 24 = 8(z - 3)$$
$$5z^2 - 15z = 5z(z - 3)$$

Factor.

$$\text{LCD} = 8 \cdot 5z \cdot (z - 3)$$
$$= 40z(z - 3)$$

(e) $m^2 + 5m + 6$, $m^2 + 4m + 4$, $2m^2 + 4m - 6$

$$m^2 + 5m + 6 = (m + 3)(m + 2)$$
$$m^2 + 4m + 4 = (m + 2)^2$$
$$2m^2 + 4m - 6 = 2(m + 3)(m - 1)$$

Factor. Notice that $(m + 2)$ is squared in the second denominator.

Use each factor, raised to its *greatest* power in any denominator. This means that $(m + 2)^2$ must be used.

$$\text{LCD} = 2(m + 3)(m + 2)^2(m - 1)$$

NOW TRY

OBJECTIVE 3 Add and subtract rational expressions with different denominators.

Recall how we add fractions with different denominators.

Example: $\frac{7}{15} + \frac{5}{12}$ As found in **Objective 2**, the LCD for 15 and 12 is 60.

$\frac{4}{4}$ and $\frac{5}{5}$ are forms of 1, the identity element of multiplication.

$$= \frac{7 \cdot 4}{15 \cdot 4} + \frac{5 \cdot 5}{12 \cdot 5}$$ Fundamental property

$$= \frac{28}{60} + \frac{25}{60}$$ Write each fraction with the common denominator.

$$= \frac{28 + 25}{60} \qquad \text{Add the numerators. Keep the common denominator.}$$

$$= \frac{53}{60}$$

We add and subtract rational expressions with different denominators similarly.

NOW TRY EXERCISE 3

Add or subtract as indicated.

(a) $\dfrac{2}{3x} + \dfrac{7}{4x}$

(b) $\dfrac{3}{z} - \dfrac{6}{z - 5}$

CLASSROOM EXAMPLE 3

Add or subtract as indicated.

(a) $\dfrac{6}{m} + \dfrac{1}{4m}$

(b) $\dfrac{2}{y} - \dfrac{1}{y + 4}$

Answers: (a) $\dfrac{25}{4m}$ (b) $\dfrac{y + 8}{y(y + 4)}$

EXAMPLE 3 Adding and Subtracting Rational Expressions (Different Denominators)

Add or subtract as indicated.

(a) $\dfrac{5}{2p} + \dfrac{3}{8p}$ The LCD for $2p$ and $8p$ is $8p$.

$$= \frac{5 \cdot 4}{2p \cdot 4} + \frac{3}{8p} \qquad \text{Fundamental property}$$

$$= \frac{20}{8p} + \frac{3}{8p} \qquad \text{Write the first fraction with the common denominator.}$$

$$= \frac{23}{8p} \qquad \text{Add the numerators. Keep the common denominator.}$$

(b) $\dfrac{6}{r} - \dfrac{5}{r - 3}$ The LCD is $r(r - 3)$.

$$= \frac{6(r - 3)}{r(r - 3)} - \frac{r \cdot 5}{r(r - 3)} \qquad \text{Fundamental property}$$

$$= \frac{6r - 18}{r(r - 3)} - \frac{5r}{r(r - 3)} \qquad \text{Distributive and commutative properties}$$

$$= \frac{6r - 18 - 5r}{r(r - 3)} \qquad \text{Subtract the numerators. Keep the common denominator.}$$

$$= \frac{r - 18}{r(r - 3)} \qquad \text{Combine like terms in the numerator.}$$

NOW TRY

TEACHING TIP

One difficulty with these problems is keeping track of the steps involved. Try these basic steps:

1. Write down the *original* problem.
2. Multiply the numerators and denominators in the original problem by the appropriate expression to get the LCD (*fraction building*).
3. Combine the numerators, and then put the result over the LCD (*the new problem*).
4. Factor the numerator if possible, and write the fraction in lowest terms.

! CAUTION Sign errors occur easily when a rational expression with two or more terms in the numerator is being subtracted. ***The subtraction sign must be distributed to every term in the numerator of the fraction that follows it.*** Study **Example 4** carefully.

CLASSROOM EXAMPLE 4

Subtract.

(a) $\dfrac{5x + 7}{2x + 7} - \dfrac{-x - 14}{2x + 7}$

(b) $\dfrac{2}{r - 2} - \dfrac{r + 3}{r - 1}$

Answers: (a) 3

(b) $\dfrac{-r^2 + r + 4}{(r - 2)(r - 1)}$

EXAMPLE 4 Subtracting Rational Expressions

Subtract.

(a) $\dfrac{7x}{3x + 1} - \dfrac{x - 2}{3x + 1}$

The denominators are the same. ***The subtraction sign must be applied to both terms in the numerator of the second rational expression.***

$$\frac{7x}{3x + 1} - \frac{x - 2}{3x + 1}$$

Use parentheses to avoid errors.

$$= \frac{7x - (x - 2)}{3x + 1} \qquad \text{Subtract the numerators. Keep the common denominator.}$$

NOW TRY ANSWERS

3. (a) $\dfrac{29}{12x}$ (b) $\dfrac{-3z - 15}{z(z - 5)}$

NOW TRY EXERCISE 4

Subtract.

(a) $\frac{18x+7}{4x+5} - \frac{2x-13}{4x+5}$

(b) $\frac{5}{x-3} - \frac{5}{x+3}$

TEACHING TIP
To avoid sign errors, emphasize the correct use of the distributive property in **Example 4.**

$$= \frac{7x - x + 2}{3x+1} \quad \text{Distributive property}$$

Be careful with signs.

$$= \frac{6x+2}{3x+1} \quad \text{Combine like terms in the numerator.}$$

$$= \frac{2(3x+1)}{3x+1} \quad \text{Factor the numerator.}$$

$$= 2 \quad \text{Fundamental property}$$

(b) $\frac{1}{q-1} - \frac{1}{q+1}$ The LCD is $(q-1)(q+1)$.

$$= \frac{1(q+1)}{(q-1)(q+1)} - \frac{1(q-1)}{(q+1)(q-1)} \quad \text{Fundamental property}$$

$$= \frac{(q+1)-(q-1)}{(q-1)(q+1)} \quad \text{Subtract the numerators.}$$

$$= \frac{q+1-q+1}{(q-1)(q+1)} \quad \text{Distributive property}$$

Be careful with signs.

$$= \frac{2}{(q-1)(q+1)} \quad \text{Combine like terms in the numerator.}$$

NOW TRY

NOW TRY EXERCISE 5

Add.

$\frac{t}{t-9} + \frac{2}{9-t}$

CLASSROOM EXAMPLE 5
Add.

$\frac{2}{x-3} + \frac{1}{3-x}$

Answer: $\frac{1}{x-3}$, or $\frac{-1}{3-x}$

TEACHING TIP
Show students the various forms of the correct answer in **Example 5.** Explain that this situation occurs often in work with rational expressions.

EXAMPLE 5 Adding Rational Expressions (Denominators Are Opposites)

Add.

$$\frac{y}{y-2} + \frac{8}{2-y} \quad \leftarrow \text{Denominators are opposites.}$$

$$= \frac{y}{y-2} + \frac{8(-1)}{(2-y)(-1)} \quad \text{Multiply the second expression by 1 in the form } \tfrac{-1}{-1}.$$

$$= \frac{y}{y-2} + \frac{-8}{y-2} \quad \text{The LCD is } y-2.$$

$$= \frac{y-8}{y-2} \quad \text{Add the numerators.}$$

We could use $2-y$ as the common denominator and rewrite the first expression.

$$\frac{y}{y-2} + \frac{8}{2-y}$$

$$= \frac{y(-1)}{(y-2)(-1)} + \frac{8}{2-y} \quad \text{Multiply the first expression by 1 in the form } \tfrac{-1}{-1}.$$

$$= \frac{-y}{2-y} + \frac{8}{2-y} \quad \text{The LCD is } 2-y.$$

$$= \frac{-y+8}{2-y} \quad \text{Add the numerators.}$$

$$= \frac{8-y}{2-y}$$

This is an equivalent form of the answer in red above. Multiply it by $\frac{-1}{-1}$ to see this.

NOW TRY

NOW TRY ANSWERS

4. (a) 4 (b) $\frac{30}{(x-3)(x+3)}$

5. $\frac{t-2}{t-9}$, or $\frac{2-t}{9-t}$

NOW TRY EXERCISE 6

Add and subtract as indicated.

$$\frac{6}{y}-\frac{1}{y-3}+\frac{3}{y^2-3y}$$

CLASSROOM EXAMPLE 6

Add and subtract as indicated.

$$\frac{4}{x-5}+\frac{-2}{x}-\frac{10}{x^2-5x}$$

Answer: $\frac{2}{x-5}$

EXAMPLE 6 Adding and Subtracting Three Rational Expressions

Add and subtract as indicated.

$$\frac{3}{x-2}+\frac{5}{x}-\frac{6}{x^2-2x}$$

$$=\frac{3}{x-2}+\frac{5}{x}-\frac{6}{x(x-2)}$$ Factor the third denominator.

$$=\frac{3x}{x(x-2)}+\frac{5(x-2)}{x(x-2)}-\frac{6}{x(x-2)}$$ The LCD is $x(x-2)$; Fundamental property

$$=\frac{3x+5(x-2)-6}{x(x-2)}$$ Add and subtract the numerators.

$$=\frac{3x+5x-10-6}{x(x-2)}$$ Distributive property

$$=\frac{8x-16}{x(x-2)}$$ Combine like terms in the numerator.

$$=\frac{8(x-2)}{x(x-2)}$$ Factor the numerator.

$$=\frac{8}{x}$$ Fundamental property

NOW TRY

NOW TRY EXERCISE 7

Subtract.

$$\frac{t-1}{t^2-2t-8}-\frac{2t+3}{t^2+3t+2}$$

CLASSROOM EXAMPLE 7

Subtract.

$$\frac{-a}{a^2+3a-4}-\frac{4a}{a^2+7a+12}$$

Answer:

$$\frac{-5a^2+a}{(a+4)(a-1)(a+3)}$$

EXAMPLE 7 Subtracting Rational Expressions

Subtract.

$$\frac{m+4}{m^2-2m-3}-\frac{2m-3}{m^2-5m+6}$$

$$=\frac{m+4}{(m-3)(m+1)}-\frac{2m-3}{(m-3)(m-2)}$$ Factor each denominator.

$$=\frac{(m+4)(m-2)}{(m-3)(m+1)(m-2)}-\frac{(2m-3)(m+1)}{(m-3)(m-2)(m+1)}$$ Fundamental property

The LCD is $(m-3)(m+1)(m-2)$.

$$=\frac{(m+4)(m-2)-(2m-3)(m+1)}{(m-3)(m+1)(m-2)}$$ Subtract the numerators.

$$=\frac{m^2+2m-8-(2m^2-m-3)}{(m-3)(m+1)(m-2)}$$ Note the careful use of parentheses. Multiply in the numerator.

$$=\frac{m^2+2m-8-2m^2+m+3}{(m-3)(m+1)(m-2)}$$ Be careful with signs. Distributive property

$$=\frac{-m^2+3m-5}{(m-3)(m+1)(m-2)}$$ Combine like terms in the numerator.

If we try to factor the numerator, we find that this expression is in lowest terms.

NOW TRY

NOW TRY ANSWERS

6. $\frac{5}{y}$

7. $\frac{-t^2+5t+11}{(t+2)(t-4)(t+1)}$

NOW TRY EXERCISE 8

Add.

$$\frac{2}{m^2 - 6m + 9} + \frac{4}{m^2 + m - 12}$$

CLASSROOM EXAMPLE 8

Add.

$$\frac{4}{p^2 - 6p + 9} + \frac{1}{p^2 + 2p - 15}$$

Answer: $\frac{5p + 17}{(p - 3)^2(p + 5)}$

NOW TRY ANSWER

8. $\frac{6m - 4}{(m - 3)^2(m + 4)}$

EXAMPLE 8 Adding Rational Expressions

Add.

$$\frac{5}{x^2 + 10x + 25} + \frac{2}{x^2 + 7x + 10}$$

$$= \frac{5}{(x + 5)^2} + \frac{2}{(x + 5)(x + 2)}$$ Factor each denominator.

$$= \frac{5(x + 2)}{(x + 5)^2(x + 2)} + \frac{2(x + 5)}{(x + 5)(x + 2)(x + 5)}$$ The LCD is $(x + 5)^2(x + 2)$; Fundamental property

$$= \frac{5(x + 2) + 2(x + 5)}{(x + 5)^2(x + 2)}$$ Add the numerators.

$$= \frac{5x + 10 + 2x + 10}{(x + 5)^2(x + 2)}$$ Distributive property

$$= \frac{7x + 20}{(x + 5)^2(x + 2)}$$ Combine like terms in the numerator.

NOW TRY

6.2 Exercises

FOR EXTRA HELP

MyMathLab®

▶ Complete solution available in MyMathLab

Concept Check *As review, add or subtract the rational numbers as indicated. Write answers in lowest terms.*

1. $\frac{8}{15} + \frac{4}{15}$
2. $\frac{5}{16} + \frac{9}{16}$
3. $\frac{5}{6} - \frac{8}{9}$
4. $\frac{3}{4} - \frac{5}{6}$
5. $\frac{5}{18} + \frac{7}{12}$
6. $\frac{3}{10} + \frac{7}{15}$

Add or subtract as indicated. ***See Example 1.***

7. $\frac{7}{t} + \frac{2}{t}$
8. $\frac{5}{r} + \frac{9}{r}$
9. ▶ $\frac{6x}{7} + \frac{y}{7}$
10. $\frac{12t}{5} + \frac{s}{5}$
11. $\frac{11}{5x} - \frac{1}{5x}$
12. $\frac{7}{4y} - \frac{3}{4y}$
13. $\frac{9}{4x^3} - \frac{17}{4x^3}$
14. $\frac{6}{5y^4} - \frac{21}{5y^4}$
15. $\frac{5x + 4}{6x + 5} + \frac{x + 1}{6x + 5}$
16. $\frac{6y + 12}{4y + 3} + \frac{2y - 6}{4y + 3}$
17. $\frac{x^2}{x + 5} - \frac{25}{x + 5}$
18. $\frac{y^2}{y + 6} - \frac{36}{y + 6}$
19. $\frac{-3p + 7}{p^2 + 7p + 12} + \frac{8p + 13}{p^2 + 7p + 12}$
20. $\frac{5x + 6}{x^2 + x - 20} + \frac{4 - 3x}{x^2 + x - 20}$
21. $\frac{a^3}{a^2 + ab + b^2} - \frac{b^3}{a^2 + ab + b^2}$
22. $\frac{p^3}{p^2 - pq + q^2} + \frac{q^3}{p^2 - pq + q^2}$

Suppose that the given expressions are denominators of rational expressions. Find the least common denominator for each group. ***See Example 2.***

23. ▶ $18x^2y^3, \quad 24x^4y^5$
24. $24a^3b^4, \quad 18a^5b^2$
25. $z - 2, \quad z$
26. $x + 3, \quad x$
27. $2y + 8, \quad y + 4$
28. $3r - 21, \quad r - 7$

Answers

1. $\frac{4}{5}$ 2. $\frac{7}{8}$ 3. $-\frac{1}{18}$ 4. $-\frac{1}{12}$ 5. $\frac{31}{36}$ 6. $\frac{23}{30}$ 7. $\frac{9}{t}$ 8. $\frac{14}{r}$ 9. $\frac{6x + y}{7}$ 10. $\frac{12t + s}{5}$ 11. $\frac{2}{x}$ 12. $\frac{1}{y}$ 13. $-\frac{2}{x^3}$ 14. $-\frac{3}{y^4}$ 15. 1 16. 2 17. $x - 5$ 18. $y - 6$ 19. $\frac{5}{p + 3}$ 20. $\frac{2}{x - 4}$ 21. $a - b$ 22. $p + q$ 23. $72x^4y^5$ 24. $72a^5b^4$ 25. $z(z - 2)$ 26. $x(x + 3)$ 27. $2(y + 4)$ 28. $3(r - 7)$ 29. $(x + 9)^2(x - 9)$ 30. $(y - 4)^2(y + 4)$ 31. $(m + n)(m - n)$ 32. $(r + s)(r - s)$

33. $x(x-4)(x+1)$

34. $y(y-2)(y-6)$

35. $(t+5)(t-2)(2t-3)$

36. $(s+1)(s-4)(3s-2)$

37. $2y(y+3)(y-3)$

38. $9x(x+2)(x-2)$

39. $2(x+2)^2(x-3)$

40. $3(a-b)^2(a+2b)$

41. The expression $\frac{x-4x-1}{x+2}$ is incorrect. The third term in the numerator should be $+1$, because the $-$ sign should be distributed over both $4x$ and -1. The answer should be $\frac{-3x+1}{x+2}$.

42. Yes, the expressions are equivalent. Multiplying $\frac{3}{5-y}$ by 1 in the form $\frac{-1}{-1}$ gives $\frac{-3}{y-5}$.

43. $\frac{31}{3t}$ **44.** $\frac{29}{4x}$

45. $\frac{5-22x}{12x^2y}$ **46.** $\frac{7-4a^2b}{18a^3b^2}$

47. $\frac{16b+9a^2}{60a^4b^6}$ **48.** $\frac{15y^3+10x}{36x^5y^5}$

49. $\frac{4pr+3sq^3}{14p^4q^4}$ **50.** $\frac{12t+5a^4b^4s}{27a^8b^7}$

51. $\frac{a^2b^5-2ab^6+3}{a^5b^7}$

52. $\frac{5t^6u^2-3t^5+6u^8}{t^{10}u^9}$

53. $\frac{1}{x(x-1)}$ **54.** $\frac{2x+3}{x(x-3)}$

55. $\frac{5a^2-7a}{(a+1)(a-3)}$

56. $\frac{5x^2-2x}{(x+4)(x-7)}$

57. 4 **58.** 3

59. 3 **60.** 2

61. $\frac{3}{x-4}$, or $\frac{-3}{4-x}$

62. $\frac{2}{2-t}$, or $\frac{-2}{t-2}$

63. $\frac{w+z}{w-z}$, or $\frac{-w-z}{z-w}$

64. $\frac{a+b}{a-b}$, or $\frac{-a-b}{b-a}$

65. $\frac{-2}{(x+1)(x-1)}$

66. $\frac{-4}{(x-1)(x+1)}$

67. $\frac{2(2x-1)}{x-1}$

68. $\frac{-x^2+x-30}{(x+3)(x-3)}$

29. $x^2-81,\quad x^2+18x+81$

30. $y^2-16,\quad y^2-8y+16$

31. $m+n,\quad m-n,\quad m^2-n^2$

32. $r+s,\quad r-s,\quad r^2-s^2$

33. $x^2-3x-4,\quad x+x^2$

34. $y^2-8y+12,\quad y^2-6y$

35. $2t^2+7t-15,\quad t^2+3t-10$

36. $s^2-3s-4,\quad 3s^2+s-2$

37. $2y+6,\quad y^2-9,\quad y$

38. $9x+18,\quad x^2-4,\quad x$

39. $2x-6,\quad x^2-x-6,\quad (x+2)^2$

40. $3a-3b,\quad a^2+ab-2b^2,\quad (a-b)^2$

41. ***Concept Check*** Consider the following *incorrect* work.

$$\frac{x}{x+2}-\frac{4x-1}{x+2}$$
$$=\frac{x-4x-1}{x+2}$$
$$=\frac{-3x-1}{x+2}$$

WHAT WENT WRONG?

42. ***Concept Check*** One student added two rational expressions and obtained the answer

$$\frac{3}{5-y}.$$

Another student obtained the answer

$$\frac{-3}{y-5}$$

for the same problem. Can both answers be correct? Why?

Add or subtract as indicated. ***See Examples 3–6.***

43. $\frac{8}{t}+\frac{7}{3t}$

44. $\frac{5}{x}+\frac{9}{4x}$

45. $\frac{5}{12x^2y}-\frac{11}{6xy}$

46. $\frac{7}{18a^3b^2}-\frac{2}{9ab}$

47. $\frac{4}{15a^4b^5}+\frac{3}{20a^2b^6}$

48. $\frac{5}{12x^5y^2}+\frac{5}{18x^4y^5}$

49. $\frac{2r}{7p^3q^4}+\frac{3s}{14p^4q}$

50. $\frac{4t}{9a^8b^7}+\frac{5s}{27a^4b^3}$

51. $\frac{1}{a^3b^2}-\frac{2}{a^4b}+\frac{3}{a^5b^7}$

52. $\frac{5}{t^4u^7}-\frac{3}{t^5u^9}+\frac{6}{t^{10}u}$

53. $\frac{1}{x-1}-\frac{1}{x}$

54. $\frac{3}{x-3}-\frac{1}{x}$

55. $\frac{3a}{a+1}+\frac{2a}{a-3}$

56. $\frac{2x}{x+4}+\frac{3x}{x-7}$

57. $\frac{11x-13}{2x-3}-\frac{3x-1}{2x-3}$

58. $\frac{13x-5}{4x-1}-\frac{x-2}{4x-1}$

59. $\frac{17y+3}{9y+7}-\frac{-10y-18}{9y+7}$

60. $\frac{4x+1}{3x+2}-\frac{-2x-3}{3x+2}$

61. $\frac{2}{4-x}+\frac{5}{x-4}$

62. $\frac{3}{2-t}+\frac{1}{t-2}$

63. $\frac{w}{w-z}-\frac{z}{z-w}$

64. $\frac{a}{a-b}-\frac{b}{b-a}$

65. $\frac{1}{x+1}-\frac{1}{x-1}$

66. $\frac{-2}{x-1}+\frac{2}{x+1}$

67. $\frac{4x}{x-1}-\frac{2}{x+1}-\frac{4}{x^2-1}$

68. $\frac{4}{x+3}-\frac{x}{x-3}-\frac{18}{x^2-9}$

69. $\frac{15}{y^2+3y}+\frac{2}{y}+\frac{5}{y+3}$

70. $\frac{7}{t-2}-\frac{6}{t^2-2t}-\frac{3}{t}$

71. $\frac{5}{x-2}+\frac{1}{x}+\frac{2}{x^2-2x}$

72. $\frac{5x}{x-3}+\frac{2}{x}+\frac{6}{x^2-3x}$

69. $\frac{7}{y}$
70. $\frac{4}{t-2}$
71. $\frac{6}{x-2}$
72. $\frac{5x+2}{x-3}$
73. $\frac{3x-2}{x-1}$
74. $\frac{6x+5}{x+3}$
75. $\frac{4x-7}{x^2-x+1}$
76. $\frac{5x-18}{x^2-2x+4}$
77. $\frac{2x+1}{x}$
78. $\frac{4x-1}{x+5}$
79. $\frac{4p^2-21p+29}{(p-2)^2}$
80. $\frac{-54r^2+42r}{(3r-1)^2}$
81. $\frac{x}{(x-2)^2(x-3)}$
82. $\frac{5m}{(m-2)^2(m+3)}$
83. $\frac{2x^2+24xy}{(x+2y)(x-y)(x+6y)}$
84. $\frac{18x^2-28xy+2y^2}{(3x+4y)(2x-y)(3x-4y)}$
85. $\frac{2x^2+21xy-10y^2}{(x+2y)(x-y)(x+6y)}$
86. $\frac{16x^2-12xy-18y^2}{(3x+4y)(2x-y)(3x-4y)}$
87. $\frac{3r-2s}{(2r-s)(3r-s)}$
88. $\frac{7y^2+11yz-y-2z}{(y+z)(y-z)(y+2z)}$
89. $\frac{10x+23}{(x+2)^2(x+3)}$
90. $\frac{3x-1}{(x+3)^2(x+1)}$
91. (a) $C(x)=\frac{10x}{49(101-x)}$ (b) 3.23 thousand dollars
92. (a) $C(x)=\frac{666.7x-6.6x^2}{(102-x)(100-x)}$ (b) first model: 127 thousand dollars; second model: 88 thousand dollars; average model: 108 thousand dollars (c) Both are 108 thousand dollars.

73. $\frac{3x}{x+1}+\frac{4}{x-1}-\frac{6}{x^2-1}$

74. $\frac{5x}{x+3}+\frac{x+2}{x}-\frac{6}{x^2+3x}$

75. $\frac{4}{x+1}+\frac{1}{x^2-x+1}-\frac{12}{x^3+1}$

76. $\frac{5}{x+2}+\frac{2}{x^2-2x+4}-\frac{60}{x^3+8}$

77. $\frac{2x+4}{x+3}+\frac{3}{x}-\frac{6}{x^2+3x}$

78. $\frac{4x+1}{x+5}-\frac{2}{x}+\frac{10}{x^2+5x}$

79. $\frac{3}{(p-2)^2}-\frac{5}{p-2}+4$

80. $\frac{8}{(3r-1)^2}+\frac{2}{3r-1}-6$

Add or subtract as indicated. ***See Examples 7 and 8.***

81. $\frac{3}{x^2-5x+6}-\frac{2}{x^2-4x+4}$

82. $\frac{2}{m^2-4m+4}+\frac{3}{m^2+m-6}$

83. $\frac{5x}{x^2+xy-2y^2}-\frac{3x}{x^2+5xy-6y^2}$

84. $\frac{6x}{6x^2+5xy-4y^2}-\frac{2y}{9x^2-16y^2}$

85. $\frac{5x-y}{x^2+xy-2y^2}-\frac{3x+2y}{x^2+5xy-6y^2}$

86. $\frac{6x+5y}{6x^2+5xy-4y^2}-\frac{x+2y}{9x^2-16y^2}$

87. $\frac{r+s}{3r^2+2rs-s^2}-\frac{s-r}{6r^2-5rs+s^2}$

88. $\frac{3y}{y^2+yz-2z^2}+\frac{4y-1}{y^2-z^2}$

89. $\frac{3}{x^2+4x+4}+\frac{7}{x^2+5x+6}$

90. $\frac{5}{x^2+6x+9}-\frac{2}{x^2+4x+3}$

Work each problem.

91. A **concours d'elegance** is a competition in which a maximum of 100 points is awarded to a car on the basis of its general attractiveness. The function

$$C(x)=\frac{1010}{49(101-x)}-\frac{10}{49}$$

approximates the cost, in thousands of dollars, of restoring a car so that it will win x points.

(a) Simplify the expression for $C(x)$ by performing the indicated subtraction.

(b) Use the simplified expression to determine how much it would cost to win 95 points. (Round to two decimal places.)

92. A **cost-benefit model** expresses cost of an undertaking in terms of benefits received. One such model gives cost in thousands of dollars to remove x percent of a pollutant as

$$C(x)=\frac{6.7x}{100-x}.$$

Another model produces the relationship

$$C(x)=\frac{6.5x}{102-x}.$$

(a) What is the cost found by averaging the two models? (*Hint:* The average of two quantities is half their sum.)

(b) Using the two given models and the answer to part (a), find the cost to the nearest thousand dollars to remove 95% $(x=95)$ of the pollutant.

(c) Average the two costs in part (b) from the given models, to the nearest thousand dollars. Compare this result with the cost obtained using the average of the two models.

6.3 Complex Fractions

OBJECTIVES

1. Simplify complex fractions by simplifying the numerator and denominator (Method 1).
2. Simplify complex fractions by multiplying by a common denominator (Method 2).
3. Compare the two methods of simplifying complex fractions.
4. Simplify rational expressions with negative exponents.

VOCABULARY

☐ complex fraction

TEACHING TIP
Have students tell whether the numerators and denominators of the complex fractions at the top of this page are single fractions.

CLASSROOM EXAMPLE 1
Use Method 1 to simplify each complex fraction.

(a) $\dfrac{\frac{y+2}{y}}{\frac{y-2}{3y}}$ (b) $\dfrac{4-\frac{3}{x}}{5-\frac{1}{x}}$

Answers: (a) $\frac{3(y+2)}{y-2}$ (b) $\frac{4x-3}{5x-1}$

A **complex fraction** is a quotient having a fraction in the numerator, denominator, or both.

$$\frac{\frac{2}{3}}{\frac{4}{9}}, \quad \frac{1+\frac{1}{x}}{2}, \quad \frac{\frac{4}{y}}{6-\frac{3}{y}}, \quad \text{and} \quad \frac{\frac{m^2-9}{m+1}}{\frac{m+3}{m^2-1}} \qquad \text{Complex fractions}$$

OBJECTIVE 1 Simplify complex fractions by simplifying the numerator and denominator (Method 1).

Simplifying a Complex Fraction (Method 1)

Step 1 Simplify the numerator and denominator separately.

Step 2 Divide by multiplying the numerator by the reciprocal of the denominator.

Step 3 Simplify the resulting fraction if possible.

In Step 2, we are treating the complex fraction as a quotient of two rational expressions and dividing. ***In order to perform this step, both the numerator and denominator must be single fractions.***

$$\frac{\frac{q-5}{8}}{\frac{q+5}{3}} \quad \begin{matrix}\leftarrow \text{Single fraction}\\ \leftarrow \text{Single fraction}\end{matrix} \qquad \frac{\frac{1}{x}+x}{\frac{x^2+1}{8}} \quad \begin{matrix}\leftarrow \text{Not a single fraction}\\ \leftarrow \text{Single fraction}\end{matrix} \qquad \frac{6+\frac{3}{x}}{\frac{x}{4}+\frac{7}{8}} \quad \begin{matrix}\leftarrow \text{Not a single fraction}\\ \leftarrow \text{Not a single fraction}\end{matrix}$$

EXAMPLE 1 Simplifying Complex Fractions (Method 1)

Use Method 1 to simplify each complex fraction.

(a) $\dfrac{\frac{x+1}{x}}{\frac{x-1}{2x}}$ Both the numerator and the denominator are single fractions. Each is already simplified. (Step 1)

$= \dfrac{x+1}{x} \div \dfrac{x-1}{2x}$ Write as a division problem.

$= \dfrac{x+1}{x} \cdot \dfrac{2x}{x-1}$ Multiply by the reciprocal of $\frac{x-1}{2x}$. (Step 2)

$= \dfrac{2x(x+1)}{x(x-1)}$ Multiply.

$= \dfrac{2(x+1)}{x-1}$ Simplify. (Step 3)

NOW TRY EXERCISE 1
Use Method 1 to simplify each complex fraction.

(a) $\dfrac{\frac{t+4}{3t}}{\frac{2t+1}{9t}}$ (b) $\dfrac{5-\frac{2}{y}}{4+\frac{1}{y}}$

(b) $\dfrac{2+\frac{1}{y}}{3-\frac{2}{y}}$ The numerator and denominator are *not* single fractions. Simplify them separately. (Step 1)

$= \dfrac{\frac{2y}{y}+\frac{1}{y}}{\frac{3y}{y}-\frac{2}{y}}$ Prepare to write the numerator and denominator as single fractions.

$= \dfrac{\frac{2y+1}{y}}{\frac{3y-2}{y}}$ The numerator is a single fraction. So is the denominator.

$= \dfrac{2y+1}{y} \div \dfrac{3y-2}{y}$ Write as a division problem.

$= \dfrac{2y+1}{y} \cdot \dfrac{y}{3y-2}$ Multiply by the reciprocal of $\frac{3y-2}{y}$. (Step 2)

$= \dfrac{2y+1}{3y-2}$ Multiply and simplify. (Step 3) NOW TRY

OBJECTIVE 2 Simplify complex fractions by multiplying by a common denominator (Method 2).

This method uses the identity property for multiplication.

Simplifying a Complex Fraction (Method 2)

Step 1 Multiply the numerator and denominator of the complex fraction by the least common denominator of the fractions in the numerator and the fractions in the denominator of the complex fraction.

Step 2 Simplify the resulting fraction if possible.

CLASSROOM EXAMPLE 2
Use Method 2 to simplify each complex fraction.

(a) $\dfrac{4-\frac{3}{x}}{5-\frac{1}{x}}$ (b) $\dfrac{3y+\frac{4}{y+1}}{2y-\frac{3}{y}}$

Answers: (a) $\dfrac{4x-3}{5x-1}$

(b) $\dfrac{3y^3+3y^2+4y}{2y^3+2y^2-3y-3}$

EXAMPLE 2 **Simplifying Complex Fractions (Method 2)**

Use Method 2 to simplify each complex fraction.

(a) $\dfrac{2+\frac{1}{y}}{3-\frac{2}{y}}$ This is the same fraction as in **Example 1(b).** Compare the solution methods.

$= \dfrac{\left(2+\frac{1}{y}\right)\cdot y}{\left(3-\frac{2}{y}\right)\cdot y}$ The LCD of all the fractions is y. Multiply the numerator and denominator by y because $\frac{y}{y}=1$. (Step 1)

$= \dfrac{2\cdot y+\frac{1}{y}\cdot y}{3\cdot y-\frac{2}{y}\cdot y}$ Distributive property (Step 2)

$= \dfrac{2y+1}{3y-2}$ Multiply. We obtain the same answer as in **Example 1(b).**

NOW TRY ANSWERS

1. (a) $\frac{3(t+4)}{2t+1}$ (b) $\frac{5y-2}{4y+1}$

NOW TRY EXERCISE 2

Use Method 2 to simplify each complex fraction.

(a) $\dfrac{5-\dfrac{2}{y}}{4+\dfrac{1}{y}}$ **(b)** $\dfrac{x+\dfrac{5}{x}}{4x-\dfrac{1}{x-2}}$

(b)
$$\frac{2p+\dfrac{5}{p-1}}{3p-\dfrac{2}{p}}$$

$$=\frac{\left(2p+\dfrac{5}{p-1}\right)\cdot p(p-1)}{\left(3p-\dfrac{2}{p}\right)\cdot p(p-1)}$$ Multiply the numerator and denominator by the LCD, $p(p-1)$. (Step 1)

$$=\frac{2p[p(p-1)]+\dfrac{5}{p-1}\cdot p(p-1)}{3p[p(p-1)]-\dfrac{2}{p}\cdot p(p-1)}$$ Distributive property (Step 2)

$$=\frac{2p[p(p-1)]+5p}{3p[p(p-1)]-2(p-1)}$$ Multiply.

$$=\frac{2p[p^2-p]+5p}{3p[p^2-p]-2p+2}$$ Distributive property

This rational expression is in lowest terms.

$$=\frac{2p^3-2p^2+5p}{3p^3-3p^2-2p+2}$$ Distributive property again

NOW TRY

OBJECTIVE 3 Compare the two methods of simplifying complex fractions.

Some students prefer one method over the other, while other students feel comfortable with both methods and rely on practice with many examples to determine which method they will use on a particular problem.

CLASSROOM EXAMPLE 3

Use both Method 1 and Method 2 to simplify each complex fraction.

(a) $\dfrac{\dfrac{5}{y+2}}{\dfrac{-3}{y^2-4}}$ (b) $\dfrac{\dfrac{1}{a}-\dfrac{1}{b}}{\dfrac{1}{a^2}-\dfrac{1}{b^2}}$

Answers: (a) $\dfrac{5(y-2)}{-3}$ (b) $\dfrac{ab}{b+a}$

EXAMPLE 3 Simplifying Complex Fractions (Both Methods)

Use both Method 1 and Method 2 to simplify each complex fraction.

Method 1

(a)
$$\frac{\dfrac{2}{x-3}}{\dfrac{5}{x^2-9}}$$

$$=\frac{\dfrac{2}{x-3}}{\dfrac{5}{(x-3)(x+3)}}$$

$$=\frac{2}{x-3}\div\frac{5}{(x-3)(x+3)}$$

$$=\frac{2}{x-3}\cdot\frac{(x-3)(x+3)}{5}$$

$$=\frac{2(x+3)}{5}$$

Method 2

(a)
$$\frac{\dfrac{2}{x-3}}{\dfrac{5}{x^2-9}}$$

$$=\frac{\dfrac{2}{x-3}}{\dfrac{5}{(x-3)(x+3)}}$$

$$=\frac{\dfrac{2}{x-3}\cdot(x-3)(x+3)}{\dfrac{5}{(x-3)(x+3)}\cdot(x-3)(x+3)}$$

$$=\frac{2(x+3)}{5}$$

NOW TRY ANSWERS

2. (a) $\dfrac{5y-2}{4y+1}$

(b) $\dfrac{x^3-2x^2+5x-10}{4x^3-8x^2-x}$

NOW TRY EXERCISE 3

Use both Method 1 and Method 2 to simplify each complex fraction.

(a) $\dfrac{\frac{1}{p-6}}{\frac{5}{p^2-36}}$ (b) $\dfrac{\frac{1}{m^2}-\frac{1}{n^2}}{\frac{1}{m}+\frac{1}{n}}$

Method 1

(b) $\dfrac{\frac{1}{x}+\frac{1}{y}}{\frac{1}{x^2}-\frac{1}{y^2}}$

$$= \frac{\frac{y}{xy}+\frac{x}{xy}}{\frac{y^2}{x^2y^2}-\frac{x^2}{x^2y^2}}$$

$$= \frac{\frac{y+x}{xy}}{\frac{y^2-x^2}{x^2y^2}}$$

$$= \frac{y+x}{xy} \div \frac{y^2-x^2}{x^2y^2}$$

$$= \frac{y+x}{xy} \cdot \frac{x^2y^2}{(y-x)(y+x)}$$

$$= \frac{xy}{y-x}$$

Method 2

(b) $\dfrac{\frac{1}{x}+\frac{1}{y}}{\frac{1}{x^2}-\frac{1}{y^2}}$

$$= \frac{\left(\frac{1}{x}+\frac{1}{y}\right)\cdot x^2y^2}{\left(\frac{1}{x^2}-\frac{1}{y^2}\right)\cdot x^2y^2}$$

$$= \frac{\left(\frac{1}{x}\right)x^2y^2+\left(\frac{1}{y}\right)x^2y^2}{\left(\frac{1}{x^2}\right)x^2y^2-\left(\frac{1}{y^2}\right)x^2y^2}$$

$$= \frac{xy^2+x^2y}{y^2-x^2}$$

$$= \frac{xy(y+x)}{(y+x)(y-x)}$$

$$= \frac{xy}{y-x}$$

NOW TRY

OBJECTIVE 4 Simplify rational expressions with negative exponents.

We begin by rewriting the expressions with only positive exponents.

CLASSROOM EXAMPLE 4

Simplify each expression, using only positive exponents in the answer.

(a) $\dfrac{a^{-2}+b^{-1}}{a^{-1}-5b^{-3}}$ (b) $\dfrac{x^{-3}+2y^{-1}}{y+2x^3}$

Answers: (a) $\dfrac{b^3+a^2b^2}{ab^3-5a^2}$

(b) $\dfrac{1}{x^3y}$

EXAMPLE 4 Simplifying Rational Expressions with Negative Exponents

Simplify each expression, using only positive exponents in the answer.

(a) $\dfrac{m^{-1}+p^{-2}}{2m^{-2}-p^{-1}}$ $\qquad a^{-n}=\frac{1}{a^n}$ (Section 4.1)

$$= \frac{\frac{1}{m}+\frac{1}{p^2}}{\frac{2}{m^2}-\frac{1}{p}}$$

Write with positive exponents. $2m^{-2} = 2 \cdot m^{-2} = \frac{2}{1} \cdot \frac{1}{m^2} = \frac{2}{m^2}$

The base of $2m^{-2}$ is m, not $2m$: $2m^{-2} = \frac{2}{m^2}$.

$$= \frac{m^2p^2\left(\frac{1}{m}+\frac{1}{p^2}\right)}{m^2p^2\left(\frac{2}{m^2}-\frac{1}{p}\right)}$$

Simplify by Method 2. Multiply the numerator and denominator by the LCD, m^2p^2.

$$= \frac{m^2p^2\cdot\frac{1}{m}+m^2p^2\cdot\frac{1}{p^2}}{m^2p^2\cdot\frac{2}{m^2}-m^2p^2\cdot\frac{1}{p}}$$

Distributive property

$$= \frac{mp^2+m^2}{2p^2-m^2p}$$

Write in lowest terms.

TEACHING TIP

A common mistake is for students to write $a^{-1}+b^{-1}$ as $\frac{1}{a+b}$. Point out that

$$a^{-1}+b^{-1} \neq (a+b)^{-1}.$$

Remind them also that a term such as $2y^{-1}$ is equal to $\frac{2}{y}$, *not* $\frac{1}{2y}$.

NOW TRY ANSWERS

3. (a) $\frac{p+6}{5}$ (b) $\frac{n-m}{mn}$

NOW TRY EXERCISE 4

Simplify each expression, using only positive exponents in the answer.

(a) $\dfrac{r^{-2} - s^{-1}}{4r^{-1} + s^{-2}}$

(b) $\dfrac{2y^{-1} - 3y^{-2}}{y^{-2} + 3x^{-1}}$

(b) $\dfrac{x^{-2} - 2y^{-1}}{y - 2x^2}$

$$= \frac{\frac{1}{x^2} - \frac{2}{y}}{y - 2x^2}$$ Write with positive exponents.

The 2 does *not* go in the denominator of this fraction.

$$= \frac{\left(\frac{1}{x^2} - \frac{2}{y}\right) x^2y}{(y - 2x^2)\, x^2y}$$ Use Method 2. Multiply by the LCD, x^2y.

$$= \frac{y - 2x^2}{(y - 2x^2)x^2y}$$ Use the distributive property in the numerator.

$$= \frac{1}{x^2y}$$ Write in lowest terms.

Remember to write 1 in the numerator.

NOW TRY

NOW TRY ANSWERS

4. (a) $\dfrac{s^2 - r^2s}{4rs^2 + r^2}$ (b) $\dfrac{2xy - 3x}{x + 3y^2}$

6.3 Exercises

FOR EXTRA HELP

MyMathLab®

Complete solution available in MyMathLab

1. complex; numerator; both
2. single; reciprocal; denominator
3. LCD; identity
4. (a) $\frac{1}{20}$ (b) -2
5. $\frac{1}{6}$
6. $\frac{1}{10}$
7. $\frac{9}{5}$
8. $\frac{15}{8}$
9. $\frac{4}{15}$
10. 1
11. $\frac{7}{17}$
12. $-\frac{16}{15}$
13. $\frac{2x}{x-1}$
14. $\frac{4t}{t+4}$
15. $\frac{2(k+1)}{3k-1}$
16. $\frac{2(1-r)}{1+r}$

Concept Check *Fill in each blank with the correct response.*

1. A(n) ________ fraction is a quotient that has a fraction in the ________, denominator, or ________.

2. To use Method 1 for simplifying a complex fraction, both the numerator and denominator must be ________ fractions. Divide by multiplying the numerator by the ________ of the ________, and then simplify.

3. To use Method 2 for simplifying a complex fraction, multiply both the numerator and denominator by the ________ of all the fractions in the numerator and denominator. This is an application of the ________ property for multiplication.

4. ***Concept Check*** Find the slope of the line that passes through each pair of points. (*Hint:* This will involve simplifying complex fractions. Recall that slope $m = \frac{y_2 - y_1}{x_2 - x_1}$.)

(a) $\left(-\frac{5}{2}, \frac{1}{6}\right)$ and $\left(\frac{5}{3}, \frac{3}{8}\right)$ (b) $\left(-\frac{5}{6}, -\frac{1}{2}\right)$ and $\left(-\frac{1}{3}, -\frac{3}{2}\right)$

Concept Check *Simplify.*

5. $\dfrac{\frac{2}{3}}{4}$ **6.** $\dfrac{\frac{3}{5}}{6}$ **7.** $\dfrac{\frac{3}{4}}{\frac{5}{12}}$ **8.** $\dfrac{\frac{5}{6}}{\frac{4}{9}}$

9. $\dfrac{\frac{5}{9} - \frac{1}{3}}{\frac{2}{3} + \frac{1}{6}}$ **10.** $\dfrac{\frac{7}{8} - \frac{3}{2}}{-\frac{1}{4} - \frac{3}{8}}$ **11.** $\dfrac{2 - \frac{1}{4}}{\frac{5}{4} + 3}$ **12.** $\dfrac{\frac{4}{3} - 2}{1 - \frac{3}{8}}$

Use either method to simplify each complex fraction. ***See Examples 1–3.***

13. $\dfrac{\frac{12}{x-1}}{\frac{6}{x}}$ **14.** $\dfrac{\frac{24}{t+4}}{\frac{6}{t}}$ **15.** $\dfrac{\frac{k+1}{2k}}{\frac{3k-1}{4k}}$ **16.** $\dfrac{\frac{1-r}{4r}}{\frac{1+r}{8r}}$

17. $\dfrac{5x^2}{9z^3}$ 18. $\dfrac{2}{3yx}$

19. $\dfrac{1+x}{-1+x}$ 20. $\dfrac{2-k}{2+k}$

21. $\dfrac{6x+1}{7x-3}$ 22. $\dfrac{4p-1}{9p+5}$

23. $\dfrac{y+x}{y-x}$ 24. $\dfrac{s-t}{s+t}$

25. $4x$ 26. $5y$

27. $x+4y$ 28. $2t+3s$

29. $\dfrac{y+4}{2}$ 30. $\dfrac{t-7}{3}$

31. $\dfrac{a+b}{ab}$ 32. $\dfrac{y-x}{xy}$

33. xy 34. rs

35. $\dfrac{3y}{2}$ 36. $\dfrac{3p^2-2p}{3p-10}$

37. $\dfrac{x^2+5x+4}{x^2+5x+10}$

38. $\dfrac{y^2-2y-3}{y^2+y-1}$

39. $\dfrac{x^2y^2}{y^2+x^2}$

40. $\dfrac{p^2q^2}{q^2-p^2}$, or $\dfrac{p^2q^2}{(q-p)(q+p)}$

41. $\dfrac{y^2+x^2}{xy^2+x^2y}$, or $\dfrac{y^2+x^2}{xy(y+x)}$

42. $\dfrac{xy}{y+x}$

43. $\dfrac{p^2+k}{p^2-3k}$ 44. $\dfrac{y-x^2}{y+4x^2}$

45. $\dfrac{1}{2xy}$ 46. $\dfrac{b+2a}{3a^2b^2}$

47. $\dfrac{m^2+6m-4}{m(m-1)}$

48. $\dfrac{m^2-m-2}{m(m-1)}$

49. $\dfrac{m^2+6m-4}{m^2-m-2}$

50. $m(m-1)$

51. $\dfrac{m^2+6m-4}{m^2-m-2}$

52. Answers will vary.

---

17. $\dfrac{\frac{4z^2x^4}{9}}{\frac{12x^2z^5}{15}}$ **18.** $\dfrac{\frac{3y^2x^3}{8}}{\frac{9y^3x^4}{16}}$ **19.** $\dfrac{\frac{1}{x}+1}{-\frac{1}{x}+1}$ **20.** $\dfrac{\frac{2}{k}-1}{\frac{2}{k}+1}$

21. $\dfrac{6+\frac{1}{x}}{7-\frac{3}{x}}$ **22.** $\dfrac{4-\frac{1}{p}}{9+\frac{5}{p}}$ **23.** $\dfrac{\frac{3}{x}+\frac{3}{y}}{\frac{3}{x}-\frac{3}{y}}$ **24.** $\dfrac{\frac{4}{t}-\frac{4}{s}}{\frac{4}{t}+\frac{4}{s}}$

25. $\dfrac{\frac{8x-24y}{10}}{\frac{x-3y}{5x}}$ **26.** $\dfrac{\frac{20x-10y}{12y}}{\frac{2x-y}{6y^2}}$ **27.** $\dfrac{\frac{x^2-16y^2}{xy}}{\frac{1}{y}-\frac{4}{x}}$ **28.** $\dfrac{\frac{4t^2-9s^2}{st}}{\frac{2}{s}-\frac{3}{t}}$

29. $\dfrac{\frac{6}{y-4}}{\frac{12}{y^2-16}}$ **30.** $\dfrac{\frac{8}{t+7}}{\frac{24}{t^2-49}}$ **31.** $\dfrac{\frac{1}{b^2}-\frac{1}{a^2}}{\frac{1}{b}-\frac{1}{a}}$ **32.** $\dfrac{\frac{1}{x^2}-\frac{1}{y^2}}{\frac{1}{x}+\frac{1}{y}}$

33. $\dfrac{x+y}{\frac{1}{y}+\frac{1}{x}}$ **34.** $\dfrac{s-r}{\frac{1}{r}-\frac{1}{s}}$ **35.** $\dfrac{y-\frac{y-3}{3}}{\frac{4}{9}+\frac{2}{3y}}$

36. $\dfrac{p-\frac{p+2}{4}}{\frac{3}{4}-\frac{5}{2p}}$ **37.** $\dfrac{\frac{x+2}{x}+\frac{1}{x+2}}{\frac{5}{x}+\frac{x}{x+2}}$ **38.** $\dfrac{\frac{y+3}{y}-\frac{4}{y-1}}{\frac{y}{y-1}+\frac{1}{y}}$

Simplify each expression, using only positive exponents in your answer. ***See Example 4.***

39. $\dfrac{1}{x^{-2}+y^{-2}}$ **40.** $\dfrac{1}{p^{-2}-q^{-2}}$ **41.** $\dfrac{x^{-2}+y^{-2}}{x^{-1}+y^{-1}}$ **42.** $\dfrac{x^{-1}-y^{-1}}{x^{-2}-y^{-2}}$

43. $\dfrac{k^{-1}+p^{-2}}{k^{-1}-3p^{-2}}$ **44.** $\dfrac{x^{-2}-y^{-1}}{x^{-2}+4y^{-1}}$ **45.** $\dfrac{x^{-1}+2y^{-1}}{2y+4x}$ **46.** $\dfrac{a^{-2}-4b^{-2}}{3b-6a}$

RELATING CONCEPTS For Individual or Group Work **(Exercises 47–52)**

Simplifying a complex fraction by Method 1 is a good way to review the methods of adding, subtracting, multiplying, and dividing rational expressions. Method 2 gives a good review of the fundamental property of rational numbers.

Refer to the following complex fraction, and ***work Exercises 47–52 in order.***

$$\dfrac{\frac{4}{m}+\frac{m+2}{m-1}}{\frac{m+2}{m}-\frac{2}{m-1}}$$

47. Add the fractions in the numerator.

48. Subtract as indicated in the denominator.

49. Divide the answer from **Exercise 47** by the answer from **Exercise 48.**

50. Go back to the original complex fraction and find the LCD of all denominators.

51. Multiply the numerator and denominator of the complex fraction by the answer from **Exercise 50.**

52. The answers for **Exercises 49 and 51** should be the same. Which method do you prefer? Explain why.

6.4 Equations with Rational Expressions and Graphs

OBJECTIVES

1. Determine the domain of the variable in a rational equation.
2. Solve rational equations.
3. Recognize the graph of a rational function.

VOCABULARY

- ☐ domain of the variable in a rational equation
- ☐ proposed solution
- ☐ extraneous solution
- ☐ discontinuous
- ☐ reciprocal function
- ☐ vertical asymptote
- ☐ horizontal asymptote

NOW TRY EXERCISE 1

Give the domain of the variable in each equation.

(a) $\frac{1}{3x} - \frac{3}{4x} = \frac{1}{3}$

(b) $\frac{1}{x-7} + \frac{2}{x+7} = \frac{14}{x^2-49}$

CLASSROOM EXAMPLE 1

Give the domain of the variable in each equation.

(a) $\frac{3}{x} + \frac{1}{4} = \frac{9}{4x}$

(b) $\frac{2}{x^2-4} + \frac{1}{x+2} = \frac{1}{x-2}$

Answers:

(a) $\{x \mid x \text{ is a real number}, x \neq 0\}$

(b) $\{x \mid x \text{ is a real number}, x \neq \pm 2\}$

NOW TRY ANSWERS

1. (a) $\{x \mid x \text{ is a real number}, x \neq 0\}$
   (b) $\{x \mid x \text{ is a real number}, x \neq \pm 7\}$

OBJECTIVE 1 Determine the domain of the variable in a rational equation.

Recall that the domain of a rational expression is the set of all possible values of the variable. (We also refer to this as "the domain of the variable.") Any value that makes the denominator 0 is excluded.

The **domain of the variable in a rational equation** is the intersection (overlap) of the domains of the rational expressions in the equation. When we solve rational equations, knowing the domain of the variable allows us to determine whether a **proposed solution**—that is, a value of the variable that *appears* to be a solution—must be discarded.

EXAMPLE 1 Determining Domains of Variables

Give the domain of the variable in each equation.

(a) $\frac{2}{x} - \frac{3}{2} = \frac{7}{2x}$

The domains of the three rational expressions in the equation are, in order,

$\{x \mid x \text{ is a real number}, x \neq 0\}$,

$\{x \mid x \text{ is a real number}\}$,

and $\{x \mid x \text{ is a real number}, x \neq 0\}$.

Set-builder notation (We could use interval notation.)

The intersection of these three domains is $\{x \mid x \text{ is a real number}, x \neq 0\}$.

(b) $\frac{2}{x-3} - \frac{3}{x+3} = \frac{12}{x^2-9}$

The domains of the three expressions are, respectively,

$\{x \mid x \text{ is a real number}, x \neq 3\}$,

$\{x \mid x \text{ is a real number}, x \neq -3\}$,

and $\{x \mid x \text{ is a real number}, x \neq \pm 3\}$.

$\pm$ is read "positive or negative," or "plus or minus."

The domain of the variable is the intersection of the three domains.

$$\{x \mid x \text{ is a real number}, x \neq \pm 3\}$$

NOW TRY

OBJECTIVE 2 Solve rational equations.

To solve rational equations, we usually multiply all terms by the least common denominator to clear the fractions. ***We can do this only with an equation*** because we want to solve for a variable (not with an expression, where we are carrying out operations).

Solving an Equation with Rational Expressions

Step 1 **Determine the domain of the variable.**

Step 2 **Multiply each side of the equation by the LCD** to clear the fractions.

Step 3 **Solve** the resulting equation.

Step 4 **Check** that each proposed solution is in the domain, and discard any values that are not. Check the remaining proposed solution(s) in the original equation.

NOW TRY EXERCISE 2

Solve.

$$\frac{1}{3x} - \frac{3}{4x} = \frac{1}{3}$$

CLASSROOM EXAMPLE 2

Solve.

$$-\frac{3}{20} + \frac{2}{x} = \frac{5}{4x}$$

Answer: $\{5\}$

TEACHING TIP

Remind students that we keep the common denominator in our answer when adding or subtracting fractions. However, when solving an equation, we eliminate the denominators by multiplying each term by the LCD.

Show examples contrasting the two methods. See the examples given with the Summary Exercises that follow this section.

TEACHING TIP

Be aware that students will discard a good solution because of an arithmetic error in checking their answer. Tell them a solution does not check if it produces 0 in any denominator of the original equation.

CLASSROOM EXAMPLE 3

Solve.

$$\frac{3}{x+1} = \frac{1}{x-1} - \frac{2}{x^2-1}$$

Answer: $\varnothing$

NOW TRY ANSWER

2. $\left\{-\frac{5}{4}\right\}$

EXAMPLE 2 Solving a Rational Equation

Solve $\frac{2}{x} - \frac{3}{2} = \frac{7}{2x}$.

Step 1 The domain, which excludes 0, was found in **Example 1(a).**

Step 2 $$2x\left(\frac{2}{x} - \frac{3}{2}\right) = 2x\left(\frac{7}{2x}\right) \quad \text{Multiply by the LCD, } 2x.$$

Step 3 $$2x\left(\frac{2}{x}\right) - 2x\left(\frac{3}{2}\right) = 2x\left(\frac{7}{2x}\right) \quad \text{Distributive property}$$

$$4 - 3x = 7 \quad \text{Multiply.}$$

$$-3x = 3 \quad \text{Subtract 4.}$$

This proposed solution is in the domain.

$$x = -1 \quad \text{Divide by } -3.$$

Step 4 CHECK (Don't forget this step.)

$$\frac{2}{x} - \frac{3}{2} = \frac{7}{2x} \quad \text{Original equation}$$

$$\frac{2}{-1} - \frac{3}{2} \stackrel{?}{=} \frac{7}{2(-1)} \quad \text{Let } x = -1.$$

$$-\frac{7}{2} = -\frac{7}{2} \; \checkmark \quad \text{True}$$

A true statement results, so the solution set is $\{-1\}$. NOW TRY

CAUTION When each side of an equation is multiplied by a *variable* expression, the resulting "solutions" may not satisfy the original equation. ***We must either determine and observe the domain or check all proposed solutions in the original equation. It is wise to do both.***

EXAMPLE 3 Solving a Rational Equation with No Solution

Solve $\frac{2}{x-3} - \frac{3}{x+3} = \frac{12}{x^2-9}$.

Step 1 The domain, which excludes ± 3, was found in **Example 1(b).**

Step 2 Factor $x^2 - 9$. Then multiply each side by the LCD, $(x+3)(x-3)$.

$$(x+3)(x-3)\left(\frac{2}{x-3} - \frac{3}{x+3}\right) = (x+3)(x-3)\left[\frac{12}{(x+3)(x-3)}\right]$$

Step 3 $$(x+3)(x-3)\left(\frac{2}{x-3}\right) - (x+3)(x-3)\left(\frac{3}{x+3}\right)$$

$$= (x+3)(x-3)\left[\frac{12}{(x+3)(x-3)}\right]$$

Distributive property

$$2(x+3) - 3(x-3) = 12 \quad \text{Multiply.}$$

$$2x + 6 - 3x + 9 = 12 \quad \text{Distributive property}$$

$$-x + 15 = 12 \quad \text{Combine like terms.}$$

$$\text{Proposed solution} \rightarrow x = 3 \quad \text{Subtract 15. Multiply by } -1.$$

NOW TRY EXERCISE 3

Solve.

$$\frac{1}{t-7}+\frac{2}{t+7}=\frac{14}{t^2-49}$$

Step 4 The proposed solution, 3, is not in the domain. Such solutions are **extraneous** and must be rejected. Substituting 3 into the original equation shows why.

CHECK
$$\frac{2}{x-3}-\frac{3}{x+3}=\frac{12}{x^2-9}$$ Original equation

$$\frac{2}{3-3}-\frac{3}{3+3}\stackrel{?}{=}\frac{12}{3^2-9}$$ Let $x=3$.

$$\frac{2}{0}-\frac{3}{6}\stackrel{?}{=}\frac{12}{0}$$ Division by 0 is undefined.

The equation has no solution. The solution set is $\varnothing$. NOW TRY

NOW TRY EXERCISE 4

Solve.

$$\frac{2}{t^2-2t-8}-\frac{4}{t^2+6t+8}=\frac{2}{t^2-16}$$

CLASSROOM EXAMPLE 4

Solve.

$$\frac{4}{x^2+x-6}-\frac{1}{x^2-4}=\frac{2}{x^2+5x+6}$$

Answer: $\{-9\}$

EXAMPLE 4 Solving a Rational Equation

Solve $\frac{3}{p^2+p-2}-\frac{1}{p^2-1}=\frac{7}{2(p^2+3p+2)}$.

Factor each denominator to find the domain and the LCD.

$$\frac{3}{(p-1)(p+2)}-\frac{1}{(p+1)(p-1)}=\frac{7}{2(p+2)(p+1)}$$ Factor the denominators.

The domain excludes 1, -2, and -1. Multiply each side of the equation by the LCD, $2(p-1)(p+2)(p+1)$.

$$2(p-1)(p+2)(p+1)\left[\frac{3}{(p-1)(p+2)}-\frac{1}{(p+1)(p-1)}\right]=2(p-1)(p+2)(p+1)\left[\frac{7}{2(p+2)(p+1)}\right]$$

$$2\cdot 3(p+1)-2(p+2)=7(p-1)$$ Distributive property

$$6p+6-2p-4=7p-7$$ $2\cdot 3=6$; Distributive property

$$4p+2=7p-7$$ Combine like terms.

$$9=3p$$ Subtract $4p$. Add 7.

Proposed solution → $3=p$ Divide by 3.

CHECK 3 is in the domain. Substitute it in the original equation to check.

$$\frac{3}{p^2+p-2}-\frac{1}{p^2-1}=\frac{7}{2(p^2+3p+2)}$$ Original equation

$$\frac{3}{3^2+3-2}-\frac{1}{3^2-1}\stackrel{?}{=}\frac{7}{2(3^2+3\cdot 3+2)}$$ Let $p=3$.

$$\frac{3}{10}-\frac{1}{8}\stackrel{?}{=}\frac{7}{40}$$ Work in the denominators.

$\frac{3}{10}-\frac{1}{8}=\frac{12}{40}-\frac{5}{40}$

$$\frac{7}{40}=\frac{7}{40}\ \checkmark$$ True

The solution set is $\{3\}$. NOW TRY

NOW TRY ANSWERS

3. $\varnothing$ 4. $\{5\}$

NOW TRY EXERCISE 5

Solve.
$$\frac{2x}{x-2}=\frac{-3}{x}+\frac{4}{x-2}$$

CLASSROOM EXAMPLE 5

Solve.
$$\frac{2}{x+3}-\frac{1}{x-1}=\frac{-x^2-3x}{x^2+2x-3}$$

Answer: $\{-5\}$

EXAMPLE 5 Solving a Rational Equation

Solve $\frac{2}{3x+1}=\frac{1}{x}-\frac{6x}{3x+1}$.

We must exclude $-\frac{1}{3}$ and 0 from the domain.

$$x(3x+1)\left(\frac{2}{3x+1}\right)=x(3x+1)\left(\frac{1}{x}-\frac{6x}{3x+1}\right)$$ Multiply by the LCD, $x(3x+1)$.

$$x(3x+1)\left(\frac{2}{3x+1}\right)=x(3x+1)\left(\frac{1}{x}\right)-x(3x+1)\left(\frac{6x}{3x+1}\right)$$ Distributive property

This is a quadratic equation.

$$2x=3x+1-6x^2$$ Multiply.

$$6x^2-x-1=0$$ Standard form

$$(3x+1)(2x-1)=0$$ Factor.

$$3x+1=0 \quad\text{or}\quad 2x-1=0$$ Zero-factor property

Proposed solutions → $x=-\frac{1}{3}$ or $x=\frac{1}{2}$ Solve each equation.

Because $-\frac{1}{3}$ is not in the domain, it is not a solution. Check that the solution set is $\left\{\frac{1}{2}\right\}$.

NOW TRY

OBJECTIVE 3 Recognize the graph of a rational function.

A function defined by a quotient of polynomials is a **rational function.** Because one or more values of x may be excluded from the domain of a rational function, their graphs are often **discontinuous**—that is, there will be one or more breaks in the graph.

One simple rational function is the **reciprocal function** $f(x)=\frac{1}{x}$. The domain of this function includes all real numbers except 0. Thus, this function pairs every real number except 0 with its reciprocal.

The closer negative values of x are to 0, the less ("more negative") y is. → ← The closer positive values of x are to 0, the greater y is.

x	−3	−2	−1	−0.5	−0.25	−0.1	0.1	0.25	0.5	1	2	3
y	$-\frac{1}{3}$	$-\frac{1}{2}$	−1	−2	−4	−10	10	4	2	1	$\frac{1}{2}$	$\frac{1}{3}$

Plotting the points from the table, we obtain the graph in FIGURE 2.

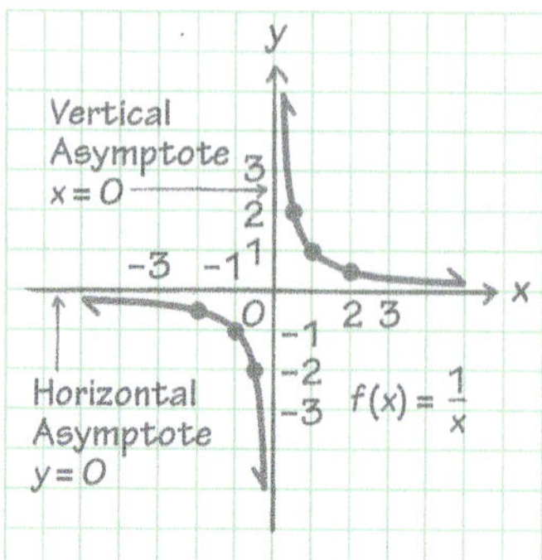

FIGURE 2

Reciprocal function

$$f(x)=\frac{1}{x}$$

Domain: $\{x\,|\,x \text{ is a real number}, x\neq 0\}$

Range: $\{y\,|\,y \text{ is a real number}, y\neq 0\}$

NOW TRY ANSWER

5. $\left\{-\frac{3}{2}\right\}$

CLASSROOM EXAMPLE 6
Graph the rational function, and give the equations of the vertical and horizontal asymptotes.

$$f(x) = \frac{2}{x+3}$$

Answer:

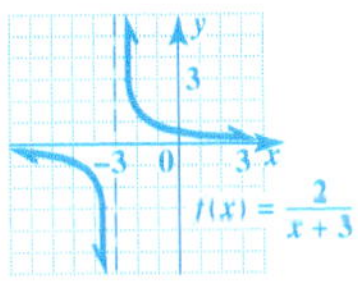

vertical asymptote: $x = -3$;
horizontal asymptote: $y = 0$

Because the domain of $f(x) = \frac{1}{x}$ includes all real numbers except 0, there is no point on the graph where $x = 0$. The vertical line with equation $x = 0$ is a **vertical asymptote.** The horizontal line with equation $y = 0$ is a **horizontal asymptote.** The graph approaches these asymptotes but does not cross them.

NOTE In general, the following are true of a rational function.

1. If the y-values approach ∞ or $-\infty$ as the x-values approach a real number a, the **vertical line $x = a$ is a vertical asymptote** of the graph.
2. If the y-values approach a real number b as $|x|$ increases without bound, the **horizontal line $y = b$ is a horizontal asymptote** of the graph.

NOW TRY EXERCISE 6
Graph the rational function, and give the equations of the vertical and horizontal asymptotes.

$$f(x) = \frac{1}{x+1}$$

EXAMPLE 6 Graphing a Rational Function

Graph the rational function, and give the equations of the vertical and horizontal asymptotes.

$$g(x) = \frac{-2}{x-3}$$

Some ordered pairs that belong to the function are listed in the table.

x	−2	−1	0	1	2	2.5	2.75	3.25	3.5	4	5	6	7
y	$\frac{2}{5}$	$\frac{1}{2}$	$\frac{2}{3}$	1	2	4	8	−8	−4	−2	−1	$-\frac{2}{3}$	$-\frac{1}{2}$

There is no point on the graph, shown in **FIGURE 3**, for $x = 3$, because 3 is excluded from the domain. The dashed line

$$x = 3 \text{ represents the vertical asymptote.}$$

It is not part of the graph. The graph gets closer to the vertical asymptote as the x-values get closer to 3. Again,

$$y = 0 \text{ is a horizontal asymptote.}$$

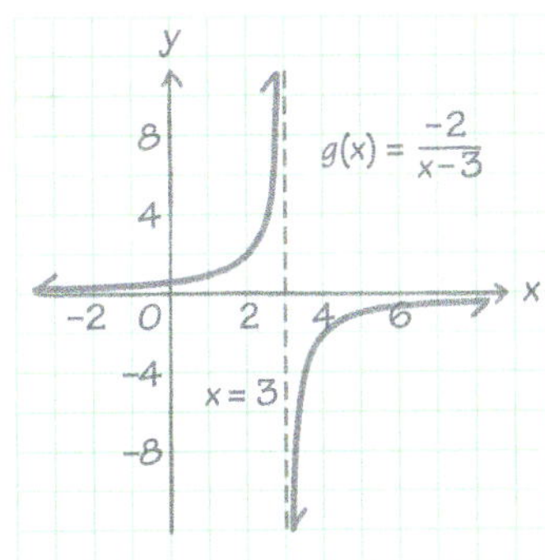

FIGURE 3

NOW TRY

NOW TRY ANSWER

6.

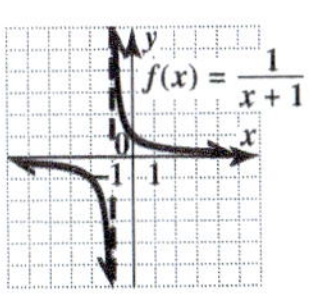

vertical asymptote: $x = -1$;
horizontal asymptote: $y = 0$

6.4 Exercises

FOR EXTRA HELP

MyMathLab®

Complete solution available in MyMathLab

1. (a) equation (b) expression (c) expression (d) equation
2. "Solve" refers to finding the solution set of an equation. This is an expression. "Solve" should be replaced by "Simplify" or "Add."

1. *Concept Check* Decide whether each of the following is an *expression* or an *equation.*

(a) $\frac{2}{x} = \frac{4}{3x} + \frac{1}{3}$ **(b)** $\frac{4}{3x} + \frac{1}{3}$ **(c)** $\frac{5}{x+1} - \frac{2}{x-1}$ **(d)** $\frac{5}{x+1} - \frac{2}{x-1} = \frac{4}{x^2-1}$

2. *Concept Check* What is wrong with the following problem?

"Solve $\frac{2x+1}{3x-4} + \frac{1}{2x+3}$."

In Exercises 3–14, we give the domains using set-builder notation.

3. $\{x \mid x \text{ is a real number}, x \neq 0\}$
4. $\{x \mid x \text{ is a real number}, x \neq 0\}$
5. $\{x \mid x \text{ is a real number}, x \neq -1, 2\}$
6. $\{x \mid x \text{ is a real number}, x \neq -4, 9\}$
7. $\left\{x \mid x \text{ is a real number}, x \neq 4, \frac{7}{2}\right\}$
8. $\left\{x \mid x \text{ is a real number}, x \neq -\frac{3}{2}, \frac{1}{3}\right\}$
9. $\left\{x \mid x \text{ is a real number}, x \neq -\frac{7}{4}, 0, \frac{13}{6}\right\}$
10. $\left\{x \mid x \text{ is a real number}, x \neq \frac{5}{3}, 0, -\frac{13}{4}\right\}$
11. $\{x \mid x \text{ is a real number}, x \neq \pm 4\}$
12. $\{x \mid x \text{ is a real number}, x \neq \pm 5\}$
13. $\{x \mid x \text{ is a real number}, x \neq 0, 1, -3, 2\}$
14. $\{x \mid x \text{ is a real number}, x \neq 0, -1, -5, 7\}$

15. $\{1\}$ 16. $\{-3\}$
17. $\{-6, 4\}$ 18. $\{-5, -3\}$
19. $\{-7, 3\}$ 20. $\{-4, 6\}$
21. $\left\{-\frac{7}{12}\right\}$ 22. $\{-11\}$
23. $\varnothing$ 24. $\varnothing$
25. $\{-3\}$ 26. $\{1\}$
27. $\{0\}$ 28. $\{0\}$
29. $\varnothing$ 30. $\varnothing$
31. $\{5\}$ 32. $\{5\}$
33. $\varnothing$ 34. $\varnothing$
35. $\left\{\frac{27}{56}\right\}$ 36. $\left\{-\frac{15}{2}\right\}$
37. $\{-3, -1\}$ 38. $\{-6, 3\}$
39. $\varnothing$ 40. $\varnothing$
41. $\{-10\}$ 42. $\{-2\}$
43. $\{-1\}$

Give the domain of the variable in each equation. ***See Example 1.***

3. $\frac{1}{3x} + \frac{1}{2x} = \frac{x}{3}$
4. $\frac{5}{6x} - \frac{8}{2x} = \frac{x}{4}$
5. $\frac{1}{x+1} - \frac{1}{x-2} = 0$
6. $\frac{3}{x+4} - \frac{2}{x-9} = 0$
7. $\frac{3x+1}{x-4} = \frac{6x+5}{2x-7}$
8. $\frac{4x-1}{2x+3} = \frac{12x-25}{6x-2}$
9. $\frac{6}{4x+7} - \frac{3}{x} = \frac{5}{6x-13}$
10. $\frac{4}{3x-5} + \frac{2}{x} = \frac{9}{4x+13}$
11. $\frac{1}{x^2-16} - \frac{2}{x-4} = \frac{1}{x+4}$
12. $\frac{2}{x^2-25} - \frac{1}{x+5} = \frac{1}{x-5}$
13. $\frac{2}{x^2-x} + \frac{1}{x+3} = \frac{4}{x-2}$
14. $\frac{3}{x^2+x} - \frac{1}{x+5} = \frac{2}{x-7}$

Solve each equation. ***See Examples 2–5.***

15. $\frac{3}{4x} = \frac{5}{2x} - \frac{7}{4}$
16. $\frac{2}{3x} = \frac{6}{5x} + \frac{8}{45}$
17. $x - \frac{24}{x} = -2$
18. $p + \frac{15}{p} = -8$
19. $\frac{x}{4} - \frac{21}{4x} = -1$
20. $\frac{x}{2} - \frac{12}{x} = 1$
21. $\frac{x-4}{x+6} = \frac{2x+3}{2x-1}$
22. $\frac{5x-8}{x+2} = \frac{5x-1}{x+3}$
23. $\frac{3x+1}{x-4} = \frac{6x+5}{2x-7}$
24. $\frac{4x-1}{2x+3} = \frac{12x-25}{6x-2}$
25. $\frac{1}{y-1} + \frac{5}{12} = \frac{-2}{3y-3}$
26. $\frac{4}{m+2} - \frac{11}{9} = \frac{1}{3m+6}$
27. $\frac{7}{6x+3} - \frac{1}{3} = \frac{2}{2x+1}$
28. $\frac{3}{4m+2} = \frac{17}{2} - \frac{7}{2m+1}$
29. $\frac{6}{x-4} + \frac{5}{x} = \frac{-20}{x^2-4x}$
30. $\frac{7}{x-4} + \frac{3}{x} = \frac{-12}{x^2-4x}$
31. $\frac{3}{x+2} - \frac{2}{x^2-4} = \frac{1}{x-2}$
32. $\frac{3}{x-2} + \frac{21}{x^2-4} = \frac{14}{x+2}$
33. $\frac{1}{y+2} + \frac{3}{y+7} = \frac{5}{y^2+9y+14}$
34. $\frac{1}{t+3} + \frac{4}{t+5} = \frac{2}{t^2+8t+15}$
35. $\frac{9}{x} + \frac{4}{6x-3} = \frac{2}{6x-3}$
36. $\frac{5}{n} + \frac{4}{6-3n} = \frac{2n}{6-3n}$
37. $\frac{1}{x-2} + \frac{1}{4} = \frac{1}{4(x^2-4)}$
38. $\frac{1}{x+4} + \frac{1}{3} = \frac{-10}{3(x^2-16)}$
39. $\frac{6}{w+3} + \frac{-7}{w-5} = \frac{-48}{w^2-2w-15}$
40. $\frac{2}{r-5} + \frac{3}{2r+1} = \frac{22}{2r^2-9r-5}$
41. $\frac{x}{x-3} + \frac{4}{x+3} = \frac{18}{x^2-9}$
42. $\frac{2x}{x-3} + \frac{4}{x+3} = \frac{-24}{x^2-9}$
43. $\frac{1}{x+4} + \frac{x}{x-4} = \frac{-8}{x^2-16}$

44. $\{-2\}$

45. $\{13\}$ **46.** $\{15\}$

47. $\{x \mid x \neq \pm 3\}$

48. $\left\{x \mid x \neq \pm \frac{3}{2}\right\}$

49.

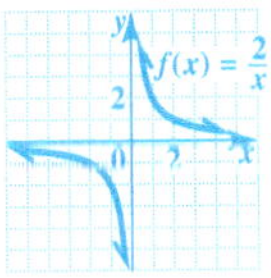

$x = 0; y = 0$

50.

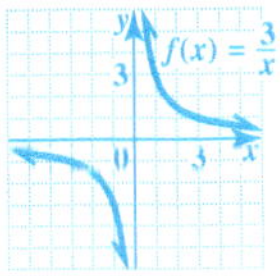

$x = 0; y = 0$

51.

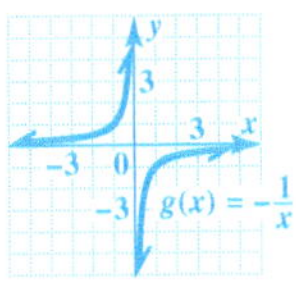

$x = 0; y = 0$

52.

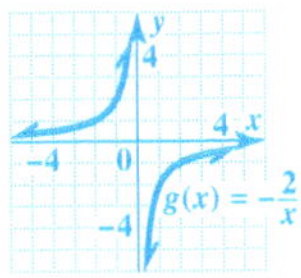

$x = 0; y = 0$

53.

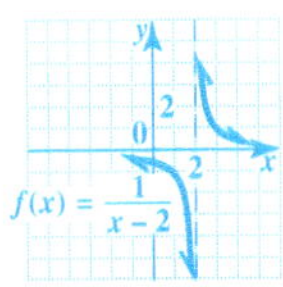

$x = 2; y = 0$

54.

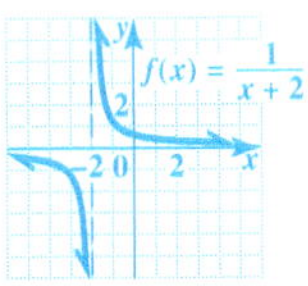

$x = -2; y = 0$

55. **(a)** 0 **(b)** 1.6 **(c)** 4.1 **(d)** The waiting time also increases.

56. **(a)** 90% **(b)** 5 **(c)** No, $x \neq 0$ because 0 is not in the domain of this function.

44. $\dfrac{5}{x-4} - \dfrac{3}{x-1} = \dfrac{x^2-1}{x^2-5x+4}$

45. $\dfrac{2}{k^2+k-6} + \dfrac{1}{k^2-k-2} = \dfrac{4}{k^2+4k+3}$

46. $\dfrac{5}{p^2+3p+2} - \dfrac{3}{p^2-4} = \dfrac{1}{p^2-p-2}$

47. $\dfrac{5x+14}{x^2-9} = \dfrac{-2x^2-5x+2}{x^2-9} + \dfrac{2x+4}{x-3}$

48. $\dfrac{4x-7}{4x^2-9} = \dfrac{-2x^2+5x-4}{4x^2-9} + \dfrac{x+1}{2x+3}$

Graph each rational function. Give the equations of the vertical and horizontal asymptotes. See Example 6.

49. $f(x) = \dfrac{2}{x}$ **50.** $f(x) = \dfrac{3}{x}$ **51.** $g(x) = -\dfrac{1}{x}$

52. $g(x) = -\dfrac{2}{x}$ **53.** $f(x) = \dfrac{1}{x-2}$ **54.** $f(x) = \dfrac{1}{x+2}$

Solve each problem.

55. The average number of vehicles waiting in line to enter a parking area is modeled by the function

$$W(x) = \frac{x^2}{2(1-x)},$$

where x is a quantity between 0 and 1 known as the **traffic intensity.** (*Source:* Mannering, F., and W. Kilareski, *Principles of Highway Engineering and Traffic Control,* John Wiley and Sons, © 2007.) For each traffic intensity, find the average number of vehicles waiting (to the nearest tenth).

(a) 0.1 **(b)** 0.8 **(c)** 0.9

(d) What happens to waiting time as traffic intensity increases?

56. The percent of deaths caused by smoking is modeled by the rational function

$$P(x) = \frac{x-1}{x},$$

where x is the number of times a smoker is more likely to die of lung cancer than a nonsmoker. This is the **incidence rate.** (*Source:* Walker, A., *Observation and Inference: An Introduction to the Methods of Epidemiology,* University of Michigan, © 1991.) For example, $x = 10$ means that a smoker is 10 times more likely than a nonsmoker to die of lung cancer.

(a) Find $P(x)$ if $x = 10$.

(b) For what values of x is $P(x) = 80\%$? (*Hint:* Change 80% to a decimal.)

(c) Can the incidence rate equal 0? Explain.

57. **(a)** 500 ft **(b)** It decreases.
58. **(a)** 140 hundreds of gallons, or 14,000 gal **(b)** It decreases.

57. The force required to keep a 2000-lb car going 30 mph from skidding on a curve, where r is the radius of the curve in feet, is given by

$$F(r) = \frac{225{,}000}{r}.$$

(a) What radius must a curve have if a force of 450 lb is needed to keep the car from skidding?

(b) As the radius of the curve is lengthened, how is the force affected?

58. The amount of heating oil produced (in gallons per day) by an oil refinery is modeled by the rational function

$$f(x) = \frac{125{,}000 - 25x}{125 + 2x},$$

where x is the amount of gasoline produced (in hundreds of gallons per day). Suppose the refinery must produce 300 gal of heating oil per day to meet the needs of its customers.

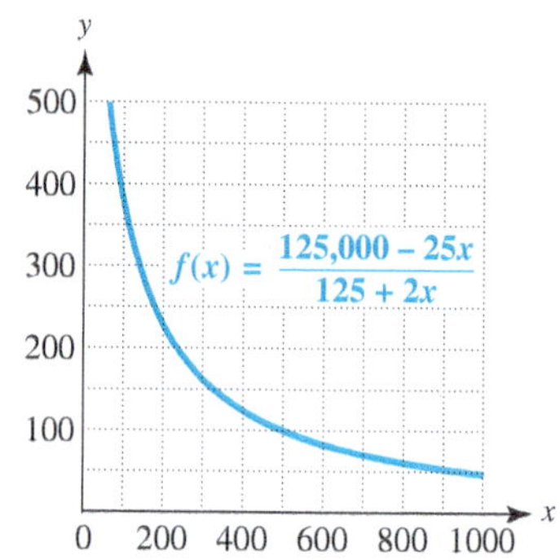

(a) How much gasoline will be produced per day?

(b) A graph of f, for $x > 0$, is shown in the figure. Use it to decide what happens to the amount of gasoline (x) produced as the amount of heating oil (y) produced increases.

SUMMARY EXERCISES Simplifying Rational Expressions vs. Solving Rational Equations

A common student error is to confuse an equation, such as $\frac{x}{2} + \frac{x}{3} = -5$, with an expression involving an operation, such as $\frac{x}{2} + \frac{x}{3}$. ***Equations are solved for a numerical answer, while problems involving operations result in simplified expressions.***

Solving an Equation	**Simplifying an Expression Involving an Operation**
Solve: $\frac{x}{2} + \frac{x}{3} = -5$ Look for the equality symbol.	***Add:*** $\frac{x}{2} + \frac{x}{3}$ There is no equality symbol.
Multiply each side by the LCD, 6.	Write both fractions with the LCD, 6.
$6\left(\frac{x}{2} + \frac{x}{3}\right) = 6(-5)$	$\frac{x}{2} + \frac{x}{3}$
$6\left(\frac{x}{2}\right) + 6\left(\frac{x}{3}\right) = 6(-5)$	$= \frac{x \cdot 3}{2 \cdot 3} + \frac{x \cdot 2}{3 \cdot 2}$
$3x + 2x = -30$	$= \frac{3x}{6} + \frac{2x}{6}$
$5x = -30$	$= \frac{3x + 2x}{6}$
$x = -6$	$= \frac{5x}{6}$
Check that the solution set is $\{-6\}$.	

Identify each exercise as an expression *or an* equation. *Then simplify the expression by performing the indicated operation, or solve the equation, as appropriate.*

1. $\frac{x}{2} - \frac{x}{4} = 5$

2. $\frac{4x-20}{x^2-25} \cdot \frac{(x+5)^2}{10}$

3. $\frac{6}{7x} - \frac{4}{x}$

4. $\frac{\frac{1}{x}+\frac{1}{y}}{\frac{1}{x}-\frac{1}{y}}$

5. $\frac{5}{7t} = \frac{52}{7} - \frac{3}{t}$

6. $\frac{x-5}{3} + \frac{1}{3} = \frac{x-2}{5}$

7. $\frac{7}{6x} + \frac{5}{8x}$

8. $\frac{4}{x} - \frac{8}{x+1} = 0$

9. $\frac{\frac{6}{x+1}-\frac{1}{x}}{\frac{2}{x}-\frac{4}{x+1}}$

10. $\frac{8}{r+2} - \frac{7}{4r+8}$

11. $\frac{x}{x+y} + \frac{2y}{x-y}$

12. $\frac{3p^2-6p}{p+5} \div \frac{p^2-4}{8p+40}$

13. $\frac{x-2}{9} \cdot \frac{5}{8-4x}$

14. $\frac{a-4}{3} + \frac{11}{6} = \frac{a+1}{2}$

15. $\frac{b^2+b-6}{b^2+2b-8} \cdot \frac{b^2+8b+16}{3b+12}$

16. $\frac{10z^2-5z}{3z^3-6z^2} \div \frac{2z^2+5z-3}{z^2+z-6}$

17. $\frac{5}{x^2-2x} - \frac{3}{x^2-4}$

18. $\frac{3}{t-1} + \frac{1}{t} = \frac{7}{2}$

19. $\frac{\frac{5}{x}-\frac{3}{y}}{\frac{9x^2-25y^2}{x^2y}}$

20. $\frac{-2}{a^2+2a-3} - \frac{5}{3-3a} = \frac{4}{3a+9}$

21. $\frac{4y^2-13y+3}{2y^2-9y+9} \div \frac{4y^2+11y-3}{6y^2-5y-6}$

22. $\frac{8}{3k+9} - \frac{8}{15} = \frac{2}{5k+15}$

23. $\frac{3r}{r-2} = 1 + \frac{6}{r-2}$

24. $\frac{6z^2-5z-6}{6z^2+5z-6} \cdot \frac{12z^2-17z+6}{12z^2-z-6}$

25. $\frac{-1}{3-x} - \frac{2}{x-3}$

26. $\frac{\frac{t}{4}-\frac{1}{t}}{1+\frac{t+4}{t}}$

27. $\frac{2}{y+1} - \frac{3}{y^2-y-2} = \frac{3}{y-2}$

28. $\frac{7}{2x^2-8x} + \frac{3}{x^2-16}$

29. $\frac{3}{y-3} - \frac{3}{y^2-5y+6} = \frac{2}{y-2}$

30. $\frac{2k+\frac{5}{k-1}}{3k-\frac{2}{k}}$

1. equation; $\{20\}$
2. expression; $\frac{2(x+5)}{5}$
3. expression; $-\frac{22}{7x}$
4. expression; $\frac{y+x}{y-x}$
5. equation; $\left\{\frac{1}{2}\right\}$
6. equation; $\{7\}$
7. expression; $\frac{43}{24x}$
8. equation; $\{1\}$
9. expression; $\frac{5x-1}{-2x+2}$, or $\frac{5x-1}{-2(x-1)}$
10. expression; $\frac{25}{4(r+2)}$
11. expression; $\frac{x^2+xy+2y^2}{(x+y)(x-y)}$
12. expression; $\frac{24p}{p+2}$
13. expression; $-\frac{5}{36}$
14. equation; $\{0\}$
15. expression; $\frac{b+3}{3}$
16. expression; $\frac{5}{3z}$
17. expression; $\frac{2x+10}{x(x-2)(x+2)}$
18. equation; $\left\{\frac{1}{7}, 2\right\}$
19. expression; $\frac{-x}{3x+5y}$
20. equation; $\{-13\}$
21. expression; $\frac{3y+2}{y+3}$
22. equation; $\left\{\frac{5}{4}\right\}$
23. equation; $\emptyset$
24. expression; $\frac{2z-3}{2z+3}$
25. expression; $\frac{-1}{x-3}$, or $\frac{1}{3-x}$
26. expression; $\frac{t-2}{8}$
27. equation; $\{-10\}$
28. expression; $\frac{13x+28}{2x(x+4)(x-4)}$
29. equation; $\emptyset$
30. expression; $\frac{k(2k^2-2k+5)}{(k-1)(3k^2-2)}$

6.5 Applications of Rational Expressions

OBJECTIVES

1. Find the value of an unknown variable in a formula.
2. Solve a formula for a specified variable.
3. Solve applications using proportions.
4. Solve applications about distance, rate, and time.
5. Solve applications about work rates.

VOCABULARY

☐ ratio
☐ proportion

NOW TRY EXERCISE 1

Use the formula in **Example 1** to find f if $p = 50$ cm and $q = 10$ cm.

CLASSROOM EXAMPLE 1

Use the formula in **Example 1** to find p if $f = 15$ cm and $q = 25$ cm.

Answer: $\frac{75}{2}$ cm

NOW TRY EXERCISE 2

Solve for m.

$$\frac{1}{m} - \frac{2}{n} = 5$$

CLASSROOM EXAMPLE 2

Solve for q.

$$\frac{3}{p} + \frac{3}{q} = \frac{5}{r}$$

Answer: $q = \frac{3rp}{5p - 3r}$

NOW TRY ANSWERS

1. $\frac{25}{3}$ cm 2. $m = \frac{n}{5n + 2}$

OBJECTIVE 1 Find the value of an unknown variable in a formula.

Formulas may contain rational expressions, such as $t = \frac{d}{r}$ and $\frac{1}{f} = \frac{1}{p} + \frac{1}{q}$.

EXAMPLE 1 Finding the Value of a Variable in a Formula

In physics, the focal length f of a lens is given by the formula

$$\frac{1}{f} = \frac{1}{p} + \frac{1}{q}.$$

In the formula, p is the distance from the object to the lens and q is the distance from the lens to the image. See **FIGURE 4**. Find q if $p = 20$ cm and $f = 10$ cm.

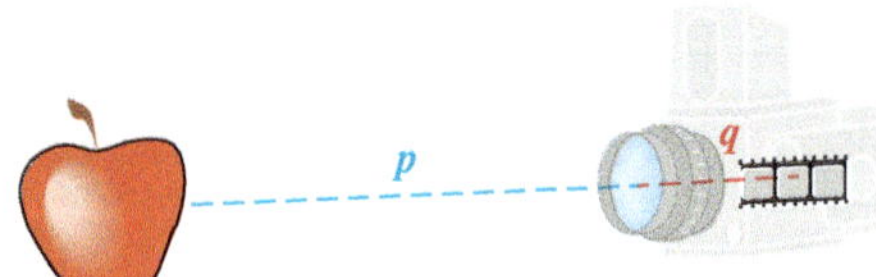

Focal Length of Camera Lens

FIGURE 4

$$\frac{1}{f} = \frac{1}{p} + \frac{1}{q}$$

$$\frac{1}{10} = \frac{1}{20} + \frac{1}{q}$$ Let $f = 10$, $p = 20$. (Solve this equation for q.)

$$20q \cdot \frac{1}{10} = 20q\left(\frac{1}{20} + \frac{1}{q}\right)$$ Multiply by the LCD, $20q$.

$$20q \cdot \frac{1}{10} = 20q\left(\frac{1}{20}\right) + 20q\left(\frac{1}{q}\right)$$ Distributive property

$$2q = q + 20$$ Multiply.

$$q = 20$$ Subtract q.

The distance from the lens to the image is 20 cm. NOW TRY

OBJECTIVE 2 Solve a formula for a specified variable.

We solve for a specified variable by isolating it on one side of the equality symbol.

EXAMPLE 2 Solving a Formula for a Specified Variable

Solve $\frac{1}{f} = \frac{1}{p} + \frac{1}{q}$ for p.

$$\frac{1}{f} = \frac{1}{p} + \frac{1}{q}$$

$$fpq \cdot \frac{1}{f} = fpq\left(\frac{1}{p} + \frac{1}{q}\right)$$ To clear the fractions, multiply by the LCD, fpq.

$$pq = fq + fp$$ Multiply; distributive property (We want the two terms with p on the same side.)

$$pq - fp = fq$$ Subtract fp.

$$p(q - f) = fq$$ Factor out p. (This is a key step.)

$$p = \frac{fq}{q - f}$$ Divide by $q - f$. NOW TRY

NOW TRY EXERCISE 3

Solve for N.

$$\frac{NR}{N-n}=t$$

CLASSROOM EXAMPLE 3

Solve for R.

$$A=\frac{Rr}{R+r}$$

Answer: $R=\frac{-Ar}{A-r}$, or $R=\frac{Ar}{r-A}$

EXAMPLE 3 Solving a Formula for a Specified Variable

Solve $I=\frac{nE}{R+nr}$ for n.

$$I=\frac{nE}{R+nr}$$

$$(R+nr)I=(R+nr)\frac{nE}{R+nr}$$ To clear the fractions, multiply by $R+nr$.

$$RI+nrI=nE$$ Distributive property on the left

Write the n-terms on the ***same*** side in preparation for factoring.

$$RI=nE-nrI$$ Subtract nrI.

$$RI=n(E-rI)$$ Factor out n.

$$\frac{RI}{E-rI}=n, \quad \text{or} \quad n=\frac{RI}{E-rI}$$ Divide by $E-rI$. Interchange sides.

NOW TRY

CAUTION Refer to the steps in **Examples 2 and 3** that factor out the desired variable. ***This variable must be a factor on only one side of the equation.*** Then each side can be divided by the remaining factor in the last step.

OBJECTIVE 3 Solve applications using proportions.

A **ratio** is a comparison of two quantities. The ratio of a to b may be written in any of the following ways.

$$a \text{ to } b, \quad a:b, \quad \text{or} \quad \frac{a}{b}$$ Ratio of a to b

Ratios are usually written as quotients in algebra. A **proportion** is a statement that two ratios are equal.

$$\frac{a}{b}=\frac{c}{d}$$ Proportion

CLASSROOM EXAMPLE 4

In a recent year, approximately 9.8% (that is, 9.8 of every 100) of the 74,165,000 children under 18 yr of age in the United States had no health insurance. How many such children were uninsured? (*Source:* U.S. Census Bureau.)

Answer: 7,268,170

EXAMPLE 4 Solving a Proportion

In 2011, about 17 of every 100 Americans had no health insurance. The population at that time was about 312 million. How many million Americans had no health insurance? (*Source:* U.S. Census Bureau.)

Step 1 **Read** the problem.

Step 2 **Assign a variable.**

Let x = the number (in millions) who had no health insurance.

Step 3 **Write an equation.** We set up a proportion. The ratio 17 to 100 should equal the ratio x to 312.

$$\frac{17}{100}=\frac{x}{312}$$ Write a proportion.

NOW TRY ANSWER

3. $N=\frac{nt}{t-R}$

NOW TRY EXERCISE 4

In 2012, about 15 of every 100 Americans lived in poverty. The population at that time was about 314 million. How many Americans lived in poverty? (*Source:* U.S. Census Bureau.)

NOW TRY EXERCISE 5

Clayton's SUV uses 28 gal of gasoline to drive 500 mi. He has 10 gal of gasoline in the SUV, and he still needs to drive 400 mi. If the car continues to use gasoline at the same rate, how many more gallons will he need?

CLASSROOM EXAMPLE 5

Lauren's car uses 15 gal of gasoline to drive 390 mi. She has 6 gal of gasoline in the car, and she still needs to drive 800 mi. If the car continues to use gasoline at the same rate, how many more gallons will she need (to the nearest tenth)?

Answer: 24.8 more gallons

Step 4 **Solve.** $31{,}200\left(\frac{17}{100}\right) = 31{,}200\left(\frac{x}{312}\right)$ Multiply by a common denominator, $100 \cdot 312 = 31{,}200$.

$5304 = 100x$ Simplify.

$53.04 = x$ Divide by 100.

Step 5 **State the answer.** 53.04 million Americans had no health insurance.

Step 6 **Check** that the ratio of 53.04 million to 312 million equals $\frac{17}{100}$.

$\frac{53.04}{312} = \frac{17}{100}$ Use a calculator to divide 53.04 by 312. A true statement results. NOW TRY

EXAMPLE 5 Solving a Proportion Involving Rates

Jodie's car uses 10 gal of gasoline to travel 210 mi. She has 5 gal of gasoline in the car. She still needs to drive 640 mi. If the car continues to use gasoline at the same rate, how many more gallons will she need?

Step 1 **Read** the problem.

Step 2 **Assign a variable.**

Let $x =$ the additional number of gallons of gas needed.

Step 3 **Write an equation.** We set up a proportion.

$$\frac{10}{210} = \frac{5+x}{640} \quad \begin{matrix}\leftarrow \text{gallons} \\ \leftarrow \text{miles}\end{matrix}$$

(gallons → 10, miles → 210)

Step 4 **Solve.** We could multiply both sides by the LCD $10 \cdot 21 \cdot 64$. Instead we use an alternative method that involves **cross products.**

For $\frac{a}{b} = \frac{c}{d}$ to be true, the cross products ad and bc must be equal.

$a \cdot d = b \cdot c$

$10 \cdot 640 = 210(5 + x)$ If $\frac{a}{b} = \frac{c}{d}$, then $ad = bc$.

$6400 = 1050 + 210x$ Multiply; distributive property

$5350 = 210x$ Subtract 1050.

$25.5 = x$ Divide by 210. Round to the nearest tenth.

Step 5 **State the answer.** Jodie will need 25.5 more gallons of gas.

Step 6 **Check.** 25.5 more gallons plus the 5 that she has equals 30.5 gal. From Step 3, the ratio of 10 to 210 must equal the ratio of 30.5 to 640.

$\frac{10}{210} \approx 0.048$ and $\frac{30.5}{640} \approx 0.048$ Use a calculator.

Because the ratios are approximately equal, the answer is correct.

NOW TRY

OBJECTIVE 4 Solve applications about distance, rate, and time.

When the distance formula $d = rt$ is solved for r or t, rational expressions result.

Rate is the ratio of distance to time, or $r = \frac{d}{t}$.

Time is the ratio of distance to rate, or $t = \frac{d}{r}$.

NOW TRY ANSWERS

4. 47.1 million

5. 12.4 more gallons

NOW TRY EXERCISE 6

A small fishing boat travels 36 mi against the current in a river in the same time that it travels 44 mi with the current. If the rate of the boat in still water is 20 mph, find the rate of the current.

CLASSROOM EXAMPLE 6

A plane travels 100 mi against the wind in the same time that it takes to travel 120 mi with the wind. The wind speed is 20 mph. Find the rate of the plane in still air.

Answer: 220 mph

TEACHING TIP

In a problem like the one in **Example 6,** the equation for a distance-rate-time problem will come from the *time* portion of the table. The equation will involve adding, subtracting, or setting the two time quantities equal to each other.

Show students how to form the equation from a literal translation of the time expressions in the problem.

EXAMPLE 6 Solving a Distance, Rate, Time Problem

A paddle wheeler travels 10 mi against the current in a river in the same time that it travels 15 mi with the current. If the rate of the current is 3 mph, find the rate of the boat in still water.

Step 1 **Read** the problem.

Step 2 **Assign a variable.** Let x = the rate of the boat in still water.

Traveling ***against*** the current slows the boat down, so the rate of the boat is the ***difference*** between its rate in still water and the rate of the current—that is, $(x - 3)$ mph.

Traveling ***with*** the current speeds the boat up, so the rate of the boat is the ***sum*** of its rate in still water and the rate of the current—that is, $(x + 3)$ mph.

Thus, $x - 3$ = the rate of the boat *against* the current,

and $x + 3$ = the rate of the boat *with* the current.

Because the time is the same going against the current as with the current, find time in terms of distance and rate for each situation. Against the current, the distance is 10 mi and the rate is $(x - 3)$ mph.

$$t = \frac{d}{r} = \frac{10}{x - 3} \quad \text{Time } against \text{ the current}$$

With the current, the distance is 15 mi and the rate is $(x + 3)$ mph.

$$t = \frac{d}{r} = \frac{15}{x + 3} \quad \text{Time } with \text{ the current}$$

	Distance	Rate	Time
Against Current	10	$x - 3$	$\frac{10}{x-3}$
With Current	15	$x + 3$	$\frac{15}{x+3}$

Summarize the information in a table.

Times are equal.

Step 3 **Write an equation.** Use the fact that the times are equal.

$$\frac{10}{x - 3} = \frac{15}{x + 3}$$

To solve, we can multiply by the LCD or use cross products.

Step 4 **Solve.**

$$(x + 3)(x - 3)\left(\frac{10}{x - 3}\right) = (x + 3)(x - 3)\left(\frac{15}{x + 3}\right) \quad \text{Multiply by the LCD, } (x + 3)(x - 3).$$

$$10(x + 3) = 15(x - 3) \quad \text{Multiply.}$$

$$10x + 30 = 15x - 45 \quad \text{Distributive property}$$

$$75 = 5x \quad \text{Subtract } 10x. \text{ Add 45.}$$

$$15 = x \quad \text{Divide by 5.}$$

Step 5 **State the answer.** The rate of the boat in still water is 15 mph.

Step 6 **Check** the answer: $\frac{10}{15 - 3} = \frac{15}{15 + 3}$ is true.

NOW TRY

NOW TRY ANSWER

6. 2 mph

CLASSROOM EXAMPLE 7
Dona drove 300 mi north from San Antonio, mostly on the freeway. She usually averaged 55 mph, but an accident slowed her rate through Dallas to 15 mph. If her trip took 6 hr, how many miles did she drive at the reduced rate?

Answer: $11\frac{1}{4}$ mi

EXAMPLE 7 Solving a Distance, Rate, Time Problem

At O'Hare International Airport in Chicago, Cheryl and Bill are walking to the gate at the same rate to catch their flight to Denver. Bill wants a window seat, so he steps onto the moving sidewalk and continues to walk while Cheryl uses the stationary sidewalk. If the sidewalk moves at 1 m per sec and Bill saves 50 sec covering the 300-m distance, what is their walking rate?

Step 1 **Read** the problem. We must find their walking rate.

Step 2 **Assign a variable.**

Let x = their walking rate in meters per second.

Thus, Cheryl travels at a rate of x meters per second and Bill travels at a rate of $(x + 1)$ meters per second. Express their times in terms of the known distances and the variable rates, as in **Example 6.** Cheryl travels 300 m at a rate of x meters per second.

$$t = \frac{d}{r} = \frac{300}{x} \quad \text{Cheryl's time}$$

Bill travels 300 m at a rate of $(x + 1)$ meters per second.

$$t = \frac{d}{r} = \frac{300}{x+1} \quad \text{Bill's time}$$

	Distance	Rate	Time
Cheryl	300	x	$\frac{300}{x}$
Bill	300	$x+1$	$\frac{300}{x+1}$

← Bill's time is 50 sec less than Cheryl's time.

Step 3 **Write an equation.** Use the times from the table.

Bill's time is Cheryl's time less 50 sec

$$\frac{300}{x+1} = \frac{300}{x} - 50$$

We *cannot* solve using cross products because there are two terms on the right.

Step 4 **Solve.**

$$x(x+1)\left(\frac{300}{x+1}\right) = x(x+1)\left(\frac{300}{x} - 50\right) \quad \text{Multiply by the LCD, } x(x+1).$$

$$x(x+1)\left(\frac{300}{x+1}\right) = x(x+1)\left(\frac{300}{x}\right) - x(x+1)(50) \quad \text{Distributive property}$$

$$300x = 300(x+1) - 50x(x+1) \quad \text{Multiply.}$$

$$300x = 300x + 300 - 50x^2 - 50x \quad \text{Distributive property}$$

$$50x^2 + 50x - 300 = 0 \quad \text{Standard form}$$

$$x^2 + x - 6 = 0 \quad \text{Divide by 50.}$$

$$(x+3)(x-2) = 0 \quad \text{Factor.}$$

$$x + 3 = 0 \quad \text{or} \quad x - 2 = 0 \quad \text{Zero-factor property}$$

$$x = -3 \quad \text{or} \quad x = 2 \quad \text{Solve each equation.}$$

Discard the negative solution because rate (speed) cannot be negative.

NOW TRY EXERCISE 7
James and Pat are driving from Atlanta to Jacksonville, a distance of 310 mi. James, whose average rate is 5 mph faster than Pat's, will drive the first 130 mi of the trip and then Pat will drive the rest of the way to their destination. If the total driving time is 5 hr, determine the average rate of each driver.

Step 5 **State the answer.** Their walking rate is 2 m per sec.

Step 6 **Check** the solution in the words of the original problem. NOW TRY

CAUTION We cannot solve the rational equation in **Example 7,**

$$\frac{300}{x+1} = \frac{300}{x} - 50,$$ To solve, multiply by the LCD.

using cross products. ***That method can be used only when there is a single rational expression on each side,*** such as in the equation in **Example 6,**

$$\frac{10}{x-3} = \frac{15}{x+3}.$$ To solve, multiply by the LCD *or* use cross products.

OBJECTIVE 5 Solve applications about work rates.

PROBLEM-SOLVING HINT If the letters r, t, and A represent the rate at which work is done, the time required, and the amount of work accomplished, respectively, then $\boldsymbol{A = rt}$. Notice the similarity to the distance formula, $d = rt$.

Amount of work can be measured in terms of jobs accomplished. Thus, if 1 job is completed, then $A = 1$, and the formula gives the rate as

$$1 = rt, \quad \text{or} \quad r = \frac{1}{t}.$$

To solve a work problem, we use the following fact to express all rates of work.

Rate of Work

If a job can be accomplished in t units of time, then the rate of work is

$$\frac{1}{t} \textbf{ job per unit of time.}$$

CLASSROOM EXAMPLE 8
Stan needs 45 min to do the dishes, while Bobbie can do them in 30 min. How long will it take them if they work together?

Answer: 18 min

EXAMPLE 8 **Solving a Work Problem**

Letitia and Kareem are working on a neighborhood cleanup. Kareem can clean up all the trash in the area in 7 hr, while Letitia can do the same job in 5 hr. How long will it take them if they work together?

Let $x =$ the number of hours it will take the two people working together.

We use $A = rt$. Because $A = 1$, the rate for each person will be $\frac{1}{t}$, where t is the time it takes the person to complete the job alone. Kareem can clean up all the trash in 7 hr, so his rate is $\frac{1}{7}$ of the job per hour. Letitia's rate is $\frac{1}{5}$ of the job per hour.

TEACHING TIP
Point out to students that the variable is in the numerator of the equation for an example like this.

$$\frac{1}{7}x = \frac{1}{7} \cdot \frac{x}{1} = \frac{x}{7}$$

	Rate	*Time Working Together*	*Fractional Part of the Job Done*
Kareem	$\frac{1}{7}$	x	$\frac{1}{7}x$
Letitia	$\frac{1}{5}$	x	$\frac{1}{5}x$

NOW TRY ANSWER
7. 60 mph; 65 mph

NOW TRY EXERCISE 8

Using her zero-turn riding lawn mower, Gina can mow her lawn in 2 hr. Using a traditional mower, Matt can mow the same lawn in 3 hr. How long will it take them to mow the lawn working together?

Part done by Kareem	+	part done by Letitia	is	1 whole job.	
$\frac{1}{7}x$	+	$\frac{1}{5}x$	=	1	Together they complete 1 job. The sum of the fractional parts must equal 1.

$$35\left(\frac{1}{7}x + \frac{1}{5}x\right) = 35 \cdot 1 \quad \text{Multiply by the LCD, 35.}$$

$$5x + 7x = 35 \quad \text{Distributive property}$$

$$12x = 35 \quad \text{Combine like terms.}$$

$$x = \frac{35}{12} \quad \text{Divide by 12.}$$

Working together, Kareem and Letitia can do the entire job in $\frac{35}{12}$ hr, or 2 hr, 55 min. Check this result in the original problem. NOW TRY

TEACHING TIP

You may want to work **Classroom Example 8** using both methods to show students that they yield the same answer.

NOTE There is another way to approach problems about rates of work. In **Example 8,** x represents the number of hours it will take the two people working together to complete the entire job. In 1 hr, $\frac{1}{x}$ of the entire job will be completed, so in 1 hr, Kareem completes $\frac{1}{7}$ of the job and Letitia completes $\frac{1}{5}$ of the job.

$$\frac{1}{7} + \frac{1}{5} = \frac{1}{x} \quad \text{The sum of their rates equals } \frac{1}{x}.$$

When each side of this equation is multiplied by $35x$, the result is $5x + 7x = 35$, the same equation we obtained in **Example 8** in the third line from the bottom. Thus, the same solution results using either approach.

NOW TRY ANSWER

8. $\frac{6}{5}$ hr, or 1 hr, 12 min

6.5 Exercises

FOR EXTRA HELP MyMathLab®

Complete solution available in MyMathLab

1. A 2. B
3. D 4. D

Concept Check *Each exercise presents a familiar formula. Give the letter of the choice that is an equivalent form of the formula.*

1. $p = br$ (percent)

A. $b = \frac{p}{r}$ **B.** $r = \frac{b}{p}$ **C.** $b = \frac{r}{p}$ **D.** $p = \frac{r}{b}$

2. $V = LWH$ (geometry)

A. $H = \frac{LW}{V}$ **B.** $L = \frac{V}{WH}$ **C.** $L = \frac{WH}{V}$ **D.** $W = \frac{H}{VL}$

3. $m = \frac{F}{a}$ (physics)

A. $a = mF$ **B.** $F = \frac{m}{a}$ **C.** $F = \frac{a}{m}$ **D.** $F = ma$

4. $I = \frac{E}{R}$ (electricity)

A. $R = \frac{I}{E}$ **B.** $R = IE$ **C.** $E = \frac{I}{R}$ **D.** $E = RI$

5. 24 6. 65.625
7. $\frac{25}{4}$ 8. about 1.349
9. Multiply each side by $a - b$.
10. Factor out r on the left.
11. $G = \frac{Fd^2}{Mm}$ 12. $M = \frac{Fd^2}{Gm}$
13. $a = \frac{bc}{c + b}$
14. $b = \frac{ac}{c - a}$, or $b = \frac{-ac}{a - c}$
15. $v = \frac{PVt}{pT}$ 16. $T = \frac{PVt}{pv}$
17. $r = \frac{nE - IR}{In}$, or $r = \frac{IR - nE}{-In}$
18. $V = at + v$
19. $b = \frac{2\mathcal{A}}{h} - B$, or $b = \frac{2\mathcal{A} - hB}{h}$
20. $n = \frac{2S}{(a + \ell)d}$
21. $r = \frac{eR}{E - e}$ 22. $x = \frac{ay - z}{1 + y}$
23. $R = \frac{D}{1 - DT}$, or $R = \frac{-D}{DT - 1}$
24. $D = \frac{R}{1 + RT}$
25. 21 girls, 7 boys
26. \$0.72

Solve each problem. ***See Example 1.***

5. In work with electric circuits, the formula

$$\frac{1}{a} = \frac{1}{b} + \frac{1}{c}$$

occurs. Find b if $a = 8$ and $c = 12$.

6. A gas law in chemistry says that

$$\frac{PV}{T} = \frac{pv}{t}.$$

Suppose that $T = 300$, $t = 350$, $V = 9$, $P = 50$, and $v = 8$. Find p.

7. A formula from anthropology says that

$$c = \frac{100b}{L}.$$

Find L if $c = 80$ and $b = 5$.

8. The gravitational force between two masses is given by

$$F = \frac{GMm}{d^2}.$$

Find M if $F = 10$, $G = 6.67 \times 10^{-11}$, $m = 1$, and $d = 3 \times 10^{-6}$.

9. *Concept Check* To solve the following equation for a, what is the first step?

$$m = \frac{ab}{a - b}$$

10. *Concept Check* To solve the following equation for r, what is the first step?

$$rp - rq = p + q$$

Solve each formula for the specified variable. ***See Examples 2 and 3.***

11. $F = \frac{GMm}{d^2}$ for G (physics) f

12. $F = \frac{GMm}{d^2}$ for M (physics)

13. $\frac{1}{a} = \frac{1}{b} + \frac{1}{c}$ for a (electricity)

14. $\frac{1}{a} = \frac{1}{b} + \frac{1}{c}$ for b (electricity)

15. $\frac{PV}{T} = \frac{pv}{t}$ for v (chemistry)

16. $\frac{PV}{T} = \frac{pv}{t}$ for T (chemistry)

17. $I = \frac{nE}{R + nr}$ for r (engineering)

18. $a = \frac{V - v}{t}$ for V (physics)

19. $\mathcal{A} = \frac{1}{2}h(b + B)$ for b (mathematics)

20. $S = \frac{n}{2}(a + \ell)d$ for n (mathematics)

21. $\frac{E}{e} = \frac{R + r}{r}$ for r (engineering)

22. $y = \frac{x + z}{a - x}$ for x (mathematics)

23. $D = \frac{R}{1 + RT}$ for R (banking)

24. $R = \frac{D}{1 - DT}$ for D (banking)

Concept Check *Use proportions to solve each problem mentally.*

25. In a mathematics class, 3 of every 4 students are girls. If there are 28 students in the class, how many are girls? How many are boys?

26. In a certain southern state, sales tax on a purchase of \$1.50 is \$0.12. What is the sales tax on a purchase of \$9.00?

27. 1.75 in. **28.** 16.25 in.
29. 5.4 in. **30.** 3.0 in.
31. 7.6 in. **32.** 15.6 in.
33. 56 teachers **34.** 61 games
35. 210 deer **36.** 630 deer

The water content of snow is affected by the temperature, wind speed, and other factors present when the snow is falling. The average snow-to-liquid ratio is 10 in. *of snow to* 1 in. *of liquid precipitation. This means that if* 10 in. *of snow fell and was melted, it would produce* 1 in. *of liquid precipitation in a rain gauge.* (*Source:* www.theweatherprediction.com)

Use a proportion to solve each problem. ***See Examples 4 and 5.***

27. A dry snow might have a snow-to-liquid ratio of 18 to 1. Using this ratio, how much liquid precipitation would be produced by 31.5 in. of snow?

28. A wet, sticky snow good for making a snow man might have a snow-to-liquid ratio of 5 to 1. How many inches of fresh snow would produce 3.25 in. of liquid precipitation using this ratio?

Solve each problem. Give answers to the nearest tenth if an approximation is needed (unless specified otherwise). ***See Examples 4 and 5.***

29. On a U.S. map, the distance between Seattle and Durango is 4.125 in. The two cities are actually 1238 mi apart. On this same map, what would be the distance between Chicago and El Paso, two cities that are actually 1606 mi apart? (*Source:* Universal Map Atlas.)

30. On a U.S. map, the distance between Reno and Phoenix is 2.5 in. The two cities are actually 768 mi apart. On this same map, what would be the distance between St. Louis and Jacksonville, two cities that are actually 919 mi apart? (*Source:* Universal Map Atlas.)

31. On a world globe, the distance between New York and Cairo, two cities that are actually 5619 mi apart, is 8.5 in. On this same globe, how far apart are Madrid and Rio de Janeiro, two cities that are actually 5045 mi apart? (*Sources:* Author's globe, *World Almanac and Book of Facts.*)

32. On a world globe, the distance between San Francisco and Melbourne, two cities that are actually 7856 mi apart, is 11.875 in. On this same globe, how far apart are Mexico City and Singapore, two cities that are actually 10,327 mi apart? (*Sources:* Author's globe, *World Almanac and Book of Facts.*)

33. During a recent school year, the average ratio of teachers to students in public elementary and secondary schools was approximately 1 to 15. If a public school had 840 students, how many teachers would be at the school if this ratio was valid there? (*Source:* U.S. National Center for Education Statistics.)

34. At one point during the 2013–2014 NBA Season, the Indiana Pacers had won 48 of their 65 regular season games. If the team continued to win the same fraction of its games, how many games would the Pacers win for the complete 82-game season? Round the answer to the nearest whole number. (*Source:* www.nba.com)

35. To estimate the deer population of a forest preserve, wildlife biologists caught, tagged, and then released 42 deer. A month later, they returned and caught a sample of 75 deer and found that 15 of them were tagged. Based on this experiment, how many deer lived in the forest preserve?

36. Suppose that in the experiment in **Exercise 35,** only 5 of the previously tagged deer were collected in the sample of 75. What would be the estimate of the deer population?

37. 25,000 fish **38.** 20,000 fish
39. 6.6 more gallons
40. 7.7 more gallons

41. $x = \frac{7}{2}$; $AC = 8$; $DF = 12$

42. $y = 2$; side $NP = 8$; side $MP = \frac{16}{3}$; side $RT = 6$; side $QT = 4$

43. 2.4 mL **44.** 0.75 mL

45. $\frac{1}{3}$ job per hr **46.** 700 mi

37. Biologists tagged 500 fish in a lake on January 1. On February 1, they returned and collected a random sample of 400 fish, 8 of which had been previously tagged. On the basis of this experiment, how many fish does the lake have?

38. Suppose that in the experiment of **Exercise 37,** 10 of the previously tagged fish were collected on February 1. What would be the estimate of the fish population?

39. Bruce Johnston's Shelby Cobra uses 5 gal of gasoline to drive 156 mi. He has 3 gal of gasoline in the car, and he still needs to drive 300 mi. If the car continues to use gasoline at the same rate, how many more gallons will he need?

40. Mike Love's T-bird uses 6 gal of gasoline to drive 141 mi. He has 4 gal of gasoline in the car, and he still needs to drive 275 mi. If the car continues to use gasoline at the same rate, how many more gallons will he need?

In geometry, two triangles with corresponding angle measures equal, called ***similar triangles,*** *have corresponding sides proportional. For example, in the figure, angle* A = *angle* D, *angle* B = *angle* E, *and angle* C = *angle* F, *so the triangles are similar. Then the following ratios of corresponding sides are equal.*

$$\frac{4}{6} = \frac{6}{9} = \frac{2x + 1}{2x + 5}$$

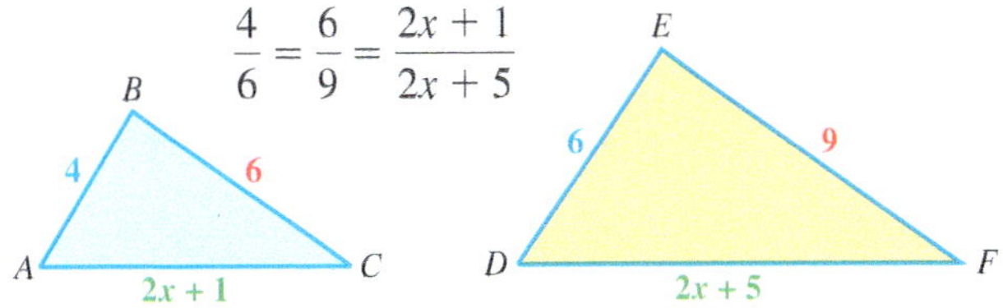

41. Solve for x, using the given proportion to find the lengths of the third sides of the triangles.

42. Suppose the following triangles are similar. Find y and the lengths of the two unknown sides of each triangle.

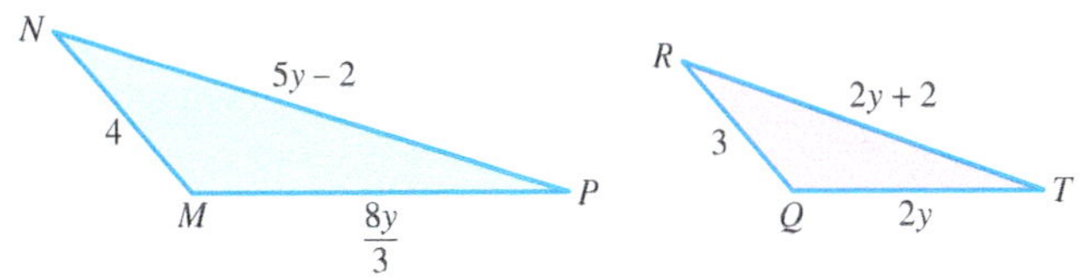

Nurses use proportions to determine the amount of a drug to administer when the dose is measured in milligrams but the drug is packaged in a diluted form in milliliters. (Source: Hoyles, C., R. Noss, and S. Pozzi, "Proportional Reasoning in Nursing Practice," *Journal for Research in Mathematics Education.)*

For example, to find the number of milliliters of fluid needed to administer 300 mg *of a drug that comes packaged as* 120 mg *in* 2 mL *of fluid, a nurse sets up the proportion*

$$\frac{120 \text{ mg}}{2 \text{ mL}} = \frac{300 \text{ mg}}{x \text{ mL}},$$

where x *represents the amount to administer in milliliters. Use this method to find the correct dose for each prescription.*

43. 120 mg of amikacin packaged as 100 mg in 2-mL vials

44. 1.5 mg of morphine packaged as 20-mg ampules diluted in 10 mL of fluid

Concept Check *Solve each problem mentally.*

45. If Marin can mow her yard in 3 hr, what is her rate (in terms of the proportion of the job per hour)?

46. A van traveling from Atlanta to Detroit averages 50 mph and takes 14 hr to make the trip. How far is it from Atlanta to Detroit?

47. 3 mph 48. 3 mph
49. 10 mph 50. 10 mph
51. 1020 mi 52. 2.4 mi
53. 1750 mi 54. 270 mi
55. 190 mi 56. 200 mi
57. $6\frac{2}{3}$ min 58. $3\frac{1}{13}$ hr

Solve each problem. ***See Examples 6 and 7.***

47. Kellen's boat travels 12 mph. Find the rate of the current of the river if she can travel 6 mi upstream in the same amount of time she can go 10 mi downstream. (Let x = the rate of the current.)

	Distance	Rate	Time
Downstream	10	$12 + x$	
Upstream	6	$12 - x$	

48. Kasey can travel 8 mi upstream in the same time it takes her to go 12 mi downstream. Her boat goes 15 mph in still water. What is the rate of the current? (Let x = the rate of the current.)

	Distance	Rate	Time
Downstream			
Upstream		$15 - x$	

49. In his boat, Sheldon can travel 30 mi downstream in the same time that it takes to travel 10 mi upstream. If the rate of the current is 5 mph, find the rate of the boat in still water.

50. In his boat, Leonard can travel 24 mi upstream in the same time that it takes to travel 36 mi downstream. If the rate of the current is 2 mph, find the rate of the boat in still water.

51. On his drive from Montpelier, Vermont, to Columbia, South Carolina, Victor averaged 51 mph. If he had been able to average 60 mph, he would have reached his destination 3 hr earlier. What is the driving distance between Montpelier and Columbia?

	Distance	Rate	Time
Actual Trip			
Alternative Trip	x	60	

52. Kelli lives in an off-campus apartment. When she rides her bike to campus, she arrives 36 min faster than when she walks. If her average walking rate is 3 mph and her average biking rate is 12 mph, how far is it from her apartment to campus?

	Distance	Rate	Time
Bike	x	12	
Walk			

53. A plane averaged 500 mph on a trip going east, but only 350 mph on the return trip. The total flying time in both directions was 8.5 hr. What was the one-way distance?

54. Mary drove from her apartment to her parents' house. She averaged 60 mph because traffic was light. On the return trip, which took 1.5 hr longer, she averaged only 45 mph on the same route. What is the distance between Mary's apartment and her parents' house?

55. On the first part of a trip to Carmel traveling on the freeway, Marge averaged 60 mph. On the rest of the trip, which was 10 mi longer than the first part, she averaged 50 mph. Find the total distance to Carmel if the second part of the trip took 30 min more than the first part.

56. During the first part of a trip on the highway, Jim and Annie averaged 60 mph. In Houston, traffic caused them to average only 30 mph. The distance they drove in Houston was 100 mi less than their distance on the highway. What was their total driving distance if they spent 50 min more on the highway than they did in Houston?

Solve each problem. ***See Example 8.***

57. Butch and Peggy want to pick up the mess that their grandson, Grant, has made in his playroom. Butch could do it in 15 min working alone. Peggy, working alone, could clean it in 12 min. How long will it take them if they work together?

	Rate	Time Working Together	Fractional Part of the Job Done
Butch	$\frac{1}{15}$	x	
Peggy	$\frac{1}{12}$	x	

58. Lou can groom a customer's dogs in 8 hr, but it takes his business partner, Janet, only 5 hr to groom the same dogs. How long will it take them to groom the dogs if they work together?

	Rate	Time Working Together	Fractional Part of the Job Done
Lou	$\frac{1}{8}$	x	
Janet	$\frac{1}{5}$	x	

59. 30 hr
60. $7\frac{1}{2}$ hr
61. $2\frac{1}{3}$ hr
62. 20 min
63. 20 hr
64. 36 hr
65. $2\frac{4}{5}$ hr
66. 240 min, or 4 hr

59. Jerry and Kuba are laying a hardwood floor. Working alone, Jerry can do the job in 20 hr. If the two of them work together, they can complete the job in 12 hr. How long would it take Kuba to lay the floor working alone? (Let $x =$ the time it would take Kuba working alone.)

	Rate	Time Working Together	Fractional Part of the Job Done
Jerry		12	
Kuba		12	

60. Mrs. Disher can grade a set of tests in 5 hr working alone. If her student teacher Mr. Howes helps her, it will take 3 hr to grade the tests. How long would it take Mr. Howes to grade the tests if he worked alone? (Let $x =$ the time it would take Mr. Howes working alone.)

	Rate	Time Working Together	Fractional Part of the Job Done
Mrs. Disher		3	
Mr. Howes		3	

61. Dixie can paint a room in 3 hr working alone. Trixie can paint the same room in 6 hr working alone. How long after Dixie starts to paint the room will it be finished if Trixie joins her 1 hr later?

62. Howard can wash a car in 30 min working alone. Raj can do the same job in 45 min working alone. How long after Howard starts to wash the car will it be finished if Raj joins him 5 min later?

63. If a vat of acid can be filled by an inlet pipe in 10 hr and emptied by an outlet pipe in 20 hr, how long will it take to fill the vat if both pipes are open?

64. A winery has a vat to hold Zinfandel. An inlet pipe can fill the vat in 9 hr, while an outlet pipe can empty it in 12 hr. How long will it take to fill the vat if both the outlet and the inlet pipes are open?

65. Suppose that Hortense and Mort can clean their entire house in 7 hr, while their toddler, Mimi, can mess it up in only 2 hr. If Hortense and Mort clean the house while Mimi is at her grandma's and then start cleaning up after Mimi the minute she gets home, how long does it take from the time Mimi gets home until the whole place is a shambles?

66. An inlet pipe can fill an artificial lily pond in 60 min, while an outlet pipe can empty it in 80 min. Through an error, both pipes are left open. How long will it take for the pond to fill?

6.6 Variation

OBJECTIVES

1. Write an equation expressing direct variation.
2. Find the constant of variation, and solve direct variation problems.
3. Solve inverse variation problems.
4. Solve joint variation problems.
5. Solve combined variation problems.

Functions in which *y depends on a multiple of x* or *y depends on a number divided by x* occur in business, mathematics, and the physical sciences and are explored in this section.

OBJECTIVE 1 Write an equation expressing direct variation.

The circumference of a circle is given by the formula $C = 2\pi r$, where r is the radius of the circle. See **FIGURE 5**. The circumference is always a constant multiple of the radius—that is, C is always found by multiplying r by the constant 2π.

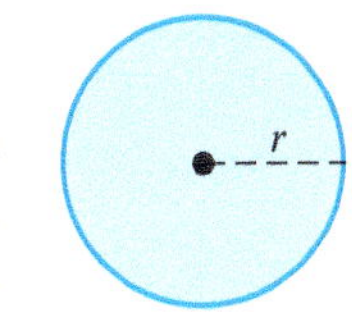

$C = 2\pi r$

FIGURE 5

As the ***radius increases,*** the ***circumference increases.***

As the ***radius decreases,*** the ***circumference decreases.***

Because of these relationships, the circumference is said to *vary directly* as the radius.

VOCABULARY

- ☐ direct variation
- ☐ constant of variation
- ☐ inverse variation
- ☐ joint variation
- ☐ combined variation

Direct Variation

y* varies directly as *x if there exists a real number k such that

$$y = kx.$$

y is said to be *proportional to* x. The number k is the **constant of variation.**

TEACHING TIP
Some simple examples of variation include the following:

Direct variation: the farther down one pushes on a car's gas pedal, the faster the car goes.

Inverse variation: the farther down one pushes on a car's brake pedal, the slower the car goes.

In direct variation, for $k > 0$, as the value of x increases, the value of y increases. Similarly, as x decreases, y decreases.

OBJECTIVE 2 Find the constant of variation, and solve direct variation problems.

The direct variation equation $y = kx$ defines a linear function, where the constant of variation k is the slope of the line. For example, the following equation describes the cost y to buy x gallons of gasoline.

$$y = 3.50x$$ See Section 2.3, Example 7.

The cost varies directly as, or is proportional to, the number of gallons purchased.

As the *number of gallons increases,* the *cost increases.*

As the *number of gallons decreases,* the *cost decreases.*

The constant of variation k is 3.50, the cost of 1 gal of gasoline.

NOW TRY EXERCISE 1
One week Morgan sold 8 dozen eggs for \$20. How much does she charge for one dozen eggs?

EXAMPLE 1 Solving a Direct Variation Problem

Eva is paid an hourly wage. One week she worked 43 hr and was paid \$795.50. How much does she earn per hour?

Let $h =$ the number of hours she works

and $P =$ her corresponding pay.

Write a variation equation.

k represents Eva's hourly wage. →	$P = kh$	P varies directly as h.
	$795.50 = 43k$	Substitute 795.50 for P and 43 for h.
This is the constant of variation. →	$k = 18.50$	Use a calculator.

Her hourly wage is \$18.50, and P and h are related by the equation

$$P = 18.50h.$$

We can use this equation to find her pay for any number of hours worked.

NOW TRY

CLASSROOM EXAMPLE 1
If 7 kg of steak cost \$45.50, how much will 1 kg of steak cost?

Answer: \$6.50

CLASSROOM EXAMPLE 2
It costs \$52 to use 800 kilowatt-hours of electricity. How much will 650 kilowatt-hours cost?

Answer: \$42.25

EXAMPLE 2 Solving a Direct Variation Problem

Hooke's law for an elastic spring states that the distance a spring stretches is directly proportional to the force applied. If a force of 150 newtons* stretches a certain spring 8 cm, how much will a force of 400 newtons stretch the spring? See **FIGURE 6.**

Let $d =$ the distance the spring stretches

and $f =$ the force applied.

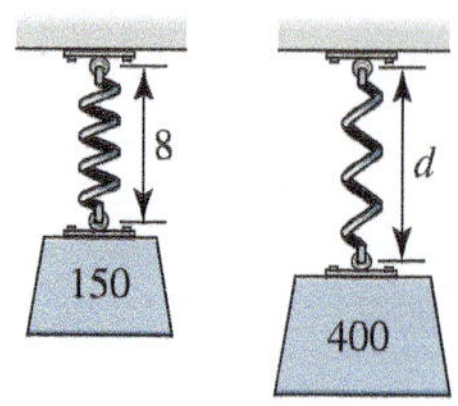

FIGURE 6

NOW TRY ANSWER
1. \$2.50

*A newton is a unit of measure of force used in physics.

NOW TRY EXERCISE 2

For a constant height, the area of a parallelogram is directly proportional to its base. If the area is 20 cm^2 when the base is 4 cm, find the area when the base is 7 cm.

Then $d = kf$ for some constant k. A force of 150 newtons stretches the spring 8 cm, so we use these values to find k.

$$d = kf \qquad \text{Variation equation}$$

Solve for k.

$$8 = k \cdot 150 \qquad \text{Let } d = 8 \text{ and } f = 150.$$

$$k = \frac{8}{150} \qquad \text{Solve for } k.$$

$$k = \frac{4}{75} \qquad \text{Write in lowest terms.}$$

Now we rewrite the variation equation $d = kf$ using $\frac{4}{75}$ for k.

$$d = \frac{4}{75}f \qquad \text{Here, } k = \frac{4}{75}.$$

For a force of 400 newtons, substitute 400 for f.

$$d = \frac{4}{75}(400) = \frac{64}{3} \qquad \text{Let } f = 400.$$

The spring will stretch $\frac{64}{3}$ cm, or $21\frac{1}{3}$ cm, if a force of 400 newtons is applied.

NOW TRY

TEACHING TIP

Students should learn to write these problems in key words and then translate into a variation equation. Thus, *distance varies directly as the square of time* in Example 3 becomes

$$d = kt^2.$$

Proceed to solve, using the steps discussed here.

Solving a Variation Problem

Step 1 Write a variation equation.

Step 2 Substitute the initial values and solve for k.

Step 3 Rewrite the variation equation with the value of k from Step 2.

Step 4 Substitute the remaining values, solve for the unknown, and find the required answer.

One variable can be proportional to a power of another variable.

Direct Variation as a Power

y* varies directly as the *n*th power of *x if there exists a real number k such that

$$y = kx^n.$$

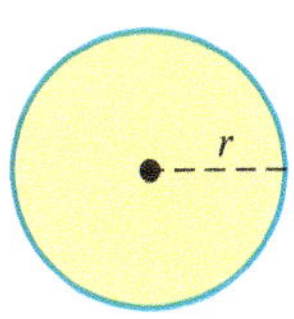

$\mathcal{A} = \pi r^2$

FIGURE 7

An example of direct variation as a power is the formula for the area of a circle, $\mathcal{A} = \pi r^2$. See **FIGURE 7**. Here, π is the constant of variation, and the area varies directly as the *square* of the radius.

CLASSROOM EXAMPLE 3

Suppose y varies directly as the cube of x, and $y = 24$ when $x = 2$. Find y when $x = 4$.

Answer: 192

EXAMPLE 3 Solving a Direct Variation Problem

The distance a body falls from rest varies directly as the square of the time it falls (disregarding air resistance). If a skydiver falls 64 ft in 2 sec, how far will she fall in 8 sec?

Step 1 Let $d =$ the distance the skydiver falls

and $t =$ the time it takes to fall.

Then d is a function of t for some constant k.

$$d = kt^2 \qquad d \text{ varies directly as the square of } t.$$

NOW TRY ANSWER

2. 35 cm^2

NOW TRY EXERCISE 3

Suppose y varies directly as the square of x, and $y = 200$ when $x = 5$. Find y when $x = 7$.

Step 2 To find the value of k, use the fact that the skydiver falls 64 ft in 2 sec.

$$d = kt^2 \quad \text{Variation equation}$$

$$64 = k(2)^2 \quad \text{Let } d = 64 \text{ and } t = 2.$$

$$k = 16 \quad \text{Find } k.$$

Step 3 Now we rewrite the variation equation $d = kt^2$ using 16 for k.

$$d = 16t^2 \quad \text{Here, } k = 16.$$

Step 4 Let $t = 8$ to find the number of feet the skydiver will fall in 8 sec.

$$d = 16(8)^2 = 1024 \quad \text{Let } t = 8.$$

The skydiver will fall 1024 ft in 8 sec. NOW TRY

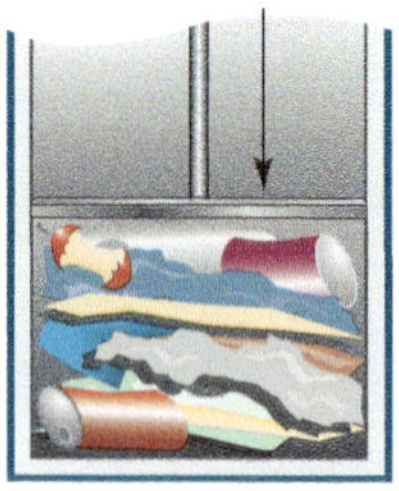

As pressure on trash increases, volume of trash decreases.

FIGURE 8

OBJECTIVE 3 Solve inverse variation problems.

Another type of variation is *inverse variation.* ***With inverse variation, for k > 0, as one variable increases, the other variable decreases.***

For example, in a closed space, volume decreases as pressure increases, which can be illustrated by a trash compactor. See FIGURE 8. As the compactor presses down, the pressure on the trash increases, and in turn, the trash occupies a smaller space.

TEACHING TIP

In function terminology, *direct variation problems* require that the independent variable be in the *numerator*.

Inverse variation problems require that the independent variable be in the *denominator*.

Inverse Variation

y* varies inversely as *x if there exists a real number k such that

$$y = \frac{k}{x}.$$

y* varies inversely as the *n*th power of *x if there exists a real number k such that

$$y = \frac{k}{x^n}.$$

TEACHING TIP

Work **Exercises 1–10** in the accompanying exercise set orally as a class to help students distinguish between direct and inverse variation.

The inverse variation equation defines a rational function. Another example of inverse variation comes from the distance formula.

$$d = rt \quad \text{Distance formula}$$

$$t = \frac{d}{r} \quad \text{Divide each side by } r.$$

Here, t (time) varies inversely as r (rate or speed), with d (distance) serving as the constant of variation. For example, if the distance between Chicago and Des Moines is 300 mi, then

$$t = \frac{300}{r}.$$

The values of r and t might be any of the following.

$r = 50, t = 6$
$r = 60, t = 5$
$r = 75, t = 4$
As r increases, t decreases.

$r = 30, t = 10$
$r = 25, t = 12$
$r = 20, t = 15$
As r decreases, t increases.

If we *increase* the rate (speed) at which we drive, time *decreases.* If we *decrease* the rate (speed) at which we drive, time *increases.*

NOW TRY ANSWER

3. 392

NOW TRY EXERCISE 4

For a constant area, the height of a triangle varies inversely as the base. If the height is 7 cm when the base is 8 cm, find the height when the base is 14 cm.

CLASSROOM EXAMPLE 4

The current in a simple electrical circuit varies inversely as the resistance. If the current is 80 amps when the resistance is 10 ohms, find the current when the resistance is 16 ohms.

Answer: 50 amps

EXAMPLE 4 Solving an Inverse Variation Problem

In the manufacture of a phone-charging device, the cost of producing the device varies inversely as the number produced. If 10,000 units are produced, the cost is \$2 per unit. Find the cost per unit to produce 25,000 units.

Let $x =$ the number of units produced

and $c =$ the cost per unit.

Here, as production increases, cost decreases, and as production decreases, cost increases. We write a variation equation using the variables c and x and the constant k.

$$c = \frac{k}{x} \qquad c \text{ varies inversely as } x.$$

To find k, we replace c with 2 and x with 10,000.

$$2 = \frac{k}{10{,}000} \qquad \text{Substitute in the variation equation.}$$

$$20{,}000 = k \qquad \text{Multiply by 10,000.}$$

Thus, the variation equation $c = \frac{k}{x}$ becomes $c = \frac{20{,}000}{x}$. When $x = 25{,}000$,

$$c = \frac{20{,}000}{25{,}000} = 0.80. \qquad \text{Let } x = 25{,}000.$$

The cost per unit to make 25,000 units is \$0.80. NOW TRY

NOW TRY EXERCISE 5

The weight of an object above Earth varies inversely as the square of its distance from the center of Earth. If an object weighs 150 lb on the surface of Earth, and the radius of Earth is about 3960 mi, how much does it weigh when it is 1000 mi above Earth's surface? Round to the nearest pound.

CLASSROOM EXAMPLE 5

Suppose p varies inversely as the cube of q, and $p = 100$ when $q = 3$. Find p when $q = 5$.

Answer: $\frac{108}{5}$

EXAMPLE 5 Solving an Inverse Variation Problem

The weight of an object above Earth varies inversely as the square of its distance from the center of Earth. A space shuttle in an elliptical orbit has a maximum distance from the center of Earth (**apogee**) of 6700 mi. Its minimum distance from the center of Earth (**perigee**) is 4090 mi. See **FIGURE 9**. If an astronaut in the shuttle weighs 57 lb at its apogee, what does the astronaut weigh at its perigee?

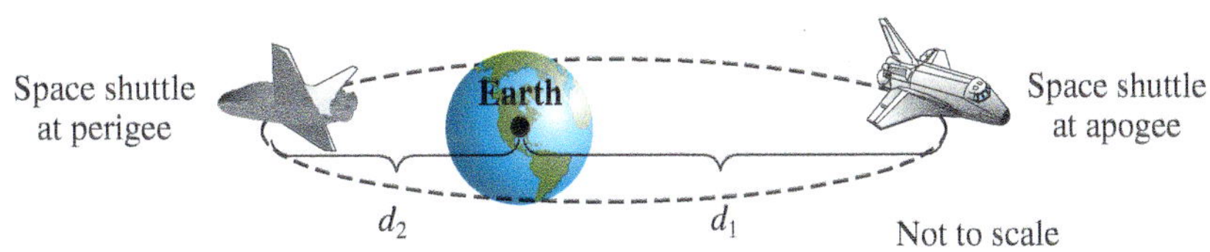

FIGURE 9

Let $w =$ the weight and $d =$ the distance from the center of Earth, for some constant k. We write a variation equation using these variables.

$$w = \frac{k}{d^2} \qquad w \text{ varies inversely as the square of } d.$$

At the apogee, the astronaut weighs 57 lb, and the distance from the center of Earth is 6700 mi. Use these values to find k.

$$57 = \frac{k}{(6700)^2} \qquad \text{Let } w = 57 \text{ and } d = 6700.$$

$$k = 57(6700)^2 \qquad \text{Solve for } k.$$

Substitute $k = 57(6700)^2$ and $d = 4090$ to find the weight at the perigee.

$$w = \frac{57(6700)^2}{(4090)^2} \approx 153 \text{ lb} \qquad \text{Use a calculator. Round to the nearest pound.}$$

NOW TRY

NOW TRY ANSWERS

4. 4 cm **5.** 96 lb

OBJECTIVE 4 Solve joint variation problems.

If one variable varies directly as the *product* of several other variables (perhaps raised to powers), the first variable is said to *vary jointly* as the others.

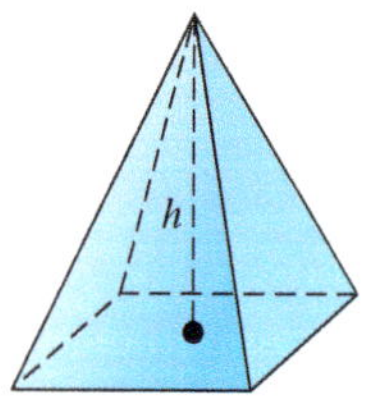

B = area of the base

FIGURE 10

Joint Variation

y* varies jointly as *x* and *z if there exists a real number k such that

$$y = kxz.$$

An example of joint variation is the formula for the volume of a right pyramid, $V = \frac{1}{3}Bh$. See **FIGURE 10**. Here, $\frac{1}{3}$ is the constant of variation, and the volume varies jointly as the area of the base and the height.

NOW TRY EXERCISE 6

The volume of a right pyramid varies jointly as the height and the area of the base. If the volume is 100 ft^3 when the area of the base is 30 ft^2 and the height is 10 ft, find the volume when the area of the base is 90 ft^2 and the height is 20 ft.

EXAMPLE 6 Solving a Joint Variation Problem

The interest on a loan or an investment is given by the formula $I = prt$. Here, for a given principal p, the interest earned, I, varies jointly as the interest rate r and the time t the principal is left earning interest. If an investment earns \$100 interest at 5% for 2 yr, how much interest will the same principal earn at 4.5% for 3 yr?

We use the formula $I = prt$, where p is the constant of variation because it is the same for both investments.

$$I = prt \quad \text{Here, } p \text{ is the constant of variation.}$$

Solve for p.

$$100 = p(0.05)(2) \quad \text{Let } I = 100,\ r = 0.05, \text{ and } t = 2.$$

$$100 = 0.1p \quad \text{Multiply.}$$

$$p = 1000 \quad \text{Divide by 0.1. Rewrite.}$$

Now we find I when $p = 1000$, $r = 0.045$, and $t = 3$.

$$I = 1000(0.045)(3) = 135 \quad \text{Let } p = 1000,\ r = 0.045, \text{ and } t = 3.$$

The interest will be \$135.

NOW TRY

CLASSROOM EXAMPLE 6

Suppose x varies jointly as y and z^2, and $x = 231$ when $y = 3$ and $z = 2$. Find x when $y = 5$ and $z = 4$.

Answer: 1540

CAUTION Note that *and* in the expression "y varies directly as x *and* z" translates as a product in the variation equation $y = kxz$. The word *and* does **not** indicate addition here.

OBJECTIVE 5 Solve combined variation problems.

There are combinations of direct and inverse variation, called **combined variation.**

EXAMPLE 7 Solving a Combined Variation Problem

Body mass index (BMI) is used to assess whether a person's weight is healthy. A BMI from 19 through 25 is considered desirable. BMI varies directly as an individual's weight in pounds and inversely as the square of their height in inches.

A person who weighs 118 lb and is 64 in. tall has a BMI of 20. (BMI is rounded to the nearest whole number.) Find the BMI of a man who weighs 165 lb and is 70 in. tall. (*Source: Washington Post.*)

Let B = BMI, w = weight, and h = height. Write a variation equation.

$$B = \frac{kw}{h^2}$$

kw ← BMI varies directly as the weight.

h^2 ← BMI varies inversely as the square of the height.

CLASSROOM EXAMPLE 7

Suppose z varies jointly as x and y^2 and inversely as w. Also, $z = \frac{3}{8}$ when $x = 2$, $y = 3$, and $w = 12$. Find z, when $x = 4$, $y = 1$, and $w = 6$.

Answer: $\frac{1}{6}$

NOW TRY ANSWER

6. 600 ft^3

To find k, let $B = 20$, $w = 118$, and $h = 64$.

$$20 = \frac{k(118)}{64^2} \qquad B = \frac{kw}{h^2}$$

$$k = \frac{20(64^2)}{118} \qquad \text{Multiply by } 64^2. \text{ Divide by } 118.$$

$$k = 694 \qquad \text{Use a calculator. Round to the nearest whole number.}$$

Now find B when $k = 694$, $w = 165$, and $h = 70$.

$$B = \frac{694(165)}{70^2} = 23 \qquad \text{Round to the nearest whole number.}$$

The man's BMI is 23.

NOW TRY

NOW TRY EXERCISE 7

In statistics, the sample size used to estimate a population mean varies directly as the variance and inversely as the square of the maximum error of the estimate. If the sample size is 200 when the variance is 25 m^2 and the maximum error of the estimate is 0.5 m, find the sample size when the variance is 25 m^2 and the maximum error of the estimate is 0.1 m.

NOW TRY ANSWER

7. 5000

6.6 Exercises

FOR EXTRA HELP

MyMathLab®

Complete solution available in MyMathLab

1. increases; decreases
2. decreases; increases
3. direct
4. direct
5. direct
6. inverse
7. inverse
8. direct
9. inverse
10. direct
11. inverse
12. inverse
13. direct
14. direct
15. joint
16. joint
17. combined
18. combined
19. The perimeter of a square varies directly as the length of its side.
20. The diameter of a circle varies directly as the length of its radius.

Concept Check *Fill in each blank with the correct response.*

1. For $k > 0$, if y varies directly as x, then when x increases, y ______, and when x decreases, y ______.

2. For $k > 0$, if y varies inversely as x, then when x increases, y ______, and when x decreases, y ______.

Concept Check *Use personal experience or intuition to determine whether the situation suggests* direct *or* inverse *variation.*

3. The number of movie tickets purchased and the total price for the tickets
4. The rate and the distance traveled by a pickup truck in 3 hr
5. The amount of pressure put on the accelerator of a car and the speed of the car
6. The percentage off an item that is on sale and the price of the item
7. Your age and the probability that you believe in the tooth fairy
8. The surface area of a balloon and its diameter
9. The demand for an item and the price of the item
10. The number of hours worked by an hourly worker and the amount of money earned

Concept Check *Determine whether each equation represents* direct, inverse, joint, *or* combined *variation.*

11. $y = \frac{3}{x}$
12. $y = \frac{8}{x}$
13. $y = 10x^2$
14. $y = 2x^3$
15. $y = 3xz^4$
16. $y = 6x^3z^2$
17. $y = \frac{4x}{wz}$
18. $y = \frac{6x}{st}$

Concept Check *Write each formula using the "language" of variation. For example, the formula for the circumference of a circle, $C = 2\pi r$, can be written as*

"The circumference of a circle varies directly as the length of its radius."

19. $P = 4s$, where P is the perimeter of a square with side of length s
20. $d = 2r$, where d is the diameter of a circle with radius r

21. $S = 4\pi r^2$, where S is the surface area of a sphere with radius r

22. $V = \frac{4}{3}\pi r^3$, where V is the volume of a sphere with radius r

23. $\mathcal{A} = \frac{1}{2}bh$, where $\mathcal{A}$ is the area of a triangle with base b and height h

24. $V = \frac{1}{3}\pi r^2 h$, where V is the volume of a cone with radius r and height h

25. *Concept Check* What is the constant of variation in each of the variation equations in **Exercises 19–24?**

26. *Concept Check* What is meant by the constant of variation in a direct variation problem? If we were to graph the linear equation $y = kx$ for some nonnegative constant k, what role would k play in the graph?

Write a variation equation for each situation. Use k as the constant of variation. ***See Examples 1–6.***

27. A varies directly as b.

28. W varies directly as f.

29. h varies inversely as t.

30. p varies inversely as s.

31. M varies directly as the square of d.

32. P varies inversely as the cube of x.

33. I varies jointly as g and h.

34. C varies jointly as a and the square of b.

Solve each problem. ***See Examples 1–7.***

35. If x varies directly as y, and $x = 9$ when $y = 3$, find x when $y = 12$.

36. If x varies directly as y, and $x = 10$ when $y = 7$, find y when $x = 50$.

37. If a varies directly as the square of b, and $a = 4$ when $b = 3$, find a when $b = 2$.

38. If h varies directly as the square of m, and $h = 15$ when $m = 5$, find h when $m = 7$.

39. If z varies inversely as w, and $z = 10$ when $w = 0.5$, find z when $w = 8$.

40. If t varies inversely as s, and $t = 3$ when $s = 5$, find s when $t = 5$.

41. If m varies inversely as p^2, and $m = 20$ when $p = 2$, find m when $p = 5$.

42. If a varies inversely as b^2, and $a = 48$ when $b = 4$, find a when $b = 7$.

43. p varies jointly as q and r^2, and $p = 200$ when $q = 2$ and $r = 3$. Find p when $q = 5$ and $r = 2$.

44. f varies jointly as g^2 and h, and $f = 50$ when $g = 4$ and $h = 2$. Find f when $g = 3$ and $h = 6$.

Solve each problem. ***See Examples 1–7.***

45. Ben bought 8.5 gal of gasoline and paid $33.32. What is the price of gasoline per gallon?

46. Sara gives horseback rides at Shadow Mountain Ranch. A 2.5-hr ride costs $50.00. What is the price per hour?

47. The weight of an object on Earth is directly proportional to the weight of that same object on the moon. A 200-lb astronaut would weigh 32 lb on the moon. How much would a 50-lb dog weigh on the moon?

48. The pressure exerted by a certain liquid at a given point is directly proportional to the depth of the point beneath the surface of the liquid. The pressure at 30 m is 80 newtons. What pressure is exerted at 50 m?

Answers

21. The surface area of a sphere varies directly as the square of its radius.
22. The volume of a sphere varies directly as the cube of its radius.
23. The area of a triangle varies jointly as the lengths of its base and height.
24. The volume of a cone varies jointly as the square of its radius and its height.
25. 4; 2; 4π; $\frac{4}{3}\pi$; $\frac{1}{2}$; $\frac{1}{3}\pi$
26. The constant of variation is the coefficient of x. In the graph of the linear equation $y = kx$, k represents the slope of the line.
27. $A = kb$
28. $W = kf$
29. $h = \frac{k}{t}$
30. $p = \frac{k}{s}$
31. $M = kd^2$
32. $P = \frac{k}{x^3}$
33. $I = kgh$
34. $C = kab^2$
35. 36
36. 35
37. $\frac{16}{9}$
38. $\frac{147}{5}$
39. 0.625
40. 3
41. $\frac{16}{5}$
42. $\frac{768}{49}$
43. $222\frac{2}{9}$
44. $84\frac{3}{8}$
45. $3.92
46. $20
47. 8 lb
48. $133\frac{1}{3}$ newtons

49. 450 cm^3
50. 120 newtons
51. 256 ft
52. 800 gal
53. $106\frac{2}{3}$ mph
54. 15 ft
55. 100 cycles per sec
56. $13\frac{1}{3}$ amps
57. $21\frac{1}{3}$ foot-candles
58. $71\frac{1}{9}$ lb
59. \$420

49. The volume of a can of tomatoes is directly proportional to the height of the can. If the volume of the can is 300 cm^3 when its height is 10.62 cm, find the volume to the nearest whole number of a can with height 15.92 cm.

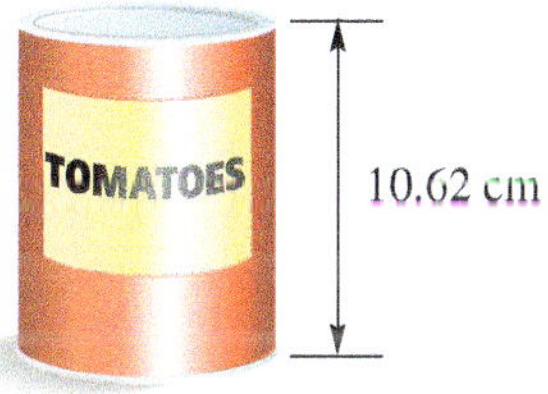

50. The force required to compress a spring is directly proportional to the change in length of the spring. If a force of 20 newtons is required to compress a certain spring 2 cm, how much force is required to compress the spring from 20 cm to 8 cm?

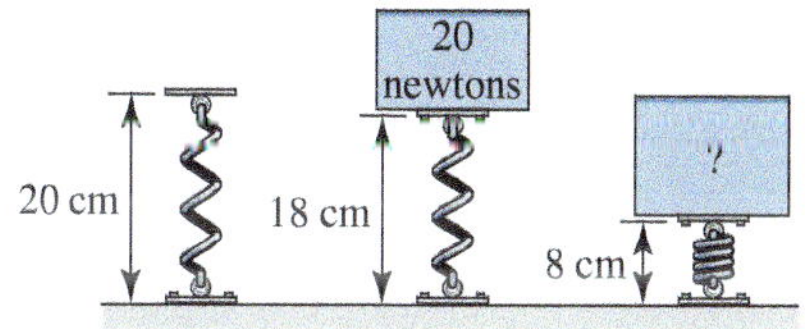

51. For a body falling freely from rest (disregarding air resistance), the distance the body falls varies directly as the square of the time. If an object is dropped from the top of a tower 576 ft high and hits the ground in 6 sec, how far did it fall in the first 4 sec?

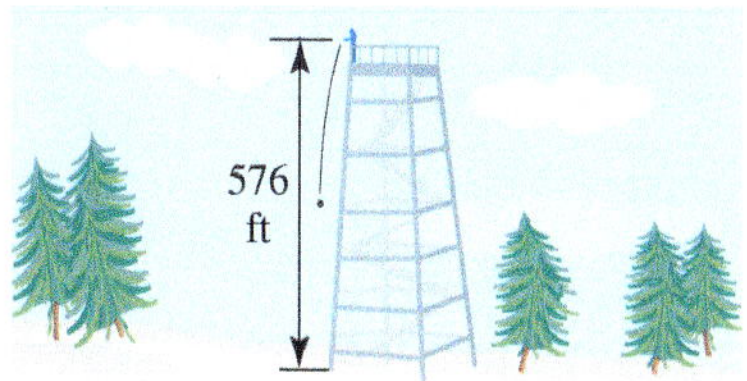

52. The amount of water emptied by a pipe varies directly as the square of the diameter of the pipe. For a certain constant water flow, a pipe emptying into a canal will allow 200 gal of water to escape in an hour. The diameter of the pipe is 6 in. How much water would a 12-in. pipe empty into the canal in an hour, assuming the same water flow?

53. Over a specified distance, rate varies inversely with time. If a Dodge Viper on a test track goes a certain distance in one-half minute at 160 mph, what rate is needed to go the same distance in three-fourths minute?

54. For a constant area, the length of a rectangle varies inversely as the width. The length of a rectangle is 27 ft when the width is 10 ft. Find the width of a rectangle with the same area if the length is 18 ft.

55. The frequency of a vibrating string varies inversely as its length. That is, a longer string vibrates fewer times in a second than a shorter string. Suppose a piano string 2 ft long vibrates 250 cycles per sec. What frequency would a string 5 ft long have?

56. The current in a simple electrical circuit varies inversely as the resistance. If the current is 20 amps when the resistance is 5 ohms, find the current when the resistance is 7.5 ohms.

57. The amount of light (measured in foot-candles) produced by a light source varies inversely as the square of the distance from the source. If the illumination produced 1 m from a light source is 768 foot-candles, find the illumination produced 6 m from the same source.

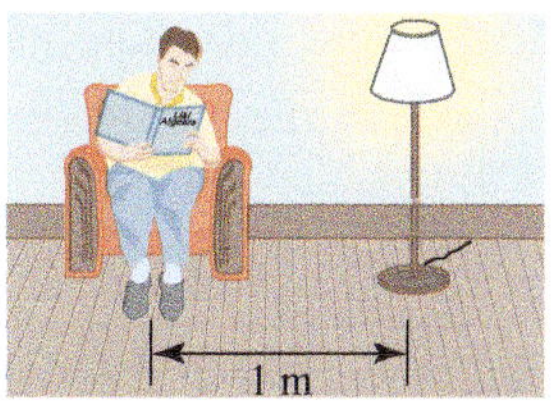

58. The force with which Earth attracts an object above Earth's surface varies inversely as the square of the distance of the object from the center of Earth. If an object 4000 mi from the center of Earth is attracted with a force of 160 lb, find the force of attraction if the object were 6000 mi from the center of Earth.

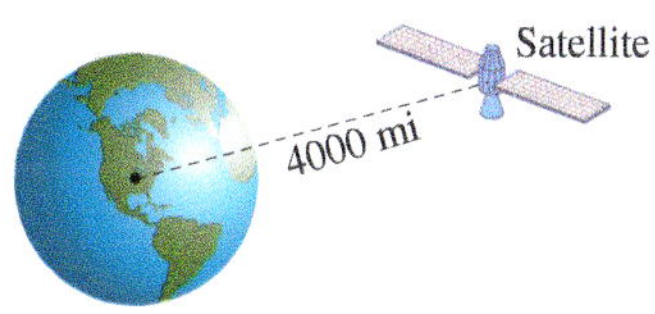

59. For a given interest rate, simple interest varies jointly as principal and time. If \$2000 left in an account for 4 yr earned interest of \$280, how much interest would be earned in 6 yr?

60. 8.5
61. 11.8 lb
62. 6.2 lb
63. 448.1 lb
64. $\frac{8}{9}$ metric ton
65. 68,600 calls
66. 1.105 L
67. Answers will vary.
68. 480 kg

60. The collision impact of an automobile varies jointly as its mass and the square of its speed. Suppose a 2000-lb car traveling at 55 mph has a collision impact of 6.1. What is the collision impact (to the nearest tenth) of the same car at 65 mph?

61. The weight of a bass varies jointly as its girth and the square of its length. (**Girth** is the distance around the body of the fish.) A prize-winning bass weighed in at 22.7 lb and measured 36 in. long with a 21-in. girth. How much (to the nearest tenth of a pound) would a bass 28 in. long with an 18-in. girth weigh?

62. The weight of a trout varies jointly as its length and the square of its girth. One angler caught a trout that weighed 10.5 lb and measured 26 in. long with an 18-in. girth. Find the weight (to the nearest tenth of a pound) of a trout that is 22 in. long with a 15-in. girth.

63. The force needed to keep a car from skidding on a curve varies inversely as the radius of the curve and jointly as the weight of the car and the square of the speed. If 242 lb of force keeps a 2000-lb car from skidding on a curve of radius 500 ft at 30 mph, what force (to the nearest tenth of a pound) would keep the same car from skidding on a curve of radius 750 ft at 50 mph?

64. The maximum load that a cylindrical column with a circular cross section can hold varies directly as the fourth power of the diameter of the cross section and inversely as the square of the height. A 9-m column 1 m in diameter will support 8 metric tons. How many metric tons can be supported by a column 12 m high and $\frac{2}{3}$ m in diameter?

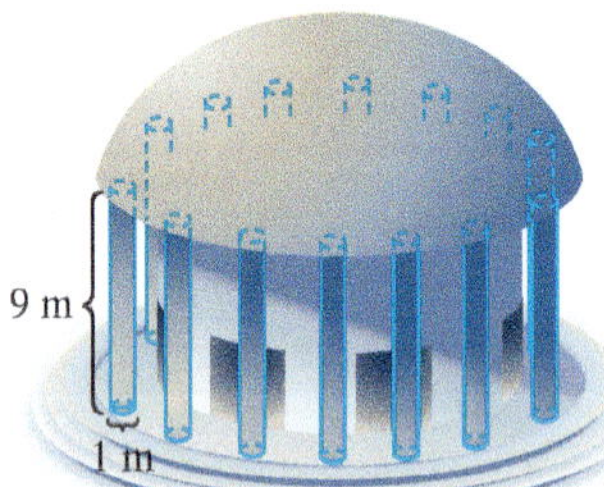

Load = 8 metric tons

65. The number of long-distance phone calls between two cities during a certain period varies jointly as the populations of the cities, p_1 and p_2, and inversely as the distance between them. If 80,000 calls are made between two cities 400 mi apart, with populations of 70,000 and 100,000, how many calls (to the nearest hundred) are made between cities with populations of 50,000 and 75,000 that are 250 mi apart?

66. The volume of gas varies inversely as the pressure and directly as the temperature. (Temperature must be measured in *Kelvin* (K), a unit of measurement used in physics.) If a certain gas occupies a volume of 1.3 L at 300 K and a pressure of 18 newtons, find the volume at 340 K and a pressure of 24 newtons.

67. A body mass index from 27 through 29 carries a slight risk of weight-related health problems, while one of 30 or more indicates a great increase in risk. Use your own height and weight and the information in **Example 7** to determine your BMI and whether you are at risk.

68. The maximum load of a horizontal beam that is supported at both ends varies directly as the width and the square of the height and inversely as the length between the supports. A beam 6 m long, 0.1 m wide, and 0.06 m high supports a load of 360 kg. What is the maximum load supported by a beam 16 m long, 0.2 m wide, and 0.08 m high?

Chapter 6 Summary

Key Terms

6.1
rational expression
rational function
reciprocal

6.2
least common denominator (LCD)

6.3
complex fraction

6.4
domain of the variable in a rational equation
proposed solution
extraneous solution
discontinuous
reciprocal function
vertical asymptote
horizontal asymptote

6.5
ratio
proportion

6.6
direct variation
constant of variation
inverse variation
joint variation
combined variation

Test Your Word Power

See how well you have learned the vocabulary in this chapter.

1. A **rational expression** is
- **A.** an algebraic expression made up of a term or the sum of a finite number of terms with real coefficients and integer exponents
- **B.** a polynomial equation of degree 2
- **C.** a quotient with one or more fractions in the numerator, denominator, or both
- **D.** a quotient of two polynomials with denominator not 0.

2. In a given set of fractions, the **least common denominator** is
- **A.** the smallest denominator of all the denominators
- **B.** the smallest expression that is divisible by all the denominators
- **C.** the largest integer that evenly divides the numerator and denominator of all the fractions
- **D.** the largest denominator of all the denominators.

3. A **complex fraction** is
- **A.** an algebraic expression made up of a term or the sum of a finite number of terms with real coefficients and integer exponents
- **B.** a polynomial equation of degree 2
- **C.** a quotient with one or more fractions in the numerator, denominator, or both
- **D.** a quotient of two polynomials with denominator not 0.

4. A **ratio**
- **A.** compares two quantities using a quotient
- **B.** says that two quotients are equal
- **C.** is a product of two quantities
- **D.** is a difference of two quantities.

5. A **proportion**
- **A.** compares two quantities using a quotient
- **B.** says that two quotients are equal
- **C.** is a product of two quantities
- **D.** is a difference of two quantities.

ANSWERS

1. D; *Examples:* $-\frac{3}{4y^2}, \frac{5x^3}{x+2}, \frac{a+3}{a^2-4a-5}$ **2.** B; *Example:* The LCD of $\frac{1}{x}, \frac{2}{3}$, and $\frac{5}{x+1}$ is $3x(x+1)$. **3.** C; *Examples:* $\frac{\frac{2}{3}}{\frac{4}{7}}, \frac{x-\frac{1}{x}}{x+\frac{1}{y}}, \frac{\frac{2}{a+1}}{a^2-1}$

4. A; *Example:* $\frac{7 \text{ in.}}{12 \text{ in.}}$ compares two quantities. **5.** B; *Example:* The proportion $\frac{2}{3} = \frac{8}{12}$ states that the two ratios are equal.

Quick Review

CONCEPTS | **EXAMPLES**

6.1 Rational Expressions and Functions; Multiplying and Dividing

Rational Function

A rational function is defined by a quotient of polynomials and has the form

$$f(x) = \frac{P(x)}{Q(x)}, \quad \text{where } Q(x) \neq 0.$$

Its domain includes all real numbers except those that make $Q(x)$ equal to 0.

Give the domain of the rational function.

$$f(x) = \frac{2x + 1}{3x + 6}$$

Solve $3x + 6 = 0$ to find $x = -2$. This value must be excluded from the domain. The domain can be written

$$\{x \mid x \text{ is a real number, } x \neq -2\} \quad \text{or} \quad (-\infty, -2) \cup (-2, \infty).$$

Fundamental Property of Rational Numbers

If $\frac{a}{b}$ is a rational number and if c is any nonzero real number, then

$$\frac{a}{b} = \frac{ac}{bc}.$$

$$\frac{3}{4} = \frac{3 \cdot 5}{4 \cdot 5} = \frac{15}{20}$$ $\frac{3}{4}$ and $\frac{15}{20}$ are equivalent.

Writing a Rational Expression in Lowest Terms

Step 1 Factor both numerator and denominator to find their greatest common factor (GCF).

Step 2 Apply the fundamental property. Divide out common factors.

Write in lowest terms.

$$\frac{2x + 8}{x^2 - 16}$$

$$= \frac{2(x + 4)}{(x - 4)(x + 4)} \quad \text{Factor.}$$

$$= \frac{2}{x - 4} \quad \text{Fundamental property}$$

Multiplying Rational Expressions

Step 1 Factor numerators and denominators.

Step 2 Apply the fundamental property.

Step 3 Multiply remaining factors in the numerators and in the denominators. Leave the denominator in factored form.

Step 4 Check that the product is in lowest terms.

Multiply.

$$\frac{x^2 + 2x + 1}{x^2 - 1} \cdot \frac{5}{3x + 3}$$

$$= \frac{(x + 1)^2}{(x - 1)(x + 1)} \cdot \frac{5}{3(x + 1)} \quad \text{Factor.}$$

$$= \frac{5}{3(x - 1)} \quad \text{Fundamental property; Multiply.}$$

Dividing Rational Expressions

Multiply the first rational expression (the dividend) by the reciprocal of the second expression (the divisor).

Divide.

$$\frac{2x + 5}{x - 3} \div \frac{2x^2 + 3x - 5}{x^2 - 9}$$

$$= \frac{2x + 5}{x - 3} \cdot \frac{x^2 - 9}{2x^2 + 3x - 5} \quad \text{Multiply by the reciprocal.}$$

$$= \frac{2x + 5}{x - 3} \cdot \frac{(x + 3)(x - 3)}{(2x + 5)(x - 1)} \quad \text{Factor.}$$

$$= \frac{x + 3}{x - 1} \quad \text{Fundamental property; Multiply.}$$

CONCEPTS	EXAMPLES

6.2 Adding and Subtracting Rational Expressions

Adding or Subtracting Rational Expressions

Step 1 **If the denominators are the same,** add or subtract the numerators. Place the result over the common denominator.

If the denominators are different, write all rational expressions with the LCD. Then add or subtract the numerators, and place the result over the common denominator.

Step 2 Make sure that the answer is in lowest terms.

Subtract.

$$\frac{1}{x+6} - \frac{3}{x+2} \quad \text{The LCD is } (x+6)(x+2).$$

$$= \frac{x+2}{(x+6)(x+2)} - \frac{3(x+6)}{(x+6)(x+2)}$$

$$= \frac{x+2-3(x+6)}{(x+6)(x+2)} \quad \text{Subtract the numerators.}$$

$$= \frac{x+2-3x-18}{(x+6)(x+2)} \quad \text{Distributive property}$$

$$= \frac{-2x-16}{(x+6)(x+2)} \quad \text{Combine like terms.}$$

6.3 Complex Fractions

Simplifying a Complex Fraction

Method 1

Step 1 Simplify the numerator and denominator separately.

Step 2 Divide by multiplying the numerator by the reciprocal of the denominator.

Step 3 Simplify the resulting fraction if possible.

Method 2

Step 1 Multiply the numerator and denominator of the complex fraction by the least common denominator of all the fractions appearing in the complex fraction.

Step 2 Simplify the resulting fraction if possible.

Simplify using both methods.

Method 1

$$\frac{\frac{1}{x^2} - \frac{1}{y^2}}{\frac{1}{x} + \frac{1}{y}}$$

$$= \frac{\frac{y^2}{x^2y^2} - \frac{x^2}{x^2y^2}}{\frac{y}{xy} + \frac{x}{xy}}$$

$$= \frac{\frac{y^2-x^2}{x^2y^2}}{\frac{y+x}{xy}}$$

$$= \frac{y^2-x^2}{x^2y^2} \div \frac{y+x}{xy}$$

$$= \frac{(y+x)(y-x)}{x^2y^2} \cdot \frac{xy}{y+x}$$

$$= \frac{y-x}{xy}$$

Method 2

$$\frac{\frac{1}{x^2} - \frac{1}{y^2}}{\frac{1}{x} + \frac{1}{y}}$$

$$= \frac{x^2y^2\left(\frac{1}{x^2} - \frac{1}{y^2}\right)}{x^2y^2\left(\frac{1}{x} + \frac{1}{y}\right)}$$

$$= \frac{y^2-x^2}{xy^2+x^2y}$$

$$= \frac{(y-x)(y+x)}{xy(y+x)}$$

$$= \frac{y-x}{xy}$$

CONCEPTS	EXAMPLES

6.4 Equations with Rational Expressions and Graphs

Solving an Equation with Rational Expressions

Step 1 Determine the domain of the variable.

Step 2 Multiply each side of the equation by the LCD to clear the fractions.

Step 3 Solve the resulting equation.

Step 4 Check that each proposed solution is in the domain, and discard any values that are not. Check the remaining proposed solution(s) in the original equation.

Solve.

$$\frac{3x+2}{x-2} + \frac{2}{x(x-2)} = \frac{-1}{x}$$ Note that 0 and 2 are excluded from the domain.

$$x(3x+2) + 2 = -(x-2)$$ Multiply by the LCD, $x(x-2)$.

$$3x^2 + 2x + 2 = -x + 2$$ Distributive property

$$3x^2 + 3x = 0$$ Add x. Subtract 2.

$$3x(x+1) = 0$$ Factor.

$$3x = 0 \quad \text{or} \quad x + 1 = 0$$ Zero-factor property

$$x = 0 \quad \text{or} \quad x = -1$$ Solve each equation.

Of the two proposed solutions, 0 must be discarded because it is not in the domain. A check confirms that the solution set is $\{-1\}$.

Graphing a Rational Function

The graph of a rational function (written in lowest terms) may have one or more breaks. At such points, the graph will approach an asymptote.

Graph $f(x) = \frac{1}{x+2}$.

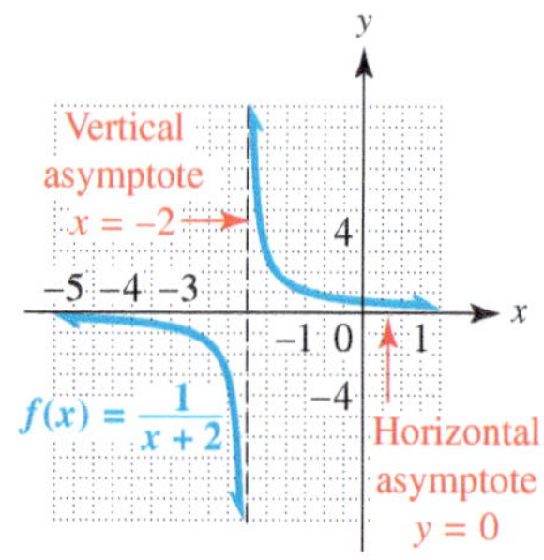

6.5 Applications of Rational Expressions

Solving a Distance, Rate, Time Problem

Use the formula

$$d = rt$$

or one of its equivalents,

$$t = \frac{d}{r} \quad \text{or} \quad r = \frac{d}{t}.$$

Solving a Work Problem

Use the fact that if a complete job is done in t units of time, then the rate of work is $\frac{1}{t}$ job per unit of time.

Solve.

A canal has a current of 2 mph. Find the rate of Amy's boat in still water if it travels 11 mi downstream in the same time that it travels 8 mi upstream.

Let x = the rate of the boat in still water.

	Distance	*Rate*	*Time*
Downstream	11	$x+2$	$\frac{11}{x+2}$
Upstream	8	$x-2$	$\frac{8}{x-2}$

The times are equal.

$$\frac{11}{x+2} = \frac{8}{x-2}$$ Write an equation. Use $t = \frac{d}{r}$.

The LCD is $(x+2)(x-2)$.

$$11(x-2) = 8(x+2)$$ Multiply by the LCD.

$$11x - 22 = 8x + 16$$ Distributive property

$$3x = 38$$ Subtract $8x$. Add 22.

$$x = 12\frac{2}{3}$$ Divide by 3.

The rate in still water is $12\frac{2}{3}$ mph.

CONCEPTS	EXAMPLES
6.6 Variation Let k be a real number.	
If $y = kx^n$, then y varies directly as x^n.	The area of a circle varies directly as the square of the radius. $\mathcal{A} = kr^2$ Here, $k = \pi$.
If $y = \frac{k}{x^n}$, then y varies inversely as x^n.	Pressure varies inversely as volume. $p = \frac{k}{V}$
If $y = kxz$, then y varies jointly as x and z.	For a given principal, interest varies jointly as interest rate and time. $I = krt$ k is the given principal.

Chapter 6 Review Exercises

6.1 *Give the domain of each rational function using* **(a)** *set-builder notation and* **(b)** *interval notation.*

1. $f(x) = \frac{-7}{3x + 18}$ **2.** $f(x) = \frac{5x + 17}{x^2 - 7x + 10}$ **3.** $f(x) = \frac{9}{x^2 - 18x + 81}$

Write each rational expression in lowest terms.

4. $\frac{12x^2 + 6x}{24x + 12}$ **5.** $\frac{25m^2 - n^2}{25m^2 - 10mn + n^2}$ **6.** $\frac{r - 2}{4 - r^2}$

Multiply or divide as indicated.

7. $\frac{(2y + 3)^2}{5y} \cdot \frac{15y^3}{4y^2 - 9}$ **8.** $\frac{w^2 - 16}{w} \cdot \frac{3}{4 - w}$

9. $\frac{z^2 - z - 6}{z - 6} \cdot \frac{z^2 - 6z}{z^2 + 2z - 15}$ **10.** $\frac{m^3 - n^3}{m^2 - n^2} \div \frac{m^2 + mn + n^2}{m + n}$

6.2 *Suppose that the given expressions are denominators of rational expressions. Find the least common denominator for each group.*

11. $32b^3$, $24b^5$ **12.** $9r^2$, $3r + 1$, 9

13. $6x^2 + 13x - 5$, $9x^2 + 9x - 4$ **14.** $3x - 12$, $x^2 - 2x - 8$, $x^2 - 8x + 16$

Add or subtract as indicated.

15. $\frac{5}{3x^6y^5} - \frac{8}{9x^4y^7}$ **16.** $\frac{5y + 13}{y + 1} - \frac{1 - 7y}{y + 1}$

17. $\frac{6}{5a + 10} + \frac{7}{6a + 12}$ **18.** $\frac{3r}{10r^2 - 3rs - s^2} + \frac{2r}{2r^2 + rs - s^2}$

6.3 *Simplify each expression.*

19. $\frac{\frac{3}{t} + 2}{\frac{4}{t} - 7}$ **20.** $\frac{\frac{2}{m - 3n}}{\frac{1}{3n - m}}$ **21.** $\frac{\frac{3}{p} - \frac{2}{q}}{\frac{9q^2 - 4p^2}{qp}}$ **22.** $\frac{x^{-2} - y^{-2}}{x^{-1} - y^{-1}}$

Answers

1. (a) $\{x \mid x \text{ is a real number}, x \neq -6\}$ (b) $(-\infty, -6) \cup (-6, \infty)$
2. (a) $\{x \mid x \text{ is a real number}, x \neq 2, 5\}$ (b) $(-\infty, 2) \cup (2, 5) \cup (5, \infty)$
3. (a) $\{x \mid x \text{ is a real number}, x \neq 9\}$ (b) $(-\infty, 9) \cup (9, \infty)$
4. $\frac{x}{2}$
5. $\frac{5m + n}{5m - n}$ 6. $\frac{-1}{2 + r}$
7. $\frac{3y^2(2y + 3)}{2y - 3}$
8. $\frac{-3(w + 4)}{w}$
9. $\frac{z(z + 2)}{z + 5}$ 10. 1
11. $96b^5$ 12. $9r^2(3r + 1)$
13. $(3x - 1)(2x + 5)(3x + 4)$
14. $3(x - 4)^2(x + 2)$
15. $\frac{15y^2 - 8x^2}{9x^6y^7}$ 16. 12
17. $\frac{71}{30(a + 2)}$
18. $\frac{13r^2 + 5rs}{(5r + s)(2r - s)(r + s)}$
19. $\frac{3 + 2t}{4 - 7t}$ 20. -2
21. $\frac{1}{3q + 2p}$ 22. $\frac{y + x}{xy}$

23. $\{-3\}$ **24.** $\{-2\}$

25. $\{0\}$ **26.** $\varnothing$

27. **(a)** equation; $\{-24\}$

(b) expression; $\frac{24+x}{6x}$

28. Although her algebra was correct, 3 is not a solution because it is not in the domain of the variable in the equation. Thus, $\varnothing$ is correct.

29. C; $x = 0$; $y = 0$

30.

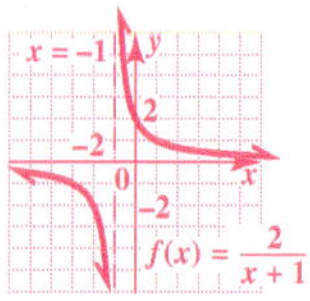

$x = -1$; $y = 0$

31. $\frac{15}{2}$ **32.** 2

33. $m = \frac{Fd^2}{GM}$

34. $M = \frac{m\mu}{\nu - \mu}$

35. 6000 passenger-km per day

6.4 *Solve each equation.*

23. $\frac{1}{t+4} + \frac{1}{2} = \frac{3}{2t+8}$

24. $\frac{-5m}{m+1} + \frac{m}{3m+3} = \frac{56}{6m+6}$

25. $\frac{2}{k-1} - \frac{4k+1}{k^2-1} = \frac{-1}{k+1}$

26. $\frac{5}{x+2} + \frac{3}{x+3} = \frac{x}{x^2+5x+6}$

27. Decide whether each of the following is an *expression* or an *equation*. Simplify the one that is an expression, and solve the one that is an equation.

(a) $\frac{4}{x} + \frac{1}{2} = \frac{1}{3}$ **(b)** $\frac{4}{x} + \frac{1}{2} - \frac{1}{3}$

28. A student obtained $x = 3$ as her final step when solving the equation

$$\frac{3}{x-3} - \frac{2}{x-2} = \frac{3}{x^2-5x+6}.$$

The answer in the back of the book was "$\varnothing$." She was sure that all her algebraic work was correct. ***WHAT WENT WRONG?***

29. ***Concept Check*** Which graph has vertical and horizontal asymptotes? What are their equations?

A.

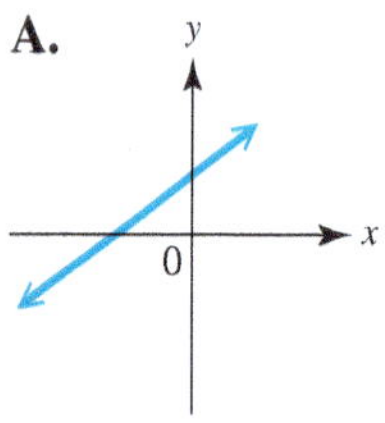

B.

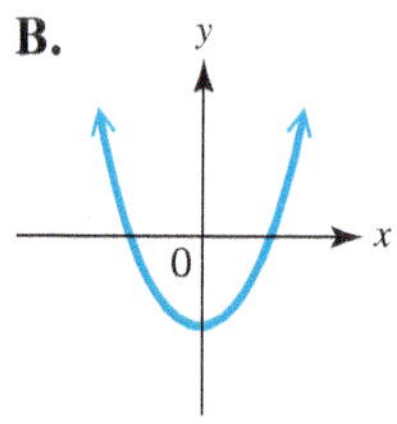

C.

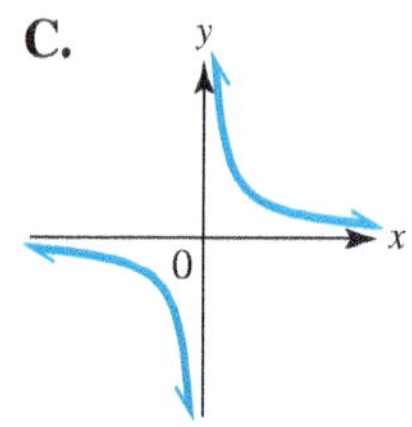

D.

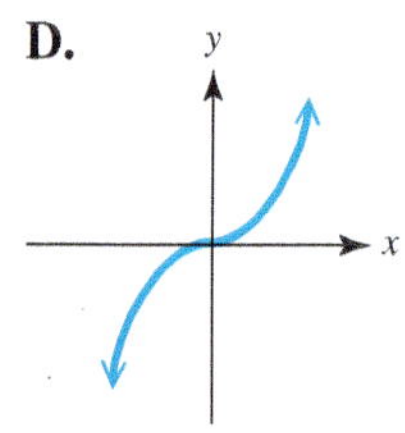

30. Graph the following rational function. Give the equations of its vertical and horizontal asymptotes.

$$f(x) = \frac{2}{x+1}$$

6.5 *Solve each problem.*

31. A law from physics has the form

$$\frac{1}{a} = \frac{1}{b} + \frac{1}{c}.$$

Find a if $b = 30$ and $c = 10$.

32. In banking and finance, the formula

$$P = \frac{M}{1+RT}$$

gives present value at simple interest. If $P = 600$, $M = 750$, and $R = 0.125$, find T.

Solve each formula for the specified variable.

33. $F = \frac{GMm}{d^2}$ for m (physics)

34. $\mu = \frac{M\nu}{M+m}$ for M (electronics)

Solve each problem.

35. An article in *Scientific American* predicted that, in the year 2050, about 23,200 of the 58,000 passenger-km per day in North America will be provided by high-speed trains. If the traffic volume in a typical region of North America is 15,000, how many passenger-kilometers per day will high-speed trains provide there? (*Source:* Schafer, A., and D. Victor, "The Past and Future of Global Mobility," *Scientific American*.)

36. 16 km per hr
37. $4\frac{4}{5}$ min 38. $3\frac{3}{5}$ hr
39. C 40. 430 mm
41. 5.59 vibrations per sec
42. 22.5 ft^3

36. A river has a current of 4 km per hr. Find the rate of Herby's boat in still water if it travels 40 km downstream in the same time that it takes to travel 24 km upstream.

	d	*r*	*t*
Upstream	24	$x-4$	
Downstream	40		

37. A sink can be filled by a cold-water tap in 8 min and by a hot-water tap in 12 min. How long would it take to fill the sink with both taps open?

	Rate	*Time Working Together*	*Fractional Part of the Job Done*
Cold		x	
Hot		x	

38. Jane and Jessica need to sort a pile of bottles at the recycling center. Working alone, Jane could do the entire job in 9 hr, while Jessica could do the entire job in 6 hr. How long will it take them if they work together?

6.6

39. In which one of the following does y vary inversely as x?

A. $y = 2x$ **B.** $y = \frac{x}{3}$ **C.** $y = \frac{3}{x}$ **D.** $y = x^2$

Solve each problem.

40. For a particular camera, the viewing distance varies directly as the amount of enlargement. A picture that is taken with this camera and enlarged 5 times should be viewed from a distance of 250 mm. Suppose a print 8.6 times the size of the negative is made. From what distance should it be viewed?

41. The frequency (number of vibrations per second) of a vibrating guitar string varies inversely as its length. That is, a longer string vibrates fewer times in a second than a shorter string. Suppose a guitar string 0.65 m long vibrates 4.3 times per sec. What frequency would a string 0.5 m long have?

42. The volume of a rectangular box of a given height is proportional to its width and length. A box with width 2 ft and length 4 ft has volume 12 ft^3. Find the volume of a box with the same height that is 3 ft wide and 5 ft long.

Chapter 6 Mixed Review Exercises

1. $\frac{1}{x-2y}$ 2. $\frac{x+5}{x+2}$
3. $\frac{6m+5}{3m^2}$
4. $\frac{11}{3-x}$, or $\frac{-11}{x-3}$
5. $\frac{x^2-6}{2(2x+1)}$ 6. $\frac{3-5x}{6x+1}$
7. $\frac{1}{3}$ 8. $\frac{s^2+t^2}{st(s-t)}$

Write each rational expression in lowest terms.

1. $\dfrac{x+2y}{x^2-4y^2}$

2. $\dfrac{x^2+2x-15}{x^2-x-6}$

Perform the indicated operations.

3. $\dfrac{2}{m} + \dfrac{5}{3m^2}$

4. $\dfrac{9}{3-x} - \dfrac{2}{x-3}$

5. $\dfrac{\frac{-3}{x} + \frac{x}{2}}{1 + \frac{x+1}{x}}$

6. $\dfrac{\frac{3}{x} - 5}{6 + \frac{1}{x}}$

7. $\dfrac{4y+16}{30} \div \dfrac{2y+8}{5}$

8. $\dfrac{t^{-2}+s^{-2}}{t^{-1}-s^{-1}}$

9. $\dfrac{k-3}{36k^2+6k+1}$

10. $\dfrac{x(9x+1)}{3x+1}$

11. $\dfrac{5a^2+4ab+12b^2}{(a+3b)(a-2b)(a+b)}$

12. $\dfrac{acd+b^2d+bc^2}{bcd}$

13. $\left\{\dfrac{1}{3}\right\}$ 14. $\left\{-\dfrac{14}{3}\right\}$

15. $\{1, 4\}$

16. $r=\dfrac{AR}{R-A}$, or $r=\dfrac{-AR}{A-R}$

17. (a) 8.32 mm (b) 44.87 diopters

18. \$53.17

19. 12 ft² 20. $4\frac{1}{2}$ mi

9. $\dfrac{k^2-6k+9}{1-216k^3} \cdot \dfrac{6k^2+17k-3}{9-k^2}$

10. $\dfrac{9x^2+46x+5}{3x^2-2x-1} \div \dfrac{x^2+11x+30}{x^3+5x^2-6x}$

11. $\dfrac{4a}{a^2-ab-2b^2} - \dfrac{6b-a}{a^2+4ab+3b^2}$

12. $\dfrac{a}{b}+\dfrac{b}{c}+\dfrac{c}{d}$

Solve each equation.

13. $\dfrac{x+3}{x^2-5x+4} - \dfrac{1}{x} = \dfrac{2}{x^2-4x}$

14. $\dfrac{3x}{x-4} + \dfrac{2}{x} = \dfrac{48}{x^2-4x}$

15. $1 - \dfrac{5}{r} = \dfrac{-4}{r^2}$

16. $A = \dfrac{Rr}{R+r}$ for r

Solve each problem.

17. The strength of a contact lens is given in units called diopters and also in millimeters of arc. As the diopters increase, the millimeters of arc decrease. The rational function

$$f(x) = \frac{337}{x}$$

relates the arc measurement $f(x)$ to the diopter measurement x. (*Source:* Bausch and Lomb.)

(a) To the nearest hundredth, what arc measurement will correspond to 40.5-diopter lenses?

(b) A lens with an arc measurement of 7.51 mm will provide what diopter strength, to the nearest hundredth?

18. At a certain gasoline station, 3 gal of unleaded gasoline cost \$12.27. How much would 13 gal of the same gasoline cost?

19. The area of a triangle varies jointly as the lengths of the base and height. A triangle with base 10 ft and height 4 ft has area 20 ft². Find the area of a triangle with base 3 ft and height 8 ft.

20. If Dr. Dawson rides his bike to his office, he averages 12 mph. If he drives his car, he averages 36 mph. His time driving is $\frac{1}{4}$ hr less than his time riding his bike. How far is his office from home?

Chapter 6 Test

FOR EXTRA HELP *Step-by-step test solutions are found on the Chapter Test Prep Videos available in* MyMathLab® *or on* YouTube.

View the complete solutions to all Chapter Test exercises in MyMathLab.

[6.1]

1. (a) $\{x \mid x$ is a real number, $x \neq -2, \frac{4}{3}\}$
(b) $(-\infty, -2) \cup \left(-2, \frac{4}{3}\right) \cup \left(\frac{4}{3}, \infty\right)$

2. $\dfrac{2x-5}{x(3x-1)}$

3. $\dfrac{3(x+3)}{4}$ 4. $\dfrac{y+4}{y-5}$

5. -2 6. $\dfrac{x+5}{x}$

[6.2]

7. $t^2(t+3)(t-2)$

1. Give the domain of the following rational function using **(a)** set-builder notation and **(b)** interval notation.

$$f(x) = \frac{x+3}{3x^2+2x-8}$$

2. Write $\dfrac{6x^2-13x-5}{9x^3-x}$ in lowest terms.

Multiply or divide as indicated.

3. $\dfrac{(x+3)^2}{4} \cdot \dfrac{6}{2x+6}$

4. $\dfrac{y^2-16}{y^2-25} \cdot \dfrac{y^2+2y-15}{y^2-7y+12}$

5. $\dfrac{3-t}{5} \div \dfrac{t-3}{10}$

6. $\dfrac{x^2-9}{x^3+3x^2} \div \dfrac{x^2+x-12}{x^3+9x^2+20x}$

7. Find the least common denominator for the group of denominators.

$$t^2+t-6, \quad t^2+3t, \quad t^2$$

8. $\dfrac{7-2t}{6t^2}$

9. $\dfrac{9a+5b}{21a^5b^3}$

10. $\dfrac{11x+21}{(x-3)^2(x+3)}$

11. $\dfrac{4}{x+2}$

[6.3]

12. $\dfrac{72}{11}$

13. $-\dfrac{1}{a+b}$

14. $\dfrac{2y^2+x^2}{xy(y-x)}$

[6.4]

15. (a) expression; $\dfrac{11(x-6)}{12}$

(b) equation; $\{6\}$

16. $\left\{\dfrac{1}{2}\right\}$

17. $\{5\}$

18. $\ell = \dfrac{2S}{n} - a$, or $\ell = \dfrac{2S - na}{n}$

19.

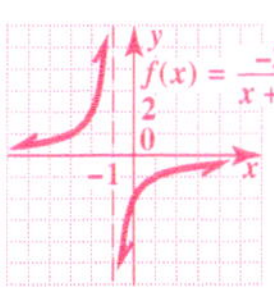

$x=-1$; $y=0$

[6.5]

20. $3\dfrac{3}{14}$ hr

21. 15 mph

22. 48,000 fish

23. (a) 3 units (b) 0

[6.6]

24. 200 amps

25. 0.8 lb

Add or subtract as indicated.

8. $\dfrac{7}{6t^2} - \dfrac{1}{3t}$

9. $\dfrac{3}{7a^4b^3} + \dfrac{5}{21a^5b^2}$

10. $\dfrac{9}{x^2-6x+9} + \dfrac{2}{x^2-9}$

11. $\dfrac{6}{x+4} + \dfrac{1}{x+2} - \dfrac{3x}{x^2+6x+8}$

Simplify each expression.

12. $\dfrac{\frac{12}{r+4}}{\frac{11}{6r+24}}$

13. $\dfrac{\frac{1}{a}-\frac{1}{b}}{\frac{a}{b}-\frac{b}{a}}$

14. $\dfrac{2x^{-2}+y^{-2}}{x^{-1}-y^{-1}}$

15. Decide whether each of the following is an *expression* or an *equation*. Simplify the one that is an expression, and solve the one that is an equation.

(a) $\dfrac{2x}{3} + \dfrac{x}{4} - \dfrac{11}{2}$

(b) $\dfrac{2x}{3} + \dfrac{x}{4} = \dfrac{11}{2}$

Solve each equation.

16. $\dfrac{1}{x} - \dfrac{4}{3x} = \dfrac{1}{x-2}$

17. $\dfrac{y}{y+2} - \dfrac{1}{y-2} = \dfrac{8}{y^2-4}$

18. Solve for the variable ℓ in this formula from mathematics: $S = \dfrac{n}{2}(a+\ell)$.

19. Graph the rational function $f(x) = \frac{-2}{x+1}$. Give the equations of its vertical and horizontal asymptotes.

Solve each problem.

20. Chris can do a job in 9 hr, while Dana can do the same job in 5 hr. How long would it take them to do the job if they worked together?

21. The rate of the current in a stream is 3 mph. Nana's boat can travel 36 mi downstream in the same time that it takes to travel 24 mi upstream. Find the rate of her boat in still water.

22. Biologists collected a sample of 600 fish from West Okoboji Lake on May 1 and tagged each of them. When they returned on June 1, a new sample of 800 fish was collected and 10 of these had been previously tagged. Use this experiment to determine the approximate fish population of West Okoboji Lake.

23. In biology, the function

$$g(x) = \frac{5x}{2+x}$$

gives the growth rate g of a population for x units of available food. (*Source:* Smith, J. Maynard, *Models in Ecology,* Cambridge University Press.)

(a) What amount of food (in appropriate units) would produce a growth rate of 3 units of growth per unit of food?

(b) What is the growth rate if no food is available?

24. The current in a simple electrical circuit is inversely proportional to the resistance. If the current is 80 amps when the resistance is 30 ohms, find the current when the resistance is 12 ohms.

25. The force of the wind blowing on a vertical surface varies jointly as the area of the surface and the square of the velocity. If a wind blowing at 40 mph exerts a force of 50 lb on a surface of 500 ft^2, how much force will a wind of 80 mph place on a surface of 2 ft^2?

Chapters R–6 Cumulative Review Exercises

1. Evaluate $|2x| + 3y - z^3$ for $x = -4$, $y = 3$, and $z = 6$.

Solve each equation or inequality.

2. $7(2x + 3) - 4(2x + 1) = 2(x + 1)$

3. $|6x - 8| - 4 = 0$

4. $\frac{2}{3}y + \frac{5}{12}y \le 20$

5. $|3x + 2| \ge 4$

Solve each problem.

6. Abushieba invested some money at 4% interest and twice as much at 3% interest. His interest for the first year was \$400. How much did he invest at each rate?

7. A triangle has area 42 m^2. The base is 14 m long. Find the height of the triangle.

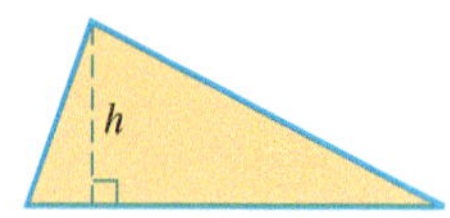

14 m

8. Graph $-4x + 2y = 8$ and give the intercepts.

Find the slope of each line described.

9. Through $(-5, 8)$ and $(-1, 2)$

10. Perpendicular to $4x - 3y = 12$

11. Write an equation of the line in **Exercise 9.** Give the equation in the form $y = mx + b$.

Graph each inequality or compound inequality.

12. $2x + 5y > 10$

13. $x - y \ge 3$ and $3x + 4y \le 12$

14. Decide whether the relation $y = -\sqrt{x + 2}$ is a function, and give its domain and range.

15. Suppose that $y = f(x)$ and $5x - 3y = 8$.

(a) Find an equation that defines $f(x)$. That is, $f(x) =$ ________.

(b) Find $f(1)$.

Solve each system.

16. $4x - y = -7$
$5x + 2y = 1$

17. $x + y - 2z = -1$
$2x - y + z = -6$
$3x + 2y - 3z = -3$

18. $x + 2y + z = 5$
$x - y + z = 3$
$2x + 4y + 2z = 11$

19. Simplify $\left(\frac{m^{-4}n^2}{m^2n^{-3}}\right) \cdot \left(\frac{m^5n^{-1}}{m^{-2}n^5}\right)$. Write the answer with only positive exponents. Assume that all variables represent nonzero real numbers.

Perform the indicated operations.

20. $(3y^2 - 2y + 6) - (-y^2 + 5y + 12)$

21. $(4f + 3)(3f - 1)$

22. $\left(\frac{1}{4}x + 5\right)^2$

23. $(3x^3 + 13x^2 - 17x - 7) \div (3x + 1)$

24. Find each of the following for the polynomial functions

$$f(x) = x^2 + 2x - 3, \quad g(x) = 2x^3 - 3x^2 + 4x - 1, \quad \text{and} \quad h(x) = x^2.$$

(a) $(f + g)(x)$ **(b)** $(g - f)(x)$ **(c)** $(f + g)(-1)$ **(d)** $(f \circ h)(x)$

[R.3] **1.** -199

[1.1] **2.** $\left\{-\frac{15}{4}\right\}$

[1.7] **3.** $\left\{\frac{2}{3}, 2\right\}$

[1.5] **4.** $\left(-\infty, \frac{240}{13}\right]$

[1.7] **5.** $(-\infty, -2] \cup \left[\frac{2}{3}, \infty\right)$

[1.3] **6.** \$4000 at 4%; \$8000 at 3%
7. 6 m

[2.1] **8.** x-intercept: $(-2, 0)$; y-intercept: $(0, 4)$

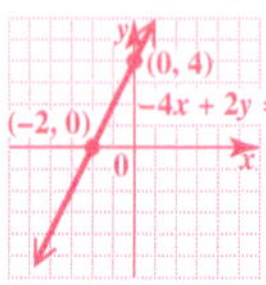

[2.2] **9.** $-\frac{3}{2}$ **10.** $-\frac{3}{4}$

[2.3] **11.** $y = -\frac{3}{2}x + \frac{1}{2}$

[2.4] **12.**

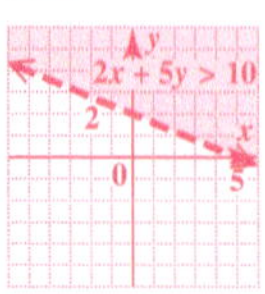

13.

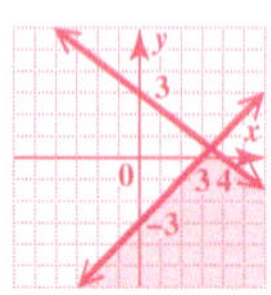

$x - y \ge 3$ and $3x + 4y \le 12$

[2.5] **14.** function; domain: $[-2, \infty)$; range: $(-\infty, 0]$

[2.6] **15. (a)** $\frac{5}{3}x - \frac{8}{3}$ **(b)** -1

[3.1] **16.** $\{(-1, 3)\}$

[3.2] **17.** $\{(-2, 3, 1)\}$ **18.** $\varnothing$

[4.1] **19.** $\frac{m}{n}$

[4.2] **20.** $4y^2 - 7y - 6$

Factor each polynomial completely.

25. $2x^2 - 13x - 45$

26. $100t^4 - 25$

27. $8p^3 + 125$

28. Write $\dfrac{y^2 - 16}{y^2 - 8y + 16}$ in lowest terms.

Perform the indicated operations. Express answers in lowest terms.

29. $\dfrac{2a^2}{a+b} \cdot \dfrac{a-b}{4a}$

30. $\dfrac{x^2 - 9}{2x + 4} \div \dfrac{x^3 - 27}{4}$

31. $\dfrac{x+4}{x-2} + \dfrac{2x-10}{x-2}$

Solve each equation.

32. $3x^2 + 4x = 7$

33. $\dfrac{-3x}{x+1} + \dfrac{4x+1}{x} = \dfrac{-3}{x^2 + x}$

Solve each problem.

34. Machine A can complete a certain job in 2 hr. To speed up the work, Machine B, which can complete the job alone in 3 hr, is brought in to help. How long will it take the two machines to complete the job working together?

35. The cost of a pizza varies directly as the square of its radius. If a pizza with a 7-in. radius costs $6.00, how much should a pizza with a 9-in. radius cost?

[4.4]
21. $12f^2 + 5f - 3$
22. $\frac{1}{16}x^2 + \frac{5}{2}x + 25$

[4.5]
23. $x^2 + 4x - 7$

[4.3]
24. **(a)** $2x^3 - 2x^2 + 6x - 4$
(b) $2x^3 - 4x^2 + 2x + 2$
(c) -14
(d) $x^4 + 2x^2 - 3$

[5.2]
25. $(2x + 5)(x - 9)$

[5.3]
26. $25(2t^2 + 1)(2t^2 - 1)$
27. $(2p + 5)(4p^2 - 10p + 25)$

[6.1]
28. $\frac{y+4}{y-4}$
29. $\frac{a(a-b)}{2(a+b)}$
30. $\frac{2(x+3)}{(x+2)(x^2+3x+9)}$

[6.2]
31. 3

[5.5]
32. $\left\{-\frac{7}{3}, 1\right\}$

[6.4]
33. $\{-4\}$

[6.5]
34. $\frac{6}{5}$ hr

[6.6]
35. $9.92

7 Roots, Radicals, and Root Functions

The formula for calculating the distance one can see to the horizon from the top of a tall building involves a *square root radical,* one of the topics covered in this chapter.

7.1 Radical Expressions and Graphs

OBJECTIVES

1. Find roots of numbers.
2. Find principal roots.
3. Graph functions defined by radical expressions.
4. Find *n*th roots of *n*th powers.
5. Use a calculator to find roots.

VOCABULARY

☐ radicand
☐ index (order)
☐ radical
☐ principal root
☐ radical expression
☐ square root function
☐ cube root function

NOW TRY EXERCISE 1

Find each root.

(a) $\sqrt[3]{1000}$ (b) $\sqrt[4]{625}$

(c) $\sqrt[4]{\frac{1}{256}}$ (d) $\sqrt[3]{0.027}$

CLASSROOM EXAMPLE 1

Find each root.

(a) $\sqrt[3]{27}$ (b) $\sqrt[3]{216}$

(c) $\sqrt[4]{256}$ (d) $\sqrt[5]{243}$

(e) $\sqrt[4]{\frac{16}{81}}$ (f) $\sqrt[3]{0.064}$

Answers: (a) 3 (b) 6 (c) 4 (d) 3 (e) $\frac{2}{3}$ (f) 0.4

NOW TRY ANSWERS

1. (a) 10 (b) 5 (c) $\frac{1}{4}$ (d) 0.3

OBJECTIVE 1 Find roots of numbers.

Recall that $6^2 = 36$. We say that "6 *squared* is 36." The opposite (or inverse) of *squaring* a number is taking its *square root.*

It is customary to write $\sqrt{\ }$ rather than $\sqrt[2]{\ }$.

$$\sqrt{36} = 6, \quad \text{because} \quad 6^2 = 36.$$

We extend this discussion to *cube roots* $\sqrt[3]{\ }$, *fourth roots* $\sqrt[4]{\ }$, and higher roots.

Meaning of $\sqrt[n]{a}$

The *n*th root of *a*, written $\sqrt[n]{a}$**,** is a number whose *n*th power equals *a*. That is,

$$\sqrt[n]{a} = b \quad \textbf{means} \quad b^n = a.$$

The number *a* is the **radicand,** *n* is the **index,** or **order,** and the expression $\sqrt[n]{a}$ is a **radical.**

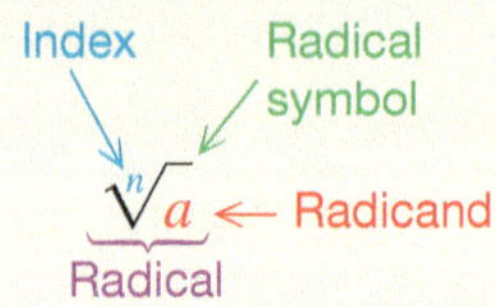

EXAMPLE 1 Simplifying Higher Roots

Find each root.

(a) $\sqrt[3]{64} = 4$, because $4^3 = 64$.

(b) $\sqrt[3]{125} = 5$, because $5^3 = 125$.

(c) $\sqrt[4]{16} = 2$, because $2^4 = 16$.

(d) $\sqrt[5]{32} = 2$, because $2^5 = 32$.

(e) $\sqrt[3]{\frac{8}{27}} = \frac{2}{3}$, because $\left(\frac{2}{3}\right)^3 = \frac{8}{27}$.

(f) $\sqrt[4]{0.0016} = 0.2$, because $(0.2)^4 = 0.0016$. NOW TRY

OBJECTIVE 2 Find principal roots.

If *n* is even, positive numbers have two *n*th roots. For example, both 4 and −4 are square roots of 16, and 2 and −2 are fourth roots of 16. For $a > 0$, $\sqrt[n]{a}$ represents the positive root, called the **principal root,** and $-\sqrt[n]{a}$ represents the negative root. For all *n*, $\sqrt[n]{0} = 0$.

***n*th Root**

Case 1 If *n* is ***even*** and *a* is ***positive or 0,*** then

$\sqrt[n]{a}$ represents the **principal *n*th root** of *a*,

and $-\sqrt[n]{a}$ represents the **negative *n*th root** of *a*.

Case 2 If *n* is ***even*** and *a* is ***negative,*** then

$\sqrt[n]{a}$ is not a real number.

Case 3 If *n* is ***odd,*** then

there is exactly one real *n*th root of *a*, written $\sqrt[n]{a}$.

If n is even, then the two nth roots of a are often written together as $\pm\sqrt[n]{a}$, with $\pm$ read "positive or negative," or "plus or minus."

NOW TRY EXERCISE 2

Find each root.

(a) $-\sqrt{25}$ **(b)** $\sqrt[4]{-625}$

(c) $\sqrt[5]{-32}$ **(d)** $\sqrt[3]{64}$

CLASSROOM EXAMPLE 2

Find each root.

(a) $\sqrt{36}$ (b) $-\sqrt{36}$

(c) $\sqrt[4]{16}$ (d) $-\sqrt[4]{16}$

(e) $\sqrt[4]{-16}$ (f) $\sqrt[5]{243}$

(g) $\sqrt[5]{-243}$

Answers: (a) 6 (b) -6

(c) 2 (d) -2

(e) $\sqrt[4]{-16}$ is not a real number.

(f) 3 (g) -3

EXAMPLE 2 Finding Roots

Find each root.

(a) $\sqrt{100} = 10$ (Case 1)

Because the radicand, 100, is *positive,* there are two square roots: 10 and -10. We want the principal square root, which is 10.

(b) $-\sqrt{100} = -10$ (Case 1)

Here, we want the negative square root, -10.

(c) $\sqrt[4]{81} = 3$ Principal 4th root (Case 1)

(d) $-\sqrt[4]{81} = -3$ Negative 4th root (Case 1)

(e) $\sqrt[4]{-81}$ (Case 2)

The index is *even* and the radicand is *negative,* so $\sqrt[4]{-81}$ is not a real number.

(f) $\sqrt[3]{8} = 2$, because $2^3 = 8$. (Case 3)

(g) $\sqrt[3]{-8} = -2$, because $(-2)^3 = -8$. (Case 3)

In parts (f) and (g), the index is *odd.* Each radical represents exactly one nth root (regardless of whether the radicand is positive, negative, or 0). NOW TRY

TEACHING TIP

At this point, it may be a good idea to review function notation.

For example, if $f(x) = \sqrt{x}$, then

$f(1) = \sqrt{1} = 1,$

$f(16) = \sqrt{16} = 4,$

$f(100) = \sqrt{100} = 10,$

and so on.

OBJECTIVE 3 Graph functions defined by radical expressions.

A **radical expression** is an algebraic expression that contains radicals.

$$3 - \sqrt{x}, \quad \sqrt[3]{x}, \quad \text{and} \quad \sqrt{2x - 1} \qquad \text{Radical expressions}$$

In earlier chapters, we graphed functions defined by polynomial and rational expressions. Now we examine the graphs of functions defined by the basic radical expressions, such as $f(x) = \sqrt{x}$ and $f(x) = \sqrt[3]{x}$.

FIGURE 1 shows the graph of the **square root function,**

$$f(x) = \sqrt{x},$$

together with a table of selected points. Only nonnegative values can be used for x, so the domain is $[0, \infty)$. Because $\sqrt{x}$ is the principal square root of x, it always has a nonnegative value, so the range is also $[0, \infty)$.

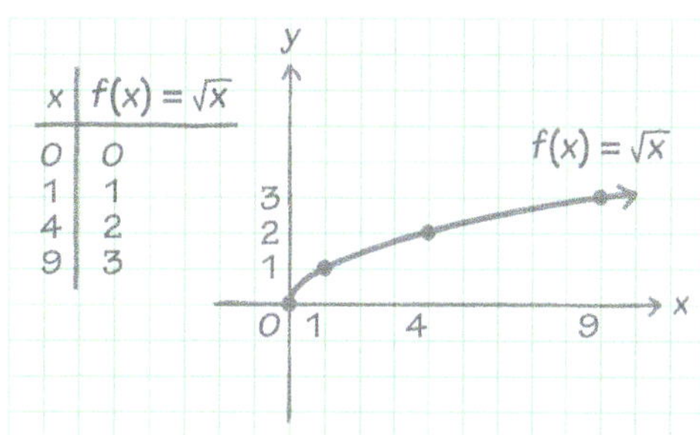

Square root function

$f(x) = \sqrt{x}$

Domain: $[0, \infty)$

Range: $[0, \infty)$

FIGURE 1

NOW TRY ANSWERS

2. **(a)** -5

(b) It is not a real number.

(c) -2 **(d)** -4

CLASSROOM EXAMPLE 3

Graph each function, and give its domain and range.

(a) $f(x) = \sqrt{x+2}$

(b) $f(x) = \sqrt[3]{x} - 1$

Answers:

(a)

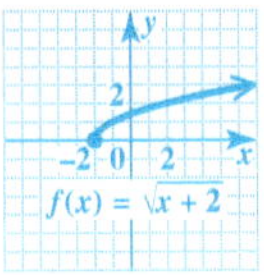

domain: $[-2, \infty)$; range: $[0, \infty)$

(b)

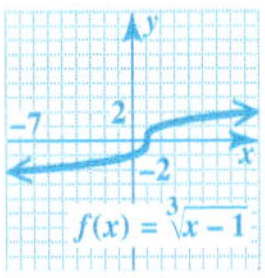

domain: $(-\infty, \infty)$; range: $(-\infty, \infty)$

FIGURE 2 shows the graph of the **cube root function**

$$f(x) = \sqrt[3]{x}.$$

Any real number (positive, negative, or 0) can be used for x in the cube root function, so $\sqrt[3]{x}$ can be positive, negative, or 0. Thus, both the domain and the range of the cube root function are $(-\infty, \infty)$.

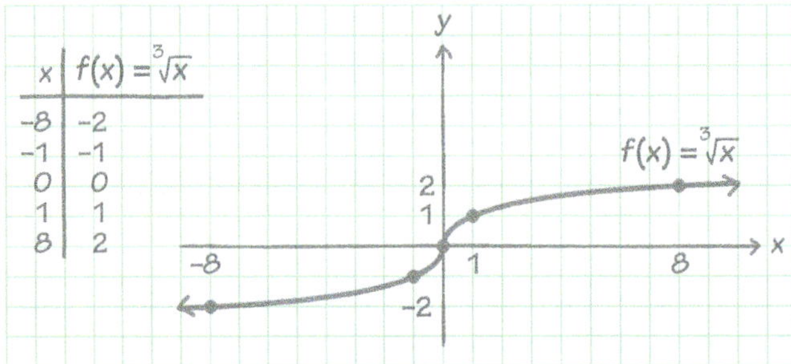

FIGURE 2

Cube root function

$f(x) = \sqrt[3]{x}$

Domain: $(-\infty, \infty)$

Range: $(-\infty, \infty)$

NOW TRY EXERCISE 3

Graph each function, and give its domain and range.

(a) $f(x) = \sqrt{x+1}$

(b) $f(x) = \sqrt[3]{x} - 1$

EXAMPLE 3 Graphing Functions Defined with Radicals

Graph each function, and give its domain and range.

(a) $f(x) = \sqrt{x-3}$

Create a table of values as given with the graph in **FIGURE 3**. The x-values were chosen in such a way that the function values are all integers. For the radicand to be nonnegative, we must have

$$x - 3 \geq 0, \quad \text{or} \quad x \geq 3.$$

Therefore, the domain of this function is $[3, \infty)$. Function values are positive or 0, so the range is $[0, \infty)$.

x	$f(x) = \sqrt{x-3}$
3	$\sqrt{3-3} = 0$
4	$\sqrt{4-3} = 1$
7	$\sqrt{7-3} = 2$

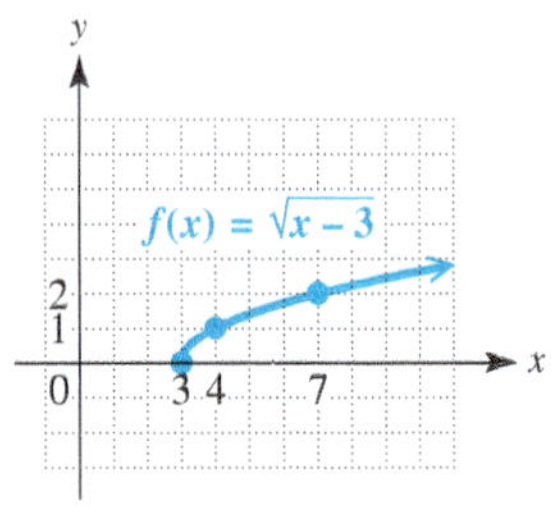

FIGURE 3

This graph is shifted 3 units to the right compared to the graph of $y = \sqrt{x}$.

(b) $f(x) = \sqrt[3]{x} + 2$

See **FIGURE 4**. Both the domain and the range are $(-\infty, \infty)$.

x	$f(x) = \sqrt[3]{x} + 2$
−8	$\sqrt[3]{-8} + 2 = 0$
−1	$\sqrt[3]{-1} + 2 = 1$
0	$\sqrt[3]{0} + 2 = 2$
1	$\sqrt[3]{1} + 2 = 3$
8	$\sqrt[3]{8} + 2 = 4$

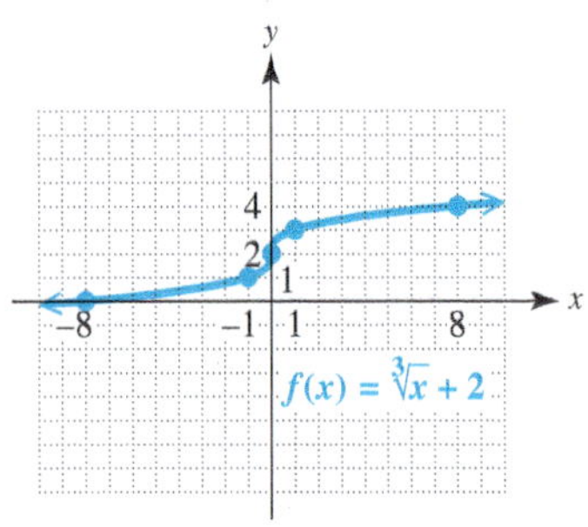

FIGURE 4

This graph is shifted 2 units up compared to the graph of $y = \sqrt[3]{x}$.

NOW TRY

NOW TRY ANSWERS

3. (a)

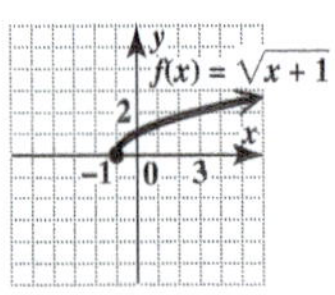

domain: $[-1, \infty)$; range: $[0, \infty)$

(b)

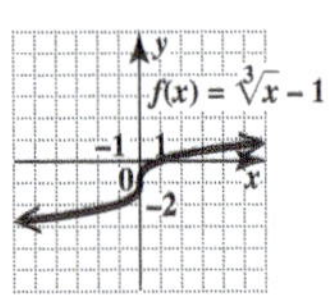

domain: $(-\infty, \infty)$; range: $(-\infty, \infty)$

OBJECTIVE 4 Find *n*th roots of *n*th powers.

Consider the expression $\sqrt{a^2}$. At first glance, we might think that it is equivalent to a. However, this is not necessarily true. For example, consider the following.

If $a = 6$, then $\sqrt{a^2} = \sqrt{6^2} = \sqrt{36} = 6$.

If $a = -6$, then $\sqrt{a^2} = \sqrt{(-6)^2} = \sqrt{36} = 6$. ← Instead of -6, we get 6, the *absolute value* of -6.

The symbol $\sqrt{a^2}$ represents the *nonnegative* square root, so we express $\sqrt{a^2}$ with absolute value bars as $|a|$ because a may be a negative number.

CLASSROOM EXAMPLE 4

Find each square root.

(a) $\sqrt{15^2}$ (b) $\sqrt{(-12)^2}$

(c) $\sqrt{y^2}$ (d) $\sqrt{(-y)^2}$

Answers: (a) 15 (b) 12 (c) $|y|$ (d) $|y|$

Meaning of $\sqrt{a^2}$

For any real number a, $\sqrt{a^2} = |a|$.

That is, the principal square root of a^2 is the absolute value of a.

NOW TRY EXERCISE 4

Find each square root.

(a) $\sqrt{11^2}$ (b) $\sqrt{(-11)^2}$

(c) $\sqrt{z^2}$ (d) $\sqrt{(-z)^2}$

EXAMPLE 4 Simplifying Square Roots Using Absolute Value

Find each square root.

(a) $\sqrt{7^2} = |7| = 7$ (b) $\sqrt{(-7)^2} = |-7| = 7$

(c) $\sqrt{k^2} = |k|$ (d) $\sqrt{(-k)^2} = |-k| = |k|$ NOW TRY

We can generalize this idea to any *n*th root.

CLASSROOM EXAMPLE 5

Simplify each root.

(a) $\sqrt[4]{(-5)^4}$ (b) $\sqrt[5]{(-5)^5}$

(c) $-\sqrt[6]{(-3)^6}$ (d) $-\sqrt[4]{m^8}$

(e) $\sqrt[3]{x^{24}}$ (f) $\sqrt[6]{y^{18}}$

Answers: (a) 5 (b) -5 (c) -3 (d) $-m^2$ (e) x^8 (f) $|y^3|$

Meaning of $\sqrt[n]{a^n}$

If ***n* is an *even* positive integer,** then $\sqrt[n]{a^n} = |a|$.

If ***n* is an *odd* positive integer,** then $\sqrt[n]{a^n} = a$.

That is, use the absolute value symbol when n is even. Absolute value is not used when n is odd.

NOW TRY EXERCISE 5

Simplify each root.

(a) $\sqrt[8]{(-2)^8}$ (b) $\sqrt[3]{(-9)^3}$

(c) $-\sqrt[4]{(-10)^4}$ (d) $-\sqrt{m^8}$

(e) $\sqrt[3]{x^{18}}$ (f) $\sqrt[4]{t^{20}}$

EXAMPLE 5 Simplifying Higher Roots Using Absolute Value

Simplify each root.

(a) $\sqrt[6]{(-3)^6} = |-3| = 3$ *n* is even. Use absolute value.

(b) $\sqrt[5]{(-4)^5} = -4$ *n* is odd.

(c) $-\sqrt[4]{(-9)^4} = -|-9| = -9$ *n* is even. Use absolute value.

(d) $-\sqrt{m^4} = -|m^2| = -m^2$ For all m, $|m^2| = m^2$.

No absolute value bars are needed here, because m^2 is nonnegative for any real number value of m.

(e) $\sqrt[3]{a^{12}} = a^4$, because $a^{12} = (a^4)^3$.

(f) $\sqrt[4]{x^{12}} = |x^3|$

We use absolute value to guarantee that the result is not negative (because x^3 is negative when x is negative). If desired, $|x^3|$ can be written as $x^2 \cdot |x|$. NOW TRY

NOW TRY ANSWERS

4. (a) 11 (b) 11 (c) $|z|$ (d) $|z|$

5. (a) 2 (b) -9 (c) -10 (d) $-m^4$ (e) x^6 (f) $|t^5|$

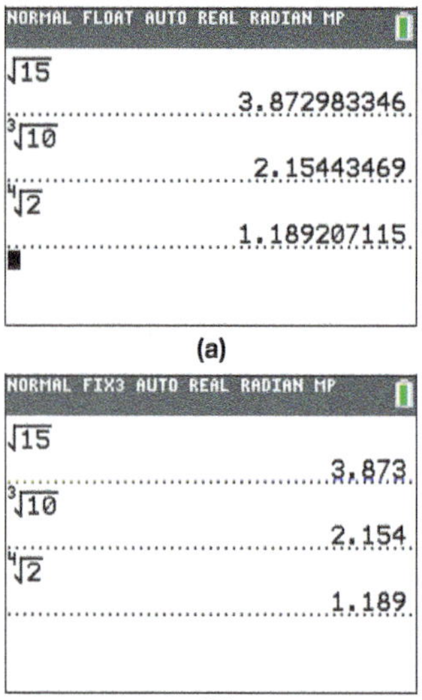

(a)

(b)

FIGURE 5

OBJECTIVE 5 Use a calculator to find roots.

While numbers such as $\sqrt{9}$ and $\sqrt[3]{-8}$ are rational, radicals are often irrational numbers. To find approximations of such radicals, we usually use a scientific or graphing calculator. For example,

$$\sqrt{15} \approx 3.872983346, \quad \sqrt[3]{10} \approx 2.15443469, \quad \text{and} \quad \sqrt[4]{2} \approx 1.189207115,$$

where the symbol $\approx$ means "is approximately equal to." In this book, we often show approximations rounded to three decimal places. Thus,

$$\sqrt{15} \approx 3.873, \quad \sqrt[3]{10} \approx 2.154, \quad \text{and} \quad \sqrt[4]{2} \approx 1.189.$$

FIGURE 5 shows how the preceding approximations are displayed on a TI-83/84 Plus graphing calculator. In **FIGURE 5(a)**, eight or nine decimal places are shown, while in **FIGURE 5(b)**, the number of decimal places is fixed at three.

There is a simple way to check that a calculator approximation is "in the ballpark." For example, because 16 is a little larger than 15, $\sqrt{16} = 4$ should be a little larger than $\sqrt{15}$. Thus, 3.873 is reasonable as an approximation for $\sqrt{15}$.

NOTE The methods for finding approximations differ among makes and models of calculators. ***Always consult your owner's manual for keystroke instructions.*** Be aware that graphing calculators often differ from scientific calculators in the order in which keystrokes are made.

CLASSROOM EXAMPLE 6

Use a calculator to approximate each radical to three decimal places.

(a) $\sqrt{17}$ (b) $-\sqrt{362}$

(c) $\sqrt[3]{9482}$ (d) $\sqrt[4]{6825}$

Answers: (a) 4.123 (b) −19.026 (c) 21.166 (d) 9.089

EXAMPLE 6 Finding Approximations for Roots

Use a calculator to verify that each approximation is correct.

(a) $\sqrt{39} \approx 6.245$ **(b)** $-\sqrt{72} \approx -8.485$

(c) $\sqrt[3]{93} \approx 4.531$ **(d)** $\sqrt[4]{39} \approx 2.499$

NOW TRY

NOW TRY EXERCISE 6

Use a calculator to approximate each radical to three decimal places.

(a) $-\sqrt{92}$ **(b)** $\sqrt[4]{39}$

(c) $\sqrt[5]{33}$

EXAMPLE 7 Using Roots to Calculate Resonant Frequency

In electronics, the resonant frequency f of a circuit may be found using the formula

$$f = \frac{1}{2\pi\sqrt{LC}},$$

where f is in cycles per second, L is in henrys, and C is in farads. (Henrys and farads are units of measure in electronics.) Find the resonant frequency f if $L = 5 \times 10^{-4}$ henry and $C = 3 \times 10^{-10}$ farad. Give the answer to the nearest thousand.

Find the value of f when $L = 5 \times 10^{-4}$ and $C = 3 \times 10^{-10}$.

$$f = \frac{1}{2\pi\sqrt{LC}} \quad \text{Given formula}$$

$$f = \frac{1}{2\pi\sqrt{(5 \times 10^{-4})(3 \times 10^{-10})}} \quad \text{Substitute for } L \text{ and } C.$$

$$f \approx 411{,}000 \quad \text{Use a calculator.}$$

The resonant frequency f is approximately 411,000 cycles per sec. NOW TRY

NOW TRY EXERCISE 7

Use the formula in **Example 7** to approximate f to the nearest thousand if

$L = 7 \times 10^{-5}$

and $C = 3 \times 10^{-9}$.

CLASSROOM EXAMPLE 7

Use the formula in Example 7 to approximate f to the nearest thousand if

$L = 6 \times 10^{-5}$

and $C = 4 \times 10^{-9}$.

Answer: 325,000 cycles per sec

NOW TRY ANSWERS

6. **(a)** −9.592 **(b)** 2.499 **(c)** 2.012

7. 347,000 cycles per sec

NOTE The expression in the second-to-last line of **Example 7** can be difficult to compute on a calculator. There are several ways to approach it, but here is one way that works nicely. Use a calculator to verify the following steps.

Step 1 Calculate the product under the radical in the denominator. The result is

$$1.5 \times 10^{-13}.$$

Step 2 Use the square root function ($\sqrt{x}$) to find the square root of this number. The result is approximately

$$3.872983346 \times 10^{-7}.$$

Step 3 Multiply by 2π, using the π function of the calculator. The result is approximately

$$2.433467206 \times 10^{-6}.$$

Step 4 Now use the reciprocal function (labeled $\frac{1}{x}$, or x^{-1}) of the calculator. The result should be 410936.296, which, rounded to the nearest thousand, is 411,000.

If the numerator had not been 1, and perhaps an expression to evaluate, one approach would be to save the result found in Step 3 in the calculator memory, evaluate the numerator, and then divide by the number saved in the memory.

7.1 Exercises

FOR EXTRA HELP MyMathLab®

Complete solution available in MyMathLab

1. E 2. F
3. D 4. B
5. C 6. E
7. C 8. B
9. C 10. D
11. (a) It is not a real number. (b) negative (c) 0

Concept Check *Match each expression from Column I with the equivalent choice from Column II. Answers may be used once, more than once, or not at all.*

I		II	
1. $-\sqrt{16}$	**2.** $\sqrt{-16}$	**A.** 3	**B.** -2
3. $\sqrt[3]{-27}$	**4.** $\sqrt[5]{-32}$	**C.** 2	**D.** -3
5. $\sqrt[4]{16}$	**6.** $-\sqrt[3]{64}$	**E.** -4	**F.** Not a real number

Concept Check *Choose the closest approximation of each square root. Do not use a calculator.*

7. $\sqrt{123.5}$

A. 9 **B.** 10 **C.** 11 **D.** 12

8. $\sqrt{67.8}$

A. 7 **B.** 8 **C.** 9 **D.** 10

Concept Check *Refer to the figure to answer each question.*

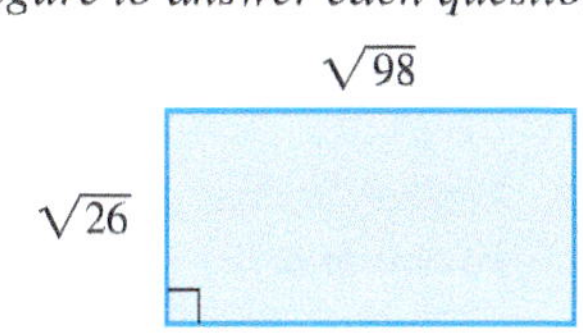

9. Which one of the following is the best estimate of its area?

A. 2500 **B.** 250 **C.** 50 **D.** 100

10. Which one of the following is the best estimate of its perimeter?

A. 15 **B.** 250 **C.** 100 **D.** 30

11. *Concept Check* Consider the expression $-\sqrt{-a}$. Decide whether it is positive, negative, 0, or not a real number in each case.

(a) $a > 0$ **(b)** $a < 0$ **(c)** $a = 0$

12. ***Concept Check*** If n is odd, under what conditions is $\sqrt[n]{a}$ the following?

(a) positive **(b)** negative **(c)** 0

Find each root. ***See Examples 1 and 2.***

13. $-\sqrt{81}$ **14.** $-\sqrt{121}$ **15.** $\sqrt[3]{216}$ **16.** $\sqrt[3]{343}$

17. $\sqrt[3]{-64}$ **18.** $\sqrt[3]{-125}$ **19.** $-\sqrt[3]{512}$ **20.** $-\sqrt[3]{1000}$

21. $\sqrt[4]{1296}$ **22.** $\sqrt[4]{625}$ **23.** $-\sqrt[4]{16}$ **24.** $-\sqrt[4]{256}$

25. $\sqrt[4]{-625}$ **26.** $\sqrt[4]{-256}$ **27.** $\sqrt[6]{64}$ **28.** $\sqrt[6]{729}$

29. $\sqrt[6]{-32}$ **30.** $\sqrt[8]{-1}$ **31.** $\sqrt{\frac{64}{81}}$ **32.** $\sqrt{\frac{100}{9}}$

33. $\sqrt[3]{\frac{64}{27}}$ **34.** $\sqrt[4]{\frac{81}{16}}$ **35.** $-\sqrt[6]{\frac{1}{64}}$ **36.** $-\sqrt[5]{\frac{1}{32}}$

37. $-\sqrt[3]{-27}$ **38.** $-\sqrt[3]{-64}$ **39.** $\sqrt{0.25}$ **40.** $\sqrt{0.36}$

41. $-\sqrt{0.49}$ **42.** $-\sqrt{0.81}$ **43.** $\sqrt[3]{0.001}$ **44.** $\sqrt[3]{0.125}$

Graph each function, and give its domain and range. ***See Example 3.***

45. $f(x) = \sqrt{x+3}$ **46.** $f(x) = \sqrt{x-5}$ **47.** $f(x) = \sqrt{x} - 2$

48. $f(x) = \sqrt{x} + 4$ **49.** $f(x) = \sqrt[3]{x} - 3$ **50.** $f(x) = \sqrt[3]{x} + 1$

51. $f(x) = \sqrt[3]{x-3}$ **52.** $f(x) = \sqrt[3]{x+1}$

Simplify each root. ***See Examples 4 and 5.***

53. $\sqrt{12^2}$ **54.** $\sqrt{19^2}$ **55.** $\sqrt{(-10)^2}$ **56.** $\sqrt{(-13)^2}$

57. $\sqrt[6]{(-2)^6}$ **58.** $\sqrt[6]{(-4)^6}$ **59.** $\sqrt[5]{(-9)^5}$ **60.** $\sqrt[5]{(-8)^5}$

61. $-\sqrt[6]{(-5)^6}$ **62.** $-\sqrt[6]{(-7)^6}$ **63.** $\sqrt{x^2}$ **64.** $-\sqrt{x^2}$

65. $\sqrt{(-z)^2}$ **66.** $\sqrt{(-q)^2}$ **67.** $\sqrt[3]{x^3}$ **68.** $-\sqrt[3]{x^3}$

69. $\sqrt[3]{x^{15}}$ **70.** $\sqrt[3]{m^9}$ **71.** $\sqrt[6]{x^{30}}$ **72.** $\sqrt[4]{k^{20}}$

Find a decimal approximation for each radical. Round answers to three decimal places. ***See Example 6.***

73. $\sqrt{9483}$ **74.** $\sqrt{6825}$ **75.** $\sqrt{284.361}$ **76.** $\sqrt{846.104}$

77. $-\sqrt{82}$ **78.** $-\sqrt{91}$ **79.** $\sqrt[3]{423}$ **80.** $\sqrt[3]{555}$

81. $\sqrt[4]{100}$ **82.** $\sqrt[4]{250}$ **83.** $\sqrt[5]{23.8}$ **84.** $\sqrt[5]{98.4}$

Solve each problem. ***See Example 7.***

85. Use the formula in **Example 7** to calculate the resonant frequency of a circuit to the nearest thousand if $L = 7.237 \times 10^{-5}$ henry and $C = 2.5 \times 10^{-10}$ farad.

12. (a) a must be positive $(a > 0)$.
(b) a must be negative $(a < 0)$.
(c) a must be 0 $(a = 0)$.

13. -9 **14.** -11
15. 6 **16.** 7
17. -4 **18.** -5
19. -8 **20.** -10
21. 6 **22.** 5
23. -2 **24.** -4
25. It is not a real number.
26. It is not a real number.
27. 2 **28.** 3
29. It is not a real number.
30. It is not a real number.
31. $\frac{8}{9}$ **32.** $\frac{10}{3}$
33. $\frac{4}{3}$ **34.** $\frac{3}{2}$
35. $-\frac{1}{2}$ **36.** $-\frac{1}{2}$
37. 3 **38.** 4
39. 0.5 **40.** 0.6
41. -0.7 **42.** -0.9
43. 0.1 **44.** 0.5

In Exercises 45–52, we give the domain and then the range.

45.
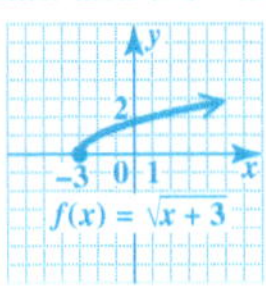

$[-3, \infty)$; $[0, \infty)$

46.
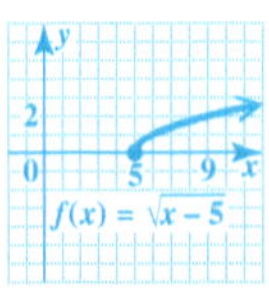

$[5, \infty)$; $[0, \infty)$

47.
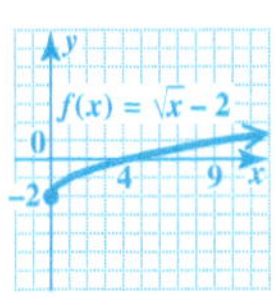

$[0, \infty)$; $[-2, \infty)$

48.
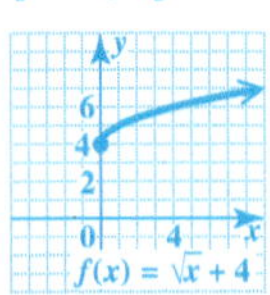

$[0, \infty)$; $[4, \infty)$

49.
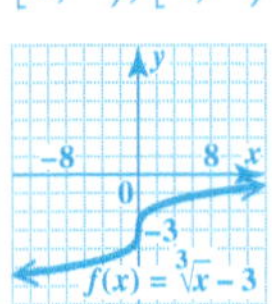

$(-\infty, \infty)$; $(-\infty, \infty)$

50.

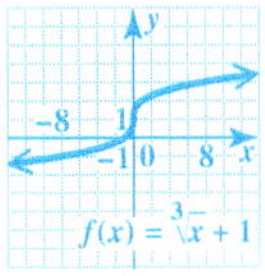

$(-\infty, \infty)$; $(-\infty, \infty)$

51.

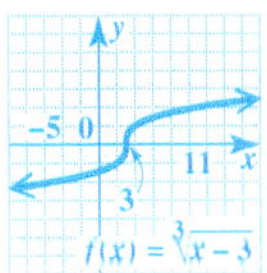

$(-\infty, \infty)$; $(-\infty, \infty)$

52.

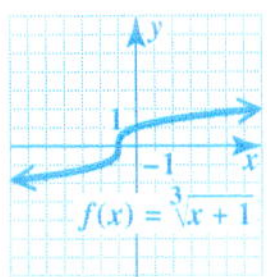

$(-\infty, \infty)$; $(-\infty, \infty)$

53. 12 **54.** 19
55. 10 **56.** 13
57. 2 **58.** 4
59. −9 **60.** −8
61. −5 **62.** −7
63. $|x|$ **64.** $-|x|$
65. $|z|$ **66.** $|q|$
67. x **68.** $-x$
69. x^5 **70.** m^3
71. $|x|^5$ (or $|x^5|$)
72. $|k|^5$ (or $|k^5|$)

73. 97.381 **74.** 82.614
75. 16.863 **76.** 29.088
77. −9.055 **78.** −9.539
79. 7.507 **80.** 8.218
81. 3.162 **82.** 3.976
83. 1.885 **84.** 2.504

85. 1,183,000 cycles per sec
86. 74 in.; 6.2 ft
87. 10 mi
88. 1.6 sec
89. 392,000 mi^2
90. 330 m^2
91. **(a)** 1.732 amps
(b) 2.236 amps
92. 24,900 ft^2

86. The threshold weight T for a person is the weight above which the risk of death increases greatly. The threshold weight in pounds for men aged 40–49 is related to height h in inches by the formula

$$h = 12.3\sqrt[3]{T}.$$

What height corresponds to a threshold weight of 216 lb for a 43-year-old man? Round the answer to the nearest inch and then to the nearest tenth of a foot.

87. According to an article in *The World Scanner Report*, the distance D, in miles, to the horizon from an observer's point of view over water or "flat" earth is given by

$$D = \sqrt{2H},$$

where H is the height of the point of view, in feet. If a person whose eyes are 6 ft above ground level is standing at the top of a hill 44 ft above "flat" earth, approximately how far to the horizon will she be able to see?

88. The time t in seconds for one complete swing of a simple pendulum, where L is the length of the pendulum in feet, and g, the acceleration due to gravity, is about 32 ft per sec^2, is

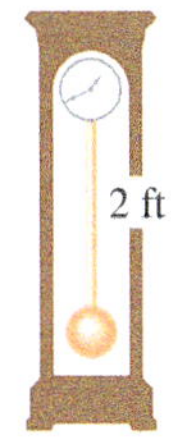

$$t = 2\pi\sqrt{\frac{L}{g}}.$$

Find the time of a complete swing of a 2-ft pendulum to the nearest tenth of a second.

89. **Heron's formula** gives a method of finding the area of a triangle if the lengths of its sides are known. Suppose that a, b, and c are the lengths of the sides. Let s denote one-half of the perimeter of the triangle (called the **semiperimeter**)—that is, $s = \frac{1}{2}(a + b + c)$. Then the area of the triangle is

$$\mathcal{A} = \sqrt{s(s-a)(s-b)(s-c)}.$$

Find the area of the Bermuda Triangle, to the nearest thousand square miles, if the "sides" of this triangle measure approximately 850 mi, 925 mi, and 1300 mi.

90. Use Heron's formula from **Exercise 89** to find the area of a triangle with sides of lengths $a = 11$ m, $b = 60$ m, and $c = 61$ m.

91. The coefficient of self-induction L (in henrys), the energy P stored in an electronic circuit (in joules), and the current I (in amps) are related by this formula.

$$I = \sqrt{\frac{2P}{L}}$$

(a) Find I if $P = 120$ and $L = 80$. **(b)** Find I if $P = 100$ and $L = 40$.

92. The Vietnam Veterans Memorial in Washington, DC, is in the shape of an unenclosed isosceles triangle with equal sides of length 246.75 ft. If the triangle were enclosed, the third side would have length 438.14 ft. Use Heron's formula from **Exercise 89** to find the area of this enclosure to the nearest hundred square feet. (*Source:* Information pamphlet obtained at the Vietnam Veterans Memorial.)

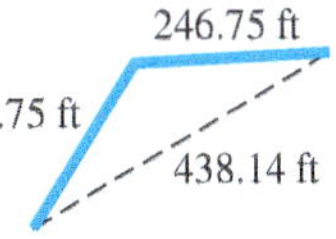

Not to scale

7.2 Rational Exponents

OBJECTIVES

1. Use exponential notation for *n*th roots.
2. Define and use expressions of the form $a^{m/n}$.
3. Convert between radicals and rational exponents.
4. Use the rules for exponents with rational exponents.

CLASSROOM EXAMPLE 1

Evaluate each exponential.

(a) $32^{1/5}$ (b) $64^{1/2}$

(c) $-81^{1/4}$ (d) $(-81)^{1/4}$

(e) $(-64)^{1/3}$ (f) $\left(\frac{1}{27}\right)^{1/3}$

Answers: (a) 2 (b) 8 (c) -3

(d) It is not a real number.

(e) -4 (f) $\frac{1}{3}$

NOW TRY EXERCISE 1

Evaluate each exponential.

(a) $81^{1/2}$ (b) $125^{1/3}$

(c) $-625^{1/4}$ (d) $(-625)^{1/4}$

(e) $(-125)^{1/3}$ (f) $\left(\frac{1}{16}\right)^{1/4}$

TEACHING TIP

Another approach to working problems like those in **Example 1** is to rewrite the base as an exponential so that the product of the exponents is 1. For example,

$$64^{1/3} = (4^3)^{1/3} = 4^1 = 4.$$

Also, try writing $a^{m/n}$ as

$$a^{\text{power } m/\text{index } n}$$

to reinforce the meaning of a rational exponent.

NOW TRY ANSWERS

1. (a) 9 (b) 5 (c) -5
   (d) It is not a real number.
   (e) -5 (f) $\frac{1}{2}$

OBJECTIVE 1 Use exponential notation for *n*th roots.

Consider the product $(3^{1/2})^2 = 3^{1/2} \cdot 3^{1/2}$. We can simplify this product as follows.

$$\begin{aligned}(3^{1/2})^2 &= 3^{1/2} \cdot 3^{1/2} \\ &= 3^{1/2+1/2} && \text{Product rule: } a^m \cdot a^n = a^{m+n} \\ &= 3^1 && \text{Add exponents.} \\ &= 3 && a^1 = a\end{aligned}$$

Also, by definition,

$$(\sqrt{3})^2 = \sqrt{3} \cdot \sqrt{3} = 3.$$

Because both $(3^{1/2})^2$ and $(\sqrt{3})^2$ are equal to 3, it seems reasonable to define

$$3^{1/2} = \sqrt{3}.$$

This suggests the following generalization.

Meaning of $a^{1/n}$

If $\sqrt[n]{a}$ is a real number, then $\quad \boldsymbol{a^{1/n} = \sqrt[n]{a}}.$

Examples: $4^{1/2} = \sqrt{4}$, $8^{1/3} = \sqrt[3]{8}$, and $16^{1/4} = \sqrt[4]{16}$

Notice that the denominator of the rational exponent is the index of the radical.

EXAMPLE 1 Evaluating Exponentials of the Form $a^{1/n}$

Evaluate each exponential.

(a) $64^{1/3} = \sqrt[3]{64} = 4$ (The denominator is the index, or root.)

(b) $100^{1/2} = \sqrt{100} = 10$ (The denominator is the index, or root. $\sqrt{\ }$ means $\sqrt[2]{\ }$.)

(c) $-256^{1/4} = -\sqrt[4]{256} = -4$

(d) $(-256)^{1/4} = \sqrt[4]{-256}$ is not a real number because the radicand, -256, is negative and the index is even.

(e) $(-32)^{1/5} = \sqrt[5]{-32} = -2$

(f) $\left(\frac{1}{8}\right)^{1/3} = \sqrt[3]{\frac{1}{8}} = \frac{1}{2}$

NOW TRY

CAUTION Notice the distinction between **Examples 1(c) and (d).** The radical in part (c) is the ***negative fourth root of a positive number,*** while the radical in part (d) is the ***principal fourth root of a negative number, which is not a real number.***

OBJECTIVE 2 Define and use expressions of the form $a^{m/n}$.

We know that $8^{1/3} = \sqrt[3]{8}$. We can define a number like $8^{2/3}$, where the numerator of the exponent is not 1. For past rules of exponents to be valid,

$$8^{2/3} = 8^{(1/3)2} = (8^{1/3})^2.$$

Because $8^{1/3} = \sqrt[3]{8}$,

$$8^{2/3} = \left(\sqrt[3]{8}\right)^2 = 2^2 = 4.$$

Generalizing from this example, we define $a^{m/n}$ as follows.

Meaning of $a^{m/n}$

If m and n are positive integers with m/n in lowest terms, then

$$a^{m/n} = \left(a^{1/n}\right)^m,$$

provided that $a^{1/n}$ is a real number. If $a^{1/n}$ is not a real number, then $a^{m/n}$ is not a real number.

NOW TRY EXERCISE 2

Evaluate each exponential.

(a) $32^{2/5}$ **(b)** $8^{5/3}$

(c) $-100^{3/2}$ **(d)** $(-121)^{3/2}$

(e) $(-125)^{4/3}$

CLASSROOM EXAMPLE 2

Evaluate each exponential.

(a) $25^{3/2}$ (b) $27^{2/3}$

(c) $-16^{3/2}$ (d) $(-64)^{2/3}$

(e) $(-36)^{3/2}$

Answers: (a) 125 (b) 9

(c) -64 (d) 16

(e) It is not a real number.

EXAMPLE 2 Evaluating Exponentials of the Form $a^{m/n}$

Evaluate each exponential.

Think: $36^{1/2} = \sqrt{36} = 6$

(a) $36^{3/2} = (36^{1/2})^3 = 6^3 = 216$

Think: $125^{1/3} = \sqrt[3]{125} = 5$

(b) $125^{2/3} = (125^{1/3})^2 = 5^2 = 25$

Be careful. The base is 4.

(c) $-4^{5/2} = -(4^{5/2}) = -(4^{1/2})^5 = -(2)^5 = -32$

Because the base here is 4, the negative sign is *not* affected by the exponent.

(d) $(-27)^{2/3} = [(-27)^{1/3}]^2 = (-3)^2 = 9$

Notice in part (c) that we first evaluate the exponential and then find its negative. In part (d), the $-$ sign is part of the base, -27.

(e) $(-100)^{3/2} = [(-100)^{1/2}]^3$, which is not a real number, because

$$(-100)^{1/2}, \quad \text{or} \quad \sqrt{-100}, \quad \text{is not a real number.}$$

NOW TRY

Recall from **Section 4.1** that for any natural number n,

$$a^{-n} = \frac{1}{a^n} \quad (\text{where } a \neq 0).$$

When a rational exponent is negative, this earlier interpretation of negative exponents is applied.

Meaning of $a^{-m/n}$

If $a^{m/n}$ is a real number, then

$$a^{-m/n} = \frac{1}{a^{m/n}} \quad (\text{where } a \neq 0).$$

NOW TRY ANSWERS

2. (a) 4 (b) 32 (c) -1000
(d) It is not a real number.
(e) 625

NOW TRY EXERCISE 3

Evaluate each exponential.

(a) $243^{-3/5}$ **(b)** $4^{-5/2}$

(c) $\left(\frac{216}{125}\right)^{-2/3}$

CLASSROOM EXAMPLE 3

Evaluate each exponential.

(a) $81^{-3/4}$ (b) $36^{-3/2}$

(c) $\left(\frac{64}{25}\right)^{-3/2}$

Answers: (a) $\frac{1}{27}$ (b) $\frac{1}{216}$

(c) $\frac{125}{512}$

EXAMPLE 3 Evaluating Exponentials of the Form $a^{-m/n}$

Evaluate each exponential.

(a) $16^{-3/4} = \frac{1}{16^{3/4}} = \frac{1}{(16^{1/4})^3} = \frac{1}{(\sqrt[4]{16})^3} = \frac{1}{2^3} = \frac{1}{8}$

The denominator of 3/4 is the index and the numerator is the exponent.

(b) $25^{-3/2} = \frac{1}{25^{3/2}} = \frac{1}{(25^{1/2})^3} = \frac{1}{(\sqrt{25})^3} = \frac{1}{5^3} = \frac{1}{125}$

(c) $\left(\frac{8}{27}\right)^{-2/3} = \frac{1}{\left(\frac{8}{27}\right)^{2/3}} = \frac{1}{\left(\sqrt[3]{\frac{8}{27}}\right)^2} = \frac{1}{\left(\frac{2}{3}\right)^2} = \frac{1}{\frac{4}{9}} = \frac{9}{4}$

$\frac{1}{\frac{4}{9}} = 1 \div \frac{4}{9} = 1 \cdot \frac{9}{4}$

We can also use the rule $\left(\frac{b}{a}\right)^{-m} = \left(\frac{a}{b}\right)^{m}$ here, as follows.

$$\left(\frac{8}{27}\right)^{-2/3} = \left(\frac{27}{8}\right)^{2/3} = \left(\sqrt[3]{\frac{27}{8}}\right)^2 = \left(\frac{3}{2}\right)^2 = \frac{9}{4}$$ The result is the same.

Take the reciprocal only of the base, *not* the exponent.

NOW TRY

TEACHING TIP

Emphasize that a negative exponent does not necessarily lead to a negative result. For example,

$8^{-1/3} = \frac{1}{2}$, not -2.

CAUTION Be careful to distinguish between exponential expressions like the following.

$16^{-1/4}$, which equals $\frac{1}{2}$, $-16^{1/4}$, which equals -2, and $-16^{-1/4}$, which equals $-\frac{1}{2}$

A negative exponent does not necessarily lead to a negative result. Negative exponents lead to reciprocals, which may be positive.

We obtain an alternative definition of $a^{m/n}$ using the power rule for exponents differently than in the earlier definition. If all indicated roots are real numbers, then

$$a^{m/n} = a^{m(1/n)} = (a^m)^{1/n}, \quad \text{so} \quad a^{m/n} = (a^m)^{1/n}.$$

Alternative Meaning of $a^{m/n}$

If all indicated roots are real numbers, then

$$a^{m/n} = (a^{1/n})^m = (a^m)^{1/n}.$$

We can now evaluate an expression such as $27^{2/3}$ in two ways.

$$27^{2/3} = (27^{1/3})^2 = 3^2 = 9$$

or $$27^{2/3} = (27^2)^{1/3} = 729^{1/3} = 9$$ The result is the same.

In most cases, it is easier to use $(a^{1/n})^m$.

NOW TRY ANSWERS

3. (a) $\frac{1}{27}$ (b) $\frac{1}{32}$ (c) $\frac{25}{36}$

CLASSROOM EXAMPLE 4

Write each exponential as a radical. Assume that all variables represent positive real numbers.

(a) $19^{1/2}$ (b) $11^{3/4}$

(c) $14x^{2/3}$ (d) $5x^{3/5} - (2x)^{3/5}$

(e) $x^{-5/7}$ (f) $(x^3 + y^3)^{1/3}$

In parts (g)–(i), write each radical as an exponential. Simplify. Assume that all variables represent positive real numbers.

(g) $\sqrt{37}$ (h) $\sqrt[4]{9^8}$

(i) $\sqrt[8]{z^8}$

Answers: (a) $\sqrt{19}$

(b) $(\sqrt[4]{11})^3$ (c) $14(\sqrt[3]{x})^2$

(d) $5(\sqrt[5]{x})^3 - (\sqrt[5]{2x})^3$

(e) $\frac{1}{(\sqrt[7]{x})^5}$ (f) $\sqrt[3]{x^3 + y^3}$

(g) $37^{1/2}$ (h) 9^2, or 81

(i) z

NOW TRY EXERCISE 4

Write each exponential as a radical. Assume that all variables represent positive real numbers.

(a) $21^{1/2}$ **(b)** $17^{5/4}$

(c) $4t^{3/5} + (4t)^{2/3}$

(d) $w^{-2/5}$ **(e)** $(a^2 - b^2)^{1/4}$

In parts (f)–(h), write each radical as an exponential. Simplify. Assume that all variables represent positive real numbers.

(f) $\sqrt[3]{15}$ **(g)** $\sqrt[4]{4^2}$

(h) $\sqrt[4]{x^4}$

NOW TRY ANSWERS

4. (a) $\sqrt{21}$ **(b)** $(\sqrt[4]{17})^5$

(c) $4(\sqrt[5]{t})^3 + (\sqrt[3]{4t})^2$

(d) $\frac{1}{(\sqrt[5]{w})^2}$ **(e)** $\sqrt[4]{a^2 - b^2}$

(f) $15^{1/3}$ **(g)** 2 **(h)** x

Radical Form of $a^{m/n}$

If all indicated roots are real numbers, then

$$a^{m/n} = \sqrt[n]{a^m} = (\sqrt[n]{a})^m.$$

That is, raise a to the mth power and then take the nth root, or take the nth root of a and then raise to the mth power.

For example, $8^{2/3} = \sqrt[3]{8^2} = \sqrt[3]{64} = 4$, and $8^{2/3} = (\sqrt[3]{8})^2 = 2^2 = 4$,

so $$8^{2/3} = \sqrt[3]{8^2} = (\sqrt[3]{8})^2.$$

OBJECTIVE 3 Convert between radicals and rational exponents.

Using the definition of rational exponents, we can simplify many problems involving radicals by converting the radicals to numbers with rational exponents. After simplifying, we can convert the answer back to radical form if required.

NOTE The ability to convert between radicals and rational exponents is important in the study of exponential and logarithmic functions later in this book.

EXAMPLE 4 Converting between Rational Exponents and Radicals

Write each exponential as a radical. Assume that all variables represent positive real numbers. Use the definition that takes the root first.

(a) $13^{1/2} = \sqrt{13}$ **(b)** $6^{3/4} = (\sqrt[4]{6})^3$ **(c)** $9m^{5/8} = 9(\sqrt[8]{m})^5$

(d) $6x^{2/3} - (4x)^{3/5} = 6(\sqrt[3]{x})^2 - (\sqrt[5]{4x})^3$

(e) $r^{-2/3} = \frac{1}{r^{2/3}} = \frac{1}{(\sqrt[3]{r})^2}$

(f) $(a^2 + b^2)^{1/2} = \sqrt{a^2 + b^2}$ ($\sqrt{a^2 + b^2} \neq a + b$)

In parts (g)–(i), write each radical as an exponential. Simplify. Assume that all variables represent positive real numbers.

(g) $\sqrt{10} = 10^{1/2}$ **(h)** $\sqrt[4]{3^8} = 3^{8/4} = 3^2 = 9$

(i) $\sqrt[6]{z^6} = z$ because z is positive.

NOW TRY

NOTE In **Example 4(i),** it is not necessary to use absolute value bars because the directions specifically state that the variable represents a positive real number. The absolute value of the positive real number z is z itself, so the result is simply z.

OBJECTIVE 4 Use the rules for exponents with rational exponents.

The definition of rational exponents allows us to apply the rules for exponents from **Section 4.1.**

Rules for Rational Exponents

Let r and s be rational numbers. For all real numbers a and b for which the indicated expressions exist, the following hold true.

$$a^r \cdot a^s = a^{r+s} \qquad a^{-r} = \frac{1}{a^r} \qquad \frac{a^r}{a^s} = a^{r-s} \qquad \left(\frac{a}{b}\right)^{-r} = \frac{b^r}{a^r}$$

$$(a^r)^s = a^{rs} \qquad (ab)^r = a^r b^r \qquad \left(\frac{a}{b}\right)^r = \frac{a^r}{b^r} \qquad a^{-r} = \left(\frac{1}{a}\right)^r$$

CLASSROOM EXAMPLE 5

Simplify each expression. Write answers with only positive exponents. Assume that all variables represent positive real numbers.

(a) $3^{1/2} \cdot 3^{1/3}$ (b) $\frac{7^{2/3}}{7^{4/3}}$

(c) $\left(\frac{a^{1/3}b^{2/3}}{b}\right)^6$ (d) $\left(\frac{a^3b^{-4}}{a^{-2}b^{1/5}}\right)^{-1/2}$

(e) $r^{2/5}(r^{3/5} + r^{8/5})$

Answers: (a) $3^{5/6}$ (b) $\frac{1}{7^{2/3}}$

(c) $\frac{a^2}{b^2}$ (d) $\frac{b^{21/10}}{a^{5/2}}$ (e) $r + r^2$

TEACHING TIP

Stress that if a problem is given in exponential form, like those in Example 5, students should apply the rules for exponents rather than complicate things by trying to convert the expressions to radical form.

CLASSROOM EXAMPLE 6

Write each radical as an exponential, and then simplify. Leave answers in exponential form. Assume that all variables represent positive real numbers.

(a) $\sqrt[4]{x^3} \cdot \sqrt[5]{x}$ (b) $\frac{\sqrt{x^5}}{\sqrt[3]{x}}$

(c) $\sqrt[3]{\sqrt[6]{x}}$

Answers: (a) $x^{19/20}$ (b) $x^{13/6}$

(c) $x^{1/18}$

EXAMPLE 5 Applying Rules for Rational Exponents

Simplify each expression. Write answers with only positive exponents. Assume that all variables represent positive real numbers.

(a) $2^{1/2} \cdot 2^{1/4}$

$= 2^{1/2+1/4}$ Product rule

$= 2^{3/4}$ Add exponents.

(b) $\frac{5^{2/3}}{5^{7/3}}$

$= 5^{2/3-7/3}$ Quotient rule

$= 5^{-5/3}$ Subtract exponents.

$= \frac{1}{5^{5/3}}$ Definition of negative exponent

(c) $\frac{(x^{1/2}y^{2/3})^4}{y}$

$= \frac{(x^{1/2})^4(y^{2/3})^4}{y}$ Power rule

$= \frac{x^2y^{8/3}}{y^1}$ Power rule; $y = y^1$

$= x^2y^{8/3-1}$ Quotient rule

$= x^2y^{5/3}$ $\frac{8}{3} - 1 = \frac{8}{3} - \frac{3}{3} = \frac{5}{3}$

(d) $\left(\frac{x^4y^{-6}}{x^{-2}y^{1/3}}\right)^{-2/3}$

$= \frac{(x^4)^{-2/3}(y^{-6})^{-2/3}}{(x^{-2})^{-2/3}(y^{1/3})^{-2/3}}$ Power rule

$= \frac{x^{-8/3}y^4}{x^{4/3}y^{-2/9}}$ Power rule

$= x^{-8/3-4/3}y^{4-(-2/9)}$ Quotient rule

$= x^{-4}y^{38/9}$ Use parentheses to avoid errors. $4 - \left(-\frac{2}{9}\right) = \frac{36}{9} + \frac{2}{9} = \frac{38}{9}$

$= \frac{y^{38/9}}{x^4}$ Definition of negative exponent

The same result is obtained if we simplify within the parentheses first, as follows.

NOW TRY EXERCISE 5

Simplify each expression. Write answers with only positive exponents. Assume that all variables represent positive real numbers.

(a) $5^{1/4} \cdot 5^{2/3}$ **(b)** $\frac{9^{3/5}}{9^{7/5}}$

(c) $\frac{(r^{2/3}t^{1/4})^8}{t}$

(d) $\left(\frac{2x^{1/2}y^{-2/3}}{x^{-3/5}y^{-1/5}}\right)^{-3}$

(e) $y^{2/3}(y^{1/3} + y^{5/3})$

$$\left(\frac{x^4y^{-6}}{x^{-2}y^{1/3}}\right)^{-2/3}$$

$= (x^{4-(-2)}y^{-6-1/3})^{-2/3}$ Quotient rule

$= (x^6y^{-19/3})^{-2/3}$ $-6 - \frac{1}{3} = -\frac{18}{3} - \frac{1}{3} = -\frac{19}{3}$

$= (x^6)^{-2/3}(y^{-19/3})^{-2/3}$ Power rule

$= x^{-4}y^{38/9}$ Power rule

$= \frac{y^{38/9}}{x^4}$ Definition of negative exponent; The result is the same.

(e) $m^{3/4}(m^{5/4} - m^{1/4})$

$= m^{3/4}(m^{5/4}) - m^{3/4}(m^{1/4})$ Distributive property

Do not make the common mistake of multiplying exponents in the first step.

$= m^{3/4+5/4} - m^{3/4+1/4}$ Product rule

$= m^{8/4} - m^{4/4}$ Add exponents.

$= m^2 - m$ Write the exponents in lowest terms.

NOW TRY

NOW TRY EXERCISE 6

Write each radical as an exponential, and then simplify. Leave answers in exponential form. Assume that all variables represent positive real numbers.

(a) $\sqrt[5]{y^3} \cdot \sqrt[3]{y}$ **(b)** $\frac{\sqrt[4]{y^3}}{\sqrt{y^5}}$

(c) $\sqrt{\sqrt[3]{y}}$

EXAMPLE 6 Applying Rules for Rational Exponents

Write each radical as an exponential, and then simplify. Leave answers in exponential form. Assume that all variables represent positive real numbers.

(a) $\sqrt[3]{x^2} \cdot \sqrt[4]{x}$

$= x^{2/3} \cdot x^{1/4}$ Convert to rational exponents.

$= x^{2/3+1/4}$ Product rule

$= x^{8/12+3/12}$ Write exponents with a common denominator.

$= x^{11/12}$ Add exponents.

(b) $\frac{\sqrt{x^3}}{\sqrt[3]{x^2}}$

$= \frac{x^{3/2}}{x^{2/3}}$ Convert to rational exponents.

$= x^{3/2-2/3}$ Quotient rule

$= x^{5/6}$ $\frac{3}{2} - \frac{2}{3} = \frac{9}{6} - \frac{4}{6} = \frac{5}{6}$

(c) $\sqrt{\sqrt[4]{z}}$

$= \sqrt{z^{1/4}}$ Convert the inside radical to a rational exponent.

$= (z^{1/4})^{1/2}$ Convert the square root to a rational exponent.

$= z^{1/8}$ Power rule

NOW TRY

NOW TRY ANSWERS

5. (a) $5^{11/12}$ (b) $\frac{1}{9^{4/5}}$ (c) $r^{16/3}t$ (d) $\frac{y^{7/5}}{8x^{33/10}}$ (e) $y + y^{7/3}$

6. (a) $y^{14/15}$ (b) $\frac{1}{y^{7/4}}$ (c) $y^{1/6}$

7.2 Exercises

FOR EXTRA HELP

MyMathLab®

Complete solution available in MyMathLab

1. C **2.** E
3. A **4.** J
5. H **6.** G
7. B **8.** F
9. D **10.** I

11. 13 **12.** 11
13. 9 **14.** 8
15. 2 **16.** 5
17. $\frac{8}{9}$ **18.** $\frac{2}{3}$
19. -3 **20.** -2
21. It is not a real number.
22. It is not a real number.
23. 1000 **24.** 512
25. 27 **26.** 36
27. -1024 **28.** -8
29. 16 **30.** 9
31. $\frac{1}{8}$ **32.** $\frac{1}{81}$
33. $\frac{1}{512}$ **34.** $\frac{1}{729}$
35. $\frac{9}{25}$ **36.** $\frac{25}{16}$
37. $\frac{27}{8}$ **38.** $\frac{32}{243}$
39. $\sqrt{10}$ **40.** $\sqrt{3}$
41. $(\sqrt[4]{8})^3$ **42.** $(\sqrt[3]{7})^2$
43. $(\sqrt[8]{9q})^5 - (\sqrt[3]{2x})^2$
44. $(\sqrt[4]{3p})^3 + \sqrt[3]{4x}$
45. $\frac{1}{(\sqrt{2m})^3}$ **46.** $\frac{1}{(\sqrt[5]{5y})^3}$
47. $(\sqrt[3]{2y+x})^2$
48. $(\sqrt{r+2z})^3$
49. $\frac{1}{(\sqrt[3]{3m^4+2k^2})^2}$
50. $\frac{1}{(\sqrt[6]{5x^2+3z^3})^5}$
51. 64 **52.** 3125
53. 64 **54.** 36
55. x^{10} **56.** r^{25}
57. $\sqrt[6]{x^5}$ **58.** $\sqrt[20]{y^{13}}$
59. $\sqrt[15]{t^8}$ **60.** $\sqrt[12]{w^7}$
61. 9 **62.** 36
63. 4 **64.** 25
65. y **66.** r

Concept Check *Match each expression from Column I with the equivalent choice from Column II.*

I		II	
1. $3^{1/2}$	**2.** $(-27)^{1/3}$	**A.** -4	**B.** 8
3. $-16^{1/2}$	**4.** $(-25)^{1/2}$	**C.** $\sqrt{3}$	**D.** $-\sqrt{6}$
5. $(-32)^{1/5}$	**6.** $(-32)^{2/5}$	**E.** -3	**F.** $\sqrt{6}$
7. $4^{3/2}$	**8.** $6^{2/4}$	**G.** 4	**H.** -2
9. $-6^{2/4}$	**10.** $36^{0.5}$	**I.** 6	**J.** Not a real number

Evaluate each exponential. ***See Examples 1–3.***

11. $169^{1/2}$ **12.** $121^{1/2}$ **13.** $729^{1/3}$ **14.** $512^{1/3}$

15. $16^{1/4}$ **16.** $625^{1/4}$ **17.** $\left(\frac{64}{81}\right)^{1/2}$ **18.** $\left(\frac{8}{27}\right)^{1/3}$

19. $(-27)^{1/3}$ **20.** $(-32)^{1/5}$ **21.** $(-144)^{1/2}$ **22.** $(-36)^{1/2}$

23. $100^{3/2}$ **24.** $64^{3/2}$ **25.** $81^{3/4}$ **26.** $216^{2/3}$

27. $-16^{5/2}$ **28.** $-32^{3/5}$ **29.** $(-8)^{4/3}$ **30.** $(-243)^{2/5}$

31. $32^{-3/5}$ **32.** $27^{-4/3}$ **33.** $64^{-3/2}$ **34.** $81^{-3/2}$

35. $\left(\frac{125}{27}\right)^{-2/3}$ **36.** $\left(\frac{64}{125}\right)^{-2/3}$ **37.** $\left(\frac{16}{81}\right)^{-3/4}$ **38.** $\left(\frac{729}{64}\right)^{-5/6}$

Write each exponential as a radical. Assume that all variables represent positive real numbers. ***See Example 4.***

39. $10^{1/2}$ **40.** $3^{1/2}$ **41.** $8^{3/4}$

42. $7^{2/3}$ **43.** $(9q)^{5/8} - (2x)^{2/3}$ **44.** $(3p)^{3/4} + (4x)^{1/3}$

45. $(2m)^{-3/2}$ **46.** $(5y)^{-3/5}$ **47.** $(2y + x)^{2/3}$

48. $(r + 2z)^{3/2}$ **49.** $(3m^4 + 2k^2)^{-2/3}$ **50.** $(5x^2 + 3z^3)^{-5/6}$

Write each radical as an exponential. Simplify. Assume that all variables represent positive real numbers. ***See Example 4.***

51. $\sqrt{2^{12}}$ **52.** $\sqrt{5^{10}}$ **53.** $\sqrt[3]{4^9}$ **54.** $\sqrt[4]{6^8}$ **55.** $\sqrt{x^{20}}$

56. $\sqrt{r^{50}}$ **57.** $\sqrt[3]{x} \cdot \sqrt{x}$ **58.** $\sqrt[4]{y} \cdot \sqrt[5]{y^2}$ **59.** $\frac{\sqrt[3]{t^4}}{\sqrt[5]{t^4}}$ **60.** $\frac{\sqrt[4]{w^3}}{\sqrt[6]{w}}$

Simplify each expression. Write answers with only positive exponents. Assume that all variables represent positive real numbers. ***See Example 5.***

61. $3^{1/2} \cdot 3^{3/2}$ **62.** $6^{4/3} \cdot 6^{2/3}$ **63.** $\frac{64^{5/3}}{64^{4/3}}$

64. $\frac{125^{7/3}}{125^{5/3}}$ **65.** $y^{7/3} \cdot y^{-4/3}$ **66.** $r^{-8/9} \cdot r^{17/9}$

67. $x^{2/3} \cdot x^{-1/4}$ **68.** $x^{2/5} \cdot x^{-1/3}$ **69.** $\dfrac{k^{1/3}}{k^{2/3} \cdot k^{-1}}$

70. $\dfrac{z^{3/4}}{z^{5/4} \cdot z^{-2}}$ **71.** $\dfrac{(x^{1/4}y^{2/5})^{20}}{x^2}$ **72.** $\dfrac{(r^{1/5}s^{2/3})^{15}}{r^2}$

73. $\dfrac{(x^{2/3})^2}{(x^2)^{7/3}}$ **74.** $\dfrac{(p^3)^{1/4}}{(p^{5/4})^2}$ **75.** $\dfrac{m^{3/4}n^{-1/4}}{(m^2n)^{1/2}}$

76. $\dfrac{(a^2b^5)^{-1/4}}{(a^{-3}b^2)^{1/6}}$ **77.** $\dfrac{p^{1/5}p^{7/10}p^{1/2}}{(p^3)^{-1/5}}$ **78.** $\dfrac{z^{1/3}z^{-2/3}z^{1/6}}{(z^{-1/6})^3}$

79. $\left(\dfrac{b^{-3/2}}{c^{-5/3}}\right)^2(b^{-1/4}c^{-1/3})^{-1}$ **80.** $\left(\dfrac{m^{-2/3}}{a^{-3/4}}\right)^4(m^{-3/8}a^{1/4})^{-2}$ **81.** $\left(\dfrac{p^{-1/4}q^{-3/2}}{3^{-1}p^{-2}q^{-2/3}}\right)^{-2}$

82. $\left(\dfrac{2^{-2}w^{-3/4}x^{-5/8}}{w^{3/4}x^{-1/2}}\right)^{-3}$ **83.** $p^{2/3}(p^{1/3} + 2p^{4/3})$ **84.** $z^{5/8}(3z^{5/8} + 5z^{11/8})$

85. $k^{1/4}(k^{3/2} - k^{1/2})$ **86.** $r^{3/5}(r^{1/2} + r^{3/4})$ **87.** $6a^{7/4}(a^{-7/4} + 3a^{-3/4})$

88. $4m^{5/3}(m^{-2/3} - 4m^{-5/3})$ **89.** $-5x^{7/6}(x^{5/6} - x^{-1/6})$ **90.** $-8y^{11/7}(y^{3/7} - y^{-4/7})$

Write each radical as an exponential, and then simplify. Leave answers in exponential form. Assume that all variables represent positive numbers. ***See Example 6.***

91. $\sqrt[5]{x^3} \cdot \sqrt[4]{x}$ **92.** $\sqrt[6]{y^5} \cdot \sqrt[3]{y^2}$ **93.** $\dfrac{\sqrt{x^5}}{\sqrt{x^8}}$ **94.** $\dfrac{\sqrt[3]{k^5}}{\sqrt[3]{k^7}}$

95. $\sqrt{y} \cdot \sqrt[3]{yz}$ **96.** $\sqrt[3]{xz} \cdot \sqrt{z}$ **97.** $\sqrt[4]{\sqrt[3]{m}}$ **98.** $\sqrt[3]{\sqrt{k}}$

99. $\sqrt{\sqrt{\sqrt{x}}}$ **100.** $\sqrt{\sqrt{\sqrt{\sqrt{x}}}}$ **101.** $\sqrt{\sqrt[3]{\sqrt[4]{x}}}$ **102.** $\sqrt[3]{\sqrt[5]{\sqrt{y}}}$

Concept Check *Work each problem.*

103. Replace a with 3 and b with 4 to show that, in general,

$$\sqrt{a^2 + b^2} \neq a + b.$$

104. Suppose someone claims that $\sqrt[n]{a^n + b^n}$ must equal $a + b$, because when $a = 1$ and $b = 0$, a true statement results:

$$\sqrt[n]{a^n + b^n} = \sqrt[n]{1^n + 0^n} = \sqrt[n]{1^n} = 1 = 1 + 0 = a + b.$$

Explain why this is faulty reasoning.

Solve each problem.

105. Meteorologists can determine the duration of a storm using the function

$$T(d) = 0.07d^{3/2},$$

where d is the diameter of the storm in miles and T is the time in hours. Find the duration of a storm with a diameter of 16 mi. Round the answer to the nearest tenth of an hour.

106. The threshold weight t, in pounds, for a person is the weight above which the risk of death increases greatly. The threshold weight in pounds for men aged 40–49 is related to height H in inches by the function

$$H(t) = (1860.867t)^{1/3}.$$

What height corresponds to a threshold weight of 200 lb for a 46-yr-old man? Round the answer to the nearest inch and then to the nearest tenth of a foot.

67. $x^{5/12}$ **68.** $x^{1/15}$
69. $k^{2/3}$ **70.** $z^{3/2}$
71. x^3y^8 **72.** rs^{10}
73. $\dfrac{1}{x^{10/3}}$ **74.** $\dfrac{1}{p^{7/4}}$
75. $\dfrac{1}{m^{1/4}n^{3/4}}$ **76.** $\dfrac{1}{b^{19/12}}$
77. p^2 **78.** $z^{1/3}$
79. $\dfrac{c^{11/3}}{b^{11/4}}$ **80.** $\dfrac{a^{5/2}}{m^{23/12}}$
81. $\dfrac{q^{5/3}}{9p^{7/2}}$ **82.** $64w^{9/2}x^{3/8}$
83. $p + 2p^2$ **84.** $3z^{5/4} + 5z^2$
85. $k^{7/4} - k^{3/4}$ **86.** $r^{11/10} + r^{27/20}$
87. $6 + 18a$ **88.** $4m - 16$
89. $-5x^2 + 5x$ **90.** $-8y^2 + 8y$
91. $x^{17/20}$ **92.** $y^{3/2}$
93. $\dfrac{1}{x^{3/2}}$ **94.** $\dfrac{1}{k^{2/3}}$
95. $y^{5/6}z^{1/3}$ **96.** $x^{1/3}z^{5/6}$
97. $m^{1/12}$ **98.** $k^{1/6}$
99. $x^{1/8}$ **100.** $x^{1/16}$
101. $x^{1/24}$ **102.** $y^{1/30}$
103. $\sqrt{a^2 + b^2} = \sqrt{3^2 + 4^2} = 5$; $a + b = 3 + 4 = 7$; $5 \neq 7$
104. The statement is true for this particular choice of values for a and b. However, it is not true *in general*. For example, let $a = 3$, $b = 4$, and $n = 2$.
105. 4.5 hr
106. 72 in.; 6.0 ft

*The **windchill factor** is a measure of the cooling effect that the wind has on a person's skin. It calculates the equivalent cooling temperature if there were no wind. The National Weather Service uses the formula*

$$\text{Windchill temperature} = 35.74 + 0.6215T - 35.75V^{4/25} + 0.4275TV^{4/25},$$

where T is the temperature in °F and V is the wind speed in miles per hour, to calculate windchill. The chart gives the windchill factor for various wind speeds and temperatures at which frostbite is a risk, and how quickly it may occur.

Temperature (°F)

Wind speed (mph) / Calm	40	30	20	10	0	−10	−20	−30	−40
5	36	25	13	1	−11	−22	−34	−46	−57
10	34	21	9	−4	−16	−28	−41	−53	−66
15	32	19	6	−7	−19	−32	−45	−58	−71
20	30	17	4	−9	−22	−35	−48	−61	−74
25	29	16	3	−11	−24	−37	−51	−64	−78
30	28	15	1	−12	−26	−39	−53	−67	−80
35	28	14	0	−14	−27	−41	−55	−69	−82
40	27	13	−1	−15	−29	−43	−57	−71	−84

Frostbites times: 30 minutes 10 minutes 5 minutes

Source: National Oceanic and Atmospheric Administration, National Weather Service.

Use the formula and a calculator to determine the windchill to the nearest tenth of a degree, given the following conditions. Compare answers with the appropriate entries in the table.

107. 30°F, 15-mph wind

108. 10°F, 30-mph wind

109. 20°F, 20-mph wind

110. 40°F, 10-mph wind

107. 19.0°; The table gives 19°.
108. −12.3°; The table gives −12°.
109. 4.2°; The table gives 4°.
110. 33.6°; The table gives 34°.

7.3 Simplifying Radicals, the Distance Formula, and Circles

OBJECTIVES

1. Use the product rule for radicals.
2. Use the quotient rule for radicals.
3. Simplify radicals.
4. Simplify products and quotients of radicals with different indexes.
5. Use the Pythagorean theorem.
6. Use the distance formula.
7. Find an equation of a circle given its center and radius.

VOCABULARY

- ☐ hypotenuse
- ☐ legs (of a right triangle)
- ☐ circle
- ☐ center
- ☐ radius

OBJECTIVE 1 Use the product rule for radicals.

Consider the expressions $\sqrt{36 \cdot 4}$ and $\sqrt{36} \cdot \sqrt{4}$. Are they equal?

$$\sqrt{36 \cdot 4} = \sqrt{144} = 12$$

$$\sqrt{36} \cdot \sqrt{4} = 6 \cdot 2 = 12$$

The result is the same.

This is an example of the **product rule for radicals.**

Product Rule for Radicals

If $\sqrt[n]{a}$ and $\sqrt[n]{b}$ are real numbers and n is a natural number, then the following holds.

$$\sqrt[n]{a} \cdot \sqrt[n]{b} = \sqrt[n]{ab}$$

That is, the product of two nth roots is the nth root of the product.

We justify the product rule using the rules for rational exponents. Because $\sqrt[n]{a} = a^{1/n}$ and $\sqrt[n]{b} = b^{1/n}$,

$$\sqrt[n]{a} \cdot \sqrt[n]{b} = a^{1/n} \cdot b^{1/n} = (ab)^{1/n} = \sqrt[n]{ab}.$$

CAUTION *Use the product rule only when the radicals have the same index.*

NOW TRY EXERCISE 1

Multiply. Assume that all variables represent positive real numbers.

(a) $\sqrt{7} \cdot \sqrt{11}$

(b) $\sqrt{2mn} \cdot \sqrt{15}$

EXAMPLE 1 Using the Product Rule

Multiply. Assume that all variables represent positive real numbers.

(a) $\sqrt{5} \cdot \sqrt{7} = \sqrt{5 \cdot 7} = \sqrt{35}$

(b) $\sqrt{11} \cdot \sqrt{p} = \sqrt{11p}$

(c) $\sqrt{7} \cdot \sqrt{11xyz} = \sqrt{77xyz}$

NOW TRY

NOW TRY EXERCISE 2

Multiply. Assume that all variables represent positive real numbers.

(a) $\sqrt[3]{4} \cdot \sqrt[3]{5}$

(b) $\sqrt[4]{5t} \cdot \sqrt[4]{6r^3}$

(c) $\sqrt[7]{20x} \cdot \sqrt[7]{3xy^3}$

(d) $\sqrt[3]{5} \cdot \sqrt[4]{9}$

EXAMPLE 2 Using the Product Rule

Multiply. Assume that all variables represent positive real numbers.

(a) $\sqrt[3]{3} \cdot \sqrt[3]{12} = \sqrt[3]{3 \cdot 12} = \sqrt[3]{36}$ Remember to write the index.

(b) $\sqrt[4]{8y} \cdot \sqrt[4]{3r^2} = \sqrt[4]{24yr^2}$

(c) $\sqrt[6]{10m^4} \cdot \sqrt[6]{5m} = \sqrt[6]{50m^5}$

(d) $\sqrt[4]{2} \cdot \sqrt[5]{2}$ This product cannot be simplified using the product rule for radicals because the indexes (4 and 5) are different.

NOW TRY

CLASSROOM EXAMPLE 1

Multiply. Assume that all variables represent positive real numbers.

(a) $\sqrt{5} \cdot \sqrt{13}$ (b) $\sqrt{7} \cdot \sqrt{xy}$

Answers: (a) $\sqrt{65}$ (b) $\sqrt{7xy}$

CLASSROOM EXAMPLE 2

Multiply. Assume that all variables represent positive real numbers.

(a) $\sqrt[3]{2} \cdot \sqrt[3]{7}$

(b) $\sqrt[6]{8r^2} \cdot \sqrt[6]{2r^3}$

(c) $\sqrt[5]{9y^2x} \cdot \sqrt[5]{8xy^2}$

(d) $\sqrt{7} \cdot \sqrt[3]{5}$

Answers: (a) $\sqrt[3]{14}$

(b) $\sqrt[6]{16r^5}$ (c) $\sqrt[5]{72y^4x^2}$

(d) This expression cannot be simplified using the product rule.

CLASSROOM EXAMPLE 3 is on the next page.

OBJECTIVE 2 Use the quotient rule for radicals.

The **quotient rule for radicals** is similar to the product rule.

Quotient Rule for Radicals

If $\sqrt[n]{a}$ and $\sqrt[n]{b}$ are real numbers, and n is a natural number, then the following holds.

$$\sqrt[n]{\frac{a}{b}} = \frac{\sqrt[n]{a}}{\sqrt[n]{b}} \quad \textbf{(where } b \neq 0\textbf{)}$$

That is, the nth root of a quotient is the quotient of the nth roots.

EXAMPLE 3 Using the Quotient Rule

Simplify. Assume that all variables represent positive real numbers.

(a) $\sqrt{\frac{16}{25}} = \frac{\sqrt{16}}{\sqrt{25}} = \frac{4}{5}$

(b) $\sqrt{\frac{7}{36}} = \frac{\sqrt{7}}{\sqrt{36}} = \frac{\sqrt{7}}{6}$

(c) $\sqrt[3]{-\frac{8}{125}} = \sqrt[3]{\frac{-8}{125}} = \frac{\sqrt[3]{-8}}{\sqrt[3]{125}} = \frac{-2}{5} = -\frac{2}{5}$ $\quad \frac{-a}{b} = -\frac{a}{b}$

(d) $\sqrt[3]{\frac{7}{216}} = \frac{\sqrt[3]{7}}{\sqrt[3]{216}} = \frac{\sqrt[3]{7}}{6}$

(e) $\sqrt[5]{\frac{x}{32}} = \frac{\sqrt[5]{x}}{\sqrt[5]{32}} = \frac{\sqrt[5]{x}}{2}$

NOW TRY ANSWERS

1. (a) $\sqrt{77}$ (b) $\sqrt{30mn}$
2. (a) $\sqrt[3]{20}$ (b) $\sqrt[4]{30tr^3}$
   (c) $\sqrt[7]{60x^2y^3}$
   (d) This expression cannot be simplified using the product rule.

NOW TRY EXERCISE 3

Simplify. Assume that all variables represent positive real numbers.

(a) $\sqrt{\frac{49}{36}}$ (b) $\sqrt{\frac{5}{144}}$

(c) $\sqrt[3]{-\frac{27}{1000}}$ (d) $\sqrt[4]{\frac{t}{16}}$

(e) $-\sqrt[5]{\frac{m^{15}}{243}}$

CLASSROOM EXAMPLE 3

Simplify. Assume that all variables represent positive real numbers.

(a) $\sqrt{\frac{100}{81}}$ (b) $\sqrt{\frac{11}{25}}$

(c) $\sqrt[3]{\frac{18}{125}}$ (d) $\sqrt{\frac{y^8}{16}}$

(e) $-\sqrt[3]{\frac{x^2}{r^{12}}}$

Answers: (a) $\frac{10}{9}$ (b) $\frac{\sqrt{11}}{5}$ (c) $\frac{\sqrt[3]{18}}{5}$ (d) $\frac{y^4}{4}$ (e) $-\frac{\sqrt[3]{x^2}}{r^4}$

TEACHING TIP

Display a partial list of squares, cubes, and fourth powers to help in simplifying.

TEACHING TIP

Show in **Example 4(b)** that, while $\sqrt{108}$ can be simplified by first factoring it as $\sqrt{9 \cdot 12}$, another step will be required because 12 has the perfect square factor 4. Explain that it is advantageous to choose the *greatest* perfect square factor to start.

NOW TRY ANSWERS

3. (a) $\frac{7}{6}$ (b) $\frac{\sqrt{5}}{12}$ (c) $-\frac{3}{10}$ (d) $\frac{\sqrt[4]{t}}{2}$ (e) $-\frac{m^3}{3}$

(f) $-\sqrt[3]{\frac{m^6}{125}} = -\frac{\sqrt[3]{m^6}}{\sqrt[3]{125}} = -\frac{m^2}{5}$ Think: $\sqrt[3]{m^6} = m^{6/3} = m^2$

NOW TRY

OBJECTIVE 3 Simplify radicals.

We use the product and quotient rules to simplify radicals. A radical is **simplified** if the following four conditions are met.

Conditions for a Simplified Radical

1. The radicand has no factor raised to a power greater than or equal to the index.
2. The radicand has no fractions.
3. No denominator contains a radical.
4. Exponents in the radicand and the index of the radical have greatest common factor 1.

Examples: $\sqrt{22}$, $\sqrt{15xy}$, $\sqrt[3]{18}$, $\frac{\sqrt[4]{m^3}}{m}$ These radicals are simplified.

$\sqrt{28}$, $\sqrt[3]{\frac{3}{5}}$, $\frac{7}{\sqrt{7}}$, $\sqrt[3]{r^{12}}$ These radicals are not simplified. Each violates one of the above conditions.

EXAMPLE 4 Simplifying Roots of Numbers

Simplify.

(a) $\sqrt{24}$

Check to see whether 24 is divisible by a perfect square (the square of a natural number) such as 4, 9, 16, … . The greatest perfect square that divides into 24 is 4.

$$\sqrt{24}$$

$= \sqrt{4 \cdot 6}$ Factor; 4 is a perfect square.

$= \sqrt{4} \cdot \sqrt{6}$ Product rule

$= 2\sqrt{6}$ $\sqrt{4} = 2$

(b) $\sqrt{108}$

As shown on the left, the number 108 is divisible by the perfect square 36. If this perfect square is not immediately clear, try factoring 108 into its prime factors, as shown on the right.

$\sqrt{108}$

$= \sqrt{36 \cdot 3}$ Factor.

$= \sqrt{36} \cdot \sqrt{3}$ Product rule

$= 6\sqrt{3}$ $\sqrt{36} = 6$

$\sqrt{108}$

$= \sqrt{2^2 \cdot 3^3}$

$= \sqrt{2^2 \cdot 3^2 \cdot 3}$ $a^3 = a^2 \cdot a$

$= \sqrt{2^2} \cdot \sqrt{3^2} \cdot \sqrt{3}$ Product rule

$= 2 \cdot 3 \cdot \sqrt{3}$ $\sqrt{2^2} = 2$, $\sqrt{3^2} = 3$

$= 6\sqrt{3}$ Multiply.

(c) $\sqrt{10}$ No perfect square (other than 1) divides into 10, so $\sqrt{10}$ cannot be simplified further.

NOW TRY EXERCISE 4

Simplify.

(a) $\sqrt{50}$ (b) $\sqrt{192}$

(c) $\sqrt{42}$ (d) $\sqrt[3]{108}$

(e) $-\sqrt[4]{80}$

CLASSROOM EXAMPLE 4

Simplify.

(a) $\sqrt{32}$ (b) $\sqrt{300}$

(c) $\sqrt{35}$ (d) $\sqrt[3]{54}$

(e) $\sqrt[4]{243}$

Answers: (a) $4\sqrt{2}$ (b) $10\sqrt{3}$

(c) $\sqrt{35}$ cannot be simplified further.

(d) $3\sqrt[3]{2}$ (e) $3\sqrt[4]{3}$

(d) $\sqrt[3]{16}$

The greatest perfect *cube* that divides into 16 is 8, so factor 16 as $8 \cdot 2$.

Remember to write the index.

$$\sqrt[3]{16}$$

$$= \sqrt[3]{8 \cdot 2} \quad \text{8 is a perfect cube.}$$

$$= \sqrt[3]{8} \cdot \sqrt[3]{2} \quad \text{Product rule}$$

$$= 2\sqrt[3]{2} \quad \sqrt[3]{8} = 2$$

(e)

$$-\sqrt[4]{162}$$

$$= -\sqrt[4]{81 \cdot 2} \quad \text{81 is a perfect 4th power.}$$

Remember the negative sign in each line.

$$= -\sqrt[4]{81} \cdot \sqrt[4]{2} \quad \text{Product rule}$$

$$= -3\sqrt[4]{2} \quad \sqrt[4]{81} = 3$$

NOW TRY

CAUTION ***Be careful with which factors belong outside the radical symbol and which belong inside.*** Note on the right in **Example 4(b)** how $2 \cdot 3$ is written outside because $\sqrt{2^2} = 2$ and $\sqrt{3^2} = 3$, while the remaining 3 is left inside the radical.

NOW TRY EXERCISE 5

Simplify. Assume that all variables represent positive real numbers.

(a) $\sqrt{36x^5}$ (b) $\sqrt{32m^5n^4}$

(c) $\sqrt[3]{-125k^3p^7}$

(d) $-\sqrt[4]{162x^7y^8}$

CLASSROOM EXAMPLE 5

Simplify. Assume that all variables represent positive real numbers.

(a) $\sqrt{25p^7}$ (b) $\sqrt{72y^3x}$

(c) $\sqrt[3]{-27y^7x^5z^6}$ (d) $-\sqrt[4]{32a^5b^7}$

Answers: (a) $5p^3\sqrt{p}$

(b) $6y\sqrt{2yx}$

(c) $-3y^2xz^2\sqrt[3]{yx^2}$

(d) $-2ab\sqrt[4]{2ab^3}$

EXAMPLE 5 Simplifying Radicals Involving Variables

Simplify. Assume that all variables represent positive real numbers.

(a) $\sqrt{16m^3}$

$$= \sqrt{16m^2 \cdot m} \quad \text{Factor.}$$

$$= \sqrt{16m^2} \cdot \sqrt{m} \quad \text{Product rule}$$

$$= 4m\sqrt{m} \quad \text{Take the square root.}$$

Absolute value bars are not needed around the m in color because all the variables represent *positive* real numbers.

(b) $\sqrt{200k^7q^8}$

$$= \sqrt{10^2 \cdot 2 \cdot (k^3)^2 \cdot k \cdot (q^4)^2} \quad \text{Factor into perfect squares.}$$

$$= 10k^3q^4\sqrt{2k} \quad \text{Take the square root.}$$

(c) $\sqrt[3]{-8x^4y^5}$

$$= \sqrt[3]{(-8x^3y^3)(xy^2)} \quad \text{Choose } -8x^3y^3 \text{ as the perfect cube that divides into } -8x^4y^5.$$

$$= \sqrt[3]{-8x^3y^3} \cdot \sqrt[3]{xy^2} \quad \text{Product rule}$$

$$= -2xy\sqrt[3]{xy^2} \quad \text{Take the cube root.}$$

(d) $-\sqrt[4]{32y^9}$

$$= -\sqrt[4]{(16y^8)(2y)} \quad 16y^8 \text{ is the greatest 4th power that divides into } 32y^9.$$

$$= -\sqrt[4]{16y^8} \cdot \sqrt[4]{2y} \quad \text{Product rule}$$

$$= -2y^2\sqrt[4]{2y} \quad \text{Take the fourth root.}$$

NOW TRY

NOW TRY ANSWERS

4. (a) $5\sqrt{2}$ (b) $8\sqrt{3}$
 (c) $\sqrt{42}$ cannot be simplified further.
 (d) $3\sqrt[3]{4}$ (e) $-2\sqrt[4]{5}$
5. (a) $6x^2\sqrt{x}$ (b) $4m^2n^2\sqrt{2m}$
 (c) $-5kp^2\sqrt[3]{p}$ (d) $-3xy^2\sqrt[4]{2x^3}$

TEACHING TIP
Warn students to remember to write the root index after simplifying a radical with an index greater than 2.

NOTE From **Example 5,** we see that if a variable is raised to a power with an exponent divisible by 2, it is a perfect square. If it is raised to a power with an exponent divisible by 3, it is a perfect cube. ***In general, if it is raised to a power with an exponent divisible by n, it is a perfect nth power.***

The conditions for a simplified radical given earlier state that an exponent in the radicand and the index of the radical should have greatest common factor 1.

NOW TRY EXERCISE 6
Simplify. Assume that all variables represent positive real numbers.
(a) $\sqrt[6]{7^2}$ **(b)** $\sqrt[6]{y^4}$

CLASSROOM EXAMPLE 6
Simplify. Assume that all variables represent positive real numbers.
(a) $\sqrt[12]{2^3}$ (b) $\sqrt[6]{t^2}$
Answers: (a) $\sqrt[4]{2}$ (b) $\sqrt[3]{t}$

EXAMPLE 6 Simplifying Radicals Using Lesser Indexes

Simplify. Assume that all variables represent positive real numbers.

(a) $\sqrt[9]{5^6}$

We write this radical using rational exponents and then write the exponent in lowest terms. We then express the answer as a radical.

$$\sqrt[9]{5^6} = (5^6)^{1/9} = 5^{6/9} = 5^{2/3} = \sqrt[3]{5^2}, \quad \text{or} \quad \sqrt[3]{25}$$

(b) $\sqrt[4]{p^2} = (p^2)^{1/4} = p^{2/4} = p^{1/2} = \sqrt{p}$ (Recall the assumption that $p > 0$.)

NOW TRY

These examples suggest the following rule.

Meaning of $\sqrt[kn]{a^{km}}$

If m is an integer, n and k are natural numbers, and all indicated roots exist, then the following holds.

$$\sqrt[kn]{a^{km}} = \sqrt[n]{a^m}$$

CLASSROOM EXAMPLE 7
Simplify $\sqrt{5} \cdot \sqrt[3]{4}$.
Answer: $\sqrt[6]{2000}$

OBJECTIVE 4 Simplify products and quotients of radicals with different indexes.

We multiply and divide radicals with different indexes using rational exponents.

NOW TRY EXERCISE 7
Simplify $\sqrt[3]{3} \cdot \sqrt{6}$.

EXAMPLE 7 Multiplying Radicals with Different Indexes

Simplify $\sqrt{7} \cdot \sqrt[3]{2}$.

Because the different indexes, 2 and 3, have a least common multiple of 6, use rational exponents to write each radical as a sixth root.

$$\sqrt{7} = 7^{1/2} = 7^{3/6} = \sqrt[6]{7^3} = \sqrt[6]{343}$$

$$\sqrt[3]{2} = 2^{1/3} = 2^{2/6} = \sqrt[6]{2^2} = \sqrt[6]{4}$$

Now we can multiply.

$$\sqrt{7} \cdot \sqrt[3]{2}$$

$$= \sqrt[6]{343} \cdot \sqrt[6]{4} \qquad \text{Substitute; } \sqrt{7} = \sqrt[6]{343},\ \sqrt[3]{2} = \sqrt[6]{4}$$

$$= \sqrt[6]{1372} \qquad \text{Product rule}$$

NOW TRY

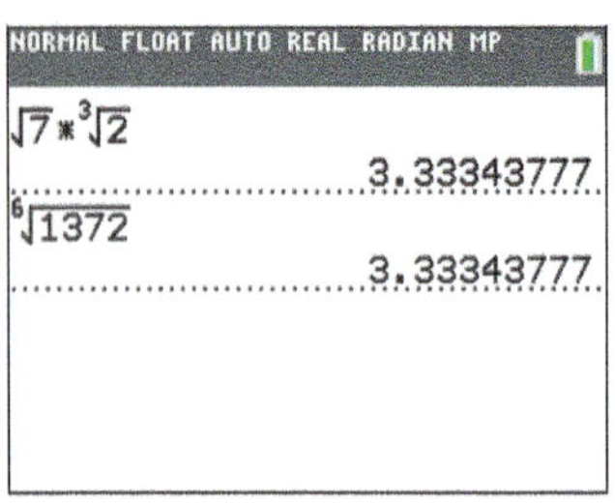

FIGURE 6

Results such as the one in **Example 7** can be supported with a calculator, as shown in **FIGURE 6**. Notice that the calculator gives the same *approximation* for the initial product and the final radical that we obtained.

NOW TRY ANSWERS
6. (a) $\sqrt[3]{7}$ **(b)** $\sqrt[3]{y^2}$
7. $\sqrt[6]{1944}$

CAUTION The computation in **FIGURE 6** is not *proof* that the two expressions are equal. The algebra in **Example 7,** however, is valid proof of their equality.

TEACHING TIP
Illustrate the equation $a^2 + b^2 = c^2$ with some familiar Pythagorean triples: 3-4-5, 5-12-13, and 7-24-25, for example.

OBJECTIVE 5 Use the Pythagorean theorem.

The **Pythagorean theorem** provides an equation that relates the lengths of the three sides of a right triangle.

Pythagorean Theorem

If a and b are the lengths of the shorter sides of a right triangle and c is the length of the longest side, then the following holds.

$$a^2 + b^2 = c^2$$

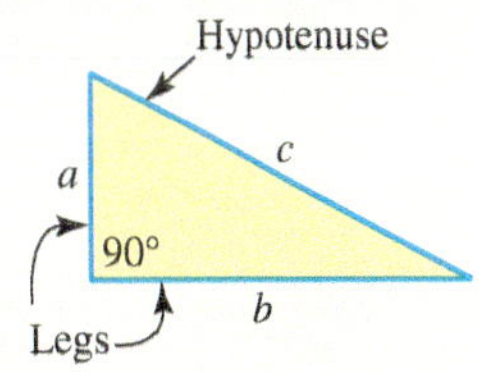

The two shorter sides are the **legs** of the triangle, and the longest side is the **hypotenuse.** The hypotenuse is the side opposite the right angle.

$$\textbf{leg}^2 + \textbf{leg}^2 = \textbf{hypotenuse}^2$$

CLASSROOM EXAMPLE 8
Find the length of the unknown side in each triangle.

(a)

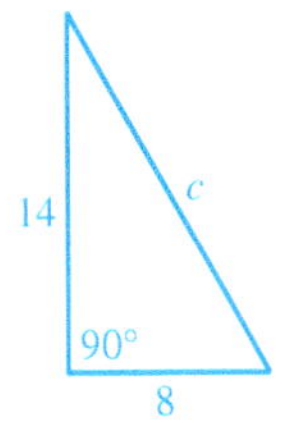

(b)

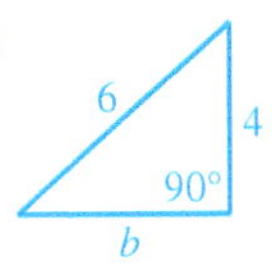

Answers: (a) $2\sqrt{65}$
(b) $2\sqrt{5}$

In **Section 8.1** we will see that an equation such as $x^2 = 7$ has two solutions: $\sqrt{7}$ (the principal, or positive, square root of 7) and $-\sqrt{7}$. Similarly, $c^2 = 52$ has two solutions, $\pm\sqrt{52} = \pm 2\sqrt{13}$. In applications we often choose only the principal (positive) square root.

NOW TRY EXERCISE 8
Find the length of the unknown side in each triangle.

(a)

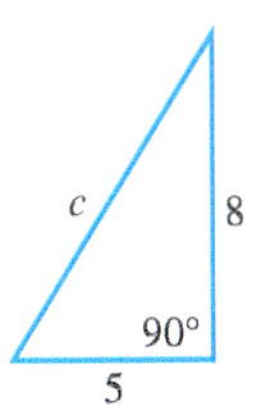

(b)

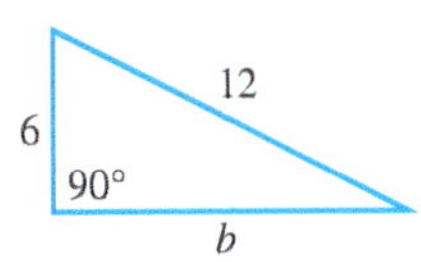

EXAMPLE 8 Using the Pythagorean Theorem

Find the length of the unknown side of the triangle in **FIGURE 7**.

Substitute carefully.

$a^2 + b^2 = c^2$	Pythagorean theorem
$4^2 + 6^2 = c^2$	Let $a = 4$ and $b = 6$.
$16 + 36 = c^2$	Apply the exponents.
$c^2 = 52$	Add. Interchange sides.
$c = \sqrt{52}$	Choose the principal root.
$c = \sqrt{4 \cdot 13}$	Factor.
$c = \sqrt{4} \cdot \sqrt{13}$	Product rule
$c = 2\sqrt{13}$	Simplify.

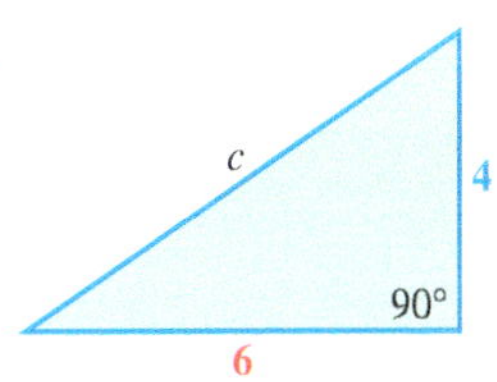

FIGURE 7

The length of the hypotenuse is $2\sqrt{13}$. NOW TRY

CAUTION In the equation $a^2 + b^2 = c^2$, be sure that the length of the hypotenuse is substituted for c and the lengths of the legs are substituted for a and b.

NOW TRY ANSWERS
8. **(a)** $\sqrt{89}$ **(b)** $6\sqrt{3}$

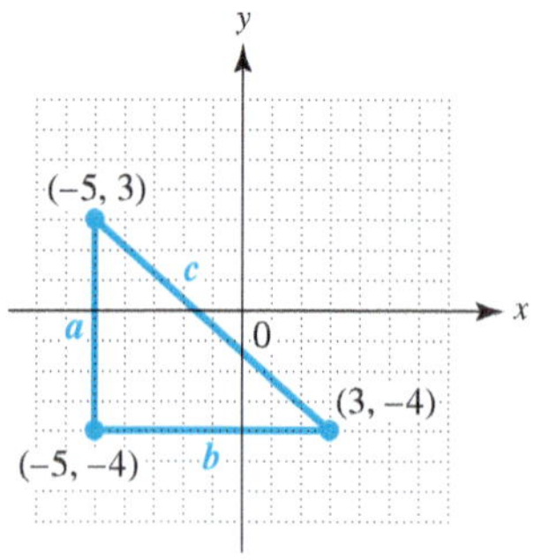

FIGURE 8

OBJECTIVE 6 Use the distance formula.

The *distance formula* allows us to find the distance between two points in the coordinate plane, or the length of the line segment joining those two points.

FIGURE 8 shows the points $(3, -4)$ and $(-5, 3)$. The vertical line through $(-5, 3)$ and the horizontal line through $(3, -4)$ intersect at the point $(-5, -4)$. Thus, the point $(-5, -4)$ becomes the vertex of the right angle in a right triangle.

By the Pythagorean theorem, the square of the length of the hypotenuse c of the right triangle in **FIGURE 8** is equal to the sum of the squares of the lengths of the two legs a and b.

$$a^2 + b^2 = c^2$$

The length a is the difference between the y-coordinates of the endpoints. The x-coordinate of both points in **FIGURE 8** is -5, so the side is vertical, and we can find a by finding the difference of the y-coordinates. We subtract -4 from 3 to obtain a positive value for a.

$$a = 3 - (-4) = 7$$

Similarly, we find b by subtracting -5 from 3.

$$b = 3 - (-5) = 8$$

Now substitute these values into the equation.

$$c^2 = a^2 + b^2$$

$$c^2 = 7^2 + 8^2 \quad \text{Let } a = 7 \text{ and } b = 8.$$

$$c^2 = 49 + 64 \quad \text{Apply the exponents.}$$

$$c^2 = 113 \quad \text{Add.}$$

$$c = \sqrt{113} \quad \text{Choose the principal root.}$$

We choose the principal root because here the distance cannot be negative. Therefore, the distance between $(-5, 3)$ and $(3, -4)$ is $\sqrt{113}$.

NOTE It is customary to leave the distance in simplified radical form. Do not use a calculator to find an approximation unless specifically directed to do so.

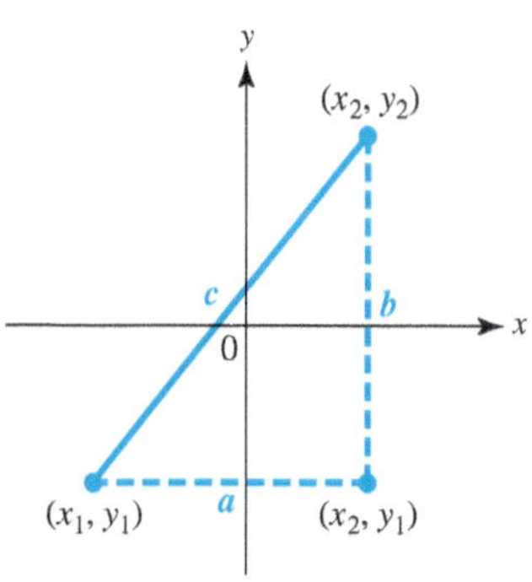

FIGURE 9

This result can be generalized. **FIGURE 9** shows the two points (x_1, y_1) and (x_2, y_2). The distance a between (x_1, y_1) and (x_2, y_1) is given by

$$a = |x_2 - x_1|,$$

and the distance b between (x_2, y_2) and (x_2, y_1) is given by

$$b = |y_2 - y_1|.$$

From the Pythagorean theorem, we obtain the following.

$$c^2 = a^2 + b^2$$

$$c^2 = (x_2 - x_1)^2 + (y_2 - y_1)^2 \quad \text{For all real numbers } a,\ |a|^2 = a^2.$$

Choosing the principal square root gives the **distance formula.** In this formula, we use d (to denote distance) rather than c.

Distance Formula

The distance d between the points (x_1, y_1) and (x_2, y_2) is given by the following.

$$d = \sqrt{(x_2 - x_1)^2 + (y_2 - y_1)^2}$$

NOW TRY EXERCISE 9
Find the distance between the points $(-4, -3)$ and $(-8, 6)$.

CLASSROOM EXAMPLE 9
Find the distance between each pair of points.
(a) $(2, -1)$ and $(5, 3)$
(b) $(-3, 2)$ and $(0, -4)$
Answers: (a) 5 (b) $3\sqrt{5}$

EXAMPLE 9 **Using the Distance Formula**

Find the distance between the points $(-3, 5)$ and $(6, 4)$.

We arbitrarily choose to let $(x_1, y_1) = (-3, 5)$ and $(x_2, y_2) = (6, 4)$.

$d = \sqrt{(x_2 - x_1)^2 + (y_2 - y_1)^2}$ Distance formula

$d = \sqrt{[6 - (-3)]^2 + (4 - 5)^2}$ Let $x_2 = 6$, $y_2 = 4$, $x_1 = -3$, $y_1 = 5$.

Substitute carefully.

$d = \sqrt{9^2 + (-1)^2}$

$d = \sqrt{82}$ Leave in radical form.

The distance is $\sqrt{82}$.

NOW TRY

OBJECTIVE 7 Find an equation of a circle given its center and radius.

A **circle** is the set of all points in a plane that lie a fixed distance from a fixed point. The fixed point is the **center,** and the fixed distance is the **radius.** We use the distance formula to find an equation of a circle.

NOW TRY EXERCISE 10
Find an equation of the circle with radius 6 and center $(0, 0)$, and graph it.

CLASSROOM EXAMPLE 10
Find an equation of the circle with radius 4 and center $(0, 0)$, and graph it .
Answer: $x^2 + y^2 = 16$

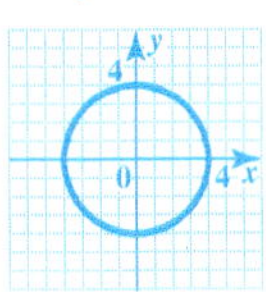

EXAMPLE 10 **Finding an Equation of a Circle and Graphing It**

Find an equation of the circle with radius 3 and center $(0, 0)$, and graph it.

If the point (x, y) is on the circle, then the distance from (x, y) to the center $(0, 0)$ is the radius 3.

$\sqrt{(x_2 - x_1)^2 + (y_2 - y_1)^2} = d$ Distance formula

$\sqrt{(x - 0)^2 + (y - 0)^2} = 3$ Let $x_1 = 0$, $y_1 = 0$, and $d = 3$.

$x^2 + y^2 = 9$ Square each side.

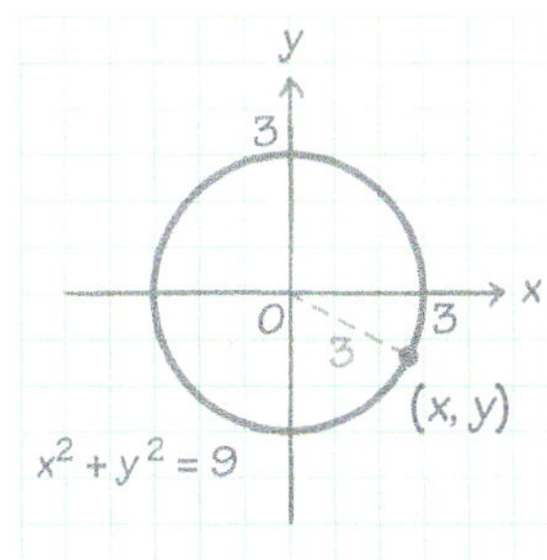

FIGURE 10

An equation of this circle is $x^2 + y^2 = 9$. The graph is shown in FIGURE 10.

NOW TRY

A circle may not be centered at the origin, as seen in the next example.

NOW TRY ANSWERS

9. $\sqrt{97}$

10. $x^2 + y^2 = 36$

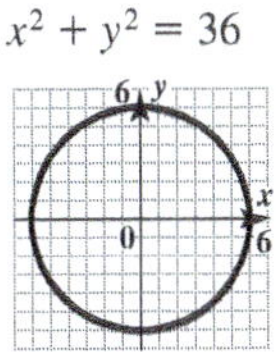

EXAMPLE 11 **Finding an Equation of a Circle and Graphing It**

Find an equation of the circle with center $(4, -3)$ and radius 5, and graph it.

$\sqrt{(x_2 - x_1)^2 + (y_2 - y_1)^2} = d$ Distance formula

$\sqrt{(x - 4)^2 + [y - (-3)]^2} = 5$ Let $x_1 = 4$, $y_1 = -3$, and $d = 5$.

$(x - 4)^2 + (y + 3)^2 = 25$ Square each side.

NOW TRY EXERCISE 11

Find an equation of the circle with center at $(-2, 2)$ and radius 3, and graph it.

CLASSROOM EXAMPLE 11

Find an equation of the circle with center at $(3, -2)$ and radius 4, and graph it.

Answer:
$(x-3)^2 + (y+2)^2 = 16$

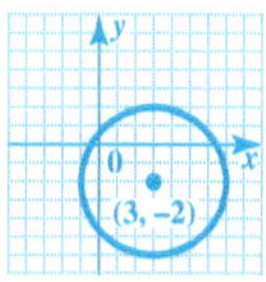

To graph the circle

$$(x-4)^2 + (y+3)^2 = 25,$$

plot the center $(4, -3)$, and then, because the radius is 5, move 5 units right, left, up, and down from the center, plotting the points

$$(9, -3), \quad (-1, -3), \quad (4, 2), \quad \text{and} \quad (4, -8).$$

Draw a smooth curve through these four points, sketching one quarter of the circle at a time. See **FIGURE 11.**

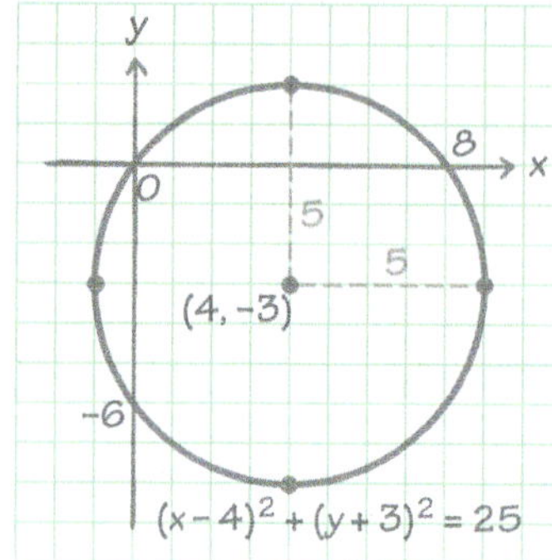

FIGURE 11

NOW TRY

Examples 10 and 11 suggest the form of an equation of a circle with radius r and center (h, k). If (x, y) is a point on the circle, then the distance from the center (h, k) to the point (x, y) is r. By the distance formula,

$$\sqrt{(x-h)^2 + (y-k)^2} = r.$$

Squaring both sides gives the **center-radius form** of the equation of a circle.

CLASSROOM EXAMPLE 12

Find an equation of the circle with center $(2, -1)$ and radius $\sqrt{10}$.

Answer:
$(x-2)^2 + (y+1)^2 = 10$

Equation of a Circle (Center-Radius Form)

An equation of a circle with radius r and center (h, k) is given by the following.

$$(x-h)^2 + (y-k)^2 = r^2$$

NOW TRY EXERCISE 12

Find an equation of the circle with center $(-5, 4)$ and radius $\sqrt{6}$.

EXAMPLE 12 Using the Center-Radius Form of the Equation of a Circle

Find an equation of the circle with center $(-1, 2)$ and radius $\sqrt{7}$.

$$(x-h)^2 + (y-k)^2 = r^2 \qquad \text{Center-radius form}$$

$$[x-(-1)]^2 + (y-2)^2 = (\sqrt{7})^2 \qquad \text{Let } h = -1,\ k = 2, \text{ and } r = \sqrt{7}.$$

Pay attention to signs here.

$$(x+1)^2 + (y-2)^2 = 7 \qquad \text{Simplify; } (\sqrt{a})^2 = a$$

NOW TRY

NOW TRY ANSWERS

11. $(x+2)^2 + (y-2)^2 = 9$

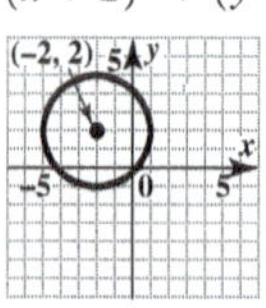

12. $(x+5)^2 + (y-4)^2 = 6$

NOTE If a circle has its center at the origin (0, 0) and radius r, then its equation is as follows.

$$(x-0)^2 + (y-0)^2 = r^2 \qquad \text{Let } h = 0,\ k = 0 \text{ in the center-radius form.}$$

$$x^2 + y^2 = r^2 \qquad \text{See **Example 10.**}$$

7.3 Exercises

FOR EXTRA HELP

MyMathLab®

Complete solution available in MyMathLab

1. D 2. D
3. B 4. A
5. D 6. D

Concept Check *Choose the correct response.*

1. Which is the greatest perfect square factor of 128?

A. 12 **B.** 16 **C.** 32 **D.** 64

2. Which is the greatest perfect cube factor of $81a^7$?

A. $8a^3$ **B.** $27a^3$ **C.** $81a^6$ **D.** $27a^6$

7. Because there are only two factors of $\sqrt[3]{x}$, $\sqrt[3]{x} \cdot \sqrt[3]{x} = (\sqrt[3]{x})^2$, or $\sqrt[3]{x^2}$.
8. Because there are only two factors, $\sqrt[4]{x} \cdot \sqrt[4]{x} = (\sqrt[4]{x})^2 = x^{2/4} = x^{1/2} = \sqrt{x}$, for $x \geq 0$.

9. $\sqrt{9}$, or 3 10. $\sqrt{25}$, or 5
11. $\sqrt{36}$, or 6 12. $\sqrt{36}$, or 6
13. $\sqrt{30}$ 14. $\sqrt{30}$
15. $\sqrt{14x}$ 16. $\sqrt{23t}$
17. $\sqrt{42pqr}$ 18. $\sqrt{35xt}$
19. $\sqrt[3]{10}$ 20. $\sqrt[3]{18}$
21. $\sqrt[3]{14xy}$ 22. $\sqrt[3]{36xy}$
23. $\sqrt[4]{33}$ 24. $\sqrt[4]{54}$
25. $\sqrt[4]{6x^3}$ 26. $\sqrt[4]{18y^3}$
27. This expression cannot be simplified by the product rule.
28. This expression cannot be simplified by the product rule.

29. $\frac{8}{11}$ 30. $\frac{4}{7}$
31. $\frac{\sqrt{3}}{5}$ 32. $\frac{\sqrt{13}}{7}$
33. $\frac{\sqrt{x}}{5}$ 34. $\frac{\sqrt{k}}{10}$
35. $\frac{p^3}{9}$ 36. $\frac{w^5}{6}$
37. $-\frac{3}{4}$ 38. $-\frac{6}{5}$
39. $\frac{\sqrt[3]{r^2}}{2}$ 40. $\frac{\sqrt[3]{t}}{5}$
41. $-\frac{3}{x}$ 42. $-\frac{5}{y}$
43. $\frac{1}{x^3}$ 44. $\frac{2}{y^4}$

45. $2\sqrt{3}$ 46. $3\sqrt{2}$
47. $12\sqrt{2}$ 48. $6\sqrt{2}$
49. $-4\sqrt{2}$ 50. $-4\sqrt{3}$
51. $-2\sqrt{7}$ 52. $-2\sqrt{6}$
53. This radical cannot be simplified further.
54. This radical cannot be simplified further.
55. $4\sqrt[3]{2}$ 56. $2\sqrt[3]{3}$
57. $-2\sqrt[3]{2}$ 58. $-5\sqrt[3]{2}$
59. $2\sqrt[3]{5}$ 60. $5\sqrt[3]{3}$
61. $-4\sqrt[4]{2}$ 62. $-5\sqrt[4]{2}$
63. $2\sqrt[5]{2}$ 64. $2\sqrt[5]{4}$
65. $-3\sqrt[5]{2}$ 66. $-4\sqrt[5]{2}$
67. $2\sqrt[6]{2}$ 68. $3\sqrt[6]{2}$

3. Which radical can be simplified?

A. $\sqrt{21}$ **B.** $\sqrt{48}$ **C.** $\sqrt[3]{12}$ **D.** $\sqrt[4]{10}$

4. Which radical *cannot* be simplified?

A. $\sqrt[3]{30}$ **B.** $\sqrt[3]{27a^2b}$ **C.** $\sqrt{\frac{25}{81}}$ **D.** $\frac{2}{\sqrt{7}}$

5. Which one of the following is *not* equal to $\sqrt{\frac{1}{2}}$? (Do not use calculator approximations.)

A. $\sqrt{0.5}$ **B.** $\sqrt{\frac{2}{4}}$ **C.** $\sqrt{\frac{3}{6}}$ **D.** $\frac{\sqrt{4}}{\sqrt{16}}$

6. Which one of the following is *not* equal to $\sqrt[3]{\frac{2}{5}}$? (Do not use calculator approximations.)

A. $\sqrt[3]{\frac{6}{15}}$ **B.** $\frac{\sqrt[3]{50}}{5}$ **C.** $\frac{\sqrt[3]{10}}{\sqrt[3]{25}}$ **D.** $\frac{\sqrt[3]{10}}{5}$

7. Why is $\sqrt[3]{x} \cdot \sqrt[3]{x}$ not equal to x? What is it equal to?

8. Why is $\sqrt[4]{x} \cdot \sqrt[4]{x}$ not equal to x? Explain why it *is* equal to $\sqrt{x}$, for $x \geq 0$.

Multiply, if possible, using the product rule. Assume that all variables represent positive real numbers. ***See Examples 1 and 2.***

9. $\sqrt{3} \cdot \sqrt{3}$ **10.** $\sqrt{5} \cdot \sqrt{5}$ **11.** $\sqrt{18} \cdot \sqrt{2}$ **12.** $\sqrt{12} \cdot \sqrt{3}$

13. $\sqrt{5} \cdot \sqrt{6}$ **14.** $\sqrt{10} \cdot \sqrt{3}$ **15.** $\sqrt{14} \cdot \sqrt{x}$ **16.** $\sqrt{23} \cdot \sqrt{t}$

17. $\sqrt{14} \cdot \sqrt{3pqr}$ **18.** $\sqrt{7} \cdot \sqrt{5xt}$ **19.** $\sqrt[3]{2} \cdot \sqrt[3]{5}$ **20.** $\sqrt[3]{3} \cdot \sqrt[3]{6}$

21. $\sqrt[3]{7x} \cdot \sqrt[3]{2y}$ **22.** $\sqrt[3]{9x} \cdot \sqrt[3]{4y}$ **23.** $\sqrt[4]{11} \cdot \sqrt[4]{3}$ **24.** $\sqrt[4]{6} \cdot \sqrt[4]{9}$

25. $\sqrt[4]{2x} \cdot \sqrt[4]{3x^2}$ **26.** $\sqrt[4]{3y^2} \cdot \sqrt[4]{6y}$ **27.** $\sqrt[3]{7} \cdot \sqrt[4]{3}$ **28.** $\sqrt[5]{8} \cdot \sqrt[6]{12}$

Simplify. Assume that all variables represent positive real numbers. ***See Example 3.***

29. $\sqrt{\frac{64}{121}}$ **30.** $\sqrt{\frac{16}{49}}$ **31.** $\sqrt{\frac{3}{25}}$ **32.** $\sqrt{\frac{13}{49}}$

33. $\sqrt{\frac{x}{25}}$ **34.** $\sqrt{\frac{k}{100}}$ **35.** $\sqrt{\frac{p^6}{81}}$ **36.** $\sqrt{\frac{w^{10}}{36}}$

37. $\sqrt[3]{-\frac{27}{64}}$ **38.** $\sqrt[3]{-\frac{216}{125}}$ **39.** $\sqrt[3]{\frac{r^2}{8}}$ **40.** $\sqrt[3]{\frac{t}{125}}$

41. $-\sqrt[4]{\frac{81}{x^4}}$ **42.** $-\sqrt[4]{\frac{625}{y^4}}$ **43.** $\sqrt[5]{\frac{1}{x^{15}}}$ **44.** $\sqrt[5]{\frac{32}{y^{20}}}$

Simplify. ***See Example 4.***

45. $\sqrt{12}$ **46.** $\sqrt{18}$ **47.** $\sqrt{288}$ **48.** $\sqrt{72}$ **49.** $-\sqrt{32}$

50. $-\sqrt{48}$ **51.** $-\sqrt{28}$ **52.** $-\sqrt{24}$ **53.** $\sqrt{30}$ **54.** $\sqrt{46}$

55. $\sqrt[3]{128}$ **56.** $\sqrt[3]{24}$ **57.** $\sqrt[3]{-16}$ **58.** $\sqrt[3]{-250}$ **59.** $\sqrt[3]{40}$

60. $\sqrt[3]{375}$ **61.** $-\sqrt[4]{512}$ **62.** $-\sqrt[4]{1250}$ **63.** $\sqrt[5]{64}$ **64.** $\sqrt[5]{128}$

65. $-\sqrt[5]{486}$ **66.** $-\sqrt[5]{2048}$ **67.** $\sqrt[6]{128}$ **68.** $\sqrt[6]{1458}$

69. $6k\sqrt{2}$ **70.** $3m\sqrt{2}$
71. $12xy^4\sqrt{xy}$ **72.** $13s^2t^5\sqrt{s}$
73. $11x^3$ **74.** $16z^6$
75. $-3t^4$ **76.** $-4y^6$
77. $-10m^4z^2$ **78.** $-5t^3s^{10}$
79. $5a^2b^3c^4$ **80.** $6y^5x^2z$
81. $\frac{1}{2}r^2t^5$ **82.** $\frac{3}{4}t^3u^2$
83. $5x\sqrt{2x}$ **84.** $10z\sqrt{3z}$
85. $-10r^5\sqrt{5r}$ **86.** $-10p^6\sqrt{2p}$
87. $x^3y^4\sqrt{13x}$ **88.** $k^4p^7\sqrt{23k}$
89. $2z^2w^3$ **90.** $4a^5b^4$
91. $-2zt^2\sqrt[3]{2z^2t}$
92. $-3mn^3\sqrt[3]{3mn}$
93. $3x^3y^4$ **94.** $3t^2u^7$
95. $-3r^3s^2\sqrt[4]{2r^3s^2}$
96. $-2km^2\sqrt[4]{2km^2}$
97. $\frac{y^5\sqrt{y}}{6}$ **98.** $\frac{v^6\sqrt{v}}{7}$
99. $\frac{x^5\sqrt[3]{x}}{3}$ **100.** $\frac{y^5\sqrt[3]{y^2}}{5}$

101. $4\sqrt{3}$ **102.** $5\sqrt{2}$
103. $\sqrt{5}$ **104.** $\sqrt{2}$
105. $x^2\sqrt{x}$ **106.** $x^3\sqrt[3]{x^2}$

107. $\sqrt[6]{432}$ **108.** $\sqrt[6]{5400}$
109. $\sqrt[12]{6912}$
110. $\sqrt[35]{2,573,571,875}$
111. $\sqrt[6]{x^5}$ **112.** $\sqrt[12]{y^7}$

113. 5 **114.** 13
115. $8\sqrt{2}$ **116.** $3\sqrt{5}$
117. $2\sqrt{14}$ **118.** $6\sqrt{2}$

119. 13 **120.** 10
121. $9\sqrt{2}$ **122.** $2\sqrt{10}$
123. $\sqrt{17}$ **124.** $\sqrt{37}$
125. 5 **126.** 13

Simplify. Assume that all variables represent positive real numbers. **See Example 5.**

69. $\sqrt{72k^2}$ **70.** $\sqrt{18m^2}$ **71.** $\sqrt{144x^3y^9}$

72. $\sqrt{169s^5t^{10}}$ **73.** $\sqrt{121x^6}$ **74.** $\sqrt{256z^{12}}$

75. $-\sqrt[3]{27t^{12}}$ **76.** $-\sqrt[3]{64y^{18}}$ **77.** $-\sqrt{100m^8z^4}$

78. $-\sqrt{25t^6s^{20}}$ **79.** $-\sqrt[3]{-125a^6b^9c^{12}}$ **80.** $-\sqrt[3]{-216y^{15}x^6z^3}$

81. $\sqrt[4]{\frac{1}{16}r^8t^{20}}$ **82.** $\sqrt[4]{\frac{81}{256}t^{12}u^8}$ **83.** $\sqrt{50x^3}$ **84.** $\sqrt{300z^3}$

85. $-\sqrt{500r^{11}}$ **86.** $-\sqrt{200p^{13}}$ **87.** $\sqrt{13x^7y^8}$ **88.** $\sqrt{23k^9p^{14}}$

89. $\sqrt[3]{8z^6w^9}$ **90.** $\sqrt[3]{64a^{15}b^{12}}$ **91.** $\sqrt[3]{-16z^5t^7}$ **92.** $\sqrt[3]{-81m^4n^{10}}$

93. $\sqrt[4]{81x^{12}y^{16}}$ **94.** $\sqrt[4]{81t^8u^{28}}$ **95.** $-\sqrt[4]{162r^{15}s^{10}}$ **96.** $-\sqrt[4]{32k^5m^{10}}$

97. $\sqrt{\frac{y^{11}}{36}}$ **98.** $\sqrt{\frac{v^{13}}{49}}$ **99.** $\sqrt[3]{\frac{x^{16}}{27}}$ **100.** $\sqrt[3]{\frac{y^{17}}{125}}$

Simplify. Assume that $x \geq 0$. **See Example 6.**

101. $\sqrt[4]{48^2}$ **102.** $\sqrt[4]{50^2}$ **103.** $\sqrt[4]{25}$

104. $\sqrt[6]{8}$ **105.** $\sqrt[10]{x^{25}}$ **106.** $\sqrt[12]{x^{44}}$

Simplify by first writing the radicals as radicals with the same index. Then multiply. Assume that all variables represent positive real numbers. **See Example 7.**

107. $\sqrt[3]{4} \cdot \sqrt{3}$ **108.** $\sqrt[3]{5} \cdot \sqrt{6}$ **109.** $\sqrt[4]{3} \cdot \sqrt[3]{4}$

110. $\sqrt[5]{7} \cdot \sqrt[7]{5}$ **111.** $\sqrt{x} \cdot \sqrt[3]{x}$ **112.** $\sqrt[3]{y} \cdot \sqrt[4]{y}$

Find the length of the unknown side in each right triangle. Simplify answers if possible. **See Example 8.**

113.
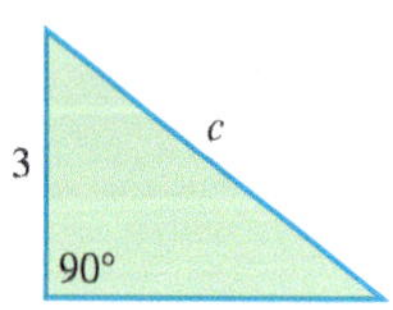

114.
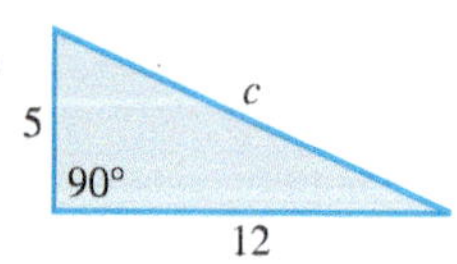

115.
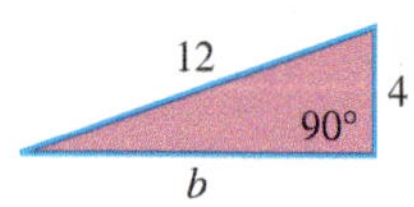

116.
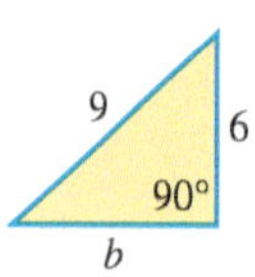

117.
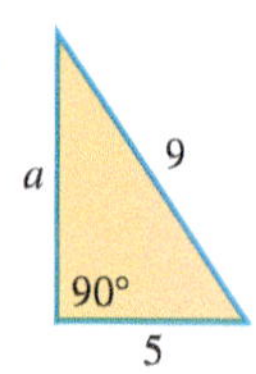

118.
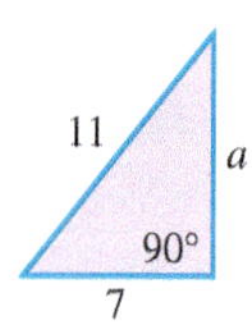

Find the distance between each pair of points. **See Example 9.**

119. $(6, 13)$ and $(1, 1)$ **120.** $(8, 13)$ and $(2, 5)$

121. $(-6, 5)$ and $(3, -4)$ **122.** $(-1, 5)$ and $(-7, 7)$

123. $(-8, 2)$ and $(-4, 1)$ **124.** $(-1, 2)$ and $(5, 3)$

125. $(4.7, 2.3)$ and $(1.7, -1.7)$ **126.** $(-2.9, 18.2)$ and $(2.1, 6.2)$

127. $6\sqrt{2}$ **128.** $\sqrt{103}$

129. $\sqrt{5y^2 - 2xy + x^2}$

130. $\sqrt{c^2 - 2cd + 5d^2}$

131. (a) B (b) C (c) D (d) A

132. The thumbtack acts as the center and the length of the string acts as the radius.

133. $x^2 + y^2 = 144$

134. $x^2 + y^2 = 81$

135. $(x + 4)^2 + (y - 3)^2 = 4$

136. $(x - 5)^2 + (y + 2)^2 = 16$

137. $(x + 8)^2 + (y + 5)^2 = 5$

138. $(x + 12)^2 + (y - 13)^2 = 7$

139.

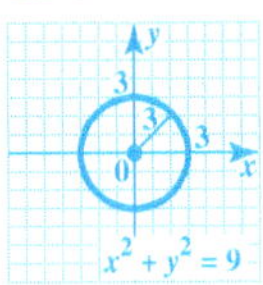

center: $(0, 0)$

140.

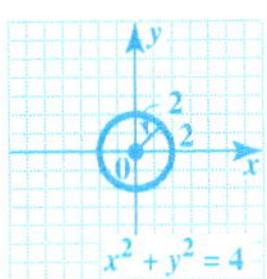

center: $(0, 0)$

141.

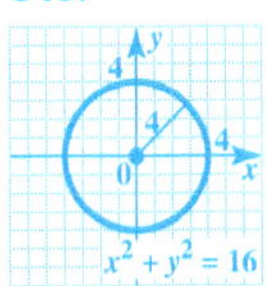

center: $(0, 0)$

142.

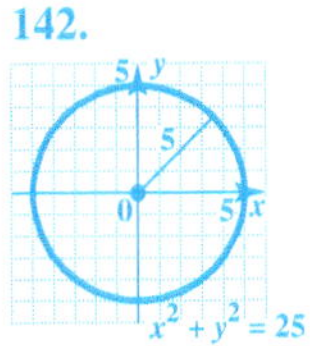

center: $(0, 0)$

143.

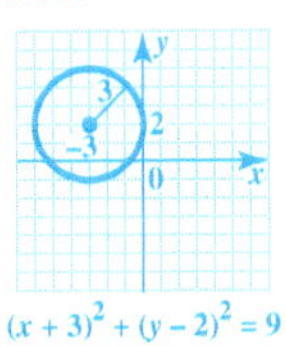

center: $(-3, 2)$

144.

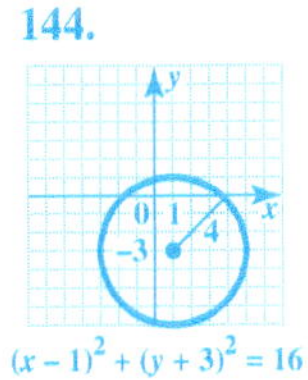

center: $(1, -3)$

145.

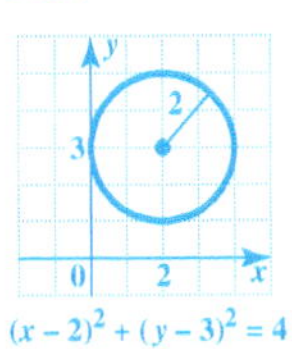

center: $(2, 3)$

146.

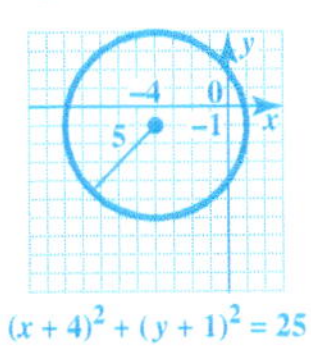

center: $(-4, -1)$

147. $2\sqrt{106} + 4\sqrt{2}$

148. $5 + \sqrt{58} + \sqrt{13}$

127. $(\sqrt{2}, \sqrt{6})$ and $(-2\sqrt{2}, 4\sqrt{6})$

128. $(\sqrt{7}, 9\sqrt{3})$ and $(-\sqrt{7}, 4\sqrt{3})$

129. $(x + y, y)$ and $(x - y, x)$

130. $(c, c - d)$ and $(d, c + d)$

Concept Check *Work each problem.*

131. Match each equation with the correct graph.

(a) $(x - 3)^2 + (y - 2)^2 = 25$

(b) $(x - 3)^2 + (y + 2)^2 = 25$

(c) $(x + 3)^2 + (y - 2)^2 = 25$

(d) $(x + 3)^2 + (y + 2)^2 = 25$

A.

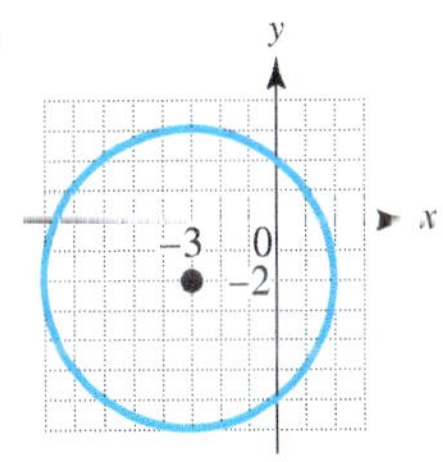

B.

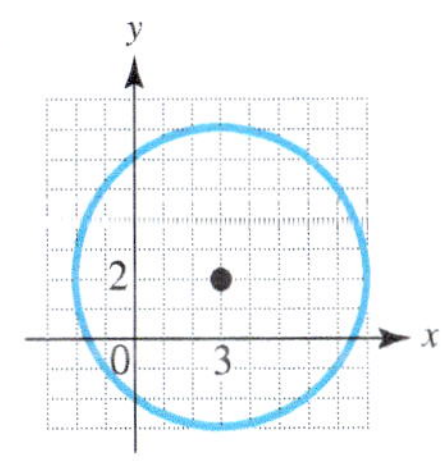

C.

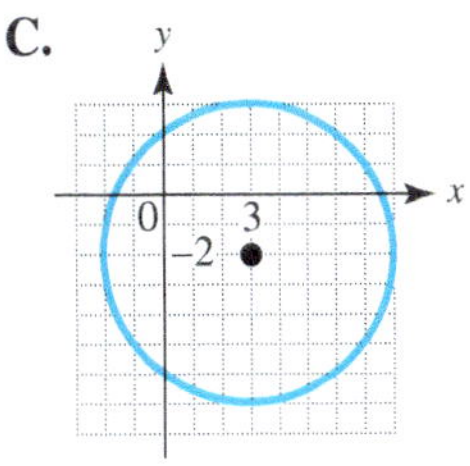

D.

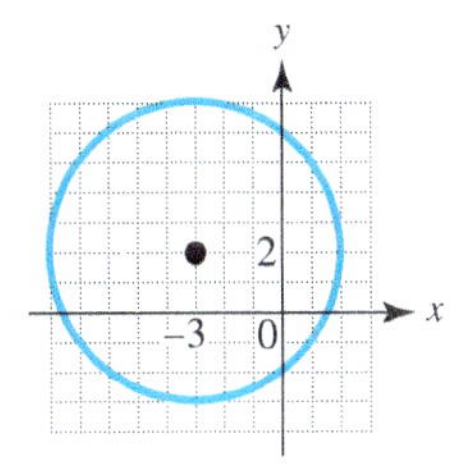

132. A circle can be drawn on a piece of posterboard by fastening one end of a string with a thumbtack, pulling the string taut with a pencil, and tracing a curve, as shown in the figure. Explain why this method works.

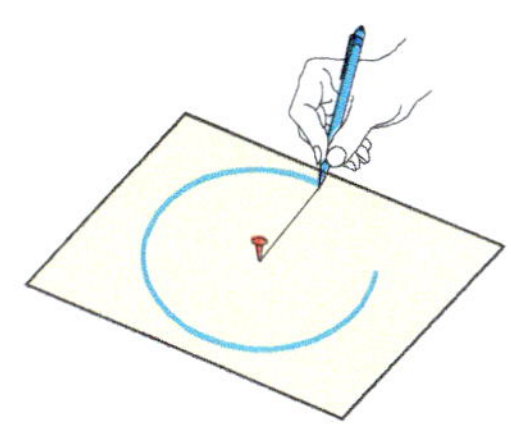

Find the equation of a circle satisfying the given conditions. ***See Examples 10–12.***

133. Center: $(0, 0)$; radius: 12

134. Center: $(0, 0)$; radius: 9

135. Center: $(-4, 3)$; radius: 2

136. Center: $(5, -2)$; radius: 4

137. Center: $(-8, -5)$; radius: $\sqrt{5}$

138. Center: $(-12, 13)$; radius: $\sqrt{7}$

Graph each circle. Identify the center. ***See Examples 10–12.***

139. $x^2 + y^2 = 9$

140. $x^2 + y^2 = 4$

141. $x^2 + y^2 = 16$

142. $x^2 + y^2 = 25$

143. $(x + 3)^2 + (y - 2)^2 = 9$

144. $(x - 1)^2 + (y + 3)^2 = 16$

145. $(x - 2)^2 + (y - 3)^2 = 4$

146. $(x + 4)^2 + (y + 1)^2 = 25$

Extending Skills *Find the perimeter of each triangle.* (*Hint: For Exercise 147, use* $\sqrt{k} + \sqrt{k} = 2\sqrt{k}$.)

147.

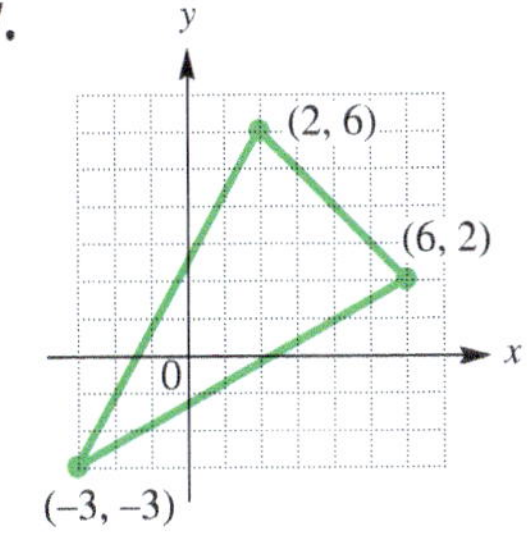

148.

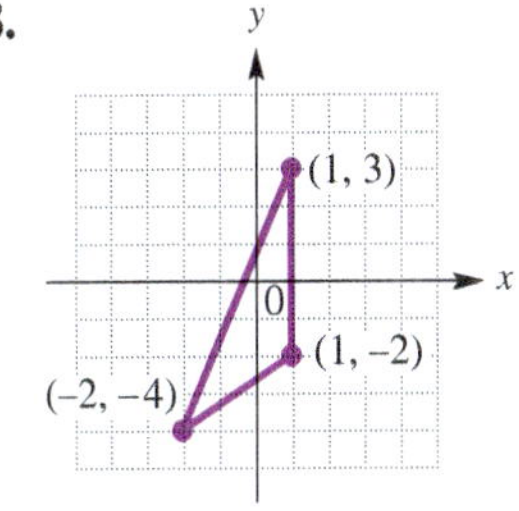

149. **(a)** $d = 1.224\sqrt{h}$ **(b)** 15.3 mi

150. 12.5 mi

151. 27.0 in.

152. $\sqrt{29}$ ft; 5.4 ft

153. 581

154. $8\sqrt{5}$ ft; 17.9 ft

Solve each problem.

149. Ted is lucky enough to own a condominium in a high rise building on the shore of Lake Michigan, and has a beautiful view of the lake from his window. He lives on the 14th floor, which is 150 ft above the ground. He discovered that he can find the number of miles to the horizon by multiplying 1.224 by the square root of his eye level in feet from the ground. Ted's eyes are 6 ft above his floor.

(a) Write a formula that could be used to calculate the distance d in miles to the horizon from a height h in feet from the ground.

(b) Use the formula from part (a) to calculate the distance, to the nearest tenth of a mile, that Ted can see to the horizon from his condominium window.

150. Refer to **Exercise 149.** Ted's neighbor Sheri lives on a floor that is 100 ft above the ground. Assuming that her eyes are 5 ft above the ground, to the nearest tenth of a mile, how far can she see to the horizon?

151. A television has a rectangular screen with a 21.7-in. width. Its height is 16 in. What is the measure of the diagonal of the screen, to the nearest tenth of an inch? (*Source:* Actual measurements of the author's television.)

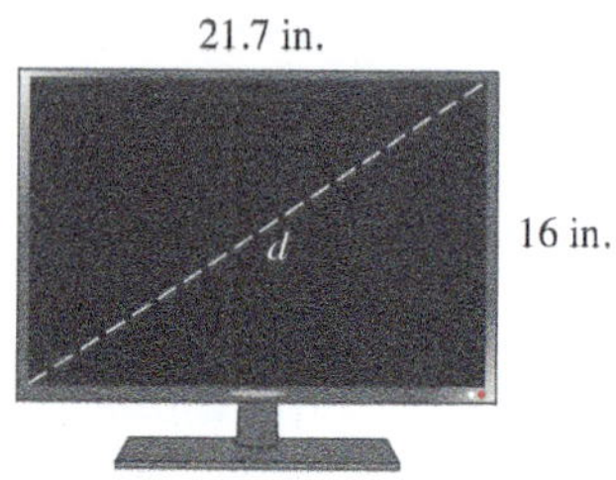

152. The length of the diagonal of a box is given by

$$D = \sqrt{L^2 + W^2 + H^2},$$

where L, W, and H are, respectively, the length, width, and height of the box. Find the length of the diagonal D of a box that is 4 ft long, 2 ft wide, and 3 ft high. Give the exact value, and then round to the nearest tenth of a foot.

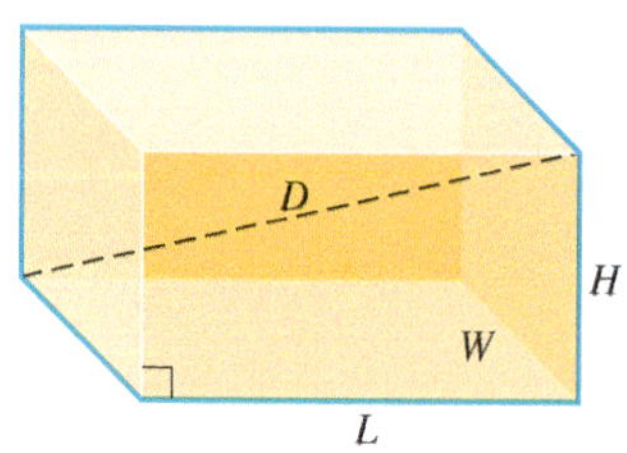

153. In the study of sound, one version of the law of tensions is

$$f_1 = f_2\sqrt{\frac{F_1}{F_2}}.$$

If $F_1 = 300$, $F_2 = 60$, and $f_2 = 260$, find f_1 to the nearest unit.

154. The illumination I, in foot-candles, produced by a light source is related to the distance d, in feet, from the light source by the equation

$$d = \sqrt{\frac{k}{I}},$$

where k is a constant. If $k = 640$, how far from the light source will the illumination be 2 foot-candles? Give the exact value, and then round to the nearest tenth of a foot.

7.4 Adding and Subtracting Radical Expressions

OBJECTIVE

1 Simplify radical expressions involving addition and subtraction.

TEACHING TIP
Explain how adding and subtracting "like radicals" is done in exactly the same way as adding and subtracting like terms. For example,

$$5\sqrt{6}+7\sqrt{6}=12\sqrt{6},$$

in the same way that

$$5x+7x=12x.$$

NOW TRY EXERCISE 1
Add or subtract to simplify each radical expression.

(a) $\sqrt{12}+\sqrt{75}$

(b) $-\sqrt{63t}+3\sqrt{28t}, \quad t \geq 0$

(c) $6\sqrt{7}-2\sqrt{3}$

CLASSROOM EXAMPLE 1
Add or subtract to simplify each radical expression.

(a) $3\sqrt{5}+7\sqrt{5}$

(b) $2\sqrt{11}-\sqrt{11}+3\sqrt{44}$

(c) $5\sqrt{12y}+6\sqrt{75y}, \quad y \geq 0$

(d) $9\sqrt{5}-4\sqrt{10}$

Answers: (a) $10\sqrt{5}$
(b) $7\sqrt{11}$ (c) $40\sqrt{3y}$
(d) The expression cannot be simplified further.

OBJECTIVE 1 Simplify radical expressions involving addition and subtraction.

Expressions such as $4\sqrt{2}+3\sqrt{2}$ and $2\sqrt{3}-5\sqrt{3}$ can be simplified using the distributive property.

$$4\sqrt{2}+3\sqrt{2}$$
$$=(4+3)\sqrt{2}=7\sqrt{2}$$

This is similar to simplifying $4x+3x$ to $7x$.

$$2\sqrt{3}-5\sqrt{3}$$
$$=(2-5)\sqrt{3}=-3\sqrt{3}$$

This is similar to simplifying $2x-5x$ to $-3x$.

CAUTION *Only radical expressions with the same index and the same radicand may be combined.*

EXAMPLE 1 Adding and Subtracting Radicals

Add or subtract to simplify each radical expression.

(a) $3\sqrt{24}+\sqrt{54}$ — Simplify each individual radical.

$=3\sqrt{4}\cdot\sqrt{6}+\sqrt{9}\cdot\sqrt{6}$ — Product rule

$=3\cdot 2\sqrt{6}+3\sqrt{6}$ — $\sqrt{4}=2$; $\sqrt{9}=3$

$=6\sqrt{6}+3\sqrt{6}$ — Multiply.

$=(6+3)\sqrt{6}$ — Distributive property

$=9\sqrt{6}$ — Add.

(b) $2\sqrt{20x}-\sqrt{45x}, \quad x \geq 0$

$=2\sqrt{4}\cdot\sqrt{5x}-\sqrt{9}\cdot\sqrt{5x}$ — Product rule

$=2\cdot 2\sqrt{5x}-3\sqrt{5x}$ — $\sqrt{4}=2$; $\sqrt{9}=3$

$=4\sqrt{5x}-3\sqrt{5x}$ — Multiply.

$=\sqrt{5x}$ — Combine like terms. ($(4-3)\sqrt{5x}=1\sqrt{5x}$, or $\sqrt{5x}$)

(c) $2\sqrt{3}-4\sqrt{5}$ The radicands differ and are already simplified, so $2\sqrt{3}-4\sqrt{5}$ cannot be simplified further.

NOW TRY

CAUTION *The root of a sum does not equal the sum of the roots.* For example,

$$\sqrt{9+16} \neq \sqrt{9}+\sqrt{16}$$

because $\sqrt{9+16}=\sqrt{25}=5$, but $\sqrt{9}+\sqrt{16}=3+4=7$.

NOW TRY ANSWERS

1. (a) $7\sqrt{3}$ (b) $3\sqrt{7t}$
(c) The expression cannot be simplified further.

NOW TRY EXERCISE 2

Add or subtract to simplify each radical expression. Assume that all variables represent positive real numbers.

(a) $3\sqrt[3]{2000} - 4\sqrt[3]{128}$

(b) $5\sqrt[4]{a^5b^3} + \sqrt[4]{81ab^7}$

(c) $\sqrt[3]{128t^4} - 2\sqrt{72t^3}$

CLASSROOM EXAMPLE 2

Add or subtract to simplify each radical expression. Assume that all variables represent positive real numbers.

(a) $-2\sqrt[4]{32} - 7\sqrt[4]{162}$

(b) $\sqrt[3]{p^4q^7} - \sqrt[3]{64pq}$

(c) $6\sqrt[3]{16z^7} + 4\sqrt{200z^5}$

Answers: (a) $-25\sqrt[4]{2}$

(b) $(pq^2 - 4)\sqrt[3]{pq}$

(c) $12z^2\sqrt[3]{2z} + 40z^2\sqrt{2z}$

EXAMPLE 2 Adding and Subtracting Radicals with Higher Indexes

Add or subtract to simplify each radical expression. Assume that all variables represent positive real numbers.

(a) $2\sqrt[3]{16} - 5\sqrt[3]{54}$ (Remember to write the index with each radical.)

$= 2\sqrt[3]{8 \cdot 2} - 5\sqrt[3]{27 \cdot 2}$ Factor.

$= 2\sqrt[3]{8} \cdot \sqrt[3]{2} - 5\sqrt[3]{27} \cdot \sqrt[3]{2}$ Product rule

$= 2 \cdot 2 \cdot \sqrt[3]{2} - 5 \cdot 3 \cdot \sqrt[3]{2}$ Find the cube roots.

$= 4\sqrt[3]{2} - 15\sqrt[3]{2}$ Multiply.

$= (4 - 15)\sqrt[3]{2}$ Distributive property

$= -11\sqrt[3]{2}$ Combine like terms.

(b) $2\sqrt[3]{x^2y} + \sqrt[3]{8x^5y^4}$

$= 2\sqrt[3]{x^2y} + \sqrt[3]{(8x^3y^3)x^2y}$ Factor.

$= 2\sqrt[3]{x^2y} + \sqrt[3]{8x^3y^3} \cdot \sqrt[3]{x^2y}$ Product rule

$= 2\sqrt[3]{x^2y} + 2xy\sqrt[3]{x^2y}$ Find the cube root.

$= (2 + 2xy)\sqrt[3]{x^2y}$ Distributive property (This result cannot be simplified further.)

(c) $5\sqrt{4x^3} + 3\sqrt[3]{64x^4}$ (Be careful. The indexes are different.)

$= 5\sqrt{4x^2 \cdot x} + 3\sqrt[3]{64x^3 \cdot x}$ Factor.

$= 5\sqrt{4x^2} \cdot \sqrt{x} + 3\sqrt[3]{64x^3} \cdot \sqrt[3]{x}$ Product rule

$= 5 \cdot 2x\sqrt{x} + 3 \cdot 4x\sqrt[3]{x}$ (Keep track of the indexes.)

$= 10x\sqrt{x} + 12x\sqrt[3]{x}$ These cannot be combined.

NOW TRY

CLASSROOM EXAMPLE 3

Perform the indicated operations. Assume that all variables represent positive real numbers.

(a) $2\sqrt{\frac{32}{36}} + 2\frac{\sqrt{27}}{\sqrt{108}}$

(b) $\sqrt{\frac{80}{y^4}} + \sqrt{\frac{81}{y^{10}}}$

Answers: (a) $\frac{4\sqrt{2}+3}{3}$

(b) $\frac{4y^3\sqrt{5}+9}{y^5}$

EXAMPLE 3 Adding and Subtracting Radicals with Fractions

Perform the indicated operations. Assume that all variables represent positive real numbers.

(a) $2\sqrt{\frac{75}{16}} + 4\frac{\sqrt{8}}{\sqrt{32}}$

$= 2\frac{\sqrt{25 \cdot 3}}{\sqrt{16}} + 4\frac{\sqrt{4 \cdot 2}}{\sqrt{16 \cdot 2}}$ Quotient rule; factor.

$= 2\left(\frac{5\sqrt{3}}{4}\right) + 4\left(\frac{2\sqrt{2}}{4\sqrt{2}}\right)$ Product rule; find the square roots.

$= \frac{5\sqrt{3}}{2} + 2$ Multiply; $\frac{\sqrt{2}}{\sqrt{2}} = 1$.

NOW TRY ANSWERS

2. (a) $14\sqrt[3]{2}$

(b) $(5a + 3b)\sqrt[4]{ab^3}$

(c) $4t\sqrt[3]{2t} - 12t\sqrt{2t}$

NOW TRY EXERCISE 3

Perform the indicated operations. Assume that all variables represent positive real numbers.

(a) $5\frac{\sqrt{5}}{\sqrt{45}} - 4\sqrt{\frac{28}{9}}$

(b) $6\sqrt[3]{\frac{16}{x^{12}}} + 7\sqrt[3]{\frac{9}{x^9}}$

$$= \frac{5\sqrt{3}}{2} + \frac{4}{2}$$ Write with a common denominator; $2 = \frac{4}{2}$

$$= \frac{5\sqrt{3}+4}{2}$$ $\frac{a}{c} + \frac{b}{c} = \frac{a+b}{c}$

(b) $$10\sqrt[3]{\frac{5}{x^6}} - 3\sqrt[3]{\frac{4}{x^9}}$$

$$= 10\frac{\sqrt[3]{5}}{\sqrt[3]{x^6}} - 3\frac{\sqrt[3]{4}}{\sqrt[3]{x^9}}$$ Quotient rule

$$= \frac{10\sqrt[3]{5}}{x^2} - \frac{3\sqrt[3]{4}}{x^3}$$ Simplify denominators.

$$= \frac{10\sqrt[3]{5}\cdot x}{x^2\cdot x} - \frac{3\sqrt[3]{4}}{x^3}$$ Write with a common denominator.

This equals x^3 so there is a common denominator.

$$= \frac{10x\sqrt[3]{5} - 3\sqrt[3]{4}}{x^3}$$ Subtract fractions.

NOW TRY

NOW TRY ANSWERS

3. (a) $\frac{5-8\sqrt{7}}{3}$

(b) $\frac{12\sqrt[3]{2}+7x\sqrt[3]{9}}{x^4}$

7.4 Exercises

FOR EXTRA HELP

MyMathLab®

Complete solution available in MyMathLab

Concept Check *Choose the correct response.*

1. Which sum can be simplified without first simplifying the individual radical expressions?

A. $\sqrt{50} + \sqrt{32}$ **B.** $3\sqrt{6} + 9\sqrt{6}$

C. $\sqrt[3]{32} - \sqrt[3]{108}$ **D.** $\sqrt[5]{6} - \sqrt[5]{192}$

2. Which difference can be simplified without first simplifying the individual radical expressions?

A. $\sqrt{81} - \sqrt{18}$ **B.** $\sqrt[3]{8} - \sqrt[3]{16}$

C. $4\sqrt[3]{7} - 9\sqrt[3]{7}$ **D.** $\sqrt{75} - \sqrt{12}$

3. *Concept Check* Even though the indexes of the terms are not equal, the sum

$$\sqrt{64} + \sqrt[3]{125} + \sqrt[4]{16}$$

can be simplified easily. What is this sum? Why can these terms be easily combined?

4. *Concept Check* On an algebra quiz, Erin gave the difference $28 - 4\sqrt{2}$ as $24\sqrt{2}$. Her teacher did not give her any credit for this answer. **WHAT WENT WRONG?**

Add or subtract to simplify each radical expression. Assume that all variables represent positive real numbers. ***See Examples 1 and 2.***

5. $\sqrt{36} - \sqrt{100}$ **6.** $\sqrt{25} - \sqrt{81}$ **7.** $-2\sqrt{48} + 3\sqrt{75}$

8. $4\sqrt{32} - 2\sqrt{8}$ **9.** $\sqrt[3]{16} + 4\sqrt[3]{54}$ **10.** $3\sqrt[3]{24} - 2\sqrt[3]{192}$

1. B 2. C

3. 15; Each radical expression simplifies to a whole number.

4. We cannot group 28 − 4 here. The terms 28 and $4\sqrt{2}$ are not like terms and cannot be combined. The difference cannot be combined.

5. −4 6. −4

7. $7\sqrt{3}$ 8. $12\sqrt{2}$

9. $14\sqrt[3]{2}$ 10. $-2\sqrt[3]{3}$

11. $5\sqrt[4]{2}$ **12.** $\sqrt[4]{5}$
13. $24\sqrt{2}$ **14.** $13\sqrt{2}$
15. The expression cannot be simplified further.
16. The expression cannot be simplified further.
17. $20\sqrt{5}$ **18.** $3\sqrt{6}$
19. $4\sqrt{2x}$ **20.** $-3\sqrt{2k}$
21. $-11m\sqrt{2}$ **22.** $-49p\sqrt{3}$
23. $7\sqrt[3]{2}$ **24.** $53\sqrt[3]{3}$
25. $2\sqrt[3]{x}$ **26.** $18\sqrt[3]{2m}$
27. $-7\sqrt[3]{x^2y}$ **28.** $-5\sqrt[3]{x^2y^2}$
29. $-x\sqrt[3]{xy^2}$ **30.** $2q^2\sqrt[3]{5q}$
31. $19\sqrt[4]{2}$ **32.** $16\sqrt[4]{2}$
33. $x\sqrt[4]{xy}$ **34.** $-m^2p\sqrt[4]{mp^2}$
35. $9\sqrt[4]{2a^3}$ **36.** $17\sqrt[4]{3x^3}$
37. $(4+3xy)\sqrt[3]{xy^2}$
38. $(5+3st)\sqrt[4]{s^3t}$
39. $4t\sqrt[3]{3st}-3s\sqrt{3st}$
40. $5a^2b^2\sqrt{5ab}+5ab\sqrt[3]{ab}$
41. $4x\sqrt[3]{x}+6x\sqrt[4]{x}$
42. $12m\sqrt[3]{m}+15m\sqrt[4]{m}$

43. $2\sqrt{2}-2$ **44.** $4\sqrt{3}-3$
45. $\frac{5\sqrt{5}}{6}$ **46.** $\frac{14\sqrt{3}}{9}$
47. $\frac{7\sqrt{2}}{6}$ **48.** $\sqrt{3}$
49. $\frac{5\sqrt{2}}{3}$ **50.** $\frac{\sqrt{3}}{2}$
51. $5\sqrt{2}+4$ **52.** $12\sqrt{2}+7$
53. $\frac{5+3x}{x^4}$ **54.** $\frac{10y^3+9}{y^5}$
55. $\frac{m\sqrt[3]{m^2}}{2}$ **56.** $-\frac{2a\sqrt[4]{a}}{3}$
57. $\frac{3x\sqrt[3]{2}-4\sqrt[3]{5}}{x^3}$
58. $\frac{-4t\sqrt[3]{4}+3\sqrt[3]{9}}{t^4}$

11. $\sqrt[4]{32}+3\sqrt[4]{2}$ **12.** $\sqrt[4]{405}-2\sqrt[4]{5}$

13. $6\sqrt{18}-\sqrt{32}+2\sqrt{50}$ **14.** $5\sqrt{8}+3\sqrt{72}-3\sqrt{50}$

15. $5\sqrt{6}+2\sqrt{10}$ **16.** $3\sqrt{11}-5\sqrt{13}$

17. $2\sqrt{5}+3\sqrt{20}+4\sqrt{45}$ **18.** $5\sqrt{54}-2\sqrt{24}-2\sqrt{96}$

19. $\sqrt{72x}-\sqrt{8x}$ **20.** $\sqrt{18k}-\sqrt{72k}$

21. $3\sqrt{72m^2}-5\sqrt{32m^2}-3\sqrt{18m^2}$ **22.** $9\sqrt{27p^2}-14\sqrt{108p^2}+2\sqrt{48p^2}$

23. $2\sqrt[3]{16}+\sqrt[3]{54}$ **24.** $15\sqrt[3]{81}+4\sqrt[3]{24}$

25. $2\sqrt[3]{27x}-2\sqrt[3]{8x}$ **26.** $6\sqrt[3]{128m}-3\sqrt[3]{16m}$

27. $3\sqrt[3]{x^2y}-5\sqrt[3]{8x^2y}$ **28.** $3\sqrt[3]{x^2y^2}-2\sqrt[3]{64x^2y^2}$

29. $3x\sqrt[3]{xy^2}-2\sqrt[3]{8x^4y^2}$ **30.** $6q^2\sqrt[3]{5q}-2q\sqrt[3]{40q^4}$

31. $5\sqrt[4]{32}+3\sqrt[4]{162}$ **32.** $2\sqrt[4]{512}+4\sqrt[4]{32}$

33. $3\sqrt[4]{x^5y}-2x\sqrt[4]{xy}$ **34.** $2\sqrt[4]{m^9p^6}-3m^2p\sqrt[4]{mp^2}$

35. $2\sqrt[4]{32a^3}+5\sqrt[4]{2a^3}$ **36.** $5\sqrt[4]{243x^3}+2\sqrt[4]{3x^3}$

37. $\sqrt[3]{64xy^2}+\sqrt[3]{27x^4y^5}$ **38.** $\sqrt[4]{625s^3t}+\sqrt[4]{81s^7t^5}$

39. $\sqrt[3]{192st^4}-\sqrt{27s^3t}$ **40.** $\sqrt{125a^5b^5}+\sqrt[3]{125a^4b^4}$

41. $2\sqrt[3]{8x^4}+3\sqrt[4]{16x^5}$ **42.** $3\sqrt[3]{64m^4}+5\sqrt[4]{81m^5}$

Perform the indicated operations. Assume that all variables represent positive real numbers. ***See Example 3.***

43. $\sqrt{8}-\frac{\sqrt{64}}{\sqrt{16}}$ **44.** $\sqrt{48}-\frac{\sqrt{81}}{\sqrt{9}}$ **45.** $\frac{2\sqrt{5}}{3}+\frac{\sqrt{5}}{6}$

46. $\frac{4\sqrt{3}}{3}+\frac{2\sqrt{3}}{9}$ **47.** $\sqrt{\frac{8}{9}}+\sqrt{\frac{18}{36}}$ **48.** $\sqrt{\frac{12}{16}}+\sqrt{\frac{48}{64}}$

49. $\frac{\sqrt{32}}{3}+\frac{2\sqrt{2}}{3}-\frac{\sqrt{2}}{\sqrt{9}}$ **50.** $\frac{\sqrt{27}}{2}-\frac{3\sqrt{3}}{2}+\frac{\sqrt{3}}{\sqrt{4}}$ **51.** $3\sqrt{\frac{50}{9}}+8\frac{\sqrt{2}}{\sqrt{8}}$

52. $5\sqrt{\frac{288}{25}}+21\frac{\sqrt{2}}{\sqrt{18}}$ **53.** $\sqrt{\frac{25}{x^8}}+\sqrt{\frac{9}{x^6}}$ **54.** $\sqrt{\frac{100}{y^4}}+\sqrt{\frac{81}{y^{10}}}$

55. $3\sqrt[3]{\frac{m^5}{27}}-2m\sqrt[3]{\frac{m^2}{64}}$ **56.** $2a\sqrt[4]{\frac{a}{16}}-5a\sqrt[4]{\frac{a}{81}}$

57. $3\sqrt[3]{\frac{2}{x^6}}-4\sqrt[3]{\frac{5}{x^9}}$ **58.** $-4\sqrt[3]{\frac{4}{t^9}}+3\sqrt[3]{\frac{9}{t^{12}}}$

59. **(a)** $\sqrt{7}$
(b) 2.645751311
(c) 2.645751311
(d) equal

60. **(a)** 9; $\sqrt{65}$; no
(b) 5; $\sqrt[3]{65}$; no
(c) $\sqrt[n]{a+b}$

61. A; 42 m
62. C

63. $(12\sqrt{5}+5\sqrt{3})$ in.
64. $24\sqrt{3}$ m
65. $(24\sqrt{2}+12\sqrt{3})$ in.
66. $36\sqrt{3}$ in.^{2}

59. *Concept Check* Consider the expression

$$\sqrt{63}+\sqrt{112}-\sqrt{252}.$$

(a) Simplify this expression using the methods of this section.

(b) Use a calculator to approximate the given expression.

(c) Use a calculator to approximate the simplified expression in part (a).

(d) Complete the following: Assuming the work in part (a) is correct, the approximations in parts (b) and (c) should be (*equal*/*unequal*).

60. *Concept Check* Let $a=1$ and let $b=64$.

(a) Evaluate $\sqrt{a}+\sqrt{b}$. Then find $\sqrt{a+b}$. Are they equal?

(b) Evaluate $\sqrt[3]{a}+\sqrt[3]{b}$. Then find $\sqrt[3]{a+b}$. Are they equal?

(c) Complete the following: In general,

$$\sqrt[n]{a}+\sqrt[n]{b}\neq \underline{\hspace{6cm}},$$

based on the observations in parts (a) and (b) of this exercise.

Solve each problem.

61. A rectangular yard has a length of $\sqrt{192}$ m and a width of $\sqrt{48}$ m. Choose the best estimate of its dimensions. Then estimate the perimeter.

A. 14 m by 7 m **B.** 5 m by 7 m **C.** 14 m by 8 m **D.** 15 m by 8 m

62. If the sides of a triangle are $\sqrt{65}$ in., $\sqrt{35}$ in., and $\sqrt{26}$ in., which one of the following is the best estimate of its perimeter?

A. 20 in. **B.** 26 in. **C.** 19 in. **D.** 24 in.

Solve each problem. Give answers as simplified radical expressions.

63. Find the perimeter of the triangle.

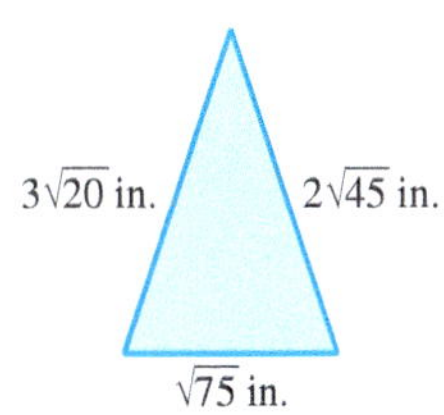

64. Find the perimeter of the rectangle.

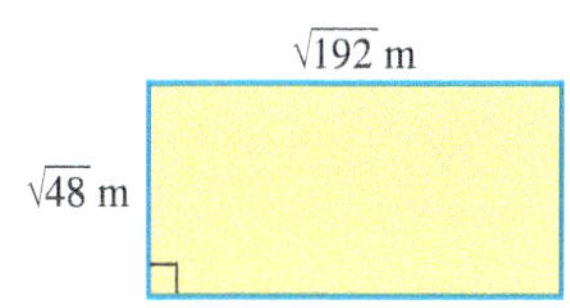

65. What is the perimeter of the computer graphic?

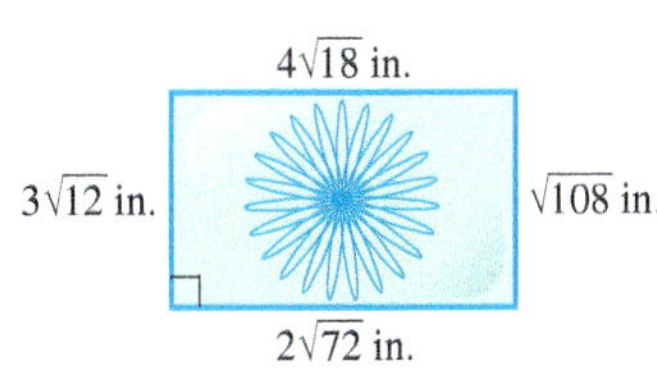

66. Find the area of the trapezoid.

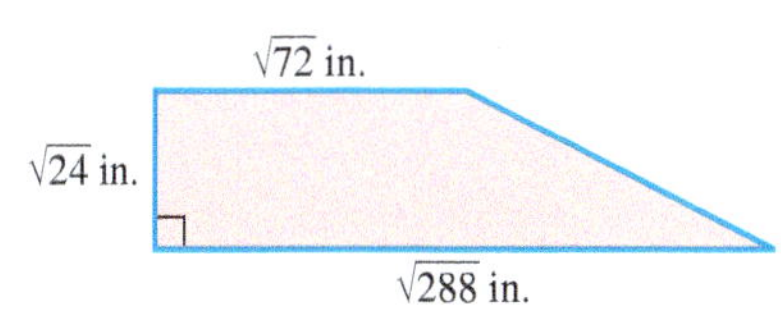

7.5 Multiplying and Dividing Radical Expressions

OBJECTIVES

1. Multiply radical expressions.
2. Rationalize denominators with one radical term.
3. Rationalize denominators with binomials involving radicals.
4. Write radical quotients in lowest terms.

VOCABULARY

☐ conjugates

NOW TRY EXERCISE 1

Multiply.

(a) $\sqrt{10}(4+\sqrt{7})$

(b) $3(\sqrt{20}-\sqrt{45})$

CLASSROOM EXAMPLE 1

Multiply.

(a) $\sqrt{11}(5-\sqrt{2})$

(b) $7(\sqrt{54}-\sqrt{24})$

Answers: (a) $5\sqrt{11}-\sqrt{22}$

(b) $7\sqrt{6}$

TEACHING TIP

Review the FOIL method with polynomials from **Section 4.4** using these examples.

$(5x+3)(6x+1)$

$(x+3y)(x-3y)$

$(7x-3)^2$

Review: $\sqrt{5}\cdot\sqrt{5}=5$

$\sqrt{5}\cdot\sqrt{6}=\sqrt{30}$

$2\sqrt{3}\cdot 4\sqrt{2}=8\sqrt{6}$

NOW TRY ANSWERS

1. (a) $4\sqrt{10}+\sqrt{70}$

(b) $-3\sqrt{5}$

OBJECTIVE 1 Multiply radical expressions.

The distributive property may be used when multiplying radical expressions.

EXAMPLE 1 Using the Distributive Property with Radicals

Multiply.

(a) $\sqrt{5}(2+\sqrt{6})$

$= \sqrt{5}\cdot 2+\sqrt{5}\cdot\sqrt{6}$ Distributive property: $a(b+c)=ab+ac$

$= 2\sqrt{5}+\sqrt{30}$ Commutative property; product rule

(b) $4(\sqrt{12}-\sqrt{27})$

$= 4\sqrt{12}-4\sqrt{27}$ Distributive property

$= 4\sqrt{4\cdot 3}-4\sqrt{9\cdot 3}$ Factor the radicands so that one factor is a perfect square.

$= 4\cdot 2\sqrt{3}-4\cdot 3\sqrt{3}$ $\sqrt{4}=2$; $\sqrt{9}=3$

$= 8\sqrt{3}-12\sqrt{3}$ Multiply.

$= -4\sqrt{3}$ $8\sqrt{3}-12\sqrt{3}=(8-12)\sqrt{3}$

NOW TRY

We multiply binomial expressions involving radicals using the FOIL method from **Section 4.4.** Recall that the acronym **FOIL** refers to the positions of the terms. We multiply the **F**irst terms, **O**uter terms, **I**nner terms, and **L**ast terms of the binomials.

EXAMPLE 2 Multiplying Binomials Involving Radical Expressions

Multiply, using the FOIL method.

(a) $(\sqrt{5}+3)(\sqrt{6}+1)$

First Outer Inner Last

$= \sqrt{5}\cdot\sqrt{6}+\sqrt{5}\cdot 1+3\cdot\sqrt{6}+3\cdot 1$

$= \sqrt{30}+\sqrt{5}+3\sqrt{6}+3$ This result cannot be simplified further.

(b) $(7-\sqrt{3})(\sqrt{5}+\sqrt{2})$

F O I L

$= 7\sqrt{5}+7\sqrt{2}-\sqrt{3}\cdot\sqrt{5}-\sqrt{3}\cdot\sqrt{2}$

$= 7\sqrt{5}+7\sqrt{2}-\sqrt{15}-\sqrt{6}$ Product rule

(c) $(\sqrt{10}+\sqrt{3})(\sqrt{10}-\sqrt{3})$

$= \sqrt{10}\cdot\sqrt{10}-\sqrt{10}\cdot\sqrt{3}+\sqrt{10}\cdot\sqrt{3}-\sqrt{3}\cdot\sqrt{3}$ FOIL method

$= 10-3$ Product rule; $-\sqrt{30}+\sqrt{30}=0$

$= 7$ Subtract.

NOW TRY EXERCISE 2

Multiply, using the FOIL method.

(a) $(8-\sqrt{5})(9-\sqrt{2})$

(b) $(\sqrt{7}+\sqrt{5})(\sqrt{7}-\sqrt{5})$

(c) $(\sqrt{15}-4)^2$

(d) $(8+\sqrt[3]{5})(8-\sqrt[3]{5})$

(e) $(\sqrt{m}-\sqrt{n})(\sqrt{m}+\sqrt{n})$ $(m \geq 0 \text{ and } n \geq 0)$

TEACHING TIP

Refer to **Example 2(d)** and explain why

$16-6\sqrt{7} \neq 10\sqrt{7}.$

CLASSROOM EXAMPLE 2

Multiply, using the FOIL method.

(a) $(2+\sqrt{3})(1+\sqrt{5})$

(b) $(4+\sqrt{5})(4-\sqrt{5})$

(c) $(\sqrt{13}-2)^2$

(d) $(4+\sqrt[3]{7})(4-\sqrt[3]{7})$

(e) $(\sqrt{r}+\sqrt{s})(\sqrt{r}-\sqrt{s})$ $(r \geq 0 \text{ and } s \geq 0)$

Answers:

(a) $2+2\sqrt{5}+\sqrt{3}+\sqrt{15}$

(b) 11 (c) $17-4\sqrt{13}$

(d) $16-\sqrt[3]{49}$ (e) $r-s$

NOW TRY ANSWERS

2. (a) $72-8\sqrt{2}-9\sqrt{5}+\sqrt{10}$

(b) 2 (c) $31-8\sqrt{15}$

(d) $64-\sqrt[3]{25}$ (e) $m-n$

The product $(\sqrt{10}+\sqrt{3})(\sqrt{10}-\sqrt{3})=(\sqrt{10})^2-(\sqrt{3})^2$ in part (c) is a difference of squares.

$$(x+y)(x-y)=x^2-y^2 \qquad \text{Here, } x=\sqrt{10} \text{ and } y=\sqrt{3}.$$

(d) $(\sqrt{7}-3)^2$

$=(\sqrt{7}-3)(\sqrt{7}-3)$ — $a^2=a \cdot a$

$=\sqrt{7}\cdot\sqrt{7}-3\sqrt{7}-3\sqrt{7}+3\cdot 3$ — FOIL method

$=7-6\sqrt{7}+9$ — Multiply. Combine like terms.

$=16-6\sqrt{7}$ — Add.

Be careful. These terms cannot be combined.

(e) $(5-\sqrt[3]{3})(5+\sqrt[3]{3})$

Remember to write the index 3 in *each* radical.

$=5\cdot 5+5\sqrt[3]{3}-5\sqrt[3]{3}-\sqrt[3]{3}\cdot\sqrt[3]{3}$ — FOIL method

$=25-\sqrt[3]{3^2}$ — Multiply. Combine like terms.

$=25-\sqrt[3]{9}$ — Apply the exponent.

(f) $(\sqrt{k}+\sqrt{y})(\sqrt{k}-\sqrt{y})$

$=(\sqrt{k})^2-(\sqrt{y})^2$ — Difference of squares

$=k-y \quad (k \geq 0 \text{ and } y \geq 0)$

NOW TRY

NOTE In **Example 2(d),** we could have used the formula for the square of a binomial to obtain the same result.

$(\sqrt{7}-3)^2$

$=(\sqrt{7})^2-2(\sqrt{7})(3)+3^2$ — $(x-y)^2=x^2-2xy+y^2$

$=7-6\sqrt{7}+9$ — Apply the exponents. Multiply.

$=16-6\sqrt{7}$ — Add.

OBJECTIVE 2 Rationalize denominators with one radical term.

A simplified radical expression has no radical in the denominator. The origin of this agreement no doubt occurred before the days of high-speed calculation, when computation was a tedious process performed by hand.

Consider the expression $\frac{1}{\sqrt{2}}$. To find a decimal approximation by hand, it is necessary to divide 1 by a decimal approximation for $\sqrt{2}$, such as 1.414. It is much easier if the divisor is a whole number. This can be accomplished by multiplying $\frac{1}{\sqrt{2}}$ by 1 in the form $\frac{\sqrt{2}}{\sqrt{2}}$. ***Multiplying by 1 in any form does not change the value of the original expression.***

$$\frac{1}{\sqrt{2}}\cdot\frac{\sqrt{2}}{\sqrt{2}}=\frac{\sqrt{2}}{2} \qquad \text{Multiply by 1; } \frac{\sqrt{2}}{\sqrt{2}}=1$$

Now the computation requires dividing 1.414 by 2 to obtain 0.707, which is easier.

With current technology, either form $\frac{1}{\sqrt{2}}$ or $\frac{\sqrt{2}}{2}$ can be approximated with the same number of keystrokes. See **FIGURE 12**, which shows how a calculator gives the same approximation for both forms of the expression.

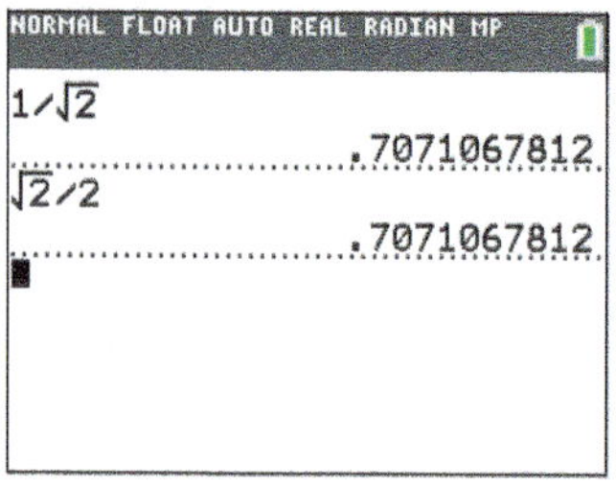

FIGURE 12

CLASSROOM EXAMPLE 3

Rationalize each denominator.

(a) $\frac{5}{\sqrt{11}}$ (b) $\frac{5\sqrt{6}}{\sqrt{5}}$

(c) $\frac{-8}{\sqrt{18}}$

Answers: (a) $\frac{5\sqrt{11}}{11}$ (b) $\sqrt{30}$

(c) $\frac{-4\sqrt{2}}{3}$

Rationalizing the Denominator

A common way of "standardizing" the form of a radical expression is to have the denominator contain no radicals. The process of removing radicals from a denominator so that the denominator contains only rational numbers is called **rationalizing the denominator.** This is done by multiplying by a form of 1.

NOW TRY EXERCISE 3

Rationalize each denominator.

(a) $\frac{8}{\sqrt{13}}$ **(b)** $\frac{9\sqrt{7}}{\sqrt{3}}$

(c) $\frac{-10}{\sqrt{20}}$

EXAMPLE 3 Rationalizing Denominators with Square Roots

Rationalize each denominator.

(a) $\frac{3}{\sqrt{7}}$

Multiply the numerator and denominator by $\sqrt{7}$. This is, in effect, multiplying by 1.

$$\frac{3}{\sqrt{7}} = \frac{3 \cdot \sqrt{7}}{\sqrt{7} \cdot \sqrt{7}} = \frac{3\sqrt{7}}{7}$$

In the denominator, $\sqrt{7} \cdot \sqrt{7} = \sqrt{7 \cdot 7} = \sqrt{49} = 7$. The final denominator is now a rational number.

TEACHING TIP

Show that rationalizing directly and simplifying the denominator first yield the same result. Explain that radicals can often be simplified in several equally correct ways.

(b) $\frac{5\sqrt{2}}{\sqrt{5}} = \frac{5\sqrt{2} \cdot \sqrt{5}}{\sqrt{5} \cdot \sqrt{5}} = \frac{5\sqrt{10}}{5} = \sqrt{10}$

(c) $\frac{-6}{\sqrt{12}}$

Less work is involved if the radical in the denominator is simplified first.

$$\frac{-6}{\sqrt{12}} = \frac{-6}{\sqrt{4 \cdot 3}} = \frac{-6}{2\sqrt{3}} = \frac{-3}{\sqrt{3}}$$

Now we rationalize the denominator.

$$\frac{-3}{\sqrt{3}} = \frac{-3 \cdot \sqrt{3}}{\sqrt{3} \cdot \sqrt{3}} = \frac{-3\sqrt{3}}{3} = -\sqrt{3}$$

NOW TRY

NOW TRY ANSWERS

3. (a) $\frac{8\sqrt{13}}{13}$ **(b)** $3\sqrt{21}$

(c) $-\sqrt{5}$

NOW TRY EXERCISE 4
Simplify each radical. In part (b), $y > 0$.

(a) $-\sqrt{\dfrac{27}{80}}$ (b) $\sqrt{\dfrac{48x^8}{y^3}}$

CLASSROOM EXAMPLE 4
Simplify each radical. In part (b), $y > 0$.

(a) $-\sqrt{\dfrac{8}{45}}$ (b) $\sqrt{\dfrac{200k^6}{y^7}}$

Answers: (a) $-\dfrac{2\sqrt{10}}{15}$

(b) $\dfrac{10k^3\sqrt{2y}}{y^4}$

TEACHING TIP
Show in **Example 4(b)** that, alternatively, multiplying directly by

$$\frac{\sqrt{p^5}}{\sqrt{p^5}}$$

will yield the same result after much simplification.

CLASSROOM EXAMPLE 5
Simplify.

(a) $\sqrt[3]{\dfrac{15}{32}}$

(b) $\sqrt[4]{\dfrac{6y}{w^2}}$ $(y \geq 0,\ w > 0)$

Answers: (a) $\dfrac{\sqrt[3]{30}}{4}$

(b) $\dfrac{\sqrt[4]{6yw^2}}{w}$

EXAMPLE 4 Rationalizing Denominators in Roots of Fractions

Simplify each radical. In part (b), $p > 0$.

(a) $-\sqrt{\dfrac{18}{125}}$

$= -\dfrac{\sqrt{18}}{\sqrt{125}}$ Quotient rule

$= -\dfrac{\sqrt{9 \cdot 2}}{\sqrt{25 \cdot 5}}$ Factor.

$= -\dfrac{3\sqrt{2}}{5\sqrt{5}}$ Product rule

$= -\dfrac{3\sqrt{2} \cdot \sqrt{5}}{5\sqrt{5} \cdot \sqrt{5}}$ Multiply by $\dfrac{\sqrt{5}}{\sqrt{5}}$.

$= -\dfrac{3\sqrt{10}}{5 \cdot 5}$ Product rule

$= -\dfrac{3\sqrt{10}}{25}$ Multiply.

(b) $\sqrt{\dfrac{50m^4}{p^5}}$

$= \dfrac{\sqrt{50m^4}}{\sqrt{p^5}}$ Quotient rule

$= \dfrac{\sqrt{25m^4 \cdot 2}}{\sqrt{p^4 \cdot p}}$ Factor.

$= \dfrac{5m^2\sqrt{2}}{p^2\sqrt{p}}$ Product rule

$= \dfrac{5m^2\sqrt{2} \cdot \sqrt{p}}{p^2\sqrt{p} \cdot \sqrt{p}}$ Multiply by $\dfrac{\sqrt{p}}{\sqrt{p}}$.

$= \dfrac{5m^2\sqrt{2p}}{p^2 \cdot p}$ Product rule

$= \dfrac{5m^2\sqrt{2p}}{p^3}$ Multiply.

NOW TRY

EXAMPLE 5 Rationalizing Denominators with Higher Roots

Simplify.

(a) $\sqrt[3]{\dfrac{27}{16}}$

Use the quotient rule, and simplify the numerator and denominator.

$$\sqrt[3]{\frac{27}{16}} = \frac{\sqrt[3]{27}}{\sqrt[3]{16}} = \frac{3}{\sqrt[3]{8} \cdot \sqrt[3]{2}} = \frac{3}{2\sqrt[3]{2}}$$

Because $2 \cdot 4 = 8$ is a perfect cube, multiply the numerator and denominator by $\sqrt[3]{4}$.

$\dfrac{3}{2\sqrt[3]{2}}$ $\sqrt[3]{\dfrac{27}{16}} = \dfrac{3}{2\sqrt[3]{2}}$ from above

$= \dfrac{3 \cdot \sqrt[3]{4}}{2\sqrt[3]{2} \cdot \sqrt[3]{4}}$ Multiply by $\sqrt[3]{4}$ in numerator and denominator. This will give $\sqrt[3]{8} = 2$ in the denominator.

$= \dfrac{3\sqrt[3]{4}}{2\sqrt[3]{8}}$ Multiply.

$= \dfrac{3\sqrt[3]{4}}{2 \cdot 2}$ $\sqrt[3]{8} = 2$

$= \dfrac{3\sqrt[3]{4}}{4}$ Multiply.

NOW TRY ANSWERS

4. (a) $-\dfrac{3\sqrt{15}}{20}$ (b) $\dfrac{4x^4\sqrt{3y}}{y^2}$

NOW TRY EXERCISE 5

Simplify.

(a) $\sqrt[3]{\frac{8}{81}}$

(b) $\sqrt[4]{\frac{7x}{y}}$ $(x \geq 0, y > 0)$

(b) $\sqrt[4]{\frac{5x}{z}}$

$= \frac{\sqrt[4]{5x}}{\sqrt[4]{z}}$ Quotient rule

$= \frac{\sqrt[4]{5x}}{\sqrt[4]{z}} \cdot \frac{\sqrt[4]{z^3}}{\sqrt[4]{z^3}}$ Multiply by 1.

$\sqrt[4]{z} \cdot \sqrt[4]{z^3}$ will give $\sqrt[4]{z^4}$.

$= \frac{\sqrt[4]{5xz^3}}{\sqrt[4]{z^4}}$ Product rule

$= \frac{\sqrt[4]{5xz^3}}{z}$ $(x \geq 0, z > 0)$

NOW TRY

CAUTION In **Example 5(a),** a typical error is to multiply the numerator and denominator by $\sqrt[3]{2}$, forgetting that $\sqrt[3]{2} \cdot \sqrt[3]{2} = \sqrt[3]{2^2}$, which does ***not*** equal 2. We need ***three*** factors of 2 to obtain 2^3 under the radical.

$$\sqrt[3]{2} \cdot \sqrt[3]{2} \cdot \sqrt[3]{2} = \sqrt[3]{2^3} \quad \text{which does equal} \quad 2.$$

TEACHING TIP

In **Example 6(a),** show how trying to rationalize with $\sqrt{2}$ or $1 + \sqrt{2}$ will not eliminate radicals in the denominator.

OBJECTIVE 3 Rationalize denominators with binomials involving radicals.

Recall the special product

$$(x + y)(x - y) = x^2 - y^2.$$

To rationalize a denominator that contains a binomial expression (one that contains exactly two terms) involving radicals, such as

$$\frac{3}{1 + \sqrt{2}},$$

we must use *conjugates.* The conjugate of $1 + \sqrt{2}$ is $1 - \sqrt{2}$. In general,

$$x + y \text{ and } x - y \text{ are } \textbf{conjugates.}$$

Specifically, if a and b represent nonnegative rational numbers, the product

$$(\sqrt{a} + \sqrt{b})(\sqrt{a} - \sqrt{b}) \quad \text{produces the rational number} \quad a - b.$$

Rationalizing a Binomial Denominator

Whenever a radical expression has a sum or difference with square root radicals in the denominator, rationalize the denominator by multiplying both the numerator and denominator by the conjugate of the denominator.

NOW TRY ANSWERS

5. (a) $\frac{2\sqrt[3]{9}}{9}$ (b) $\frac{\sqrt[4]{7xy^3}}{y}$

NOW TRY EXERCISE 6

Rationalize each denominator.

(a) $\frac{4}{1+\sqrt{3}}$ (b) $\frac{4}{5+\sqrt{7}}$

(c) $\frac{\sqrt{3}+\sqrt{7}}{\sqrt{5}-\sqrt{2}}$

(d) $\frac{8}{\sqrt{3x}-\sqrt{y}}$ $(3x \neq y, x > 0, y > 0)$

CLASSROOM EXAMPLE 6

Rationalize each denominator.

(a) $\frac{4}{2-\sqrt{3}}$

(b) $\frac{7}{\sqrt{2}+\sqrt{13}}$

(c) $\frac{\sqrt{3}+\sqrt{5}}{\sqrt{2}-\sqrt{7}}$

(d) $\frac{2}{\sqrt{k}+\sqrt{z}}$ $(k \neq z, k > 0, z > 0)$

Answers:

(a) $8+4\sqrt{3}$

(b) $\frac{-7(\sqrt{2}-\sqrt{13})}{11}$

(c) $\frac{-(\sqrt{6}+\sqrt{21}+\sqrt{10}+\sqrt{35})}{5}$

(d) $\frac{2(\sqrt{k}-\sqrt{z})}{k-z}$

TEACHING TIP

Many students feel uncomfortable about answers that are not "simple numbers" like 2, −6, $\frac{1}{2}$, and so on. Warn them that many of the answers in the examples and exercises in this section are messier than usual.

NOW TRY ANSWERS

6. (a) $-2+2\sqrt{3}$

(b) $\frac{2(5-\sqrt{7})}{9}$

(c) $\frac{\sqrt{15}+\sqrt{6}+\sqrt{35}+\sqrt{14}}{3}$

(d) $\frac{8(\sqrt{3x}+\sqrt{y})}{3x-y}$

EXAMPLE 6 Rationalizing Binomial Denominators

Rationalize each denominator.

(a) $\frac{3}{1+\sqrt{2}}$

Again, we are multiplying by a form of 1.

$= \frac{3(1-\sqrt{2})}{(1+\sqrt{2})(1-\sqrt{2})}$ Multiply the numerator and denominator by $1-\sqrt{2}$, the conjugate of the denominator.

$= \frac{3(1-\sqrt{2})}{-1}$ The denominator is now a rational number.

$(1+\sqrt{2})(1-\sqrt{2})$
$= 1^2 - (\sqrt{2})^2$
$= 1 - 2$, or -1

$= \frac{3}{-1}(1-\sqrt{2})$ $\frac{a \cdot b}{c} = \frac{a}{c} \cdot b$

$= -3(1-\sqrt{2})$ Simplify.

$= -3+3\sqrt{2}$ Distributive property

Either of these forms is correct.

(b) $\frac{5}{4-\sqrt{3}}$

$= \frac{5(4+\sqrt{3})}{(4-\sqrt{3})(4+\sqrt{3})}$ Multiply the numerator and denominator by $4+\sqrt{3}$.

$= \frac{5(4+\sqrt{3})}{16-3}$ Multiply in the denominator.

$= \frac{5(4+\sqrt{3})}{13}$ Subtract in the denominator.

Notice that we leave the numerator in factored form. This makes it easier to determine whether the expression is written in lowest terms.

(c) $\frac{\sqrt{2}-\sqrt{3}}{\sqrt{5}+\sqrt{3}}$

$= \frac{(\sqrt{2}-\sqrt{3})(\sqrt{5}-\sqrt{3})}{(\sqrt{5}+\sqrt{3})(\sqrt{5}-\sqrt{3})}$ Multiply the numerator and denominator by $\sqrt{5}-\sqrt{3}$.

$= \frac{\sqrt{10}-\sqrt{6}-\sqrt{15}+3}{5-3}$ Multiply.

$= \frac{\sqrt{10}-\sqrt{6}-\sqrt{15}+3}{2}$ Subtract in the denominator.

(d) $\frac{3}{\sqrt{5m}-\sqrt{p}}$ $(5m \neq p, m > 0, p > 0)$

$= \frac{3(\sqrt{5m}+\sqrt{p})}{(\sqrt{5m}-\sqrt{p})(\sqrt{5m}+\sqrt{p})}$ Multiply the numerator and denominator by $\sqrt{5m}+\sqrt{p}$.

$= \frac{3(\sqrt{5m}+\sqrt{p})}{5m-p}$ Multiply in the denominator. NOW TRY

OBJECTIVE 4 Write radical quotients in lowest terms.

NOW TRY EXERCISE 7

Write each quotient in lowest terms.

(a) $\dfrac{15 - 6\sqrt{2}}{18}$

(b) $\dfrac{15k + \sqrt{50k^2}}{20k} \quad (k > 0)$

TEACHING TIP

Expressions like the one in **Example 7(a)** occur when the quadratic formula is used in the next chapter. Errors often happen when students try to write in lowest terms before factoring. Warn them here (and whenever the opportunity arises) that we simplify to lowest terms *only* after the numerator and denominator have been factored.

CLASSROOM EXAMPLE 7

Write each quotient in lowest terms.

(a) $\dfrac{24 - 36\sqrt{7}}{16}$

(b) $\dfrac{2x + \sqrt{32x^2}}{6x} \quad (x > 0)$

Answers: (a) $\dfrac{6 - 9\sqrt{7}}{4}$

(b) $\dfrac{1 + 2\sqrt{2}}{3}$

NOW TRY ANSWERS

7. (a) $\dfrac{5 - 2\sqrt{2}}{6}$ **(b)** $\dfrac{3 + \sqrt{2}}{4}$

EXAMPLE 7 Writing Radical Quotients in Lowest Terms

Write each quotient in lowest terms.

(a) $\dfrac{6 + 2\sqrt{5}}{4}$

$= \dfrac{2(3 + \sqrt{5})}{2 \cdot 2}$ Factor the numerator and denominator. (This is a key step.)

$= \dfrac{3 + \sqrt{5}}{2}$ Divide out the common factor.

Here is an alternative method for writing this expression in lowest terms.

$$\frac{6 + 2\sqrt{5}}{4} = \frac{6}{4} + \frac{2\sqrt{5}}{4} = \frac{3}{2} + \frac{\sqrt{5}}{2} = \frac{3 + \sqrt{5}}{2}$$

(b) $\dfrac{5y - \sqrt{8y^2}}{6y} \quad (y > 0)$

$= \dfrac{5y - 2y\sqrt{2}}{6y}$ $\sqrt{8y^2} = \sqrt{4y^2 \cdot 2} = 2y\sqrt{2}$

$= \dfrac{y(5 - 2\sqrt{2})}{6y}$ Factor the numerator.

$= \dfrac{5 - 2\sqrt{2}}{6}$ Divide out the common factor.

NOW TRY

CAUTION *Be careful to factor before writing a quotient in lowest terms.*

7.5 Exercises

FOR EXTRA HELP

MyMathLab®

Complete solution available in MyMathLab

1. E 2. C 3. A 4. F 5. D 6. B

Concept Check *Match each part of a rule for a special product in Column I with the other part in Column II. Assume that A and B represent positive real numbers.*

I	II
1. $(A + \sqrt{B})(A - \sqrt{B})$	**A.** $A - B$
2. $(\sqrt{A} + B)(\sqrt{A} - B)$	**B.** $A + 2B\sqrt{A} + B^2$
3. $(\sqrt{A} + \sqrt{B})(\sqrt{A} - \sqrt{B})$	**C.** $A - B^2$
4. $(\sqrt{A} + \sqrt{B})^2$	**D.** $A - 2\sqrt{AB} + B$
5. $(\sqrt{A} - \sqrt{B})^2$	**E.** $A^2 - B$
6. $(\sqrt{A} + B)^2$	**F.** $A + 2\sqrt{AB} + B$

7. $3\sqrt{6}+2\sqrt{3}$
8. $5\sqrt{10}-\sqrt{30}$
9. $20\sqrt{2}$ 10. $14\sqrt{2}$
11. -2 12. -22
13. -1 14. -7
15. 6 16. 15
17. $\sqrt{6}-\sqrt{2}+\sqrt{3}-1$
18. $\sqrt{15}-2\sqrt{3}+3\sqrt{5}-6$
19. $\sqrt{22}+\sqrt{55}-\sqrt{14}-\sqrt{35}$
20. $\sqrt{39}+\sqrt{143}-\sqrt{21}-\sqrt{77}$
21. $8-\sqrt{15}$ 22. $-19+\sqrt{77}$
23. $9+4\sqrt{5}$ 24. $12-2\sqrt{11}$
25. $26-2\sqrt{105}$
26. $8-4\sqrt{3}$
27. $4-\sqrt[3]{36}$ 28. $\sqrt[3]{9}-36$
29. 10 30. 2
31. $6x+3\sqrt{x}-2\sqrt{5x}-\sqrt{5}$
32. $4p-36\sqrt{p}+\sqrt{7p}-9\sqrt{7}$
33. $9r-s$ 34. $k-16m$
35. $4\sqrt[3]{4y^2}-19\sqrt[3]{2y}-5$
36. $15\sqrt[3]{3z^2}-3\sqrt[3]{9z}-14$
37. $3x-4$ 38. $6y-16$
39. $4x-y$ 40. $p-25s$
41. $2\sqrt{6}-1$ 42. $-2\sqrt{10}+4$
43. $\sqrt{7}$ 44. $\sqrt{11}$
45. $5\sqrt{3}$ 46. $2\sqrt{6}$
47. $\frac{\sqrt{6}}{2}$ 48. $\frac{\sqrt{42}}{6}$
49. $\frac{9\sqrt{15}}{5}$ 50. $\frac{3\sqrt{22}}{11}$
51. $-\frac{7\sqrt{3}}{12}$ 52. $-\frac{5\sqrt{6}}{12}$
53. $\frac{\sqrt{14}}{2}$ 54. $\frac{\sqrt{30}}{3}$
55. $-\frac{\sqrt{14}}{10}$ 56. $-\frac{\sqrt{39}}{15}$
57. $\frac{2\sqrt{6x}}{x}$ 58. $\frac{2\sqrt{13y}}{y}$
59. $\frac{-8\sqrt{3k}}{k}$ 60. $\frac{-4\sqrt{13m}}{m}$
61. $\frac{-5m^2\sqrt{6mn}}{n^2}$
62. $\frac{-7r\sqrt{2rs}}{s^3}$

Multiply, and then simplify. Assume that all variables represent positive real numbers. ***See Examples 1 and 2.***

7. $\sqrt{6}(3+\sqrt{2})$ **8.** $\sqrt{10}(5-\sqrt{3})$ **9.** $5(\sqrt{72}-\sqrt{8})$

10. $7(\sqrt{50}-\sqrt{18})$ **11.** $(\sqrt{7}+3)(\sqrt{7}-3)$ **12.** $(\sqrt{3}-5)(\sqrt{3}+5)$

13. $(\sqrt{2}-\sqrt{3})(\sqrt{2}+\sqrt{3})$ **14.** $(\sqrt{7}+\sqrt{14})(\sqrt{7}-\sqrt{14})$

15. $(\sqrt{8}-\sqrt{2})(\sqrt{8}+\sqrt{2})$ **16.** $(\sqrt{20}-\sqrt{5})(\sqrt{20}+\sqrt{5})$

17. $(\sqrt{2}+1)(\sqrt{3}-1)$ **18.** $(\sqrt{3}+3)(\sqrt{5}-2)$

19. $(\sqrt{11}-\sqrt{7})(\sqrt{2}+\sqrt{5})$ **20.** $(\sqrt{13}-\sqrt{7})(\sqrt{3}+\sqrt{11})$

21. $(2\sqrt{3}+\sqrt{5})(3\sqrt{3}-2\sqrt{5})$ **22.** $(\sqrt{7}-\sqrt{11})(2\sqrt{7}+3\sqrt{11})$

23. $(\sqrt{5}+2)^2$ **24.** $(\sqrt{11}-1)^2$

25. $(\sqrt{21}-\sqrt{5})^2$ **26.** $(\sqrt{6}-\sqrt{2})^2$

27. $(2+\sqrt[3]{6})(2-\sqrt[3]{6})$ **28.** $(\sqrt[3]{3}+6)(\sqrt[3]{3}-6)$

29. $(2+\sqrt[3]{2})(4-2\sqrt[3]{2}+\sqrt[3]{4})$ **30.** $(\sqrt[3]{3}-1)(\sqrt[3]{9}+\sqrt[3]{3}+1)$

31. $(3\sqrt{x}-\sqrt{5})(2\sqrt{x}+1)$ **32.** $(4\sqrt{p}+\sqrt{7})(\sqrt{p}-9)$

33. $(3\sqrt{r}-\sqrt{s})(3\sqrt{r}+\sqrt{s})$ **34.** $(\sqrt{k}+4\sqrt{m})(\sqrt{k}-4\sqrt{m})$

35. $(\sqrt[3]{2y}-5)(4\sqrt[3]{2y}+1)$ **36.** $(\sqrt[3]{9z}-2)(5\sqrt[3]{9z}+7)$

37. $(\sqrt{3x}+2)(\sqrt{3x}-2)$ **38.** $(\sqrt{6y}-4)(\sqrt{6y}+4)$

39. $(2\sqrt{x}+\sqrt{y})(2\sqrt{x}-\sqrt{y})$ **40.** $(\sqrt{p}+5\sqrt{s})(\sqrt{p}-5\sqrt{s})$

41. $[(\sqrt{2}+\sqrt{3})-\sqrt{6}][(\sqrt{2}+\sqrt{3})+\sqrt{6}]$

42. $[(\sqrt{5}-\sqrt{2})-\sqrt{3}][(\sqrt{5}-\sqrt{2})+\sqrt{3}]$

Rationalize each denominator. Assume that all variables represent positive real numbers. ***See Examples 3 and 4.***

43. $\frac{7}{\sqrt{7}}$ **44.** $\frac{11}{\sqrt{11}}$ **45.** $\frac{15}{\sqrt{3}}$ **46.** $\frac{12}{\sqrt{6}}$ **47.** $\frac{\sqrt{3}}{\sqrt{2}}$

48. $\frac{\sqrt{7}}{\sqrt{6}}$ **49.** $\frac{9\sqrt{3}}{\sqrt{5}}$ **50.** $\frac{3\sqrt{2}}{\sqrt{11}}$ **51.** $\frac{-7}{\sqrt{48}}$ **52.** $\frac{-5}{\sqrt{24}}$

53. $\sqrt{\frac{7}{2}}$ **54.** $\sqrt{\frac{10}{3}}$ **55.** $-\sqrt{\frac{7}{50}}$ **56.** $-\sqrt{\frac{13}{75}}$ **57.** $\sqrt{\frac{24}{x}}$

58. $\sqrt{\frac{52}{y}}$ **59.** $\frac{-8\sqrt{3}}{\sqrt{k}}$ **60.** $\frac{-4\sqrt{13}}{\sqrt{m}}$ **61.** $-\sqrt{\frac{150m^5}{n^3}}$ **62.** $-\sqrt{\frac{98r^3}{s^5}}$

63. $\frac{12x^3\sqrt{2xy}}{y^5}$ 64. $\frac{11t^4\sqrt{2tu}}{u^6}$

65. $\frac{5\sqrt{2my}}{y^2}$ 66. $\frac{2\sqrt{5rm}}{m^2}$

67. $-\frac{4k\sqrt{3z}}{z}$ 68. $-\frac{5m\sqrt{3mp}}{p}$

69. $\frac{\sqrt[3]{18}}{3}$ 70. $\frac{\sqrt[3]{100}}{5}$

71. $\frac{\sqrt[3]{12}}{3}$ 72. $\frac{\sqrt[3]{20}}{4}$

73. $\frac{\sqrt[3]{18}}{4}$ 74. $\frac{\sqrt[3]{30}}{3}$

75. $-\frac{\sqrt[3]{2pr}}{r}$ 76. $-\frac{\sqrt[3]{6xy}}{y}$

77. $\frac{x^2\sqrt[3]{y^2}}{y}$ 78. $\frac{m^3\sqrt[3]{q^2}}{q}$

79. $\frac{2\sqrt[4]{x^3}}{x}$ 80. $\frac{3\sqrt[4]{y^3}}{y}$

81. $\frac{\sqrt[4]{2yz^3}}{z}$ 82. $\frac{\sqrt[4]{7ts^2}}{s}$

83. $\frac{3(4-\sqrt{5})}{11}$ 84. $\frac{4(5-\sqrt{6})}{19}$

85. $\frac{6\sqrt{2}+4}{7}$ 86. $\frac{3\sqrt{3}+3}{2}$

87. $\frac{2(3\sqrt{5}-2\sqrt{3})}{33}$

88. $\frac{3\sqrt{2}+2\sqrt{7}}{10}$

89. $2\sqrt{3}+\sqrt{10}-3\sqrt{2}-\sqrt{15}$

90. $\sqrt{15}+\sqrt{10}+3\sqrt{2}+2\sqrt{3}$

91. $\sqrt{m}-2$ 92. $\sqrt{r}+3$

93. $\frac{4(\sqrt{x}+2\sqrt{y})}{x-4y}$

94. $\frac{5(3\sqrt{r}-\sqrt{s})}{9r-s}$

95. $\frac{x-2\sqrt{xy}+y}{x-y}$

96. $\frac{a+2\sqrt{ab}+b}{a-b}$

97. $\frac{5\sqrt{k}(2\sqrt{k}-\sqrt{q})}{4k-q}$

98. $\frac{3\sqrt{x}(\sqrt{x}+2\sqrt{y})}{x-4y}$

63. $\sqrt{\frac{288x^7}{y^9}}$ **64.** $\sqrt{\frac{242t^9}{u^{11}}}$ **65.** $\frac{5\sqrt{2m}}{\sqrt{y^3}}$

66. $\frac{2\sqrt{5r}}{\sqrt{m^3}}$ **67.** $-\sqrt{\frac{48k^2}{z}}$ **68.** $-\sqrt{\frac{75m^3}{p}}$

Simplify. Assume that all variables represent positive real numbers. ***See Example 5.***

69. $\sqrt[3]{\frac{2}{3}}$ **70.** $\sqrt[3]{\frac{4}{5}}$ **71.** $\sqrt[3]{\frac{4}{9}}$ **72.** $\sqrt[3]{\frac{5}{16}}$ **73.** $\sqrt[3]{\frac{9}{32}}$

74. $\sqrt[3]{\frac{10}{9}}$ **75.** $-\sqrt[3]{\frac{2p}{r^2}}$ **76.** $-\sqrt[3]{\frac{6x}{y^2}}$ **77.** $\sqrt[3]{\frac{x^6}{y}}$ **78.** $\sqrt[3]{\frac{m^9}{q}}$

79. $\sqrt[4]{\frac{16}{x}}$ **80.** $\sqrt[4]{\frac{81}{y}}$ **81.** $\sqrt[4]{\frac{2y}{z}}$ **82.** $\sqrt[4]{\frac{7t}{s^2}}$

Rationalize each denominator. Assume that all variables represent positive real numbers and no denominators are 0. ***See Example 6.***

83. $\frac{3}{4+\sqrt{5}}$ **84.** $\frac{4}{5+\sqrt{6}}$ **85.** $\frac{\sqrt{8}}{3-\sqrt{2}}$

86. $\frac{\sqrt{27}}{3-\sqrt{3}}$ **87.** $\frac{2}{3\sqrt{5}+2\sqrt{3}}$ **88.** $\frac{-1}{3\sqrt{2}-2\sqrt{7}}$

89. $\frac{\sqrt{2}-\sqrt{3}}{\sqrt{6}-\sqrt{5}}$ **90.** $\frac{\sqrt{5}+\sqrt{6}}{\sqrt{3}-\sqrt{2}}$ **91.** $\frac{m-4}{\sqrt{m}+2}$

92. $\frac{r-9}{\sqrt{r}-3}$ **93.** $\frac{4}{\sqrt{x}-2\sqrt{y}}$ **94.** $\frac{5}{3\sqrt{r}+\sqrt{s}}$

95. $\frac{\sqrt{x}-\sqrt{y}}{\sqrt{x}+\sqrt{y}}$ **96.** $\frac{\sqrt{a}+\sqrt{b}}{\sqrt{a}-\sqrt{b}}$ **97.** $\frac{5\sqrt{k}}{2\sqrt{k}+\sqrt{q}}$ **98.** $\frac{3\sqrt{x}}{\sqrt{x}-2\sqrt{y}}$

Write each quotient in lowest terms. Assume that all variables represent positive real numbers. ***See Example 7.***

99. $\frac{30-20\sqrt{6}}{10}$ **100.** $\frac{24+12\sqrt{5}}{12}$ **101.** $\frac{3-3\sqrt{5}}{3}$ **102.** $\frac{-5+5\sqrt{2}}{5}$

103. $\frac{16-4\sqrt{8}}{12}$ **104.** $\frac{12-9\sqrt{72}}{18}$ **105.** $\frac{6p+\sqrt{24p^3}}{3p}$ **106.** $\frac{11y-\sqrt{242y^5}}{22y}$

Extending Skills *Rationalize each denominator. Assume that all radicals represent real numbers and no denominators are 0.*

107. $\frac{3}{\sqrt{x+y}}$ **108.** $\frac{5}{\sqrt{m-n}}$ **109.** $\frac{p}{\sqrt{p+2}}$ **110.** $\frac{q}{\sqrt{5+q}}$

99. $3-2\sqrt{6}$ 100. $2+\sqrt{5}$

101. $1-\sqrt{5}$ 102. $-1+\sqrt{2}$

103. $\frac{4-2\sqrt{2}}{3}$ 104. $\frac{2-9\sqrt{2}}{3}$

105. $\frac{6+2\sqrt{6p}}{3}$ 106. $\frac{1-y\sqrt{2y}}{2}$

107. $\frac{3\sqrt{x+y}}{x+y}$ 108. $\frac{5\sqrt{m-n}}{m-n}$

109. $\frac{p\sqrt{p+2}}{p+2}$ 110. $\frac{q\sqrt{5+q}}{5+q}$

111. Each expression is approximately equal to 0.2588190451.

112. Each expression is approximately equal to -3.732050808.

113. $\frac{33}{8(6+\sqrt{3})}$

114. $\frac{11}{2(2\sqrt{5}+3)}$

115. $\frac{4x-y}{3x(2\sqrt{x}+\sqrt{y})}$

116. $\frac{p-9q}{4q(\sqrt{p}+3\sqrt{q})}$

Solve each problem.

111. The following expression occurs in a standard problem in trigonometry.

$$\frac{1}{\sqrt{2}}\cdot\frac{\sqrt{3}}{2}-\frac{1}{\sqrt{2}}\cdot\frac{1}{2}$$

Show that it simplifies to $\frac{\sqrt{6}-\sqrt{2}}{4}$. Then verify, using a calculator approximation.

112. The following expression occurs in a standard problem in trigonometry.

$$\frac{\sqrt{3}+1}{1-\sqrt{3}}$$

Show that it simplifies to $-2-\sqrt{3}$. Then verify, using a calculator approximation.

Extending Skills *In calculus, it is sometimes desirable to* ***rationalize the numerator.*** *For example, to rationalize the numerator of*

$$\frac{6-\sqrt{2}}{4},$$

we multiply the numerator and the denominator by the conjugate of the numerator.

$$\frac{6-\sqrt{2}}{4}=\frac{(6-\sqrt{2})(6+\sqrt{2})}{4(6+\sqrt{2})}=\frac{36-2}{4(6+\sqrt{2})}=\frac{34}{4(6+\sqrt{2})}=\frac{17}{2(6+\sqrt{2})}$$

Rationalize each numerator. Assume that all variables represent positive real numbers.

113. $\frac{6-\sqrt{3}}{8}$ **114.** $\frac{2\sqrt{5}-3}{2}$ **115.** $\frac{2\sqrt{x}-\sqrt{y}}{3x}$ **116.** $\frac{\sqrt{p}-3\sqrt{q}}{4q}$

SUMMARY EXERCISES Performing Operations with Radicals and Rational Exponents

Conditions for a Simplified Radical

1. The radicand has no factor raised to a power greater than or equal to the index.
2. The radicand has no fractions.
3. No denominator contains a radical.
4. Exponents in the radicand and the index of the radical have greatest common factor 1.

Concept Check *Give the reason why each radical is not simplified.*

1. $\sqrt{\frac{2}{5}}$ **2.** $\sqrt[15]{x^5}$ **3.** $\frac{5}{\sqrt[3]{10}}$ **4.** $\sqrt[3]{x^5y^6}$

Perform all indicated operations, and express each answer in simplest form with positive exponents. Assume that all variables represent positive real numbers.

5. $6\sqrt{10}-12\sqrt{10}$ **6.** $\sqrt{7}(\sqrt{7}-\sqrt{2})$ **7.** $(1-\sqrt{3})(2+\sqrt{6})$

8. $\sqrt{50}-\sqrt{98}+\sqrt{72}$ **9.** $(3\sqrt{5}+2\sqrt{7})^2$ **10.** $\frac{-3}{\sqrt{6}}$

1. The radicand is a fraction, $\frac{2}{5}$.
2. The exponent in the radicand and the index of the radical have greatest common factor 5.
3. The denominator contains a radical, $\sqrt[3]{10}$.
4. The radicand has two factors, x and y, that are raised to powers greater than the index, 3.

5. $-6\sqrt{10}$ 6. $7-\sqrt{14}$

7. $2+\sqrt{6}-2\sqrt{3}-3\sqrt{2}$

8. $4\sqrt{2}$

9. $73+12\sqrt{35}$

10. $\frac{-\sqrt{6}}{2}$

11. $4(\sqrt{7}-\sqrt{5})$
12. $-3+2\sqrt{2}$
13. -44
14. $\frac{\sqrt{x}+\sqrt{5}}{x-5}$
15. $2abc^3\sqrt[3]{b^2}$
16. $5\sqrt[3]{3}$
17. $3(\sqrt{5}-2)$
18. $\frac{\sqrt{15x}}{5x}$
19. $\frac{8}{5}$
20. $\frac{\sqrt{2}}{8}$
21. $-\sqrt[3]{100}$
22. $11+2\sqrt{30}$
23. $-3\sqrt{3x}$
24. $52-30\sqrt{3}$
25. $\frac{\sqrt[3]{117}}{9}$
26. $3\sqrt{2}+\sqrt{15}+\sqrt{42}+\sqrt{35}$
27. $2\sqrt[4]{27}$
28. $\frac{1+\sqrt[3]{3}+\sqrt[3]{9}}{-2}$
29. $\frac{x\sqrt[3]{x^2}}{y}$
30. $-4\sqrt{3}-3$
31. $xy^{6/5}$
32. $x^{10}y$
33. $\frac{1}{25x^2}$
34. $7+4\cdot 3^{1/2}$, or $7+4\sqrt{3}$
35. $3\sqrt[3]{2x^2}$
36. -2
37. (a) 8 (b) $\{-8, 8\}$
38. (a) 10 (b) $\{-10, 10\}$
39. (a) $\{-4, 4\}$ (b) -4
40. (a) $\{-5, 5\}$ (b) -5

11. $\dfrac{8}{\sqrt{7}+\sqrt{5}}$ **12.** $\dfrac{1-\sqrt{2}}{1+\sqrt{2}}$ **13.** $(\sqrt{5}+7)(\sqrt{5}-7)$

14. $\dfrac{1}{\sqrt{x}-\sqrt{5}}$, $x \neq 5$ **15.** $\sqrt[3]{8a^3b^5c^9}$ **16.** $\dfrac{15}{\sqrt[3]{9}}$

17. $\dfrac{3}{\sqrt{5}+2}$ **18.** $\sqrt{\dfrac{3}{5x}}$ **19.** $\dfrac{16\sqrt{3}}{5\sqrt{12}}$

20. $\dfrac{2\sqrt{25}}{8\sqrt{50}}$ **21.** $\dfrac{-10}{\sqrt[3]{10}}$ **22.** $\dfrac{\sqrt{6}+\sqrt{5}}{\sqrt{6}-\sqrt{5}}$

23. $\sqrt{12x}-\sqrt{75x}$ **24.** $(5-3\sqrt{3})^2$ **25.** $\sqrt[3]{\dfrac{13}{81}}$

26. $\dfrac{\sqrt{3}+\sqrt{7}}{\sqrt{6}-\sqrt{5}}$ **27.** $\dfrac{6}{\sqrt[4]{3}}$ **28.** $\dfrac{1}{1-\sqrt[3]{3}}$

29. $\sqrt[3]{\dfrac{x^2y}{x^{-3}y^4}}$ **30.** $\sqrt{12}-\sqrt{108}-\sqrt[3]{27}$ **31.** $\dfrac{x^{-2/3}y^{4/5}}{x^{-5/3}y^{-2/5}}$

32. $\left(\dfrac{x^{3/4}y^{2/3}}{x^{1/3}y^{5/8}}\right)^{24}$ **33.** $(125x^3)^{-2/3}$ **34.** $\dfrac{4^{1/2}+3^{1/2}}{4^{1/2}-3^{1/2}}$

35. $\sqrt[3]{16x^2}-\sqrt[3]{54x^2}+\sqrt[3]{128x^2}$ **36.** $(1-\sqrt[3]{3})(1+\sqrt[3]{3}+\sqrt[3]{9})$

Students often have trouble distinguishing between the following two types of problems.

Simplifying a Radical Involving a Square Root	**Solving an Equation Using Square Roots**
Exercise: Simplify $\sqrt{25}$.	*Exercise:* Solve $x^2 = 25$.
Answer: 5	*Answer:* $\{-5, 5\}$
In this situation, $\sqrt{25}$ represents the positive square root of 25, namely 5.	In this situation, $x^2 = 25$ has two solutions, the negative square root of 25 or the positive square root of 25: -5, 5.

In Exercises 37–40, provide the appropriate responses.

37. (a) Simplify $\sqrt{64}$.
(b) Solve $x^2 = 64$.

38. (a) Simplify $\sqrt{100}$.
(b) Solve $x^2 = 100$.

39. (a) Solve $x^2 = 16$.
(b) Simplify $-\sqrt{16}$.

40. (a) Solve $x^2 = 25$.
(b) Simplify $-\sqrt{25}$.

7.6 Solving Equations with Radicals

OBJECTIVES

1. Solve radical equations using the power rule.
2. Solve radical equations with indexes greater than 2.
3. Use the power rule to solve a formula for a specified variable.

VOCABULARY

- ☐ radical equation
- ☐ proposed solution
- ☐ extraneous solution

OBJECTIVE 1 Solve radical equations using the power rule.

An equation that includes one or more radical expressions with a variable is called a **radical equation.**

$$\sqrt{x-4}=8, \quad \sqrt{5x+12}=3\sqrt{2x-1}, \quad \text{and} \quad \sqrt[3]{6+x}=27 \qquad \text{Radical equations}$$

Solving radical equations involves a process that we have not yet seen, and it requires careful application. Notice that the equation $x = 1$ has only one solution. Its solution set is $\{1\}$. If we square both sides of this equation, we get $x^2 = 1$. This new equation has *two* solutions: -1 and 1. The solution of the original equation is also a solution of the equation following squaring. However, that equation has another solution, -1, that is *not* a solution of the original equation.

When solving equations with radicals, we use this idea of raising both sides to a power, which is an application of the **power rule.**

Power Rule for Solving an Equation with Radicals

If both sides of an equation are raised to the same power, all solutions of the original equation are also solutions of the new equation.

The power rule does not say that all solutions of the new equation are solutions of the original equation. They may or may not be. A value of the variable that appears to be a solution is a **proposed solution.** Such solutions that do not satisfy the original equation are **extraneous solutions.** They must be rejected.

CLASSROOM EXAMPLE 1

Solve $\sqrt{5x+1}=4$.

Answer: $\{3\}$

! CAUTION *When the power rule is used to solve an equation, every solution of the new equation must be checked in the original equation.*

NOW TRY EXERCISE 1

Solve $\sqrt{9x+7}=5$.

EXAMPLE 1 Using the Power Rule

Solve $\sqrt{3x+4}=8$.

$(\sqrt{3x+4})^2 = 8^2$	Use the power rule and square each side.
$3x+4=64$	Apply the exponents.
$3x=60$	Subtract 4.
$x=20$	Divide by 3.

$(\sqrt{a})^2 = \sqrt{a}\cdot\sqrt{a} = a$

TEACHING TIP

Review this rule: For $x \geq 0$,

$$(\sqrt{x})^2 = \sqrt{x}\cdot\sqrt{x} = x.$$

That is, when squaring the square root of a number, the result is the number.

CHECK

$\sqrt{3x+4}=8$	Original equation
$\sqrt{3\cdot 20+4}\stackrel{?}{=}8$	Let $x = 20$.
$\sqrt{64}\stackrel{?}{=}8$	Simplify.
$8=8$ ✓	True

Because 20 satisfies the *original* equation, the solution set is $\{20\}$. NOW TRY

NOW TRY ANSWER

1. $\{2\}$

Solving an Equation with Radicals

Step 1 **Isolate the radical.** Make sure that one radical term is alone on one side of the equation.

Step 2 **Apply the power rule.** Raise each side of the equation to a power that is the same as the index of the radical.

Step 3 **Solve** the resulting equation. If it still contains a radical, repeat Steps 1 and 2.

Step 4 **Check** all proposed solutions in the original equation. Discard any values that are not solutions of the original equation.

NOW TRY EXERCISE 2

Solve $\sqrt{3x+4}+5=0.$

CLASSROOM EXAMPLE 2

Solve $\sqrt{5x+3}+2=0.$

Answer: $\varnothing$

TEACHING TIP

Emphasize that after raising both sides of an equation to a power, it is essential to check *all* proposed solutions, as some may not be actual solutions. Emphasize that solutions must be checked in the *original* equation.

EXAMPLE 2 Using the Power Rule

Solve $\sqrt{5x-1}+3=0.$

Step 1 $\sqrt{5x-1}=-3$ To isolate the radical on one side, subtract 3 from each side.

Step 2 $\left(\sqrt{5x-1}\right)^2=(-3)^2$ Square each side.

Step 3 $5x-1=9$ Apply the exponents.

$5x=10$ Add 1.

$x=2$ Divide by 5.

Step 4 **CHECK** $\sqrt{5x-1}+3=0$ Original equation

$\sqrt{5\cdot 2-1}+3\stackrel{?}{=}0$ Let $x=2$.

$3+3=0$ False

Be sure to check the proposed solution.

This false result shows that the *proposed* solution 2 is *not* a solution of the original equation. It is extraneous. The solution set is $\varnothing$.

NOW TRY

NOTE We could have determined after Step 1 that the equation in **Example 2** has no solution because the expression on the left cannot be negative.

The next examples involve finding the square of a binomial. Recall the rule from **Section 4.4.**

$$(x+y)^2=x^2+2xy+y^2$$

CLASSROOM EXAMPLE 3

Solve $\sqrt{5-x}=x+1.$

Answer: $\{1\}$

EXAMPLE 3 Using the Power Rule (Squaring a Binomial)

Solve $\sqrt{4-x}=x+2.$

Step 1 The radical is isolated on the left side of the equation.

Step 2 Square each side. The square of $x+2$ is $(x+2)^2=x^2+2(x)(2)+4.$

$$\left(\sqrt{4-x}\right)^2=(x+2)^2$$

Remember the middle term.

$$4-x=x^2+4x+4$$

Twice the product of 2 and x

NOW TRY ANSWER

2. $\varnothing$

NOW TRY EXERCISE 3

Solve $\sqrt{16 - x} = x + 4$.

Step 3 The new equation is quadratic, so write it in standard form.

$$x^2 + 5x = 0 \quad \text{Standard form of } 4 - x = x^2 + 4x + 4$$

$$x(x + 5) = 0 \quad \text{Factor.}$$

Set *each* factor equal to 0.

$$x = 0 \quad \text{or} \quad x + 5 = 0 \quad \text{Zero-factor property}$$

$$x = -5 \quad \text{Solve for } x.$$

Step 4 Check each proposed solution in the original equation.

CHECK

$$\sqrt{4 - x} = x + 2 \qquad\qquad \sqrt{4 - x} = x + 2$$

$$\sqrt{4 - 0} \stackrel{?}{=} 0 + 2 \quad \text{Let } x = 0. \qquad \sqrt{4 - (-5)} \stackrel{?}{=} -5 + 2 \quad \text{Let } x = -5.$$

$$\sqrt{4} \stackrel{?}{=} 2 \qquad\qquad \sqrt{9} \stackrel{?}{=} -3$$

$$2 = 2 \checkmark \quad \text{True} \qquad\qquad 3 = -3 \quad \text{False}$$

The number 0 is a solution, but the proposed solution -5 is extraneous. The solution set is $\{0\}$.

NOW TRY

NOW TRY EXERCISE 4

Solve

$\sqrt{x^2 - 3x + 18} = x + 3$.

CLASSROOM EXAMPLE 4

Solve $\sqrt{1 - 2x - x^2} = x + 1$.

Answer: $\{0\}$

EXAMPLE 4 Using the Power Rule (Squaring a Binomial)

Solve $\sqrt{x^2 - 4x + 9} = x - 1$.

Squaring gives $(x - 1)^2 = x^2 - 2(x)(1) + 1^2$ on the right.

$$\left(\sqrt{x^2 - 4x + 9}\right)^2 = (x - 1)^2 \quad \text{Remember the middle term.}$$

$$x^2 - 4x + 9 = x^2 - 2x + 1$$

Twice the product of x and -1

$$-2x = -8 \quad \text{Subtract } x^2 \text{ and 9. Add } 2x.$$

$$x = 4 \quad \text{Divide by } -2.$$

CHECK

$$\sqrt{x^2 - 4x + 9} = x - 1 \quad \text{Original equation}$$

$$\sqrt{4^2 - 4 \cdot 4 + 9} \stackrel{?}{=} 4 - 1 \quad \text{Let } x = 4.$$

$$3 = 3 \checkmark \quad \text{True}$$

The solution set is $\{4\}$.

NOW TRY

CLASSROOM EXAMPLE 5

Solve $\sqrt{2x + 3} + \sqrt{x + 1} = 1$.

Answer: $\{-1\}$

TEACHING TIP

A common error occurs when students try to square both sides by squaring term by term. Remind students again that the right side of the equation is a binomial and must be squared by using this rule.

$$(x + y)^2 = x^2 + 2xy + y^2$$

EXAMPLE 5 Using the Power Rule (Squaring Twice)

Solve $\sqrt{5x + 6} + \sqrt{3x + 4} = 2$.

Isolate one radical on one side of the equation by subtracting $\sqrt{3x + 4}$ from each side.

$$\sqrt{5x + 6} = 2 - \sqrt{3x + 4} \quad \text{Subtract } \sqrt{3x + 4}.$$

$$\left(\sqrt{5x + 6}\right)^2 = \left(2 - \sqrt{3x + 4}\right)^2 \quad \text{Square each side.}$$

$$5x + 6 = 4 - 4\sqrt{3x + 4} + (3x + 4) \quad \text{Be careful here.}$$

Remember the middle term. Twice the product of 2 and $-\sqrt{3x + 4}$

The equation still contains a radical, so isolate the radical term on the right and square both sides again.

NOW TRY ANSWERS

3. $\{0\}$ **4.** $\{1\}$

NOW TRY EXERCISE 5

Solve

$\sqrt{3x+1} - \sqrt{x+4} = 1.$

$$5x + 6 = 4 - 4\sqrt{3x+4} + 3x + 4 \quad \text{Result after squaring}$$

$$5x + 6 = 8 - 4\sqrt{3x+4} + 3x \quad \text{Combine like terms.}$$

$$2x - 2 = -4\sqrt{3x+4} \quad \text{Subtract 8 and } 3x.$$

Divide *each* term by 2.

$$x - 1 = -2\sqrt{3x+4} \quad \text{Divide by 2.}$$

$$(x-1)^2 = \left(-2\sqrt{3x+4}\right)^2 \quad \text{Square each side again.}$$

$$x^2 - 2x + 1 = (-2)^2\left(\sqrt{3x+4}\right)^2 \quad \text{On the right, } (ab)^2 = a^2b^2.$$

$$x^2 - 2x + 1 = 4(3x+4) \quad \text{Apply the exponents.}$$

$$x^2 - 2x + 1 = 12x + 16 \quad \text{Distributive property}$$

$$x^2 - 14x - 15 = 0 \quad \text{Standard form}$$

$$(x-15)(x+1) = 0 \quad \text{Factor.}$$

$$x - 15 = 0 \quad \text{or} \quad x + 1 = 0 \quad \text{Zero-factor property}$$

$$x = 15 \quad \text{or} \quad x = -1 \quad \text{Solve each equation.}$$

CHECK First check the proposed solution $x = 15$.

$$\sqrt{5x+6} + \sqrt{3x+4} = 2 \quad \text{Original equation}$$

$$\sqrt{5(15)+6} + \sqrt{3(15)+4} \stackrel{?}{=} 2 \quad \text{Let } x = 15.$$

$$\sqrt{81} + \sqrt{49} \stackrel{?}{=} 2 \quad \text{Simplify.}$$

$$9 + 7 \stackrel{?}{=} 2 \quad \text{Take square roots.}$$

$$16 = 2 \quad \text{False}$$

Now check the proposed solution -1.

$$\sqrt{5x+6} + \sqrt{3x+4} = 2 \quad \text{Original equation}$$

$$\sqrt{5(-1)+6} + \sqrt{3(-1)+4} \stackrel{?}{=} 2 \quad \text{Let } x = -1.$$

$$\sqrt{1} + \sqrt{1} \stackrel{?}{=} 2 \quad \text{Evaluate the radicands.}$$

$$1 + 1 \stackrel{?}{=} 2 \quad \text{Take square roots.}$$

$$2 = 2 \quad ✓ \text{ True}$$

The proposed solution -1 is valid, but 15 is extraneous and must be rejected. Thus, the solution set is $\{-1\}$.

NOW TRY 5

OBJECTIVE 2 Solve radical equations with indexes greater than 2.

CLASSROOM EXAMPLE 6

Solve $\sqrt[3]{2x+7} = \sqrt[3]{3x-2}$.

Answer: $\{9\}$

EXAMPLE 6 Using the Power Rule for a Power Greater Than 2

Solve $\sqrt[3]{z+5} = \sqrt[3]{2z-6}$.

$$\left(\sqrt[3]{z+5}\right)^3 = \left(\sqrt[3]{2z-6}\right)^3 \quad \text{Cube each side.}$$

$$z + 5 = 2z - 6 \quad \text{Apply the exponents.}$$

$$11 = z \quad \text{Subtract } z. \text{ Add 6.}$$

NOW TRY ANSWER

5. $\{5\}$

NOW TRY EXERCISE 6

Solve $\sqrt[3]{4x-5} = \sqrt[3]{3x+2}$.

CHECK

$$\sqrt[3]{z+5} = \sqrt[3]{2z-6} \quad \text{Original equation}$$

$$\sqrt[3]{11+5} \stackrel{?}{=} \sqrt[3]{2 \cdot 11 - 6} \quad \text{Let } z = 11.$$

$$\sqrt[3]{16} = \sqrt[3]{16} \ \checkmark \quad \text{True}$$

The solution set is $\{11\}$.

NOW TRY

OBJECTIVE 3 Use the power rule to solve a formula for a specified variable.

NOW TRY EXERCISE 7

Solve the formula for a.

$$x = \sqrt{\frac{y+2}{a}}$$

CLASSROOM EXAMPLE 7

Solve the formula for R.

$$Z = \sqrt{\frac{R}{T}}$$

Answer: $R = TZ^2$

EXAMPLE 7 Solving a Formula from Electronics for a Variable

An important property of a radio-frequency transmission line is its **characteristic impedance,** represented by Z and measured in ohms. If L and C are the inductance and capacitance, respectively, per unit of length of the line, then these quantities are related by the formula $Z = \sqrt{\frac{L}{C}}$. Solve this formula for C.

$$Z = \sqrt{\frac{L}{C}}$$

Our goal is to isolate C on one side of the equality symbol.

$$Z^2 = \left(\sqrt{\frac{L}{C}}\right)^2 \quad \text{Square each side.}$$

$$Z^2 = \frac{L}{C} \quad (\sqrt{a})^2 = a$$

$$CZ^2 = L \quad \text{Multiply by } C.$$

$$C = \frac{L}{Z^2} \quad \text{Divide by } Z^2.$$

NOW TRY

NOW TRY ANSWERS

6. $\{7\}$ 7. $a = \frac{y+2}{x^2}$

7.6 Exercises

FOR EXTRA HELP

MyMathLab®

Complete solution available in MyMathLab

Concept Check *Check each equation to see if the given value for x is a solution.*

1. $\sqrt{3x+18} - x = 0$
   (a) 6 (b) -3

2. $\sqrt{3x-3} - x + 1 = 0$
   (a) 1 (b) 4

3. $\sqrt{x+2} - \sqrt{9x-2} = -2\sqrt{x-1}$
   (a) 2 (b) 7

4. $\sqrt{8x-3} - 2x = 0$
   (a) $\frac{3}{2}$ (b) $\frac{1}{2}$

5. *Concept Check* Is 9 a solution of the following equation?
$$\sqrt{x} = -3$$
If not, can there be a solution of this equation?

6. *Concept Check* Before even attempting to solve
$$\sqrt{3x+18} = x,$$
how can we be sure that the equation cannot have a negative solution?

Solve each equation. ***See Examples 1–4.***

7. $\sqrt{x-2} = 3$

8. $\sqrt{x+1} = 7$

9. $\sqrt{6k-1} = 1$

10. $\sqrt{7x-3} = 6$

11. $\sqrt{4r+3} + 1 = 0$

12. $\sqrt{5k-3} + 2 = 0$

Answers:

1. (a) yes (b) no
2. (a) yes (b) yes
3. (a) yes (b) no
4. (a) yes (b) yes
5. No. There is no solution. The radical expression, which is nonnegative, cannot equal a negative number.
6. Because the radical on the left side cannot be negative, and it must equal x, x cannot be negative.
7. $\{11\}$ 8. $\{48\}$
9. $\left\{\frac{1}{3}\right\}$ 10. $\left\{\frac{39}{7}\right\}$
11. $\emptyset$ 12. $\emptyset$

13. $\{5\}$ 14. $\{24\}$
15. $\{18\}$ 16. $\{20\}$
17. $\{5\}$ 18. $\{7\}$
19. $\{4\}$ 20. $\{16\}$
21. $\{17\}$ 22. $\{20\}$
23. $\{5\}$ 24. $\{6\}$
25. $\emptyset$ 26. $\emptyset$
27. $\{0\}$ 28. $\{1\}$
29. $\{0\}$ 30. $\{2\}$
31. $\emptyset$ 32. $\emptyset$
33. $\{1\}$ 34. $\{8\}$

35. It is incorrect to just square each term. The right side should be $(8-x)^2 = 64 - 16x + x^2$. The correct first step is $3x + 4 = 64 - 16x + x^2$, and the solution set is $\{4\}$.

36. It is incorrect to just square each term. First, we must isolate a radical, and then square both sides. One way to do this is to write $\sqrt{5x+6} = 3 + \sqrt{x+3}$, and then square to obtain $5x + 6 = 9 + 6\sqrt{x+3} + x + 3$. The solution set is $\{6\}$.

37. $\{1\}$ 38. $\{9\}$
39. $\{-1\}$ 40. $\{-4\}$
41. $\{14\}$ 42. $\{-4\}$
43. $\{8\}$ 44. $\{5\}$
45. $\{0\}$ 46. $\{-2\}$
47. $\emptyset$ 48. $\emptyset$
49. $\{7\}$ 50. $\{3\}$
51. $\{7\}$ 52. $\{6\}$
53. $\{4, 20\}$ 54. $\{-1, 1\}$
55. $\emptyset$ 56. $\emptyset$
57. $\left\{\frac{5}{4}\right\}$ 58. $\left\{\frac{2}{9}\right\}$
59. $\{9, 17\}$ 60. $\{-2, -1\}$
61. $\left\{\frac{1}{4}, 1\right\}$ 62. $\{0, 3\}$

13. $\sqrt{3x+1} - 4 = 0$
14. $\sqrt{5x+1} - 11 = 0$
15. $4 - \sqrt{x-2} = 0$
16. $9 - \sqrt{4x+1} = 0$
17. $\sqrt{9x-4} = \sqrt{8x+1}$
18. $\sqrt{4x-2} = \sqrt{3x+5}$
19. $2\sqrt{x} = \sqrt{3x+4}$
20. $2\sqrt{x} = \sqrt{5x-16}$
21. $3\sqrt{x-1} = 2\sqrt{2x+2}$
22. $5\sqrt{4x+1} = 3\sqrt{10x+25}$
23. $x = \sqrt{x^2+4x-20}$
24. $x = \sqrt{x^2-3x+18}$
25. $x = \sqrt{x^2+3x+9}$
26. $x = \sqrt{x^2-4x-8}$
27. $\sqrt{9-x} = x+3$
28. $\sqrt{5-x} = x+1$
29. $\sqrt{k^2+2k+9} = k+3$
30. $\sqrt{x^2-3x+3} = x-1$
31. $\sqrt{x^2+12x-4} = x-4$
32. $\sqrt{x^2-15x+15} = x-5$
33. $\sqrt{r^2+9r+15} - r - 4 = 0$
34. $\sqrt{m^2+3m+12} - m - 2 = 0$

35. *Concept Check* In solving the equation $\sqrt{3x+4} = 8 - x$, a student wrote the following for her first step. **WHAT WENT WRONG?** Solve the given equation correctly.

$$3x + 4 = 64 + x^2$$

36. *Concept Check* In solving the equation $\sqrt{5x+6} - \sqrt{x+3} = 3$, a student wrote the following for his first step. **WHAT WENT WRONG?** Solve the given equation correctly.

$$(5x+6) + (x+3) = 9$$

Solve each equation. ***See Examples 5 and 6.***

37. $\sqrt[3]{2x+5} = \sqrt[3]{6x+1}$
38. $\sqrt[3]{p+5} = \sqrt[3]{2p-4}$
39. $\sqrt[3]{x^2+5x+1} = \sqrt[3]{x^2+4x}$
40. $\sqrt[3]{r^2+2r+8} = \sqrt[3]{r^2+3r+12}$
41. $\sqrt[3]{2m-1} = \sqrt[3]{m+13}$
42. $\sqrt[3]{2k-11} = \sqrt[3]{5k+1}$
43. $\sqrt[4]{x+12} = \sqrt[4]{3x-4}$
44. $\sqrt[4]{z+11} = \sqrt[4]{2z+6}$
45. $\sqrt[3]{x-8} + 2 = 0$
46. $\sqrt[3]{r+1} + 1 = 0$
47. $\sqrt[4]{2k-5} + 4 = 0$
48. $\sqrt[4]{8z-3} + 2 = 0$
49. $\sqrt{k+2} - \sqrt{k-3} = 1$
50. $\sqrt{r+6} - \sqrt{r-2} = 2$
51. $\sqrt{2r+11} - \sqrt{5r+1} = -1$
52. $\sqrt{3x-2} - \sqrt{x+3} = 1$
53. $\sqrt{3p+4} - \sqrt{2p-4} = 2$
54. $\sqrt{4x+5} - \sqrt{2x+2} = 1$
55. $\sqrt{3-3p} - 3 = \sqrt{3p+2}$
56. $\sqrt{4x+7} - 4 = \sqrt{4x-1}$
57. $\sqrt{2\sqrt{x+11}} = \sqrt{4x+2}$
58. $\sqrt{1+\sqrt{24-10x}} = \sqrt{3x+5}$

Extending Skills *For each equation, write the expressions with rational exponents as radical expressions, and then solve, using the procedures explained in this section.*

59. $(2x-9)^{1/2} = 2 + (x-8)^{1/2}$
60. $(3w+7)^{1/2} = 1 + (w+2)^{1/2}$
61. $(2w-1)^{2/3} - w^{1/3} = 0$
62. $(x^2-2x)^{1/3} - x^{1/3} = 0$

63. $L = CZ^2$ 64. $\mathcal{A} = \pi r^2$
65. $K = \frac{V^2 m}{2}$ 66. $m = \frac{2K}{V^2}$
67. $M = \frac{r^2 F}{m}$ 68. $F = \frac{Mm}{r^2}$
69. $r = \frac{a}{4\pi^2 N^2}$
70. (a) 62.5 m (b) 155.1 m

Solve each formula for the indicated variable. ***See Example 7.*** *(Source: Cooke, N., and J. Orleans,* Mathematics Essential to Electricity and Radio, *McGraw-Hill.)*

63. $Z = \sqrt{\frac{L}{C}}$ for L **64.** $r = \sqrt{\frac{\mathcal{A}}{\pi}}$ for $\mathcal{A}$ **65.** $V = \sqrt{\frac{2K}{m}}$ for K

66. $V = \sqrt{\frac{2K}{m}}$ for m **67.** $r = \sqrt{\frac{Mm}{F}}$ for M **68.** $r = \sqrt{\frac{Mm}{F}}$ for F

The formula

$$N = \frac{1}{2\pi}\sqrt{\frac{a}{r}}$$

is used to find the rotational rate N of a space station. Here, a is the acceleration and r represents the radius of the space station, in meters. To find the value of r that will make N simulate the effect of gravity on Earth, the equation must be solved for r, using the required value of N. (Source: Kastner, B., Space Mathematics, *NASA.)*

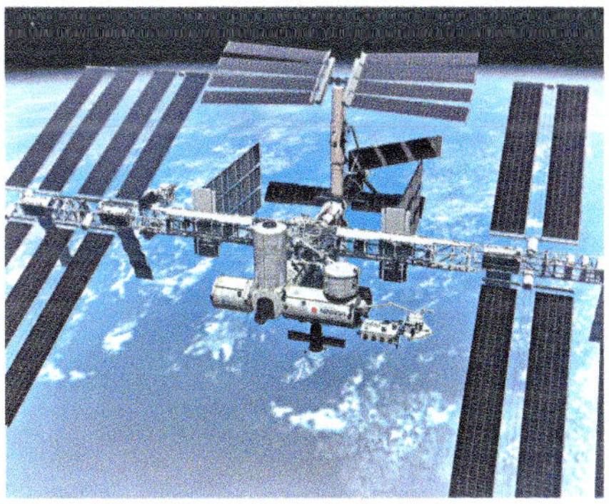

69. Solve the equation for r.

70. (a) Approximate the value of r so that $N = 0.063$ rotation per sec if $a = 9.8$ m per sec^2.

(b) Approximate the value of r so that $N = 0.04$ rotation per sec if $a = 9.8$ m per sec^2.

7.7 Complex Numbers

OBJECTIVES

1. Simplify numbers of the form $\sqrt{-b}$, where $b > 0$.
2. Recognize subsets of the complex numbers.
3. Add and subtract complex numbers.
4. Multiply complex numbers.
5. Divide complex numbers.
6. Simplify powers of i.

VOCABULARY

- ☐ complex number
- ☐ real part
- ☐ imaginary part
- ☐ pure imaginary number
- ☐ nonreal complex number
- ☐ complex conjugates

OBJECTIVE 1 Simplify numbers of the form $\sqrt{-b}$, where $b > 0$.

The equation $x^2 + 1 = 0$ has no real number solution because any solution must be a number whose square is -1. In the set of real numbers, all squares are nonnegative numbers because the product of two positive numbers or two negative numbers is positive and $0^2 = 0$. To provide a solution of the equation

$$x^2 + 1 = 0,$$

we introduce a new number i.

Imaginary Unit *i*

The **imaginary unit *i*** is defined as follows.

$$i = \sqrt{-1}, \quad \text{and thus} \quad i^2 = -1$$

That is, i is the principal square root of -1.

We can use this definition to define any square root of a negative real number.

Meaning of $\sqrt{-b}$

For any positive real number b, $\sqrt{-b} = i\sqrt{b}$.

NOW TRY EXERCISE 1

Write each number as a product of a real number and i.

(a) $\sqrt{-49}$ (b) $-\sqrt{-121}$

(c) $\sqrt{-3}$ (d) $\sqrt{-32}$

CLASSROOM EXAMPLE 1

Write each number as a product of a real number and i.

(a) $\sqrt{-25}$ (b) $-\sqrt{-81}$

(c) $\sqrt{-7}$ (d) $\sqrt{-44}$

Answers: (a) $5i$ (b) $-9i$

(c) $i\sqrt{7}$ (d) $2i\sqrt{11}$

EXAMPLE 1 Simplifying Square Roots of Negative Numbers

Write each number as a product of a real number and i.

(a) $\sqrt{-100} = i\sqrt{100} = 10i$

(b) $-\sqrt{-36} = -i\sqrt{36} = -6i$

(c) $\sqrt{-2} = i\sqrt{2}$

(d) $\sqrt{-54} = i\sqrt{54} = i\sqrt{9 \cdot 6} = 3i\sqrt{6}$

NOW TRY

CAUTION It is easy to mistake $\sqrt{2}i$ for $\sqrt{2i}$, with the i under the radical. For this reason, we usually write $\sqrt{2}i$ as $i\sqrt{2}$, as in the definition of $\sqrt{-b}$.

When finding a product such as $\sqrt{-4} \cdot \sqrt{-9}$, we cannot use the product rule for radicals because it applies only to *nonnegative* radicands.

For this reason, we change $\sqrt{-b}$ to the form $i\sqrt{b}$ before performing any multiplications or divisions.

NOW TRY EXERCISE 2

Multiply.

(a) $\sqrt{-4} \cdot \sqrt{-16}$

(b) $\sqrt{-5} \cdot \sqrt{-11}$

(c) $\sqrt{-3} \cdot \sqrt{-12}$

(d) $\sqrt{13} \cdot \sqrt{-2}$

TEACHING TIP

Keep $i^2 = -1$ and $i = \sqrt{-1}$ visible during the discussion of examples throughout this section.

CLASSROOM EXAMPLE 2

Multiply.

(a) $\sqrt{-16} \cdot \sqrt{-25}$

(b) $\sqrt{-6} \cdot \sqrt{-5}$

(c) $\sqrt{-8} \cdot \sqrt{-6}$

(d) $\sqrt{-5} \cdot \sqrt{7}$

Answers: (a) -20 (b) $-\sqrt{30}$

(c) $-4\sqrt{3}$ (d) $i\sqrt{35}$

EXAMPLE 2 Multiplying Square Roots of Negative Numbers

Multiply.

(a) $\sqrt{-4} \cdot \sqrt{-9}$

First write all square roots in terms of i.

$= i\sqrt{4} \cdot i\sqrt{9}$ — $\sqrt{-b} = i\sqrt{b}$

$= i \cdot 2 \cdot i \cdot 3$ — Take square roots.

$= 6i^2$ — Multiply.

$= 6(-1)$ — Substitute -1 for i^2.

$= -6$

(b) $\sqrt{-3} \cdot \sqrt{-7}$

$= i\sqrt{3} \cdot i\sqrt{7}$ — $\sqrt{-b} = i\sqrt{b}$

$= i^2\sqrt{3 \cdot 7}$ — Product rule

$= (-1)\sqrt{21}$ — Substitute -1 for i^2.

$= -\sqrt{21}$ — $(-1)a = -a$

(c) $\sqrt{-2} \cdot \sqrt{-8}$

$= i\sqrt{2} \cdot i\sqrt{8}$ — $\sqrt{-b} = i\sqrt{b}$

$= i^2\sqrt{2 \cdot 8}$ — Product rule

$= (-1)\sqrt{16}$ — $i^2 = -1$

$= -4$ — Take the square root.

(d) $\sqrt{-5} \cdot \sqrt{6}$

$= i\sqrt{5} \cdot \sqrt{6}$

$= i\sqrt{30}$

NOW TRY

CAUTION Using the product rule for radicals *before* using the definition of $\sqrt{-b}$ gives an *incorrect* answer. **Example 2(a)** shows that

$\sqrt{-4} \cdot \sqrt{-9} = -6,$ — Correct (**Example 2(a)**)

but $\sqrt{-4(-9)} = \sqrt{36} = 6.$ — Incorrect

Thus $\sqrt{-4} \cdot \sqrt{-9} \neq \sqrt{-4(-9)}.$

NOW TRY ANSWERS

1. (a) $7i$ (b) $-11i$ (c) $i\sqrt{3}$ (d) $4i\sqrt{2}$
2. (a) -8 (b) $-\sqrt{55}$ (c) -6 (d) $i\sqrt{26}$

NOW TRY EXERCISE 3

Divide.

(a) $\dfrac{\sqrt{-72}}{\sqrt{-8}}$ (b) $\dfrac{\sqrt{-48}}{\sqrt{3}}$

CLASSROOM EXAMPLE 3

Divide.

(a) $\dfrac{\sqrt{-80}}{\sqrt{-5}}$ (b) $\dfrac{\sqrt{-40}}{\sqrt{10}}$

Answers: (a) 4 (b) $2i$

EXAMPLE 3 Dividing Square Roots of Negative Numbers

Divide.

(a) $\dfrac{\sqrt{-75}}{\sqrt{-3}}$

$= \dfrac{i\sqrt{75}}{i\sqrt{3}}$ First write all square roots in terms of i.

$= \sqrt{\dfrac{75}{3}}$ $\frac{i}{i} = 1$; Quotient rule

$= \sqrt{25}$ Divide.

$= 5$

(b) $\dfrac{\sqrt{-32}}{\sqrt{8}}$

$= \dfrac{i\sqrt{32}}{\sqrt{8}}$ $\sqrt{-32} = i\sqrt{32}$

$= i\sqrt{\dfrac{32}{8}}$ Quotient rule

$= i\sqrt{4}$ Divide.

$= 2i$

NOW TRY

OBJECTIVE 2 Recognize subsets of the complex numbers.

A new set of numbers, the *complex numbers,* are defined as follows.

TEACHING TIP

Show examples of real numbers and pure imaginary numbers written in standard form. For example,

$3 = 3 + 0i$

and $4i = 0 + 4i$.

TEACHING TIP

Refer students back to the diagram of real numbers in Section R.1.

Complex Number

If a and b are real numbers, then any number of the form $\boldsymbol{a + bi}$ is a **complex number.** In the complex number $a + bi$, the number a is the **real part** and b is the **imaginary part.***

For a complex number $a + bi$, if $b = 0$, then $a + bi = a$, which is a real number.

Thus, the set of real numbers is a subset of the set of complex numbers.

If $a = 0$ and $b \neq 0$, the complex number is a **pure imaginary number.** For example, $3i$ is a pure imaginary number. A number such as $7 + 2i$ is a **nonreal complex number.**

A complex number written in the form $a + bi$ is in **standard form.** In this section, most answers will be given in standard form, but if a or b is 0, we consider answers such as a or bi to be in standard form.

The relationships among the various sets of numbers are shown in **FIGURE 13.**

TEACHING TIP

Relate the examples and exercises in this section to our work with simplifying algebraic expressions from earlier chapters. However, be sure to emphasize that $i^2 = -1$. Stress that when i^2 appears, simply replace it with -1.

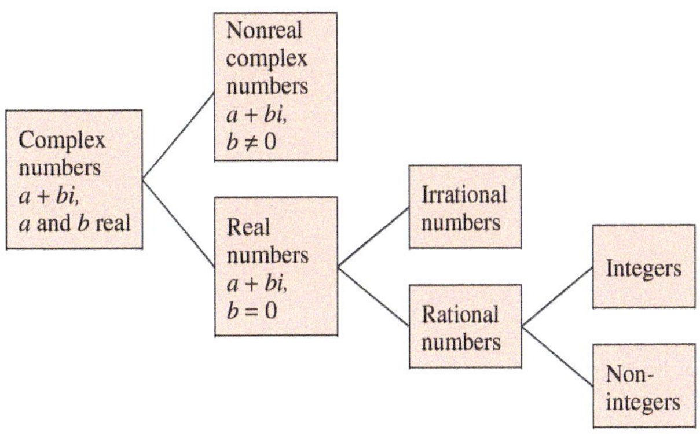

FIGURE 13

NOW TRY ANSWERS

3. (a) 3 (b) $4i$

*Some texts define bi as the imaginary part of the complex number $a + bi$.

OBJECTIVE 3 Add and subtract complex numbers.

The commutative, associative, and distributive properties for real numbers are also valid for complex numbers. ***Thus, to add complex numbers, we add their real parts and add their imaginary parts.***

NOW TRY EXERCISE 4

Add.

(a) $(-3+2i)+(4+7i)$

(b) $(5-i)+(-3+3i)+(6-4i)$

EXAMPLE 4 Adding Complex Numbers

Add.

(a) $(2+3i)+(6+4i)$

$= (2+6)+(3+4)i$ Commutative, associative, and distributive properties

$= 8+7i$ Add real parts. Add imaginary parts.

(b) $(4+2i)+(3-i)+(-6+3i)$

$= [4+3+(-6)]+[2+(-1)+3]i$ Associative property

$= 1+4i$ Add real parts. Add imaginary parts. NOW TRY

CLASSROOM EXAMPLE 4

Add.

(a) $(-1-8i)+(9-3i)$

(b) $(-3+2i)+(1-3i)+(-7-5i)$

Answers: (a) $8-11i$ (b) $-9-6i$

To subtract complex numbers, we subtract their real parts and subtract their imaginary parts.

NOW TRY EXERCISE 5

Subtract.

(a) $(7+10i)-(3+5i)$

(b) $(5-2i)-(9-7i)$

(c) $(-1+12i)-(-1-i)$

EXAMPLE 5 Subtracting Complex Numbers

Subtract.

(a) $(6+5i)-(3+2i)$

$= (6-3)+(5-2)i$ Properties of real numbers

$= 3+3i$ Subtract real parts. Subtract imaginary parts.

(b) $(7-3i)-(8-6i)$

$= (7-8)+[-3-(-6)]i$

$= -1+3i$

(c) $(-9+4i)-(-9+8i)$

$= (-9+9)+(4-8)i$

$= 0-4i$ Be careful.

$= -4i$ NOW TRY

CLASSROOM EXAMPLE 5

Subtract.

(a) $(-1+2i)-(4+i)$

(b) $(8-5i)-(12-3i)$

(c) $(-10+6i)-(-10+10i)$

Answers: (a) $-5+i$ (b) $-4-2i$ (c) $-4i$

OBJECTIVE 4 Multiply complex numbers.

We multiply complex numbers in the same way that we multiply polynomials.

CLASSROOM EXAMPLE 6

Multiply.

(a) $6i(4+3i)$

(b) $(6-4i)(2+4i)$

(c) $(3+2i)(3+4i)$

Answers: (a) $-18+24i$ (b) $28+16i$ (c) $1+18i$

EXAMPLE 6 Multiplying Complex Numbers

Multiply.

(a) $4i(2+3i)$

$= 4i(2)+4i(3i)$ Distributive property

$= 8i+12i^2$ Multiply.

$= 8i+12(-1)$ Substitute -1 for i^2.

$= -12+8i$ Standard form

NOW TRY ANSWERS

4. (a) $1+9i$ (b) $8-2i$

5. (a) $4+5i$ (b) $-4+5i$ (c) $13i$

NOW TRY EXERCISE 6
Multiply.
(a) $8i(3 - 5i)$
(b) $(7 - 2i)(4 + 3i)$

TEACHING TIP
Remind students to combine the real terms in their results.

(b) $(3 + 5i)(4 - 2i)$

$= \underbrace{3(4)}_{\text{First}} + \underbrace{3(-2i)}_{\text{Outer}} + \underbrace{5i(4)}_{\text{Inner}} + \underbrace{5i(-2i)}_{\text{Last}}$ Use the FOIL method.

$= 12 - 6i + 20i - 10i^2$ Multiply.

$= 12 + 14i - 10(-1)$ Add imaginary parts. Substitute -1 for i^2.

$= 12 + 14i + 10$ Multiply.

$= 22 + 14i$ Add real parts.

(c) $(2 + 3i)(1 - 5i)$

$= 2(1) + 2(-5i) + 3i(1) + 3i(-5i)$ FOIL method

$= 2 - 10i + 3i - 15i^2$ Multiply.

$= 2 - 7i - 15(-1)$ $i^2 = -1$ (Use parentheses around -1 to avoid errors.)

$= 2 - 7i + 15$ Multiply.

$= 17 - 7i$ Add real parts. NOW TRY

The two complex numbers $a + bi$ and $a - bi$ are **complex conjugates,** or simply *conjugates,* of each other. ***The product of a complex number and its conjugate is always a real number,*** as shown here.

$$(a + bi)(a - bi)$$

$= a^2 - abi + abi - b^2i^2$ FOIL method

$= a^2 - b^2(-1)$ Combine like terms; $i^2 = -1$

$= a^2 + b^2$ (The product eliminates i.)

For example, $(3 + 7i)(3 - 7i) = 3^2 + 7^2 = 9 + 49 = 58.$

OBJECTIVE 5 Divide complex numbers.

CLASSROOM EXAMPLE 7
Find each quotient.
(a) $\frac{23 - i}{3 - i}$ (b) $\frac{5 - i}{i}$
Answers: (a) $7 + 2i$
(b) $-1 - 5i$

EXAMPLE 7 Dividing Complex Numbers

Find each quotient.

(a) $\frac{8 + 9i}{5 + 2i}$

$= \frac{(8 + 9i)(5 - 2i)}{(5 + 2i)(5 - 2i)}$ Multiply numerator and denominator by $5 - 2i$, the conjugate of the denominator.

$= \frac{40 - 16i + 45i - 18i^2}{5^2 + 2^2}$ In the denominator, $(a + bi)(a - bi) = a^2 + b^2$.

$= \frac{40 + 29i - 18(-1)}{25 + 4}$ In the numerator, add imaginary parts; $i^2 = -1$

$= \frac{58 + 29i}{29}$ Multiply. Add real parts. Add in the denominator.

$= \frac{29(2 + i)}{29}$ Factor the numerator. (Factor first. Then divide out the common factor.)

$= 2 + i$ Lowest terms

TEACHING TIP
Call students' attention to the final step in the work for **Example 7(a).** *Factor first, and then simplify.*

NOW TRY ANSWERS
6. (a) $40 + 24i$ **(b)** $34 + 13i$

NOW TRY EXERCISE 7

Find each quotient.

(a) $\frac{4+2i}{1+3i}$ **(b)** $\frac{5-4i}{i}$

(b) $\frac{1+i}{i}$

$= \frac{(1+i)(-i)}{i(-i)}$ Multiply numerator and denominator by $-i$, the conjugate of i.

$= \frac{-i-i^2}{-i^2}$ Use the distributive property in the numerator. Multiply in the denominator.

$= \frac{-i-(-1)}{-(-1)}$ Substitute -1 for i^2. Use parentheses to avoid errors.

$= \frac{-i+1}{1}$

$= 1-i$ $\quad \frac{a}{1} = a$ NOW TRY

TEACHING TIP
Explain that a larger positive power of *i* can also be simplified by dividing the exponent by 4 and observing the remainder.

OBJECTIVE 6 Simplify powers of *i*.

Because i^2 is defined to be -1, we can find greater powers of i as shown in the following examples.

$$i^3 = i \cdot i^2 = i(-1) = -i \qquad i^6 = i^2 \cdot i^4 = (-1) \cdot 1 = -1$$
$$i^4 = i^2 \cdot i^2 = (-1)(-1) = 1 \qquad i^7 = i^3 \cdot i^4 = (-i) \cdot 1 = -i$$
$$i^5 = i \cdot i^4 = i \cdot 1 = i \qquad i^8 = i^4 \cdot i^4 = 1 \cdot 1 = 1$$

Notice that the powers of i rotate through the four numbers i, -1, $-i$, and 1. Greater powers of i can be simplified using the fact that $i^4 = 1$.

CLASSROOM EXAMPLE 8
Find each power of *i*.
(a) i^{28} (b) i^{19}
(c) i^{-9} (d) i^{-22}
Answers: (a) 1 (b) $-i$
(c) $-i$ (d) -1

EXAMPLE 8 Simplifying Powers of *i*

Find each power of *i*.

(a) $i^{12} = (i^4)^3 = 1^3 = 1$ $\quad i^4 = 1$

(b) $i^{39} = i^{36} \cdot i^3 = (i^4)^9 \cdot i^3 = 1^9 \cdot (-i) = -i$

(c) $i^{-2} = \frac{1}{i^2} = \frac{1}{-1} = -1$

(d) $i^{-1} = \frac{1}{i} = \frac{1(-i)}{i(-i)} = \frac{-i}{-i^2} = \frac{-i}{-(-1)} = \frac{-i}{1} = -i$ NOW TRY

NOW TRY EXERCISE 8

Find each power of *i*.

(a) i^{16} **(b)** i^{21}
(c) i^{-6} **(d)** i^{-13}

NOW TRY ANSWERS
7. **(a)** $1-i$ **(b)** $-4-5i$
8. **(a)** 1 **(b)** i **(c)** -1 **(d)** $-i$

7.7 Exercises

FOR EXTRA HELP MyMathLab®

Complete solution available in MyMathLab

Concept Check *List all of the following sets to which each number belongs. A number may belong to more than one set.*

real numbers pure imaginary numbers nonreal complex numbers complex numbers

1. $3+5i$ **2.** $-7i$ **3.** $\sqrt{2}$

4. $\frac{13}{3}$ **5.** $\sqrt{-49}$ **6.** $-\sqrt{-8}$

1. nonreal complex, complex
2. pure imaginary, nonreal complex, complex
3. real, complex
4. real, complex
5. pure imaginary, nonreal complex, complex
6. pure imaginary, nonreal complex, complex

Concept Check *Decide whether each expression is equal to 1, -1, i, or $-i$.*

7. $\sqrt{-1}$ **8.** $-\sqrt{-1}$ **9.** i^2 **10.** $-i^2$ **11.** $\frac{1}{i}$ **12.** $(-i)^2$

Write each number as a product of a real number and i. Simplify all radical expressions. **See Example 1.**

13. $\sqrt{-169}$
14. $\sqrt{-225}$
15. $-\sqrt{-144}$
16. $-\sqrt{-196}$
17. $\sqrt{-5}$
18. $\sqrt{-21}$
19. $\sqrt{-48}$
20. $\sqrt{-96}$

Multiply or divide as indicated. **See Examples 2 and 3.**

21. $\sqrt{-7} \cdot \sqrt{-15}$
22. $\sqrt{-3} \cdot \sqrt{-19}$
23. $\sqrt{-4} \cdot \sqrt{-25}$
24. $\sqrt{-9} \cdot \sqrt{-81}$
25. $\sqrt{-3} \cdot \sqrt{11}$
26. $\sqrt{-10} \cdot \sqrt{2}$
27. $\dfrac{\sqrt{300}}{\sqrt{-100}}$
28. $\dfrac{\sqrt{-40}}{\sqrt{-10}}$
29. $\dfrac{\sqrt{-75}}{\sqrt{3}}$
30. $\dfrac{\sqrt{-160}}{\sqrt{10}}$
31. $\dfrac{-\sqrt{-64}}{\sqrt{-16}}$
32. $\dfrac{-\sqrt{-100}}{\sqrt{-25}}$

Add or subtract as indicated. Give answers in standard form. **See Examples 4 and 5.**

33. $(3 + 2i) + (-4 + 5i)$
34. $(7 + 15i) + (-11 + 14i)$
35. $(5 - i) + (-5 + i)$
36. $(-2 + 6i) + (2 - 6i)$
37. $(4 + i) - (-3 - 2i)$
38. $(9 + i) - (3 + 2i)$
39. $(-3 - 4i) - (-1 - 4i)$
40. $(-2 - 3i) - (-5 - 3i)$
41. $(-4 + 11i) + (-2 - 4i) + (7 + 6i)$
42. $(-1 + i) + (2 + 5i) + (3 + 2i)$
43. $[(7 + 3i) - (4 - 2i)] + (3 + i)$
44. $[(7 + 2i) + (-4 - i)] - (2 + 5i)$

Concept Check *Fill in the blank with the correct response.*

45. Because $(4 + 2i) - (3 + i) = 1 + i$, using the definition of subtraction we can check this to find that

$$(1 + i) + (3 + i) = \_\_\_\_\_.$$

46. Because $\frac{-5}{2 - i} = -2 - i$, using the definition of division we can check this to find that

$$(-2 - i)(2 - i) = \_\_\_\_\_.$$

Multiply. **See Example 6.**

47. $(3i)(27i)$
48. $(5i)(125i)$
49. $(-8i)(-2i)$
50. $(-32i)(-2i)$
51. $5i(-6 + 2i)$
52. $3i(4 + 9i)$
53. $(4 + 3i)(1 - 2i)$
54. $(7 - 2i)(3 + i)$
55. $(4 + 5i)^2$
56. $(3 + 2i)^2$
57. $2i(-4 - i)^2$
58. $3i(-3 - i)^2$
59. $(12 + 3i)(12 - 3i)$
60. $(6 + 7i)(6 - 7i)$
61. $(4 + 9i)(4 - 9i)$
62. $(7 + 2i)(7 - 2i)$
63. $(1 + i)^2(1 - i)^2$
64. $(2 - i)^2(2 + i)^2$

Concept Check *Answer each of the following.*

65. Let a and b represent real numbers.
(a) What is the conjugate of $a + bi$?
(b) If we multiply $a + bi$ by its conjugate, we obtain _____ + _____, which is always a real number.

7. i 8. $-i$ 9. -1 10. 1 11. $-i$ 12. -1 13. $13i$ 14. $15i$ 15. $-12i$ 16. $-14i$ 17. $i\sqrt{5}$ 18. $i\sqrt{21}$ 19. $4i\sqrt{3}$ 20. $4i\sqrt{6}$ 21. $-\sqrt{105}$ 22. $-\sqrt{57}$ 23. -10 24. -27 25. $i\sqrt{33}$ 26. $2i\sqrt{5}$ 27. $\sqrt{3}$ 28. 2 29. $5i$ 30. $4i$ 31. -2 32. -2 33. $-1 + 7i$ 34. $-4 + 29i$ 35. 0 36. 0 37. $7 + 3i$ 38. $6 - i$ 39. -2 40. 3 41. $1 + 13i$ 42. $4 + 8i$ 43. $6 + 6i$ 44. $1 - 4i$ 45. $4 + 2i$ 46. -5 47. -81 48. -625 49. -16 50. -64 51. $-10 - 30i$ 52. $-27 + 12i$ 53. $10 - 5i$ 54. $23 + i$ 55. $-9 + 40i$ 56. $5 + 12i$ 57. $-16 + 30i$ 58. $-18 + 24i$ 59. 153 60. 85 61. 97 62. 53 63. 4 64. 25 65. (a) $a - bi$ (b) a^2; b^2

66. By what complex number should we multiply the numerator and denominator of $\frac{2 + i\sqrt{2}}{2 - i\sqrt{2}}$ to write the quotient in standard form?

A. $\sqrt{2}$ B. $i\sqrt{2}$ C. $2 + i\sqrt{2}$ D. $2 - i\sqrt{2}$

Find each quotient. ***See Example 7.***

67. $\frac{2}{1 - i}$ 68. $\frac{2}{1 + i}$ 69. $\frac{8i}{2 + 2i}$ 70. $\frac{-8i}{1 + i}$

71. $\frac{-7 + 4i}{3 + 2i}$ 72. $\frac{-38 - 8i}{7 + 3i}$ 73. $\frac{2 - 3i}{2 + 3i}$ 74. $\frac{-1 + 5i}{3 + 2i}$

75. $\frac{3 + i}{i}$ 76. $\frac{5 - i}{i}$ 77. $\frac{3 - i}{-i}$ 78. $\frac{5 + i}{-i}$

Find each power of i. ***See Example 8.***

79. i^{18} 80. i^{26} 81. i^{89} 82. i^{48} 83. i^{38}

84. i^{102} 85. i^{43} 86. i^{83} 87. i^{-5} 88. i^{-17}

Concept Check *Provide a short response in Exercises 89 and 90.*

89. A student simplified i^{-18} as shown. Provide mathematical justification for this correct work.

$$i^{-18} = i^{-18} \cdot i^{20} = i^{-18+20} = i^2 = -1$$

90. Explain why the following two expressions must be equal. (Do not actually perform the computation.)

$$(46 + 25i)(3 - 6i) \quad \text{and} \quad (46 + 25i)(3 - 6i)i^{12}$$

Ohm's law *for the current I in a circuit with voltage E, resistance R, capacitive reactance X_c, and inductive reactance X_L is*

$$I = \frac{E}{R + (X_L - X_c)i}.$$

Use this law to work each exercise.

91. Find I if $E = 2 + 3i$, $R = 5$, $X_L = 4$, and $X_c = 3$.

92. Find E if $I = 1 - i$, $R = 2$, $X_L = 3$, and $X_c = 1$.

Answers

66. C

67. $1 + i$ 68. $1 - i$
69. $2 + 2i$ 70. $-4 - 4i$
71. $-1 + 2i$ 72. $-5 + i$
73. $-\frac{5}{13} - \frac{12}{13}i$
74. $\frac{7}{13} + \frac{17}{13}i$
75. $1 - 3i$ 76. $-1 - 5i$
77. $1 + 3i$ 78. $-1 + 5i$

79. -1 80. -1
81. i 82. 1
83. -1 84. -1
85. $-i$ 86. $-i$
87. $-i$ 88. $-i$

89. Because $i^{20} = (i^4)^5 = 1^5 = 1$, the student multiplied by 1, which is justified by the identity property for multiplication.
90. $i^{12} = (i^4)^3 = 1^3 = 1$, so by the identity property for multiplication, the two products must be equal.

91. $\frac{1}{2} + \frac{1}{2}i$
92. 4

Chapter 7 Summary

Key Terms

7.1
radicand
index (order)
radical
principal root
radical expression
square root function
cube root function

7.3
hypotenuse
legs (of a right triangle)
circle
center
radius

7.5
conjugates

7.6
radical equation
proposed solution
extraneous solution

7.7
complex number
real part
imaginary part
pure imaginary number
nonreal complex number
complex conjugates

New Symbols

$\sqrt{}$	radical symbol	$\pm$	"positive or negative" or "plus or minus"	$a^{1/n}$	a to the power $\frac{1}{n}$	i	imaginary unit
$\sqrt[n]{a}$	radical; principal nth root of a	$\approx$	is approximately equal to	$a^{m/n}$	a to the power $\frac{m}{n}$		

Test Your Word Power

See how well you have learned the vocabulary in this chapter.

1. A **radicand** is
   A. the index of a radical
   B. the number or expression under the radical symbol
   C. the positive root of a number
   D. the radical symbol.

2. The **Pythagorean theorem** states that, in a right triangle,
   A. the sum of the measures of the angles is 180°
   B. the sum of the lengths of the two shorter sides equals the length of the longest side
   C. the longest side is opposite the right angle
   D. the square of the length of the longest side equals the sum of the squares of the lengths of the two shorter sides.

3. A **hypotenuse** is
   A. either of the two shorter sides of a triangle
   B. the shortest side of a triangle
   C. the side opposite the right angle in a triangle
   D. the longest side in any triangle.

4. **Rationalizing the denominator** is the process of
   A. eliminating fractions from a radical expression
   B. changing the denominator of a fraction from a radical to a rational number
   C. clearing a radical expression of radicals
   D. multiplying radical expressions.

5. An **extraneous solution** is a value
   A. that does not satisfy the original equation
   B. that makes an equation true
   C. that makes an expression equal 0
   D. that checks in the original equation.

6. A **complex number** is
   A. a real number that includes a complex fraction
   B. a zero multiple of i
   C. a number of the form $a + bi$, where a and b are real numbers
   D. the square root of -1.

ANSWERS

1. B; *Example:* In $\sqrt{3xy}$, $3xy$ is the radicand. **2.** D; *Example:* In a right triangle where $a = 6$, $b = 8$, and $c = 10$, $6^2 + 8^2 = 10^2$.
3. C; *Example:* In a right triangle where the sides measure 9, 12, and 15 units, the hypotenuse is the side opposite the right angle, with measure 15 units. **4.** B; *Example:* To rationalize the denominator of $\frac{5}{\sqrt{3}+1}$, multiply both the numerator and denominator by $\sqrt{3} - 1$ to obtain $\frac{5(\sqrt{3}-1)}{2}$.
5. A; *Example:* The proposed solution 2 is extraneous in $\sqrt{5x-1} + 3 = 0$. **6.** C; *Examples:* -5 (or $-5 + 0i$), $7i$ (or $0 + 7i$), $\sqrt{2} - 4i$

Quick Review

CONCEPTS	EXAMPLES

7.1 Radical Expressions and Graphs

$\sqrt[n]{a} = b$ means $b^n = a$.

$\sqrt[n]{a}$ is the **principal nth root** of a.

$\sqrt[n]{a^n} = |a|$ if n is even. $\sqrt[n]{a^n} = a$ if n is odd.

The two square roots of 64 are $\sqrt{64} = 8$ (the principal square root) and $-\sqrt{64} = -8$.

$$\sqrt[4]{(-2)^4} = |-2| = 2 \qquad \sqrt[3]{-27} = -3$$

Functions Defined by Radical Expressions

The square root function

$$f(x) = \sqrt{x}$$

and the cube root function

$$f(x) = \sqrt[3]{x}$$

are two important functions defined by radical expressions.

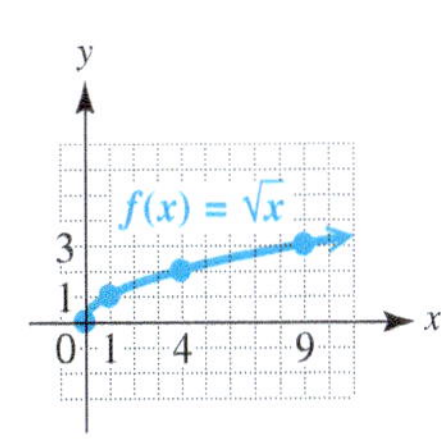

Square root function

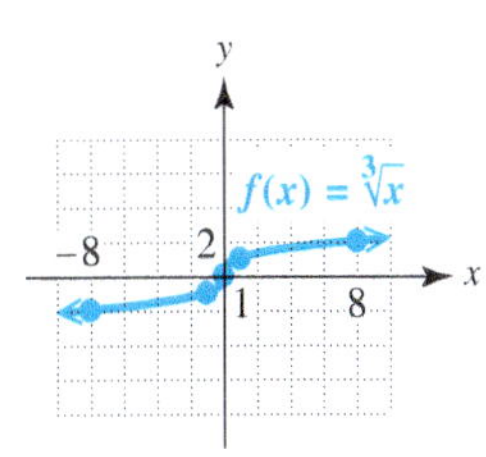

Cube root function

CONCEPTS | **EXAMPLES**

7.2 Rational Exponents

$a^{1/n} = \sqrt[n]{a}$ whenever $\sqrt[n]{a}$ exists.

If m and n are positive integers with $\frac{m}{n}$ in lowest terms, then $a^{m/n} = (a^{1/n})^m$, provided that $a^{1/n}$ is a real number.

All of the usual definitions and rules for exponents are valid for rational exponents.

Apply the rules for rational exponents.

$$81^{1/2} = \sqrt{81} = 9 \qquad -64^{1/3} = -\sqrt[3]{64} = -4$$

$$8^{5/3} = (8^{1/3})^5 = 2^5 = 32 \qquad (y^{2/5})^{10} = y^4$$

Write with positive exponents.

$$5^{-1/2} \cdot 5^{1/4} = 5^{-1/2+1/4} = 5^{-1/4} = \frac{1}{5^{1/4}}$$

$$\frac{x^{-1/3}}{x^{-1/2}} = x^{-1/3-(-1/2)} = x^{-1/3+1/2} = x^{1/6}, \quad x > 0$$

7.3 Simplifying Radicals, the Distance Formula, and Circles

Product and Quotient Rules for Radicals

If $\sqrt[n]{a}$ and $\sqrt[n]{b}$ are real numbers and n is a natural number, then the following hold.

$$\sqrt[n]{a} \cdot \sqrt[n]{b} = \sqrt[n]{ab} \quad \text{and} \quad \sqrt[n]{\frac{a}{b}} = \frac{\sqrt[n]{a}}{\sqrt[n]{b}} \quad (\text{where } b \neq 0)$$

Simplify.

$$\sqrt{3} \cdot \sqrt{7} = \sqrt{21} \qquad \sqrt[5]{x^3y} \cdot \sqrt[5]{xy^2} = \sqrt[5]{x^4y^3}$$

$$\frac{\sqrt{x^5}}{\sqrt{x^4}} = \sqrt{\frac{x^5}{x^4}} = \sqrt{x}, \quad x > 0$$

Conditions for a Simplified Radical

1. The radicand has no factor raised to a power greater than or equal to the index.
2. The radicand has no fractions.
3. No denominator contains a radical.
4. Exponents in the radicand and the index of the radical have greatest common factor 1.

$$\sqrt{18} = \sqrt{9 \cdot 2} = 3\sqrt{2}$$

$$\sqrt[3]{54x^5y^3} = \sqrt[3]{27x^3y^3 \cdot 2x^2} = 3xy\sqrt[3]{2x^2}$$

$$\sqrt{\frac{7}{4}} = \frac{\sqrt{7}}{\sqrt{4}} = \frac{\sqrt{7}}{2}$$

$$\sqrt[9]{x^3} = x^{3/9} = x^{1/3}, \quad \text{or} \quad \sqrt[3]{x}$$

Pythagorean Theorem

If a and b are the lengths of the shorter sides of a right triangle and c is the length of the longest side, then the following holds.

$$a^2 + b^2 = c^2$$

Find b for the triangle in the figure.

$10^2 + b^2 = (2\sqrt{61})^2$	Let a = 10, c = $2\sqrt{61}$.
$b^2 = 4(61) - 100$	Square and then subtract 100.
$b^2 = 144$	Simplify.
$b = 12$	$12^2 = 144$, and $12 > 0$.

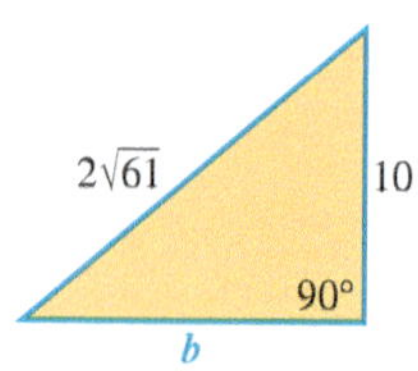

Distance Formula

The distance d between the points (x_1, y_1) and (x_2, y_2) is given by the following.

$$d = \sqrt{(x_2 - x_1)^2 + (y_2 - y_1)^2}$$

Find the distance between $(3, -2)$ and $(-1, 1)$.

$\sqrt{(-1-3)^2 + [1-(-2)]^2}$	Substitute.
$= \sqrt{(-4)^2 + 3^2}$	Subtract.
$= \sqrt{16+9}$	Square.
$= \sqrt{25}$	Add.
$= 5$	Take the square root.

CONCEPTS | **EXAMPLES**

Circles

The circle with radius r and center at (h, k) has an equation of the form

$$(x - h)^2 + (y - k)^2 = r^2.$$

The circle with equation $(x + 2)^2 + (y - 3)^2 = 25$, which can be written

$$[x - (-2)]^2 + (y - 3)^2 = 5^2,$$

has center $(-2, 3)$ and radius 5.

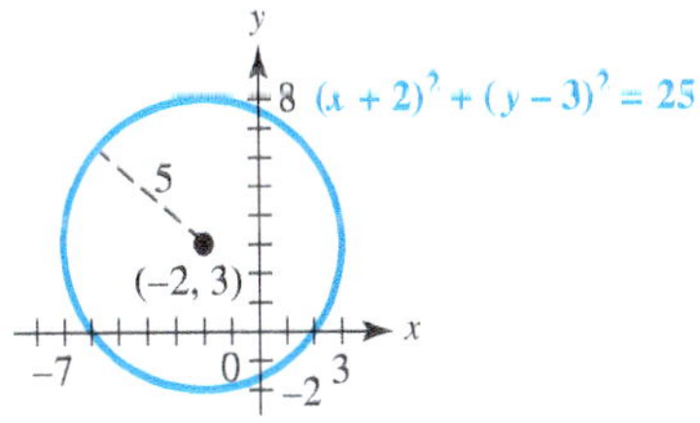

7.4 Adding and Subtracting Radical Expressions

Only radical expressions with the same index and the same radicand may be combined.

Simplify.

$$\begin{aligned} &2\sqrt{28} - 3\sqrt{63} + 8\sqrt{112} \\ &= 2\sqrt{4 \cdot 7} - 3\sqrt{9 \cdot 7} + 8\sqrt{16 \cdot 7} \\ &= 2 \cdot 2\sqrt{7} - 3 \cdot 3\sqrt{7} + 8 \cdot 4\sqrt{7} \\ &= 4\sqrt{7} - 9\sqrt{7} + 32\sqrt{7} \\ &= (4 - 9 + 32)\sqrt{7} \\ &= 27\sqrt{7} \end{aligned}$$

$\sqrt{15} + \sqrt{30}$
$\sqrt{3} + \sqrt[3]{9}$ } Cannot be simplified further

7.5 Multiplying and Dividing Radical Expressions

Multiply binomial radical expressions by using the FOIL method. Special products from **Section 4.4** may apply.

Perform the operations and simplify.

$$\begin{aligned} &(\sqrt{2} + \sqrt{7})(\sqrt{3} - \sqrt{6}) \\ &= \sqrt{6} - 2\sqrt{3} + \sqrt{21} - \sqrt{42} \qquad \sqrt{12} = 2\sqrt{3} \end{aligned}$$

$$\begin{aligned} &(\sqrt{5} - \sqrt{10})(\sqrt{5} + \sqrt{10}) \\ &= 5 - 10 \\ &= -5 \end{aligned}$$

$$\begin{aligned} &(\sqrt{3} - \sqrt{2})^2 \\ &= 3 - 2\sqrt{3} \cdot \sqrt{2} + 2 \\ &= 5 - 2\sqrt{6} \end{aligned}$$

Rationalize the denominator by multiplying both the numerator and the denominator by the same expression, one that will yield a rational number in the final denominator.

$$\frac{\sqrt{7}}{\sqrt{5}} = \frac{\sqrt{7} \cdot \sqrt{5}}{\sqrt{5} \cdot \sqrt{5}} = \frac{\sqrt{35}}{5}$$

$$\begin{aligned} \frac{4}{\sqrt{5} - \sqrt{2}} &= \frac{4(\sqrt{5} + \sqrt{2})}{(\sqrt{5} - \sqrt{2})(\sqrt{5} + \sqrt{2})} \\ &= \frac{4(\sqrt{5} + \sqrt{2})}{5 - 2} = \frac{4(\sqrt{5} + \sqrt{2})}{3} \end{aligned}$$

To write a radical quotient in lowest terms, factor the numerator and denominator and then divide out any common factor(s).

$$\frac{5 + 15\sqrt{6}}{10} = \frac{5(1 + 3\sqrt{6})}{5 \cdot 2} = \frac{1 + 3\sqrt{6}}{2}$$

CONCEPTS	EXAMPLES

7.6 Solving Equations with Radicals

Solving an Equation with Radicals

Step 1 Isolate one radical on one side of the equation.

Step 2 Raise both sides of the equation to a power that is the same as the index of the radical.

Step 3 Solve the resulting equation. If it still contains a radical, repeat Steps 1 and 2.

Step 4 Check all proposed solutions in the *original* equation. Discard any values that are not solutions of the original equation.

Solve $\sqrt{2x+3} - x = 0$.

$$\sqrt{2x+3} = x \quad \text{Add } x.$$
$$\left(\sqrt{2x+3}\right)^2 = x^2 \quad \text{Square each side.}$$
$$2x + 3 = x^2 \quad \text{Apply the exponents.}$$
$$x^2 - 2x - 3 = 0 \quad \text{Standard form}$$
$$(x-3)(x+1) = 0 \quad \text{Factor.}$$
$$x - 3 = 0 \quad \text{or} \quad x + 1 = 0 \quad \text{Zero-factor property}$$
$$x = 3 \quad \text{or} \quad x = -1 \quad \text{Solve each equation.}$$

A check shows that 3 is a solution, but -1 is extraneous. The solution set is $\{3\}$.

7.7 Complex Numbers

$i = \sqrt{-1}$, where $i^2 = -1$.

For any positive number b, $\sqrt{-b} = i\sqrt{b}$.

To multiply radicals with negative radicands, first change each factor to the form $i\sqrt{b}$ and then multiply. The same procedure applies to quotients.

Express each in $a + bi$ form.

$$\sqrt{-25} = i\sqrt{25} = 5i$$
$$\sqrt{-3} \cdot \sqrt{-27}$$
$$= i\sqrt{3} \cdot i\sqrt{27} \quad \sqrt{-b} = i\sqrt{b}$$
$$= i^2\sqrt{81}$$
$$= -1 \cdot 9 \quad i^2 = -1$$
$$= -9$$
$$\frac{\sqrt{-18}}{\sqrt{-2}} = \frac{i\sqrt{18}}{i\sqrt{2}} = \sqrt{\frac{18}{2}} = \sqrt{9} = 3$$

Adding and Subtracting Complex Numbers
Add (or subtract) the real parts and add (or subtract) the imaginary parts.

Perform the operations.

$$(5 + 3i) + (8 - 7i) = 13 - 4i \qquad (5 + 3i) - (8 - 7i) = -3 + 10i$$

Multiplying Complex Numbers
Multiply complex numbers using the FOIL method.

Multiply. $(2 + i)(5 - 3i)$

$$= 10 - 6i + 5i - 3i^2 \quad \text{Use the FOIL method.}$$
$$= 10 - i - 3(-1) \quad i^2 = -1$$
$$= 10 - i + 3 \quad \text{Multiply.}$$
$$= 13 - i \quad \text{Add real parts.}$$

Dividing Complex Numbers
Divide complex numbers by multiplying the numerator and the denominator by the conjugate of the denominator.

Divide. $\frac{20}{3+i}$

$$= \frac{20(3-i)}{(3+i)(3-i)} \quad \text{Multiply both numerator and denominator by the conjugate of the denominator.}$$
$$= \frac{20(3-i)}{9-i^2} \quad (a+b)(a-b) = a^2 - b^2$$
$$= \frac{20(3-i)}{10} \quad i^2 = -1 \text{ and } 9 - (-1) = 10.$$
$$= 2(3-i) \quad \text{Divide out the common factor, 10.}$$
$$= 6 - 2i \quad \text{Distributive property}$$

Chapter 7 Review Exercises

7.1 *Find each root.*

1. $\sqrt{1764}$ **2.** $-\sqrt{289}$ **3.** $\sqrt[3]{216}$

4. $\sqrt[3]{-125}$ **5.** $-\sqrt[3]{27}$ **6.** $\sqrt[5]{-32}$

7. Under what conditions is $\sqrt[n]{a}$ not a real number?

Simplify each root so that no radicals appear. Assume that x represents any real number.

8. $\sqrt{x^2}$ **9.** $-\sqrt{x^2}$ **10.** $\sqrt[3]{x^3}$

Find a decimal approximation for each radical. Round answers to three decimal places.

11. $-\sqrt{47}$ **12.** $\sqrt[3]{-129}$ **13.** $\sqrt[4]{605}$

14. $\sqrt[4]{500}$ **15.** $-\sqrt[3]{500}$ **16.** $-\sqrt{28}$

Graph each function and give its domain and range.

17. $f(x) = \sqrt{x-1}$ **18.** $f(x) = \sqrt[3]{x} + 4$

7.2

19. Fill in the blanks with the correct responses: One way to evaluate $8^{2/3}$ is to first find the ______ root of ______, which is ______. Then raise that result to the ______ power, to obtain an answer of ______. Therefore, $8^{2/3}$ = ______.

20. Which one of the following is a positive number?

A. $(-27)^{2/3}$ **B.** $(-64)^{5/3}$ **C.** $(-100)^{1/2}$ **D.** $(-32)^{1/5}$

21. If a is a negative number and n is odd, then what must be true about m for $a^{m/n}$ to be **(a)** positive **(b)** negative?

22. If a is negative and n is even, is $a^{1/n}$ a real number?

Simplify. If the expression does not represent a real number, say so.

23. $49^{1/2}$ **24.** $-121^{1/2}$ **25.** $16^{5/4}$ **26.** $-8^{2/3}$

27. $-\left(\frac{36}{25}\right)^{3/2}$ **28.** $\left(-\frac{1}{8}\right)^{-5/3}$ **29.** $\left(\frac{81}{10{,}000}\right)^{-3/4}$ **30.** $(-16)^{3/4}$

Write each exponential as a radical.

31. $(m + 3n)^{1/2}$ **32.** $(3a + b)^{-5/3}$

Write each radical as an exponential.

33. $\sqrt{7^9}$ **34.** $\sqrt[5]{p^4}$

Simplify each expression. Write answers with only positive exponents. Assume that all variables represent positive real numbers.

35. $5^{1/4} \cdot 5^{7/4}$ **36.** $\frac{96^{2/3}}{96^{-1/3}}$ **37.** $\frac{(a^{1/3})^4}{a^{2/3}}$

38. $\frac{y^{-1/3} \cdot y^{5/6}}{y}$ **39.** $\left(\frac{z^{-1}x^{-3/5}}{2^{-2}z^{-1/2}x}\right)^{-1}$ **40.** $r^{-1/2}(r + r^{3/2})$

---

1. 42 **2.** −17
3. 6 **4.** −5
5. −3 **6.** −2
7. $\sqrt[n]{a}$ is not a real number if n is even and a is negative.
8. $|x|$
9. $-|x|$ **10.** x
11. −6.856 **12.** −5.053
13. 4.960 **14.** 4.729
15. −7.937 **16.** −5.292
17.
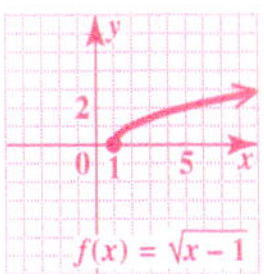

domain: $[1, \infty)$; range: $[0, \infty)$
18.
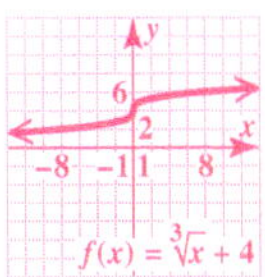

domain: $(-\infty, \infty)$; range: $(-\infty, \infty)$
19. cube (third); 8; 2; second; 4; 4
20. A
21. **(a)** m must be even. **(b)** m must be odd.
22. no
23. 7 **24.** −11
25. 32 **26.** −4
27. $-\frac{216}{125}$ **28.** −32
29. $\frac{1000}{27}$
30. It is not a real number.
31. $\sqrt{m+3n}$
32. $\frac{1}{(\sqrt[3]{3a+b})^5}$, or $\frac{1}{\sqrt[3]{(3a+b)^5}}$
33. $7^{9/2}$ **34.** $p^{4/5}$
35. 5^2, or 25 **36.** 96
37. $a^{2/3}$ **38.** $\frac{1}{y^{1/2}}$
39. $\frac{z^{1/2}x^{8/5}}{4}$ **40.** $r^{1/2} + r$

41. $s^{1/2}$ **42.** $r^{3/2}$
43. $p^{1/2}$ **44.** $k^{9/4}$
45. $m^{13/3}$ **46.** $z^{1/12}$
47. $x^{1/8}$ **48.** $x^{1/15}$
49. $x^{1/36}$
50. The product rule for exponents applies only if the bases are the same.

51. $\sqrt{66}$ **52.** $\sqrt{5r}$
53. $\sqrt[3]{30}$ **54.** $\sqrt[4]{21}$
55. $2\sqrt{5}$ **56.** $5\sqrt{3}$
57. $-5\sqrt{5}$ **58.** $-3\sqrt[3]{4}$
59. $10y^3\sqrt{y}$ **60.** $4pq^2\sqrt[3]{p}$
61. $3a^2b\sqrt[3]{4a^2b^2}$
62. $2r^2t\sqrt[3]{79r^2t}$
63. $\frac{y\sqrt{y}}{12}$ **64.** $\frac{m^5}{3}$
65. $\frac{\sqrt[3]{r^2}}{2}$ **66.** $\frac{a^2\sqrt[4]{a}}{3}$
67. $\sqrt{15}$ **68.** $p\sqrt{p}$
69. $\sqrt[12]{2000}$ **70.** $\sqrt[10]{x^7}$
71. 10 **72.** $\sqrt{197}$
73. $x^2 + y^2 = 121$
74. $(x + 2)^2 + (y - 4)^2 = 9$
75. $(x + 1)^2 + (y + 3)^2 = 25$
76. $(x - 4)^2 + (y - 2)^2 = 36$

77.
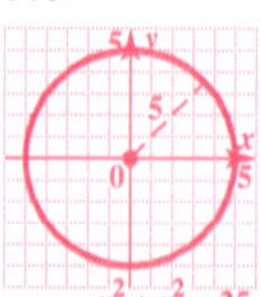
$x^2 + y^2 = 25$
center: (0, 0)

78.
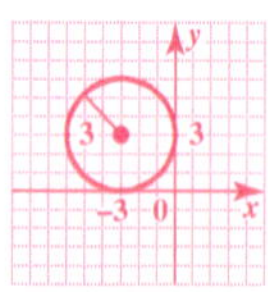
$(x + 3)^2 + (y - 3)^2 = 9$
center: (−3, 3)

79.
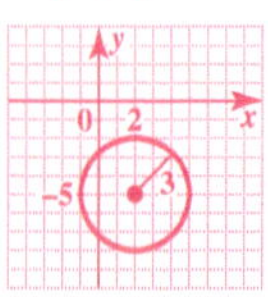
$(x - 2)^2 + (y + 5)^2 = 9$
center: (2, −5)

80. It is impossible for the sum of the squares of two real numbers to be negative.

81. $-11\sqrt{2}$ **82.** $23\sqrt{5}$
83. $7\sqrt{3y}$ **84.** $26m\sqrt{6m}$
85. $19\sqrt[3]{2}$ **86.** $-8\sqrt[4]{2}$

87. $1 - \sqrt{3}$ **88.** 2
89. $9 - 7\sqrt{2}$ **90.** $15 - 2\sqrt{26}$
91. 29
92. $2\sqrt[3]{2y^2} + 2\sqrt[3]{4y} - 3$

Write each radical as an exponential and then simplify. Leave answers in exponential form. Assume that all variables represent positive real numbers.

41. $\sqrt[8]{s^4}$ **42.** $\sqrt[6]{r^9}$ **43.** $\frac{\sqrt{p^5}}{p^2}$

44. $\sqrt[4]{k^3} \cdot \sqrt{k^3}$ **45.** $\sqrt[3]{m^5} \cdot \sqrt[3]{m^8}$ **46.** $\sqrt[4]{\sqrt[3]{z}}$

47. $\sqrt{\sqrt{\sqrt{x}}}$ **48.** $\sqrt[3]{\sqrt[5]{x}}$ **49.** $\sqrt{\sqrt[6]{\sqrt[3]{x}}}$

50. The product rule does not apply to $3^{1/4} \cdot 2^{1/5}$. Why?

7.3 *Simplify. Assume that all variables represent positive real numbers.*

51. $\sqrt{6} \cdot \sqrt{11}$ **52.** $\sqrt{5} \cdot \sqrt{r}$ **53.** $\sqrt[3]{6} \cdot \sqrt[3]{5}$ **54.** $\sqrt[4]{7} \cdot \sqrt[4]{3}$

55. $\sqrt{20}$ **56.** $\sqrt{75}$ **57.** $-\sqrt{125}$ **58.** $\sqrt[3]{-108}$

59. $\sqrt{100y^7}$ **60.** $\sqrt[3]{64p^4q^6}$ **61.** $\sqrt[3]{108a^8b^5}$ **62.** $\sqrt[3]{632r^8t^4}$

63. $\sqrt{\frac{y^3}{144}}$ **64.** $\sqrt[3]{\frac{m^{15}}{27}}$ **65.** $\sqrt[3]{\frac{r^2}{8}}$ **66.** $\sqrt[4]{\frac{a^9}{81}}$

Simplify each radical expression.

67. $\sqrt[6]{15^3}$ **68.** $\sqrt[4]{p^6}$ **69.** $\sqrt[3]{2} \cdot \sqrt[4]{5}$ **70.** $\sqrt{x} \cdot \sqrt[5]{x}$

71. Find the length x of the hypotenuse of the right triangle shown.

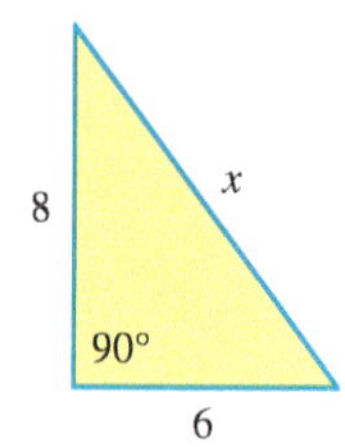

72. Find the distance between the points $(-4, 7)$ and $(10, 6)$.

Find the equation of a circle satisfying the given conditions.

73. Center $(0, 0)$, $r = 11$ **74.** Center $(-2, 4)$, $r = 3$

75. Center $(-1, -3)$, $r = 5$ **76.** Center $(4, 2)$, $r = 6$

Graph each circle. Identify the center.

77. $x^2 + y^2 = 25$ **78.** $(x + 3)^2 + (y - 3)^2 = 9$ **79.** $(x - 2)^2 + (y + 5)^2 = 9$

80. Why does the equation $x^2 + y^2 = -1$ have no points on its graph?

7.4 *Perform the indicated operations. Assume that all variables represent positive real numbers.*

81. $2\sqrt{8} - 3\sqrt{50}$ **82.** $8\sqrt{80} - 3\sqrt{45}$ **83.** $-\sqrt{27y} + 2\sqrt{75y}$

84. $2\sqrt{54m^3} + 5\sqrt{96m^3}$ **85.** $3\sqrt[3]{54} + 5\sqrt[3]{16}$ **86.** $-6\sqrt[4]{32} + \sqrt[4]{512}$

7.5 *Multiply, and then simplify.*

87. $(\sqrt{3} + 1)(\sqrt{3} - 2)$ **88.** $(\sqrt{7} + \sqrt{5})(\sqrt{7} - \sqrt{5})$

89. $(3\sqrt{2} + 1)(2\sqrt{2} - 3)$ **90.** $(\sqrt{13} - \sqrt{2})^2$

91. $(\sqrt[3]{2} + 3)(\sqrt[3]{4} - 3\sqrt[3]{2} + 9)$ **92.** $(\sqrt[3]{4y} - 1)(\sqrt[3]{4y} + 3)$

93. $\frac{\sqrt{30}}{5}$ 94. $-3\sqrt{6}$
95. $\frac{3\sqrt{7py}}{y}$ 96. $\frac{\sqrt{22}}{4}$
97. $-\frac{\sqrt[3]{45}}{5}$ 98. $\frac{3m\sqrt[3]{4n}}{n^2}$
99. $\frac{\sqrt{2}-\sqrt{7}}{-5}$ 100. $\frac{5(\sqrt{6}+3)}{3}$
101. $\frac{1-\sqrt{5}}{4}$ 102. $\frac{-6+\sqrt{3}}{2}$
103. $\{2\}$ 104. $\{6\}$
105. $\{0, 5\}$ 106. $\{9\}$
107. $\{3\}$ 108. $\{7\}$
109. $\left\{-\frac{1}{2}\right\}$ 110. $\{14\}$
111. $\varnothing$ 112. $\{7\}$
113. $H = \sqrt{L^2 - W^2}$
114. 7.9 ft
115. $4i$ 116. $10i\sqrt{2}$
117. $-10 - 2i$ 118. $14 + 7i$
119. $-\sqrt{35}$ 120. -45
121. 3 122. $5 + i$
123. $32 - 24i$ 124. $1 - i$
125. $-i$ 126. 1
127. -1 128. 1

Rationalize each denominator. Assume that all variables represent positive real numbers.

93. $\frac{\sqrt{6}}{\sqrt{5}}$ **94.** $\frac{-6\sqrt{3}}{\sqrt{2}}$ **95.** $\frac{3\sqrt{7p}}{\sqrt{y}}$ **96.** $\sqrt{\frac{11}{8}}$

97. $-\sqrt[3]{\frac{9}{25}}$ **98.** $\sqrt[3]{\frac{108m^3}{n^5}}$ **99.** $\frac{1}{\sqrt{2}+\sqrt{7}}$ **100.** $\frac{-5}{\sqrt{6}-3}$

Write each quotient in lowest terms.

101. $\frac{2-2\sqrt{5}}{8}$ **102.** $\frac{-18+\sqrt{27}}{6}$

7.6 *Solve each equation.*

103. $\sqrt{8x+9} = 5$ **104.** $\sqrt{2x-3} - 3 = 0$

105. $\sqrt{7x+1} = x + 1$ **106.** $3\sqrt{x} = \sqrt{10x-9}$

107. $\sqrt{x^2+3x+7} = x + 2$ **108.** $\sqrt{x+2} - \sqrt{x-3} = 1$

109. $\sqrt[3]{5x-1} = \sqrt[3]{3x-2}$ **110.** $\sqrt[3]{1-2x} - \sqrt[3]{-x-13} = 0$

111. $\sqrt[4]{x-1} + 2 = 0$ **112.** $\sqrt[4]{x+7} = \sqrt[4]{2x}$

Carpenters stabilize wall frames with a diagonal brace, as shown in the figure. The length of the brace is given by

$$L = \sqrt{H^2 + W^2}.$$

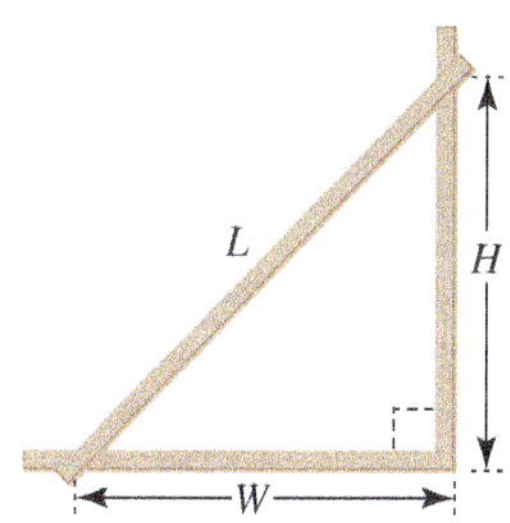

113. Solve this formula for H.

114. If the bottom of the brace is attached 9 ft from the corner and the brace is 12 ft long, how far up the corner post should it be nailed? Give the answer to the nearest tenth of a foot.

7.7 *Write each number as a product of a real number and i. Simplify.*

115. $\sqrt{-16}$ **116.** $\sqrt{-200}$

Perform the indicated operations. Give answers in standard form.

117. $(-2 + 5i) + (-8 - 7i)$ **118.** $(5 + 4i) - (-9 - 3i)$

119. $\sqrt{-5} \cdot \sqrt{-7}$ **120.** $\sqrt{-25} \cdot \sqrt{-81}$

121. $\frac{\sqrt{-72}}{\sqrt{-8}}$ **122.** $(2 + 3i)(1 - i)$

123. $(6 - 2i)^2$ **124.** $\frac{3-i}{2+i}$

Find each power of i.

125. i^{11} **126.** i^{36} **127.** i^{-10} **128.** i^{-8}

Chapter 7 Mixed Review Exercises

Simplify. Assume that all variables represent positive real numbers.

1. $-\sqrt[4]{256}$ **2.** $1000^{-2/3}$ **3.** $\dfrac{z^{-1/5} \cdot z^{3/10}}{z^{7/10}}$ **4.** $\sqrt[3]{54z^9t^8}$

5. $\sqrt{-49}$ **6.** $\dfrac{-1}{\sqrt{12}}$ **7.** $\sqrt[3]{\dfrac{12}{25}}$ **8.** i^{-1000}

9. $-5\sqrt{18} + 12\sqrt{72}$ **10.** $(4 - 9i) + (-1 + 2i)$ **11.** $\dfrac{\sqrt{50}}{\sqrt{-2}}$

12. $\dfrac{3 + \sqrt{54}}{6}$ **13.** $(3 + 2i)^2$ **14.** $8\sqrt[3]{x^3y^2} - 2x\sqrt[3]{y^2}$

Solve each equation.

15. $\sqrt{x + 4} = x - 2$ **16.** $\sqrt[3]{2x - 9} = \sqrt[3]{5x + 3}$

17. $\sqrt{6 + 2x} - 1 = \sqrt{7 - 2x}$ **18.** $\sqrt{7x + 11} - 5 = 0$

19. $\sqrt{6x + 2} - \sqrt{5x + 3} = 0$ **20.** $\sqrt{3 + 5x} - \sqrt{x + 11} = 0$

21. $\sqrt{11 + 2x} + 1 = \sqrt{5x + 1}$ **22.** $\sqrt{5x + 6} - \sqrt{x + 3} = 3$

Answers:

1. -4 2. $\frac{1}{100}$
3. $\frac{1}{z^{3/5}}$ 4. $3z^3t^2\sqrt[3]{2t^2}$
5. $7i$ 6. $-\frac{\sqrt{3}}{6}$
7. $\frac{\sqrt[3]{60}}{5}$ 8. 1
9. $57\sqrt{2}$ 10. $3 - 7i$
11. $-5i$ 12. $\frac{1 + \sqrt{6}}{2}$
13. $5 + 12i$ 14. $6x\sqrt[3]{y^2}$
15. $\{5\}$ 16. $\{-4\}$
17. $\left\{\frac{3}{2}\right\}$ 18. $\{2\}$
19. $\{1\}$ 20. $\{2\}$
21. $\{7\}$ 22. $\{6\}$

Chapter 7 Test

FOR EXTRA HELP *Step-by-step test solutions are found on the Chapter Test Prep Videos available in* MyMathLab® *or on* YouTube.

View the complete solutions to all Chapter Test exercises in MyMathLab.

Evaluate.

1. $-\sqrt{841}$ **2.** $\sqrt[3]{-512}$ **3.** $125^{1/3}$

Find a decimal approximation for each radical. Round answers to three decimal places.

4. $\sqrt{478}$ **5.** $\sqrt[3]{-832}$

6. Graph the function $f(x) = \sqrt{x + 6}$, and give the domain and range.

Simplify. Assume that all variables represent positive real numbers.

7. $\left(\dfrac{16}{25}\right)^{-3/2}$ **8.** $(-64)^{-4/3}$ **9.** $\dfrac{3^{2/5}x^{-1/4}y^{2/5}}{3^{-8/5}x^{7/4}y^{1/10}}$

10. $\left(\dfrac{x^{-4}y^{-6}}{x^{-2}y^3}\right)^{-2/3}$ **11.** $7^{3/4} \cdot 7^{-1/4}$ **12.** $\sqrt[3]{a^4} \cdot \sqrt[3]{a^7}$

13. Find the exact length of side b in the figure shown at the right.

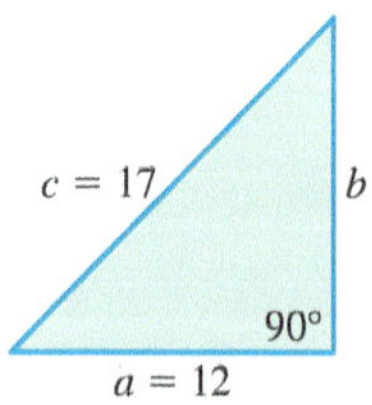

14. Find the distance between the points $(-4, 2)$ and $(2, 10)$.

15. Write an equation of a circle with center $(-4, 6)$ and radius 5.

16. Graph the circle $(x - 2)^2 + (y + 3)^2 = 16$. Identify the center.

Simplify. Assume that all variables represent positive real numbers.

17. $\sqrt{54x^5y^6}$ **18.** $\sqrt[4]{32a^7b^{13}}$

Answers:

[7.1]
1. -29 2. -8

[7.2]
3. 5

[7.1]
4. 21.863
5. -9.405
6. domain: $[-6, \infty)$; range: $[0, \infty)$

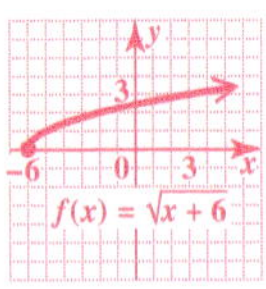

[7.2]
7. $\frac{125}{64}$ 8. $\frac{1}{256}$
9. $\frac{9y^{3/10}}{x^2}$ 10. $x^{4/3}y^6$
11. $7^{1/2}$, or $\sqrt{7}$

[7.3]
12. $a^3\sqrt[3]{a^2}$, or $a^{11/3}$
13. $\sqrt{145}$ 14. 10
15. $(x + 4)^2 + (y - 6)^2 = 25$

16. 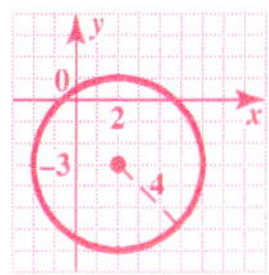

$(x-2)^2+(y+3)^2=16$

center: $(2, -3)$; radius: 4

17. $3x^2y^3\sqrt{6x}$ 18. $2ab^3\sqrt[4]{2a^3b}$

19. $\sqrt[6]{200}$

[7.4]

20. $26\sqrt{5}$ 21. $(2ts-3t^2)\sqrt[3]{2s^2}$

[7.5]

22. $66+\sqrt{5}$ 23. $23-4\sqrt{15}$

24. $-\frac{\sqrt{10}}{4}$ 25. $\frac{2\sqrt[3]{25}}{5}$

26. $-2(\sqrt{7}-\sqrt{5})$

27. $3+\sqrt{6}$

[7.6]

28. (a) 59.8

(b) $T=\frac{V_0^2-V^2}{-V^2k}$, or $T=\frac{V^2-V_0^2}{V^2k}$

29. $\{-1\}$ 30. $\{3\}$

31. $\{-3\}$

[7.7]

32. $-5-8i$ 33. $-2+16i$

34. $3+4i$ 35. i

36. (a) true (b) true (c) false (d) true

19. $\sqrt{2}\cdot\sqrt[3]{5}$ (Express as a radical.)

20. $3\sqrt{20}-5\sqrt{80}+4\sqrt{500}$

21. $\sqrt[3]{16t^3s^5}-\sqrt[3]{54t^6s^2}$

22. $(7\sqrt{5}+4)(2\sqrt{5}-1)$

23. $(\sqrt{3}-2\sqrt{5})^2$ **24.** $\frac{-5}{\sqrt{40}}$ **25.** $\frac{2}{\sqrt[3]{5}}$ **26.** $\frac{-4}{\sqrt{7}+\sqrt{5}}$

27. Write $\frac{6+\sqrt{24}}{2}$ in lowest terms.

28. The following formula from physics relates the velocity V of sound to the temperature T.

$$V=\frac{V_0}{\sqrt{1-kT}}$$

(a) Approximate V to the nearest tenth if $V_0=50$, $k=0.01$, and $T=30$.

(b) Solve the formula for T.

Solve each equation.

29. $\sqrt[3]{5x}=\sqrt[3]{2x-3}$ **30.** $\sqrt{x+6}=9-2x$

31. $\sqrt{x+4}-\sqrt{1-x}=-1$

In Exercises 32–34, perform the indicated operations. Give the answers in standard form.

32. $(-2+5i)-(3+6i)-7i$ **33.** $(1+5i)(3+i)$ **34.** $\frac{7+i}{1-i}$

35. Simplify i^{37}.

36. Answer *true* or *false* to each of the following.

(a) $i^2=-1$ **(b)** $i=\sqrt{-1}$ **(c)** $i=-1$ **(d)** $\sqrt{-3}=i\sqrt{3}$

Chapters R–7 Cumulative Review Exercises

[R.3]

1. 1 2. $-\frac{14}{9}$

[1.1]

3. $\{-4\}$ 4. $\{-12\}$

5. $\{6\}$

[1.7] 6. $\left\{-\frac{10}{3}, 1\right\}$ [1.5] 7. $(-6, \infty)$

[1.3]

8. 36 nickels; 64 quarters

9. $2\frac{2}{39}$ L

[2.1]

10. 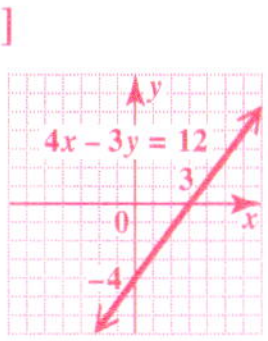

Evaluate each expression for $a=-3$, $b=5$, and $c=-4$.

1. $|2a^2-3b+c|$ **2.** $\frac{(a+b)(a+c)}{3b-6}$

Solve each equation.

3. $3(x+2)-4(2x+3)=-3x+2$ **4.** $\frac{1}{3}x+\frac{1}{4}(x+8)=x+7$

5. $0.04x+0.06(100-x)=5.88$ **6.** $|6x+7|=13$

7. Find the solution set of $-5-3(x-2)<11-2(x+2)$. Write it in interval notation.

Solve each problem.

8. A piggy bank has 100 coins, all of which are nickels and quarters. The total value of the money is \$17.80. How many of each denomination are there in the bank?

9. How many liters of pure alcohol must be mixed with 40 L of 18% alcohol to obtain a 22% alcohol solution?

10. Graph the equation $4x-3y=12$.

11. Find the slope of the line passing through the points $(-4, 6)$ and $(2, -3)$. Then find the equation of the line and write it in the form $y=mx+b$.

12. If $f(x) = 3x - 7$, find $f(-10)$.

Solve each system.

13. $3x - y = 23$
 $2x + 3y = 8$

14. $x + y + z = 1$
 $x - y - z = -3$
 $x + y - z = -1$

15. Several years ago, sending five 2-oz letters and three 3-oz letters by first-class mail would have cost \$5.39. Sending three 2-oz letters and five 3-oz letters would have cost \$5.73. What was the rate for one 2-oz letter and one 3-oz letter? (*Source:* U.S. Postal Service.)

Perform the indicated operations.

16. $(3k^3 - 5k^2 + 8k - 2) - (4k^3 + 11k + 7) + (2k^2 - 5k)$

17. $(8x - 7)(x + 3)$

18. $\dfrac{6y^4 - 3y^3 + 5y^2 + 6y - 9}{2y + 1}$

Factor each polynomial completely.

19. $2p^2 - 5pq + 3q^2$

20. $3k^4 + k^2 - 4$

21. $x^3 + 512$

Solve by the zero-factor property.

22. $2x^2 + 11x + 15 = 0$

23. $5x(x - 1) = 2(1 - x)$

24. What is the domain of $f(x) = \dfrac{4}{x^2 - 9}$? Use set-builder notation.

Perform each operation and express the answer in lowest terms.

25. $\dfrac{y^2 + y - 12}{y^3 + 9y^2 + 20y} \div \dfrac{y^2 - 9}{y^3 + 3y^2}$

26. $\dfrac{1}{x + y} + \dfrac{3}{x - y}$

Simplify.

27. $\dfrac{\frac{-6}{x - 2}}{\frac{8}{3x - 6}}$

28. $\dfrac{\frac{1}{a} - \frac{1}{b}}{\frac{a}{b} - \frac{b}{a}}$

29. $\dfrac{x^{-1}}{y - x^{-1}}$

30. Solve the equation $\dfrac{x + 1}{x - 3} = \dfrac{4}{x - 3} + 6$.

31. Danielle can ride her bike 4 mph faster than her husband, Richard. If Danielle can ride 48 mi in the same time that Richard can ride 24 mi, what are their rates?

Write each expression in simplest form, using only positive exponents. Assume that all variables represent positive real numbers.

32. $27^{-2/3}$

33. $\sqrt[3]{16x^2y} \cdot \sqrt[3]{3x^3y}$

34. $\sqrt{50} + \sqrt{8}$

35. $\dfrac{1}{\sqrt{10} - \sqrt{8}}$

36. Find the distance between the points $(-4, 4)$ and $(-2, 9)$.

37. Solve the equation $\sqrt{3x - 8} = x - 2$.

38. Express $\frac{6 - 2i}{1 - i}$ in standard form.

[2.2, 2.3]
11. $-\frac{3}{2}$; $y = -\frac{3}{2}x$

[2.6]
12. -37

[3.1]
13. $\{(7, -2)\}$

[3.2]
14. $\{(-1, 1, 1)\}$

[3.3]
15. 2-oz letter: \$0.61; 3-oz letter: \$0.78

[4.2]
16. $-k^3 - 3k^2 - 8k - 9$

[4.4]
17. $8x^2 + 17x - 21$

[4.5]
18. $3y^3 - 3y^2 + 4y + 1 + \frac{-10}{2y + 1}$

[5.2, 5.3]
19. $(2p - 3q)(p - q)$
20. $(3k^2 + 4)(k - 1)(k + 1)$
21. $(x + 8)(x^2 - 8x + 64)$

[5.5]
22. $\left\{-3, -\frac{5}{2}\right\}$
23. $\left\{-\frac{2}{5}, 1\right\}$

[6.1]
24. $\{x \mid x \text{ is a real number}, x \neq \pm 3\}$
25. $\frac{y}{y + 5}$

[6.2]
26. $\frac{4x + 2y}{(x + y)(x - y)}$

[6.3]
27. $-\frac{9}{4}$
28. $\frac{-1}{a + b}$
29. $\frac{1}{xy - 1}$

[6.4]
30. $\varnothing$

[6.5]
31. Danielle: 8 mph; Richard: 4 mph

[7.2]
32. $\frac{1}{9}$

[7.3]
33. $2x\sqrt[3]{6x^2y^2}$

[7.4]
34. $7\sqrt{2}$

[7.5]
35. $\frac{\sqrt{10} + 2\sqrt{2}}{2}$

[7.3]
36. $\sqrt{29}$

[7.6]
37. $\{3, 4\}$

[7.7]
38. $4 + 2i$

8 Quadratic Equations, Inequalities, and Functions

Quadratic functions, one of the topics of this chapter, have graphs that are *parabolas*. Cross sections of telescopes, satellite dishes, and automobile headlights form parabolas, as do the cables that support suspension bridges.

8.1 The Square Root Property and Completing the Square

OBJECTIVES

1. Review the zero-factor property.
2. Learn the square root property.
3. Solve quadratic equations of the form $(ax + b)^2 = c$ by extending the square root property.
4. Solve quadratic equations by completing the square.
5. Solve quadratic equations with solutions that are not real numbers.

VOCABULARY

☐ quadratic equation
☐ second-degree equation

Recall from **Section 5.5** that a *quadratic equation* is defined as follows.

Quadratic Equation

A **quadratic equation** (in x here) can be written in the form

$$ax^2 + bx + c = 0,$$

where a, b, and c are real numbers and $a \neq 0$. The given form is called **standard form.**

Examples: $4x^2 + 4x - 5 = 0$ and $3x^2 = 4x - 8$ Quadratic equations (The first equation is in standard form.)

A quadratic equation is a **second-degree equation**—that is, an equation with a squared variable term and no terms of greater degree.

OBJECTIVE 1 Review the zero-factor property.

In **Section 5.5** we used the zero-factor property to solve quadratic equations.

Zero-Factor Property

If two numbers have a product of 0, then at least one of the numbers must be 0. **That is, if $ab = 0$, then $a = 0$ or $b = 0$.**

NOW TRY EXERCISE 1
Solve $2x^2 + 5x - 12 = 0$.

CLASSROOM EXAMPLE 1
Solve $2x^2 - 3x = -1$.
Answer: $\left\{\frac{1}{2}, 1\right\}$

EXAMPLE 1 Using the Zero-Factor Property

Solve $3x^2 - 5x - 28 = 0$.

This trinomial can be factored. The quadratic equation must be in standard form.

$$3x^2 - 5x - 28 = 0$$

$$(3x + 7)(x - 4) = 0 \quad \text{Factor.}$$

$$3x + 7 = 0 \quad \text{or} \quad x - 4 = 0 \quad \text{Zero-factor property}$$

$$3x = -7 \quad \text{or} \quad x = 4 \quad \text{Solve each equation.}$$

$$x = -\frac{7}{3}$$

To check, substitute each value in the original equation. The solution set is $\left\{-\frac{7}{3}, 4\right\}$.

NOW TRY

TEACHING TIP
Explain to students that until now, all quadratic equations have been carefully selected so that they could be solved using the zero-factor property. The methods in the following sections will enable us to solve *all* quadratic equations.

OBJECTIVE 2 Learn the square root property.

Not every quadratic equation can be solved using the zero-factor property. Other methods of solving quadratic equations are based on the following property.

Square Root Property

If x and k are complex numbers and $x^2 = k$, then

$$x = \sqrt{k} \quad \text{or} \quad x = -\sqrt{k}.$$

NOW TRY ANSWER
1. $\left\{-4, \frac{3}{2}\right\}$

TEACHING TIP
Tell students to expect a maximum of two solutions in the solution set of a quadratic equation.
Caution students about the common error of forgetting to write the negative solution in solving equations such as the one in Example 2(a).

These steps justify the square root property.

$$x^2 = k$$

$$x^2 - k = 0 \quad \text{Subtract } k.$$

$$(x - \sqrt{k})(x + \sqrt{k}) = 0 \quad \text{Factor.}$$

$$x - \sqrt{k} = 0 \quad \text{or} \quad x + \sqrt{k} = 0 \quad \text{Zero-factor property}$$

$$\boldsymbol{x = \sqrt{k}} \quad \text{or} \quad \boldsymbol{x = -\sqrt{k}} \quad \text{Solve each equation.}$$

CAUTION If $k \neq 0$, then using the square root property always produces *two* square roots, one positive and one negative.

NOW TRY EXERCISE 2
Solve each equation.
(a) $t^2 = 10$
(b) $2x^2 - 90 = 0$

CLASSROOM EXAMPLE 2
Solve each equation.
(a) $m^2 = 64$
(b) $3x^2 - 54 = 0$
Answers: (a) $\{8, -8\}$
(b) $\{3\sqrt{2}, -3\sqrt{2}\}$

EXAMPLE 2 Using the Square Root Property

Solve each equation.

(a) $x^2 = 5$

By the square root property, if $x^2 = 5$, then

$$x = \sqrt{5} \quad \text{or} \quad x = -\sqrt{5}.$$

Don't forget the negative solution.

These solutions are irrational numbers. The solution set is $\{\sqrt{5}, -\sqrt{5}\}$.

(b)

$$4x^2 - 48 = 0$$

$$4x^2 = 48 \quad \text{Add 48.}$$

$$x^2 = 12 \quad \text{Divide by 4.}$$

Don't stop here. Simplify the radicals.

$$x = \sqrt{12} \quad \text{or} \quad x = -\sqrt{12} \quad \text{Square root property}$$

$$x = 2\sqrt{3} \quad \text{or} \quad x = -2\sqrt{3} \quad \sqrt{12} = \sqrt{4} \cdot \sqrt{3} = 2\sqrt{3}$$

CHECK $4x^2 - 48 = 0$ Original equation

$$4(2\sqrt{3})^2 - 48 \stackrel{?}{=} 0 \quad \text{Let } x = 2\sqrt{3}.$$

$$4(12) - 48 \stackrel{?}{=} 0$$

$$48 - 48 \stackrel{?}{=} 0$$

$$0 = 0 \; ✓ \; \text{True}$$

$(2\sqrt{3})^2 = 2^2 \cdot (\sqrt{3})^2$

$$4(-2\sqrt{3})^2 - 48 \stackrel{?}{=} 0 \quad \text{Let } x = -2\sqrt{3}.$$

$$4(12) - 48 \stackrel{?}{=} 0$$

$$48 - 48 \stackrel{?}{=} 0$$

$$0 = 0 \; ✓ \; \text{True}$$

The solution set is $\{2\sqrt{3}, -2\sqrt{3}\}$.

(c)

$$3x^2 + 5 = 11$$

$$3x^2 = 6 \quad \text{Subtract 5.}$$

$$x^2 = 2 \quad \text{Divide by 3.}$$

$$x = \sqrt{2} \quad \text{or} \quad x = -\sqrt{2} \quad \text{Square root property}$$

The solution set is $\{\sqrt{2}, -\sqrt{2}\}$. NOW TRY

NOTE Using the symbol $\pm$ (read "positive or negative," or "plus or minus"), the solutions in **Example 2** could be written $\pm\sqrt{5}$, $\pm 2\sqrt{3}$, and $\pm\sqrt{2}$.

NOW TRY ANSWERS
2. (a) $\{\sqrt{10}, -\sqrt{10}\}$
(b) $\{3\sqrt{5}, -3\sqrt{5}\}$

NOW TRY EXERCISE 3
Tim is dropping roofing nails from the top of a roof 25 ft high into a large bucket on the ground. Use the formula in **Example 3** to determine how long it will take a nail dropped from 25 ft to hit the bottom of the bucket.

CLASSROOM EXAMPLE 3
An expert marksman can hold a silver dollar at forehead level, drop it, draw his gun, and shoot the coin as it passes waist level. If the coin falls about 4 ft, use the formula in **Example 3** to find the time that elapses between the dropping of the coin and the shot.

Answer: 0.5 sec

EXAMPLE 3 Using the Square Root Property in an Application

Galileo Galilei developed a formula for freely falling objects described by

$$d = 16t^2,$$

where d is the distance in feet that an object falls (disregarding air resistance) in t seconds, regardless of weight.

If the Leaning Tower of Pisa is about 180 ft tall, use Galileo's formula to determine how long it would take an object dropped from the top of the tower to fall to the ground. (*Source:* www.brittanica.com)

Galileo Galilei (1564–1642)

$$d = 16t^2 \quad \text{Galileo's formula}$$

$$180 = 16t^2 \quad \text{Let } d = 180.$$

$$11.25 = t^2 \quad \text{Divide by 16.}$$

$$t = \sqrt{11.25} \quad \text{or} \quad t = -\sqrt{11.25} \quad \text{Square root property}$$

Time cannot be negative, so we discard $-\sqrt{11.25}$. Using a calculator,

$$\sqrt{11.25} = 3.4 \quad \text{so} \quad t = 3.4. \quad \text{Round to the nearest tenth.}$$

The object would fall to the ground in 3.4 sec. NOW TRY

CAUTION When solving applications, a solution of a quadratic equation may not satisfy the physical requirements as in **Example 3.** Reject such solutions as answers.

OBJECTIVE 3 Solve quadratic equations of the form $(ax + b)^2 = c$ by extending the square root property.

NOW TRY EXERCISE 4
Solve $(x + 3)^2 = 25$.

CLASSROOM EXAMPLE 4
Solve $(x - 3)^2 = 16$.

Answer: $\{-1, 7\}$

EXAMPLE 4 Extending the Square Root Property

Solve $(x - 5)^2 = 36$.

$$\underset{x^2}{(x - 5)^2} = \underset{k}{36} \quad \text{Substitute } (x-5)^2 \text{ for } x^2 \text{ and 36 for } k \text{ in the square root property.}$$

$$x - 5 = \sqrt{36} \quad \text{or} \quad x - 5 = -\sqrt{36} \quad \text{Square root property}$$

$$x - 5 = 6 \quad \text{or} \quad x - 5 = -6 \quad \text{Take square roots.}$$

$$x = 11 \quad \text{or} \quad x = -1 \quad \text{Add 5.}$$

CHECK $(x - 5)^2 = 36$ Original equation

$(11 - 5)^2 \stackrel{?}{=} 36$ Let $x = 11$.	$(-1 - 5)^2 \stackrel{?}{=} 36$ Let $x = -1$.
$6^2 \stackrel{?}{=} 36$	$(-6)^2 \stackrel{?}{=} 36$
$36 = 36$ ✓ True	$36 = 36$ ✓ True

Both values satisfy the original equation. The solution set is $\{-1, 11\}$. NOW TRY

NOW TRY ANSWERS
3. 1.25 sec
4. $\{-8, 2\}$

NOW TRY EXERCISE 5
Solve $(5x - 4)^2 = 27$.

CLASSROOM EXAMPLE 5
Solve $(3x + 1)^2 = 2$.

Answer: $\left\{\frac{-1 \pm \sqrt{2}}{3}\right\}$

EXAMPLE 5 Extending the Square Root Property

Solve $(2x - 3)^2 = 18$.

$$2x - 3 = \sqrt{18} \quad \text{or} \quad 2x - 3 = -\sqrt{18} \qquad \text{Square root property}$$

$$2x = 3 + \sqrt{18} \quad \text{or} \quad 2x = 3 - \sqrt{18} \qquad \text{Add 3.}$$

$$x = \frac{3 + \sqrt{18}}{2} \quad \text{or} \quad x = \frac{3 - \sqrt{18}}{2} \qquad \text{Divide by 2.}$$

$$x = \frac{3 + 3\sqrt{2}}{2} \quad \text{or} \quad x = \frac{3 - 3\sqrt{2}}{2} \qquad \sqrt{18} = \sqrt{9 \cdot 2} = 3\sqrt{2}$$

We show the check for the first value. The check for the other value is similar.

CHECK

$$(2x - 3)^2 = 18 \qquad \text{Original equation}$$

$$\left[2\left(\frac{3 + 3\sqrt{2}}{2}\right) - 3\right]^2 \stackrel{?}{=} 18 \qquad \text{Let } x = \frac{3 + 3\sqrt{2}}{2}.$$

$$(3 + 3\sqrt{2} - 3)^2 \stackrel{?}{=} 18 \qquad \text{Multiply.}$$

$$(3\sqrt{2})^2 \stackrel{?}{=} 18 \qquad \text{Simplify.}$$

$$18 = 18 \checkmark \qquad \text{True}$$

The $\pm$ symbol denotes two solutions.

The solution set is $\left\{\frac{3 + 3\sqrt{2}}{2}, \frac{3 - 3\sqrt{2}}{2}\right\}$, abbreviated $\left\{\frac{3 \pm 3\sqrt{2}}{2}\right\}$. NOW TRY

TEACHING TIP
For some classes, it may be desirable to cover completing the square in a second class session.

TEACHING TIP
Have students practice finding the constant to be added to obtain a perfect square trinomial in each case.

$x^2 + 4x + \underline{?}$ [4]

$t^2 - 2t + \underline{?}$ [1]

$m^2 + 5m + \underline{?}$ $\left[\frac{25}{4}\right]$

$x^2 - \frac{2}{3}x + \underline{?}$ $\left[\frac{1}{9}\right]$

OBJECTIVE 4 Solve quadratic equations by completing the square.

We can use the square root property to solve *any* quadratic equation by writing it in the form

$$\text{Square of a binomial} \rightarrow (x + k)^2 = n. \leftarrow \text{Constant}$$

That is, we must write the left side of the equation as a perfect square trinomial that can be factored as $(x + k)^2$, the square of a binomial, and the right side must be a constant. This process is called **completing the square.**

Recall that the perfect square trinomial

$$x^2 + 10x + 25 \quad \text{can be factored as} \quad (x + 5)^2.$$

In the trinomial, the coefficient of x (the first-degree term) is 10 and the constant term is 25. If we take half of 10 and square it, we obtain the constant term, 25.

Coefficient of x ↓ Constant ↓

$$\left[\frac{1}{2}(10)\right]^2 = 5^2 = 25$$

Similarly, in $x^2 + 12x + 36$, $\left[\frac{1}{2}(12)\right]^2 = 6^2 = 36$,

and in $m^2 - 6m + 9$, $\left[\frac{1}{2}(-6)\right]^2 = (-3)^2 = 9$.

This relationship is true in general and is the idea behind completing the square.

NOW TRY ANSWER
5. $\left\{\frac{4 \pm 3\sqrt{3}}{5}\right\}$

NOW TRY EXERCISE 6

Solve $x^2 + 6x - 2 = 0$.

CLASSROOM EXAMPLE 6

Solve $x^2 - 2x - 10 = 0$.

Answer: $\{1 \pm \sqrt{11}\}$

EXAMPLE 6 Solving a Quadratic Equation by Completing the Square ($a = 1$)

Solve $x^2 + 8x + 10 = 0$.

The trinomial on the left is nonfactorable, so this quadratic equation cannot be solved using the zero-factor property. It is not in the correct form to solve using the square root property. We can solve it by completing the square.

$$x^2 + 8x + 10 = 0 \quad \text{Original equation}$$

Only terms with variables remain on the left side.

$$x^2 + 8x = -10 \quad \text{Subtract 10.}$$

$$x^2 + 8x + \underline{?} = -10 \quad \text{We must add a constant.}$$

Needs to be a perfect square trinomial

Take half the coefficient of the first-degree term, $8x$, and square the result.

$$\left[\frac{1}{2}(8)\right]^2 = 4^2 = 16 \longleftarrow \text{Desired constant}$$

We add this constant, 16, to *each* side of the equation and continue solving as shown.

$$x^2 + 8x + 16 = -10 + 16 \quad \text{Add 16 to each side.}$$

This is a key step.

$$(x + 4)^2 = 6 \quad \text{Factor on the left. Add on the right.}$$

$$x + 4 = \sqrt{6} \quad \text{or} \quad x + 4 = -\sqrt{6} \quad \text{Square root property}$$

$$x = -4 + \sqrt{6} \quad \text{or} \quad x = -4 - \sqrt{6} \quad \text{Add } -4.$$

CHECK

$$x^2 + 8x + 10 = 0 \quad \text{Original equation}$$

$$\left(-4 + \sqrt{6}\right)^2 + 8\left(-4 + \sqrt{6}\right) + 10 \stackrel{?}{=} 0 \quad \text{Let } x = -4 + \sqrt{6}.$$

Remember the middle term when squaring $-4 + \sqrt{6}$.

$$16 - 8\sqrt{6} + 6 - 32 + 8\sqrt{6} + 10 \stackrel{?}{=} 0$$

$$0 = 0 \quad \checkmark \quad \text{True}$$

The check of $-4 - \sqrt{6}$ is similar. The solution set is

$$\left\{-4 + \sqrt{6}, -4 - \sqrt{6}\right\}, \quad \text{or} \quad \left\{-4 \pm \sqrt{6}\right\}.$$

NOW TRY

TEACHING TIP

Emphasize that students need to know the mechanics presented in this section to understand material presented later in the book (e.g., finding the vertex of a parabola).

TEACHING TIP

In Step 4, remind students to take the square root of both sides and remind them to use the $\pm$ sign on the right side. Then they should solve for the unknown and write the result in lowest terms.

NOW TRY ANSWER

6. $\{-3 \pm \sqrt{11}\}$

Completing the Square to Solve $ax^2 + bx + c = 0$ (Where $a \neq 0$)

Step 1 **Be sure the second-degree term has coefficient 1.**
- If the coefficient of the second-degree term is 1, go to Step 2.
- If it is not 1, but some other nonzero number a, divide each side of the equation by a.

Step 2 **Write the equation in correct form.** Make sure that all variable terms are on one side of the equality symbol and the constant term is on the other side.

Step 3 **Complete the square.**
- Take half the coefficient of the first-degree term, and square it.
- Add the square to each side of the equation.
- Factor the variable side, which should be a perfect square trinomial, as the square of a binomial. Combine terms on the other side.

Step 4 **Solve** the equation using the square root property.

NOW TRY EXERCISE 7

Solve $x^2 + x - 3 = 0$.

CLASSROOM EXAMPLE 7

Solve $x^2 + 3x - 1 = 0$.

Answer: $\left\{\frac{-3 \pm \sqrt{13}}{2}\right\}$

EXAMPLE 7 Solving a Quadratic Equation by Completing the Square ($a = 1$)

Solve $x^2 + 5x - 1 = 0$.

The coefficient of the second-degree term is 1, so we begin with Step 2.

Step 2 $x^2 + 5x = 1$ Add 1 to each side.

Step 3 Take half the coefficient of the first-degree term and square the result.

$$\left[\frac{1}{2}(5)\right]^2 = \left(\frac{5}{2}\right)^2 = \frac{25}{4}$$

$$x^2 + 5x + \frac{25}{4} = 1 + \frac{25}{4} \quad \text{Add the square to each side of the equation.}$$

$$\left(x + \frac{5}{2}\right)^2 = \frac{29}{4} \quad \text{Factor on the left. Add on the right.}$$

Step 4 $x + \frac{5}{2} = \sqrt{\frac{29}{4}}$ or $x + \frac{5}{2} = -\sqrt{\frac{29}{4}}$ Square root property

$$x + \frac{5}{2} = \frac{\sqrt{29}}{2} \quad \text{or} \quad x + \frac{5}{2} = -\frac{\sqrt{29}}{2} \qquad \sqrt{\frac{a}{b}} = \frac{\sqrt{a}}{\sqrt{b}}$$

$$x = -\frac{5}{2} + \frac{\sqrt{29}}{2} \quad \text{or} \quad x = -\frac{5}{2} - \frac{\sqrt{29}}{2} \qquad \text{Add } -\tfrac{5}{2}.$$

$$x = \frac{-5 + \sqrt{29}}{2} \quad \text{or} \quad x = \frac{-5 - \sqrt{29}}{2} \qquad \frac{a}{c} \pm \frac{b}{c} = \frac{a \pm b}{c}$$

The solution set is $\left\{\frac{-5 \pm \sqrt{29}}{2}\right\}$.

NOW TRY

CLASSROOM EXAMPLE 8

Solve $3x^2 + 6x - 2 = 0$.

Answer: $\left\{\frac{-3 \pm \sqrt{15}}{3}\right\}$

EXAMPLE 8 Solving a Quadratic Equation by Completing the Square ($a \neq 1$)

Solve $2x^2 - 4x - 5 = 0$.

Divide each side by 2 to obtain 1 as the coefficient of the second-degree term.

$$x^2 - 2x - \frac{5}{2} = 0 \qquad \text{Divide by 2. (Step 1)}$$

$$x^2 - 2x = \frac{5}{2} \qquad \text{Add } \tfrac{5}{2}. \text{ (Step 2)}$$

$$\left[\frac{1}{2}(-2)\right]^2 = (-1)^2 = 1 \qquad \text{Complete the square. (Step 3)}$$

$$x^2 - 2x + 1 = \frac{5}{2} + 1 \qquad \text{Add 1 to each side.}$$

$$(x - 1)^2 = \frac{7}{2} \qquad \text{Factor on the left. Add on the right.}$$

$$x - 1 = \sqrt{\frac{7}{2}} \quad \text{or} \quad x - 1 = -\sqrt{\frac{7}{2}} \qquad \text{Square root property (Step 4)}$$

$$x = 1 + \sqrt{\frac{7}{2}} \quad \text{or} \quad x = 1 - \sqrt{\frac{7}{2}} \qquad \text{Add 1.}$$

$$x = 1 + \frac{\sqrt{14}}{2} \quad \text{or} \quad x = 1 - \frac{\sqrt{14}}{2} \qquad \sqrt{\frac{7}{2}} = \frac{\sqrt{7}}{\sqrt{2}} = \frac{\sqrt{7}}{\sqrt{2}} \cdot \frac{\sqrt{2}}{\sqrt{2}} = \frac{\sqrt{14}}{2}$$

TEACHING TIP

After working through several examples, have students, either individually or in pairs, work through an exercise from the exercise set. Refer them to the steps listed in the box.

NOW TRY ANSWER

7. $\left\{\frac{-1 \pm \sqrt{13}}{2}\right\}$

NOW TRY EXERCISE 8

Solve $3x^2 + 12x - 5 = 0$.

Add the two terms in each solution as follows.

$$1 + \frac{\sqrt{14}}{2} = \frac{2}{2} + \frac{\sqrt{14}}{2} = \frac{2 + \sqrt{14}}{2}$$

$$1 - \frac{\sqrt{14}}{2} = \frac{2}{2} - \frac{\sqrt{14}}{2} = \frac{2 - \sqrt{14}}{2}$$

$1 = \frac{2}{2}$

The solution set is $\left\{\frac{2 \pm \sqrt{14}}{2}\right\}$. NOW TRY

OBJECTIVE 5 Solve quadratic equations with solutions that are not real numbers.

If $k < 0$ in the equation $x^2 = k$, then there will be two nonreal complex solutions.

NOW TRY EXERCISE 9

Solve each equation.

(a) $t^2 = -24$

(b) $(x + 4)^2 = -36$

(c) $x^2 + 8x + 21 = 0$

CLASSROOM EXAMPLE 9

Solve each equation.

(a) $x^2 = -17$

(b) $(x + 5)^2 = -100$

(c) $5x^2 - 15x + 12 = 0$

Answers: (a) $\{\pm i\sqrt{17}\}$

(b) $\{-5 \pm 10i\}$

(c) $\left\{\frac{3}{2} \pm \frac{\sqrt{15}}{10} i\right\}$

EXAMPLE 9 Solving Quadratic Equations (Nonreal Complex Solutions)

Solve each equation.

(a) $x^2 = -15$

$x = \sqrt{-15}$ or $x = -\sqrt{-15}$ Square root property

$x = i\sqrt{15}$ or $x = -i\sqrt{15}$ $\sqrt{-a} = i\sqrt{a}$

The solution set is $\{i\sqrt{15}, -i\sqrt{15}\}$, or $\{\pm i\sqrt{15}\}$.

(b) $(x + 2)^2 = -16$

$x + 2 = \sqrt{-16}$ or $x + 2 = -\sqrt{-16}$ Square root property

$x + 2 = 4i$ or $x + 2 = -4i$ $\sqrt{-16} = 4i$

$x = -2 + 4i$ or $x = -2 - 4i$ Add -2.

The solution set is $\{-2 + 4i, -2 - 4i\}$, or $\{-2 \pm 4i\}$.

(c) $x^2 + 2x + 7 = 0$

$x^2 + 2x = -7$ Subtract 7.

$x^2 + 2x + 1 = -7 + 1$ $\left[\frac{1}{2}(2)\right]^2 = 1$; Add 1 to each side.

$(x + 1)^2 = -6$ Factor on the left. Add on the right.

$x + 1 = \sqrt{-6}$ or $x + 1 = -\sqrt{-6}$ Square root property

$x + 1 = i\sqrt{6}$ or $x + 1 = -i\sqrt{6}$ $\sqrt{-6} = i\sqrt{6}$

$x = -1 + i\sqrt{6}$ or $x = -1 - i\sqrt{6}$ Add -1.

The solution set is $\{-1 + i\sqrt{6}, -1 - i\sqrt{6}\}$, or $\{-1 \pm i\sqrt{6}\}$. NOW TRY

NOW TRY ANSWERS

8. $\left\{\frac{-6 \pm \sqrt{51}}{3}\right\}$

9. **(a)** $\{\pm 2i\sqrt{6}\}$

(b) $\{-4 \pm 6i\}$

(c) $\{-4 \pm i\sqrt{5}\}$

8.1 Exercises

FOR EXTRA HELP

MyMathLab®

Complete solution available in MyMathLab

Concept Check *Answer each question.*

1. Which of the following are quadratic equations?

A. $x + 2 = 0$ **B.** $x^2 - 8x + 16 = 0$ **C.** $2t^2 - 5t = 3$ **D.** $x^3 + x^2 + 4 = 0$

2. Which quadratic equation identified in **Exercise 1** is in standard form?

3. A student incorrectly solved $x^2 - x - 2 = 5$ as follows. ***WHAT WENT WRONG?***

$$x^2 - x - 2 = 5$$
$$(x-2)(x+1) = 5 \quad \text{Factor.}$$
$$x - 2 = 5 \quad \text{or} \quad x + 1 = 5 \quad \text{Zero-factor property}$$
$$x = 7 \quad \text{or} \quad x = 4 \quad \text{Solve each equation.}$$

4. A student was asked to solve the quadratic equation $x^2 = 16$ and did not get full credit for the solution set $\{4\}$. ***WHAT WENT WRONG?***

Use the zero-factor property to solve each equation. (Hint: In Exercises 9 and 10, write the equation in standard form first.) ***See Example 1.***

5. $x^2 + 3x + 2 = 0$
6. $x^2 + 8x + 15 = 0$
7. $3x^2 + 8x - 3 = 0$
8. $2x^2 + x - 6 = 0$
9. $2x^2 = 9x - 4$
10. $5x^2 = 11x - 2$

Use the square root property to solve each equation. ***See Examples 2, 4, and 5.***

11. $x^2 = 81$
12. $x^2 = 225$
13. $x^2 = 17$
14. $x^2 = 19$
15. $x^2 = 32$
16. $x^2 = 54$
17. $x^2 - 20 = 0$
18. $p^2 - 50 = 0$
19. $3x^2 - 72 = 0$
20. $5z^2 - 200 = 0$
21. $2x^2 + 7 = 61$
22. $3x^2 + 8 = 80$
23. $3x^2 - 10 = 86$
24. $4x^2 - 7 = 65$
25. $(x + 2)^2 = 25$
26. $(t + 8)^2 = 9$
27. $(x - 6)^2 = 49$
28. $(x - 4)^2 = 64$
29. $(x - 4)^2 = 3$
30. $(x + 3)^2 = 11$
31. $(t + 5)^2 = 48$
32. $(m - 6)^2 = 27$
33. $(3x - 1)^2 = 7$
34. $(2x - 5)^2 = 10$
35. $(4p + 1)^2 = 24$
36. $(5t + 2)^2 = 12$
37. $(2 - 5t)^2 = 12$
38. $(1 - 4p)^2 = 24$

Use Galileo's formula to solve each problem. Round answers to the nearest tenth. ***See Example 3.***

39. The sculpture of American presidents at Mount Rushmore National Memorial is 500 ft above the valley floor. How long would it take a rock dropped from the top of the sculpture to fall to the ground? (*Source:* www.travelsd.com)

40. The Gateway Arch in St. Louis, Missouri, is 630 ft tall. How long would it take an object dropped from the top of the arch to fall to the ground? (*Source:* www.gatewayarch.com)

1. B, C 2. B
3. The zero-factor property requires a product equal to 0. The first step should have been to rewrite the equation with 0 on one side.
4. The equation is also true for $x = -4$. The solution set is $\{-4, 4\}$, or $\{\pm 4\}$.
5. $\{-2, -1\}$ 6. $\{-5, -3\}$
7. $\left\{-3, \frac{1}{3}\right\}$ 8. $\left\{-2, \frac{3}{2}\right\}$
9. $\left\{\frac{1}{2}, 4\right\}$ 10. $\left\{\frac{1}{5}, 2\right\}$
11. $\{\pm 9\}$ 12. $\{\pm 15\}$
13. $\{\pm\sqrt{17}\}$ 14. $\{\pm\sqrt{19}\}$
15. $\{\pm 4\sqrt{2}\}$ 16. $\{\pm 3\sqrt{6}\}$
17. $\{\pm 2\sqrt{5}\}$ 18. $\{\pm 5\sqrt{2}\}$
19. $\{\pm 2\sqrt{6}\}$ 20. $\{\pm 2\sqrt{10}\}$
21. $\{\pm 3\sqrt{3}\}$ 22. $\{\pm 2\sqrt{6}\}$
23. $\{\pm 4\sqrt{2}\}$ 24. $\{\pm 3\sqrt{2}\}$
25. $\{-7, 3\}$ 26. $\{-11, -5\}$
27. $\{-1, 13\}$ 28. $\{-4, 12\}$
29. $\{4 \pm \sqrt{3}\}$ 30. $\{-3 \pm \sqrt{11}\}$
31. $\{-5 \pm 4\sqrt{3}\}$
32. $\{6 \pm 3\sqrt{3}\}$
33. $\left\{\frac{1 \pm \sqrt{7}}{3}\right\}$
34. $\left\{\frac{5 \pm \sqrt{10}}{2}\right\}$
35. $\left\{\frac{-1 \pm 2\sqrt{6}}{4}\right\}$
36. $\left\{\frac{-2 \pm 2\sqrt{3}}{5}\right\}$
37. $\left\{\frac{2 \pm 2\sqrt{3}}{5}\right\}$
38. $\left\{\frac{1 \pm 2\sqrt{6}}{4}\right\}$
39. 5.6 sec 40. 6.3 sec
41. Solve $(2x + 1)^2 = 5$ by the square root property. Solve $x^2 + 4x = 12$ by completing the square.
42. Divide each side by 2.

41. *Concept Check* Which one of the two equations

$$(2x+1)^2 = 5 \quad \text{and} \quad x^2 + 4x = 12,$$

is more suitable for solving by the square root property? Which one is more suitable for solving by completing the square?

42. *Concept Check* According to the procedure described in this section, what would be the first step in solving $2x^2 + 8x = 9$ by completing the square?

Concept Check Find the constant that must be added to make each expression a perfect square trinomial. Then factor the trinomial.

43. $x^2 + 6x +$ ____ It factors as ______.

44. $x^2 + 14x +$ ____ It factors as ______.

45. $p^2 - 12p +$ ____ It factors as ______.

46. $x^2 - 20x +$ ____ It factors as ______.

47. $q^2 + 9q +$ ____ It factors as ______.

48. $t^2 + 3t +$ ____ It factors as ______.

Determine the number that will complete the square to solve each equation, after the constant term has been written on the right side and the coefficient of the second-degree term is 1. Do not actually solve. ***See Examples 6–8.***

49. $x^2 + 4x - 2 = 0$ **50.** $t^2 + 2t - 1 = 0$ **51.** $x^2 + 10x + 18 = 0$

52. $x^2 + 8x + 11 = 0$ **53.** $3w^2 - w - 24 = 0$ **54.** $4z^2 - z - 39 = 0$

Solve each equation by completing the square. Use the results of ***Exercises 49–54*** *to solve Exercises 57–62.* ***See Examples 6–8.***

55. $x^2 - 2x - 24 = 0$ **56.** $m^2 - 4m - 32 = 0$ **57.** $x^2 + 4x - 2 = 0$

58. $t^2 + 2t - 1 = 0$ **59.** $x^2 + 10x + 18 = 0$ **60.** $x^2 + 8x + 11 = 0$

61. $3w^2 - w = 24$ **62.** $4z^2 - z = 39$ **63.** $x^2 + 7x - 1 = 0$

64. $x^2 + 13x - 3 = 0$ **65.** $2k^2 + 5k - 2 = 0$ **66.** $3r^2 + 2r - 2 = 0$

67. $5x^2 - 10x + 2 = 0$ **68.** $2x^2 - 16x + 25 = 0$ **69.** $9x^2 - 24x = -13$

70. $25n^2 - 20n = 1$ **71.** $z^2 - \frac{4}{3}z = -\frac{1}{9}$ **72.** $p^2 - \frac{8}{3}p = -1$

73. $0.1x^2 - 0.2x - 0.1 = 0$ (*Hint*: First clear the decimals.)

74. $0.1p^2 - 0.4p + 0.1 = 0$ (*Hint*: First clear the decimals.)

Find the nonreal complex solutions of each equation. ***See Example 9.***

75. $x^2 = -12$ **76.** $x^2 = -18$ **77.** $(r-5)^2 = -4$

78. $(t+6)^2 = -9$ **79.** $(6x-1)^2 = -8$ **80.** $(4m-7)^2 = -27$

81. $m^2 + 4m + 13 = 0$ **82.** $t^2 + 6t + 10 = 0$ **83.** $3r^2 + 4r + 4 = 0$

84. $4x^2 + 5x + 5 = 0$ **85.** $-m^2 - 6m - 12 = 0$ **86.** $-x^2 - 5x - 10 = 0$

Extending Skills Solve for x. Assume that a and b represent positive real numbers.

87. $x^2 - b = 0$ **88.** $x^2 = 4b$ **89.** $4x^2 = b^2 + 16$

90. $9x^2 - 25a = 0$ **91.** $(5x - 2b)^2 = 3a$ **92.** $x^2 - a^2 - 36 = 0$

43. 9; $(x+3)^2$ **44.** 49; $(x+7)^2$
45. 36; $(p-6)^2$
46. 100; $(x-10)^2$
47. $\frac{81}{4}$; $\left(q+\frac{9}{2}\right)^2$
48. $\frac{9}{4}$; $\left(t+\frac{3}{2}\right)^2$
49. 4 **50.** 1
51. 25 **52.** 16
53. $\frac{1}{36}$ **54.** $\frac{1}{64}$
55. $\{-4, 6\}$ **56.** $\{-4, 8\}$
57. $\{-2 \pm \sqrt{6}\}$
58. $\{-1 \pm \sqrt{2}\}$
59. $\{-5 \pm \sqrt{7}\}$
60. $\{-4 \pm \sqrt{5}\}$
61. $\left\{-\frac{8}{3}, 3\right\}$ **62.** $\left\{-3, \frac{13}{4}\right\}$
63. $\left\{\frac{-7 \pm \sqrt{53}}{2}\right\}$
64. $\left\{\frac{-13 \pm \sqrt{181}}{2}\right\}$
65. $\left\{\frac{-5 \pm \sqrt{41}}{4}\right\}$
66. $\left\{\frac{-1 \pm \sqrt{7}}{3}\right\}$
67. $\left\{\frac{5 \pm \sqrt{15}}{5}\right\}$
68. $\left\{\frac{8 \pm \sqrt{14}}{2}\right\}$
69. $\left\{\frac{4 \pm \sqrt{3}}{3}\right\}$
70. $\left\{\frac{2 \pm \sqrt{5}}{5}\right\}$
71. $\left\{\frac{2 \pm \sqrt{3}}{3}\right\}$
72. $\left\{\frac{4 \pm \sqrt{7}}{3}\right\}$
73. $\{1 \pm \sqrt{2}\}$ **74.** $\{2 \pm \sqrt{3}\}$
75. $\{\pm 2i\sqrt{3}\}$ **76.** $\{\pm 3i\sqrt{2}\}$
77. $\{5 \pm 2i\}$ **78.** $\{-6 \pm 3i\}$
79. $\left\{\frac{1}{6} \pm \frac{\sqrt{2}}{3}i\right\}$
80. $\left\{\frac{7}{4} \pm \frac{3\sqrt{3}}{4}i\right\}$
81. $\{-2 \pm 3i\}$ **82.** $\{-3 \pm i\}$
83. $\left\{-\frac{2}{3} \pm \frac{2\sqrt{2}}{3}i\right\}$
84. $\left\{-\frac{5}{8} \pm \frac{\sqrt{55}}{8}i\right\}$

85. $\{-3 \pm i\sqrt{3}\}$
86. $\left\{-\frac{5}{2} \pm \frac{\sqrt{15}}{2}i\right\}$
87. $\{\pm\sqrt{b}\}$
88. $\{\pm 2\sqrt{b}\}$
89. $\left\{\pm\frac{\sqrt{b^2+16}}{2}\right\}$
90. $\left\{\pm\frac{5\sqrt{a}}{3}\right\}$
91. $\left\{\frac{2b \pm \sqrt{3a}}{5}\right\}$
92. $\{\pm\sqrt{a^2+36}\}$
93. x^2 94. x
95. $6x$ 96. 1
97. 9
98. $(x+3)^2$, or $x^2 + 6x + 9$

RELATING CONCEPTS For Individual or Group Work **(Exercises 93–98)**

The Greeks had a method of completing the square geometrically in which they literally changed a figure into a square. For example, to complete the square for $x^2 + 6x$, we begin with a square of side x, as in the figure on the top. We add three rectangles of width 1 to the right side and the bottom to get a region with area $x^2 + 6x$. To fill in the corner (complete the square), we must add nine 1-by-1 squares as shown.

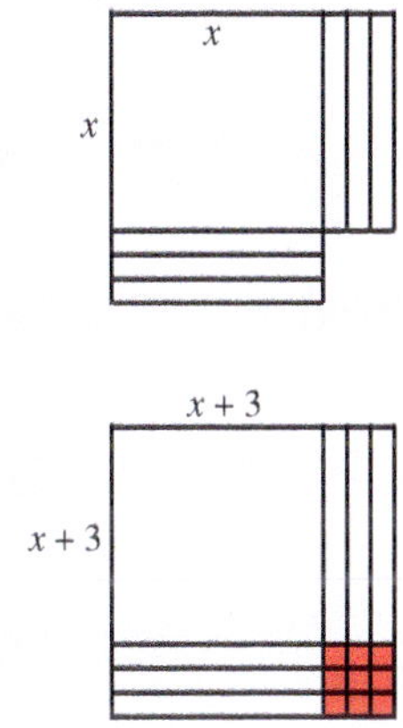

Work Exercises 93–98 in order.

93. What is the area of the original square?

94. What is the area of each strip?

95. What is the total area of the six strips?

96. What is the area of each small square in the corner of the second figure?

97. What is the total area of the small squares?

98. What is the area of the new "complete" square?

8.2 The Quadratic Formula

OBJECTIVES

1. Derive the quadratic formula.
2. Solve quadratic equations using the quadratic formula.
3. Use the discriminant to determine number and type of solutions.

VOCABULARY

☐ quadratic formula
☐ discriminant

TEACHING TIP
Going through the derivation of the quadratic formula is a good review of the previous section. Students should be able to follow the individual steps (although they may not be able to present the derivation in its entirety).

In this section, we complete the square to solve the general quadratic equation

$$ax^2 + bx + c = 0,$$

where a, b, and c are complex numbers and $a \neq 0$. The solution of this general equation gives a formula for finding the solution of *any* specific quadratic equation.

OBJECTIVE 1 Derive the quadratic formula.

We solve $ax^2 + bx + c = 0$ by completing the square (where $a > 0$) as follows.

$$ax^2 + bx + c = 0$$ Use the steps given in **Section 8.1.**

$$x^2 + \frac{b}{a}x + \frac{c}{a} = 0$$ Divide by a. (Step 1)

$$x^2 + \frac{b}{a}x = -\frac{c}{a}$$ Subtract $\frac{c}{a}$. (Step 2)

$$\left[\frac{1}{2}\left(\frac{b}{a}\right)\right]^2 = \left(\frac{b}{2a}\right)^2 = \frac{b^2}{4a^2}$$ Complete the square. (Step 3)

$$x^2 + \frac{b}{a}x + \frac{b^2}{4a^2} = -\frac{c}{a} + \frac{b^2}{4a^2}$$ Add $\frac{b^2}{4a^2}$ to each side.

$$\left(x + \frac{b}{2a}\right)^2 = \frac{b^2}{4a^2} + \frac{-c}{a}$$ Factor on the left. Rearrange the terms on the right.

$$\left(x + \frac{b}{2a}\right)^2 = \frac{b^2}{4a^2} + \frac{-4ac}{4a^2}$$ Write with a common denominator.

$$\left(x + \frac{b}{2a}\right)^2 = \frac{b^2 - 4ac}{4a^2}$$ Add fractions.

$$x + \frac{b}{2a} = \sqrt{\frac{b^2 - 4ac}{4a^2}} \quad \text{or} \quad x + \frac{b}{2a} = -\sqrt{\frac{b^2 - 4ac}{4a^2}}$$ Square root property (Step 4)

We can simplify $\sqrt{\frac{b^2 - 4ac}{4a^2}}$ as $\frac{\sqrt{b^2 - 4ac}}{\sqrt{4a^2}}$, or $\frac{\sqrt{b^2 - 4ac}}{2a}$.

The right side of each equation can be expressed as follows.

$$x + \frac{b}{2a} = \frac{\sqrt{b^2 - 4ac}}{2a} \quad \text{or} \quad x + \frac{b}{2a} = \frac{-\sqrt{b^2 - 4ac}}{2a}$$

$$x = \frac{-b}{2a} + \frac{\sqrt{b^2 - 4ac}}{2a} \quad \text{or} \quad x = \frac{-b}{2a} - \frac{\sqrt{b^2 - 4ac}}{2a}$$

If $a < 0$, the same two solutions are obtained.

$$\mathbf{x = \frac{-b + \sqrt{b^2 - 4ac}}{2a}} \quad \text{or} \quad \mathbf{x = \frac{-b - \sqrt{b^2 - 4ac}}{2a}}$$

This result is the **quadratic formula,** which is abbreviated as follows.

TEACHING TIP

When using the quadratic formula, after writing the equation so that one side is equal to 0, have students circle the coefficients representing *a*, *b*, and *c*. Doing so ensures that the negative sign is not omitted when these coefficients are substituted into the quadratic formula.

Quadratic Formula

The solutions of the equation $ax^2 + bx + c = 0$ (where $a \neq 0$) are given by

$$\mathbf{x = \frac{-b \pm \sqrt{b^2 - 4ac}}{2a}}.$$

CAUTION *In the quadratic formula, the square root is added to or subtracted from the value of $-b$ before dividing by 2a.*

OBJECTIVE 2 Solve quadratic equations using the quadratic formula.

CLASSROOM EXAMPLE 1

Solve $4x^2 - 11x - 3 = 0$.

Answer: $\left\{-\frac{1}{4}, 3\right\}$

EXAMPLE 1 Using the Quadratic Formula (Rational Solutions)

Solve $6x^2 - 5x - 4 = 0$.

This equation is in standard form, so we identify the values of *a*, *b*, and *c*. Here *a*, the coefficient of the second-degree term, is 6, and *b*, the coefficient of the first-degree term, is -5. The constant *c* is -4. Now substitute into the quadratic formula.

TEACHING TIP

To avoid possible sign errors, have students evaluate b^2 and $-4ac$, and then write these numbers in their proper places under the radical symbol.

$$x = \frac{-b \pm \sqrt{b^2 - 4ac}}{2a}$$ Quadratic formula

$$x = \frac{-(-5) \pm \sqrt{(-5)^2 - 4(6)(-4)}}{2(6)}$$ $a = 6, b = -5, c = -4$

Use parentheses and substitute carefully to avoid errors.

$$x = \frac{5 \pm \sqrt{25 + 96}}{12}$$

$$x = \frac{5 \pm \sqrt{121}}{12}$$ Simplify the radical.

$$x = \frac{5 \pm 11}{12}$$ Take the square root.

NOW TRY EXERCISE 1
Solve $2x^2 + 3x - 20 = 0$.

TEACHING TIP
Remind students to carry out the arithmetic in the case of two rational solutions.

There are two values represented, one from the + sign and one from the − sign.

$$x = \frac{5 + 11}{12} = \frac{16}{12} = \frac{4}{3} \quad \text{or} \quad x = \frac{5 - 11}{12} = \frac{-6}{12} = -\frac{1}{2}$$

Check each value in the original equation. The solution set is $\left\{-\frac{1}{2}, \frac{4}{3}\right\}$.

NOW TRY

NOTE We could have factored the trinomial and then used the zero-factor property to solve the equation in **Example 1.**

$6x^2 - 5x - 4 = 0$	See **Example 1.**
$(3x - 4)(2x + 1) = 0$	Factor.
$3x - 4 = 0$ or $2x + 1 = 0$	Zero-factor property
$3x = 4$ or $2x = -1$	Solve each equation.
$x = \frac{4}{3}$ or $x = -\frac{1}{2}$	Same solutions as in **Example 1**

When solving a quadratic equation, it is a good idea to try to factor the trinomial first. If it can be factored, then apply the zero-factor property. If it cannot be factored or if factoring is difficult, then use the quadratic formula.

NOW TRY EXERCISE 2
Solve $3x^2 + 1 = -5x$.

CLASSROOM EXAMPLE 2
Solve $2x^2 + 19 = 14x$.

Answer: $\left\{\frac{7 \pm \sqrt{11}}{2}\right\}$

TEACHING TIP
Stress that the values of a, b, and c are determined only *after* the quadratic equation is written in standard form.

EXAMPLE 2 Using the Quadratic Formula (Irrational Solutions)

Solve $4x^2 = 8x - 1$.

Write the equation in standard form as $4x^2 - 8x + 1 = 0$. ← This is a key step.

$$x = \frac{-b \pm \sqrt{b^2 - 4ac}}{2a} \quad \text{Quadratic formula}$$

$$x = \frac{-(-8) \pm \sqrt{(-8)^2 - 4(4)(1)}}{2(4)} \quad a = 4,\ b = -8,\ c = 1$$

$$x = \frac{8 \pm \sqrt{64 - 16}}{8} \quad \text{Simplify in the numerator and denominator.}$$

$$x = \frac{8 \pm \sqrt{48}}{8} \quad \text{Subtract under the radical.}$$

$$x = \frac{8 \pm 4\sqrt{3}}{8} \quad \sqrt{48} = \sqrt{16} \cdot \sqrt{3} = 4\sqrt{3}$$

$$x = \frac{4(2 \pm \sqrt{3})}{4(2)} \quad \text{Factor.}$$

$$x = \frac{2 \pm \sqrt{3}}{2} \quad \text{Divide out the common factor 4 to write in lowest terms.}$$

The solution set is $\left\{\frac{2 \pm \sqrt{3}}{2}\right\}$.

NOW TRY

NOW TRY ANSWERS

1. $\left\{-4, \frac{5}{2}\right\}$

2. $\left\{\frac{-5 \pm \sqrt{13}}{6}\right\}$

TEACHING TIP
Remind students to *factor first, divide out second* when writing answers in lowest terms.

CAUTION

1. ***Before solving, every quadratic equation must be expressed in standard form*** $ax^2 + bx + c = 0$, whether we use the zero-factor property or the quadratic formula.
2. ***When writing solutions in lowest terms, be sure to factor first. Then divide out the common factor.*** See the last two steps in **Example 2.**

NOW TRY EXERCISE 3
Solve $(x + 5)(x - 1) = -18$.

CLASSROOM EXAMPLE 3
Solve $(x + 5)(x + 1) = 10x$.

Answer: $\{2 \pm i\}$

TEACHING TIP
Have students work several problems individually or in groups. Include problems with different types of solutions.

EXAMPLE 3 Using the Quadratic Formula (Nonreal Complex Solutions)

Solve $(9x + 3)(x - 1) = -8$.

This is a quadratic equation—when the first terms $9x$ and x are multiplied, we get a second-degree term, $9x^2$. We must write the equation in standard form.

$$(9x + 3)(x - 1) = -8$$

$$9x^2 - 6x - 3 = -8 \quad \text{Multiply.}$$

$$\text{Standard form} \rightarrow 9x^2 - 6x + 5 = 0 \quad \text{Add 8.}$$

From the equation $9x^2 - 6x + 5 = 0$, we identify $a = 9$, $b = -6$, and $c = 5$.

$$x = \frac{-b \pm \sqrt{b^2 - 4ac}}{2a} \quad \text{Quadratic formula}$$

$$x = \frac{-(-6) \pm \sqrt{(-6)^2 - 4(9)(5)}}{2(9)} \quad \text{Substitute.}$$

$$x = \frac{6 \pm \sqrt{-144}}{18} \quad \text{Simplify.}$$

$$x = \frac{6 \pm 12i}{18} \quad \sqrt{-144} = 12i$$

$$x = \frac{6(1 \pm 2i)}{6(3)} \quad \text{Factor.}$$

$$x = \frac{1 \pm 2i}{3} \quad \text{Divide out the common factor 6 to write in lowest terms.}$$

$$x = \frac{1}{3} \pm \frac{2}{3}i \quad \text{Standard form } a + bi \text{ for a complex number}$$

The solution set is $\left\{\frac{1}{3} \pm \frac{2}{3}i\right\}$. NOW TRY

OBJECTIVE 3 Use the discriminant to determine number and type of solutions.

The solutions of the quadratic equation $ax^2 + bx + c = 0$ are given by

$$x = \frac{-b \pm \sqrt{b^2 - 4ac}}{2a}. \quad \leftarrow \text{Discriminant}$$

The expression under the radical symbol, $b^2 - 4ac$, is called the **discriminant** because it distinguishes among the number of solutions—one or two—and the type of solutions—rational, irrational, or nonreal complex—of a quadratic equation.

NOW TRY ANSWER
3. $\{-2 \pm 3i\}$

TEACHING TIP
Give an example illustrating each case in the table.

$15x^2 + 11x = 14$
[discriminant 961; two rational solutions]

$2x^2 + 3x = 4$
[discriminant 41; two irrational solutions]

$x^2 + 20x + 100 = 0$
[discriminant 0; one rational solution]

$2x^2 + 3x + 4 = 0$
[discriminant -23; two nonreal complex solutions]

Using the Discriminant

If a, b, and c are integers in a quadratic equation $ax^2 + bx + c = 0$, then the discriminant $b^2 - 4ac$ can be used to determine the number and type of solutions of the equation as follows.

Discriminant	*Number and Type of Solutions*
Positive, and the square of an integer	Two rational solutions
Positive, but not the square of an integer	Two irrational solutions
Zero	One rational solution
Negative	Two nonreal complex solutions

We can also use the discriminant to help decide how to solve a quadratic equation.

If a, b, and c are integers and the discriminant is a perfect square (including 0), then the equation can be solved using the zero-factor property. Otherwise, the quadratic formula should be used.

CLASSROOM EXAMPLE 4
Find the discriminant. Use it to predict the number and type of solutions for each equation. Tell whether the equation can be solved using the zero-factor property or whether the quadratic formula should be used.

(a) $10x^2 - x - 2 = 0$
(b) $3x^2 - x = 7$
(c) $16x^2 + 25 = 40x$

Answers:
(a) 81; two rational solutions; zero-factor property
(b) 85; two irrational solutions; quadratic formula
(c) 0; one rational solution; zero-factor property

EXAMPLE 4 Using the Discriminant

Find the discriminant. Use it to predict the number and type of solutions for each equation. Tell whether the equation can be solved using the zero-factor property or whether the quadratic formula should be used.

(a) $6x^2 - x - 15 = 0$

First identify the values of a, b, and c. Because $-x = -1x$, the value of b is -1. We find the discriminant by evaluating $b^2 - 4ac$.

$b^2 - 4ac$ — Use parentheses and substitute carefully.

$= (-1)^2 - 4(6)(-15)$ — $a = 6$, $b = -1$, $c = -15$ (all integers)

$= 1 + 360$ — Apply the exponent. Multiply.

$= 361$ — Add.

$= 19^2$, which is a perfect square.

The discriminant 361 is a perfect square, so referring to the table we see that there will be two rational solutions. We can solve using the zero-factor property.

TEACHING TIP
Have students, either individually or in groups, solve the equations in **Example 4** using the information gained from the discriminant.

(b) $3x^2 - 4x = 5$

Write in standard form as $3x^2 - 4x - 5 = 0$.

$b^2 - 4ac$ — Discriminant

$= (-4)^2 - 4(3)(-5)$ — $a = 3$, $b = -4$, $c = -5$ (all integers)

$= 16 + 60$ — Apply the exponent. Multiply.

$= 76$ — Add.

Because 76 is positive but *not* the square of an integer, the equation will have two irrational solutions. We solve using the quadratic formula.

NOW TRY EXERCISE 4

Find the discriminant. Use it to predict the number and type of solutions for each equation. Tell whether the equation can be solved using the zero-factor property or whether the quadratic formula should be used.

(a) $8x^2 - 6x - 5 = 0$

(b) $9x^2 = 24x - 16$

(c) $3x^2 + 2x = -1$

NOW TRY ANSWERS

4. (a) 196; two rational solutions; zero-factor property
(b) 0; one rational solution; zero-factor property
(c) -8; two nonreal complex solutions; quadratic formula

(c) $4x^2 + x + 1 = 0$

$x = 1x$, so $b = 1$.

$b^2 - 4ac$	Discriminant
$= 1^2 - 4(4)(1)$	$a = 4$, $b = 1$, $c = 1$ (all integers)
$= 1 - 16$	Apply the exponent. Multiply.
$= -15$	Subtract.

Because the discriminant is negative, there will be two nonreal complex solutions. We solve using the quadratic formula.

(d) $4x^2 + 9 = 12x$ Write in standard form as $4x^2 - 12x + 9 = 0$.

$b^2 - 4ac$	Discriminant
$= (-12)^2 - 4(4)(9)$	$a = 4$, $b = -12$, $c = 9$ (all integers)
$= 144 - 144$	Apply the exponent. Multiply.
$= 0$	Subtract.

The discriminant is 0, so there is only one rational solution. We solve using the zero-factor property.

NOW TRY

8.2 Exercises

FOR EXTRA HELP

MyMathLab®

Complete solution available in MyMathLab

Concept Check *Answer each question.*

1. The documentation for an early version of Microsoft *Word* for Windows used the following for the quadratic formula. Was this correct? If not, correct it.

$$x = -b \pm \frac{\sqrt{b^2 - 4ac}}{2a}$$ Correct or incorrect?

2. One patron wrote the quadratic formula, as shown here, on a wall at the Cadillac Bar in Houston, Texas. Was this correct? If not, correct it.

$$x = \frac{-b\sqrt{b^2 - 4ac}}{2a}$$ Correct or incorrect?

3. A student solved $5x^2 - 5x + 1 = 0$ incorrectly as follows. ***WHAT WENT WRONG?***

$$x = \frac{-(-5) \pm \sqrt{(-5)^2 - 4(5)(1)}}{2(5)}$$

$$x = \frac{5 \pm \sqrt{5}}{10}$$

$$x = \frac{1}{2} \pm \sqrt{5} \quad \text{Solution set: } \left\{\frac{1}{2} \pm \sqrt{5}\right\}$$

4. A student incorrectly claimed that the equation $2x^2 - 5 = 0$ cannot be solved using the quadratic formula because there is no first-degree x-term. ***WHAT WENT WRONG?*** Give the values of a, b, and c for this equation.

Use the quadratic formula to solve each equation. (All solutions for these equations are real numbers.) ***See Examples 1 and 2.***

5. $x^2 - 8x + 15 = 0$ **6.** $x^2 + 3x - 28 = 0$ **7.** $2x^2 + 4x + 1 = 0$

1. No. The fraction bar should extend under the term $-b$. The correct formula is $x = \frac{-b \pm \sqrt{b^2 - 4ac}}{2a}$.
2. No. The patron forgot the $\pm$ sign in the numerator. (See the correct formula in the Exercise 1 answer.)
3. The last step is wrong. Because 5 is not a common factor in the numerator, the fraction cannot be simplified. The solution set is $\left\{\frac{5 \pm \sqrt{5}}{10}\right\}$.
4. No. The quadratic formula can be used to solve *any* quadratic equation. Here, the quadratic formula can be used with $a = 2$, $b = 0$, and $c = -5$.
5. $\{3, 5\}$ 6. $\{-7, 4\}$
7. $\left\{\frac{-2 \pm \sqrt{2}}{2}\right\}$
8. $\left\{\frac{-3 \pm \sqrt{17}}{4}\right\}$
9. $\left\{\frac{1 \pm \sqrt{3}}{2}\right\}$ 10. $\left\{\frac{-1 \pm \sqrt{2}}{3}\right\}$
11. $\{5 \pm \sqrt{7}\}$ 12. $\{1 \pm \sqrt{5}\}$
13. $\left\{\frac{-1 \pm \sqrt{2}}{2}\right\}$ 14. $\left\{\frac{1 \pm 2\sqrt{5}}{2}\right\}$

15. $\left\{\frac{-1 \pm \sqrt{7}}{3}\right\}$ **16.** $\left\{\frac{13 \pm \sqrt{163}}{3}\right\}$

17. $\{1 \pm \sqrt{5}\}$ **18.** $\left\{\frac{-1 \pm \sqrt{7}}{6}\right\}$

19. $\left\{\frac{-2 \pm \sqrt{10}}{2}\right\}$

20. $\left\{\frac{-3 \pm \sqrt{21}}{3}\right\}$

21. $\{-1 \pm 3\sqrt{2}\}$ **22.** $\{3 \pm \sqrt{17}\}$

23. $\left\{\frac{1 \pm \sqrt{29}}{2}\right\}$ **24.** $\left\{\frac{3 \pm \sqrt{73}}{2}\right\}$

25. $\left\{\frac{-4 \pm \sqrt{91}}{3}\right\}$

26. $\left\{\frac{-3 \pm \sqrt{33}}{2}\right\}$

27. $\left\{\frac{-3 \pm \sqrt{57}}{8}\right\}$

28. $\left\{\frac{5 \pm \sqrt{41}}{8}\right\}$

29. $\left\{\frac{3}{2} \pm \frac{\sqrt{15}}{2}i\right\}$ **30.** $\left\{\frac{5}{2} \pm \frac{\sqrt{55}}{2}i\right\}$

31. $\{3 \pm i\sqrt{5}\}$ **32.** $\{-2 \pm i\sqrt{7}\}$

33. $\left\{\frac{1}{2} \pm \frac{\sqrt{6}}{2}i\right\}$ **34.** $\left\{\frac{1}{3} \pm \frac{\sqrt{6}}{3}i\right\}$

35. $\left\{-\frac{2}{3} \pm \frac{\sqrt{2}}{3}i\right\}$

36. $\left\{-\frac{3}{4} \pm \frac{\sqrt{7}}{4}i\right\}$

37. $\left\{\frac{1}{2} \pm \frac{1}{4}i\right\}$ **38.** $\left\{\frac{2}{3} \pm \frac{1}{3}i\right\}$

39. 0; B; zero-factor property
40. 0; B; zero-factor property
41. 8; C; quadratic formula
42. 180; C; quadratic formula
43. 49; A; zero-factor property
44. 64; A; zero-factor property
45. −80; D; quadratic formula
46. −2304; D; quadratic formula

47. (a) 25; zero-factor property; $\left\{-3, -\frac{4}{3}\right\}$
(b) 44; quadratic formula; $\left\{\frac{7 \pm \sqrt{11}}{2}\right\}$

48. (a) $\left\{-\frac{7}{5}\right\}$ **(b)** $\left\{\frac{7}{2}\right\}$
(c) $\left\{-\frac{1}{3}, 2\right\}$ **(d)** $\left\{-\frac{1}{2}, \frac{3}{2}\right\}$

49. −10 or 10 **50.** −14 or 14
51. 16 **52.** 9
53. 25 **54.** 9
55. $b = \frac{44}{5}$; $\frac{3}{10}$ **56.** $c = 2$; 2

8. $2x^2 + 3x - 1 = 0$ **9.** $2x^2 - 2x = 1$ **10.** $9x^2 + 6x = 1$

11. $x^2 + 18 = 10x$ **12.** $x^2 - 4 = 2x$ **13.** $4x^2 + 4x - 1 = 0$

14. $4r^2 - 4r - 19 = 0$ **15.** $2 - 2x = 3x^2$ **16.** $26r - 2 = 3r^2$

17. $\frac{x^2}{4} - \frac{x}{2} = 1$ **18.** $p^2 + \frac{p}{3} = \frac{1}{6}$ **19.** $-2t(t + 2) = -3$

20. $-3x(x + 2) = -4$ **21.** $(r - 3)(r + 5) = 2$ **22.** $(x + 1)(x - 7) = 1$

23. $(x + 2)(x - 3) = 1$ **24.** $(x - 5)(x + 2) = 6$ **25.** $p = \frac{5(5 - p)}{3(p + 1)}$

26. $x = \frac{2(x + 3)}{x + 5}$ **27.** $(2x + 1)^2 = x + 4$ **28.** $(2x - 1)^2 = x + 2$

Use the quadratic formula to solve each equation. (All solutions for these equations are nonreal complex numbers.) ***See Example 3.***

29. $x^2 - 3x + 6 = 0$ **30.** $x^2 - 5x + 20 = 0$ **31.** $r^2 - 6r + 14 = 0$

32. $t^2 + 4t + 11 = 0$ **33.** $4x^2 - 4x = -7$ **34.** $9x^2 - 6x = -7$

35. $x(3x + 4) = -2$ **36.** $z(2z + 3) = -2$

37. $(2x - 1)(8x - 4) = -1$ **38.** $(x - 1)(9x - 3) = -2$

Find the discriminant. Use it to determine whether the solutions for each equation are

A. *two rational numbers* **B.** *one rational number*
C. *two irrational numbers* **D.** *two nonreal complex numbers.*

Tell whether the equation can be solved using the zero-factor property or whether the quadratic formula should be used. Do not actually solve. ***See Example 4.***

39. $25x^2 + 70x + 49 = 0$ **40.** $4x^2 - 28x + 49 = 0$ **41.** $x^2 + 4x + 2 = 0$

42. $9x^2 - 12x - 1 = 0$ **43.** $3x^2 = 5x + 2$ **44.** $4x^2 = 4x + 3$

45. $3m^2 - 10m + 15 = 0$ **46.** $18x^2 + 60x + 82 = 0$

47. Find the discriminant for each quadratic equation. Use it to tell whether the equation can be solved using the zero-factor property or whether the quadratic formula should be used. Then solve each equation.

(a) $3x^2 + 13x = -12$ **(b)** $2x^2 + 19 = 14x$

48. Based on the answers in **Exercises 39–46,** solve the equation given in each exercise.

(a) Exercise 39 **(b) Exercise 40** **(c) Exercise 43** **(d) Exercise 44**

Extending Skills *Find the value of a, b, or c so that each equation will have exactly one rational solution. (Hint: The discriminant must equal 0 for an equation to have one rational solution.)*

49. $p^2 + bp + 25 = 0$ **50.** $r^2 - br + 49 = 0$ **51.** $am^2 + 8m + 1 = 0$

52. $at^2 + 24t + 16 = 0$ **53.** $9x^2 - 30x + c = 0$ **54.** $4m^2 + 12m + c = 0$

55. One solution of $4x^2 + bx - 3 = 0$ is $-\frac{5}{2}$. Find b and the other solution.

56. One solution of $3x^2 - 7x + c = 0$ is $\frac{1}{3}$. Find c and the other solution.

8.3 Equations Quadratic in Form

OBJECTIVES

1. Solve an equation with fractions by writing it in quadratic form.
2. Use quadratic equations to solve applied problems.
3. Solve an equation with radicals by writing it in quadratic form.
4. Solve an equation that is quadratic in form by substitution.

VOCABULARY

☐ quadratic in form

NOW TRY EXERCISE 1

Solve $\frac{2}{x} + \frac{3}{x+2} = 1.$

CLASSROOM EXAMPLE 1

Solve $\frac{4}{x-1} + 9 = -\frac{7}{x}.$

Answer: $\left\{-1, \frac{7}{9}\right\}$

TEACHING TIP

Advise students to try to factor *first*. Point out that the quadratic formula will *always* work in case one has difficulty with a factorable trinomial.

CLASSROOM EXAMPLE 2

In $1\frac{3}{4}$ hr Cody rows his boat 5 mi upriver and comes back. The rate of the current is 3 mph. How fast does Cody row?

Answer: 7 mph

NOW TRY ANSWER

1. $\{-1, 4\}$

OBJECTIVE 1 Solve an equation with fractions by writing it in quadratic form.

A variety of nonquadratic equations can be written in the form of a quadratic equation and solved using the methods of this chapter.

EXAMPLE 1 Solving a Rational Equation That Leads to a Quadratic Equation

Solve $\frac{1}{x} + \frac{1}{x-1} = \frac{7}{12}.$

Clear fractions by multiplying each side by the least common denominator, $12x(x-1)$. (The domain is $\{x \mid x \text{ is a real number}, x \neq 0, 1\}$.)

$$12x(x-1)\left(\frac{1}{x} + \frac{1}{x-1}\right) = 12x(x-1)\left(\frac{7}{12}\right) \quad \text{Multiply by the LCD.}$$

$$12x(x-1)\frac{1}{x} + 12x(x-1)\frac{1}{x-1} = 12x(x-1)\frac{7}{12} \quad \text{Distributive property}$$

$$12(x-1) + 12x = 7x(x-1) \quad \text{Multiply.}$$

$$12x - 12 + 12x = 7x^2 - 7x \quad \text{Distributive property}$$

$$24x - 12 = 7x^2 - 7x \quad \text{Combine like terms.}$$

This trinomial is factorable.

$$7x^2 - 31x + 12 = 0 \quad \text{Standard form}$$

$$(7x-3)(x-4) = 0 \quad \text{Factor.}$$

$$7x - 3 = 0 \quad \text{or} \quad x - 4 = 0 \quad \text{Zero-factor property}$$

$$7x = 3 \quad \text{or} \quad x = 4 \quad \text{Solve each equation.}$$

$$x = \frac{3}{7}$$

These values are in the domain. Check them in the original equation. The solution set is $\left\{\frac{3}{7}, 4\right\}$.

NOW TRY

OBJECTIVE 2 Use quadratic equations to solve applied problems.

Some distance-rate-time (or motion) problems lead to quadratic equations.

EXAMPLE 2 Solving a Motion Problem

A riverboat for tourists averages 12 mph in still water. It takes the boat 1 hr, 4 min to travel 6 mi upstream and return. Find the rate of the current.

Step 1 **Read** the problem carefully.

Step 2 **Assign a variable.** Let x = the rate of the current.

The current slows down the boat as it travels upstream, so the rate of the boat traveling upstream is its rate in still water *less* the rate of the current, or $(12 - x)$ mph. See **FIGURE 1** on the next page.

NOW TRY EXERCISE 2
A small fishing boat averages 18 mph in still water. It takes the boat $\frac{9}{10}$ hr to travel 8 mi upstream and return. Find the rate of the current.

Riverboat traveling upstream—the current slows it down.

FIGURE 1

Similarly, the current speeds up the boat as it travels downstream, so its rate downstream is $(12 + x)$ mph. Thus,

$$12 - x = \text{the rate upstream in miles per hour,}$$

and

$$12 + x = \text{the rate downstream in miles per hour.}$$

	d	*r*	*t*
Upstream	6	$12 - x$	$\frac{6}{12 - x}$
Downstream	6	$12 + x$	$\frac{6}{12 + x}$

Complete a table. Use the distance formula, $d = rt$, solved for time t, $t = \frac{d}{r}$, to write expressions for t.

Step 3 **Write an equation.** We use the total time, 1 hr, 4 min, written as a fraction.

$$1 + \frac{4}{60} = 1 + \frac{1}{15} = \frac{16}{15} \text{ hr} \qquad \text{Total time}$$

The time upstream plus the time downstream equals $\frac{16}{15}$ hr.

$$\underset{\downarrow}{\text{Time upstream}} + \underset{\downarrow}{\text{Time downstream}} = \underset{\downarrow}{\text{Total time}}$$

$$\frac{6}{12 - x} + \frac{6}{12 + x} = \frac{16}{15}$$

Step 4 **Solve** the equation. The LCD is $15(12 - x)(12 + x)$.

$$15(12 - x)(12 + x)\left(\frac{6}{12 - x} + \frac{6}{12 + x}\right)$$

$$= 15(12 - x)(12 + x)\left(\frac{16}{15}\right) \qquad \text{Multiply by the LCD.}$$

$$15(12 + x) \cdot 6 + 15(12 - x) \cdot 6 = 16(12 - x)(12 + x) \qquad \text{Distributive property; Multiply.}$$

$$90(12 + x) + 90(12 - x) = 16(144 - x^2) \qquad \text{Multiply.}$$

$$1080 + 90x + 1080 - 90x = 2304 - 16x^2 \qquad \text{Distributive property}$$

$$2160 = 2304 - 16x^2 \qquad \text{Combine like terms.}$$

$$16x^2 = 144 \qquad \text{Add } 16x^2\text{. Subtract 2160.}$$

$$x^2 = 9 \qquad \text{Divide by 16.}$$

$$x = 3 \quad \text{or} \quad x = -3 \qquad \text{Square root property}$$

Step 5 **State the answer.** The current rate cannot be -3, so the answer is 3 mph.

Step 6 **Check** that this value satisfies the original problem. NOW TRY

TEACHING TIP
Alternatively, a student may wish to write the equation $x^2 = 9$ in Example 2 as

$$x^2 - 9 = 0,$$

then factor on the left and solve using the zero-factor property.

PROBLEM-SOLVING HINT Recall from **Section 6.5** that a person's work rate is $\frac{1}{t}$ part of the job per hour, where t is the time in hours required to do the complete job. Thus, the part of the job the person will do in x hours is $\frac{1}{t}x$.

NOW TRY ANSWER
2. 2 mph

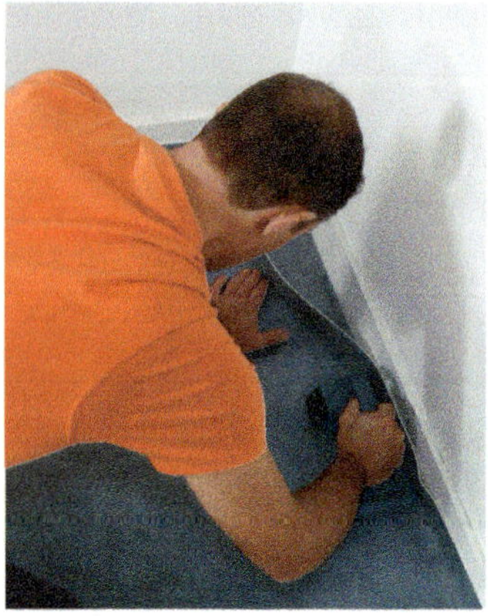

CLASSROOM EXAMPLE 3
Two chefs are preparing a banquet. One chef could prepare the banquet in 2 hr less time than the other. Together, they complete the job in 5 hr. How long (to the nearest tenth) would it take the faster chef working alone?

Answer: 9.1 hr

EXAMPLE 3 Solving a Work Problem

It takes two carpet layers 4 hr to carpet a room. If each worked alone, one of them could do the job in 1 hr less time than the other. How long would it take each carpet layer to complete the job alone?

Step 1 **Read** the problem again. There will be two answers.

Step 2 **Assign a variable.**

Let x = the number of hours for the slower carpet layer to complete the job.

Then $x - 1$ = the number of hours for the faster carpet layer to complete the job.

The slower worker's rate is $\frac{1}{x}$, and the faster worker's rate is $\frac{1}{x-1}$. Together they can do the job in 4 hr. Complete a table as shown.

	Rate	Time Working Together	Fractional Part of the Job Done
Slower Worker	$\frac{1}{x}$	4	$\frac{1}{x}(4)$
Faster Worker	$\frac{1}{x-1}$	4	$\frac{1}{x-1}(4)$

Sum is 1 whole job.

Step 3 **Write an equation.**

Part done by slower worker + Part done by faster worker = 1 whole job

$$\frac{4}{x} + \frac{4}{x-1} = 1$$

Step 4 **Solve** the equation from Step 3.

$$x(x-1)\left(\frac{4}{x} + \frac{4}{x-1}\right) = x(x-1)(1) \quad \text{Multiply by the LCD, } x(x-1).$$

$$4(x-1) + 4x = x(x-1) \quad \text{Distributive property}$$

$$4x - 4 + 4x = x^2 - x \quad \text{Distributive property}$$

$$x^2 - 9x + 4 = 0 \quad \text{Standard form}$$

The trinomial on the left cannot be factored, so the equation cannot be solved using the zero-factor property. We use the quadratic formula.

$$x = \frac{-b \pm \sqrt{b^2 - 4ac}}{2a} \quad \text{Quadratic formula}$$

$$x = \frac{-(-9) \pm \sqrt{(-9)^2 - 4(1)(4)}}{2(1)} \quad a = 1, b = -9, c = 4$$

$$x = \frac{9 \pm \sqrt{65}}{2} \quad \text{Simplify.}$$

$$x = \frac{9 + \sqrt{65}}{2} = 8.5 \quad \text{or} \quad x = \frac{9 - \sqrt{65}}{2} = 0.5 \quad \text{Use a calculator. Round to the nearest tenth.}$$

NOW TRY EXERCISE 3

Two electricians are running wire to finish a basement. One electrician could finish the job in 2 hr less time than the other. Together, they complete the job in 6 hr. How long (to the nearest tenth) would it take the slower electrician to complete the job alone?

Step 5 **State the answer.** Only the solution 8.5 makes sense in the original problem.

If $x = 0.5$, then $\quad x - 1 = 0.5 - 1 = -0.5$, Time cannot be negative.

which cannot represent the time for the faster worker. The slower worker could do the job in 8.5 hr and the faster in $8.5 - 1 = 7.5$ hr.

Step 6 **Check** that these results satisfy the original problem. NOW TRY

NOW TRY EXERCISE 4

Solve each equation.

(a) $x = \sqrt{9x - 20}$

(b) $x + \sqrt{x} = 20$

CLASSROOM EXAMPLE 4

Solve $2x = \sqrt{x} + 1$.

Answer: $\{1\}$

TEACHING TIP

As with rational equations, answers to equations with radicals must be checked. Remind students that the square root of a negative number is a pure imaginary number.

OBJECTIVE 3 Solve an equation with radicals by writing it in quadratic form.

EXAMPLE 4 Solving Radical Equations That Lead to Quadratic Equations

Solve each equation.

(a) $x = \sqrt{6x - 8}$

This equation is not quadratic. However, squaring each side of the equation gives a quadratic equation that can be solved using the zero-factor property.

$$x^2 = \left(\sqrt{6x - 8}\right)^2 \quad \text{Square each side.}$$

$$x^2 = 6x - 8 \quad (\sqrt{a})^2 = a$$

This trinomial is factorable.

$$x^2 - 6x + 8 = 0 \quad \text{Standard form}$$

$$(x - 4)(x - 2) = 0 \quad \text{Factor.}$$

$$x - 4 = 0 \quad \text{or} \quad x - 2 = 0 \quad \text{Zero-factor property}$$

$$x = 4 \quad \text{or} \quad x = 2 \quad \text{Proposed solutions}$$

Squaring each side of a radical equation can introduce extraneous solutions. ***All proposed solutions must be checked in the original (not the squared) equation.***

CHECK

$x = \sqrt{6x - 8}$		$x = \sqrt{6x - 8}$	
$4 \stackrel{?}{=} \sqrt{6(4) - 8}$	Let $x = 4$.	$2 \stackrel{?}{=} \sqrt{6(2) - 8}$	Let $x = 2$.
$4 \stackrel{?}{=} \sqrt{16}$		$2 \stackrel{?}{=} \sqrt{4}$	
$4 = 4$ ✓	True	$2 = 2$ ✓	True

Both solutions check, so the solution set is $\{2, 4\}$.

(b)

$$x + \sqrt{x} = 6$$

$$\sqrt{x} = 6 - x \quad \text{Isolate the radical on one side.}$$

$$\left(\sqrt{x}\right)^2 = (6 - x)^2 \quad \text{Square each side.}$$

$$x = 36 - 12x + x^2 \quad (a - b)^2 = a^2 - 2ab + b^2$$

$$x^2 - 13x + 36 = 0 \quad \text{Standard form}$$

$$(x - 4)(x - 9) = 0 \quad \text{Factor.}$$

$$x - 4 = 0 \quad \text{or} \quad x - 9 = 0 \quad \text{Zero-factor property}$$

$$x = 4 \quad \text{or} \quad x = 9 \quad \text{Proposed solutions}$$

CHECK

$x + \sqrt{x} = 6$		$x + \sqrt{x} = 6$	
$4 + \sqrt{4} \stackrel{?}{=} 6$	Let $x = 4$.	$9 + \sqrt{9} \stackrel{?}{=} 6$	Let $x = 9$.
$6 = 6$ ✓	True	$12 = 6$	False

Only the solution 4 checks, so the solution set is $\{4\}$. NOW TRY

NOW TRY ANSWERS

3. 13.1 hr

4. (a) $\{4, 5\}$ (b) $\{16\}$

TEACHING TIP
Show several specific examples of *quadratic forms* such as

$3x^4 + 4x^2 + 2$

and $5x^6 + 2x^3 - 7$.

OBJECTIVE 4 Solve an equation that is quadratic in form by substitution.

A nonquadratic equation that can be written in the form

$$au^2 + bu + c = 0,$$

for $a \neq 0$ and an algebraic expression u, is **quadratic in form.**

Many equations that are quadratic in form can be solved more easily by defining and substituting a "temporary" variable u for an expression involving the variable in the original equation.

NOW TRY EXERCISE 5
Define a variable u in terms of x, and write each equation in the quadratic form $au^2 + bu + c = 0$.

(a) $x^4 - 10x^2 + 9 = 0$

(b) $6(x + 2)^2 - 11(x + 2) + 4 = 0$

EXAMPLE 5 Defining Substitution Variables

Define a variable u in terms of x, and write each equation in the quadratic form $au^2 + bu + c = 0$.

(a) $x^4 - 13x^2 + 36 = 0$

Look at the two terms involving the variable x, ignoring their coefficients. Try to find one variable expression that is the square of the other. Here $x^4 = (x^2)^2$, so we can define $u = x^2$, and rewrite the original equation as a quadratic equation in u.

$$u^2 - 13u + 36 = 0 \qquad \text{Here, } u = x^2.$$

(b) $2(4x - 3)^2 + 7(4x - 3) + 5 = 0$

Because this equation involves both $(4x - 3)^2$ and $(4x - 3)$, we let $u = 4x - 3$.

$$2u^2 + 7u + 5 = 0 \qquad \text{Here, } u = 4x - 3.$$

(c) $2x^{2/3} - 11x^{1/3} + 12 = 0$

We apply a power rule for exponents, $(a^m)^n = a^{mn}$ **(Section 4.1).** Because $(x^{1/3})^2 = x^{2/3}$, we define $u = x^{1/3}$ and write the original equation as follows.

$$2u^2 - 11u + 12 = 0 \qquad \text{Here, } u = x^{1/3}.$$

NOW TRY

CLASSROOM EXAMPLE 5
Define a variable u in terms of x, and write each equation in the quadratic form $au^2 + bu + c = 0$.

(a) $2x^4 + 5x^2 - 12 = 0$

(b) $2(x + 5)^2 - 7(x + 5) + 6 = 0$

(c) $x^{4/3} - 8x^{2/3} + 16 = 0$

Answers:

(a) $u = x^2$; $2u^2 + 5u - 12 = 0$

(b) $u = x + 5$; $2u^2 - 7u + 6 = 0$

(c) $u = x^{2/3}$; $u^2 - 8u + 16 = 0$

EXAMPLE 6 Solving Equations That Are Quadratic in Form

Solve each equation.

(a)

$x^4 - 13x^2 + 36 = 0$		See Example 5(a).
$(x^2)^2 - 13x^2 + 36 = 0$		$x^4 = (x^2)^2$
Quadratic in form → $u^2 - 13u + 36 = 0$		Let $u = x^2$.
$(u - 4)(u - 9) = 0$		Factor.
$u - 4 = 0$	or $u - 9 = 0$	Zero-factor property
Don't stop here. → $u = 4$	or $u = 9$	Solve.
$x^2 = 4$	or $x^2 = 9$	Substitute x^2 for u.
$x = \pm 2$	or $x = \pm 3$	Square root property

Each value can be verified by substituting it into the original equation for x. The equation $x^4 - 13x^2 + 36 = 0$ is a fourth-degree equation and has four solutions, $-3, -2, 2, 3$.* The solution set is abbreviated $\{\pm 2, \pm 3\}$.

CLASSROOM EXAMPLE 6
Solve each equation.

(a) $9x^4 - 37x^2 + 4 = 0$

(b) $x^4 - 4x^2 = -2$

Answers:

(a) $\left\{\pm\frac{1}{3}, \pm 2\right\}$

(b) $\left\{\pm\sqrt{2 + \sqrt{2}}, \pm\sqrt{2 - \sqrt{2}}\right\}$

NOW TRY ANSWERS

5. (a) $u = x^2$; $u^2 - 10u + 9 = 0$
   (b) $u = x + 2$; $6u^2 - 11u + 4 = 0$

*In general, an equation in which an nth-degree polynomial equals 0 has n complex solutions, although some of them may be repeated.

NOW TRY EXERCISE 6

Solve each equation.

(a) $x^4 - 17x^2 + 16 = 0$

(b) $x^4 + 4 = 8x^2$

TEACHING TIP

Point out that we can obtain *up to* four distinct solutions to a fourth-degree equation.

(b)

$$4x^4 + 1 = 5x^2$$

$$4(x^2)^2 + 1 = 5x^2 \qquad x^4 = (x^2)^2$$

$$4u^2 + 1 = 5u \qquad \text{Let } u = x^2.$$

$$4u^2 - 5u + 1 = 0 \qquad \text{Standard form}$$

$$(4u - 1)(u - 1) = 0 \qquad \text{Factor.}$$

$$4u - 1 = 0 \quad \text{or} \quad u - 1 = 0 \qquad \text{Zero-factor property}$$

$$u = \frac{1}{4} \quad \text{or} \quad u = 1 \qquad \text{Solve.}$$

This is a key step.

$$x^2 = \frac{1}{4} \quad \text{or} \quad x^2 = 1 \qquad \text{Substitute } x^2 \text{ for } u.$$

$$x = \pm\frac{1}{2} \quad \text{or} \quad x = \pm 1 \qquad \text{Square root property}$$

Check that the solution set is $\left\{\pm\frac{1}{2}, \pm 1\right\}$.

(c)

$$x^4 = 6x^2 - 3$$

$$x^4 - 6x^2 + 3 = 0 \qquad \text{Standard form}$$

$$(x^2)^2 - 6x^2 + 3 = 0 \qquad x^4 = (x^2)^2$$

$$u^2 - 6u + 3 = 0 \qquad \text{Let } u = x^2.$$

The trinomial on the left is nonfactorable, so we cannot solve the equation using the zero-factor property. To solve, we use the quadratic formula.

$$u = \frac{-(-6) \pm \sqrt{(-6)^2 - 4(1)(3)}}{2(1)} \qquad a = 1, b = -6, c = 3$$

$$u = \frac{6 \pm \sqrt{24}}{2} \qquad \text{Simplify.}$$

$$u = \frac{6 \pm 2\sqrt{6}}{2} \qquad \sqrt{24} = \sqrt{4} \cdot \sqrt{6} = 2\sqrt{6}$$

$$u = \frac{2(3 \pm \sqrt{6})}{2} \qquad \text{Factor.}$$

$$u = 3 \pm \sqrt{6} \qquad \text{Divide out the common factor 2.}$$

$$x^2 = 3 + \sqrt{6} \quad \text{or} \quad x^2 = 3 - \sqrt{6} \qquad \text{Substitute } x^2 \text{ for } u.$$

Find *both* square roots in each case.

$$x = \pm\sqrt{3 + \sqrt{6}} \quad \text{or} \quad x = \pm\sqrt{3 - \sqrt{6}} \qquad \text{Square root property}$$

The solution set contains four numbers and is written

$$\left\{\pm\sqrt{3 + \sqrt{6}}, \pm\sqrt{3 - \sqrt{6}}\right\}.$$

NOW TRY

NOW TRY ANSWERS

6. (a) $\{\pm 1, \pm 4\}$

(b) $\left\{\pm\sqrt{4 + 2\sqrt{3}}, \pm\sqrt{4 - 2\sqrt{3}}\right\}$

NOTE The quadratic expressions in equations like those in **Examples 6(a) and (b)** can be factored directly.

$$x^4 - 13x^2 + 36 = 0 \quad \text{Example 6(a) equation}$$
$$(x^2 - 9)(x^2 - 4) = 0 \quad \text{Factor.}$$
$$(x + 3)(x - 3)(x + 2)(x - 2) = 0 \quad \text{Factor again.}$$

Using the zero-factor property gives the same solutions that we obtained in **Example 6(a).** Equations that include nonfactorable quadratic expressions (as in **Example 6(c)**) must be solved using substitution and the quadratic formula.

TEACHING TIP
Stress that u is just a temporary variable that helps us solve the given equation. As in any equation, we must solve for the variable in the *original* equation.

Solving an Equation That Is Quadratic in Form by Substitution

Step 1 **Define a temporary variable u,** based on the relationship between the variable expressions in the given equation. Substitute u in the original equation and rewrite the equation in the form $au^2 + bu + c = 0$.

Step 2 **Solve the quadratic equation obtained in Step 1** either by factoring the trinomial and applying the zero-factor property or by using the quadratic formula.

Step 3 **Replace u with the expression it defined in Step 1.**

Step 4 **Solve the resulting equations for the original variable.**

Step 5 **Check** all solutions by substituting them in the original equation.

CLASSROOM EXAMPLE 7
Solve each equation.
(a) $5(x + 3)^2 + 9(x + 3) = 2$
(b) $4x^{2/3} = 3x^{1/3} + 1$

Answers: (a) $\left\{-5, -\frac{14}{5}\right\}$
(b) $\left\{-\frac{1}{64}, 1\right\}$

TEACHING TIP
Although substitution is the method shown here, one could multiply out the expression on the left in **Example 7(a),** then factor directly and solve using the zero-factor property.

EXAMPLE 7 Solving Equations That Are Quadratic in Form

Solve each equation.

(a) $2(4x - 3)^2 + 7(4x - 3) + 5 = 0$

Step 1 Because of the repeated quantity $4x - 3$, substitute u for $4x - 3$.

$$2(4x - 3)^2 + 7(4x - 3) + 5 = 0 \quad \text{See Example 5(b).}$$
$$2u^2 + 7u + 5 = 0 \quad \text{Let } u = 4x - 3.$$

Step 2
$$(2u + 5)(u + 1) = 0 \quad \text{Factor.}$$
$$2u + 5 = 0 \quad \text{or} \quad u + 1 = 0 \quad \text{Zero-factor property}$$

Don't stop here.
$$u = -\frac{5}{2} \quad \text{or} \quad u = -1 \quad \text{Solve for } u.$$

Step 3
$$4x - 3 = -\frac{5}{2} \quad \text{or} \quad 4x - 3 = -1 \quad \text{Substitute } 4x - 3 \text{ for } u.$$

Step 4
$$4x = \frac{1}{2} \quad \text{or} \quad 4x = 2 \quad \text{Solve for } x.$$
$$x = \frac{1}{8} \quad \text{or} \quad x = \frac{1}{2}$$

Step 5 Check that the solution set of the original equation is $\left\{\frac{1}{8}, \frac{1}{2}\right\}$.

NOW TRY EXERCISE 7

Solve each equation.

(a) $6(x-4)^2 + 11(x-4) - 10 = 0$

(b) $2x^{2/3} - 7x^{1/3} + 3 = 0$

(b) $2x^{2/3} - 11x^{1/3} + 12 = 0$

Step 1 Because $x^{2/3} = (x^{1/3})^2$, we substitute u for $x^{1/3}$.

$$2x^{2/3} - 11x^{1/3} + 12 = 0 \quad \text{See Example 5(c).}$$

$$2(x^{1/3})^2 - 11x^{1/3} + 12 = 0 \quad x^{2/3} = (x^{1/3})^2$$

$$2u^2 - 11u + 12 = 0 \quad \text{Let } u = x^{1/3}.$$

Step 2

$$(2u-3)(u-4) = 0 \quad \text{Factor.}$$

$$2u - 3 = 0 \quad \text{or} \quad u - 4 = 0 \quad \text{Zero-factor property}$$

$$u = \frac{3}{2} \quad \text{or} \quad u = 4 \quad \text{Solve for } u.$$

Step 3

$$x^{1/3} = \frac{3}{2} \quad \text{or} \quad x^{1/3} = 4 \quad \text{Substitute } x^{1/3} \text{ for } u.$$

Step 4

$$(x^{1/3})^3 = \left(\frac{3}{2}\right)^3 \quad \text{or} \quad (x^{1/3})^3 = 4^3 \quad \text{Cube each side.}$$

$$x = \frac{27}{8} \quad \text{or} \quad x = 64 \quad \text{Apply the exponents.}$$

Step 5 Because the original equation involves variables with rational exponents, check that neither of these solutions is extraneous. The solution set is $\left\{\frac{27}{8}, 64\right\}$.

NOW TRY

CAUTION A common error when solving problems like those in **Examples 6 and 7** is to stop too soon. ***Once we have solved for u, we must remember to substitute and solve for the values of the original variable.***

NOW TRY ANSWERS

7. (a) $\left\{\frac{3}{2}, \frac{14}{3}\right\}$ (b) $\left\{\frac{1}{8}, 27\right\}$

8.3 Exercises

FOR EXTRA HELP

MyMathLab®

Complete solution available in MyMathLab

Concept Check *Based on the discussion and examples of this section, give the first step to solve each equation. Do not actually solve.*

1. $\frac{14}{x} = x - 5$

2. $\sqrt{1+x} + x = 5$

3. $(x^2+x)^2 - 8(x^2+x) + 12 = 0$

4. $3x = \sqrt{16 - 10x}$

5. *Concept Check* Study this incorrect "solution." ***WHAT WENT WRONG?***

$$x = \sqrt{3x+4} \quad \text{Square each side.}$$

$$x^2 = 3x + 4$$

$$x^2 - 3x - 4 = 0$$

$$(x-4)(x+1) = 0$$

$$x - 4 = 0 \quad \text{or} \quad x + 1 = 0$$

$$x = 4 \quad \text{or} \quad x = -1$$

Solution set: $\{4, -1\}$

6. *Concept Check* Study this incorrect "solution." ***WHAT WENT WRONG?***

$$2(x-1)^2 - 3(x-1) + 1 = 0 \quad \text{Let}$$

$$2u^2 - 3u + 1 = 0 \quad u = x - 1.$$

$$(2u-1)(u-1) = 0$$

$$2u - 1 = 0 \quad \text{or} \quad u - 1 = 0$$

$$u = \frac{1}{2} \quad \text{or} \quad u = 1$$

Solution set: $\left\{\frac{1}{2}, 1\right\}$

1. Multiply by the LCD, x.
2. Isolate the radical term on one side.
3. Substitute a variable for $x^2 + x$.
4. Square each side.
5. The proposed solution -1 does not check. The solution set is $\{4\}$.
6. The solutions given are for u. Each must be set equal to $x - 1$ and solved for x. The correct solution set is $\left\{\frac{3}{2}, 2\right\}$.

7. $\{-2, 7\}$ 8. $\{-6, -2\}$
9. $\{-4, 7\}$ 10. $\left\{-\frac{1}{4}, 2\right\}$
11. $\left\{-\frac{2}{3}, 1\right\}$ 12. $\{-3, 1\}$
13. $\left\{-\frac{14}{17}, 5\right\}$ 14. $\left\{-8, \frac{9}{11}\right\}$
15. $\left\{-\frac{11}{7}, 0\right\}$ 16. $\left\{0, \frac{59}{13}\right\}$
17. $\left\{\frac{-1 \pm \sqrt{13}}{2}\right\}$
18. $\left\{\frac{-1 \pm \sqrt{193}}{12}\right\}$
19. $\left\{-\frac{8}{3}, -1\right\}$ 20. $\left\{-\frac{7}{3}, \frac{1}{3}\right\}$
21. $\left\{\frac{2 \pm \sqrt{22}}{3}\right\}$
22. $\left\{\frac{3 \pm \sqrt{3}}{3}\right\}$
23. $\left\{\frac{-1 \pm \sqrt{5}}{4}\right\}$
24. $\left\{\frac{5 \pm \sqrt{5}}{6}\right\}$
25. (a) $(20 - t)$ mph (b) $(20 + t)$ mph
26. (a) $\frac{1}{m}$ job per hr (b) $\frac{2}{m}$ job
27. the rate of her boat in still water; $x - 5$; $x + 5$; row 1 of table: $15, x - 5, \frac{15}{x-5}$; row 2 of table: $15, x + 5, \frac{15}{x+5}$; $\frac{15}{x-5} + \frac{15}{x+5} = 4$; 10 mph
28. $x - 2$; row 1 of table: $\frac{1}{x-2}, 2, \frac{2}{x-2}$; row 2 of table: $\frac{1}{x}, 2, \frac{2}{x}$; $\frac{2}{x-2} + \frac{2}{x} = 1$; Jaime: 5.2 hr; Carlos: 3.2 hr

Solve each equation. Check the solutions. ***See Example 1.***

7. $\frac{14}{x} = x - 5$

8. $\frac{-12}{x} = x + 8$

9. $1 - \frac{3}{x} - \frac{28}{x^2} = 0$

10. $4 - \frac{7}{r} - \frac{2}{r^2} = 0$

11. $3 - \frac{1}{t} = \frac{2}{t^2}$

12. $1 + \frac{2}{x} = \frac{3}{x^2}$

13. $\frac{1}{x} + \frac{2}{x+2} = \frac{17}{35}$

14. $\frac{2}{m} + \frac{3}{m+9} = \frac{11}{4}$

15. $\frac{2}{x+1} + \frac{3}{x+2} = \frac{7}{2}$

16. $\frac{4}{3-p} + \frac{2}{5-p} = \frac{26}{15}$

17. $\frac{3}{2x} - \frac{1}{2(x+2)} = 1$

18. $\frac{4}{3x} - \frac{1}{2(x+1)} = 1$

19. $3 = \frac{1}{t+2} + \frac{2}{(t+2)^2}$

20. $1 + \frac{2}{3z+2} = \frac{15}{(3z+2)^2}$

21. $\frac{6}{p} = 2 + \frac{p}{p+1}$

22. $\frac{x}{2-x} + \frac{2}{x} = 5$

23. $1 - \frac{1}{2x+1} - \frac{1}{(2x+1)^2} = 0$

24. $1 - \frac{1}{3x-2} - \frac{1}{(3x-2)^2} = 0$

Concept Check *Answer each question.*

25. A boat travels 20 mph in still water, and the rate of the current is t mph.

(a) What is the rate of the boat when it travels upstream?

(b) What is the rate of the boat when it travels downstream?

26. It takes m hours to grade a set of papers.

(a) What is the grader's rate (in job per hour)?

(b) How much of the job will the grader do in 2 hr?

Solve each problem. Round answers to the nearest tenth. ***See Examples 2 and 3.***

27. In 4 hr, Kerrie can travel 15 mi upriver and come back. The rate of the current is 5 mph. Find the rate of her boat in still water.

Let x = ______________.

The rate traveling upriver (*against* the current) is _____ mph.

The rate traveling back downriver (*with* the current) is _____ mph.

Complete the table.

	d	*r*	*t*
Up			
Down			

Write an equation, and complete the solution.

28. Carlos can complete a certain lab test in 2 hr less time than Jaime can. If they can finish the job together in 2 hr, how long would it take each of them working alone?

Let x = Jaime's time alone (in hours).

Then _____ = Carlos' time alone (in hours).

Complete the table.

	Rate	*Time Working Together*	*Fractional Part of the Job Done*
Carlos			
Jaime			

Write an equation, and complete the solution.

29. 25 mph **30.** 255 mph
31. 50 mph **32.** 80 km per hr
33. 3.6 hr **34.** 11.1 hr
35. Rusty: 25.0 hr; Nancy: 23.0 hr

29. On a windy day William found that he could travel 16 mi downstream and then 4 mi back upstream at top speed in a total of 48 min. What was the top speed of William's boat if the rate of the current was 15 mph? (Let x represent the rate of the boat in still water.)

	d	r	t
Upstream	4	$x - 15$	
Downstream	16		

30. Vera flew for 6 hr at a constant rate. She traveled 810 mi with the wind, then turned around and traveled 720 mi against the wind. The wind speed was a constant 15 mph. Find the rate of the plane.

	d	r	t
With Wind	810		
Against Wind	720		

31. The distance from Jackson to Lodi is about 40 mi, as is the distance from Lodi to Manteca. Adrian drove from Jackson to Lodi, stopped in Lodi for a high-energy drink, and then drove on to Manteca at 10 mph faster. Driving time for the entire trip was 88 min. Find her rate from Jackson to Lodi. (*Source: State Farm Road Atlas.*)

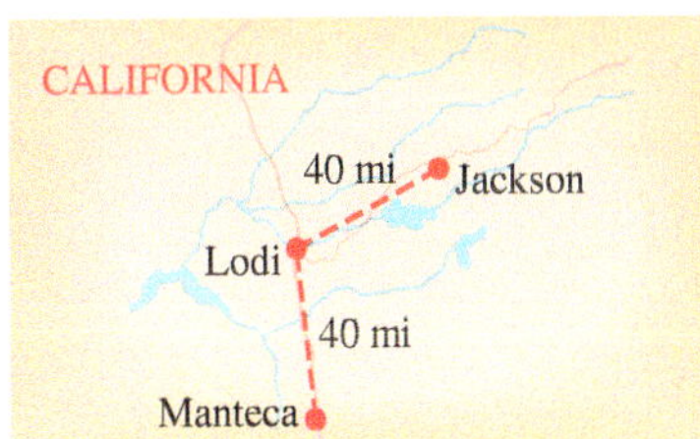

32. Medicine Hat and Cranbrook are 300 km apart. Steve rides his Harley 20 km per hr faster than Mohammad rides his Yamaha. Find Steve's average rate if he travels from Cranbrook to Medicine Hat in $1\frac{1}{4}$ hr less time than Mohammad. (*Source: State Farm Road Atlas.*)

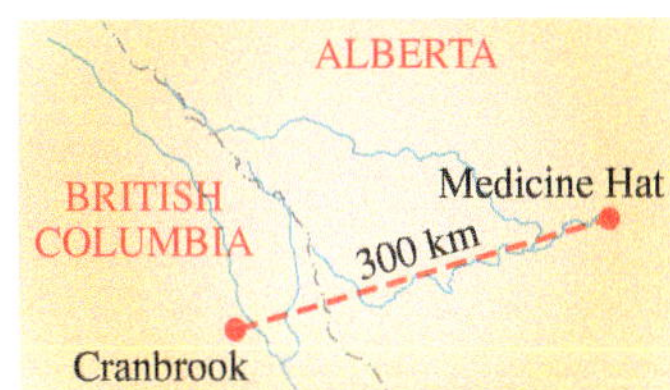

33. Working together, two people can cut a large lawn in 2 hr. One person can do the job alone in 1 hr less time than the other. How long (to the nearest tenth) would it take the faster worker to do the job? (Let x represent the time of the faster worker.)

	Rate	Time Working Together	Fractional Part of the Job Done
Faster Worker	$\frac{1}{x}$	2	
Slower Worker		2	

34. Working together, two people can clean an office building in 5 hr. One person takes 2 hr longer than the other to clean the building alone. How long (to the nearest tenth) would it take the slower worker to clean the building alone? (Let x represent the time of the slower worker.)

	Rate	Time Working Together	Fractional Part of the Job Done
Faster Worker			
Slower Worker	$\frac{1}{x}$		

35. Rusty and Nancy are planting flowers. Working alone, Rusty would take 2 hr longer than Nancy to plant the flowers. Working together, they do the job in 12 hr. How long (to the nearest tenth) would it have taken each person working alone?

36. Joel: 2.6 hr; Noel: 3.6 hr
37. 3 hr; 6 hr **38.** 9 min
39. $\{2, 5\}$ **40.** $\{1, 4\}$
41. $\{3\}$ **42.** $\left\{\frac{1}{2}\right\}$
43. $\left\{\frac{8}{9}\right\}$ **44.** $\left\{\frac{3}{4}\right\}$
45. $\{9\}$ **46.** $\{16\}$
47. $\left\{\frac{2}{5}\right\}$ **48.** $\left\{\frac{5}{6}\right\}$
49. $\{-2\}$ **50.** $\{-1\}$
51. $\{\pm 2, \pm 5\}$ **52.** $\{\pm 1, \pm 6\}$
53. $\left\{\pm 1, \pm\frac{3}{2}\right\}$ **54.** $\left\{\pm 1, \pm\frac{4}{3}\right\}$
55. $\{\pm 2, \pm 2\sqrt{3}\}$
56. $\{\pm 3, \pm 2\sqrt{2}\}$
57. $\{-6, -5\}$ **58.** $\{-1, 8\}$
59. $\left\{-\frac{16}{3}, -2\right\}$ **60.** $\{-4, 1\}$
61. $\{-8, 1\}$ **62.** $\{-1, 27\}$
63. $\{-64, 27\}$ **64.** $\left\{-\frac{512}{27}, 27\right\}$
65. $\left\{\pm 1, \pm\frac{27}{8}\right\}$ **66.** $\left\{\pm 1, \pm\frac{64}{27}\right\}$
67. $\left\{-\frac{1}{3}, \frac{1}{6}\right\}$ **68.** $\left\{-\frac{4}{3}, \frac{1}{2}\right\}$
69. $\left\{-\frac{1}{2}, 3\right\}$ **70.** $\left\{\frac{2}{3}, 2\right\}$
71. $\left\{\pm\frac{\sqrt{6}}{3}, \pm\frac{1}{2}\right\}$
72. $\left\{\frac{-4 \pm \sqrt{41}}{2}\right\}$
73. $\{3, 11\}$ **74.** $\{8\}$
75. $\{25\}$ **76.** $\{-3, -2, 1, 2\}$
77. $\left\{-\sqrt[3]{5}, -\frac{\sqrt[3]{4}}{2}\right\}$
78. $\left\{-4, -\frac{1}{2}\right\}$
79. $\left\{\frac{4}{3}, \frac{9}{4}\right\}$ **80.** $\{1, \sqrt[3]{9}\}$
81. $\left\{\pm\frac{\sqrt{9+\sqrt{65}}}{2}, \pm\frac{\sqrt{9-\sqrt{65}}}{2}\right\}$
82. $\left\{\pm\frac{\sqrt{11+\sqrt{89}}}{4}, \pm\frac{\sqrt{11-\sqrt{89}}}{4}\right\}$
83. $\left\{\pm 1, \pm\frac{\sqrt{6}}{2}i\right\}$
84. $\left\{\pm\frac{1}{2}i, \pm i\right\}$

36. Joel can work through a stack of invoices in 1 hr less time than Noel can. Working together they take $1\frac{1}{2}$ hr. How long (to the nearest tenth) would it take each person working alone?

37. Two pipes together can fill a tank in 2 hr. One of the pipes, used alone, takes 3 hr longer than the other to fill the tank. How long would each pipe take to fill the tank alone?

38. A washing machine can be filled in 6 min if both the hot and cold water taps are fully opened. Filling the washer with hot water alone takes 9 min longer than filling it with cold water alone. How long does it take to fill the washer with cold water?

Solve each equation. Check the solutions. ***See Example 4.***

39. $x = \sqrt{7x - 10}$ **40.** $z = \sqrt{5z - 4}$ **41.** $2x = \sqrt{11x + 3}$

42. $4x = \sqrt{6x + 1}$ **43.** $3x = \sqrt{16 - 10x}$ **44.** $4t = \sqrt{8t + 3}$

45. $t + \sqrt{t} = 12$ **46.** $p - 2\sqrt{p} = 8$ **47.** $x = \sqrt{\frac{6 - 13x}{5}}$

48. $r = \sqrt{\frac{20 - 19r}{6}}$ **49.** $-x = \sqrt{\frac{8 - 2x}{3}}$ **50.** $-x = \sqrt{\frac{3x + 7}{4}}$

Solve each equation. Check the solutions. ***See Examples 5–7.***

51. $x^4 - 29x^2 + 100 = 0$ **52.** $x^4 - 37x^2 + 36 = 0$ **53.** $4q^4 - 13q^2 + 9 = 0$

54. $9x^4 - 25x^2 + 16 = 0$ **55.** $x^4 + 48 = 16x^2$ **56.** $z^4 + 72 = 17z^2$

57. $(x + 3)^2 + 5(x + 3) + 6 = 0$ **58.** $(x - 4)^2 + (x - 4) - 20 = 0$

59. $3(m + 4)^2 - 8 = 2(m + 4)$ **60.** $(t + 5)^2 + 6 = 7(t + 5)$

61. $x^{2/3} + x^{1/3} - 2 = 0$ **62.** $x^{2/3} - 2x^{1/3} - 3 = 0$

63. $r^{2/3} + r^{1/3} - 12 = 0$ **64.** $3x^{2/3} - x^{1/3} - 24 = 0$

65. $4x^{4/3} - 13x^{2/3} + 9 = 0$ **66.** $9t^{4/3} - 25t^{2/3} + 16 = 0$

67. $2 + \frac{5}{3x - 1} = \frac{-2}{(3x - 1)^2}$ **68.** $3 - \frac{7}{2p + 2} = \frac{6}{(2p + 2)^2}$

69. $2 - 6(z - 1)^{-2} = (z - 1)^{-1}$ **70.** $3 - 2(x - 1)^{-1} = (x - 1)^{-2}$

The following exercises are not grouped by type. Solve each equation. (Exercises 83 and 84 require knowledge of complex numbers.) ***See Examples 1 and 4–7.***

71. $12x^4 - 11x^2 + 2 = 0$ **72.** $\left(x - \frac{1}{2}\right)^2 + 5\left(x - \frac{1}{2}\right) - 4 = 0$

73. $\sqrt{2x + 3} = 2 + \sqrt{x - 2}$ **74.** $\sqrt{m + 1} = -1 + \sqrt{2m}$

75. $2(1 + \sqrt{r})^2 = 13(1 + \sqrt{r}) - 6$ **76.** $(x^2 + x)^2 + 12 = 8(x^2 + x)$

77. $2m^6 + 11m^3 + 5 = 0$ **78.** $8x^6 + 513x^3 + 64 = 0$

79. $6 = 7(2w - 3)^{-1} + 3(2w - 3)^{-2}$ **80.** $x^6 - 10x^3 = -9$

81. $2x^4 - 9x^2 = -2$ **82.** $8x^4 + 1 = 11x^2$

83. $2x^4 + x^2 - 3 = 0$ **84.** $4x^4 + 5x^2 + 1 = 0$

SUMMARY EXERCISES Applying Methods for Solving Quadratic Equations

TEACHING TIP
Spend some lecture time briefly reviewing the methods outlined in the table. Emphasize that the two methods most often used are the *zero-factor property* if the trinomial can be factored (look for this first) and the *quadratic formula* (if we cannot factor).

1. square root property
2. zero-factor property
3. quadratic formula
4. quadratic formula
5. zero-factor property
6. square root property

7. $\{\pm\sqrt{7}\}$ 8. $\left\{-\frac{3}{2}, \frac{5}{3}\right\}$
9. $\{-3 \pm \sqrt{5}\}$ 10. $\{-2, 8\}$
11. $\left\{-\frac{3}{2}, 4\right\}$ 12. $\left\{-3, \frac{1}{3}\right\}$
13. $\left\{\frac{2 \pm \sqrt{2}}{2}\right\}$ 14. $\{\pm 2i\sqrt{3}\}$
15. $\left\{\frac{1}{2}, 2\right\}$ 16. $\{\pm 1, \pm 3\}$
17. $\left\{\frac{-3 \pm 2\sqrt{2}}{2}\right\}$
18. $\left\{\frac{4}{5}, 3\right\}$
19. $\{\pm\sqrt{2}, \pm\sqrt{7}\}$
20. $\left\{\frac{1 \pm \sqrt{5}}{4}\right\}$
21. $\left\{-\frac{1}{2} \pm \frac{\sqrt{3}}{2}i\right\}$
22. $\left\{-\frac{\sqrt[3]{175}}{5}, 1\right\}$
23. $\left\{\frac{3}{2}\right\}$ 24. $\left\{\frac{2}{3}\right\}$
25. $\{\pm 6\sqrt{2}\}$ 26. $\left\{-\frac{2}{3}, 2\right\}$
27. $\{-4, 9\}$ 28. $\{\pm 13\}$
29. $\left\{1 \pm \frac{\sqrt{3}}{3}i\right\}$ 30. $\{3\}$
31. $\left\{\frac{1}{6} \pm \frac{\sqrt{47}}{6}i\right\}$
32. $\left\{-\frac{1}{3}, \frac{1}{6}\right\}$

We have introduced four methods for solving quadratic equations written in standard form $ax^2 + bx + c = 0$.

Method	*Advantages*	*Disadvantages*
Zero-factor property	This is usually the fastest method.	Not all trinomials are factorable. Some factorable trinomials are difficult to factor.
Square root property	This is the simplest method for solving equations of the form $(ax + b)^2 = c$.	Few equations are given in this form.
Completing the square	This method can always be used, although most people prefer the quadratic formula.	It requires more steps than other methods.
Quadratic formula	This method can always be used.	Sign errors are common when evaluating $\sqrt{b^2 - 4ac}$.

Concept Check *Decide whether the* zero-factor property, *the* square root property, *or the* quadratic formula *is most appropriate for solving each quadratic equation. Do not actually solve.*

1. $(2x + 3)^2 = 4$ **2.** $4x^2 - 3x = 1$ **3.** $x^2 + 5x - 8 = 0$

4. $2x^2 + 3x = 1$ **5.** $3x^2 = 2 - 5x$ **6.** $x^2 = 5$

Solve each quadratic equation by the method of your choice.

7. $p^2 = 7$ **8.** $6x^2 - x - 15 = 0$ **9.** $n^2 + 6n + 4 = 0$

10. $(x - 3)^2 = 25$ **11.** $\frac{5}{x} + \frac{12}{x^2} = 2$ **12.** $3x^2 = 3 - 8x$

13. $2r^2 - 4r + 1 = 0$ ***14.** $x^2 = -12$ **15.** $x\sqrt{2} = \sqrt{5x - 2}$

16. $x^4 - 10x^2 + 9 = 0$ **17.** $(2x + 3)^2 = 8$ **18.** $\frac{2}{x} + \frac{1}{x - 2} = \frac{5}{3}$

19. $t^4 + 14 = 9t^2$ **20.** $8x^2 - 4x = 2$ ***21.** $z^2 + z + 1 = 0$

22. $5x^6 + 2x^3 - 7 = 0$ **23.** $4t^2 - 12t + 9 = 0$ **24.** $x\sqrt{3} = \sqrt{2 - x}$

25. $r^2 - 72 = 0$ **26.** $-3x^2 + 4x = -4$ **27.** $x^2 - 5x - 36 = 0$

28. $w^2 = 169$ ***29.** $3p^2 = 6p - 4$ **30.** $z = \sqrt{\frac{5z + 3}{2}}$

***31.** $\frac{4}{r^2} + 3 = \frac{1}{r}$ **32.** $2(3x - 1)^2 + 5(3x - 1) = -2$

*This exercise requires knowledge of complex numbers.

8.4 Formulas and Further Applications

OBJECTIVES

1. Solve formulas involving squares and square roots for specified variables.
2. Solve applied problems using the Pythagorean theorem.
3. Solve applied problems using area formulas.
4. Solve applied problems using quadratic functions as models.

NOW TRY EXERCISE 1

Solve each formula for the specified variable. Keep $\pm$ in the answer in part (a).

(a) $n = \dfrac{ab}{E^2}$ for E

(b) $S = \sqrt{\dfrac{pq}{n}}$ for p

CLASSROOM EXAMPLE 1

Solve each formula for the specified variable. Keep $\pm$ in the answer in part (a).

(a) $\mathscr{A} = \pi r^2$ for r

(b) $s = 30\sqrt{\dfrac{a}{p}}$ for a

Answers: (a) $r = \dfrac{\pm\sqrt{\mathscr{A}\pi}}{\pi}$

(b) $a = \dfrac{ps^2}{900}$

CLASSROOM EXAMPLE 2 is on the next page.

NOW TRY ANSWERS

1. (a) $E = \dfrac{\pm\sqrt{abn}}{n}$

(b) $p = \dfrac{nS^2}{q}$

OBJECTIVE 1 Solve formulas involving squares and square roots for specified variables.

EXAMPLE 1 Solving for Specified Variables

Solve each formula for the specified variable. Keep $\pm$ in the answer in part (a).

(a) $w = \dfrac{kFr}{v^2}$ for v

$$w = \frac{kFr}{v^2}$$ The goal is to isolate v on one side.

$$v^2 w = kFr \quad \text{Multiply by } v^2.$$

$$v^2 = \frac{kFr}{w} \quad \text{Divide by } w.$$

$$v = \pm\sqrt{\frac{kFr}{w}} \quad \text{Square root property}$$ Include both positive and negative roots.

$$v = \frac{\pm\sqrt{kFr}}{\sqrt{w}} \cdot \frac{\sqrt{w}}{\sqrt{w}} \quad \text{Rationalize the denominator.}$$

$$v = \frac{\pm\sqrt{kFrw}}{w} \quad \sqrt{a}\cdot\sqrt{b} = \sqrt{ab};\ \sqrt{a}\cdot\sqrt{a} = a$$

(b) $d = \sqrt{\dfrac{4\mathscr{A}}{\pi}}$ for $\mathscr{A}$

$$d = \sqrt{\frac{4\mathscr{A}}{\pi}}$$ The goal is to isolate $\mathscr{A}$ on one side.

$$d^2 = \frac{4\mathscr{A}}{\pi} \quad \text{Square each side.}$$

$$\pi d^2 = 4\mathscr{A} \quad \text{Multiply by } \pi.$$

$$\frac{\pi d^2}{4} = \mathscr{A}, \quad \text{or} \quad \mathscr{A} = \frac{\pi d^2}{4} \quad \text{Divide by 4. Interchange sides.}$$

NOW TRY

EXAMPLE 2 Solving for a Specified Variable

Solve $s = 2t^2 + kt$ for t.

Because the given equation has terms with t^2 and t, write it in standard form $ax^2 + bx + c = 0$, with t as the variable instead of x.

$$s = 2t^2 + kt$$

$$0 = 2t^2 + kt - s \quad \text{Subtract } s.$$

$$2t^2 + kt - s = 0 \quad \text{Standard form}$$

NOW TRY EXERCISE 2
Solve for r.

$$r^2 + 9r = -c$$

To solve $2t^2 + kt - s = 0$, use the quadratic formula with $a = 2$, $b = k$, and $c = -s$.

$$t = \frac{-k \pm \sqrt{k^2 - 4(2)(-s)}}{2(2)} \quad \text{Substitute.}$$

$$t = \frac{-k \pm \sqrt{k^2 + 8s}}{4} \quad \text{Simplify.}$$

The solutions are $t = \dfrac{-k + \sqrt{k^2 + 8s}}{4}$ and $t = \dfrac{-k - \sqrt{k^2 + 8s}}{4}$. NOW TRY

OBJECTIVE 2 Solve applied problems using the Pythagorean theorem.

The Pythagorean theorem, represented by the equation

$$a^2 + b^2 = c^2,$$

is illustrated in **FIGURE 2** and was introduced in **Section 7.3.** It is used to solve applications involving right triangles.

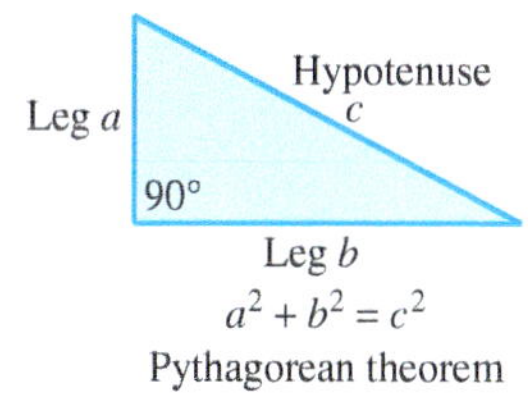

FIGURE 2

NOW TRY EXERCISE 3
Matt is building a new barn, with length 10 ft more than width. While determining the footprint of the barn, he measured the diagonal as 50 ft. What will be the dimensions of the barn?

EXAMPLE 3 Using the Pythagorean Theorem

Two cars left an intersection at the same time, one heading due north, the other due west. Some time later, they were exactly 100 mi apart. The car headed north had gone 20 mi farther than the car headed west. How far had each car traveled?

Step 1 **Read** the problem carefully.

Step 2 **Assign a variable.**

Let x = the distance traveled by the car headed west.

Then $x + 20$ = the distance traveled by the car headed north.

See **FIGURE 3**. The cars are 100 mi apart, so the hypotenuse of the right triangle equals 100.

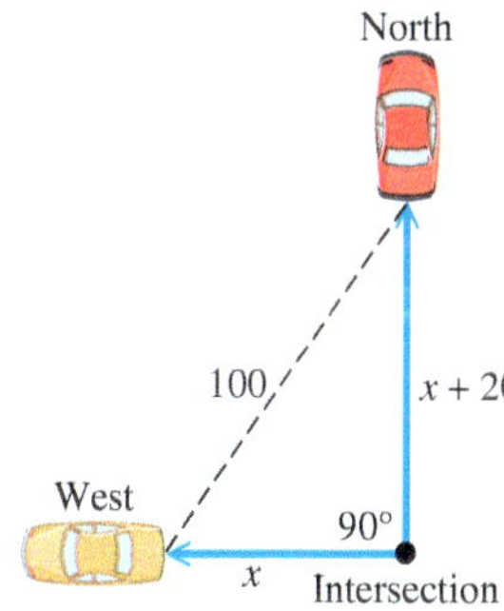

FIGURE 3

Step 3 **Write an equation.** Use the Pythagorean theorem.

$$a^2 + b^2 = c^2 \quad \text{Pythagorean theorem}$$

$$x^2 + (x + 20)^2 = 100^2 \quad \text{See FIGURE 3.}$$

$(x + y)^2 = x^2 + 2xy + y^2$

Step 4 **Solve.**

$$x^2 + x^2 + 40x + 400 = 10{,}000 \quad \text{Square the binomial.}$$

$$2x^2 + 40x - 9600 = 0 \quad \text{Standard form}$$

$$x^2 + 20x - 4800 = 0 \quad \text{Divide by 2.}$$

$$(x + 80)(x - 60) = 0 \quad \text{Factor.}$$

$$x + 80 = 0 \quad \text{or} \quad x - 60 = 0 \quad \text{Zero-factor property}$$

$$x = -80 \quad \text{or} \quad x = 60 \quad \text{Solve for } x.$$

Step 5 **State the answer.** Distance cannot be negative, so discard the negative solution. The required distances are 60 mi and $60 + 20 = 80$ mi.

Step 6 **Check.** Here $60^2 + 80^2 = 100^2$, so the answer is correct. NOW TRY

NOW TRY ANSWERS

2. $r = \dfrac{-9 \pm \sqrt{81 - 4c}}{2}$

3. 30 ft by 40 ft

NOW TRY EXERCISE 4

A football practice field is 30 yd wide and 40 yd long. A strip of grass sod of uniform width is to be placed around the perimeter of the practice field. There is enough money budgeted for 296 sq yd of sod. How wide will the strip be?

CLASSROOM EXAMPLE 4

Suppose the pool in Example 4 is 20 ft by 40 ft and there is enough seed to cover 700 ft^2. How wide should the grass strip be?

Answer: 5 ft

OBJECTIVE 3 Solve applied problems using area formulas.

EXAMPLE 4 Solving an Area Problem

A rectangular reflecting pool in a park is 20 ft wide and 30 ft long. The gardener wants to plant a strip of grass of uniform width around the edge of the pool. She has enough seed to cover 336 ft^2. How wide will the strip be?

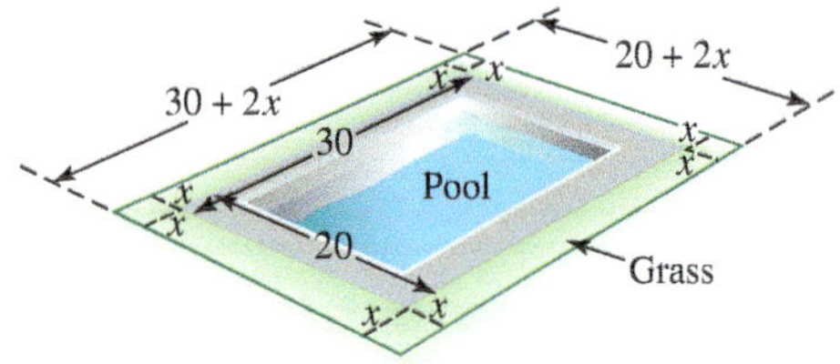

FIGURE 4

Step 1 **Read** the problem carefully.

Step 2 **Assign a variable.** The pool is shown in FIGURE 4.

Let x = the unknown width of the grass strip.

Then $20 + 2x$ = the width of the large rectangle (the width of the pool plus two grass strips),

and $30 + 2x$ = the length of the large rectangle.

Step 3 **Write an equation.** Refer to FIGURE 4.

$(30 + 2x)(20 + 2x)$ Area of large rectangle (length · width)

$30 \cdot 20$, or 600 Area of pool (in square feet)

The area of the large rectangle minus the area of the pool should equal 336 ft^2, the area of the grass strip.

Area of large rectangle − Area of pool = Area of grass

$$(30 + 2x)(20 + 2x) - 600 = 336$$

Step 4 **Solve.**

$$600 + 60x + 40x + 4x^2 - 600 = 336 \quad \text{Multiply.}$$
$$4x^2 + 100x - 336 = 0 \quad \text{Standard form}$$
$$x^2 + 25x - 84 = 0 \quad \text{Divide by 4.}$$
$$(x + 28)(x - 3) = 0 \quad \text{Factor.}$$
$$x + 28 = 0 \quad \text{or} \quad x - 3 = 0 \quad \text{Zero-factor property}$$
$$x = -28 \quad \text{or} \quad x = 3 \quad \text{Solve for } x.$$

Step 5 **State the answer.** The width cannot be −28 ft, so the grass strip should be 3 ft wide.

Step 6 **Check.** If $x = 3$, we can find the area of the large rectangle (which includes the grass strip).

$$(30 + 2 \cdot 3)(20 + 2 \cdot 3) = 36 \cdot 26 = 936 \text{ ft}^2 \quad \text{Area of pool and strip}$$

The area of the pool is $30 \cdot 20 = 600 \text{ ft}^2$. So, the area of the grass strip is

$$936 - 600 = 336 \text{ ft}^2, \quad \text{as required.}$$

NOW TRY

NOW TRY ANSWER

4. 2 yd

OBJECTIVE 4 Solve applied problems using quadratic functions as models.

Some applied problems can be modeled by *quadratic functions,* which for real numbers a, b, and c, can be written in the form

$$f(x) = ax^2 + bx + c \quad (\text{where } a \neq 0).$$

NOW TRY EXERCISE 5

If an object is projected upward from the top of a 120-ft building at 60 ft per sec, its position (in feet above the ground) is given by

$$s(t) = -16t^2 + 60t + 120,$$

where t is time in seconds after it was projected. When does it hit the ground (to the nearest tenth)?

CLASSROOM EXAMPLE 5

A ball is projected upward from the ground. Its distance in feet from the ground at t seconds is

$$s(t) = -16t^2 + 64t.$$

At what times (to the nearest tenth) will the ball be 32 ft from the ground?

Answer: at 0.6 sec and at 3.4 sec

EXAMPLE 5 Solving an Applied Problem Using a Quadratic Function

If an object is projected upward from the top of a 144-ft building at 112 ft per sec, its position (in feet above the ground) is given by

$$s(t) = -16t^2 + 112t + 144,$$

where t is time in seconds after it was projected. When does it hit the ground?

When the object hits the ground, its distance above the ground is 0. We must find the value of t that makes $s(t) = 0$.

$$0 = -16t^2 + 112t + 144 \qquad \text{Let } s(t) = 0.$$

$$0 = t^2 - 7t - 9 \qquad \text{Divide by } -16.$$

$$t = \frac{-(-7) \pm \sqrt{(-7)^2 - 4(1)(-9)}}{2(1)} \qquad \text{Substitute into the quadratic formula. Let } a = 1,\ b = -7, \text{ and } c = -9.$$

$$t = \frac{7 \pm \sqrt{85}}{2} = \frac{7 \pm 9.2}{2} \qquad \text{Use a calculator. Round to the nearest tenth.}$$

The two solutions are

$$t = 8.1 \quad \text{or} \quad t = -1.1.$$

Time cannot be negative, so we discard the negative solution. The object hits the ground 8.1 sec after it is projected. NOW TRY

CLASSROOM EXAMPLE 6

Refer to Example 6.

(a) Use the model to approximate the CPI for 2000, to the nearest whole number.

(b) In what year did the CPI reach 150? (Round down for the year.)

Answers: (a) 175 (b) 1994

EXAMPLE 6 Using a Quadratic Function to Model the CPI

The Consumer Price Index (CPI) is used to measure trends in prices for a "basket" of goods purchased by typical American families. This index uses a base period of 1982–1984, which means that the index number for that period is 100. The quadratic function

$$f(x) = -0.0003x^2 + 4.55x + 83.7$$

approximates the CPI for the years 1980–2012, where x is the number of years that have elapsed since 1980. (*Source:* Bureau of Labor Statistics.)

(a) Use the model to approximate the CPI for 2012.

For 2012, $x = 2012 - 1980 = 32$, so we find $f(32)$.

$$f(x) = -0.0003x^2 + 4.55x + 83.7 \qquad \text{Given model}$$

$$f(32) = -0.0003(32)^2 + 4.55(32) + 83.7 \qquad \text{Let } x = 32.$$

$$f(32) = 229 \qquad \text{Nearest whole number}$$

According to the model, the CPI for 2012 was 229.

NOW TRY ANSWER

5. 5.2 sec after it is projected

NOW TRY EXERCISE 6

Refer to **Example 6.**

(a) Use the model to approximate the CPI for 2010, to the nearest whole number.

(b) In what year did the CPI reach 175? (Round down for the year.)

(b) In what year did the CPI reach 200?
Find the value of x that makes $f(x) = 200$.

$$f(x) = -0.0003x^2 + 4.55x + 83.7 \quad \text{Given model}$$

$$200 = -0.0003x^2 + 4.55x + 83.7 \quad \text{Let } f(x) = 200.$$

$$0 = -0.0003x^2 + 4.55x - 116.3 \quad \text{Subtract 200.}$$

$$x = \frac{-4.55 \pm \sqrt{4.55^2 - 4(-0.0003)(-116.3)}}{2(-0.0003)} \quad \text{Use } a = -0.0003,\ b = 4.55, \text{ and } c = -116.3 \text{ in the quadratic formula.}$$

$$x = 25.6 \quad \text{or} \quad x = 15{,}141.1 \quad \text{Use a calculator. Round to the nearest tenth.}$$

Rounding the first solution 25.6 down, the CPI first reached 200 in

$$1980 + 25 = 2005.$$

(Reject the solution $x = 15{,}141.1$, as this corresponds to a totally unreasonable year.)

NOW TRY

NOW TRY ANSWERS

6. (a) 220 **(b)** 2000

8.4 Exercises

FOR EXTRA HELP

MyMathLab®

Complete solution available in MyMathLab

Concept Check *Answer each question.*

1. In solving a formula that has the specified variable in the denominator, what is the first step?

2. What is the first step in solving a formula like $gw^2 = 2r$ for w?

3. What is the first step in solving a formula like $gw^2 = kw + 24$ for w?

4. Why is it particularly important to check all proposed solutions to an applied problem against the information in the original problem?

For each triangle, solve for m in terms of the other variables (where $m > 0$).

5.

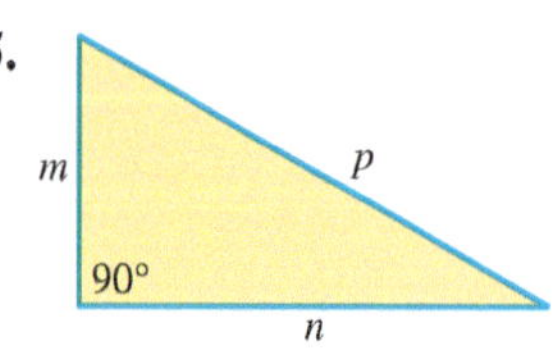

6.

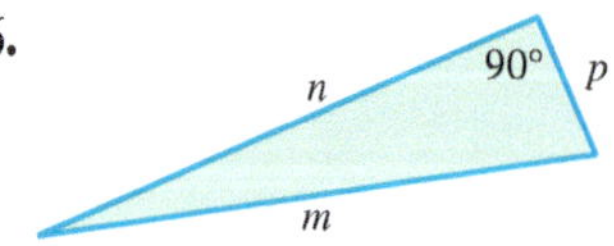

Solve each formula for the specified variable. (Leave ± in the answers.) ***See Examples 1 and 2.***

7. $d = kt^2$ for t

8. $S = 6e^2$ for e

9. $S = 4\pi r^2$ for r

10. $s = kwd^2$ for d

11. $I = \dfrac{ks}{d^2}$ for d

12. $R = \dfrac{k}{d^2}$ for d

13. $F = \dfrac{kA}{v^2}$ for v

14. $L = \dfrac{kd^4}{h^2}$ for h

15. $V = \dfrac{1}{3}\pi r^2 h$ for r

16. $V = \pi r^2 h$ for r

17. $At^2 + Bt = -C$ for t

18. $S = 2\pi rh + \pi r^2$ for r

19. $D = \sqrt{kh}$ for h

20. $F = \dfrac{k}{\sqrt{d}}$ for d

21. $p = \sqrt{\dfrac{k\ell}{g}}$ for ℓ

22. $p = \sqrt{\dfrac{k\ell}{g}}$ for g

1. Find a common denominator, and then multiply both sides by the common denominator.

2. Solve for w^2 by dividing each side by g.

3. Write it in standard form (with 0 on one side, in decreasing powers of w).

4. In an applied problem, often one answer will not satisfy the original statement. For example, a negative answer may not be possible.

5. $m = \sqrt{p^2 - n^2}$

6. $m = \sqrt{p^2 + n^2}$

7. $t = \dfrac{\pm\sqrt{dk}}{k}$ **8.** $e = \dfrac{\pm\sqrt{6S}}{6}$

9. $r = \dfrac{\pm\sqrt{S\pi}}{2\pi}$ **10.** $d = \dfrac{\pm\sqrt{skw}}{kw}$

11. $d = \dfrac{\pm\sqrt{skI}}{I}$ **12.** $d = \dfrac{\pm\sqrt{kR}}{R}$

13. $v = \dfrac{\pm\sqrt{kAF}}{F}$

14. $h = \dfrac{\pm d^2\sqrt{kL}}{L}$

15. $r = \dfrac{\pm\sqrt{3\pi Vh}}{\pi h}$

16. $r = \frac{\pm\sqrt{V\pi h}}{\pi h}$

17. $t = \frac{-B \pm \sqrt{B^2 - 4AC}}{2A}$

18. $r = \frac{-\pi h \pm \sqrt{\pi^2 h^2 + \pi S}}{\pi}$

19. $h = \frac{D^2}{k}$ **20.** $d = \frac{k^2}{F^2}$

21. $\ell = \frac{p^2 g}{k}$ **22.** $g = \frac{k\ell}{p^2}$

23. $R = \frac{E^2 - 2pr \pm E\sqrt{E^2 - 4pr}}{2p}$

24. $S = -\frac{t}{3}$ or $S = \frac{t}{2}$

25. $r = \frac{5pc}{4}$ or $r = -\frac{2pc}{3}$

26. $t = \frac{-v \pm \sqrt{v^2 + 2gS}}{g}$

27. $I = \frac{-cR \pm \sqrt{c^2R^2 - 4cL}}{2cL}$

28. $I = \frac{E \pm \sqrt{E^2 - 4RP}}{2R}$

29. 7.9, 8.9, 11.9
30. 2.3, 5.3, 5.8
31. eastbound ship: 80 mi; southbound ship: 150 mi
32. 120 ft
33. 8 in., 15 in., 17 in.
34. 9 ft, 12 ft, 15 ft
35. length: 24 ft; width: 10 ft
36. 5 ft

Extending Skills *Solve each equation for the specified variable. (Leave ± in the answers.)*

23. $p = \frac{E^2R}{(r+R)^2}$ for R (where $E > 0$)

24. $S(6S - t) = t^2$ for S

25. $10p^2c^2 + 7pcr = 12r^2$ for r

26. $S = vt + \frac{1}{2}gt^2$ for t

27. $LI^2 + RI + \frac{1}{c} = 0$ for I

28. $P = EI - RI^2$ for I

Solve each problem. When appropriate, round answers to the nearest tenth. ***See Example 3.***

29. Find the lengths of the sides of the triangle.

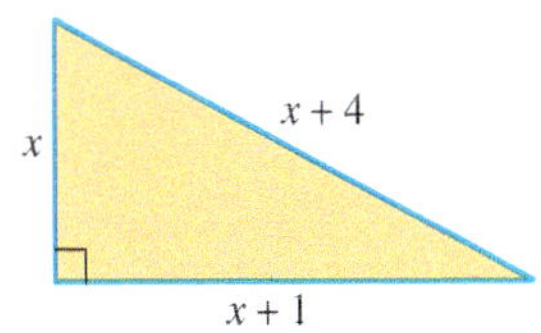

30. Find the lengths of the sides of the triangle.

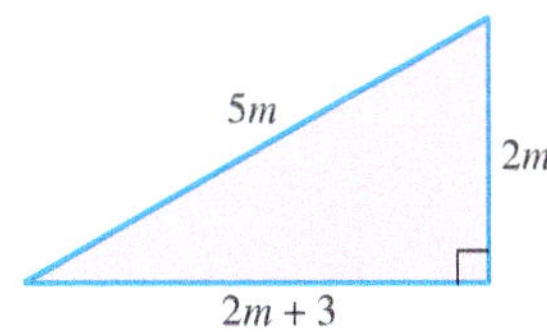

31. Two ships leave port at the same time, one heading due south and the other heading due east. Several hours later, they are 170 mi apart. If the ship traveling south traveled 70 mi farther than the other ship, how many miles did they each travel?

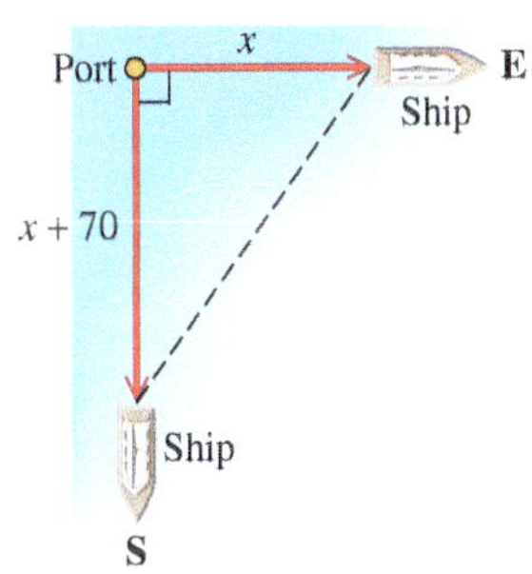

32. Deborah is flying a kite that is 30 ft farther above her hand than its horizontal distance from her. The string from her hand to the kite is 150 ft long. How high is the kite?

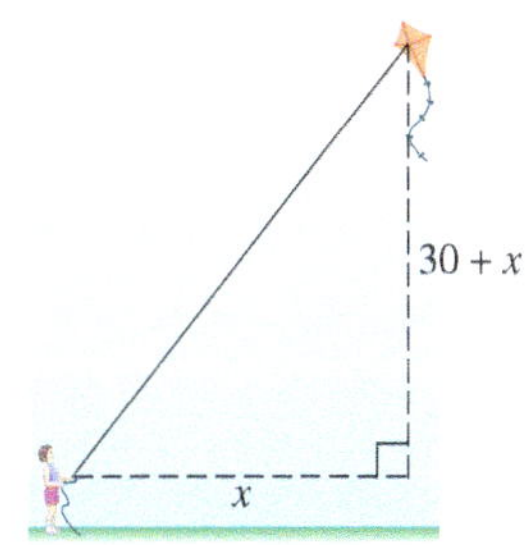

33. A game board is in the shape of a right triangle. The hypotenuse is 2 in. longer than the longer leg, and the longer leg is 1 in. less than twice as long as the shorter leg. How long is each side of the game board?

34. Manuel is planting a vegetable garden in the shape of a right triangle. The longer leg is 3 ft longer than the shorter leg, and the hypotenuse is 3 ft longer than the longer leg. Find the lengths of the three sides of the garden.

35. The diagonal of a rectangular rug measures 26 ft, and the length is 4 ft more than twice the width. Find the length and width of the rug.

36. A 13-ft ladder is leaning against a house. The distance from the bottom of the ladder to the house is 7 ft less than the distance from the top of the ladder to the ground. How far is the bottom of the ladder from the house?

37. 2 ft **38.** 1 ft
39. 7 m by 12 m
40. length: 26 m; width: 16 m
41. 20 in. by 12 in.
42. 21 in. by 23 in.
43. 1 sec and 8 sec
44. 9 sec after it is projected

Solve each problem. ***See Example 4.***

37. A club swimming pool is 30 ft wide and 40 ft long. The club members want an exposed aggregate border in a strip of uniform width around the pool. They have enough material for 296 ft^2. How wide can the strip be?

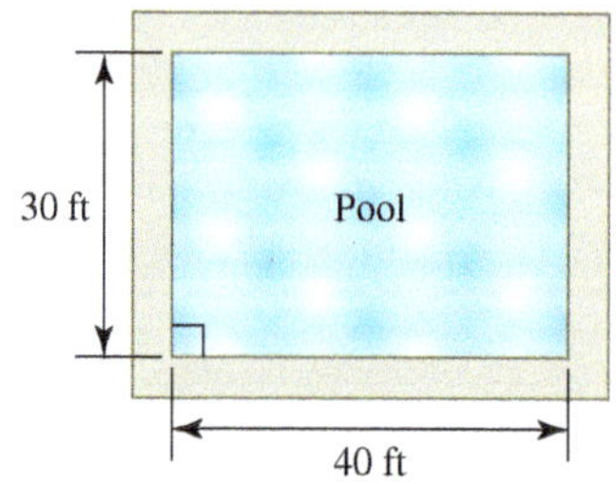

38. Lyudmila wants to buy a rug for a room that is 20 ft long and 15 ft wide. She wants to leave an even strip of flooring uncovered around the edges of the room. How wide a strip will she have if she buys a rug with an area of 234 ft^2?

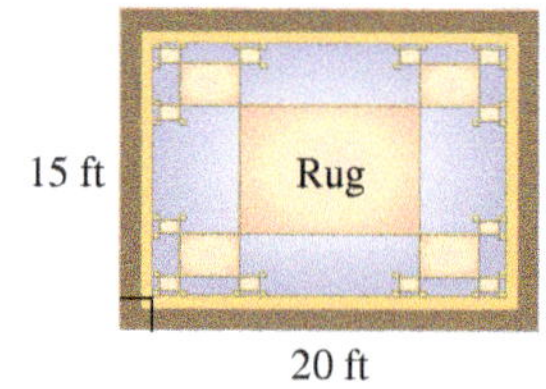

39. A rectangle has a length 2 m less than twice its width. When 5 m are added to the width, the resulting figure is a square with an area of 144 m^2. Find the dimensions of the original rectangle.

40. Mariana's backyard measures 20 m by 30 m. She wants to put a flower garden in the middle of the yard, leaving a strip of grass of uniform width around the flower garden. Mariana must have 184 m^2 of grass. Under these conditions, what will the length and width of the garden be?

41. A rectangular piece of sheet metal has a length that is 4 in. less than twice the width. A square piece 2 in. on a side is cut from each corner. The sides are then turned up to form an uncovered box of volume 256 $in.^3$. Find the length and width of the original piece of metal.

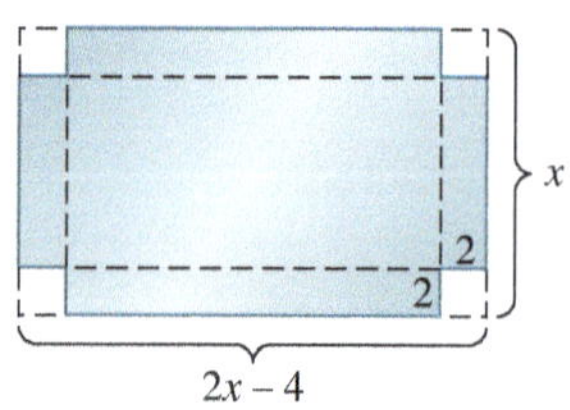

42. A rectangular piece of cardboard is 2 in. longer than it is wide. A square piece 3 in. on a side is cut from each corner. The sides are then turned up to form an uncovered box of volume 765 $in.^3$. Find the dimensions of the original piece of cardboard.

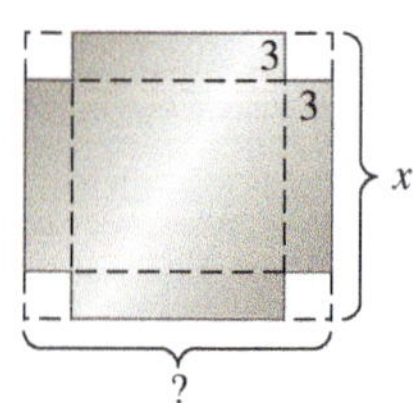

Solve each problem. When appropriate, round answers to the nearest tenth. ***See Example 5.***

43. An object is projected directly upward from the ground. After t seconds its distance in feet above the ground is

$$s(t) = 144t - 16t^2.$$

After how many seconds will the object be 128 ft above the ground? (*Hint:* Look for a common factor before solving the equation.)

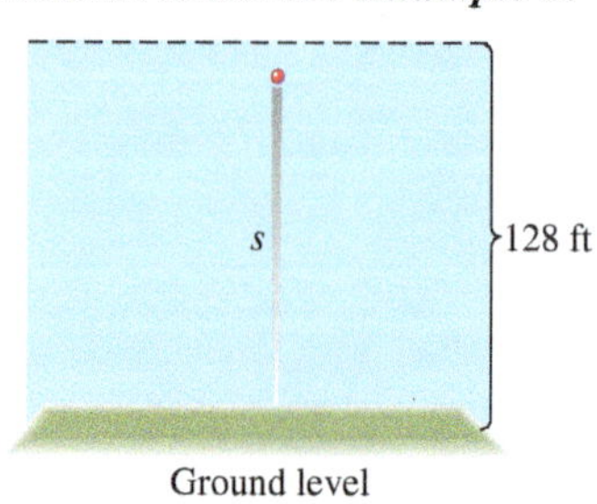

44. When does the object in **Exercise 43** strike the ground?

45. 2.4 sec and 5.6 sec
46. 3.7 sec and 9.3 sec
47. 9.2 sec 48. 11.1 sec

49. It reaches its *maximum* height at 5 sec because this is the only time it reaches 400 ft.
50. Because the discriminant is negative, the ball never reaches a height of 425 ft.

51. $1.50 52. $19.59
53. 0.035, or 3.5%
54. 0.045, or 4.5%

45. A ball is projected upward from the ground. Its distance in feet from the ground in t seconds is given by

$$s(t) = -16t^2 + 128t.$$

At what times will the ball be 213 ft from the ground?

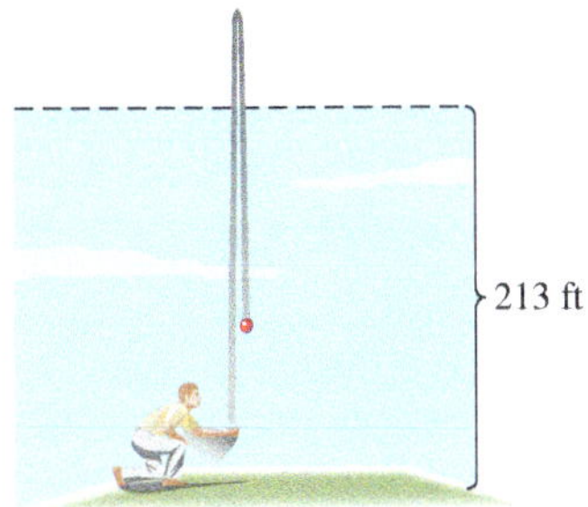

46. A toy rocket is launched from ground level. Its distance in feet from the ground in t seconds is given by

$$s(t) = -16t^2 + 208t.$$

At what times will the rocket be 550 ft from the ground?

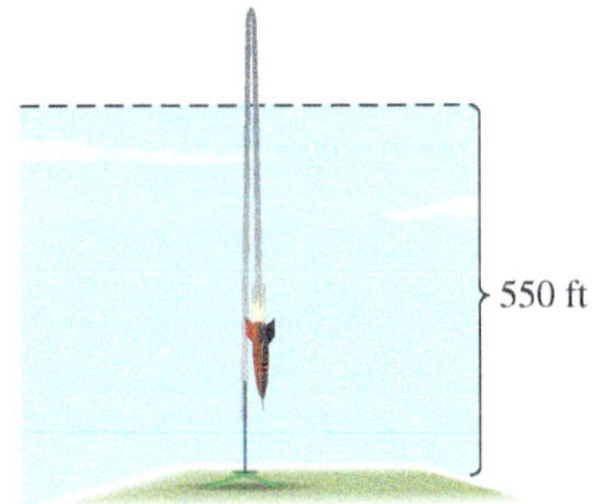

47. The following function gives the distance in feet a car going approximately 68 mph will skid in t seconds.

$$D(t) = 13t^2 - 100t$$

Find the time it would take for the car to skid 180 ft.

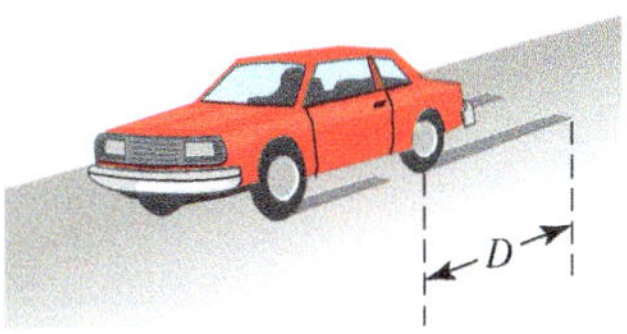

48. Refer to the function in **Exercise 47.** Find the time it would take for the car to skid 500 ft.

A ball is projected upward from ground level, and its distance in feet from the ground in t seconds is given by

$$s(t) = -16t^2 + 160t.$$

49. After how many seconds does the ball reach a height of 400 ft? Describe in words its position at this height.

50. After how many seconds does the ball reach a height of 425 ft? Interpret the mathematical result here.

Extending Skills *Solve each problem using a quadratic equation.*

51. A certain bakery has found that the daily demand for blueberry muffins is $\frac{6000}{p}$, where p is the price of a muffin in cents. The daily supply is $3p - 410$. Find the price at which supply and demand are equal.

52. In one area the demand for Blu-ray discs is $\frac{1900}{P}$ per day, where P is the price in dollars per disc. The supply is $5P - 1$ per day. At what price, to the nearest cent, does supply equal demand?

53. The formula $A = P(1 + r)^2$ gives the amount A in dollars that P dollars will grow to in 2 yr at interest rate r (where r is given as a decimal), using compound interest. What interest rate will cause $2000 to grow to $2142.45 in 2 yr?

54. Use the formula $A = P(1 + r)^2$ to find the interest rate r at which a principal P of $10,000 will increase to $10,920.25 in 2 yr.

55. 5.5 m per sec
56. 2.1 m per sec

57. 5 or 14 58. 4

59. (a) $560 billion
(b) $560 billion; They are the same.
60. (a) $480 billion
(b) $480 billion; They are the same.
61. 2008 62. 2004

William Froude was a 19th century naval architect who used the following expression, known as the ***Froude number,*** *in shipbuilding.*

$$\frac{v^2}{g\ell}$$

This expression was also used by R. McNeill Alexander in his research on dinosaurs. (Source: "How Dinosaurs Ran," Scientific American.*)*

Use this expression to find the value of v (in meters per second), given $g = 9.8$ m per sec^2. (Round to the nearest tenth.)

55. Rhinoceros: $\ell = 1.2$;
Froude number $= 2.57$

56. Triceratops: $\ell = 2.8$;
Froude number $= 0.16$

Recall that corresponding sides of similar triangles are proportional. Use this fact to find the lengths of the indicated sides of each pair of similar triangles. Check all possible solutions in both triangles. Sides of a triangle cannot be negative (and are not drawn to scale here).

57. Side AC

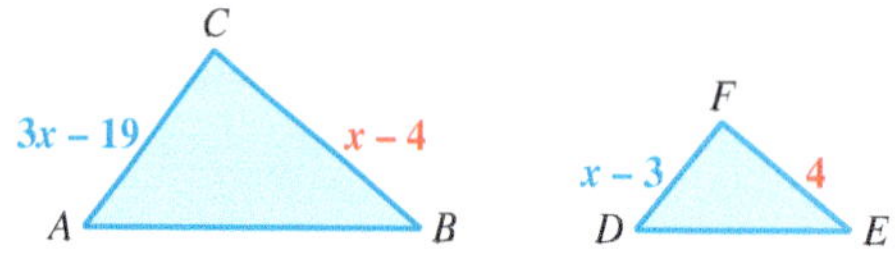

58. Side RQ

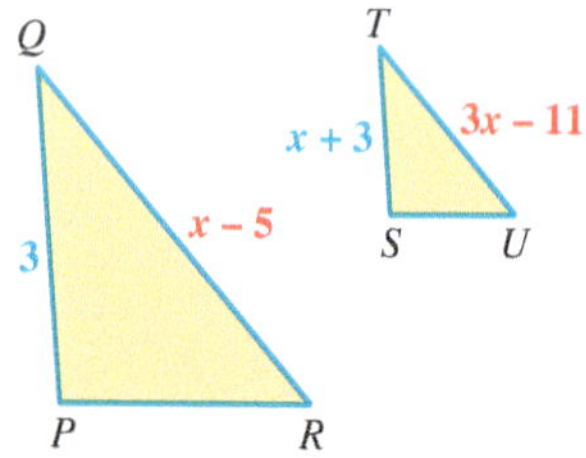

Total spending (in billions of dollars) in the United States from all sources on physician and clinical services for the years 2000–2012 are shown in the bar graph and can be modeled by the quadratic function

$$f(x) = -0.2901x^2 + 25.90x + 291.6.$$

Here, $x = 0$ represents 2000, $x = 2$ represents 2002, and so on. Use the graph and the model to work Exercises 59–62. ***See Example 6.***

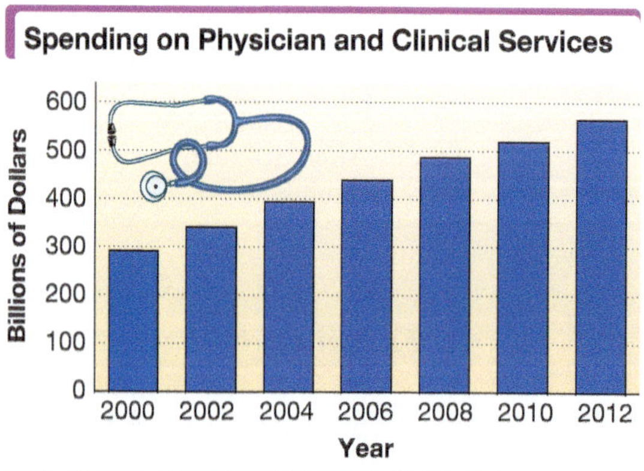

Source: Centers for Medicare and Medicaid Services.

59. Approximate spending on physician and clinical services in 2012 to the nearest $10 billion using **(a)** the graph and **(b)** the model. How do the two approximations compare?

60. Repeat **Exercise 59** for the year 2008.

61. According to the model, in what year did spending on physician and clinical services first exceed $500 billion? (Round down for the year.)

62. Repeat **Exercise 61** for $400 billion.

8.5 Graphs of Quadratic Functions

OBJECTIVES

1. Graph a quadratic function.
2. Graph parabolas with horizontal and vertical shifts.
3. Use the coefficient of x^2 to predict the shape and direction in which a parabola opens.
4. Find a quadratic function to model data.

VOCABULARY

- ☐ parabola
- ☐ vertex
- ☐ axis of symmetry (axis)
- ☐ quadratic function

TEACHING TIP

As you graph parabolas with students, have them give the vertex, axis, domain, and range.

OBJECTIVE 1 Graph a quadratic function.

FIGURE 5 gives a graph of the simplest *quadratic function* $y = x^2$. This graph is a **parabola.** The point $(0, 0)$, the lowest point on the curve, is the **vertex** of the parabola. The vertical line through the vertex is the **axis of symmetry,** or simply the **axis,** of the parabola. Here its equation is $x = 0$. A parabola is **symmetric about its axis**—that is, if the graph were folded along the axis, the two portions of the curve would coincide.

x	y
−2	4
−1	1
0	0
1	1
2	4

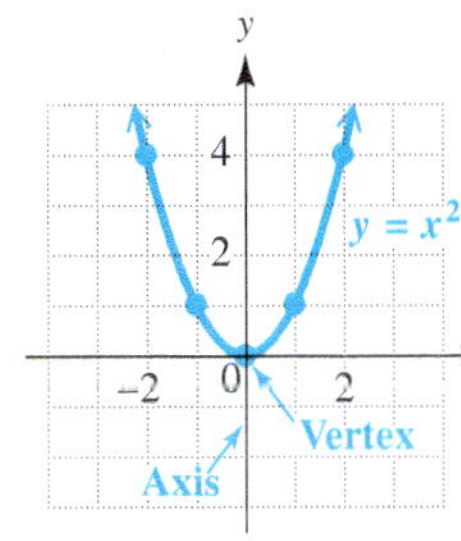

FIGURE 5

As **FIGURE 5** suggests, x can be any real number, so the domain of the function $y = x^2$, written in interval notation, is $(-\infty, \infty)$. Values of y are always nonnegative, so the range is $[0, \infty)$.

Quadratic Function

A function that can be written in the form

$$f(x) = ax^2 + bx + c$$

for real numbers a, b, and c, where $a \neq 0$, is a **quadratic function.**

The graph of any quadratic function is a parabola with a vertical axis.

NOTE We use the variable y and function notation $f(x)$ interchangeably. Although we use the letter f most often to name quadratic functions, other letters can be used. We use the capital letter F to distinguish between different parabolas graphed on the same coordinate axes.

Parabolas have a special reflecting property that makes them useful in the design of telescopes, radar equipment, solar furnaces, and automobile headlights. (See **FIGURE 6.**)

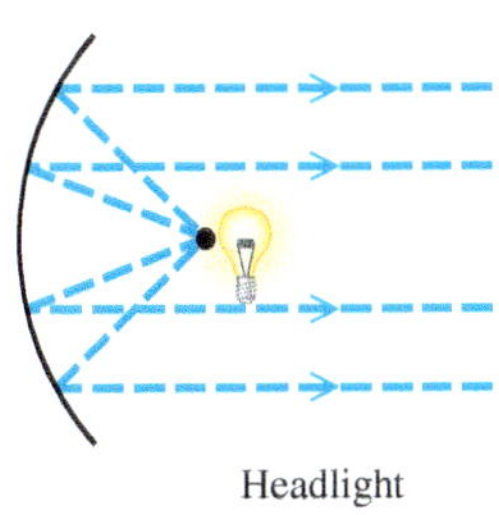

FIGURE 6

OBJECTIVE 2 Graph parabolas with horizontal and vertical shifts.

Parabolas need not have their vertices at the origin, as does the graph of $f(x) = x^2$.

NOW TRY EXERCISE 1

Graph $f(x) = x^2 - 3$. Give the vertex, axis, domain, and range.

CLASSROOM EXAMPLE 1

Graph $f(x) = x^2 + 3$. Give the vertex, axis, domain, and range.

Answer:

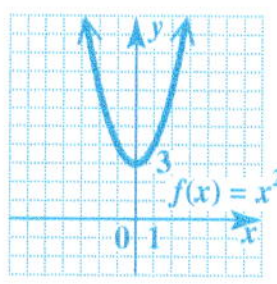

vertex: $(0, 3)$; axis: $x = 0$; domain: $(-\infty, \infty)$; range: $[3, \infty)$

EXAMPLE 1 Graphing a Parabola (Vertical Shift)

Graph $F(x) = x^2 - 2$.

The graph of $F(x) = x^2 - 2$ has the same shape as that of $f(x) = x^2$ but is *shifted*, or *translated*, 2 units down, with vertex $(0, -2)$. Every function value is 2 less than the corresponding function value of $f(x) = x^2$. Plotting points on both sides of the vertex gives the graph in **FIGURE 7**.

x	$f(x) = x^2$	$F(x) = x^2 - 2$
−2	4	2
−1	1	−1
0	0	−2
1	1	−1
2	4	2

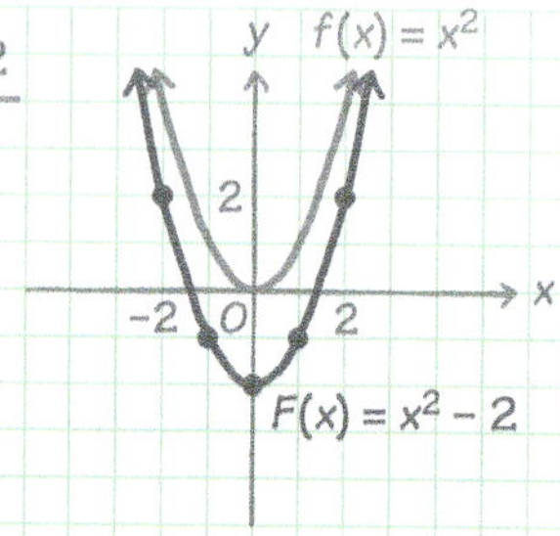

$F(x) = x^2 - 2$
Vertex: $(0, -2)$
Axis of symmetry: $x = 0$
Domain: $(-\infty, \infty)$
Range: $[-2, \infty)$
The graph of $f(x) = x^2$ is shown for comparison.

FIGURE 7

This parabola is symmetric about its axis $x = 0$, so the plotted points are "mirror images" of each other. Because x can be any real number, the domain is $(-\infty, \infty)$. The value of y (or $F(x)$) is always greater than or equal to -2, so the range is $[-2, \infty)$.

NOW TRY

CLASSROOM EXAMPLE 2

Graph $f(x) = (x + 2)^2$. Give the vertex, axis, domain, and range.

Answer:

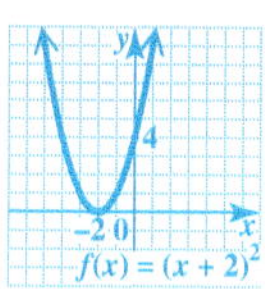

vertex: $(-2, 0)$; axis: $x = -2$; domain: $(-\infty, \infty)$; range: $[0, \infty)$

Vertical Shift of a Parabola

The graph of $F(x) = x^2 + k$ is a parabola.

- The graph has the same shape as the graph of $f(x) = x^2$.
- The parabola is shifted k units up if $k > 0$ and $|k|$ units down if $k < 0$.
- The vertex of the parabola is $(0, k)$.

NOW TRY EXERCISE 2

Graph $f(x) = (x + 1)^2$. Give the vertex, axis, domain, and range.

EXAMPLE 2 Graphing a Parabola (Horizontal Shift)

Graph $F(x) = (x - 2)^2$.

If $x = 2$, then $F(x) = 0$, giving the vertex $(2, 0)$. The graph of $F(x) = (x - 2)^2$ has the same shape as that of $f(x) = x^2$ but is shifted 2 units to the right. Plotting points on one side of the vertex, and using symmetry about the axis $x = 2$ to find corresponding points on the other side, gives the graph in **FIGURE 8**.

x	$F(x) = (x - 2)^2$
0	4
1	1
2	0
3	1
4	4

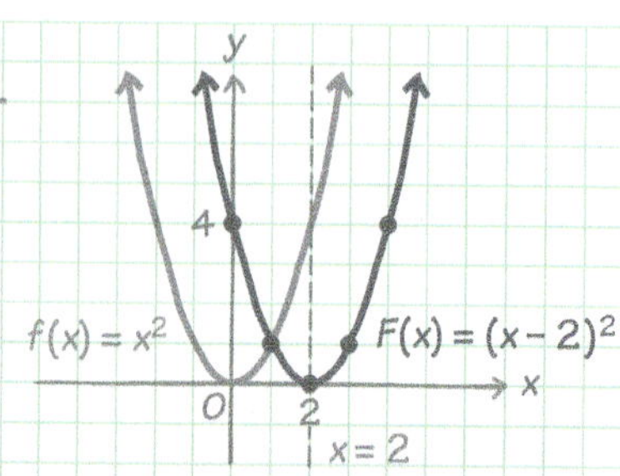

$F(x) = (x - 2)^2$
Vertex: $(2, 0)$
Axis of symmetry: $x = 2$
Domain: $(-\infty, \infty)$
Range: $[0, \infty)$

FIGURE 8

NOW TRY

NOW TRY ANSWERS

1.

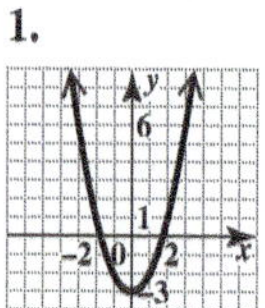

$f(x) = x^2 - 3$

vertex: $(0, -3)$; axis: $x = 0$; domain: $(-\infty, \infty)$; range: $[-3, \infty)$

2.

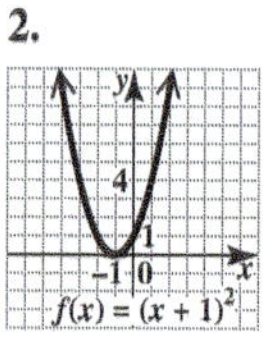

vertex: $(-1, 0)$; axis: $x = -1$; domain: $(-\infty, \infty)$; range: $[0, \infty)$

Horizontal Shift of a Parabola

The graph of $F(x) = (x - h)^2$ is a parabola.

- The graph has the same shape as the graph of $f(x) = x^2$.
- The parabola is shifted h units to the right if $h > 0$ and $|h|$ units to the left if $h < 0$.
- The vertex of the parabola is $(h, 0)$.

CAUTION ***Errors frequently occur when horizontal shifts are involved.*** To determine the direction and magnitude of a horizontal shift, find the value that causes the expression $x - h$ to equal 0, as shown below.

$F(x) = (x - 5)^2$	$F(x) = (x + 5)^2$
Because **+5** causes $x - 5$ to equal 0, the graph of $F(x)$ illustrates a shift of **5 units to the right.**	Because **−5** causes $x + 5$ to equal 0, the graph of $F(x)$ illustrates a shift of **5 units to the left.**

NOW TRY EXERCISE 3

Graph $f(x) = (x + 1)^2 - 2$. Give the vertex, axis, domain, and range.

CLASSROOM EXAMPLE 3

Graph $f(x) = (x - 2)^2 + 1$. Give the vertex, axis, domain, and range.

Answer:

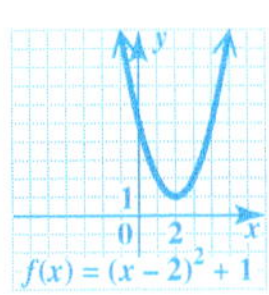

vertex: $(2, 1)$; axis: $x = 2$; domain: $(-\infty, \infty)$; range: $[1, \infty)$

EXAMPLE 3 Graphing a Parabola (Horizontal and Vertical Shifts)

Graph $F(x) = (x + 3)^2 - 2$.

This graph has the same shape as that of $f(x) = x^2$, but is shifted 3 units to the left (because $x + 3 = 0$ if $x = -3$) and 2 units down (because of the negative sign in -2). See **FIGURE 9**.

x	$F(x)$
−5	2
−4	−1
−3	−2
−2	−1
−1	2

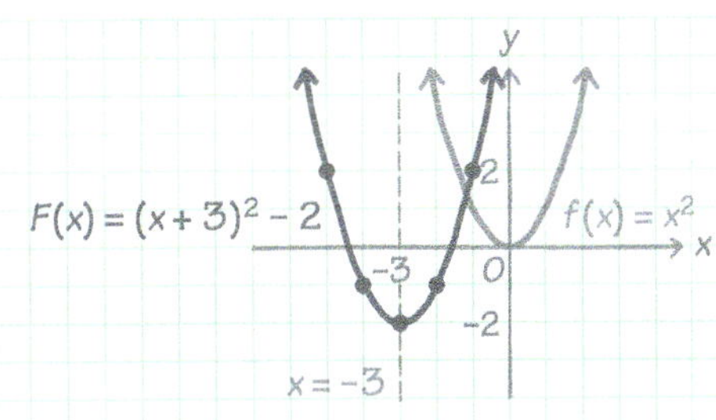

$F(x) = (x + 3)^2 - 2$
Vertex: $(-3, -2)$
Axis of symmetry: $x = -3$
Domain: $(-\infty, \infty)$
Range: $[-2, \infty)$

FIGURE 9

NOW TRY

Vertex and Axis of a Parabola

The graph of $F(x) = (x - h)^2 + k$ is a parabola.

- The graph has the same shape as the graph of $f(x) = x^2$.
- The vertex of the parabola is (h, k).
- The axis of symmetry is the vertical line $x = h$.

NOW TRY ANSWER

3.

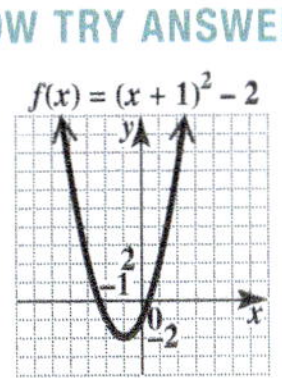

vertex: $(-1, -2)$; axis: $x = -1$; domain: $(-\infty, \infty)$; range: $[-2, \infty)$

OBJECTIVE 3 Use the coefficient of x^2 to predict the shape and direction in which a parabola opens.

Not all parabolas open up, and not all parabolas have the same shape as the graph of $f(x) = x^2$.

NOW TRY EXERCISE 4
Graph $f(x) = -3x^2$. Give the vertex, axis, domain, and range.

CLASSROOM EXAMPLE 4
Graph $f(x) = -2x^2 - 3$. Give the vertex, axis, domain, and range.

Answer:

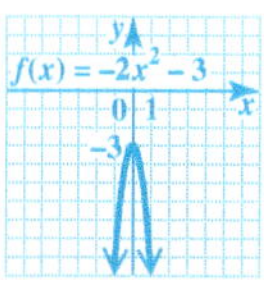

vertex: $(0, -3)$; axis: $x = 0$;
domain: $(-\infty, \infty)$; range: $(-\infty, -3]$

EXAMPLE 4 Graphing a Parabola That Opens Down

Graph $f(x) = -\frac{1}{2}x^2$.

This parabola is shown in **FIGURE 10**. The coefficient $-\frac{1}{2}$ affects the shape of the graph—the $\frac{1}{2}$ makes the parabola wider $\left(\text{because the values of } \frac{1}{2}x^2 \text{ increase more slowly than those of } x^2\right)$, and the negative sign makes the parabola open down. The graph is not shifted in any direction. Unlike the parabolas graphed in **Examples 1–3**, the vertex $(0, 0)$ has the *greatest* function value of any point on the graph.

x	$f(x)$
-2	-2
-1	$-\frac{1}{2}$
0	0
1	$-\frac{1}{2}$
2	-2

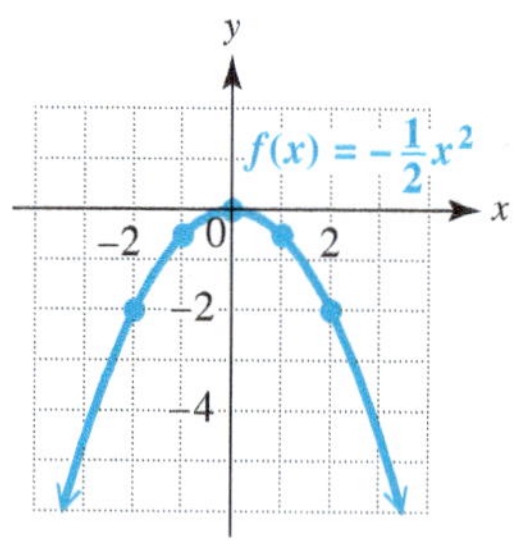

$f(x) = -\frac{1}{2}x^2$
Vertex: $(0, 0)$
Axis of symmetry: $x = 0$
Domain: $(-\infty, \infty)$
Range: $(-\infty, 0]$

FIGURE 10

NOW TRY

CLASSROOM EXAMPLE 5
Graph $f(x) = \frac{1}{2}(x - 2)^2 + 1$. Give the vertex, axis, domain, and range.

Answer:

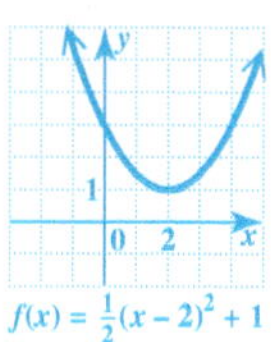

vertex: $(2, 1)$; axis: $x = 2$;
domain: $(-\infty, \infty)$; range: $[1, \infty)$

General Characteristics of $F(x) = a(x - h)^2 + k$ (Where $a \neq 0$)

1. The graph of the quadratic function
$$F(x) = a(x - h)^2 + k \quad (\text{where } a \neq 0)$$
is a parabola with vertex (h, k) and vertical line $x = h$ as axis of symmetry.
2. The graph opens up if $a > 0$ and down if $a < 0$.
3. The graph is wider than that of $f(x) = x^2$ if $0 < |a| < 1$.
The graph is narrower than that of $f(x) = x^2$ if $|a| > 1$.

NOW TRY EXERCISE 5
Graph $f(x) = 2(x - 1)^2 + 2$. Give the vertex, axis, domain, and range.

EXAMPLE 5 Using the General Characteristics to Graph a Parabola

Graph $F(x) = -2(x + 3)^2 + 4$.

The parabola opens down (because $a < 0$) and is narrower than the graph of $f(x) = x^2$ because $|-2| = 2$ and $2 > 1$. This causes values of $F(x)$ to decrease more quickly than those of $f(x) = -x^2$. This parabola has vertex $(-3, 4)$, as shown in **FIGURE 11**. To complete the graph, we plotted the ordered pairs $(-4, 2)$ and, by symmetry, $(-2, 2)$. Symmetry can be used to find additional ordered pairs that satisfy the equation.

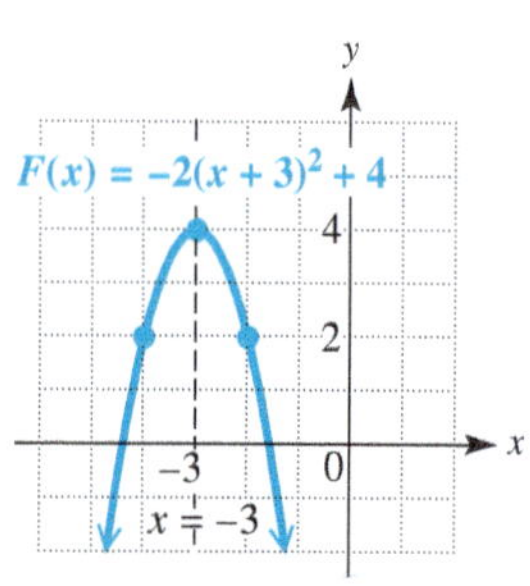

$F(x) = -2(x + 3)^2 + 4$
Vertex: $(-3, 4)$
Axis of symmetry: $x = -3$
Domain: $(-\infty, \infty)$
Range: $(-\infty, 4]$

FIGURE 11

NOW TRY

NOW TRY ANSWERS

4.

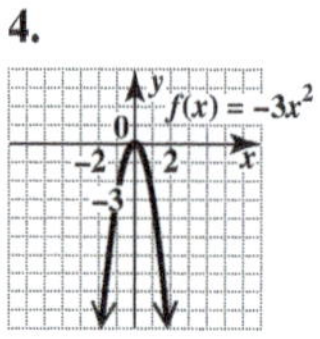

vertex: $(0, 0)$;
axis: $x = 0$;
domain: $(-\infty, \infty)$;
range: $(-\infty, 0]$

5.

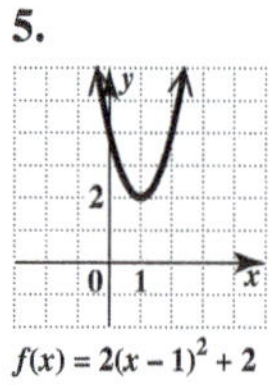

vertex: $(1, 2)$;
axis: $x = 1$;
domain: $(-\infty, \infty)$;
range: $[2, \infty)$

TEACHING TIP
Have students decide whether each parabola opens up or down and whether it is wider or narrower than the graph of $f(x) = x^2$.

(a) $f(x) = -\frac{2}{3}x^2$

(b) $f(x) = \frac{3}{4}x^2 + 1$

(c) $f(x) = -2x^2 - 3$

(d) $f(x) = 3x^2 + 2$

Answers:
(a) down; wider
(b) up; wider
(c) down; narrower
(d) up; narrower

CLASSROOM EXAMPLE 6
Using the points $(0, 5900)$, $(4, 7300)$, and $(8, 7300)$, find another quadratic model for the data on higher-order multiple births in **Example 6.**

Answer:
$y = -43.75x^2 + 525x + 5900$

OBJECTIVE 4 Find a quadratic function to model data.

EXAMPLE 6 Modeling the Number of Multiple Births

The number of higher-order multiple births (triplets or more) in the United States is shown in the table. Here, x represents the number of years since 1996 and y represents the number of higher-order multiple births (to the nearest hundred).

Year	*x*	*y*
1996	0	5900
2000	4	7300
2002	6	7400
2004	8	7300
2006	10	6500
2008	12	6300
2010	14	5500
2012	16	4900

Source: National Center for Health Statistics.

Find a quadratic function that models the data.

A scatter diagram of the ordered pairs (x, y) is shown in **FIGURE 12**. The general shape suggested by the scatter diagram indicates that a parabola should approximate these points, as shown by the dashed curve in **FIGURE 13**. The equation for such a parabola would have a negative coefficient for x^2 because the graph opens down.

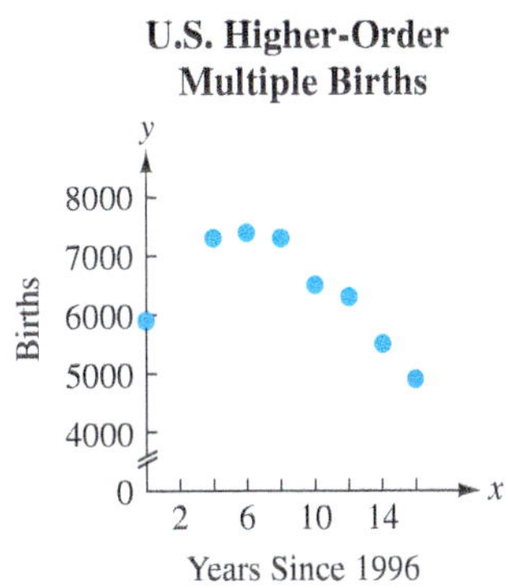

FIGURE 12

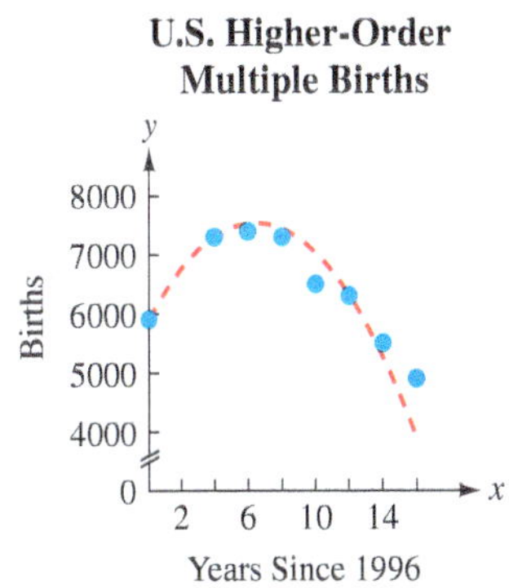

FIGURE 13

To find a quadratic function of the form

$$y = ax^2 + bx + c$$

that models, or *fits*, these data, we choose three representative ordered pairs from the table and use them to write a system of three equations.

$$(0, 5900), \quad (6, 7400), \quad \text{and} \quad (12, 6300) \qquad \text{Three ordered pairs } (x, y)$$

We substitute the x- and y-values from each ordered pair into the quadratic form $y = ax^2 + bx + c$ to obtain three equations.

$$a(0)^2 + b(0) + c = 5900 \xrightarrow{\text{Simplify.}} c = 5900 \quad (1)$$

$$a(6)^2 + b(6) + c = 7400 \longrightarrow 36a + 6b + c = 7400 \quad (2)$$

$$a(12)^2 + b(12) + c = 6300 \longrightarrow 144a + 12b + c = 6300 \quad (3)$$

NOW TRY EXERCISE 6
Using the points (0, 5900), (4, 7300), and (12, 6300), find another quadratic model for the data on higher-order multiple births in **Example 6.** (Round values of a and b to the nearest tenth.)

To find the values of a, b, and c, we solve this system of three equations in three variables using the methods of **Section 3.2.** From equation (1), $c = 5900$, so we substitute 5900 for c in equations (2) and (3) to obtain two equations in two variables.

$$36a + 6b + 5900 = 7400 \xrightarrow{\text{Subtract 5900.}} 36a + 6b = 1500 \quad (4)$$
$$144a + 12b + 5900 = 6300 \longrightarrow 144a + 12b = 400 \quad (5)$$

We eliminate b from this system of equations in two variables by multiplying equation (4) by -2 and adding the results to equation (5).

$$-72a - 12b = -3000 \quad \text{Multiply equation (4) by } -2.$$
$$144a + 12b = 400 \quad (5)$$
$$72a = -2600 \quad \text{Add.}$$
$$a = -36.1 \quad \text{Use a calculator. Round to one decimal place.}$$

We substitute -36.1 for a in equation (4) to find that $b = 466.6$. (Substituting in equation (5) will give $b = 466.5$ due to rounding procedures.) Using the values we found for a, b, and c, the model is

$$y = \underset{a}{-36.1}x^2 + \underset{b}{466.6}x + \underset{c}{5900}.$$

NOW TRY

NOTE If we had chosen three different ordered pairs of data in **Example 6,** a slightly different, though similar, model would result. (See **Now Try Exercise 6.**)

The *quadratic regression* feature on a graphing calculator can also be used to generate the quadratic model that best fits given data. See your owner's manual for details.

NOW TRY ANSWER
6. $y = -39.6x^2 + 508.4x + 5900$ (Answers may vary slightly due to rounding.)

8.5 Exercises

FOR EXTRA HELP MyMathLab®

Complete solution available in MyMathLab

1. (a) B (b) C (c) A (d) D

Concept Check *Match each quadratic function in parts (a)–(d) with its graph from choices A–D.*

1. (a) $f(x) = (x + 2)^2 - 1$

(b) $f(x) = (x + 2)^2 + 1$

(c) $f(x) = (x - 2)^2 - 1$

(d) $f(x) = (x - 2)^2 + 1$

A.
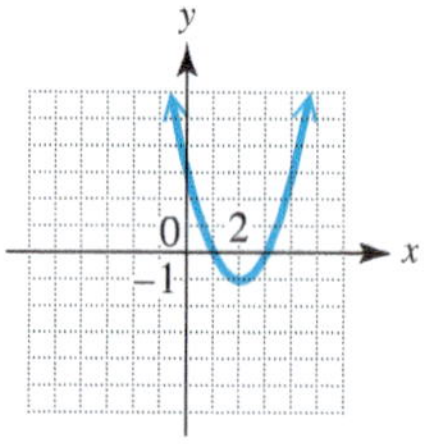

B.
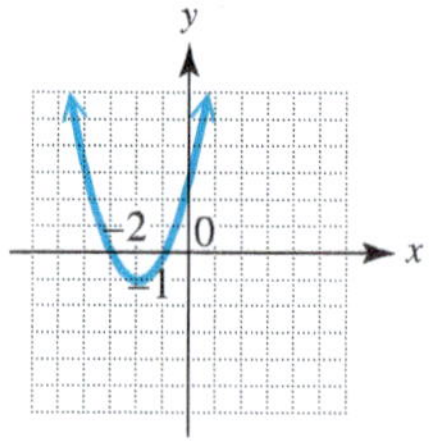

C.
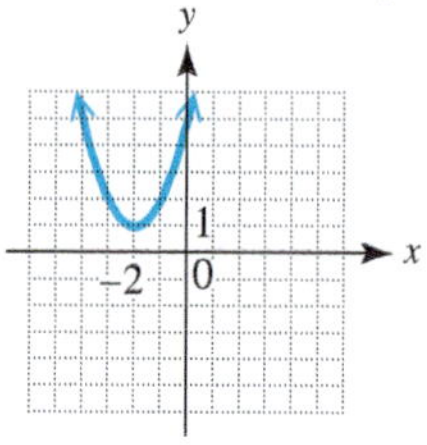

D.
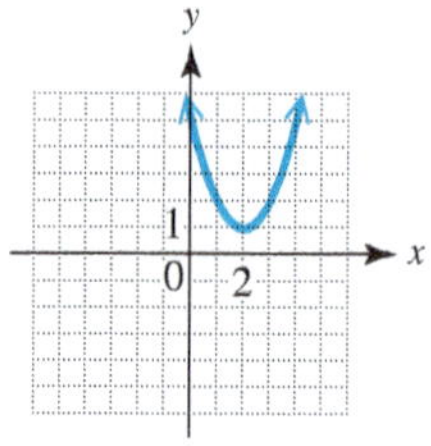

2. (a) D (b) C (c) B (d) A

3. (a) D (b) B (c) C (d) A

4. (a) I (b) IV (c) II (d) III
(e) narrower (f) wider
(g) up (h) down

5. (0, 0) 6. (0, 0)
7. (0, 4) 8. (0, −4)
9. (1, 0) 10. (−3, 0)
11. (−3, −4) 12. (−5, −8)
13. (5, 6) 14. (2, 1)

15. down; wider
16. down; narrower
17. up; narrower
18. up; wider
19. down; narrower
20. down; wider

21.

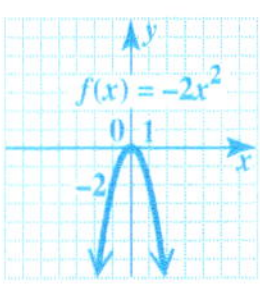

vertex: (0, 0); axis: $x = 0$; domain: $(-\infty, \infty)$; range: $(-\infty, 0]$

22.

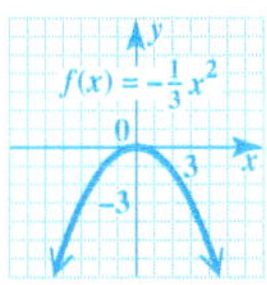

vertex: (0, 0); axis: $x = 0$; domain: $(-\infty, \infty)$; range: $(-\infty, 0]$

23.

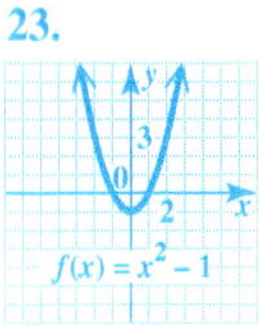

vertex: (0, −1); axis: $x = 0$; domain: $(-\infty, \infty)$; range: $[-1, \infty)$

24.

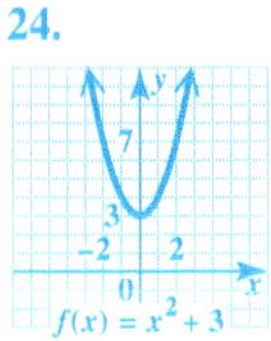

vertex: (0, 3); axis: $x = 0$; domain: $(-\infty, \infty)$; range: $[3, \infty)$

25.

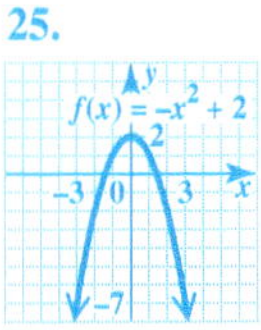

vertex: (0, 2); axis: $x = 0$; domain: $(-\infty, \infty)$; range: $(-\infty, 2]$

26.

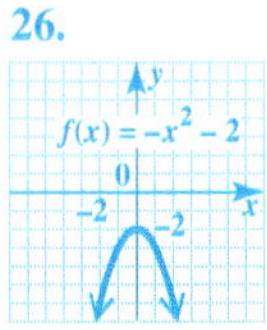

vertex: (0, −2); axis: $x = 0$; domain: $(-\infty, \infty)$; range: $(-\infty, -2]$

27.

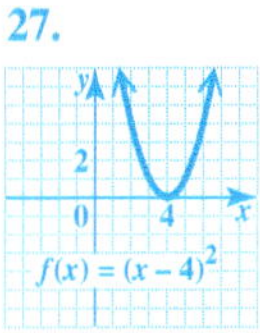

vertex: (4, 0); axis: $x = 4$; domain: $(-\infty, \infty)$; range: $[0, \infty)$

28.

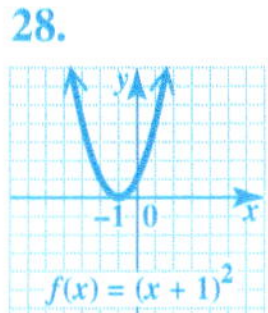

vertex: (−1, 0); axis: $x = -1$; domain: $(-\infty, \infty)$; range: $[0, \infty)$

2. (a) $f(x) = -x^2 + 2$

(b) $f(x) = -x^2 - 2$

(c) $f(x) = -(x + 2)^2$

(d) $f(x) = -(x - 2)^2$

A.

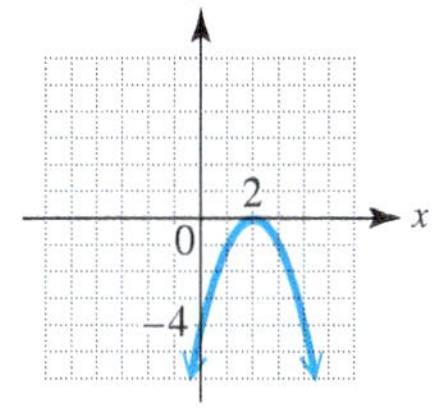

B.

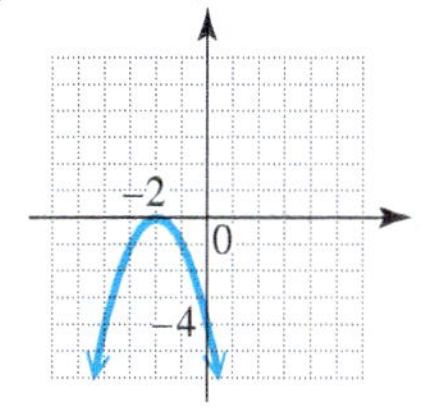

C.

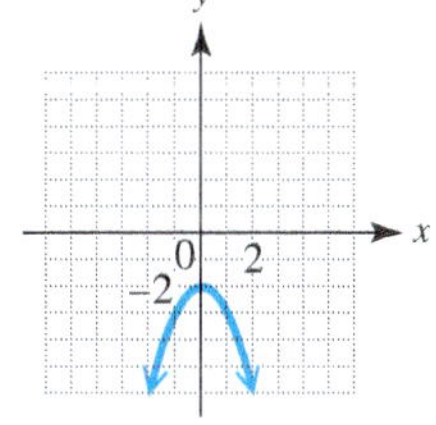

D.

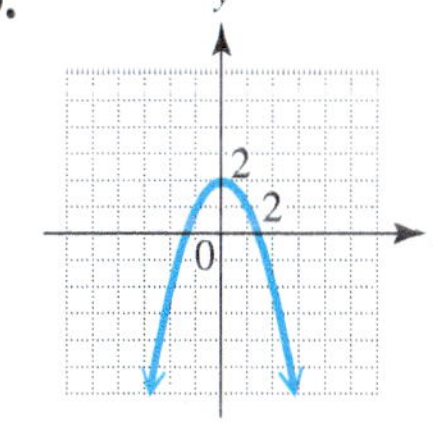

3. ***Concept Check*** Match each quadratic function with the description of the parabola that is its graph.

(a) $f(x) = (x - 4)^2 - 2$ A. Vertex (2, −4), opens down

(b) $f(x) = (x - 2)^2 - 4$ B. Vertex (2, −4), opens up

(c) $f(x) = -(x - 4)^2 - 2$ C. Vertex (4, −2), opens down

(d) $f(x) = -(x - 2)^2 - 4$ D. Vertex (4, −2), opens up

4. ***Concept Check*** For $f(x) = a(x - h)^2 + k$, in what quadrant is the vertex if the values of h and k are as follows?

(a) $h > 0, k > 0$ (b) $h > 0, k < 0$ (c) $h < 0, k > 0$ (d) $h < 0, k < 0$

Consider the value of a, and make the correct choice.

(e) If $|a| > 1$, then the graph is (*narrower* / *wider*) than the graph of $f(x) = x^2$.

(f) If $0 < |a| < 1$, then the graph is (*narrower* / *wider*) than the graph of $f(x) = x^2$.

(g) If $a > 0$, then the graph opens (*up* / *down*).

(h) If $a < 0$, then the graph opens (*up* / *down*).

Identify the vertex of each parabola. ***See Examples 1–4.***

5. $f(x) = -3x^2$ 6. $f(x) = \frac{1}{2}x^2$ 7. $f(x) = x^2 + 4$ 8. $f(x) = x^2 - 4$

9. $f(x) = (x - 1)^2$ 10. $f(x) = (x + 3)^2$ 11. $f(x) = (x + 3)^2 - 4$

12. $f(x) = (x + 5)^2 - 8$ 13. $f(x) = -(x - 5)^2 + 6$ 14. $f(x) = -(x - 2)^2 + 1$

For each quadratic function, tell whether the graph opens up or down and whether the graph is wider, narrower, or the same shape as the graph of $f(x) = x^2$. ***See Examples 4 and 5.***

15. $f(x) = -\frac{2}{5}x^2$ 16. $f(x) = -2x^2$

17. $f(x) = 3x^2 + 1$ 18. $f(x) = \frac{2}{3}x^2 - 4$

19. $f(x) = -4(x + 2)^2 + 5$ 20. $f(x) = -\frac{1}{3}(x + 6)^2 + 3$

Graph each parabola. Give the vertex, axis of symmetry, domain, and range. ***See Examples 1–5.***

21. $f(x) = -2x^2$

22. $f(x) = -\frac{1}{3}x^2$

23. $f(x) = x^2 - 1$

24. $f(x) = x^2 + 3$

25. $f(x) = -x^2 + 2$

26. $f(x) = -x^2 - 2$

27. $f(x) = (x - 4)^2$

28. $f(x) = (x + 1)^2$

29. $f(x) = (x + 2)^2 - 1$

30. $f(x) = (x - 1)^2 + 2$

31. $f(x) = 2(x - 2)^2 - 4$

32. $f(x) = 3(x - 2)^2 + 1$

33. $f(x) = -2(x + 3)^2 + 4$

34. $f(x) = -2(x - 2)^2 - 3$

35. $f(x) = -\frac{1}{2}(x + 1)^2 + 2$

36. $f(x) = -\frac{2}{3}(x + 2)^2 + 1$

37. $f(x) = 2(x - 2)^2 - 3$

38. $f(x) = \frac{4}{3}(x - 3)^2 - 2$

In Exercises 39–44, decide whether a linear function *or a* quadratic function *would be a more appropriate model for each set of graphed data. If linear, tell whether the slope should be* positive *or* negative. *If quadratic, tell whether the coefficient of x^2 should be* positive *or* negative. ***See Example 6.***

39. **Time Spent Playing Video Games**

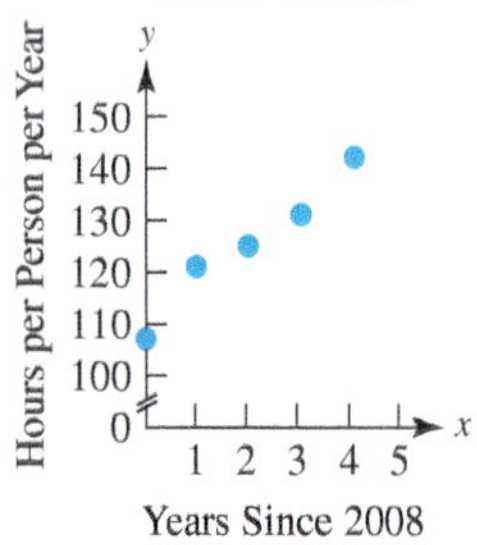

Source: www.statisca.com

40. **Average Daily Volume of First-Class Mail**

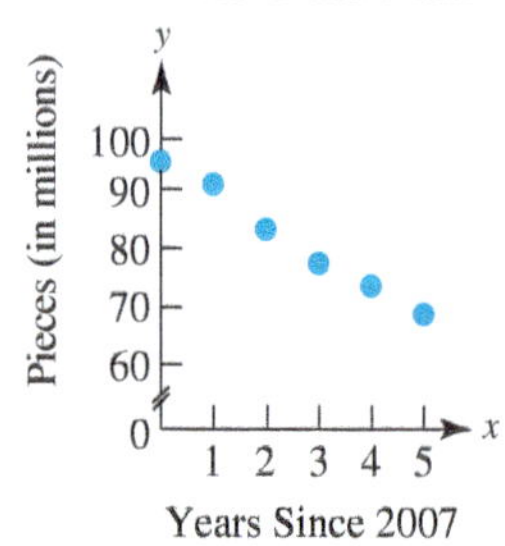

Source: U.S. Postal Service.

41. **Food Assistance Spending in Iowa**

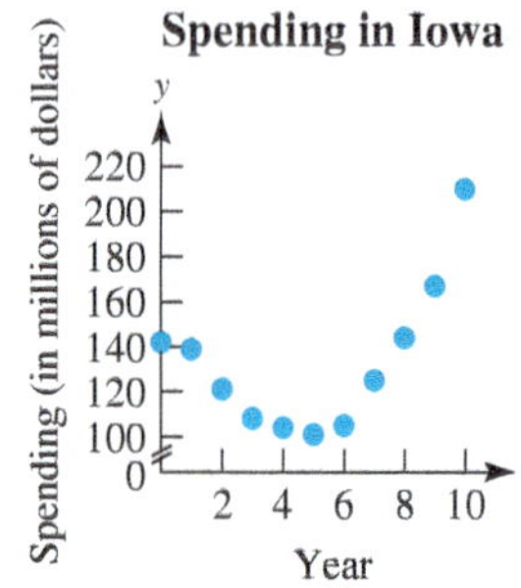

Source: Iowa Department of Human Services.

42. **U.S. Foreign-Born Population**

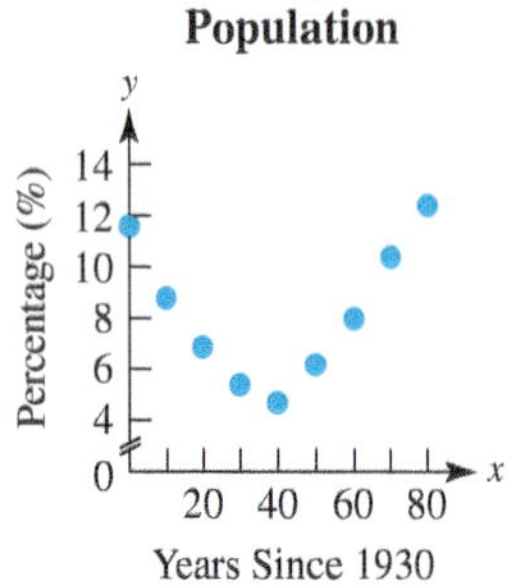

Source: U.S. Census Bureau.

43. **High School Students Who Smoke**

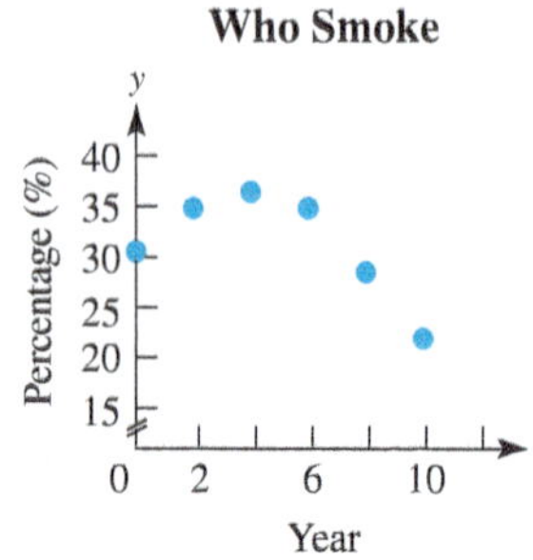

Source: www.cdc.gov

44. **Social Security Assets***

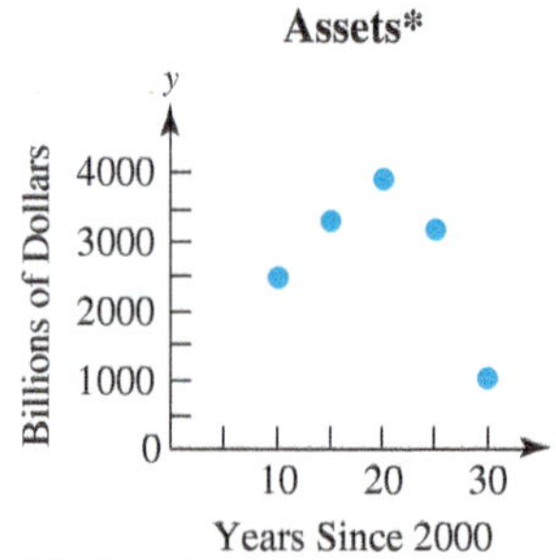

*Projected
Source: Social Security Administration.

29.

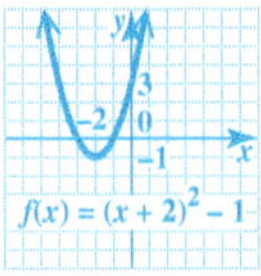

vertex: $(-2, -1)$; axis: $x = -2$; domain: $(-\infty, \infty)$; range: $[-1, \infty)$

30.

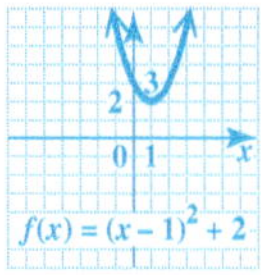

vertex: $(1, 2)$; axis: $x = 1$; domain: $(-\infty, \infty)$; range: $[2, \infty)$

31.

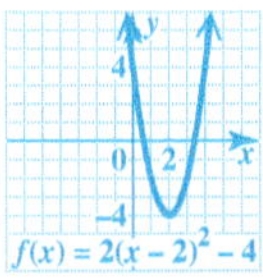

vertex: $(2, -4)$; axis: $x = 2$; domain: $(-\infty, \infty)$; range: $[-4, \infty)$

32.

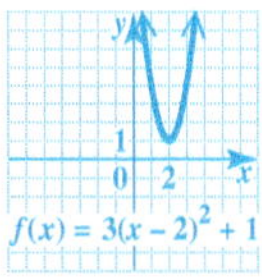

vertex: $(2, 1)$; axis: $x = 2$; domain: $(-\infty, \infty)$; range: $[1, \infty)$

33.

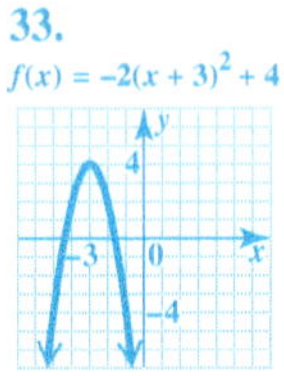

vertex: $(-3, 4)$; axis: $x = -3$; domain: $(-\infty, \infty)$; range: $(-\infty, 4]$

34.

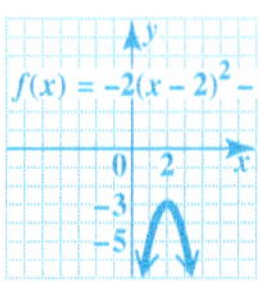

vertex: $(2, -3)$; axis: $x = 2$; domain: $(-\infty, \infty)$; range: $(-\infty, -3]$

35.

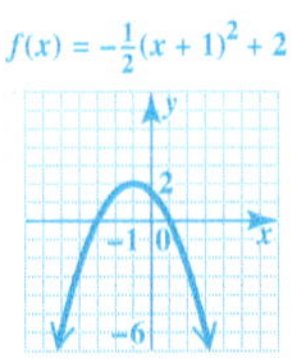

vertex: $(-1, 2)$; axis: $x = -1$; domain: $(-\infty, \infty)$; range: $(-\infty, 2]$

36.

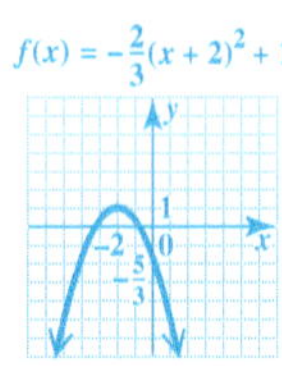

vertex: $(-2, 1)$; axis: $x = -2$; domain: $(-\infty, \infty)$; range: $(-\infty, 1]$

37.

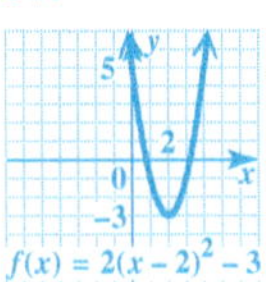

vertex: $(2, -3)$; axis: $x = 2$; domain: $(-\infty, \infty)$; range: $[-3, \infty)$

38.

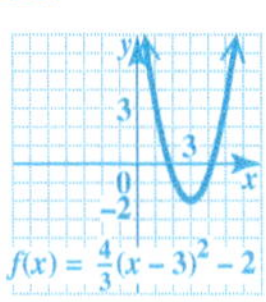

vertex: $(3, -2)$; axis: $x = 3$; domain: $(-\infty, \infty)$; range: $[-2, \infty)$

39. linear function; positive
40. linear function; negative
41. quadratic function; positive
42. quadratic function; positive
43. quadratic function; negative
44. quadratic function; negative

45. (a)

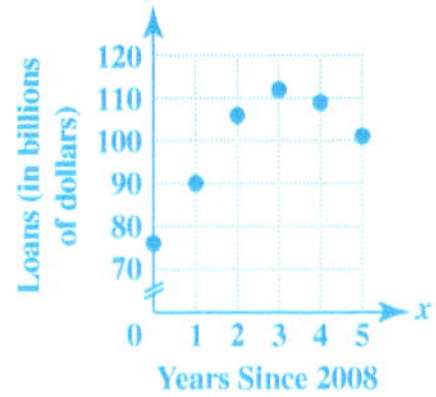

(b) quadratic function

(c) negative

(d) $y = -3.5x^2 + 22.5x + 76$

(e) 2010: \$107 billion; 2012: \$110 billion; The model approximates the data fairly well.

46. (a)

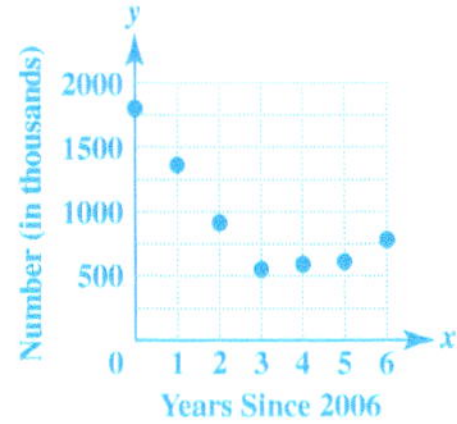

(b) quadratic function

(c) positive

(d) $y = 66.25x^2 - 567.5x + 1800$

(e) 2008: 930 thousand; 2011: 619 thousand; The model approximates the data fairly well.

Solve each problem. ***See Example 6.***

45. Federal student loans (in billions of dollars) are shown in the table, where x represents the number of years since 2008 and y represents total student loans.

Year	x	y
2008	0	76
2009	1	90
2010	2	106
2011	3	112
2012	4	109
2013	5	101

Source: The College Board.

(a) Use the ordered pairs (x, y) to make a scatter diagram of the data.

(b) Would a linear or quadratic function better model the data?

(c) Should the coefficient a of x^2 in a quadratic model $y = ax^2 + bx + c$ be positive or negative?

(d) Use the ordered pairs $(0, 76)$, $(3, 112)$, and $(5, 101)$ to find a quadratic function that models the data.

(e) Use the model from part (d) to approximate total federal student loans (in billions of dollars) for 2010 and 2012. How well does the model approximate the actual data from the table?

46. The number (in thousands) of new, privately owned housing units started in the United States is shown in the table. Here x represents years since 2006 and y represents total housing starts.

Year	x	y
2006	0	1800
2007	1	1360
2008	2	910
2009	3	550
2010	4	590
2011	5	610
2012	6	780

Source: U.S. Census Bureau.

(a) Use the ordered pairs (x, y) to make a scatter diagram of the data.

(b) Would a linear or quadratic function better model the data?

(c) Should the coefficient a of x^2 in a quadratic model $y = ax^2 + bx + c$ be positive or negative?

(d) Use the ordered pairs $(0, 1800)$, $(4, 590)$, and $(6, 780)$ to find a quadratic function that models the data.

(e) Use the model from part (d) to approximate the number of housing starts for 2008 and 2011 to the nearest thousand. How well does the model approximate the actual data from the table?

8.6 More about Parabolas and Their Applications

OBJECTIVES

1. Find the vertex of a vertical parabola.
2. Graph a quadratic function.
3. Use the discriminant to find the number of x-intercepts of a parabola with a vertical axis.
4. Use quadratic functions to solve problems involving maximum or minimum value.
5. Graph parabolas with horizontal axes.

NOW TRY EXERCISE 1

Find the vertex of the graph of

$$f(x) = x^2 + 2x - 8.$$

NOW TRY EXERCISE 2

Find the vertex of the graph of

$$f(x) = -4x^2 + 16x - 10.$$

CLASSROOM EXAMPLE 1

Find the vertex of the graph of

$$f(x) = x^2 + 4x - 9.$$

Answer: $(-2, -13)$

CLASSROOM EXAMPLE 2

Find the vertex of the graph of

$$f(x) = 2x^2 - 4x + 1.$$

Answer: $(1, -1)$

NOW TRY ANSWERS

1. $(-1, -9)$ **2.** $(2, 6)$

OBJECTIVE 1 Find the vertex of a vertical parabola.

When the equation of a parabola is given in the form $f(x) = ax^2 + bx + c$, there are two ways to locate the vertex.

1. Complete the square. (See **Examples 1 and 2.**)

2. Use a formula derived by completing the square. (See **Example 3.**)

EXAMPLE 1 Completing the Square to Find the Vertex ($a = 1$)

Find the vertex of the graph of $f(x) = x^2 - 4x + 5$.

We can express $x^2 - 4x + 5$ in the form $(x - h)^2 + k$ by completing the square on $x^2 - 4x$, as in **Section 8.1.** The process is slightly different here because we want to keep $f(x)$ alone on one side of the equation. Instead of adding the appropriate number to each side, we *add and subtract* it on the right.

$$f(x) = x^2 - 4x + 5$$

$$f(x) = (x^2 - 4x \qquad) + 5 \quad \text{Group the variable terms.}$$

$$\left[\tfrac{1}{2}(-4)\right]^2 = (-2)^2 = 4 \quad \text{Square half the coefficient of the first-degree term.}$$

This is equivalent to adding 0.

$$f(x) = (x^2 - 4x + 4 - 4) + 5 \quad \text{Add and subtract 4.}$$

$$f(x) = (x^2 - 4x + 4) - 4 + 5 \quad \text{Bring } -4 \text{ outside the parentheses.}$$

$$f(x) = (x - 2)^2 + 1 \quad \text{Factor. Combine like terms.}$$

The vertex of this parabola is $(2, 1)$. NOW TRY

EXAMPLE 2 Completing the Square to Find the Vertex ($a \neq 1$)

Find the vertex of the graph of $f(x) = -3x^2 + 6x - 1$.

Because the x^2-term has a coefficient other than 1, we factor that coefficient out of the first two terms before completing the square.

$$f(x) = -3x^2 + 6x - 1$$

$$f(x) = (-3x^2 + 6x) - 1 \quad \text{Group the variable terms.}$$

$$f(x) = -3(x^2 - 2x) - 1 \quad \text{Factor out } -3.$$

$$f(x) = -3(x^2 - 2x \qquad) - 1 \quad \text{Prepare to complete the square.}$$

$$\left[\tfrac{1}{2}(-2)\right]^2 = (-1)^2 = 1 \quad \text{Square half the coefficient of the first-degree term.}$$

$$f(x) = -3(x^2 - 2x + 1 - 1) - 1 \quad \text{Add and subtract 1.}$$

Now bring −1 outside the parentheses. Be sure to multiply it by −3.

This is a key step.

$$f(x) = -3(x^2 - 2x + 1) + (-3)(-1) - 1 \quad \text{Distributive property}$$

$$f(x) = -3(x^2 - 2x + 1) + 3 - 1 \quad \text{Multiply.}$$

$$f(x) = -3(x - 1)^2 + 2 \quad \text{Factor. Combine like terms.}$$

The vertex is $(1, 2)$. NOW TRY

TEACHING TIP
Work through an example showing the "completing the square" process. Then show the vertex formula

$$\left(\frac{-b}{2a}, \frac{4ac-b^2}{4a}\right).$$

Substitute the values for a, b, and c into this formula to verify the result.

We can complete the square to derive a formula for the vertex of the graph of the quadratic function $f(x) = ax^2 + bx + c$ (where $a \neq 0$).

$f(x) = ax^2 + bx + c$	Standard form
$f(x) = (ax^2 + bx) + c$	Group the terms with x.
$f(x) = a\left(x^2 + \frac{b}{a}x \qquad \right) + c$	Factor a from the first two terms.
$\left[\frac{1}{2}\left(\frac{b}{a}\right)\right]^2 = \left(\frac{b}{2a}\right)^2 = \frac{b^2}{4a^2}$	Square half the coefficient of the first-degree term.
$f(x) = a\left(x^2 + \frac{b}{a}x + \frac{b^2}{4a^2} - \frac{b^2}{4a^2}\right) + c$	Add and subtract $\frac{b^2}{4a^2}$.
$f(x) = a\left(x^2 + \frac{b}{a}x + \frac{b^2}{4a^2}\right) + a\left(-\frac{b^2}{4a^2}\right) + c$	Distributive property
$f(x) = a\left(x^2 + \frac{b}{a}x + \frac{b^2}{4a^2}\right) - \frac{b^2}{4a} + c$	$-\frac{ab^2}{4a^2} = -\frac{b^2}{4a}$
$f(x) = a\left(x + \frac{b}{2a}\right)^2 + \frac{4ac - b^2}{4a}$	Factor. Rewrite terms with a common denominator.
$f(x) = a\left[x - \underbrace{\left(\frac{-b}{2a}\right)}_{h}\right]^2 + \underbrace{\frac{4ac - b^2}{4a}}_{k}$	$f(x) = a(x-h)^2 + k$ The vertex (h, k) can be expressed in terms of a, b, and c.

The expression for k can be found by replacing x with $\frac{-b}{2a}$. Using function notation, if $y = f(x)$, then the y-value of the vertex is $f\left(\frac{-b}{2a}\right)$.

Vertex Formula

The graph of the quadratic function $f(x) = ax^2 + bx + c$ (where $a \neq 0$) has vertex

$$\left(\frac{-b}{2a}, f\left(\frac{-b}{2a}\right)\right).$$

The axis of symmetry of the parabola is the line having equation

$$x = \frac{-b}{2a}.$$

CLASSROOM EXAMPLE 3
Use the vertex formula to find the vertex of the graph of

$$f(x) = -2x^2 + 3x - 1.$$

Answer: $\left(\frac{3}{4}, \frac{1}{8}\right)$

EXAMPLE 3 Using the Formula to Find the Vertex

Use the vertex formula to find the vertex of the graph of $f(x) = x^2 - x - 6$.

The x-coordinate of the vertex of the parabola is given by $\frac{-b}{2a}$.

$$\frac{-b}{2a} = \frac{-(-1)}{2(1)} = \frac{1}{2} \leftarrow x\text{-coordinate of vertex}$$

$a = 1$, $b = -1$, and $c = -6$.

NOW TRY EXERCISE 3
Use the vertex formula to find the vertex of the graph of

$f(x) = 3x^2 - 2x + 8.$

The y-coordinate of the vertex of $f(x) = x^2 - x - 6$ is $f\left(\frac{-b}{2a}\right) = f\left(\frac{1}{2}\right)$.

$$f\left(\frac{1}{2}\right) = \left(\frac{1}{2}\right)^2 - \frac{1}{2} - 6 = \frac{1}{4} - \frac{1}{2} - 6 = -\frac{25}{4} \leftarrow y\text{-coordinate of vertex}$$

The vertex is $\left(\frac{1}{2}, -\frac{25}{4}\right)$.

NOW TRY

OBJECTIVE 2 Graph a quadratic function.

Graphing a Quadratic Function $y = f(x)$

Step 1 **Determine whether the graph opens up or down.**

- If $a > 0$, then the parabola opens up.
- If $a < 0$, then it opens down.

Step 2 **Find the vertex.** Use the vertex formula or complete the square.

Step 3 **Find any intercepts.**

- To find the x-intercepts (if any), solve $f(x) = 0$.
- To find the y-intercept, evaluate $f(0)$.

Step 4 **Complete the graph.** Plot the points found so far. Find and plot additional points as needed, using symmetry about the axis.

TEACHING TIP
Point out that a minimum of two points on either side of the vertex is needed for an accurate graph.

CLASSROOM EXAMPLE 4
Graph the quadratic function

$f(x) = x^2 - 6x + 5.$

Give the vertex, axis, domain, and range.

Answer:

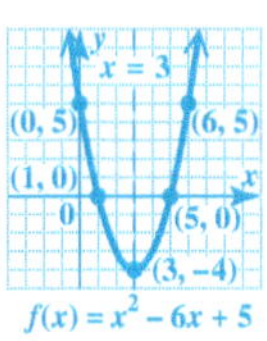

vertex: $(3, -4)$; axis: $x = 3$;
domain: $(-\infty, \infty)$;
range: $[-4, \infty)$

EXAMPLE 4 Graphing a Quadratic Function

Graph the quadratic function $f(x) = x^2 - x - 6$.

Step 1 From the equation, $a = 1$, so the graph of the function opens up.

Step 2 The vertex, $\left(\frac{1}{2}, -\frac{25}{4}\right)$, was found in **Example 3** using the vertex formula.

Step 3 Find any intercepts. The vertex, $\left(\frac{1}{2}, -\frac{25}{4}\right)$, is in quadrant IV and the graph opens up, so there will be two x-intercepts. Let $f(x) = 0$ and solve.

$$f(x) = x^2 - x - 6$$

$$0 = x^2 - x - 6 \quad \text{Let } f(x) = 0.$$

$$0 = (x - 3)(x + 2) \quad \text{Factor.}$$

$$x - 3 = 0 \quad \text{or} \quad x + 2 = 0 \quad \text{Zero-factor property}$$

$$x = 3 \quad \text{or} \quad x = -2 \quad \text{Solve each equation.}$$

The x-intercepts are $(3, 0)$ and $(-2, 0)$.
Find the y-intercept by evaluating $f(0)$.

$$f(x) = x^2 - x - 6$$

$$f(0) = 0^2 - 0 - 6 \quad \text{Let } x = 0.$$

$$f(0) = -6 \quad \text{Apply the exponent. Subtract.}$$

The y-intercept is $(0, -6)$.

NOW TRY ANSWER
3. $\left(\frac{1}{3}, \frac{23}{3}\right)$

NOW TRY EXERCISE 4

Graph the quadratic function

$$f(x) = x^2 + 2x - 3.$$

Give the vertex, axis, domain, and range.

Step 4 Plot the points found so far and additional points as needed using symmetry about the axis, $x = \frac{1}{2}$. The graph is shown in **FIGURE 14**.

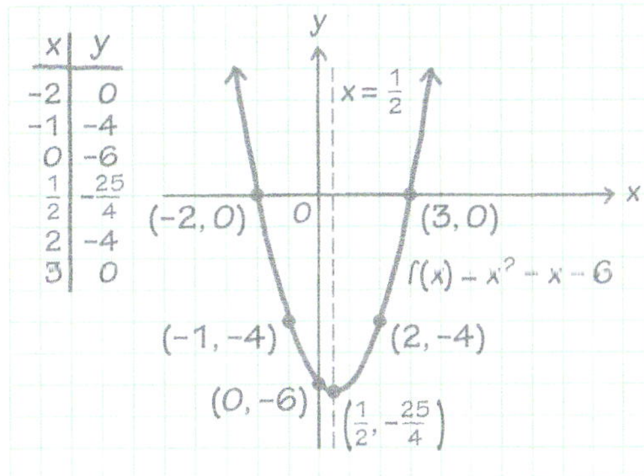

$f(x) = x^2 - x - 6$

Vertex: $\left(\frac{1}{2}, -\frac{25}{4}\right)$

Axis of symmetry: $x = \frac{1}{2}$

Domain: $(-\infty, \infty)$

Range: $\left[-\frac{25}{4}, \infty\right)$

FIGURE 14

NOW TRY

OBJECTIVE 3 Use the discriminant to find the number of *x*-intercepts of a parabola with a vertical axis.

Recall from **Section 8.2** that

$$b^2 - 4ac \qquad \text{Discriminant}$$

is the *discriminant* of the quadratic equation $ax^2 + bx + c = 0$ and that we can use it to determine the number of real solutions of a quadratic equation.

In a similar way, we can use the discriminant of a quadratic *function* to determine the number of *x*-intercepts of its graph. The three possibilities are shown in **FIGURE 15**.

1. If the discriminant is positive, the parabola will have two *x*-intercepts.
2. If the discriminant is 0, there will be only one *x*-intercept, and it will be the vertex of the parabola.
3. If the discriminant is negative, the graph will have no *x*-intercepts.

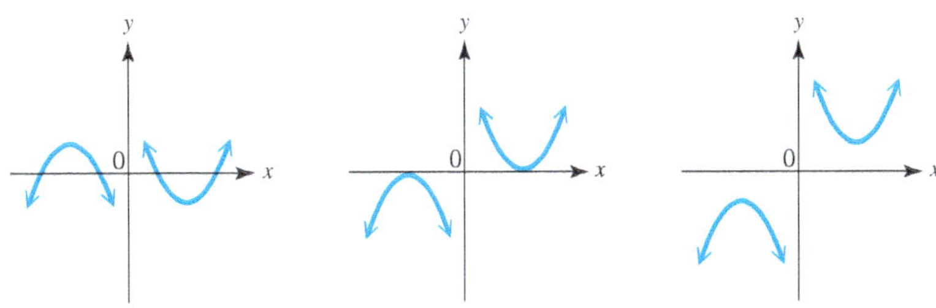

$b^2 - 4ac > 0$	$b^2 - 4ac = 0$	$b^2 - 4ac < 0$
Two *x*-intercepts	One *x*-intercept	No *x*-intercepts

FIGURE 15

CLASSROOM EXAMPLE 5

Find the discriminant and use it to determine the number of *x*-intercepts of the graph of each quadratic function.

(a) $f(x) = -3x^2 - x + 2$

(b) $f(x) = x^2 - x + 1$

(c) $f(x) = x^2 - 8x + 16$

Answers: (a) 25; two (b) -3; none (c) 0; one

EXAMPLE 5 Using the Discriminant to Determine Number of *x*-Intercepts

Find the discriminant and use it to determine the number of *x*-intercepts of the graph of each quadratic function.

(a) $f(x) = 2x^2 + 3x - 5$

$$b^2 - 4ac \qquad \text{Discriminant}$$
$$= 3^2 - 4(2)(-5) \qquad a = 2, b = 3, c = -5$$
$$= 9 - (-40) \qquad \text{Apply the exponent. Multiply.}$$
$$= 49 \qquad \text{Subtract.}$$

Because the discriminant is positive, the parabola has two *x*-intercepts.

NOW TRY ANSWER

4.

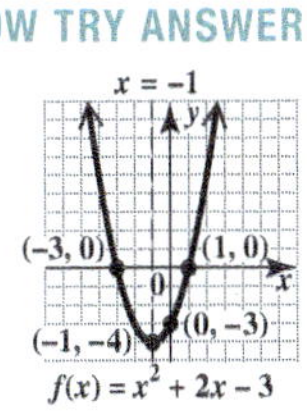

vertex: $(-1, -4)$; axis: $x = -1$; domain: $(-\infty, \infty)$; range: $[-4, \infty)$

NOW TRY EXERCISE 5
Find the discriminant and use it to determine the number of x-intercepts of the graph of each quadratic function.

(a) $f(x) = -2x^2 + 3x - 2$
(b) $f(x) = 3x^2 + 2x - 1$
(c) $f(x) = 4x^2 - 12x + 9$

(b) $f(x) = -3x^2 - 1$

$$b^2 - 4ac \qquad \text{Discriminant}$$
$$= 0^2 - 4(-3)(-1) \qquad a = -3,\ b = 0,\ c = -1$$
$$= 0 - 12 \qquad \text{Apply the exponent. Multiply.}$$
$$= -12 \qquad \text{Subtract.}$$

The discriminant is negative, so the graph has no x-intercepts.

(c) $f(x) = 9x^2 + 6x + 1$

$$b^2 - 4ac \qquad \text{Discriminant}$$
$$= 6^2 - 4(9)(1) \qquad a = 9,\ b = 6,\ c = 1$$
$$= 36 - 36 \qquad \text{Apply the exponent. Multiply.}$$
$$= 0 \qquad \text{Subtract.}$$

Because the discriminant is 0, the parabola has only one x-intercept (its vertex).

NOW TRY

OBJECTIVE 4 Use quadratic functions to solve problems involving maximum or minimum value.

The vertex of the graph of a quadratic function is either the highest or the lowest point on the parabola. It provides the following information.

1. The y-value of the vertex gives the maximum or minimum value of y.
2. The x-value tells where the maximum or minimum occurs.

PROBLEM-SOLVING HINT In many applied problems we must find the greatest or least value of some quantity. When we can express that quantity in terms of a quadratic function, the value of k in the vertex (h, k) gives that optimum value.

CLASSROOM EXAMPLE 6
Solve the problem in Example 6 if the farmer has only 100 ft of fencing.
Answer: The field should be 25 ft by 50 ft with maximum area 1250 ft².

EXAMPLE 6 Finding the Maximum Area of a Rectangular Region

A farmer has 120 ft of fencing to enclose a rectangular area next to a building. (See **FIGURE 16**.) Find the maximum area he can enclose and the dimensions of the field when the area is maximized.

FIGURE 16

Let x = the width of the field.

$$x + x + \text{length} = 120 \qquad \text{Sum of the sides is 120 ft.}$$
$$2x + \text{length} = 120 \qquad \text{Combine like terms.}$$
$$\text{length} = 120 - 2x \qquad \text{Subtract } 2x.$$

NOW TRY ANSWERS
5. (a) -7; none (b) 16; two (c) 0; one

NOW TRY EXERCISE 6
Solve the problem in **Example 6** if the farmer has only 80 ft of fencing.

The area $\mathcal{A}(x)$ is given by the product of the length and width.

$$\mathcal{A}(x) = (120 - 2x)x \quad \text{Area = length} \cdot \text{width}$$

$$\mathcal{A}(x) = 120x - 2x^2 \quad \text{Distributive property}$$

To determine the maximum area, use the vertex formula to find the vertex of the parabola $\mathcal{A}(x) = 120x - 2x^2$. Write the equation in standard form.

$$\mathcal{A}(x) = -2x^2 + 120x \quad a = -2,\ b = 120,\ c = 0$$

Then $$x = \frac{-b}{2a} = \frac{-120}{2(-2)} = \frac{-120}{-4} = 30,$$

and $$\mathcal{A}(30) = -2(30)^2 + 120(30) = -2(900) + 3600 = 1800.$$

The graph is a parabola that opens down. Its vertex is $(30, 1800)$. The maximum area will be 1800 ft^2 when x, the width, is 30 ft and the length is

$$120 - 2(30) = 60 \text{ ft.}$$

NOW TRY

CLASSROOM EXAMPLE 7
A toy rocket is launched from the ground so that its distance in feet above the ground after t seconds is

$$s(t) = -16t^2 + 208t.$$

Find the maximum height it reaches and the number of seconds it takes to reach that height.

Answer: 676 ft; 6.5 sec

CAUTION ***Be careful when interpreting the meanings of the coordinates of the vertex.*** The first coordinate, x, gives the value for which the *function value*, y or $f(x)$, is a maximum or a minimum.

Read a problem carefully to determine whether to find the value of the independent variable, the function value, or both.

NOW TRY EXERCISE 7
A stomp rocket is launched from the ground with an initial velocity of 48 ft per sec so that its distance in feet above the ground after t seconds is

$$s(t) = -16t^2 + 48t.$$

Find the maximum height attained by the rocket and the number of seconds it takes to reach that height.

EXAMPLE 7 Finding the Maximum Height Attained by a Projectile

If air resistance is neglected, a projectile on Earth shot straight upward with an initial velocity of 40 m per sec will be at a height s in meters given by

$$s(t) = -4.9t^2 + 40t,$$

where t is the number of seconds elapsed after projection. After how many seconds will it reach its maximum height, and what is this maximum height?

For this function, $a = -4.9$, $b = 40$, and $c = 0$. Use the vertex formula.

$$t = \frac{-b}{2a} = \frac{-40}{2(-4.9)} \approx 4.1 \quad \text{Use a calculator. Round to the nearest tenth.}$$

This indicates that the maximum height is attained at 4.1 sec. To find this maximum height, calculate $s(4.1)$.

$$s(t) = -4.9t^2 + 40t$$

$$s(4.1) = -4.9(4.1)^2 + 40(4.1) \quad \text{Let } t = 4.1.$$

$$s(4.1) \approx 81.6 \quad \text{Use a calculator. Round to the nearest tenth.}$$

The projectile will attain a maximum height of 81.6 m at 4.1 sec. NOW TRY

OBJECTIVE 5 Graph parabolas with horizontal axes.

If x and y are interchanged in the equation

$$y = ax^2 + bx + c, \quad \text{the equation becomes} \quad x = ay^2 + by + c.$$

Because of the interchange of the roles of x and y, these parabolas are horizontal (with horizontal lines as axes of symmetry).

NOW TRY ANSWERS
6. The field should be 20 ft by 40 ft with maximum area 800 ft^2.
7. 36 ft; 1.5 sec

Graph of a Horizontal Parabola

The graph of $x = ay^2 + by + c$ or $x = a(y - k)^2 + h$ is a parabola.

- The vertex of the parabola is (h, k).
- The axis of symmetry is the horizontal line $y = k$.
- The graph opens to the right if $a > 0$ and to the left if $a < 0$.

NOW TRY EXERCISE 8

Graph $x = (y + 2)^2 - 1$. Give the vertex, axis, domain, and range.

CLASSROOM EXAMPLE 8

Graph $x = (y + 1)^2 - 4$. Give the vertex, axis, domain, and range.

Answer:

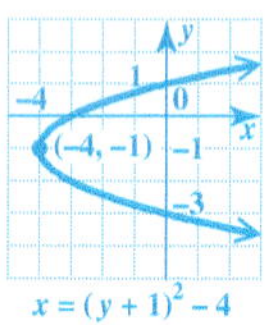

vertex: $(-4, -1)$; axis: $y = -1$; domain: $[-4, \infty)$; range: $(-\infty, \infty)$

EXAMPLE 8 Graphing a Horizontal Parabola ($a = 1$)

Graph $x = (y - 2)^2 - 3$. Give the vertex, axis, domain, and range.

This graph has its vertex at $(-3, 2)$ because the roles of x and y are interchanged. It opens to the right (the positive x-direction) because $a = 1$ and $1 > 0$, and has the same shape as $y = x^2$ (but situated horizontally). Plotting a few additional points gives the graph shown in **FIGURE 17**.

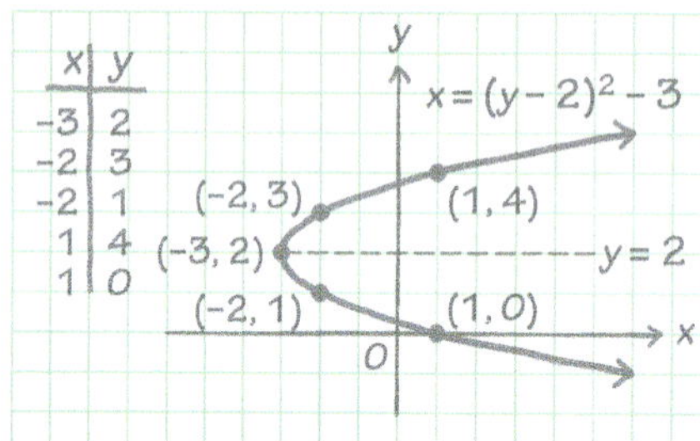

$x = (y - 2)^2 - 3$
Vertex: $(-3, 2)$
Axis of symmetry: $y = 2$
Domain: $[-3, \infty)$
Range: $(-\infty, \infty)$

FIGURE 17

NOW TRY

NOW TRY EXERCISE 9

Graph $x = -3y^2 - 6y - 5$. Give the vertex, axis, domain, and range.

CLASSROOM EXAMPLE 9

Graph $x = -y^2 + 2y + 5$. Give the vertex, axis, domain, and range.

Answer:

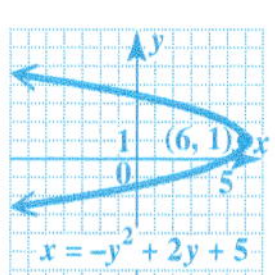

vertex: $(6, 1)$; axis: $y = 1$; domain: $(-\infty, 6]$; range: $(-\infty, \infty)$

EXAMPLE 9 Graphing a Horizontal Parabola ($a \neq 1$)

Graph $x = -2y^2 + 4y - 3$. Give the vertex, axis, domain, and range.

$$x = -2y^2 + 4y - 3$$

$$x = (-2y^2 + 4y) - 3 \quad \text{Group the variable terms.}$$

$$x = -2(y^2 - 2y) - 3 \quad \text{Factor out } -2.$$

$$\left[\tfrac{1}{2}(-2)\right]^2 = (-1)^2 = 1$$

$$x = -2(y^2 - 2y + 1 - 1) - 3 \quad \text{Complete the square within the parentheses. Add and subtract 1.}$$

$$x = -2(y^2 - 2y + 1) + (-2)(-1) - 3 \quad \text{Distributive property}$$

Be careful here.

$$x = -2(y - 1)^2 - 1 \quad \text{Factor. Simplify.}$$

Because of the negative coefficient -2 in $x = -2(y - 1)^2 - 1$, the graph opens to the left (the negative x-direction). The graph is narrower than the graph of $y = x^2$ because $|-2| = 2$, and $2 > 1$. See **FIGURE 18**.

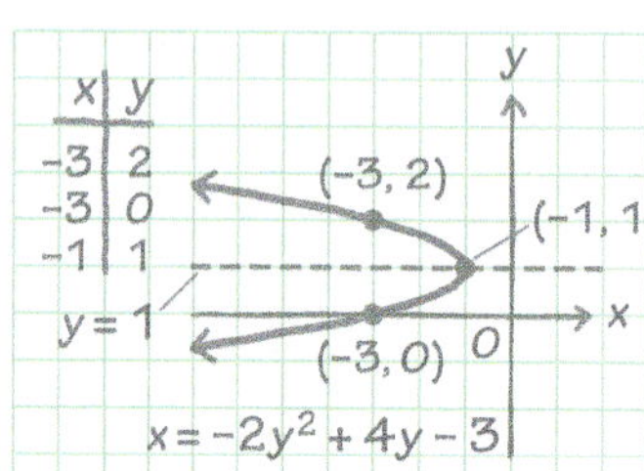

$x = -2y^2 + 4y - 3$
Vertex: $(-1, 1)$
Axis of symmetry: $y = 1$
Domain: $(-\infty, -1]$
Range: $(-\infty, \infty)$

FIGURE 18

NOW TRY

NOW TRY ANSWERS

8.

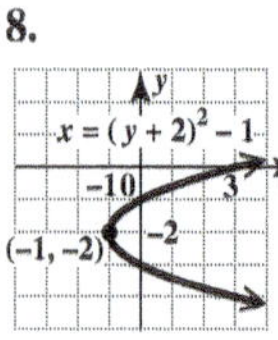

vertex: $(-1, -2)$; axis: $y = -2$; domain: $[-1, \infty)$; range: $(-\infty, \infty)$

9.

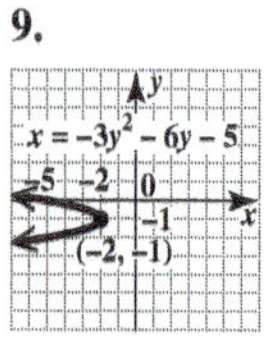

vertex: $(-2, -1)$; axis: $y = -1$; domain: $(-\infty, -2]$; range: $(-\infty, \infty)$

TEACHING TIP
Perhaps mention that the vertex for a quadratic relation of the form

$$x = ay^2 + by + c$$

is given by

$$\left(\frac{4ac - b^2}{4a}, \frac{-b}{2a}\right).$$

! CAUTION ***Only quadratic equations solved for y (whose graphs are vertical parabolas) are examples of functions.*** The horizontal parabolas in **Examples 8 and 9** are *not* graphs of functions because they do not satisfy the conditions of the vertical line test.

Summary of Graphs of Parabolas

Equation	*Graph*
$y = ax^2 + bx + c$ or $y = a(x - h)^2 + k$	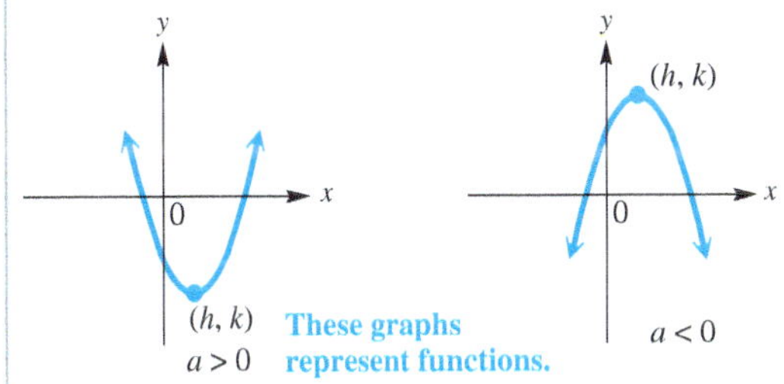
$x = ay^2 + by + c$ or $x = a(y - k)^2 + h$	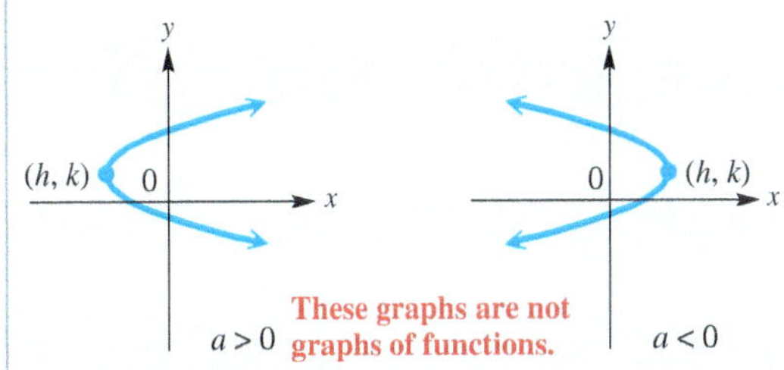

8.6 Exercises

FOR EXTRA HELP

MyMathLab®

Complete solution available in MyMathLab

1. If x is squared, it has a vertical axis. If y is squared, it has a horizontal axis.
2. A parabola with a horizontal axis fails the vertical line test.
3. Find the discriminant of the function. If it is positive, there are two x-intercepts. If it is 0, there is one x-intercept (at the vertex), and if it is negative, there is no x-intercept.
4. none
5. A, D are vertical parabolas. B, C are horizontal parabolas.
6. A, D
7. $(-4, -6)$ 8. $(-5, -2)$
9. $(1, -3)$ 10. $(2, 4)$
11. $\left(-\frac{1}{2}, -\frac{29}{4}\right)$ 12. $\left(\frac{1}{2}, \frac{19}{4}\right)$

Concept Check *Answer each question.*

1. How can we determine just by looking at the equation of a parabola whether it has a vertical or a horizontal axis?
2. Why can't the graph of a quadratic function be a parabola with a horizontal axis?
3. How can we determine the number of x-intercepts of the graph of a quadratic function without graphing the function?
4. If the vertex of the graph of a quadratic function is $(1, -3)$, and the graph opens down, how many x-intercepts does the graph have?
5. Which equations have a graph that is a vertical parabola? A horizontal parabola?

   A. $y = -x^2 + 20x + 80$ **B.** $x = 2y^2 + 6y + 5$
   C. $x + 1 = (y + 2)^2$ **D.** $f(x) = (x - 4)^2$
6. Which of the equations in **Exercise 5** represent functions?

Find the vertex of each parabola. ***See Examples 1–3.***

7. $f(x) = x^2 + 8x + 10$
8. $f(x) = x^2 + 10x + 23$
9. $f(x) = -2x^2 + 4x - 5$
10. $f(x) = -3x^2 + 12x - 8$
11. $f(x) = x^2 + x - 7$
12. $f(x) = x^2 - x + 5$

13. $(-1, 3)$; up; narrower; no x-intercepts
14. $(1, 1)$; up; narrower; no x-intercepts
15. $\left(\frac{5}{2}, \frac{37}{4}\right)$; down; same shape; two x-intercepts
16. $\left(\frac{7}{2}, \frac{57}{4}\right)$; down; same shape; two x-intercepts
17. $(-3, -9)$; to the right; wider
18. $(-55, -10)$; to the right; wider

19. F
20. A
21. C
22. B
23. D
24. E

25. $f(x) = x^2 + 8x + 10$

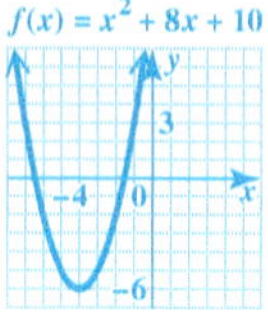

vertex: $(-4, -6)$; axis: $x = -4$; domain: $(-\infty, \infty)$; range: $[-6, \infty)$

26. $f(x) = x^2 + 10x + 23$

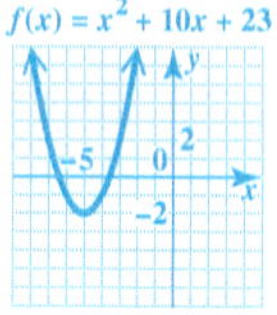

vertex: $(-5, -2)$; axis: $x = -5$; domain: $(-\infty, \infty)$; range: $[-2, \infty)$

27.

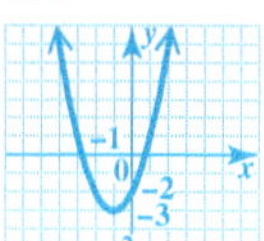

vertex: $(-1, -3)$; axis: $x = -1$; domain: $(-\infty, \infty)$; range: $[-3, \infty)$

28.

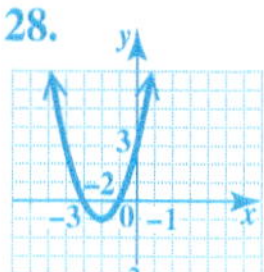

vertex: $(-2, -1)$; axis: $x = -2$; domain: $(-\infty, \infty)$; range: $[-1, \infty)$

29. $f(x) = -2x^2 + 4x - 5$

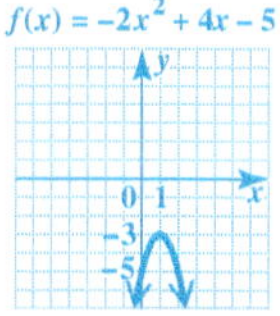

vertex: $(1, -3)$; axis: $x = 1$; domain: $(-\infty, \infty)$; range: $(-\infty, -3]$

30. $f(x) = -3x^2 + 12x - 8$

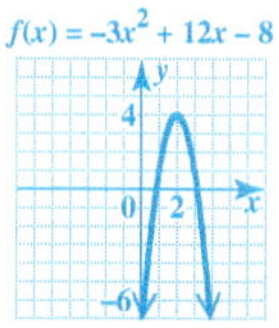

vertex: $(2, 4)$; axis: $x = 2$; domain: $(-\infty, \infty)$; range: $(-\infty, 4]$

31.

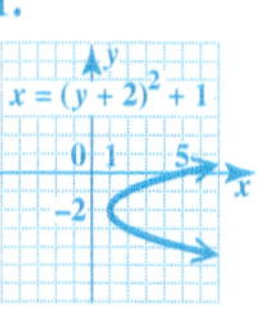

vertex: $(1, -2)$; axis: $y = -2$; domain: $[1, \infty)$; range: $(-\infty, \infty)$

32.

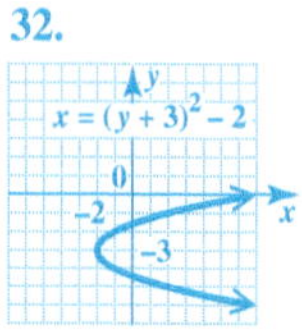

vertex: $(-2, -3)$; axis: $y = -3$; domain: $[-2, \infty)$; range: $(-\infty, \infty)$

Find the vertex of each parabola. For each equation, decide whether the graph opens up, down, to the left, *or* to the right, *and whether it is* wider, narrower, *or the* same shape *as the graph of* $y = x^2$. *If it is a parabola with a vertical axis of symmetry, find the discriminant and use it to determine the number of x-intercepts.* ***See Examples 1–3, 5, 8, and 9.***

13. $f(x) = 2x^2 + 4x + 5$
14. $f(x) = 3x^2 - 6x + 4$
15. $f(x) = -x^2 + 5x + 3$
16. $f(x) = -x^2 + 7x + 2$
17. $x = \frac{1}{3}y^2 + 6y + 24$
18. $x = \frac{1}{2}y^2 + 10y - 5$

Concept Check *Match each equation in Exercises 19–24 with its graph in choices A–F.*

19. $y = 2x^2 + 4x - 3$
20. $y = -x^2 + 3x + 5$
21. $y = -\frac{1}{2}x^2 - x + 1$
22. $x = y^2 + 6y + 3$
23. $x = -y^2 - 2y + 4$
24. $x = 3y^2 + 6y + 5$

A.

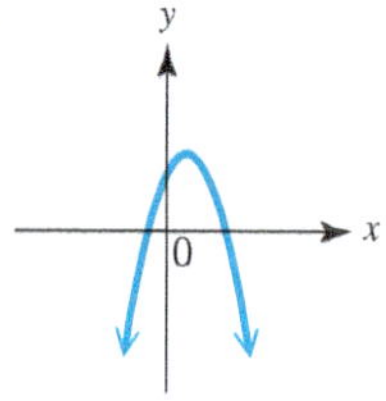

B.

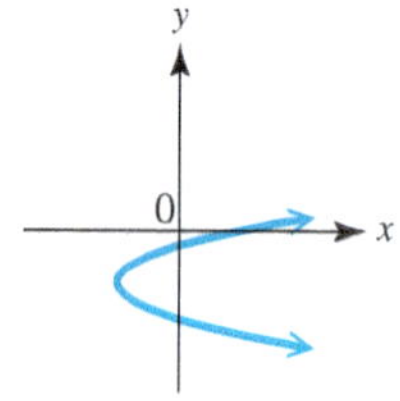

C.

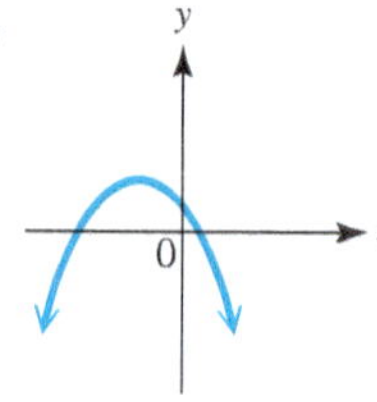

D.

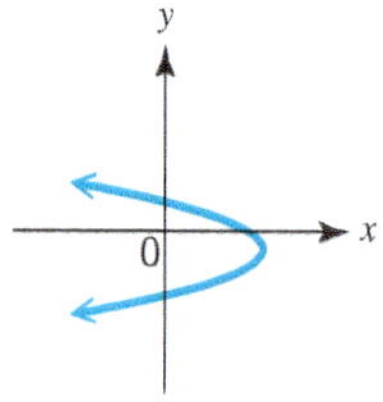

E.

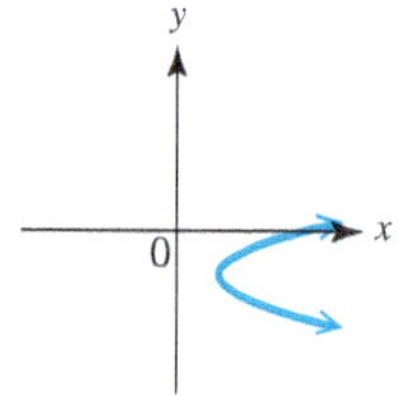

F.

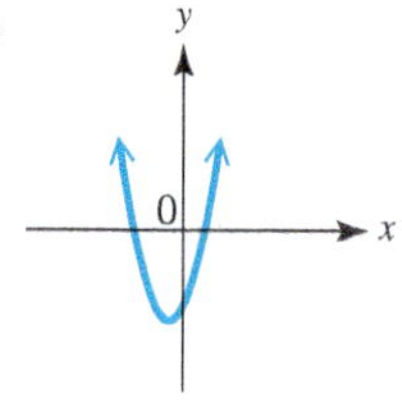

Graph each parabola. (Use the results of ***Exercises 7–10*** *in Exercises 25, 26, 29, and 30.) Give the vertex, axis of symmetry, domain, and range.* ***See Examples 4, 8, and 9.***

25. $f(x) = x^2 + 8x + 10$
26. $f(x) = x^2 + 10x + 23$
27. $f(x) = x^2 + 2x - 2$
28. $f(x) = x^2 + 4x + 3$
29. $f(x) = -2x^2 + 4x - 5$
30. $f(x) = -3x^2 + 12x - 8$
31. $x = (y + 2)^2 + 1$
32. $x = (y + 3)^2 - 2$
33. $x = -(y - 3)^2 - 1$
34. $x = -(y - 2)^2 + 4$
35. $x = -\frac{1}{5}y^2 + 2y - 4$
36. $x = -\frac{1}{2}y^2 - 4y - 6$
37. $x = 3y^2 + 12y + 5$
38. $x = 4y^2 + 16y + 11$

Solve each problem. ***See Examples 6 and 7.***

39. Find the pair of numbers whose sum is 40 and whose product is a maximum. (*Hint:* Let x and $40 - x$ represent the two numbers.)

40. Find the pair of numbers whose sum is 60 and whose product is a maximum.

33.

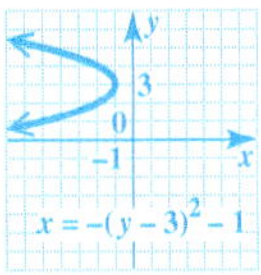

vertex: $(-1, 3)$;
axis: $y = 3$;
domain: $(-\infty, -1]$;
range: $(-\infty, \infty)$

34.

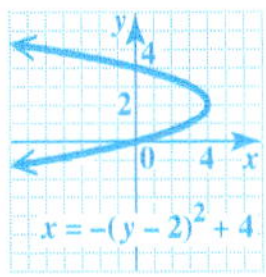

vertex: $(4, 2)$;
axis: $y = 2$;
domain: $(-\infty, 4]$;
range: $(-\infty, \infty)$

35.

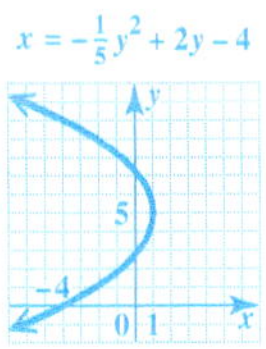

vertex: $(1, 5)$;
axis: $y = 5$;
domain: $(-\infty, 1]$;
range: $(-\infty, \infty)$

36.

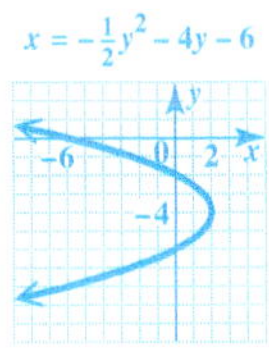

vertex: $(2, -4)$;
axis: $y = -4$;
domain: $(-\infty, 2]$;
range: $(-\infty, \infty)$

37.

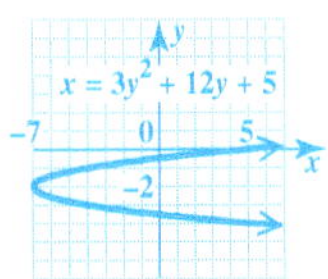

vertex: $(-7, -2)$;
axis: $y = -2$;
domain: $[-7, \infty)$;
range: $(-\infty, \infty)$

38.

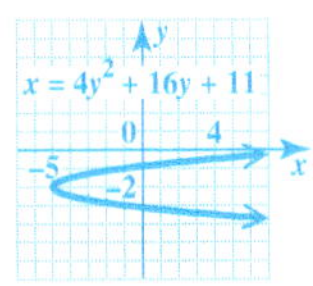

vertex: $(-5, -2)$;
axis: $y = -2$;
domain: $[-5, \infty)$;
range: $(-\infty, \infty)$

39. 20 and 20

40. 30 and 30

41. 140 ft by 70 ft; 9800 ft^2

42. 50 ft by 25 ft; 1250 ft^2

43. 2 sec; 65 ft

44. 10 hr; 180 students

45. 20 units; \$210

46. 35 units; \$275

41. Polk Community College wants to construct a rectangular parking lot on land bordered on one side by a highway. It has 280 ft of fencing that is to be used to fence off the other three sides. What should be the dimensions of the lot if the enclosed area is to be a maximum? What is the maximum area?

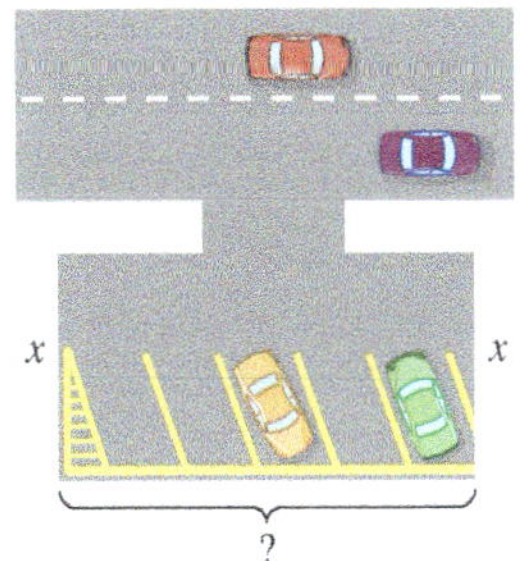

42. Bonnie has 100 ft of fencing material to enclose a rectangular exercise run for her dog. One side of the run will border her house, so she will only need to fence three sides. What dimensions will give the enclosure the maximum area? What is the maximum area?

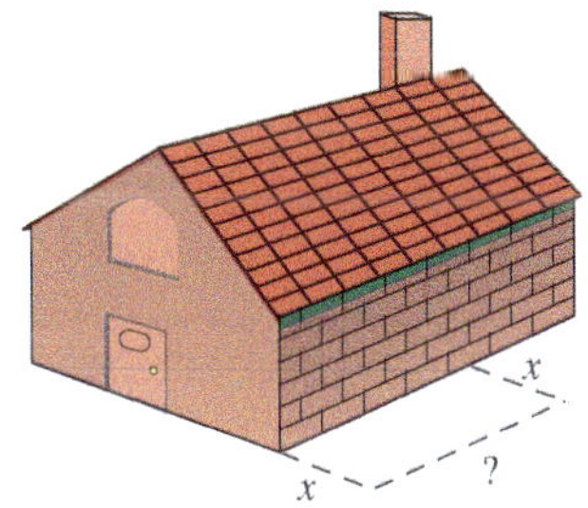

43. Two physics students from American River College find that when a bottle of California wine is shaken several times, held upright, and uncorked, its cork travels according to the function

$$s(t) = -16t^2 + 64t + 1,$$

where s is its height in feet above the ground t seconds after being released. After how many seconds will it reach its maximum height? What is the maximum height?

44. Professor Barbu has found that the number of students attending his intermediate algebra class is approximated by

$$S(x) = -x^2 + 20x + 80,$$

where x is the number of hours that the Campus Center is open daily. Find the number of hours that the center should be open so that the number of students attending class is a maximum. What is this maximum number of students?

45. Klaus has a taco stand. He has found that his daily costs are approximated by

$$C(x) = x^2 - 40x + 610,$$

where $C(x)$ is the cost, in dollars, to sell x units of tacos. Find the number of units of tacos he should sell to minimize his costs. What is the minimum cost?

46. Mohammad has a frozen yogurt cart. His daily costs are approximated by

$$C(x) = x^2 - 70x + 1500,$$

where $C(x)$ is the cost, in dollars, to sell x units of frozen yogurt. Find the number of units of frozen yogurt he must sell to minimize his costs. What is the minimum cost?

47. If an object on Earth is projected upward with an initial velocity of 32 ft per sec, then its height after t seconds is given by

$$s(t) = -16t^2 + 32t.$$

Find the maximum height attained by the object and the number of seconds it takes to hit the ground.

48. A projectile on Earth is fired straight upward so that its distance (in feet) above the ground t seconds after firing is given by

$$s(t) = -16t^2 + 400t.$$

Find the maximum height it reaches and the number of seconds it takes to reach that height.

49. The percent of the U.S. population that was foreign-born during the years 1930–2010 can be modeled by the quadratic function

$$f(x) = 0.0043x^2 - 0.3245x + 11.53,$$

where $x = 0$ represents 1930, $x = 10$ represents 1940, and so on. (*Source:* U.S. Census Bureau.)

(a) The coefficient of x^2 in the model is positive, so the graph of this quadratic function is a parabola that opens up. Will the y-value of the vertex of this graph be a maximum or a minimum?

(b) According to the model, in what year during this period was the percent of foreign-born population a minimum? (Round down for the year.) Use the actual x-value of the vertex, to the nearest tenth, to find this percent, also to the nearest tenth.

50. The percent of births in the United States to teenage mothers during the years 2005–2012 can be modeled by the quadratic function

$$f(x) = -0.3661x^2 + 1.565x + 39.21,$$

where $x = 0$ represents 2005, $x = 1$ represents 2006, and so on. (*Source:* CDC.)

(a) The coefficient of x^2 in the model is negative, so the graph of this quadratic function is a parabola that opens down. Will the y-value of the vertex of this graph be a maximum or a minimum?

(b) According to the model, in what year during this period was the percent of births in the United States to teenage mothers a maximum? (Round down for the year.) Use the actual x-value of the vertex, to the nearest tenth, to find this percent, also to the nearest tenth.

The graph shows how Social Security trust fund assets are expected to change, and suggests that a quadratic function would be a good fit to the data. The data are approximated by the function

$$f(x) = -20.57x^2 + 758.9x - 3140.$$

In the model, $x = 10$ *represents 2010,* $x = 15$ *represents 2015, and so on, and* $f(x)$ *is in billions of dollars.*

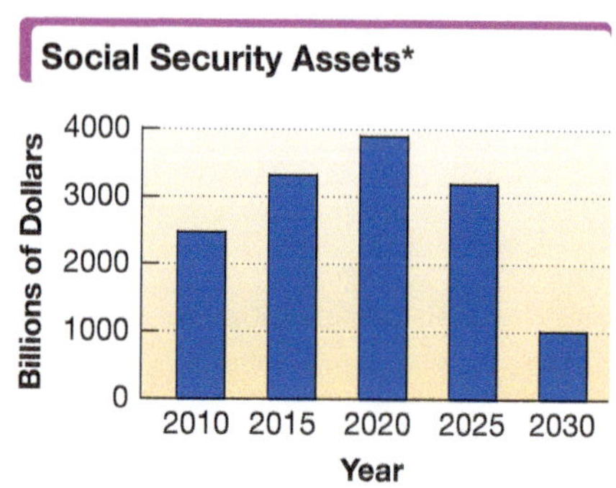

*Projected

Source: Social Security Administration.

51. How could we have predicted that this quadratic model would have a negative coefficient for x^2, based only on the graph shown?

52. Algebraically determine the vertex of the graph, with coordinates to four significant digits. Interpret the answer as it applies to this application.

53. (a) $R(x) = (100 - x)(200 + 4x)$
$= 20{,}000 + 200x - 4x^2$

(b)
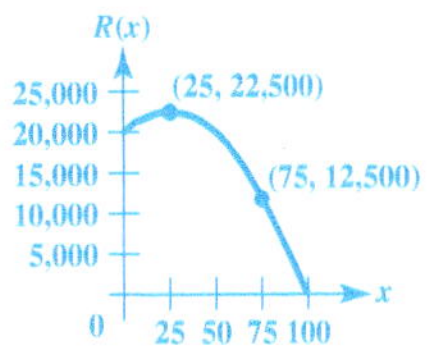

(c) 25 (d) \$22,500

54. (a) $R(x) = (42 - x)(48 + 2x)$
$= -2x^2 + 36x + 2016$

(b)
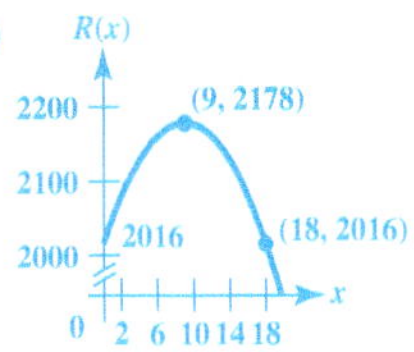

(c) 9 (d) \$2178

55. center: $(-2, -3)$; $r = 2$
56. center: $(4, 6)$; $r = 7$
57. center: $(-5, 7)$; $r = 9$
58. center: $(1, -2)$; $r = 3$

Extending Skills *In each problem, find the following.*

(a) A function $R(x)$ that describes the total revenue received

(b) The graph of the function from part (a)

(c) The number of unsold seats that will produce the maximum revenue

(d) The maximum revenue

53. A charter flight charges a fare of \$200 per person, plus \$4 per person for each unsold seat on the plane. The plane holds 100 passengers. Let x represent the number of unsold seats. (*Hint:* To find $R(x)$, multiply the number of people flying, $100 - x$, by the price per ticket, $200 + 4x$.)

54. A charter bus charges a fare of \$48 per person, plus \$2 per person for each unsold seat on the bus. The bus has 42 seats. Let x represent the number of unsold seats. (*Hint:* To find $R(x)$, multiply the number riding, $42 - x$, by the price per ticket, $48 + 2x$.)

Extending Skills *In this section, we completed the square to find the vertex of a parabola in the form $f(x) = ax^2 + bx + c$. (See **Examples 1 and 2.**) Recall from **Section 7.3** that an equation of a circle with center (h, k) and radius r is*

$$(x - h)^2 + (y - k)^2 = r^2. \quad \text{Center-radius form of the equation of a circle}$$

Consider the equation $x^2 + y^2 + 2x + 6y - 15 = 0$, which is also the equation of a circle. We can write this equation in center-radius form by completing the squares on x and y.

$$x^2 + y^2 + 2x + 6y - 15 = 0$$

$$x^2 + y^2 + 2x + 6y = 15 \quad \text{Transform so that the constant is on the right.}$$

$$(x^2 + 2x \qquad) + (y^2 + 6y \qquad) = 15 \quad \text{Write in anticipation of completing the square.}$$

$$\left[\tfrac{1}{2}(2)\right]^2 = 1 \qquad \left[\tfrac{1}{2}(6)\right]^2 = 9$$

$$(x^2 + 2x + 1) + (y^2 + 6y + 9) = 15 + 1 + 9 \quad \text{Complete the squares on both } x \text{ and } y\text{. Add 1 and 9 on } both \text{ sides of the equation.}$$

$$(x + 1)^2 + (y + 3)^2 = 25 \quad \text{Factor on the left. Add on the right.}$$

$$[x - (-1)]^2 + [y - (-3)]^2 = 5^2 \quad \text{Write in center-radius form.}$$

The final equation shows that the circle has center $(-1, -3)$ and radius 5.
Find the center and radius of each circle.

55. $x^2 + y^2 + 4x + 6y + 9 = 0$

56. $x^2 + y^2 - 8x - 12y + 3 = 0$

57. $x^2 + y^2 + 10x - 14y - 7 = 0$

58. $x^2 + y^2 - 2x + 4y - 4 = 0$

8.7 Polynomial and Rational Inequalities

OBJECTIVES

1. Solve quadratic inequalities.
2. Solve polynomial inequalities of degree 3 or greater.
3. Solve rational inequalities.

VOCABULARY

☐ quadratic inequality
☐ rational inequality

OBJECTIVE 1 Solve quadratic inequalities.

We can combine methods of solving linear inequalities and methods of solving quadratic equations to solve *quadratic inequalities.*

Quadratic Inequality

A **quadratic inequality** (in x here) can be written in the form

$$ax^2 + bx + c < 0, \qquad ax^2 + bx + c > 0,$$
$$ax^2 + bx + c \le 0, \quad \text{or} \quad ax^2 + bx + c \ge 0,$$

where a, b, and c are real numbers and $a \ne 0$.

One way to solve a quadratic inequality involves graphing the related quadratic function. This method is justified because the graph is *continuous*—that is, it has no breaks.

NOW TRY EXERCISE 1

Use the graph to solve each quadratic inequality.

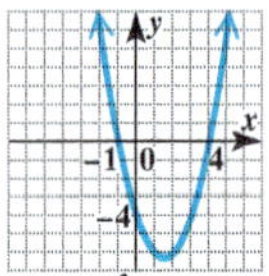

$f(x) = x^2 - 3x - 4$

(a) $x^2 - 3x - 4 > 0$

(b) $x^2 - 3x - 4 < 0$

CLASSROOM EXAMPLE 1

Use the graph to solve each quadratic inequality.

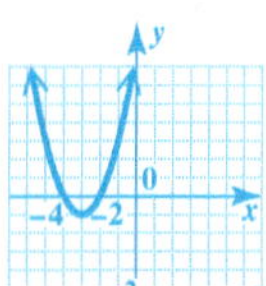

$f(x) = x^2 + 6x + 8$

(a) $x^2 + 6x + 8 > 0$

(b) $x^2 + 6x + 8 < 0$

Answers:

(a) $(-\infty, -4) \cup (-2, \infty)$

(b) $(-4, -2)$

EXAMPLE 1 Solving Quadratic Inequalities by Graphing

Solve each inequality.

(a) $x^2 - x - 12 > 0$

We graph the related quadratic function $f(x) = x^2 - x - 12$. We are particularly interested in the x-intercepts, which are found as in **Section 8.6** by letting $f(x) = 0$ and solving the following quadratic equation.

$$x^2 - x - 12 = 0 \quad \text{Let } f(x) = 0.$$

$$(x - 4)(x + 3) = 0 \quad \text{Factor.}$$

$$x - 4 = 0 \quad \text{or} \quad x + 3 = 0 \quad \text{Zero-factor property}$$

$$x = 4 \quad \text{or} \quad x = -3 \leftarrow \text{The } x\text{-intercepts are } (4, 0) \text{ and } (-3, 0).$$

The graph opens up because the coefficient of x^2 is positive. See **FIGURE 19(a)**. Notice that x-values less than -3 or greater than 4 result in y-values *greater than* 0. Thus, the solution set of $x^2 - x - 12 > 0$, written in interval notation, is

$$(-\infty, -3) \cup (4, \infty).$$

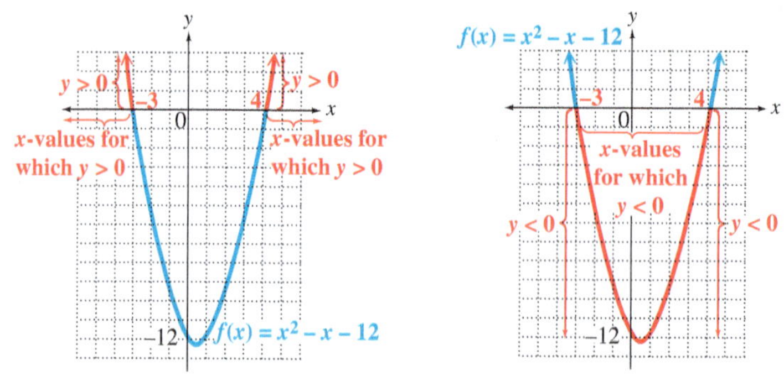

The graph is *above* the x-axis for $(-\infty, -3) \cup (4, \infty)$.

(a)

The graph is *below* the x-axis for $(-3, 4)$.

(b)

FIGURE 19

(b) $x^2 - x - 12 < 0$

We want values of y that are *less than* 0. See **FIGURE 19(b)**. Notice from the graph that x-values between -3 and 4 result in y-values less than 0. Thus, the solution set of $x^2 - x - 12 < 0$, written in interval notation, is $(-3, 4)$.

NOW TRY

NOTE If the inequalities in **Example 1** had used $\geq$ and $\leq$, the solution sets would have included the x-values of the intercepts, which make the quadratic expression equal to 0. They would have been written in interval notation as

$$(-\infty, -3] \cup [4, \infty) \quad \text{and} \quad [-3, 4].$$

Square brackets would indicate that the endpoints -3 and 4 are *included* in the solution sets.

NOW TRY ANSWERS

1. **(a)** $(-\infty, -1) \cup (4, \infty)$

(b) $(-1, 4)$

Another method for solving a quadratic inequality uses the basic ideas of **Example 1** without actually graphing the related quadratic function.

CLASSROOM EXAMPLE 2

Solve and graph the solution set.

$$2x^2 + 3x - 2 \geq 0$$

Answer: $(-\infty, -2] \cup \left[\frac{1}{2}, \infty\right)$

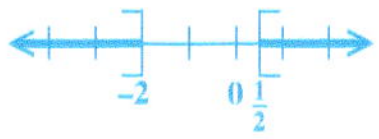

EXAMPLE 2 Solving a Quadratic Inequality Using Test Values

Solve and graph the solution set of $x^2 - x - 12 > 0$.

Solve the quadratic equation $x^2 - x - 12 = 0$ as in **Example 1(a).**

$$x^2 - x - 12 = 0 \quad \text{Let } f(x) = 0.$$

$$(x - 4)(x + 3) = 0 \quad \text{Factor.}$$

$$x - 4 = 0 \quad \text{or} \quad x + 3 = 0 \quad \text{Zero-factor property}$$

$$x = 4 \quad \text{or} \quad x = -3 \quad \text{Solve each equation.}$$

The numbers 4 and −3 divide a number line into Intervals A, B, and C, as shown in **FIGURE 20.** ***Be careful to put the lesser number on the left.***

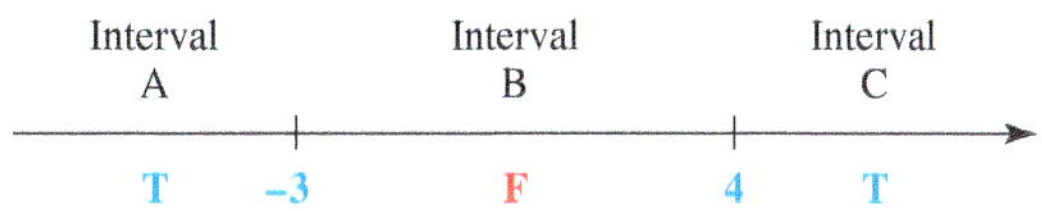

FIGURE 20

Notice the similarity between **FIGURE 20** and the x-axis with intercepts (−3, 0) and (4, 0) in **FIGURE 19(a).**

The numbers 4 and −3 are the only values that make the quadratic expression $x^2 - x - 12$ equal to 0. All other numbers make the expression either positive or negative. The sign of the expression can change from positive to negative or from negative to positive only at a number that makes it 0.

Therefore, if one number in an interval satisfies the inequality, then all numbers in that interval will satisfy the inequality.

TEACHING TIP

In **Example 2,** point out that the solutions to the quadratic *equation* indicate the values that make the left side equal to 0. The solutions to the quadratic *inequality* give values of the variable that cause the left side to become positive.

To see if the numbers in Interval A satisfy the inequality, choose any number from Interval A in **FIGURE 20** (that is, any number less than −3). We choose −5. Substitute this test value for x in the original inequality $x^2 - x - 12 > 0$.

$$x^2 - x - 12 > 0 \quad \text{Original inequality}$$

$$(-5)^2 - (-5) - 12 \stackrel{?}{>} 0 \quad \text{Let } x = -5.$$

Use parentheses to avoid sign errors.

$$25 + 5 - 12 \stackrel{?}{>} 0 \quad \text{Simplify.}$$

$$18 > 0 \checkmark \quad \text{True}$$

Because −5 satisfies the inequality, *all* numbers from Interval A are solutions.

Now try 0 from Interval B.

$$x^2 - x - 12 > 0 \quad \text{Original inequality}$$

$$0^2 - 0 - 12 \stackrel{?}{>} 0 \quad \text{Let } x = 0.$$

$$-12 > 0 \quad \text{False}$$

The numbers in Interval B are *not* solutions.

Now try 5 from Interval C.

$$x^2 - x - 12 > 0 \quad \text{Original inequality}$$

$$5^2 - 5 - 12 \stackrel{?}{>} 0 \quad \text{Let } x = 5.$$

$$8 > 0 \quad \text{True}$$

All numbers from Interval C are solutions.

NOW TRY EXERCISE 2

Solve and graph the solution set.

$$x^2 + 2x - 8 > 0$$

TEACHING TIP

Warn students to pay attention to the symbols at the endpoints (brackets or parentheses), carefully graph the solution set (including the arrows), and properly write the solution set in interval notation.

Based on these results (shown by the colored letters in **FIGURE 20**), the solution set includes all numbers in Intervals A and C, as shown in **FIGURE 21**. The solution set is written in interval notation as

$$(-\infty, -3) \cup (4, \infty).$$

−3 and 4 are *not* included because the symbol $>$ does not include equality.

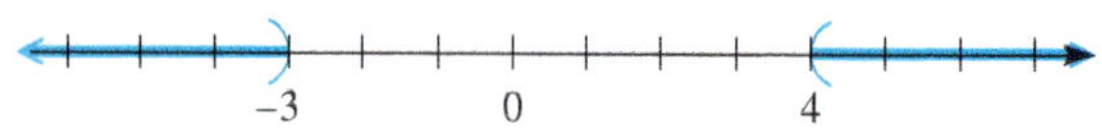

FIGURE 21

This agrees with the solution set found in **Example 1(a)**. NOW TRY

Solving a Quadratic Inequality

Step 1 **Write the inequality as an equation and solve it.**

Step 2 **Use the solutions from Step 1 to determine intervals.** Graph the values found in Step 1 on a number line. These values divide the number line into intervals.

Step 3 **Find the intervals that satisfy the inequality.** Substitute a test value from each interval into the original inequality to determine the intervals that satisfy the inequality. All numbers in those intervals are in the solution set. A graph of the solution set will usually look like one of these.

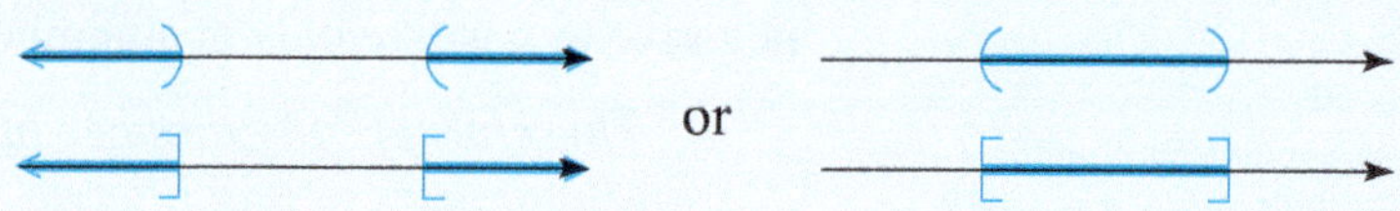

Step 4 **Consider the endpoints separately.** The values from Step 1 are included in the solution set if the inequality symbol is $\leq$ or $\geq$. They are not included if it is $<$ or $>$.

CLASSROOM EXAMPLE 3

Solve and graph the solution set.

$$2x^2 - 9x \leq 18$$

Answer: $\left[-\frac{3}{2}, 6\right]$

EXAMPLE 3 Solving a Quadratic Inequality

Step 1 Solve and graph the solution set of $2x^2 + 5x \leq 12$.

$2x^2 + 5x = 12$	Related quadratic equation
$2x^2 + 5x - 12 = 0$	Standard form
$(2x - 3)(x + 4) = 0$	Factor.
$2x - 3 = 0$ or $x + 4 = 0$	Zero-factor property
$x = \frac{3}{2}$ or $x = -4$	Solve each equation.

Step 2 The numbers $\frac{3}{2}$ and -4 divide a number line into three intervals. See **FIGURE 22.**

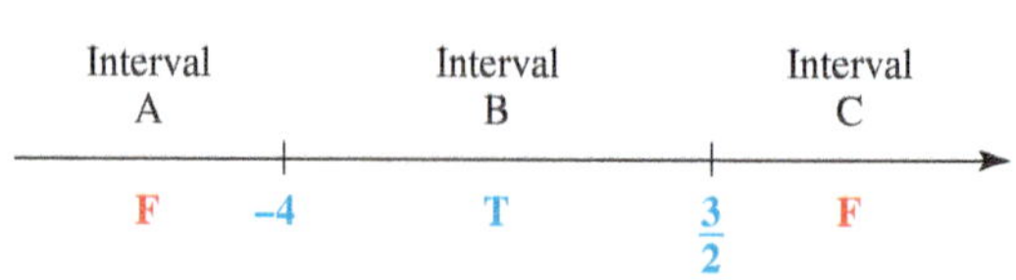

FIGURE 22

NOW TRY ANSWER

2. $(-\infty, -4) \cup (2, \infty)$

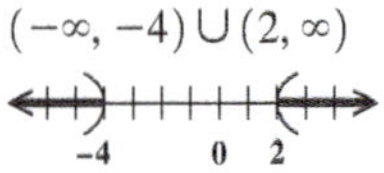

NOW TRY EXERCISE 3
Solve and graph the solution set.

$$3x^2 - 11x \leq 4$$

Steps 3 and 4 Substitute a test value from each interval in the *original* inequality $2x^2 + 5x \leq 12$ to determine which intervals satisfy the inequality.

Interval	*Test Value*	*Test of Inequality*	*True or False?*
A	-5	$25 \leq 12$	F
B	0	$0 \leq 12$	T
C	2	$18 \leq 12$	F

We use a table to organize this information. (Verify it.)

The numbers in Interval B are solutions. See **FIGURE 23**. The solution set is the interval

$$\left[-4, \frac{3}{2}\right].$$

-4 and $\frac{3}{2}$ are included because the symbol $\leq$ includes equality.

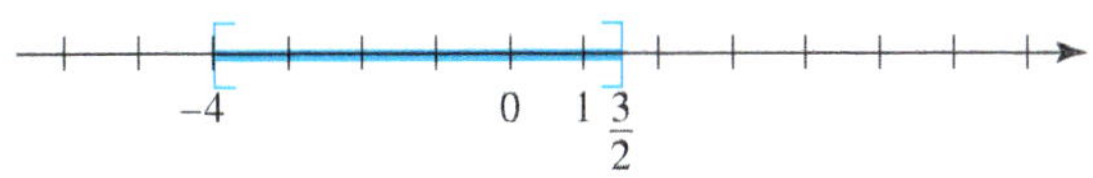

FIGURE 23

NOW TRY

NOW TRY EXERCISE 4
Solve each inequality.
(a) $(4x - 1)^2 > -3$
(b) $(4x - 1)^2 < -3$

CLASSROOM EXAMPLE 4
Solve each inequality.
(a) $(3x - 2)^2 > -2$
(b) $(3x - 2)^2 < -2$
Answers: (a) $(-\infty, \infty)$ (b) $\emptyset$

EXAMPLE 4 Solving Special Cases

Solve each inequality.

(a) $(2x - 3)^2 > -1$

Because $(2x - 3)^2$ is never negative, it is always greater than -1. Thus, the solution set for $(2x - 3)^2 > -1$ is the set of all real numbers, $(-\infty, \infty)$.

(b) $(2x - 3)^2 < -1$

Using similar reasoning as in part (a), there is no solution for this inequality. The solution set is $\emptyset$.

NOW TRY

OBJECTIVE 2 Solve polynomial inequalities of degree 3 or greater.

CLASSROOM EXAMPLE 5
Solve and graph the solution set.

$$(2x + 1)(3x - 1)(x + 4) > 0$$

Answer: $\left(-4, -\frac{1}{2}\right) \cup \left(\frac{1}{3}, \infty\right)$

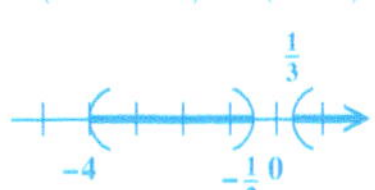

EXAMPLE 5 Solving a Third-Degree Polynomial Inequality

Solve and graph the solution set of $(x - 1)(x + 2)(x - 4) \leq 0$.

This is a *cubic* (third-degree) inequality rather than a quadratic inequality, but it can be solved using the preceding method by extending the zero-factor property to more than two factors. (Step 1)

$$(x - 1)(x + 2)(x - 4) = 0 \qquad \text{Set the factored polynomial } \textit{equal} \text{ to 0.}$$

$$x - 1 = 0 \quad \text{or} \quad x + 2 = 0 \quad \text{or} \quad x - 4 = 0 \qquad \text{Zero-factor property}$$

$$x = 1 \quad \text{or} \quad x = -2 \quad \text{or} \quad x = 4 \qquad \text{Solve each equation.}$$

Locate the numbers -2, 1, and 4 on a number line, as in **FIGURE 24**, to determine the Intervals A, B, C, and D. (Step 2)

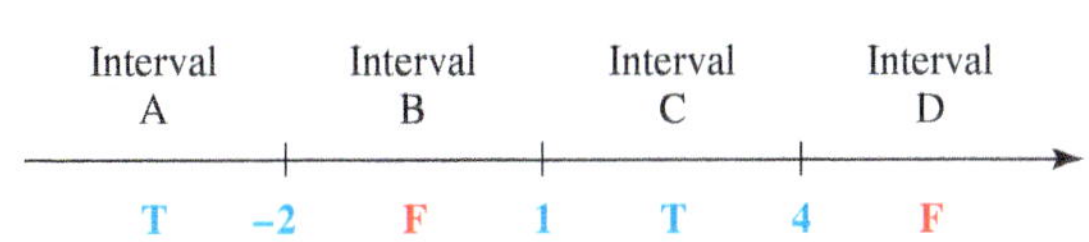

FIGURE 24

NOW TRY ANSWERS

3. $\left[-\frac{1}{3}, 4\right]$

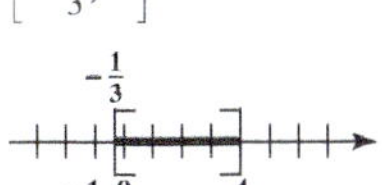

4. **(a)** $(-\infty, \infty)$ **(b)** $\emptyset$

NOW TRY EXERCISE 5
Solve and graph the solution set.

$(x + 4)(x - 3)(2x + 1) \leq 0$

Substitute a test value from each interval in the *original* inequality to determine which intervals satisfy $(x - 1)(x + 2)(x - 4) \leq 0$. (Step 3)

Interval	*Test Value*	*Test of Inequality*	*True or False?*
A	-3	$-28 \leq 0$	T
B	0	$8 \leq 0$	F
C	2	$-8 \leq 0$	T
D	5	$28 \leq 0$	F

The numbers in Intervals A and C are in the solution set, which is written as the interval $(-\infty, -2] \cup [1, 4]$, and graphed in **FIGURE 25**. The three endpoints are included because the inequality symbol $\leq$ includes equality. (Step 4)

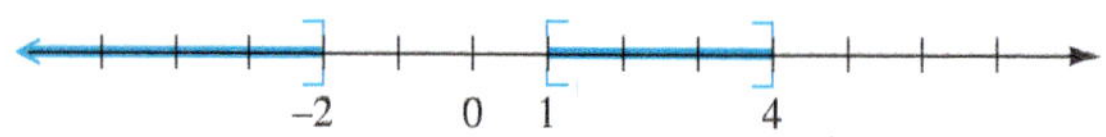

FIGURE 25

NOW TRY

OBJECTIVE 3 Solve rational inequalities.

Rational inequalities involve rational expressions and are solved similarly.

Solving a Rational Inequality

Step 1 **Write the inequality so that 0 is on one side** and there is a single fraction on the other side.

Step 2 **Determine the values that make the numerator or denominator equal to 0.**

Step 3 **Divide a number line into intervals.** Use the values from Step 2.

Step 4 **Find the intervals that satisfy the inequality.** Test a value from each interval by substituting it into the *original* inequality.

Step 5 **Consider the endpoints separately.** Exclude any values that make the denominator 0.

CLASSROOM EXAMPLE 6
Solve and graph the solution set.

$$\frac{2}{x-4} < 3$$

Answer: $(-\infty, 4) \cup \left(\frac{14}{3}, \infty\right)$

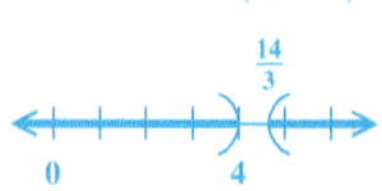

EXAMPLE 6 Solving a Rational Inequality

Solve and graph the solution set of $\frac{-1}{x-3} > 1$.

Write the inequality so that 0 is on one side. (Step 1)

$$\frac{-1}{x-3} - 1 > 0 \quad \text{Subtract 1.}$$

$$\frac{-1}{x-3} - \frac{x-3}{x-3} > 0 \quad \text{Use } x - 3 \text{ as the common denominator.}$$

Be careful with signs.

$$\frac{-1 - x + 3}{x-3} > 0 \quad \text{Write the left side as a single fraction.}$$

$$\frac{-x+2}{x-3} > 0 \quad \text{Combine like terms in the numerator.}$$

NOW TRY ANSWER

5. $(-\infty, -4] \cup \left[-\frac{1}{2}, 3\right]$

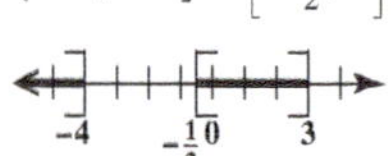

NOW TRY EXERCISE 6

Solve and graph the solution set.

$$\frac{3}{x+1} > 4$$

The sign of $\frac{-x+2}{x-3}$ will change from positive to negative or negative to positive only at those values that make the numerator or denominator 0. The number 2 makes the numerator 0, and 3 makes the denominator 0. (Step 2) These two numbers, 2 and 3, divide a number line into three intervals. See **FIGURE 26**. (Step 3)

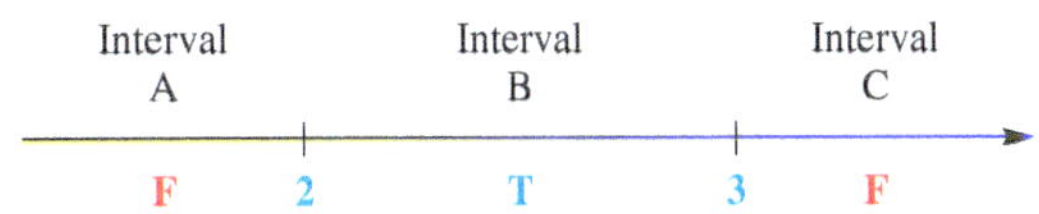

FIGURE 26

Substituting a test value from each interval in the *original* inequality, $\frac{-1}{x-3} > 1$, gives the results shown in the table. (Step 4)

Interval	*Test Value*	*Test of Inequality*	*True or False?*
A	0	$\frac{1}{3} > 1$	F
B	2.5	$2 > 1$	T
C	4	$-1 > 1$	F

The numbers in Interval B are solutions, so the solution set is the interval $(2, 3)$. This interval does not include 3 because it makes the denominator in the original inequality 0. The number 2 is not included either because the inequality symbol $>$ does not include equality. (Step 5) See **FIGURE 27**.

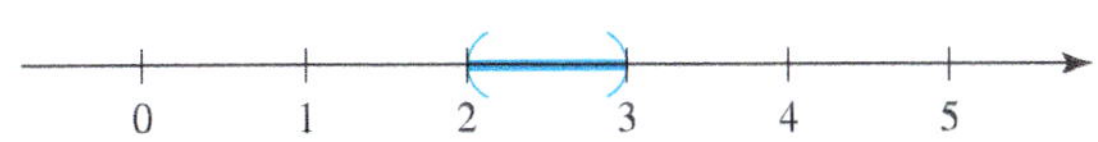

FIGURE 27

NOW TRY

CAUTION ***When solving a rational inequality, any number that makes the denominator 0 must be excluded from the solution set.***

CLASSROOM EXAMPLE 7

Solve and graph the solution set.

$$\frac{x+2}{x-1} \leq 5$$

Answer: $(-\infty, 1) \cup \left[\frac{7}{4}, \infty\right)$

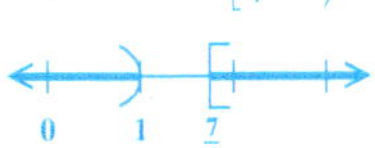

EXAMPLE 7 Solving a Rational Inequality

Solve and graph the solution set of $\frac{x-2}{x+2} \leq 2$.

Write the inequality so that 0 is on one side. (Step 1)

$$\frac{x-2}{x+2} - 2 \leq 0 \quad \text{Subtract 2.}$$

$$\frac{x-2}{x+2} - \frac{2(x+2)}{x+2} \leq 0 \quad \text{Use } x+2 \text{ as the common denominator.}$$

Be careful with signs.

$$\frac{x-2-2x-4}{x+2} \leq 0 \quad \text{Write as a single fraction.}$$

$$\frac{-x-6}{x+2} \leq 0 \quad \text{Combine like terms in the numerator.}$$

NOW TRY ANSWER

6. $\left(-1, -\frac{1}{4}\right)$

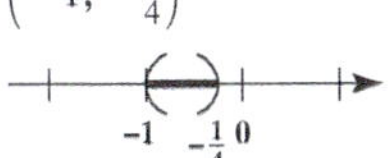

NOW TRY EXERCISE 7
Solve and graph the solution set.

$$\frac{x-3}{x+3} \le 2$$

Because $\frac{-x-6}{x+2} \le 0$, the number -6 makes the numerator 0, and -2 makes the denominator 0. (Step 2) These two numbers determine three intervals on a number line. See **FIGURE 28.** (Step 3)

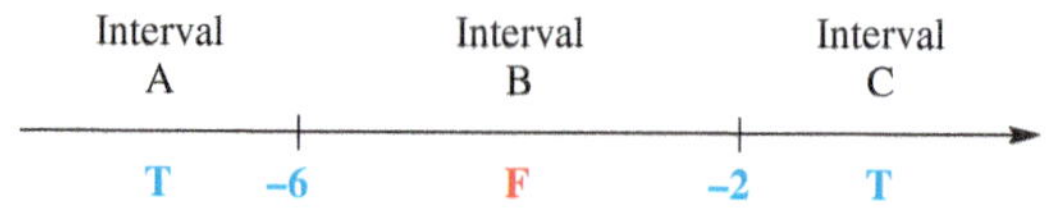

FIGURE 28

Substitute a test value from each interval in the *original* inequality $\frac{x-2}{x+2} \le 2$. (Step 4)

Interval	*Test Value*	*Test of Inequality*	*True or False?*
A	-8	$\frac{5}{3} \le 2$	T
B	-4	$3 \le 2$	F
C	0	$-1 \le 2$	T

The numbers in Intervals A and C are solutions. The solution set is the interval

$$(-\infty, -6] \cup (-2, \infty).$$

The number -6 satisfies the original inequality, but -2 does not because it makes the denominator 0. (Step 5) See **FIGURE 29.**

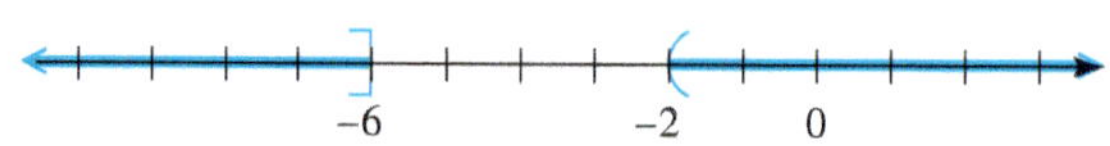

FIGURE 29

NOW TRY

NOW TRY ANSWER

7. $(-\infty, -9] \cup (-3, \infty)$

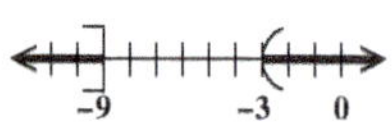

8.7 Exercises

FOR EXTRA HELP

MyMathLab®

Complete solution available in MyMathLab

1. *Concept Check* The solution set of the inequality $x^2 + x - 12 < 0$ is the interval $(-4, 3)$. Without actually performing any work, give the solution set of the inequality

$$x^2 + x - 12 \ge 0.$$

2. Explain how to determine whether to include or exclude endpoints when solving a quadratic or higher-degree inequality.

In each exercise, the graph of a quadratic function f is given. Use the graph to find the solution set of each equation or inequality. ***See Example 1.***

3. **(a)** $x^2 - 4x + 3 = 0$
(b) $x^2 - 4x + 3 > 0$
(c) $x^2 - 4x + 3 < 0$

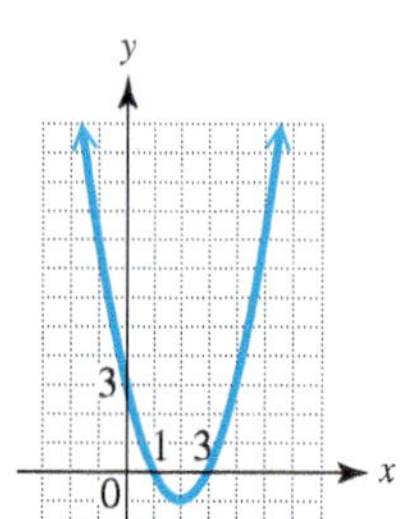

$f(x) = x^2 - 4x + 3$

4. **(a)** $3x^2 + 10x - 8 = 0$
(b) $3x^2 + 10x - 8 \ge 0$
(c) $3x^2 + 10x - 8 < 0$

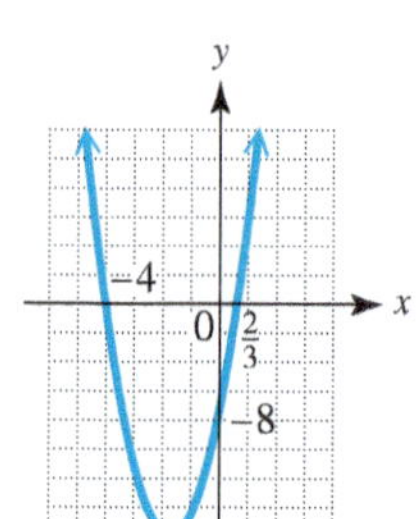

$f(x) = 3x^2 + 10x - 8$

Answers

1. $(-\infty, -4] \cup [3, \infty)$
2. Include the endpoints if the symbol is $\ge$ or $\le$. Exclude the endpoints if the symbol is $>$ or $<$.
3. **(a)** $\{1, 3\}$ **(b)** $(-\infty, 1) \cup (3, \infty)$ **(c)** $(1, 3)$
4. **(a)** $\left\{-4, \frac{2}{3}\right\}$ **(b)** $(-\infty, -4] \cup \left[\frac{2}{3}, \infty\right)$ **(c)** $\left(-4, \frac{2}{3}\right)$
5. **(a)** $\{-2, 5\}$ **(b)** $[-2, 5]$ **(c)** $(-\infty, -2] \cup [5, \infty)$
6. **(a)** $\left\{-3, \frac{5}{2}\right\}$ **(b)** $\left[-3, \frac{5}{2}\right]$ **(c)** $(-\infty, -3] \cup \left[\frac{5}{2}, \infty\right)$

5. **(a)** $-x^2 + 3x + 10 = 0$
(b) $-x^2 + 3x + 10 \geq 0$
(c) $-x^2 + 3x + 10 \leq 0$

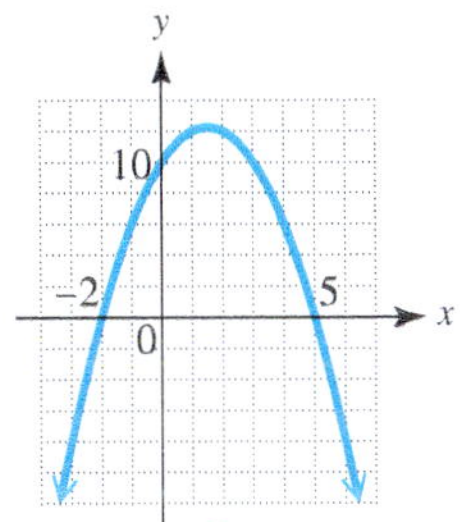

$f(x) = -x^2 + 3x + 10$

6. **(a)** $-2x^2 - x + 15 = 0$
(b) $-2x^2 - x + 15 \geq 0$
(c) $-2x^2 - x + 15 \leq 0$

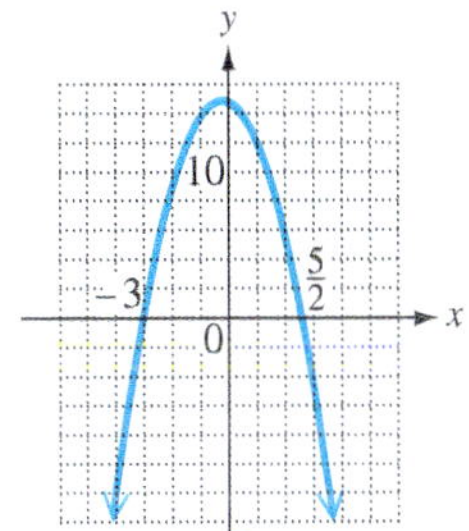

$f(x) = -2x^2 - x + 15$

Solve each inequality, and graph the solution set. ***See Examples 2 and 3.*** *(Hint: In Exercises 23 and 24, use the quadratic formula.)*

7. $(x+1)(x-5) > 0$ **8.** $(x+6)(x-2) > 0$ **9.** $(x+4)(x-6) < 0$

10. $(x+4)(x-8) < 0$ **11.** $x^2 - 4x + 3 \geq 0$ **12.** $x^2 - 3x - 10 \geq 0$

13. $10x^2 + 9x \geq 9$ **14.** $3x^2 + 10x \geq 8$ **15.** $4x^2 - 9 \leq 0$

16. $9x^2 - 25 \leq 0$ **17.** $6x^2 + x \geq 1$ **18.** $4x^2 + 7x \geq -3$

19. $z^2 - 4z \geq 0$ **20.** $x^2 + 2x < 0$ **21.** $3x^2 - 5x \leq 0$

22. $2z^2 + 3z > 0$ **23.** $x^2 - 6x + 6 \geq 0$ **24.** $3x^2 - 6x + 2 \leq 0$

Solve each inequality. ***See Example 4.***

25. $(4-3x)^2 \geq -2$ **26.** $(7-6x)^2 \geq -1$ **27.** $(3x+5)^2 \leq -4$

28. $(8x+5)^2 \leq -5$ **29.** $(2x+5)^2 < 0$ **30.** $(3x-7)^2 < 0$

31. $(5x-1)^2 \geq 0$ **32.** $(4x+1)^2 \geq 0$

Solve each inequality, and graph the solution set. ***See Example 5.***

33. $(x-1)(x-2)(x-4) < 0$ **34.** $(2x+1)(3x-2)(4x+7) < 0$

35. $(x-4)(2x+3)(3x-1) \geq 0$ **36.** $(x+2)(4x-3)(2x+7) \geq 0$

Solve each inequality, and graph the solution set. ***See Examples 6 and 7.***

37. $\frac{x-1}{x-4} > 0$ **38.** $\frac{x+1}{x-5} > 0$ **39.** $\frac{2x+3}{x-5} \leq 0$ **40.** $\frac{3x+7}{x-3} \leq 0$

41. $\frac{8}{x-2} \geq 2$ **42.** $\frac{20}{x-1} \geq 1$ **43.** $\frac{3}{2x-1} < 2$ **44.** $\frac{6}{x-1} < 1$

45. $\frac{x-3}{x+2} \geq 2$ **46.** $\frac{m+4}{m+5} \geq 2$ **47.** $\frac{x-8}{x-4} < 3$ **48.** $\frac{2t-3}{t+1} > 4$

49. $\frac{4k}{2k-1} < k$ **50.** $\frac{r}{r+2} < 2r$ **51.** $\frac{2x-3}{x^2+1} \geq 0$

52. $\frac{9x-8}{4x^2+25} < 0$ **53.** $\frac{(3x-5)^2}{x+2} > 0$ **54.** $\frac{(5x-3)^2}{2x+1} \leq 0$

We do not give answer graphs for Exercises 7–54.

7. $(-\infty, -1) \cup (5, \infty)$
8. $(-\infty, -6) \cup (2, \infty)$
9. $(-4, 6)$ **10.** $(-4, 8)$
11. $(-\infty, 1] \cup [3, \infty)$
12. $(-\infty, -2] \cup [5, \infty)$
13. $\left(-\infty, -\frac{3}{2}\right] \cup \left[\frac{3}{5}, \infty\right)$
14. $(-\infty, -4] \cup \left[\frac{2}{3}, \infty\right)$
15. $\left[-\frac{3}{2}, \frac{3}{2}\right]$ **16.** $\left[-\frac{5}{3}, \frac{5}{3}\right]$
17. $\left(-\infty, -\frac{1}{2}\right] \cup \left[\frac{1}{3}, \infty\right)$
18. $(-\infty, -1] \cup \left[-\frac{3}{4}, \infty\right)$
19. $(-\infty, 0] \cup [4, \infty)$
20. $(-2, 0)$
21. $\left[0, \frac{5}{3}\right]$
22. $\left(-\infty, -\frac{3}{2}\right) \cup (0, \infty)$
23. $(-\infty, 3-\sqrt{3}] \cup [3+\sqrt{3}, \infty)$
24. $\left[\frac{3-\sqrt{3}}{3}, \frac{3+\sqrt{3}}{3}\right]$
25. $(-\infty, \infty)$ **26.** $(-\infty, \infty)$
27. $\emptyset$ **28.** $\emptyset$
29. $\emptyset$ **30.** $\emptyset$
31. $(-\infty, \infty)$ **32.** $(-\infty, \infty)$
33. $(-\infty, 1) \cup (2, 4)$
34. $\left(-\infty, -\frac{7}{4}\right) \cup \left(-\frac{1}{2}, \frac{2}{3}\right)$
35. $\left[-\frac{3}{2}, \frac{1}{3}\right] \cup [4, \infty)$
36. $\left[-\frac{7}{2}, -2\right] \cup \left[\frac{3}{4}, \infty\right)$
37. $(-\infty, 1) \cup (4, \infty)$
38. $(-\infty, -1) \cup (5, \infty)$
39. $\left[-\frac{3}{2}, 5\right)$ **40.** $\left[-\frac{7}{3}, 3\right)$
41. $(2, 6]$ **42.** $(1, 21]$
43. $\left(-\infty, \frac{1}{2}\right) \cup \left(\frac{5}{4}, \infty\right)$
44. $(-\infty, 1) \cup (7, \infty)$
45. $[-7, -2)$ **46.** $[-6, -5)$
47. $(-\infty, 2) \cup (4, \infty)$
48. $\left(-\frac{7}{2}, -1\right)$
49. $\left(0, \frac{1}{2}\right) \cup \left(\frac{5}{2}, \infty\right)$ **50.** $\left(-2, -\frac{3}{2}\right) \cup (0, \infty)$ **51.** $\left[\frac{3}{2}, \infty\right)$ **52.** $\left(-\infty, \frac{8}{9}\right)$
53. $\left(-2, \frac{5}{3}\right) \cup \left(\frac{5}{3}, \infty\right)$ **54.** $\left(-\infty, -\frac{1}{2}\right) \cup \left\{\frac{3}{5}\right\}$

Chapter 8 Summary

Key Terms

8.1
quadratic equation
second-degree equation

8.2
quadratic formula
discriminant

8.3
quadratic in form

8.5
parabola
vertex
axis of symmetry (axis)
quadratic function

8.7
quadratic inequality
rational inequality

Test Your Word Power

See how well you have learned the vocabulary in this chapter.

1. The **quadratic formula** is
A. a formula to find the number of solutions of a quadratic equation
B. a formula to find the type of solutions of a quadratic equation
C. the standard form of a quadratic equation
D. a general formula for solving any quadratic equation.

2. A **quadratic function** is a function that can be written in the form
A. $f(x) = mx + b$, for real numbers m and b
B. $f(x) = \frac{P(x)}{Q(x)}$, where $Q(x) \neq 0$
C. $f(x) = ax^2 + bx + c$, for real numbers a, b, and c $(a \neq 0)$
D. $f(x) = \sqrt{x}$, for $x \geq 0$.

3. A **parabola** is the graph of
A. any equation in two variables
B. a linear equation
C. an equation of degree 3
D. a quadratic equation in two variables.

4. The **vertex** of a parabola is
A. the point where the graph intersects the y-axis
B. the point where the graph intersects the x-axis
C. the lowest point on a parabola that opens up or the highest point on a parabola that opens down
D. the origin.

5. The **axis** of a parabola is
A. either the x-axis or the y-axis
B. the vertical line (of a vertical parabola) or the horizontal line (of a horizontal parabola) through the vertex
C. the lowest or highest point on the graph of a parabola
D. a line through the origin.

6. A parabola is **symmetric about its axis** because
A. its graph is near the axis
B. its graph is identical on each side of the axis
C. its graph looks different on each side of the axis
D. its graph intersects the axis.

ANSWERS

1. D; *Example:* The solutions of $ax^2 + bx + c = 0$ $(a \neq 0)$ are given by $x = \frac{-b \pm \sqrt{b^2 - 4ac}}{2a}$. **2.** C; *Examples:* $f(x) = x^2 - 2$, $f(x) = (x + 4)^2 + 1$, $f(x) = x^2 - 4x + 5$ **3.** D; *Examples:* See the figures in the Quick Review for **Sections 8.5 and 8.6.** **4.** C; *Example:* The graph of $y = (x + 3)^2$ has vertex $(-3, 0)$, which is the lowest point on the graph. **5.** B; *Example:* The axis of $y = (x + 3)^2$ is the vertical line $x = -3$. **6.** B; *Example:* Because the graph of $y = (x + 3)^2$ is symmetric about its axis $x = -3$, the points $(-2, 1)$ and $(-4, 1)$ are on the graph.

Quick Review

CONCEPTS	EXAMPLES
8.1 The Square Root Property and Completing the Square **Square Root Property** If x and k are complex numbers and $x^2 = k$, then $\mathbf{x = \sqrt{k}}$ or $\mathbf{x = -\sqrt{k}}$.	Solve $(x - 1)^2 = 8$. $x - 1 = \sqrt{8}$ or $x - 1 = -\sqrt{8}$ $x = 1 + 2\sqrt{2}$ or $x = 1 - 2\sqrt{2}$ The solution set is $\{1 + 2\sqrt{2}, 1 - 2\sqrt{2}\}$, or $\{1 \pm 2\sqrt{2}\}$.

CONCEPTS	EXAMPLES

Completing the Square

To solve $ax^2 + bx + c = 0$ (where $a \neq 0$), follow these steps.

Step 1 If $a \neq 1$, divide each side by a.

Step 2 Write the equation with the variable terms on one side and the constant on the other.

Step 3 Complete the square.

- Take half the coefficient of x and square it.
- Add the square to each side.
- Factor the perfect square trinomial, and write it as the square of a binomial. Combine terms on the other side.

Step 4 Use the square root property to solve.

Solve $2x^2 - 4x - 18 = 0$.

$x^2 - 2x - 9 = 0$ Divide by 2.

$x^2 - 2x = 9$ Add 9.

$\left[\frac{1}{2}(-2)\right]^2 = (-1)^2 = 1$

$x^2 - 2x + 1 = 9 + 1$ Add 1.

$(x - 1)^2 = 10$ Factor. Add.

$x - 1 = \sqrt{10}$ or $x - 1 = -\sqrt{10}$ Square root property

$x = 1 + \sqrt{10}$ or $x = 1 - \sqrt{10}$

The solution set is $\{1 + \sqrt{10}, 1 - \sqrt{10}\}$, or $\{1 \pm \sqrt{10}\}$.

8.2 The Quadratic Formula

Quadratic Formula

The solutions of $ax^2 + bx + c = 0$ (where $a \neq 0$) are given by

$$x = \frac{-b \pm \sqrt{b^2 - 4ac}}{2a}.$$

Solve $3x^2 + 5x + 2 = 0$.

$x = \frac{-5 \pm \sqrt{5^2 - 4(3)(2)}}{2(3)}$ $a = 3, b = 5, c = 2$

$x = \frac{-5 \pm 1}{6}$ Simplify.

$x = -\frac{2}{3}$ or $x = -1$ Two solutions, one from + and one from −

The solution set is $\left\{-1, -\frac{2}{3}\right\}$.

The Discriminant

The discriminant $b^2 - 4ac$ of $ax^2 + bx + c = 0$ (where a, b, and c are integers) can be used to determine the number and type of solutions.

Discriminant	*Number and Type of Solutions*
Positive, and the square of an integer	Two rational solutions
Positive, but not the square of an integer	Two irrational solutions
Zero	One rational solution
Negative	Two nonreal complex solutions

For $x^2 + 3x - 10 = 0$, the discriminant is

$b^2 - 4ac$

$= 3^2 - 4(1)(-10)$ $a = 1, b = 3, c = -10$

$= 49.$ Two rational solutions

8.3 Equations Quadratic in Form

A nonquadratic equation that can be written in the form $au^2 + bu + c = 0$, for $a \neq 0$ and an algebraic expression u, is quadratic in form.

Solving an Equation Quadratic in Form (Substitution)

Step 1 Define a temporary variable u.

Step 2 Solve the quadratic equation obtained in Step 1.

Step 3 Replace u with the expression it defined.

Step 4 Solve the resulting equations for the original variable.

Step 5 Check all solutions in the original equation.

Solve $3(x + 5)^2 + 7(x + 5) + 2 = 0$.

$3u^2 + 7u + 2 = 0$ Let $u = x + 5$.

$(3u + 1)(u + 2) = 0$ Factor.

$3u + 1 = 0$ or $u + 2 = 0$ Zero-factor property

$u = -\frac{1}{3}$ or $u = -2$ Solve for u.

$x + 5 = -\frac{1}{3}$ or $x + 5 = -2$ Replace u with $x + 5$.

$x = -\frac{16}{3}$ or $x = -7$ Subtract 5.

Check that the solution set is $\left\{-7, -\frac{16}{3}\right\}$.

CONCEPTS | **EXAMPLES**

8.4 Formulas and Further Applications

Solving a Formula for a Squared Variable

Case 1 **If the variable appears only to the second power:** Isolate the squared variable on one side of the equation, and then use the square root property.

Case 2 **If the variable appears to the first and second powers:** Write the equation in standard form, and then use the quadratic formula.

Solve $A = \frac{2mp}{r^2}$ for r. (***Case 1***)

$r^2A = 2mp$ Multiply by r^2.

$r^2 = \frac{2mp}{A}$ Divide by A.

$r = \pm\sqrt{\frac{2mp}{A}}$ Square root property

$r = \frac{\pm\sqrt{2mpA}}{A}$ Rationalize denominator.

Solve $x^2 + rx = t$ for x. (***Case 2***)

$x^2 + rx - t = 0$ Standard form

$x = \frac{-r \pm \sqrt{r^2 - 4(1)(-t)}}{2(1)}$

$a = 1, b = r, c = -t$

$x = \frac{-r \pm \sqrt{r^2 + 4t}}{2}$

Simplify.

8.5 Graphs of Quadratic Functions

1. The graph of the quadratic function

$$F(x) = a(x - h)^2 + k \quad (\text{where } a \neq 0)$$

is a parabola with vertex (h, k) and the vertical line $x = h$ as axis of symmetry.

2. The graph opens up if $a > 0$ and down if $a < 0$.

3. The graph is wider than the graph of $f(x) = x^2$ if $0 < |a| < 1$ and narrower if $|a| > 1$.

Graph $f(x) = -(x + 3)^2 + 1$.

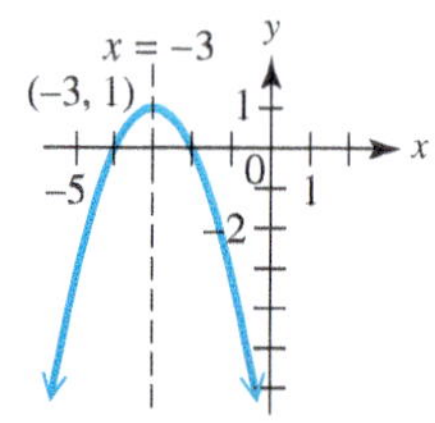

The graph opens down because $a < 0$.

Vertex: $(-3, 1)$

Axis of symmetry: $x = -3$

Domain: $(-\infty, \infty)$

Range: $(-\infty, 1]$

8.6 More about Parabolas and Their Applications

The vertex of the graph of

$$f(x) = ax^2 + bx + c$$

(where $a \neq 0$) may be found by completing the square or using the vertex formula $\left(\frac{-b}{2a}, f\left(\frac{-b}{2a}\right)\right)$.

Graphing a Quadratic Function

Step 1 Determine whether the graph opens up or down.

Step 2 Find the vertex.

Step 3 Find any intercepts.

Step 4 Find and plot additional points as needed.

Horizontal Parabolas

The graph of

$$x = ay^2 + by + c \quad \text{or} \quad x = a(y - k)^2 + h$$

is a horizontal parabola with vertex (h, k) and the horizontal line $y = k$ as axis of symmetry. The graph opens to the right if $a > 0$ and to the left if $a < 0$.

Horizontal parabolas do not represent functions.

Graph $f(x) = x^2 + 4x + 3$.

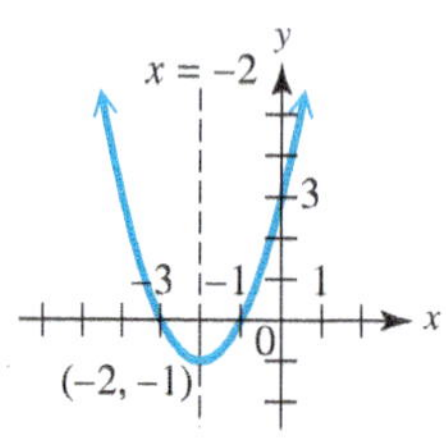

The graph opens up because $a > 0$.

Vertex: $(-2, -1)$

The solutions of $x^2 + 4x + 3 = 0$ are -1 and -3, so the x-intercepts are $(-1, 0)$ and $(-3, 0)$.

$f(0) = 3$, so the y-intercept is $(0, 3)$.

Axis of symmetry: $x = -2$

Domain: $(-\infty, \infty)$

Range: $[-1, \infty)$

Graph $x = 2y^2 + 6y + 5$.

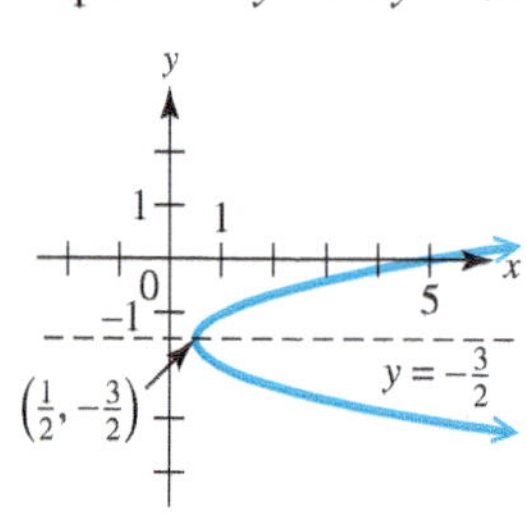

The graph opens to the right because $a > 0$.

Vertex: $\left(\frac{1}{2}, -\frac{3}{2}\right)$

Axis of symmetry: $y = -\frac{3}{2}$

Domain: $\left[\frac{1}{2}, \infty\right)$

Range: $(-\infty, \infty)$

CONCEPTS | **EXAMPLES**

8.7 Polynomial and Rational Inequalities

Solving a Quadratic (or Higher-Degree Polynomial) Inequality

Step 1 Write the inequality as an equation and solve.

Solve $2x^2 + 5x + 2 < 0$.

$$2x^2 + 5x + 2 = 0 \quad \text{Related equation}$$

$$(2x + 1)(x + 2) = 0 \quad \text{Factor.}$$

$$2x + 1 = 0 \quad \text{or} \quad x + 2 = 0 \quad \text{Zero-factor property}$$

$$x = -\frac{1}{2} \quad \text{or} \quad x = -2 \quad \text{Solve each equation.}$$

Step 2 Use the values found in Step 1 to divide a number line into intervals.

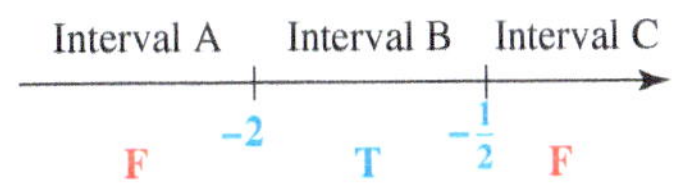

Intervals: $(-\infty, -2)$, $\left(-2, -\frac{1}{2}\right)$, $\left(-\frac{1}{2}, \infty\right)$

Step 3 Substitute a test value from each interval into the *original* inequality to determine the intervals that belong to the solution set.

Test values: -3 (Interval A), -1 (Interval B), 0 (Interval C)

$x = -3$ makes the original inequality false, $x = -1$ makes it true, and $x = 0$ makes it false.

Step 4 Consider the endpoints separately.

The solution set is the interval $\left(-2, -\frac{1}{2}\right)$. The endpoints are not included because the symbol $<$ does not include equality.

Solving a Rational Inequality

Step 1 Write the inequality so that 0 is on one side and there is a single fraction on the other side.

Solve $\frac{x}{x+2} \geq 4$.

$$\frac{x}{x+2} - 4 \geq 0 \quad \text{Subtract 4.}$$

$$\frac{x}{x+2} - \frac{4(x+2)}{x+2} \geq 0 \quad \text{Write with a common denominator.}$$

$$\frac{-3x - 8}{x+2} \geq 0 \quad \text{Subtract fractions.}$$

Step 2 Determine the values that make the numerator or denominator 0.

$-\frac{8}{3}$ makes the numerator 0, and -2 makes the denominator 0.

Step 3 Use the values from Step 2 to divide a number line into intervals.

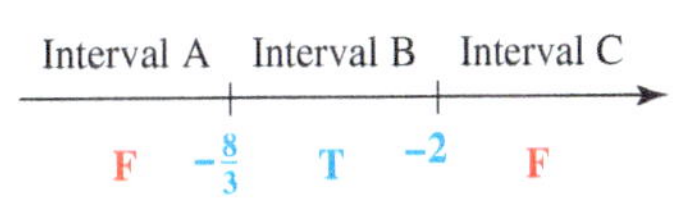

Intervals: $\left(-\infty, -\frac{8}{3}\right)$, $\left(-\frac{8}{3}, -2\right)$, $(-2, \infty)$

Step 4 Substitute a test value from each interval into the *original* inequality to determine the intervals that belong to the solution set.

Test values: -4 from Interval A makes the original inequality false, $-\frac{7}{3}$ from Interval B makes it true, and 0 from Interval C makes it false.

Step 5 Consider the endpoints separately.

The solution set is the interval $\left[-\frac{8}{3}, -2\right)$. The endpoint -2 is not included because it makes the denominator 0.

Chapter 8 Review Exercises

Answers

1. $\{\pm 11\}$
2. $\{\pm\sqrt{3}\}$
3. $\left\{-\frac{15}{2}, \frac{5}{2}\right\}$
4. $\left\{\frac{2}{3} \pm \frac{5}{3}i\right\}$
5. $\{-2 \pm \sqrt{19}\}$
6. $\left\{\frac{1}{2}, 1\right\}$
7. By the square root property, the first step should be $x = \sqrt{12}$ or $x = -\sqrt{12}$. The solution set is $\{\pm 2\sqrt{3}\}$.
8. 5.9 sec
9. $\left\{-\frac{7}{2}, 3\right\}$
10. $\left\{\frac{-5 \pm \sqrt{53}}{2}\right\}$
11. $\left\{\frac{1 \pm \sqrt{41}}{2}\right\}$
12. $\left\{-\frac{3}{4} \pm \frac{\sqrt{23}}{4}i\right\}$
13. $\left\{\frac{2}{3} \pm \frac{\sqrt{2}}{3}i\right\}$
14. $\left\{\frac{-7 \pm \sqrt{37}}{2}\right\}$
15. 17; C
16. 64; A
17. −92; D
18. 0; B
19. $\left\{-\frac{5}{2}, 3\right\}$
20. $\left\{-\frac{1}{2}, 1\right\}$
21. $\{-4\}$
22. $\left\{-\frac{11}{6}, -\frac{19}{12}\right\}$
23. $\left\{-\frac{343}{8}, 64\right\}$
24. $\{\pm 1, \pm 3\}$
25. 7 mph
26. 40 mph
27. 4.6 hr
28. Zoran: 2.6 hr; Claude: 3.6 hr

8.1 *Solve each equation using the square root property or completing the square.*

1. $t^2 = 121$ **2.** $p^2 = 3$ **3.** $(2x + 5)^2 = 100$

***4.** $(3x - 2)^2 = -25$ **5.** $x^2 + 4x = 15$ **6.** $2x^2 - 3x = -1$

7. A student gave the following incorrect "solution." **WHAT WENT WRONG?**

$$x^2 = 12$$
$$x = \sqrt{12} \qquad \text{Square root property}$$
$$x = 2\sqrt{3} \qquad \text{Solution set: } \{2\sqrt{3}\}$$

8. The High Roller in Las Vegas, the world's largest observation wheel as of 2014, has a height of about 168 m. Use the metric version of Galileo's formula,

$$d = 4.9t^2 \quad \text{(where } d \text{ is in meters),}$$

to find how long it would take a wallet dropped from the top of the High Roller to reach the ground. Round the answer to the nearest tenth of a second. (*Source:* www.caesars.com)

8.2 *Solve each equation using the quadratic formula.*

9. $2x^2 + x - 21 = 0$ **10.** $x^2 + 5x = 7$ **11.** $(t + 3)(t - 4) = -2$

***12.** $2x^2 + 3x + 4 = 0$ ***13.** $3p^2 = 2(2p - 1)$ **14.** $x(2x - 7) = 3x^2 + 3$

Find the discriminant and use it to predict whether the solutions to each equation are

A. *two rational numbers* **B.** *one rational number*

C. *two irrational numbers* **D.** *two nonreal complex numbers.*

15. $x^2 + 5x + 2 = 0$ **16.** $4t^2 = 3 - 4t$

17. $4x^2 = 6x - 8$ **18.** $9z^2 + 30z + 25 = 0$

8.3 *Solve each equation. Check the solutions.*

19. $\frac{15}{x} = 2x - 1$ **20.** $\frac{1}{n} + \frac{2}{n+1} = 2$

21. $-2r = \sqrt{\frac{48 - 20r}{2}}$ **22.** $8(3x + 5)^2 + 2(3x + 5) - 1 = 0$

23. $2x^{2/3} - x^{1/3} - 28 = 0$ **24.** $p^4 - 10p^2 + 9 = 0$

Solve each problem. Round answers to the nearest tenth, as necessary.

25. Bahaa paddled a canoe 20 mi upstream, then paddled back. If the rate of the current was 3 mph and the total trip took 7 hr, what was Bahaa's rate?

26. Carol Ann drove 8 mi to pick up a friend, and then drove 11 mi to a mall at a rate 15 mph faster. If Carol Ann's total travel time was 24 min, what was her rate on the trip to pick up her friend?

27. An old machine processes a batch of checks in 1 hr more time than a new one. How long would it take the old machine to process a batch of checks that the two machines together process in 2 hr?

28. Zoran can process a stack of invoices 1 hr faster than Claude can. Working together, they take 1.5 hr. How long would it take each person working alone?

---

*This exercise requires knowledge of complex numbers.

8.4 *Solve each formula for the specified variable. (Give answers with ±.)*

29. $k = \dfrac{rF}{wv^2}$ for v **30.** $p = \sqrt{\dfrac{yz}{6}}$ for y **31.** $mt^2 = 3mt + 6$ for t

Solve each problem. Round answers to the nearest tenth, as necessary.

32. A large machine requires a part in the shape of a right triangle with a hypotenuse 9 ft less than twice the length of the longer leg. The shorter leg must be $\frac{3}{4}$ the length of the longer leg. Find the lengths of the three sides of the part.

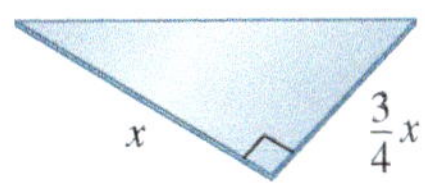

33. A square has an area of 256 cm^2. If the same amount is removed from one dimension and added to the other, the resulting rectangle has an area 16 cm^2 less. Find the dimensions of the rectangle.

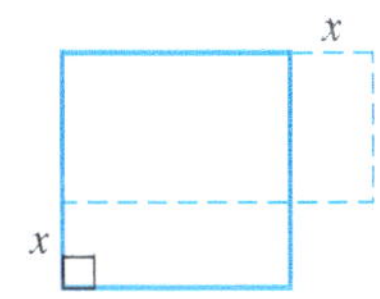

34. Allen wants to buy a mat for a photograph that measures 14 in. by 20 in. He wants to have an even border around the picture when it is mounted on the mat. If the area of the mat he chooses is 352 in.^{2}, how wide will the border be?

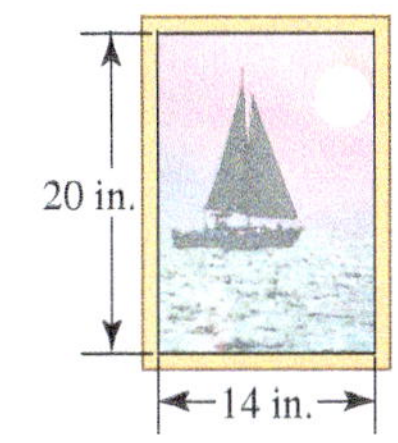

35. If a square piece of cardboard has 3-in. squares cut from its corners and then has the flaps folded up to form an open-top box, the volume of the box is given by the formula

$$V = 3(x - 6)^2,$$

where x is the length of each side of the original piece of cardboard in inches. What original length would yield a box with volume 432 in.^{3}?

36. A searchlight moves horizontally back and forth along a wall. The distance of the light from a starting point at t minutes is given by the quadratic function

$$f(t) = 100t^2 - 300t.$$

How long will it take before the light returns to the starting point?

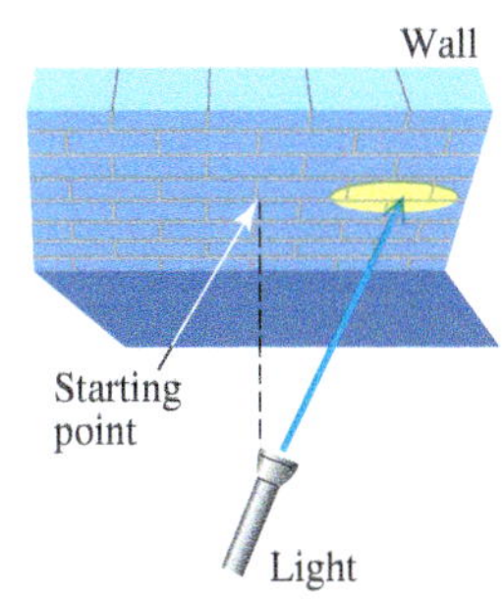

8.5, 8.6 *Identify the vertex of each parabola.*

37. $f(x) = -(x - 1)^2$ **38.** $f(x) = (x - 3)^2 + 7$

39. $x = (y - 3)^2 - 4$ **40.** $y = -3x^2 + 4x - 2$

Graph each parabola. Give the vertex, axis of symmetry, domain, and range.

41. $y = 2(x - 2)^2 - 3$ **42.** $f(x) = -2x^2 + 8x - 5$

43. $x = 2(y + 3)^2 - 4$ **44.** $x = -\frac{1}{2}y^2 + 6y - 14$

Solve each problem.

45. The height (in feet) of a projectile t seconds after being fired from Earth into the air is given by

$$f(t) = -16t^2 + 160t.$$

Find the number of seconds required for the projectile to reach maximum height. What is the maximum height?

Answers

29. $v = \dfrac{\pm\sqrt{rFkw}}{kw}$

30. $y = \dfrac{6p^2}{z}$

31. $t = \dfrac{3m \pm \sqrt{9m^2 + 24m}}{2m}$

32. 9 ft, 12 ft, 15 ft

33. 12 cm by 20 cm

34. 1 in.

35. 18 in.

36. 3 min

37. (1, 0) **38.** (3, 7)

39. (−4, 3) **40.** $\left(\frac{2}{3}, -\frac{2}{3}\right)$

41.

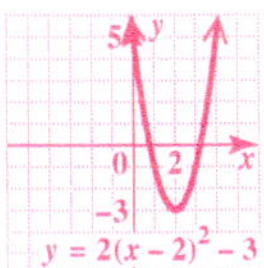

vertex: (2, −3); axis: $x = 2$; domain: $(-\infty, \infty)$; range: $[-3, \infty)$

42.

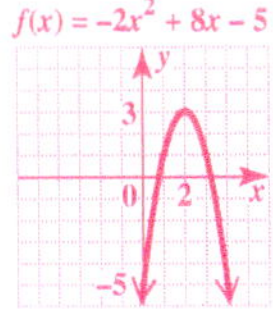

vertex: (2, 3); axis: $x = 2$; domain: $(-\infty, \infty)$; range: $(-\infty, 3]$

43.

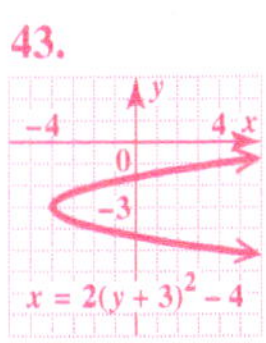

vertex: (−4, −3); axis: $y = -3$; domain: $[-4, \infty)$; range: $(-\infty, \infty)$

44.

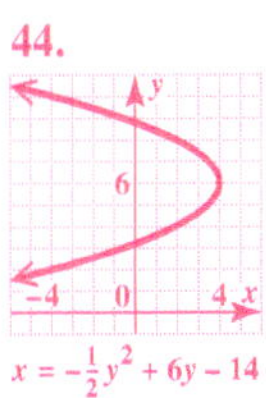

vertex: (4, 6); axis: $y = 6$; domain: $(-\infty, 4]$; range: $(-\infty, \infty)$

45. 5 sec; 400 ft

46. length: 50 m; width: 50 m; maximum area: 2500 m^2

47. $\left(-\infty, -\frac{3}{2}\right) \cup (4, \infty)$

48. [−4, 3]

46. Find the length and width of a rectangle having a perimeter of 200 m if the area is to be a maximum. What is the maximum area?

8.7 *Solve each inequality, and graph the solution set.*

47. $(x-4)(2x+3)>0$

48. $x^2+x\le 12$

49. $(x+2)(x-3)(x+5)\le 0$

50. $(4x+3)^2\le -4$

51. $\dfrac{6}{2z-1}<2$

52. $\dfrac{3t+4}{t-2}\le 1$

49. $(-\infty,-5]\cup[-2,3]$

50. $\varnothing$

51. $\left(-\infty,\frac{1}{2}\right)\cup(2,\infty)$

52. $[-3,2)$

Chapter 8 Mixed Review Exercises

Solve each equation or inequality.

1. $V=r^2+R^2h$ for R

***2.** $3t^2-6t=-4$

3. $(3x+11)^2=7$

4. $S=\dfrac{Id^2}{k}$ for d

5. $(8x-7)^2\ge -1$

6. $2x-\sqrt{x}=6$

7. $x^4-8x^2=-1$

8. $\dfrac{-2}{x+5}\le -5$

9. $6+\dfrac{15}{s^2}=-\dfrac{19}{s}$

10. $(x^2-2x)^2=11(x^2-2x)-24$

11. $(r-1)(2r+3)(r+6)<0$

Work each problem.

12. Match each equation in parts (a)–(f) with the figure that most closely resembles its graph in choices A–F.

(a) $g(x)=x^2-5$ **(b)** $h(x)=-x^2+4$ **(c)** $F(x)=(x-1)^2$

(d) $G(x)=(x+1)^2$ **(e)** $H(x)=(x-1)^2+1$ **(f)** $K(x)=(x+1)^2+1$

A.

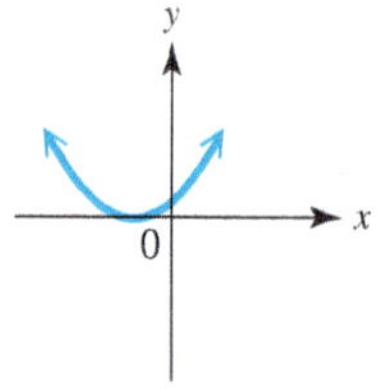

B.

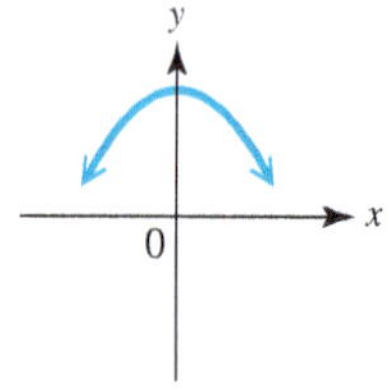

C.

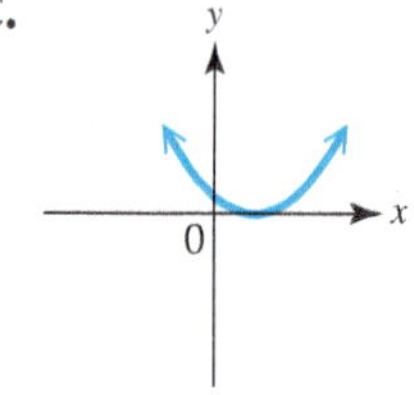

D.

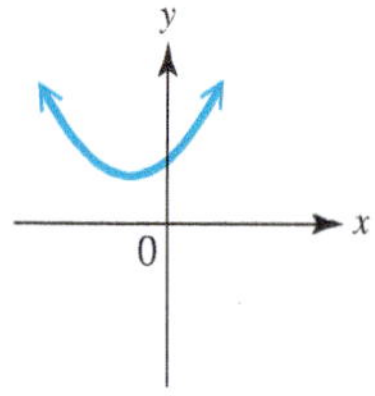

E.

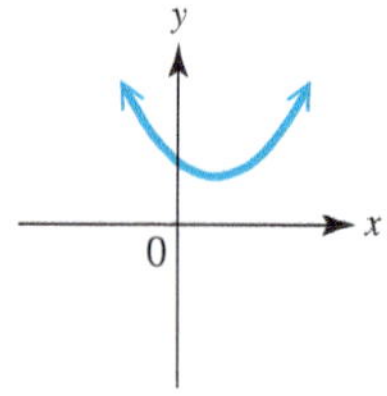

F.

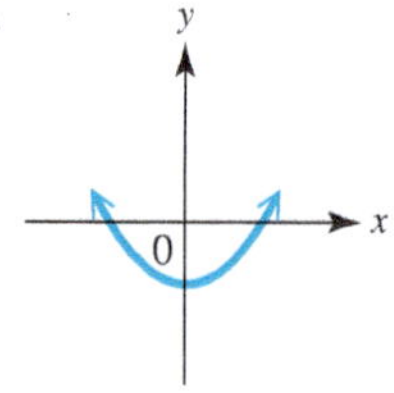

13. Graph $f(x)=4x^2+4x-2$. Give the vertex, axis of symmetry, domain, and range.

14. In 4 hr, Rajeed can travel 15 mi upriver and come back. The rate of the current is 5 mph. Find the rate of the boat in still water.

15. Two pieces of a large wooden puzzle fit together to form a rectangle with length 1 cm less than twice the width. The diagonal, where the two pieces meet, is 2.5 cm in length. Find the length and width of the rectangle.

1. $R=\dfrac{\pm\sqrt{Vh-r^2h}}{h}$

2. $\left\{1\pm\dfrac{\sqrt{3}}{3}i\right\}$

3. $\left\{\dfrac{-11\pm\sqrt{7}}{3}\right\}$

4. $d=\dfrac{\pm\sqrt{SkI}}{I}$

5. $(-\infty,\infty)$

6. $\{4\}$

7. $\{\pm\sqrt{4+\sqrt{15}},\pm\sqrt{4-\sqrt{15}}\}$

8. $\left(-5,-\dfrac{23}{5}\right]$

9. $\left\{-\dfrac{5}{3},-\dfrac{3}{2}\right\}$

10. $\{-2,-1,3,4\}$

11. $(-\infty,-6)\cup\left(-\dfrac{3}{2},1\right)$

12. (a) F (b) B (c) C (d) A (e) E (f) D

13.

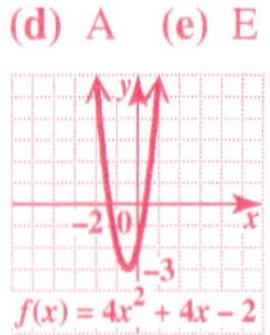

vertex: $\left(-\frac{1}{2},-3\right)$; axis: $x=-\frac{1}{2}$; domain: $(-\infty,\infty)$; range: $[-3,\infty)$

14. 10 mph

15. length: 2 cm; width: 1.5 cm

*This exercise requires knowledge of complex numbers.

Chapter 8 Test

FOR EXTRA HELP *Step-by-step test solutions are found on the Chapter Test Prep Videos available in* MyMathLab® *or on* YouTube.

Solve each equation using the square root property or completing the square.

1. $t^2 = 54$ **2.** $(7x + 3)^2 = 25$ **3.** $x^2 + 2x = 4$

Solve each equation using the quadratic formula.

4. $2x^2 - 3x - 1 = 0$ ***5.** $3t^2 - 4t = -5$ **6.** $3x = \sqrt{\frac{9x + 2}{2}}$

***7.** If k is a negative number, then which one of the following equations will have two nonreal complex solutions?

A. $x^2 = 4k$ **B.** $x^2 = -4k$ **C.** $(x + 2)^2 = -k$ **D.** $x^2 + k = 0$

8. What is the discriminant for $2x^2 - 8x - 3 = 0$? How many and what type of solutions does this equation have? (Do not actually solve.)

Solve each equation by any method.

9. $3 - \frac{16}{x} - \frac{12}{x^2} = 0$ **10.** $4x^2 + 7x - 3 = 0$

11. $9x^4 + 4 = 37x^2$ **12.** $12 = (2n + 1)^2 + (2n + 1)$

13. Solve $S = 4\pi r^2$ for r. (Leave $\pm$ in your answer.)

Solve each problem.

14. Terry and Callie do word processing. For a certain prospectus, Callie can prepare it 2 hr faster than Terry can. If they work together, they can do the entire prospectus in 5 hr. How long will it take each of them working alone to prepare the prospectus? Round answers to the nearest tenth of an hour.

15. Qihong paddled a canoe 10 mi upstream and then paddled back to the starting point. If the rate of the current was 3 mph and the entire trip took $3\frac{1}{2}$ hr, what was Qihong's rate?

16. Endre has a pool 24 ft long and 10 ft wide. He wants to construct a concrete walk around the pool. If he plans for the walk to be of uniform width and cover 152 ft², what will the width of the walk be?

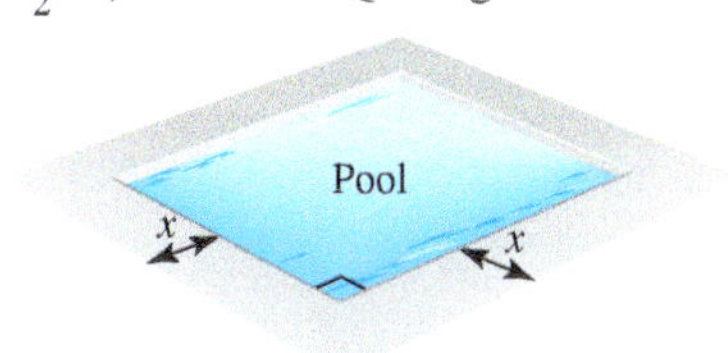

17. At a point 30 m from the base of a tower, the distance to the top of the tower is 2 m more than twice the height of the tower. Find the height of the tower.

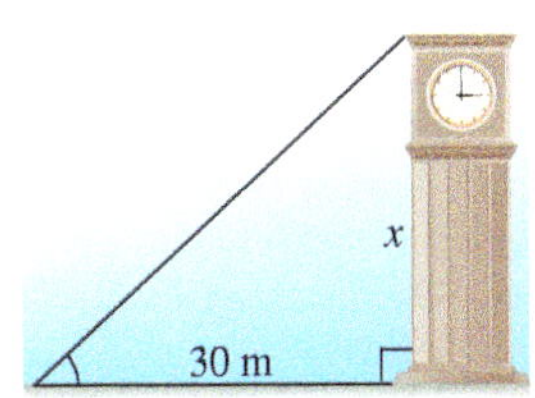

18. Which one of the following figures most closely resembles the graph of $f(x) = a(x - h)^2 + k$ if $a < 0$, $h > 0$, and $k < 0$?

A.
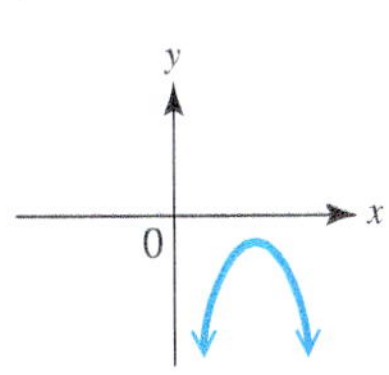

B.
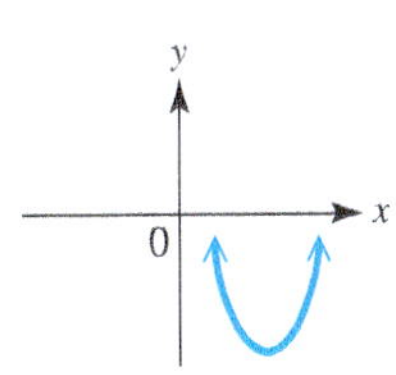

C.
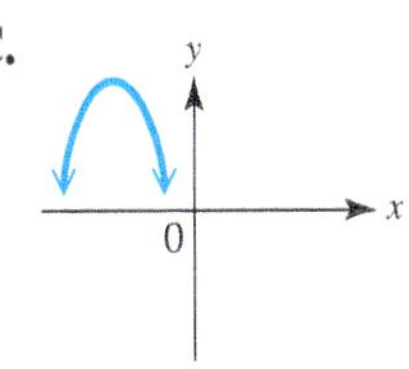

D.
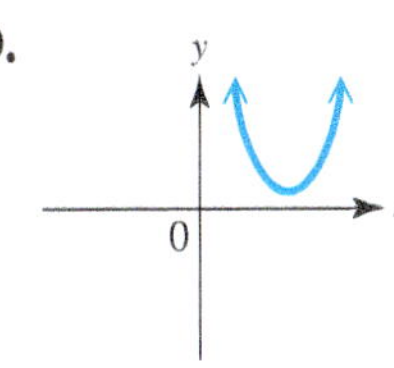

*This exercise requires knowledge of complex numbers.

View the complete solutions to all Chapter Test exercises in MyMathLab.

[8.1]
1. $\{\pm 3\sqrt{6}\}$ 2. $\left\{-\frac{8}{7}, \frac{2}{7}\right\}$
3. $\{-1 \pm \sqrt{5}\}$

[8.2]
4. $\left\{\frac{3 \pm \sqrt{17}}{4}\right\}$
5. $\left\{\frac{2}{3} \pm \frac{\sqrt{11}}{3}i\right\}$

[8.3]
6. $\left\{\frac{2}{3}\right\}$

[8.1]
7. A

[8.2]
8. discriminant: 88; There are two irrational solutions.

[8.1–8.3]
9. $\left\{-\frac{2}{3}, 6\right\}$
10. $\left\{\frac{-7 \pm \sqrt{97}}{8}\right\}$
11. $\left\{\pm\frac{1}{3}, \pm 2\right\}$ 12. $\left\{-\frac{5}{2}, 1\right\}$

[8.4]
13. $r = \frac{\pm\sqrt{\pi S}}{2\pi}$

[8.3]
14. Terry: 11.1 hr; Callie: 9.1 hr
15. 7 mph

[8.4]
16. 2 ft 17. 16 m

[8.5, 8.6]
18. A

19.
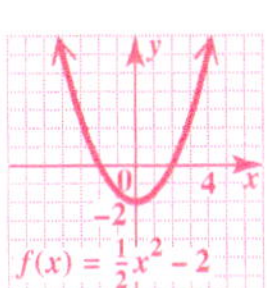

vertex: $(0, -2)$; axis: $x = 0$; domain: $(-\infty, \infty)$; range: $[-2, \infty)$

20. $f(x) = -x^2 + 4x - 1$
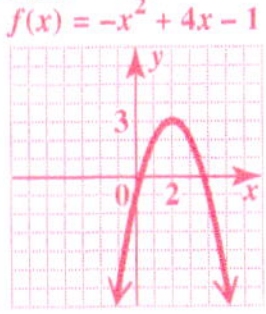

vertex: $(2, 3)$; axis: $x = 2$; domain: $(-\infty, \infty)$; range: $(-\infty, 3]$

21.

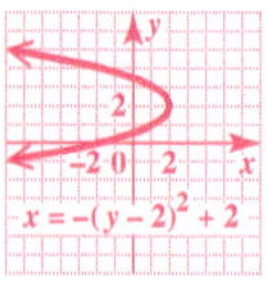

vertex: $(2, 2)$; axis: $y = 2$; domain: $(-\infty, 2]$; range: $(-\infty, \infty)$

22. 160 ft by 320 ft

[8.7]

23. $(-\infty, -5) \cup \left(\frac{3}{2}, \infty\right)$

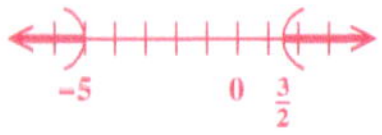

24. $(-\infty, 4) \cup [9, \infty)$

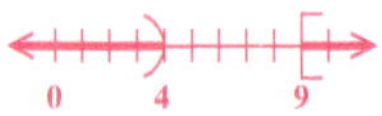

Graph each parabola. Identify the vertex, axis of symmetry, domain, and range.

19. $f(x) = \frac{1}{2}x^2 - 2$ **20.** $f(x) = -x^2 + 4x - 1$ **21.** $x = -(y - 2)^2 + 2$

22. Houston Community College is planning to construct a rectangular parking lot on land bordered on one side by a highway. The plan is to use 640 ft of fencing to fence off the other three sides. What should the dimensions of the lot be if the enclosed area is to be a maximum?

Solve each inequality, and graph the solution set.

23. $2x^2 + 7x > 15$ **24.** $\frac{5}{t - 4} \le 1$

Chapters R–8 Cumulative Review Exercises

[R.1, 7.7]

1. (a) $-2, 0, 7$

(b) $-\frac{7}{3}, -2, 0, 0.7, 7, \frac{32}{3}$

(c) All are real except $\sqrt{-8}$.

(d) All are complex numbers.

[1.1] **2.** $\left\{\frac{4}{5}\right\}$ **[1.7]** **3.** $\left\{\frac{11}{10}, \frac{7}{2}\right\}$

[7.6] **4.** $\left\{\frac{2}{3}\right\}$ **[6.4]** **5.** $\varnothing$

[8.1, 8.2]

6. $\left\{\frac{7 \pm \sqrt{177}}{4}\right\}$

[8.3]

7. $\{\pm 1, \pm 2\}$

[1.5] **8.** $[1, \infty)$ **[1.7]** **9.** $\left[2, \frac{8}{3}\right]$

[8.7]

10. $(1, 3)$ **11.** $(-2, 1)$

[2.1, 2.5]

12.

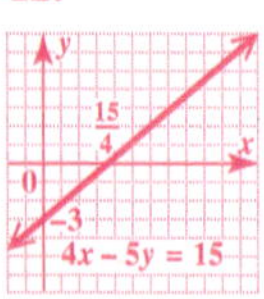

function; domain: $(-\infty, \infty)$; range: $(-\infty, \infty)$; $f(x) = \frac{4}{5}x - 3$

[2.4, 2.5]

13.

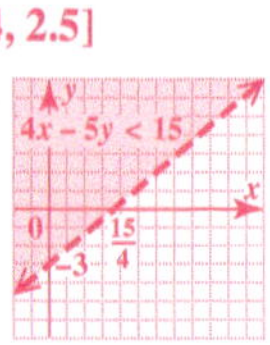

not a function

1. Let $S = \left\{-\frac{7}{3}, -2, -\sqrt{3}, 0, 0.7, \sqrt{12}, \sqrt{-8}, 7, \frac{32}{3}\right\}$. List the elements of S that are elements of each set.

(a) Integers **(b)** Rational numbers **(c)** Real numbers **(d)** Complex numbers

Solve each equation or inequality.

2. $7 - (4 + 3t) + 2t = -6(t - 2) - 5$ **3.** $|6x - 9| = |-4x + 2|$

4. $2x = \sqrt{\frac{5x + 2}{3}}$ **5.** $\frac{3}{x - 3} - \frac{2}{x - 2} = \frac{3}{x^2 - 5x + 6}$

6. $(r - 5)(2r + 3) = 1$ **7.** $x^4 - 5x^2 + 4 = 0$

8. $-2x + 4 \le -x + 3$ **9.** $|3x - 7| \le 1$

10. $x^2 - 4x + 3 < 0$ **11.** $\frac{3}{p + 2} > 1$

Graph each relation. Decide whether or not y can be expressed as a function f of x, and if so, give its domain and range, and write using function notation.

12. $4x - 5y = 15$ **13.** $4x - 5y < 15$ **14.** $y = -2(x - 1)^2 + 3$

15. Find the slope and intercepts of the line with equation $-2x + 7y = 16$.

16. Write an equation for the specified line. Express each equation in slope-intercept form.

(a) Through $(2, -3)$ and parallel to the line with equation $5x + 2y = 6$

(b) Through $(-4, 1)$ and perpendicular to the line with equation $5x + 2y = 6$

Solve each system of equations.

17. $2x - 4y = 10$
$9x + 3y = 3$

18. $x + y + 2z = 3$
$-x + y + z = -5$
$2x + 3y - z = -8$

19. Two top-grossing superhero films are *Marvel's The Avengers* and *Spiderman 2*. The two films together grossed \$1143 million. *Spiderman 2* grossed \$139 million less than *Marvel's The Avengers*. How much did each film gross? (*Source:* Box Office Mojo.)

[8.5]

14. 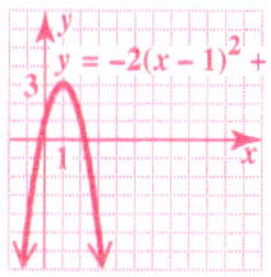

function;
domain: $(-\infty, \infty)$;
range: $(-\infty, 3]$;
$f(x) = -2(x-1)^2 + 3$

[2.2]

15. $m = \frac{2}{7}$; x-intercept: $(-8, 0)$; y-intercept: $\left(0, \frac{16}{7}\right)$

[2.3]

16. (a) $y = -\frac{5}{2}x + 2$

(b) $y = \frac{2}{5}x + \frac{13}{5}$

[3.1] 17. $\{(1, -2)\}$

[3.2] 18. $\{(3, -4, 2)\}$

[3.3]

19. *Marvel's The Avengers:* \$641 million; *Spiderman 2:* \$502 million

[4.1]

20. $\frac{x^8}{y^4}$ 21. $\frac{4}{xy^2}$

[4.4]

22. $\frac{4}{9}t^2 + 12t + 81$

[4.5]

23. $4x^2 - 6x + 11 + \frac{4}{x+2}$

[5.1–5.3]

24. $(4m-3)(6m+5)$
25. $(2x+3y)(4x^2-6xy+9y^2)$
26. $(3x-5y)^2$

[6.1] 27. $-\frac{5}{18}$

[6.2] 28. $-\frac{8}{x}$

[6.3] 29. $\frac{r-s}{r}$

[7.3] 30. $\frac{3\sqrt[3]{4}}{4}$

[7.5]

31. $\sqrt{7} + \sqrt{5}$

[8.4]

32. southbound car: 57 mi; eastbound car: 76 mi

Write with positive exponents only. Assume that variables represent positive real numbers.

20. $\left(\frac{x^{-3}y^2}{x^5y^{-2}}\right)^{-1}$

21. $\frac{(4x^{-2})^2(2y^3)}{8x^{-3}y^5}$

Perform the indicated operations.

22. $\left(\frac{2}{3}t + 9\right)^2$

23. Divide $4x^3 + 2x^2 - x + 26$ by $x + 2$.

Factor completely.

24. $24m^2 + 2m - 15$

25. $8x^3 + 27y^3$

26. $9x^2 - 30xy + 25y^2$

Simplify. Express each answer in lowest terms. Assume denominators are nonzero.

27. $\frac{5x+2}{-6} \div \frac{15x+6}{5}$

28. $\frac{3}{2-x} - \frac{5}{x} + \frac{6}{x^2-2x}$

29. $\frac{\frac{r}{s} - \frac{s}{r}}{\frac{r}{s} + 1}$

Simplify each radical expression.

30. $\sqrt[3]{\frac{27}{16}}$

31. $\frac{2}{\sqrt{7} - \sqrt{5}}$

32. Two cars left an intersection at the same time, one heading due south and the other due east. Later they were exactly 95 mi apart. The car heading east had gone 38 mi less than twice as far as the car heading south. How far had each car traveled?

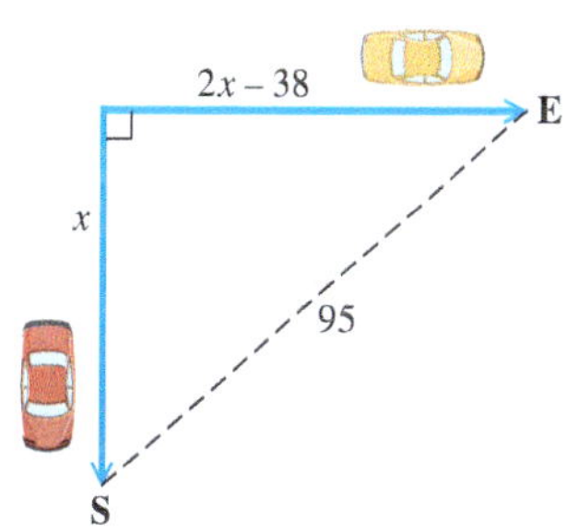

9 Inverse, Exponential, and Logarithmic Functions

The magnitudes of earthquakes, intensities of sounds, and population growth and decay are some examples of applications of *exponential* and *logarithmic functions*.

9.1 Inverse Functions

OBJECTIVES

1. Decide whether a function is one-to-one and, if it is, find its inverse.
2. Use the horizontal line test to determine whether a function is one-to-one.
3. Find the equation of the inverse of a function.
4. Graph f^{-1}, given the graph of f.

VOCABULARY

- ☐ one-to-one function
- ☐ inverse of a function

In this chapter we study two important types of functions, *exponential* and *logarithmic*. These functions are related: They are *inverses* of one another.

OBJECTIVE 1 Decide whether a function is one-to-one and, if it is, find its inverse.

Suppose we define the function

$$G = \{(-2, 2), (-1, 1), (0, 0), (1, 3), (2, 5)\}.$$

We can form another set of ordered pairs from G by interchanging the x- and y-values of each pair in G. We can call this set F, so

$$F = \{(2, -2), (1, -1), (0, 0), (3, 1), (5, 2)\}.$$

To show that these two sets are related as just described, F is called the *inverse* of G. For a function f to have an inverse, f must be a *one-to-one function.*

One-to-One Function

In a **one-to-one function,** each x-value corresponds to only one y-value, and each y-value corresponds to only one x-value.

The function shown in FIGURE 1(a) is not one-to-one because the y-value 7 corresponds to *two* x-values, 2 and 3. That is, the ordered pairs $(2, 7)$ and $(3, 7)$ both belong to the function. The function in FIGURE 1(b) is one-to-one.

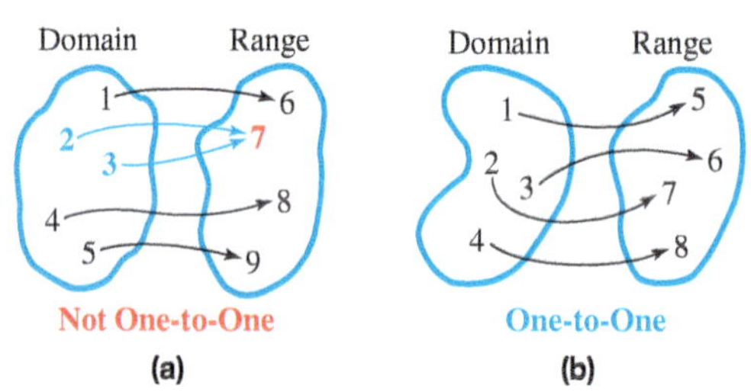

FIGURE 1

The *inverse* of any one-to-one function f is found by interchanging the components of the ordered pairs of f. The inverse of f is written f^{-1}. Read f^{-1} as **"the inverse of f"** or **"f-inverse."**

CAUTION The symbol $f^{-1}(x)$ does ***not*** represent $\frac{1}{f(x)}$.

Inverse of a Function

The **inverse** of a one-to-one function f, written f^{-1}, is the set of all ordered pairs of the form (y, x), where (x, y) belongs to f. ***The inverse is formed by interchanging x and y, so the domain of f becomes the range of f^{-1} and the range of f becomes the domain of f^{-1}.***

For inverses f and f^{-1}, it follows that for all x in their domains,

$$(f \circ f^{-1})(x) = x \quad \text{and} \quad (f^{-1} \circ f)(x) = x.$$

NOW TRY EXERCISE 1

Decide whether each function is one-to-one. If it is, find the inverse.

(a) $F = \{(-1, -2), (0, 0)$
$(1, -2), (2, -8)\}$

(b) $G = \{(0, 0), (1, 1),$
$(4, 2), (9, 3)\}$

(c) A Norwegian physiologist has developed a rule for predicting running times based on the time to run 5 km (5K). An example for one runner is shown here. (*Source:* Stephen Seiler, Agder College, Kristiansand, Norway.)

Distance	*Time*
1.5K	4:22
3K	9:18
5K	16:00
10K	33:40

CLASSROOM EXAMPLE 1

Decide whether each function is one-to-one. If it is, find the inverse.

(a) $F = \{(2, 5), (3, 6), (4, 8), (8, 7)\}$

(b) $G = \{(0, 3), (-1, 2), (1, 3)\}$

Answers: (a) one-to-one;
$F^{-1} = \{(5, 2), (6, 3), (8, 4), (7, 8)\}$
(b) not one-to-one

NOW TRY ANSWERS

1. (a) not one-to-one
(b) one-to-one;
$G^{-1} = \{(0, 0), (1, 1), (2, 4), (3, 9)\}$

(c)

Time	*Distance*
4:22	1.5K
9:18	3K
16:00	5K
33:40	10K

EXAMPLE 1 Finding Inverses of One-to-One Functions

Decide whether each function is one-to-one. If it is, find the inverse.

(a) $F = \{(-2, 1), (-1, 0), (0, 1), (1, 2), (2, 2)\}$

Each x-value in f corresponds to just one y-value. However, the y-value 1 corresponds to two x-values, -2 and 0. Also, the y-value 2 corresponds to both 1 and 2. Because some y-values correspond to more than one x-value, F is not one-to-one and does not have an inverse.

(b) $G = \{(3, 1), (0, 2), (2, 3), (4, 0)\}$

Every x-value in G corresponds to only one y-value, and every y-value corresponds to only one x-value, so G is a one-to-one function. The inverse function is found by interchanging the x- and y-values in each ordered pair.

$$G^{-1} = \{(1, 3), (2, 0), (3, 2), (0, 4)\}$$

The domain and range of G become the range and domain, respectively, of G^{-1}.

(c) The table shows the number of days in which the air in the Los Angeles–Long Beach–Santa Ana metropolitan area failed to meet acceptable air-quality standards for the years 2005–2010.

Year	*Number of Days Exceeding Standards*
2005	29
2006	26
2007	18
2008	31
2009	18
2010	4

Source: U.S. Environmental Protection Agency.

Let f be the function defined in the table, with the years forming the domain and the number of days exceeding the air-quality standards forming the range. Then f is not one-to-one because in two different years (2007 and 2009) the number of days with unacceptable air quality was the same, 18.

NOW TRY

OBJECTIVE 2 Use the horizontal line test to determine whether a function is one-to-one.

By graphing a function and observing the graph, we can use the *horizontal line test* to tell whether the function is one-to-one.

Horizontal Line Test

A function is one-to-one if every horizontal line intersects the graph of the function at most once.

The horizontal line test follows from the definition of a one-to-one function. Any two points that lie on the same horizontal line have the same y-coordinate. No two ordered pairs that belong to a one-to-one function may have the same y-coordinate. Therefore, no horizontal line will intersect the graph of a one-to-one function more than once.

EXAMPLE 2 Using the Horizontal Line Test

Use the horizontal line test to determine whether each graph is the graph of a one-to-one function.

(a)

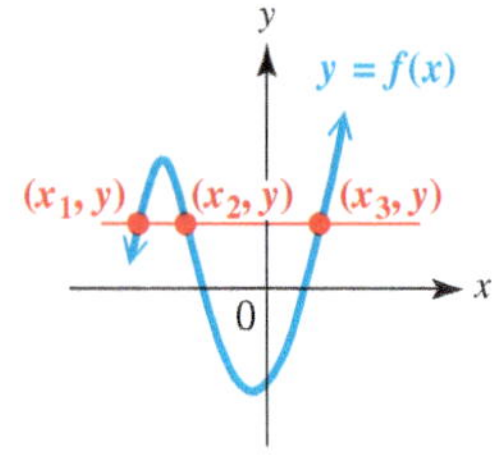

FIGURE 2

Because a horizontal line intersects the graph in more than one point (actually three points), the function in **FIGURE 2** is not one-to-one.

(b)

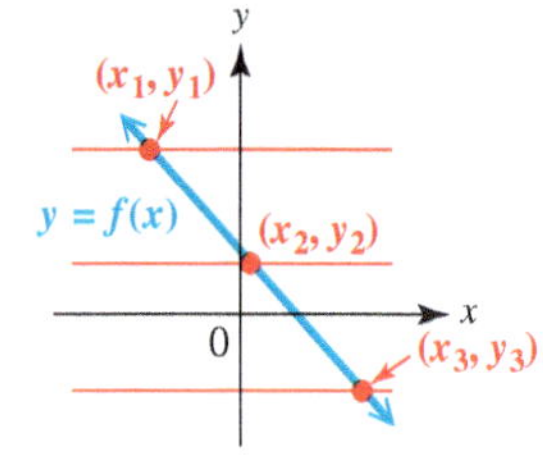

FIGURE 3

Every horizontal line will intersect the graph in exactly one point. The function in **FIGURE 3** is one-to-one.

NOW TRY

NOW TRY EXERCISE 2

Use the horizontal line test to determine whether each graph is the graph of a one-to-one function.

(a)

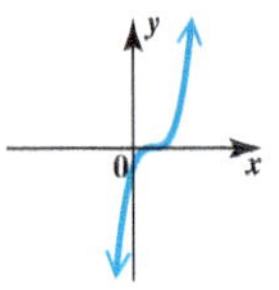

(b)

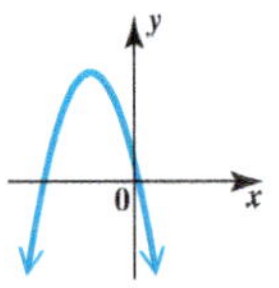

CLASSROOM EXAMPLE 2

Use the horizontal line test to determine whether each graph is the graph of a one-to-one function.

(a)

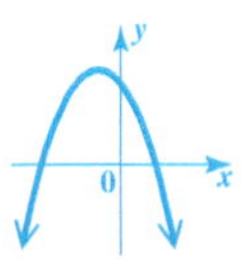

(b)

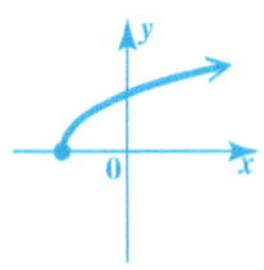

Answers:

(a) not one-to-one

(b) one-to-one

OBJECTIVE 3 Find the equation of the inverse of a function.

The inverse of a one-to-one function is found by interchanging the x- and y-values of each of its ordered pairs. The equation of the inverse of a function $y = f(x)$ is found in the same way.

Finding the Equation of the Inverse of $y = f(x)$

For a one-to-one function f defined by an equation $y = f(x)$, find the defining equation of the inverse as follows.

Step 1 Interchange x and y.

Step 2 Solve for y.

Step 3 Replace y with $f^{-1}(x)$.

CLASSROOM EXAMPLE 3

Decide whether each equation defines a one-to-one function. If so, find the equation that defines the inverse.

(a) $f(x) = 3x - 4$

(b) $f(x) = x^3 + 1$

(c) $f(x) = (x - 3)^2$

Answers:

(a) one-to-one function; $f^{-1}(x) = \frac{x+4}{3}$, or $f^{-1}(x) = \frac{1}{3}x + \frac{4}{3}$

(b) one-to-one function; $f^{-1}(x) = \sqrt[3]{x-1}$

(c) not a one-to-one function

EXAMPLE 3 Finding Equations of Inverses

Decide whether each equation defines a one-to-one function. If so, find the equation that defines the inverse.

(a) $f(x) = 2x + 5$

The graph of $y = 2x + 5$ is a nonvertical line, so by the horizontal line test, f is a one-to-one function. To find the inverse, follow the steps.

$$y = 2x + 5 \quad \text{Let } y = f(x).$$

$$x = 2y + 5 \quad \text{Interchange } x \text{ and } y. \text{ (Step 1)}$$

$$2y = x - 5 \quad \text{Solve for } y. \text{ (Step 2)}$$

$$y = \frac{x-5}{2}$$

$$f^{-1}(x) = \frac{x-5}{2} \quad \text{Replace } y \text{ with } f^{-1}(x). \text{ (Step 3)}$$

This equation can be written as follows.

$$f^{-1}(x) = \frac{x}{2} - \frac{5}{2}, \quad \text{or} \quad f^{-1}(x) = \frac{1}{2}x - \frac{5}{2} \qquad \frac{a-b}{c} = \frac{a}{c} - \frac{b}{c}$$

NOW TRY ANSWERS

2. **(a)** one-to-one
(b) not one-to-one

NOW TRY EXERCISE 3

Decide whether each equation defines a one-to-one function. If so, find the equation that defines the inverse.

(a) $f(x) = 5x - 7$

(b) $f(x) = (x + 1)^2$

(c) $f(x) = x^3 - 4$

TEACHING TIP

To emphasize the relationship between inverse functions, for $f(x) = 2x + 5$ and $x = 3$, find

$$y = 2 \cdot 3 + 5 = 11.$$

Now, for $f^{-1}(x) = \frac{x-5}{2}$ and $x = 11$, find

$$y = \frac{11 - 5}{2} = 3.$$

This shows numerically that when the resulting y-value is substituted in $f^{-1}(x)$, we obtain the original x-value.

Thus, f^{-1} is a linear function. In the function

$$f(x) = 2x + 5,$$

the range value is found by starting with a value of x, multiplying by 2, and adding 5. In the equation for the inverse

$$f^{-1}(x) = \frac{x-5}{2}, \quad \text{One form of } f^{-1}(x)$$

we *subtract* 5, and then *divide* by 2. This shows how an inverse is used to "undo" what a function does to the variable x.

(b) $y = x^2 + 2$

This equation has a vertical parabola as its graph, so some horizontal lines will intersect the graph at two points. For example, both $x = 3$ and $x = -3$ correspond to $y = 11$. Because of the x^2-term, there are many pairs of x-values that correspond to the same y-value. This means that the function $y = x^2 + 2$ is not one-to-one and does not have an inverse.

Applying the steps for finding the equation of an inverse leads to the following.

$$y = x^2 + 2$$

$$x = y^2 + 2 \quad \text{Interchange } x \text{ and } y.$$

$$y^2 = x - 2 \quad \text{Solve for } y.$$

$$y = \pm\sqrt{x-2} \quad \text{Square root property}$$

The last step shows that there are two y-values for each choice of $x > 2$, so we again see that the given function is not one-to-one. It does not have an inverse.

(c) $f(x) = (x - 2)^3$

A cubing function like this is one-to-one. (See **Section 4.3.**)

$$f(x) = (x-2)^3$$

$$y = (x-2)^3 \quad \text{Replace } f(x) \text{ with } y.$$

$$x = (y-2)^3 \quad \text{Interchange } x \text{ and } y.$$

$$\sqrt[3]{x} = \sqrt[3]{(y-2)^3} \quad \text{Take the cube root on each side.}$$

$$\sqrt[3]{x} = y - 2 \quad \sqrt[3]{a^3} = a$$

$$y = \sqrt[3]{x} + 2 \quad \text{Solve for } y.$$

$$f^{-1}(x) = \sqrt[3]{x} + 2 \quad \text{Replace } y \text{ with } f^{-1}(x).$$

NOW TRY

CLASSROOM EXAMPLE 4

Find $f^{-1}(x)$ for

$$f(x) = \frac{x-4}{x+2}, \quad x \neq -2.$$

Answer:

$$f^{-1}(x) = \frac{-2x-4}{x-1}, \quad x \neq 1$$

EXAMPLE 4 Using Factoring to Find an Inverse

It is shown in standard college algebra texts that

$$f(x) = \frac{x+1}{x-2}, \quad x \neq 2$$

is one-to-one. Find $f^{-1}(x)$.

$$f(x) = \frac{x+1}{x-2}, \quad x \neq 2 \quad \text{Given equation}$$

$$y = \frac{x+1}{x-2}, \quad x \neq 2 \quad \text{Replace } f(x) \text{ with } y.$$

NOW TRY ANSWERS

3. **(a)** one-to-one function; $f^{-1}(x) = \frac{x+7}{5}$, or $f^{-1}(x) = \frac{1}{5}x + \frac{7}{5}$

(b) not a one-to-one function

(c) one-to-one function; $f^{-1}(x) = \sqrt[3]{x+4}$

NOW TRY EXERCISE 4

Find $f^{-1}(x)$ for

$$f(x) = \frac{x+3}{x-4}, \quad x \neq 4.$$

Interchange x and y to find the inverse. Then solve for y.

$$x = \frac{y+1}{y-2}, \quad y \neq 2$$

$$x(y-2) = y+1 \qquad \text{Multiply by } y-2.$$

$$xy - 2x = y + 1 \qquad \text{Distributive property}$$

$$xy - y = 2x + 1 \qquad \text{Subtract } y. \text{ Add } 2x.$$

$$y(x-1) = 2x+1 \qquad \text{Factor out } y.$$

$$y = \frac{2x+1}{x-1} \qquad \text{Divide by } x-1.$$

Replace y with $f^{-1}(x)$ and note the restriction that $x \neq 1$.

$$f^{-1}(x) = \frac{2x+1}{x-1}, \quad x \neq 1$$

NOW TRY

OBJECTIVE 4 Graph f^{-1}, given the graph of f.

One way to graph the inverse of a function f whose equation is given follows.

Graphing the Inverse

Step 1 Find several ordered pairs that belong to f.

Step 2 Interchange x and y to obtain ordered pairs that belong to f^{-1}.

Step 3 Plot those points, and sketch the graph of f^{-1} through them.

TEACHING TIP

Point out that the graphs for $f(x)$ and $f^{-1}(x)$ in **Example 5** intersect along the line $y = x$ at the x-value for which $f(x) = f^{-1}(x)$.

We can also select points on the graph of f and *use symmetry* to find corresponding points on the graph of f^{-1}.

For example, suppose the point (a, b) shown in **FIGURE 4** belongs to a one-to-one function f. Then the point (b, a) belongs to f^{-1}. The line segment connecting (a, b) and (b, a) is perpendicular to, and cut in half by, the line $y = x$. The points (a, b) and (b, a) are "mirror images" of each other with respect to $y = x$.

We can find the graph of f^{-1} from the graph of f by locating the mirror image of each point in f with respect to the line $y = x$.

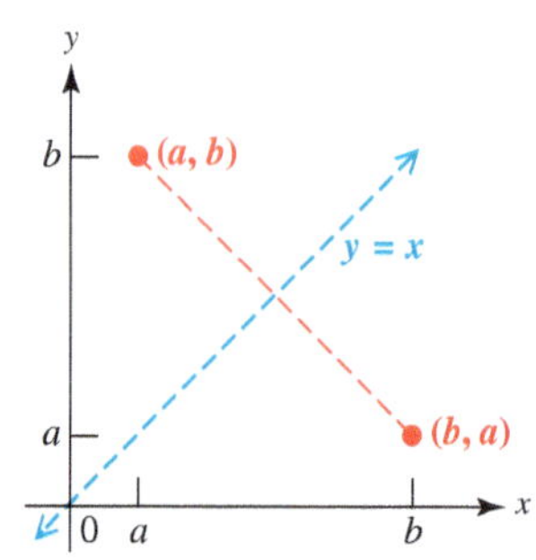

FIGURE 4

CLASSROOM EXAMPLE 5

Use the fact that if (a, b) lies on the graph of f, than (b, a) lies on the graph of f^{-1} to graph the inverse of the function f shown.

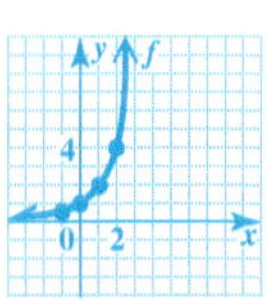

Answer:

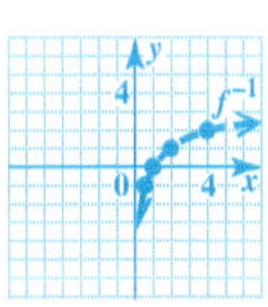

EXAMPLE 5 Graphing the Inverse

Use the fact that if (a, b) lies on the graph of f, then (b, a) lies on the graph of f^{-1}.

(a) **FIGURE 5(a)** on the next page shows the graph of a one-to-one function f. Sketch the graph of f^{-1}.

The points $\left(-1, \frac{1}{2}\right)$, $(0, 1)$, $(1, 2)$, and $(2, 4)$ lie on the graph of f. To find the graph of f^{-1}, plot the points

$$\left(\frac{1}{2}, -1\right), (1, 0), (2, 1), \text{ and } (4, 2).$$

NOW TRY ANSWER

4. $f^{-1}(x) = \frac{4x+3}{x-1}, \quad x \neq 1$

NOW TRY EXERCISE 5

Use the fact that if (a, b) lies on the graph of f, than (b, a) lies on the graph of f^{-1} to graph the inverse of the function f shown.

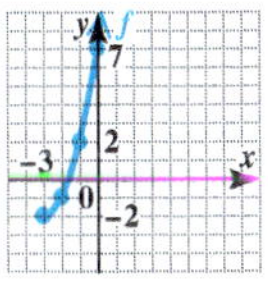

Join the points with the same shape curve as seen for f. See **FIGURE 5(b)**.

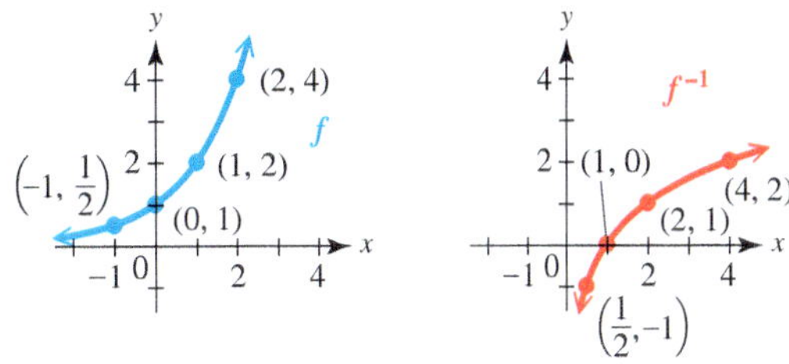

(a) (b)

FIGURE 5

(b) Graph the inverse of each function f (shown in blue) in **FIGURE 6**.

Each inverse is shown in red. In both cases, the graph of f^{-1} is a reflection of the graph of f across the line $y = x$.

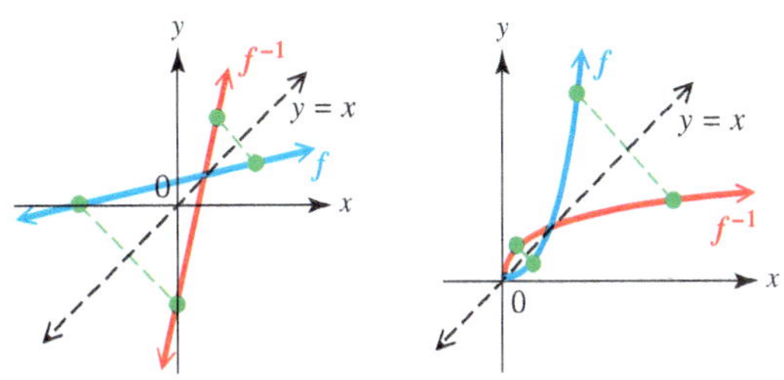

FIGURE 6

NOW TRY

NOW TRY ANSWER

5.

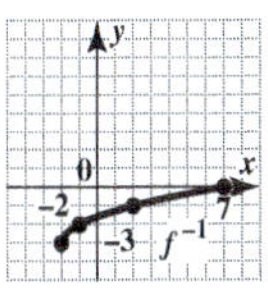

9.1 Exercises

FOR EXTRA HELP

MyMathLab®

Complete solution available in MyMathLab

1. B
2. Choice A is one-to-one.
   In choice B, $f(1) = f(-1) = 1$, and so $f(x) = x^2$ is not one-to-one.
   In choice C, $f(2) = f(-2) = 2$, and so $f(x) = |x|$ is not one-to-one.
   In choice D, $f(-2) = f(4) = -9$, and so $f(x) = -x^2 + 2x - 1$ is also not one-to-one.
3. A
4. D

Concept Check *In Exercises 1–4, choose the correct response from the given list.*

1. If a function is made up of ordered pairs in such a way that the same y-value appears in a correspondence with two different x-values, then

A. the function is one-to-one
B. the function is not one-to-one
C. its graph does not pass the vertical line test
D. it has an inverse function associated with it.

2. Which equation defines a one-to-one function? Explain why the others are not, using specific examples.

A. $f(x) = x$ **B.** $f(x) = x^2$ **C.** $f(x) = |x|$ **D.** $f(x) = -x^2 + 2x - 1$

3. Only one of the graphs illustrates a one-to-one function. Which one is it? **(See Example 2.)**

A.

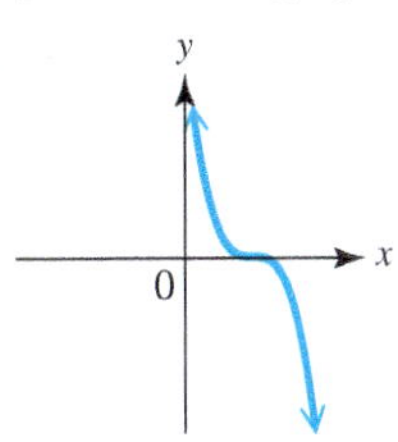

B.

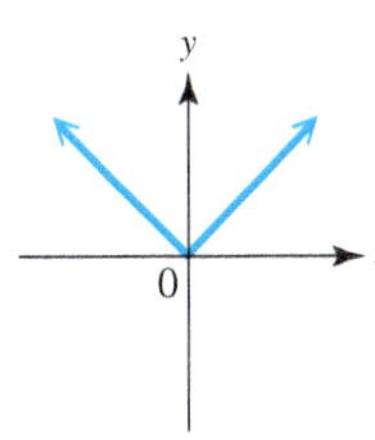

C.

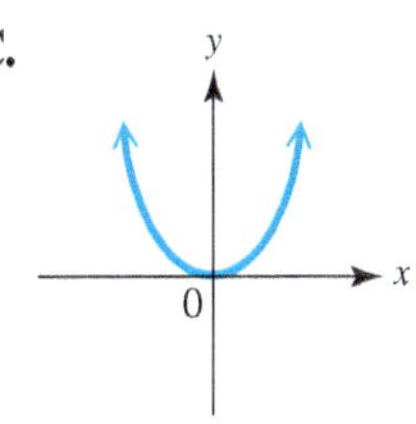

D.

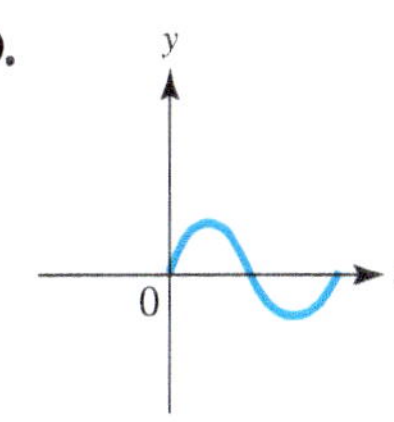

4. If a function f is one-to-one and the point (p, q) lies on the graph of f, then which point *must* lie on the graph of f^{-1}?

A. $(-p, q)$ **B.** $(-q, -p)$ **C.** $(p, -q)$ **D.** (q, p)

5. This function is one-to-one.
6. This function is one-to-one.
7. Yes, it is one-to-one. By adding 1 to 1058, two distances would be the same, so the function would not be one-to-one.
8. This function is not one-to-one because two sodas in the list have 41 mg of caffeine.

9. $\{(6,3),(10,2),(12,5)\}$

10. $\left\{(3,-1),(5,0),(0,5),\left(-\frac{1}{2},7\right)\right\}$

11. not one-to-one
12. not one-to-one
13. $f^{-1}(x)=x-3$
14. $f^{-1}(x)=x-8$
15. $f^{-1}(x)=-2x-4$
16. $f^{-1}(x)=-4x-32$
17. $f^{-1}(x)=\frac{x-4}{2}$, or $f^{-1}(x)=\frac{1}{2}x-2$
18. $f^{-1}(x)=\frac{x-1}{3}$, or $f^{-1}(x)=\frac{1}{3}x-\frac{1}{3}$

Answer each question.

5. The table shows trans fat content in a fast-food product in various countries, based on type of frying oil used. If the set of countries is the domain and the set of trans fat percentages is the range of a function, is it one-to-one? Why or why not?

Country	*Percentage of Trans Fat in McDonald's Chicken*
Scotland	14
United States	11
Peru	9
Poland	8
Russia	5
Denmark	1

Source: New England Journal of Medicine.

6. The table shows the top world oil exporters in thousands of barrels. If the set of countries is the domain and the set of numbers of barrels exported is the range of a function, is it one-to-one? If not, explain why.

Country	*Number of Barrels Exported (in thousands)*
Saudi Arabia	8865
Russia	7201
United Arab Emirates	2595
Kuwait	2414
Nigeria	2254
Iraq	2235

Source: U.S. Energy Information Administration.

7. The road mileage between Denver, Colorado, and several selected U.S. cities is shown in the table. If we consider this as a function that pairs each city with a distance, is it a one-to-one function? How could we change the answer to this question by adding 1 mile to one of the distances shown?

City	*Distance to Denver (in miles)*
Atlanta	1398
Dallas	781
Indianapolis	1058
Kansas City, MO	600
Los Angeles	1059
San Francisco	1235

8. The table lists caffeine amounts in several popular 12-oz sodas, If the set of sodas is the domain and the set of caffeine amounts is the range of a function, is it one-to-one? If not, explain why.

Soda	*Caffeine (in mg)*
Mountain Dew	55
Diet Coke	45
Dr. Pepper	41
Sunkist Orange Soda	41
Diet Pepsi-Cola	36
Coca-Cola Classic	34

Source: National Soft Drink Association.

19. $f^{-1}(x) = x^2 + 3, \quad x \geq 0$
20. $f^{-1}(x) = x^2 - 2, \quad x \geq 0$
21. not one-to-one
22. not one-to-one
23. $f^{-1}(x) = \sqrt[3]{x + 4}$
24. $f^{-1}(x) = \sqrt[3]{x - 5}$
25. $f^{-1}(x) = \dfrac{-2x + 4}{x - 1}, \quad x \neq 1$
26. $f^{-1}(x) = \dfrac{-5x + 3}{x - 1}, \quad x \neq 1$
27. $f^{-1}(x) = \dfrac{-5x - 2}{x - 4}, \quad x \neq 4$
28. $f^{-1}(x) = \dfrac{-4x - 10}{x - 5}, \quad x \neq 5$
29. $f^{-1}(x) = \dfrac{5x + 1}{2x + 2}, \quad x \neq -1$
30. $f^{-1}(x) = \dfrac{4x + 2}{3x + 3}, \quad x \neq -1$
31. (a) 8 (b) 3
32. (a) 16 (b) 4
33. (a) 1 (b) 0
34. (a) $\frac{1}{4}$ (b) -2
35. (a) one-to-one
(b) 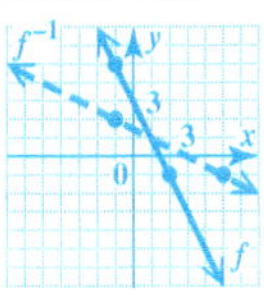
36. (a) one-to-one
(b) 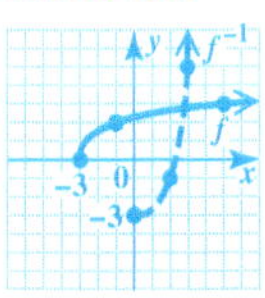
37. (a) not one-to-one
38. (a) not one-to-one
39. (a) one-to-one
(b) 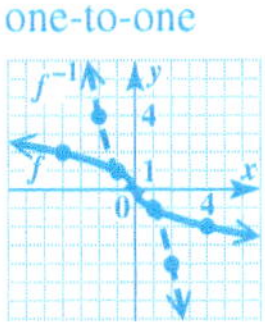
40. (a) one-to-one
(b) 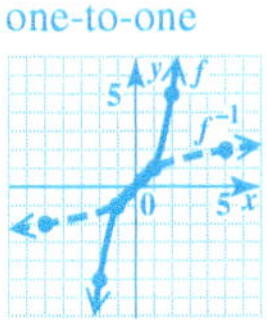

If the function is one-to-one, find its inverse. ***See Examples 1–3.***

9. $\{(3, 6), (2, 10), (5, 12)\}$

10. $\left\{(-1, 3), (0, 5), (5, 0), \left(7, -\frac{1}{2}\right)\right\}$

11. $\{(-1, 3), (2, 7), (4, 3), (5, 8)\}$

12. $\{(-8, 6), (-4, 3), (0, 6), (5, 10)\}$

13. $f(x) = x + 3$

14. $f(x) = x + 8$

15. $f(x) = -\frac{1}{2}x - 2$

16. $f(x) = -\frac{1}{4}x - 8$

17. $f(x) = 2x + 4$

18. $f(x) = 3x + 1$

19. $f(x) = \sqrt{x - 3}, \quad x \geq 3$

20. $f(x) = \sqrt{x + 2}, \quad x \geq -2$

21. $f(x) = 3x^2 + 2$

22. $f(x) = 4x^2 - 1$

23. $f(x) = x^3 - 4$

24. $f(x) = x^3 + 5$

Each function is one-to-one. Find its inverse. ***See Example 4.***

25. $f(x) = \dfrac{x + 4}{x + 2}, \quad x \neq -2$

26. $f(x) = \dfrac{x + 3}{x + 5}, \quad x \neq -5$

27. $f(x) = \dfrac{4x - 2}{x + 5}, \quad x \neq -5$

28. $f(x) = \dfrac{5x - 10}{x + 4}, \quad x \neq -4$

29. $f(x) = \dfrac{-2x + 1}{2x - 5}, \quad x \neq \dfrac{5}{2}$

30. $f(x) = \dfrac{-3x + 2}{3x - 4}, \quad x \neq \dfrac{4}{3}$

Concept Check *Let $f(x) = 2^x$. We will see in the next section that this function is one-to-one. Find each value, always working part (a) before part (b).*

31. (a) $f(3)$ (b) $f^{-1}(8)$

32. (a) $f(4)$ (b) $f^{-1}(16)$

33. (a) $f(0)$ (b) $f^{-1}(1)$

34. (a) $f(-2)$ (b) $f^{-1}\left(\frac{1}{4}\right)$

The graphs of some functions are given in Exercises 35–40.

(a) Use the horizontal line test to determine whether the function is one-to-one.

(b) If the function is one-to-one, then graph the inverse of the function. ***See Example 5.***

35. 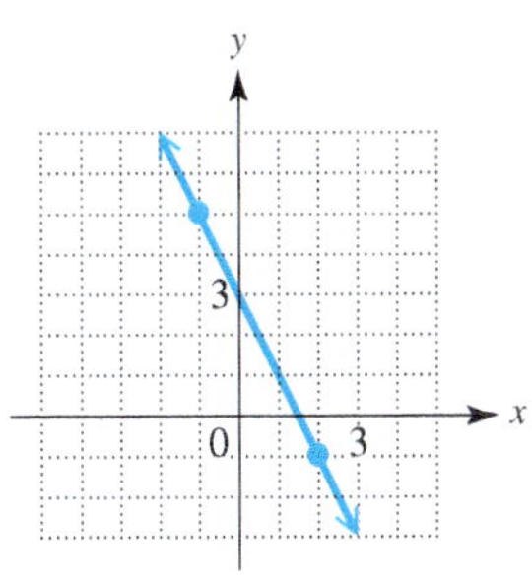

36. 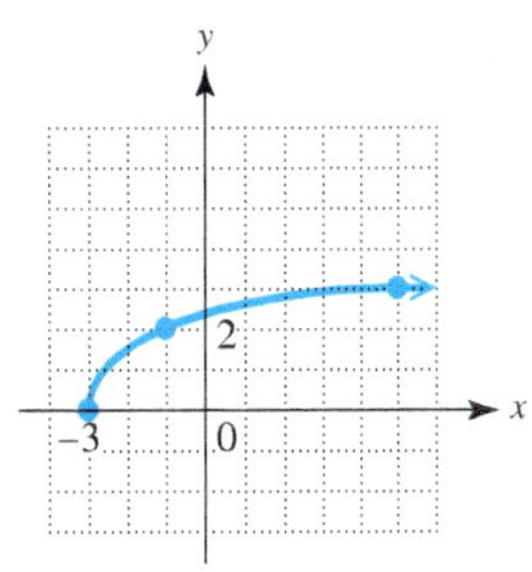

37. 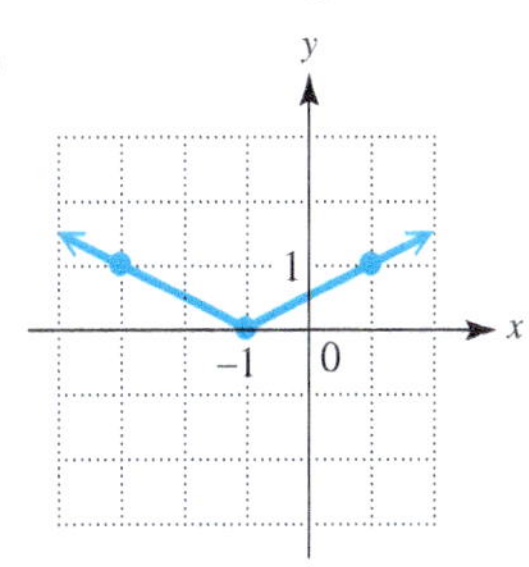

38. 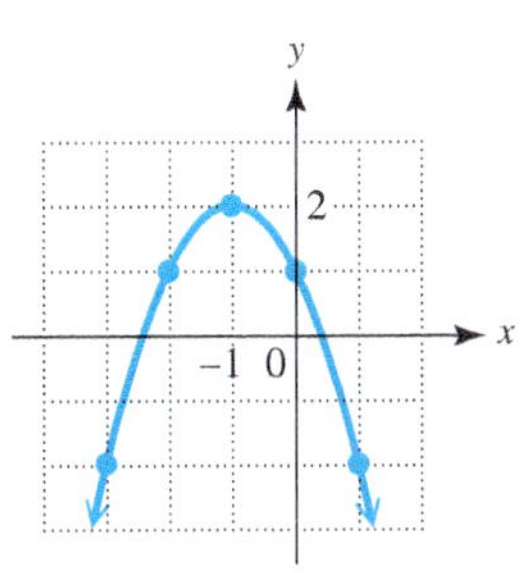

39. 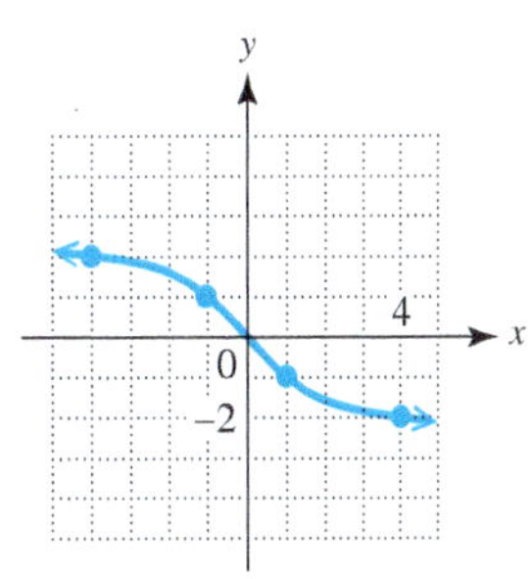

40. 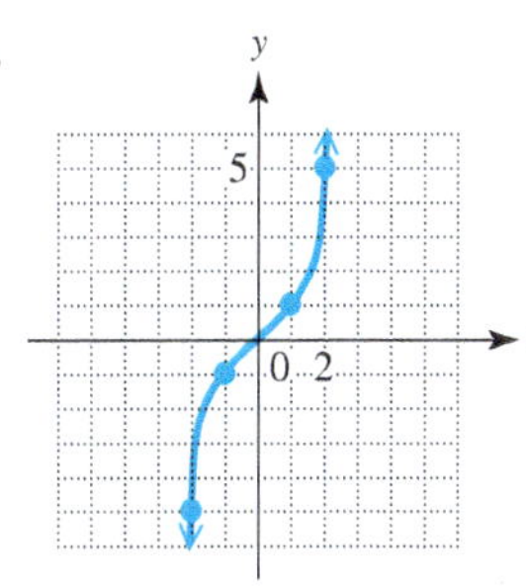

41.

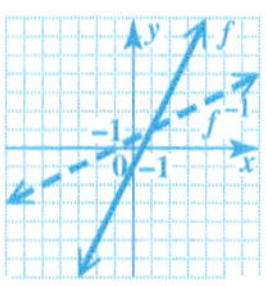

42.

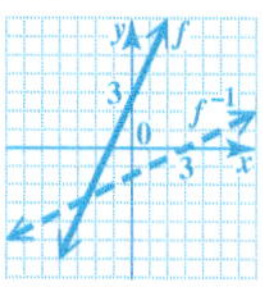

43.

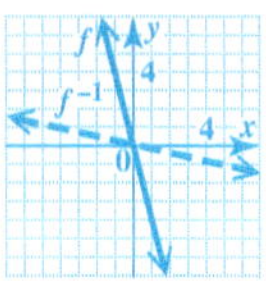

44.

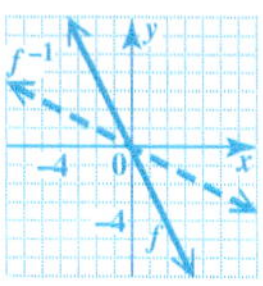

45.

x	$f(x)$
0	0
1	1
4	2

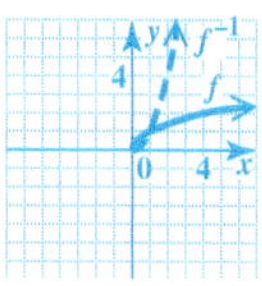

46.

x	$f(x)$
0	0
1	−1
4	−2

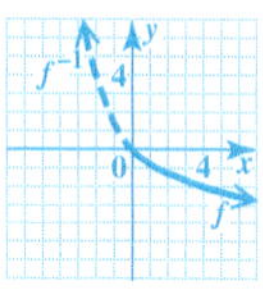

47.

x	$f(x)$
−1	−3
0	−2
1	−1
2	6

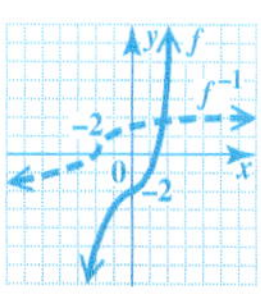

48.

x	$f(x)$
−2	−5
−1	2
0	3
1	4

49. $f^{-1}(x) = \frac{x+5}{4}$, or $f^{-1}(x) = \frac{1}{4}x + \frac{5}{4}$

50. MY GRAPHING CALCULATOR IS THE GREATEST THING SINCE SLICED BREAD.

51. If the function were not one-to-one, there would be ambiguity in some of the characters, as they could represent more than one letter.

52. Answers will vary. For example, Jane Doe is

1004 5 2748 129 68 3379 129.

Each function in Exercises 41–48 is one-to-one. Graph the function as a solid line (or curve) and then graph its inverse on the same set of axes as a dashed line (or curve). In Exercises 45–48, complete the table so that graphing the function will be easier. ***See Example 5.***

41. $f(x) = 2x - 1$ **42.** $f(x) = 2x + 3$ **43.** $f(x) = -4x$ **44.** $f(x) = -2x$

45. $f(x) = \sqrt{x}$, $x \geq 0$

x	$f(x)$
0	
1	
4	

46. $f(x) = -\sqrt{x}$, $x \geq 0$

x	$f(x)$
0	
1	
4	

47. $f(x) = x^3 - 2$

x	$f(x)$
−1	
0	
1	
2	

48. $f(x) = x^3 + 3$

x	$f(x)$
−2	
−1	
0	
1	

RELATING CONCEPTS For Individual or Group Work (Exercises 49–52)

Inverse functions can be used to send and receive coded information. A simple example might use the function $f(x) = 2x + 5$. (Note that it is one-to-one.) Suppose that each letter of the alphabet is assigned a numerical value according to its position, as follows.

A	1	G	7	L	12	Q	17	V	22
B	2	H	8	M	13	R	18	W	23
C	3	I	9	N	14	S	19	X	24
D	4	J	10	O	15	T	20	Y	25
E	5	K	11	P	16	U	21	Z	26
F	6								

This is an Enigma machine, used by the Germans in World War II to send coded messages.

Using the function, the word ALGEBRA *would be encoded as*

7 29 19 15 9 41 7,

because

$$f(\text{A}) = f(1) = 2(1) + 5 = 7, \quad f(\text{L}) = f(12) = 2(12) + 5 = 29, \quad \textit{and so on.}$$

The message would then be decoded by using the inverse of f, which is $f^{-1}(x) = \frac{x-5}{2}$ $\left(\text{or } f^{-1}(x) = \frac{1}{2}x - \frac{5}{2}\right)$. For example,

$$f^{-1}(7) = \frac{7-5}{2} = 1 = \text{A}, \quad f^{-1}(29) = \frac{29-5}{2} = 12 = \text{L}, \quad \textit{and so on.}$$

Work Exercises 49–52 in order.

49. Suppose that you are an agent for a detective agency. Today's function for your code is $f(x) = 4x - 5$. Find the rule for f^{-1} algebraically.

50. You receive the following coded message today. (Read across from left to right.)

47 95 23 67 −1 59 27 31 51 23 7 −1 43 7 79 43 −1 75 55 67
31 71 75 27 15 23 67 15 −1 75 15 71 75 75 27 31 51
23 71 31 51 7 15 71 43 31 7 15 11 3 67 15 −1 11

Use the letter/number assignment described earlier to decode the message.

51. Why is a one-to-one function essential in this encoding/decoding process?

52. Use $f(x) = x^3 + 4$ to encode your name, using the letter/number assignment described earlier.

9.2 Exponential Functions

OBJECTIVES

1. Use a calculator to find approximations of exponentials.
2. Define and graph exponential functions.
3. Solve exponential equations of the form $a^x = a^k$ for x.
4. Use exponential functions in applications involving growth or decay.

VOCABULARY

- □ exponential function with base *a*
- □ asymptote
- □ exponential equation

OBJECTIVE 1 Use a calculator to find approximations of exponentials.

In **Section 7.2** we showed how to evaluate 2^x for rational values of x.

$$2^3 = 8, \quad 2^{-1} = \frac{1}{2}, \quad 2^{1/2} = \sqrt{2}, \quad \text{and} \quad 2^{3/4} = \sqrt[4]{2^3} = \sqrt[4]{8}$$

Examples of 2^x for rational x

In more advanced courses it is shown that 2^x exists for all real number values of x, both rational and irrational. We can use a calculator to find approximations of exponential expressions that are not easily determined.

EXAMPLE 1 Approximating Using a Calculator

Use a calculator to find an approximation to three decimal places for each exponential expression.

(a) $2^{1.6}$ **(b)** $2^{-1.3}$ **(c)** $2^{1/3}$

FIGURE 7 shows how a TI-84 calculator approximates these values. The display shows more decimal places than we usually need, so we round to three decimal places as directed.

$$2^{1.6} \approx 3.031, \quad 2^{-1.3} \approx 0.406, \quad 2^{1/3} \approx 1.260$$

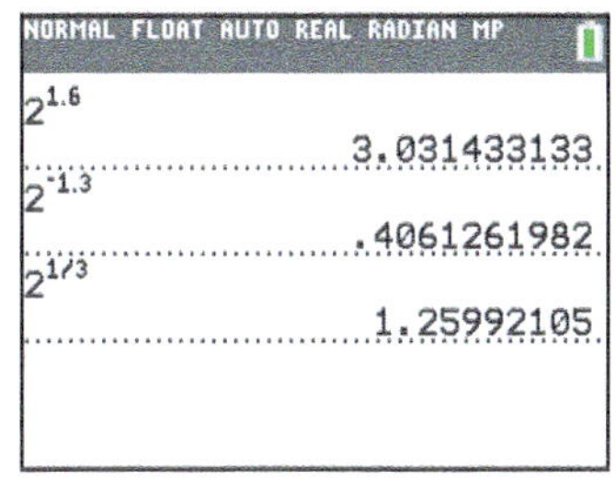

FIGURE 7

NOW TRY

NOW TRY EXERCISE 1

Use a calculator to find an approximation to three decimal places for each exponential expression.

(a) $2^{3.1}$ **(b)** $2^{-1.7}$ **(c)** $2^{1/4}$

CLASSROOM EXAMPLE 1

Use a calculator to find an approximation to three decimal places for each exponential expression.

(a) $2^{4.5}$ (b) $2^{-1.1}$ (c) $2^{1/5}$

Answers: (a) 22.627 (b) 0.467 (c) 1.149

OBJECTIVE 2 Define and graph exponential functions.

The following definition of an exponential function assumes that a^x exists for all real numbers x.

Exponential Function

For $a > 0$, $a \neq 1$, and all real numbers x,

$$f(x) = a^x$$

defines the **exponential function with base *a*.**

TEACHING TIP

A calculator with the following keys will be essential in this chapter.

y^x, 10^x, $\log x$,

e^x, $\ln x$

Be sure that students can locate them on their calculators.

NOTE ***The two restrictions on the value of a in the definition of an exponential function $f(x) = a^x$ are important.***

1. The restriction $a > 0$ is necessary so that the function can be defined for all real numbers x. Letting a be negative ($a = -2$, for instance) and letting $x = \frac{1}{2}$ would give the expression $(-2)^{1/2}$, which is not real.
2. The restriction $a \neq 1$ is necessary because 1 raised to any power is equal to 1, resulting in the linear function $f(x) = 1$.

NOW TRY ANSWERS

1. (a) 8.574 (b) 0.308 (c) 1.189

When graphing an exponential function of the form $f(x) = a^x$, pay particular attention to whether $a > 1$ or $0 < a < 1$.

NOW TRY EXERCISE 2

Graph $f(x) = 4^x$.

CLASSROOM EXAMPLE 2

Graph $f(x) = 10^x$.

Answer:

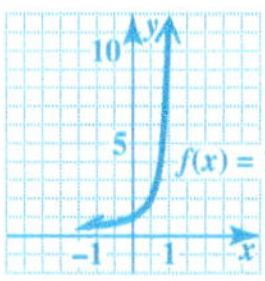

TEACHING TIP

When graphing these exponential functions, take care to point out to students that the x-axis is an *asymptote.*

EXAMPLE 2 Graphing an Exponential Function ($a > 1$)

Graph $f(x) = 2^x$. Then compare it to the graph of $F(x) = 5^x$.

Choose some values of x, and find the corresponding values of $f(x)$. Plotting these points and drawing a smooth curve through them gives the darker graph shown in **FIGURE 8**. This graph is typical of the graph of an exponential function of the form $f(x) = a^x$, where $a > 1$.

The larger the value of a, the faster the graph rises.

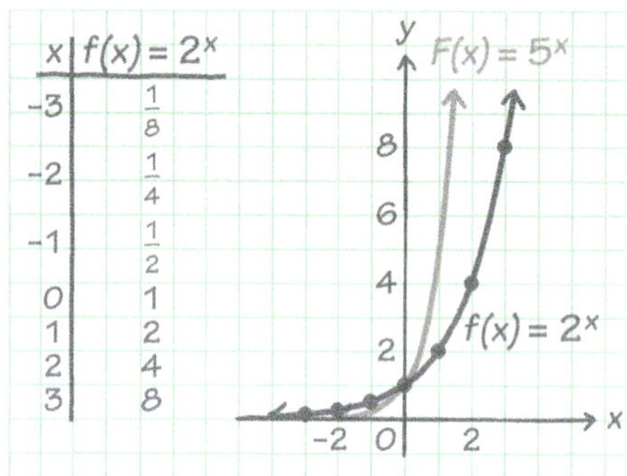

Exponential function with base $a > 1$

Domain: $(-\infty, \infty)$

Range: $(0, \infty)$

y-intercept: $(0, 1)$

The function is one-to-one, and its graph rises from left to right.

FIGURE 8

CLASSROOM EXAMPLE 3

Graph $g(x) = \left(\frac{1}{4}\right)^x$.

Answer:

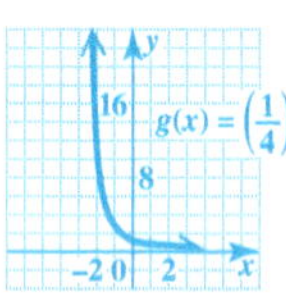

The vertical line test assures us that the graphs in **FIGURE 8** represent functions. **FIGURE 8** also shows an important characteristic of exponential functions with $a > 1$:

As x gets larger, y increases at a faster and faster rate. NOW TRY

CAUTION The graph of this exponential function *approaches* the x-axis, but does ***not*** touch it.

NOW TRY EXERCISE 3

Graph $g(x) = \left(\frac{1}{10}\right)^x$.

EXAMPLE 3 Graphing an Exponential Function ($0 < a < 1$)

Graph $g(x) = \left(\frac{1}{2}\right)^x$.

Find some points on the graph. The graph in **FIGURE 9** is similar to that of $f(x) = 2^x$ (**FIGURE 8**) except that here ***as x gets larger, y decreases.*** This graph is typical of the graph of an exponential function of the form $f(x) = a^x$, where $0 < a < 1$.

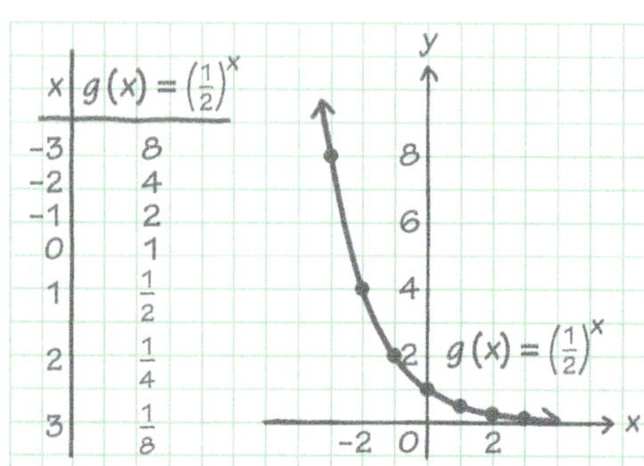

Exponential function with base $0 < a < 1$

Domain: $(-\infty, \infty)$

Range: $(0, \infty)$

y-intercept: $(0, 1)$

The function is one-to-one, and its graph falls from left to right.

FIGURE 9

NOW TRY

NOW TRY ANSWERS

2.

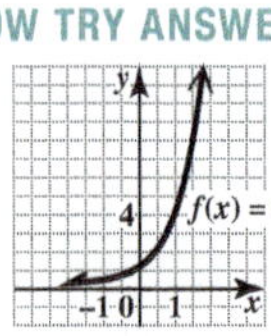

3.

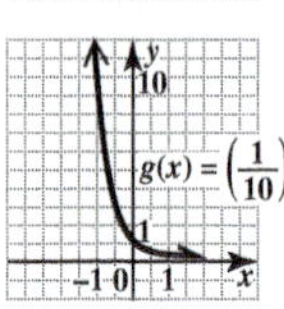

TEACHING TIP
Stress the importance of the information given in the box.
Point out that because the range of $f(x) = a^x$ is $(0, \infty)$, $f(x)$ *never* equals 0.

Characteristics of the Graph of $f(x) = a^x$

1. The graph contains the point $(0, 1)$, which is its y-intercept.

2. The function is one-to-one.

- When $a > 1$, the graph will *rise* from left to right. (See **FIGURE 8**.)
- When $0 < a < 1$, the graph will *fall* from left to right. (See **FIGURE 9**.)

In both cases, the graph goes from the second quadrant to the first.

3. The graph will approach the x-axis, but never touch it. (From **Section 6.4**, recall that such a line is an **asymptote.**)

4. The domain is $(-\infty, \infty)$, and the range is $(0, \infty)$.

NOW TRY EXERCISE 4
Graph $f(x) = 4^{2x-1}$.

CLASSROOM EXAMPLE 4
Graph $f(x) = 2^{4x-3}$.

Answer:

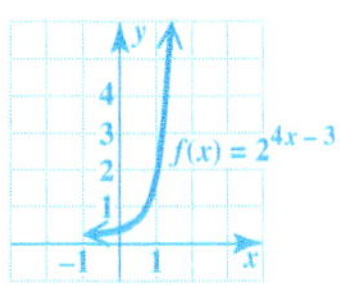

EXAMPLE 4 Graphing a More Complicated Exponential Function

Graph $f(x) = 3^{2x-4}$.

Find some ordered pairs. We let $x = 0$ and $x = 2$ and find values of $f(x)$, or y.

$$y = 3^{2(0)-4} \quad \text{Let } x = 0.$$
$$y = 3^{-4}$$
$$y = \frac{1}{81} \quad a^{-n} = \frac{1}{a^n}$$

$$y = 3^{2(2)-4} \quad \text{Let } x = 2.$$
$$y = 3^0$$
$$y = 1 \quad a^0 = 1$$

These ordered pairs, $\left(0, \frac{1}{81}\right)$ and $(2, 1)$, along with the other ordered pairs shown in the table, lead to the graph in **FIGURE 10**.

x	y
0	$\frac{1}{81}$
1	$\frac{1}{9}$
2	1
3	9

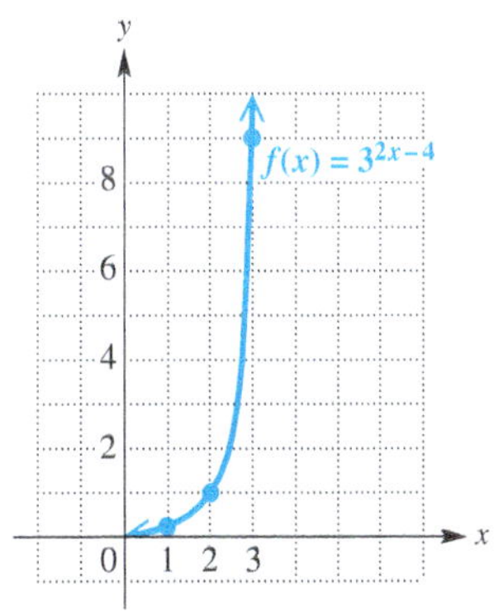

FIGURE 10

The graph is similar to the graph of $f(x) = 3^x$ except that it is shifted to the right and rises more rapidly.

NOW TRY

TEACHING TIP
It helps students to remember this property if it is stated in words:
"If two exponential expressions with the same base are equal, then the exponents must be equal."
(An exception to this occurs when the base is 1, which is not allowed as an exponential base.)

OBJECTIVE 3 Solve exponential equations of the form $a^x = a^k$ for x.

Until this chapter, we have solved only equations that had the variable as a base, like $x^2 = 8$. In these equations, all exponents have been constants. An **exponential equation** is an equation that has a variable in an exponent, such as

$$9^x = 27.$$

We can use the following property to solve certain exponential equations.

Property for Solving an Exponential Equation

For $a > 0$ and $a \neq 1$, if $a^x = a^y$ then $x = y$.

This property would not necessarily be true if $a = 1$.

NOW TRY ANSWER

4.

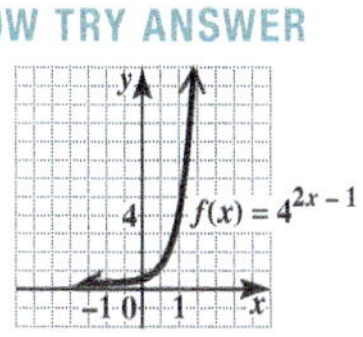

NOW TRY EXERCISE 5
Solve the equation.
$$8^x = 16$$

CLASSROOM EXAMPLE 5
Solve the equation.
$$25^x = 125$$
Answer: $\left\{\frac{3}{2}\right\}$

CLASSROOM EXAMPLE 6
Solve each equation.
(a) $25^{x-2} = 125^x$
(b) $4^x = \frac{1}{32}$
(c) $\left(\frac{3}{4}\right)^x = \frac{16}{9}$
Answers: (a) $\{-4\}$ (b) $\left\{-\frac{5}{2}\right\}$
(c) $\{-2\}$

NOW TRY ANSWER
5. $\left\{\frac{4}{3}\right\}$

Solving an Exponential Equation

Step 1 **Each side must have the same base.** If the two sides of the equation do not have the same base, express each as a power of the same base if possible.

Step 2 **Simplify exponents** if necessary, using the rules of exponents.

Step 3 **Set exponents equal** using the property given in this section.

Step 4 **Solve** the equation obtained in Step 3.

NOTE The steps used in **Examples 5 and 6** cannot be applied to an equation like $3^x = 12$ because Step 1 cannot easily be done. We solve such equations in **Section 9.6.**

EXAMPLE 5 Solving an Exponential Equation

Solve the equation $9^x = 27$.

$$9^x = 27$$

$$(3^2)^x = 3^3 \quad \text{Write with the same base; } 9 = 3^2 \text{ and } 27 = 3^3. \text{ (Step 1)}$$

$$3^{2x} = 3^3 \quad \text{Power rule for exponents (Step 2)}$$

$$2x = 3 \quad \text{If } a^x = a^y, \text{ then } x = y. \text{ (Step 3)}$$

$$x = \frac{3}{2} \quad \text{Solve for } x. \text{ (Step 4)}$$

CHECK Substitute $\frac{3}{2}$ for x.

$$9^x = 9^{3/2} = (9^{1/2})^3 = 3^3 = 27 \quad ✓ \text{ True}$$

The solution set is $\left\{\frac{3}{2}\right\}$. **NOW TRY**

EXAMPLE 6 Solving Exponential Equations

Solve each equation.

(a) $4^{3x-1} = 16^{x+2}$

Be careful multiplying the exponents.

$$4^{3x-1} = (4^2)^{x+2} \quad \text{Write with the same base; } 16 = 4^2.$$

$$4^{3x-1} = 4^{2x+4} \quad \text{Power rule for exponents}$$

$$3x - 1 = 2x + 4 \quad \text{Set exponents equal.}$$

$$x = 5 \quad \text{Subtract } 2x. \text{ Add 1.}$$

CHECK $4^{3x-1} = 16^{x+2}$

$$4^{3(5)-1} \stackrel{?}{=} 16^{5+2} \quad \text{Substitute. Let } x = 5.$$

$$4^{14} \stackrel{?}{=} 16^7 \quad \text{Perform the operations in the exponents.}$$

$$4^{14} \stackrel{?}{=} (4^2)^7 \quad 16 = 4^2$$

$$4^{14} = 4^{14} \quad ✓ \text{ True}$$

The solution set is $\{5\}$.

NOW TRY EXERCISE 6

Solve each equation.

(a) $3^{2x-1} = 27^{x+4}$

(b) $5^x = \frac{1}{625}$

(c) $\left(\frac{2}{7}\right)^x = \frac{343}{8}$

(b) $6^x = \frac{1}{216}$

$6^x = \frac{1}{6^3}$ $216 = 6^3$

$6^x = 6^{-3}$ Write with the same base; $\frac{1}{6^3} = 6^{-3}$.

$x = -3$ Set exponents equal.

CHECK $6^x = 6^{-3} = \frac{1}{6^3} = \frac{1}{216}$ ✓   Substitute -3 for x. True.

The solution set is $\{-3\}$.

(c) $\left(\frac{2}{3}\right)^x = \frac{9}{4}$

$\left(\frac{2}{3}\right)^x = \left(\frac{4}{9}\right)^{-1}$ $\frac{9}{4} = \left(\frac{4}{9}\right)^{-1}$

$\left(\frac{2}{3}\right)^x = \left[\left(\frac{2}{3}\right)^2\right]^{-1}$ Write with the same base.

$\left(\frac{2}{3}\right)^x = \left(\frac{2}{3}\right)^{-2}$ Power rule for exponents

$x = -2$ Set exponents equal.

Check that the solution set is $\{-2\}$.   NOW TRY

OBJECTIVE 4 Use exponential functions in applications involving growth or decay.

CLASSROOM EXAMPLE 7

The function

$f(x) = 0.07414(1.004267)^x$

approximates the carbon dioxide concentration in parts per million, where x is the year (based on data from 1959 to 2013). Use this function and a calculator to approximate the concentration of carbon dioxide, to the nearest unit, for the year 2012. How well does the function approximate the actual figure of 394?

Answer: 390; This is slightly below the actual figure.

EXAMPLE 7 Solving an Application Involving Exponential Growth

The graph in **FIGURE 11** shows the concentration of carbon dioxide (in parts per million) in the air. This concentration is increasing exponentially.

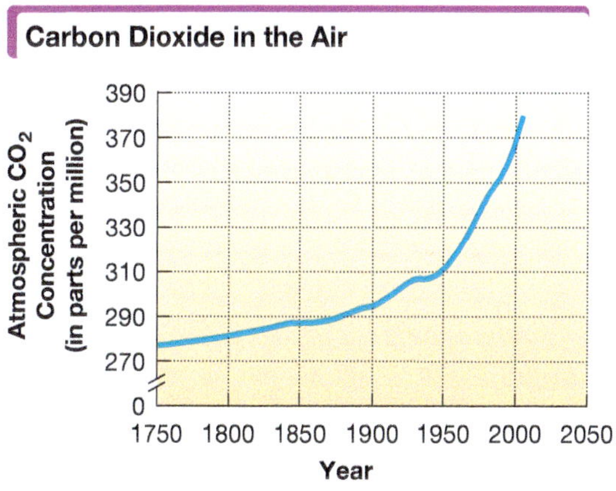

Source: *Sacramento Bee;* National Oceanic and Atmospheric Administration.

FIGURE 11

NOW TRY ANSWERS

6. **(a)** $\{-13\}$ **(b)** $\{-4\}$ **(c)** $\{-3\}$

NOW TRY EXERCISE 7

Use the function in **Example 7** to approximate the carbon dioxide concentration in 2000, to the nearest unit.

The data in the preceding graph are approximated by the function

$$f(x) = 266(1.001)^x,$$

where x is number of years since 1750. Use this function and a calculator to approximate the concentration of carbon dioxide in parts per million, to the nearest unit, for each year.

(a) 1900

Because x represents number of years since 1750, $x = 1900 - 1750 = 150$.

$$f(x) = 266(1.001)^x \quad \text{Given function}$$
$$f(150) = 266(1.001)^{150} \quad \text{Let } x = 150.$$
$$f(150) \approx 309 \quad \text{Evaluate with a calculator.}$$

The concentration in 1900 was 309 parts per million.

(b) 1950

$$f(x) = 266(1.001)^x \quad \text{Given function}$$
$$f(200) = 266(1.001)^{200} \quad x = 1950 - 1750 = 200$$
$$f(200) \approx 325 \quad \text{Evaluate with a calculator.}$$

The concentration in 1950 was 325 parts per million. NOW TRY

NOW TRY EXERCISE 8

Use the function in **Example 8** to approximate the pressure at 6000 m, to the nearest unit.

CLASSROOM EXAMPLE 8

Use the function in **Example 8** to approximate the pressure at 8000 m, to the nearest unit.

Answer: 355 millibars

EXAMPLE 8 Applying an Exponential Decay Function

The atmospheric pressure (in millibars) at a given altitude x, in meters, can be approximated by the function

$$f(x) = 1038(1.000134)^{-x}, \quad \text{for values of } x \text{ between 0 and 10,000.}$$

Because the base is greater than 1 and the coefficient of x in the exponent is negative, function values decrease as x increases. This means that as altitude increases, atmospheric pressure decreases. (*Source:* Miller, A. and J. Thompson, *Elements of Meteorology*, Fourth Edition, Charles E. Merrill Publishing Company.)

(a) According to this function, what is the pressure at ground level?

$$f(x) = 1038(1.000134)^{-x} \quad \text{Given function}$$
$$f(0) = 1038(1.000134)^{-0} \quad \text{Let } x = 0.$$
$$= 1038(1) \quad a^0 = 1$$
$$= 1038$$

At ground level, $x = 0$.

The pressure is 1038 millibars.

(b) What is the pressure at 5000 m, to the nearest unit?

$$f(x) = 1038(1.000134)^{-x} \quad \text{Given function}$$
$$f(5000) = 1038(1.000134)^{-5000} \quad \text{Let } x = 5000.$$
$$f(5000) \approx 531 \quad \text{Evaluate with a calculator.}$$

The pressure is 531 millibars. NOW TRY

NOTE The function in **Example 8** is equivalent to

$$f(x) = 1038\left(\frac{1}{1.000134}\right)^x.$$

In this form, the base a satisfies the condition $0 < a < 1$.

NOW TRY ANSWERS

7. 342 parts per million

8. 465 millibars

9.2 Exercises

FOR EXTRA HELP

MyMathLab®

Complete solution available in MyMathLab

1. C
2. A
3. rises
4. falls
5. A
6. The graph of an exponential function is a smooth curve that rises or falls depending on the base. If the base is between 0 and 1, it falls, and if it is greater than 1, it rises. For example, the graph of $f(x) = 3^x$ rises from left to right, has (0, 1) as its y-intercept, and has the x-axis as horizontal asymptote. The domain is $(-\infty, \infty)$, and the range is $(0, \infty)$.
7. 3.732
8. 6.498
9. 0.344
10. 0.272
11. 1.995
12. 3.162
13. 1.587
14. 1.431
15. 0.192
16. 0.072
17. 73.517
18. 6.964

19.

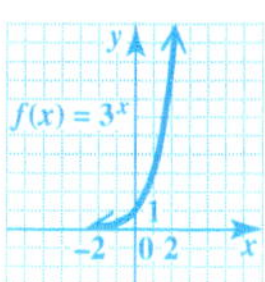

20.

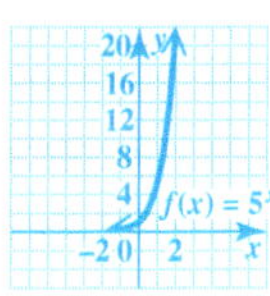

21.

$g(x) = \left(\frac{1}{3}\right)^x$

22.

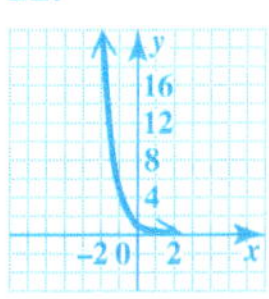

$g(x) = \left(\frac{1}{5}\right)^x$

23.

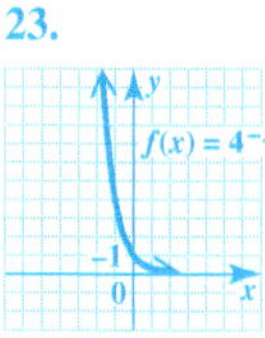

24.

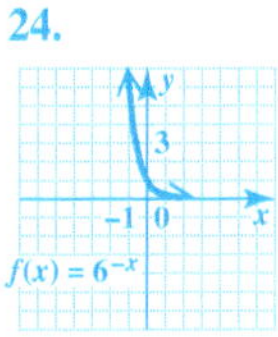

25.

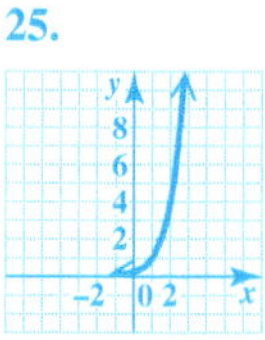

$f(x) = 2^{2x-2}$

26.

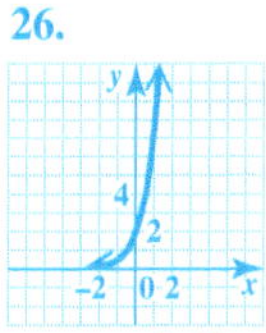

$f(x) = 2^{2x+1}$

27. {2}
28. {2}
29. $\left\{\frac{3}{2}\right\}$
30. $\left\{\frac{2}{3}\right\}$
31. {7}
32. {4}
33. {−3}
34. {−4}
35. {−1}
36. {−1}
37. {−3}
38. {−3}

Concept Check *Provide the answers in Exercises 1–6.*

1. Which point lies on the graph of $f(x) = 3^x$?

A. (1, 0) **B.** (3, 1) **C.** (0, 1) **D.** $\left(\sqrt{3}, \frac{1}{3}\right)$

2. Which statement is true?

A. The point $\left(\frac{1}{2}, \sqrt{5}\right)$ lies on the graph of $f(x) = 5^x$.

B. $f(x) = 2^x$ is not a one-to-one function.

C. The y-intercept of the graph of $f(x) = 10^x$ is (0, 10).

D. The graph of $y = 4^x$ rises at a faster rate than the graph of $y = 10^x$.

3. For an exponential function $f(x) = a^x$, if $a > 1$, then the graph (*rises*/*falls*) from left to right.

4. For an exponential function $f(x) = a^x$, if $0 < a < 1$, then the graph (*rises*/*falls*) from left to right.

5. The asymptote of the graph of $f(x) = a^x$

A. is the x-axis **B.** is the y-axis

C. has equation $x = 1$ **D.** has equation $y = 1$.

6. Describe the characteristics of the graph of an exponential function. Use the exponential function $f(x) = 3^x$ and the words *asymptote*, *domain*, and *range* in the explanation.

Use a calculator to find an approximation to three decimal places for each exponential expression. ***See Example 1.***

7. $2^{1.9}$ **8.** $2^{2.7}$ **9.** $2^{-1.54}$ **10.** $2^{-1.88}$

11. $10^{0.3}$ **12.** $10^{0.5}$ **13.** $4^{1/3}$ **14.** $6^{1/5}$

15. $\left(\frac{1}{3}\right)^{1.5}$ **16.** $\left(\frac{1}{3}\right)^{2.4}$ **17.** $\left(\frac{1}{4}\right)^{-3.1}$ **18.** $\left(\frac{1}{4}\right)^{-1.4}$

Graph each exponential function. ***See Examples 2–4.***

19. $f(x) = 3^x$ **20.** $f(x) = 5^x$ **21.** $g(x) = \left(\frac{1}{3}\right)^x$ **22.** $g(x) = \left(\frac{1}{5}\right)^x$

23. $f(x) = 4^{-x}$ **24.** $f(x) = 6^{-x}$ **25.** $f(x) = 2^{2x-2}$ **26.** $f(x) = 2^{2x+1}$

Solve each equation. ***See Examples 5 and 6.***

27. $6^x = 36$ **28.** $8^x = 64$ **29.** $100^x = 1000$

30. $8^x = 4$ **31.** $16^{2x+1} = 64^{x+3}$ **32.** $9^{2x-8} = 27^{x-4}$

33. $5^x = \frac{1}{125}$ **34.** $3^x = \frac{1}{81}$ **35.** $5^x = 0.2$

36. $10^x = 0.1$ **37.** $\left(\frac{3}{2}\right)^x = \frac{8}{27}$ **38.** $\left(\frac{4}{3}\right)^x = \frac{27}{64}$

39. 100 g **40.** 31.25 g
41. 0.30 g
42.

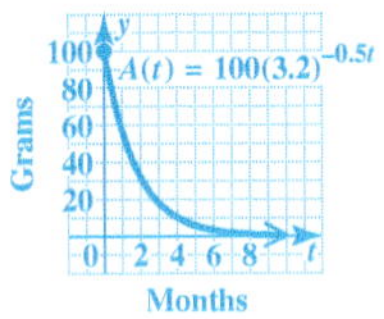

43. (a) 0.6°C (b) 0.3°C
44. (a) 0.9°C (b) 0.4°C
45. (a) 1.4°C (b) 0.5°C
46. (a) 3.2°C (b) 0.7°C

47. (a) 2.403 million
(b) 32.971 million
(c) 188.962 million

48. (a) 115.54 million tons
(b) 140.84 million tons
(c) 171.68 million tons
(d) The model gives 231.06 million tons. The actual value is slightly greater.

The amount of radioactive material in an ore sample is given by the function

$$A(t) = 100(3.2)^{-0.5t},$$

where A(t) is the amount present, in grams, of the sample t months after the initial measurement.

Use this function for Exercises 39–42.

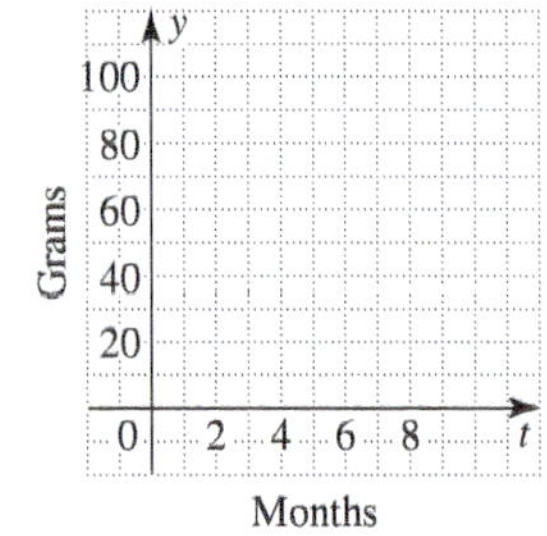

39. How much radioactive material was present at the initial measurement? (*Hint:* $t = 0$.)

40. How much, to the nearest hundredth, was present 2 months later?

41. How much, to the nearest hundredth, was present 10 months later?

42. Graph the function on the axes as shown.

A major scientific periodical published an article in 1990 dealing with the problem of global warming. The article was accompanied by a graph that illustrated two possible scenarios.

(a) *The warming might be modeled by an exponential function of the form*

$$f(x) = (1.046 \times 10^{-38})(1.0444^x).$$

(b) *The warming might be modeled by a linear function of the form*

$$g(x) = 0.009x - 17.67.$$

In both cases, x represents the year, and the function value represents the increase in degrees Celsius due to the warming. Use these functions to approximate the increase in temperature for each year, to the nearest tenth.

43. 2000 **44.** 2010 **45.** 2020 **46.** 2040

Solve each problem. ***See Examples 7 and 8.***

47. The average number of Facebook users in millions at the end of each year from 2004 to 2013 can be modeled by the exponential function

$$f(x) = 2.403(2.394)^x,$$

where $x = 0$ represents 2004, $x = 1$ represents 2005, and so on. Use this model to approximate the number of Facebook users in each year, to the nearest thousandth. (*Source:* http://statista.com)

(a) 2004 **(b)** 2007 **(c)** 2009

48. Based on figures from 1960 through 2010, the municipal solid waste in millions of tons in the United States can be modeled by the exponential function

$$f(x) = 94.78(1.02)^x,$$

where $x = 0$ corresponds to 1960, $x = 10$ corresponds to 1970, and so on. Use this model to approximate the municipal solid waste generated in each year, to the nearest hundredth. (*Source:* U.S. Environmental Protection Agency.)

(a) 1970 **(b)** 1980 **(c)** 1990

(d) In 2005, the actual number was 252.7 million tons. How does this compare to the number that the model gives?

49. (a) \$5000 (b) \$2973
(c) \$1768
(d)

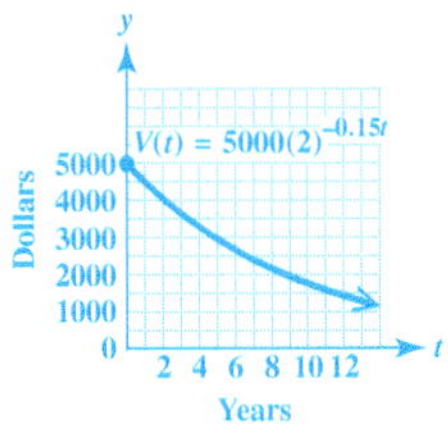

50. (a) 6.67 yr after it was purchased
(b) 13.33 yr after it was purchased

49. A small business estimates that the value $V(t)$ of a copy machine is decreasing according to the function

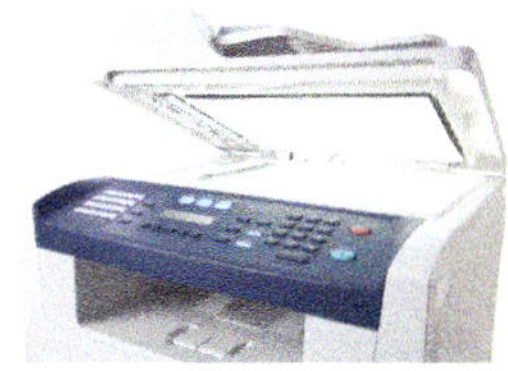

$$V(t) = 5000(2)^{-0.15t},$$

where t is the number of years that have elapsed since the machine was purchased, and $V(t)$ is in dollars.

(a) What was the original value of the machine?

(b) What is the value of the machine 5 yr after purchase, to the nearest dollar?

(c) What is the value of the machine 10 yr after purchase, to the nearest dollar?

(d) Graph the function.

50. Refer to the function in **Exercise 49.**

(a) When will the value of the machine be \$2500? (*Hint:* Let $V(t) = 2500$, divide both sides by 5000, and use the method of **Example 5.**)

(b) When will the value of the machine be \$1250?

9.3 Logarithmic Functions

OBJECTIVES

1. Define a logarithm.
2. Convert between exponential and logarithmic forms, and evaluate logarithms.
3. Solve logarithmic equations of the form $\log_a b = k$ for a, b, or k.
4. Use the definition of logarithm to simplify logarithmic expressions.
5. Define and graph logarithmic functions.
6. Use logarithmic functions in applications involving growth or decay.

VOCABULARY

☐ logarithm
☐ logarithmic equation
☐ logarithmic function with base a

OBJECTIVE 1 Define a logarithm.

The graph of $y = 2^x$ is the curve shown in blue in **FIGURE 12**. Because $y = 2^x$ defines a one-to-one function, it has an inverse. Interchanging x and y gives

$$x = 2^y, \quad \text{the inverse of} \quad y = 2^x.$$

Roles of x and y are interchanged.

As we saw in **Section 9.1,** the graph of the inverse is found by reflecting the graph of $y = 2^x$ across the line $y = x$. The graph of $x = 2^y$ is shown as a red curve in **FIGURE 12**.

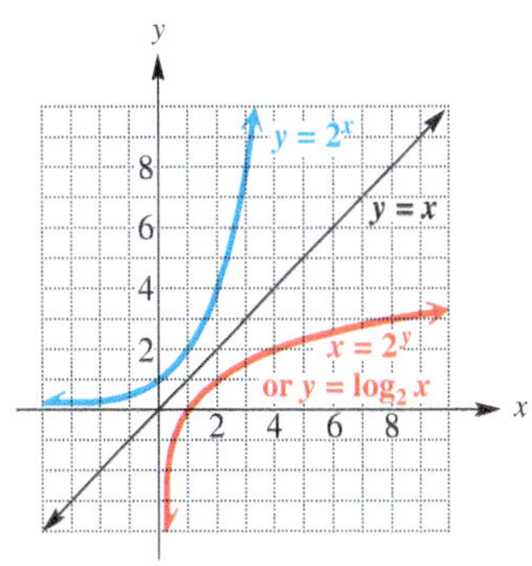

FIGURE 12

We can also write the equation of the red curve using a new notation that involves the concept of *logarithm*.

Logarithm

For all positive numbers a, where $a \neq 1$, and all positive numbers x,

$$y = \log_a x \quad \text{means the same as} \quad x = a^y.$$

The abbreviation **log** is used for the word **logarithm.** Read **$\log_a x$** as **"the logarithm of x with base a"** or **"the base a logarithm of x."** To remember the location of the base and the exponent in each form, refer to the following diagrams.

Logarithmic form	Exponential form
Logarithmic form: $y = \log_a x$ (Exponent: y; Base: a)	Exponential form: $x = a^y$ (Exponent: y; Base: a)

CLASSROOM EXAMPLE 1
Fill in the blanks with the equivalent forms.

Exponential Form	*Logarithmic Form*
$2^5 = 32$	______
$100^{1/2} = 10$	______
______	$\log_8 4 = \frac{2}{3}$
______	$\log_6 \frac{1}{1296} = -4$

Answers: $\log_2 32 = 5$; $\log_{100} 10 = \frac{1}{2}$; $8^{2/3} = 4$; $6^{-4} = \frac{1}{1296}$

Meaning of $\log_a x$

A logarithm is an exponent. ***The expression $\log_a x$ represents the exponent to which the base a must be raised to obtain x.***

OBJECTIVE 2 Convert between exponential and logarithmic forms, and evaluate logarithms.

We can use the definition of logarithm to carry out these conversions.

NOW TRY EXERCISE 1

(a) Write $6^3 = 216$ in logarithmic form.

(b) Write $\log_{64} 4 = \frac{1}{3}$ in exponential form.

EXAMPLE 1 Converting between Exponential and Logarithmic Forms

The table shows several pairs of equivalent forms.

Exponential Form	*Logarithmic Form*
$3^2 = 9$	$\log_3 9 = 2$
$\left(\frac{1}{5}\right)^{-2} = 25$	$\log_{1/5} 25 = -2$
$10^5 = 100{,}000$	$\log_{10} 100{,}000 = 5$
$4^{-3} = \frac{1}{64}$	$\log_4 \frac{1}{64} = -3$

$y = \log_a x$ means $x = a^y$.

NOW TRY

CLASSROOM EXAMPLE 2
Use a calculator to approximate each logarithm to four decimal places.

(a) $\log_2 13$ (b) $\log_6 18$
(c) $\log_{1/4} 20$ (d) $\log_{10} 50$

Answers: (a) 3.7004 (b) 1.6131 (c) −2.1610 (d) 1.6990

EXAMPLE 2 Evaluating Logarithms

Use a calculator to approximate each logarithm to four decimal places.

(a) $\log_2 5$ **(b)** $\log_3 12$ **(c)** $\log_{1/2} 12$ **(d)** $\log_{10} 20$

FIGURE 13 shows how a TI-84 Plus calculator approximates the logarithms in parts (a)–(c).

TEACHING TIP
The function for finding a logarithm on the TI-84 Plus, as shown in **FIGURES 13 and 14**, is found in the MATH menu, option A.

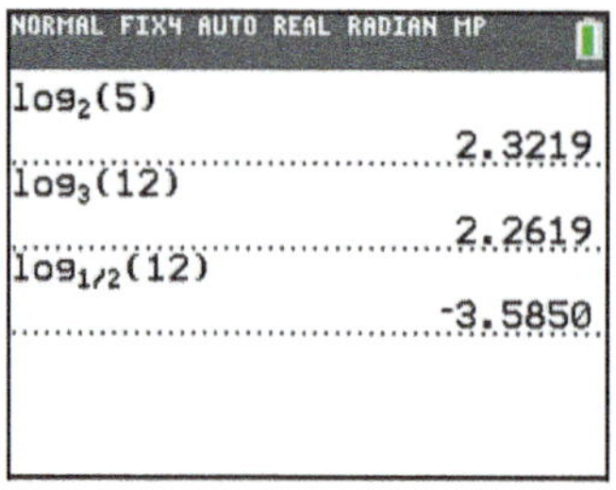

FIGURE 13

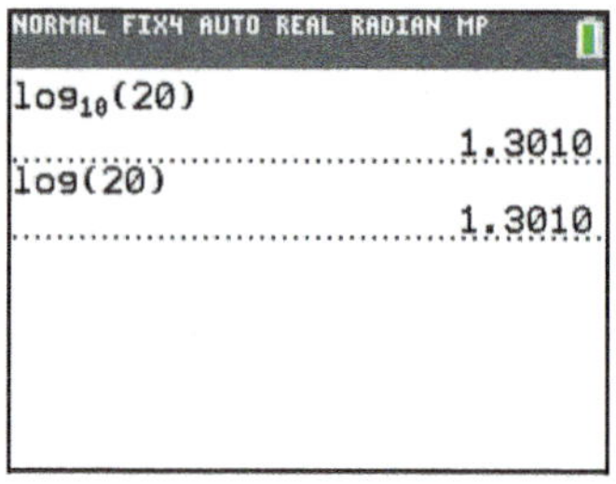

FIGURE 14

FIGURE 14 shows the approximation for the expression $\log_{10} 20$ in part (d). Notice that the second display, which indicates log 20 (with no base shown), gives the same result. We shall see in a later section that when no base is indicated, the base is understood to be 10. A base 10 logarithm is a *common logarithm.*

NOW TRY ANSWERS

1. **(a)** $\log_6 216 = 3$
   (b) $64^{1/3} = 4$

NOW TRY EXERCISE 2
Use a calculator to approximate each logarithm to four decimal places.
(a) $\log_2 7$ (b) $\log_5 8$
(c) $\log_{1/3} 12$ (d) $\log_{10} 18$

To four decimal places, the logarithms in parts (a)–(d) are as follows.

$$\log_2 5 \approx 2.3219, \qquad \log_3 12 \approx 2.2619,$$
$$\log_{1/2} 12 \approx -3.5850, \qquad \log_{10} 20 \approx 1.3010$$

NOW TRY

NOTE In **Section 9.5** we will introduce another method of calculating logarithms like those in **Example 2(a)–(c)** using the *change-of-base rule*.

OBJECTIVE 3 Solve logarithmic equations of the form $\log_a b = k$ for a, b, or k.

A **logarithmic equation** is an equation with a logarithm in at least one term.

CLASSROOM EXAMPLE 3
Solve each equation.
(a) $\log_3 x = -3$
(b) $\log_{1/3} (2x + 5) = 2$
(c) $\log_x 5 = 4$
(d) $\log_{36} \sqrt[4]{6} = x$
Answers: (a) $\left\{\frac{1}{27}\right\}$ (b) $\left\{-\frac{22}{9}\right\}$
(c) $\sqrt[4]{5}$ (d) $\left\{\frac{1}{8}\right\}$

EXAMPLE 3 Solving Logarithmic Equations

Solve each equation.

(a) $\log_4 x = -2$

By the definition of logarithm, $\log_4 x = -2$ is equivalent to $x = 4^{-2}$.

$$x = 4^{-2} = \frac{1}{16}$$

The solution set is $\left\{\frac{1}{16}\right\}$.

(b) $\log_{1/2} (3x + 1) = 2$

$$3x + 1 = \left(\frac{1}{2}\right)^2 \qquad \text{Write in exponential form.}$$

This is a key step.

$$3x + 1 = \frac{1}{4} \qquad \text{Apply the exponent.}$$

$$12x + 4 = 1 \qquad \text{Multiply each term by 4.}$$

$$12x = -3 \qquad \text{Subtract 4.}$$

$$x = -\frac{1}{4} \qquad \text{Divide by 12. Write in lowest terms.}$$

CHECK $\log_{1/2} (3x + 1) = 2$

$$\log_{1/2} \left[3\left(-\frac{1}{4}\right) + 1\right] \stackrel{?}{=} 2 \qquad \text{Let } x = -\tfrac{1}{4}.$$

$$\log_{1/2} \frac{1}{4} \stackrel{?}{=} 2 \qquad \text{Simplify within parentheses.}$$

$$\left(\frac{1}{2}\right)^2 \stackrel{?}{=} \frac{1}{4} \qquad \text{Write in exponential form.}$$

$$\frac{1}{4} = \frac{1}{4} \qquad \checkmark \text{ True}$$

The solution set is $\left\{-\frac{1}{4}\right\}$.

NOW TRY ANSWERS
2. (a) 2.8074 (b) 1.2920
(c) −2.2619 (d) 1.2553

NOW TRY EXERCISE 3

Solve each equation.

(a) $\log_2 x = -5$

(b) $\log_{3/2}(2x - 1) = 3$

(c) $\log_x 10 = 2$

(d) $\log_{125} \sqrt[3]{5} = x$

(c)

$$\log_x 3 = 2$$

$$x^2 = 3 \quad \text{Write in exponential form.}$$

$$x = \pm\sqrt{3} \quad \text{Take square roots.}$$

Be careful here. $-\sqrt{3}$ is extraneous.

Only the *principal* square root satisfies the equation because the base must be a positive number.

CHECK

$$\log_x 3 = 2$$

$$\log_{\sqrt{3}} 3 \stackrel{?}{=} 2 \quad \text{Let } x = \sqrt{3}.$$

$$(\sqrt{3})^2 \stackrel{?}{=} 3 \quad \text{Write in exponential form.}$$

$$3 = 3 \quad ✓ \text{ True}$$

The solution set is $\{\sqrt{3}\}$.

(d)

$$\log_{49} \sqrt[3]{7} = x$$

$$49^x = \sqrt[3]{7} \quad \text{Write in exponential form.}$$

$$(7^2)^x = 7^{1/3} \quad \text{Write with the same base.}$$

$$7^{2x} = 7^{1/3} \quad \text{Power rule for exponents}$$

$$2x = \frac{1}{3} \quad \text{Set exponents equal.}$$

$$x = \frac{1}{6} \quad \text{Divide by 2 } \left(\text{which is the same as multiplying by } \tfrac{1}{2}\right).$$

Check to verify that the solution set is $\left\{\frac{1}{6}\right\}$. NOW TRY

CLASSROOM EXAMPLE 4

Use the special properties to evaluate each expression.

(a) $\log_{12} 12$ (b) $\log_{\sqrt{3}} \sqrt{3}$

(c) $\log_3 1$ (d) $\log_{0.4} 1$

(e) $\log_4 4^7$ (f) $5^{\log_5 6}$

(g) $\log_4 64$

Answers: (a) 1 (b) 1 (c) 0 (d) 0 (e) 7 (f) 6 (g) 3

OBJECTIVE 4 Use the definition of logarithm to simplify logarithmic expressions.

The definition of logarithm allows us to state several special properties.

Special Properties of Logarithms

For any positive real number b, where $b \neq 1$, the following hold true.

$$\log_b b^r = r \qquad b^{\log_b r} = r \quad (\text{where } r > 0)$$

$$\log_b b = 1 \qquad \log_b 1 = 0$$

NOW TRY EXERCISE 4

Use the special properties to evaluate each expression.

(a) $\log_{10} 10$ **(b)** $\log_8 1$

(c) $\log_{0.1} 1$ **(d)** $\log_3 3^9$

(e) $5^{\log_5 3}$ **(f)** $\log_3 81$

EXAMPLE 4 Using Properties of Logarithms

Use the special properties to evaluate.

(a) $\log_7 7 = 1$ $\quad \log_b b = 1$

(b) $\log_{\sqrt{2}} \sqrt{2} = 1$

(c) $\log_9 1 = 0$ $\quad \log_b 1 = 0$

(d) $\log_{0.2} 1 = 0$

(e) $\log_2 2^6 = 6$ $\quad \log_b b^r = r$

(f) $\log_3 3^{-2.5} = -2.5$

(g) $4^{\log_4 9} = 9$ $\quad b^{\log_b r} = r$

(h) $10^{\log_{10} 13} = 13$

(i) $\log_2 32$

We know that $32 = 2^5$, so we replace 32 with 2^5 and use the first property above.

$$\log_2 32 = \log_2 2^5 = 5$$

NOW TRY

NOW TRY ANSWERS

3. (a) $\left\{\frac{1}{32}\right\}$ (b) $\left\{\frac{35}{16}\right\}$ (c) $\{\sqrt{10}\}$ (d) $\left\{\frac{1}{9}\right\}$

4. (a) 1 (b) 0 (c) 0 (d) 9 (e) 3 (f) 4

OBJECTIVE 5 Define and graph logarithmic functions.

Logarithmic Function

If a and x are positive numbers, where $a \neq 1$, then

$$g(x) = \log_a x$$

defines the **logarithmic function with base a.**

NOW TRY EXERCISE 5

Graph $f(x) = \log_6 x$.

CLASSROOM EXAMPLE 5

Graph $f(x) = \log_4 x$.

Answer:

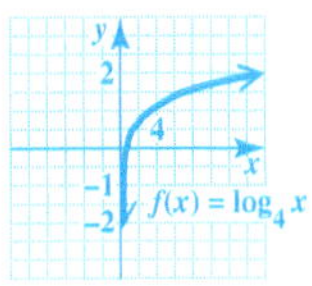

EXAMPLE 5 Graphing a Logarithmic Function ($a > 1$)

Graph $f(x) = \log_2 x$.

By writing $y = f(x) = \log_2 x$ in exponential form as

$$x = 2^y,$$

we can identify ordered pairs that satisfy the equation. It is easier to choose values for y and find the corresponding values of x. Plotting the points in the table of ordered pairs and connecting them with a smooth curve gives the graph in **FIGURE 15.** This graph is typical of logarithmic functions with base $a > 1$.

$x = 2^y$	y
$\frac{1}{4}$	-2
$\frac{1}{2}$	-1
1	0
2	1
4	2

Be careful to write the x- and y-values in the correct order.

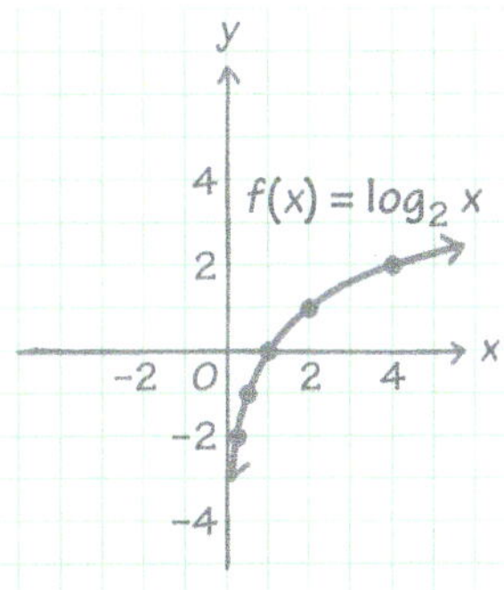

FIGURE 15

Logarithmic function with base $a > 1$

Domain: $(0, \infty)$

Range: $(-\infty, \infty)$

x-intercept: $(1, 0)$

The function is one-to-one, and its graph rises from left to right.

NOW TRY

CLASSROOM EXAMPLE 6

Graph $g(x) = \log_{1/10} x$.

Answer:

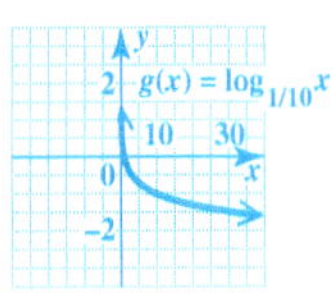

NOW TRY EXERCISE 6

Graph $g(x) = \log_{1/4} x$.

EXAMPLE 6 Graphing a Logarithmic Function ($0 < a < 1$)

Graph $g(x) = \log_{1/2} x$.

We write $y = g(x) = \log_{1/2} x$ in exponential form as

$$x = \left(\frac{1}{2}\right)^y,$$

then choose values for y and find the corresponding values of x. Plotting these points and connecting them with a smooth curve gives the graph in **FIGURE 16.** This graph is typical of logarithmic functions with base $0 < a < 1$.

$x = \left(\frac{1}{2}\right)^y$	y
$\frac{1}{4}$	2
$\frac{1}{2}$	1
1	0
2	-1
4	-2

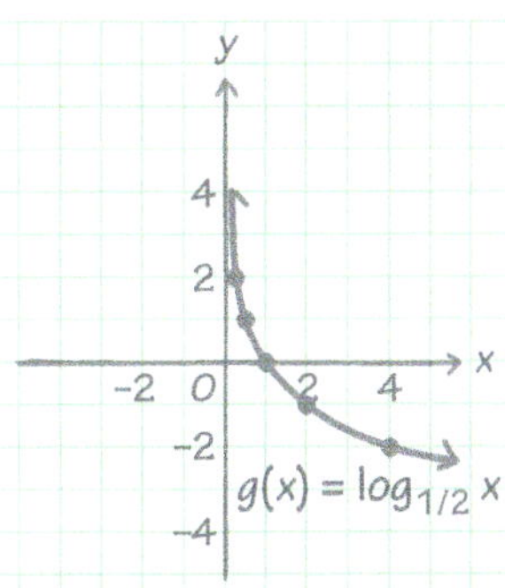

FIGURE 16

Logarithmic function with base $0 < a < 1$

Domain: $(0, \infty)$

Range: $(-\infty, \infty)$

x-intercept: $(1, 0)$

The function is one-to-one, and its graph falls from left to right.

NOW TRY

NOW TRY ANSWERS

5.

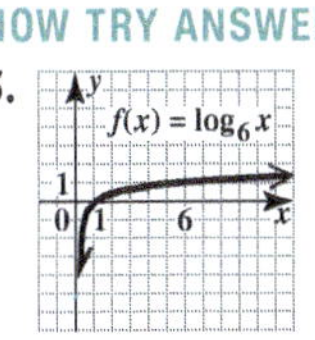

6.

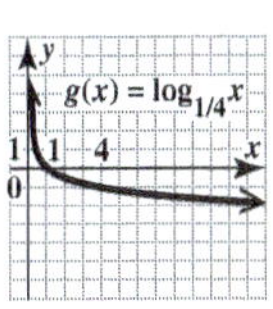

TEACHING TIP
To emphasize that the domain of the logarithmic function is strictly positive, have students try to find

$\log 0$ and $\log(-2)$

on their calculators and comment on the results.

Characteristics of the Graph of $g(x) = \log_a x$

1. The graph contains the point $(1, 0)$, which is its x-intercept.
2. The function is one-to-one.
   - When $a > 1$, the graph will *rise* from left to right, from the fourth quadrant to the first. (See **FIGURE 15.**)
   - When $0 < a < 1$, the graph will *fall* from left to right, from the first quadrant to the fourth. (See **FIGURE 16.**)
3. The graph will approach the y-axis, but never touch it. (The y-axis is an asymptote.)
4. The domain is $(0, \infty)$, and the range is $(-\infty, \infty)$.

NOTE See the similar box titled

"Characteristics of the Graph of $f(x) = a^x$"

in **Section 9.2.** Compare the four characteristics one by one to see how the concepts of inverse functions, introduced in **Section 9.1,** are illustrated by these two classes of functions.

OBJECTIVE 6 Use logarithmic functions in applications involving growth or decay.

NOW TRY EXERCISE 7
Suppose the gross national product (GNP) of a small country (in millions of dollars) is approximated by

$$G(t) = 15.0 + 2.00 \log_{10} t,$$

where t is time in years since 2013. Approximate to the nearest tenth the GNP for each value of t.

(a) $t = 1$ **(b)** $t = 10$

EXAMPLE 7 Solving an Application of a Logarithmic Function

The barometric pressure in inches of mercury at a distance of x miles from the eye of a typical hurricane can be approximated by

$$f(x) = 27 + 1.105 \log_{10}(x + 1).$$

(*Source:* Miller, A. and R. Anthes, *Meteorology,* Fifth Edition, Charles E. Merrill Publishing Company.) Approximate the pressure 9 mi from the eye of the hurricane.

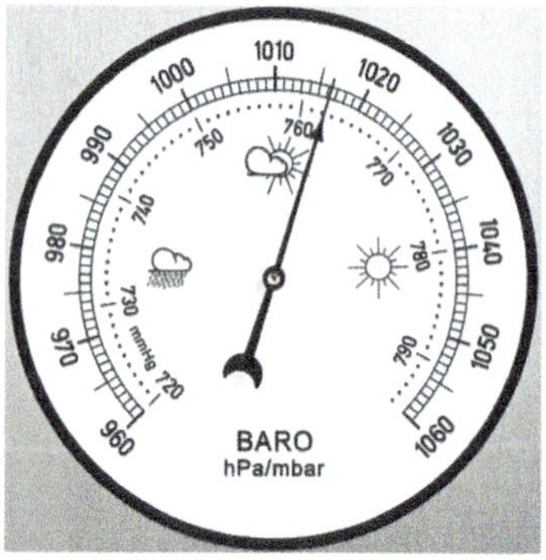

Let $x = 9$, and find $f(9)$.

$$f(x) = 27 + 1.105 \log_{10}(x + 1)$$

$$f(9) = 27 + 1.105 \log_{10}(9 + 1) \quad \text{Let } x = 9.$$

$$f(9) = 27 + 1.105 \log_{10} 10 \quad \text{Add inside parentheses.}$$

$$f(9) = 27 + 1.105(1) \quad \log_{10} 10 = 1$$

$$f(9) = 28.105 \quad \text{Add.}$$

The pressure 9 mi from the eye of the hurricane is 28.105 in. NOW TRY

CLASSROOM EXAMPLE 7
A population of mites in a laboratory is growing according to the function

$$P(t) = 80 \log_{10}(t + 10),$$

where t is the number of days after a study is begun.

(a) Find the number of mites at the beginning of the study.
(b) Find the number present after 90 days.

Answers: (a) 80 (b) 160

NOW TRY ANSWERS
7. (a) \$15.0 million
(b) \$17.0 million

9.3 Exercises

FOR EXTRA HELP

MyMathLab®

Complete solution available in MyMathLab

1. (a) B (b) E (c) D (d) F (e) A (f) C
2. (a) C (b) F (c) B (d) A (e) E (f) D
3. $(0, \infty)$; $(-\infty, \infty)$
4. $f(x) = 3^x$
5. $\log_4 1024 = 5$
6. $\log_3 729 = 6$
7. $\log_{1/2} 8 = -3$
8. $\log_{1/6} 216 = -3$
9. $\log_{10} 0.001 = -3$
10. $\log_{36} 6 = \frac{1}{2}$
11. $\log_{625} 5 = \frac{1}{4}$
12. $\log_{343} 7 = \frac{1}{3}$
13. $\log_8 \frac{1}{4} = -\frac{2}{3}$
14. $\log_{16} \frac{1}{8} = -\frac{3}{4}$
15. $\log_5 1 = 0$
16. $\log_7 1 = 0$
17. $4^3 = 64$
18. $2^9 = 512$
19. $10^{-4} = \frac{1}{10{,}000}$
20. $100^1 = 100$
21. $6^0 = 1$
22. $\pi^0 = 1$
23. $9^{1/2} = 3$
24. $64^{1/6} = 2$
25. $\left(\frac{1}{4}\right)^{1/2} = \frac{1}{2}$
26. $\left(\frac{1}{8}\right)^{1/3} = \frac{1}{2}$
27. $5^{-1} = 5^{-1}$
28. $10^{-2} = 10^{-2}$
29. (a) C (b) B (c) B (d) C
30. By using the word "radically," the teacher meant for him to consider roots. Because 3 is the square (2nd) root of 9, $\log_9 3 = \frac{1}{2}$.

1. *Concept Check* Match each logarithmic equation in Column I with the corresponding exponential equation in Column II.

I	II
(a) $\log_{1/3} 3 = -1$	**A.** $8^{1/3} = \sqrt[3]{8}$
(b) $\log_5 1 = 0$	**B.** $\left(\frac{1}{3}\right)^{-1} = 3$
(c) $\log_2 \sqrt{2} = \frac{1}{2}$	**C.** $4^1 = 4$
(d) $\log_{10} 1000 = 3$	**D.** $2^{1/2} = \sqrt{2}$
(e) $\log_8 \sqrt[3]{8} = \frac{1}{3}$	**E.** $5^0 = 1$
(f) $\log_4 4 = 1$	**F.** $10^3 = 1000$

2. *Concept Check* Match each logarithm in Column I with its corresponding value in Column II.

I	II
(a) $\log_4 16$	**A.** -2
(b) $\log_3 81$	**B.** -1
(c) $\log_3 \left(\frac{1}{3}\right)$	**C.** 2
(d) $\log_{10} 0.01$	**D.** 0
(e) $\log_5 \sqrt{5}$	**E.** $\frac{1}{2}$
(f) $\log_{13} 1$	**F.** 4

3. *Concept Check* The domain of $f(x) = a^x$ is $(-\infty, \infty)$, while the range is $(0, \infty)$. Therefore because $g(x) = \log_a x$ is the inverse of f, the domain of g is ________, while the range of g is ________.

4. *Concept Check* The graphs of both $f(x) = 3^x$ and $g(x) = \log_3 x$ rise from left to right. Which one rises at a faster rate?

Write in logarithmic form. ***See Example 1.***

5. $4^5 = 1024$ **6.** $3^6 = 729$ **7.** $\left(\frac{1}{2}\right)^{-3} = 8$ **8.** $\left(\frac{1}{6}\right)^{-3} = 216$

9. $10^{-3} = 0.001$ **10.** $36^{1/2} = 6$ **11.** $\sqrt[4]{625} = 5$ **12.** $\sqrt[3]{343} = 7$

13. $8^{-2/3} = \frac{1}{4}$ **14.** $16^{-3/4} = \frac{1}{8}$ **15.** $5^0 = 1$ **16.** $7^0 = 1$

Write in exponential form. ***See Example 1.***

17. $\log_4 64 = 3$ **18.** $\log_2 512 = 9$ **19.** $\log_{10} \frac{1}{10{,}000} = -4$

20. $\log_{100} 100 = 1$ **21.** $\log_6 1 = 0$ **22.** $\log_\pi 1 = 0$

23. $\log_9 3 = \frac{1}{2}$ **24.** $\log_{64} 2 = \frac{1}{6}$ **25.** $\log_{1/4} \frac{1}{2} = \frac{1}{2}$

26. $\log_{1/8} \frac{1}{2} = \frac{1}{3}$ **27.** $\log_5 5^{-1} = -1$ **28.** $\log_{10} 10^{-2} = -2$

29. *Concept Check* Match each logarithm in Column I with its value in Column II.

I	II
(a) $\log_8 8$	**A.** -1
(b) $\log_{16} 1$	**B.** 0
(c) $\log_{0.3} 1$	**C.** 1
(d) $\log_{\sqrt{7}} \sqrt{7}$	**D.** 0.1

30. *Concept Check* When a student asked his teacher to explain how to evaluate

$$\log_9 3$$

without showing any work, his teacher told him to "Think radically." Explain what the teacher meant by this hint.

31. 3.1699 **32.** 3.9069
33. 1.7959 **34.** 2.0244
35. −1.7925 **36.** −2.0478
37. −1.5850 **38.** −2.8074
39. 1.9243 **40.** 2.1004
41. 1.6990 **42.** 1.9542

43. $\left\{\frac{1}{3}\right\}$ **44.** $\left\{\frac{1}{3}\right\}$
45. $\{81\}$ **46.** $\{25\}$
47. $\left\{\frac{1}{5}\right\}$ **48.** $\left\{\frac{1}{2}\right\}$
49. $\{1\}$ **50.** $\{1\}$
51. $\{x \mid x > 0, x \neq 1\}$
52. $\{x \mid x > 0, x \neq 1\}$
53. $\{5\}$ **54.** $\{10\}$
55. $\left\{\frac{5}{3}\right\}$ **56.** $\left\{\frac{3}{4}\right\}$
57. $\{4\}$ **58.** $\{9\}$
59. $\left\{\frac{3}{2}\right\}$ **60.** $\left\{\frac{3}{2}\right\}$
61. $\{30\}$ **62.** $\{37\}$

63. 1 **64.** 1
65. 0 **66.** 0
67. 9 **68.** 6
69. −1 **70.** −6
71. 9 **72.** 3
73. 5 **74.** 11
75. 6 **76.** 7
77. 4 **78.** 3
79. −1 **80.** −1
81. $\frac{1}{3}$ **82.** $\frac{1}{3}$

83.

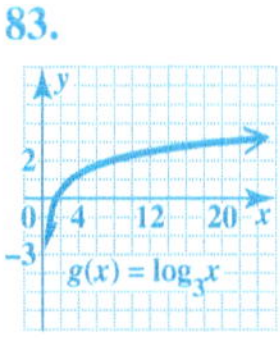

84.

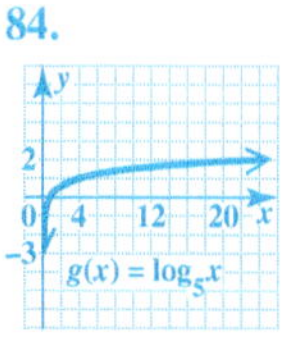

85.

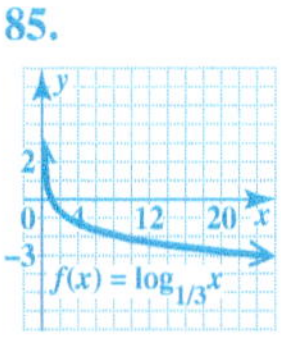

86.

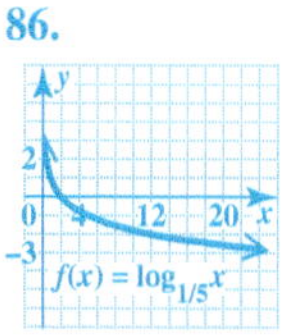

87.

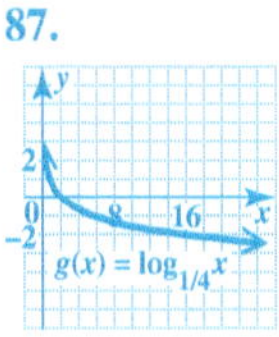

88.

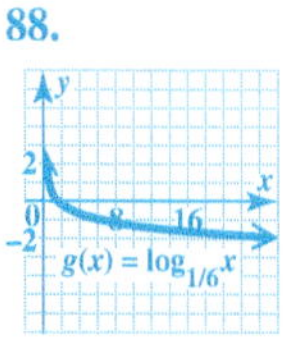

89. Every power of 1 is equal to 1, and thus it cannot be used as a base.
90. Answers will vary.

91. 8 **92.** 16 **93.** 24
94. $f(0) = 8$, $f(10) = 16$, and $f(60) = 24$, so the points $(0, 8)$, $(10, 16)$, and $(60, 24)$ all lie on the graph of f.

Use a calculator to approximate each logarithm to four decimal places. **See Example 2.**

31. $\log_2 9$ **32.** $\log_2 15$ **33.** $\log_5 18$ **34.** $\log_5 26$

35. $\log_{1/4} 12$ **36.** $\log_{1/5} 27$ **37.** $\log_2 \left(\frac{1}{3}\right)$ **38.** $\log_2 \left(\frac{1}{7}\right)$

39. $\log_{10} 84$ **40.** $\log_{10} 126$ **41.** $\log 50$ **42.** $\log 90$

Solve each equation. **See Example 3.**

43. $x = \log_{27} 3$ **44.** $x = \log_{125} 5$ **45.** $\log_x 9 = \frac{1}{2}$ **46.** $\log_x 5 = \frac{1}{2}$

47. $\log_x 125 = -3$ **48.** $\log_x 64 = -6$ **49.** $\log_{12} x = 0$ **50.** $\log_4 x = 0$

51. $\log_x x = 1$ **52.** $\log_x 1 = 0$ **53.** $\log_x \frac{1}{25} = -2$

54. $\log_x \frac{1}{10} = -1$ **55.** $\log_8 32 = x$ **56.** $\log_{81} 27 = x$

57. $\log_\pi \pi^4 = x$ **58.** $\log_{\sqrt{2}} (\sqrt{2})^9 = x$ **59.** $\log_6 \sqrt{216} = x$

60. $\log_4 \sqrt{64} = x$ **61.** $\log_4 (2x + 4) = 3$ **62.** $\log_3 (2x + 7) = 4$

Use the special properties of logarithms to evaluate each expression. **See Example 4.**

63. $\log_3 3$ **64.** $\log_8 8$ **65.** $\log_5 1$ **66.** $\log_{12} 1$

67. $\log_4 4^9$ **68.** $\log_5 5^6$ **69.** $\log_2 2^{-1}$ **70.** $\log_4 4^{-6}$

71. $6^{\log_6 9}$ **72.** $12^{\log_{12} 3}$ **73.** $8^{\log_8 5}$ **74.** $5^{\log_5 11}$

75. $\log_2 64$ **76.** $\log_2 128$ **77.** $\log_3 81$ **78.** $\log_3 27$

79. $\log_4 \left(\frac{1}{4}\right)$ **80.** $\log_6 \left(\frac{1}{6}\right)$ **81.** $\log_6 \sqrt[3]{6}$ **82.** $\log_9 \sqrt[3]{9}$

If (p, q) is on the graph of $f(x) = a^x$ (for $a > 0$ and $a \neq 1$), then (q, p) is on the graph of $f^{-1}(x) = \log_a x$. Use this fact, and refer to the graphs required in **Exercises 19–24** *of* **Section 9.2** *to graph each logarithmic function.* **See Examples 5 and 6.**

83. $g(x) = \log_3 x$ **84.** $g(x) = \log_5 x$ **85.** $f(x) = \log_{1/3} x$ **86.** $f(x) = \log_{1/5} x$

87. $g(x) = \log_{1/4} x$ $\left(\textit{Hint: } 4^{-x} = \left(\frac{1}{4}\right)^x.\right)$ **88.** $g(x) = \log_{1/6} x$ $\left(\textit{Hint: } 6^{-x} = \left(\frac{1}{6}\right)^x.\right)$

89. ***Concept Check*** Explain why 1 is not allowed as a base for a logarithmic function.

90. ***Concept Check*** Compare the summary of facts about the graph of $f(x) = a^x$ in **Section 9.2** with the similar summary of facts about the graph of $g(x) = \log_a x$ in this section. Make a list of the facts that reinforce the concept that f and g are inverse functions.

Use the graph at the right to predict the value of $f(t)$ for the given value of t.

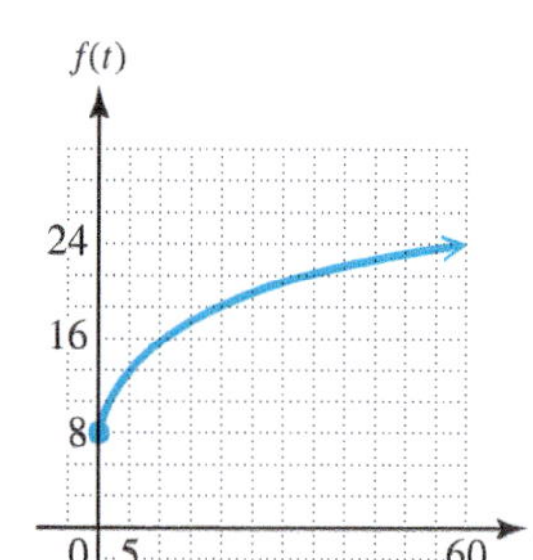

91. $t = 0$ **92.** $t = 10$ **93.** $t = 60$

94. Show that the points determined in **Exercises 91–93** lie on the graph of

$$f(t) = 8 \log_5 (2t + 5).$$

95. (a) 130 thousand units
(b) 190 thousand units
(c)

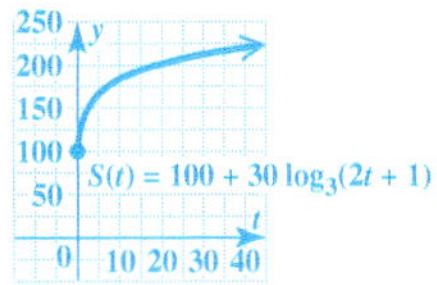

96. (a) 6 (b) 12 (c) 18
(d)

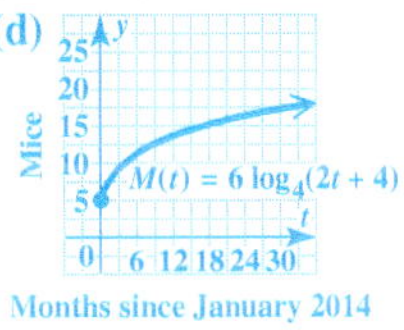

97. about 4 times as powerful
98. about 316 times as powerful

Solve each problem. ***See Example 7.***

95. Sales (in thousands of units) of a new product are approximated by the function

$$S(t) = 100 + 30 \log_3 (2t + 1),$$

where t is the number of years after the product is introduced.

(a) What were the sales, to the nearest unit, after 1 yr?

(b) What were the sales, to the nearest unit, after 13 yr?

(c) Graph $y = S(t)$.

96. A study showed that the number of mice in an old abandoned house was approximated by the function

$$M(t) = 6 \log_4 (2t + 4),$$

where t is measured in months and $t = 0$ corresponds to January 2014. Find the number of mice in the house for each month.

(a) January 2014 **(b)** July 2014 **(c)** July 2016 **(d)** Graph $y = M(t)$.

The ***Richter scale*** *is used to measure the intensity of earthquakes. The Richter scale rating of an earthquake of intensity x is given by*

$$R = \log_{10} \frac{x}{x_0},$$

where x_0 is the intensity of an earthquake of a certain (small) size. The figure here shows Richter scale ratings for selected Southern California earthquakes with magnitudes greater than 4.7.

Southern California Earthquakes
(with magnitudes greater than 4.7)

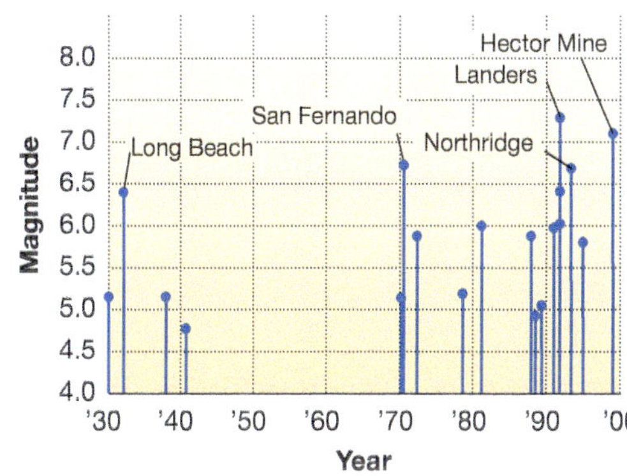

Source: Caltech; U.S. Geological Survey.

97. The Northridge earthquake had a Richter scale rating of 6.7. The Landers earthquake had a rating of 7.3. How much more powerful was the Landers quake than the Northridge quake?

98. Compare the smallest rated earthquake in the figure (at 4.8) with the Landers quake (at 7.3). How much more powerful was the Landers quake?

9.4 Properties of Logarithms

OBJECTIVES

1. Use the product rule for logarithms.
2. Use the quotient rule for logarithms.
3. Use the power rule for logarithms.
4. Use properties to write alternative forms of logarithmic expressions.

Logarithms were used as an aid to numerical calculation for several hundred years. Today the widespread use of calculators has made the use of logarithms for calculation obsolete. However, logarithms are still very important in applications and in further work in mathematics.

OBJECTIVE 1 Use the product rule for logarithms.

One way in which logarithms simplify problems is by changing a problem of multiplication into one of addition. We know that $\log_2 4 = 2$, $\log_2 8 = 3$, and $\log_2 32 = 5$.

$$\log_2 32 = \log_2 4 + \log_2 8 \qquad 5 = 2 + 3$$

$$\log_2 (4 \cdot 8) = \log_2 4 + \log_2 8 \qquad 32 = 4 \cdot 8$$

This is an example of the following rule.

Product Rule for Logarithms

If x, y, and b are positive real numbers, where $b \neq 1$, then the following holds true.

$$\log_b xy = \log_b x + \log_b y$$

That is, the logarithm of a product is the sum of the logarithms of the factors.

NOTE The word statement of the product rule can be restated by replacing "logarithm" with "exponent." The rule then becomes the familiar rule for multiplying exponential expressions:

The *exponent* of a product is the sum of the *exponents* of the factors.

To prove this rule, let $m = \log_b x$ and $n = \log_b y$, and recall that

$$\log_b x = m \quad \text{means} \quad b^m = x \quad \text{and} \quad \log_b y = n \quad \text{means} \quad b^n = y.$$

Now consider the product xy.

$xy = b^m \cdot b^n$ Substitute.

$xy = b^{m+n}$ Product rule for exponents

$\log_b xy = m + n$ Convert to logarithmic form.

$\log_b xy = \log_b x + \log_b y$ Substitute.

The last statement is the result we wished to prove.

CLASSROOM EXAMPLE 1
Use the product rule to rewrite each logarithm.
(a) $\log_4 (3 \cdot 7)$
(b) $\log_8 10 + \log_8 3$
(c) $\log_8 8k$, $k > 0$
(d) $\log_5 m^2$, $m > 0$

Answers: (a) $\log_4 3 + \log_4 7$
(b) $\log_8 30$ (c) $1 + \log_8 k$
(d) $2 \log_5 m$

NOW TRY EXERCISE 1
Use the product rule to rewrite each logarithm.
(a) $\log_{10} (7 \cdot 9)$
(b) $\log_5 11 + \log_5 8$
(c) $\log_5 (5x)$, $x > 0$
(d) $\log_2 t^3$, $t > 0$

EXAMPLE 1 Using the Product Rule

Use the product rule to rewrite each logarithm. Assume $x > 0$.

(a) $\log_5 (6 \cdot 9)$

$= \log_5 6 + \log_5 9$ Product rule

(b) $\log_7 8 + \log_7 12$

$= \log_7 (8 \cdot 12)$ Product rule

$= \log_7 96$ Multiply.

(c) $\log_3 (3x)$

$= \log_3 3 + \log_3 x$ Product rule

$= 1 + \log_3 x$ $\log_3 3 = 1$

(d) $\log_4 x^3$

$= \log_4 (x \cdot x \cdot x)$ $x^3 = x \cdot x \cdot x$

$= \log_4 x + \log_4 x + \log_4 x$ Product rule

$= 3 \log_4 x$ Combine like terms.

NOW TRY

TEACHING TIP
Caution students that

$$\frac{\log_b x}{\log_b y} \neq \log_b x - \log_b y.$$

NOW TRY ANSWERS
1. (a) $\log_{10} 7 + \log_{10} 9$
(b) $\log_5 88$
(c) $1 + \log_5 x$
(d) $3 \log_2 t$

OBJECTIVE 2 Use the quotient rule for logarithms.

The rule for division is similar to the rule for multiplication.

Quotient Rule for Logarithms

If x, y, and b are positive real numbers, where $b \neq 1$, then the following holds true.

$$\log_b \frac{x}{y} = \log_b x - \log_b y$$

That is, the logarithm of a quotient is the difference of the logarithm of the numerator and the logarithm of the denominator.

The proof of this rule is similar to the proof of the product rule.

NOW TRY EXERCISE 2

Use the quotient rule to rewrite each logarithm.

(a) $\log_{10} \frac{7}{9}$

(b) $\log_4 x - \log_4 12, \quad x > 0$

(c) $\log_5 \frac{25}{27}$

CLASSROOM EXAMPLE 2

Use the quotient rule to rewrite each logarithm.

(a) $\log_7 \frac{9}{4}$

(b) $\log_3 p - \log_3 q,\ p > 0,\ q > 0$

(c) $\log_4 \frac{3}{16}$

Answers: (a) $\log_7 9 - \log_7 4$
(b) $\log_3 \frac{p}{q}$ (c) $\log_4 3 - 2$

EXAMPLE 2 Using the Quotient Rule

Use the quotient rule to rewrite each logarithm. Assume $x > 0$.

(a) $\log_4 \frac{7}{9}$

$= \log_4 7 - \log_4 9$ Quotient rule

(b) $\log_5 6 - \log_5 x$

$= \log_5 \frac{6}{x}$ Quotient rule

(c) $\log_3 \frac{27}{5}$

$= \log_3 27 - \log_3 5$ Quotient rule

$= 3 - \log_3 5$ $\log_3 27 = 3$

(d) $\log_6 28 - \log_6 7$

$= \log_6 \frac{28}{7}$ Quotient rule

$= \log_6 4$ $\frac{28}{7} = 4$ NOW TRY

CAUTION ***There is no property of logarithms to rewrite the logarithm of a sum or difference.*** For example, we ***cannot*** write $\log_b (x + y)$ in terms of $\log_b x$ and $\log_b y$. Also,

$$\log_b \frac{x}{y} \neq \frac{\log_b x}{\log_b y}.$$

OBJECTIVE 3 Use the power rule for logarithms.

An exponential expression such as

$$2^3 \quad \text{means} \quad 2 \cdot 2 \cdot 2.$$

The base is used as a factor 3 times. Similarly, the product rule can be extended to rewrite the logarithm of a power as the product of the exponent and the logarithm of the base.

$\log_5 2^3$

$= \log_5 (2 \cdot 2 \cdot 2)$

$= \log_5 2 + \log_5 2 + \log_5 2$

$= 3 \log_5 2$

$\log_2 7^4$

$= \log_2 (7 \cdot 7 \cdot 7 \cdot 7)$

$= \log_2 7 + \log_2 7 + \log_2 7 + \log_2 7$

$= 4 \log_2 7$

Furthermore, we saw in **Example 1(d)** that $\log_4 x^3 = 3 \log_4 x$. These examples suggest the following rule.

TEACHING TIP

Be sure that students can distinguish between the two expressions

$\log_b x^r$ and $(\log_b x)^r$.

The power rule applies only to the first of the two expressions.

Power Rule for Logarithms

If x and b are positive real numbers, where $b \neq 1$, and if r is any real number, then the following holds true.

$$\log_b x^r = r \log_b x$$

That is, the logarithm of a number to a power equals the exponent times the logarithm of the number.

Examples: $\log_b m^5 = 5 \log_b m$ and $\log_3 5^4 = 4 \log_3 5$

NOW TRY ANSWERS

2. (a) $\log_{10} 7 - \log_{10} 9$
(b) $\log_4 \frac{x}{12}$
(c) $2 - \log_5 27$

To prove the power rule, let $\log_b x = m$.

$b^m = x$	Convert to exponential form.
$(b^m)^r = x^r$	Raise to the power r.
$b^{mr} = x^r$	Power rule for exponents
$\log_b x^r = rm$	Convert to logarithmic form; commutative property
$\log_b x^r = r \log_b x$	$m = \log_b x$ from above

This is the statement to be proved.

TEACHING TIP
For a review of rational exponents, refer students to **Section 7.2.**

As a special case of the power rule, let $r = \frac{1}{p}$, so

$$\log_b \sqrt[p]{x} = \log_b x^{1/p} = \frac{1}{p} \log_b x.$$

For example, using this result, with $x > 0$,

$$\log_b \sqrt[5]{x} = \log_b x^{1/5} = \frac{1}{5} \log_b x \quad \text{and} \quad \log_b \sqrt[3]{x^4} = \log_b x^{4/3} = \frac{4}{3} \log_b x.$$

Another special case is as follows.

$$\log_b \frac{1}{x} = \log_b x^{-1} = -\log_b x$$

NOW TRY EXERCISE 3
Use the power rule to rewrite each logarithm. Assume $a > 0$, $x > 0$, and $a \neq 1$.

(a) $\log_7 5^3$ **(b)** $\log_a \sqrt{10}$

(c) $\log_3 \sqrt[4]{x^3}$ **(d)** $\log_4 \frac{1}{x^5}$

EXAMPLE 3 Using the Power Rule

Use the power rule to rewrite each logarithm. Assume $b > 0$, $x > 0$, and $b \neq 1$.

(a) $\log_5 4^2$

$= 2 \log_5 4$

(b) $\log_b x^5$

$= 5 \log_b x$

(c) $\log_b \sqrt{7}$

$= \log_b 7^{1/2}$ $\sqrt{x} = x^{1/2}$

$= \frac{1}{2} \log_b 7$ Power rule

(d) $\log_2 \sqrt[5]{x^2}$

$= \log_2 x^{2/5}$ $\sqrt[5]{x^2} = x^{2/5}$

$= \frac{2}{5} \log_2 x$ Power rule

(e) $\log_3 \frac{1}{x^4}$

$= \log_3 x^{-4}$ Definition of negative exponent

$= -4 \log_3 x$ Power rule

NOW TRY

CLASSROOM EXAMPLE 3
Use the power rule to rewrite each logarithm. Assume $a > 0$, $b > 0$, $x > 0$, $a \neq 1$, and $b \neq 1$.

(a) $\log_3 5^2$ (b) $\log_a x^4$

(c) $\log_b \sqrt{8}$ (d) $\log_2 \sqrt[7]{x^4}$

Answers: (a) $2 \log_3 5$ (b) $4 \log_a x$ (c) $\frac{1}{2} \log_b 8$ (d) $\frac{4}{7} \log_2 x$

We summarize the properties of logarithms from the previous section and this one.

Properties of Logarithms

If x, y, and b are positive real numbers, where $b \neq 1$, and r is any real number, then the following hold true.

Special Properties (from Section 9.3)	$\log_b b^r = r$	$b^{\log_b r} = r$ (where $r > 0$)
Product Rule	$\log_b xy = \log_b x + \log_b y$	
Quotient Rule	$\log_b \frac{x}{y} = \log_b x - \log_b y$	
Power Rule	$\log_b x^r = r \log_b x$	

NOW TRY ANSWERS

3. **(a)** $3 \log_7 5$ **(b)** $\frac{1}{2} \log_a 10$ **(c)** $\frac{3}{4} \log_3 x$ **(d)** $-5 \log_4 x$

OBJECTIVE 4 Use properties to write alternative forms of logarithmic expressions.

EXAMPLE 4 Writing Logarithms in Alternative Forms

Use properties of logarithms to rewrite each expression if possible. Assume that all variables represent positive real numbers.

(a) $\log_4 4x^3$

$= \log_4 4 + \log_4 x^3$ Product rule

$= 1 + 3 \log_4 x$ $\log_4 4 = 1$; power rule

(b) $\log_7 \sqrt{\dfrac{m}{n}}$

$= \log_7 \left(\dfrac{m}{n}\right)^{1/2}$ Write the radical expression with a rational exponent.

$= \dfrac{1}{2} \log_7 \dfrac{m}{n}$ Power rule

$= \dfrac{1}{2}(\log_7 m - \log_7 n)$ Quotient rule

(c) $\log_5 \dfrac{a^2}{bc}$

$= \log_5 a^2 - \log_5 bc$ Quotient rule

$= 2 \log_5 a - \log_5 bc$ Power rule

$= 2 \log_5 a - (\log_5 b + \log_5 c)$ Product rule

$= 2 \log_5 a - \log_5 b - \log_5 c$

Parentheses are necessary here.

(d) $4 \log_b m - \log_b n, \quad b \neq 1$

$= \log_b m^4 - \log_b n$ Power rule

$= \log_b \dfrac{m^4}{n}$ Quotient rule

(e) $\log_b (x + 1) + \log_b (2x + 1) - \dfrac{2}{3} \log_b x, \quad b \neq 1$

$= \log_b (x + 1) + \log_b (2x + 1) - \log_b x^{2/3}$ Power rule

$= \log_b \dfrac{(x + 1)(2x + 1)}{x^{2/3}}$ Product and quotient rules

$= \log_b \dfrac{2x^2 + 3x + 1}{x^{2/3}}$ Multiply in the numerator.

(f) $\log_8 (2p + 3r)$ cannot be rewritten using the properties of logarithms. ***There is no property of logarithms to rewrite the logarithm of a sum.*** NOW TRY

In the next example, we use numerical values for $\log_2 5$ and $\log_2 3$. While we use the equality symbol to give these values, they are actually approximations because most logarithms of this type are irrational numbers. ***We use = with the understanding that the values are correct to four decimal places.***

NOW TRY EXERCISE 4

Use properties of logarithms to rewrite each expression if possible. Assume that all variables represent positive real numbers.

(a) $\log_3 9z^4$

(b) $\log_6 \sqrt{\dfrac{n}{3m}}$

(c) $\log_2 x + 3 \log_2 y - \log_2 z$

(d) $\log_5 (x + 10) + \log_5 (x - 10) - \dfrac{3}{5} \log_5 x, \quad x > 10$

(e) $\log_7 (49 + 2x)$

CLASSROOM EXAMPLE 4

Use properties of logarithms to rewrite each expression if possible. Assume that all variables represent positive real numbers.

(a) $\log_6 36m^5$

(b) $\log_2 \sqrt{9z}$

(c) $\log_b \dfrac{8r^2}{m - 1}, \quad m > 1, b \neq 1$

(d) $2 \log_b x + 3 \log_b y, \quad b \neq 1$

(e) $\log_4 (3x + y)$

Answers: (a) $2 + 5 \log_6 m$

(b) $\log_2 3 + \frac{1}{2} \log_2 z$

(c) $\log_b 8 + 2 \log_b r - \log_b (m - 1)$

(d) $\log_b x^2y^3$

(e) cannot be rewritten

TEACHING TIP

Use exercises similar to **Examples 3 and 4** to orally check students' understanding of the rules in this section.

NOW TRY ANSWERS

4. (a) $2 + 4 \log_3 z$

(b) $\dfrac{1}{2}(\log_6 n - \log_6 3 - \log_6 m)$

(c) $\log_2 \dfrac{xy^3}{z}$ **(d)** $\log_5 \dfrac{x^2 - 100}{x^{3/5}}$

(e) cannot be rewritten

NOW TRY EXERCISE 5

Given that $\log_2 7 = 2.8074$ and $\log_2 10 = 3.3219$, use properties of logarithms to evaluate each expression.

(a) $\log_2 70$ **(b)** $\log_2 0.7$

(c) $\log_2 49$

CLASSROOM EXAMPLE 5

Use the values given in Example 5 and properties of logarithms to evaluate each expression.

(a) $\log_2 25$ (b) $\log_2 \frac{5}{3}$

(c) $\log_2 75$

Answers: (a) 4.6438 (b) 0.7369 (c) 6.2288

EXAMPLE 5 Using the Properties of Logarithms with Numerical Values

Given that $\log_2 5 = 2.3219$ and $\log_2 3 = 1.5850$, use properties of logarithms to evaluate each expression.

(a) $\log_2 15$

$= \log_2 (3 \cdot 5)$	Factor 15.
$= \log_2 3 + \log_2 5$	Product rule
$= 1.5850 + 2.3219$	Substitute the given values.
$= 3.9069$	Add.

(b) $\log_2 0.6$

$= \log_2 \frac{3}{5}$	$0.6 = \frac{6}{10} = \frac{3}{5}$
$= \log_2 3 - \log_2 5$	Quotient rule
$= 1.5850 - 2.3219$	Substitute the given values.
$= -0.7369$	Subtract.

(c) $\log_2 27$

$= \log_2 3^3$	Write 27 as a power of 3.
$= 3 \log_2 3$	Power rule
$= 3(1.5850)$	Substitute the given value.
$= 4.7550$	Multiply.

NOW TRY

NOW TRY EXERCISE 6

Decide whether each statement is *true* or *false*.

(a) $\log_2 16 + \log_2 16 = \log_2 32$

(b) $(\log_2 4)(\log_3 9) = \log_6 36$

CLASSROOM EXAMPLE 6

Decide whether each statement is *true* or *false*.

(a) $\log_6 36 - \log_6 6 = \log_6 30$

(b) $\log_4 (\log_2 16) = \frac{\log_6 6}{\log_6 36}$

Answers: (a) false (b) false

EXAMPLE 6 Deciding Whether Statements about Logarithms Are True

Decide whether each statement is *true* or *false*.

(a) $\log_2 8 - \log_2 4 = \log_2 4$

Evaluate each side.

$\log_2 8 - \log_2 4$	Left side	$\log_2 4$	Right side
$= \log_2 2^3 - \log_2 2^2$	Write 8 and 4 as powers of 2.	$= \log_2 2^2$	Write 4 as a power of 2.
$= 3 - 2$	$\log_a a^x = x$	$= 2$	$\log_a a^x = x$
$= 1$	Subtract.		

The statement is false because $1 \neq 2$.

(b) $\log_3 (\log_2 8) = \frac{\log_7 49}{\log_8 64}$

$\log_3 (\log_2 8)$	Left side	$\frac{\log_7 49}{\log_8 64}$	Right side
$= \log_3 (\log_2 2^3)$	Write 8 as a power of 2.	$= \frac{\log_7 7^2}{\log_8 8^2}$	Write 49 and 64 using exponents.
$= \log_3 3$	$\log_a a^x = x$	$= \frac{2}{2}$	$\log_a a^x = x$
$= 1$	$3 = 3^1$	$= 1$	Simplify.

The statement is true because $1 = 1$.

NOW TRY

NOW TRY ANSWERS

5. **(a)** 6.1293 **(b)** −0.5145 **(c)** 5.6148

6. **(a)** false **(b)** false

9.4 Exercises

FOR EXTRA HELP

MyMathLab®

Complete solution available in MyMathLab

1. false; $\log_b x + \log_b y = \log_b xy$
2. true
3. true
4. false; $\log_b x^r = r \log_b x$
5. In the notation $\log_a (x + y)$, the parentheses do not indicate multiplication. They indicate that $x + y$ is the result of raising a to some power.
6. We cannot apply the product rule because $\log_b xy = \log_b x + \log_b y$ only if x and y are positive numbers.
7. $\log_{10} 7 + \log_{10} 8$
8. $\log_{10} 7 - \log_{10} 8$
9. 4
10. $6 \log_{10} 3$
11. 9
12. 4
13. $\log_7 4 + \log_7 5$
14. $\log_8 9 + \log_8 11$
15. $\log_5 8 - \log_5 3$
16. $\log_3 7 - \log_3 5$
17. $2 \log_4 6$
18. $4 \log_5 7$
19. $\frac{1}{3}\log_3 4 - 2\log_3 x - \log_3 y$
20. $\frac{1}{3}\log_7 13 - \log_7 p - 2\log_7 q$
21. $\frac{1}{2}\log_3 x + \frac{1}{2}\log_3 y - \frac{1}{2}\log_3 5$
22. $\frac{1}{2}\log_6 p + \frac{1}{2}\log_6 q - \frac{1}{2}\log_6 7$
23. $\frac{1}{3}\log_2 x + \frac{1}{5}\log_2 y - 2\log_2 r$
24. $\frac{1}{4}\log_4 z + \frac{1}{5}\log_4 w - 2\log_4 s$
25. $\log_b xy$
26. $\log_b wz$
27. $\log_a \frac{m}{n}$
28. $\log_b \frac{x}{y}$
29. $\log_a \frac{rt^3}{s}$
30. $\log_a \frac{pr^2}{q}$
31. $\log_a \frac{125}{81}$
32. $\log_a \frac{125}{3}$

Concept Check *Decide whether each statement of a logarithmic property is* true *or* false. *If it is false, correct it by changing the right side of the equation.*

1. $\log_b x + \log_b y = \log_b (x + y)$

2. $\log_b \frac{x}{y} = \log_b x - \log_b y$

3. $\log_b b^x = x$

4. $\log_b x^r = \log_b rx$

5. *Concept Check* A student erroneously wrote $\log_a (x + y) = \log_a x + \log_a y$. When his teacher explained that this was indeed wrong, the student claimed that he had used the distributive property. **WHAT WENT WRONG?**

6. *Concept Check* Consider the following "proof" that $\log_2 16$ does not exist.

$$\log_2 16$$
$$= \log_2 (-4)(-4)$$
$$= \log_2 (-4) + \log_2 (-4)$$

The logarithm of a negative number is not defined, so the final step cannot be evaluated. Thus $\log_2 16$ does not exist. **WHAT WENT WRONG?**

Use the indicated rule of logarithms to complete each equation. ***See Examples 1–3.***

7. $\log_{10} (7 \cdot 8) =$ __________ (product rule)

8. $\log_{10} \frac{7}{8} =$ __________ (quotient rule)

9. $3^{\log_3 4} =$ __________ (special property)

10. $\log_{10} 3^6 =$ __________ (power rule)

11. $\log_3 3^9 =$ __________ (special property)

12. $\log_3 9^2 =$ __________ (special property)

Use properties of logarithms to express each logarithm as a sum or difference of logarithms, or as a single number if possible. Assume that all variables represent positive real numbers. ***See Examples 1–4.***

13. $\log_7 (4 \cdot 5)$

14. $\log_8 (9 \cdot 11)$

15. $\log_5 \frac{8}{3}$

16. $\log_3 \frac{7}{5}$

17. $\log_4 6^2$

18. $\log_5 7^4$

19. $\log_3 \frac{\sqrt[3]{4}}{x^2 y}$

20. $\log_7 \frac{\sqrt[3]{13}}{pq^2}$

21. $\log_3 \sqrt{\frac{xy}{5}}$

22. $\log_6 \sqrt{\frac{pq}{7}}$

23. $\log_2 \frac{\sqrt[3]{x} \cdot \sqrt[5]{y}}{r^2}$

24. $\log_4 \frac{\sqrt[4]{z} \cdot \sqrt[5]{w}}{s^2}$

Use properties of logarithms to write each expression as a single logarithm. Assume that all variables are defined in such a way that the variable expressions are positive, and bases are positive numbers not equal to 1. ***See Examples 1–4.***

25. $\log_b x + \log_b y$

26. $\log_b w + \log_b z$

27. $\log_a m - \log_a n$

28. $\log_b x - \log_b y$

29. $(\log_a r - \log_a s) + 3 \log_a t$

30. $(\log_a p - \log_a q) + 2 \log_a r$

31. $3 \log_a 5 - 4 \log_a 3$

32. $3 \log_a 5 - \frac{1}{2} \log_a 9$

33. $\log_{10}(x^2 + 8x + 15)$
34. $\log_{10}(x^2 + 10x + 24)$
35. $\log_p \frac{x^3 y^{1/2}}{z^{3/2} a^3}$
36. $\log_b \frac{x^{1/3} y^{2/3}}{s^{3/4} t^{2/3}}$
37. 1.2552
38. 0.6020
39. −0.6532
40. 0.6532
41. 1.5562
42. 2.2094
43. 0.2386
44. 0.0602
45. 0.4771
46. −0.9542
47. 4.7710
48. 5.7190
49. false
50. false
51. true
52. true
53. true
54. true
55. false
56. false

33. $\log_{10}(x + 3) + \log_{10}(x + 5)$ **34.** $\log_{10}(x + 4) + \log_{10}(x + 6)$

35. $3 \log_p x + \frac{1}{2} \log_p y - \frac{3}{2} \log_p z - 3 \log_p a$

36. $\frac{1}{3} \log_b x + \frac{2}{3} \log_b y - \frac{3}{4} \log_b s - \frac{2}{3} \log_b t$

To four decimal places, the values of $\log_{10} 2$ *and* $\log_{10} 9$ *are*

$$\log_{10} 2 = 0.3010 \quad \text{and} \quad \log_{10} 9 = 0.9542.$$

Use these values and properties of logarithms to evaluate each expression. DO NOT USE A CALCULATOR. ***See Example 5.***

37. $\log_{10} 18$ **38.** $\log_{10} 4$ **39.** $\log_{10} \frac{2}{9}$ **40.** $\log_{10} \frac{9}{2}$

41. $\log_{10} 36$ **42.** $\log_{10} 162$ **43.** $\log_{10} \sqrt[4]{9}$ **44.** $\log_{10} \sqrt[5]{2}$

45. $\log_{10} 3$ **46.** $\log_{10} \frac{1}{9}$ **47.** $\log_{10} 9^5$ **48.** $\log_{10} 2^{19}$

Decide whether each statement is true *or* false. ***See Example 6.***

49. $\log_2(8 + 32) = \log_2 8 + \log_2 32$ **50.** $\log_2(64 - 16) = \log_2 64 - \log_2 16$

51. $\log_3 7 + \log_3 7^{-1} = 0$ **52.** $\log_3 49 + \log_3 49^{-1} = 0$

53. $\log_6 60 - \log_6 10 = 1$ **54.** $\log_3 8 + \log_3 \frac{1}{8} = 0$

55. $\frac{\log_{10} 7}{\log_{10} 14} = \frac{1}{2}$ **56.** $\frac{\log_{10} 10}{\log_{10} 100} = \frac{1}{10}$

9.5 Common and Natural Logarithms

OBJECTIVES

1. Evaluate common logarithms using a calculator.
2. Use common logarithms in applications.
3. Evaluate natural logarithms using a calculator.
4. Use natural logarithms in applications.
5. Use the change-of-base rule.

VOCABULARY

☐ common logarithm
☐ natural logarithm

Logarithms are important in many applications in biology, engineering, economics, and social science. In this section we find numerical approximations for logarithms. Traditionally, base 10 logarithms were used most often because our number system is base 10. Logarithms to base 10 are **common logarithms,** and

$$\mathbf{\log_{10} x} \text{ is abbreviated as } \mathbf{\log x},$$

where the base is understood to be 10.

OBJECTIVE 1 Evaluate common logarithms using a calculator.

In **Example 1,** we give the results of evaluating some common logarithms using a calculator with a LOG key. Consult your calculator manual to see how to use this key.

EXAMPLE 1 Evaluating Common Logarithms

Using a calculator, evaluate each logarithm to four decimal places.

(a) $\log 327.1$ **(b)** $\log 437{,}000$

(c) $\log 0.0615$ **(d)** $\log 10^{6.1988}$

NOW TRY EXERCISE 1

Using a calculator, evaluate each logarithm to four decimal places.

(a) log 115

(b) log 539,000

(c) log 0.023

(d) $\log 10^{12.2139}$

CLASSROOM EXAMPLE 1

Using a calculator, evaluate each logarithm to four decimal places.

(a) log 41,600

(b) log 43.5

(c) log 0.442

(d) $\log 10^{6.4123}$

Answers: (a) 4.6191 (b) 1.6385 (c) −0.3546 (d) 6.4123

FIGURE 17 shows how a graphing calculator displays these common logarithms to four decimal places.

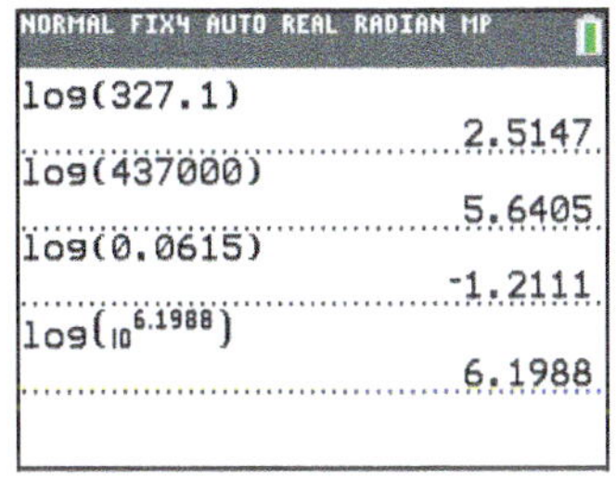

FIGURE 17

NOW TRY

In **Example 1(c),** log 0.0615 ≈ −1.2111, which is a negative result. ***The common logarithm of a number between 0 and 1 is always negative*** because the logarithm is the exponent on 10 that produces the number. In this case, we have

$$10^{-1.2111} \approx 0.0615.$$

If the exponent (the logarithm) were positive, the result would be greater than 1 because $10^0 = 1$. The graph in **FIGURE 18** illustrates these concepts.

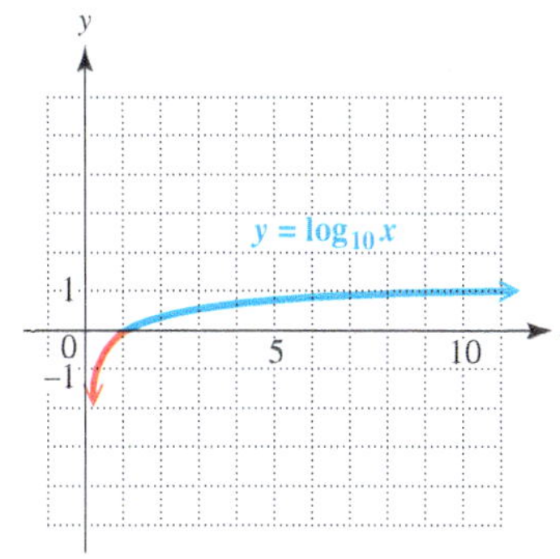

FIGURE 18

OBJECTIVE 2 Use common logarithms in applications.

In chemistry, pH is a measure of the acidity or alkalinity of a solution. Pure water, for example, has pH 7. In general, acids have pH numbers less than 7, and alkaline solutions have pH values greater than 7, as shown in **FIGURE 19**.

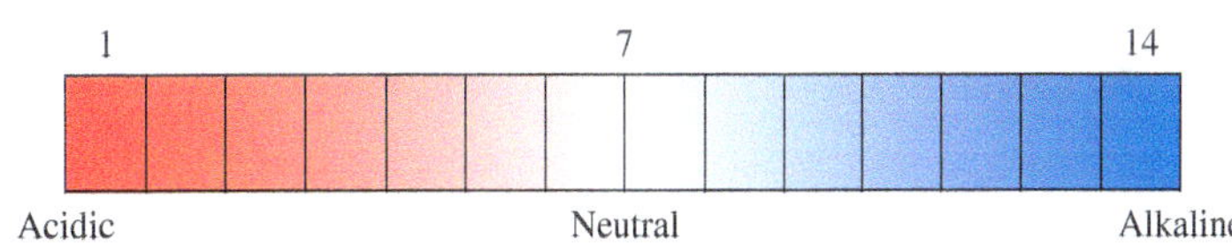

FIGURE 19 pH Scale

The **pH** of a solution is defined as

$$\mathbf{pH} = -\log\left[H_3O^+\right],$$

where $[H_3O^+]$ is the hydronium ion concentration in moles per liter. ***It is customary to round pH values to the nearest tenth.***

CLASSROOM EXAMPLE 2

Find the pH of water with a hydronium ion concentration of

$$1.2 \times 10^{-3}.$$

If it is taken from a wetland, is the wetland a rich fen, a poor fen, or a bog?

Answer: 2.9; bog

EXAMPLE 2 Using pH in an Application

Wetlands are classified as *bogs, fens, marshes,* and *swamps,* on the basis of pH values. A pH value between 6.0 and 7.5, such as that of Summerby Swamp in Michigan's Hiawatha National Forest, indicates that the wetland is a "rich fen." When the pH is between 3.0 and 6.0, the wetland is a "poor fen," and if the pH falls to 3.0 or less, it is a "bog." (*Source:* Mohlenbrock, R., "Summerby Swamp, Michigan," *Natural History.*)

NOW TRY ANSWERS

1. (a) 2.0607 (b) 5.7316 (c) −1.6383 (d) 12.2139

NOW TRY EXERCISE 2
Water taken from a wetland has a hydronium ion concentration of

$$3.4 \times 10^{-5}.$$

Find the pH value for the water and classify the wetland as a rich fen, a poor fen, or a bog.

Suppose that the hydronium ion concentration of a sample of water from a wetland is 6.3×10^{-3}. How would this wetland be classified?

$\text{pH} = -\log(6.3 \times 10^{-3})$	$\text{pH} = -\log[H_3O^+]$
$\text{pH} = -(\log 6.3 + \log 10^{-3})$	Product rule
$\text{pH} = -[0.7993 - 3(1)]$	Use a calculator to find log 6.3.
$\text{pH} = -0.7993 + 3$	Distributive property
$\text{pH} \approx 2.2$	Add.

The pH is less than 3.0, so the wetland is a bog. NOW TRY

NOW TRY EXERCISE 3
Find the hydronium ion concentration of a solution with pH 2.6.

CLASSROOM EXAMPLE 3
Find the hydronium ion concentration of a solution with pH 4.6.

Answer: 2.5×10^{-5}

EXAMPLE 3 Finding Hydronium Ion Concentration

Find the hydronium ion concentration of drinking water with pH 6.5.

$\text{pH} = -\log[H_3O^+]$	
$6.5 = -\log[H_3O^+]$	Let pH = 6.5.
$\log[H_3O^+] = -6.5$	Multiply by −1. Interchange sides.
$[H_3O^+] = 10^{-6.5}$	Write in exponential form, base 10.
$[H_3O^+] \approx 3.2 \times 10^{-7}$	Evaluate with a calculator. NOW TRY

The loudness of sound is measured in a unit called a **decibel,** abbreviated **dB.** To measure with this unit, we first assign an intensity of I_0 to a very faint sound, called the **threshold sound.** If a particular sound has intensity I, then the decibel level of this louder sound is

$$\mathbf{D = 10 \log\left(\frac{I}{I_0}\right).}$$

Any sound over 85 dB exceeds what hearing experts consider safe. Permanent hearing damage can be suffered at levels above 150 dB.

Loudness of Common Sounds

Decibel Level	*Example*
60	Normal conversation
90	Rush hour traffic, lawn mower
100	Garbage truck, chain saw, pneumatic drill
120	Rock concert, thunderclap
140	Gunshot blast, jet engine
180	Rocket launching pad

Source: Deafness Research Foundation.

CLASSROOM EXAMPLE 4
Find the decibel level to the nearest whole number of a whisper with intensity I of $115I_0$.

Answer: 21 dB

NOW TRY EXERCISE 4
Find the decibel level to the nearest whole number of the sound from a jet engine with intensity I of

$$6.312 \times 10^{13} I_0.$$

EXAMPLE 4 Measuring the Loudness of Sound

If music delivered through Bluetooth headphones has intensity I of $3.162 \times 10^{11} I_0$, find the average decibel level. (*Source:* CNET.)

$$D = 10 \log\left(\frac{I}{I_0}\right)$$

$$D = 10 \log\left(\frac{3.162 \times 10^{11} I_0}{I_0}\right) \quad \text{Substitute the given value for } I.$$

$$D = 10 \log(3.162 \times 10^{11})$$

$$D \approx 115 \quad \text{Evaluate with a calculator. Round to the nearest unit.}$$

NOW TRY

NOW TRY ANSWERS
2. 4.5; poor fen
3. 2.5×10^{-3}
4. 138 dB

Leonhard Euler (1707–1783)

The number *e* is named after Euler.

OBJECTIVE 3 Evaluate natural logarithms using a calculator.

Logarithms used in applications are often **natural logarithms,** which have as base the number ***e*.** The letter e was chosen to honor Leonhard Euler, who published extensive results on the number in 1748. It is an irrational number, so its decimal expansion never terminates and never repeats.

One way to see how e appears in an exponential situation involves calculating the values of

$$\left(1+\frac{1}{x}\right)^x \quad \text{as } x \text{ gets larger without bound.}$$

The table shows these values for $x = 1, 10, 100, 1000, 10{,}000,$ and $100{,}000$.

x	$\left(1+\frac{1}{x}\right)^x$
1	2
10	2.59374246
100	2.704813829
1000	2.716923932
10,000	2.718145927
100,000	2.718268237

These approximations are found using a calculator.

It appears that as x gets larger without bound, $\left(1+\frac{1}{x}\right)^x$ approaches some number. This number is e.

Approximation for e

$$e \approx 2.718281828$$

A scientific or graphing calculator with an e^x key can approximate powers of e. See **FIGURE 20**.

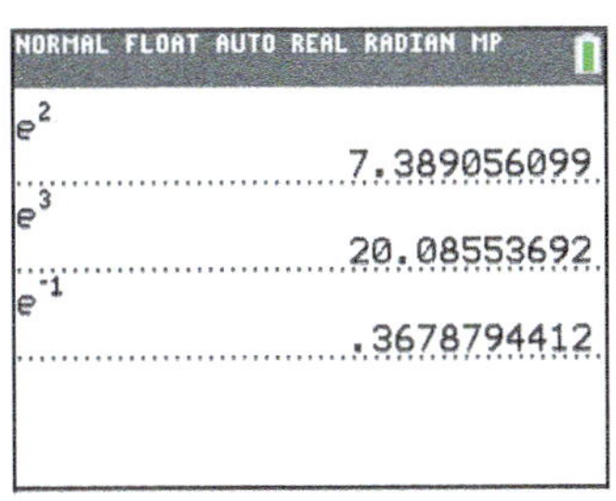

FIGURE 20

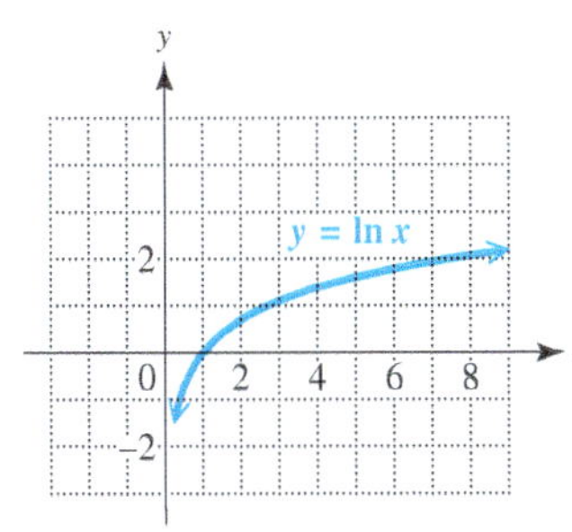

FIGURE 21

Logarithms with base e are called natural logarithms because they occur in natural situations that involve growth or decay.

The base e logarithm of x is written $\mathbf{\ln x}$ (read "**el en *x***").

The graph of $y = \ln x$ is given in **FIGURE 21.**

A calculator key labeled LN is used to evaluate natural logarithms. Consult your calculator manual to see how to use this key.

NOW TRY EXERCISE 5

Using a calculator, evaluate each logarithm to four decimal places.

(a) $\ln 0.26$ **(b)** $\ln 12$

(c) $\ln 150$ **(d)** $\ln e^{5.8321}$

CLASSROOM EXAMPLE 5

Using a calculator, evaluate each logarithm to four decimal places.

(a) $\ln 0.01$ (b) $\ln 27$

(c) $\ln 529$ (d) $\ln e^{-1.2136}$

Answers: (a) -4.6052 (b) 3.2958

(c) 6.2710 (d) -1.2136

EXAMPLE 5 Evaluating Natural Logarithms

Using a calculator, evaluate each logarithm to four decimal places.

(a) $\ln 0.5841 \approx -0.5377$ **(b)** $\ln 192.7 \approx 5.2611$

(c) $\ln 10.84 \approx 2.3832$ **(d)** $\ln e^{4.6832} \approx 4.6832$

FIGURE 22 shows how a graphing calculator displays these natural logarithms to four decimal places. As with common logarithms, ***a number between 0 and 1 has a negative natural logarithm.*** See part (a), where $\ln 0.5841$ is negative.

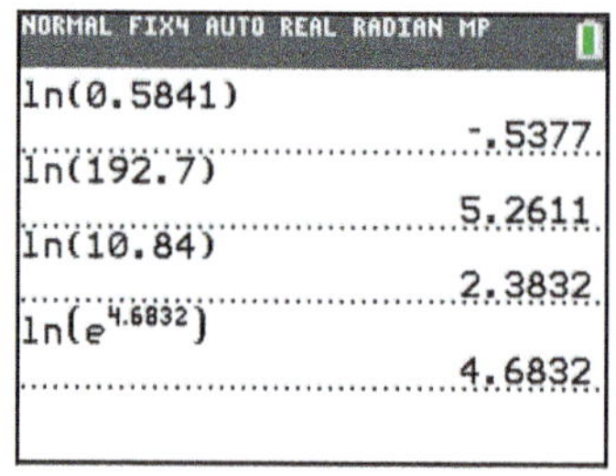

FIGURE 22

NOW TRY

OBJECTIVE 4 Use natural logarithms in applications.

NOW TRY EXERCISE 6

Use the logarithmic function in **Example 6** to approximate the altitude when atmospheric pressure is 600 millibars. Round to the nearest hundred.

CLASSROOM EXAMPLE 6

Use the logarithmic function in **Example 6** to approximate the altitude when atmospheric pressure is 700 millibars. Round to the nearest hundred.

Answer: 2700 m

EXAMPLE 6 Applying a Natural Logarithmic Function

The altitude in meters that corresponds to an atmospheric pressure of x millibars is given by the logarithmic function

$$f(x) = 51{,}600 - 7457 \ln x.$$

(*Source:* Miller, A. and J. Thompson, *Elements of Meteorology,* Fourth Edition, Charles E. Merrill Publishing Company.) Use this function to find the altitude when atmospheric pressure is 400 millibars. Round to the nearest hundred.

Let $x = 400$ and substitute in the expression for $f(x)$.

$$f(x) = 51{,}600 - 7457 \ln x$$

$$f(400) = 51{,}600 - 7457 \ln 400 \quad \text{Let } x = 400.$$

$$f(400) \approx 6900 \quad \text{Evaluate with a calculator.}$$

Atmospheric pressure is 400 millibars at 6900 m. NOW TRY

NOTE In **Example 6,** the final answer was obtained using a calculator *without* rounding the intermediate values. In general, it is best to wait until the final step to round the answer. Otherwise, a buildup of round-off error may cause the final answer to have an incorrect final decimal place digit or digits.

OBJECTIVE 5 Use the change-of-base rule.

In **Example 2** of **Section 9.3** we illustrated how the TI-84 Plus calculator evaluates logarithms for any base. If a calculator does not have this function, we calculate such logarithms using the change-of-base rule, which allows us to convert logarithms from one base to another.

NOW TRY ANSWERS

5. (a) -1.3471 **(b)** 2.4849

(c) 5.0106 **(d)** 5.8321

6. 3900 m

TEACHING TIP
Have students compare the results of applying the change-of-base rule shown in **FIGURE 23** below to those in **FIGURE 13** in **Section 9.3** to see how they correspond.

Change-of-Base Rule

If $a > 0$, $a \neq 1$, $b > 0$, $b \neq 1$, and $x > 0$, then the following holds true.

$$\log_a x = \frac{\log_b x}{\log_b a}$$

NOTE Any positive number other than 1 can be used for base b in the change-of-base rule. Usually the only practical bases are e and 10 because calculators generally have dedicated keys for these two bases.

TEACHING TIP
Caution students that they must use the (LOG) key once in the numerator and also once in the denominator when applying the change-of-base rule. That is,

$$\frac{\log_b x}{\log_b a} \neq \log_b x - \log_b a.$$

To derive the change-of-base rule, let $\log_a x = m$.

$$\log_a x = m$$

$$a^m = x \quad \text{Convert to exponential form.}$$

$$\log_b a^m = \log_b x \quad \text{Take the logarithm on each side.}$$

$$m \log_b a = \log_b x \quad \text{Power rule}$$

$$(\log_a x)(\log_b a) = \log_b x \quad \text{Substitute for } m.$$

$$\log_a x = \frac{\log_b x}{\log_b a} \quad \text{Divide by } \log_b a.$$

This last statement is the change-of-base rule.

NOW TRY EXERCISE 7
Use a calculator and the change-of-base rule to approximate each logarithm to four decimal places.
(a) $\log_2 7$ **(b)** $\log_5 8$
(c) $\log_{1/3} 12$

EXAMPLE 7 Using the Change-of-Base Rule

Use a calculator and the change-of-base rule to approximate each logarithm to four decimal places.

(a) $\log_2 5$ **(b)** $\log_3 12$ **(c)** $\log_{1/2} 12$

See **FIGURE 23**. To four decimal places, the logarithms in parts (a)–(c) are as follows.

$$\log_2 5 \approx 2.3219, \quad \log_3 12 \approx 2.2619, \quad \log_{1/2} 12 \approx -3.5850$$

CLASSROOM EXAMPLE 7
Use a calculator and the change-of-base rule to approximate each logarithm to four decimal places.
(a) $\log_2 13$ (b) $\log_6 18$
(c) $\log_{1/4} 20$
Answers: (a) 3.7004
(b) 1.6131 (c) -2.1610

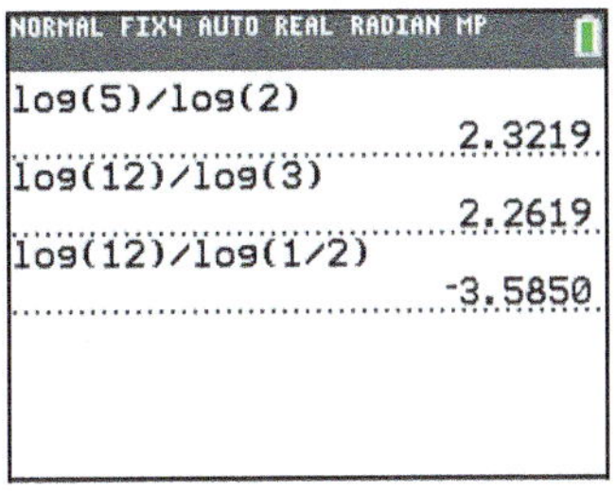

FIGURE 23

Compare these results to **Example 2(a)–(c)** of **Section 9.3**. NOW TRY

NOTE Either common or natural logarithms can be used when applying the change-of-base rule. Verify that the same results are found in **Example 7** for natural logarithms.

NOW TRY ANSWERS
7. (a) 2.8074 (b) 1.2920
(c) -2.2619

9.5 Exercises

FOR EXTRA HELP MyMathLab®

Complete solution available in MyMathLab

1. C 2. A
3. C 4. B
5. 2 6. 2
7. 1.6335 8. 1.9912
9. 2.5164 10. 2.6601
11. −1.4868 12. −0.7592
13. 9.6421 14. 3.1112
15. 9.6776 16. 4.3284
17. 2.0592 18. 2.1187
19. −2.8896 20. −3.8304
21. 5.9613 22. 6.8486
23. −11.4007 24. −1.4724
25. 4.1506 26. 5.0096
27. 2.3026 28. 0.4343
29. 40 30. 45
31. (a) 2.552424846 (b) 1.552424846 (c) 0.552424846 (d) The whole number parts will vary, but the decimal parts are the same.
32. Answers will vary. Suppose the last name is Miller, with $k = 6$. (a) 0.7781512504 (b) 6 (c) Because $y = \log x$ and $y = 10^x$ are inverses, we will always obtain k as our answer.
33. poor fen 34. poor fen
35. bog 36. bog
37. rich fen 38. rich fen

Concept Check *Choose the correct response.*

1. What is the base in the expression $\log x$?

A. x **B.** 1 **C.** 10 **D.** e

2. What is the base in the expression $\ln x$?

A. e **B.** 1 **C.** 10 **D.** x

3. Given that $10^0 = 1$ and $10^1 = 10$, between what two consecutive integers is the value of $\log 6.3$?

A. 6 and 7 **B.** 10 and 11 **C.** 0 and 1 **D.** −1 and 0

4. Given that $e^1 \approx 2.718$ and $e^2 \approx 7.389$, between what two consecutive integers is the value of $\ln 6.3$?

A. 6 and 7 **B.** 1 and 2 **C.** 2 and 3 **D.** 0 and 1

5. *Concept Check* Without using a calculator, give the value of $\log 10^2$.

6. *Concept Check* Without using a calculator, give the value of $\ln e^2$.

You will need a calculator for most of the remaining exercises in this set.

Evaluate each logarithm to four decimal places when appropriate. ***See Examples 1 and 5.***

7. $\log 43$ **8.** $\log 98$ **9.** $\log 328.4$

10. $\log 457.2$ **11.** $\log 0.0326$ **12.** $\log 0.1741$

13. $\log 10^{9.6421}$ **14.** $\log 10^{3.1112}$ **15.** $\log (4.76 \times 10^9)$

16. $\log (2.13 \times 10^4)$ **17.** $\ln 7.84$ **18.** $\ln 8.32$

19. $\ln 0.0556$ **20.** $\ln 0.0217$ **21.** $\ln 388.1$

22. $\ln 942.6$ **23.** $\ln e^{-11.4007}$ **24.** $\ln e^{-1.4724}$

25. $\ln (8.59 \times e^2)$ **26.** $\ln (7.46 \times e^3)$ **27.** $\ln 10$

28. $\log e$ **29.** $10 \ln e^4$ **30.** $15 \ln e^3$

31. *Concept Check* Use a calculator to find approximations of each logarithm.

(a) $\log 356.8$ **(b)** $\log 35.68$ **(c)** $\log 3.568$

(d) Observe the answers and make a conjecture concerning the decimal values of the common logarithms of numbers greater than 1 that have the same digits.

32. *Concept Check* Let k represent the number of letters in your last name.

(a) Use a calculator to find $\log k$.

(b) Raise 10 to the power indicated by the number found in part (a). What is the result?

(c) Use the concepts of **Section 9.1** and explain why we obtained the answer found in part (b). Would it matter what number we used for k to observe the same result?

Suppose that water from a wetland area is sampled and found to have the given hydronium ion concentration. Is the wetland a rich fen, *a* poor fen, *or a* bog? ***See Example 2.***

33. 3.1×10^{-5} **34.** 2.5×10^{-5} **35.** 2.5×10^{-2}

36. 3.6×10^{-2} **37.** 2.7×10^{-7} **38.** 2.5×10^{-7}

39. 11.6 40. 8.4
41. 4.3 42. 5.9
43. 4.0×10^{-8} 44. 1.0×10^{-2}
45. 4.0×10^{-6} 46. 2.5×10^{-5}
47. (a) 142 dB (b) 126 dB (c) 120 dB
48. (a) 35.0 yr (b) 14.2 yr (c) 9.0 yr
49. (a) 800 yr (b) 5200 yr (c) 11,500 yr
50. every 2 hr

Find the pH of the substance with the given hydronium ion concentration. ***See Example 2.***

39. Ammonia, 2.5×10^{-12}

40. Sodium bicarbonate, 4.0×10^{-9}

41. Grapes, 5.0×10^{-5}

42. Tuna, 1.3×10^{-6}

Find the hydronium ion concentration of the substance with the given pH. ***See Example 3.***

43. Human blood plasma, 7.4

44. Human gastric contents, 2.0

45. Spinach, 5.4

46. Bananas, 4.6

Solve each problem. ***See Examples 4 and 6.***

47. Managements of sports stadiums and arenas often encourage fans to make as much noise as possible. Find the average decibel level

$$D = 10 \log \left(\frac{I}{I_0} \right)$$

for each venue with the given intensity I.

(a) NFL fans, Kansas City Chiefs at Arrowhead Stadium:

$I = (1.58 \times 10^{14}) I_0$ (*Source:* www.guinessworldrecords.com)

(b) NBA fans, Sacramento Kings at Sleep Train Arena:

$I = (3.9 \times 10^{12}) I_0$ (*Source:* www.guinessworldrecords.com)

(c) MLB fans, Baltimore Orioles at Camden Yards:

$I = (1.1 \times 10^{12}) I_0$ (*Source:* www.baltimoresportsreport.com)

48. The time t in years for an amount increasing at a rate of r (in decimal form) to double is given by

$$t(r) = \frac{\ln 2}{\ln (1 + r)}.$$

This is the **doubling time.** Find the doubling time to the nearest tenth for an investment at each interest rate.

(a) 2% (or 0.02) **(b)** 5% (or 0.05) **(c)** 8% (or 0.08)

49. The number of years, $N(x)$, since two independently evolving languages split off from a common ancestral language is approximated by

$$N(x) = -5000 \ln x,$$

where x is the percent of words (in decimal form) from the ancestral language common to both languages now. Find the number of years (to the nearest hundred years) since the split for each percent of common words.

(a) 85% (or 0.85) **(b)** 35% (or 0.35) **(c)** 10% (or 0.10)

50. The concentration of a drug injected into the bloodstream decreases with time. The intervals of time T when the drug should be administered are given by

$$T = \frac{1}{k} \ln \frac{C_2}{C_1},$$

where k is a constant determined by the drug in use, C_2 is the concentration at which the drug is harmful, and C_1 is the concentration below which the drug is ineffective. (*Source:* Horelick, B. and S. Koont, "Applications of Calculus to Medicine: Prescribing Safe and Effective Dosage," *UMAP Module 202.*) Thus, if $T = 4$, the drug should be administered every 4 hr. For a certain drug, $k = \frac{1}{3}$, $C_2 = 5$, and $C_1 = 2$. How often should the drug be administered? (*Hint:* Round down.)

51. (a) 17.8 million
(b) 2005

52. (a) 55% (b) 90%

53. (a) \$54 per ton
(b) If $x = 0$, then

$$\ln(1 - x) = \ln 1 = 0,$$

so $T(x)$ would be negative. If $x = 1$, then

$$\ln(1 - x) = \ln 0,$$

but the domain of $\ln x$ is $(0, \infty)$.

54. (a) 5.6 yr
(b) $x > 0$ and $\frac{87 - x}{63}$ is positive and in the domain of the function only if $24 < x < 87$.

55. 2.2619 56. 2.0850
57. 0.6826 58. 0.7124
59. 0.3155 60. 0.2994
61. 0.8736 62. 2.0115
63. 2.4849 64. 2.7081
65. 0.4421 66. 0.4796

67. In each case, the product is 1. $\log_a b$ and $\log_b a$ are reciprocals.

68. $\log_3 \sqrt{2} = \log_3 2^{1/2} = \frac{1}{2}\log_3 2$. A calculator gives the same approximation for $\log_3 \sqrt{2}$ and $\frac{1}{2}\log_3 2$: 0.3155.

51. The growth of outpatient surgeries at hospitals is approximated by

$$f(x) = 10.6049 + 2.3556 \ln x,$$

where x is the number of years since 1990, and $f(x)$ is in millions. (*Source:* American Hospital Association.)

(a) What does this model give for the number of outpatient surgeries in 2011?

(b) According to this model, when did outpatient surgeries reach 17,000,000? (*Hint:* Substitute for $f(x)$, and then write the equation in exponential form to solve it.)

52. In the central Sierra Nevada of California, the percent of moisture that falls as snow rather than rain is approximated by

$$f(x) = 86.3 \ln x - 680,$$

where x is the altitude in feet.

(a) What percent of the moisture at 5000 ft falls as snow?

(b) What percent at 7500 ft falls as snow?

53. The **cost-benefit equation**

$$T(x) = -0.642 - 189 \ln(1 - x)$$

describes the approximate tax $T(x)$, in dollars per ton, that would result in an $x\%$ (in decimal form) reduction in carbon dioxide emissions.

(a) What tax will reduce emissions 25%?

(b) Explain why the equation is not valid for $x = 0$ or $x = 1$.

54. The age in years of a female blue whale of length x in feet is approximated by

$$f(x) = -2.57 \ln\left(\frac{87 - x}{63}\right).$$

(a) How old is a female blue whale that measures 80 ft?

(b) The equation that defines this function has domain $24 < x < 87$. Explain why.

Use the change-of-base rule (with either common or natural logarithms) to approximate each logarithm to four decimal places. ***See Example 7.***

55. $\log_3 12$ **56.** $\log_4 18$ **57.** $\log_5 3$

58. $\log_7 4$ **59.** $\log_3 \sqrt{2}$ **60.** $\log_6 \sqrt[3]{5}$

61. $\log_\pi e$ **62.** $\log_\pi 10$ **63.** $\log_e 12$

64. $\log_e 15$ **65.** $\log_{12} 3$ **66.** $\log_{18} 4$

67. *Concept Check* Multiply the results in **Exercises 55 and 65.** Then do the same for **Exercises 56 and 66.** Make a conjecture about the relationship between

$$\log_a b \quad \text{and} \quad \log_b a.$$

68. *Concept Check* Apply the power rule to the expression in **Exercise 59.** Write the result, and then verify that it is equal to $\log_3 \sqrt{2}$.

9.6 Exponential and Logarithmic Equations; Further Applications

OBJECTIVES

1. Solve equations involving variables in the exponents.
2. Solve equations involving logarithms.
3. Solve applications of compound interest.
4. Solve applications involving base e exponential growth and decay.

VOCABULARY

- ☐ compound interest
- ☐ continuous compounding

We solved exponential and logarithmic equations in **Sections 9.2 and 9.3.** General methods for solving these equations depend on the following properties.

Properties for Solving Exponential and Logarithmic Equations

For all real numbers $b > 0$, $b \neq 1$, and any real numbers x and y, the following hold true.

1. If $x = y$, then $b^x = b^y$.

2. If $b^x = b^y$, then $x = y$. (We used this property in **Section 9.2.**)

3. If $x = y$, and $x > 0$, $y > 0$, then $\log_b x = \log_b y$.

4. If $x > 0$, $y > 0$, and $\log_b x = \log_b y$, then $x = y$.

OBJECTIVE 1 Solve equations involving variables in the exponents.

In **Examples 1 and 2,** we use Property 3.

NOW TRY EXERCISE 1
Solve the equation. Approximate the solution to three decimal places.

$$5^x = 20$$

CLASSROOM EXAMPLE 1
Solve the equation. Approximate the solution to three decimal places.

$$10^x = 4$$

Answer: $\{0.602\}$

EXAMPLE 1 Solving an Exponential Equation

Solve $3^x = 12$. Approximate the solution to three decimal places.

$$3^x = 12$$

$$\log 3^x = \log 12 \quad \text{Property 3 (common logs)}$$

$$x \log 3 = \log 12 \quad \text{Power rule}$$

$$\text{Exact solution} \longrightarrow x = \frac{\log 12}{\log 3} \quad \text{Divide by } \log 3.$$

$$\text{Decimal approximation} \longrightarrow x \approx 2.262 \quad \text{Use a calculator.}$$

CHECK $3^x = 3^{2.262} \approx 12$ ✓ True
Evaluate with a calculator.

The solution set is $\{2.262\}$. **NOW TRY**

CAUTION Be careful: $\frac{\log 12}{\log 3}$ is ***not*** equal to log 4. Check to see that

$$\log 4 \approx 0.6021, \quad \text{but} \quad \frac{\log 12}{\log 3} \approx 2.262.$$

NOTE In **Example 1,** we used Property 3 with common logarithms. We could just as easily have used natural logarithms. Verify that

$$\frac{\ln 12}{\ln 3} = \frac{\log 12}{\log 3} \approx 2.262.$$

When an exponential equation has e as the base, as in the next example, it is easiest to use base e (natural) logarithms.

NOW TRY ANSWER
1. $\{1.861\}$

NOW TRY EXERCISE 2
Solve $e^{0.12x} = 10$. Approximate the solution to three decimal places.

CLASSROOM EXAMPLE 2
Solve $e^{0.02x} = 192$. Approximate the solution to three decimal places.

Answer: $\{262.875\}$

EXAMPLE 2 Solving an Exponential Equation with Base e

Solve $e^{0.003x} = 40$. Approximate the solution to three decimal places.

$$\ln e^{0.003x} = \ln 40 \quad \text{Property 3 (natural logs)}$$
$$0.003x \ln e = \ln 40 \quad \text{Power rule}$$
$$0.003x = \ln 40 \quad \ln e = \ln e^1 = 1$$
$$x = \frac{\ln 40}{0.003} \quad \text{Divide by 0.003.}$$
$$x \approx 1229.626 \quad \text{Evaluate with a calculator.}$$

Check that $e^{0.003(1229.626)} \approx 40$. The solution set is $\{1229.626\}$. NOW TRY

NOW TRY EXERCISE 3
Solve $3^{-x+5} = 8$. Approximate the solution to three decimal places.

CLASSROOM EXAMPLE 3
Solve $5^{-x+7} = 9$. Approximate the solution to three decimal places.

Answer: $\{5.635\}$

EXAMPLE 3 Solving an Exponential Equation

Solve $7^{-x+4} = 17$. Approximate the solution to three decimal places.

$$7^{-x+4} = 17$$
$$\log 7^{-x+4} = \log 17 \quad \text{Property 3 (common logs)}$$
$$(-x + 4) \log 7 = \log 17 \quad \text{Power rule}$$
$$-x \log 7 + 4 \log 7 = \log 17 \quad \text{Distributive property}$$
$$-x \log 7 = \log 17 - 4 \log 7 \quad \text{Subtract 4 log 7.}$$
$$x = \frac{\log 17 - 4 \log 7}{-\log 7} \quad \text{Divide by } -\log 7.$$
$$x \approx 2.544 \quad \text{Evaluate with a calculator.}$$

Check that $7^{-2.544+4} \approx 17$. The solution set is $\{2.544\}$. NOW TRY

General Method for Solving an Exponential Equation

Take logarithms having the same base on both sides and then use the power rule of logarithms or the special property $\log_b b^x = x$. (See **Examples 1–3.**)

As a special case, if both sides can be written as exponentials with the same base, do so, and set the exponents equal. (See **Section 9.2.**)

OBJECTIVE 2 Solve equations involving logarithms.

NOW TRY EXERCISE 4
Solve $\log_6 (2x + 4) = 2$.

CLASSROOM EXAMPLE 4
Solve $\log_5 (3x + 2) = 3$.

Answer: $\{41\}$

EXAMPLE 4 Solving a Logarithmic Equation

Solve $\log_3 (4x + 1) = 4$.

$$\log_3 (4x + 1) = 4$$
$$4x + 1 = 3^4 \quad \text{Convert to exponential form.}$$
$$4x + 1 = 81 \quad 3^4 = 81$$
$$4x = 80 \quad \text{Subtract 1.}$$
$$x = 20 \quad \text{Divide by 4.}$$

Check to see that $\log_3 [4(20) + 1] = \log_3 81 = 4$ is true. The solution set is $\{20\}$.

NOW TRY

NOW TRY ANSWERS
2. $\{19.188\}$ 3. $\{3.107\}$
4. $\{16\}$

NOW TRY EXERCISE 5
Solve $\log_5 (x-1)^3 = 2$. Give the exact solution.

CLASSROOM EXAMPLE 5
Solve $\log_3 (x+1)^5 = 3$. Give the exact solution.

Answer: $\{-1+\sqrt[5]{27}\}$

EXAMPLE 5 Solving a Logarithmic Equation

Solve $\log_2 (x+5)^3 = 4$. Give the exact solution.

$$\log_2 (x+5)^3 = 4$$

$$(x+5)^3 = 2^4 \quad \text{Convert to exponential form.}$$

$$(x+5)^3 = 16 \quad 2^4 = 16$$

$$x+5 = \sqrt[3]{16} \quad \text{Take the cube root on each side.}$$

$$x = -5+\sqrt[3]{16} \quad \text{Add } -5.$$

$$x = -5+2\sqrt[3]{2} \quad \sqrt[3]{16} = \sqrt[3]{8\cdot 2} = \sqrt[3]{8}\cdot\sqrt[3]{2} = 2\sqrt[3]{2}$$

CHECK

$$\log_2 (x+5)^3 = 4 \quad \text{Original equation}$$

$$\log_2 \left(-5+2\sqrt[3]{2}+5\right)^3 \stackrel{?}{=} 4 \quad \text{Let } x = -5+2\sqrt[3]{2}.$$

$$\log_2 \left(2\sqrt[3]{2}\right)^3 \stackrel{?}{=} 4 \quad \text{Work inside the parentheses.}$$

$$\log_2 16 \stackrel{?}{=} 4 \quad \left(2\sqrt[3]{2}\right)^3 = 2^3\left(\sqrt[3]{2}\right)^3 = 8\cdot 2 = 16$$

$$2^4 \stackrel{?}{=} 16 \quad \text{Write in exponential form.}$$

$$16 = 16 \checkmark \quad \text{True}$$

A true statement results, so the solution set is $\{-5+2\sqrt[3]{2}\}$. NOW TRY

CAUTION Recall that the domain of $f(x) = \log_b x$ is $(0, \infty)$. ***For this reason, always check that each proposed solution of an equation with logarithms yields only logarithms of positive numbers in the original equation.***

CLASSROOM EXAMPLE 6
Solve.

$\log_8 (2x+5) + \log_8 3 = \log_8 33$

Answer: $\{3\}$

EXAMPLE 6 Solving a Logarithmic Equation

Solve $\log_2 (x+1) - \log_2 x = \log_2 7$.

$$\log_2 (x+1) - \log_2 x = \log_2 7$$

Transform the left side to an expression with only *one* logarithm.

$$\log_2 \frac{x+1}{x} = \log_2 7 \quad \text{Quotient rule}$$

$$\frac{x+1}{x} = 7 \quad \text{Property 4}$$

$$x+1 = 7x \quad \text{Multiply by } x.$$

$$1 = 6x \quad \text{Subtract } x.$$

$$\frac{1}{6} = x \quad \text{Divide by 6.}$$

This proposed solution must be checked.

We cannot take the logarithm of a *nonpositive* number, so both $x+1$ and x must be positive here. If $x = \frac{1}{6}$, then this condition is satisfied.

TEACHING TIP
Before working an example like **Examples 6 and 7**, do the following.

1. Review the product, quotient, and power rules for logarithms.
2. Tell students that a logarithmic equation cannot be solved with *two* logarithms on one side of the equation. The equation must be transformed so that one side contains an expression with only *one* logarithm.
3. Explain that there may be extraneous solutions.

NOW TRY ANSWER

5. $\{1+\sqrt[3]{25}\}$

NOW TRY EXERCISE 6
Solve.

$\log_4 (2x + 13) - \log_4 (x + 1) = \log_4 10$

CHECK

$$\log_2 (x + 1) - \log_2 x = \log_2 7 \quad \text{Original equation}$$

$$\log_2 \left(\frac{1}{6} + 1\right) - \log_2 \frac{1}{6} \stackrel{?}{=} \log_2 7 \quad \text{Let } x = \tfrac{1}{6}.$$

$$\log_2 \frac{7}{6} - \log_2 \frac{1}{6} \stackrel{?}{=} \log_2 7 \quad \text{Add.}$$

$$\log_2 \frac{\frac{7}{6}}{\frac{1}{6}} \stackrel{?}{=} \log_2 7 \quad \text{Quotient rule}$$

$\frac{\frac{7}{6}}{\frac{1}{6}} = \frac{7}{6} \div \frac{1}{6} = \frac{7}{6} \cdot \frac{6}{1} = 7$

$$\log_2 7 = \log_2 7 \quad ✓ \quad \text{True}$$

A true statement results, so the solution set is $\left\{\frac{1}{6}\right\}$. **NOW TRY**

NOW TRY EXERCISE 7
Solve.

$\log_4 (x + 2) + \log_4 2x = 2$

CLASSROOM EXAMPLE 7
Solve.

$\log x + \log (x + 15) = 2$

Answer: $\{5\}$

EXAMPLE 7 Solving a Logarithmic Equation

Solve $\log x + \log (x - 21) = 2$.

$$\log x + \log (x - 21) = 2$$

$$\log x(x - 21) = 2 \quad \text{Product rule}$$

The base is 10.

$$x(x - 21) = 10^2 \quad \text{Write in exponential form.}$$

$$x^2 - 21x = 100 \quad \text{Distributive property; Multiply.}$$

$$x^2 - 21x - 100 = 0 \quad \text{Standard form}$$

$$(x - 25)(x + 4) = 0 \quad \text{Factor.}$$

$$x - 25 = 0 \quad \text{or} \quad x + 4 = 0 \quad \text{Zero-factor property}$$

$$x = 25 \quad \text{or} \quad x = -4 \quad \text{Proposed solutions}$$

The value -4 must be rejected as a solution because it leads to the logarithm of a negative number in the original equation.

$$\log (-4) + \log (-4 - 21) = 2 \quad \text{The left side is undefined.}$$

Check that the only solution is 25, so the solution set is $\{25\}$. **NOW TRY**

CAUTION ***Do not reject a potential solution just because it is nonpositive. Reject any value that leads to the logarithm of a nonpositive number.***

Solving a Logarithmic Equation

Step 1 **Transform the equation so that a single logarithm appears on one side.** Use the product rule or quotient rule of logarithms to do this.

Step 2 **Do one of the following.**

(a) Use Property 4.

If $\log_b x = \log_b y$, then $x = y$. (See **Example 6.**)

(b) Write the equation in exponential form.

If $\log_b x = k$, then $x = b^k$. (See **Examples 5 and 7.**)

NOW TRY ANSWERS

6. $\left\{\frac{3}{8}\right\}$ **7.** $\{2\}$

OBJECTIVE 3 Solve applications of compound interest.

We have solved simple interest problems using the formula $I = prt$. In most cases, interest paid or charged is **compound interest** (interest paid on both principal and interest). The formula for compound interest is an application of exponential functions. ***In this book, monetary amounts are given to the nearest cent.***

Compound Interest Formula (for a Finite Number of Periods)

If a principal of P dollars is deposited at an annual rate of interest r compounded (paid) n times per year, then the account will contain

$$A = P\left(1 + \frac{r}{n}\right)^{nt}$$

dollars after t years. (In this formula, r is expressed as a decimal.)

NOW TRY EXERCISE 8

How much money will there be in an account at the end of 10 yr if \$10,000 is deposited at 2.5% compounded monthly?

CLASSROOM EXAMPLE 8

Find the value of \$2000 deposited at 4% compounded quarterly for 10 yr.

Answer: \$2977.73

EXAMPLE 8 Solving a Compound Interest Problem for *A*

How much money will there be in an account at the end of 5 yr if \$1000 is deposited at 3% compounded quarterly? (Assume no withdrawals are made.)

Because interest is compounded quarterly, $n = 4$.

$$A = P\left(1 + \frac{r}{n}\right)^{nt} \quad \text{Compound interest formula}$$

$$A = 1000\left(1 + \frac{0.03}{4}\right)^{4 \cdot 5} \quad \text{Substitute } P = 1000,\ r = 0.03 \text{ (because } 3\% = 0.03),\ n = 4, \text{ and } t = 5.$$

$$A = 1000(1.0075)^{20} \quad \text{Simplify.}$$

$$A = 1161.18 \quad \text{Evaluate with a calculator.}$$

The account will contain \$1161.18. NOW TRY

NOW TRY EXERCISE 9

Find the number of years, to the nearest hundredth, it will take for money deposited in an account paying 4% interest compounded quarterly to double.

CLASSROOM EXAMPLE 9

Find the number of years, to the nearest hundredth, it will take for money deposited in an account paying 4% interest compounded semiannually to double.

Answer: 17.50 yr

EXAMPLE 9 Solving a Compound Interest Problem for *t*

Suppose inflation is averaging 3% per year. To the nearest hundredth of a year, how long will it take for prices to double? (This is the **doubling time** of the money.)

We want the number of years t for P dollars to grow to $2P$ dollars at 3% per year.

$$A = P\left(1 + \frac{r}{n}\right)^{nt} \quad \text{Compound interest formula}$$

$$2P = P\left(1 + \frac{0.03}{1}\right)^{1t} \quad \text{Let } A = 2P,\ r = 0.03, \text{ and } n = 1.$$

$$2 = (1.03)^t \quad \text{Divide by } P. \text{ Simplify.}$$

$$\log 2 = \log (1.03)^t \quad \text{Property 3}$$

$$\log 2 = t \log (1.03) \quad \text{Power rule}$$

$$t = \frac{\log 2}{\log 1.03} \quad \text{Interchange sides. Divide by } \log 1.03.$$

$$t \approx 23.45 \quad \text{Evaluate with a calculator.}$$

Prices will double in 23.45 yr. To check, verify that $1.03^{23.45} \approx 2$. NOW TRY

NOW TRY ANSWERS

8. \$12,836.92

9. 17.42 yr

TEACHING TIP
For the continuous compounding formula, refer to P as the amount at "present" and A as the amount "after." It should make sense to students that e^{rt} is a multiplier.

Interest can be compounded over various time periods per year, including

annually, semiannually, quarterly, daily, and so on.

The number of compounding periods can get larger and larger. If the value of n increases without bound, we have an example of **continuous compounding.** The formula for continuous compounding is derived in advanced courses, and is an example of exponential growth involving the number e.

Continuous Compound Interest Formula

If a principal of P dollars is deposited at an annual rate of interest r compounded continuously for t years, the final amount A on deposit is given by

$$A = Pe^{rt}.$$

NOW TRY EXERCISE 10
Suppose that \$4000 is invested at 3% interest for 2 yr.

(a) How much will the investment grow to if it is compounded continuously?

(b) How long would it take for the original investment to double? Round to the nearest tenth.

CLASSROOM EXAMPLE 10
Suppose that \$2000 is invested at 5% interest for 10 yr.

(a) How much will the investment grow to if it is compounded continuously?

(b) How long would it take for the original investment to double? Round to the nearest tenth.

Answers: (a) \$3297.44
(b) 13.9 yr

EXAMPLE 10 Solving a Continuous Compound Interest Problem

In **Example 8** we found that \$1000 invested for 5 yr at 3% interest compounded quarterly would grow to \$1161.18.

(a) How much would this investment grow to if it is compounded continuously?

$A = Pe^{rt}$ Continuous compounding formula

$A = 1000e^{0.03(5)}$ Let $P = 1000$, $r = 0.03$, and $t = 5$.

$A = 1000e^{0.15}$ Multiply in the exponent.

$A = 1161.83$ Evaluate with a calculator.

The investment will grow to \$1161.83 (which is \$0.65 more than the amount in **Example 8** when interest was compounded quarterly).

(b) How long would it take for the initial investment amount to triple? Round to the nearest tenth.

We must find the value of t that will cause A to be $3(\$1000) = \$3000$.

$A = Pe^{rt}$ Continuous compounding formula

$3000 = 1000e^{0.03t}$ Let $A = 3P = 3000$, $P = 1000$, and $r = 0.03$.

$3 = e^{0.03t}$ Divide by 1000.

$\ln 3 = \ln e^{0.03t}$ Take natural logarithms.

$\ln 3 = 0.03t$ $\ln e^k = k$

$t = \dfrac{\ln 3}{0.03}$ Divide by 0.03. Interchange sides.

$t \approx 36.6$ Evaluate with a calculator.

It would take 36.6 yr for the original investment to triple. NOW TRY

NOW TRY ANSWERS
10. (a) \$4247.35
(b) 23.1 yr

OBJECTIVE 4 Solve applications involving base e exponential growth and decay.

When situations involve growth or decay of a population, the amount or number of some quantity present at time t can be approximated by

$$f(t) = y_0 e^{kt}.$$

In this equation, y_0 is the amount or number present at time $t = 0$ and k is a constant.

The continuous compounding of money is an example of exponential growth. In **Example 11,** we investigate exponential decay.

EXAMPLE 11 Solving an Application Involving Exponential Decay

Carbon 14 is a radioactive form of carbon that is found in all living plants and animals. After death, radioactive carbon 14 disintegrates according to the function

$$f(t) = y_0 e^{-0.000121t},$$

where t is time in years, $f(t)$ is the amount of the sample at time t, and y_0 is the initial amount present at $t = 0$.

(a) If an initial sample contains $y_0 = 10$ g of carbon 14, how many grams, to the nearest hundredth, will be present after 3000 yr?

Let $y_0 = 10$ and $t = 3000$ in the formula, and evaluate with a calculator.

$$f(3000) = 10e^{-0.000121(3000)} \approx 6.96 \text{ g}$$

(b) How long would it take to the nearest year for the initial sample to decay to half of its original amount? (This is the **half-life.**)

$f(t) = y_0 e^{-0.000121t}$	Exponential decay formula
$5 = 10e^{-0.000121t}$	Let $y_0 = 10$ and $f(t) = \frac{1}{2}(10) = 5$.
$\frac{1}{2} = e^{-0.000121t}$	Divide by 10.
$\ln \frac{1}{2} = -0.000121t$	Take natural logarithms; $\ln e^k = k$.
$t = \dfrac{\ln \frac{1}{2}}{-0.000121}$	Divide by -0.000121. Interchange sides.
$t \approx 5728$	Evaluate with a calculator.

The half-life is 5728 yr. NOW TRY

NOW TRY EXERCISE 11

Radium 226 decays according to the function

$$f(t) = y_0 e^{-0.00043t},$$

where t is time in years.

(a) If an initial sample contains $y_0 = 4.5$ g of radium 226, how many grams, to the nearest tenth, will be present after 150 yr?

(b) What is the half-life of radium 226? Round to the nearest year.

CLASSROOM EXAMPLE 11

Radioactive strontium decays according to the function

$$f(t) = y_0 e^{-0.0239t},$$

where t is time in years.

(a) If an initial sample contains $y_0 = 12$ g of radioactive strontium, how many grams, to the nearest tenth, will be present after 35 yr?

(b) What is the half-life of radioactive strontium? Round to the nearest year.

Answers: (a) 5.20 g (b) 29 yr

NOW TRY ANSWERS

11. (a) 4.2 g **(b)** 1612 yr

9.6 Exercises

FOR EXTRA HELP

MyMathLab®

Complete solution available in MyMathLab

Many of the problems in these exercises require a calculator with logarithm capability.

Concept Check *Tell whether common logarithms or natural logarithms would be a better choice to use for solving each equation. Do not actually solve.*

1. $10^{0.0025x} = 75$ **2.** $10^{3x+1} = 13$

3. $e^{x-2} = 24$ **4.** $e^{-0.28x} = 30$

1. common logarithms
2. common logarithms
3. natural logarithms
4. natural logarithms

5. $\{0.827\}$ **6.** $\{0.792\}$
7. $\{0.833\}$ **8.** $\{-0.725\}$
9. $\{1.201\}$ **10.** $\{0.497\}$
11. $\{2.269\}$ **12.** $\{-13.257\}$
13. $\{15.967\}$ **14.** $\{-1.741\}$
15. $\{-6.067\}$ **16.** $\{-4.607\}$
17. $\{261.291\}$ **18.** $\{566.866\}$
19. $\{-10.718\}$ **20.** $\{-18.892\}$
21. $\{3\}$ **22.** $\{2\}$
23. $\{5.879\}$ **24.** $\{43.301\}$
25. $\{-\pi\}$, or $\{-3.142\}$
26. $\left\{\frac{\pi}{2}\right\}$, or $\{1.571\}$
27. $\{1\}$ **28.** $\left\{\frac{2}{3}\right\}$
29. $\{4\}$ **30.** $\{23\}$
31. $\left\{\frac{2}{3}\right\}$ **32.** $\left\{\frac{133}{12}\right\}$
33. $\left\{\frac{33}{2}\right\}$ **34.** $\left\{\frac{17}{2}\right\}$
35. $\{-1 + \sqrt[3]{49}\}$
36. $\{3 + 4\sqrt[3]{4}\}$
37. $\{\pm 3\}$ **38.** $\{\pm 5\}$
39. 2 cannot be a solution because

$$\log(2-3) = \log(-1),$$

and -1 is not in the domain of $\log x$.

40. -4 is a solution because it leads to a *positive* value of $3 - x$:

$$3 - (-4) = 7.$$

The number -4 is in the domain of $\log(3 - x)$.

41. $\left\{\frac{1}{3}\right\}$ **42.** $\left\{\frac{3}{2}\right\}$
43. $\{2\}$ **44.** $\{4\}$
45. $\varnothing$ **46.** $\varnothing$
47. $\{8\}$ **48.** $\{1\}$
49. $\left\{\frac{4}{3}\right\}$ **50.** $\left\{\frac{1}{2}\right\}$
51. $\{8\}$ **52.** $\{4\}$
53. \$2539.47; 10.2 yr
54. \$3828.78; 14.7 yr

Solve each equation. Approximate solutions to three decimal places. **See Examples 1 and 3.**

5. $7^x = 5$ **6.** $4^x = 3$ **7.** $9^{-x+2} = 13$

8. $6^{-x+1} = 22$ **9.** $3^{2x} = 14$ **10.** $5^{3x} = 11$

11. $2^{x+3} = 5^x$ **12.** $6^{x+3} = 4^x$ **13.** $2^{x+3} = 3^{x-4}$

14. $4^{x-2} = 5^{3x+2}$ **15.** $4^{2x+3} = 6^{x-1}$ **16.** $3^{2x+1} = 5^{x-1}$

Solve each equation. Use natural logarithms. When appropriate, approximate solutions to three decimal places. **See Example 2.**

17. $e^{0.012x} = 23$ **18.** $e^{0.006x} = 30$ **19.** $e^{-0.205x} = 9$

20. $e^{-0.103x} = 7$ **21.** $\ln e^{3x} = 9$ **22.** $\ln e^{2x} = 4$

23. $\ln e^{0.45x} = \sqrt{7}$ **24.** $\ln e^{0.04x} = \sqrt{3}$ **25.** $\ln e^{-x} = \pi$

26. $\ln e^{2x} = \pi$ **27.** $e^{\ln 2x} = e^{\ln(x+1)}$ **28.** $e^{\ln(6-x)} = e^{\ln(4+2x)}$

Solve each equation. Give exact solutions. **See Examples 4 and 5.**

29. $\log_4(2x + 8) = 2$ **30.** $\log_5(5x + 10) = 3$

31. $\log_3(6x + 5) = 2$ **32.** $\log_5(12x - 8) = 3$

33. $\log_2(2x - 1) = 5$ **34.** $\log_6(4x + 2) = 2$

35. $\log_7(x + 1)^3 = 2$ **36.** $\log_4(x - 3)^3 = 4$

37. $\log_2(x^2 + 7) = 4$ **38.** $\log_6(x^2 + 11) = 2$

39. *Concept Check* Suppose that in solving a logarithmic equation having the term $\log(x - 3)$, we obtain the proposed solution 2. We know that our algebraic work is correct, so we give $\{2\}$ as the solution set. **WHAT WENT WRONG?**

40. *Concept Check* Suppose that in solving a logarithmic equation having the term $\log(3 - x)$, we obtain the proposed solution -4. We know that our algebraic work is correct, so we reject -4 and give $\varnothing$ as the solution set. **WHAT WENT WRONG?**

Solve each equation. Give exact solutions. **See Examples 6 and 7.**

41. $\log(6x + 1) = \log 3$ **42.** $\log(7 - 2x) = \log 4$

43. $\log_5(3t + 2) - \log_5 t = \log_5 4$ **44.** $\log_2(t + 5) - \log_2(t - 1) = \log_2 3$

45. $\log 4x - \log(x - 3) = \log 2$ **46.** $\log(-x) + \log 3 = \log(2x - 15)$

47. $\log_2 x + \log_2(x - 7) = 3$ **48.** $\log(2x - 1) + \log 10x = \log 10$

49. $\log 5x - \log(2x - 1) = \log 4$ **50.** $\log_3 x + \log_3(2x + 5) = 1$

51. $\log_2 x + \log_2(x - 6) = 4$ **52.** $\log_2 x + \log_2(x + 4) = 5$

Solve each problem. **See Examples 8–10.**

53. How much money will there be in an account at the end of 6 yr if \$2000 is deposited at 4% compounded quarterly? (Assume no withdrawals are made.) To one decimal place, how long will it take for the account to grow to \$3000?

54. How much money will there be in an account at the end of 7 yr if \$3000 is deposited at 3.5% compounded quarterly? (Assume no withdrawals are made.) To one decimal place, how long will it take for the account to grow to \$5000?

55. What will be the amount A in an account with initial principal \$4000 if interest is compounded continuously at an annual rate of 3.5% for 6 yr? To one decimal place, how long will it take for the initial amount to double?

56. Refer to the first question in **Exercise 54.** Does the money grow to a greater value under those conditions, or when invested for 7 yr at 3% compounded continuously?

57. Find the amount of money in an account after 12 yr if \$5000 is deposited at 7% annual interest compounded as follows.

(a) Annually **(b)** Semiannually **(c)** Quarterly

(d) Daily (Use $n = 365$.) **(e)** Continuously

58. Find the amount of money in an account after 8 yr if \$4500 is deposited at 6% annual interest compounded as follows.

(a) Annually **(b)** Semiannually **(c)** Quarterly

(d) Daily (Use $n = 365$.) **(e)** Continuously

59. How much money must be deposited today to amount to \$1850 in 40 yr at 6.5% compounded continuously?

60. How much money must be deposited today to amount to \$1000 in 10 yr at 5% compounded continuously?

Solve each problem. ***See Example 11.***

61. Revenues of software publishers in the United States for the years 2004–2009 can be modeled by the function

$$S(x) = 114{,}711e^{0.0467x},$$

where $x = 0$ represents 2004, $x = 1$ represents 2005, and so on, and $S(x)$ is in millions of dollars. Approximate, to the nearest unit, revenue for 2007. (*Source:* U.S. Census Bureau.)

62. Based on selected figures obtained during the years 1970–2011, the total number of bachelor's degrees earned in the United States can be modeled by the function

$$D(x) = 794{,}383e^{0.0174x},$$

where $x = 0$ corresponds to 1970, $x = 10$ corresponds to 1980, and so on. Approximate, to the nearest unit, the number of bachelor's degrees earned in 2010. (*Source:* U.S. National Center for Education Statistics.)

63. Suppose that the amount, in grams, of plutonium 241 present in a given sample is determined by the function

$$A(t) = 2.00e^{-0.053t},$$

where t is measured in years. Approximate the amount present, to the nearest hundredth, in the sample after the given number of years.

(a) 4 **(b)** 10 **(c)** 20 **(d)** What was the initial amount present?

64. Suppose that the amount, in grams, of radium 226 present in a given sample is determined by the function

$$A(t) = 3.25e^{-0.00043t},$$

where t is measured in years. Approximate the amount present, to the nearest hundredth, in the sample after the given number of years.

(a) 20 **(b)** 100 **(c)** 500 **(d)** What was the initial amount present?

55. \$4934.71; 19.8 yr
56. The account described in Exercise 54 (at 3.5% compounded quarterly) will yield an amount which is \$127.75 greater, so it is a better deal.
57. (a) \$11,260.96 (b) \$11,416.64 (c) \$11,497.99 (d) \$11,580.90 (e) \$11,581.83
58. (a) \$7172.32 (b) \$7221.18 (c) \$7246.46 (d) \$7272.05 (e) \$7272.33
59. \$137.41 60. \$606.53
61. \$131,962 million
62. 1,593,305 degrees
63. (a) 1.62 g (b) 1.18 g (c) 0.69 g (d) 2.00 g
64. (a) 3.22 g (b) 3.11 g (c) 2.62 g (d) 3.25 g

65. (a) 179.73 g (b) 21.66 yr
66. (a) 1.76 units (b) 5.55 hr

65. A sample of 400 g of lead 210 decays to polonium 210 according to the function

$$A(t) = 400e^{-0.032t},$$

where t is time in years. Approximate answers to the nearest hundredth.

(a) How much lead will be left in the sample after 25 yr?

(b) How long will it take the initial sample to decay to half of its original amount?

66. The concentration of a drug in a person's system decreases according to the function

$$C(t) = 2e^{-0.125t},$$

where $C(t)$ is in appropriate units, and t is in hours. Approximate answers to the nearest hundredth.

(a) How much of the drug will be in the system after 1 hr?

(b) How long will it take for the concentration to be half of its original amount?

Chapter 9 Summary

Key Terms

9.1
one-to-one function
inverse of a function

9.2
exponential function with base a
asymptote
exponential equation

9.3
logarithm
logarithmic equation
logarithmic function with base a

9.5
common logarithm
natural logarithm

9.6
compound interest
continuous compounding

New Symbols

$f^{-1}(x)$	inverse of $f(x)$
$\log_a x$	logarithm of x with base a
$\log x$	common (base 10) logarithm of x
$\ln x$	natural (base e) logarithm of x
e	a constant, approximately 2.718281828

Test Your Word Power

See how well you have learned the vocabulary in this chapter.

1. In a **one-to-one function**
   - **A.** each x-value corresponds to only one y-value
   - **B.** each x-value corresponds to one or more y-values
   - **C.** each x-value is the same as each y-value
   - **D.** each x-value corresponds to only one y-value and each y-value corresponds to only one x-value.

2. If f is a one-to-one function, then the **inverse** of f is
   - **A.** the set of all solutions of f
   - **B.** the set of all ordered pairs formed by interchanging the coordinates of the ordered pairs of f
   - **C.** the set of all ordered pairs that are the opposite (negative) of the coordinates of the ordered pairs of f
   - **D.** an equation involving an exponential expression.

3. An **exponential function** is a function defined by an expression of the form
   - **A.** $f(x) = ax^2 + bx + c$ for real numbers a, b, c $(a \neq 0)$
   - **B.** $f(x) = \log_a x$ for positive numbers a and x $(a \neq 1)$
   - **C.** $f(x) = a^x$ for all real numbers x $(a > 0, a \neq 1)$
   - **D.** $f(x) = \sqrt{x}$ for $x \geq 0$.

4. An **asymptote** is
   A. a line that a graph intersects just once
   B. a line that the graph of a function more and more closely approaches as the x-values increase or decrease without bound
   C. the x-axis or y-axis
   D. a line about which a graph is symmetric.

5. A **logarithmic function** is a function that is defined by an expression of the form
   A. $f(x) = ax^2 + bx + c$ for real numbers a, b, c $(a \neq 0)$
   B. $f(x) = \log_a x$ for positive numbers a and x $(a \neq 1)$
   C. $f(x) = a^x$ for all real numbers x $(a > 0, a \neq 1)$
   D. $f(x) = \sqrt{x}$ for $x \geq 0$.

6. A **logarithm** is
   A. an exponent
   B. a base
   C. an equation
   D. a polynomial.

ANSWERS

1. D; *Example:* The function $f = \{(0, 2), (1, -1), (3, 5), (-2, 3)\}$ is one-to-one. **2.** B; *Example:* The inverse of the one-to-one function f defined in Answer 1 is $f^{-1} = \{(2, 0), (-1, 1), (5, 3), (3, -2)\}$. **3.** C; *Examples:* $f(x) = 4^x$, $g(x) = \left(\frac{1}{2}\right)^x$ **4.** B; *Example:* The graph of $f(x) = 2^x$ has the x-axis $(y = 0)$ as an asymptote. **5.** B; *Examples:* $f(x) = \log_3 x$, $g(x) = \log_{1/3} x$ **6.** A; *Example:* $\log_a x$ is the exponent to which a must be raised to obtain x; $\log_3 9 = 2$ because $3^2 = 9$.

Quick Review

CONCEPTS | **EXAMPLES**

9.1 Inverse Functions

Horizontal Line Test

A function is one-to-one if every horizontal line intersects the graph of the function at most once.

Inverse Functions

For a one-to-one function $y = f(x)$, the equation that defines the inverse function f^{-1} is found as follows.

Step 1 Interchange x and y.

Step 2 Solve for y.

Step 3 Replace y with $f^{-1}(x)$.

In general, the graph of f^{-1} is the mirror image of the graph of f with respect to the line $y = x$. If (a, b) lies on the graph of f, then (b, a) lies on the graph of f^{-1}.

Find f^{-1} if $f(x) = 2x - 3$.

The graph of f is a nonhorizontal (slanted) straight line, so f is one-to-one by the horizontal line test.

Find $f^{-1}(x)$ as follows.

$$f(x) = 2x - 3$$

$$y = 2x - 3 \quad \text{Let } y = f(x).$$

$$x = 2y - 3 \quad \text{Interchange } x \text{ and } y.$$

$$x + 3 = 2y \quad \text{Solve for } y.$$

$$y = \frac{x + 3}{2}$$

$$f^{-1}(x) = \frac{1}{2}x + \frac{3}{2} \quad \text{Replace } y \text{ with } f^{-1}(x).$$

The graphs of a function f and its inverse f^{-1} are shown here.

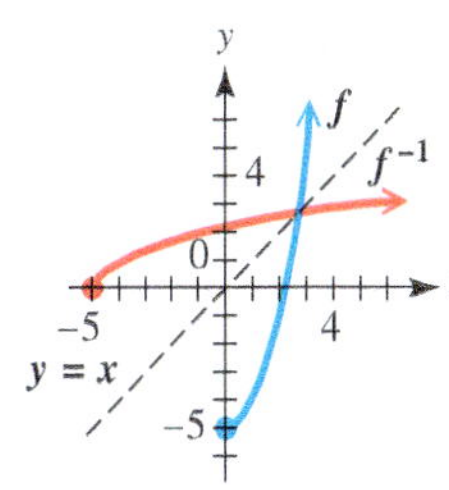

CONCEPTS | **EXAMPLES**

9.2 Exponential Functions

For $a > 0$, $a \neq 1$, and all real numbers x,

$$f(x) = a^x$$

defines the exponential function with base a.

Graph of $f(x) = a^x$

1. The graph contains the point $(0, 1)$, which is its y-intercept.
2. When $a > 1$, the graph rises from left to right.
   When $0 < a < 1$, the graph falls from left to right.
3. The x-axis is an asymptote.
4. The domain is $(-\infty, \infty)$, and the range is $(0, \infty)$.

$f(x) = 3^x$ defines the exponential function with base 3.

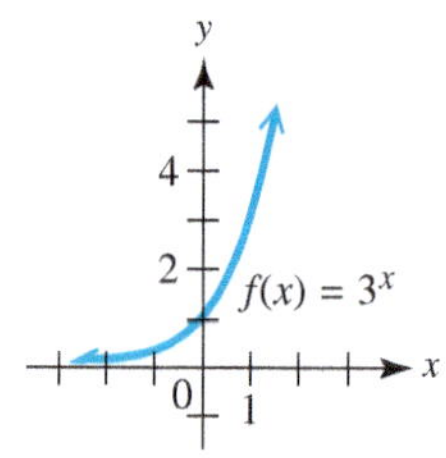

9.3 Logarithmic Functions

For all positive real numbers a, where $a \neq 1$, and all positive real numbers x,

$$y = \log_a x \quad \text{means} \quad x = a^y.$$

Special Properties of Logarithms

For $b > 0$, where $b \neq 1$, the following hold true.

$$\log_b b^r = r \qquad b^{\log_b r} = r \quad (\text{where } r > 0)$$

$$\log_b b = 1 \qquad \log_b 1 = 0$$

If a and x are positive real numbers, where $a \neq 1$, then

$$g(x) = \log_a x$$

defines the logarithmic function with base a.

Graph of $g(x) = \log_a x$

1. The graph contains the point $(1, 0)$, which is its x-intercept.
2. When $a > 1$, the graph rises from left to right.
   When $0 < a < 1$, the graph falls from left to right.
3. The y-axis is an asymptote.
4. The domain is $(0, \infty)$, and the range is $(-\infty, \infty)$.

$y = \log_2 x$ means $x = 2^y$.

$\log_5 5^6 = 6$ $\quad$ $4^{\log_4 6} = 6$

$\log_3 3 = 1$ $\quad$ $\log_5 1 = 0$

$g(x) = \log_3 x$ defines the logarithmic function with base 3.

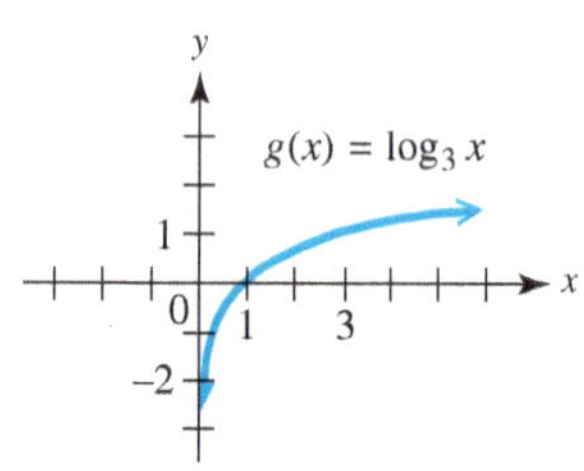

9.4 Properties of Logarithms

If x, y, and b are positive real numbers, where $b \neq 1$, and r is any real number, then the following hold.

Special Properties (repeated)

$$\log_b b^r = r \quad \text{and} \quad b^{\log_b r} = r \quad (\text{where } r > 0)$$

Product Rule $\quad \log_b xy = \log_b x + \log_b y$

Quotient Rule $\quad \log_b \frac{x}{y} = \log_b x - \log_b y$

Power Rule $\quad \log_b x^r = r \log_b x$

$6^{\log_6 10} = 10$ $\quad$ $\log_3 3^4 = 4$ $\quad$ Special properties

$\log_2 3m = \log_2 3 + \log_2 m$ $\quad$ Product rule

$\log_5 \frac{9}{4} = \log_5 9 - \log_5 4$ $\quad$ Quotient rule

$\log_{10} 2^3 = 3 \log_{10} 2$ $\quad$ Power rule

CONCEPTS | **EXAMPLES**

9.5 Common and Natural Logarithms

Common logarithms (base 10) are used in applications such as pH, sound level, and intensity of an earthquake. Use the (LOG) key of a calculator to evaluate common logarithms.

Use the formula $\mathbf{pH} = -\mathbf{log}\,[\mathbf{H_3O^+}]$ to find the pH (to one decimal place) of grapes with hydronium ion concentration 5.0×10^{-5}.

$\text{pH} = -\log\,(5.0 \times 10^{-5})$ Substitute.

$\text{pH} = -(\log 5.0 + \log 10^{-5})$ Property of logarithms

$\text{pH} \approx 4.3$ Evaluate with a calculator.

Natural logarithms (base *e*) are found in formulas for applications of growth and decay, such as continuous compounding of money, decay of chemical compounds, and biological growth. Use the (LN) key or both the (INV) and (e^x) keys to evaluate natural logarithms.

Use the formula for doubling time (in years)

$$t(r) = \frac{\ln 2}{\ln\,(1 + r)}$$

to find doubling time to the nearest tenth of a year, at a rate of 4%.

$$t(0.04) = \frac{\ln 2}{\ln\,(1 + 0.04)} \approx 17.7$$ Let $r = 0.04$, and evaluate with a calculator.

The doubling time is 17.7 yr.

Change-of-Base Rule

If $a > 0$, $a \neq 1$, $b > 0$, $b \neq 1$, and $x > 0$, then the following holds true.

$$\log_a x = \frac{\log_b x}{\log_b a}$$

Use the change-of-base rule to approximate $\log_3 17$.

$$\log_3 17 = \frac{\ln 17}{\ln 3} = \frac{\log 17}{\log 3} \approx 2.5789$$

9.6 Exponential and Logarithmic Equations; Further Applications

To solve exponential equations, use these properties (where $b > 0$, $b \neq 1$).

1. If $b^x = b^y$, then $x = y$.

Solve. $2^{3x} = 2^5$

$3x = 5$ Set exponents equal.

$x = \frac{5}{3}$ Divide by 3.

The solution set is $\left\{\frac{5}{3}\right\}$.

2. If $x = y$, $x > 0$, $y > 0$, then $\log_b x = \log_b y$.

Solve. $5^x = 8$

$\log 5^x = \log 8$ Take common logarithms.

$x \log 5 = \log 8$ Power rule

$x = \frac{\log 8}{\log 5} \approx 1.292$ Divide by log 5.

The solution set is $\{1.292\}$.

To solve logarithmic equations, use these properties, where $b > 0$, $b \neq 1$, $x > 0$, and $y > 0$. First use the properties of **Section 9.4,** if necessary, to write the equation in the proper form.

1. If $\log_b x = \log_b y$, then $x = y$.

Solve. $\log_3 2x = \log_3\,(x + 1)$

$2x = x + 1$

$x = 1$ Subtract x.

This value checks, so the solution set is $\{1\}$.

CONCEPTS	EXAMPLES
2. If $\log_b x = k$, then $x = b^k$.	Solve. $\log_2 (3x - 1) = 4$ $3x - 1 = 2^4$ Exponential form $3x - 1 = 16$ Apply the exponent. $3x = 17$ Add 1. $x = \frac{17}{3}$ Divide by 3.
Always check any proposed solutions in logarithmic equations.	This value checks, so the solution set is $\left\{\frac{17}{3}\right\}$.

Chapter 9 Review Exercises

1. not one-to-one
2. one-to-one
3. $f^{-1}(x) = \frac{x - 7}{-3}$, or $f^{-1}(x) = -\frac{1}{3}x + \frac{7}{3}$
4. $f^{-1}(x) = \frac{x^3 + 4}{6}$
5. not one-to-one
6. This function is not one-to-one because two states in the list have minimum wage $8.00.

7.

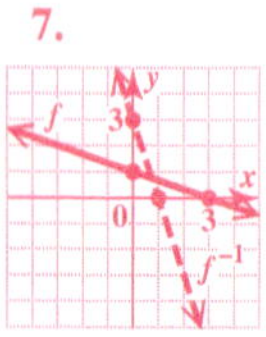

8.

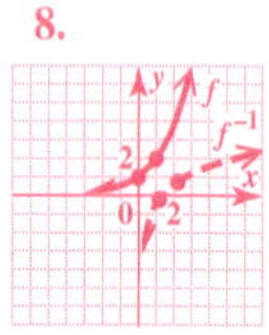

9. 172.466 10. 0.034
11. 0.079

9.1 *Determine whether each graph is the graph of a one-to-one function.*

1.

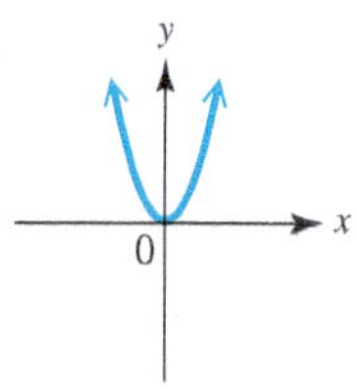

2.

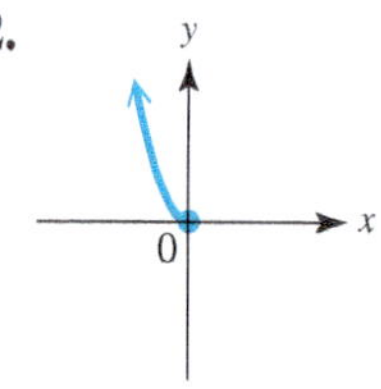

Determine whether each function is one-to-one. If it is, find its inverse.

3. $f(x) = -3x + 7$ **4.** $f(x) = \sqrt[3]{6x - 4}$ **5.** $f(x) = -x^2 + 3$

6. The table lists basic minimum wages in several states. If the set of states is the domain and the set of wage amounts is the range of a function, is it one-to-one? If not, explain why.

State	Minimum Wage (in dollars)
Washington	9.04
Nevada	8.25
California	8.00
Massachusetts	8.00
Ohio	7.70
Iowa	7.25

Source: U.S. Department of Labor.

Each function graphed is one-to-one. Graph its inverse.

7.

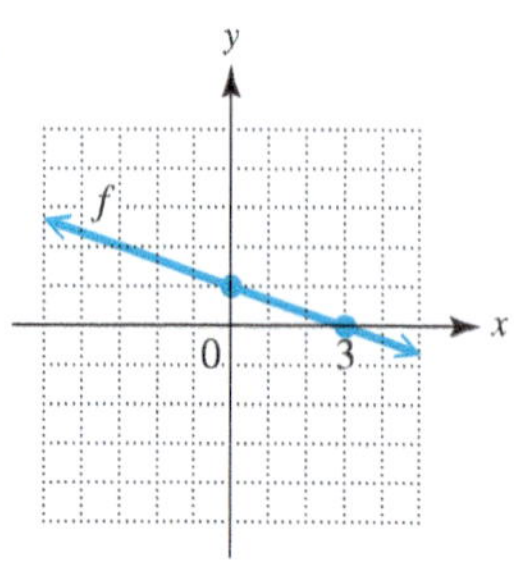

8.

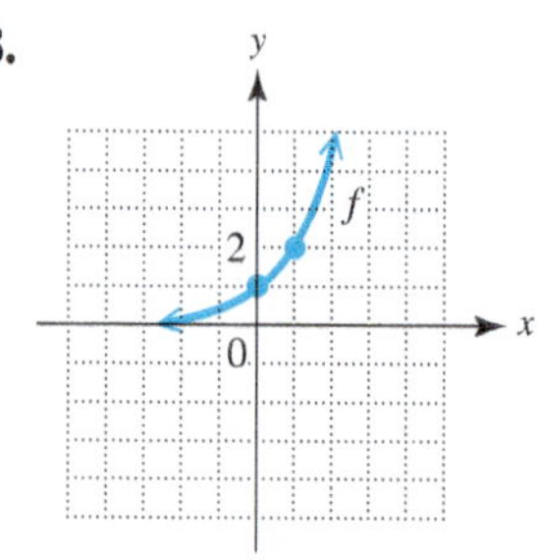

9.2 *Use a calculator to find an approximation to three decimal places for each exponential expression.*

9. $5^{3.2}$ **10.** $\left(\frac{1}{5}\right)^{2.1}$ **11.** $8.3^{-1.2}$

Graph each exponential function.

12. $f(x) = 3^x$ **13.** $f(x) = \left(\frac{1}{3}\right)^x$ **14.** $f(x) = 2^{2x+3}$

Solve each equation.

15. $5^{2x+1} = 25$ **16.** $4^{3x} = 8^{x+4}$ **17.** $\left(\frac{1}{27}\right)^{x-1} = 9^{2x}$

18. Sulfur dioxide emissions in the United States, in millions of tons, from 1980 through 2012 can be approximated by the exponential function

$$f(x) = 30.7032(1.0403)^{-x},$$

where $x = 0$ corresponds to 1980, $x = 5$ to 1985, and so on. Use this function to approximate emissions, to the nearest million tons, for each year. (*Source:* EPA.)

(a) 1985 **(b)** 1995 **(c)** 2005

9.3 *Use a calculator to approximate each logarithm to four decimal places.*

19. $\log_2 6$ **20.** $\log_7 3$ **21.** $\log_{10} 55$

Graph each logarithmic function.

22. $g(x) = \log_3 x$ (*Hint:* See **Exercise 12.**) **23.** $g(x) = \log_{1/3} x$ (*Hint:* See **Exercise 13.**)

24. Evaluate without using a calculator.

(a) $4^{\log_4 12}$ **(b)** $\log_9 9^{13}$ **(c)** $\log_5 625$

Solve each equation.

25. $\log_8 64 = x$ **26.** $\log_2 \sqrt{8} = x$ **27.** $\log_x \left(\frac{1}{49}\right) = -2$

28. $\log_4 x = \frac{3}{2}$ **29.** $\log_k 4 = 1$ **30.** $\log_6 x = -2$

A company has found that total sales, in thousands of dollars, are given by the function

$$S(x) = 100 \log_2 (x + 2),$$

where x is the number of weeks after a major advertising campaign was introduced.

31. What were total sales 6 weeks after the campaign was introduced?

32. Graph the function.

9.4 *Use properties of logarithms to express each logarithm as a sum or difference of logarithms. Assume that all variables represent positive real numbers.*

33. $\log_2 3xy^2$ **34.** $\log_4 \frac{\sqrt{x} \cdot w^2}{z}$

Use properties of logarithms to write each expression as a single logarithm. Assume that all variables represent positive real numbers, $b \neq 1$.

35. $\log_b 3 + \log_b x - 2 \log_b y$ **36.** $\log_3 (x + 7) - \log_3 (4x + 6)$

9.5 *Evaluate each logarithm to four decimal places.*

37. $\log 28.9$ **38.** $\log 0.257$ **39.** $\ln 28.9$ **40.** $\ln 0.257$

12.

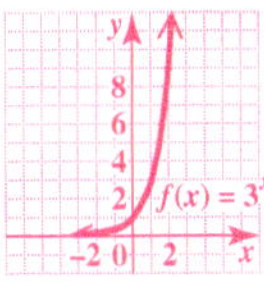

13.

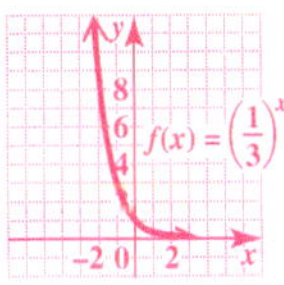

14.

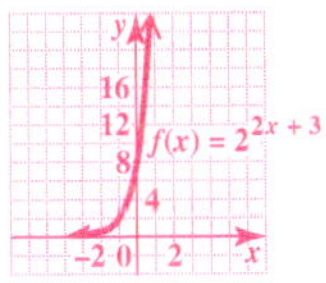

15. $\left\{\frac{1}{2}\right\}$ **16.** $\{4\}$

17. $\left\{\frac{3}{7}\right\}$

18. (a) 25 million tons **(b)** 17 million tons **(c)** 11 million tons

19. 2.5850 **20.** 0.5646

21. 1.7404

22.

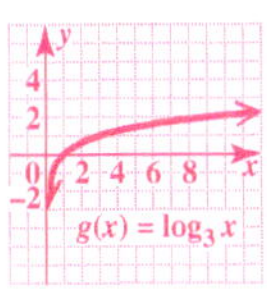

23.

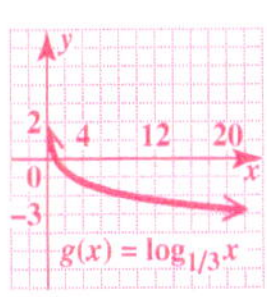

24. (a) 12 **(b)** 13 **(c)** 4

25. $\{2\}$ **26.** $\left\{\frac{3}{2}\right\}$

27. $\{7\}$ **28.** $\{8\}$

29. $\{4\}$ **30.** $\left\{\frac{1}{36}\right\}$

31. \$300,000

32.

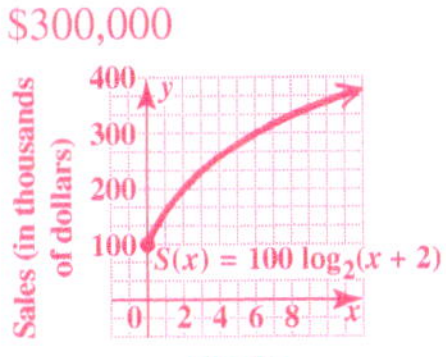

33. $\log_2 3 + \log_2 x + 2 \log_2 y$

34. $\frac{1}{2} \log_4 x + 2 \log_4 w - \log_4 z$

35. $\log_b \frac{3x}{y^2}$ **36.** $\log_3 \left(\frac{x+7}{4x+6}\right)$

37. 1.4609 **38.** -0.5901

39. 3.3638 **40.** -1.3587

41. 0.9251 42. 1.7925
43. 6.4 44. 8.4
45. 2.5×10^{-5}
46. Magnitude 1 is about 6.3 times as intense as magnitude 3.
47. (a) 18 yr (b) 12 yr
(c) 7 yr (d) 6 yr
(e) Each comparison shows approximately the same number. For example, in part (a) the doubling time is 18 yr (rounded) and $\frac{72}{4} = 18$. Thus, the formula $t = \frac{72}{100r}$ (called the *rule of 72*) is an excellent approximation of the doubling time formula.
48. $\{2.042\}$
49. $\{4.907\}$ 50. $\{18.310\}$
51. $\left\{\frac{1}{9}\right\}$ 52. $\{-6 + \sqrt[3]{25}\}$
53. $\{2\}$ 54. $\left\{\frac{3}{8}\right\}$
55. $\{4\}$ 56. $\{1\}$
57. When the power rule was applied in the second step, the domain was changed from $\{x \mid x \neq 0\}$ to $\{x \mid x > 0\}$. The valid solution -10 was "lost." The solution set is $\{\pm 10\}$.
58. D
59. \$24,403.80

Use the change-of-base rule (with either common or natural logarithms) to approximate each logarithm to four decimal places.

41. $\log_{16} 13$ **42.** $\log_4 12$

Find the pH *of each substance with the given hydronium ion concentration.*

43. Milk, 4.0×10^{-7} **44.** Crackers, 3.8×10^{-9}

45. If orange juice has pH 4.6, what is its hydronium ion concentration?

46. The magnitude of a star is given by the equation

$$M = 6 - 2.5 \log \frac{I}{I_0},$$

where I_0 is the measure of the faintest star and I is the actual intensity of the star being measured. The dimmest stars are of magnitude 6, and the brightest are of magnitude 1. Determine the ratio of intensities between stars of magnitude 1 and 3.

47. Section 9.5, Exercise 48 introduced the doubling function

$$t(r) = \frac{\ln 2}{\ln (1 + r)},$$

that gives the number of years required to double money when it is invested at interest rate r (in decimal form) compounded annually. To the nearest year, how long does it take to double money at each rate?

(a) 4% **(b)** 6% **(c)** 10% **(d)** 12%

(e) Compare each answer in parts (a)–(d) with the following numbers. Explain.

$$\frac{72}{4}, \quad \frac{72}{6}, \quad \frac{72}{10}, \quad \frac{72}{12}$$

9.6 *Solve each equation. Approximate solutions to three decimal places.*

48. $3^x = 9.42$ **49.** $2^{x-1} = 15$ **50.** $e^{0.06x} = 3$

Solve each equation. Give exact solutions.

51. $\log_3 (9x + 8) = 2$ **52.** $\log_5 (x + 6)^3 = 2$

53. $\log_3 (x + 2) - \log_3 x = \log_3 2$ **54.** $\log (2x + 3) = 1 + \log x$

55. $\log_4 x + \log_4 (8 - x) = 2$ **56.** $\log_2 x + \log_2 (x + 15) = \log_2 16$

57. Consider the following "solution" of the equation $\log x^2 = 2$. **WHAT WENT WRONG?** Give the correct solution set.

$\log x^2 = 2$	Original equation
$2 \log x = 2$	Power rule for logarithms
$\log x = 1$	Divide each side by 2.
$x = 10^1$	Write in exponential form.
$x = 10$	$10^1 = 10$

Solution set: $\{10\}$

58. Which is *not* a representation of the solution of $7^x = 23$?

A. $\frac{\log 23}{\log 7}$ **B.** $\frac{\ln 23}{\ln 7}$ **C.** $\log_7 23$ **D.** $\log_{23} 7$

Solve each problem. Use a calculator as necessary.

59. If \$20,000 is deposited at 4% annual interest compounded quarterly, how much will be in the account after 5 yr, assuming no withdrawals are made?

60. $11,190.72

61. Plan A is better because it would pay $2.92 more.

62. 13.9 days

63. $4267 **64.** 11%

60. How much will $10,000 compounded continuously at 3.75% annual interest amount to in 3 yr?

61. Which is a better plan?

Plan A: Invest $1000 at 4% compounded quarterly for 3 yr

Plan B: Invest $1000 at 3.9% compounded monthly for 3 yr

62. What is the half-life of a radioactive substance that decays according to the function

$Q(t) = A_0 e^{-0.05t}$, where t is in days, to the nearest tenth?

A machine purchased for business use ***depreciates,*** *or loses value, over a period of years. The value of the machine at the end of its useful life is its* ***scrap value.*** *By one method of depreciation, the scrap value, S, is given by*

$$S = C(1 - r)^n,$$

where C is original cost, n is useful life in years, and r is the constant percent of depreciation.

63. Find the scrap value of a machine costing $30,000, having a useful life of 12 yr and a constant annual rate of depreciation of 15%. Round to the nearest dollar.

64. A machine has a "half-life" of 6 yr. Find the constant annual rate of depreciation, to the nearest percent.

Chapter 9 Mixed Review Exercises

1. 7 **2.** 36
3. 4 **4.** e
5. −5 **6.** 5.4

7. {72} **8.** {3}

9. $\left\{\frac{1}{9}\right\}$ **10.** $\left\{\frac{4}{3}\right\}$

11. {3} **12.** {0.348}

13. {0} **14.** $\left\{\frac{1}{8}\right\}$

15. $\left\{\frac{11}{3}\right\}$ **16.** {−2, −1}

17. 67%

18. (a) 1.209061955
(b) 7
(c) {1.209061955}

19. Answers will vary. Suppose the name is Jeffery Cole, with $m = 7$ and $n = 4$.
(a) $\log_7 4$ is the exponent to which 7 must be raised to obtain 4.
(b) 0.7124143742
(c) 4

Evaluate.

1. $\log_2 128$ **2.** $5^{\log_5 36}$ **3.** $e^{\ln 4}$

4. $10^{\log e}$ **5.** $\log_3 3^{-5}$ **6.** $\ln e^{5.4}$

Solve each equation.

7. $\log_3(x + 9) = 4$ **8.** $\ln e^x = 3$ **9.** $\log_x \frac{1}{81} = 2$

10. $27^x = 81$ **11.** $2^{2x-3} = 8$ **12.** $8^{3x} = 5^{x+1}$

13. $5^{x+2} = 25^{2x+1}$ **14.** $\log_3 (x + 1) - \log_3 x = 2$

15. $\log (3x - 1) = \log 10$ **16.** $\ln (x^2 + 3x + 4) = \ln 2$

Solve each problem.

17. Recall from **Exercise 49** in **Section 9.5** that the number of years, $N(x)$, since two independently evolving languages split off from a common ancestral language is approximated by

$$N(x) = -5000 \ln x,$$

where x is the percent of words from the ancestral language common to both languages now. Find x to the nearest percent if the split occurred 2000 yr ago.

18. To solve the equation $5^x = 7$, we must find the exponent to which 5 must be raised in order to obtain 7. This is $\log_5 7$.

(a) Use the change-of-base rule and a calculator to find $\log_5 7$.

(b) Raise 5 to the number found in part (a). What is the result?

(c) Using as many decimal places as the calculator gives, write the solution set of $5^x = 7$.

19. Let m be the number of letters in your first name, and let n be the number of letters in your last name.

(a) Explain what $\log_m n$ means. **(b)** Use a calculator to find $\log_m n$.

(c) Raise m to the power indicated by the number found in part (b). What is the result?

20. (a) 0.325 **(b)** 0.673

20. One measure of the diversity of the species in an ecological community is the **index of diversity,** the logarithmic expression

$$-(p_1 \ln p_1 + p_2 \ln p_2 + \cdots + p_n \ln p_n),$$

where $p_1, p_2, \ldots, p_n$ are the proportions of a sample belonging to each of n species in the sample. (*Source:* Ludwig, J. and J. Reynolds, *Statistical Ecology: A Primer on Methods and Computing,* New York, John Wiley and Sons.) Approximate the index of diversity to the nearest thousandth if a sample of 100 from a community produces the following numbers.

(a) 90 of one species, 10 of another **(b)** 60 of one species, 40 of another

Chapter 9 Test

FOR EXTRA HELP *Step-by-step test solutions are found on the Chapter Test Prep Videos available in* MyMathLab® *or on* YouTube.

View the complete solutions to all Chapter Test exercises in MyMathLab.

[9.1]
1. (a) not one-to-one
(b) one-to-one
2. $f^{-1}(x) = x^3 - 7$
3.

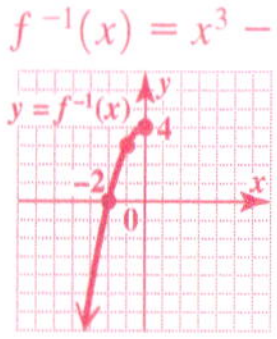

[9.2]
4.

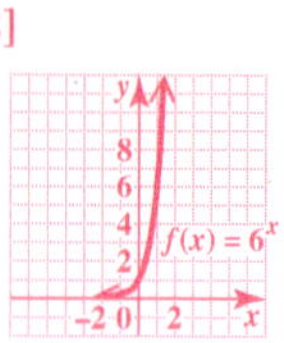

[9.3]
5.

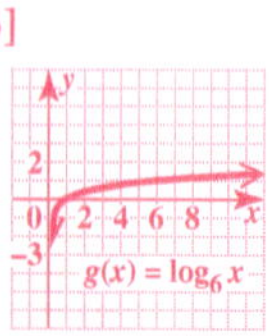

6. 9
7. 6 **8.** 0

[9.2]
9. $\{-4\}$ **10.** $\left\{-\frac{13}{3}\right\}$
11. (a) 775 millibars
(b) 265 millibars

[9.3]
12. $\log_4 0.0625 = -2$
13. $7^2 = 49$ **14.** $\{32\}$
15. $\left\{\frac{1}{2}\right\}$ **16.** $\{2\}$
17. 5; 2; fifth; 32

[9.4]
18. $2 \log_3 x + \log_3 y$
19. $\frac{1}{2} \log_5 x - \log_5 y - \log_5 z$

1. Decide whether each function is one-to-one.

(a) $f(x) = x^2 + 9$

(b)

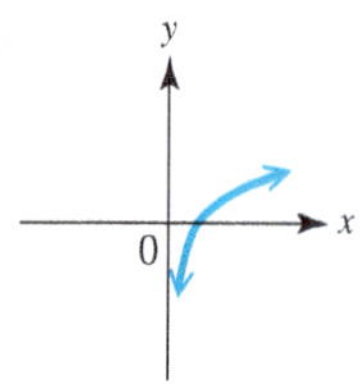

2. Find $f^{-1}(x)$ for the one-to-one function

$$f(x) = \sqrt[3]{x + 7}.$$

3. Graph the inverse given the graph of $y = f(x)$.

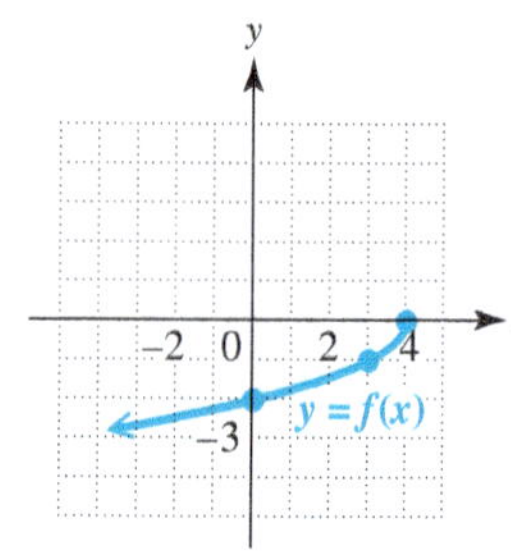

Graph each function.

4. $f(x) = 6^x$ **5.** $g(x) = \log_6 x$

Use the special properties of logarithms to evaluate each expression.

6. $7^{\log_7 9}$ **7.** $\log_3 3^6$ **8.** $\log_5 1$

Solve each equation. Give exact solutions.

9. $5^x = \frac{1}{625}$ **10.** $2^{3x-7} = 8^{2x+2}$

11. The atmospheric pressure (in millibars) at a given altitude x (in meters) is approximated by

$$f(x) = 1013e^{-0.0001341x}.$$

Use this function to approximate the atmospheric pressure at each altitude.

(a) 2000 m **(b)** 10,000 m

12. Write in logarithmic form: $4^{-2} = 0.0625$.

13. Write in exponential form: $\log_7 49 = 2$.

Solve each equation.

14. $\log_{1/2} x = -5$ **15.** $x = \log_9 3$ **16.** $\log_x 16 = 4$

17. Fill in the blanks with the correct responses: The value of $\log_2 32$ is _____. This means that if we raise _____ to the _____ power, the result is _____.

Use properties of logarithms to write each expression as a sum or difference of logarithms. Assume that variables represent positive real numbers.

18. $\log_3 x^2 y$ **19.** $\log_5 \left(\frac{\sqrt{x}}{yz}\right)$

Use properties of logarithms to write each expression as a single logarithm. Assume that variables represent positive real numbers, $b \neq 1$.

20. $3 \log_b s - \log_b t$

21. $\frac{1}{4} \log_b r + 2 \log_b s - \frac{2}{3} \log_b t$

Evaluate each logarithm to four decimal places.

22. $\log 23.1$

23. $\ln 0.82$

24. Use the change-of-base rule to express $\log_3 19$ as described.

(a) in terms of common logarithms

(b) in terms of natural logarithms

(c) approximated to four decimal places

25. Solve $3^x = 78$. Approximate the solution to three decimal places.

26. Solve $\log_8 (x + 5) + \log_8 (x - 2) = 1$.

27. Suppose that \$10,000 is invested at 4.5% annual interest, compounded quarterly. How much will be in the account in 5 yr if no money is withdrawn?

28. Suppose that \$15,000 is invested at 5% annual interest, compounded continuously.

(a) How much will be in the account in 5 yr if no money is withdrawn?

(b) To one decimal place, how long will it take for the initial principal to double?

20. $\log_b \frac{s^3}{t}$

21. $\log_b \frac{r^{1/4} s^2}{t^{2/3}}$

[9.5]

22. 1.3636

23. -0.1985

24. (a) $\frac{\log 19}{\log 3}$ (b) $\frac{\ln 19}{\ln 3}$ (c) 2.6801

[9.6]

25. $\{3.966\}$ **26.** $\{3\}$

27. \$12,507.51

28. (a) \$19,260.38 (b) 13.9 yr

Chapters R–9 Cumulative Review Exercises

Let $S = \left\{-\frac{9}{4}, -2, -\sqrt{2}, 0, 0.6, \sqrt{11}, \sqrt{-8}, 6, \frac{30}{3}\right\}$. *List the elements of S that are members of each set.*

1. Integers

2. Rational numbers

3. Irrational numbers

Simplify each expression.

4. $|-8| + 6 - |-2| - (-6 + 2)$

5. $2(-5) + (-8)(4) - (-3)$

Solve each equation or inequality.

6. $7 - (3 + 4x) + 2x = -5(x - 1) - 3$

7. $2x + 2 \leq 5x - 1$

8. $|2x - 5| = 9$

9. $|4x + 2| > 10$

Graph.

10. $5x + 2y = 10$

11. $-4x + y \leq 5$

12. The graph indicates the number of international travelers to the United States from 2006 through 2010.

(a) Is this the graph of a function?

(b) What is the slope of the line in the graph? Interpret the slope in the context of international travelers to the United States.

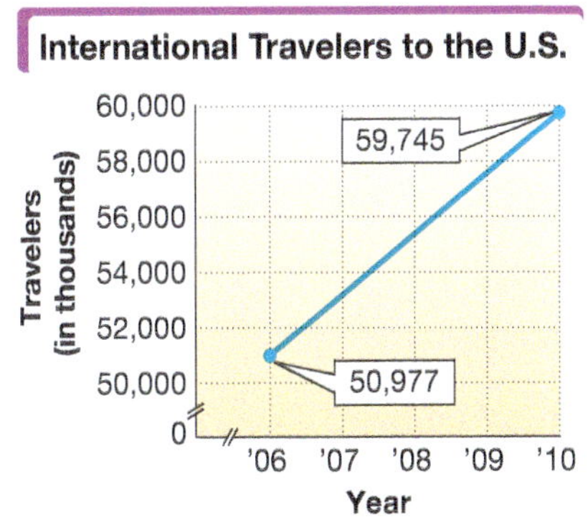

Source: U.S. Department of Commerce.

[R.1]

1. $-2, 0, 6, \frac{30}{3}$ (or 10)

2. $-\frac{9}{4}, -2, 0, 0.6, 6, \frac{30}{3}$ (or 10)

3. $-\sqrt{2}, \sqrt{11}$

[R.2, R.3]

4. 16 **5.** -39

[1.1] **6.** $\left\{-\frac{2}{3}\right\}$ [1.5] **7.** $[1, \infty)$

[1.7]

8. $\{-2, 7\}$

9. $(-\infty, -3) \cup (2, \infty)$

[2.1] **10.**

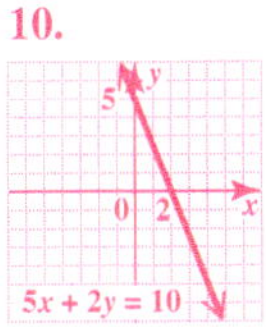

[2.4] **11.**

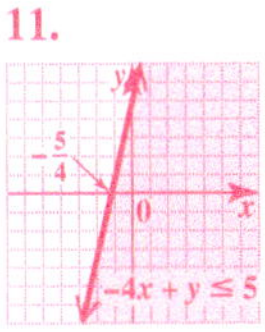

[2.2, 2.5]

12. (a) yes

(b) 2192; The number of travelers increased by an average of 2192 thousand per year during these years.

[2.3]
13. $y = \frac{3}{4}x - \frac{19}{4}$

[3.1, 3.2]
14. $\{(4, 2)\}$ 15. $\{(1, -1, 4)\}$

[3.3]
16. 6 lb

[4.4]
17. $6p^2 + 7p - 3$ 18. $16k^2 - 24k + 9$

[4.2]
19. $-5m^3 + 2m^2 - 7m + 4$

[4.5]
20. $5x^2 - 2x + 8$

[5.1]
21. $x(8 + x^2)$

[5.2]
22. $(3y - 2)(8y + 3)$
23. $z(5z + 1)(z - 4)$

[5.3]
24. $(4a + 5b^2)(4a - 5b^2)$
25. $(2c + d)(4c^2 - 2cd + d^2)$
26. $(4r + 7q)^2$

[4.1]
27. $-\frac{1875p^{13}}{8}$

[6.1]
28. $\frac{x + 5}{x + 4}$

[6.2]
29. $\frac{-3k - 19}{(k + 3)(k - 2)}$

[7.3]
30. $12\sqrt{2}$

[7.4]
31. $-27\sqrt{2}$

[7.6]
32. $\{0, 4\}$

[7.7]
33. 41

[8.1, 8.2]
34. $\left\{\frac{1 \pm \sqrt{13}}{6}\right\}$

[8.7]
35. $(-\infty, -4) \cup (2, \infty)$

[8.3]
36. $\{\pm 1, \pm 2\}$

[9.2]
37. $\{-1\}$

[8.5]
38.

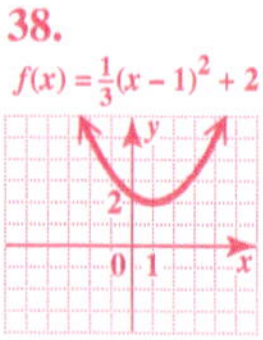

[9.2]
39.

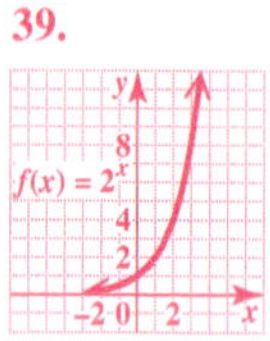

[9.3]
40.

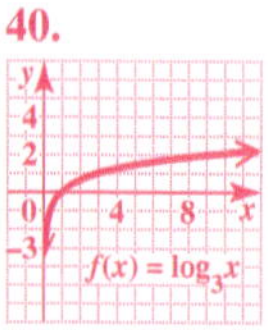

[9.4]
41. $3 \log x + \frac{1}{2} \log y - \log z$

[9.6]
42. (a) 25,000 (b) 30,500
(c) 37,300 (d) in 3.5 hr, or at 3:30 P.M.

13. Find an equation of the line passing through $(5, -1)$ and parallel to the line with equation $3x - 4y = 12$. Write the equation in slope-intercept form.

Solve each system.

14. $5x - 3y = 14$
$2x + 5y = 18$

15. $x + 2y + 3z = 11$
$3x - y + z = 8$
$2x + 2y - 3z = -12$

16. Candy worth \$1.00 per lb is to be mixed with 10 lb of candy worth \$1.96 per lb to obtain a mixture that will be sold for \$1.60 per lb. How many pounds of the \$1.00 candy should be used?

Number of Pounds	Price per Pound (in dollars)	Value (in dollars)
x	1.00	$1x$
	1.60	

Perform the indicated operations.

17. $(2p + 3)(3p - 1)$

18. $(4k - 3)^2$

19. $(3m^3 + 2m^2 - 5m) - (8m^3 + 2m - 4)$

20. Divide $15x^3 - x^2 + 22x + 8$ by $3x + 1$.

Factor.

21. $8x + x^3$

22. $24y^2 - 7y - 6$

23. $5z^3 - 19z^2 - 4z$

24. $16a^2 - 25b^4$

25. $8c^3 + d^3$

26. $16r^2 + 56rq + 49q^2$

Perform the indicated operations.

27. $\frac{(5p^3)^4(-3p^7)}{2p^2(4p^4)}$

28. $\frac{x^2 - 9}{x^2 + 7x + 12} \div \frac{x - 3}{x + 5}$

29. $\frac{2}{k + 3} - \frac{5}{k - 2}$

Simplify.

30. $\sqrt{288}$

31. $2\sqrt{32} - 5\sqrt{98}$

32. Solve $\sqrt{2x + 1} - \sqrt{x} = 1$.

33. Multiply $(5 + 4i)(5 - 4i)$.

Solve each equation or inequality.

34. $3x^2 - x - 1 = 0$

35. $x^2 + 2x - 8 > 0$

36. $x^4 - 5x^2 + 4 = 0$

37. $5^{x+3} = \left(\frac{1}{25}\right)^{3x+2}$

Graph.

38. $f(x) = \frac{1}{3}(x - 1)^2 + 2$

39. $f(x) = 2^x$

40. $f(x) = \log_3 x$

41. Use properties of logarithms to write the following as a sum or difference of logarithms. Assume that variables represent positive real numbers.

$$\log \frac{x^3\sqrt{y}}{z}$$

42. Let the number of bacteria present in a certain culture be given by

$$B(t) = 25{,}000e^{0.2t},$$

where t is time measured in hours, and $t = 0$ corresponds to noon. Approximate, to the nearest hundred, the number of bacteria present at each time.

(a) noon **(b)** 1 P.M. **(c)** 2 P.M.

(d) When will the population double?

APPENDIX Synthetic Division

OBJECTIVES

1. Use synthetic division to divide by a polynomial of the form $x - k$.
2. Use the remainder theorem to evaluate a polynomial.
3. Use the remainder theorem to decide whether a given number is a solution of an equation.

TEACHING TIP
Help students identify the pattern: Numbers *above* the line are added, a number *below* the line is multiplied by the synthetic divisor, then moved to the right above the line.

OBJECTIVE 1 Use synthetic division to divide by a polynomial of the form $x - k$.

If a polynomial in x is divided by a binomial of the form $x - k$, a shortcut method called **synthetic division** can be used. Consider the following.

Polynomial Division

$$
\begin{array}{r}
3x^2 + 9x + 25 \\
x - 3\overline{)3x^3 + 0x^2 - 2x + 5} \\
\underline{3x^3 - 9x^2} \\
9x^2 - 2x \\
\underline{9x^2 - 27x} \\
25x + 5 \\
\underline{25x - 75} \\
80
\end{array}
$$

Synthetic Division

$$
\begin{array}{rrrrr}
 & & 3 & 9 & 25 \\
1 - 3\,\overline{)} & 3 & 0 & -2 & 5 \\
 & \underline{3} & \underline{-9} & & \\
 & & 9 & -2 & \\
 & & \underline{9} & \underline{-27} & \\
 & & & 25 & 5 \\
 & & & \underline{25} & \underline{-75} \\
 & & & & 80
\end{array}
$$

On the right above, exactly the same division is shown written without the variables. This is why it is *essential* to use 0 as a placeholder in synthetic division. All the numbers in color on the right are repetitions of the numbers directly above them, so we omit them, as shown on the left below.

$$
\begin{array}{rrrrr}
 & & 3 & 9 & 25 \\
1 - 3\,\overline{)} & 3 & 0 & -2 & 5 \\
 & & \underline{-9} & & \\
 & & 9 & -2 & \\
 & & & \underline{-27} & \\
 & & & 25 & 5 \\
 & & & & \underline{-75} \\
 & & & & 80
\end{array}
\qquad
\begin{array}{rrrrr}
 & & 3 & 9 & 25 \\
1 - 3\,\overline{)} & 3 & 0 & -2 & 5 \\
 & & \underline{-9} & & \\
 & & 9 & & \\
 & & & \underline{-27} & \\
 & & & 25 & \\
 & & & & \underline{-75} \\
 & & & & 80
\end{array}
$$

The numbers in color on the left are again repetitions of the numbers directly above them. They too are omitted, as shown on the right above. If we bring the 3 in the dividend down to the beginning of the bottom row, the top row can be omitted because it duplicates the bottom row.

$$
\begin{array}{rrrrr}
1 - 3\,\overline{)} & 3 & 0 & -2 & 5 \\
 & & -9 & -27 & -75 \\
\hline
 & 3 & 9 & 25 & 80
\end{array}
$$

We omit the 1 at the upper left—it represents $1x$, which will always be the first term in the divisor. To simplify the arithmetic, we replace subtraction in the second row by addition. To compensate, we change the -3 at the upper left to its additive inverse, 3.

$$\begin{array}{r|rrrrl}
\text{Additive inverse} \rightarrow 3 & 3 & 0 & -2 & 5 & \\
 & & 9 & 27 & 75 & \leftarrow \text{Signs changed} \\
\hline
 & 3 & 9 & 25 & 80 & \leftarrow \text{Remainder}
\end{array}$$

The quotient is read from the bottom row.

$$3x^2 + 9x + 25 + \frac{80}{x-3}$$

The first three numbers in the bottom row are the coefficients of the quotient polynomial with degree 1 less than the degree of the dividend. The last number gives the remainder.

Synthetic division is used only when dividing a polynomial by a binomial of the form x − k.

NOW TRY EXERCISE 1

Use synthetic division to divide $4x^3 + 18x^2 + 19x + 7$ by $x + 3$.

CLASSROOM EXAMPLE 1

Use synthetic division to divide $2x^3 + 13x^2 + 30x + 25$ by $x + 3$.

Answer: $2x^2 + 7x + 9 - \frac{2}{x+3}$

EXAMPLE 1 Using Synthetic Division

Use synthetic division to divide $5x^2 + 16x + 15$ by $x + 2$.

We change $x + 2$ into the form $x - k$ by writing it as

$$x + 2 = x - (-2), \quad \text{where } k = -2.$$

Now we write the coefficients of $5x^2 + 16x + 15$, placing -2 to the left.

$$x + 2 \text{ leads to } -2. \rightarrow \begin{array}{r|rrr} -2 & 5 & 16 & 15 \end{array} \leftarrow \text{Coefficients}$$

$$\begin{array}{r|rrr}
-2 & 5 & 16 & 15 \\
 & & -10 & \\
\hline
 & 5 & &
\end{array}$$

Bring down the 5, and multiply: $-2 \cdot 5 = -10$.

$$\begin{array}{r|rrr}
-2 & 5 & 16 & 15 \\
 & & -10 & -12 \\
\hline
 & 5 & 6 &
\end{array}$$

Add 16 and -10, getting 6, and multiply 6 and -2 to get -12.

$$\begin{array}{r|rrr}
-2 & 5 & 16 & 15 \\
 & & -10 & -12 \\
\hline
 & 5 & 6 & 3
\end{array}$$

Add 15 and -12, getting 3.

Read the result from the bottom row. 3 ← Remainder

Remember that a fraction bar means division.

$$\frac{5x^2 + 16x + 15}{x+2} = 5x + 6 + \frac{3}{x+2}$$

NOW TRY

CLASSROOM EXAMPLE 2

Use synthetic division to divide $3x^3 - 2x + 21$ by $x + 2$.

Answer: $3x^2 - 6x + 10 + \frac{1}{x+2}$

NOW TRY EXERCISE 2

Use synthetic division to divide $-3x^4 + 13x^3 - 6x^2 + 31$ by $x - 4$.

EXAMPLE 2 Using Synthetic Division with a Missing Term

Use synthetic division to divide $-4x^5 + x^4 + 6x^3 + 2x^2 + 50$ by $x - 2$.

$$\begin{array}{r|rrrrrr}
2 & -4 & 1 & 6 & 2 & 0 & 50 \\
 & & -8 & -14 & -16 & -28 & -56 \\
\hline
 & -4 & -7 & -8 & -14 & -28 & -6
\end{array}$$

Use the steps given above, first inserting a 0 for the missing x-term.

Read the result from the bottom row.

$$\frac{-4x^5 + x^4 + 6x^3 + 2x^2 + 50}{x-2} = -4x^4 - 7x^3 - 8x^2 - 14x - 28 + \frac{-6}{x-2}$$

NOW TRY

NOW TRY ANSWERS

1. $4x^2 + 6x + 1 + \frac{4}{x+3}$
2. $-3x^3 + x^2 - 2x - 8 + \frac{-1}{x-4}$

OBJECTIVE 2 Use the remainder theorem to evaluate a polynomial.

We can use synthetic division to evaluate polynomials. In the synthetic division of **Example 2,** where the polynomial was divided by $x - 2$, the remainder was -6.

Replacing x in the polynomial with 2 gives the following.

$$-4x^5 + x^4 + 6x^3 + 2x^2 + 50$$

$$= -4 \cdot 2^5 + 2^4 + 6 \cdot 2^3 + 2 \cdot 2^2 + 50 \quad \text{Replace } x \text{ with 2.}$$

$$= -4 \cdot 32 + 16 + 6 \cdot 8 + 2 \cdot 4 + 50 \quad \text{Evaluate the powers.}$$

$$= -128 + 16 + 48 + 8 + 50 \quad \text{Multiply.}$$

$$= -6 \quad \text{Add.}$$

TEACHING TIP
In **Example 3,** evaluate $P(-2)$ using direct substitution methods for comparison.

This number, -6, is the same number as the remainder. Dividing by $x - 2$ produced a remainder equal to the result when x is replaced with 2. This always happens, as the following **remainder theorem** states. This result is proved in more advanced courses.

CLASSROOM EXAMPLE 3
Let $P(x) = x^3 - 5x^2 + 7x - 3$. Use the remainder theorem to evaluate $P(-2)$.

Answer: -45

Remainder Theorem

If the polynomial $P(x)$ is divided by $x - k$, then the remainder is equal to $P(k)$.

NOW TRY EXERCISE 3
Let $P(x) = 3x^3 - 2x^2 + 5x + 30$. Use the remainder theorem to evaluate $P(-2)$.

EXAMPLE 3 Using the Remainder Theorem

Let $P(x) = 2x^3 - 5x^2 - 3x + 11$. Use the remainder theorem to evaluate $P(-2)$.

Divide $P(x)$ by $x - (-2)$, using synthetic division.

Value of k →	-2)	2	-5	-3	11	
			-4	18	-30	
		2	-9	15	-19	← Remainder

Thus, $P(-2) = -19$. NOW TRY

OBJECTIVE 3 Use the remainder theorem to decide whether a given number is a solution of an equation.

NOW TRY EXERCISE 4
Use the remainder theorem to decide whether -4 is a solution of the equation.

$$5x^3 + 19x^2 - 2x + 8 = 0$$

EXAMPLE 4 Using the Remainder Theorem

Use the remainder theorem to decide whether -5 is a solution of the equation.

$$2x^4 + 12x^3 + 6x^2 - 5x + 75 = 0$$

If synthetic division gives a remainder of 0, then -5 is a solution. Otherwise, it is not.

Proposed solution →	-5)	2	12	6	-5	75	
			-10	-10	20	-75	
		2	2	-4	15	0	← Remainder

CLASSROOM EXAMPLE 4
Use the remainder theorem to decide whether 2 is a solution of the equation.

$$3x^3 - 11x^2 + 17x - 14 = 0$$

Answer: yes

Because the remainder is 0, the polynomial has value 0 when $k = -5$. So -5 is a solution of the given equation. NOW TRY

The synthetic division in **Example 4** shows that $x - (-5)$ divides the polynomial with 0 remainder. Thus $x - (-5) = x + 5$ is a *factor* of the polynomial.

$$2x^4 + 12x^3 + 6x^2 - 5x + 75 \quad \text{factors as} \quad (x + 5)(2x^3 + 2x^2 - 4x + 15).$$

The second factor is the quotient polynomial in the last row of the synthetic division.

NOW TRY ANSWERS
3. -12 **4.** yes

Exercises

FOR EXTRA HELP MyMathLab®

Concept Check *Complete the statement.*

1. The factored form of $2x^3 + x^2 - x + 10$ is $(x + 2)(2x^2 - 3x + 5)$. The remainder theorem assures us that when $2x^3 + x^2 - x + 10$ is divided by $x + 2$, the remainder is ______.

2. See **Exercise 1.** Given $P(x) = 2x^3 + x^2 - x + 10$, we can determine that $P(-2) =$ ______.

Use synthetic division to find each quotient. ***See Examples 1 and 2.***

3. $\dfrac{x^2 - 6x + 5}{x - 1}$ **4.** $\dfrac{x^2 - 4x - 21}{x + 3}$ **5.** $\dfrac{4m^2 + 19m - 5}{m + 5}$

6. $\dfrac{3x^2 - 5x - 12}{x - 3}$ **7.** $\dfrac{2a^2 + 8a + 13}{a + 2}$ **8.** $\dfrac{4y^2 - 5y - 20}{y - 4}$

9. $(p^2 - 3p + 5) \div (p + 1)$ **10.** $(z^2 + 4z - 6) \div (z - 5)$

11. $\dfrac{4a^3 - 3a^2 + 2a - 3}{a - 1}$ **12.** $\dfrac{5p^3 - 6p^2 + 3p + 14}{p + 1}$

13. $(x^5 - 2x^3 + 3x^2 - 4x - 2) \div (x - 2)$

14. $(2y^5 - 5y^4 - 3y^2 - 6y - 23) \div (y - 3)$

15. $(-4r^6 - 3r^5 - 3r^4 + 5r^3 - 6r^2 + 3r + 3) \div (r - 1)$

16. $(2t^6 - 3t^5 + 2t^4 - 5t^3 + 6t^2 - 3t - 2) \div (t - 2)$

17. $(-3y^5 + 2y^4 - 5y^3 - 6y^2 - 1) \div (y + 2)$

18. $(m^6 + 2m^4 - 5m + 11) \div (m - 2)$

Use the remainder theorem to evaluate $P(k)$. ***See Example 3.***

19. $P(x) = 2x^3 - 4x^2 + 5x - 3$; $k = 2$ **20.** $P(x) = x^3 + 3x^2 - x + 5$; $k = -1$

21. $P(x) = -x^3 - 5x^2 - 4x - 2$; $k = -4$ **22.** $P(x) = -x^3 + 5x^2 - 3x + 4$; $k = 3$

23. $P(x) = 2x^3 - 4x^2 + 5x - 33$; $k = 3$ **24.** $P(x) = x^3 - 3x^2 + 4x - 4$; $k = 2$

Use the remainder theorem to decide whether the given number is a solution of the equation. ***See Example 4.***

25. $x^3 - 2x^2 - 3x + 10 = 0$; $x = -2$ **26.** $x^3 - 3x^2 - x + 10 = 0$; $x = -2$

27. $3x^3 + 2x^2 - 2x + 11 = 0$; $x = -2$ **28.** $3x^3 + 10x^2 + 3x - 9 = 0$; $x = -2$

29. $2x^3 - x^2 - 13x + 24 = 0$; $x = -3$ **30.** $5x^3 + 22x^2 + x - 28 = 0$; $x = -4$

31. $x^4 + 2x^3 - 3x^2 + 8x = 8$; $x = -2$ **32.** $x^4 - x^3 - 6x^2 + 5x = -10$; $x = -2$

RELATING CONCEPTS For Individual or Group Work **(Exercises 33–38)**

We can show a connection between dividing one polynomial by another and factoring the first polynomial. Let $P(x) = 2x^2 + 5x - 12$. ***Work Exercises 33–38 in order.***

33. Factor $P(x)$. **34.** Solve $P(x) = 0$. **35.** Evaluate $P(-4)$. **36.** Evaluate $P\left(\frac{3}{2}\right)$.

37. Complete the sentence: If $P(a) = 0$, then $x -$ ______ is a factor of $P(x)$.

38. Use the conclusion reached in **Exercise 37** to decide whether $x - 3$ is a factor of $Q(x) = 3x^3 - 4x^2 - 17x + 6$. Factor $Q(x)$ completely.

Answers

1. 0
2. 0
3. $x - 5$
4. $x - 7$
5. $4m - 1$
6. $3x + 4$
7. $2a + 4 + \dfrac{5}{a + 2}$
8. $4y + 11 + \dfrac{24}{y - 4}$
9. $p - 4 + \dfrac{9}{p + 1}$
10. $z + 9 + \dfrac{39}{z - 5}$
11. $4a^2 + a + 3$
12. $5p^2 - 11p + 14$
13. $x^4 + 2x^3 + 2x^2 + 7x + 10 + \dfrac{18}{x - 2}$
14. $2y^4 + y^3 + 3y^2 + 6y + 12 + \dfrac{13}{y - 3}$
15. $-4r^5 - 7r^4 - 10r^3 - 5r^2 - 11r - 8 + \dfrac{-5}{r - 1}$
16. $2t^5 + t^4 + 4t^3 + 3t^2 + 12t + 21 + \dfrac{40}{t - 2}$
17. $-3y^4 + 8y^3 - 21y^2 + 36y - 72 + \dfrac{143}{y + 2}$
18. $m^5 + 2m^4 + 6m^3 + 12m^2 + 24m + 43 + \dfrac{97}{m - 2}$
19. 7
20. 8
21. -2
22. 13
23. 0
24. 0
25. yes
26. no
27. no
28. no
29. yes
30. yes
31. no
32. yes
33. $(2x - 3)(x + 4)$
34. $\left\{-4, \frac{3}{2}\right\}$
35. 0
36. 0
37. a
38. Yes, $x - 3$ is a factor; $Q(x) = (x - 3)(3x - 1)(x + 2)$

ANSWERS TO SELECTED EXERCISES

In this section, we provide the answers that we think most students will obtain when they work the exercises using the methods explained in the text. If your answer does not look exactly like the one given here, it is not necessarily wrong. In many cases, there are equivalent forms of the answer that are correct. For example, if the answer section shows $\frac{3}{4}$ and your answer is 0.75, you have obtained the right answer, but written it in a different (yet equivalent) form. Unless the directions specify otherwise, 0.75 is just as valid an answer as $\frac{3}{4}$.

In general, if your answer does not agree with the one given in the text, see whether it can be transformed into the other form. If it can, then it is the correct answer. If you still have doubts, consult your instructor.

R REVIEW OF THE REAL NUMBER SYSTEM

Section R.1 (pages 9–12)

1. yes **3.** $\{1, 2, 3, 4, 5\}$ **5.** $\{5, 6, 7, 8, \dots\}$ **7.** $\{\dots, -1, 0, 1, 2, 3, 4\}$ **9.** $\{10, 12, 14, 16, \dots\}$ **11.** $\{-4, 4\}$ **13.** $\varnothing$

In Exercises 15 and 17, we give one possible answer.

15. $\{x \mid x$ is an even natural number less than or equal to $8\}$
17. $\{x \mid x$ is a positive multiple of $4\}$

19. 

21. 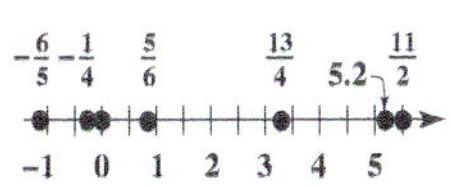

23. **(a)** 8, 13, $\frac{75}{5}$ (or 15) **(b)** 0, 8, 13, $\frac{75}{5}$ **(c)** -9, 0, 8, 13, $\frac{75}{5}$ **(d)** -9, -0.7, 0, $\frac{6}{7}$, $4.\overline{6}$, 8, $\frac{21}{2}$, 13, $\frac{75}{5}$ **(e)** $-\sqrt{6}$, $\sqrt{7}$ **(f)** All are real numbers. **25.** false; Some are whole numbers, but negative integers are not. **27.** false; No irrational number is an integer. **29.** true **31.** true **33.** true **35.** **(a)** A **(b)** A **(c)** B **(d)** B **37.** **(a)** -6 **(b)** 6 **39.** **(a)** 12 **(b)** 12 **41.** **(a)** $-\frac{6}{5}$ **(b)** $\frac{6}{5}$ **43.** 8 **45.** $\frac{3}{2}$ **47.** -5 **49.** -2 **51.** -4.5 **53.** 5 **55.** 6 **57.** 0 **59.** **(a)** Las Vegas; The population increased by 45.4%. **(b)** Detroit; The population decreased by 3.6%. **61.** Pacific Ocean, Indian Ocean, Caribbean Sea, South China Sea, Gulf of California **63.** true **65.** true **67.** false **69.** true **71.** true **73.** $2 < 6$ **75.** $4 > -9$ **77.** $-10 < -5$ **79.** $x > 0$ **81.** $7 > y$ **83.** $5 \geq 5$ **85.** $3t - 4 \leq 10$ **87.** $5x + 3 \neq 0$ **89.** $t \leq -3$ **91.** $3x < 4$ **93.** $-6 < 10$; true **95.** $10 \geq 10$; true **97.** $-3 \geq -3$; true **99.** $-8 > -6$; false **101.** $5 \geq -5$; true **103.** Iowa (IA), Ohio (OH), Pennsylvania (PA) **105.** $x < y$

Section R.2 (pages 19–23)

1. positive; $18 + 6 = 24$ **3.** greater; $-14 + 9 = -5$ **5.** the number with greater absolute value is subtracted from the one with lesser absolute value; $5 - 12 = -7$ **7.** additive inverses; $4 + (-4) = 0$ **9.** negative; $-5(15) = -75$ **11.** -19 **13.** 9 **15.** $-\frac{19}{12}$ **17.** -1.85 **19.** -11 **21.** 21 **23.** -13 **25.** -10.18 **27.** $\frac{67}{30}$ **29.** 14 **31.** -5 **33.** -6 **35.** -11 **37.** 16 **39.** -4 **41.** 8 **43.** 3.018 **45.** $-\frac{7}{4}$ **47.** $-\frac{7}{8}$ **49.** 1 **51.** 6 **53.** $\frac{13}{2}$, or $6\frac{1}{2}$ **55.** It is true for multiplication (and division). It is false for addition and false for subtraction when the number to be subtracted has the lesser absolute value. A more precise statement is, "The product or quotient of two negative numbers is positive." **57.** -35 **59.** 40 **61.** 2 **63.** -12 **65.** $\frac{6}{5}$ **67.** 1 **69.** 0 **71.** 0 **73.** -7 **75.** 6 **77.** -4 **79.** 0 **81.** undefined **83.** $\frac{25}{102}$ **85.** $-\frac{9}{13}$ **87.** -2.1 **89.** 10,000 **91.** $\frac{17}{18}$ **93.** $\frac{17}{36}$ **95.** $-\frac{19}{24}$ **97.** $-\frac{22}{45}$ **99.** $-\frac{2}{15}$ **101.** $\frac{3}{5}$ **103.** $-\frac{35}{27}$ **105.** $-\frac{1}{3}$ **107.** $-\frac{4}{3}$ **109.** -12.351 **111.** -15.876 **113.** -4.14 **115.** 4800 **117.** $\frac{1}{3}$ **119.** 70.21 **121.** The two quotients are reciprocals by inspection. So once the correct answer to Exercise 117 is known $\left(\frac{1}{3}\right)$, the answer to Exercise 118 is simply its reciprocal (3). **123.** 112°F **125.** \$30.13 **127.** **(a)** \$466.02 **(b)** \$190.68 **129.** **(a)** 59.53% **(b)** -36.87% **(c)** 58.64% **131.** 1995: $-$\$164 billion; 2000: \$236 billion; 2005: $-$\$318 billion; 2012: $-$\$1087 billion

Section R.3 (pages 28–31)

1. false; $-7^6 = -(7^6)$ **3.** true **5.** true **7.** true **9.** false; The base is 8. **11.** **(a)** 64 **(b)** -64 **(c)** 64 **(d)** -64 **13.** 10^4 **15.** $\left(\frac{3}{4}\right)^5$ **17.** $(-9)^3$ **19.** z^7 **21.** 0^6 **23.** 16 **25.** 0.021952 **27.** $\frac{1}{125}$ **29.** $\frac{256}{625}$ **31.** -125 **33.** 256 **35.** -729 **37.** -4096 **39.** 9 **41.** 13 **43.** -20 **45.** $\frac{10}{11}$ **47.** -0.7 **49.** not a real number **51.** **(a)** not a real number **(b)** negative **53.** 24 **55.** 15 **57.** 55 **59.** -91 **61.** -8 **63.** -48 **65.** -2 **67.** -79 **69.** -10 **71.** 2 **73.** -2 **75.** undefined **77.** -7 **79.** -1 **81.** 17 **83.** -96 **85.** $-\frac{15}{238}$ **87.** 8 **89.** $-\frac{5}{16}$ **91.** $-\frac{11}{4}$, or -2.75 **93.** $-\frac{3}{16}$ **95.** \$1572 **97.** \$3296 **99.** 0.035 **101.** **(a)** \$28.3 billion **(b)** \$42.2 billion **(c)** \$56.0 billion **(d)** The amount spent on pets almost exactly doubled from 2001 to 2013.

Section R.4 (pages 37–39)

1. B **3.** A **5.** order **7.** same; same **9.** -7 **11.** $2m + 2p$ **13.** $-12x + 12y$ **15.** $8k$ **17.** $-2r$ **19.** cannot be simplified **21.** $8a$ **23.** $-2d + f$ **25.** $x + y$ **27.** $-6y + 3$ **29.** $p + 11$ **31.** $-2k + 15$ **33.** $-3m + 2$ **35.** -1 **37.** $2p + 7$ **39.** $-6z - 39$ **41.** $(5 + 8)x = 13x$ **43.** $(5 \cdot 9)r = 45r$ **45.** $9y + 5x$ **47.** 7 **49.** 0 **51.** $8(-4) + 8x = -32 + 8x$ **53.** 0 **55.** Answers will vary. One example of commutativity is washing your face and brushing your teeth. An example of non-commutativity is putting on your socks and putting on your shoes. **57.** 1900 **59.** 75 **61.** 431 **63.** No. One example is $7 + (5 \cdot 3) = (7 + 5)(7 + 3)$, which is false. **65.** associative property **66.** associative property **67.** commutative property **68.** associative property **69.** distributive property **70.** arithmetic facts

1 LINEAR EQUATIONS, INEQUALITIES, AND APPLICATIONS

Section 1.1 (pages 50–53)

1. algebraic expression; does; is not **3.** true; solution; solution set **5.** identity; all real numbers **7.** A, C **9.** Both sides are evaluated as 30, so 6 is a solution. **11.** equation **13.** expression **15.** equation **17.** A sign error was made when the distributive property was applied. The left side of the second line should be $8x - 4x + 6$. The correct solution is 1. **19.** $\{-1\}$ **21.** $\{-4\}$ **23.** $\{-7\}$ **25.** $\{0\}$ **27.** $\{4\}$ **29.** $\{-\frac{7}{8}\}$ **31.** $\emptyset$; contradiction **33.** $\{-\frac{5}{3}\}$ **35.** $\{-\frac{1}{2}\}$ **37.** $\{2\}$ **39.** $\{-2\}$ **41.** {all real numbers}; identity **43.** $\{-1\}$ **45.** $\{7\}$ **47.** $\{2\}$ **49.** {all real numbers}; identity **51.** $\{\frac{3}{2}\}$ **53.** 12 **55. (a)** 10^2, or 100 **(b)** 10^3, or 1000 **57.** $\{-\frac{18}{5}\}$ **59.** $\{-\frac{5}{6}\}$ **61.** $\{6\}$ **63.** $\{4\}$ **65.** $\{3\}$ **67.** $\{3\}$ **69.** $\{0\}$ **71.** $\{35\}$ **73.** $\{-10\}$ **75.** $\{2000\}$ **77.** $\{25\}$ **79.** $\{40\}$ **81.** $\{3\}$

Section 1.2 (pages 60–65)

1. formula **3. (a)** 35% **(b)** 18% **(c)** 2% **(d)** 7.5% **(e)** 150%

There may be other acceptable forms of the answers in Exercises 5–35.

5. $r = \frac{I}{pt}$ **7. (a)** $b = \frac{\mathcal{A}}{h}$ **(b)** $h = \frac{\mathcal{A}}{b}$ **9.** $L = \frac{P - 2W}{2}$, or $L = \frac{P}{2} - W$ **11. (a)** $W = \frac{V}{LH}$ **(b)** $H = \frac{V}{LW}$ **13.** $r = \frac{C}{2\pi}$ **15. (a)** $b = \frac{2\mathcal{A} - hB}{h}$, or $b = \frac{2\mathcal{A}}{h} - B$ **(b)** $B = \frac{2\mathcal{A} - hb}{h}$, or $B = \frac{2\mathcal{A}}{h} - b$ **17.** $C = \frac{5}{9}(F - 32)$ **19. (a)** $x = \frac{C - B}{A}$ **(b)** $A = \frac{C - B}{x}$ **21.** $t = \frac{A - P}{Pr}$ **23.** D **25.** $y = -4x + 1$ **27.** $y = \frac{1}{2}x + 3$ **29.** $y = -\frac{4}{9}x + \frac{11}{9}$ **31.** $y = \frac{3}{2}x + \frac{5}{2}$ **33.** $y = \frac{6}{5}x - \frac{7}{5}$ **35.** $y = \frac{3}{2}x - 3$ **37.** 3.140 hr **39.** 52 mph **41.** 113°F **43.** 230 m **45. (a)** 240 in. **(b)** 480 in. **47.** 2 in. **49.** 75% water; 25% alcohol **51.** 3% **53. (a)** .558 **(b)** .481 **(c)** .438 **(d)** .315 **55.** 34% **57.** 103.4 million **59.** \$72,324 **61.** \$14,465 **63.** 12% **65.** 23.8% **67.** 49.9% **69. (a)** 0.79 m^2 **(b)** 0.96 m^2 **71. (a)** 116 mg **(b)** 141 mg

Section 1.3 (pages 73–78)

1. (a) $x + 15$ **(b)** $15 > x$ **3. (a)** $x - 8$ **(b)** $8 < x$ **5.** D **7.** $2x - 13$ **9.** $12 + 4x$ **11.** $8(x - 16)$ **13.** $\frac{3x}{10}$ **15.** $x + 6 = -31; -37$ **17.** $x - (-4x) = x + 9; \frac{9}{4}$ **19.** $14 - \frac{2}{3}x = 10; 6$ **21.** expression; $-11x + 63$ **23.** equation; $\{\frac{51}{11}\}$ **25.** expression; $\frac{1}{3}x - \frac{13}{2}$ **27.** *Step 1:* the number of patents each corporation secured; *Step 2:* patents that Samsung secured; *Step 3:* x; $x - 1414$; *Step 4:* 6457; *Step 5:* 6457; 5043; *Step 6:* 1414; IBM patents; 5043; 11,500 **29.** width: 165 ft; length: 265 ft **31.** 24.34 in. by 29.88 in. **33.** 850 mi; 925 mi; 1300 mi **35.** Exxon Mobil: \$449.9 billion; Wal-Mart: \$469.2 billion **37.** Eiffel Tower: 984 ft; Leaning Tower: 180 ft **39.** Obama: 332 votes; Romney: 206 votes **41.** 49.3% **43.** \$8265 **45.** \$49.48 **47.** \$225 **49.** \$4000 at 3%; \$8000 at 4% **51.** \$10,000 at 4.5%; \$19,000 at 3% **53.** \$24,000 **55.** 5 L **57.** 4 L **59.** 1 gal **61.** 150 lb **63.** We cannot expect the final mixture to be worth more than the more expensive of the two ingredients. **65. (a)** $800 - x$ **(b)** $800 - y$ **66. (a)** $0.05x$; $0.10(800 - x)$ **(b)** $0.05y$; $0.10(800 - y)$ **67. (a)** $0.05x + 0.10(800 - x) = 800(0.0875)$ **(b)** $0.05y + 0.10(800 - y) = 800(0.0875)$ **68. (a)** \$200 at 5%; \$600 at 10% **(b)** 200 L of 5% acid; 600 L of 10% acid **(c)** The processes are the same. The amounts of money in Problem A correspond to the amounts of solution in Problem B.

Section 1.4 (pages 83–87)

1. \$5.70 **3.** 30 mph **5.** The problem asks for the *distance* to the workplace. To find this distance, we must multiply the rate, 10 mph, by the time, $\frac{3}{4}$ hr. **7.** No, the answers must be whole numbers because they represent the numbers of coins. **9.** 17 pennies; 17 dimes; 10 quarters **11.** 23 loonies; 14 toonies **13.** 28 \$10 coins; 13 \$20 coins **15.** 872 adult tickets **17.** 8.10 m per sec **19.** 10.35 m per sec **21.** $2\frac{1}{2}$ hr **23.** 7:50 P.M. **25.** 45 mph **27.** $\frac{1}{2}$ hr **29.** 60°, 60°, 60° **31.** 40°, 45°, 95° **33.** Both measure 122°. **35.** 64°, 26° **37.** 65°, 115° **39.** 19, 20, 21 **41.** 27, 28, 29, 30 **43.** 61 yr old **45.** 28, 30, 32 **47.** 21, 23, 25 **49.** 40°, 80° **50.** 120° **51.** The sum is equal to the measure of the angle found in **Exercise 50.** **52.** The sum of the measures of angles ① and ② is equal to the measure of angle ③.

Section 1.5 (pages 99–102)

1. D **3.** B **5.** F **7. (a)** $x < 100$ **(b)** $100 \le x \le 129$ **(c)** $130 \le x \le 159$ **(d)** $160 \le x \le 189$ **(e)** $x \ge 190$

9. $[16, \infty)$

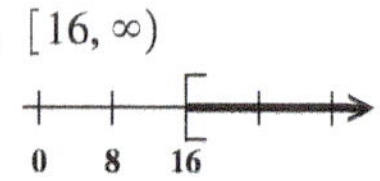

11. $(7, \infty)$

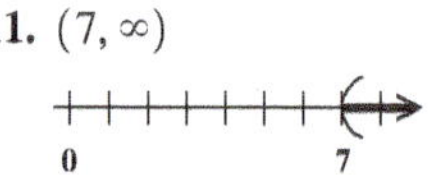

13. $(-\infty, -4)$

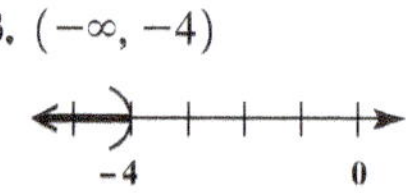

15. $(-\infty, -40]$

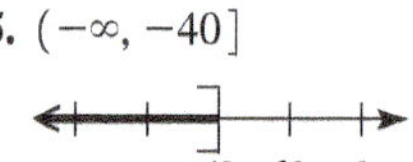

17. $(-\infty, 4]$

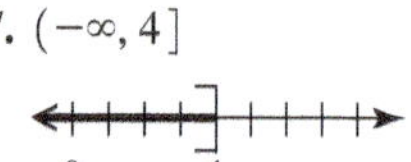

19. $(-\infty, -10]$

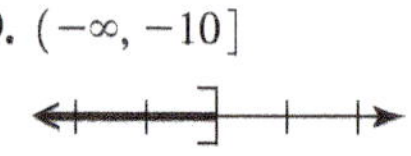

21. $(-\infty, 14)$

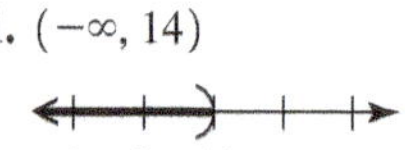

23. $\left(-\infty, -\frac{15}{2}\right)$

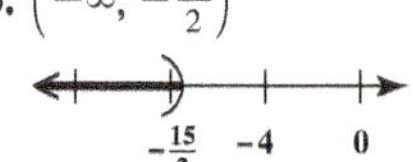

25. $\left[\frac{1}{2}, \infty\right)$

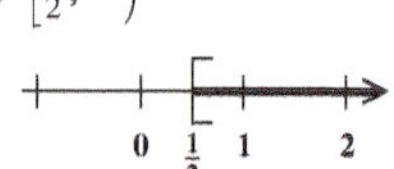

27. $[2, \infty)$

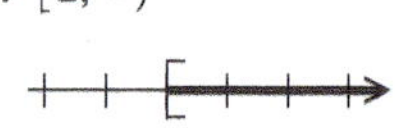

29. $(3, \infty)$

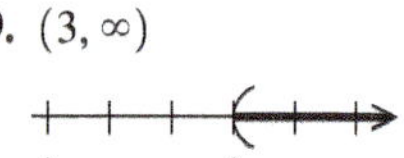

31. $(-\infty, 4)$

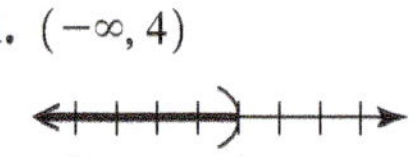

33. $\left(-\infty, \frac{23}{6}\right]$ **35.** $\left(-\infty, \frac{76}{11}\right)$

37. $(-\infty, \infty)$ **39.** $\varnothing$ **41.** A; B

43. $(1, 11)$ **45.** $[-14, 10]$

47. $[-5, 6]$ **49.** $\left[-\frac{14}{3}, 2\right]$

51. $\left(-\frac{1}{3}, \frac{1}{9}\right]$ **53.** $\left[-1, \frac{5}{2}\right]$

55. $\left[-\frac{1}{2}, \frac{35}{2}\right]$ **57.** $(0, 1)$ **59.** $(-2, 2)$ **61.** $[3, \infty)$ **63.** $[-9, \infty)$ **65.** at least 80 **67.** 26 months **69.** more than 4.4 in. **71.** **(a)** 140 to 184 lb **(b)** 107 to 141 lb **(c)** Answers will vary. **73.** 26 DVDs

Section 1.6 (pages 108–110)

1. true **3.** false; The union is $(-\infty, 7) \cup (7, \infty)$. **5.** false; The intersection is $\varnothing$. **7.** $\{1, 3, 5\}$, or B **9.** $\{4\}$, or D **11.** $\varnothing$ **13.** $\{1, 2, 3, 4, 5, 6\}$, or A

15. **17.**

19. $(-3, 2)$ **21.** $(-\infty, 2]$

23. $\varnothing$

25. $[5, 9]$ **27.** $(-3, -1)$

29. $(-\infty, 4]$ **31.**

33. **35.** $(-\infty, 8]$

37. $[-2, \infty)$ **39.** $(-\infty, \infty)$

41. $(-\infty, -5) \cup (5, \infty)$ **43.** $(-\infty, -1) \cup (2, \infty)$

45. $(-\infty, \infty)$ **47.** $[-4, -1]$ **49.** $[-9, -6]$ **51.** $(-\infty, 3)$ **53.** $[3, 9)$

55. intersection; $(-5, -1)$ **57.** union; $(-\infty, 4)$

59. union; $(-\infty, 0] \cup [2, \infty)$ **61.** intersection; $[4, 12]$

63. {Tuition and fees} **65.** {Tuition and fees, Board rates, Dormitory charges} **67.** Maria, Joe **68.** none of them **69.** none of them **70.** Luigi, Than **71.** Maria, Joe **72.** all of them

Section 1.7 (pages 118–121)

1. E; C; D; B; A **3.** **(a)** one **(b)** two **(c)** none **5.** $\{-12, 12\}$ **7.** $\{-5, 5\}$ **9.** $\{-6, 12\}$ **11.** $\{-5, 6\}$ **13.** $\left\{-3, \frac{11}{2}\right\}$ **15.** $\left\{-\frac{19}{2}, \frac{9}{2}\right\}$ **17.** $\left\{\frac{7}{3}, 3\right\}$ **19.** $\{12, 36\}$ **21.** $\{-12, 12\}$ **23.** $\{-10, -2\}$ **25.** $\left\{-\frac{32}{3}, 8\right\}$ **27.** $\{-75, 175\}$

29. $(-\infty, -3) \cup (3, \infty)$ **31.** $(-\infty, -4] \cup [4, \infty)$

33. $(-\infty, -25] \cup [15, \infty)$ **35.** $\left(-\infty, -\frac{12}{5}\right) \cup \left(\frac{8}{5}, \infty\right)$

37. $(-\infty, -2) \cup (8, \infty)$ **39.** $\left(-\infty, -\frac{9}{5}\right] \cup [3, \infty)$

41. **(a)** **(b)**

43. $[-3, 3]$ **45.** $(-4, 4)$

47. $(-25, 15)$ **49.** $\left[-\frac{12}{5}, \frac{8}{5}\right]$

51. $[-2, 8]$ **53.** $\left(-\frac{9}{5}, 3\right)$

55. $(-\infty, -5) \cup (13, \infty)$ **57.** $\left(-\frac{13}{3}, 3\right)$

59. $\{-6, -1\}$ **61.** $\left[-\frac{10}{3}, 4\right]$

63. $\left[-\frac{7}{6}, -\frac{5}{6}\right]$ **65.** $[3, 13]$

67. $\left(-\infty, \frac{1}{2}\right] \cup \left[\frac{7}{6}, \infty\right)$ **69.** $\left\{-\frac{5}{3}, \frac{11}{3}\right\}$ **71.** $(-\infty, -20) \cup (40, \infty)$

73. $\{-5, 1\}$ **75.** $\{3, 9\}$ **77.** $\{0, 20\}$ **79.** $\{-5, 5\}$
81. $\{-5, -3\}$ **83.** $(-\infty, -3) \cup (2, \infty)$ **85.** $[-10, 0]$
87. $(-\infty, 20] \cup [30, \infty)$ **89.** $\left\{-\frac{5}{3}, \frac{1}{3}\right\}$ **91.** $\{-1, 3\}$ **93.** $\left\{-3, \frac{5}{3}\right\}$
95. $\left\{-\frac{1}{3}, -\frac{1}{15}\right\}$ **97.** $\left\{-\frac{5}{4}\right\}$ **99.** $(-\infty, \infty)$ **101.** $\varnothing$ **103.** $\left\{-\frac{1}{4}\right\}$
105. $\varnothing$ **107.** $(-\infty, \infty)$ **109.** $\left\{-\frac{3}{7}\right\}$ **111.** $\left\{\frac{2}{5}\right\}$ **113.** $(-\infty, \infty)$
115. $\varnothing$ **117.** between 30.72 and 33.28 oz, inclusive **119.** between 31.2 and 32.8 oz, inclusive **121.** $(-0.05, 0.05)$
123. $(2.74975, 2.75025)$ **125.** between 6.8 and 9.8 lb
127. $|x - 1000| \le 100$; $900 \le x \le 1100$ **129.** 810.5 ft
130. Bank of America Center, Texaco Heritage Plaza **131.** Williams Tower, Bank of America Center, Texaco Heritage Plaza, Enterprise Plaza, Centerpoint Energy Plaza, Continental Center I, Fulbright Tower
132. (a) $|x - 810.5| \ge 95$ **(b)** $x \ge 905.5$ or $x \le 715.5$
(c) JPMorgan Chase Tower, Wells Fargo Plaza, One Shell Plaza
(d) It makes sense because it includes all buildings *not* listed in the answer to **Exercise 131.**

2 LINEAR EQUATIONS, GRAPHS, AND FUNCTIONS

Section 2.1 (pages 143–146)

1. origin **3.** y; x; x; y **5.** two; 2; 6 **7. (a)** x represents the year; y represents personal spending on medical care in billions of dollars. **(b)** about \$2360 billion **(c)** $(2012, 2360)$ **(d)** 2008 **9. (a)** I **(b)** III **(c)** II **(d)** IV **(e)** none **(f)** none **11. (a)** I or III **(b)** II or IV **(c)** II or IV **(d)** I or III

13.–21.

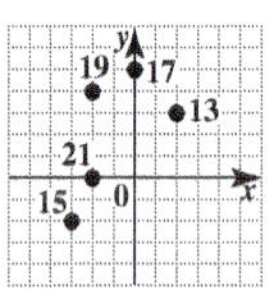

23. (a) -4; -3; -2; -1; 0
(b)

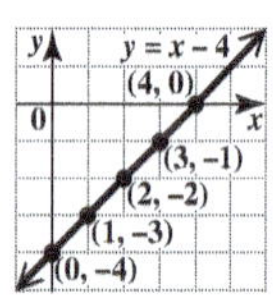

25. (a) -3; 3; 2; -1
(b)

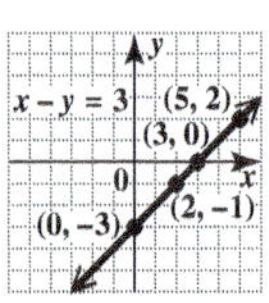

27. (a) $\frac{5}{2}$; 5; $\frac{3}{2}$; 1
(b)

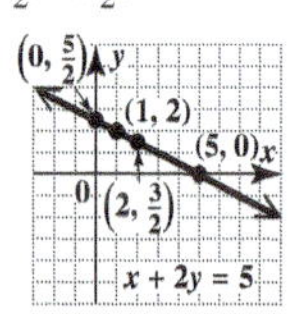

29. (a) -4; 5; $-\frac{12}{5}$; $\frac{5}{4}$
(b)

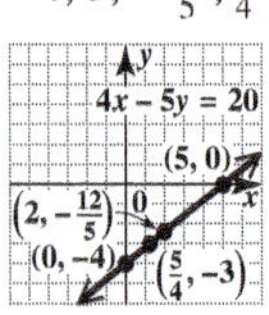

31. (a) 3; 1; -1; -3
(b)

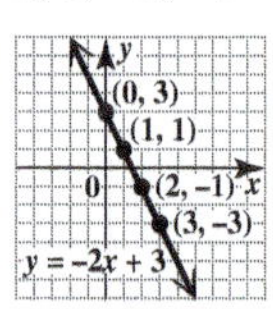

33. (a) 1 **(b)** 2 **(c)** For every increase in x by 1 unit, y increases by 2 units.

35. $(6, 0)$; $(0, 4)$

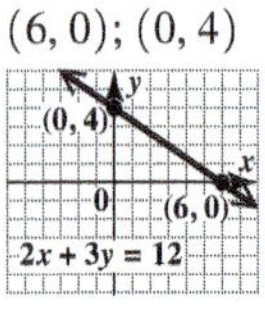

37. $(6, 0)$; $(0, -2)$

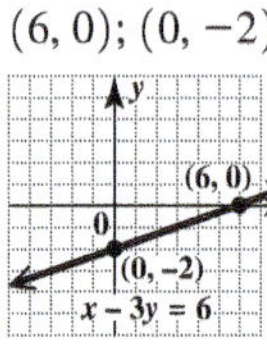

39. $(-2, 0)$; $\left(0, -\frac{5}{3}\right)$

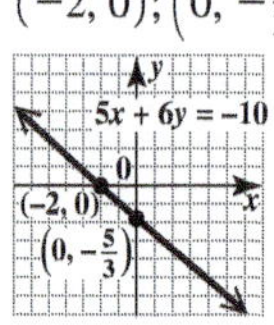

41. $\left(\frac{21}{2}, 0\right)$; $\left(0, -\frac{7}{3}\right)$

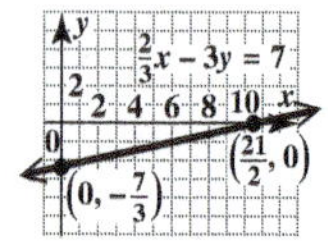

43. none; $(0, 5)$

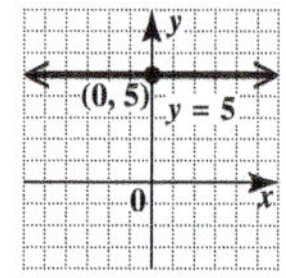

45. $(2, 0)$; none

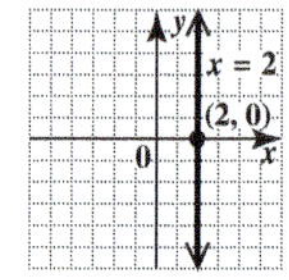

47. $(-4, 0)$; none

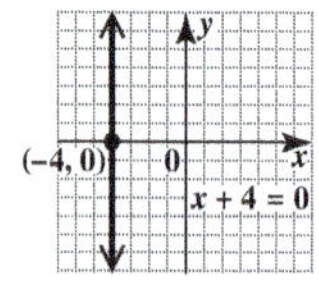

49. none; $(0, -2)$

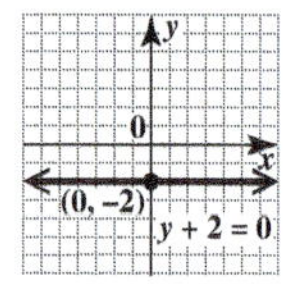

51. $(0, 0)$; $(0, 0)$

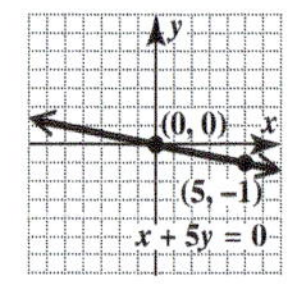

53. $(0, 0)$; $(0, 0)$

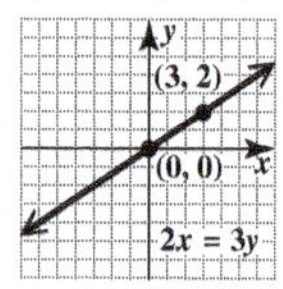

55. $(0, 0)$; $(0, 0)$

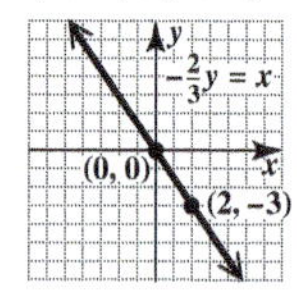

57. (a) $(-2, 0)$; $(0, 3)$
(b) B
(c)

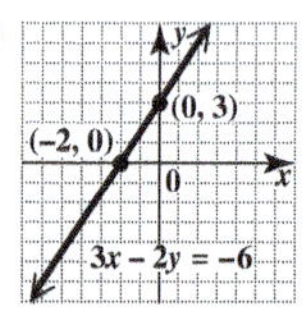

59. (a) none; $(0, -1)$
(b) A
(c)

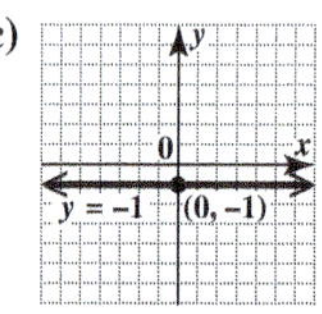

61. C **63.** B **65.** $(-5, -1)$ **67.** $\left(\frac{9}{2}, -\frac{3}{2}\right)$ **69.** $\left(0, \frac{11}{2}\right)$
71. $(2.1, 0.9)$ **73.** $(1, 1)$ **75.** $\left(-\frac{5}{12}, \frac{5}{28}\right)$ **77.** $Q(11, -4)$
79. $Q(4.5, 0.75)$

Section 2.2 (pages 156–162)

1. A, B, D, F **3.** 5 ft **5. (a)** C **(b)** A **(c)** D **(d)** B **7.** 2
9. undefined **11.** 1 **13.** -1 **15. (a)** B **(b)** C **(c)** A **(d)** D
17. 2 **19.** $\frac{5}{2}$ **21.** 0 **23.** undefined **25. (a)** 8 **(b)** rises
27. (a) $\frac{5}{6}$ **(b)** rises **29. (a)** 0 **(b)** horizontal **31. (a)** $-\frac{1}{2}$
(b) falls **33. (a)** undefined **(b)** vertical **35. (a)** -1 **(b)** falls
37. 6 **39.** -3 **41.** -2 **43.** $\frac{4}{3}$ **45.** $-\frac{5}{2}$ **47.** undefined

In part (a) of Exercises 49–53, we used the intercepts. Other points can be used.

49. (a) $(4, 0)$ and $(0, -8)$; 2 **(b)** $y = 2x - 8$; 2 **(c)** $A = 2, B = -1$; 2
51. (a) $(4, 0)$ and $(0, -3)$; $-\frac{3}{4}$ **(b)** $y = -\frac{3}{4}x + 3$; $-\frac{3}{4}$
(c) $A = 3, B = 4$; $-\frac{3}{4}$ **53. (a)** $(-3, 0)$ and $(0, -3)$; -1
(b) $y = -x - 3$; -1 **(c)** $A = 1, B = 1$; -1

55. $-\frac{1}{2}$

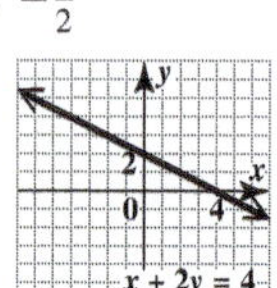

57. $\frac{5}{2}$

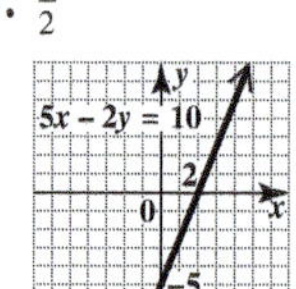

59. 4

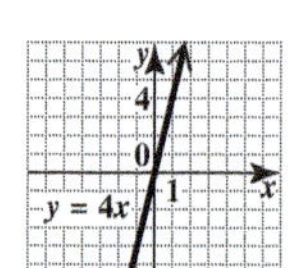

61. undefined

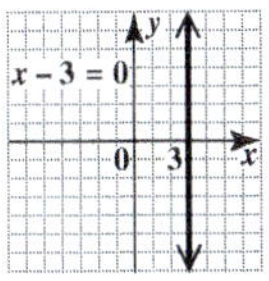

63. 0

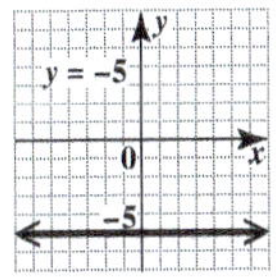

65. 0

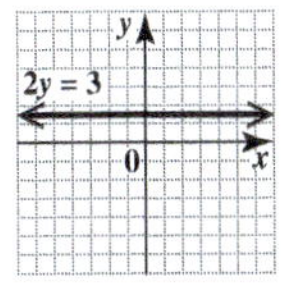

67.

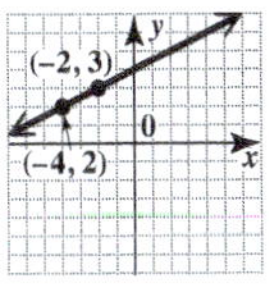

69.

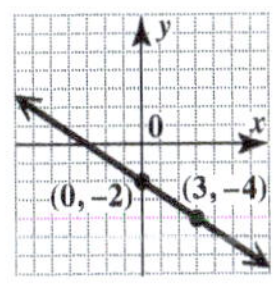

71.

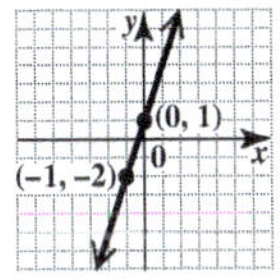

73.

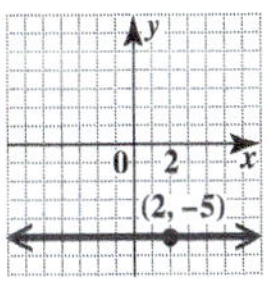

75.

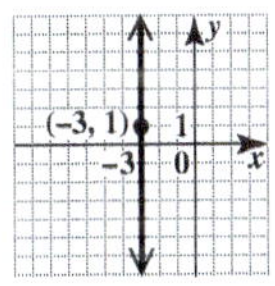

77. $-\frac{4}{9}; \frac{9}{4}$ **79.** parallel **81.** perpendicular **83.** neither **85.** parallel **87.** neither **89.** perpendicular **91.** 19.5 ft **93.** −$4000 per year; The value of the machine is decreasing $4000 per year during these years. **95.** 0% per year (or no change); The percent of pay raise is not changing—it is 3% per year during these years. **97.** **(a)** In 2012, there were 326 million wireless subscriber connections in the U.S. **(b)** 14.2 **(c)** The number of subscribers increased by an average of 14.2 million per year from 2007 to 2012. **99.** **(a)** −7 theaters per yr **(b)** The negative slope means that the number of drive-in theaters decreased by an average of 7 per year from 2005 to 2012. **101.** $0.08 per year; The price of a gallon of gasoline increased by an average of $0.08 per year from 1980 to 2012. **103.** −1859 thousand cameras per year; The number of digital cameras sold decreased by an average of 1859 thousand per year from 2010 to 2013. **105.** Because the slopes of both pairs of opposite sides are equal, the figure is a parallelogram. **107.** $\frac{1}{3}$ **108.** $\frac{1}{3}$ **109.** $\frac{1}{3}$ **110.** $\frac{1}{3} = \frac{1}{3} = \frac{1}{3}$ is true. **111.** collinear **112.** not collinear

Section 2.3 (pages 172–176)

1. A **3.** A **5.** $3x + y = 10$ **7.** A **9.** C **11.** H **13.** B **15.** $y = 5x + 15$ **17.** $y = -\frac{2}{3}x + \frac{4}{5}$ **19.** $y = x - 1$ **21.** $y = \frac{2}{5}x + 5$ **23.** $y = \frac{2}{3}x + 1$ **25.** $y = -x - 2$ **27.** $y = 2x - 4$ **29.** $y = -\frac{3}{5}x + 3$

31. **(a)** $y = x + 4$ **(b)** 1 **(c)** $(0, 4)$ **(d)**

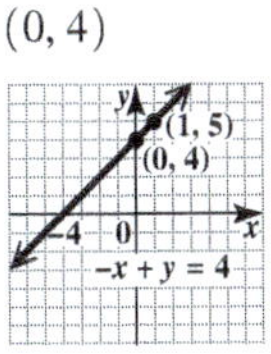

33. **(a)** $y = -\frac{6}{5}x + 6$ **(b)** $-\frac{6}{5}$ **(c)** $(0, 6)$ **(d)**

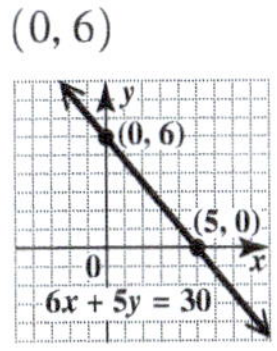

35. **(a)** $y = \frac{4}{5}x - 4$ **(b)** $\frac{4}{5}$ **(c)** $(0, -4)$ **(d)**

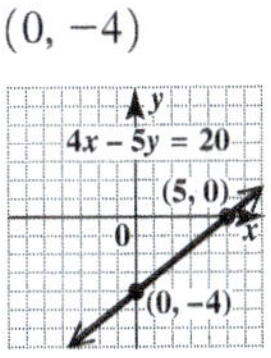

37. **(a)** $y = -\frac{1}{2}x - 2$ **(b)** $-\frac{1}{2}$ **(c)** $(0, -2)$ **(d)**

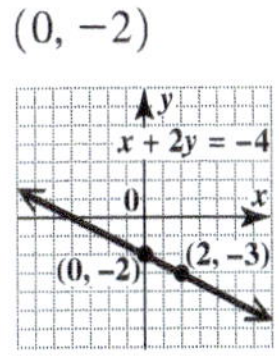

39. **(a)** $y = -2x + 18$ **(b)** $2x + y = 18$ **41.** **(a)** $y = -\frac{3}{4}x + \frac{5}{2}$ **(b)** $3x + 4y = 10$ **43.** **(a)** $y = \frac{1}{2}x + \frac{13}{2}$ **(b)** $x - 2y = -13$

45. **(a)** $y = 4x - 12$ **(b)** $4x - y = 12$ **47.** **(a)** $y = 1.4x + 4$ **(b)** $7x - 5y = -20$ **49.** $2x - y = 2$ **51.** $x + 2y = 8$ **53.** $y = 5$ **55.** $x = 7$ **57.** $y = -3$ **59.** $2x - 13y = -6$ **61.** $y = 5$ **63.** $x = 9$ **65.** $y = -\frac{3}{2}$ **67.** $y = 8$ **69.** $x = 0.5$ **71.** **(a)** $y = 3x - 19$ **(b)** $3x - y = 19$ **73.** **(a)** $y = \frac{1}{2}x - 1$ **(b)** $x - 2y = 2$ **75.** **(a)** $y = -\frac{1}{2}x + 9$ **(b)** $x + 2y = 18$ **77.** **(a)** $y = 7$ **(b)** $y = 7$ **79.** $y = 45x$; $(0, 0)$, $(5, 225)$, $(10, 450)$ **81.** $y = 3.75x$; $(0, 0)$, $(5, 18.75)$, $(10, 37.50)$ **83.** $y = 140x$; $(0, 0)$, $(5, 700)$, $(10, 1400)$ **85.** **(a)** $y = 149x + 15$ **(b)** $(5, 760)$; The cost for 5 tickets and a parking pass is $760. **(c)** $313 **87.** **(a)** $y = 41x + 99$ **(b)** $(5, 304)$; The cost for a 5-month membership is $304. **(c)** $591 **89.** **(a)** $y = 95x + 36$ **(b)** $(5, 511)$; The cost of the plan for 5 months is $511. **(c)** $2316 **91.** **(a)** $y = 6x + 30$ **(b)** $(5, 60)$; It costs $60 to rent the saw for 5 days. **(c)** 18 days **93.** **(a)** $y = -1357x + 7030$; Sales of portable media/MP3 players in the United States decreased by $1357 million per year from 2010 to 2013. **(b)** $5673 million **95.** **(a)** $y = 3.875x + 31.3$ **(b)** $73.9 billion; It is very close to the actual value. **97.** 32; 212 **98.** $(0, 32)$ and $(100, 212)$ **99.** $\frac{9}{5}$ **100.** $F = \frac{9}{5}C + 32$ **101.** $C = \frac{5}{9}(F - 32)$ **102.** 86° **103.** 10° **104.** −40°

Section 2.4 (pages 183–185)

1. **(a)** yes **(b)** yes **(c)** no **(d)** yes **3.** **(a)** no **(b)** no **(c)** no **(d)** yes **5.** solid; below **7.** dashed; above **9.** ≤ **11.** >

13.

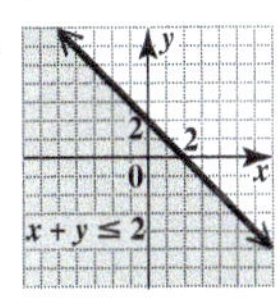

15.

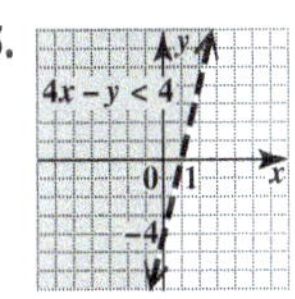

17.

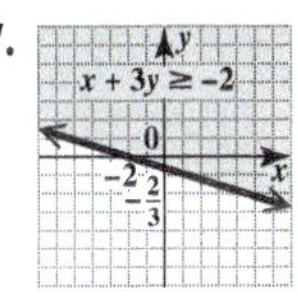

19.

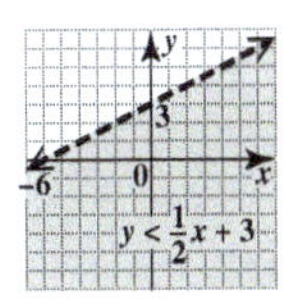

21.

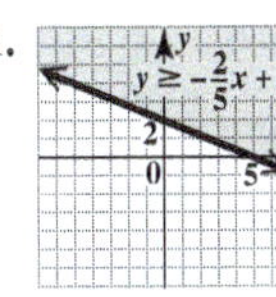

23.

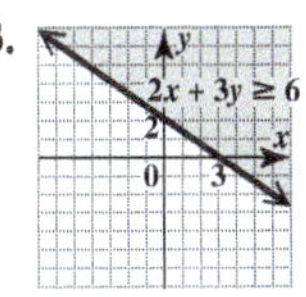

25.

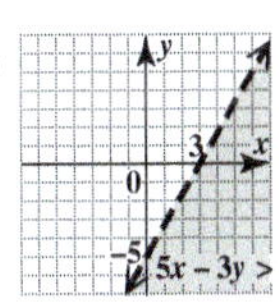

27.

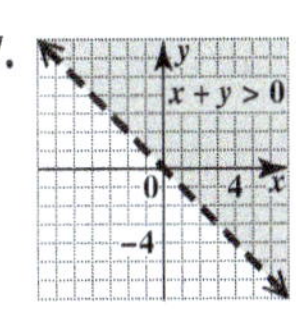

29.

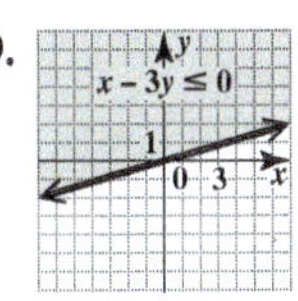

31.

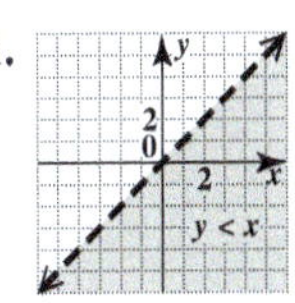

33.

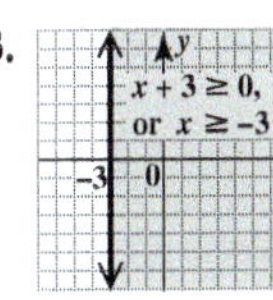

35.

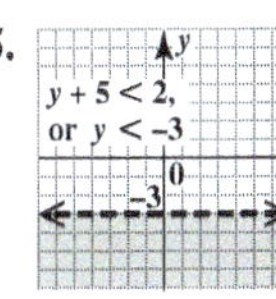

37. 2; $(0, -4)$; $2x - 4$; solid; above; ≥; $\ge 2x - 4$

39.

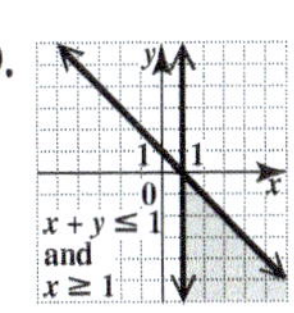

41.

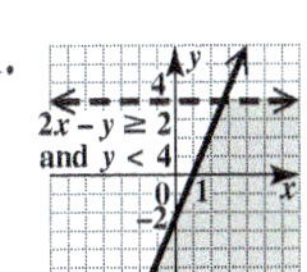

43.

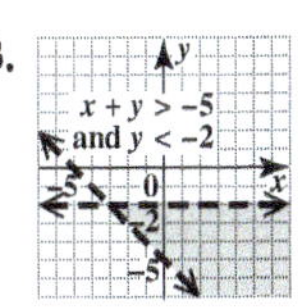

45. $-3 < x < 3$

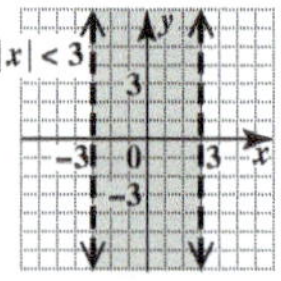

47. $-2 < x + 1 < 2$
$-3 < x < 1$

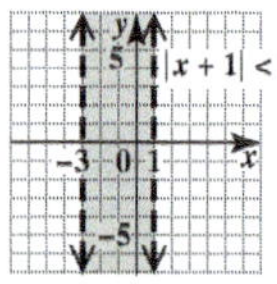

49.

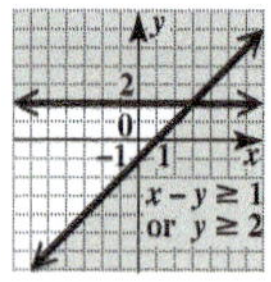

51.

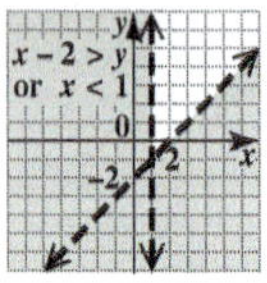

53.

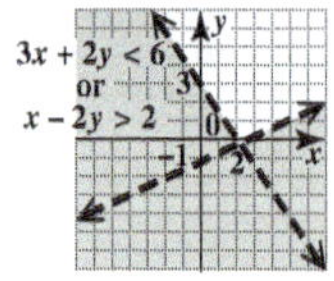

55. $x \le 200$, $x \ge 100$, $y \ge 3000$

56.

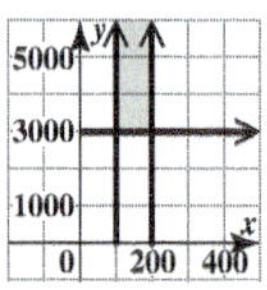

57. $C = 50x + 100y$ **58.** Some examples are $(100, 5000)$, $(150, 3000)$, and $(150, 5000)$. The corner points are $(100, 3000)$ and $(200, 3000)$.

59. The least value occurs when $x = 100$ and $y = 3000$. **60.** The company should use 100 workers and manufacture 3000 units to achieve the least possible cost.

Section 2.5 (pages 193–195)

1. relation; ordered pairs **3.** domain; range **5.** independent variable; dependent variable **7.** $\{(2, -2), (2, 0), (2, 1)\}$ **9.** $\{(1960, 0.76), (1980, 2.69), (2000, 5.39), (2013, 8.38)\}$ **11.** $\{(A, 4), (B, 3), (C, 2), (D, 1), (F, 0)\}$

In Exercises 13 and 15, answers will vary.

13.

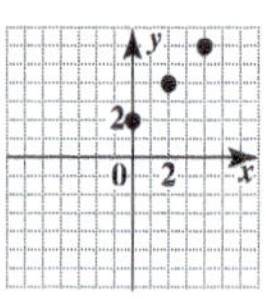

15.

x	y
−3	−4
−3	1
2	0

17. function; domain: $\{5, 3, 4, 7\}$; range: $\{1, 2, 9, 6\}$ **19.** not a function; domain: $\{2, 0\}$; range: $\{4, 2, 5\}$ **21.** function; domain: $\{-3, 4, -2\}$; range: $\{1, 7\}$ **23.** not a function; domain: $\{1, 0, 2\}$; range: $\{1, -1, 0, 4, -4\}$ **25.** function; domain: $\{2, 5, 11, 17, 3\}$; range: $\{1, 7, 20\}$ **27.** not a function; domain: $\{1\}$; range: $\{5, 2, -1, -4\}$ **29.** function; domain: $\{4, 2, 0, -2\}$; range: $\{-3\}$ **31.** function; domain: $\{-2, 0, 3\}$; range: $\{2, 3\}$ **33.** function; domain: $(-\infty, \infty)$; range: $(-\infty, \infty)$ **35.** not a function; domain: $\{-2\}$; range: $(-\infty, \infty)$ **37.** not a function; domain: $(-\infty, 0]$; range: $(-\infty, \infty)$ **39.** function; domain: $(-\infty, \infty)$; range: $(-\infty, 4]$ **41.** not a function; domain: $[-4, 4]$; range: $[-3, 3]$ **43.** not a function; domain: $(-\infty, \infty)$; range: $[2, \infty)$ **45.** function; $(-\infty, \infty)$ **47.** function; $(-\infty, \infty)$ **49.** function; $(-\infty, \infty)$ **51.** not a function; $[0, \infty)$ **53.** not a function; $(-\infty, \infty)$ **55.** function; $[0, \infty)$ **57.** function; $[3, \infty)$ **59.** function; $\left[-\frac{1}{2}, \infty\right)$ **61.** function; $(-\infty, \infty)$ **63.** function; $(-\infty, 0) \cup (0, \infty)$ **65.** function; $(-\infty, 4) \cup (4, \infty)$ **67.** function; $(-\infty, 0) \cup (0, \infty)$ **69. (a)** yes **(b)** domain: $\{2009, 2010, 2011, 2012, 2013\}$; range: $\{44.0, 43.4, 43.1, 42.9\}$ **(c)** 42.9; 2010 **(d)** Answers will vary. Two possible answers are $(2010, 43.4)$ and $(2013, 43.1)$.

Section 2.6 (pages 201–205)

1. $f(x)$; function; domain; x; f of x (or "f at x") **3.** line; -2; linear; $-2x + 4$; -2; 3; -2 **5.** 4 **7.** 13 **9.** -11 **11.** 4 **13.** -296 **15.** 3 **17.** 2.75 **19.** $-3p + 4$ **21.** $3x + 4$ **23.** $-3x - 2$ **25.** $-6t + 1$ **27.** $-\pi^2 + 4\pi + 1$ **29.** $-3x - 3h + 4$ **31.** $-\frac{p^2}{9} + \frac{4p}{3} + 1$ **33. (a)** -1 **(b)** -1 **35. (a)** 2 **(b)** 3 **37. (a)** 15 **(b)** 10 **39. (a)** 4 **(b)** 1 **41. (a)** 3 **(b)** -3 **43. (a)** -3 **(b)** 2 **45. (a)** 2 **(b)** 0 **(c)** -1 **47. (a)** $f(x) = -\frac{1}{3}x + 4$ **(b)** 3 **49. (a)** $f(x) = 3 - 2x^2$ **(b)** -15 **51. (a)** $f(x) = \frac{4}{3}x - \frac{8}{3}$ **(b)** $\frac{4}{3}$

53. domain: $(-\infty, \infty)$; range: $(-\infty, \infty)$

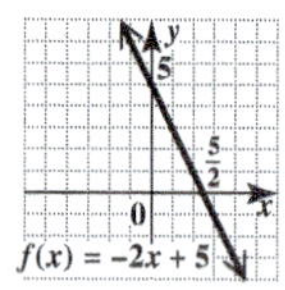

55. domain: $(-\infty, \infty)$; range: $(-\infty, \infty)$

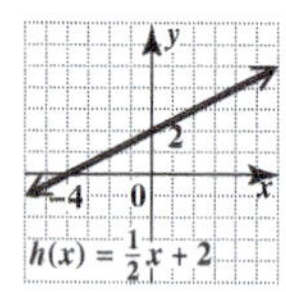

57. domain: $(-\infty, \infty)$; range: $(-\infty, \infty)$

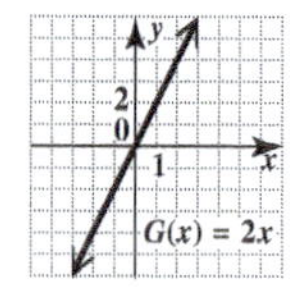

59. domain: $(-\infty, \infty)$; range: $\{-4\}$

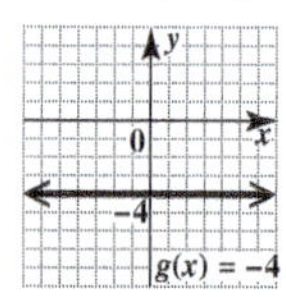

61. domain: $(-\infty, \infty)$; range: $\{0\}$

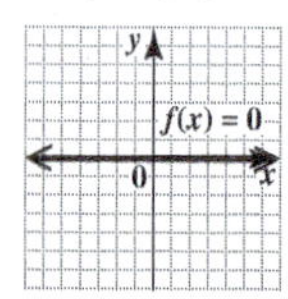

63. x-axis **65. (a)** 11.25 (dollars) **(b)** 3 is the value of the independent variable, which represents a package weight of 3 lb. $f(3)$ is the value of the dependent variable, which represents the cost to mail a 3-lb package. **(c)** \$18.75; $f(5) = 18.75$ **67. (a)** $f(x) = 12x + 100$ **(b)** 1600; The cost to print 125 t-shirts is \$1600. **(c)** 75; $f(75) = 1000$; The cost to print 75 t-shirts is \$1000. **69. (a)** 1.1 **(b)** 5 **(c)** -1.2 **(d)** $(0, 3.5)$ **(e)** $f(x) = -1.2x + 3.5$ **71. (a)** $[0, 100]$; $[0, 3000]$ **(b)** 25 hr; 25 hr **(c)** 2000 gal **(d)** $f(0) = 0$; The pool is empty at time 0. **(e)** $f(25) = 3000$; After 25 hr, there are 3000 gal of water in the pool. **73. (a)** 194.53 cm **(b)** 177.29 cm **(c)** 177.41 cm **(d)** 163.65 cm **75.** Because it falls from left to right, the slope is negative. **76.** $-\frac{3}{2}$ **77.** $-\frac{3}{2}$; $\frac{2}{3}$ **78.** $\left(\frac{7}{3}, 0\right)$ **79.** $\left(0, \frac{7}{2}\right)$ **80.** $f(x) = -\frac{3}{2}x + \frac{7}{2}$ **81.** $-\frac{17}{2}$ **82.** $\frac{23}{3}$

3 SYSTEMS OF LINEAR EQUATIONS

Section 3.1 (pages 225–230)

1. 4; -3 **3.** $\varnothing$ **5.** no **7.** D; The ordered-pair solution must be in quadrant IV because that is where the graphs of the equations intersect. **9. (a)** B **(b)** C **(c)** A **(d)** D **11.** solution **13.** not a solution

15. $\{(-2, -3)\}$

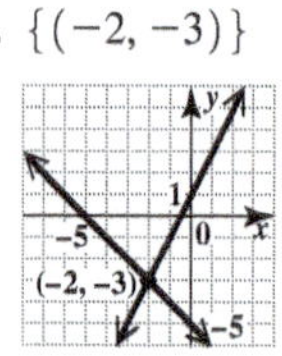

17. $\{(0, 1)\}$

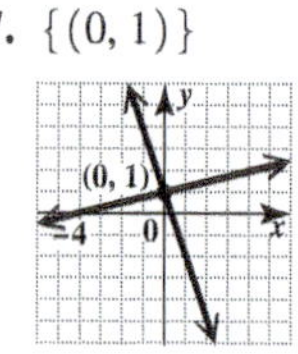

19. $\{(2, 0)\}$

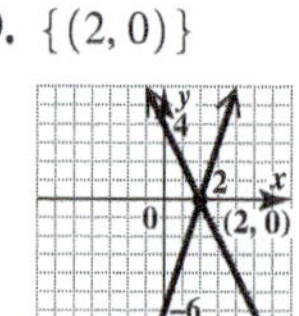

21. $\{(1, 2)\}$ **23.** $\{(2, 3)\}$ **25.** $\left\{\left(\frac{22}{9}, \frac{22}{3}\right)\right\}$ **27.** $\{(5, 4)\}$
29. $\{(1, 3)\}$ **31.** $\left\{\left(-5, -\frac{10}{3}\right)\right\}$ **33.** $\{(2, 6)\}$
35. $\{(x, y) \mid 2x - y = 0\}$; dependent equations **37.** $\varnothing$; inconsistent system
39. $\{(4, 2)\}$ **41.** $\{(-8, 4)\}$ **43.** $\{(0, 0)\}$ **45.** $\{(2, -4)\}$
47. $\{(3, -1)\}$ **49.** $\{(2, -3)\}$ **51.** $\{(x, y) \mid 7x + 2y = 6\}$; dependent equations **53.** $\left\{\left(\frac{3}{2}, -\frac{3}{2}\right)\right\}$ **55.** $\varnothing$; inconsistent system **57.** $\{(0, 0)\}$
59. $\{(0, -4)\}$ **61.** $\left\{\left(6, -\frac{5}{6}\right)\right\}$ **63.** $y = -\frac{3}{7}x + \frac{4}{7}$; $y = -\frac{3}{7}x + \frac{3}{14}$; no solution **65.** Both are $y = -\frac{2}{3}x + \frac{1}{3}$; infinitely many solutions
67. $\{(-3, 2)\}$ **69.** $\left\{\left(\frac{1}{3}, \frac{1}{2}\right)\right\}$ **71.** $\{(-4, 6)\}$
73. $\{(x, y) \mid 4x - y = -2\}$ **75.** $\left\{\left(1, \frac{1}{2}\right)\right\}$ **77.** None of the coefficients in the first system are -1 or 1, and solving by substitution would involve fractions. In the second system, we can solve for x in the first equation to avoid using fractions as coefficients.
79. **(a)** \$4 **(b)** 300 half-gallons **(c)** supply: 200 half-gallons; demand: 400 half-gallons **81.** **(a)** 2009–2011 **(b)** 2011 and 2012 **(c)** 2010; 13,000 **(d)** (2010, 13,000) **(e)** *The Big Bang Theory* viewership was increasing, *Dancing with the Stars* viewership was decreasing, and *60 Minutes* viewership remained fairly constant.
83. hiking **85.** (8.7, 35.9) **87.** $\{(2, 4)\}$ **89.** $\left\{\left(\frac{1}{2}, 2\right)\right\}$
91. $\left\{\left(\frac{c}{a}, 0\right)\right\}$ **93.** $\left\{\left(-\frac{1}{a}, -5\right)\right\}$

Section 3.2 (pages 237–239)

1. Answers will vary. Some possible answers are **(a)** two perpendicular walls and the ceiling in a normal room, **(b)** the floors of three different levels of an office building, and **(c)** three pages of a book (because they intersect in the spine). **3.** The statement means that when -1 is substituted for x, 2 is substituted for y, and 3 is substituted for z in the three equations, the resulting three statements are true. **5.** 4 **7.** $\{(3, 2, 1)\}$
9. $\{(1, 4, -3)\}$ **11.** $\{(0, 2, -5)\}$ **13.** $\{(1, 0, 3)\}$ **15.** $\left\{\left(1, \frac{3}{10}, \frac{2}{5}\right)\right\}$
17. $\left\{\left(-\frac{7}{3}, \frac{22}{3}, 7\right)\right\}$ **19.** $\{(-12, 18, 0)\}$ **21.** $\{(0.8, -1.5, 2.3)\}$
23. $\{(4, 5, 3)\}$ **25.** $\{(2, 2, 2)\}$ **27.** $\left\{\left(\frac{8}{3}, \frac{2}{3}, 3\right)\right\}$ **29.** $\{(-1, 0, 0)\}$
31. $\{(-4, 6, 2)\}$ **33.** $\{(-3, 5, -6)\}$ **35.** $\varnothing$; inconsistent system
37. $\{(x, y, z) \mid x - y + 4z = 8\}$; dependent equations **39.** $\{(3, 0, 2)\}$
41. $\{(x, y, z) \mid 2x + y - z = 6\}$; dependent equations **43.** $\{(0, 0, 0)\}$
45. $\varnothing$; inconsistent system **47.** $\{(2, 1, 5, 3)\}$ **49.** $\{(-2, 0, 1, 4)\}$

Section 3.3 (pages 250–256)

1. **(a)** 6 oz **(b)** 15 oz **(c)** 24 oz **(d)** 30 oz **3.** \$1.89x
5. **(a)** $(10 - x)$ mph **(b)** $(10 + x)$ mph **7.** wins: 92; losses: 70
9. length: 78 ft; width: 36 ft **11.** AT&T: \$126.4 billion; Verizon: \$115.8 billion **13.** $x = 40$ and $y = 50$, so the angles measure 40° and 50°.
15. hockey: \$354.84; basketball: \$315.66 **17.** ribeye: \$17.29; salmon: \$14.99 **19.** DVD: \$12.96; Blu-ray: \$19.49 **21.** 6 gal of 25%; 14 gal of 35% **23.** pure acid: 6 L; 10% acid: 48 L **25.** nuts: 14 kg; cereal: 16 kg **27.** \$1000 at 2%; \$2000 at 4% **29.** train: 60 km per hr; plane: 160 km per hr **31.** scooter: 25 mph; bicycle: 10 mph
33. boat: 21 mph; current: 3 mph **35.** \$0.75-per-lb candy: 3 lb; \$1.25-per-lb candy: 7 lb **37.** general admission: 76; with student ID: 108
39. 8 for a citron; 5 for a wood apple **41.** $x + y + z = 180$; angle measures: 70°, 30°, 80° **43.** first: 20°; second: 70°; third: 90°
45. shortest: 12 cm; middle: 25 cm; longest: 33 cm **47.** gold: 13; silver: 11; bronze: 9 **49.** general admission: 1170; courtside: 985; bench: 130 **51.** bookstore A: 140; bookstore B: 280; bookstore C: 380
53. first chemical: 50 kg; second chemical: 400 kg; third chemical: 300 kg
55. wins: 30; losses: 12; overtime losses: 6 **57.** box of fish: 8 oz; box of bugs: 2 oz; box of worms: 5 oz

4 EXPONENTS, POLYNOMIALS, AND POLYNOMIAL FUNCTIONS

Section 4.1 (pages 276–280)

1. incorrect; $(ab)^2 = a^2b^2$ **3.** incorrect; $\left(\frac{4}{a}\right)^3 = \frac{4^3}{a^3}$ **5.** correct
7. Do not multiply the bases. $4^5 \cdot 4^2 = 4^7$ **9.** 13^{12} **11.** 8^{10} **13.** x^{17}
15. $-27w^8$ **17.** $18x^3y^8$ **19.** The product rule does not apply.
21. **(a)** B **(b)** C **(c)** B **(d)** C **23.** 1 **25.** -1 **27.** 1 **29.** 2
31. 0 **33.** -2 **35.** **(a)** B **(b)** D **(c)** B **(d)** D **37.** $\frac{1}{5^4}$, or $\frac{1}{625}$
39. $\frac{1}{3^5}$, or $\frac{1}{243}$ **41.** $\frac{1}{9}$ **43.** $\frac{1}{(4x)^2}$, or $\frac{1}{16x^2}$ **45.** $\frac{4}{x^2}$ **47.** $-\frac{1}{a^3}$
49. $\frac{1}{(-a)^4}$, or $\frac{1}{a^4}$ **51.** $\frac{1}{(-3x)^3}$, or $-\frac{1}{27x^3}$ **53.** $\frac{11}{30}$ **55.** $-\frac{5}{24}$
57. $-\frac{3}{4}$ **59.** 16 **61.** $\frac{27}{4}$ **63.** 4^2, or 16 **65.** x^4 **67.** $\frac{1}{r^3}$ **69.** 6^6
71. $\frac{1}{6^{10}}$ **73.** 7^2, or 49 **75.** r^3 **77.** The quotient rule does not apply.
79. x^{18} **81.** $\frac{27}{125}$ **83.** $64t^3$ **85.** $-216x^6$ **87.** $-\frac{64m^6}{t^3}$ **89.** $\frac{s^{12}}{t^{20}}$
91. **(a)** B **(b)** D **(c)** D **(d)** B **93.** 64 **95.** $\frac{27}{8}$ **97.** $\frac{25}{16}$
99. $\frac{81}{16t^4}$ **101.** $16x^2$ **103.** $\frac{32}{x^5}$ **105.** $\frac{1}{3}$ **107.** $\frac{1}{a^5}$ **109.** 5^6 **111.** $\frac{1}{x^{12}}$
113. $\frac{1}{k^2}$ **115.** $-4r^6$ **117.** $\frac{625}{a^{10}}$ **119.** $\frac{z^4}{x^3}$ **121.** $-14k^3$ **123.** $\frac{p^4}{5}$
125. $\frac{m^8}{n^{11}}$ **127.** $\frac{1}{2pq}$ **129.** $\frac{4}{a^2}$ **131.** $\frac{1}{6y^{13}}$ **133.** $\frac{4k^5}{m^2}$ **135.** $\frac{2k^5}{3}$
137. $\frac{8}{3pq^{10}}$ **139.** $\frac{y^9}{8}$ **141.** $\frac{n^{10}}{25m^{18}}$ **143.** $-\frac{125y^3}{x^{30}}$ **145.** $\frac{4k^{17}}{125}$
147. $-\frac{3}{32m^8p^4}$ **149.** $\frac{2}{3y^4}$ **151.** $\frac{3p^8}{16q^{14}}$ **153.** after; power; a; 10^n
155. 5.3×10^2 **157.** 8.3×10^{-1} **159.** 6.92×10^{-6} **161.** -3.85×10^4
163. $\$1 \times 10^9$ (or $\$10^9$); $\$1 \times 10^{12}$ (or $\$10^{12}$); $\$3.8 \times 10^{12}$; 2.57891×10^5 **165.** 72,000 **167.** 0.00254 **169.** $-60{,}000$
171. 0.000012 **173.** 0.06 **175.** 0.0000025 **177.** 200,000 **179.** 3000
181. 7.5×10^9 **183.** 4×10^{17} **185.** **(a)** 3.174×10^8 **(b)** $\$1 \times 10^{12}$ (or $\$10^{12}$) **(c)** \$3151 **187.** 30,500, or 3.05×10^4 **189.** 300 sec
191. approximately 5.87×10^{12} mi **193.** **(a)** 5.16×10^2 **(b)** 998 mi²

Section 4.2 (pages 284–286)

1. A **3.** 7; 1 **5.** -15; 2 **7.** 1; 4 **9.** $\frac{1}{6}$; 1 **11.** 8; 0 **13.** -1; 3
15. $2x^3 - 3x^2 + x + 4$; $2x^3$; 2 **17.** $p^7 - 8p^5 + 4p^3$; p^7; 1

19. $-3m^4 - m^3 + 10$; $-3m^4$; -3 **21.** monomial; 0 **23.** binomial; 1 **25.** binomial; 8 **27.** monomial; 6 **29.** trinomial; 3 **31.** none of these; 5 **33.** $8z^4$ **35.** $7m^3$ **37.** $5x$ **39.** already simplified **41.** $-t + 13s$ **43.** $8k^2 + 2k - 7$ **45.** $-2n^4 - n^3 + n^2$ **47.** $-2ab^2 + 20a^2b$ **49.** $3m + 11$ **51.** $-p - 4$ **53.** $8x^2 + x - 2$ **55.** $-t^4 + 2t^2 - t + 5$ **57.** $5y^3 - 3y^2 + 5y + 1$ **59.** $r + 13$ **61.** $-2a^2 - 2a - 7$ **63.** $-3z^5 + z^2 + 7z$ **65.** $12p - 4$ **67.** $-9p^2 + 11p - 9$ **69.** $5a + 18$ **71.** $14m^2 - 13m + 6$ **73.** $13z^2 + 10z - 3$ **75.** $10y^3 - 7y^2 + 5y + 8$ **77.** $-5a^4 - 6a^3 + 9a^2 - 11$ **79.** $3y^2 - 4y + 2$ **81.** $-4m^2 + 4n^2 - 7n$ **83.** $y^4 - 4y^2 - 4$ **85.** $10z^2 - 16z$ **87.** $12x^2 + 8x + 5$

Section 4.3 (pages 294–297)

1. polynomial; one; terms; powers **3. (a)** -10 **(b)** 8 **(c)** -4 **5. (a)** 8 **(b)** -10 **(c)** 0 **7. (a)** 8 **(b)** 2 **(c)** 4 **9. (a)** 7 **(b)** 1 **(c)** 1 **11. (a)** 8 **(b)** 74 **(c)** 6 **13. (a)** -11 **(b)** 4 **(c)** -8 **15. (a)** 1.0 million lb **(b)** 56.3 million lb **(c)** 158.8 million lb **17. (a)** \$280.6 billion **(b)** \$568.5 billion **(c)** \$711.4 billion **19. (a)** $8x - 3$ **(b)** $2x - 17$ **21. (a)** $-x^2 + 12x - 12$ **(b)** $9x^2 + 4x + 6$ **23.** $f(x)$ and $g(x)$ can be any two polynomials that have a sum of $3x^3 - x + 3$, such as $f(x) = 3x^3 + 1$ and $g(x) = -x + 2$. **25.** $x^2 + 2x - 9$ **27.** 6 **29.** $x^2 - x - 6$ **31.** 6 **33.** -33 **35.** 0 **37.** $-\frac{9}{4}$ **39.** $-\frac{9}{2}$ **41. (a)** $P(x) = 8.49x - 50$ **(b)** \$799 **43.** B **45.** A **47.** 6 **49.** 83 **51.** 53 **53.** 13 **55.** $2x^2 + 11$ **57.** $2x - 2$ **59.** $\frac{97}{4}$ **61.** 8 **63.** 1 **65.** 9 **67.** 1 **69.** $(f \circ g)(x) = 63{,}360x$; It computes the number of inches in x miles. **71. (a)** $s = \frac{x}{4}$ **(b)** $y = \frac{x^2}{16}$ **(c)** 2.25 **73.** $(\mathcal{A} \circ r)(t) = 4\pi t^2$; This is the area of the circular layer as a function of time.

75.
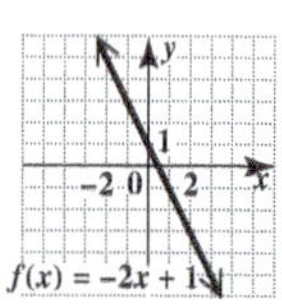

domain: $(-\infty, \infty)$; range: $(-\infty, \infty)$

77.
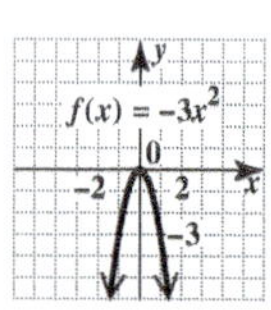

domain: $(-\infty, \infty)$; range: $(-\infty, 0]$

79.
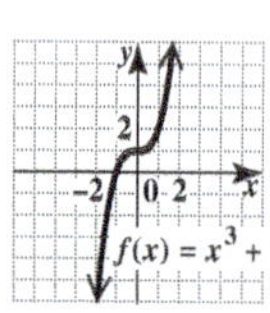

domain: $(-\infty, \infty)$; range: $(-\infty, \infty)$

Section 4.4 (pages 304–306)

1. C **3.** D **5.** $-24m^5$ **7.** $-28x^7y^4$ **9.** $-6x^2 + 15x$ **11.** $-2q^3 - 3q^4$ **13.** $18k^4 + 12k^3 + 6k^2$ **15.** $6t^3 + t^2 - 14t - 3$ **17.** $m^3 - 3m^2 - 40m$ **19.** $24z^3 - 20z^2 - 16z$ **21.** $4x^5 - 4x^4 - 24x^3$ **23.** $6y^2 + y - 12$ **25.** $25m^2 - 9n^2$ **27.** $-2b^3 + 2b^2 + 18b + 12$ **29.** $8z^4 - 14z^3 + 17z^2 + 20z - 3$ **31.** $6p^4 + p^3 + 4p^2 - 27p - 6$ **33.** $m^2 - 3m - 40$ **35.** $12k^2 + k - 6$ **37.** $15x^2 - 13x + 2$ **39.** $20x^2 + 23xy + 6y^2$ **41.** $3z^2 + zw - 4w^2$ **43.** $12c^2 + 16cd - 3d^2$ **45.** $x^2 - 81$ **47.** $4p^2 - 9$ **49.** $25m^2 - 1$ **51.** $9a^2 - 4c^2$ **53.** $16m^2 - 49n^4$ **55.** $75y^7 - 12y$ **57.** $y^2 - 10y + 25$ **59.** $x^2 + 2x + 1$ **61.** $4p^2 + 28p + 49$ **63.** $16n^2 - 24nm + 9m^2$ **65.** $k^2 - \frac{10}{7}kp + \frac{25}{49}p^2$ **67.** $0.04x^2 - 0.56xy + 1.96y^2$ **69.** $16x^2 - \frac{4}{9}$ **71.** $0.1x^2 + 0.63x - 0.13$ **73.** $3w^2 - \frac{23}{4}wz - \frac{1}{2}z^2$ **75.** $25x^2 + 10x + 1 + 60xy + 12y + 36y^2$ **77.** $4a^2 + 4ab + b^2 - 12a - 6b + 9$ **79.** $4a^2 + 4ab + b^2 - 9$ **81.** $4h^2 - 4hk + k^2 - j^2$ **83.** $y^3 + 6y^2 + 12y + 8$ **85.** $125r^3 - 75r^2s + 15rs^2 - s^3$ **87.** $q^4 - 8q^3 + 24q^2 - 32q + 16$ **89.** $6a^3 + 7a^2b + 4ab^2 + b^3$ **91.** $4z^4 - 17z^3x + 12z^2x^2 - 6zx^3 + x^4$ **93.** $m^4 - 4m^2p^2 + 4mp^3 - p^4$ **95.** $a^4b - 7a^2b^3 - 6ab^4$ **97.** 49; 25; $49 \neq 25$ **99.** 2401; 337; $2401 \neq 337$ **101.** $\frac{9}{2}x^2 - 2y^2$ **103.** $15x^2 - 2x - 24$ **105.** $10x^2 - 2x$ **107.** $2x^2 - x - 3$ **109.** $8x^3 - 27$ **111.** $2x^3 - 18x$ **113.** -20 **115.** $2x^2 - 6x$ **117.** 36 **119.** $\frac{35}{4}$ **121.** $\frac{1859}{64}$ **123.** $a - b$ **124.** $\mathcal{A} = s^2$; $(a - b)^2$ **125.** $(a - b)b$, or $ab - b^2$; $2ab - 2b^2$ **126.** b^2 **127.** a^2; a **128.** $a^2 - (2ab - 2b^2) - b^2 = a^2 - 2ab + b^2$ **129. (a)** They must be equal. **(b)** $(a - b)^2 = a^2 - 2ab + b^2$

130.
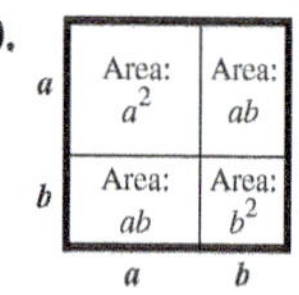

The large square is made up of two smaller squares and two congruent rectangles. The sum of the areas is $a^2 + 2ab + b^2$. Because $(a + b)^2$ must represent the same quantity, they must be equal. Thus, $(a + b)^2 = a^2 + 2ab + b^2$.

Section 4.5 (pages 311–313)

1. quotient; exponents **3.** 0 **5.** $3x^3 - 2x^2 + 1$ **7.** $3y + 4 - \frac{5}{y}$ **9.** $3 + \frac{5}{m} + \frac{6}{m^2}$ **11.** $\frac{2}{7n} - \frac{3}{2m} + \frac{9}{7mn}$ **13.** $\frac{2y}{x} + \frac{3}{4} + \frac{3w}{x}$ **15.** $r^2 - 7r + 6$ **17.** $y - 4$ **19.** $q + 8$ **21.** $t + 5$ **23.** $p - 4 + \frac{44}{p + 6}$ **25.** $m^2 + 2m - 1$ **27.** $m^2 + m + 3$ **29.** $x^2 + 2x - 3 + \frac{6}{4x + 1}$ **31.** $2x - 5 + \frac{-4x + 5}{3x^2 - 2x + 4}$ **33.** $x^2 + x + 3$ **35.** $2x^2 - x - 5$ **37.** $3x^2 + 6x + 11 + \frac{26}{x - 2}$ **39.** $2k^2 + 3k - 1$ **41.** $z^2 + 3$ **43.** $2y^2 + 2$ **45.** $p^2 + p + 1$ **47.** $x^2 - 4x + 2 + \frac{9x - 4}{x^2 + 3}$ **49.** $p^2 + \frac{5}{2}p + 2 + \frac{-1}{2p + 2}$ **51.** $\frac{3}{2}a - 10 + \frac{77}{2a + 6}$ **53.** $\frac{2}{3}x - 1$ **55.** $\frac{3}{4}a - 2 + \frac{1}{4a + 3}$ **57.** $(2p + 7)$ feet **59.** -13; -13; They are the same, which suggests that when $P(x)$ is divided by $x - r$, the result is $P(r)$. Here, $r = -1$. **61.** $5x - 1$; 0 **63.** $2x - 3$; -1 **65.** $4x^2 + 6x + 9$; $\frac{3}{2}$ **67.** $\frac{x^2 - 9}{2x}$, $x \neq 0$ **69.** $-\frac{5}{4}$ **71.** $\frac{x - 3}{2x}$, $x \neq 0$ **73.** 0 **75.** $-\frac{35}{4}$ **77.** $\frac{7}{2}$

5 FACTORING

Section 5.1 (pages 329–330)

1. C **3.** $3m$ **5.** $8xy$ **7.** $3(r + t)^2$ **9.** $12(m - 5)$ **11.** $4(1 + 5z)$ **13.** cannot be factored **15.** $8k(k^2 + 3)$ **17.** $-2p^2q^4(2p + q)$ **19.** $7x^3(1 + 5x - 2x^2)$ **21.** $2t^3(5t^2 - 1 - 2t)$ **23.** $5ac(3ac^2 - 5c + 1)$ **25.** $16zn^3(zn^3 + 4n^4 - 2z^2)$ **27.** $7ab(2a^2b + a - 3a^4b^2 + 6b^3)$

29. $(m-4)(2m+5)$ **31.** $11(2z-1)$ **33.** $(2-x)^2(1+2x)$ **35.** $(3-x)(6+2x-x^2)$ **37.** $20z(2z+1)(3z+4)$ **39.** $5(m+p)^2(m+p-2-3m^2-6mp-3p^2)$ **41.** $r(-r^2+3r+5)$; $-r(r^2-3r-5)$ **43.** $12s^4(-s+4)$; $-12s^4(s-4)$ **45.** $2x^2(-x^3+3x+2)$; $-2x^2(x^3-3x-2)$ **47.** $(m+q)(x+y)$ **49.** $(5m+n)(2+k)$ **51.** $(2-q)(2-3p)$ **53.** $(p+q)(p-4z)$ **55.** $(2x+3)(y+1)$ **57.** $(m+4)(m^2-6)$ **59.** $(a^2+b^2)(-3a+2b)$ **61.** $(y-2)(x-2)$ **63.** $(3y-2)(3y^3-4)$ **65.** $(1-a)(1-b)$ **67.** $2(m-3q)(x+y)$ **69.** $4(a^2+2b)(a-b^2)$ **71.** $y^2(2x+1)(x^2-7)$ **73.** $m^{-5}(3+m^2)$ **75.** $p^{-3}(3+2p)$

Section 5.2 (pages 336–338)

1. D **3.** C **5.** The factor $(4x+10)$ can be factored further into $2(2x+5)$, giving the *completely* factored form as $2(2x+5)(x-2)$. **7.** $x+3$ **9.** $m-7$ **11.** $r-5$ **13.** $x-2a$ **15.** $2x-3$ **17.** $2u+v$ **19.** $(y-3)(y+10)$ **21.** $(p+8)(p+7)$ **23.** prime **25.** $(a+5b)(a-7b)$ **27.** $(a-6b)(a-3b)$ **29.** prime **31.** $-(6m-5)(m+3)$ **33.** $(5x-6)(2x+3)$ **35.** $(4k+3)(5k+8)$ **37.** $(3a-2b)(5a-4b)$ **39.** $(6m-5)^2$ **41.** prime **43.** $(2xz-1)(3xz+4)$ **45.** $3(4x+5)(2x+1)$ **47.** $-5(a+6)(3a-4)$ **49.** $-11x(x-6)(x-4)$ **51.** $2xy^3(x-12y)^2$ **53.** $6a(a-3)(a+5)$ **55.** $13y(y+4)(y-1)$ **57.** $3p(2p-1)^2$ **59.** $(6p^3-r)(2p^3-5r)$ **61.** $(5k+4)(2k+1)$ **63.** $(3m+3p+5)(m+p-4)$ **65.** $(a+b)^2(a-3b)(a+2b)$ **67.** $(p+q)^2(p+3q)$ **69.** $(z-x)^2(z+2x)$ **71.** $(p^2-8)(p^2-2)$ **73.** $(2x^2+3)(x^2-6)$ **75.** $(4x^2+3)(4x^2+1)$

Section 5.3 (pages 342–344)

1. A, D **3.** B, C **5.** A sum of two squares can be factored if the binomial terms have a common factor greater than 1. **7.** $(p+4)(p-4)$ **9.** $(5x+2)(5x-2)$ **11.** $2(3a+7b)(3a-7b)$ **13.** $4(4m^2+y^2)(2m+y)(2m-y)$ **15.** $(y+z+9)(y+z-9)$ **17.** $(4+x+3y)(4-x-3y)$ **19.** $(p^2+16)(p+4)(p-4)$ **21.** $(k-3)^2$ **23.** $(2z+w)^2$ **25.** $(4m-1+n)(4m-1-n)$ **27.** $(2r-3+s)(2r-3-s)$ **29.** $(x+y-1)(x-y+1)$ **31.** $2(7m+3n)^2$ **33.** $(p+q+1)^2$ **35.** $(a-b+4)^2$ **37.** $(x-3)(x^2+3x+9)$ **39.** $(6-t)(36+6t+t^2)$ **41.** $(x+4)(x^2-4x+16)$ **43.** $(10+y)(100-10y+y^2)$ **45.** $(2x+1)(4x^2-2x+1)$ **47.** $(5x-6)(25x^2+30x+36)$ **49.** $(x-2y)(x^2+2xy+4y^2)$ **51.** $(4g-3h)(16g^2+12gh+9h^2)$ **53.** $(7p+5q)(49p^2-35pq+25q^2)$ **55.** $3(2n+3p)(4n^2-6np+9p^2)$ **57.** $(y+z+4)(y^2+2yz+z^2-4y-4z+16)$ **59.** $(m^2-5)(m^4+5m^2+25)$ **61.** $(3-10x^3)(9+30x^3+100x^6)$ **63.** $(5y^2+z)(25y^4-5y^2z+z^2)$ **65.** $(5p+2q)(25p^2-10pq+4q^2+5p-2q)$ **67.** $(3a-4b)(9a^2+12ab+16b^2+5)$ **69.** $(t-3)(2t+1)(4t^2-2t+1)$ **71.** $(8m-9n)(8m+9n-64m^2-72mn-81n^2)$ **73.** $(x^3-y^3)(x^3+y^3)$; $(x-y)(x^2+xy+y^2)(x+y)(x^2-xy+y^2)$ **74.** $(x^2+xy+y^2)(x^2-xy+y^2)$ **75.** $(x^2-y^2)(x^4+x^2y^2+y^4)$; $(x-y)(x+y)(x^4+x^2y^2+y^4)$ **76.** $x^4+x^2y^2+y^4$ **77.** The product must equal $x^4+x^2y^2+y^4$. Multiply $(x^2+xy+y^2)(x^2-xy+y^2)$ to verify this. **78.** Start by factoring as a difference of squares.

Section 5.4 (pages 347–348)

1. **(a)** B **(b)** D **(c)** A **(d)** A, C **(e)** A, B **3.** $3p^2(p-6)(p+5)$ **5.** $3pq(a+6b)(a-5b)$ **7.** prime **9.** $(6b+1)(b-3)$ **11.** $(x-10)(x^2+10x+100)$ **13.** $(p+2)(4+m)$ **15.** $9m(m-5+2m^2)$ **17.** $2(3m-10)(9m^2+30m+100)$ **19.** $(3m-5n)^2$ **21.** $(k-9)(q+r)$ **23.** $16z^2x(zx-2)$ **25.** $(x+7)(x-5)$ **27.** $(25+x^2)(5-x)(5+x)$ **29.** $(p+1)(p^2-p+1)$ **31.** $(8m+25)(8m-25)$ **33.** $6z(2z^2-z+3)$ **35.** $16(4b+5c)(4b-5c)$ **37.** $8(4+5z)(16-20z+25z^2)$ **39.** $(5r-s)(2r+5s)$ **41.** $4pq(2p+q)(3p+5q)$ **43.** $3(4k^2+9)(2k+3)(2k-3)$ **45.** $(m-n)(m^2+mn+n^2+m+n)$ **47.** $(x-2m-n)(x+2m+n)$ **49.** $6p^3(3p^2-4+2p^3)$ **51.** $2(x+4)(x-5)$ **53.** $8mn$ **55.** $2(5p+9)(5p-9)$ **57.** $4rx(3m^2+mn+10n^2)$ **59.** $(7a-4b)(3a+b)$ **61.** prime **63.** $(p+8q-5)^2$ **65.** $(7m^2+1)(3m^2-5)$ **67.** $(2r-t)(r^2-rt+19t^2)$ **69.** $(x+3)(x^2+1)(x+1)(x-1)$ **71.** $(m+n-5)(m-n+1)$

Section 5.5 (pages 355–358)

1. D; The polynomial is not factored. **3.** $\{-10, 5\}$ **5.** $\left\{-\frac{8}{3}, \frac{5}{2}\right\}$ **7.** $\{-2, 5\}$ **9.** $\{-6, -3\}$ **11.** $\left\{-\frac{1}{2}, 4\right\}$ **13.** $\left\{-\frac{1}{3}, \frac{4}{5}\right\}$ **15.** $\{-4, 0\}$ **17.** $\{0, 6\}$ **19.** $\{-2, 2\}$ **21.** $\{-3, 3\}$ **23.** $\{3\}$ **25.** $\left\{-\frac{4}{3}\right\}$ **27.** $\{-4, 2\}$ **29.** $\left\{-\frac{1}{2}, 6\right\}$ **31.** $\{-3, 4\}$ **33.** $\left\{-5, -\frac{1}{5}\right\}$ **35.** $\{1, 6\}$ **37.** $\left\{-\frac{1}{2}, 0, 5\right\}$ **39.** $\{-1, 0, 3\}$ **41.** $\left\{-\frac{4}{3}, 0, \frac{4}{3}\right\}$ **43.** $\left\{-\frac{5}{2}, -1, 1\right\}$ **45.** $\{-3, 3, 6\}$ **47.** By dividing each side by a variable expression, she lost the solution 0. The solution set is $\left\{-\frac{4}{3}, 0, \frac{4}{3}\right\}$. **49.** $\left\{-\frac{1}{2}, 6\right\}$ **51.** $\left\{-\frac{2}{3}, \frac{4}{15}\right\}$ **53.** $\left\{-\frac{3}{2}, \frac{1}{2}\right\}$ **55.** width: 16 ft; length: 20 ft **57.** base: 12 ft; height: 5 ft **59.** 50 ft by 100 ft **61.** -6 and -5 or 5 and 6 **63.** length: 15 in.; width: 9 in. **65.** 5 sec **67.** $6\frac{1}{4}$ sec **69.** L appears on both sides of the equation. **71.** $F=\frac{k}{d-D}$ **73.** $r=\frac{-2k-3y}{a-1}$, or $r=\frac{2k+3y}{1-a}$ **75.** $y=\frac{-x}{w-3}$, or $y=\frac{x}{3-w}$

6 RATIONAL EXPRESSIONS AND FUNCTIONS

Section 6.1 (pages 373–376)

1. integers; 0; rational; polynomials; 0 **3.** **(a)** $\{x \mid x \text{ is a real number}, x \neq 7\}$ **(b)** $(-\infty, 7) \cup (7, \infty)$ **5.** **(a)** $\left\{x \mid x \text{ is a real number}, x \neq -\frac{1}{7}\right\}$ **(b)** $\left(-\infty, -\frac{1}{7}\right) \cup \left(-\frac{1}{7}, \infty\right)$

7. **(a)** $\{x \mid x \text{ is a real number}, x \neq 0\}$ **(b)** $(-\infty, 0) \cup (0, \infty)$
9. **(a)** $\left\{x \mid x \text{ is a real number}, x \neq -2, \frac{3}{2}\right\}$
(b) $(-\infty, -2) \cup \left(-2, \frac{3}{2}\right) \cup \left(\frac{3}{2}, \infty\right)$ **11.** **(a)** $\{x \mid x \text{ is a real number}\}$
(b) $(-\infty, \infty)$ **13.** **(a)** $\{x \mid x \text{ is a real number}\}$ **(b)** $(-\infty, \infty)$
15. $\frac{2}{15}$ **17.** $\frac{9}{10}$ **19.** $\frac{3}{4}$ **21.** **(a)** C **(b)** A **(c)** D **(d)** B **(e)** E
(f) F **23.** B, E, F **25.** B **27.** x **29.** $\frac{x-3}{x+5}$ **31.** $\frac{x+3}{2x(x-3)}$
33. It is already in lowest terms. **35.** $\frac{6}{7}$ **37.** $\frac{z}{6}$ **39.** $\frac{t-3}{3}$ **41.** $\frac{2}{t-3}$
43. $\frac{x-3}{x+1}$ **45.** $\frac{4x+1}{4x+3}$ **47.** $a^2 - ab + b^2$ **49.** $\frac{c+6d}{c-d}$ **51.** $\frac{a+b}{a-b}$
53. -1 **55.** $-(x+y)$, or $-x-y$ **57.** $-(x+2)$, or $-x-2$
59. $-\frac{x+y}{x-y}$ (There are other answers.) **61.** $-\frac{1}{2}$ **63.** It is already in lowest terms. **65.** $\frac{x+4}{x-2}$ **67.** $\frac{2x+3}{x+2}$ **69.** $\frac{7x}{6}$ **71.** $-\frac{p+5}{2p}$ (There are other answers.) **73.** $\frac{35}{4}$ **75.** $-(z+1)$, or $-z-1$ **77.** $\frac{14x^2}{5}$
79. $\frac{-m(m+7)}{m+1}$ (There are other answers.) **81.** -2 **83.** $\frac{x+4}{x-4}$
85. $\frac{a^2+ab+b^2}{a-b}$ **87.** $\frac{2x-3}{2(x-3)}$ **89.** $\frac{a^2+2ab+4b^2}{a+2b}$ **91.** $\frac{2x+3}{2x-3}$
93. $\frac{3a^3b^2}{4}$ **95.** $\frac{27}{2mn^7}$ **97.** $\frac{k+5p}{2k+5p}$ **99.** $(k-1)(k-2)$

Section 6.2 (pages 382–384)

1. $\frac{4}{5}$ **3.** $-\frac{1}{18}$ **5.** $\frac{31}{36}$ **7.** $\frac{9}{t}$ **9.** $\frac{6x+y}{7}$ **11.** $\frac{2}{x}$ **13.** $-\frac{2}{x^3}$ **15.** 1
17. $x-5$ **19.** $\frac{5}{p+3}$ **21.** $a-b$ **23.** $72x^4y^5$ **25.** $z(z-2)$
27. $2(y+4)$ **29.** $(x+9)^2(x-9)$ **31.** $(m+n)(m-n)$
33. $x(x-4)(x+1)$ **35.** $(t+5)(t-2)(2t-3)$
37. $2y(y+3)(y-3)$ **39.** $2(x+2)^2(x-3)$ **41.** The expression $\frac{x-4x-1}{x+2}$ is incorrect. The third term in the numerator should be $+1$ because the $-$ sign should be distributed over both $4x$ and -1. The answer should be $\frac{-3x+1}{x+2}$. **43.** $\frac{31}{3t}$ **45.** $\frac{5-22x}{12x^2y}$ **47.** $\frac{16b+9a^2}{60a^4b^6}$
49. $\frac{4pr+3sq^3}{14p^4q^4}$ **51.** $\frac{a^2b^5-2ab^6+3}{a^5b^7}$ **53.** $\frac{1}{x(x-1)}$
55. $\frac{5a^2-7a}{(a+1)(a-3)}$ **57.** 4 **59.** 3 **61.** $\frac{3}{x-4}$, or $\frac{-3}{4-x}$
63. $\frac{w+z}{w-z}$, or $\frac{-w-z}{z-w}$ **65.** $\frac{-2}{(x+1)(x-1)}$ **67.** $\frac{2(2x-1)}{x-1}$
69. $\frac{7}{y}$ **71.** $\frac{6}{x-2}$ **73.** $\frac{3x-2}{x-1}$ **75.** $\frac{4x-7}{x^2-x+1}$ **77.** $\frac{2x+1}{x}$
79. $\frac{4p^2-21p+29}{(p-2)^2}$ **81.** $\frac{x}{(x-2)^2(x-3)}$
83. $\frac{2x^2+24xy}{(x+2y)(x-y)(x+6y)}$ **85.** $\frac{2x^2+21xy-10y^2}{(x+2y)(x-y)(x+6y)}$
87. $\frac{3r-2s}{(2r-s)(3r-s)}$ **89.** $\frac{10x+23}{(x+2)^2(x+3)}$
91. **(a)** $C(x) = \frac{10x}{49(101-x)}$ **(b)** 3.23 thousand dollars

Section 6.3 (pages 389–390)

1. complex; numerator; both **3.** LCD; identity **5.** $\frac{1}{6}$ **7.** $\frac{9}{5}$ **9.** $\frac{4}{15}$
11. $\frac{7}{17}$ **13.** $\frac{2x}{x-1}$ **15.** $\frac{2(k+1)}{3k-1}$ **17.** $\frac{5x^2}{9z^3}$ **19.** $\frac{1+x}{-1+x}$
21. $\frac{6x+1}{7x-3}$ **23.** $\frac{y+x}{y-x}$ **25.** $4x$ **27.** $x+4y$ **29.** $\frac{y+4}{2}$ **31.** $\frac{a+b}{ab}$
33. xy **35.** $\frac{3y}{2}$ **37.** $\frac{x^2+5x+4}{x^2+5x+10}$ **39.** $\frac{x^2y^2}{y^2+x^2}$ **41.** $\frac{y^2+x^2}{xy^2+x^2y}$, or $\frac{y^2+x^2}{xy(y+x)}$ **43.** $\frac{p^2+k}{p^2-3k}$ **45.** $\frac{1}{2xy}$ **47.** $\frac{m^2+6m-4}{m(m-1)}$
48. $\frac{m^2-m-2}{m(m-1)}$ **49.** $\frac{m^2+6m-4}{m^2-m-2}$ **50.** $m(m-1)$
51. $\frac{m^2+6m-4}{m^2-m-2}$ **52.** Answers will vary.

Section 6.4 (pages 395–398)

1. **(a)** equation **(b)** expression **(c)** expression **(d)** equation
In Exercises 3–13, we give the domains using set-builder notation.
3. $\{x \mid x \text{ is a real number}, x \neq 0\}$ **5.** $\{x \mid x \text{ is a real number}, x \neq -1, 2\}$
7. $\left\{x \mid x \text{ is a real number}, x \neq 4, \frac{7}{2}\right\}$ **9.** $\left\{x \mid x \text{ is a real number}, x \neq -\frac{7}{4}, 0, \frac{13}{6}\right\}$ **11.** $\{x \mid x \text{ is a real number}, x \neq \pm 4\}$
13. $\{x \mid x \text{ is a real number}, x \neq 0, 1, -3, 2\}$ **15.** $\{1\}$ **17.** $\{-6, 4\}$
19. $\{-7, 3\}$ **21.** $\left\{-\frac{7}{12}\right\}$ **23.** $\emptyset$ **25.** $\{-3\}$ **27.** $\{0\}$ **29.** $\emptyset$
31. $\{5\}$ **33.** $\emptyset$ **35.** $\left\{\frac{27}{56}\right\}$ **37.** $\{-3, -1\}$ **39.** $\emptyset$ **41.** $\{-10\}$
43. $\{-1\}$ **45.** $\{13\}$ **47.** $\{x \mid x \neq \pm 3\}$

49.
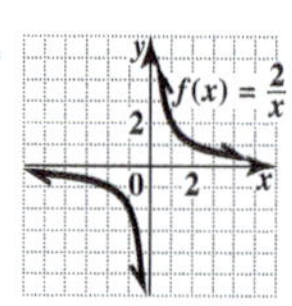

$x = 0;\ y = 0$

51.
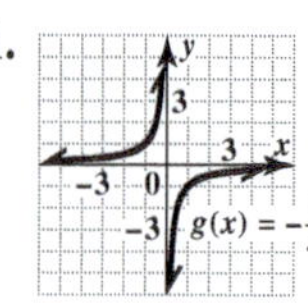

$x = 0;\ y = 0$

53.
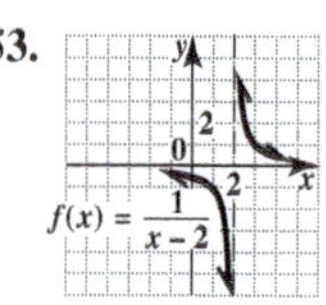

$x = 2;\ y = 0$

55. **(a)** 0 **(b)** 1.6 **(c)** 4.1 **(d)** The waiting time also increases.
57. **(a)** 500 ft **(b)** It decreases.

Section 6.5 (pages 406–411)

1. A **3.** D **5.** 24 **7.** $\frac{25}{4}$ **9.** Multiply each side by $a-b$.
11. $G = \frac{Fd^2}{Mm}$ **13.** $a = \frac{bc}{c+b}$ **15.** $v = \frac{PVt}{pT}$ **17.** $r = \frac{nE-IR}{In}$, or $r = \frac{IR-nE}{-In}$ **19.** $b = \frac{2\mathcal{A}}{h} - B$, or $b = \frac{2\mathcal{A}-hB}{h}$ **21.** $r = \frac{eR}{E-e}$
23. $R = \frac{D}{1-DT}$, or $R = \frac{-D}{DT-1}$ **25.** 21 girls, 7 boys **27.** 1.75 in.
29. 5.4 in. **31.** 7.6 in. **33.** 56 teachers **35.** 210 deer **37.** 25,000 fish
39. 6.6 more gallons **41.** $x = \frac{7}{2}$; $AC = 8$; $DF = 12$ **43.** 2.4 mL

45. $\frac{1}{3}$ job per hr **47.** 3 mph **49.** 10 mph **51.** 1020 mi **53.** 1750 mi **55.** 190 mi **57.** $6\frac{2}{3}$ min **59.** 30 hr **61.** $2\frac{1}{3}$ hr **63.** 20 hr **65.** $2\frac{4}{5}$ hr

Section 6.6 (pages 417–420)

1. increases; decreases **3.** direct **5.** direct **7.** inverse **9.** inverse **11.** inverse **13.** direct **15.** joint **17.** combined **19.** The perimeter of a square varies directly as the length of its side. **21.** The surface area of a sphere varies directly as the square of its radius. **23.** The area of a triangle varies jointly as the lengths of its base and height. **25.** $4; 2; 4\pi; \frac{4}{3}\pi; \frac{1}{2}; \frac{1}{3}\pi$ **27.** $A = kb$ **29.** $h = \frac{k}{t}$ **31.** $M = kd^2$ **33.** $I = kgh$ **35.** 36 **37.** $\frac{16}{9}$ **39.** 0.625 **41.** $\frac{16}{5}$ **43.** $222\frac{2}{9}$ **45.** \$3.92 **47.** 8 lb **49.** 450 cm^3 **51.** 256 ft **53.** $106\frac{2}{3}$ mph **55.** 100 cycles per sec **57.** $21\frac{1}{3}$ foot-candles **59.** \$420 **61.** 11.8 lb **63.** 448.1 lb **65.** 68,600 calls **67.** Answers will vary.

7 ROOTS, RADICALS, AND ROOT FUNCTIONS

Section 7.1 (pages 439–441)

1. E **3.** D **5.** C **7.** C **9.** C **11.** **(a)** It is not a real number. **(b)** negative **(c)** 0 **13.** -9 **15.** 6 **17.** -4 **19.** -8 **21.** 6 **23.** -2 **25.** It is not a real number. **27.** 2 **29.** It is not a real number. **31.** $\frac{8}{9}$ **33.** $\frac{4}{3}$ **35.** $-\frac{1}{2}$ **37.** 3 **39.** 0.5 **41.** -0.7 **43.** 0.1

In Exercises 45–51, we give the domain and then the range.

45.

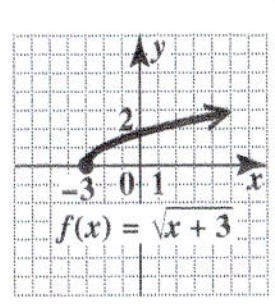

$[-3, \infty)$; $[0, \infty)$

47.

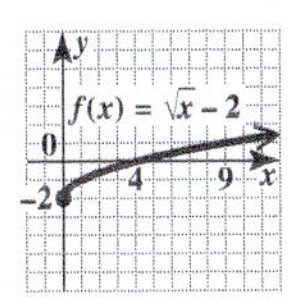

$[0, \infty)$; $[-2, \infty)$

49.

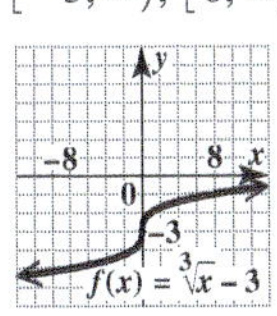

$(-\infty, \infty)$; $(-\infty, \infty)$

51.

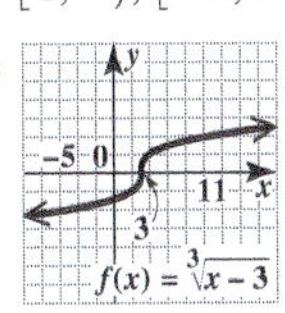

$(-\infty, \infty)$; $(-\infty, \infty)$

53. 12 **55.** 10 **57.** 2 **59.** -9 **61.** -5 **63.** $|x|$ **65.** $|z|$ **67.** x **69.** x^5 **71.** $|x|^5$ (or $|x^5|$) **73.** 97.381 **75.** 16.863 **77.** -9.055 **79.** 7.507 **81.** 3.162 **83.** 1.885 **85.** 1,183,000 cycles per sec **87.** 10 mi **89.** 392,000 mi^2 **91.** **(a)** 1.732 amps **(b)** 2.236 amps

Section 7.2 (pages 448–450)

1. C **3.** A **5.** H **7.** B **9.** D **11.** 13 **13.** 9 **15.** 2 **17.** $\frac{8}{9}$ **19.** -3 **21.** It is not a real number. **23.** 1000 **25.** 27 **27.** -1024 **29.** 16 **31.** $\frac{1}{8}$ **33.** $\frac{1}{512}$ **35.** $\frac{9}{25}$ **37.** $\frac{27}{8}$ **39.** $\sqrt{10}$ **41.** $(\sqrt[4]{8})^3$ **43.** $(\sqrt[8]{9q})^5 - (\sqrt[3]{2x})^2$ **45.** $\frac{1}{(\sqrt{2m})^3}$ **47.** $(\sqrt[3]{2y+x})^2$ **49.** $\frac{1}{(\sqrt[3]{3m^4+2k^2})^2}$ **51.** 64 **53.** 64 **55.** x^{10} **57.** $\sqrt[6]{x^5}$ **59.** $\sqrt[15]{t^8}$ **61.** 9 **63.** 4 **65.** y **67.** $x^{5/12}$ **69.** $k^{2/3}$ **71.** x^3y^8 **73.** $\frac{1}{x^{10/3}}$ **75.** $\frac{1}{m^{1/4}n^{3/4}}$ **77.** p^2 **79.** $\frac{c^{11/3}}{b^{11/4}}$ **81.** $\frac{q^{5/3}}{9p^{7/2}}$ **83.** $p + 2p^2$ **85.** $k^{7/4} - k^{3/4}$ **87.** $6 + 18a$ **89.** $-5x^2 + 5x$ **91.** $x^{17/20}$ **93.** $\frac{1}{x^{3/2}}$ **95.** $y^{5/6}z^{1/3}$ **97.** $m^{1/12}$ **99.** $x^{1/8}$ **101.** $x^{1/24}$ **103.** $\sqrt{a^2+b^2} = \sqrt{3^2+4^2} = 5$; $a + b = 3 + 4 = 7$; $5 \neq 7$ **105.** 4.5 hr **107.** 19.0°; The table gives 19°. **109.** 4.2°; The table gives 4°.

Section 7.3 (pages 458–462)

1. D **3.** B **5.** D **7.** Because there are only two factors of $\sqrt[3]{x}$, $\sqrt[3]{x} \cdot \sqrt[3]{x} = (\sqrt[3]{x})^2$, or $\sqrt[3]{x^2}$. **9.** $\sqrt{9}$, or 3 **11.** $\sqrt{36}$, or 6 **13.** $\sqrt{30}$ **15.** $\sqrt{14x}$ **17.** $\sqrt{42pqr}$ **19.** $\sqrt[3]{10}$ **21.** $\sqrt[4]{14xy}$ **23.** $\sqrt[4]{33}$ **25.** $\sqrt[4]{6x^3}$ **27.** This expression cannot be simplified by the product rule. **29.** $\frac{8}{11}$ **31.** $\frac{\sqrt{3}}{5}$ **33.** $\frac{\sqrt{x}}{5}$ **35.** $\frac{p^3}{9}$ **37.** $-\frac{3}{4}$ **39.** $\frac{\sqrt[3]{r^2}}{2}$ **41.** $-\frac{3}{x}$ **43.** $\frac{1}{x^3}$ **45.** $2\sqrt{3}$ **47.** $12\sqrt{2}$ **49.** $-4\sqrt{2}$ **51.** $-2\sqrt{7}$ **53.** This radical cannot be simplified further. **55.** $4\sqrt[3]{2}$ **57.** $-2\sqrt[3]{2}$ **59.** $2\sqrt[3]{5}$ **61.** $-4\sqrt[4]{2}$ **63.** $2\sqrt[5]{2}$ **65.** $-3\sqrt[5]{2}$ **67.** $2\sqrt[6]{2}$ **69.** $6k\sqrt{2}$ **71.** $12xy^4\sqrt{xy}$ **73.** $11x^3$ **75.** $-3t^4$ **77.** $-10m^4z^2$ **79.** $5a^2b^3c^4$ **81.** $\frac{1}{2}r^2t^5$ **83.** $5x\sqrt{2x}$ **85.** $-10r^5\sqrt{5r}$ **87.** $x^3y^4\sqrt{13x}$ **89.** $2z^2w^3$ **91.** $-2zt^2\sqrt[3]{2z^2t}$ **93.** $3x^3y^4$ **95.** $-3r^3s^2\sqrt[4]{2r^3s^2}$ **97.** $\frac{y^5\sqrt{y}}{6}$ **99.** $\frac{x^5\sqrt[3]{x}}{3}$ **101.** $4\sqrt{3}$ **103.** $\sqrt{5}$ **105.** $x^2\sqrt{x}$ **107.** $\sqrt[6]{432}$ **109.** $\sqrt[12]{6912}$ **111.** $\sqrt[6]{x^5}$ **113.** 5 **115.** $8\sqrt{2}$ **117.** $2\sqrt{14}$ **119.** 13 **121.** $9\sqrt{2}$ **123.** $\sqrt{17}$ **125.** 5 **127.** $6\sqrt{2}$ **129.** $\sqrt{5y^2 - 2xy + x^2}$ **131.** **(a)** B **(b)** C **(c)** D **(d)** A **133.** $x^2 + y^2 = 144$ **135.** $(x + 4)^2 + (y - 3)^2 = 4$ **137.** $(x + 8)^2 + (y + 5)^2 = 5$

139.

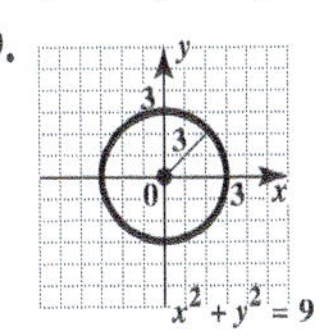

center: $(0, 0)$

141.

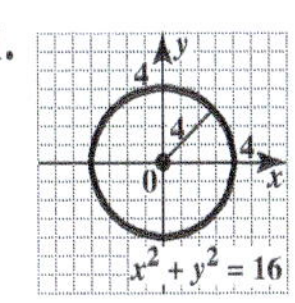

center: $(0, 0)$

143.

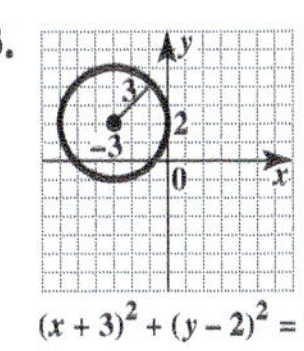

center: $(-3, 2)$

145.

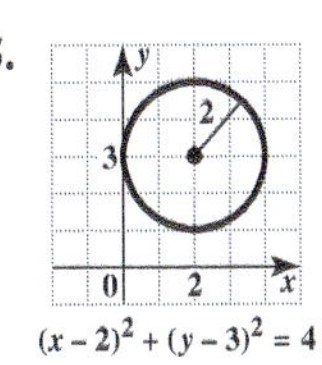

center: $(2, 3)$

147. $2\sqrt{106} + 4\sqrt{2}$ **149.** **(a)** $d = 1.224\sqrt{h}$ **(b)** 15.3 mi **151.** 27.0 in. **153.** 581

Section 7.4 (pages 465–467)

1. B **3.** 15; Each radical expression simplifies to a whole number. **5.** -4 **7.** $7\sqrt{3}$ **9.** $14\sqrt[3]{2}$ **11.** $5\sqrt[4]{2}$ **13.** $24\sqrt{2}$ **15.** The expression cannot be simplified further. **17.** $20\sqrt{5}$

19. $4\sqrt{2x}$ **21.** $-11m\sqrt{2}$ **23.** $7\sqrt[3]{2}$ **25.** $2\sqrt[3]{x}$ **27.** $-7\sqrt[4]{x^2y}$ **29.** $-x\sqrt[3]{xy^2}$ **31.** $19\sqrt[4]{2}$ **33.** $x\sqrt[4]{xy}$ **35.** $9\sqrt[4]{2a^3}$ **37.** $(4+3xy)\sqrt[3]{xy^2}$ **39.** $4t\sqrt[3]{3st}-3s\sqrt{3st}$ **41.** $4x\sqrt[3]{x}+6x\sqrt[4]{x}$ **43.** $2\sqrt{2}-2$ **45.** $\frac{5\sqrt{5}}{6}$ **47.** $\frac{7\sqrt{2}}{6}$ **49.** $\frac{5\sqrt{2}}{3}$ **51.** $5\sqrt{2}+4$ **53.** $\frac{5+3x}{x^4}$ **55.** $\frac{m\sqrt[3]{m^2}}{2}$ **57.** $\frac{3x\sqrt[3]{2}-4\sqrt[3]{5}}{x^3}$ **59.** **(a)** $\sqrt{7}$ **(b)** 2.645751311 **(c)** 2.645751311 **(d)** equal **61.** A; 42 m **63.** $(12\sqrt{5}+5\sqrt{3})$ in. **65.** $(24\sqrt{2}+12\sqrt{3})$ in.

Section 7.5 (pages 474–477)

1. E **3.** A **5.** D **7.** $3\sqrt{6}+2\sqrt{3}$ **9.** $20\sqrt{2}$ **11.** -2 **13.** -1 **15.** 6 **17.** $\sqrt{6}-\sqrt{2}+\sqrt{3}-1$ **19.** $\sqrt{22}+\sqrt{55}-\sqrt{14}-\sqrt{35}$ **21.** $8-\sqrt{15}$ **23.** $9+4\sqrt{5}$ **25.** $26-2\sqrt{105}$ **27.** $4-\sqrt[3]{36}$ **29.** 10 **31.** $6x+3\sqrt{x}-2\sqrt{5x}-\sqrt{5}$ **33.** $9r-s$ **35.** $4\sqrt[3]{4y^2}-19\sqrt[3]{2y}-5$ **37.** $3x-4$ **39.** $4x-y$ **41.** $2\sqrt{6}-1$ **43.** $\sqrt{7}$ **45.** $5\sqrt{3}$ **47.** $\frac{\sqrt{6}}{2}$ **49.** $\frac{9\sqrt{15}}{5}$ **51.** $-\frac{7\sqrt{3}}{12}$ **53.** $\frac{\sqrt{14}}{2}$ **55.** $-\frac{\sqrt{14}}{10}$ **57.** $\frac{2\sqrt{6x}}{x}$ **59.** $\frac{-8\sqrt{3k}}{k}$ **61.** $\frac{-5m^2\sqrt{6mn}}{n^2}$ **63.** $\frac{12x^3\sqrt{2xy}}{y^5}$ **65.** $\frac{5\sqrt{2my}}{y^2}$ **67.** $-\frac{4k\sqrt{3z}}{z}$ **69.** $\frac{\sqrt[3]{18}}{3}$ **71.** $\frac{\sqrt[3]{12}}{3}$ **73.** $\frac{\sqrt[3]{18}}{4}$ **75.** $-\frac{\sqrt[3]{2pr}}{r}$ **77.** $\frac{x^2\sqrt[3]{y^2}}{y}$ **79.** $\frac{2\sqrt[4]{x^3}}{x}$ **81.** $\frac{\sqrt[4]{2yz^3}}{z}$ **83.** $\frac{3(4-\sqrt{5})}{11}$ **85.** $\frac{6\sqrt{2}+4}{7}$ **87.** $\frac{2(3\sqrt{5}-2\sqrt{3})}{33}$ **89.** $2\sqrt{3}+\sqrt{10}-3\sqrt{2}-\sqrt{15}$ **91.** $\sqrt{m}-2$ **93.** $\frac{4(\sqrt{x}+2\sqrt{y})}{x-4y}$ **95.** $\frac{x-2\sqrt{xy}+y}{x-y}$ **97.** $\frac{5\sqrt{k}(2\sqrt{k}-\sqrt{q})}{4k-q}$ **99.** $3-2\sqrt{6}$ **101.** $1-\sqrt{5}$ **103.** $\frac{4-2\sqrt{2}}{3}$ **105.** $\frac{6+2\sqrt{6p}}{3}$ **107.** $\frac{3\sqrt{x+y}}{x+y}$ **109.** $\frac{p\sqrt{p+2}}{p+2}$ **111.** Each expression is approximately equal to 0.2588190451. **113.** $\frac{33}{8(6+\sqrt{3})}$ **115.** $\frac{4x-y}{3x(2\sqrt{x}+\sqrt{y})}$

Section 7.6 (pages 483–485)

1. **(a)** yes **(b)** no **3.** **(a)** yes **(b)** no **5.** No. There is no solution. The radical expression, which is nonnegative, cannot equal a negative number. **7.** $\{11\}$ **9.** $\{\frac{1}{3}\}$ **11.** $\emptyset$ **13.** $\{5\}$ **15.** $\{18\}$ **17.** $\{5\}$ **19.** $\{4\}$ **21.** $\{17\}$ **23.** $\{5\}$ **25.** $\emptyset$ **27.** $\{0\}$ **29.** $\{0\}$ **31.** $\emptyset$ **33.** $\{1\}$ **35.** It is incorrect to just square each term. The right side should be $(8-x)^2 = 64-16x+x^2$. The correct first step is $3x+4 = 64-16x+x^2$, and the solution set is $\{4\}$. **37.** $\{1\}$ **39.** $\{-1\}$ **41.** $\{14\}$ **43.** $\{8\}$ **45.** $\{0\}$ **47.** $\emptyset$ **49.** $\{7\}$ **51.** $\{7\}$ **53.** $\{4, 20\}$ **55.** $\emptyset$ **57.** $\{\frac{5}{4}\}$ **59.** $\{9, 17\}$ **61.** $\{\frac{1}{4}, 1\}$ **63.** $L = CZ^2$ **65.** $K = \frac{V^2m}{2}$ **67.** $M = \frac{r^2F}{m}$ **69.** $r = \frac{a}{4\pi^2N^2}$

Section 7.7 (pages 490–492)

1. nonreal complex, complex **3.** real, complex **5.** pure imaginary, nonreal complex, complex **7.** i **9.** -1 **11.** $-i$ **13.** $13i$ **15.** $-12i$ **17.** $i\sqrt{5}$ **19.** $4i\sqrt{3}$ **21.** $-\sqrt{105}$ **23.** -10 **25.** $i\sqrt{33}$ **27.** $\sqrt{3}$ **29.** $5i$ **31.** -2 **33.** $-1+7i$ **35.** 0 **37.** $7+3i$ **39.** -2 **41.** $1+13i$ **43.** $6+6i$ **45.** $4+2i$ **47.** -81 **49.** -16 **51.** $-10-30i$ **53.** $10-5i$ **55.** $-9+40i$ **57.** $-16+30i$ **59.** 153 **61.** 97 **63.** 4 **65.** **(a)** $a-bi$ **(b)** a^2; b^2 **67.** $1+i$ **69.** $2+2i$ **71.** $-1+2i$ **73.** $-\frac{5}{13}-\frac{12}{13}i$ **75.** $1-3i$ **77.** $1+3i$ **79.** -1 **81.** i **83.** -1 **85.** $-i$ **87.** $-i$ **89.** Because $i^{20} = (i^4)^5 = 1^5 = 1$, the student multiplied by 1, which is justified by the identity property for multiplication. **91.** $\frac{1}{2}+\frac{1}{2}i$

8 QUADRATIC EQUATIONS, INEQUALITIES, AND FUNCTIONS

Section 8.1 (pages 511–513)

1. B, C **3.** The zero-factor property requires a product equal to 0. The first step should have been to rewrite the equation with 0 on one side. **5.** $\{-2, -1\}$ **7.** $\{-3, \frac{1}{3}\}$ **9.** $\{\frac{1}{2}, 4\}$ **11.** $\{\pm 9\}$ **13.** $\{\pm\sqrt{17}\}$ **15.** $\{\pm 4\sqrt{2}\}$ **17.** $\{\pm 2\sqrt{5}\}$ **19.** $\{\pm 2\sqrt{6}\}$ **21.** $\{\pm 3\sqrt{3}\}$ **23.** $\{\pm 4\sqrt{2}\}$ **25.** $\{-7, 3\}$ **27.** $\{-1, 13\}$ **29.** $\{4\pm\sqrt{3}\}$ **31.** $\{-5\pm 4\sqrt{3}\}$ **33.** $\left\{\frac{1\pm\sqrt{7}}{3}\right\}$ **35.** $\left\{\frac{-1\pm 2\sqrt{6}}{4}\right\}$ **37.** $\left\{\frac{2\pm 2\sqrt{3}}{5}\right\}$ **39.** 5.6 sec **41.** Solve $(2x+1)^2 = 5$ by the square root property. Solve $x^2+4x = 12$ by completing the square. **43.** 9; $(x+3)^2$ **45.** 36; $(p-6)^2$ **47.** $\frac{81}{4}$; $\left(q+\frac{9}{2}\right)^2$ **49.** 4 **51.** 25 **53.** $\frac{1}{36}$ **55.** $\{-4, 6\}$ **57.** $\{-2\pm\sqrt{6}\}$ **59.** $\{-5\pm\sqrt{7}\}$ **61.** $\{-\frac{8}{3}, 3\}$ **63.** $\left\{\frac{-7\pm\sqrt{53}}{2}\right\}$ **65.** $\left\{\frac{-5\pm\sqrt{41}}{4}\right\}$ **67.** $\left\{\frac{5\pm\sqrt{15}}{5}\right\}$ **69.** $\left\{\frac{4\pm\sqrt{3}}{3}\right\}$ **71.** $\left\{\frac{2\pm\sqrt{3}}{3}\right\}$ **73.** $\{1\pm\sqrt{2}\}$ **75.** $\{\pm 2i\sqrt{3}\}$ **77.** $\{5\pm 2i\}$ **79.** $\left\{\frac{1}{6}\pm\frac{\sqrt{2}}{3}i\right\}$ **81.** $\{-2\pm 3i\}$ **83.** $\left\{-\frac{2}{3}\pm\frac{2\sqrt{2}}{3}i\right\}$ **85.** $\{-3\pm i\sqrt{3}\}$ **87.** $\{\pm\sqrt{b}\}$ **89.** $\left\{\pm\frac{\sqrt{b^2+16}}{2}\right\}$ **91.** $\left\{\frac{2b\pm\sqrt{3a}}{5}\right\}$ **93.** x^2 **94.** x **95.** $6x$ **96.** 1 **97.** 9 **98.** $(x+3)^2$, or x^2+6x+9

Section 8.2 (pages 518–519)

1. No. The fraction bar should extend under the term $-b$. The correct formula is $x = \frac{-b\pm\sqrt{b^2-4ac}}{2a}$. **3.** The last step is wrong. Because 5 is not a common factor in the numerator, the fraction cannot be simplified. The solution set is $\left\{\frac{5\pm\sqrt{5}}{10}\right\}$. **5.** $\{3, 5\}$ **7.** $\left\{\frac{-2\pm\sqrt{2}}{2}\right\}$

9. $\left\{\frac{1 \pm \sqrt{3}}{2}\right\}$ **11.** $\{5 \pm \sqrt{7}\}$ **13.** $\left\{\frac{-1 \pm \sqrt{2}}{2}\right\}$

15. $\left\{\frac{-1 \pm \sqrt{7}}{3}\right\}$ **17.** $\{1 \pm \sqrt{5}\}$ **19.** $\left\{\frac{-2 \pm \sqrt{10}}{2}\right\}$

21. $\{-1 \pm 3\sqrt{2}\}$ **23.** $\left\{\frac{1 \pm \sqrt{29}}{2}\right\}$ **25.** $\left\{\frac{-4 \pm \sqrt{91}}{3}\right\}$

27. $\left\{\frac{-3 \pm \sqrt{57}}{8}\right\}$ **29.** $\left\{\frac{3}{2} \pm \frac{\sqrt{15}}{2}i\right\}$ **31.** $\{3 \pm i\sqrt{5}\}$

33. $\left\{\frac{1}{2} \pm \frac{\sqrt{6}}{2}i\right\}$ **35.** $\left\{-\frac{2}{3} \pm \frac{\sqrt{2}}{3}i\right\}$ **37.** $\left\{\frac{1}{2} \pm \frac{1}{4}i\right\}$

39. 0; B; zero-factor property **41.** 8; C; quadratic formula **43.** 49; A; zero-factor property **45.** -80; D; quadratic formula **47.** **(a)** 25; zero-factor property; $\left\{-3, -\frac{4}{3}\right\}$ **(b)** 44; quadratic formula; $\left\{\frac{7 \pm \sqrt{11}}{2}\right\}$

49. -10 or 10 **51.** 16 **53.** 25 **55.** $b = \frac{44}{5}; \frac{3}{10}$

Section 8.3 (pages 527–530)

1. Multiply by the LCD, x. **3.** Substitute a variable for $x^2 + x$. **5.** The proposed solution -1 does not check. The solution set is $\{4\}$.

7. $\{-2, 7\}$ **9.** $\{-4, 7\}$ **11.** $\left\{-\frac{2}{3}, 1\right\}$ **13.** $\left\{-\frac{14}{17}, 5\right\}$

15. $\left\{-\frac{11}{7}, 0\right\}$ **17.** $\left\{\frac{-1 \pm \sqrt{13}}{2}\right\}$ **19.** $\left\{-\frac{8}{3}, -1\right\}$

21. $\left\{\frac{2 \pm \sqrt{22}}{3}\right\}$ **23.** $\left\{\frac{-1 \pm \sqrt{5}}{4}\right\}$ **25.** **(a)** $(20 - t)$ mph **(b)** $(20 + t)$ mph **27.** the rate of her boat in still water; $x - 5$; $x + 5$; row 1 of table: 15, $x - 5$, $\frac{15}{x-5}$; row 2 of table: 15, $x + 5$, $\frac{15}{x+5}$; $\frac{15}{x-5} + \frac{15}{x+5} = 4$; 10 mph **29.** 25 mph **31.** 50 mph

33. 3.6 hr **35.** Rusty: 25.0 hr; Nancy: 23.0 hr **37.** 3 hr; 6 hr

39. $\{2, 5\}$ **41.** $\{3\}$ **43.** $\left\{\frac{8}{9}\right\}$ **45.** $\{9\}$ **47.** $\left\{\frac{2}{5}\right\}$ **49.** $\{-2\}$

51. $\{\pm 2, \pm 5\}$ **53.** $\left\{\pm 1, \pm \frac{3}{2}\right\}$ **55.** $\{\pm 2, \pm 2\sqrt{3}\}$

57. $\{-6, -5\}$ **59.** $\left\{-\frac{16}{3}, -2\right\}$ **61.** $\{-8, 1\}$ **63.** $\{-64, 27\}$

65. $\left\{\pm 1, \pm \frac{27}{8}\right\}$ **67.** $\left\{-\frac{1}{3}, \frac{1}{6}\right\}$ **69.** $\left\{-\frac{1}{2}, 3\right\}$ **71.** $\left\{\pm \frac{\sqrt{6}}{3}, \pm \frac{1}{2}\right\}$

73. $\{3, 11\}$ **75.** $\{25\}$ **77.** $\left\{-\sqrt[3]{5}, -\frac{\sqrt[3]{4}}{2}\right\}$ **79.** $\left\{\frac{4}{3}, \frac{9}{4}\right\}$

81. $\left\{\pm \frac{\sqrt{9 + \sqrt{65}}}{2}, \pm \frac{\sqrt{9 - \sqrt{65}}}{2}\right\}$ **83.** $\left\{\pm 1, \pm \frac{\sqrt{6}}{2}i\right\}$

Section 8.4 (pages 536–540)

1. Find a common denominator, and then multiply both sides by the common denominator. **3.** Write it in standard form (with 0 on one side, in decreasing powers of w). **5.** $m = \sqrt{p^2 - n^2}$ **7.** $t = \frac{\pm\sqrt{dk}}{k}$

9. $r = \frac{\pm\sqrt{S\pi}}{2\pi}$ **11.** $d = \frac{\pm\sqrt{skI}}{I}$ **13.** $v = \frac{\pm\sqrt{kAF}}{F}$

15. $r = \frac{\pm\sqrt{3\pi Vh}}{\pi h}$ **17.** $t = \frac{-B \pm \sqrt{B^2 - 4AC}}{2A}$ **19.** $h = \frac{D^2}{k}$

21. $\ell = \frac{p^2 g}{k}$ **23.** $R = \frac{E^2 - 2pr \pm E\sqrt{E^2 - 4pr}}{2p}$

25. $r = \frac{5pc}{4}$ or $r = -\frac{2pc}{3}$ **27.** $I = \frac{-cR \pm \sqrt{c^2R^2 - 4cL}}{2cL}$

29. 7.9, 8.9, 11.9 **31.** eastbound ship: 80 mi; southbound ship: 150 mi **33.** 8 in., 15 in., 17 in. **35.** length: 24 ft; width: 10 ft **37.** 2 ft **39.** 7 m by 12 m **41.** 20 in. by 12 in. **43.** 1 sec and 8 sec **45.** 2.4 sec and 5.6 sec **47.** 9.2 sec **49.** It reaches its *maximum* height at 5 sec because this is the only time it reaches 400 ft. **51.** \$1.50 **53.** 0.035, or 3.5% **55.** 5.5 m per sec **57.** 5 or 14 **59.** **(a)** \$560 billion **(b)** \$560 billion; They are the same. **61.** 2008

Section 8.5 (pages 546–549)

1. **(a)** B **(b)** C **(c)** A **(d)** D **3.** **(a)** D **(b)** B **(c)** C **(d)** A **5.** $(0, 0)$ **7.** $(0, 4)$ **9.** $(1, 0)$ **11.** $(-3, -4)$ **13.** $(5, 6)$ **15.** down; wider **17.** up; narrower **19.** down; narrower

21.
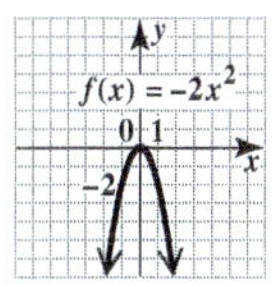

vertex: $(0, 0)$; axis: $x = 0$; domain: $(-\infty, \infty)$; range: $(-\infty, 0]$

23.
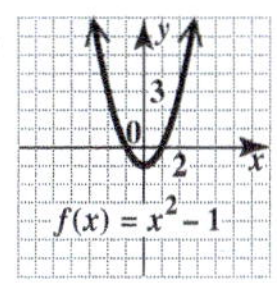

vertex: $(0, -1)$; axis: $x = 0$; domain: $(-\infty, \infty)$; range: $[-1, \infty)$

25.
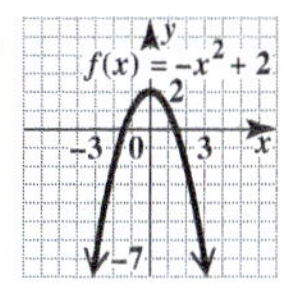

vertex: $(0, 2)$; axis: $x = 0$; domain: $(-\infty, \infty)$; range: $(-\infty, 2]$

27.
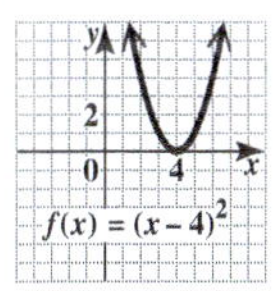

vertex: $(4, 0)$; axis: $x = 4$; domain: $(-\infty, \infty)$; range: $[0, \infty)$

29.
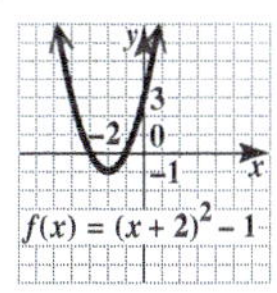

vertex: $(-2, -1)$; axis: $x = -2$; domain: $(-\infty, \infty)$; range: $[-1, \infty)$

31.
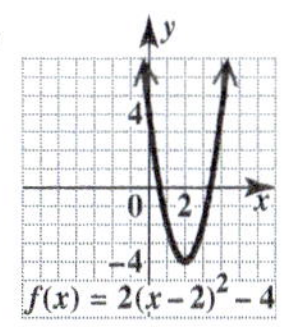

vertex: $(2, -4)$; axis: $x = 2$; domain: $(-\infty, \infty)$; range: $[-4, \infty)$

33.
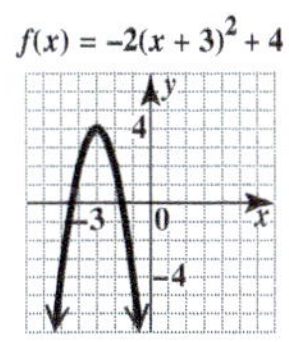

vertex: $(-3, 4)$; axis: $x = -3$; domain: $(-\infty, \infty)$; range: $(-\infty, 4]$

35.
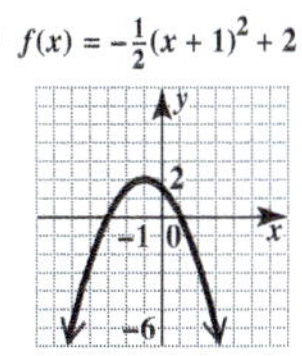

vertex: $(-1, 2)$; axis: $x = -1$; domain: $(-\infty, \infty)$; range: $(-\infty, 2]$

37.
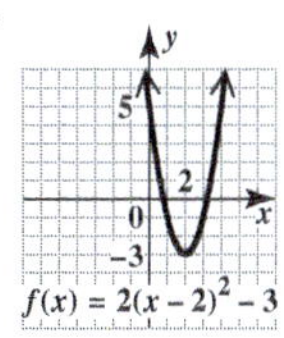

vertex: $(2, -3)$; axis: $x = 2$; domain: $(-\infty, \infty)$; range: $[-3, \infty)$

39. linear function; positive **41.** quadratic function; positive **43.** quadratic function; negative

45. **(a)**

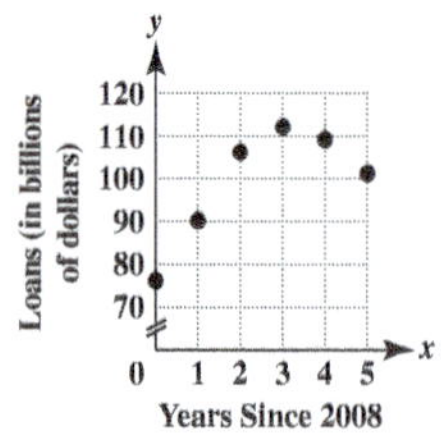

(b) quadratic function **(c)** negative **(d)** $y = -3.5x^2 + 22.5x + 76$ **(e)** 2010: $107 billion; 2012: $110 billion; The model approximates the data very well.

Section 8.6 (pages 557–561)

1. If x is squared, it has a vertical axis. If y is squared, it has a horizontal axis. **3.** Find the discriminant of the function. If it is positive, there are two x-intercepts. If it is 0, there is one x-intercept (at the vertex), and if it is negative, there is no x-intercept. **5.** A, D are vertical parabolas. B, C are horizontal parabolas. **7.** $(-4, -6)$ **9.** $(1, -3)$ **11.** $\left(-\frac{1}{2}, -\frac{29}{4}\right)$ **13.** $(-1, 3)$; up; narrower; no x-intercepts **15.** $\left(\frac{5}{2}, \frac{37}{4}\right)$; down; same shape; two x-intercepts **17.** $(-3, -9)$; to the right; wider **19.** F **21.** C **23.** D

25.

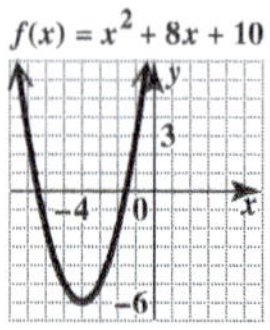

vertex: $(-4, -6)$; axis: $x = -4$; domain: $(-\infty, \infty)$; range: $[-6, \infty)$

27.

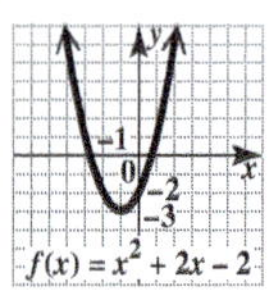

vertex: $(-1, -3)$; axis: $x = -1$; domain: $(-\infty, \infty)$; range: $[-3, \infty)$

29.

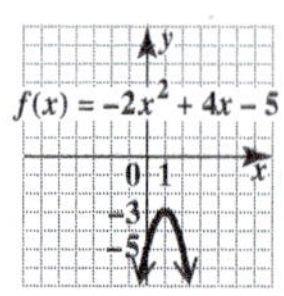

vertex: $(1, -3)$; axis: $x = 1$; domain: $(-\infty, \infty)$; range: $(-\infty, -3]$

31.

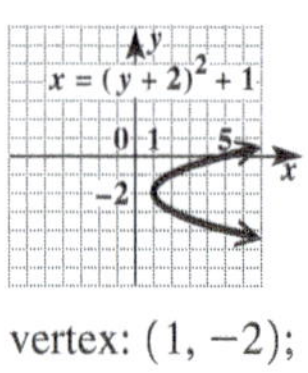

vertex: $(1, -2)$; axis: $y = -2$; domain: $[1, \infty)$; range: $(-\infty, \infty)$

33.

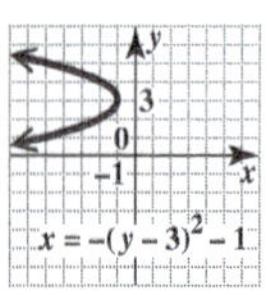

vertex: $(-1, 3)$; axis: $y = 3$; domain: $(-\infty, -1]$; range: $(-\infty, \infty)$

35.

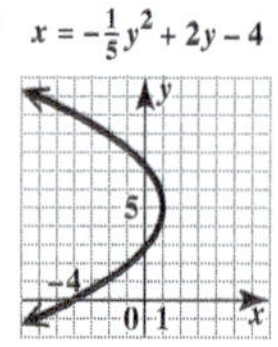

vertex: $(1, 5)$; axis: $y = 5$; domain: $(-\infty, 1]$; range: $(-\infty, \infty)$

37.

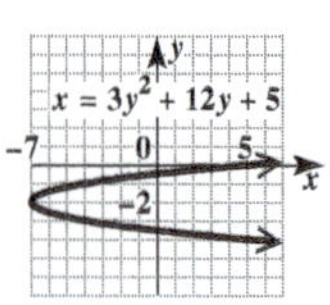

vertex: $(-7, -2)$; axis: $y = -2$; domain: $[-7, \infty)$; range: $(-\infty, \infty)$

39. 20 and 20 **41.** 140 ft by 70 ft; 9800 ft^2 **43.** 2 sec; 65 ft **45.** 20 units; $210 **47.** 16 ft; 2 sec **49.** **(a)** minimum **(b)** 1967; 5.4% **51.** The coefficient of x^2 is negative because a parabola that models the data must open down.

53. **(a)** $R(x) = (100 - x)(200 + 4x)$
$= 20{,}000 + 200x - 4x^2$

(b)

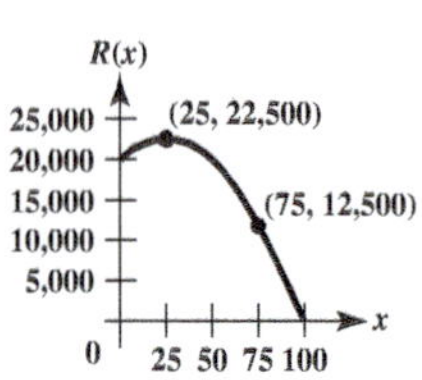

(c) 25 **(d)** $22,500 **55.** center: $(-2, -3)$; $r = 2$ **57.** center: $(-5, 7)$; $r = 9$

Section 8.7 (pages 568–569)

1. $(-\infty, -4] \cup [3, \infty)$ **3.** **(a)** $\{1, 3\}$ **(b)** $(-\infty, 1) \cup (3, \infty)$ **(c)** $(1, 3)$ **5.** **(a)** $\{-2, 5\}$ **(b)** $[-2, 5]$ **(c)** $(-\infty, -2] \cup [5, \infty)$

7. $(-\infty, -1) \cup (5, \infty)$

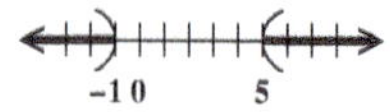

9. $(-4, 6)$

11. $(-\infty, 1] \cup [3, \infty)$

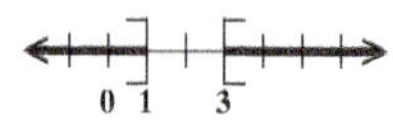

13. $\left(-\infty, -\frac{3}{2}\right] \cup \left[\frac{3}{5}, \infty\right)$

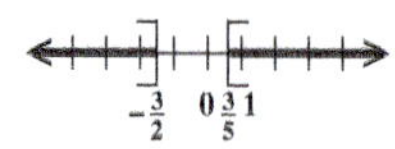

15. $\left[-\frac{3}{2}, \frac{3}{2}\right]$

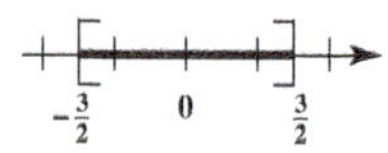

17. $\left(-\infty, -\frac{1}{2}\right] \cup \left[\frac{1}{3}, \infty\right)$

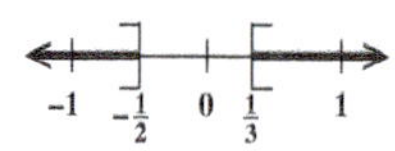

19. $(-\infty, 0] \cup [4, \infty)$

21. $\left[0, \frac{5}{3}\right]$

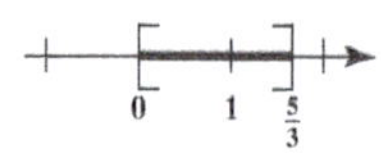

23. $(-\infty, 3 - \sqrt{3}] \cup [3 + \sqrt{3}, \infty)$

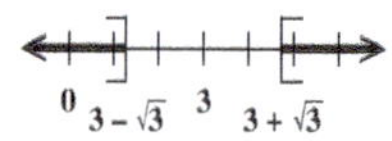

25. $(-\infty, \infty)$ **27.** $\varnothing$ **29.** $\varnothing$ **31.** $(-\infty, \infty)$

33. $(-\infty, 1) \cup (2, 4)$

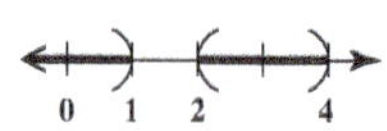

35. $\left[-\frac{3}{2}, \frac{1}{3}\right] \cup [4, \infty)$

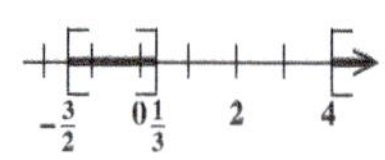

37. $(-\infty, 1) \cup (4, \infty)$

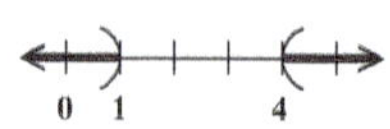

39. $\left[-\frac{3}{2}, 5\right)$

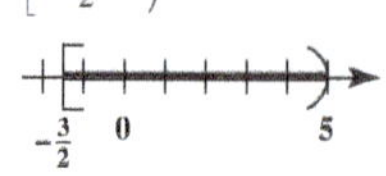

41. $(2, 6]$

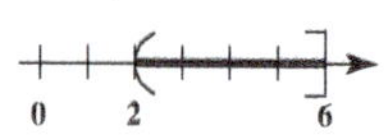

43. $\left(-\infty, \frac{1}{2}\right) \cup \left(\frac{5}{4}, \infty\right)$

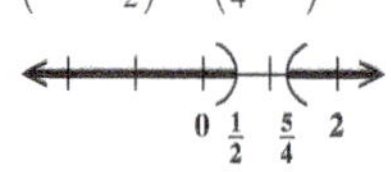

45. $[-7, -2)$

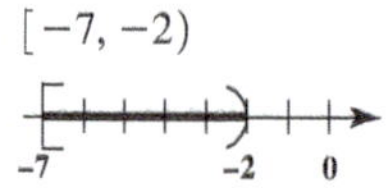

47. $(-\infty, 2) \cup (4, \infty)$

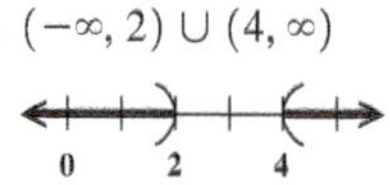

49. $\left(0, \frac{1}{2}\right) \cup \left(\frac{5}{2}, \infty\right)$

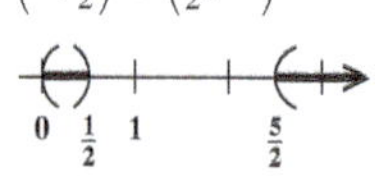

51. $\left[\frac{3}{2}, \infty\right)$

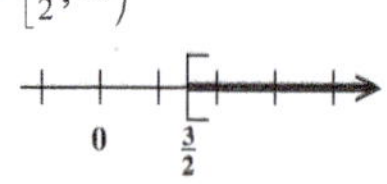

53. $\left(-2, \frac{5}{3}\right) \cup \left(\frac{5}{3}, \infty\right)$

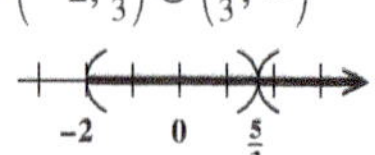

9 INVERSE, EXPONENTIAL, AND LOGARITHMIC FUNCTIONS

Section 9.1 (pages 587–590)

1. B **3.** A **5.** This function is one-to-one. **7.** Yes, it is one-to-one. By adding 1 to 1058, two distances would be the same, so the function would not be one-to-one. **9.** $\{(6, 3), (10, 2), (12, 5)\}$
11. not one-to-one **13.** $f^{-1}(x) = x - 3$ **15.** $f^{-1}(x) = -2x - 4$
17. $f^{-1}(x) = \frac{x-4}{2}$, or $f^{-1}(x) = \frac{1}{2}x - 2$ **19.** $f^{-1}(x) = x^2 + 3, x \geq 0$
21. not one-to-one **23.** $f^{-1}(x) = \sqrt[3]{x+4}$
25. $f^{-1}(x) = \frac{-2x+4}{x-1}, x \neq 1$ **27.** $f^{-1}(x) = \frac{-5x-2}{x-4}, x \neq 4$
29. $f^{-1}(x) = \frac{5x+1}{2x+2}, x \neq -1$ **31.** **(a)** 8 **(b)** 3
33. **(a)** 1 **(b)** 0
35. **(a)** one-to-one **37.** **(a)** not one-to-one **39.** **(a)** one-to-one
(b)

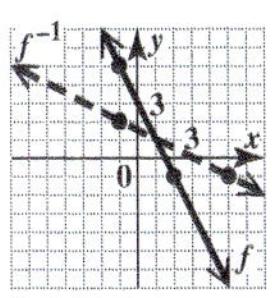

(b)

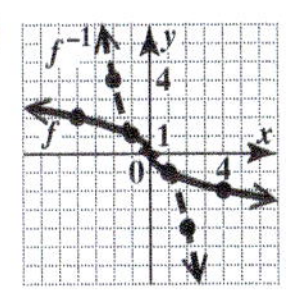

41.

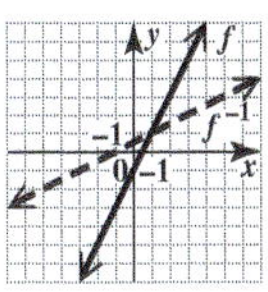

43.

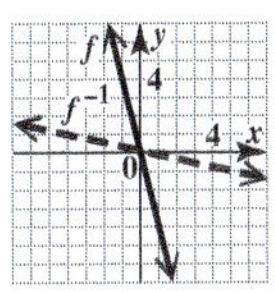

45.

x	$f(x)$
0	0
1	1
4	2

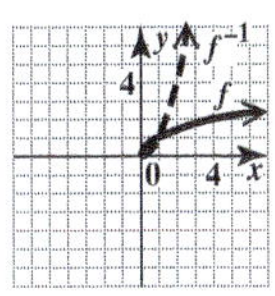

47.

x	$f(x)$
−1	−3
0	−2
1	−1
2	6

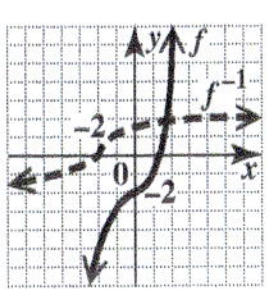

49. $f^{-1}(x) = \frac{x+5}{4}$, or $f^{-1}(x) = \frac{1}{4}x + \frac{5}{4}$

50. MY GRAPHING CALCULATOR IS THE GREATEST THING SINCE SLICED BREAD. **51.** If the function were not one-to-one, there would be ambiguity in some of the characters, as they could represent more than one letter. **52.** Answers will vary. For example, Jane Doe is 1004 5 2748 129 68 3379 129.

Section 9.2 (pages 597–599)

1. C **3.** rises **5.** A **7.** 3.732 **9.** 0.344 **11.** 1.995 **13.** 1.587
15. 0.192 **17.** 73.517 **19.**

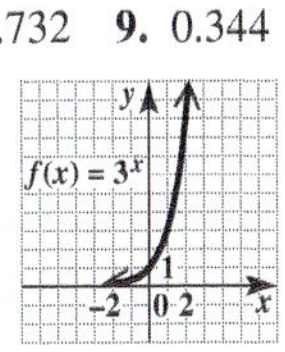

21.

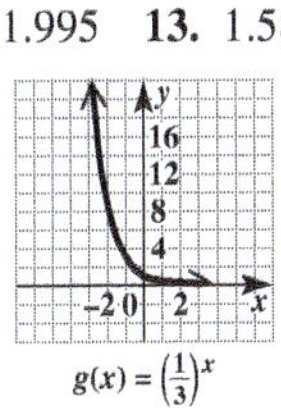

23.

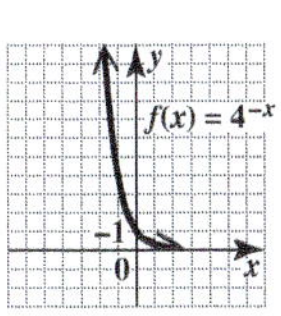

25.

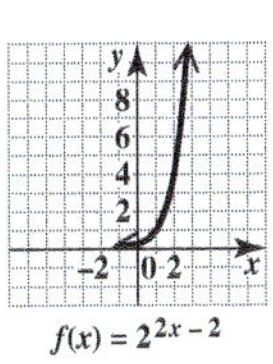

27. $\{2\}$ **29.** $\left\{\frac{3}{2}\right\}$
31. $\{7\}$ **33.** $\{-3\}$
35. $\{-1\}$ **37.** $\{-3\}$
39. 100 g **41.** 0.30 g
43. **(a)** 0.6°C **(b)** 0.3°C **45.** **(a)** 1.4°C **(b)** 0.5°C
47. **(a)** 2.403 million **(b)** 32.971 million **(c)** 188.962 million

49. **(a)** \$5000 **(b)** \$2973 **(c)** \$1768
(d)

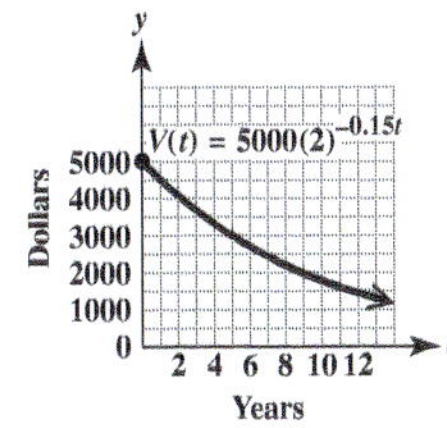

Section 9.3 (pages 605–607)

1. **(a)** B **(b)** E **(c)** D **(d)** F **(e)** A **(f)** C **3.** $(0, \infty); (-\infty, \infty)$
5. $\log_4 1024 = 5$ **7.** $\log_{1/2} 8 = -3$ **9.** $\log_{10} 0.001 = -3$
11. $\log_{625} 5 = \frac{1}{4}$ **13.** $\log_8 \frac{1}{4} = -\frac{2}{3}$ **15.** $\log_5 1 = 0$ **17.** $4^3 = 64$
19. $10^{-4} = \frac{1}{10{,}000}$ **21.** $6^0 = 1$ **23.** $9^{1/2} = 3$ **25.** $\left(\frac{1}{4}\right)^{1/2} = \frac{1}{2}$
27. $5^{-1} = 5^{-1}$ **29.** **(a)** C **(b)** B **(c)** B **(d)** C **31.** 3.1699
33. 1.7959 **35.** −1.7925 **37.** −1.5850 **39.** 1.9243 **41.** 1.6990
43. $\left\{\frac{1}{3}\right\}$ **45.** $\{81\}$ **47.** $\left\{\frac{1}{5}\right\}$ **49.** $\{1\}$ **51.** $\{x \mid x > 0, x \neq 1\}$
53. $\{5\}$ **55.** $\left\{\frac{5}{3}\right\}$ **57.** $\{4\}$ **59.** $\left\{\frac{3}{2}\right\}$ **61.** $\{30\}$ **63.** 1 **65.** 0
67. 9 **69.** −1 **71.** 9 **73.** 5 **75.** 6 **77.** 4 **79.** −1 **81.** $\frac{1}{3}$
83.

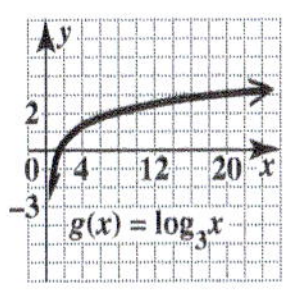

85.

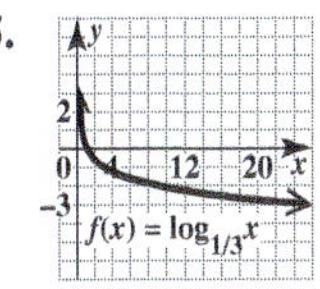

87.

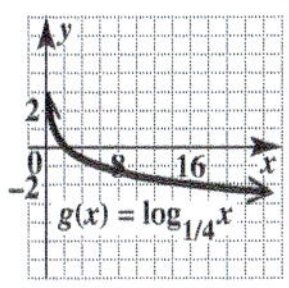

89. Every power of 1 is equal to 1, and thus it cannot be used as a base.
91. 8 **93.** 24 **95.** **(a)** 130 thousand units **(b)** 190 thousand units
(c)

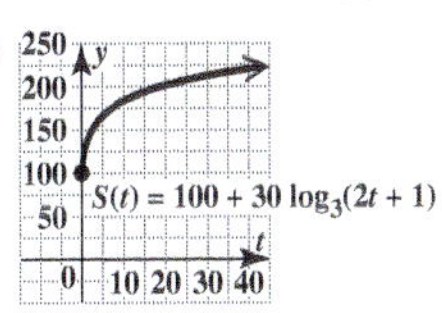

97. about 4 times as powerful

Section 9.4 (pages 613–614)

1. false; $\log_b x + \log_b y = \log_b xy$ **3.** true **5.** In the notation $\log_a(x + y)$, the parentheses do not indicate multiplication. They indicate that $x + y$ is the result of raising a to some power. **7.** $\log_{10} 7 + \log_{10} 8$
9. 4 **11.** 9 **13.** $\log_7 4 + \log_7 5$ **15.** $\log_5 8 - \log_5 3$ **17.** $2 \log_4 6$
19. $\frac{1}{3}\log_3 4 - 2\log_3 x - \log_3 y$ **21.** $\frac{1}{2}\log_3 x + \frac{1}{2}\log_3 y - \frac{1}{2}\log_3 5$
23. $\frac{1}{3}\log_2 x + \frac{1}{5}\log_2 y - 2\log_2 r$ **25.** $\log_b xy$ **27.** $\log_a \frac{m}{n}$
29. $\log_a \frac{rt^3}{s}$ **31.** $\log_a \frac{125}{81}$ **33.** $\log_{10}(x^2 + 8x + 15)$ **35.** $\log_p \frac{x^3 y^{1/2}}{z^{3/2}a^3}$
37. 1.2552 **39.** −0.6532 **41.** 1.5562 **43.** 0.2386 **45.** 0.4771
47. 4.7710 **49.** false **51.** true **53.** true **55.** false

Section 9.5 (pages 620–622)

1. C **3.** C **5.** 2 **7.** 1.6335 **9.** 2.5164 **11.** −1.4868 **13.** 9.6421
15. 9.6776 **17.** 2.0592 **19.** −2.8896 **21.** 5.9613 **23.** −11.4007
25. 4.1506 **27.** 2.3026 **29.** 40 **31.** **(a)** 2.552424846
(b) 1.552424846 **(c)** 0.552424846 **(d)** The whole number parts will vary, but the decimal parts are the same. **33.** poor fen **35.** bog

37. rich fen **39.** 11.6 **41.** 4.3 **43.** 4.0×10^{-8} **45.** 4.0×10^{-6} **47. (a)** 142 dB **(b)** 126 dB **(c)** 120 dB **49. (a)** 800 yr **(b)** 5200 yr **(c)** 11,500 yr **51. (a)** 17.8 million **(b)** 2005 **53. (a)** \$54 per ton **(b)** If $x = 0$, then $\ln(1 - x) = \ln 1 = 0$, so $T(x)$ would be negative. If $x = 1$, then $\ln(1 - x) = \ln 0$, but the domain of $\ln x$ is $(0, \infty)$. **55.** 2.2619 **57.** 0.6826 **59.** 0.3155 **61.** 0.8736 **63.** 2.4849 **65.** 0.4421 **67.** In each case, the product is 1. $\log_a b$ and $\log_b a$ are reciprocals.

Section 9.6 (pages 629–632)

1. common logarithms **3.** natural logarithms **5.** $\{0.827\}$ **7.** $\{0.833\}$ **9.** $\{1.201\}$ **11.** $\{2.269\}$ **13.** $\{15.967\}$ **15.** $\{-6.067\}$ **17.** $\{261.291\}$ **19.** $\{-10.718\}$ **21.** $\{3\}$ **23.** $\{5.879\}$ **25.** $\{-\pi\}$, or $\{-3.142\}$ **27.** $\{1\}$ **29.** $\{4\}$ **31.** $\left\{\frac{2}{3}\right\}$ **33.** $\left\{\frac{33}{2}\right\}$ **35.** $\{-1 + \sqrt[3]{49}\}$ **37.** $\{\pm 3\}$ **39.** 2 cannot be a solution because $\log(2 - 3) = \log(-1)$, and -1 is not in the domain of $\log x$. **41.** $\left\{\frac{1}{3}\right\}$ **43.** $\{2\}$ **45.** $\varnothing$ **47.** $\{8\}$ **49.** $\left\{\frac{4}{3}\right\}$ **51.** $\{8\}$ **53.** \$2539.47; 10.2 yr **55.** \$4934.71; 19.8 yr **57. (a)** \$11,260.96 **(b)** \$11,416.64 **(c)** \$11,497.99 **(d)** \$11,580.90 **(e)** \$11,581.83 **59.** \$137.41 **61.** \$131,962 million **63. (a)** 1.62 g **(b)** 1.18 g **(c)** 0.69 g **(d)** 2.00 g **65. (a)** 179.73 g **(b)** 21.66 yr

APPENDIX

Appendix (page 646)

1. 0 **3.** $x - 5$ **5.** $4m - 1$ **7.** $2a + 4 + \frac{5}{a + 2}$ **9.** $p - 4 + \frac{9}{p + 1}$ **11.** $4a^2 + a + 3$ **13.** $x^4 + 2x^3 + 2x^2 + 7x + 10 + \frac{18}{x - 2}$ **15.** $-4r^5 - 7r^4 - 10r^3 - 5r^2 - 11r - 8 + \frac{-5}{r - 1}$ **17.** $-3y^4 + 8y^3 - 21y^2 + 36y - 72 + \frac{143}{y + 2}$ **19.** 7 **21.** -2 **23.** 0 **25.** yes **27.** no **29.** yes **31.** no **33.** $(2x - 3)(x + 4)$ **34.** $\left\{-4, \frac{3}{2}\right\}$ **35.** 0 **36.** 0 **37.** a **38.** Yes, $x - 3$ is a factor; $Q(x) = (x - 3)(3x - 1)(x + 2)$

GLOSSARY

For a more complete discussion, see the section(s) in parentheses.

A

absolute value The absolute value of a number is the undirected distance between 0 and the number on a number line. (Section R.1)

absolute value equation An absolute value equation is an equation that involves the absolute value of a variable expression. (Section 1.7)

absolute value inequality An absolute value inequality is an inequality that involves the absolute value of a variable expression. (Section 1.7)

additive inverse (opposite, negative) The additive inverse of a number x, symbolized $-x$, is the number that is the same distance from 0 on the number line as x, but on the opposite side of 0. The number 0 is its own additive inverse. For all real numbers x, $x + (-x) = (-x) + x = 0$. (Section R.1)

algebraic expression An algebraic expression is a collection of constants, variables, operation symbols, and/or grouping symbols. (Sections R.3, 4.2)

asymptote A line that a graph more and more closely approaches as the graph gets farther away from the origin is an asymptote of the graph. (Sections 6.4, 9.2)

axis of symmetry (axis) The axis of symmetry of a parabola is the vertical or horizontal line (depending on the orientation of the graph) through the vertex of the parabola. (Section 8.5)

B

base The base in an exponential expression is the expression that is the repeated factor. In b^x, b is the base. (Sections R.3, 4.1)

binomial A binomial is a polynomial consisting of exactly two terms. (Section 4.2)

boundary line In the graph of an inequality, the boundary line separates the region that satisfies the inequality from the region that does not satisfy the inequality. (Section 2.4)

C

center The fixed point that is a fixed distance from all the points that form a circle is the center of the circle. (Section 7.3)

circle A circle is the set of all points in a plane that lie a fixed distance from a fixed point. (Section 7.3)

coefficient (See **numerical coefficient.**)

combined variation A relationship among variables that involves both direct and inverse variation is combined variation. (Section 6.6)

common logarithm A common logarithm is a logarithm having base 10. (Section 9.5)

complex conjugates The complex numbers $a + bi$ and $a - bi$ are complex conjugates. (Section 7.7)

complex fraction A complex fraction is a quotient with one or more fractions in the numerator, denominator, or both. (Section 6.3)

complex number A complex number is any number that can be written in the form $a + bi$, where a and b are real numbers and i is the imaginary unit. (Section 7.7)

components In an ordered pair (x, y), x and y are the components of the ordered pair. (Section 2.1)

composite function If g is a function of x, and f is a function of $g(x)$, then $f(g(x))$ defines the composite function of f and g. It is symbolized $(f \circ g)(x)$. (Section 4.3)

composition of functions The process of finding a composite function is called composition of functions. (Section 4.3)

compound inequality A compound inequality consists of two inequalities linked by a connective word such as *and* or *or*. (Section 1.6)

compound interest Compound interest is paid or charged on both principal and previous interest. (Section 9.6)

conditional equation A conditional equation is true for some replacements of the variable and false for others. (Section 1.1)

conjugates The binomials $x + y$ and $x - y$ are conjugates. (Section 7.5)

consecutive even (or odd) integers Two even (or odd) integers that differ by two are consecutive even (or odd) integers. (Section 1.4)

consecutive integers Two integers that differ by one are consecutive integers. (Section 1.4)

consistent system A system of equations with a solution is a consistent system. (Section 3.1)

constant A constant is a fixed, unchanging number. (Section R.3)

constant function A constant function is a linear function whose graph is a horizontal line. It has the form $f(x) = b$, where b is a real number. (Section 2.6)

constant of variation In the variation equations $y = kx$, $y = \frac{k}{x}$, and $y = kxz$, the real number k is the constant of variation. (Section 6.6)

continuous compounding Continuous compounding of interest involves compounding where the number of compounding periods is allowed to increase without bound, leading to a formula with the number e. (Section 9.6)

contradiction A contradiction is an equation that is never true. It has no solution. (Section 1.1)

coordinate Every point on a number line is associated with a unique real number, called the coordinate of the point. (Section R.1)

coordinates of a point The numbers in an ordered pair are the coordinates of the corresponding point in the plane. (Section 2.1)

cube root function The function $f(x) = \sqrt[3]{x}$ is the cube root function. (Section 7.1)

cubing function The polynomial function $f(x) = x^3$ is the cubing function. (Section 4.3)

D

degree of a polynomial The degree of a polynomial is the greatest degree of any of the terms in the polynomial. (Section 4.2)

degree of a term The degree of a term is the sum of the exponents on the variables in the term. (Section 4.2)

dependent equations Equations of a system that have the same graph (because they are different forms of the same equation) are dependent equations. (Section 3.1)

dependent variable In an equation relating x and y, if the value of the variable y depends on the value of the variable x, then y is the dependent variable. (Section 2.5)

descending powers A polynomial in one variable is written in descending powers of the variable if the exponents on the variables of the terms of the polynomial decrease from left to right. (Section 4.2)

difference The answer to a subtraction problem is the difference. (Section R.2)

difference of cubes The difference of cubes, $x^3 - y^3$, can be factored as $x^3 - y^3 = (x - y)(x^2 + xy + y^2)$. (Section 5.3)

difference of squares The difference of squares, $x^2 - y^2$, can be factored as the product of the sum and difference of two terms, or $x^2 - y^2 = (x + y)(x - y)$. (Section 5.3)

direct variation y varies directly as x if there exists a real number k such that $y = kx$. (Section 6.6)

discontinuous A graph that has one or more breaks due to excluded values of the domain is discontinuous. (Section 6.4)

discriminant The discriminant of the equation $ax^2 + bx + c = 0$ is the quantity $b^2 - 4ac$ under the radical symbol in the quadratic formula. (Section 8.2)

domain The set of all first components (x-values) in the ordered pairs of a relation is the domain. (Section 2.5)

domain of the variable in a rational equation The domain of the variable in a rational equation is the intersection of the domains of the rational expressions in the equation. (Section 6.4)

double solution If a quadratic equation has two solutions that are not distinct, that solution is a double solution. (Section 5.5)

E

elements (members) The elements (members) of a set are the objects that belong to the set. (Section R.1)

empty (null) set The empty (null) set, denoted by { } or $\emptyset$, is the set containing no elements. (Section R.1)

equation An equation is a statement that two quantities are equal. (Section R.1)

equivalent equations Equivalent equations are equations that have the same solution set. (Section 1.1)

equivalent inequalities Equivalent inequalities are inequalities that have the same solution set. (Section 1.5)

exponent (power) An exponent, or power, is a number that indicates how many times its base is used as a factor. In b^x, x is the exponent. (Sections R.3, 4.1)

exponential equation An exponential equation is an equation that has a variable in at least one exponent. (Section 9.2)

exponential expression A number or letter (variable) written with an exponent is an exponential expression. (Sections R.3, 4.1)

exponential function with base a An exponential function with base a is a function of the form $f(x) = a^x$, where $a > 0$ and $a \neq 1$ for all real numbers x. (Section 9.2)

extraneous solution A proposed solution to an equation that does not satisfy the original equation is an extraneous solution. (Sections 6.4 and 7.6)

F

factoring Factoring is the process of writing a polynomial as the product of two or more simpler polynomials. (Section 5.1)

factors Two numbers whose product is a third number are factors of that third number. (Section R.3)

finite set A finite set is a set where the counting of its elements comes to an end. (Section R.1)

first-degree equation A first-degree (linear) equation has no term with the variable to a power greater than 1. (Section 2.1)

focus variable In the elimination method for solving a system of equations in three variables, the first variable to be eliminated is the focus variable. (Section 3.2)

formula A formula is an equation in which variables are used to describe a relationship among several quantities. (Section 1.2)

function A function is a relation in which, for each distinct value of the first component of the ordered pairs, there is exactly one value of the second component. (Section 2.5)

G

graph of an equation The graph of an equation in two variables is the set of all points that correspond to all of the ordered pairs that satisfy the equation. (Section 2.1)

graph of a number The point on a number line that corresponds to a number is its graph. (Section R.1)

greatest common factor (GCF) The greatest common factor of a list of integers is the largest factor of all those integers. The greatest common factor of the terms of a polynomial is the largest factor of all the terms in the polynomial. (Section 5.1)

H

horizontal asymptote If the y-values of a rational function approach a real number b as $|x|$ increases without bound, then the horizontal line $y = b$ is a horizontal asymptote of the graph. (Section 6.4)

hypotenuse The hypotenuse is the longest side in a right triangle. It is the side opposite the right angle. (Section 7.3)

I

identity An identity is an equation that is true for all valid replacements of the variable. It has an infinite number of solutions. (Section 1.1)

identity element for addition (additive identity) For all real numbers a, $a + 0 = 0 + a = a$. The number 0 is the identity element for addition. (Section R.4)

identity element for multiplication (multiplicative identity) For all real numbers a, $a \cdot 1 = 1 \cdot a = a$. The number 1 is the identity element for multiplication. (Section R.4)

identity function The simplest polynomial function is the identity function, $f(x) = x$. (Section 4.3)

imaginary part The imaginary part of the complex number $a + bi$ is b. (Section 7.7)

inconsistent system A system of equations with no solution is an inconsistent system. (Section 3.1)

independent equations Equations of a system that have different graphs are independent equations. (Section 3.1)

independent variable In an equation relating x and y, if the value of the variable y depends on the value of the variable x, then x is the independent variable. (Section 2.5)

index (order) In a radical of the form $\sqrt[n]{a}$, n is the index or order. (Section 7.1)

inequality An inequality is a statement that two expressions are not equal. (Sections R.1, 1.5)

infinite set An infinite set is a set where the counting of its elements does not come to an end. (Section R.1)

integers The set of integers is $\{\ldots, -3, -2, -1, 0, 1, 2, 3, \ldots\}$. (Section R.1)

intersection The intersection of two sets A and B, symbolized $A \cap B$, is the set of elements that belong to *both A and B*. (Section 1.6)

interval An interval is a portion of a number line. (Section 1.5)

inverse of a function f The inverse of a one-to-one function f, written f^{-1}, is the set of all ordered pairs of the form (y, x) where (x, y) belongs to f. (Section 9.1)

inverse variation y varies inversely as x if there exists a real number k such that $y = \frac{k}{x}$. (Section 6.6)

irrational numbers An irrational number cannot be written as the quotient of two integers, but can be represented by a point on a number line. (Section R.1)

J

joint variation y varies jointly as x and z if there exists a real number k such that $y = kxz$. (Section 6.6)

L

leading coefficient The numerical coefficient of the leading term of a polynomial is the leading coefficient. (Section 4.2)

leading term When a polynomial is written in descending powers of the variable, the greatest-degree term is written first and is the leading term of the polynomial. (Section 4.2)

least common denominator (LCD) Given several denominators, the least multiple that is divisible by all the denominators is the least common denominator. (Section 6.2)

legs (of a right triangle) The two shorter perpendicular sides of a right triangle are its legs. (Section 7.3)

like terms Terms with exactly the same variables raised to exactly the same powers are like terms. (Sections R.4, 4.2)

linear equation in one variable A linear equation in one variable (here x) can be written in the form $Ax + B = C$, where A, B, and C are real numbers and $A \neq 0$. (Section 1.1)

linear equation in two variables A linear equation in two variables (here x and y) is an equation that can be written in the form $Ax + By = C$, where A, B, and C are real numbers and A and B are not both 0. (Section 2.1)

linear function A function f that can be written in the form $f(x) = ax + b$, where a and b are real numbers, is a linear function. The value of a is the slope m of the graph of the function. (Section 2.6)

linear inequality in one variable A linear inequality in one variable (here x) can be written in the form $Ax + B < C$ or $Ax + B > C$ (or with $\leq$ or $\geq$), where A, B, and C are real numbers and $A \neq 0$. (Section 1.5)

linear inequality in two variables A linear inequality in two variables (here x and y) can be written in the form $Ax + By < C$ or $Ax + By > C$ (or with $\leq$ or $\geq$), where A, B, and C are real numbers and A and B are not both 0. (Section 2.4)

logarithm A logarithm is an exponent. The expression $\log_a x$ represents the exponent to which the base a must be raised to obtain x. (Section 9.3)

logarithmic equation A logarithmic equation is an equation with a logarithm of a variable expression in at least one term. (Section 9.3)

logarithmic function with base a A logarithmic function with base a is a function of the form $g(x) = \log_a x$, where $a > 0$ and $a \neq 1$ for all positive real numbers x. (Section 9.3)

M

mathematical model In a real-world problem, a mathematical model is one or more equations (or inequalities) that describe the situation. (Section 1.2)

monomial A monomial is a polynomial consisting of exactly one term. (Section 4.2)

multiplicative inverse (reciprocal) The multiplicative inverse (reciprocal) of a nonzero number x, symbolized $\frac{1}{x}$, is the real number which has the property that the product of the two numbers is 1. For all nonzero real numbers x, $\frac{1}{x} \cdot x = x \cdot \frac{1}{x} = 1$. (Sections R.2, 6.1)

N

natural (counting) numbers The set of natural numbers is the set of numbers used for counting, $\{1, 2, 3, 4, \ldots\}$. (Section R.1)

natural logarithm A natural logarithm is a logarithm having base e. (Section 9.5)

negative of a polynomial The negative of a polynomial is that polynomial with the sign of every term changed. (Section 4.2)

negative square root The negative square root of a positive number x, written with a radical symbol as $-\sqrt{x}$, is the negative number which, when squared, gives x as the result. (Section R.3)

nonreal complex number If $a + bi$ has $b \neq 0$, then it is a nonreal complex number. (Section 7.7)

number line A line that has a point designated to correspond to the real number 0, and a standard unit chosen to represent the distance between 0 and 1, is a number line. All real numbers correspond to one and only one number on such a line. (Section R.1)

numerical coefficient The numerical factor in a term is the numerical coefficient, or simply, the coefficient of the other factors. (Sections R.4, 4.2)

O

one-to-one function In a one-to-one function each x-value corresponds to only one y-value, and each y-value corresponds to only one x-value. (Section 9.1)

order (See **index.**)

ordered pair An ordered pair is a pair of numbers (x, y) written within parentheses in which the order of the numbers is important. (Section 2.1)

ordered triple An ordered triple is a triple of numbers written within parentheses in the form (x, y, z). (Section 3.2)

origin The point at which the x-axis and y-axis of a rectangular coordinate system intersect is the origin. (Section 2.1)

P

parabola The graph of a quadratic function is a parabola. (Section 8.5)

percent Percent, written with the symbol %, means "per one hundred." (Section 1.2)

percent decrease If a quantity decreases from one measure to another, the amount of decrease compared to the original amount, when expressed as a percent, is the percent decrease. (Section 1.2)

percent increase If a quantity increases from one measure to another, the amount of increase compared to the original amount, when expressed as a percent, is the percent decrease. (Section 1.2)

perfect square trinomial A perfect square trinomial is a trinomial that can be factored as the square of a binomial. (Section 5.3)

plot To plot a point in a rectangular coordinate system is to locate it based on its x- and y-coordinates. (Section 2.1)

polynomial A polynomial is a term or a finite sum of terms of the form ax^n, where a is a real number, x is a variable, and the exponent n is a whole number. (Section 4.2)

polynomial function A function defined by a polynomial in one variable, consisting of one or more terms, is a polynomial function. (Section 4.3)

polynomial in x A polynomial containing only the variable x is a polynomial in x. (Section 4.2)

positive (principal) square root The positive (principal) square root of a positive number x, written with a radical symbol as $\sqrt{x}$, is the positive number which, when squared, gives x as the result. (Section R.3)

prime polynomial A prime polynomial is a polynomial that cannot be factored into factors having only integer coefficients. (Section 5.1)

principal root For even indexes, the symbols $\sqrt{\ }$, $\sqrt[4]{\ }$, $\sqrt[6]{\ }$, . . . , $\sqrt[n]{\ }$ are used for nonnegative roots, which are principal roots. (Section 7.1)

product The answer to a multiplication problem is the product. (Section R.2)

proportion A proportion is a statement that two ratios are equal. (Section 6.5)

proposed solution A value of the variable that appears to be a solution of an equation before checking is a proposed solution. (Sections 6.4 and 7.6)

pure imaginary number A complex number $a + bi$ where $a = 0$ and $b \neq 0$ is a pure imaginary number. (Section 7.7)

Q

quadrant A quadrant is one of the four regions in the plane determined by the axes in a rectangular coordinate system. (Section 2.1)

quadratic equation A quadratic equation (in x here) can be written in the form $ax^2 + bx + c = 0$, where a, b, and c are real numbers and $a \neq 0$. (Sections 5.5, 8.1)

quadratic formula The solutions of the equation $ax^2 + bx + c = 0$ (where $a \neq 0$) are given by the quadratic formula, $x = \dfrac{-b \pm \sqrt{b^2 - 4ac}}{2a}$. (Section 8.2)

quadratic function A function that can be written in the form $f(x) = ax^2 + bx + c$, for real numbers a, b, and c, where $a \neq 0$, is a quadratic function. (Section 8.5)

quadratic inequality A quadratic inequality (in x here) can be written in the form $ax^2 + bx + c < 0$ or $ax^2 + bx + c > 0$ (or with $\leq$ or $\geq$), where a, b, and c are real numbers and $a \neq 0$. (Section 8.7)

quadratic in form A nonquadratic equation that can be written in the form $au^2 + bu + c = 0$, for $a \neq 0$ and an algebraic expression u, is quadratic in form. (Section 8.3)

quotient The answer to a division problem is the quotient. (Section R.2)

R

radical An expression consisting of a radical symbol, root index, and radicand is a radical. (Section 7.1)

radical equation A radical equation is an equation with a variable in at least one radicand. (Section 7.6)

radical expression A radical expression is an algebraic expression that contains radicals. (Section 7.1)

radicand The number or expression under a radical symbol is the radicand. (Section 7.1)

radius The radius of a circle is the fixed distance between the center and any point on the circle. (Section 7.3)

range The set of all second components (y-values) in the ordered pairs of a relation is the range. (Section 2.5)

ratio A ratio is a comparison of two quantities using a quotient. (Section 6.5)

rational equation An equation that involves rational expressions is a rational equation. (Section 6.4)

rational expression The quotient of two polynomials with denominator not 0 is a rational expression. (Section 6.1)

rational function A function that is defined by a quotient of polynomials is a rational function. (Section 6.1)

rational inequality An inequality that involves rational expressions is a rational inequality. (Section 8.7)

rational numbers Rational numbers can be written as the quotient of two integers, with denominator not 0. (Section R.1)

real numbers Real numbers include all numbers that can be represented by points on a number line—that is, all rational and irrational numbers. (Section R.1)

real part The real part of the complex number $a + bi$ is a. (Section 7.7)

reciprocal (See **multiplicative inverse.**)

reciprocal function The rational function $f(x) = \frac{1}{x}$, where $x \neq 0$, is the reciprocal function. (Section 6.4)

rectangular (Cartesian) coordinate system The x-axis and y-axis placed at a right angle at their zero points form a rectangular coordinate system, also called a Cartesian coordinate system. (Section 2.1)

relation A relation is any set of ordered pairs. (Section 2.5)

rise Rise refers to the vertical change between two points on a line—that is, the change in y-values. (Section 2.2)

run Run refers to the horizontal change between two points on a line—that is, the change in x-values. (Section 2.2)

S

scatter diagram A scatter diagram is a graph of ordered pairs of data (usually real-world data). (Section 2.3)

second-degree equation (See **quadratic equation.**)

set A set is a collection of objects. (Section R.1)

signed numbers Signed numbers are numbers that can be written with a positive or negative sign. (Section R.1)

slope The ratio of the change in y to the change in x for any two points on a line is the slope of the line. (Section 2.2)

solution A solution of an equation is any replacement for the variable that makes the equation true. (Section 1.1)

solution set The solution set of an equation is the set of all solutions of the equation. (Section 1.1)

solution set of a linear system The solution set of a linear system of equations consists of all ordered pairs that satisfy all the equations of the system at the same time. (Section 3.1)

square root The inverse of squaring a number is taking its square root. That is, a number a is a square root of k if $a^2 = k$. (Section R.3)

square root function The root function $f(x) = \sqrt{x}$, where $x \geq 0$, is the square root function. (Section 7.1)

squaring function The polynomial function $f(x) = x^2$ is the squaring function. (Section 4.3)

sum The answer to an addition problem is the sum. (Section R.2)

sum of cubes The sum of cubes, $x^3 + y^3$, can be factored as $x^3 + y^3 = (x + y) \cdot (x^2 - xy + y^2)$. (Section 5.3)

system of equations A system of equations consists of two or more equations to be solved at the same time. (Section 3.1)

system of linear equations A system of linear equations consists of two or more linear equations to be solved at the same time. (Section 3.1)

T

table of ordered pairs (table of values) A table of ordered pairs (table of values) is a systematic method for listing corresponding x- and y-coordinates when graphing an equation in two variables. (Section 2.1)

term A term is a number, a variable, or the product or quotient of a number and one or more variables raised to powers. (Sections R.4, 4.2)

three-part inequality An inequality that says that one number is between two other numbers is a three-part inequality. (Section 1.5)

trinomial A trinomial is a polynomial consisting of exactly three terms. (Section 4.2)

U

union The union of two sets A and B, symbolized $A \cup B$, is the set of elements that belong to *either A or B* (or both). (Section 1.6)

unlike terms Unlike terms are terms that do not satisfy the conditions that define like terms. (Section R.4)

V

variable A variable is a symbol, usually a letter, used to represent an unknown number. (Section R.1)

vertex The point on a parabola that has the least y-value (if the parabola opens up) or the greatest y-value (if the parabola opens down) is the vertex of the parabola. (Section 8.5)

vertical asymptote If the y-values of a rational function approach $-\infty$ or ∞ as the x-values approach a real number a, then the vertical line $x = a$ is a vertical asymptote of the graph. (Section 6.4)

W

whole numbers The set of whole numbers is $\{0, 1, 2, 3, 4, \ldots\}$. (Section R.1)

working equation In the elimination method for solving a system of equations in three variables, the working equation is used twice to eliminate the focus variable. (Section 3.2)

X

***x*-axis** The horizontal number line in a rectangular coordinate system is the x-axis. (Section 2.1)

***x*-intercept** A point where a graph intersects the x-axis is an x-intercept. (Section 2.1)

Y

***y*-axis** The vertical number line in a rectangular coordinate system is the y-axis. (Section 2.1)

***y*-intercept** A point where a graph intersects the y-axis is a y-intercept. (Section 2.1)

PHOTO CREDITS

FRONT MATTER
p. vT Margaret L. Lial; **p. vB** John Hornsby

CHAPTER R
p. 22 Vukas/Fotolia; **p. 30** Terry McGinnis; **p. 31** Terry McGinnis; **p. 42** WavebreakMediaMicro/Fotolia

CHAPTER 1
p. 43 Vibe Images/Fotolia; **p. 53** Gstockstudio/Fotolia; **p. 58** John Hornsby; **p. 63** Steve Broer/Shutterstock; **p. 64** Wavebreakmedia/Shutterstock; **p. 76T** Luciano Mortula/Shutterstock; **p. 76M** Hornsbystamp; **p. 76B** Rob/Fotolia; **p. 77** Stuart Jenner/Shutterstock; **p. 84T** Pearson Education; **p. 84B** Photo Disc/Getty Images; **p. 85** Bikeriderlondon/Shutterstock; **p. 88L** ValentinValkov/Fotolia; **p. 88R** Nyul/Fotolia; **p. 101L** Jupiterimages/Goodshoot/Thinkstock; **p. 101R** John Hornsby; **p. 108** Faraways/Shutterstock; **p. 110** Michaeljung/Fotolia; **p. 117** Degtiarova Viktoriia/Shutterstock; **p. 121** Lunamarina/Fotolia; **p. 131** John Hornsby

CHAPTER 2
p. 135 Konstantin Yolshin/Fotolia; **p. 136** Library of Congress; **p. 147** Alex_SK/Fotolia; **p. 155** Joanna Zielinska/Fotolia; **p. 171** Photastic/Shutterstock; **p. 174** Holbox/Shutterstock; **p. 175** Wrangler/Shutterstock; **p. 176** Barry Blackburn/Shutterstock; **p. 177** Studio DMM Photography, Designs & Art/Shutterstock; **p. 203T** C Squared Studios/Stockbyte/Getty Images; **p. 203B** Comstock; **p. 205** Neelsky/Shutterstock

CHAPTER 3
p. 215 Filtv/Fotolia; **p. 240T** Photoestelar/Fotolia; **p. 240B** Rob Marmion/Shutterstock; **p. 247** Morgan Lane Photography/Shutterstock; **p. 251T** Nadia Zagainova/Shutterstock; **p. 251B** Joe Gough/Fotolia; **p. 254L** Hieronymus Ukkel/Fotolia; **p. 254R** Dwphotos/Shutterstock; **p. 255T** Stephen Coburn/Shutterstock; **p. 255B** Shutterstock

CHAPTER 4
p. 265 Etien/Fotolia; **p. 278** Darrin Henry/Fotolia; **p. 280T** Vovan/Shutterstock; **p. 280M** Jiawangkun/Shutterstock; **p. 280B** Multi-State Lottery Association; **p. 290** Gorbelabda/Shutterstock; **p. 294** Fair Trade USA; **p. 295** Topseller/Shutterstock; **p. 316** Lucky Dragon/Fotolia; **p. 318** Markus Bormann/Fotolia; **p. 322** National Atomic Museum Foundation

CHAPTER 5
p. 323 Perytskyy/Fotolia; **p. 354** Peter Barrett/Shutterstock

CHAPTER 6
p. 365 David Albrecht/Fotolia; **p. 384** Photoart/Fotolia; **p. 397** Lars Christensen/Shutterstock; **p. 403** Lucio/Fotolia; **p. 405** Dmitry Kalinovsky/Shutterstock; **p. 408T** John Hornsby; **p. 408M** Hurst Photo/Shutterstock; **p. 408B** James Marvin Phelps/Shutterstock; **p. 409** Scott Rothstein/Shutterstock; **p. 417** AVAVA/Shutterstock; **p. 420** John Hornsby

CHAPTER 7
p. 433 Achim Baqué/Fotolia; **p. 450** John Hornsby; **p. 485** Johnson Space Center/NASA; **p. 502** Scol22/Fotolia

CHAPTER 8
p. 503 CrackerClips/Fotolia; **p. 506** Library of Congress; **p. 511L** Jovannig/Fotolia; **p. 511R** Rudi1976/Fotolia; **p. 522** Auremar/Fotolia; **p. 535** Iofoto/Shutterstock; **p. 540** Danny M Clark/Fotolia; **p. 545** Wabkmiami/Fotolia; **p. 549T** Tyler Olson/Fotolia; **p. 549B** Racheal Grazias/Shutterstock; **p. 559** Corbis; **p. 560** Bonnie Jacobs/iStock

CHAPTER 9
p. 581 Shi Yali/Shutterstock; **p. 583** Frontpage/Shutterstock; **p. 590** Andy Crawford/Dorling Kindersley; **p. 598** Wyatt Rivard/Shutterstock; **p. 599** Aleksandr Makarenko/Shutterstock; **p. 604** Per Tillmann/Fotolia; **p. 615** Denton Rumsey/Shutterstock; **p. 616** StockLite/Shutterstock; **p. 617** Georgios Kollidas/Fotolia; **p. 621** Tawana Frink/Fotolia; **p. 622** Jose Alberto Tejo/Shutterstock; **p. 627** Alina555/iStock/360/Getty Images; **p. 631** Michaeljung/Fotolia

INDEX

Formulas

Figure	*Formulas*	*Illustration*

Square

Perimeter: $P = 4s$

Area: $\mathcal{A} = s^2$

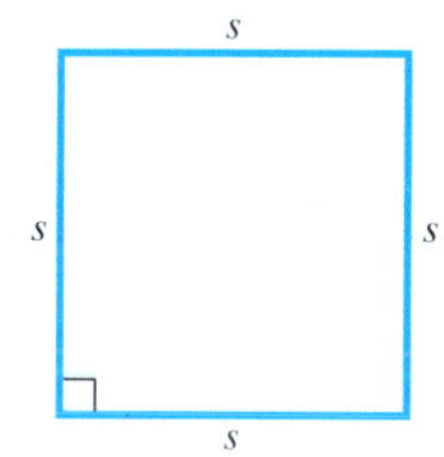

Rectangle

Perimeter: $P = 2L + 2W$

Area: $\mathcal{A} = LW$

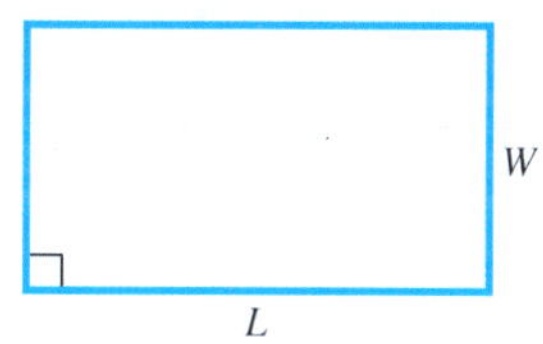

Triangle

Perimeter: $P = a + b + c$

Area: $\mathcal{A} = \frac{1}{2}bh$

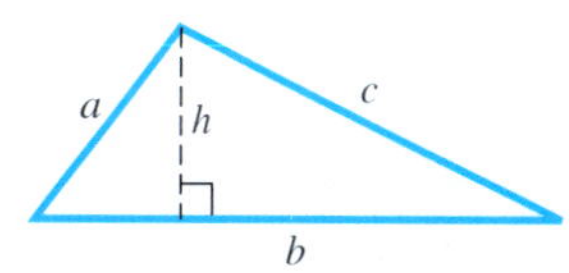

Parallelogram

Perimeter: $P = 2a + 2b$

Area: $\mathcal{A} = bh$

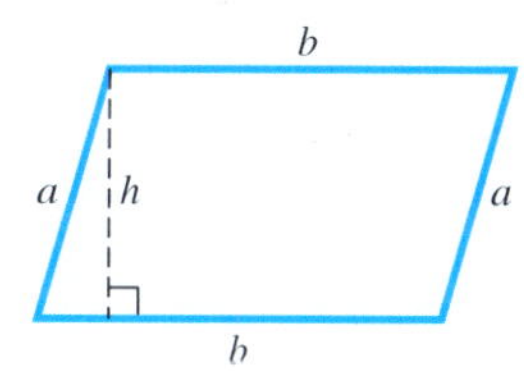

Trapezoid

Perimeter: $P = a + b + c + B$

Area: $\mathcal{A} = \frac{1}{2}h(b + B)$

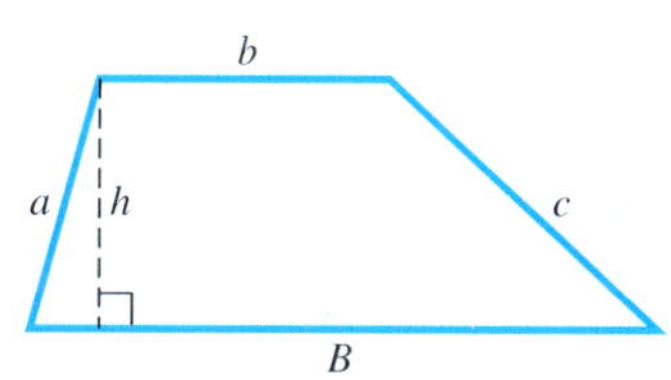

Circle

Diameter: $d = 2r$

Circumference: $C = 2\pi r$

$C = \pi d$

Area: $\mathcal{A} = \pi r^2$

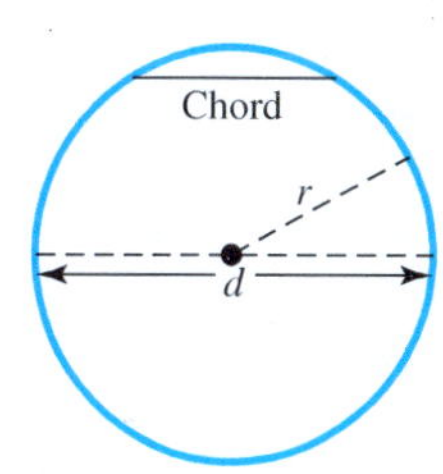

Triangles and Angles

Right Triangle

Triangle has one 90° (right) angle.

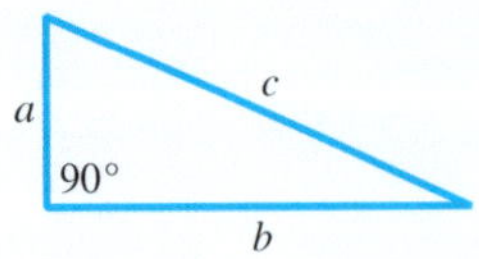

Pythagorean Theorem (for right triangles)

$a^2 + b^2 = c^2$

Right Angle

Measure is 90°.

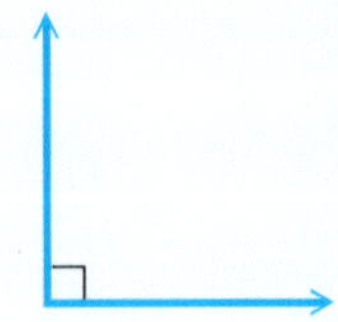

Isosceles Triangle

Two sides are equal.

$AB = BC$

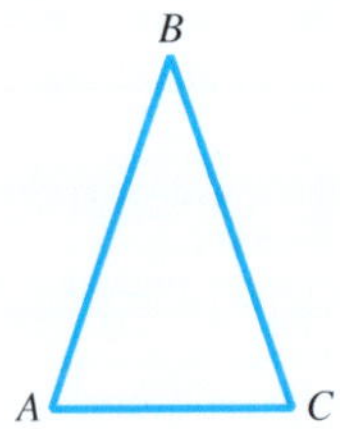

Straight Angle

Measure is 180°.

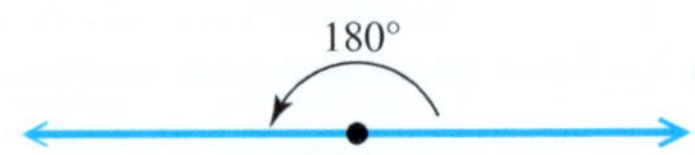

Equilateral Triangle

All sides are equal.

$AB = BC = CA$

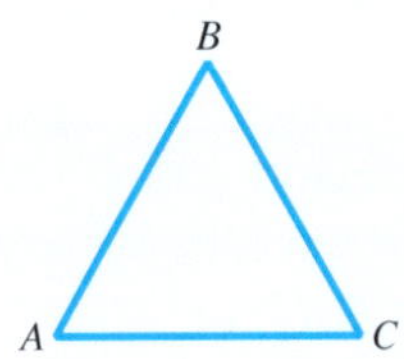

Complementary Angles

The sum of the measures of two complementary angles is 90°.

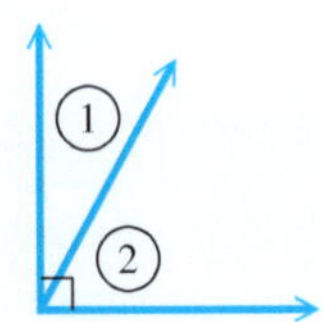

Angles ① and ② are complementary.

Sum of the Angles of Any Triangle

$A + B + C = 180°$

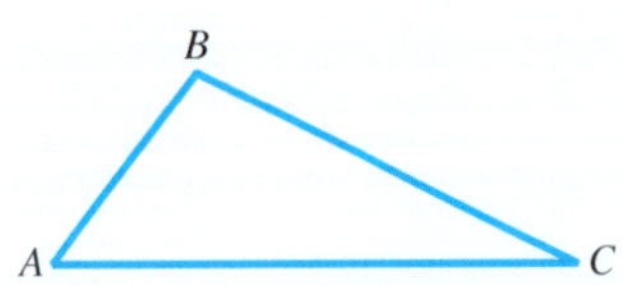

Supplementary Angles

The sum of the measures of two supplementary angles is 180°.

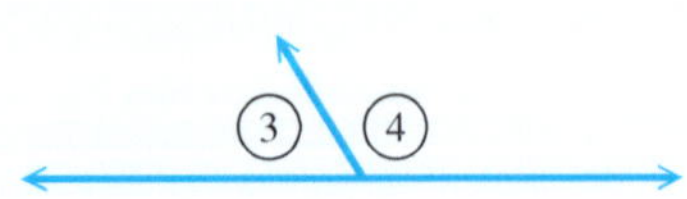

Angles ③ and ④ are supplementary.

Similar Triangles

Corresponding angles are equal. Corresponding sides are proportional.

$A = D, B = E, C = F$

$$\frac{AB}{DE} = \frac{AC}{DF} = \frac{BC}{EF}$$

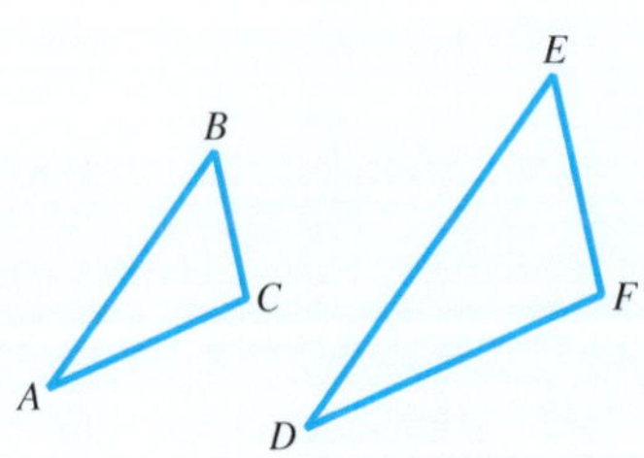

Vertical Angles

Vertical angles have equal measures.

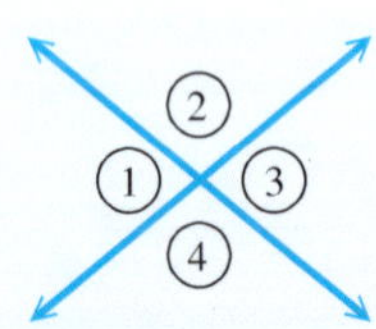

Angle ① = Angle ③

Angle ② = Angle ④